Physical Processes in Clouds and Cloud Modeling

This book presents the most comprehensive and systematic description currently available of both classical and novel theories of cloud processes, providing a much-needed link between cloud theory, observation, experimental results, and cloud modeling. The book shows why and how modern models serve as a major tool of investigation of cloud processes responsible for atmospheric phenomena, including climate change. It systematically describes classical as well as recent advancements in cloud physics, including cloud-aerosol interaction; collisions of particles in turbulent clouds; and the formation of multiphase cloud particles. As the first of its kind to serve as a practical guide for using state-of-the-art numerical cloud models, major emphasis is placed on explaining how microphysical processes are treated in modern numerical cloud-resolving models. The book will be a valuable resource for advanced students, researchers, and numerical model designers in cloud physics, atmospheric science, meteorology, and environmental science.

Dr. Alexander P. Khain is Professor in the Institute of Earth Sciences at the Hebrew University of Jerusalem. He is a renowned leading expert in developing cloud and cloud-resolving models with precise microphysics in order to investigate the physics of clouds and precipitation. He has participated in several American, European, and Asian international research projects where his advanced microphysical schemes were widely used to investigate natural and anthropogenic aerosol effects and relations between microphysics and atmospheric dynamics. He has published two books on tropical cyclones and their interaction with the ocean, and approximately 200 academic papers on cloud physics, cloud-aerosol interaction, and numerical modeling of clouds, storms, and hurricanes.

Dr. Mark Pinsky is Professor in the Institute of Earth Sciences at the Hebrew University of Jerusalem. He is a leading expert in the investigation and modeling of drop condensation/evaporation and turbulence impact on collision processes in clouds. The cloud models he has developed have enabled us – for the first time – to explain the impact of turbulence on precipitation formation as well as drizzle formation. Dr. Pinsky has participated in several joint European and American research projects in the fields of cloud physics, precipitation enhancement, and satellite and radar meteorology. He has published more than 100 academic papers on cloud physics, cloud modeling, radar meteorology, and estimation of symmetry measure in chemistry.

Physical Processes in Clouds and Cloud Modeling

ALEXANDER P. KHAIN
Hebrew University of Jerusalem

MARK PINSKY
Hebrew University of Jerusalem

CAMBRIDGE
UNIVERSITY PRESS

University Printing House, Cambridge CB2 8BS, United Kingdom

One Liberty Plaza, 20th Floor, New York, NY 10006, USA

477 Williamstown Road, Port Melbourne, VIC 3207, Australia

314–321, 3rd Floor, Plot 3, Splendor Forum, Jasola District Centre, New Delhi – 110025, India

79 Anson Road, #06–04/06, Singapore 079906

Cambridge University Press is part of the University of Cambridge.

It furthers the University's mission by disseminating knowledge in the pursuit of education, learning, and research at the highest international levels of excellence.

www.cambridge.org
Information on this title: www.cambridge.org/9780521767439
DOI: 10.1017/9781139049481

First published 2018

Printed in the United States of America by Sheridan Books, Inc.

A catalogue record for this publication is available from the British Library.

Library of Congress Cataloging-in-Publication Data
Names: Khain, A. P. (Aleksandr Pavlovich), author. | Pinsky, Mark (Senior scientist), author.
Title: Physical processes in clouds and cloud modeling / Alexander Khain (Hebrew University of Jerusalem) and Mark Pinsky (Hebrew University of Jerusalem).
Description: Cambridge, United Kingdom ; New York, NY : Cambridge University Press, 2018. | Includes bibliographical references and index.
Identifiers: LCCN 2017049967 | ISBN 9780521767439 (hardback) | ISBN 0521767431 (hardback)
Subjects: LCSH: Cloud physics–Mathematical models.
Classification: LCC QC921.5 .K43 2018 | DDC 551.57/6–dc23
LC record available at https://lccn.loc.gov/2017049967

ISBN 978-0-521-76743-9 Hardback

Contents

Preface *page* vii
List of Abbreviations ix
List of Symbols xiii

1 Clouds: Definitions and Significance 1
1.1 The Importance of Clouds 1
1.2 Clouds and Cloud-Related Phenomena 4
1.3 Investigating Clouds: The Purpose and Scope of This Book 15
References 18

2 Cloud Particles and Their Representation in Cloud Models 19
2.1 General Characteristics of Cloud Particles and Their Description in Cloud Models 19
2.2 Atmospheric Aerosols 29
2.3 Cloud Drops 39
2.4 Cloud Ice 51
References 63

3 Basic Equations 68
3.1 Thermodynamics of Dry and Moist Air 68
3.2 Budget and Motion Equations 79
3.3 Turbulence 85
3.4 Scales of Atmospheric Motions and Equation Averaging 100
3.5 Dynamic, Thermodynamic, and Kinetic Equations in Cloud Models 109
3.6 Similarity of Averaged and Non-Averaged Equations 114
References 119

4 Numerical Methods Used in Cloud Models 122
4.1 Finite-Difference Approximation and Representation of Derivatives 122
4.2 Equation of Advection. Stability and Errors of Numerical Schemes 126
4.3 Equations of Friction and Diffusion 135
4.4 Gravity and Inertia Gravity Waves 137
4.5 Numerical Schemes for Non-Hydrostatic Models 142
4.6 Comments Concerning Application of Numerical Schemes in Cloud Models 147
References 148

5 Warm Microphysical Processes 151
5.1 Droplet Nucleation 152
5.2 Condensational (Diffusional) Drop Growth 165
5.3 Parameterization of Droplet Nucleation in Cloud Models 188
5.4 Calculation of Diffusional Growth and of Supersaturation in Numerical Cloud Models 198
5.5 Drop Dynamics 207
5.6 Gravitational Drop Collisions 222
5.7 Methods Used for Solving Stochastic Collection Equation and Stochastic Breakup Equation 235
5.8 Turbulent Collisions of Drops and Their Parameterization in Cloud Models 253
5.9 Turbulent Mixing in Clouds 276
5.10 Numerical Modeling of Mixing Effects 298
5.11 Formation of DSD and Raindrops via Warm Processes 313
References 331

6 Microphysical Processes in Ice and Mixed-Phase Clouds 344
6.1 Main Ice-Related Processes and Ice Particles Description 345
6.2 Nucleation of Ice Particles 354
6.3 Phase Transformations in Mixed-Phase Clouds due to Diffusion Growth and Deposition 376
6.4 Motion of Ice Particles 397
6.5 Collisions and Coalescence in Mixed-Phase Clouds and Ice Clouds 416
6.6 Melting and Freezing 440
6.7 Dry and Wet Growth of Graupel and Hail 460
6.8 Ice Multiplication and Its Representation in Cloud Models 477
References 486

7 Modeling: A Powerful Tool for Cloud Investigation 497
7.1 Characteristics of State-of-the-Art Cloud and Cloud-Resolving Models 497

7.2 Two Methodologies in Cloud Microphysics: Bulk Schemes and Bin Schemes 505
7.3 Effects of Aerosols on the Structure and Microphysics of Clouds and Cloud Systems 536
7.4 Some Advances in Modeling Clouds and Cloud-Related Phenomena 563
7.5 Conclusions and Perspectives of Cloud Modeling 579
References 581

Appendix A Tensors 595
Appendix B Collision Efficiency between Drops and Turbulent Enhancement Factor 597
Appendix C Graupel–Drop Collision Efficiency and Kernel 609
Index 623

Preface

Clouds, to a great extent, govern the radiation balance and energy fluxes reaching Earth and have a strong impact on the hydrological cycle and freshwater distribution over the globe. Latent heat release in clouds is the energy source for atmospheric phenomena from the scale of a separate cloud to global circulation. Clouds are also a major component of hazardous atmospheric phenomena such as hurricanes, storms, and hailstorms. Changes in cloud microphysical and radiative properties are the main factors leading to climate change. All of these determine the vital impact that clouds exert on all forms of life on the planet, human life included.

Studies of clouds and cloud-related phenomena are an integral part of modern meteorology, providing data to be directly applied in weather forecasting, agriculture, environment control, air traffic control, and many other spheres of human activity.

One of the major methods of investigating clouds and cloud-related phenomena is numerical modeling. Numerical cloud models solve coupled dynamic and microphysical equations for different processes occurring in clouds. The microphysical processes characterizing transformation of cloud particles are typically accompanied by phase transitions and latent heat release determine, to a large extent, the type, dynamics, and structure of clouds.

There are several classic books on cloud physics and cloud microphysical processes, among them are the following (see the References in Chapter 1 for details): *A Short Course in Cloud Physics* by Rogers and Yau (1996); the fundamental book *Cloud Microphysics and Precipitation* by Pruppacher and Klett (1997); *Cloud and Precipitation Microphysics* by Straka (2009), describing simplified expressions for rates of microphysical processes used in bulk-parameterization; and *Thermodynamics, Kinetics, and Microphysics of Clouds* by Khvorostyanov and Curry (2014), dedicated to theoretical analysis of specific microphysical processes. These books, however, do not discuss the issues of numerical cloud model design. In the absence of detailed information, many scientists (especially students) use cloud models as "black boxes." Inevitably, analysis of model results in this case is rough and often conjectural and descriptive.

Our book is aimed at filling in this gap. Major focus is put on explaining how cloud microphysical processes are treated in modern numerical cloud models. We stress the relationship between the physics of the cloud processes and the mathematical description of these processes in cloud models. Cloud hydrometeors of different types such as liquid drops, low density ice particles such as snowflakes, and high-density particles such as hail are active participants in condensation, collision, sedimentation, and other phase-transition processes. These processes unfold in different ways (and, accordingly, are differently represented in the models) depending on specific conditions and hydrometeors. Therefore, the two largest chapters in this book are dedicated to description and modeling of warm microphysical processes occurring at positive temperatures, and mixed-phase microphysical processes (in particular, ice processes) occurring at negative temperatures.

Considerable attention is paid to the two main approaches used for description of microphysical processes in models. The bin approach focuses on the evolution of size distributions of hydrometeors. The bulk parameterization is less time-consuming, while treating only changes of several moments of these distributions. In this book, we systematically describe how cloud microphysical processes are treated in modern bin and bulk cloud models.

Along with the classic studies, we analyze numerous modern research contributions made over the past two decades that have greatly improved understanding of the cloud microphysical processes. We present the state-of-the-art view of liquid precipitation and ice precipitation formation. This book also outlines the recent progress in modeling techniques, which enabled an increase in the number of the microphysical processes described by models and also improved the representation of these processes in models of both bin and bulk types. Comparison of model results obtained by both approaches and analysis of their importance for atmospheric and cloud physics is also a distinctive feature of this book.

To make the content accessible for a wide reading audience, we included extensive background material that is intended as the reference source on a vast variety of cloud-related phenomena. Large numbers of illustrations, figures, and tables are provided to facilitate better understanding of this material. We hope that the approach undertaken in this book, which combines the theory of cloud microphysical processes and their numerical realizations in models, will prove to be helpful for students, researchers, and cloud-model designers in the fields of cloud physics, atmospheric science, and meteorology.

Work on this book took several years. We express our sincere gratitude to Dr. Elena Negnevitsky, who took over the hard work of editing. She can rightfully be considered a coauthor of this book.

We thank our former and current students Igor Sednev, Yaron Segal, Nir Benmoshe, Leehi Magaritz-Ronen, Eyal Ilotoviz, Jacob Shpund, and others for their help with model development and for performing numerical experiments. We express our deep gratitude to colleagues from different countries: Profs. Daniel Rosenfeld, Hans Pruppacher, Vaughan Phillips, Isaac Ginis, Alexei Korolev, Ilia Mazin, Klaus Beheng, Alexander Ryzhkov; Drs. Axel Seifert, Thara Prabhakaran, Mikhail Ovchinnikov, Jiwen Fan, Barry Lynn and many others for their fruitful collaboration, exchange of ideas and work in the framework of joint grants. This book would have never appeared without this invaluable collaboration. We remember with gratitude our late colleague Dr. Andrei Pokrovsky, who made an enormous contribution to the development of the Hebrew University Cloud Model.

We express our gratitude to the U.S.-Israel Binational Science Foundation, the Israel Science Foundation, the U.S. Department of Energy, the Ministry of Science, Technology and Space of Israel, and to the Hebrew University of Jerusalem for support of our scientific work.

Our thanks go to Matt Lloyd and Zoë Pruce at Cambridge University Press, and to Shaheer Husanne, Project Manager at SPi Global, for their continuous support and advice.

And of course, we thank our wives, who for many years patiently suffered our long working hours and often had to witness quite heated debates on chapters, sections, sentences, and words of the text.

Abbreviations

2DC	A Two-Dimensional Cloud probe
2DP	A Two-Dimensional Precipitation probe
ACE	Aerosol Characterization Experiment
ACTOS	Airborne Cloud Turbulence Observation System
ADI	Alternating-Direction Implicit method
AF	Adiabatic Fraction
AP	Aerosol Particle
ARM	Atmospheric Radiation Measurement
ARW	Advanced Research WRF solver
ASASP	Active-Scattering Aerosol Spectrometer Probe
ASTEX	Atlantic Stratocumulus Transition Experiment
ATSR	Along Track Scanning Radiometer
BB	Bright Band
BBC	Baltex Bridge Campaign
BIMGQ	Bin Integral Method with Gauss Quadrature
BL	Boundary Layer
BM	Bin Microphysics
BRM	Berry and Reinhardt Method
CAIAPEEX	Cloud-Aerosol Interaction and Precipitation Enhancement Experiment
CAPE	Convective Available Potential Energy
CaPE	Convection and Precipitation/ Electrification experiment
CASP	Canadian Atlantic Storms Program
CCN	Cloud Condensation Nuclei
CCOPE	Cooperative Convective Precipitation Experiment
CFAD	Contoured Frequency Altitude Diagram
CFDC	Continuous Flow Diffusion Chamber
CFL	Courant–Friedrichs–Lewy condition
CIMMS	Cooperative Institute for Mesoscale Meteorological Studies model
CKE	Collision Kinetic Energy
CN	Condensational Nuclei
COPE	Convective Precipitation Experiment
COPE-MED	Convective Precipitation Experiment - Microphysical and Entrainment Dependencies
COSMO	Consortium for Small-Scale Modeling
CPI	Cloud Particle Imager
CRM	Cloud Resolving Model
CRYSTAL-FACE	NASA Cirrus Regional Study of Tropical Anvils and Cirrus Layers - Florida Area Cirrus Experiment
CWC	Cloud Water Content
DHARMA	Distributed Hydrodynamic Aerosol and Radiative Modeling Application
DHF	Deliquescent-Heterogeneous Freezing
DNS	Direct Numerical Simulation
DSD	Drop Size Distribution
DYCOMS	Dynamics and Chemistry of Marine Stratocumulus field study
EARLINET	European Aerosol Research Lidar Network
EC	Elemental Carbon
ECHAM	Atmospheric general circulation Model
EL	Equilibrium Level
EMPM	Explicit Mixing Parcel Model
ETEM	Entity-Type Entrainment Mixing process
EUCREX	European Cloud-Radiation Experiment
FATE	First ATSR Tropical Experiment
FD	Freeing Drops
FFT	Fast Fourier Transform
FIRE	First ISCCP Regional Experiment
FSSP	Forward Scattering Spectrometer Probe
GATE	Global Atlantic Tropical Experiment
GCCN	Giant Cloud Condensation Nuclei

GCE	Goddard Cumulus Ensemble model
GCM	General (Global) Circulation Model
GT	Glaciation Time
HDI	Hydrodynamic Drop Interaction
HMT	Hydrometeorology Testbed project
HUCM	Hebrew University Cloud Model
HVPS	High Volume Precipitation Spectrometer
ICE	International Cirrus Experiment
ICE-L	Ice in Clouds Experiment – Layer clouds
ICE-T	Ice in Clouds Experiment–Tropical
IN	Ice Nuclei
INCA	Interaction with Chemistry and Aerosols model
INSPECT-1	First Ice Nuclei Spectroscopy study
INSPECTRO	Influence of clouds on the Spectral actinic flux density in the lower troposphere field campaigns
IPCC	Intergovernmental Panel on Climate Change
ISCCP	International Satellite Cloud Climatology Project
ISDAC	Indirect and Semi-Direct Aerosol Campaign
ISF	Ice Sponge Front
ITCZ	Inter-tropical Convergence Zone
IWC	Ice Water Content
IWP	Ice Water Pass
JMA-NHM	Japan Meteorological Agency Non-Hydrostatic Model
LAM	Limited Area Model
LBA-SMOCC	Large-Scale Biosphere Atmosphere Experiment in Amazonia -Smoke Aerosols, Clouds, Rainfall, and Climate: Climate Campaign
LCL	Lifting Condensation Level
LCM	Lagrangian Cloud Model
LDM	Linear Discrete Method
LE	Large Eddies
LEM	Lagrangian-Eulerian Model
LES	Large-Eddy Simulation
LFC	Level of Free Convection
LFM	Linear Flux Method
LMDzT	Laboratoire de Meteorologie Dynamique general circulation model
LSM	Land-Surface Model
LST	Local Standard Time
LWC	Liquid Water Content
LWF	Liquid Water Fraction
LWP	Liquid Water Pass
MCS	Mesoscale Convective System
MD	Mass Distribution
MM5	Fifth-Generation Penn State/NCAR Mesoscale Model
MMM	Microphysical Method of Moments
M-PACE	Mixed-Phase Arctic Cloud Experiment
MPL	Maximum Parcel Level
NASA	National Aeronautics and Space Administration
NASA FIRE	NASA FIRE Arctic Cloud Experiment
ACE	Atmospheric Chemistry Experiment
NCAR	National Center for Atmospheric Research
NCEP	National Centers for Environmental Prediction
NOAA	National Oceanic and Atmospheric Administration
OAP	Optical Array Probe
PBAP	Primary Biological Aerosols Particles
PBE	Positive Buoyant Energy
PCASP	Passive Cavity Aerosol Spectrometer Probe
PDF	Probability Distribution Function
PMS	Particle Measuring System
PRE-STORM	Preliminary Regional Experiment for Storm-scale Operational and Research Meteorology
PSD	Particle Size Distribution
PVM	Particulate Volume Monitor
RACORO	Routine AAF CLOWD Optical Radiative Observations
RAMS	Regional Atmospheric Modeling System
RDF	Radial Distribution Function
RF	Research Flight
RH	Relative Humidity
RICO	Rain in Cumulus over the Ocean field program
RK2	Runge-Kutta 2 step method
RK3	Runge-Kutta 3 step method
RMS	Root Mean Square
RSD	Raindrop Size Distribution
RWC	Rain Water Content
SAM	System for Atmospheric Modeling
SBE	Stochastic Breakup Equation
SBM	Spectral Bin Microphysics
SCE	Stochastic Collection Equation
SCMS	Small Cumulus Microphysics Study
SLL	Schumann–Ludlam Limit

SMOCC	Smoke Aerosols, Clouds, Rainfall and Climate program
SST	Sea-Surface Temperature
STD	STandard Deviation
STEPS	Severe Thunderstorm Electrification and Precipitation Study
SUCCESS	Subsonic Aircraft: Contrail and Cloud Effect Special Study
TARA	Transportable Atmospheric Radar
TAU	Tel-Aviv University model
TC	Tropical Cyclone
TEM	Trajectory Ensemble Model
TKE	Turbulent Kinetic Energy
TLS	Tethered Lifting System
TOGA	Tropical Ocean Global Atmosphere program
TOGA COARE	TOGA Coupled Ocean-Atmosphere Response Experiment
TOM	Total Organic Matter
UHSAS	Ultra-High Sensitivity Aerosol Spectrometer
UT	Universal Time
UTC	Coordinated Universal Time
UWNMS	The University of Wisconsin Non-Hydrostatic Model
WBF	Wegener–Bergeron–Findeisen process
WDM6	WRF Double-Moment 6-class Microphysics scheme
WINSIC	Water Insoluble Inorganic Carbon
WINSOC	Water Insoluble Organic Compounds
WRF	Weather Research and Forecasting model
WSF	WRF Software Framework
WSOC	Water Soluble Organic Compounds

Symbols

Symbol	Description	Units
$\mathbf{A}$, A_i	Lagrangian acceleration of the air	ms^{-2}
B_{ll}	Longitudinal correlation function of turbulent velocity	m^2s^{-2}
B_{nn}	Lateral correlation function of turbulent velocity	m^2s^{-2}
c_p	Specific heat capacity at constant pressure	$J\ kg^{-1}\ K^{-1}$
c_v	Specific heat capacity at constant volume	$J\ kg^{-1}\ K^{-1}$
c_w	Specific heat of water	$J\ kg^{-1}\ K^{-1}$
C_0	Courant number	–
C_{0_cr}	Critical Courant number	–
C_d	Drag coefficient	–
C_{ph}	Phase speed	m/s
d, D	Diameter or width of particle	m
D	Diffusivity of water vapor in air	m^2s^{-1}
Da	Damköhler number	–
e	Water vapor pressure	Nm^{-2}
e_w	Saturation vapor pressure over water	Nm^{-2}
e_i	- over ice	Nm^{-2}
E	Turbulent kinetic energy per unit mass	m^2s^{-2}
E_l	- on subgrid scale	m^2s^{-2}
$E(r_1, r_2)$	Collision efficiency	–
E_{coal}	Coalescence efficiency	–
E_T	Total coalescence energy	J
f	Coriolis parameter	–
f_w	Liquid water fraction	–
$f(m)$	Size distribution of water drops as a function of mass	$kg^{-1}m^{-3}$
$f_r(r)$	- as a function of radius	m^{-4}
$f_i(m)$	Size distribution of ice particles or ice crystals	$kg^{-1}m^{-3}$
$f_g(m)$	- of graupel	$kg^{-1}m^{-3}$
$f_h(m)$	- of hail	$kg^{-1}m^{-3}$
$f_s(m)$	- of snow	$kg^{-1}m^{-3}$
F_v	Ventilation coefficient	–
$F(k)$	Spectrum of turbulent kinetic energy	m^3s^{-2}
Fl	Flatness of velocity gradient	–
g, $\mathbf{g}$	Acceleration of gravity	ms^{-2}
$g(m)$	Mass distribution of drops	m^{-3}
h	Planck constant	J s
h	Thickness or length of particle	m
h	static energy of dry air	m^2s^{-2}
h_l	- of moist air	m^2s^{-2}
k	Boltzmann constant	JK^{-1}
κ	Heat conductivity	$W\ K^{-1}m^{-1}$
k_a	- of air	$W\ K^{-1}m^{-1}$
k_i	- of ice	$W\ K^{-1}m^{-1}$
k_w	- of water	$W\ K^{-1}m^{-1}$
k	Wave number	m^{-1}
K	Turbulent coefficient	m^2s^{-1}
$K(r_1, r_2)$	Collision kernel	m^3s^{-1}
K_{g_col}	Collection kernel	m^3s^{-1}
K_{num}	Coefficient of numerical viscosity	m^2s^{-1}
Kn	Knudsen number	–
l	Spatial scale, mixing length	m
L	External turbulent scale	m
L_w	Latent heat of water evaporation	$J\ kg^{-1}$
L_i	- of ice sublimation	$J\ kg^{-1}$
L_m	- of ice melting/freezing	$J\ kg^{-1}$
m	Mass of cloud particle, or drop	kg
m_i	of ice particle	kg
m_N	- of aerosol particle	kg
m_{is}	Wettability parameter	–
M	Mass content	$kg\ m^{-3}$
M_w	Molecular weight of water	$kg\ mol^{-1}$
M_N	- of aerosol salt	$kg\ mol^{-1}$
M^k	The moment of particle size distribution of order k	
N	Concentration of cloud particles or drops	m^{-3}

(cont.)

Symbol	Description	Units
N_i	- of ice particles	m^{-3}
N_N	- of aerosols	m^{-3}
N_{CCN}	- of cloud condensational nuclei	m^{-3}
N_{IN}	- of ice nuclei	m^{-3}
N_c	- of crystals	m^{-3}
N_r	- of raindrops	m^{-3}
N_B	Brunt–Väisälä frequency	Hz
p	Pressure of moist air	Pa
p_a	of dry air	Pa
P_r	Probability of spontaneous breakup	–
Pr	Prandtl number	–
q	Mixing ratio of water vapor	kg/kg
q_c	- of cloud droplets	kg/kg
q_l	- of liquid water	kg/kg
q_{iw}	- of ice water	kg/kg
q_r	- of rain water	kg/kg
q_S	- saturated	kg/kg
q_t	- total	kg/kg
q_v	Specific humidity	kg/kg
q_i	Ice water content (IWC)	$kg\ m^{-3}$
q_N	Aerosol mass content	$kg\ m^{-3}$
q_w	Liquid water content (LWC)	$kg\ m^{-3}$
Q	Integral of supersaturation	s
r	radius of wet particle, or liquid drop	m
r_N	- of dry aerosol particle	m
r_{cr}	- critical	m
r_d	- of water drop	m
r_i	- of ice particle	m
$\bar{r}$	Mean radius	m
r_v	Mean volume radius	m
r_{eff}	Effective radius	m
r_{mod}	Modal radius	m
R	Universal gas constant	$J\ K^{-1}mol^{-1}$
R	Specific gas constant of moist air	$J\ K^{-1}kg^{-1}$
R_a	- of dry air	$J\ K^{-1}kg^{-1}$
R_v	- of water vapor	$J\ K^{-1}kg^{-1}$
R	Potential evaporation parameter	–
Ra	Rayleigh number	–
Ra_{cr}	Critical Rayleigh number	–
Ra_t	Turbulent Rayleigh number	–
Re	Reynolds number	–
Re_{cr}	Critical Reynolds number	–
Re_λ	Taylor microscale Reynolds number	–
Re_t	Turbulent Reynolds number	–
Ri	Richardson number	–
Ri_{cr}	Critical Richardson number	–
s_N	Solubility of aerosol water solution	–
S_w	Supersaturation over water	–
S_i	- over ice	–
S_{qs}	- quasi-steady	–
S_{ij}	Velocity gradient tensor	s^{-1}
Sk	Skewness of velocity gradient	–
St	Stokes number	–
Sc	Schmidt number	–
T	Absolute temperature of air	K
T_C	Temperature of air in Celsius scale	°C
T_d	Dew point	°C
T_L	Lagrangian time scale	s
T_v	Virtual temperature	K
T_0	Triple point (273.15)	K
t	Time	s
t_f	- of freezing	s
t_{gl}	- glaciation time	s
$\mathbf{u}$, u_i	Air velocity, turbulent air velocity	ms^{-1}
V	air volume	m^3
$\mathbf{V}$, V_i	Particle velocity	ms^{-1}
V'_i	Particle velocity relative to the air	ms^{-1}
V_{imp}	Impact velocity	ms^{-1}
V_g	Sedimentation velocity of particles (terminal velocity)	ms^{-1}
V_{swept}	Swept volume	m^3s^{-1}
w	Vertical air velocity	ms^{-1}
We	Weber number	–
$\mathbf{x}$, x_i	Spatial coordinates	m
X	Best number (Davis number)	–
Z	Radar reflectivity	dBZ
β_{ext}	Optical extinction	m^{-1}
$\Gamma(p,T)$	Psychrometric correction	–
γ_a	Dry adiabatic lapse rate	Km^{-1}
γ_{ma}	Moist adiabatic lapse rate	Km^{-1}
γ_d	Aspect ratio of drop	–
γ_i	Aspect ratio of ice crystals	–
γ_F	Fourier number	–

(cont.)

Symbol	Description	Units
δ_f	Dispersion of size distribution	–
δ_s	Soluble fraction of aerosol particle	–
δ_{ij}	Kronecker symbol	–
Δt	Time step	s
Δx, Δy, Δz	Grid space lengths	m
ε	Turbulent energy dissipation rate	m^2s^{-3}
ε_{ijk}	Levy–Civita tensor	–
Φ_s	Osmotic coefficient	–
λ	Taylor microscale	m^{-1}
λ	Entrainment parameter	m
μ	Dynamic viscosity of air	kg $m^{-1}s^{-1}$
μ	Chemical potential	–
μ	Cloud fraction	$Jmol^{-1}$
μ_k, l_k	Kolmogorov microscale	m
ν	Kinematic viscosity of air	m^2s^{-1}
ν_N	Number of dissociated ions per aerosol molecular	–
Π	Exner function	–
Π_0	Reference Exner function	–
π''	Deviation of Exner function	–
θ	Potential temperature	K
θ_e	Equivalent potential temperature	K
θ_{il}	Ice-water potential temperature	K
θ_l	Liquid potential temperature	K
ρ	Density of wet air	kg m^{-3}
ρ_a	- of dry air	kg m^{-3}
ρ_{bulk}	- bulk density of particle	kg m^{-3}
ρ_v	- of water vapor (absolute humidity)	kg m^{-3}
ρ_N	- of aerosol particle	kg m^{-3}
ρ_w	- of water	kg m^{-3}
ρ_i	- of ice	kg m^{-3}
σ_w	Surface tension of water-air interface	Nm^{-1}
σ_i	- of ice-air interface	Nm^{-1}
σ_{iw}	- of water-ice interface	Nm^{-1}
σ_f	RMS size distribution width	m
τ	Lifetime	s
τ_{mix}	Characteristic mixing time	s
τ_p, τ_d	Particle or drop relaxation time	s
τ_k	Kolmogorov time scale	s
τ_{pr}	Phase relaxation time	s

1 Clouds: Definitions and Significance

Clouds are visible clusters of small water drops and ice particles. Being a widespread atmospheric phenomenon, clouds are part of most processes occurring in the atmosphere. To a great extent, they govern solar radiation and energy fluxes reaching Earth, and have a strong effect on the hydrological cycle and freshwater distribution over the globe. Clouds are the energy source for different mesoscale phenomena, including dangerous ones such as tornadoes and hurricanes. All this determines the vital impact that clouds exert on all forms of life on the planet, human life included. Studies of clouds and cloud-related phenomena are an integral part of modern meteorology, providing data to be directly applied in weather forecasting, agriculture, environment control, air traffic control, and many other spheres of human activity.

1.1 The Importance of Clouds

The Sun is the major source of energy for Earth's oceans, atmosphere, land, and biosphere. The flux of about 342 J/s of shortwave solar radiation falls upon every square meter of Earth. The difference between the total amount of energy that Earth receives from the Sun and the total amount of energy that Earth reflects and emits back into space in the form of infrared radiation determines our planet's energy budget. Earth's climate system tends toward eventually reaching a *radiation balance* between the incoming solar energy and the outgoing thermal energy. If more solar energy comes in, Earth warms up and emits more heat into space to restore the radiation balance.

Figure 1.1.1 shows the composition of the annual mean global energy balance for the decade 2000–2010. The average top-of-atmosphere (TOA) imbalance is $0.6 = 340.2 - 239.7 - 99.9\ \mathrm{Wm}^{-2}$. This small imbalance is more than two orders of magnitude smaller than each of the individual balance components. Covering a substantial portion of Earth's surface, clouds significantly influence the TOA balance. The main impact factor is a so-called cloud *albedo* effect, which is a measure of cloud reflectivity and is defined as the ratio of the reflected radiation to the incident radiation. It varies within a vast range from less than 10% to more than 90% and depends on several factors, such as liquid water content and ice content, sizes of drops and ice particles, cloud thickness, the Sun's zenith angle, etc. The smaller the drops and the

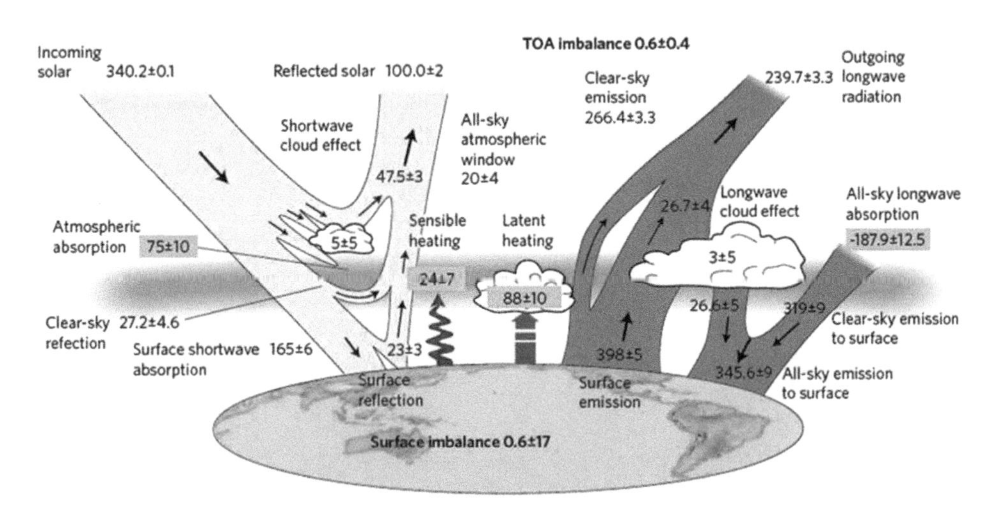

Figure 1.1.1 The annual mean global energy budget of Earth for the period 2000–2010. All the fluxes are in Wm^{-2}. The solar fluxes are marked yellow and the infrared fluxes are marked pink. The four flux quantities in the purple-shaded boxes represent the principal components of the atmospheric energy balance (from Graeme et al., 2012; reprinted with permission from Macmillan Publishers Ltd.).

larger the liquid water content, the higher the cloud albedo is, all other factors being equal. The overall reflectance of planet Earth is about 30%, meaning that about 30% of the incoming shortwave solar radiation is reflected back into space. If, hypothetically, all clouds were removed, the global albedo would decrease to about 15%. Figure 1.1.1 shows that the cloud albedo effect decreases the radiative influx by 47.5 ± 3 Wm^{-2}.

At the same time, in comparison to the clear-sky case, clouds reduce the outgoing long-wave radiation flux by approximately 26.4 ± 4 Wm^{-2} (Figure 1.1.1), creating a phenomenon known as the *greenhouse effect*. The global cloud albedo effect is significantly larger than the greenhouse effect. The net cloud-induced loss of radiation from Earth can be estimated as 21.1 ± 5 Wm^{-2}. This accounts for a net cooling effect that clouds have on Earth's climate, as illustrated in **Figure 1.1.2**. The scatter diagram of the global air temperature vs. the total cloud cover clearly demonstrates the inverse dependence between the temperature and the cloud cover with the correlation coefficient of about 0.5. A simple linear fit model suggests that an increase in the global cloud cover by 1% corresponds to a global temperature decrease of about 0.07°C. The diagram also shows that the cloud impact strongly depends on the cloud cover, which varies in the range of 63–70%.

Different types of clouds play different roles in the radiation balance. *Cirrus clouds* transmit most of the incoming shortwave radiation but trap some of the outgoing long-wave radiation. Their cloud greenhouse forcing is greater than their cloud albedo forcing, resulting in net warming of Earth. *Stratocumulus clouds* reflect much of the incoming shortwave radiation but also reemit large amounts of long-wave radiation. Their cloud albedo forcing is larger than their cloud greenhouse forcing, causing a net cooling of Earth. *Deep convective clouds* emit little long-wave radiation at the top but much of it at the bottom, and reflect much of the incoming shortwave radiation. Their cloud greenhouse and albedo forcing are both high, but are nearly in balance, causing neither warming nor cooling.

Since the atmosphere is nearly in energetic equilibrium, the net effect of different factors is substantially smaller than the value of any individual factor (or component) and smaller than errors in estimations of each individual factor. This fact indicates an almost precise radiation and energetic balance and complicates predictions of climate change, because changes result from small differences between large components of the balance. Variation of the cloud cover is one of the most important factors affecting the radiation balance. Even comparatively small changes in cloud cover may substantially affect the climate and affect climatic changes. Hence the great attention paid today to factors that can change the reflectivity properties of clouds and of the cloud cover (Rosenfeld et al., 2006).

Another important factor affecting the radiation balance is the aerosol effect on clouds. Rain and drizzle formation have a high impact, as well. Human activity can also affect the cloud cover and cloud radiation properties, thus creating anthropogenic climate changes (Ramanathan et al., 2001).

Clouds are an important component of the global *hydrological cycle* schematically shown in **Figure 1.1.3.**

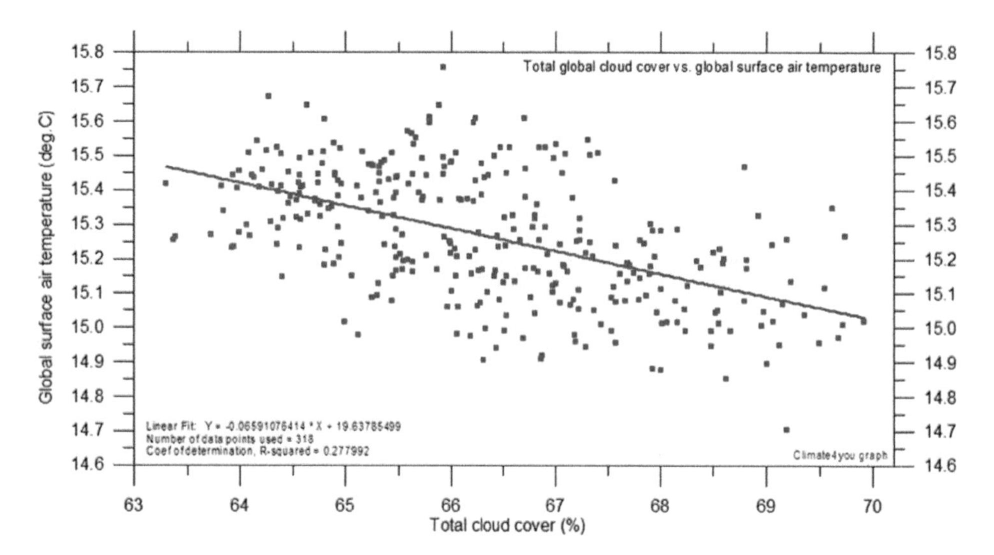

Figure 1.1.2 Scatter diagram showing the total monthly global cloud cover plotted versus the monthly global surface air temperature (diagram of Dr. Ole Humlum; published with permission of the author).

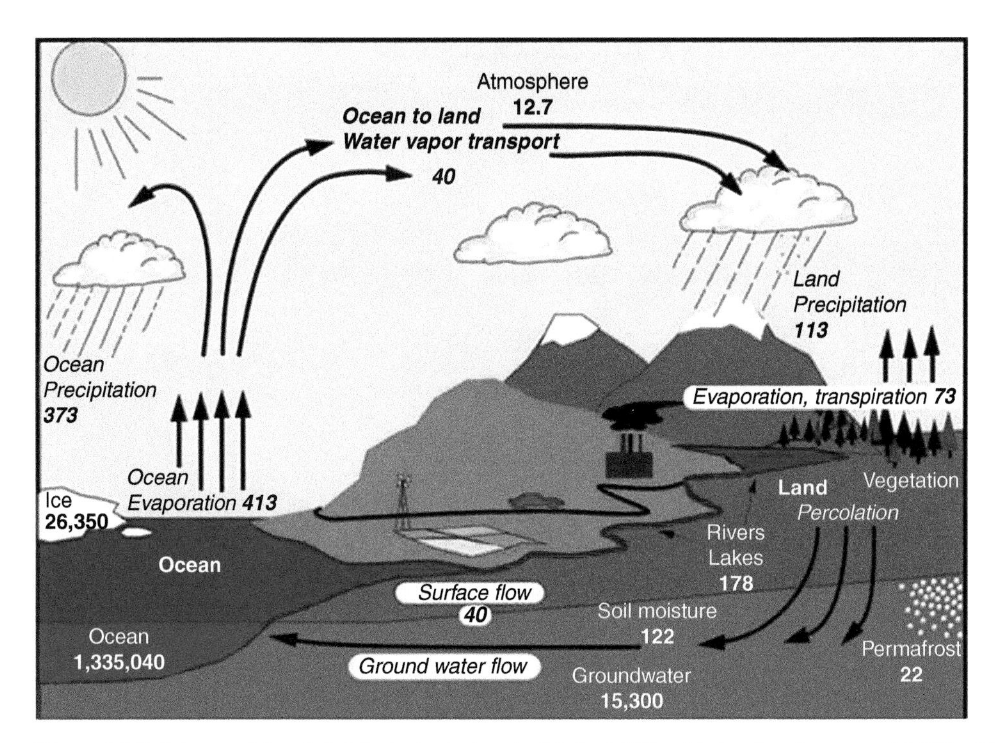

Figure 1.1.3 Scheme of the global hydrological cycle. Units: 1,000 cubic km for storage and 1,000 cubic km/year for exchanges (from Trenberth et al., 2007; courtesy of © American Meteorological Society. Used with permission).

Water vapor evaporated from Earth's surface – mostly from the ocean surface – is transported by atmospheric motions sometimes over hundreds or thousands of kilometers and then ascends. The temperature in the ascending air volumes decreases and the air becomes saturated or oversaturated. This process leads to formation of clouds containing droplets formed on small aerosol particles. Droplets grow, decreasing the amount of water vapor in the air. At freezing temperatures, ice particles form and grow, producing different types of ice hydrometeors: ice crystals, aggregates (snow), graupel, and hail. The growth of ice particles also decreases the air humidity. Thus, clouds are responsible for the amount of water vapor in the atmosphere. Collisions of cloud particles lead to formation of large drops and precipitation. The rate and type of precipitation (liquid vs. solid) depends on a cloud type and environment conditions (temperature and humidity). Water reaching the surface as precipitation may seep underground or get into rivers, where it travels large distances before evaporating. The combination of evaporation, precipitation, and transport of liquid water into (or beneath) soil and along rivers and ocean streams creates the complicated global hydrological cycle.

Since water vapor and clouds play a crucial role in the atmospheric energy balance, there is a close relationship between the climate and the hydrological cycle. The hydrological cycle is also essential in shaping Earth's environment, the availability of water being a critical factor for life as well as for many chemical reactions and transformations affecting the physical environment. Describing various components of the hydrological cycle and analyzing the mechanisms responsible for the exchange of water between different reservoirs are thus important elements of climatology.

The role of clouds is not limited to their effect on radiation and the hydrological cycle. The latent heat released during condensation of water vapor is a dominant energy source of atmospheric motions of different scales. The release of the latent heat in clouds is the main source of kinetic energy of global circulation of the atmosphere. The kinetic energy of clouds and related phenomena, such as thunderstorms, tropical cyclones, etc., is largely determined by the latent heat release in clouds. Clouds are an important energetic source of turbulent motions. The latent heat in clouds is released during condensational growth of liquid droplets and ice particles as well as during freezing of liquid droplets and melting of ice. These processes take place at microscales, the sizes of cloud particles ranging from about 1 μm to 1,000 μm, and in rare cases up to 1 cm.

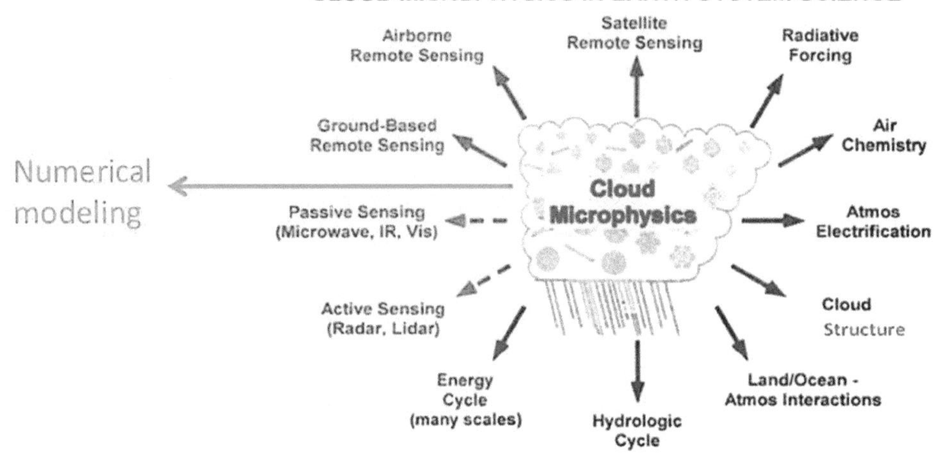

Figure 1.1.4 Links between cloud microphysics and other fields of atmospheric and earth sciences and methods of cloud investigation (from Tao and Moncrieff, 2009 with changes; courtesy of © Willey and Sons Ltd.).

Thus, atmospheric motions within a wide range from centimeters to global scales are affected by microscale processes taking place in clouds.

Cloud effects depend on cloud type, height, cloud age, cloud thickness, etc. *Cloud physics* represents a special branch of *earth science*. *Cloud microphysics* studies formation, growth, transition, conversion, and sedimentation of cloud particles (drops, ice particles, and aerosol particles). The cloud microphysical processes determine the microstructure of clouds that include the spatial-temporal field of masses and concentrations of cloud particles (hydrometeors), as well as size distributions of cloud particles belonging to different hydrometeor types. Cloud microphysics is related to many areas of earth science: atmospheric radiation (radiation forcing), hydrology, atmospheric electrification, aerosol science, and atmospheric chemistry. These links are illustrated in **Figure 1.1.4**.

1.2 Clouds and Cloud-Related Phenomena

1.2.1 Typology of Clouds

Clouds can be classified according to their geometrical (morphological) structure and (roughly) assigned to two large classes. Clouds belonging to the first class have horizontal scales far exceeding their thickness. These clouds are stratiform-like clouds. The word "stratus" comes from the Latin prefix *strato-*, meaning "layer." Stratiform-like clouds are further separated into subclasses according to the altitudes of their cloud base as low-, middle-, or high-level clouds. Each subclass is, in turn, separated into different types according to cloud forms and structures determined by their dynamical and microphysical properties. The second class includes clouds of vertical development (convective clouds) whose thickness is of the same order or even larger than their horizontal size and whose base is typically located in the boundary layer (BL). Convective clouds are driven by atmospheric instability and buoyancy forces. The class of convective clouds is separated into subclasses according to altitudes of their tops. The main subclasses of clouds are shown in **Figure 1.2.1**. Next, we briefly describe the main specific phenomenological properties of cloud types shown in Figure 1.2.1. A more detailed description and classification can be found on Cloud Atlas (www.clouds-online.com/)

Stratiform-Like Clouds

Subclasses of low-level stratiform-like clouds include four cloud types: stratus (St), stratus fractus (St fr.), stratocumulus clouds (Sc), and nimbostratus (Ns). Horizontal sizes of *St* exceed the cloud depth by several orders of magnitude. They are homogeneous in the horizontal direction and often cover the entire sky. The height of cloud base of *St* typically ranges from 0.1 to 0.7 km, and the thickness of the cloud layer ranges from 0.2 to 0.8 km. They sometimes produce drizzle, freezing drizzle, or snow (aggregates), depending on temperature. *St* form due to turbulent mixing of air in the BL, or in a manner similar to that of fog formation when the ambient air temperature decreases, thus increasing the relative humidity. Once the temperature drops below the dew point, a stratus cloud forms. *St* are essentially aboveground fog formed either through

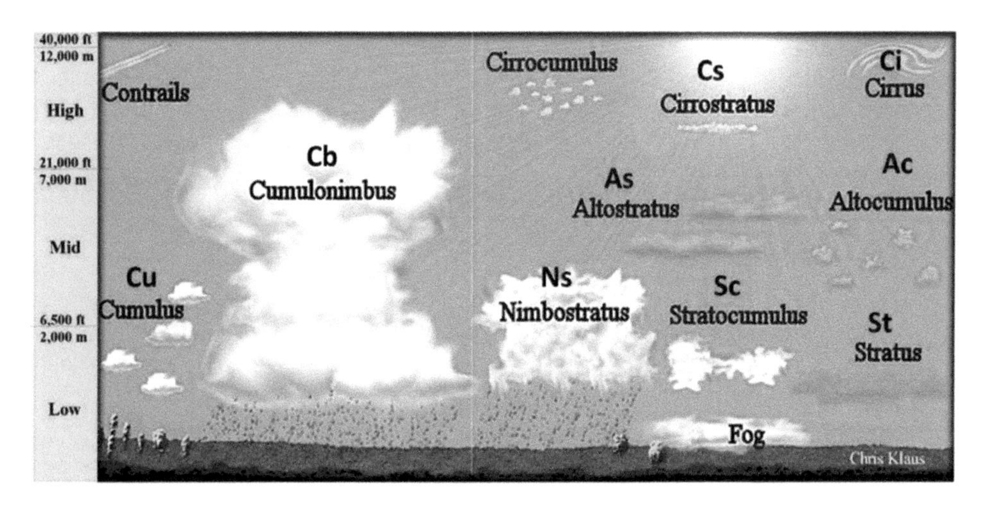

Figure 1.2.1 Classification of clouds. Clouds of vertical development (convective) are depicted on the left side of the panel; stratiform-like clouds are depicted on the right side of the panel (*The Scheme*, by Christopher Klaus, published with changes; with permission of the author).

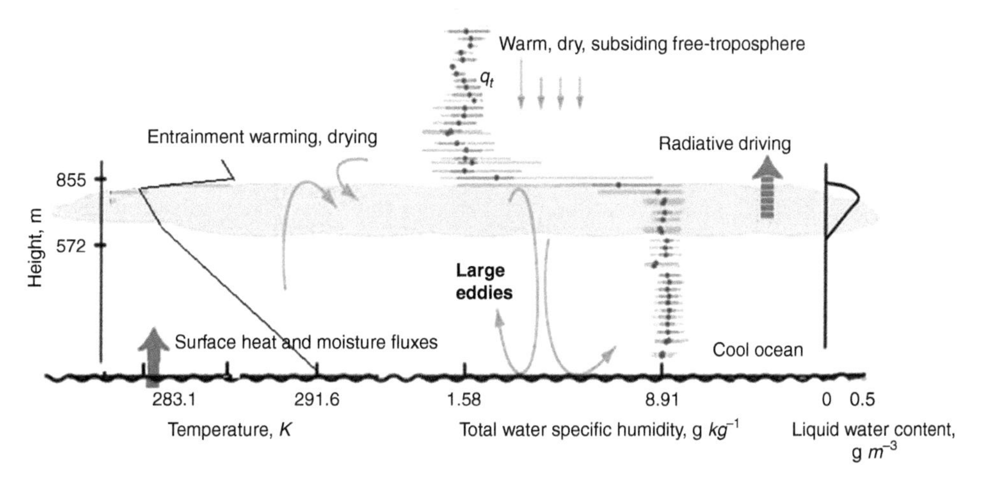

Figure 1.2.2 Conceptual scheme of a well-mixed boundary layer covered by a non-precipitating maritime *Sc* over the cool, eastern subtropical oceans (research flight near 30°N, 120°W) (with changes) (courtesy of Bjorn Stevens).

lifting of morning fog or through cold air moving at low altitudes over a region. Sometimes *St* are called "high fog." They can be composed of water droplets, supercooled water droplets, or ice crystals, depending upon the ambient temperature. *St fr.* clouds look like separate shreds with broken edges.

***Stratocumulus clouds* (*Sc*)** is a basic cloud type of low-level stratiform-like clouds. *Sc* contain elements of *Cu* that are embedded into zones of stratiform clouds. Cloud tops of *Sc* is typically less than 2.5 km. The vertical velocities are usually caused by convective cells in the BL and reach 1–2 m/s. The maximum vertical velocities often take place near the cloud base.

A conceptual scheme of a well-mixed boundary layer covered by a non-precipitating maritime *Sc* is shown in **Figure 1.2.2**.

Liquid Sc clouds develop within a neutral or slightly unstable BL, typically over relatively cool ocean surfaces. The BL is well mixed and limited from above by the stable inversion layer in which the temperature increases with height. Figure 1.2.2 shows the vertical profiles of temperature and total specific humidity q_t, i.e., the total amount of water including water vapor and liquid per kg of air.

Mixing in the vertical direction leads to linearly decreases in the temperature from the surface upward

Figure 1.2.3 Typical stratocumulus cloud (credit: dszc/Getty Images).

Figure 1.2.4 A raining nimbostratus cloud (from https://pixabay.com/en/mongolia-steppe-rain-clouds-487112/).

with a gradient of 9.8 K km^{-1} and to nearly constant values of q_t within the mixing layer of the BL below the inversion level. Sharp increase in T and decrease in q_t take place within the inversion layer. The inversion is typically supported by large-scale subsidence of dry and warm air. The depth of *Sc* is a few hundred meters. Humidity and temperature in the BL are determined to a large extent by surface fluxes. Radiative fluxes from the upper boundary of Sc lead to cooling of the upper boundary and favor the vertical mixing within the BL. Mixing in the BL is due to large eddies arising within it. In non-precipitating *Sc*, cloud droplets are smaller than 20 μm in radius. When the cloud depth is larger and exceeds about 0.3 km, and the air humidity in the BL is high, liquid water content exceeds the critical value, and drizzle (drops of about 100 μm in radius) forms and falls down. Formation of drizzle dramatically changes the cloud structure and cover and, as a result, the radiative characteristics of *Sc*. A typical *Sc* in the BL is shown in **Figure 1.2.3**.

Nimbostratus **(*Ns*)** is a low-to-middle troposphere stratiform cloud that has considerable vertical extent (2–5 km) and horizontal extent (tens to hundred kms) and produces precipitation over a large area. *Nimbo* comes from the Latin word *nimbus*, meaning "precipitation." *Ns* is a major source of precipitation. They usually have a darker color than stratus clouds due to their high moisture content. A typical *Ns* is shown in **Figure 1.2.4.** They form along a warm front or an occluded front within zones of weak updrafts. Often, *Ns* form from an altostratus cloud when it thickens and descends into lower altitudes.

Altostratus **(*As*) *and Altocumulus*** **(*Ac*)** are stratiform-like clouds of middle level. *Alto* comes from the Latin word *altus*, meaning "high." *As* look like a sheet or a layer and are generally uniform gray to bluish-green in color (**Figure 1.2.5**). *As* are lighter in color than *Ns*, but darker than cirrostratus. *As* are transparent and the sun can be seen through thin ones. They can produce light precipitation that evaporates, not reaching the surface. If the precipitation increases in persistence and intensity, the *As* may thicken into *Ns*, as previously mentioned. They most often take the form of a featureless sheet; however, they can also be fragmented. The appearance of *As* is often a sign of an upper-level warm front approaching. They may be composed of ice crystals whose sizes increase as the altitude decreases. Near cloud top, the ice crystals are largely hexagonal plates, becoming more conglomerated as the cloud descends.

Altocumulus **(*Ac*)** may appear as parallel bands (**Figure 1.2.6**) or rounded masses. Typically, a portion of an *Ac* is shaded, which distinguishes them from the high-level cirrocumulus. They usually form by convection in unstable layers at the upper levels. The presence of *Ac* during a humid summer morning is a sign of thunderstorms that may occur later in the day.

Cirrostratus **(*Cs*) *and Cirrus*** **(*Ci*)** are stratiform-like clouds of the upper level. The word *cirrus* is the Latin word for "curl". These clouds arise due to slow large-scale updrafts at the updraft velocity of the order of 1 to few cm/s in zones of atmospheric fronts. *Cs* are ice crystals stratiform clouds of high location that can appear as a striated sheet in the sky. They are transparent, so the sun or the moon are always clearly visible

Figure 1.2.5 Altostratus (credit: Wallace Garrison/Getty Images).

Figure 1.2.6 Typical *Ac* clouds (from https://pixabay.com/en/mongolia-steppe-rain-clouds-487112/).

Figure 1.2.7 A Cs (from https://pixabay.com/en/cirrostratus-skyscape-sky-cloud-246294/).

through them. A photo of a Cs cloud is shown in **Figure 1.2.7.** *Ci* consist of ice crystals of different types, the fraction of a particular type depending on the temperature. The dominating crystal types in such clouds are thick hexagonal plates and short, solid hexagonal columns.

Clouds of Vertical Development (Convective Clouds)
A specific feature of convective clouds is that they form as a result of a thermal instability in the atmosphere. Their updrafts are caused largely by the buoyancy force, which explains the term "convective clouds." The friction force in the BL leads to formation of zones of convergence and divergence. The vertical velocities arising in the convergence zone foster formation of convective clouds. The subclasses of convective clouds include cumulus (*Cu*), cumulus congestus (*TCu*) and cumulonimbus (*Cb*). The word *cumulus* comes from the Latin word meaning "heap" or "pile." Vertical sizes of *Cu* and *Cb* are of the same order as their horizontal sizes.

***Cumulus clouds* (*Cu*)** are often described as "puffy," "cotton like," or "fluffy"; their bases look flat. They are low-level clouds with altitudes generally less than 1,000 m, characterized by an unstable BL capped from above by a stable inversion layer. Sometimes single *Cu* may form, but usually they are organized into 2D rolls or 3D convective cells. The vertical velocities in *Cu*, determined by the magnitude of the BL instability, are of the order of few m/s. Typical *Cu* are shown in **Figure 1.2.8.** They contain only liquid droplets, i.e., they are liquid clouds. The microphysical processes in *Cu* are called warm microphysical processes.

Figure 1.2.8 Fair-weather cumulus clouds (from https://pixabay.com/en/cloud-blue-white-cloudscape-sky-1044223/).

Figure 1.2.9 TCu, developing into a cumulonimbus near Key Biscayne, Florida. The clear-contrast cloud boundaries indicate the liquid phase. Smoothed cloud boundaries indicate ice particles (from Houze's Cloud Atlas, http://www.atmos.washington.edu/Atlas/oro.html; courtesy of R. Houze).

***Cumulus congestus* (*TCu*).** If the atmosphere is unstable within a layer of several kilometers deep, *Cu* can further develop in the vertical direction and become *TCu*. More rarely, Sc can transform into *TCu*. The vertical velocity in *TCu* is typically of several m/s. High vertical velocities are caused by the buoyancy force, while the cloud top height is about 6 km. They contain both liquid and ice particles, thus being mixed-phase clouds (**Figure 1.2.9**). *TCu* are capable of producing severe turbulence and showers of moderate-to-heavy intensity.

***Cumulonimbus clouds* (*Cb*)** are thunderstorm clouds that form if *TCu* clouds continue to grow vertically, sometimes up to the tropopause level (12–16 km). The vertical velocity in *Cb* sometimes reaches 40–50 m/s. A single-cell *Cb* at its mature stage is shown in **Figure 1.2.10**. The *Cb* has a well-defined anvil consisting of ice crystals. The existence of the cloud anvil indicates the presence of a stable layer above the cloud, so the anvil appears in the zone of air detrainment within the cloud. *Cb* are mixed-phase clouds usually producing heavy rains, thunderstorms, hailstorms, and often lightning. The precipitation rates can reach several cm/hour and even tens of cm per hour. Hailstones may reach 5 cm in diameter.

Figure 1.2.10 Single-cell cumulonimbus (from https://pixabay.com/en/cumulus-nimbus-cloud-white-large-491106/).

Crude schemes of the microphysical structures of developing and decaying convective clouds are shown in **Figure 1.2.11.** At the developing stage, aerosols known as *cloud condensational nuclei* ascend in updrafts. The vertical motions can be caused by gravitational waves, orography, temperature inhomogeneity of the underlying surface and other factors. Clouds form due to air cooling in updrafts when the relative humidity exceeds 100%. As the relative humidity exceeds a certain critical value, part of atmospheric aerosols turns into cloud droplets (droplet nucleation). The droplets ascend, growing by condensation of water vapor. The latent heat release increases the vertical velocity. When droplets reach about 20 μm in radius, intense collisions start leading to formation of raindrops, and small raindrops ascend and freeze. In order to freeze, a raindrop should either have an immersion aerosol inside or collide with ice particles. Frozen raindrops give rise to hail particles. At cold temperatures, ice crystals are activated on insoluble aerosol particles. Other mechanisms of ice crystal nucleation also exist, e.g., ice crystals can grow by deposition of water vapor. Collisions between ice crystals lead to formation of aggregates called snow or snowflakes. When the

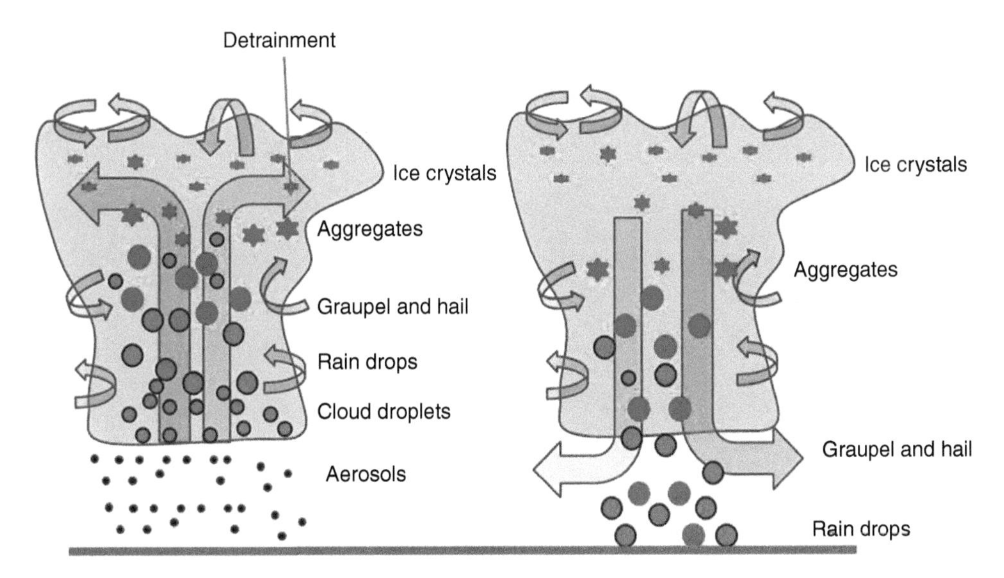

Figure 1.2.11 Schemes of the microphysical structure of a developing (left) and a decaying (right) mixed-phase convective cloud.

Figure 1.2.12 A photo of orographic clouds (from http://213.239.206.108/Bilder/HTML/dig1410_01.html; courtesy of photographer, Dr. Bernhard Mühr).

aggregates accrete a significant amount of cloud droplets, their density increases and they become graupel. Graupel and hail are the main high-density precipitating hydrometeors in *Cb*.

At the decaying stage, the cloud updrafts decrease or may even be replaced by downdrafts. Raindrops, graupel, and hail fall down. Depending on the environmental conditions, some (or all) graupel and hail melt. The largest particles that do not fully melt reach the surface.

There are clouds that typically develop in zones of air updrafts caused by hills and mountains. As the air mass ascends, temperature decreases and the relative humidity rises to 100%, creating clouds and frequent precipitation. Such clouds are called *orographic clouds* (*OC*) (**Figure 1.2.12**). Being linked to the topography,

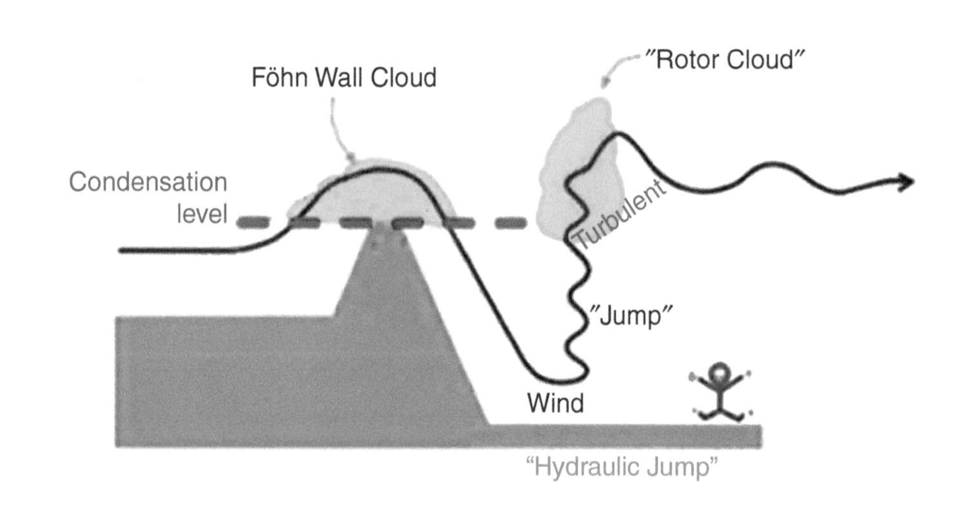

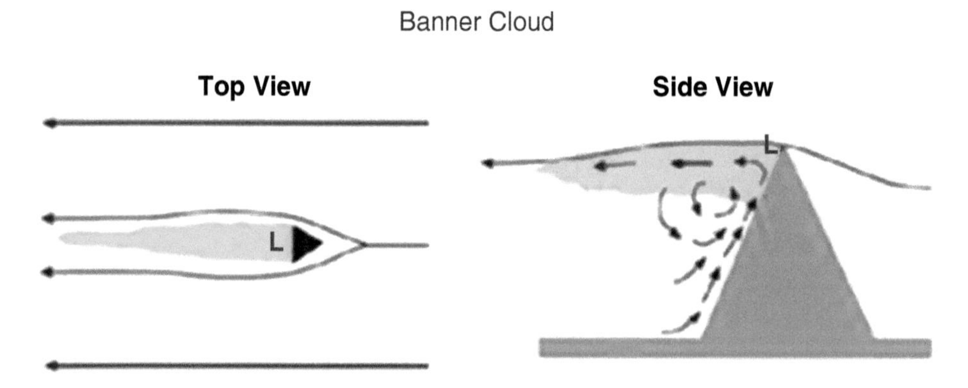

Figure 1.2.13 Structure of wind flows and related cloudiness arising over mountains (from Houze's Cloud Atlas, www.atmos.washington.edu/Atlas/oro.html; courtesy of R. Houze).

OC are usually standing clouds, even if the winds at the same height are very strong. As a result, mountains form local precipitation regimes. Typically, more precipitation falls over the upwind slope of a mountain region while precipitation amount decreases on the downwind side since precipitation on the windward side removes water from the air. Depending on stability conditions, *OC* can be both stratiform like

Table 1.2.1 Characteristics of the main cloud types in extratropical zone (from Mazin and Shmeter, 1983*; courtesy of I. Mazin).

Cloud Type	Cloud Base Height (km)	Cloud Depth (km)	Horizontal Size (km)	Phase State	Life Time	Characteristic Mean Vertical Velocities (cm/s)	Precipitation
St	0.1–0.7	0.1–1.0	10–1,000	Liquid	Day and more	1–10	No or drizzle
Sc	0.4–2.0	0.1–1.0	10–1,000	Liquid or mixed-phase	Day and more	1–10	No or drizzle
Ac	2–6	0.1–0.8	10–100	Liquid or mixed-phase	Day and more	1–10	No
Cc	6–9	0.2–1.0	10–100	Ice or mixed-phase	Day and more	1–10	No
Ns	0.1–2.0	Up to several km	100–1,000	Mixed-phase	Day and more	1–10	Rain, snow
As	3–6	Up to several km	100–1,000	Ice or mixed-phase	Day and more	1–10	Rain, snow
Cs	5–9	Up to several km	100–1,000	Ice	Day and more	1–10	No
Ci	6–10	From several tens to 1.5 km	10–1,000	Ice	Day and more	1–10	No
Cu	0.8–2.0	<3	1–5	Liquid	Tens of min.	100	No or drizzle
TCu	0.8–2.0	3–5	5–10	Liquid or mixed-phase	Tens of min.	Up to 5,000	No or not strong showers
Cb	0.4–1.5	<14	Up to 50–100	Mixed-phase	From tens of min. to several hours	Up to 5,000	Showers, hailshaft

* Values of the vertical velocity for *Cu* and *Cb* are averaged updraft velocities. Velocities of downdrafts are typically 1.5–2 times less. The data concerning the thickness of *Cu* and *Cb* are related to summertime clouds at middle latitudes; in winter, these clouds have lower thickness.

and convective, and therefore can be classified as shown in Figure 1.2.1.

OC can produce both liquid and ice precipitation. The existence of mountains leads to formation of *wave clouds*. The standing waves form in stable atmosphere over a mountain and can either form above or in the lee of the mountain. Clouds develop in updrafts and evaporate in downdrafts (**Figure 1.2.13**). Depending on the wind speed, stability of the atmosphere, and mountain height, different cloud structures can appear over mountains. In case of a moderate wind wave, the lee wave develops, as shown in Figure 1.2.13a. Under strong background wind, a strong air flow arises over the downwind slope, which may end as a "hydraulic jump" accompanied by formation of strong updraft and convective clouds (Figure 1.2.13b). In case of highly stable atmosphere, the air cannot overcome the ridge. Clouds rise over the upwind side of the mountain (Figure 1.2.13c). The air masses flow around the mountain sides.

Table 1.2.1 summarizes the characteristics of the main classes of nontropical clouds. Orographic clouds are not included as a separate cloud type, as they can be both stratiform-like and convective.

On meteorological maps, types of clouds are denoted using certain abbreviations that, alongside brief descriptions, are presented in **Table 1.2.2**.

1.2.2 Cloud Systems and Cloud-Related Phenomena

Clouds evolve during their lifetime and can transform from one class into another. Clouds often form systems known as mesoscale cloud systems (MCS). The formation of cloud systems of certain structures is often referred to as *self-organization of clouds*. Cloud systems form clusters with characteristic spatial scales of hundreds of kilometers and a characteristic lifetime of tens of hours. Many MCS have their own specific dynamics and structure. Next, several examples of cloud self-organization are presented.

Table 1.2.2 Symbols used for notation of some types of atmospheric clouds on weather maps.

Cloud Abbreviation	C_L	Description (Abridged from W.M.O. Code)	C_M	Description (Abridged from W.M.O. Code)	C_H	Description (Abridged from W.M.O. Code)
St or Fs – Stratus or Fractostratus	**1**	Cu far weather, little vertical development & flattened	**1**	Thin As (most of cloud layer semitransparent)	**1**	Filaments of Ci. or "mares tails," scattered and not increasing
Ci – Cirrus	**2**	Cu considerable development, towering with or without other Cu or SC bases at same level	**2**	Thick As greater part sufficiently dense to hide sun (or moon), or Ns	**2**	Dense Ci in patches or twisted sheaves, usually not increasing, sometimes like remains of Cb; or towers tufts
Cs – Cirrostratus	**3**	Cb with tops lacking clear-cut outlines, but distinctly not cirriform or anvil shaped; with or without Cu, Sc, St	**3**	Thin Ac, mostly semi-transparent, cloud elements not changing much at a single level	**3**	Dense Ci, often anvil-shaped derived from associated Cb
Ac – Altocumulus	**4**	Sc formed by spreading out of Cu; Cu often present also	**4**	Thin Ac in patches; cloud elements continually changing and/or occurring at more than one level	**4**	Ci, often hook-shaped gradually spreading over the sky and usually thickening as a whole
As – Altostratus	**5**	Sc not formed by spreading out of Cu	**5**	Thin Ac in bands or in a layer gradually spreading over sky and usually thickening as a whole	**5**	Ci and Cs, often in converging bands or Cs alone; generally overspreading and growing denser, the continuous layer not reaching 45 altitude
Sc – Stratocumulus	**6**	St or Fs or both, but no Fs of bad weather	**6**	Ac formed by the spreading out of Cu	**6**	Ci and Cs, often in converging bands or Cs alone; generally overspreading and growing denser; the continuous layer exceeding 45 altitude
Ns – Nimbostraus	**7**	Fs and/or Fc of bad weather (scud)	**7**	Double-layered Ac, or a thick layer of Ac, not increasing; or Ac with As and/or Ns	**7**	Veil of Cs covering the entire sky
Cu or Fc – Cumulus or Fractocumulus	**8**	Cu and Sc (not formed by spreading out of Cu) with bases at different levels	**8**	Ac in the form of Cu-shaped tufts or Ac with turrets	**8**	Cs not increasing and not covering entire sky
Cb – Cumulonimbus	**9**	Cb having a clearly fibrous (cirriform) top, often anvil-shaped, with or without Cu, Sc, ST or scud	**9**	Ac of a chaotic sky, usually at different levels; patches of dense Ci are usually present	**9**	Cc alone or Cc with some Ci or Cs but the Cc being the main cirriform cloud

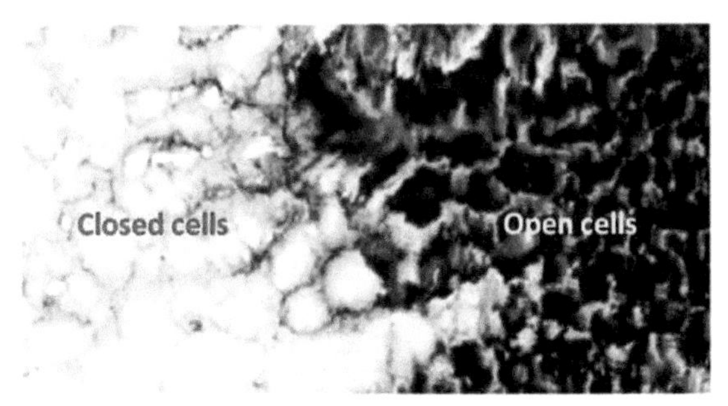

Figure 1.2.14 Convective cells. Left: cloud streets over the southern Appalachian Mountains (Seth Adams and Jim Foster, courtesy of authors). The streets are formed by small *Cu*. Right: 3D convective cells (from Rosenfeld et al., 2006).

Convective cells. Thermal instability in the BL leads to formation of convective cells consisting of zones of updrafts and downdrafts. Clouds develop in the updrafts and form different geometrical patterns depending on the shapes of convective cells. The cells can be either 2D rolls or 3D convective cells (**Figure 1.2.14**). Convective cells are referred to as closed cells if updrafts are in the center of the cells, or as open cells if there are downdrafts in the cell centers (Figure 1.2.14, right). In closed cells, the area of updrafts is larger than that of downdrafts. Since the horizontally averaged vertical velocities in the BL are close to zero, updrafts are weak, clouds in these cells are of *Sc* type, and the cloud coverage is large.

In open cells, updrafts are located along the cell periphery and cover an area lower than that covered by downdrafts. Therefore, the vertical updraft velocities are stronger than the downdraft velocities and clouds forming at the cell periphery are of *Cu* type. Open cells are mostly observed in cold-air outbreaks, when cold and dry air from continents moves over a relatively warm water surface and the BL is thermally unstable. Closed cells typically arise over cold ocean surfaces. Transition from closed cells to open ones can be a result of changes in the surface temperature, but can also be caused by microphysical processes such as drizzle formation, when droplet concentration decreases. Open cells can convert to closed ones in case of increasing aerosol concentration when droplet concentration also increases while droplet size decreases. The area covered by these droplets enlarges since they do not collide and sediment only slightly.

The updrafts in 2D convective rolls lead to formation of cloud streets (Figure1.2.14, left). The convective rolls are typically elongated along the wind direction. The presence of convective cells intensifies the vertical circulation in the BL and increases surface fluxes and wind near the surface. In the presence of the rolls, air volumes experience spiral motions in the BL, ascending within cloud streets and descending between them.

Squall lines. A common structure of MCS over both oceans and land is squall lines. As shown in Figure 1.2.11, convective clouds at the decaying stage create downdrafts that are cold due to melting ice and evaporation of rain drops. Cold air descends, creating so-called cool pools within the surface layer. The convergence of the spreading cold air and warm air in the BL leads to formation of updrafts and of new clouds. In some cases, including squall lines, the convergence caused by the downdrafts supports the parent clouds at the convective front. Convergence at low levels prevents cloud dissipation, increasing the lifetime of the convective cloud and the entire cloud system. At low levels, the convergence zone can move at velocity that differs from the wind velocity. Thus, squall lines may propagate at velocity typically different from that of the environmental wind. An example of a squall line is presented in **Figure 1.2.15.**

Schematic diagram of geometrical and microphysical structure of a squall line is shown in **Figure 1.2.16.** The scheme of a squall line shown in the diagram describes the following process. A deep convective cloud (convective region) creates strong precipitation (convective showers). The squall line is located within the vertical wind shear. As a result of air detrainment from cloud anvil (together with ice crystals and aggregates), a stratiform area forms behind the convective region (the left part of the diagram). Ice particles grow by deposition, and their collisions lead to formation of aggregates. Sedimentation of aggregates their melting and consequent drop in evaporation lead to a dramatic decrease of temperature in the stratiform zone within the BL. The divergence of the cold air in the BL supports the convective region that may shift with time at velocity

Figure 1.2.15 A squall line in Western NSW in Australia, November 2015 (from http://www.rawartists.org/djmimages; courtesy of David Metcalf).

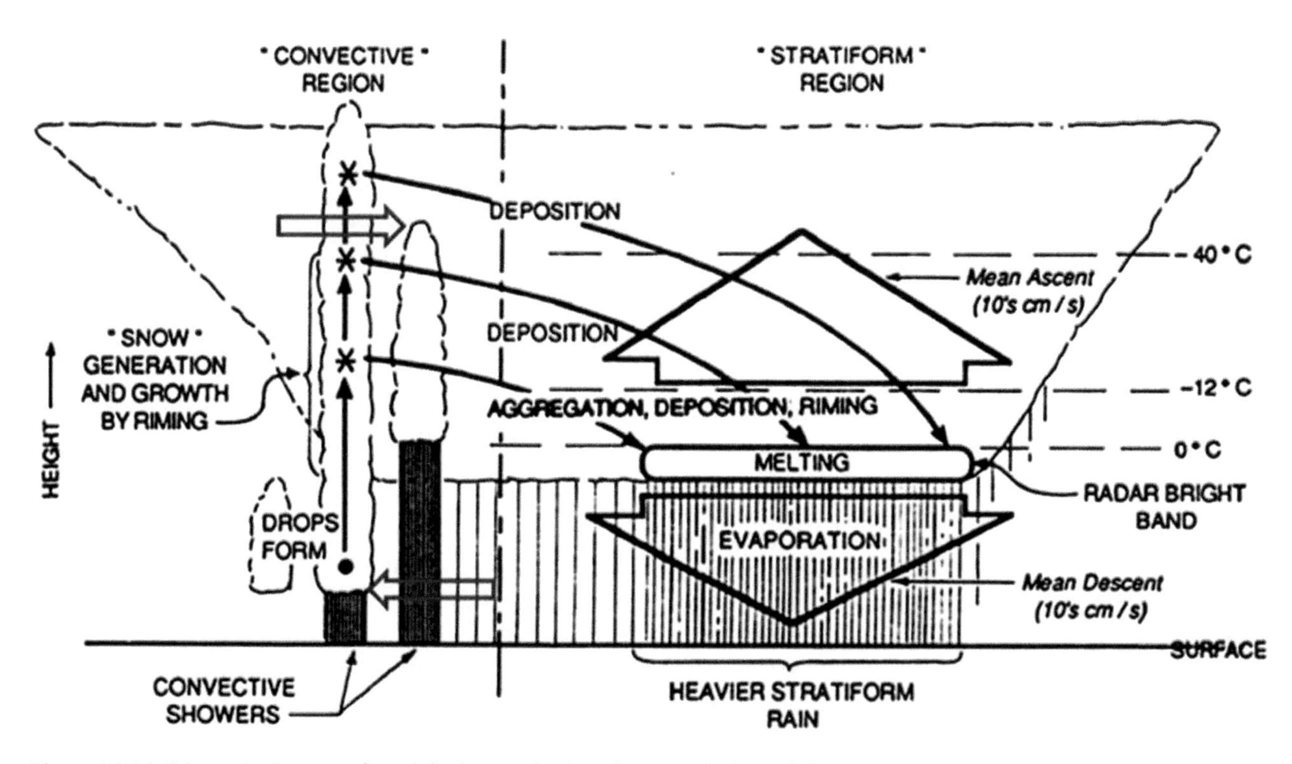

Figure 1.2.16 Schematic diagram of precipitation mechanisms for a tropical squall line. Straight solid arrows indicate convective updrafts; wide-open arrows indicate mesoscale ascent and subsidence in the stratiform region where vapor deposition and evaporation occur. Curved solid arrows indicate particle trajectories. Arrows show the wind direction relative to the moving squall line (adapted from Houze, 1989; courtesy of © Willey and Sons Ltd.).

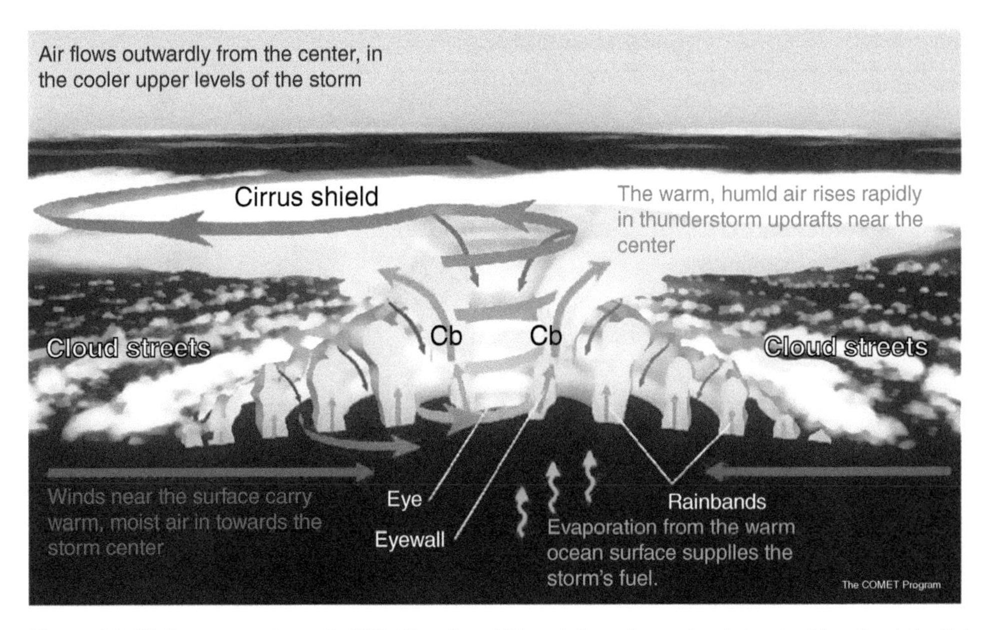

Figure 1.2.17 A cross-section of a TC. Clouds at TC periphery form cloud streets. The cloud depth increases toward TC center. Convective clouds form rainbands resembling a squall line propagating toward TC center. At a certain distance from TC center, the air ascends in deep *Cb*, forming the TC eyewall (courtesy of H. Kane and I. Ginis).

different from that of the background wind. Thus, squall lines are the result of interaction between convective and stratiform thermodynamics and the background flow dynamics.

Tropical cyclones. An important example of a cloud-related mesoscale phenomenon is tropical cyclones (TC). Intense TC are called hurricanes or typhoons. The initial mesoscale vortex in the BL leads to formation of a radial wind component directed at the minimum pressure area in the center of the vortex. The convergence of the moist air leads to formation and invigoration of convective clouds in the central zone of the vortex. At TC periphery, clouds arise in the form of cloud streets. Center rainbands form toward TC. Air influx in the BL leads to formation of deep convective clouds *Cb* in the TC eyewall. The anvils of the *Cb* form a shield of cirrus clouds. The increasing latent heat release leads to a further temperature increase and to a pressure fall in the vortex center, increasing the air influx of air and moisture in the TC BL. As a result, convection in the central zone becomes stronger, causing TC intensification. The cloud structure in a TC is illustrated in **Figure 1.2.17.**

1.3 Investigating Clouds: The Purpose and Scope of This Book

The two main methods of investigating clouds, their evolution, structure, and cloud processes are experimental studies and theoretical studies. Experimental studies include both laboratory experiments and in situ measurements in natural clouds (**Figure 1.3.1**).

In laboratory experiments, scientists investigate particular microphysical processes occurring in clouds and determine parameters associated with these processes, using various tools and techniques. For instance, efficiency of collisions between drops in laminar and turbulent flows, drop growth on aerosol particles, drop breakup due to collisions, and spontaneous breakup have been studied in wind tunnels. Fundamental results concerning the melting of ice particles and freezing of drops have been obtained in cloud chambers in laboratory conditions. Recent laboratory experiments have investigated fine microphysical processes such as clustering of cloud droplets in a 3D turbulent flow with characteristic spatial scales of concentration inhomogeneity of tenth of a centimeter. The advantages of laboratory studies are the ability to repeat the experiments under the same conditions and to control variations of external parameters. However, laboratory experiments have some drawbacks that are actually the reverse side of the advantages, since laboratory experiment conditions often differ from those in real clouds. In addition, processes in real clouds often take place at temporal and spatial scales much larger than those in the laboratory.

In view of these imperfections of laboratory cloud investigations, in situ measurements in real clouds are

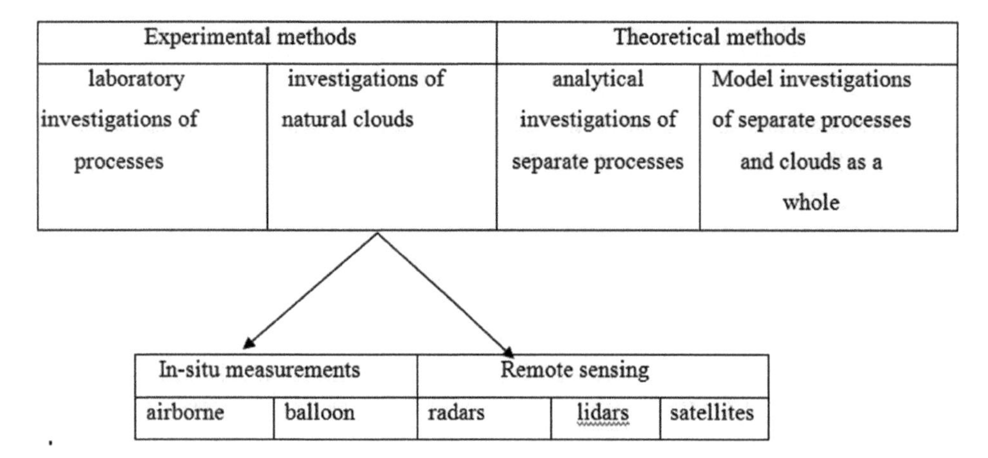

Figure 1.3.1 Methods of investigating clouds.

of primary importance. Measurements can be performed directly in clouds using probes positioned on an aircraft or a balloon, or indirectly using remote sensing instruments such as meteorological radars, satellites, and lidars. Devices installed on the ground measuring characteristics of low clouds and precipitation are also used for in situ measurements.

The advantages of in situ measurements are significant, as in most cases probe measurements of microphysical and dynamic cloud parameters enable interpretation of the results. At the same time, direct in situ measurements provide only a limited volume of information as they are conducted within relatively rare field campaigns with set time periods, seasons, and geographical locations. Some areas within deep convective clouds are inaccessible for in situ measurements due to dangers.

Remote sensing methods enable getting long-term series of measurements over large areas covered with clouds. For example, the meteorological radar network covers all the vast territory of the United States. Satellite measurements are performed all over the globe. However, since remote sensing signals form under the impact of multiple atmospheric and electromagnetic processes, the measurements are to a large extent uncertain and their interpretations are ambiguous.

Another drawback related to both direct measurements and remote sensing measurements is that measurements in real clouds cannot be repeated under the same environmental conditions. And finally, there is no measurement that could directly explain physics of a natural phenomenon or a process related to clouds. Such explanation is the goal of theoretical studies.

Theoretical studies can be roughly separated into analytical investigations and cloud investigations using numerical models. Actually, any theoretical study can be considered a model of a real process, with some simplifications. Analytical models typically consider separate processes under multiple simplified assumptions. The advantage of analytical modeling is the ability to perform a detailed analysis of a process that enables scientists to reach a comprehensive understanding of physics underlying the process under investigation. In many cases, the analytical studies reveal relationships between different cloud parameters. These data are then used for parameterization of processes in large cloud models, and guide the modeling itself and help to interpret modeling results. Yet, numerous simplifications constitute substantial limitations of the analytical methods.

Numerical models typically consider more general problems and include far more processes and governing parameters than analytical models. Equation systems formulated for a numerical model are typically much more complicated than those used in the analytical approach. As the latter usually have no analytical solutions, different numerical methods are used to obtain the solutions. Numerical cloud models solve equations, which allow the analysis of the formation of cloud microstructure, as well as the evolution of cloud particles and their distributions. The knowledge of cloud microphysical structure allows the calculation of both cloud radiative properties and cloud precipitating properties. Formation of cloud particles, their growth, and transformations are typically accompanied by phase transitions and latent heat release that determines cloud dynamics and all cloud-related phenomena. These coupled dynamic and microphysical processes are described in cloud models.

Mathematical equations underlying cloud models can be separated into three types. The first type includes equations expressing the conservation laws of heat and mass and some basic equations of hydrodynamics of moist air. These equations represent Newton's second law and the first law of thermodynamics. The second type includes microphysical equations for known physical processes, e.g., the diffusion growth equation and equation describing particle collisions (so-called Smoluchovski equation). Theories of some microphysical processes are not yet fully developed or too complicated to be applied in practice. In this case, cloud model designers use equations of the third type, namely, empiric and semi-empiric equations containing parameters established from observations. As understanding of cloud microphysical processes improves, it is possible to develop more accurate equations, gradually replacing the third type of equations by equations of the second type. All of these equations, both partial differential equations and ordinary differential equations, constitute a system of equations that is solved using numerical methods, usually the finite difference method in which partial derivatives are replaced by finite differences. This actually means replacing the original equation system by the system of algebraic equations. Sometimes, to save computer time, the original exact equations are replaced by simplified ones or by lookup tables prepared in advance.

To formulate cloud model equations containing multiple unknown parameters, results obtained in the laboratory and in situ measurements are widely used. Comparison of results of cloud modeling to those obtained via measurements and observations is a necessary condition for a model's justification. At the same time, theoretical studies and numerical modeling raise issues to be studied experimentally and thus trigger novel laboratory experiments and in situ observations. In addition, theoretical studies often determine accuracy requirements and new parameters to be measured. In combination with other investigation methods, cloud modeling has proved to be a highly effective approach.

Early numerical cloud models treated evolution of separate clouds. The increase in the computational power allows utilization of models that describe cloud formation and evolution over larger areas. Such models are known as cloud-resolving models (CRM). Further progress in weather prediction, simulation of climate, and climatic changes requires a more comprehensive representation of clouds and cloud processes in large-scale numerical atmospheric models.

There are several classical books on cloud physics and cloud microphysical processes, among them *A Short Course in Cloud Physics*, by Rogers and Yau (1989) and a fundamental book on cloud microphysics by Pruppacher and Klett (1997). Straka (2009) describes simplified expressions used for solving microphysical equations (bulk parameterization). The book by Khvorostyanov and Curry (2014) is dedicated to theoretical analysis of specific microphysical processes.

There are two major differences between this book and earlier works. First, we describe how these cloud microphysical processes are treated in modern numerical cloud models. We try to stress the relationship between the physics of the cloud processes and mathematical descriptions of these processes in cloud models. Clouds contain different cloud particles, which can be separated into cloud hydrometeors of different types, for instance, liquid droplets, low density ice particles – such as snow particles – and high density particles such as hail. Each hydrometeor type consists of particles of different sizes from submicron sizes up to a few centimeters. Precipitating and radiative cloud properties depend on the size distributions of cloud hydrometeors. In this book, we systematically consider the major modeling approaches – namely, spectral bin microphysics (SBM) – based on solving the equations for size distribution functions of cloud hydrometeors, and the bulk-parameterization approach that includes calculations of several moments of these size distributions. Second, we analyze the progress made in the understanding of the cloud microphysical processes achieved over the past two decades. We present the modern view on understanding the processes of liquid precipitation and ice precipitation formation.

Modeling is often detached from the actual physical base. Many scientists use models or different microphysical schemes (packages) such as black boxes. The analysis of model results in this case becomes rough and often remains conjectural and descriptive. We hope that the approach undertaken in this book, which combines the theory of cloud microphysical processes and their numerical realizations in models, will prove helpful for students, researchers, and cloud model designers in the fields of cloud physics, atmospheric science, and meteorology. The book contains plentiful background material and is intended as the reference source on a vast variety of cloud-related issues.

References

Graeme, L.S., J. Li, M. Wild, C.A. Clayson, N. Loeb, S. Kato, T. L'Ecuyer, P.W. Stackhouse, Jr., M. Lebsock, and T. Andrews, 2012: An update on Earth's energy balance in light of the latest global observations. *Nature Geoscience*, **5**, 691–696.

Houze, R.A., Jr., 1989: Observed structure of mesoscale convective systems and implications for large-scale heating. *Q. J. Royal Meteorol. Soc.*, **115**, 425–461.

2010: Review. Clouds in tropical cyclones. *Mon. Wea. Rev.*, **138**, 293–344.

Khvorostyanov, V.I., and J.A. Curry, 2014: *Thermodynamics, Kinetics, and Microphysics of Clouds.* Cambridge University Press, ISBN: 9781107016033.

Mazin, I.P., and S.M. Shmeter, 1983: *Clouds, Their Structure and Formation.* Gidrometizdat, 278.

Pruppacher, H.R., J.D Klett, 1997: *Microphysics of Clouds and Precipitation.* 2nd edn. Oxford Press, 914.

Ramanathan, V.C.P.J., P.J. Crutzen, J.T. Kiehl, and D. Rosenfeld, 2001: Aerosols, climate, and the hydrological cycle. *Science*, **294**, #5549, 2119–2124.

Rogers, R.R., and M.K. Yau, 1989: *A Short Course in Cloud Physics.* Pergamon Press 293.

Rosenfeld, D., Y.J. Kaufman, and I. Koren, 2006: Switching cloud cover and dynamical regimes from open to closed Benard cells in response to the suppression of precipitation by aerosols. *Atmos. Chem. Phys.*, **6**, 2503–2511.

Straka, J.M., 2009: *Cloud and Precipitation Microphysics.* Cambridge University Press, 392.

Tao, W.-K., and M.W. Moncrieff, 2009: Multiscale cloud system modeling. *Rev. Geophys.*, **47**, RG4002, doi:10.1029/2008RG000276.

Trenberth, K.E., L. Smith, T. Qian, A. Dai, and J. Fasullo, 2007: Estimates of the global water budget and its annual cycle using observational and model data. *J. Hydrometeorol.*, **8**, 758–769.

2 Cloud Particles and Their Representation in Cloud Models

2.1 General Characteristics of Cloud Particles and Their Description in Cloud Models

In the atmospheric science, atmospheric and cloud particles participating in microphysical processes are divided into three large classes: aerosols, liquid drops, and ice particles. Each class will be described in detail in Sections 2.2, 2.3, and 2.4. In this section, we present the general characteristics of the particles and methods used for their description.

Aerosol particles (or aerosols, hereafter APs) is a collective name for suspended particular matter ranging in size from small clusters with the diameter of 10 nm to giant particles of ~20 μm in diameter (Levin and Cotton, 2009). Some APs are insoluble and remain solid regardless of the humidity of the air around them. Other particles are either soluble or contain a soluble fraction. If air humidity exceeds a certain threshold known as deliquesce humidity, these APs start growing due to the condensation of water vapor and become wet APs, also known as haze particles, or simply haze. At certain humidity values, the size of haze particles tends to an equilibrium size, when the flux of water molecules onto their surface is equal to the flux of molecules evaporating from the particle surface. The equilibrium size depends on the size of the initial dry AP, as well as on its chemical composition and on the environmental humidity (see Section 5.1 for details). Typically, the mass of water in haze exceeds by many times that of dry AP, which means that haze particles represent comparatively diluted solutions.

The role of APs is manifold. These particles affect the visibility of the air, which, depending on different sizes and concentrations of APs, can vary from a few hundred meters to several hundred kilometers. APs determine the radiative parameters of the atmosphere affecting the temperature and temperature gradients, etc. Via the radiative parameters, APs affect the radiative balance and the heat balance of the atmosphere and consequently have an impact on climate. APs are important contributors to air pollution.

It is important to stress that all droplets and most ice particles in clouds form on APs. APs giving rise to droplet and ice formation are known as condensational nuclei (CN) and ice nuclei (IN), respectively. Since a significant fraction of APs is of anthropogenic nature, a natural question arises whether these man-made APs can impact precipitation amounts and rates, as well as local and even global climate. New data allow one to answer this question positively (e.g., Rosenfeld et al., 2014; Fan et al., 2015). While APs affect the precipitation and radiation properties of clouds, clouds themselves have an impact on the concentration and size of APs. Moreover, chemical processes in cloud drops lead to formation of new APs. One of the major topics in modern cloud physics, as well as in this book, is cloud–aerosol interaction.

Man-made APs used for cloud seeding to achieve rain enhancement are of special interest. For instance, submicron APs consisting of silver iodide (AgI) are used for glaciogenic seeding, leading to production of additional amounts of ice crystals in clouds. AgI particles begin to grow as ice crystals at comparatively warm temperatures (~−5°C).

Drops are liquid particles with diameters ranging from about 1 μm to ~1 cm and are traditionally classified into two groups: small drops called cloud droplets have diameters from 1 μm to ~50 μm and raindrops with diameters exceeding 50 μm. Raindrops start forming as a result of droplet collisions and continue growing as they collect cloud droplets.

Droplets form by condensation of water vapor molecules onto the surface of haze particles. The flux of molecules is determined by diffusion process. Accordingly, the condensational droplet growth is known as the diffusional growth. Since droplets contain soluble mass, they are weak solutions. Diffusional growth increases the mass of water within droplets and decreases droplet salinity. In case the APs giving rise to haze and droplet formation contain an insoluble fraction, droplets contain insoluble inclusions. Insoluble particles can also be captured by drops as a result of collisions with APs and other drops. These insoluble particles (immersion nuclei) determine the process of a drop freezing at low temperatures.

Drops with diameters below ~ 200 μm are almost spherical in shape. Rain drops of larger sizes become oblate in the vertical direction due to resistance of the environmental air during their fall. Rain drops with diameters close to ~1 cm become unstable and break up spontaneously producing several smaller raindrops. The process of spontaneous breakup determines the maximum diameter of raindrops (of ~ 1 cm).

At certain temperatures colder than −5°C ice particles originate in clouds. Ice particles form in several ways. They can form by freezing of haze and drops, as well as by condensation of water vapor molecules upon insoluble APs. In cloud physics, four large classes of ice particles are distinguished: ice crystals, aggregates (snowflakes), graupel, and hail. Each class is separated into a great number of subclasses that differ in their shape and density.

There are more than eighty different subclasses (or types) of ice crystals, which differ in their shape (Pruppacher and Klett, 1997). The sizes of ice crystals range from a few microns to a few thousand microns. The characteristic feature of ice crystals is that they typically have an ordered crystalline structure. Ice crystals are mostly nonspherical.

Aggregates (sometimes called snow) form as a result of the collection of ice crystals. Aggregates have high porosity and hence low density and can reach sizes of a few centimeters in diameter. Larger aggregates become unstable and break up. Collisions of aggregates with water drops (called the riming process) increase the rimed (high density) fraction in aggregates. Substantially rimed ice particles are known as *graupel*. Graupel particles are denser than aggregates and reach a maximum diameter of about 1 cm. There exist different shapes of graupel, but on average their shapes are closer to spherical than those of ice crystals.

Collisions of frozen drops (small spherical ice particles) with unfrozen drops, as well as collisions of graupel with liquid drops, increase particle density. Particles with density close to that of pure ice (0.91 g/cm^3) are known as *hail*. Hail can also form by freezing of rain drops. Both graupel and hail grow by the collection of liquid drops. If the amount of drops collected by a graupel (or hail) particle is significant, the release of the latent heat of fusion does not allow all water to freeze on the particle surface. This process is called wet growth. If all liquid water collected by graupel or hail freezes at the particle surface, this is dry growth. Usually, wet growth begins when the particle is quite large and collects water drops rapidly. Graupel that originates by wet growth is often assigned to hail. The shape of hail is almost spherical, with maximum diameters sometimes reaching several centimeters.

Cloud physics and, in particular, cloud modeling, faces the challenge of dealing with describing the great variety of atmospheric particles of different shapes and densities, as well as their interactions. To simplify the problem, it is accepted to introduce "virtual" or "equivalent particles" characterized by a comparatively small number of parameters and a simplified geometry. Real atmospheric particles are assigned to one of the classes of equivalent particles. Thus, the description of all the variety of atmospheric particles is reduced to the description of simplified equivalent particles.

2.1.1 Equivalent Particles

Equivalent particles are characterized by several parameters, described next.

Particle mass. The main parameter characterizing a cloud particle is its mass. The mass of a particle determines its behavior in thermodynamic and hydrodynamic processes. An equivalent particle has the same mass as a real one.

Bulk density. The second basic characteristic of a cloud particle is its bulk density ρ_{bulk}, defined as the mass-to-volume ratio. The bulk density of water and hail is equal to the density of corresponding bulk materials, i.e., the density of water and pure ice, respectively. However, many APs and ice particles have densities lower than those of the real bulk material due to the presence of some air in their capillary spaces. The bulk density of large aggregates (snowflakes) can be as low as 0.01 g/cm^3. The bulk density of graupel typically ranges from 0.2 g/cm^3 to 0.8 g/cm^3. Bulk density doesn't only depend on the air volume within particles, as different particles can contain matters of different densities. For instance, particles can contain water, soluble materials of different types, insoluble inclusions, etc. Riming, i.e., collisions between aggregates and water droplets, increases the bulk density of aggregates. Riming of aggregates may end up in the formation of graupel or hail.

Equivalent shapes. As regards describing particle shapes, several models of equivalent particles are applied depending on the particular purposes of the research. In numerical models, particles of diverse shapes are approximated by equivalent particles, which are typically axisymmetric bodies characterized by two dimensions. For instance, plate-like particles and branch-like ones (dendrites) are approximated by discs of diameter d and thickness h. Columnar particles are modeled as cylinders of length h and width d. Such particles are often approximated also as oblate or prolate spheroids. Represented by their equivalent shapes,

Table 2.1.1 The geometrical characteristics of simplified particle shapes.

Shape	Sphere	Circular Disk	Oblate Spheroid	Prolate Spheroid
	2r	h, d	h, d	h, d
Surface area	$4\pi r^2$	$\pi d\left(\frac{d}{2}+h\right)$	$\frac{\pi}{2}d^2+\frac{\pi h^2}{4\varepsilon}\ln\frac{1+\varepsilon}{1-\varepsilon}$ $\varepsilon=\sqrt{1-\frac{h^2}{d^2}}$	$\frac{\pi}{2}d^2+\frac{\pi dh}{2\varepsilon}\mathrm{asin}(\varepsilon)$ $\varepsilon=\sqrt{1-\frac{d^2}{h^2}}$
Volume	$\frac{4}{3}\pi r^3$	$\frac{\pi}{4}d^2h$	$\frac{\pi}{6}d^2h$	$\frac{\pi}{6}dh^2$
Mean volume radius	r	$\left(\frac{3}{16}d^2h\right)^{1/3}$	$\frac{1}{2}(d^2h)^{1/3}$	$\frac{1}{2}(dh^2)^{1/3}$

nonspherical particles are characterized by the aspect ratio h/d. Actually, the aspect ratio is the ratio of a particle's size along the axis of symmetry to its size along the transverse direction. The aspect ratio for discs and oblate particles is less than one, while for columns and prolate particles this ratio is larger than one. The main geometrical characteristics of discs (which can be considered as cylinders with low aspect ratios), prolate, and oblate spheroids are presented in **Table 2.1.1**. Using the empirical relationships between the main particle dimensions, it is possible to calculate the volume of the particle.

Equivalent radius. In cloud models, it is often too complicated to utilize particles with equivalent shapes characterized by two dimensions. Besides, the information about the shape of particles is often unavailable, or there is too wide a variability of particle shapes within the region of interest that is difficult to embrace. In these cases, particles are represented by equivalent spheres of the same mass as the particles under study. The radius of the equivalent spheres is designated as the equivalent radius r. The equivalent radius can be defined in several ways depending on the properties of equivalent spheres, which should be similar to those of the properties of the real particles. Sometimes, the equivalent radius of cloud particles is defined as the Stokes radius, if the particle size is determined by its fall velocity. In this case, it is assumed that the equivalent sphere falls down at the same velocity as the real particle. In some cases, when the measurements of APs and ice crystals are performed by photoelectric probes, the equivalent radius is defined as the radius of a spherical particle that disperses light in a given direction similarly to the real particle. Such definition is useful while solving problems related to calculation of radiation balances. In numerical models, the equivalent radius is typically defined as the radius of the sphere having the same mass and the same bulk density as the real particle. In this case, the equivalent radius is the mean volume radius r_v calculated using the expressions shown in Table 2.1.1.

Representation of particle properties via particle mass. In numerical models, all the geometrical parameters are represented as functions of particle masses and bulk densities using *in-situ* or laboratory measurements. A great variety of the relationships between geometrical dimensions themselves, geometrical dimensions vs. density, etc., is presented in Pruppacher and Klett (1997). We present here one example of such representation. In the Hebrew University Cloud Model (hereafter, HUCM) (e.g., Khain et al., 2004b) the thickness-diameter relationship for dendrites of Plc-s and Pld crystal types and the bulk density of dendrites are $h = 9.96\cdot 10^{-3}d^{0.415}$ and $\rho_{bulk} = 0.588\cdot d^{-0.377}$, respectively (Pruppacher and Klett, 1997) (the geometrical sizes are given in centimeters and the density in g cm^{-3}). Using these empirical expressions and formulas approximating the shape of dendrites by a cylinder (see Table 2.1.1), the geometrical parameters and the bulk density can be expressed as a function of the mass as

$$d = 14.02m^{0.491},\ h = 0.0297m^{0.204},$$
$$r_v = 1.032m^{0.395},\ \rho_{bulk} = 0.0912m^{-0.185} \quad (2.1.1a)$$

In Equation (2.1.1a), the geometrical sizes are given in centimeters, the mass in g, and the density in g cm^{-3}.

Fall velocity. An important characteristic of cloud particles is their fall velocity. Representation of particles via equivalent particles (including equivalent spheres) does not mean that the fall velocities of real particles are similar to those of equivalent ones because the equivalent particles may have different hydrodynamic properties. For instance, it is difficult to expect that a real dendrite falls at the same velocity as the equivalent sphere of the same mass and the same bulk density. Typically, empirical expressions for fall velocities are used for each type of particles. By way of example, in HUCM the fall velocity dependence of Pld dendrites (in the diameter range of $0.05\,\text{cm} < d < 0.3\,\text{cm}$) is chosen as $V_t = 4.22 \cdot 10^1 d^{0.442}$ (Pruppacher and Klett, 1997) (the diameter is in millimeters and the velocity is in cm s^{-1}). It leads to the following dependence of fall velocity on dendrite mass as (the mass is in g and the velocity is in cm s^{-1}):

$$V_t = 1.36 \cdot 10^2 m^{0.217} \quad (2.1.1b)$$

Dependencies of geometrical parameters, bulk density, and fall velocity for Pld-type dendrites on particle mass are calculated using Equations (2.1.1a, 2.1.1b) are shown in **Figure 2.1.1**.

In different cloud models, expressions such as Equations (2.1.1a and 2.1.1b) allow one to use the same mass grids (containing from several tens to a few hundred bins, or mass categories) for all the types of particles (i.e., drops, ice crystals, aggregates, graupel, and hail). Knowing the type and the mass of a particle one can calculate all the rest of its parameters required for calculation of its growth by condensation and collisions, as well as its reflectivity properties.

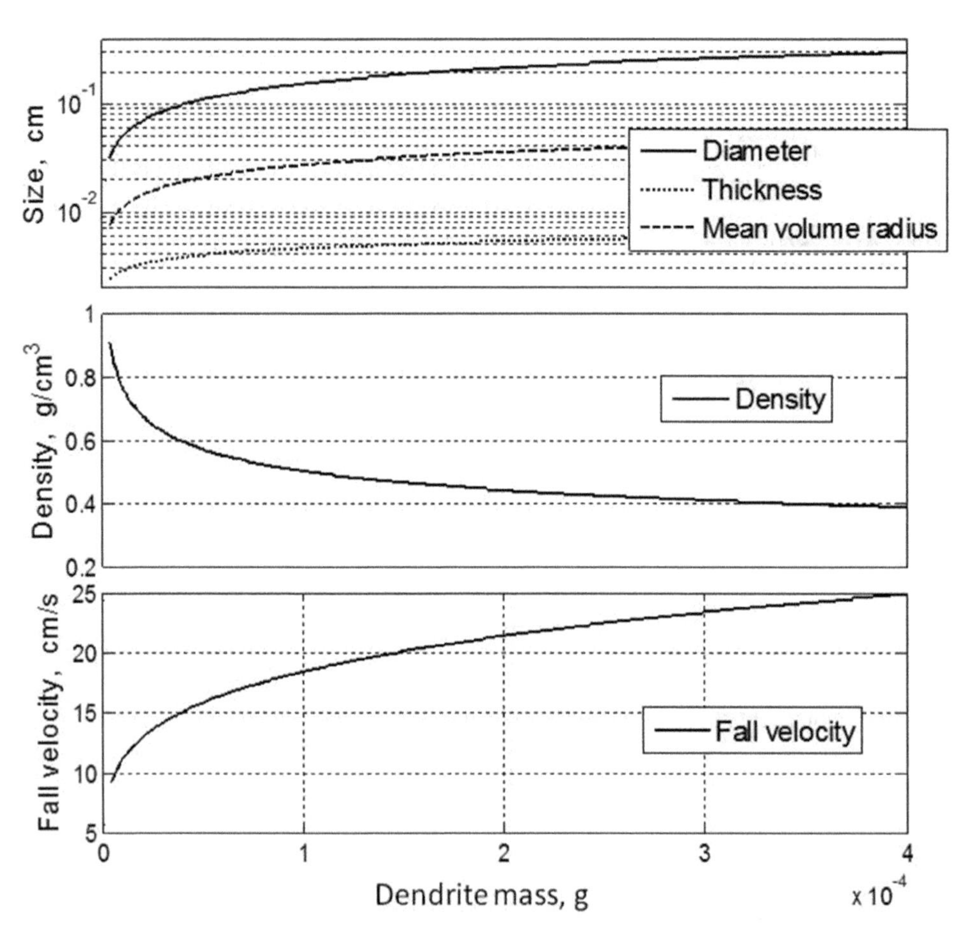

Figure 2.1.1 Dependencies of geometrical parameters, bulk density, and fall velocity for Pld-type dendrites on particle mass.

2.1.2 Number Distributions and Mass Distributions

Knowing particle size distributions (hereafter PSD) of cloud particles is of major importance while dealing with multiple issues in cloud physics and environmental science (e.g., cloud modeling, deposition calculations, and radiation models). PSDs are often referred to as size spectra of particles. Typically, PSD is defined as $f(m)$, i.e., a function of particle mass m, thus $f(m)dm$ is the number of particles per unit of volume within the mass ranging from m to $m + dm$. It is convenient to use $f(m)$ because all the parameters of particles can be expressed via the particle mass, as discussed previously. $f(m)$ is measured in units of $g^{-1}cm^{-3}$ (in the international system of units [SI], the units are $kg^{-1}m^{-3}$). Particle concentration N (sometimes called a number concentration) is then expressed as

$$N = \int_{m_{\min}}^{m_{\max}} f(m)dm \tag{2.1.2}$$

Since $f(m) = \frac{dN}{dm}$, $f(m)$ is often referred to as number distribution, or number density.

In numerical modeling, the mass density function $g(m)$ is also widely used (e.g., Berry and Reinhard, 1974). Normalization of the mass density function is performed as

$$M = \int_{m_{\min}}^{m_{\max}} g(m)dm, \tag{2.1.3}$$

where M is the mass content in units g cm^{-3} (or in SI, kg m^{-3}). The comparison of Equations (2.1.2) and (2.1.3) shows that

$$g(m) = m \cdot f(m) \tag{2.1.4}$$

Since $g(m) = \frac{dM}{dm}$, the function $g(m)$ is often referred to as mass distribution or mass density.

In addition to the PSD depending on particle mass, the PSD depending on radius r, i.e., $f_r(r)$ (or diameter d), are often used (in case of nonspherical particles, r is the equivalent radius). According to the definition of $f_r(r)$, $f_r(r)dr$ is the number of particles per unit of volume within the radii ranging from r to $r + dr$. $f_r(r)$ is measured in units of m^{-4}; if a particle's radius is measured in μm, units $cm^{-3}\mu m^{-1}$ are often used. According to the definition, particle concentration N is

$$N = \int_{r_{\min}}^{r_{\max}} f_r(r)dr, \tag{2.1.5}$$

where $r_{\min}$ and $r_{\max}$ are the minimum and maximum radii in the particle spectrum, respectively. It implies that $f_r(r) = \frac{dN}{dr}$. The relationship between $f_r(r)$ and $f(m)$ is obvious:

$$f_r(r) = f(m)\frac{dm}{dr} = \frac{3m}{r}f(m) \tag{2.1.6}$$

The equality on the right side of Equation (2.1.6) is valid in case the bulk density does not change as the particle mass varies.

Since the size of cloud particles varies by several orders of magnitude, utilization of $f_r(r)$ is not always convenient, the density distribution $\tilde{f}_r(\log_{10} r)$ is introduced with respect to the logarithm of radius, so $N = \int_{\log_{10} r_{\min}}^{\log_{10} r_{\max}} \tilde{f}_r(r)d\log_{10} r$. The relationship between these two PSDs $f_r(r)$ and $\tilde{f}_r(\log r)$ is obvious:

$$\tilde{f}_r(\log_{10} r) = \frac{dN}{d\log_{10} r} = 2.3rf_r(r) \tag{2.1.7}$$

The following important comment is required regarding the determination and interpretation of PSD. Particles in clouds are often separated by distances significantly exceeding their sizes. The distance between particles changes randomly. Thus, a PSD measured *in situ* or calculated in models should be evaluated using particle numbers large enough to be sufficient for obtaining statistically representative results. Accordingly, the air volume to be measured should be large enough to contain this number of particles. PSDs obtained within too-small air volumes may fluctuate due to the Poisson noise, which complicates the physical interpretation of PSD.

As can be seen from the definitions, PSDs are analogous to the probability distribution functions (PDFs) used in the probability theory. The main difference between PSD and PDF is the normalization condition. In the probability theory, the integral of PDF is equal to unity. The integrals of PSDs are equal either to particle concentration values or to mass values per unit of volume, making them convenient for utilization in cloud physics.

PSDs are often characterized by their moments, which have an important physical meaning in cloud physics. A moment of order k is defined as follows:!

$$M^k = \overline{r^k} = \int_{r_{\min}}^{r_{\max}} r^k f_r(r)dr \tag{2.1.8}$$

As seen from Equation (2.1.8), particle concentration is the zero-order moment of the size distribution, while the third moment is related to mass content and the sixth moment is related to radar reflectivity. In cloud physics, different quantities characterizing cloud parameters can be written using Equation (2.1.8) (**Table 2.1.2**).

The width of PSD, as well as the optical extinction, are described using the second moment. The effective radius r_{eff}, widely used for evaluation of the radiation properties of clouds and aerosols, is represented by the ratio of the third moment to the second moment. It is noteworthy that $\bar{r} \leq r_v \leq r_{eff}$.

The moments can be written also for the function $f(m)$:

$$M^k = \overline{m^k} = \int_{m_{\min}}^{m_{\max}} m^k f(m) dm \tag{2.1.9}$$

In this case, the zero moment is the concentration, the first moment is the mass content and the second moment is related to the radar reflectivity.

2.1.3 Representation of Size Distributions in Cloud Models

There are several approaches to represent PSDs in numerical models. In so-called spectrum bin microphysical models (SBM), PSDs are defined on discrete mass grids containing from several tens to several hundred mass bins. Each bin contains particles of mass m_i, where i is the number of a bin. In bulk-parameterization schemes, PSDs are approximated using the Gamma or exponential distributions. PSDs of APs are often

Table 2.1.2 Expressions for different quantities characterizing size distributions of particles used in Cloud Physics

Quantity	Expression	Units
Particle concentration	$N = \int_{r_{\min}}^{r_{\max}} f_r(r) dr$	m^{-3}
Mean radius	$\bar{r} = \frac{1}{N} \int_{r_{\min}}^{r_{\max}} r f_r(r) dr$	m
RMS size distribution width	$\sigma_f = \left[\frac{1}{N} \int_{r_{\min}}^{r_{\max}} (r - \bar{r})^2 f_r(r) dr \right]^{1/2}$	m
Dispersion of size distribution (known also as relative dispersion or coefficient of variation)	$\delta_f = \frac{\sigma_f}{\bar{r}}$	
Optical extinction	$\beta_{ext} = 2\pi \int_{r_{\min}}^{r_{\max}} r^2 f_r(r) dr$	m^{-1}
Mean volume radius	$r_v = \left(\int_{r_{\min}}^{r_{\max}} r^3 f_r(r) dr \Big/ \int_{r_{\min}}^{r_{\max}} f_r(r) dr \right)^{1/3}$	m
Effective radius	$r_{eff} = \int_{r_{\min}}^{r_{\max}} r^3 f_r(r) dr \Big/ \int_{r_{\min}}^{r_{\max}} r^2 f_r(r) dr$	m
Modal radius, r_{mod}	$\left. \frac{df_r(r)}{dr} \right\|_{r = r_{\text{mod}}} = 0$	m
Mass content	$M = \frac{4}{3} \pi \rho \int_{r_{\min}}^{r_{\max}} r^3 f_r(r) dr$	kg m^{-3}
Radar reflectivity	$Z = 198.0618 + 10 \cdot \log_{10} \int_{r_{\min}}^{r_{\max}} r^6 f_r(r) dr$	dBZ

approximated by log-normal distributions or sum of several (typically three) log-normal distributions referred to as modes of the distributions.

Representation of PSD on Regular (Fixed) and Movable Mass Grids in Models with Spectral Bin Microphysics
Sizes of cloud particles range within several orders of magnitude. Typically, it is highly important to describe the evolution of the smallest particles as accurately as possible. To achieve this, it is widely accepted to define PSDs on a logarithmically equidistant mass grid where the mass corresponding to the next bin exceeds that in the previous bin by factor α:

$$m_{i+1} = \alpha m_i \tag{2.1.10}$$

According to Equation (2.1.10), the mass of particles belonging to the i-th bin is

$$m_i = m_0 \alpha^{i-1} = m_0 \exp[(i-1)\ln\alpha], \tag{2.1.11}$$

where m_0 is the mass of the particles in the first bin. In many SBM models, $\alpha = 2$, but the value $\alpha = \sqrt{2}$ is also often used. It is clear that the relationship between the radii of particles (in case of ice particles we mean the melting radius) in the neighboring bins is

$$r_{i+1} = \alpha^{1/3} r_i, \tag{2.1.12}$$

hence

$$r_i = r_0 \exp\left[\frac{(i-1)\ln\alpha}{3}\right], \tag{2.1.13}$$

where r_0 is the radius of particles in the first bin.

Following Berry and Reinhard (1974), we introduce size distribution $\tilde{g}(\ln r)$ (where r is either the drop radius or the melting radius of the ice particle), defined on the logarithmically equidistant mass grid and obeying the following normalization condition:

$$M = \int_{\ln r_0}^{\ln r_{\max}} \tilde{g}(\ln r) d\ln r \tag{2.1.14}$$

Since $\tilde{g}(\ln r) d\ln r$ is the mass within the range of $d\ln r$ corresponding to the range of mass dm, we can obtain

$$\tilde{g}(\ln r) d\ln r = g(m)dm = mf(m)dm \tag{2.1.15}$$

Accordingly, PSDs introduced above can be expressed via $\tilde{g}(\ln r)$ as

$$g(m) = \frac{1}{3m}\tilde{g}(\ln r); \; f(m) = \frac{1}{3m^2}\tilde{g}(\ln r) \tag{2.1.16}$$

It follows from Equation (2.1.12) that on the logarithmically equidistant mass grid the grid increment is equal to constant

$$d\ln r = \ln r_{i+1} - \ln r_i = \frac{1}{3}\ln\alpha = const, \tag{2.1.17}$$

where i is the bin number. Following Berry and Reinhard (1974), we denote $G_i = \tilde{g}(\ln r_i)$. From Equation (2.1.14) and Equation (2.1.17), we get the expressions for the mass M_i and concentration N_i of the particles in the i-th mass bin

$$M_i = \frac{1}{3}\ln\alpha \cdot G_i \tag{2.1.18}$$

$$N_i = \frac{1}{3}\ln\alpha \cdot \frac{G_i}{m_i}. \tag{2.1.19}$$

Hence, the total particle mass and concentration can be represented as the sums:

$$M = \frac{1}{3}\ln\alpha \sum_i \tilde{g}(\ln r_i) = \frac{1}{3}\ln\alpha \sum_i G_i, \tag{2.1.20}$$

$$N = \frac{1}{3}\ln\alpha \cdot \sum_i \frac{G_i}{m_i}. \tag{2.1.21}$$

Representation of size distributions on discrete mass grid in the form described by Equations (2.1.18), (2.1.19), (2.1.20), and (2.1.21) is quite friendly for numerical simulations. It allows calculation of particle mass knowing particle concentration in the bin, and vice versa. This representation is widely used in SBM models for simulation of single clouds, cloud ensembles, squall lines, storms, mesoscale rain events, and hurricanes.

A comment is required concerning the choice of the number of bins in numerical cloud models. It is clear that in order to describe the shape of PSD, the number of bins should not be too small. At the same time, any substantial increase in the number of bins leads both to an increase in the computational time and to a decrease in the statistical significance of the quantities calculated in each bin. Reliable statistics can be obtained if the particle concentration within each bin obeys the condition $N_i \Delta x \Delta y \Delta z >> 1$, where Δx, Δy and Δz are distances between the neighboring grid points in a spatial mesh used in the model (see Section 5.6). It means that the number of particles belonging to a certain bin within the cloud volume should be large enough. Otherwise, particle concentration fluctuations of the Poisson noise type can arise in each bin, especially at the PSD tails. For this reason, averaging of the SD values is necessary, either over the neighboring bins or with respect to time or volumes. The latter is equivalent to SD averaging along an aircraft leg in measurements.

It is sometimes assumed that the representation of PSDs (Equations 2.1.18, 2.1.19, 2.1.20, 2.1.21) is too simplified, leading to errors in calculations of high-order

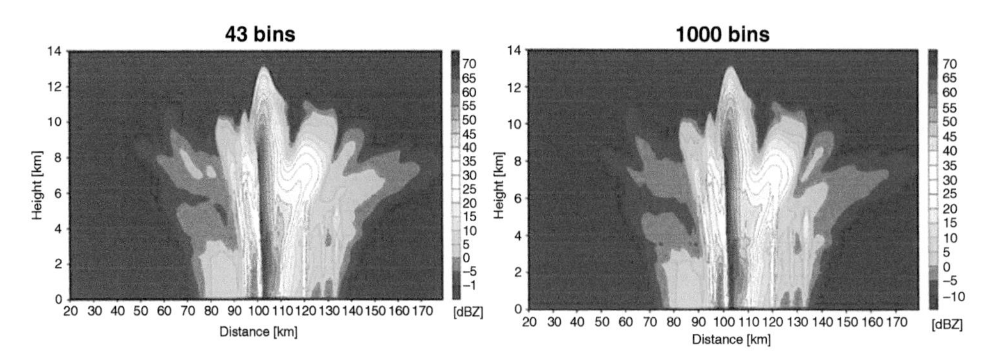

Figure 2.1.2 Fields of the radar reflectivity of a hail storm in Oklahoma, calculated using PSDs represented on the mass grid containing 43 bins (left panel) and interpolated on the grid containing 1,000 bins (right panel).

moments such as radar reflectivity, which can change within the same bin (category). Our experience does not reveal any difference in case the number of bins is high enough (several tens). For instance, **Figure 2.1.2** (left panel) shows the field of the radar reflectivity of a strong hail storm in Oklahoma. The storm was calculated using the HUCM in which SDs are represented on a mass grid containing forty three mass bins covering the size range of hail from 2 μm to ~6.5 cm (Ryzhkov et al., 2011). In order to check the response of the radar reflectivity to the grid resolution, the radar reflectivity field was calculated also on the grid containing 1,000 logarithmically equidistant bins (Figure 2.1.2, right panel). The particle concentrations in each bin of this grid were obtained by the interpolation of the PSD values calculated on a forty three-bin grid. No significant difference was found between the radar reflectivity fields calculated on the 43-bin grid and on the 1,000-bin grid.

It should be stressed that sizes of drops and ice particles change in the course of diffusional growth/evaporation of drops and depositional/sublimation of ice, as well as during collisions. Due to this fact, the particle masses obtained as a result of these processes do not form an exact logarithmic mass grid. Therefore, in order to use a regular logarithmic mass grid, one needs to perform a remapping (interpolation) of PSD onto the regular bin grid. Problems related to this remapping are discussed in Section 5.4.

Another approach to PSD representation on a discrete mass grid was applied by Pinsky and Khain (2002) and Pinsky et al. (2008) in the Lagrangian model of the atmospheric boundary layer capped by stratocumulus clouds. In this approach, all drops are assumed to be concentrated at grid points r_i (like delta functions). The normalization condition used in this representation is

$$N = \sum_i N_i; M = \sum_i m_i N_i \tag{2.1.22}$$

Such PSD representation can be considered as a discrete analog of function $f_r(r)$. The initial grid of radii (and masses) in this representation is chosen to be logarithmically equidistant. An advantage of this approach is that masses m_i of the mass grid are not fixed but may change, allowing the drop distribution to be represented on a movable mass grid. For instance, the values of masses m_i of the mass grid change according to the diffusional growth/evaporation equation. As a result, no remapping on a regular mass grid is required at the stage of diffusion growth/evaporation of droplets, which dramatically decreases the total number of remappings, which in turn reduces the remapping-induced errors in calculation of size distributions.

Another possibility of increasing the accuracy of calculations of aerosols and smallest droplets is the utilization of a high-resolution grid. In the model by Pinsky and Khain (2002); Pinsky et al. (2008); Magaritz et al. (2009) the number of bins was quite large and varied from 500 to 2,000 in different model versions. Knowing concentrations in each bin, one can calculate the number and mass distributions using Equations (2.1.4) and (2.1.6).

The Method of Moments

Another method to represent particle spectra, known as the microphysical method of moments (MMM), was developed in a number of studies (Bleck, 1970; Danielsen et al., 1972; Ochs and Yao, 1978; Enukashvily, 1980). It was further elaborated by Tzivion et al. (1987, 1989), Feingold et al. (1994), and Reisin et al. (1996). In this method, grid points m_i and m_{i+1} are regarded, respectively, as the lower and the upper boundaries of the i-th bin (category). Similar to the approaches described previously, the mass grid is logarithmically equidistant. A specific feature of this approach is that the mass grid is used for representation of the PSD moments in each category and not for PSDs. The PSD moments in each bin are defined as

$$M_i^v = \int_{m_i}^{m_{i+1}} m^v f(m) dm \quad (2.1.23)$$

Thus, in the method of moments the kinetic equations are formulated not for PSD, but for its moments in each category. Since PSD can be characterized by several moments, the equation system may include an equation for each moment. In one-moment schemes (Bleck, 1970; Danielsen et al., 1972), the first moment ($v = 1$) representing masses in the bins is calculated:

$$M_i^1 = M_i = \int_{m_i}^{m_{i+1}} mf(m) dm \quad (2.1.24)$$

In multi-moment schemes (Tzivion et al., 1987), the PSDs in each bin is characterized by several moments: concentration (Equation 2.1.25), (zero moment $v = 0$), mass (first moment) (Equation 2.1.24) and reflectivity ($v = 2$) (Equation 2.1.26):

$$M_i^0 = N_i = \int_{m_i}^{m_{i+1}} f(m) dm \quad (2.1.25)$$

$$M_i^2 = Z_i = \int_{m_i}^{m_{i+1}} m^2 f(m) dm \quad (2.1.26)$$

To calculate the integrals (Equations 2.1.24, 2.1.25, and 2.1.26), it is necessary to make some assumptions about the behavior of the PSD within each bin. Enukashvily (1980) used the linear approximation to the PSD within each category. This approach does not always satisfy the condition of positiveness throughout each category. To overcome this problem, the points where the PSD approximations are equal to zero are found in each category in the course of simulations. Integration is then performed only within the segment of the interval where the function is positive. Tzivion et al. (1987) applied a linear distribution to approximate the integrands within each category. This approach actually excludes the emergence of negative PSDs within bins.

Representation of PSDs by several moments within each bin may, potentially, be more accurate than other PSD representations described previously, since it attempts to describe the nonuniformity of PSDs within bins. However, while dealing with microphysical processes such as diffusional growth, collisions, etc., the method of moments requires a complicated formulation of kinetic equations. Moreover, independent equation systems must be solved for different moments, for instance, for masses and concentration in bins, which at least doubles the amount of computer recourses and time required as compared to the simpler and more computationally efficient method of Berry and Reinhard (1974).

Two-Dimensional Size Distributions

In some cases, particles of similar mass have different properties. For instance, drops of the same mass may have different salinity (i.e., different aerosol mass), different insoluble inclusions, different charges, etc. Graupel and snow of the same mass can contain different rimed fractions, charges, etc. To describe such cases with high accuracy, two-dimensional (or sometimes multidimensional) size distributions are used. For instance, to describe the situation when drops of the same mass have different salt mass (different salinity), Bott (2000) used 2D size distribution function $f_r(r, r_N)$, where $f_r(r, r_N)drdr_N$ is the concentration of drops within the radii interval of $[r, r + dr]$ and the aerosol radii within interval of $[r_N, r_N + dr_N]$. In this case, a 2D size grid is used, one dimension being assigned for drops and the other for aerosols. The total concentration of particles is calculated in this case as $N = \iint f_r(r, r_N)drdr_N$. A similar approach was used by Khain et al. (2004a), who used 2D size distribution of drops containing different charges in order to simulate the effects of drop charge on raindrop formation. However, utilization of two-dimensional and multidimensional size distributions is extremely time consuming, so at present only a few models of such type have been developed. Most models use averaging of drop (or ice particle) properties for particles belonging to the same mass bin, e.g., the averaged salinity of drops belonging to the same bin (Magaritz et al., 2009). Khain et al. (2011) calculated the averaged rimed mass for snow particles assigned to the same mass bin.

Flossmann et al. (1985), Flossmann and Pruppacher (1988), and Respondek et al. (1995) used the mass distribution functions for aerosol masses located both outside ($g_{APa}(m)$) and within drops ($g_{APd}(m)$) to evaluate wet removal aerosol pollutants. In these studies, drops belonging to the same bin are characterized by the averaged aerosol properties (e.g., drops belonging to the same bin have the same salinity).

Representation of Size Distributions in Models Using Bulk-Parameterization of Microphysical Processes

In SBM, masses and concentrations of particles should be calculated in each bin (category). To decrease the

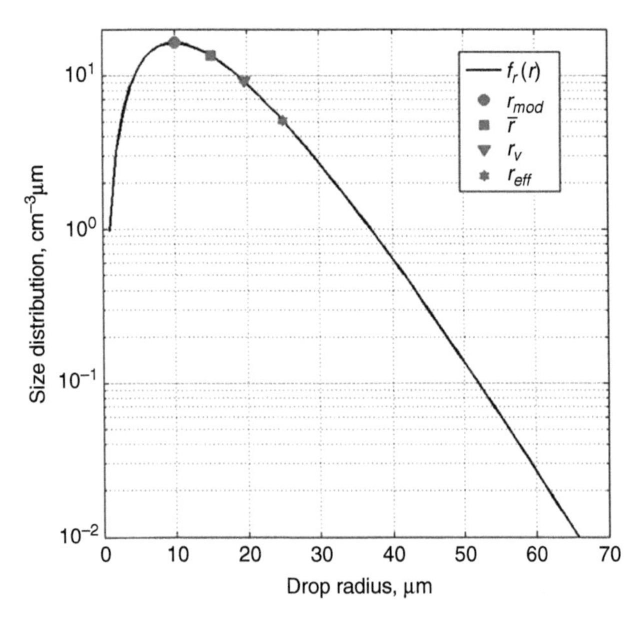

Figure 2.1.3 An example of general Gamma distribution (Equation 2.1.28) with parameters $N_0 = 300\,\mathrm{cm}^{-3}$, $\beta = 5\,\mu\mathrm{m}$, $\alpha = 3$ and $\mu = 1$. Also shown: the mean $\bar{r} = 15\,\mu\mathrm{m}$, the modal $r_{\mathrm{mod}} = 10\,\mu\mathrm{m}$, the mean volume $r_v = 19.6\,\mu\mathrm{m}$, and the effective $r_{eff} = 25\,\mu\mathrm{m}$ radii.

number of variables, in bulk-parameterization schemes, PSDs are parameterized (approximated) by Gamma distributions containing three parameters (Saleeby and Cotton, 2004):

$$f_r(r) = \frac{N_0}{\Gamma(\alpha)\beta}\left(\frac{r}{\beta}\right)^{\alpha-1}\exp\left(-\frac{r}{\beta}\right) \tag{2.1.27}$$

or the general Gamma distributions containing four parameters:

$$f_r(r) = \frac{N_0\mu}{\Gamma(\alpha)\beta}\left(\frac{r}{\beta}\right)^{\mu\alpha-1}\exp\left[-\left(\frac{r}{\beta}\right)^{\mu}\right] \tag{2.1.28}$$

These distributions are related to the number density distribution $f(m)$ and the mass density distribution $g(m)$ according to Equations (2.1.4) and (2.1.6). An example of general Gamma distribution (Equation 2.1.28) with parameters $N_0 = 300\,\mathrm{cm}^{-3}$, $\beta = 5\,\mu\mathrm{m}$, $\alpha = 3$, and $\mu = 1$ is presented in **Figure 2.1.3**, which also shows the mean $\bar{r} = 15\,\mu\mathrm{m}$, the modal $r_{\mathrm{mod}} = 10\,\mu\mathrm{m}$, the mean volume $r_v = 19.6\,\mu\mathrm{m}$, and the effective $r_{eff} = 25\,\mu\mathrm{m}$ radii.

To approximate PSDs of large particles such as raindrops, snowflakes, graupel, and hail, the exponential (Marshall-Palmer) distribution is often used:

$$f_r(r) = \frac{N_0}{\beta}\exp\left(-\frac{r}{\beta}\right) \tag{2.1.29}$$

The distribution (Equation 2.1.29) contains two parameters: N_0 and β.

Table 2.1.3 Gamma distribution
$f_r(r) = \frac{N_0}{\Gamma(\alpha)\beta}\left(\frac{r}{\beta}\right)^{\alpha-1}\exp\left(-\frac{r}{\beta}\right)$ (three parameters N_0, α, β).

Concentration	N_0
Mean radius	$\bar{r} = \beta\alpha$
RMS size distribution width	$\sigma_f = \beta\sqrt{\alpha}$
Dispersion of size distribution	$\delta_f = \alpha^{-1/2}$
Optical extinction	$\beta_{ext} = 2\pi N_0\beta^2\alpha(\alpha+1)$
Mean volume radius	$r_v = \beta\left(\prod_{k=0}^{2}(\alpha+k)\right)^{1/3}$
Effective radius	$r_{eff} = \beta(\alpha+2)$
Modal radius	$r_{\mathrm{mod}} = \beta(\alpha-1)$
Mass content	$M = \frac{4}{3}\pi\rho N_0\beta^3\prod_{k=0}^{2}(\alpha+k)$
Reflectivity	$Z = 198.0618 + 10\,\mathrm{lg}\left[N\beta^6\prod^{5}(\alpha+k)\right]$

Utilization of Equations (2.1.27), (2.1.28), and (2.1.29) is convenient from the mathematical point of view because all the moments of these distributions can be expressed using analytical expressions (see **Tables 2.1.3–2.1.5**). In addition, using PSD in the form of Equations (2.1.27)–(2.1.29) allows analytical calculation of most microphysical quantities, such as averaged bulk densities, fall velocities, etc., thus making the calculations simpler and less time-consuming. Due to their advantages, these shapes for approximating PSDs of different cloud hydrometeors are widely used (see the book by Straka (2009) for a detailed description of such calculations).

Note that along with distribution $f_r(r)$ given in Equation (2.1.28), PSD $f(m)$ is also often used in the form of generalized Gamma distribution (e.g., Seifert and Beheng, 2006)

$$f(m) = \tilde{N}_0 m^{\tilde{\nu}}\exp\left(-\tilde{\lambda}m^{\tilde{\mu}}\right) \tag{2.1.30}$$

with the normalization condition described by Equation (2.1.2). Parameters in Equation (2.1.30) have specific names. $\tilde{N}_0$ is the intercept parameter, $\tilde{\nu}$ is the shape parameter, $\tilde{\lambda}$ is the slope or scale parameter, and $\tilde{\mu}$ is the dispersion parameter. The parameters $\tilde{\nu}$ and $\tilde{\mu}$ determine the shape of $f(m)$ at very small m (initial shape parameter) and for very large m (final shape parameter), respectively. In case a one-to-one relationship exists between radius and mass as for spherical particles, both $f_r(r)$ and $f(m)$ are represented by general Gamma

Table 2.1.4 General Gamma distribution $f_r(r) = \frac{N_0 \mu}{\Gamma(\alpha)\beta}\left(\frac{r}{\beta}\right)^{\mu\alpha-1} \exp\left[-\left(\frac{r}{\beta}\right)^{\mu}\right]$ (four parameters N_0, α, β, μ).

Concentration	N_0
Mean radius	$\bar{r} = \beta \frac{\Gamma(\alpha+\mu^{-1})}{\Gamma(\alpha)}$
RMS size distribution width	$\sigma_f = \beta \frac{\sqrt{\Gamma(\alpha+2\mu^{-1})\Gamma(\alpha) - \Gamma^2(\alpha+\mu^{-1})}}{\Gamma(\alpha)}$
Dispersion of size distribution	$\delta_f = \sqrt{\frac{\Gamma(\alpha+2\mu^{-1})\Gamma(\alpha)}{\Gamma^2(\alpha+\mu^{-1})} - 1}$
Optical extinction	$\beta_{ext} = 2\pi N_0 \beta^2 \frac{\Gamma(\alpha+2\mu^{-1})}{\Gamma(\alpha)}$
Mean volume radius	$r_v = \beta\left(\frac{\Gamma(\alpha+3\mu^{-1})}{\Gamma(\alpha)}\right)^{1/3}$
Effective radius	$r_{eff} = \beta \frac{\Gamma(\alpha+3\mu^{-1})}{\Gamma(\alpha+2\mu^{-1})}$
Modal radius	$r_{\text{mod}} = \beta(\alpha - \mu^{-1})^{1/\mu}$
Mass content	$M = \frac{4}{3}\pi\rho N_0 \beta^3 \frac{\Gamma(\alpha+3\mu^{-1})}{\Gamma(\alpha)}$
Reflectivity	$Z = 198.0618 + 10\,\lg\left[N_0 \beta^6 \frac{\Gamma(\alpha+6\mu^{-1})}{\Gamma(\alpha)}\right]$

distribution. One can easily convert parameters of the respective PSDs.

Often, expressions of general Gamma distribution in the form $f(D) = N_0' D^{\nu'} \exp(-\lambda' D^{\mu'})$are used for approximation of cloud particle PSDs as a function of effective diameter D (e.g., Ferrier, 1994; Milbrandt and McTaggart-Cowan, 2010). The parameter names of this distribution are the same as $f(m)$. Using the relation $m = \frac{4}{3}\pi\rho r^3$ (ρ is bulk density of hydrometeor), one can get the following relationships between parameters of $f(D)$ and $f(m)$: $N_0' = 3\tilde{N}_0\left(\frac{1}{6}\pi\rho\right)^{\tilde{\nu}+1}$, $\nu' = 3\tilde{\nu} + 2$, $\lambda' = \tilde{\lambda}\left(\frac{1}{6}\pi\rho\right)^{\tilde{\mu}}$, and $\mu' = 3\tilde{\mu}$.

As was noted previously, log-normal distribution with three parameters is also used in cloud physics for the description of aerosol and sometimes as droplet size distribution (e.g., Pinsky et al., 2014). The log-normal distribution is described by the following equation:

$$f_r(r) = \frac{N_0}{r\sigma\sqrt{2\pi}} \exp\left\{-\frac{(\ln r - \ln R)^2}{2\sigma^2}\right\} \quad (2.1.31)$$

Table 2.1.6 presents moments of log-normal distribution used in cloud physics.

Table 2.1.5 Exponential (Marshall–Palmer) distribution $f_r(r) = \frac{N_0}{\beta} \exp\left(-\frac{r}{\beta}\right)$ (two parameters: N_0, β).

Concentration	N_0
Mean radius	$\bar{r} = \beta$
RMS size distribution width	$\sigma_f = \beta$
Dispersion of size distribution	$\delta_f = 1$
Optical extinction	$\beta_{ext} = 4\pi N_0 \beta^2$
Mean volume radius	$r_v = \sqrt[3]{6}\beta$
Effective radius	$r_{eff} = 3\beta$
Modal radius	$r_{\text{mod}} = 0$
Particle content	$M = 8\pi\rho N_0 \beta^3$
Reflectivity	$Z = 226.6351 + 10\,\lg(N\beta^6)$

Table 2.1.6 Lognormal distribution $f_r(r) = \frac{N_0}{r\sigma\sqrt{2\pi}} \exp\left\{-\frac{(\ln r - \ln R)^2}{2\sigma^2}\right\}$ (three parameters N, R, σ).

Concentration	N_0
Mean radius	$\bar{r} = R \exp\frac{\sigma^2}{2}$
RMS size distribution width	$\sigma_f = R \exp\frac{\sigma^2}{2}\left[\exp(\sigma^2) - 1\right]^{1/2}$
Dispersion of size distribution	$\delta_f = \left[\exp(\sigma^2) - 1\right]^{1/2}$
Optical extinction	$\beta_{ext} = 2\pi N_0 R^2 \exp(2\sigma^2)$
Mean volume radius	$r_v = R\exp\frac{3\sigma^2}{2}$
Effective radius	$r_{eff} = R \exp\frac{5\sigma^2}{2}$
Modal radius	$r_{\text{mod}} = R\exp(-\sigma^2)$
Mass content	$M = \frac{4}{3}\pi\rho N_0 R^3 \exp\frac{9\sigma^2}{2}$
Reflectivity	$Z = 198.0618 + 10\,\lg\left[N_0 R^6 \exp(18\sigma^2)\right]$

2.2 Atmospheric Aerosols

A detailed description of atmospheric aerosols is given in the books by Hobbs (1993); Pruppacher and Klett (1997) and Levin and Cotton (2009). In this section, we provide the reader with some basic information on APs that is essential for cloud modeling. Since APs give rise to formations of droplets and ice crystals, it is important to know size distributions and the chemical composition of APs. These parameters are often used as the initial conditions in cloud simulations. APs are affected by clouds, which makes them the focus of studies in acid rains. Some cloud chemical models are aimed, among other things, at accurate reproducing of AP formation in the atmosphere.

2.2.1 Types and Size Distributions of Atmospheric Aerosols

Atmospheric APs can be classified into subgroups based on their geographical location, size, the mechanism of formation, and chemical composition. All these aspects of classification are closely interrelated.

Geographically, APs are classified into maritime, continental, and background aerosols. According to their size, APs are separated into Aitken APs (0.001–0.1 μm), large APs (0.1–1 μm), and giant APs (>1 μm) (Hobbs, 1993). Aitken APs were named after J. Aitken, who built the first probe to measure the number of APs that can be activated to turn into droplets at very high supersaturations (Mason, 1971).

APs are classified according to their size into three ranges: 0.001–0.1 μm (the nucleation mode); 0.1–1 μm (the accumulation mode); and $\geq 1\,\mu$m (the coarse particle mode) (Whitby, 1973). The nucleation mode is produced by gas-to-particle conversion, the accumulated mode by coagulation of smaller APs alongside heterogeneous condensation, and the coarse mode by mechanical processes.

The contribution of APs to cloud processes depends on their size and chemical composition. APs can be classified according to chemical composition into the mineral APs and biogenic and organic APs. General solubility and the soluble/insoluble fractions are the most important quantities, since soluble particles give rise mostly to cloud droplets, while insoluble particles are mainly responsible for formation of ice crystals and droplet freezing. These parameters have to be taken into account in numerical cloud modeling.

The sources of APs are multiple and include seas and oceans; land surface (especially deserts and mountains); volcanoes, burning biomass (e.g., forest fires) and fuel combustion; chemical and photochemical reactions in the atmosphere, etc. Sources and spatial distribution of APs are described in detail by Hobbs (1993) and Levin and Cotton (2009). Due to microphysical processes, APs grow with time, and the size distributions of aerosols depend on the AP's age, i.e., on time elapsed since the AP's formation.

Knowledge on AP size distributions is extremely important for cloud physics and cloud modeling, as the size distributions determine the meteorological processes related to clouds and radiation. Multiple measurements in different regions of Earth's surface showed that within the radius ranges of 0.1–10 μm, the aerosol number distribution $f_r(r_N)$ obeys the power law (a fact first established by Junge (1955):

$$f_r(r_N) = A r_N^{-\alpha}, \tag{2.2.1}$$

where α is ~ 4. **Figure 2.2.1** illustrates size distributions of APs in different geographical regions with high and low AP concentrations. In certain cases, variations of α can be very significant.

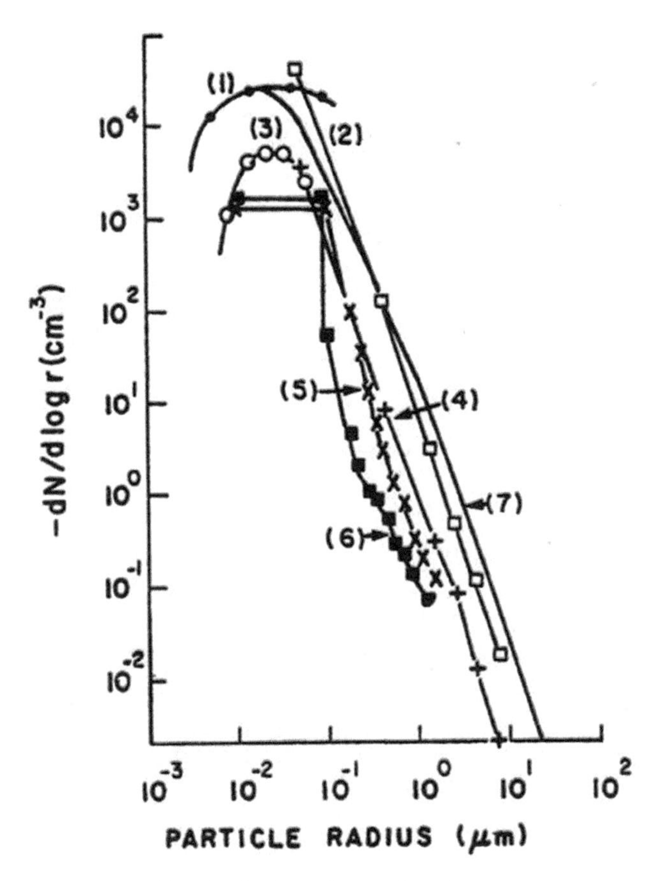

Figure 2.2.1 Size distribution of aerosol particles ($10^{-2}\,\mu\text{m} \leq r_N \leq 10\,\mu\text{m}$) in the air over various locations in Central Europe and the US Data obtained by Junge (1955): (1) Frankfurt, ion counter; (2) Frankfurt, nuclei counter and impactor; (3) Zugspitze, 3,000 m, ion counter; (4) Zugspitze, 3,000 m, nuclei counter and impactor. Data obtained by Junge (1969) using the Royco counter and nuclei counter: (5) Crater Lake, 2,200 m without subsidence; (6) Crater Lake, 2,200 m with subsidence. Data obtained by Noll and Pilat (1971); (7) Seattle, State of Washington (adopted from Pruppacher and Klett, 1997, with the permission of Springer, and from Jaenicke, 1993, courtesy of Elsevier).

AP size distributions are often approximated by lognormal size distribution Equation (2.1.31). Typically, AP size distribution is represented as a sum of three lognormal functions as

$$\tilde{f}(r_N) = \frac{dN_N}{d(\log r_N)} = \frac{1}{\sqrt{2\pi}} \exp \sum_{i=1}^{3} \frac{N_i}{\log \sigma_i} \left\{ -\frac{(\log r_N/R_i)^2}{2(\log \sigma_i)^2} \right\}, \tag{2.2.2}$$

where r_N is the particle radius (in μm), R_i is the measure of the mean i-th radius (in μm), N_i is the i-th concentration, and $\log \sigma_i$ is a measure of i-th lognormal function width. In Equation (2.2.2), nine parameters are needed to describe AP size distribution. Typically, each term of Equation (2.2.2) describes APs in certain size ranges. Each range corresponds to the specific physical mechanism of AP formation. The number of lognormal function in Equation (2.2.2) is equal, therefore, to the number of the size ranges (I, II, or III). However, sometimes size distribution of aerosol belonging to a certain size range is also approximated as a sum of several log-normal functions, as shown in **Table 2.2.1**.

The main types of tropospheric aerosols (polar, background, marine, remote continental, desert, rural, and urban) differ by geographic location, chemical composition, and size distributions. Table 2.2.1 presents the parameters of size distributions for these types of tropospheric aerosols. The size and volume distributions of APs corresponding to parameters from Table 2.2.1 are plotted in **Figure 2.2.2**. For polar AP, each size range is described by one term in Equation (2.2.2). For rural and urban APs, a sum of the three terms in Equation (2.2.2) describes only the nucleation mode (range I).

Figure 2.2.2b shows that while the total AP concentration is determined by small AP, the total AP mass is typically determined by large AP.

In numerical models, each AP range is often approximated by a certain lognormal function in Equation (2.2.2). Parameters of such functions for the main AP types are presented in **Table 2.2.2**. Comparison of Tables 2.2.2 and 2.2.1 shows that all AP types contain the coarse mode, but concentration of APs in this mode is much lower than that in the accumulation mode or in the nucleation mode. AP distributions corresponding to Table 2.2.2 are presented in **Figure 2.2.3**.

In humid air, soluble APs begin growing, and their size is determined by the environment humidity (Section 5.1). Wet APs are called haze particles. Size distributions of haze particles belonging to different AP types and calculated at RH = 100 perecent are presented in Figure 2.2.3. One can see that the dry AP is several times smaller than the corresponding haze particle.

Table 2.2.1 Parameters of aerosol size distributions approximated by the sum of the three lognormal functions (from Jaenicke, 1993, Courtesy of Elsevier).

Aerosol	Range[a]	i	N_i (cm^{-3})	R_i (μm)	$\log \sigma_i$
Polar	I	1	2.17×10^1	0.0689	0.245
	II	2	1.86×10^{-1}	0.375	0.300
	III	3	3.04×10^{-4}	4.29	0.291
Background	I	1	1.29×10^2	0.0036	0.645
	II	2	5.97×10^1	0.127	0.253
	II	3	6.35×10^1	0.259	0.425
Maritime	I	1	1.33×10^2	0.0039	0.657
	II	2	6.66×10^1	0.133	0.210
	II	3	3.06×10^0	0.29	0.396
Remote continental	I	1	3.20×10^3	0.01	0.161
	I	2	2.90×10^3	0.058	0.217
	II	3	3.00×10^{-1}	0.9	0.380
Desert dust storm	I	1	7.26×10^2	0.001	0.247
	I	2	1.14×10^3	0.0188	0.770
	III	3	1.78×10^{-1}	10.8	0.438
Rural	I	1	6.65×10^3	0.00739	0.225
	I	2	1.47×10^2	0.0269	0.557
	I	3	1.99×10^3	0.0419	0.266
Urban	I	1	9.93×10^4	0.00651	0.245
	I	2	1.11×10^3	0.00714	0.666
	I	3	3.64×10^4	0.0248	0.337

*In Table 2.2.1, ranges I, II, and III correspond to the Aitken mode (nucleation mode), accumulation mode, and the coarse (or giant) particle mode, respectively.

The AP types presented in Tables 2.2.1 and 2.2.2 have the following general characteristics.

Polar aerosols: APs located close to the surface in the Arctic and the Antarctic (Jaenicke and Schutz, 1982; Ito, 1982; Shaw, 1986; Jaenicke et al., 1992). The distribution reflects the fact that these APs are aged (low numbers of AP in the nucleation mode).

Background aerosols: APs located above cloud layers. Measurements of these APs were mostly carried out on mountain tops or within subsiding air masses typical of mid-tropospheric conditions (Hobbs, 1993).

Remote maritime aerosols: According to Hobbs (1993), maritime APs are supposedly composed of a

Table 2.2.2 Aerosol distribution parameters for four tested cases. Aerosol radii and the width of the aerosol modes are given in **μm**, concentrations within the modes are in cm^{-3} (from Ghan et al., 2011; courtesy of John Wiley & Sons, Inc.)

	Nuclei mode			Accumulation mode			Coarse mode		
	r_1	σ_1	N_1	r_2	σ_2	N_2	r_3	σ_3	N_3
Marine	0.005	1.6	340	0.035	2.0	60	0.31	2.7	3.1
Clean continental	0.008	1.6	1,000	0.034	2.1	800	0.46	2.2	0.72
Background	0.008	1.7	6,400	0.038	2.0	2,300	0.51	2.16	3.2
Urban	0.007	1.8	106,000	0.027	2.16	32,000	0.43	2.21	5.4

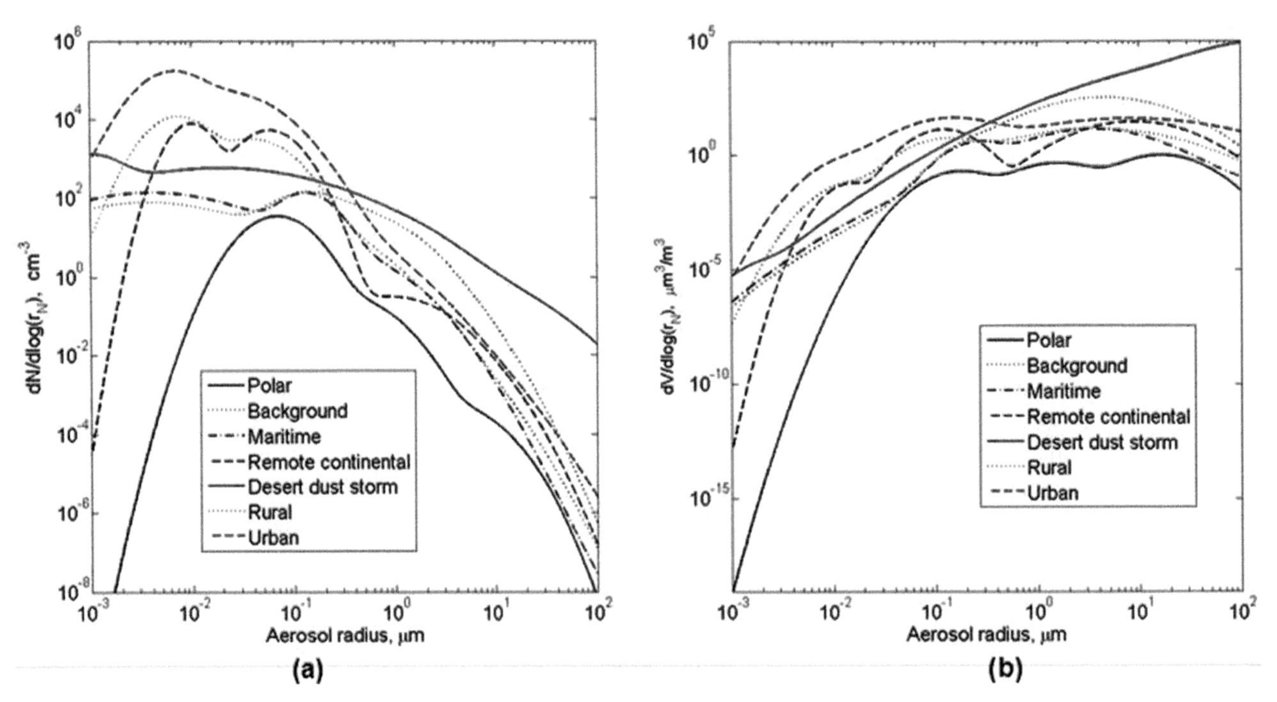

Figure 2.2.2 Model of number size distributions **(a)** and volume distributions **(b)** of the selected AP types (according to Table 2.2.1).

combination of sea salt and particulate oxidation products of biogenic dimethyl sulphide. Most submicron APs consist of sulphate and organics. The specification of the organics is still not known exactly, but apparently there are substantial amounts of both soluble and insoluble carbon. The impact of organic and inorganic fractions within APs depends both on the geography and pollution penetrating oceans from continents. Insoluble carbon species tend to be concentrated in the smallest AP sizes. This is an important fact since some fraction of these organic species acts as a surfactant, forming layers on AP surfaces and thus affecting the AP's performance during formation of drops and ice.

The sea salt AP size distributions extend well down into the submicron range (see Levin and Cotton, 2009 for particular references). Sea salt contribution is governed by wind speed. The values of parameters in Table 2.2.1 are given for the typical ambient humidity and wind speed of 5.5–7.9 m/s.

Figure 2.2.4 shows size distributions of APs within a marine boundary layer covered by stratocumulus clouds, as measured in the field experiment DYCOMS II during two research flights: RF01 (no drizzling clouds) and RF07 (light drizzle), used in a modeling study (Magaritz et al., 2009). In this case, the background wind was low (~6–7 m/s). The maximum AP radius measured in this experiment was 1.4 μm; the AP concentration was 160–180 cm^{-3}.

Typically, the concentration in submicron modes of maritime APs varies from a few tens per cm^3 to several

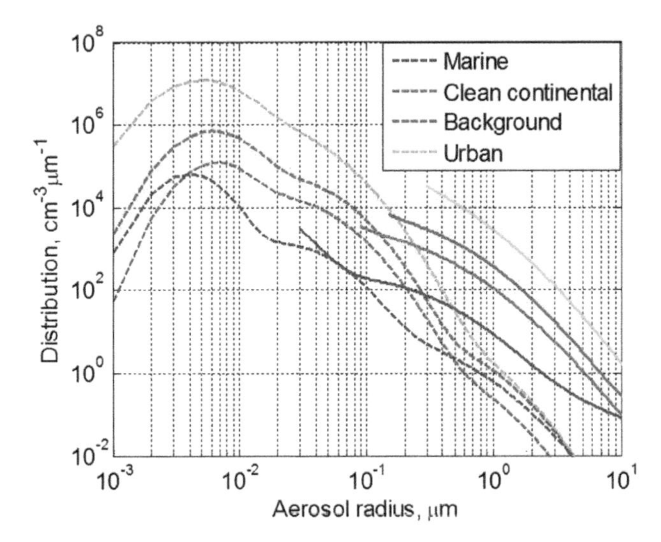

Figure 2.2.3 The distributions of dry aerosols (dashed lines) and wet aerosols (solid lines) at cloud base $T = 6°C$. Wet aerosols are haze particles (from Pinsky et al., 2014; courtesy of © John Wiley & Sons, Inc.).

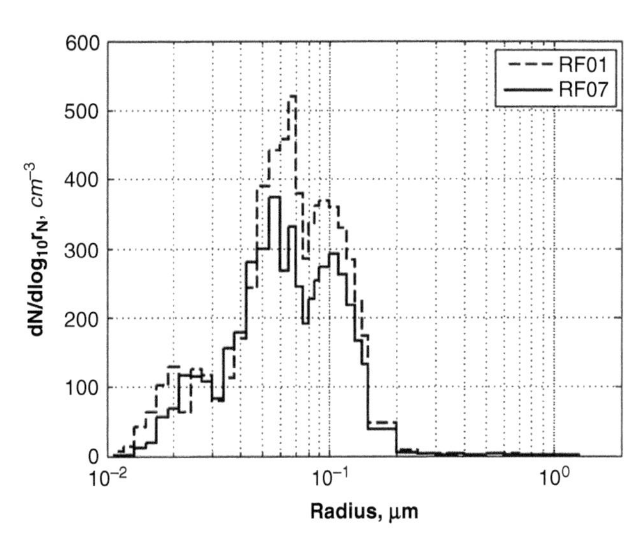

Figure 2.2.4 Size distributions of dry aerosols in RF01 and RF07 flights used in the simulations. The distributions are based on composite data obtained with a radial differential mobility analyzer and with a passive cavity aerosol spectrometer probe (from Magaritz et al., 2009; courtesy of © American Meteorological Society. Used with permission).

hundred per cm^3. The variations are especially pronounced at the tail of the distributions where the concentration of particles is low, as can be seen in **Figure 2.2.5**, showing the mass distributions of maritime APs.

Sea spray: Under strong winds (storms or tropical cyclones), sea spray forms near the surface due to breakup of waves and other mechanical processes. A specific feature of spray drops is their high salinity (about 35 g of salt per 1 kg of solution over the Atlantic Ocean), and a significant presence of large drops. The largest radii of dry aerosols in spray may exceed 50 μm, and the radius of spray drops may exceed 200 μm. The maximum size of spray drops increases with an increase in the wind speed. The mass distributions of large APs over sea at the wind speed ranging from 15 m/s to 20 m/s within the lowest 20 m-deep atmospheric layer are presented in **Figure 2.2.6**. The sizes of the particles in Figure 2.2.6 are recalculated for 80% environmental humidity. The concentration of spray drops with radii exceeding 1 μm is a few cm^{-3}.

Remote continental aerosols: These APs are located close to the surface in continental regions that are not greatly affected by human activities. However, nowadays remote continental APs consist of some natural matter (dust, biogenic material) mixed with pollution aerosols at varying levels of dilution that are of anthropogenic nature.

Desert dust-storm aerosols: These natural aerosols may be considered an extreme AP type (Hobbs, 1993). The total concentration of such aerosols is comparatively low (~1500 cm^{-3}), but their mass loading is quite high. Because the deserts cover huge areas and are often located within zones of strong winds, their potential for a long-range transport is high, and desert dust influences vast areas. For instance, the Saharan dust may affect hail storms in Florida (Van den Heever et al., 2006). Desert dust contains a small soluble fraction (~0.1) and can play the role of cloud condensational nuclei, i.e., under favorable conditions can give rise to droplet formation. The typical size distribution of desert dust APs is presented in Figure 2.2.2.

Rural aerosols: Aerosols in rural areas are mainly continental and are moderately impacted by anthropogenic factors. The results of measurements in the lower troposphere over the high plains of North America presented by Hobbs et al. (1985) are closer to the model distribution values for rural and urban aerosols given in Table 2.2.1, rather than to those for the background or remote continental aerosols.

Urban aerosols: Due to the dominant influence of emissions from industries, home heating, and traffic, aerosols over cities and other human settlements are of a special interest. Urban aerosols typically consist mostly of the Aitken particles, with some presence of giant APs, and are characterized by high AP concentration from thousands to tens of thousands of AP per cm^3. The AP size distribution is characterized by the three modes with modal diameters of 0.005–0.02 μm, 0.1 μm, and 1.0 μm,

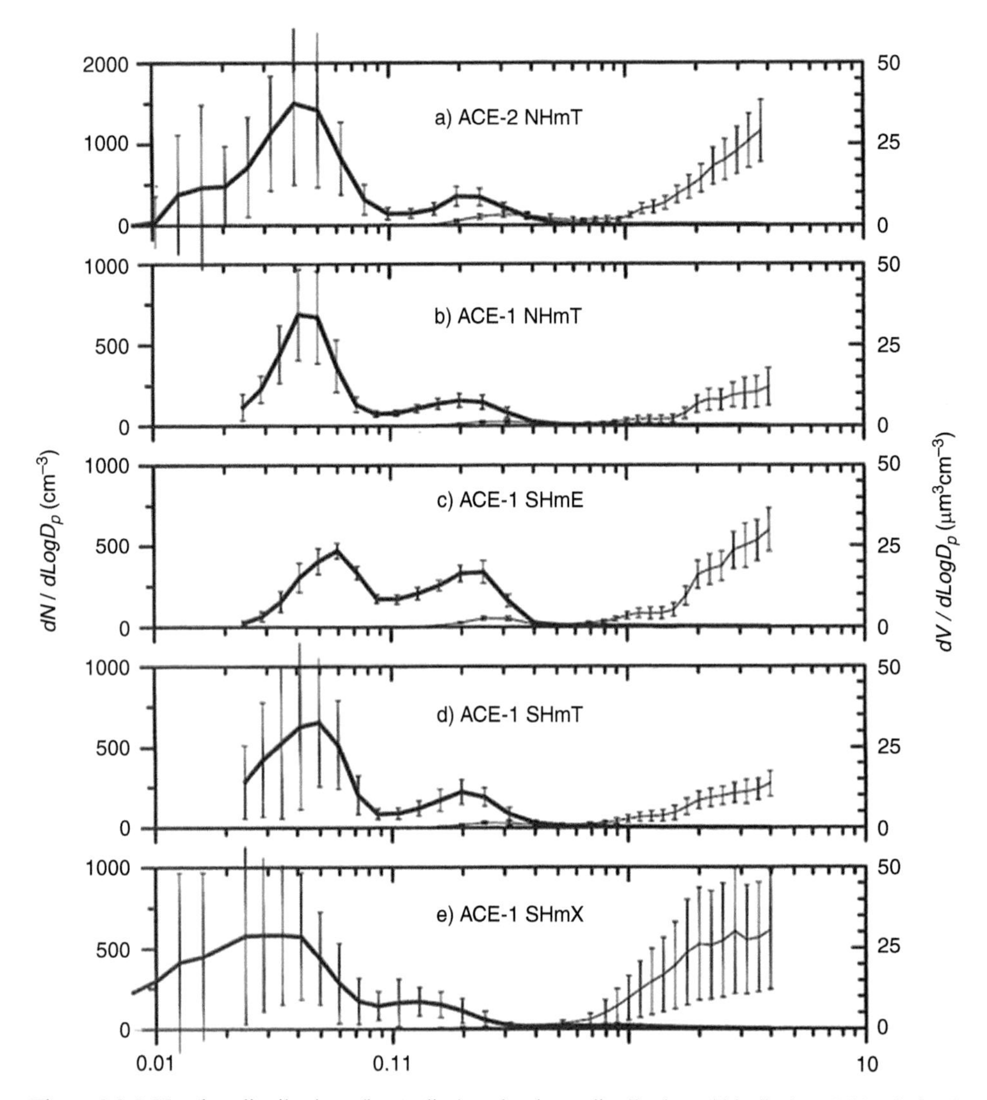

Figure 2.2.5 Number distributions (heavy line) and volume distributions (thin line) at 55% relative humidity for different air masses during the first two Aerosol Characterization Experiments (ACEs). The vertical bars represent one standard deviation of the mean number or volume in the size bin over the averaging period (23–114 hr) (from Bates et al., 2002; courtesy of © John Wiley & Sons, Inc.).

respectively. The distribution of APs within submicron size range is highly variable depending on time of day, season, etc. An example of size distribution of urban APs is presented in Figure 2.2.2. The amount of APs with radii close to 0.001 μm is often uncertain because of instrumentation problems. In Figure 2.2.2, the main difference among the curves is for the smallest and the largest particles, which probably reflects the influence of AP residence time (aerosol ages). The minimum variation of AP concentration is for AP sizes ranging from 0.1 μm to 1 μm. Particles of this size influence atmospheric visibility, causing variations up to more than two and a half orders of magnitude.

The values and distributions presented in Table 2.2.1 and Figure 2.2.2 characterize some averaged AP properties. In real conditions, the magnitudes of quantities characterizing distributions of different AP classes and modes vary widely, covering one to two orders of magnitude, as shown in **Figure 2.2.7**. Thus, when particular study cases are simulated, it is desirable to use aerosol data measured for the corresponding cases.

The chemical composition of APs at remote, rural, and polluted venues in Western Europe is presented in **Figure 2.2.8**. One can see significant fractions of carbonates, as well as insoluble fractions. The insoluble organics tend to be of smaller sizes than soluble species.

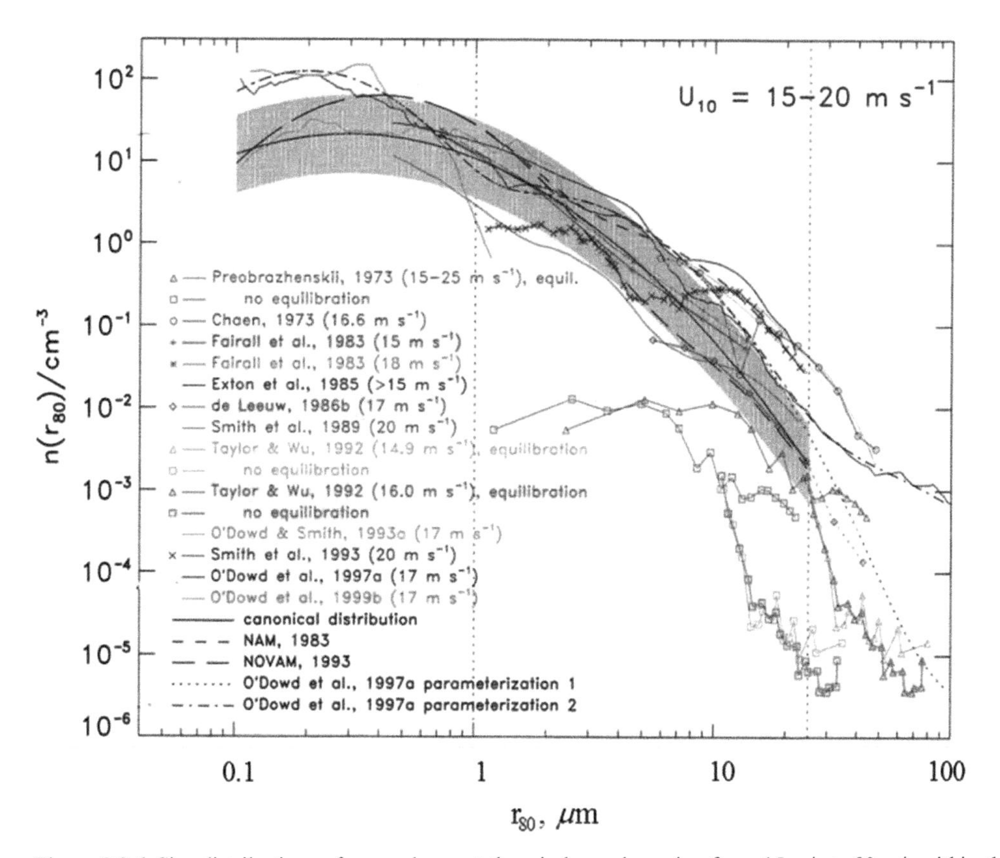

Figure 2.2.6 Size distributions of spray drops at the wind speed ranging from 15 m/s to 20 m/s within the lowest 20 m-deep atmospheric layer recalculated for 8% relative humidity. The shaded area denotes the standard deviation of the measurements; $n(r_{80}) = d \ln N / d \ln r_{80}$. Results obtained by different authors are denoted by different symbols. The references are presented in Lewis and Schwartz, 2004 (from Lewis and Schwartz, 2004; courtesy of © John Wiley & Sons, Inc.).

Biomass burning APs: Biomass burning emissions are the second AP source on a global scale (besides industrial pollution). Biomass frequently burns mostly in the tropical regions, especially in Brazil and Southern Africa. The size distributions of APs formed as a result of biomass burning contain two dominating modes: the accumulation mode centered within the range of 0.1–0.3 μm in diameter, and the coarse mode centered within the range of 4 μm in diameter. Concentrations of APs in polluted zones can exceed 10^4 cm^{-3} (Rissler et al., 2006). Biomass burning aerosols consist of carbonaceous species, a significant fraction being black carbon. About half of the organic matter is water soluble. The inorganic fraction is comparatively low. Table 2.2.3 presents the chemical composition of different AP types.

2.2.2 Vertical Profiles of AP Concentration

The vertical distribution of APs is an important property that determines the residence time in the atmosphere and long-range transport potentials of APs, as well as AP impact on climate. It is important to know the vertical distribution of AP in order to estimate lateral penetration of APs into clouds. Currently, observation data on AP vertical distributions are scarce, while data on AP size distribution and chemical composition as a function of height are actually lacking.

Typically, AP concentration is nearly constant within the well-mixed atmospheric boundary layer and decreases exponentially with height up to the level of 4 km. Above this level, AP concentration almost does not vary over height (Mazin and Schmeter, 1983). These height dependencies are often approximated as

$$N_N(z) = \begin{cases} N_1, & z < z_{mix} \\ N_1 \exp\left(-\dfrac{z - z_{mix}}{H_p}\right), & z_{mix} < z < z_h \\ N_2, & z > z_h \end{cases} \tag{2.2.3}$$

where $N_N(z)$is the total AP concentration, z_{mix} is the depth of the well-mixed layer near the surface, and H_p

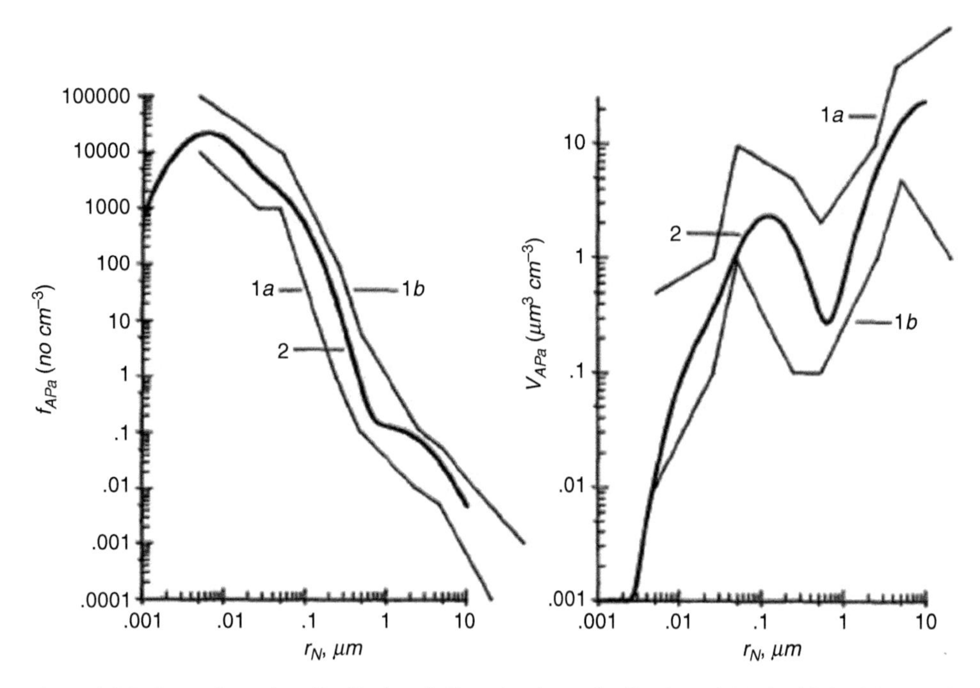

Figure 2.2.7 Aerosol number distribution (left) and volume distribution (right) for high plains as observed by Hobbs et al. (1985) (curves 1a, 1b) and as used in the model based on Equation (2.2.2) (curve 2) (from Respondek et al., 1995; courtesy of © American Meteorological Society. Used with permission).

is the characteristic spatial scale at which the AP concentration decreases by 2.7 times.

To take into account the deviation of the vertical distribution of AP concentration from the exponential distribution, Jaenicke (1993) proposes the approximation formula

$$N_N(z) = N_{N0}\left[\exp\left(-\frac{z}{|H_p|}\right) + \left(\frac{N_{Nbg}}{N_{N0}}\right)^v\right]^v, \tag{2.2.4}$$

where $v = H_p/|H_p|$, N_{N0} is the number concentration of APs at the surface and N_{Nbg} is the number concentration of the corresponding background APs aloft. The values of the parameters in Equation (2.2.4) for different AP types are presented in **Table 2.2.4**.

Table 2.2.4 shows that if parameter $p_b = 0$, the mass content M_N of APs decreases exponentially with height:

$$M_N(z) = M_{N0}\exp\left(-\frac{z}{H_{M_N}}\right) \tag{2.2.5}$$

Figure 2.2.9 shows the vertical profiles of AP concentration of different aerosol types. Interpreting these results, one should take into account that APs concentrations have a large variability. This is to a large extent due to the short residence times of small particles, whose contribution into the number concentrations is dominating. The background aerosols exhibit an approximately constant vertical profile, with concentrations within the range of 300–3,000 cm^{-3} above the mixing layer up to the tropopause. The concentrations of polar APs are more stable, with the values around 200 cm^{-3} up to the tropopause, while concentrations at the surface are often much lower. Remote continental aerosols generally show a decrease in concentration over altitude even within the mixed layer. The steepness of the decrease depends on the surface concentrations. Above the mixed layer, continental AP concentrations approach the values of the background AP concentrations.

Over the oceans, AP concentration typically almost does not change with height, while AP mass rapidly decreases with height, being determined by the largest sea salt particles whose concentrations are low. In the mixed layer, the concentration of maritime APs can increase with height (negative values of H_p in Equation. (2.2.4)). In other cases, AP concentration is higher near the surface, and consequently $H_p > 0$. Observations show fluctuations (both decreasing and increasing) in AP concentration in the vicinity of clouds.

Variability of the height dependencies of AP concentration is high. In some cases, the aerosol type is the same over the height, in other cases the type of APs changes

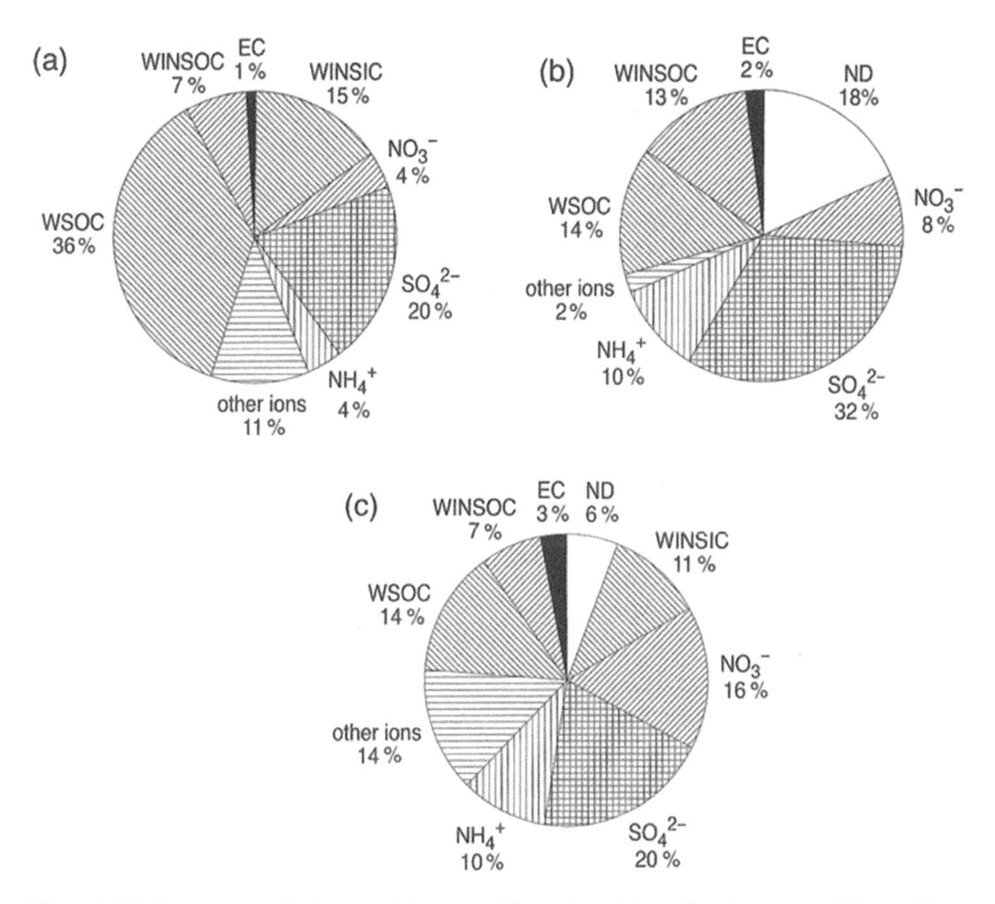

Figure 2.2.8 Average mass balance at **(a)** remote, **(b)** rural, and **(c)** polluted venues in Western Europe. ES = elemental carbon, WINSOC = water insoluble organic compounds, WINSIC = water insoluble inorganic carbon, WSOC = water soluble organic compounds, and ND = not determined (from Zappoli et al., 1999; courtesy of © Elsevier).

with height indicating different sources of APs at different altitudes. An example of the case when the AP type does not change with height is presented in **Figure 2.2.10a**, showing vertical profiles of the AP concentration as measured during 152 research flights over Beijing (China) during the period February 2005–September 2006 (Liu et al., 2009). Figure 2.2.10a shows also the AP concentration profiles averaged over all flights. One can see that the averaged AP concentration near the surface is 6600 cm^{-3}, while at the maximum it exceeds 12,000 cm^{-3}. The averaged AP concentration decreases exponentially with height as $N_N = 6612 \exp(-z/1419)$. More detailed analysis shows that all vertical profiles of the AP concentration can be separated into two groups, as shown in Figures 2.2.10b and 2.2.10c. The profiles belonging to group A were observed in unstable well-mixed BL and are characterized by nearly constant AP concentration within the BL and exponential decrease above the BL. The profiles belonging to group B occur under stable BL and are characterized by monotonic decrease in the AP concentration with height. In both cases aerosol size distributions do not change much with height, indicating that all APs were of the same type.

The vertical profiles of AP concentration in the case when the AP types are different at different heights are presented in **Figure 2.2.11**. The AP concentration was measured over the Indian Ocean not far from the southern coast of India during March 2006 (Corrigan et al., 2008). One can see that during the first half of the observed period the AP concentration decreased exponentially above the top of the BL. During the second half of the observed period the AP concentration increased above the BL and reached extremely high for the ocean values exceeding 2,000 cm^{-3}. Analysis of air trajectories showed that this increase in the AP concentration was caused by penetration of polluted air from Africa and Arabia.

The increase in the AP concentration above the BL top during Sahara dust intrusion is typical for the eastern Atlantic (Lynn et al., 2016).

Table 2.2.3 Mass concentrations (in $\mu g\ m^{-3}$) of tropospheric aerosols of various chemical compositions (from Jaenicke, 1993, courtesy of Elsevier).

Aerosol:	Urban			Remote Continental	Polar		Maritime
Location:	Tees-side England	West Covina California	Sapporo Japan	Wank Mt. Germany	Ny-Alesund Spitsbergen	Barrow Alaska	Atlantic Ocean
Year:	1967	1973	1982	1972–1982	1979	1986	pre 1975
SO_4^{2-}	13.80	16.47	2.8–5.3	2.15	2.32	1.91	2.58
NO_3^-	3.00	9.70	0.1–1.6	0.85	0.055	0.13	0.050
Cl^-	3.18	0.73	0.1–1.6	0.087	0.013	1.11	4.63
Br^-	0.07	0.53				0.05	0.015
NH_4^+	4.84	6.93	0.6–1.8	1.00	0.23	0.65	0.16
Na^+	1.18	3.10	0.3–0.9	0.047	0.042	0.68	2.91
K^+	0.44	0.90		0.045	0.023	0.38	0.11
Ca^{2+}	1.56	1.93	0.2–1.0	0.082	0.073		0.17
Mg^{2+}	0.60	1.37			0.032		0.40
Al_2O_3	3.63	6.43		0.20			
SiO_2	5.91	21.10		0.51			
Fe_2O_3	5.32	3.83		0.10	0.24		0.14
CaO	—[a]	—[a]		0.09	0.91		
PbO				0.020			
ZnO				0.020			
Cu				0.002			
Cd				0.001			
Totals[b]	43.53	75.70	32.3	48.73	3.94	4.87[c]	11.17
TOM[d]	56		2.4–16.0	4	0.14		2
Biological material[e]	17.0	22.5	7.3				

[a] All Ca are assumed to be water soluble.
[b] Totals are the individual compounds added up.
[c] The total is from Li and Winchester (1989), who also include Methanesulfonate (0.012), Formate⁻ (0.24), Acetate⁻ (0.73), Propionate⁻ (0.147), and Pyruvate⁻ (0.010).
[d] TOM is the total organic matter calculated on the basis of carbon content.
[e] Biological material is estimated for rural and urban aerosols only, using the fraction of 14.6% of the total mass.

2.2.3 Residence Time

Significant variations in tropospheric AP concentrations are related to differences in the AP residence time. The residence time τ depends on the composition of APs, their size, and geographical location. In a generalized form, τ can be represented as (Jaenicke, 1988, 1993)

$$\frac{1}{\tau} = \frac{1}{K}\left(\frac{r_N}{R}\right)^2 + \frac{1}{K}\left(\frac{r_N}{R}\right)^{-2} + \frac{1}{\tau_{wet}}, \tag{2.2.6}$$

where $K \approx 1.28 \cdot 10^8 s$, $R = 0.3\,\mu m$ is the radius of particles with the maximum residence time and τ_{wet} is the measure of the influence of wet removal (τ_{wet} is eight days in the lower troposphere and three days in the middle and upper troposphere). This expression yields the maximum residence time of about one week for particles with radius of 0.3 μm in cloudy areas, and up to several weeks in the upper atmosphere. Long residence time of particles of this size is attributed to a

Table 2.2.4 Parameters of vertical aerosol profiles[a] (from Jaenicke, 1993, courtesy of Elsevier).

Aerosol Type	Altitude (m)	Scale Height H_p (m)	Surface Value p_0	Background Value p_B
		Aerosol mass concentration ($p = M_a$, in μg m^{-3})		
Ocean	→2,400	900	16	0
Remote continental	→2,400	730	20	0
Desert	→6,000	2,000	150	0
Polar	→6,000	30,000	3	0
Background	→tropopause	∞	1	0
		Aerosol number concentration ($p = N_N$, in cm^{-3})[b]		
Ocean	→1,000	−290 to 440	10–3,000	300
Remote continental	→2,500	1,100 to 550	3,000–30,000	300
	→500	−130	10	
Polar	500→			200
	tropopause	∞	200	
Background	→tropopause	∞	300–2,000	300

[a]The parameters are defined in Equation (2.2.4). The general term p is to be replaced for N_N or M_N, in order to get number concentration or mass content, respectively. $p = p_0\left[e^{-z/|H_p|} + \left(\frac{p_b}{p_0}\right)^{v}\right]^{v}$

[b]The order of H_p values corresponds to the order of N_{N0} values. For instance, the values $H_p = -290$ m and 440 m correspond to $N_{N0} = 10\,cm^{-3}$ and $N_{N0} = 3{,}000\,cm^{-3}$, respectively.

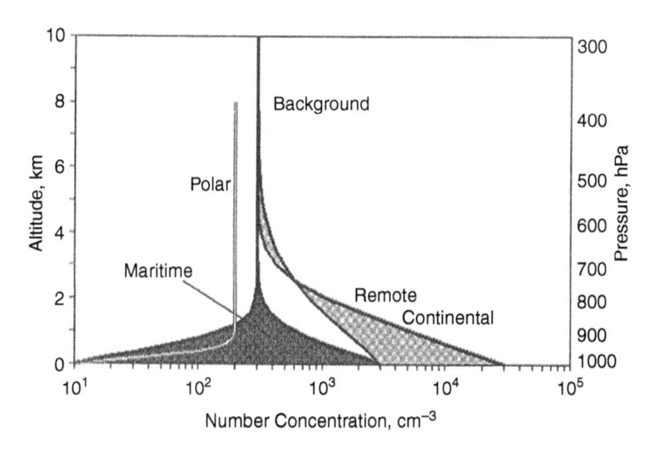

Figure 2.2.9 Vertical distribution of aerosol number concentration. The ranges of number concentrations are shown for ocean APs as well as for remote continental APs (from Jaenicke, 1993, courtesy of © Elsevier).

comparative low efficiency of wet removal process, when its efficiency is evaluated by averaging over large scales. Smaller and larger particles are suspended in the air for a much shorter time. The short residence time of smaller APs is the result of the Brownian coagulation, which shifts AP size to larger values. The short residence time of larger APs results from their higher settling velocities. Large APs are removed from the atmosphere faster than small ones. The longer residence times of APs in the upper troposphere, together with the spatial sources, appear to be the reasons for the comparatively constant background APs. Note that Equation (2.2.6) represents only the averaged values.

2.3 Cloud Drops

Drops are the major hydrometeors in clouds responsible for precipitation in warm clouds and for formation of ice particles. Drop freezing can be either spontaneous or caused by collisions with ice particles. The range of cloud drop diameters is wide, from about 1 μm to ~0.9 cm. Drops exceeding about 500 μm in radius become asymmetric, being oblate in the vertical direction due to resistance of the environmental air during their fall (**Figures 2.3.1**). The raindrop aspect ratio decreases with an increasing drop size. The dependence of aspect ratio (AR) on the equivalent drop diameter is shown in **Figure 2.3.2** and can be calculated using the formula by Brandes et al. (2002) for diameters below about 8.5 μm.

Large raindrops experience breakup that can be spontaneous (when a drop breaks up due to instability of its shape during its fall) or collisional when the breakup takes place after a collision with other particles (Low

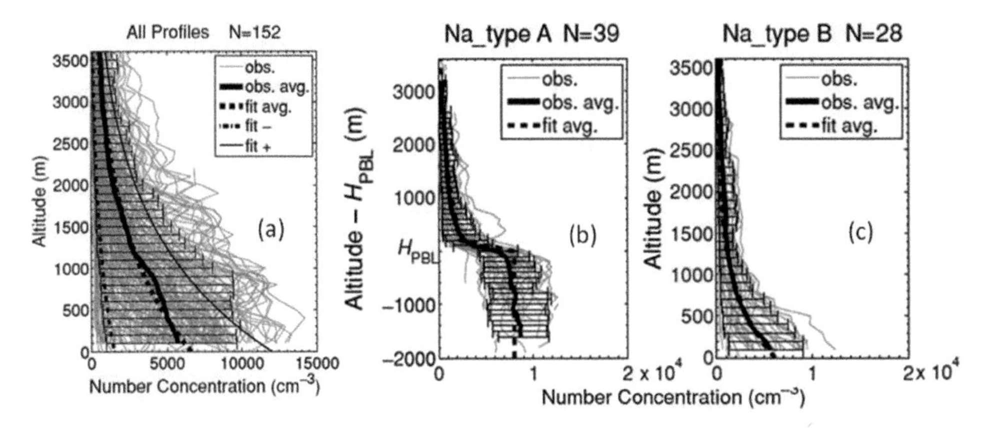

Figure 2.2.10 (a) All vertical profiles of AP concentration. The solid black line, solid grey lines, and horizontal error bars represent the average vertical profile (obs. avg.), the observed profiles (obs.), and the 10th and 90th percentiles of the AP concentration values at each altitude level, respectively. The dashed black line shows a regression curve $N_N = 6612\exp(-z/1419)$. The dash-dotted black line and the thin solid black line represent the AP concentration profiles under clean air condition and the heavy polluted condition. (b) Profiles of the AP concentration in cases of an elevated inversion layer and well mixed BL. (c) Profiles of the AP concentration in cases of relatively stable surface layers. Notations are as in the left panel (from Liu et al. 2009, courtesy of © Tellus).

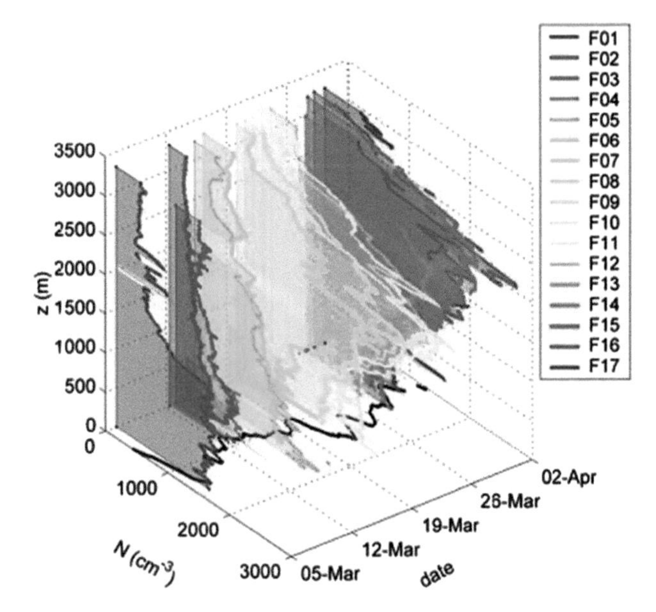

Figure 2.2.11 Vertical profiles of the AP concentration during the Maldives Autonomous Unmanned Aerial Vehicle Campaign (MAC) (March, 2006). The legend indicates flight numbers. The solid black line at $z = 0$ shows the AP concentration collected at the surface station (from Corrigan et al., 2008, courtesy of © ACP).

and List, 1982). The probability of breakup caused by either of the mechanisms increases with increasing drop size. Typically, both factors are working together, and in the presence of collisions they break up large raindrops before their size reaches the size corresponding to a spontaneous breakup. The spontaneous breakup is important when concentration of particles is low and collisions are rare. These factors determine the maximum drop size of about 0.9 cm in diameter.

The drop concentration ranges from ~50 cm^{-3} in clouds developing in very clean air observed over open oceans to ~ several thousand per cm^3 in clouds developing in polluted air. The mean separation distance between cloud drops usually exceeds 1 mm. The maximum liquid water contents are observed in convective clouds and do not exceed 4–5 g/m^3, so the mass loading is relatively low, being of the order of $1 \div 5 \times 10^{-3}$. At the same time, the mass of drops can have a mechanical impact on cloud flows as it decreases buoyancy. This effect is especially pronounced in maritime convective clouds (Emanuel, 1994).

The total range of drop sizes is divided into two subranges: drops with diameters below 50 cm^{-3} are assigned to cloud droplets, while drops of larger size are assigned to rain drops. This division is quite natural because the mass distribution $g(r)$ indicates the existence of a pronounced minimum within the radius range of 25–35 μm. **Figure 2.3.3** presents a schematic representation of the mass distribution as the function of drop radius. This minimum is caused by a rapid increase in the collision rate between drops with radii exceeding ~25 μm and smaller cloud droplets. As a result, the largest cloud droplets reaching this "triggering" radius start growing rapidly shifting toward larger sizes. The raindrop mode has a maximum that shifts toward larger sizes with time. This shift can be seen in **Figure 2.3.4**.

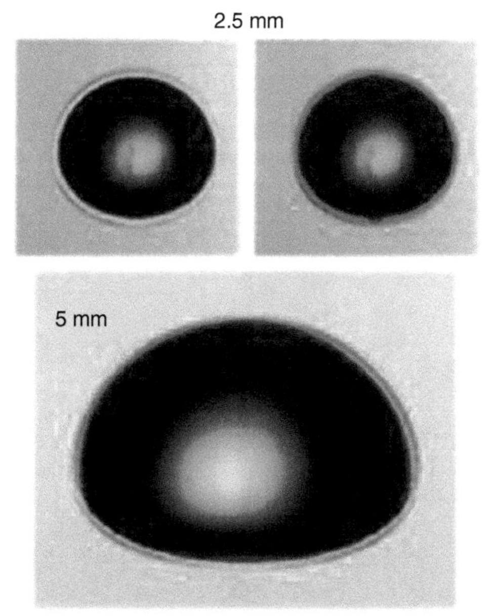

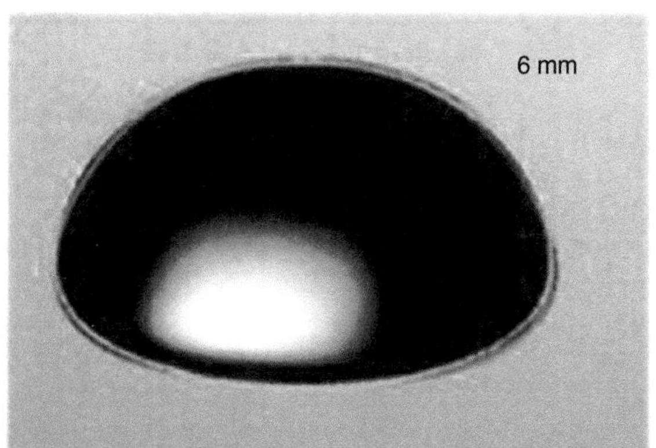

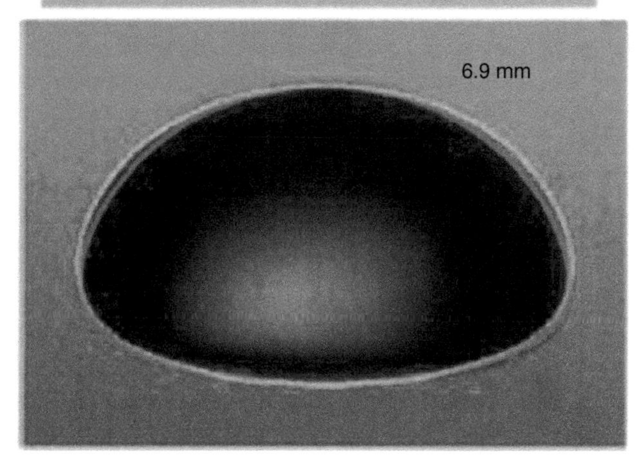

Figure 2.3.1 Shapes of drops of different diameters obtained from wind-tunnel measurements (from Thurai et al., 2009; courtesy of © American Meteorological Society. used with permission).

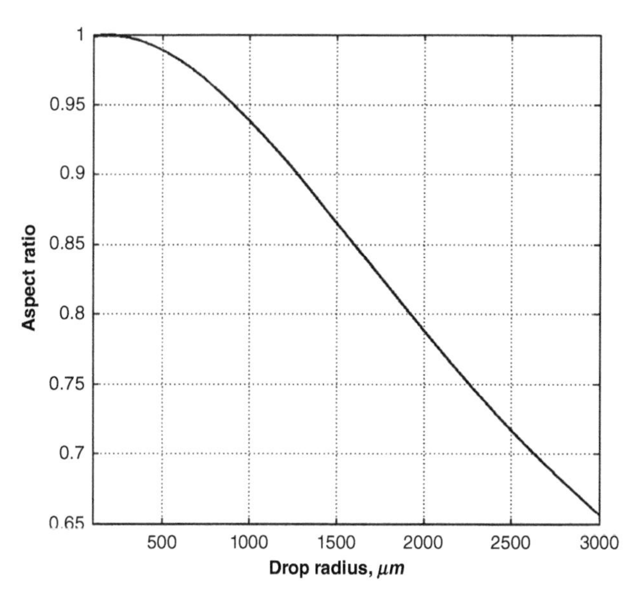

Figure 2.3.2 Dependence of the raindrop aspect ratio on the equivalent radius. Aspect ratio is calculated as $\gamma_d = 0.9951 + 0.02510D - 0.03644D^2 + 0.005303D^3 - 0.0002492D^4$, where drop diameter is in mm (Brandes et al., 2002).

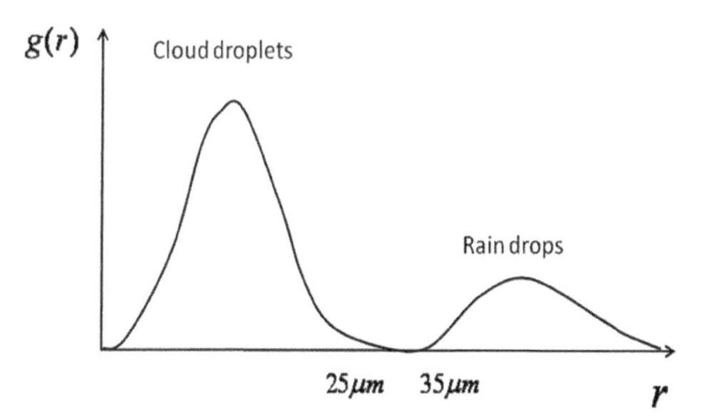

Figure 2.3.3 Schematic representation of drop mass distribution as a function of drop radius.

As can be seen in Figure 2.3.3, the drop spectra are well separated into two main modes: cloud droplets and raindrops. This fact is used in bulk parameterization schemes where cloud droplets and raindrops are treated as different hydrometeor classes (types). The ratios between masses of cloud droplets and raindrops depend on their location within a cloud and on the stage of cloud evolution. In developing clouds, droplets form near the cloud base and grow by diffusion in the cloud updrafts. At this stage, droplet spectrum contains only cloud droplets. The first raindrops with

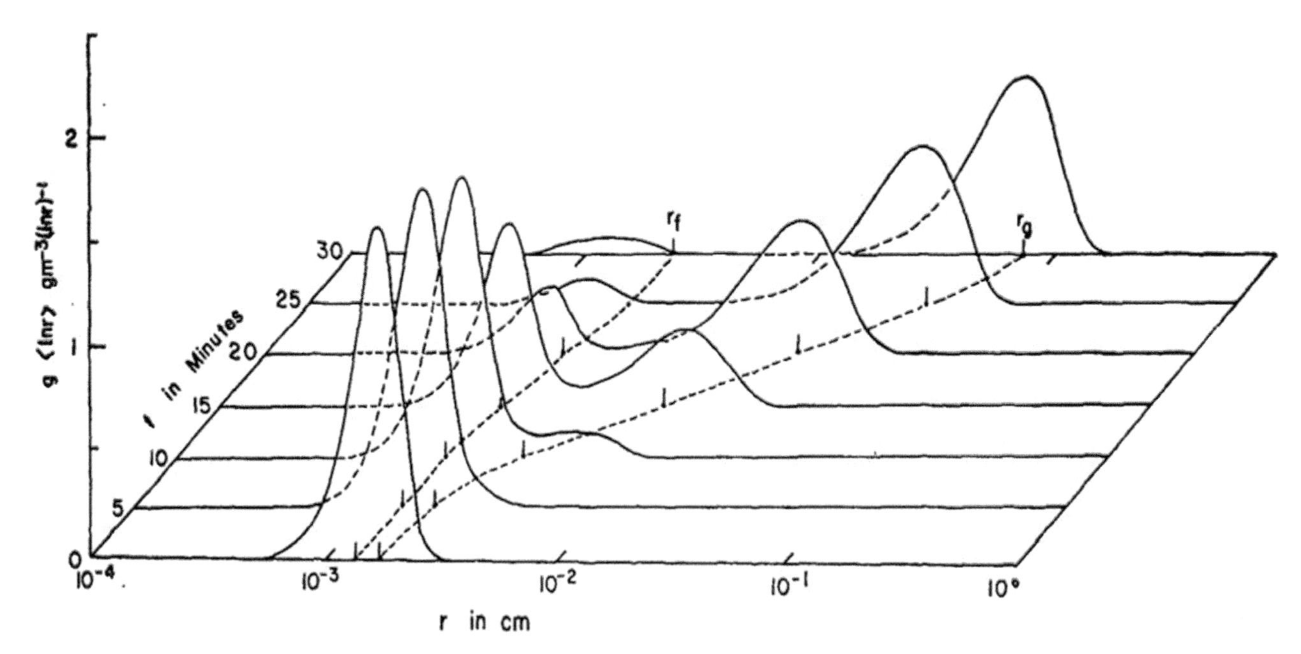

Figure 2.3.4 Evolution of DSDs from the initial droplet spectrum centered at 14-μm radius (from Berry and Reinhardt, 1974; courtesy of © American Meteorological Society. Used with permission)

radii of about 60 μm form as a result of collisions between cloud droplets; sometimes the emergence of first raindrops is identified when the radar reflectivity echo of ~−5 dBZ is first obtained. The distance above the cloud base where the first raindrops form depends on droplet concentration, size, on the intensity of in-cloud turbulence, and some other factors. The process of formation of first raindrops is analyzed in Section 5.11, where drop size distribution (DSD) formation is discussed in detail. The maximum values of droplet concentration are reached in developing clouds just before raindrops start forming. Emerging raindrops efficiently collect cloud droplets, thus decreasing their concentration. Maximum droplet concentrations are highly correlated with the concentration of soluble APs because droplets form on these APs. The relationship between the concentration and sizes of APs and the similar parameters of cloud droplets is discussed in Section 5.1. At a later stage of cloud evolution raindrops fall, efficiently collecting cloud droplets on the way. Hence, droplet concentration decreases dramatically while raindrops contribute largely to the liquid water content (LWC), which is the sum of cloud droplet water content (CWC) and raindrop water content (RWC). Below the cloud base, LWC is formed largely by raindrops.

The evolution of DSDs as a result of drop collisions is illustrated in Figure 2.3.4, which represents the solution of the stochastic equation of collisions (Section 5.6) at a given initial distribution of cloud droplets. This example illustrates the formation of raindrops by collisions within a spatially homogeneous volume. One can see that soon after the first raindrops form (~10 min in Figure 2.3.4), the process of raindrop formation accelerates rapidly, and within several minutes the majority of cloud droplets are captured by raindrops. Even when the concentration of cloud droplets is very low, the concentration of raindrops is by several orders of magnitude lower. For this reason, the raindrop mode is well seen in mass distributions (as in Figure 2.3.4). In DSD, the rain mode is much less pronounced.

The shapes of DSDs depend on the distributions of soluble APs or, more exactly, on the distributions of cloud condensation nuclei (CCN), vertical velocity, as well as on the environment conditions (the atmospheric instability, environment air humidity, and some other factors (Section 5.1). The DSDs can be different in stratiform and convective clouds. Although values of some DSD parameters are highly variable, there are some characteristics of DSDs that change relatively insignificantly. For instance, the droplet spectrum dispersion has specific changes over height: it decreases from high values (about 5) within a few tens of meters in the vicinity of cloud base to ~0.2–0.4 and then remains nearly constant at the value of ~0.2 in

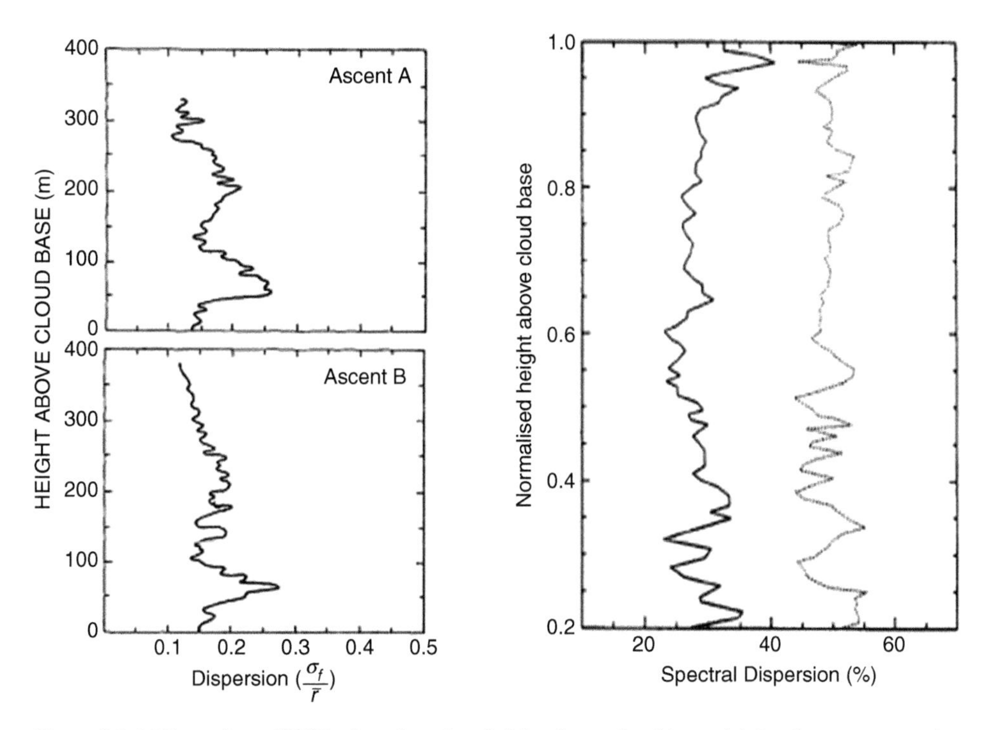

Figure 2.3.5 Dispersion of DSD plotted against height above cloud base: **(a)** for the two ascents in cumulus clouds (Politovich, 1993) and **(b)** for an ascent in maritime air mass (solid line) during FATE on November 6, 1991 and for an ascent in continental air mass (dashed line) over the North Sea on February 26, 1991 (from Martin et al., 1994; courtesy of © American Meteorological Society. Used with permission).

cumulus clouds (e.g., Politovich, 1993) and of ~0.4 in stratiform clouds (e.g., Martin et al., 1994). The typical vertical profiles of the spectrum dispersion measured in cumulus clouds and stratiform clouds are presented in **Figure 2.3.5**.

Since in most cases the mean drop radius within clouds increases with height, the fact that the DSD dispersion is nearly constant with height indicates an increase in the droplet spectrum width with height. In other words, we see the droplet spectrum broadening. Searching for an explanation for this broadening has been a major problem in the theoretical warm–rain cloud physics. The reason is that, according to the classical theory, diffusional growth of droplets leads to droplet spectrum relative dispersion, rapidly decreasing with height to ~0.01 at a few hundred meters above cloud base (Rogers and Yau, 1996) (see Section 5.2 for details). Thus, observed DSDs are always wider than DSDs that could be obtained in accordance with the classical theory of diffusional droplet growth. In this sense, we can say that all DSDs are rather wide despite the fact that the DSDs widths vary within a wide range. It should be noted that speaking about the droplet spectrum narrowing with height, predicted by the equation of diffusional growth, we mean only the narrowing in the space of droplet radii (or diameters). According to this equation, the rates of increase in squares of droplets' radii are the same for droplets of all radii, while the masses of larger droplets grow faster than that of small ones. So, the spectrum with respect to square of a droplets sizes does not become narrower, and the spectrum with respect to droplet's masses widens.

2.3.1 Examples of Droplet Size Distributions Measured *in situ*

In situ, DSD are usually measured and averaged along the airplane track within a single cloud, or more often over several clouds crossed by the airplane at a certain height. For this reason, the broadening of DSDs over height is often attributed to the averaging performed over all local DSDs that are comparatively narrow. An example of DSD broadening induced by averaging is presented in **Figure 2.3.6**.

It seems, however, that most local DSDs (measured at comparatively short intervals of a few hundred meters) are also wide, i.e., their droplet spectra

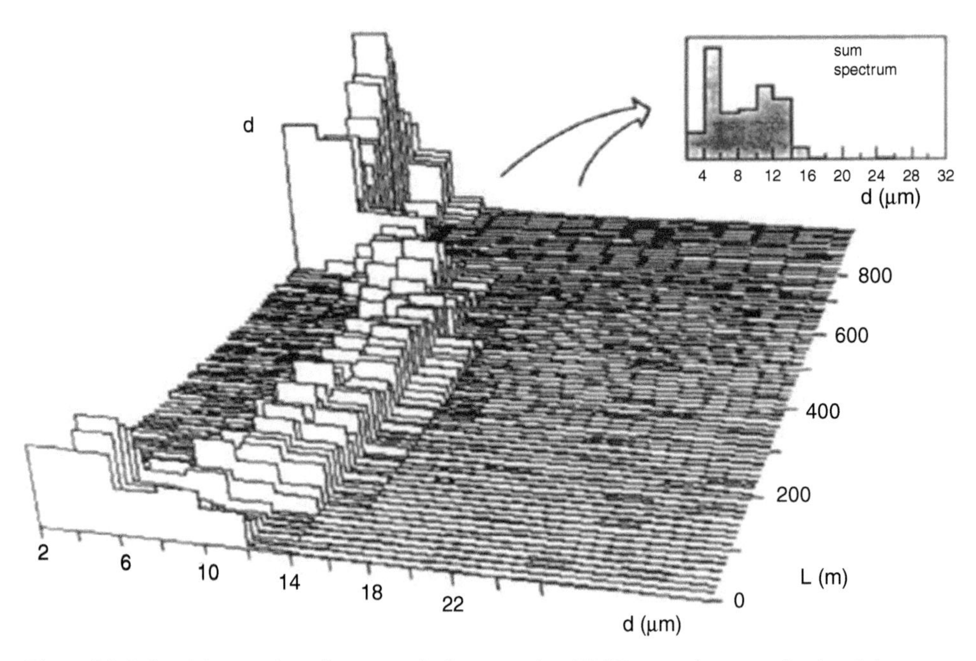

Figure 2.3.6 Spatial averaging of comparatively narrow local DSDs over the zone of unimodal spectra results in a wide bimodal spectrum (from Korolev, 1994, courtesy of © Elsevier).

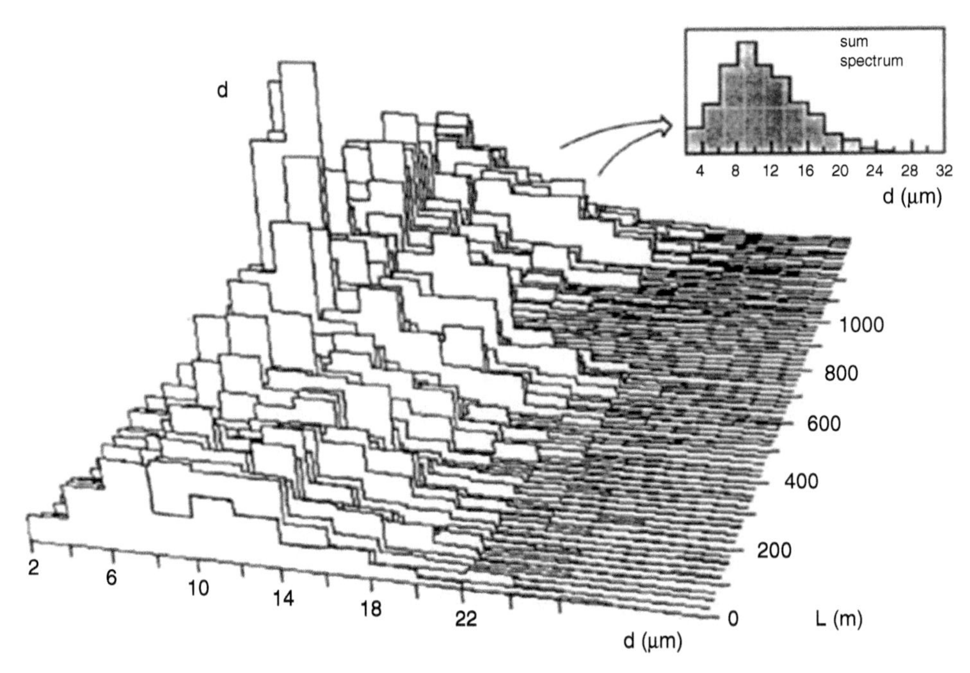

Figure 2.3.7 Spatial averaging over a zone of wide bimodal and multimodal spectra yields a wide unimodal spectrum (from Korolev, 1994, courtesy of © Elsevier).

dispersion is of 0.2–0.3. An example of a case when local DSDs are wide within stratocumulus clouds is presented in **Figure 2.3.7**.

The local DSDs can have the width of the same order of magnitude as that of spatially averaged DSDs in cumulus clouds, as illustrated in **Figure 2.3.8**. Most local DSDs shown in this figure have droplet spectra relative dispersion of 0.2–0.3. It means that spatial averaging of DSDs is not the only reason of the fact that observed DSDs are wide.

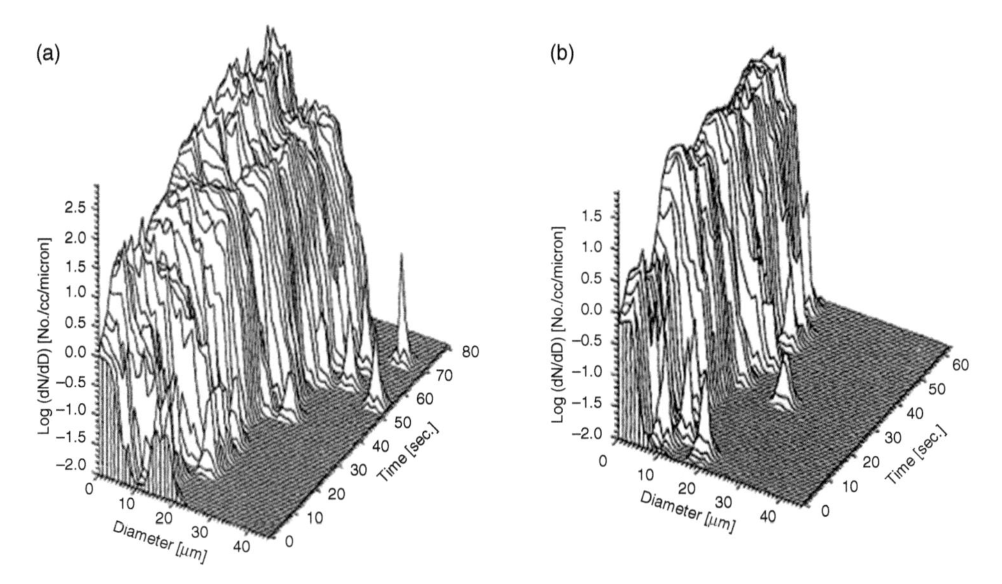

Figure 2.3.8 Cloud droplet spectra measured during a flight in the vicinity of the base of cumulus clouds **(a)** on March 6, 1990 and **(b)** on February 6, 1990 (from Levin et al., 1996; courtesy of © American Meteorological Society. Used with permission).

Another characteristic feature of DSDs is that many local DSDs are bimodal and even multimodal. The appearance of bimodal DSDs could be a result of spatial averaging of local unimodal DSDs, as shown in Figure 2.3.6. However, a frequently observed situation is presented in Figure 2.3.7, showing that local DSDs are largely bimodal or even multimodal. The averaging of such local DSDs masks the bimodality of local DSDs. The bimodality and multimodality of DSDs take place not only in stratocumulus clouds (as shown in Figure 2.3.6), but in cumulus clouds, as well. A classic example of bimodal DSDs measured in maritime cumulus clouds was reported about forty years ago by Warner (1969) (**Figure 2.3.9**). One can see two modes centered at the diameters of ~4–5 μm and 10–12 μm. It is noteworthy that the gap between the modes changes with time and height and disappears at certain heights.

Recent in situ measurements indicate that the bimodality or multimodality of DSD is quite a regular phenomenon that may take place in polluted continental clouds, as well. An example is presented in **Figure 2.3.10**, showing DSDs measured in situ during Cloud-Aerosol Interaction and Rain Enhancement experiment (CAIAPEEX) in tops of developing deep convective clouds on June 22, 2009 in Hyderabad during a monsoon in India. One can see that the shapes of DSDs change significantly with increasing height, as well as along the aircraft traverse at the same height. It should be noted that such variability of DSD shapes, as well as the existence of very small droplets in all local DSDs, were observed several kilometers above the cloud base. The physical mechanisms of formation of very small droplets, as well as bimodal and multimodal DSDs, are discussed in Section 5.11.

As was just mentioned, the DSD width varies within wide ranges. One of the main microphysical mechanisms leading to this variety in the DSD width is the effect of aerosols. Multiple observations show that the dynamical and microphysical properties of clouds arising over the sea are different from those arising over the land; hence, clouds are traditionally separated into maritime and continental types. Since the air over oceans contains more water vapor, cloud bases of maritime clouds are typically lower by 500–1,500 m than bases of continental clouds that are usually located at the heights of 1.5–2.5 km. Typically, maritime convective clouds have lower vertical velocities than continental clouds, with maximum updrafts between 5 m/s and 10 m/s (Jorgensen et al., 1985; Jorgensen and LeMone, 1989). Maximum vertical velocities in maritime deep cumulus clouds reach 20–25 m/s (Heymsfield et al., 2010). At the same time, vertical velocities in continental hailstorms can exceed 30–40 m/s. In addition to the differences in the dynamics, maritime and continental clouds have different microphysics due to different aerosol loadings in the environments. Effects of vertical velocities and aerosols on the formation of DSDs and size distributions of ice hydrometeors are discussed in Chapters 5, 6, and 7.

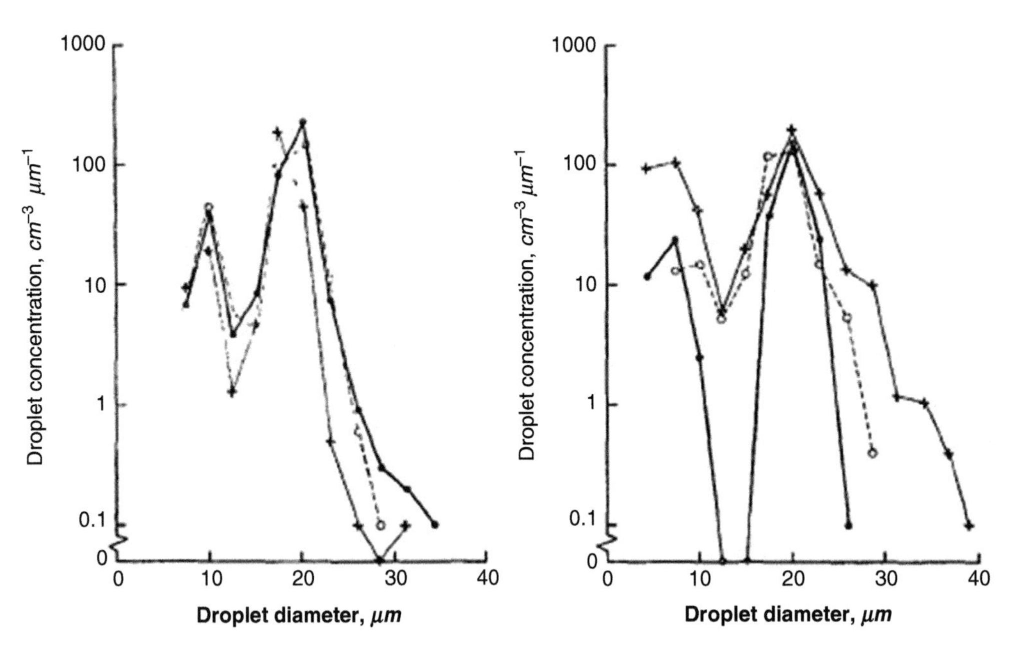

Figure 2.3.9 The detailed droplet spectra measured in maritime clouds (from Warner, 1969; courtesy of © American Meteorological Society. Used with permission).

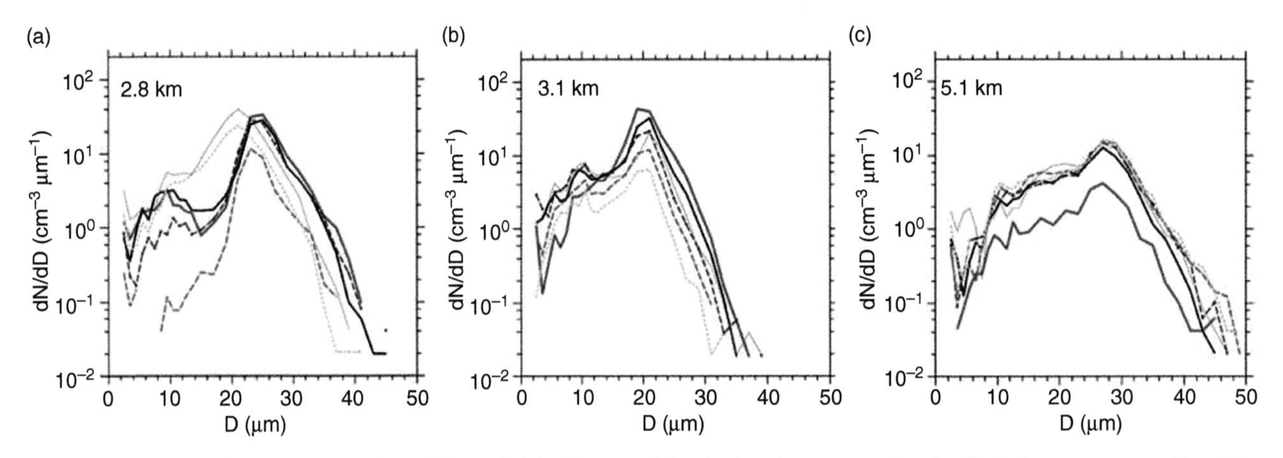

Figure 2.3.10 Plots of DSDs measured at different heights in tops of developing deep convective clouds during a monsoon (June 22, 2009). The DSDs were measured at a frequency of 1 Hz (from Prabha et al., 2011; courtesy of © American Meteorological Society. Used with permission).

The air over oceans usually contains APs in much lower concentrations than the air over continents, thus clouds developing over oceans have a lower droplet concentration than continental clouds. As a result, the terms "maritime" and "continental" are often used to stress the difference in the maxima of droplet concentrations, which is of about 50–100 cm^{-3} in maritime clouds and 500–3,000 cm^{-3} in continental clouds. However, this separation of clouds into continental and maritime based on droplet concentration only is not always valid. Penetration of continental APs (often of anthropogenic nature) to the atmosphere over oceans results in a heavy pollution of maritime clouds, making their droplet concentrations close to those in continental clouds.

An increase in AP concentration leads to a higher droplet concentration and to smaller droplet sizes. As a result, DSDs in clouds developing in polluted air

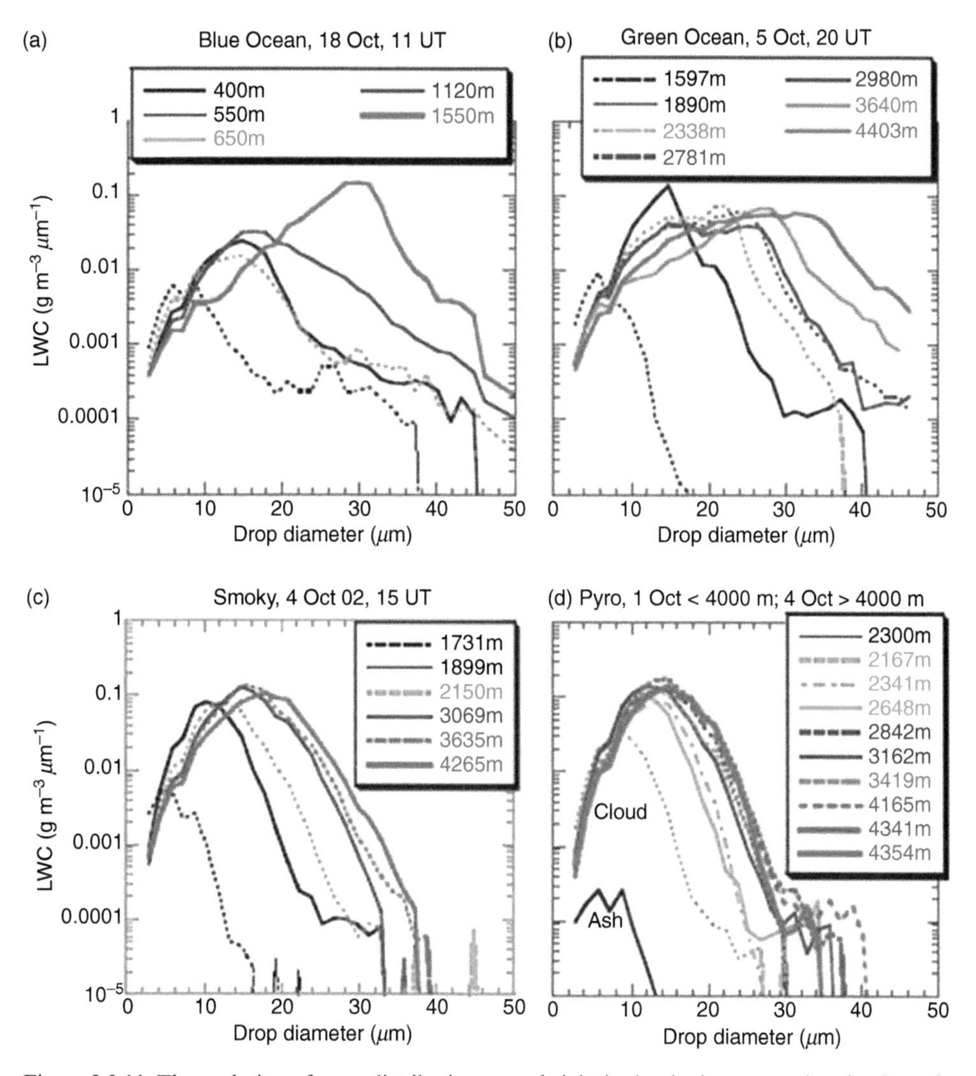

Figure 2.3.11 The evolution of mass distributions over height in developing convective clouds, at four aerosol concentration values: **A**: Blue Ocean, October 18, 2002, 11:00 UT (universal time), off the northeast Brazilian coast (4S 38W); **B**: Green Ocean, October 5, 2002, 20:00 UT, in the clean air at the western tip of the Amazon (6S 73W); **C**: Smoky clouds, October 4, 2002,15:00 UT (10S 62W) in Rondonia and **D**: Pyro-clouds, October 1, 19:00 UT (10S 56W), and on October 4, 19:00 UT (10S, 67W). The lowest DSDs in each plot represents the conditions at the cloud base, except in D, where the mass distribution for large ash particles outside the cloud is also shown (from Andreae et al., 2004; reprinted with permission from AAAS).

are typically narrower than those developing in clean air.

Aerosol effects on DSDs were investigated in detail during the LBA-SMOCC campaign in the Amazon region in 2002 (Andreae et al., 2004). These measurements were performed at low, intermediate, and very high AP concentrations. The corresponding clouds are referred to as "Blue-Ocean", "Green Ocean" and "Smoky" clouds, respectively. Extremely polluted clouds developing over forest fires are referred to as "Pyro-clouds".

Figure 2.3.11 shows droplet mass distributions (MD functions) measured in situ at different levels in developing clouds during the LBA-SMOCC campaign. Panel A shows MDs of droplets in "Blue-Ocean" clouds developing over the sea near the coastal line. The concentration of CCNs at 1% of supersaturation was found to be ~400 cm^{-3}. In this case, the maximum droplet concentration is about 100 cm^{-3}. At such low drop concentration, the DSDs become wide at small distances of less than ~1 km above the cloud base and contain large droplets able to trigger efficient collisions

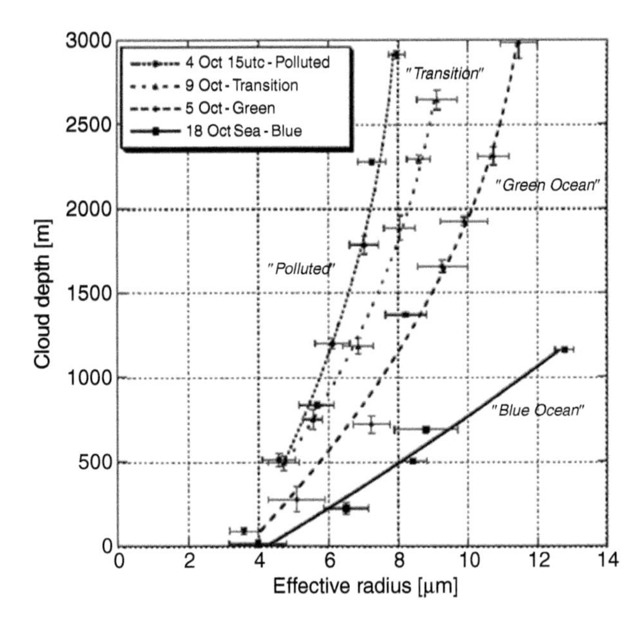

Figure 2.3.12 The vertical profiles of the effective radius in tops of convective clouds developing at different aerosol concentrations, calculated using the DSDs measured in situ in the LBA–SMOCC campaign. Notations: Blue Ocean clouds measured over the sea; Green Ocean clouds measured in comparatively clean air over jungle; transition and polluted clouds measured in zones of burning biomass, in polluted atmosphere (from Freud et al., 2008, courtesy of © ACP).

and raindrop formation. The Green Ocean clouds develop in the atmosphere containing ~1700 cm^{-3} CCN at 1% of supersaturation. The droplet concentration in these clouds is of several hundred per cm^3. The MDs measured in Green Ocean clouds over the jungle are presented in panel B. The droplet spectra are narrower than over the sea, but still wide. Panel C shows MDs in clouds growing in zones of burning biomass where CCN concentration exceeds 5,000–10,000 cm^{-3} (Smoky clouds). Droplet concentration in these clouds reaches 2,000–3,000 cm^{-3}. The droplet spectra in polluted clouds are much narrower than those developing in clean air. Panel D shows MDs in Pyro-clouds arising over forest fires, which are extremely polluted with very narrow MDs and droplet concentrations typically exceeding several thousand per cm^3.

Formation of raindrops by droplet collisions depends on collision rate, which increases with the growth of the amount of largest droplets and of the total drop mass (see Section 5.6 for details). The onset of raindrop formation is determined by emergence of the largest cloud droplets with radii exceeding ~20–25 μm. The latter often indicates that other parameters characterizing DSD (e.g., the mean radii or the effective radii) (see the definitions in Table 2.1.2) have reached certain threshold values. Observation data (Rosenfeld and Gutman, 1994; Freud et al., 2008; Gerber, 1996; Yum and Hudson, 2002; VanZanten et al., 2005; Twohy et al., 2005) and numerical results (Pinsky and Khain, 2002; Magaritz et al., 2009) indicate that first raindrops form in ascending cloud volumes of cumulus clouds if the effective radius reaches ~14–15 μm, while in stratocumulus clouds first drizzle drops form when the effective radius reaches 10–12 μm. The vertical profiles of the effective droplet radius in clouds developing at different aerosol concentrations, calculated using the DSDs measured in situ in the LBA–SMOCC campaign, are presented in **Figure 2.3.12**. Assuming these values (~14–15 μm) as the threshold values of the effective radii, Figure 2.3.12 indicates that in clouds developing in clean air the first raindrops form at lower levels above the cloud base than in polluted clouds. One can see that the effective radius in developing clouds does not reach 14–15 μm, hence the first raindrops form at higher distances from the cloud base, which are not shown in the figure.

2.3.2 Examples of Raindrop Size Distributions Measured *in situ*

The process of raindrop formation is nonstationary (Figure 2.3.4) and highly inhomogeneous in space, which indicates that raindrop size distributions (RSD) vary over height and time. There are two main physical mechanisms determining the maximum size of raindrops: collisions and breakup. Raindrops with radius below ~3 mm do not experience breakups and grow by collisions with cloud droplets and, rarer, with raindrops. Raindrops with radii exceeding 3.5 mm change their shape and break up into several fragments (Section 5.6). The growth of raindrop size by collisions depends on the amount of cloud droplets that they can collect during their fall to the surface. For instance, in warm stratocumulus clouds with the depth of 200–400 m, the maximum radius of raindrops (drizzle) does not exceed 300 μm (**Figure 2.3.13**). Raindrops forming in small warm rain cumulus clouds are larger and reach about 1.5 mm in radius. **Figure 2.3.14** shows RSDs measured in small warm maritime cumulus clouds observed in the RICO field experiment.

The maximum raindrop size in convective clouds increases with the increase of rain rate. **Figure 2.3.15** shows classical Marshall–Palmer (1948) distributions (Table 2.1.5) of raindrops measured near the surface.

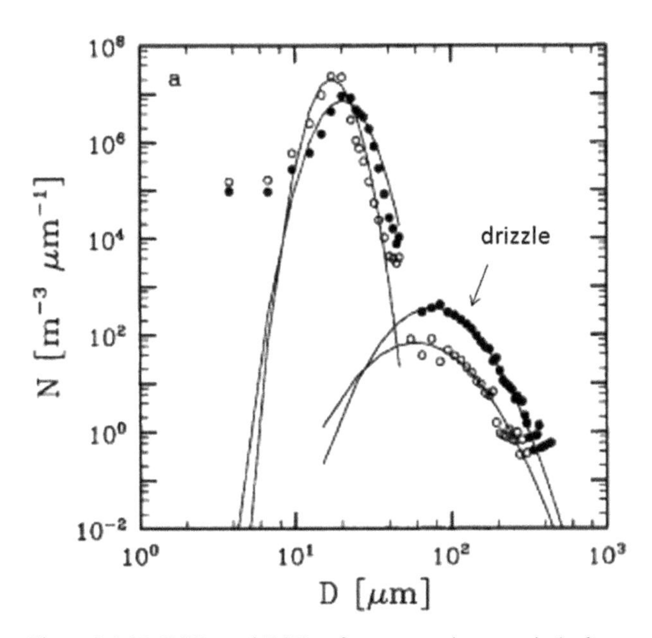

Figure 2.3.13 DSDs and RSDs of one two-minute period of heavy drizzle (closed circles) and one two-minute period of light drizzle (open circles) measured at cloud top in the RF07 research flight. The functional fits to the data are shown by solid lines (from VanZanten et al., 2005; courtesy of © American Meteorological Society. Used with permission).

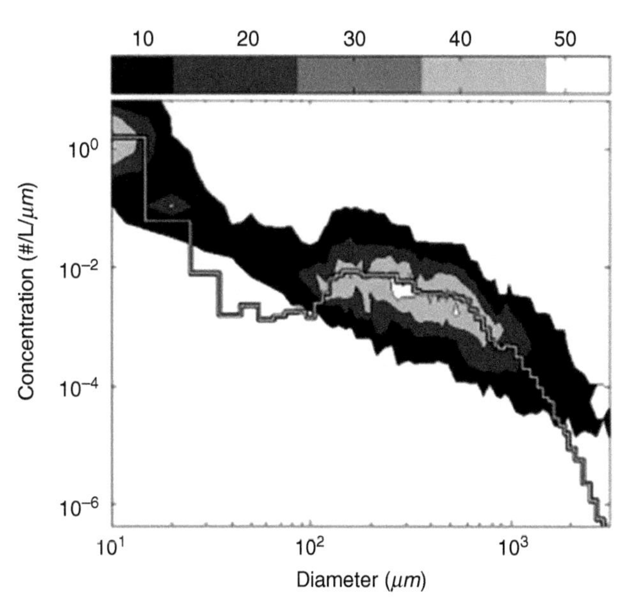

Figure 2.3.14 The mean of 237 RSDs observed at 600 ft (183m) altitude over the ocean on January 19, 2005. The contours show the number of RSDs passing through the region. Very few individual RSDs have any counts between 30 μm and 100 μm. Such RSDs do not appear in the contour plot because zero values are not included in log–log plots (from Baker et al., 2009; courtesy of © American Meteorological Society. Used with permission).

This figure also shows DSDs produced in warm stratiform clouds and in thunderstorms and averaged over many measurements. The maximum size of raindrops also depends on whether these drops form by collisions of liquid droplets or by melting of ice particles. Raindrops of deep continental clouds often form by melting ice hail and graupel. The maximum size of these particles is larger than that of raindrops; first of all, because ice particles, in contrast to raindrops, do not experience breakup (with the exception of snowflakes that can break up after reaching very large sizes). Besides, ice particles ascend to higher levels and have longer residential time in clouds. Figure 2.3.15 shows that in thunderstorms the maximum radius of raindrops may exceed 3 mm. In the case of strong hail storms when radar reflectivity exceed 65–70 dBZ, melting of large graupel or hail leads to formation of raindrops with diameters up to 0.8–0.9 cm (Ryzhkov et al., 2011).

The effect of the melting of snowflakes on the maximum drop size and on the RSD is shown in **Figure 2.3.16**, presenting the mean RSD derived from the ten-minute disdrometer samples measured in winter storms in the California coastal zone during the National Oceanic and Atmospheric Administration's (NOAA's) Hydrometeorology Test Bed (HMT) project (2003–2004), which is referred to as HMT-04. A significant

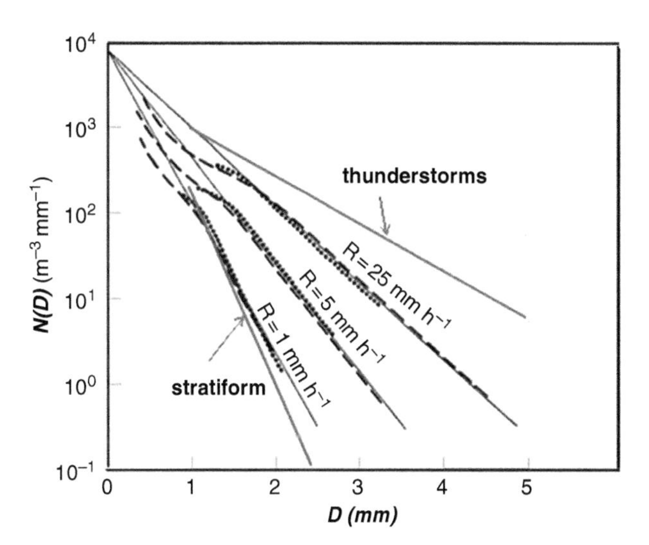

Figure 2.3.15 Averaged rain size distributions. Black lines: distributions presented by Marshall and Palmer (1948) for rain measured in Ottawa. Dashed black lines are from observations; solid lines are approximations by exponential distribution. Red and blue lines show averaged rain size distributions measured in stratiform clouds and thunderstorms, respectively. The RSDs are taken from Wallops Island data (Marshall and Palmer, 1948; Kostinski and Jameson, 1999).

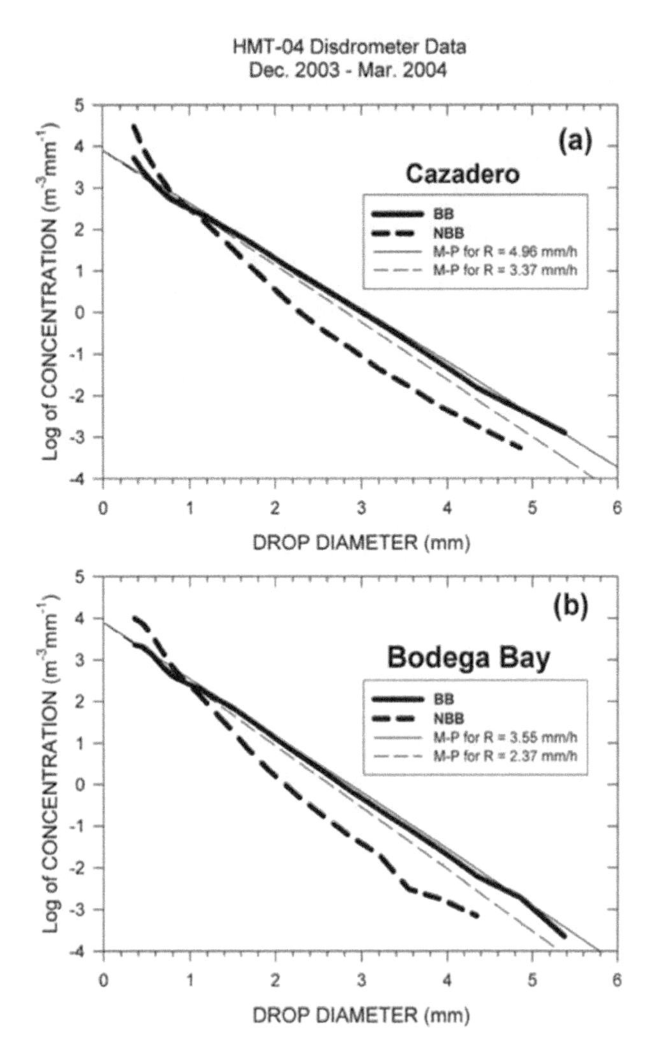

Figure 2.3.16 Mean drop size distributions at (a) Cazadero and (b) Bodega Bay derived from the 10-minute disdrometer samples measured in winter storms in the California coastal zone during NOAA's HMT project (2003–2004), HMT-04 winter season. Heavy lines represent the measurements for periods of BB (solid) and NBB (dashed). Thin straight lines represent the exponential Marshall–Palmer spectra for the same average rainfall intensities as in the HMT-04 observations, where thin solid lines correspond to the BB rain rates at each site and thin dashed lines correspond to the NBB rain rates (from Martner et al., 2008; courtesy of © American Meteorological Society. Used with permission).

fraction of the winter storms in this zone does not reveal the bright band (BB), i.e., an increase in the radar reflectivity within the melting layer caused by melting of large snowflakes. However, some storms demonstrate a pronounced BB. We can see that in the presence of BB the raindrops are larger and their amount is also larger than in case of no BB (NBB).

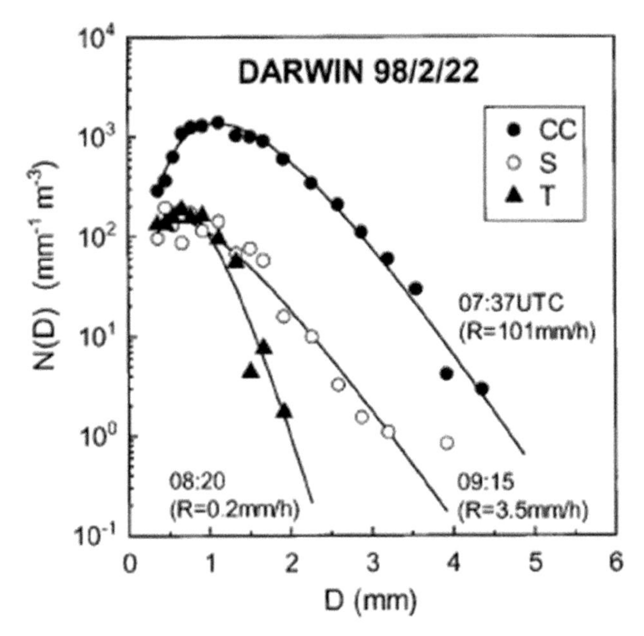

Figure 2.3.17 RSDs measured in different parts of the continental squall line (CC is the convective center; S is the stratiform region; and T is the reflectivity throughout the region) on February 22, 1998, observed in Darwin, Australia. Approximation by Gamma distribution is plotted by solid lines (from Maki et al., 2001; courtesy of © American Meteorological Society. Used with permission).

Finally, we present **Figure 2.3.17**, showing RSDs in different segments of the same squall line. One can see that RSDs are quite different in the two zones of the squall line, namely, the zone of intense convection and the stratiform zone behind the zone of intense convection. One can see that raindrops falling in the zone of intense convection reach larger sizes than those falling in the stratiform zone. It is related to the fact that the largest raindrops in the convective zone form as a result of the melting of hail particles, which can reach large sizes. In the stratiform zone hail does not form.

Figures 2.3.15, 2.3.16, and 2.3.17 show that averaged RSDs, at least for raindrops with $D > 1$ mm, are approximated by the Marshall–Palmer distributions reasonably well. However, parameters of these distributions vary significantly depending on several factors. To get better approximation of DSDs by Gamma distribution (and exponential distributions for raindrops, in a particular case), parameters of the distribution should be changed with height in certain relation, as discussed in Section 7.2.

To summarize, DSDs as well as RSDs differ significantly depending on a number of factors, and often have quite complicated shapes. The range of drop sizes may reach several orders of magnitude. The brief description presented here gives some idea of the challenges one faces while modeling DSDs. The complexity lies in reproducing the multi-aspect physical processes responsible for shaping droplet and raindrop distributions that are spatially nonuniform and time dependent.

2.4 Cloud Ice

Ice microphysics is perhaps the least understood field of cloud physics. The reason might be the high variety of cloud ice types and the complex nature of the processes leading to ice particle formation and evolution. All cloud ice particles are generally subdivided into four large classes: ice crystals, aggregates (snowflakes), graupel, and hail. Some models, however, distinguish two classes of ice particles: low bulk density ice particles (ice crystals and snow) and ice particles of high bulk density (graupel and hail).

2.4.1 Ice Crystals

Ice crystals are the dominating hydrometeor type in cirrus clouds, anvils of deep cumulus clouds, ice stratiform, and mixed-phase stratiform clouds. Radiative properties of cloudy atmosphere (e.g., albedo) are determined largely by ice crystal concentration, size distributions, and crystal shapes in cirrus clouds covering huge areas. Cirrus cloud radiative forcing, both solar and infrared, is influenced by the pattern of crystal size spectra. According to Zhang et al. (1999), the net radiative forcing for bimodal spectra is lower than that for single-modal spectra. The solar radiative forcing of cirrus clouds estimated for nonspherical ice crystals is less than that calculated if ice particles are represented as equivalent spheres. The cloud radiative forcing, solar albedo, and infrared emittance change significantly as the mean crystal size changes. In most cirrus clouds, the net cloud radiative forcing is positive, with the exception of cirrus clouds with a large number of small ice crystals (of the mean maximum dimension <30 μm) and cirrus clouds with bimodal crystal size distribution. Ice crystals play a highly important role in microphysics of mixed-phase clouds, as they participate in production of all other types of cloud ice particles. The contribution of ice crystals to winter precipitation can also be significant.

There is no unique definition of ice crystals. According to Pruppacher and Klett (1997), "If ice particles grow by diffusion of water vapor (this process is known as deposition), they are called ice crystals or snow crystals." However, many ice particles that grow by riming, i.e., by collisions with water drops, or form from small frozen drops are also assigned to the class of ice crystals. Sometimes, cloud modelers assign ice particles to ice crystals if they have the maximum linear size below ~100–150 μm and fall velocities of several tens of cm/s. According to this approach, frozen drops with radii below 100–150 μm are also assigned to ice crystals (Hall, 1980; Khain et al., 2011). At the same time, the maximum linear sizes of ice crystals in real clouds can reach several mm. Apparently, it is most appropriate to define an ice crystal as an ice particle of a specific nonspherical shape that forms as a result of depositional growth or splintering by freezing of drops or collisions between ice particles and drops.

Sometimes it is assumed that bulk density of ice crystals is low. However, this is true only for large crystals of branch-like type (snow crystals). In fact, even for these crystals the bulk density may be close to that of pure ice when their linear sizes are below several tens of μm. The bulk density then decreases during the condensational (depositional) growth. This decrease in the bulk density of ice crystals occurring when their size increases is taken into account in many cloud microphysical models (e.g., Khain et al., 2004).

As regards their origin, ice crystals can be separated into *primary* ice crystals and *secondary* ice crystals. Primary ice crystals emerge out of deposition of water vapor on haze particles and on so-called INs, which are non-soluble APs. Secondary ice crystals emerge as a result of splintering that occurs, for instance, during drop freezing or when graupel and hail collide with drops with diameters exceeding ~20 μm (Hallet and Mossop, 1974).

Primary crystals forming at temperatures over -20°C are typically single crystals, i.e., crystals of single crystal structure. From the crystallographic point of view, these crystals have one common basic shape, namely that of a six-fold (hexagonal) symmetric prism with two basal planes and six prism planes (Pruppacher and Klett, 1997). Under different environmental conditions, some faces (known as meta-stable faces) grow quickly, giving rise to the formation of the crystal's edges and corners, while other faces grow slowly and become the bounding faces of the crystal. **Figure 2.4.1** shows a dendrite crystal of a regular shape; its central part is an IN (or a small frozen droplet), which triggered the process of the crystal formation. The branches grew due to the depositional growth.

Laboratory measurements and in situ observations indicate that each ice crystal habit (a group of ice crystals having a similar shape) forms within its own ranges of temperatures and ice supersaturations. To illustrate these dependencies, habit diagrams are used (an example is shown in **Figure 2.4.2**). The temperature is a major factor that determines formation of a certain crystal type. For instance, plates form within the temperature range of 0°C to −4°C and of −8°C to −22°C, while columns form within the temperature range of −4°C to −8°C. Branch-like crystals (dendrites) tend to grow within the temperature range of −10°C to −20°C at comparatively high ice supersaturations.

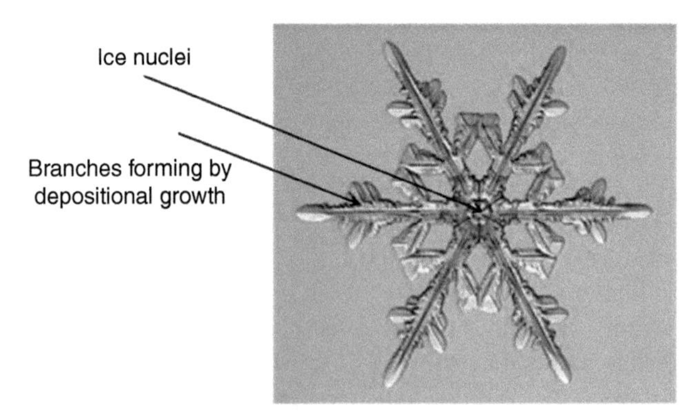

Figure 2.4.1 A dendrite crystal of a regular shape; its central part is an IN (or a small frozen drop), which triggered the crystal formation. The branches grew due to the depositional growth (from www.flickr.com/photos/klibbrecht/15781855351/in/photostream/; permission from photographer Dr. K.G. Libbrecht).

The snow crystal habit diagram does not represent the entire variety of crystal shapes. A significant fraction of ice crystals has irregular shapes and forms during diffusional growth of frozen drops. These crystals are of irregular shapes because bulges and protrusions often form in the course of freezing. Ice particles formed from single-crystalline frozen drops are likely to turn into two-layer crystals (double dendrites) (Pruppacher and Klett, 1997). Ice crystals growing around frozen drops forming their center are quite widespread and, depending on the surrounding air temperature, constitute 20–50% of all crystals. The diameter of the central frozen drop was found to range between 2.5 and 25 μm. At lower temperatures, frozen drops become polycrystalline and trigger formation of so-called spatial crystals, which are often a combination of columnar crystals. The columns may have a conical or a pyramidal shape (bullet shape) with the tip pointing toward the frozen drop in their common center (**Figure 2.4.3**).

Spatial crystals may also develop as a result of collisions of super-cooled drops with ice particles. At comparatively warm temperatures, these collisions lead to formation of single crystals with a crystallographic structure that may be the same or different from that of the colliding crystal. At cold temperatures, the

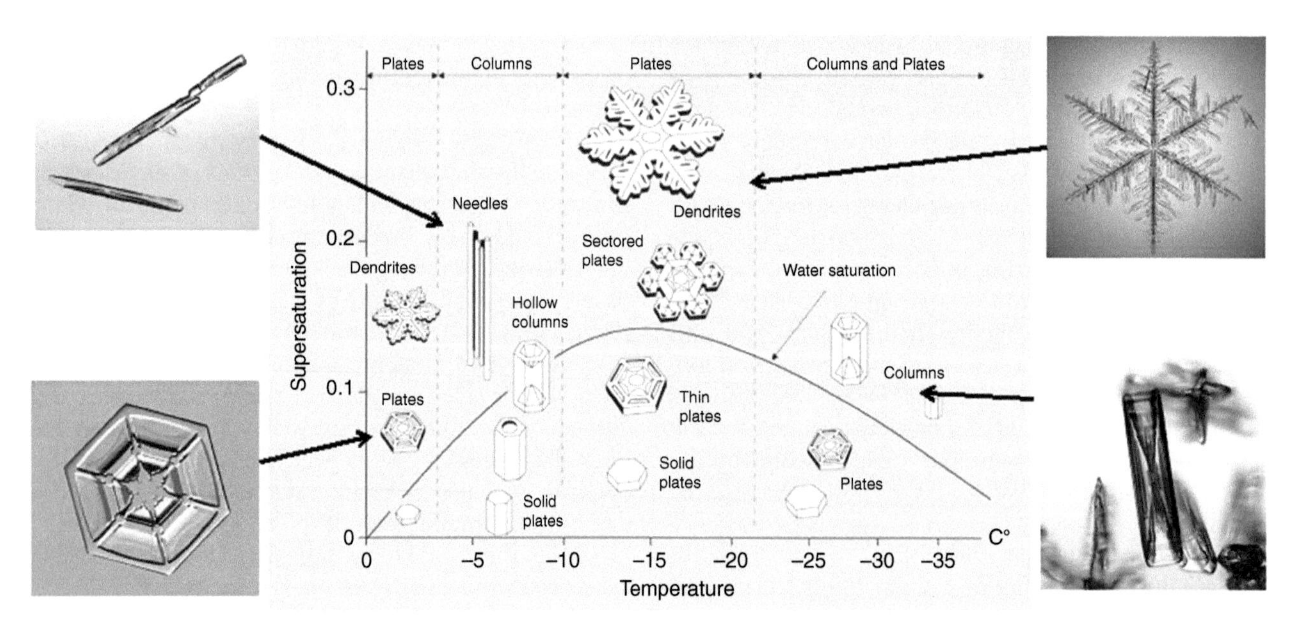

Figure 2.4.2 The snow crystal habit diagram: ice supersaturation–temperature with changes (from www.snowcrystals.com/morphology/morphology.html; permission of the author K.G. Libbrecht).

Figure 2.4.3 A bullet rosette (from www.snowcrystals.com/guide/sampler.jpg, permission from Prof. K.G. Libbrecht).

colliding drop may become a polycrystalline frozen drop that continues growing due to vapor diffusion to form snow crystals whose parts have various spatial orientations. Ice crystals of non-regular form also form by splintering during drop freezing or as a result of ice-ice collisions (Section 6.8).

Using the new laboratory and in situ measured data, Bailey and Hallett (2009) reconsidered and extended the snow crystal habit diagram presented in Figure 2.4.2. The new diagram shown in **Figure 2.4.4** covers the temperature range from -1°C down to -70°C. While retaining the well-established descriptions of habits from the previous diagrams for temperatures above -18°C, the new diagram gives more precise information for lower temperatures. The diagram shows that at low temperatures the dominating habits are composed of polycrystals of various shapes. There are two distinct habit versions: a plate-like crystal at temperatures from -20°C to -40°C, and a columnar crystal at temperatures from -40°C to -70°C. The new diagram also emphasizes the fact that most individual crystals are complex, irregular, and imperfect in appearance, in agreement with the results reported by Korolev et al. (1999, 2000). Plates and columns are the most common habits at low ice supersaturation. Many crystals have such irregular shapes that they were mistakenly identified as aggregates in past studies. Additionally, the majority of very small crystals growing at low ice supersaturation are compact faceted polycrystals and not spheroids.

Substantial changes in habits can occur due to vertical motions occurring in the course of transition from one growth condition to another. A modified crystal often "stores" the history of its nucleation and changing growth conditions. When crystals fall from one cloud region to another, their shapes change and become quite complicated, containing elements or combinations of different basic crystal shapes, such as stellar crystals with spatial plates, columns with dendrites, etc. (Pruppacher and Klett, 1997). Columns and bullet rosettes falling from zones with temperatures below -40°C into warmer regions continue growing like plates. Bullet rosettes typically become mixed rosettes with plate-like components after bullets and columns experience a reduction in the aspect ratio while growing larger and wider, sometimes becoming hollow or capped depending on ice supersaturation (Figure 2.4.4).

According to the classification suggested by Magono and Lee (1966), there are about eighty distinct types of habits given specific names. It is clear that cloud models cannot describe the whole variety of ice crystals shapes. In most models, ice crystals are approximated by "effective" spheres with certain size distributions. However, there are models taking into account several types of ice crystals (e.g., plate-like crystals, dendrites and columns) (e.g., Khain et al., 2004b). There are detailed empirical formulas relating masses and types of the crystals to their linear sizes, bulk densities, and fall velocities (Pruppacher and Klett, 1997; Mitchell and Heymsfield, 2005; Heymsfiled et al., 2007; Straka, 2009; see also Sections 6.1 and 6.4). These formulas can be used in cloud modeling.

Sometimes, there is a necessity to represent ice crystals shapes more accurately than just approximation by sphere, cylinder, disc, or ellipsoid. Wang (1997) presented simple mathematical expressions describing quite complicated shapes of crystals. For instance, he proposed describing shapes of *plates* and *dendrites* using the following formula for the distance from the center of an ice crystal to its edge:

$$r = a\left[\sin^2(n\theta)\right]^b + c, \tag{2.4.1}$$

where r and θ are the radial and the angular coordinates with respect to the center of an ice crystal, respectively. The amplitude parameter a, width parameter b, center size parameter c, and polygonaliry parameter n are adjustable to fit the shape and size of the ice crystals. The ranges of these parameters are: $a = c \div \infty$; $b = 0 \div \infty$; $c = 0 \div \infty$; $n = 0, \frac{1}{2}, 1, \frac{3}{2}, 2, \ldots$

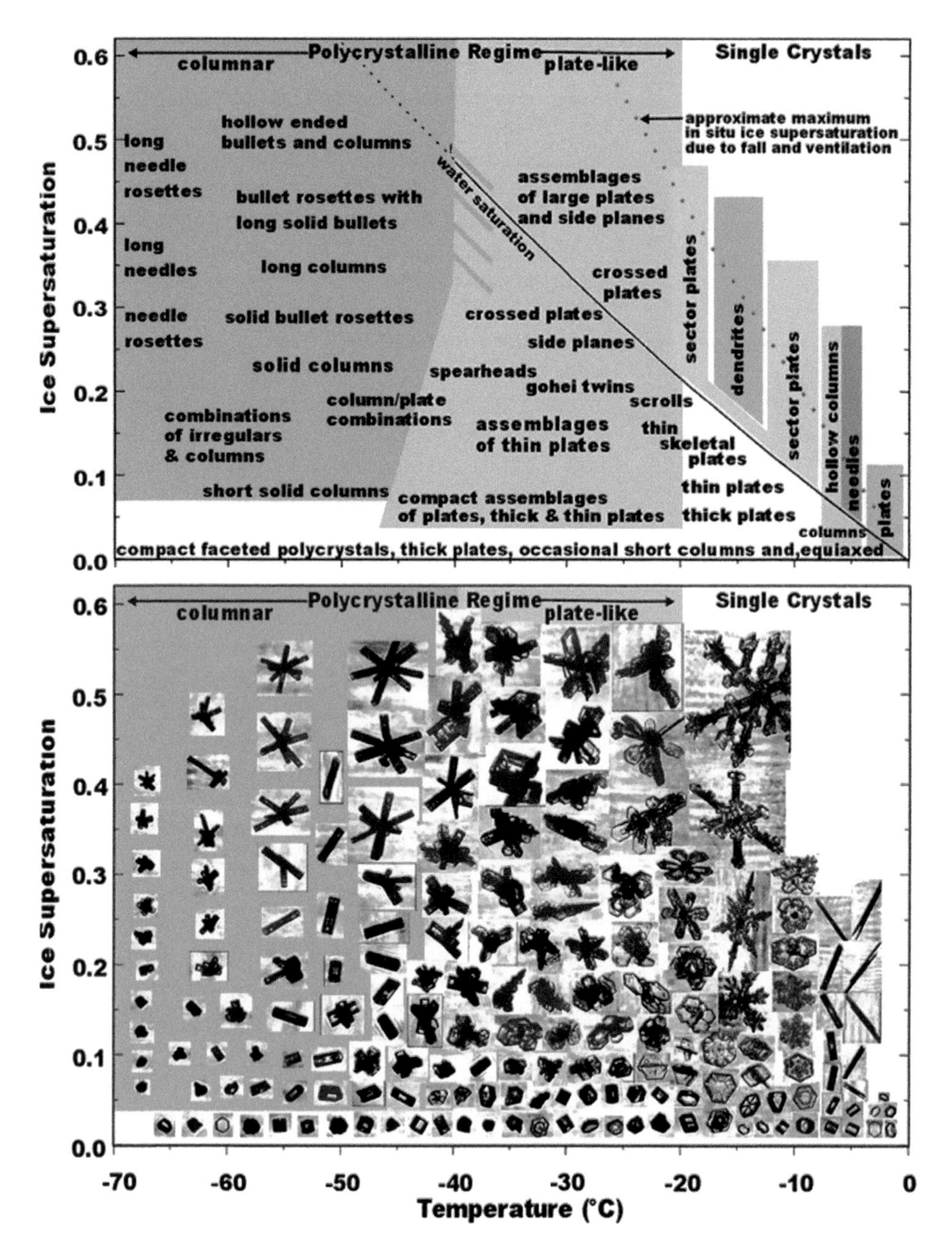

Figure 2.4.4 The snow crystal habit diagram for atmospheric ice crystals based on laboratory results and field studies (from Bailey and Hallett, 2009; courtesy of © American Meteorological Society. Used with permission).

A rectangle projection of a cylindrical column-like crystal shape of length L and diameter D is described by the formula:

$$\frac{4x^2}{L^2} + \frac{4y^2}{D^2}\left(1 + \varepsilon - \frac{4x^2}{L^2}\right) = 1, \tag{2.4.2}$$

where $\varepsilon << 1$ is a small positive number. 3D structures of ice crystals are described using formula

$$r = \left\{a\left[\cos^2(m\theta)\right]^b + c\right\}^d \left\{a'\left[\sin^2(n\varphi)\right]^{b'} + c'\right\}^{d'} \tag{2.4.3}$$

This equation is an extension of Equation (2.4.1) applied to both the θ and φ angles. The shape generated by this expression has $2mn$ branches. For example, an eight-lobe combination of bullets shown

Figure 2.4.5 Left: Different shapes of plate-like and dendrite-like crystals obtained using Equation (2.4.1) under different values of *a*, *b*, and *c*. Right: An example of an eight-lobe combination of bullets (from Wang, 1997; courtesy of © American Meteorological Society. Used with permission).

in **Figure 2.4.5** (right) can be generated using formula $r = [1 - \cos^4(2\varphi)]^{20}[1 - \sin^4(2\theta)]^{20}$.

The concentrations of ice crystals in clouds at temperatures warmer than ~−38°C vary from 0.1 L^{-1} to a few hundred per liter depending on cloud type and geographical location. In anvils of deep convective clouds at temperatures colder than ~−38°C, the concentration of ice crystals can reach a few per cm^3 and even several hundred per cm^3 (Rosenfeld and Woodley, 2000; Heymsfield et al., 2009). These ice crystals form as a result of homogeneous freezing of small cloud droplets. The range of the ice water content (IWC) in cloud anvils of deep convective clouds is between 0.1 and 0.4 gm^{-3}. IWC typically decreases with an increase in the distance from the cloud axis (Tian et al., 2010).

Parameters of ice crystal size distributions were measured in numerous field experiments (Field et al., 2007; Tian et al., 2010). **Table 2.4.1** shows the parameters of ice crystal size distributions measured in several projects such as FIRE (the First ISCCP Regional Experiment), ICE (International Cirrus Experiment), EUCREX (European Cloud and Radiation Experiment), and in other field measurements. Although these in situ-measured cirrus PSDs vary greatly from case to case, in cirrus clouds they often have the bimodal pattern alongside the unimodal pattern. In many cases, ice crystal size distributions were found to vary vertically (e.g., Heymsfield and Platt, 1984; Koch, 1996).

The crystal size spectra measured at various cloud levels during EUCREX 1994 over the polar seas (Koch, 1996) are shown in **Figure 2.4.6**. The unimodal size spectra of ice crystals were found near the cloud top, while the bimodal size spectra were observed in the lower layers of the cloud. The second maximum peak in the bimodal size spectra drifts to a larger crystal size within the lower cloud layers. The bimodal ice crystal size spectra in cirrus were also measured during FIRE II using PMS 2DC probes and an ice particle replicator (Arnott et al., 1994). The maximum equivalent diameter of these ice crystals is about ~600 μm, which corresponds to the maximum crystal dimension of 2,000 μm and, sometimes, of 5,000 μm.

2.4.2 Aggregates and Snowflakes

Aggregates form by collisions between ice crystals, and grow by collecting other ice crystals and aggregates. A picture of aggregates formed by snow crystals is presented in **Figure 2.4.7**.

The maximum size of aggregates typically ranges between 400 and 1,500 μm. Aggregates whose size exceeds this range are assigned to snowflakes. In cloud modeling, aggregates within the entire range of sizes are sometimes referred to as snow. Air temperature and crystal shape are two dominant factors in the aggregation process. Investigations of snowflake formation performed during the second Canadian Atlantic Storms Program (CASP II in January–March 1992, in severe maritime storms that lashed the coast of Newfoundland (Stewart, 1991; Stewart and Crawford, 1995)) showed

Table 2.4.1 Parameters of ice crystals measured in cirrus clouds (from Zhang et al., 1999 with changes; courtesy of Elsevier).
Notations:
R_{mini}: crystal size at the minimum concentration trough.
R_{maxi1}: crystal size at the first maximum concentration peak.
R_{maxi2}: crystal size at second maximum concentration peak.
N_{maxi1}: the first maximum in crystal concentration.
N_{maxi2}: the second maximum in crystal concentration.

Cloud Type	Instruments	Spectrum Pattern	Microphysical Features
cirrus uncinus, cirrostratus		bimodal, single-modal	range: 20–100 μm, R_{maxi2} = 500 μm
stratospheric cirrus anvil	2D grey imaging probe	tri-modal	
cirrus		single-modal, bimodal	
all types of cirrus (FIRE II)	ice particle replicator	bimodal	range: 20–450 μm R_{maxi1} = 20 μm, R_{mini} = 150 μm, R_{maxi2} = 250–300 μm, $N_{maxi1} = 7 \times 10^5$ m^{-3}, $N_{maxi2} = 5 \times 10^2$ m^{-3}
cirrus (FIRE II)	PMS 2DC, ice particle replicator	bimodal	R_{maxi1} = 20 μm, R_{mini} = 100 μm, R_{maxi2} = 200–300 μm, $N_{maxi1} = 10^5$ m^{-3}, $N_{maxi2} = 10^3$ m^{-3}
Arctic cirrus (EUCREX)	PMS OAP 2D2-C	single-modal, bimodal	range: 40–500 μm, R_{mini} = 100–200 μm, R_{maxi1} = 40 μm, R_{maxi2} = 100–300 μm, $N_{maxi1} = 10^3$ m^{-3}, $N_{maxi2} = 10^2$ m^{-3}
jet stream cirrus frontal cirrus (EUCREX)	PMS OAP, Hallet replicator	single-modal, single-modal	more ice water and larger ice crystals are found in frontal cirrus than in jet stream cirrus
contrail cirrus, natural cirrus (ICE)	PMS 2D-C	single-modal, bimodal	high ice particle number density and small particles are found in contrail cirrus; R_{maxi2} = 300–400 μm, $N_{maxi1} = 10^3$ m^{-3}, $N_{maxi2} = 10^1$ m^{-3}, R_{mini} = 200 μm

that formation of snowflakes began with intense growth of dendrites at −15°C. The diameters of snowflakes increased with decreasing height (from ~1 cm at ~2 km level to 2–4 cm at 0.6 km). The total concentration of particles decreased from about 0.5 m^{-3} to about 0.1 m^{-3}. Aggregates with sizes up to about 5 cm were observed over a 30-km region by means of the high-volume precipitation spectrometer (HVPS). The snowflakes falling to the surface had a significant fraction of melted water. **Figure 2.4.8** shows images of 4–5 cm snowflakes and 1-km averaged histograms of snowflakes size distribution observed at 0.6 km (+1°C).

Snowflakes are likely to break up after reaching the maximum size of several cm. The breakup can be

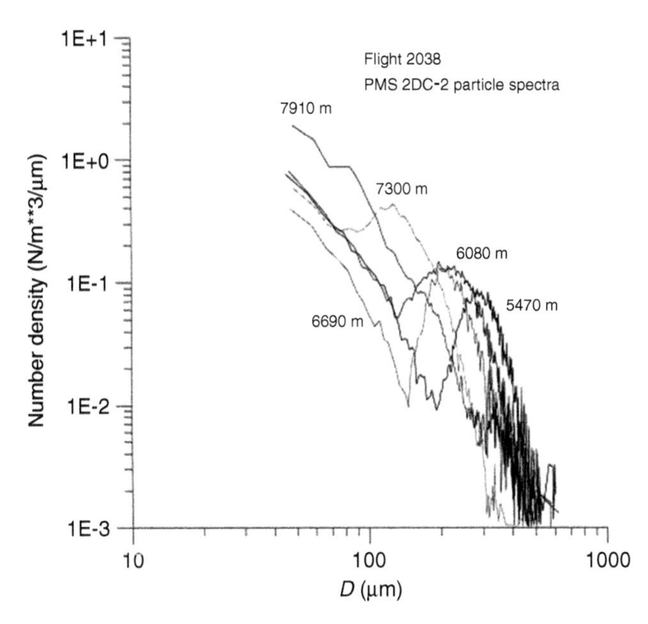

Figure 2.4.6 Bimodal ice crystal size spectra measured during EUCREX. *D* is the diameter of a sphere with an equivalent surface area (from Zhang et al., 1999; courtesy of © Elsevier).

Figure 2.4.7 The picture of aggregates formed by snow crystals (from www.its.caltech.edu/~atomic/snowcrystals/; courtesy of photographer, Prof. K.G. Libbrecht).

spontaneous or collisional. In the latter case, breakup takes place after the collision between a snowflake and a particle (likely graupel or hail) that is moving at a velocity significantly higher than the snowflake (see Section 6.8). The bulk density of snowflakes ranges between 0.005 g cm^{-3} and 0.5 g cm^{-3}, the most frequent values ranging between 0.01 and 0.2 g cm^{-3} (Pruppacher and Klett, 1997).

2.4.3 Graupel

In clouds that contain both crystals and supercooled droplets, ice crystals and aggregates collect small water droplets. **Figure 2.4.9** shows typical weakly and substantially rimed ice crystals. Intense riming increases the bulk density of crystals and aggregates and makes their shapes closer to spherical with significant porosity (**Figure 2.4.10**).

When the bulk density of rimed particles exceeds ~0.2 g cm^{-3}, they are assigned to graupel. Graupel is the most widespread type of ice hydrometeors in convective clouds, its bulk density varying from 0.2 to ~0.8 g cm^{-3}. In case graupel forms via riming of aggregates, its bulk density increases with its size. Graupel can also form by collisions between frozen drops and smaller droplets (Pruppacher and Klett, 1997). In this case, the bulk density decreases with an increasing graupel size. Graupel sizes range between 100 μm and ~0.8–1 cm in diameter. Larger graupels are assigned to the hail class. Graupel particles have different shapes depending on the shape and size of the original crystals or aggregates that started growing through riming. Crystals of different shapes experience different kinds of motions during their fall. Some crystals rotate rapidly, while others do not rotate at all or rotate slowly. Therefore, collisions between ice crystals and droplets have various impacts over particle surfaces, leading to the formation of different graupel shapes. A photo of graupel particles of different forms falling together with snow aggregates (snowflakes) is presented in **Figure 2.4.11**.

Examples of graupel size distributions measured at the height of $z = 4$ km during a storm on June 20, 1995, near Fort Collins, Colorado, are shown in **Figure 2.4.12**. One can see that graupel size distributions can be bimodal, similar to the distributions of ice crystals. These size distributions were measured during the same storm with the time increment of ~0.5 h. The comparison of these distributions demonstrates substantial variability of graupel concentrations over time and space, the range being from several tens per m^3 to several tens per liter. One can also see that graupel size distributions have deviations from the exponential patterns, i.e., from the Marshall–Palmer distributions.

2.4.4 Hail

Hail is the largest ice hydrometeor of the highest density close to that of pure ice (0.91 g cm^{-3}). Hail formation can be triggered by different mechanisms. The mechanism dominating in the tropical zone is the freezing of

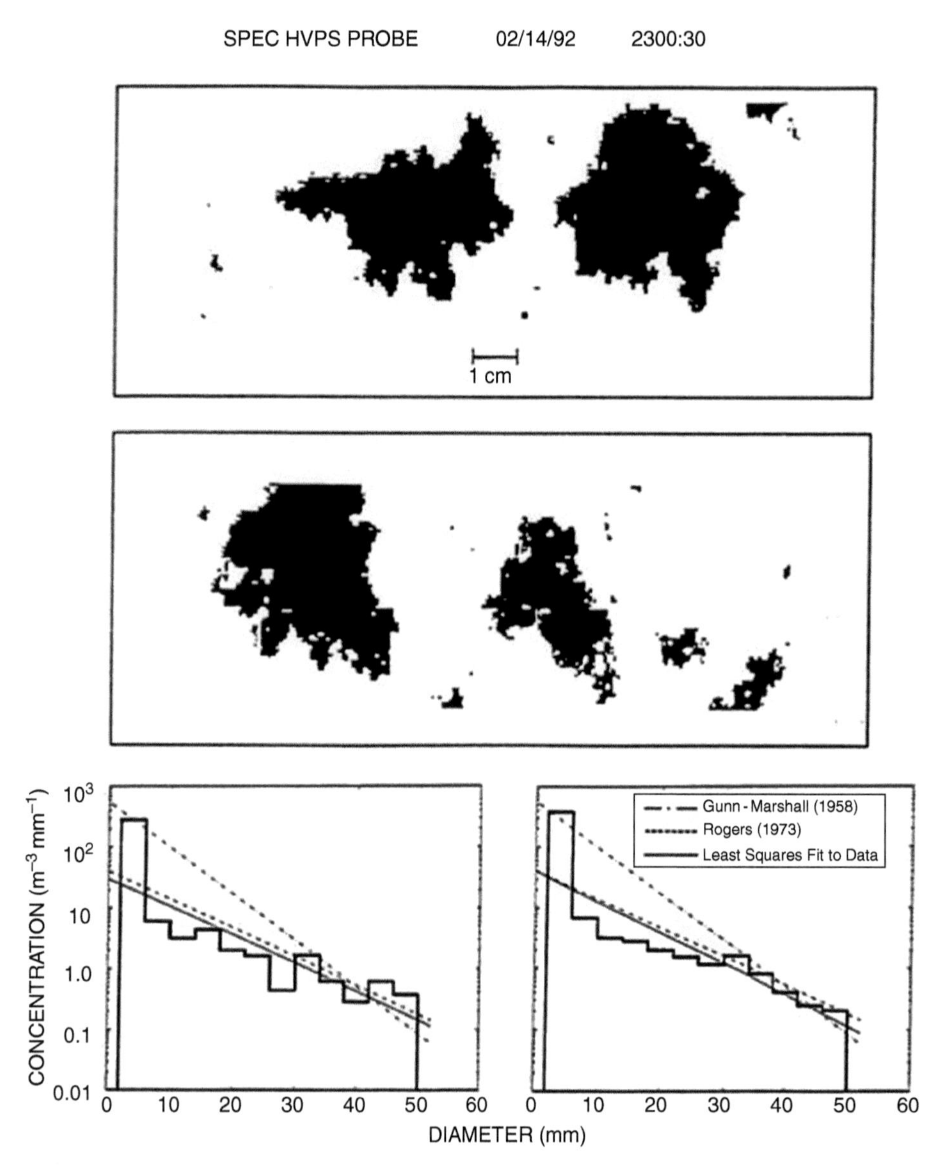

Figure 2.4.8 Images of 4–5 cm-sized snowflakes and 1 km-averaged histograms of particle size distributions observed at the height of 0.6 km at +1.0°C via two independent HVPS channels. Dashed lines show different approximations of size distributions by analytic formulas (from Lawson et al., 1998; courtesy of © American Meteorological Society).

comparatively large rain drops just above the freezing level (Takahashi, 2006). These frozen drops/hail particles do not exceed 4–5 mm in diameter. In mid-latitude hail storms, hail typically forms from a hail embryo (a frozen drop or an ice crystal) at the upper levels of deep convective clouds. The embryo grows through the collection of supercooled water at low temperatures. Hailstones form under conditions of high atmospheric instability and vertical updrafts exceeding 30–40 m/s. These strong updrafts both increase the residential time of hail growing in cloud and transport more supercooled water from below, thus intensifying the process of riming. Hail of this type forms as a result of intense riming of graupel as well as by accretion of supercooled water by frozen raindrops.

The aspect ratio of hailstones is about 0.8. In many cases, however, hailstones are nonspherical disc-shaped particles. These have preferential orientation during their fall with the small axis perpendicular to the direction of hail movement. During their fall, hailstones collect liquid supercooled water largely in their equatorial zones. As a result, the diameters of

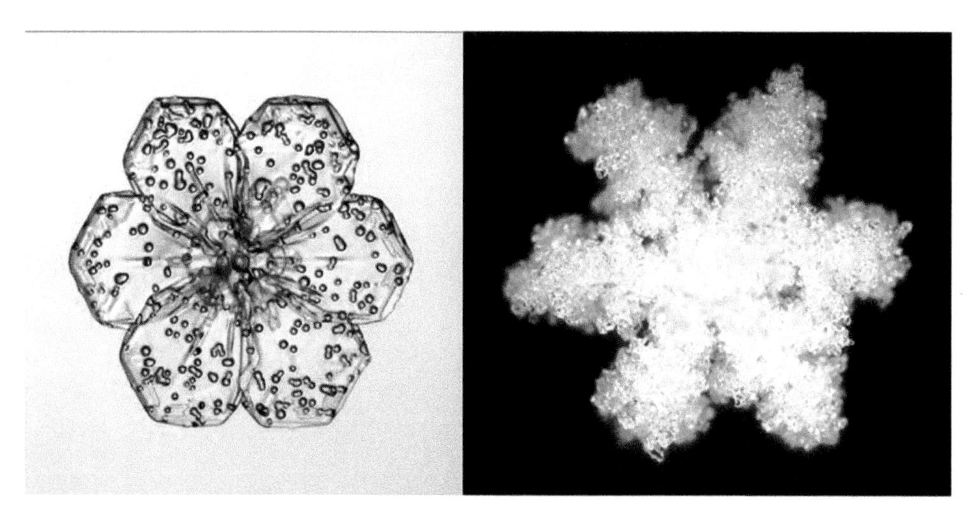

Figure 2.4.9 Pictures of weakly rimed (left) and heavily rimed (right) ice crystals. The frozen droplets seen on the surface of the crystals are much smaller than the crystals themselves (from www.its.caltech.edu/~atomic/snowcrystals/; courtesy of photographer, Prof. K.G. Libbrecht).

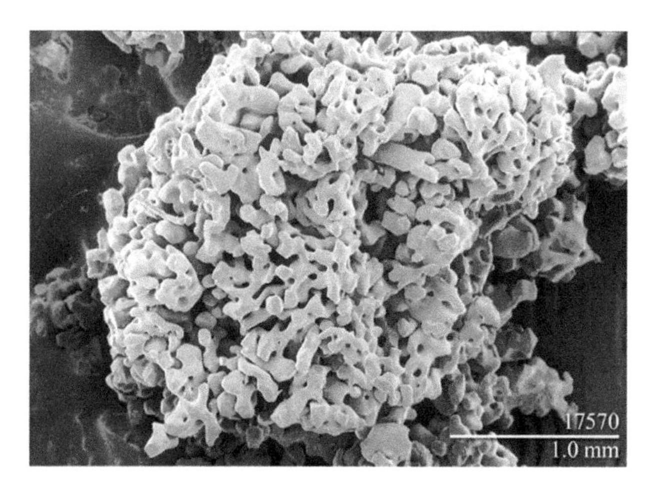

Figure 2.4.10 Heavily rimed aggregates (from http://www.snowcrystals.com/courtesy of photographer, Prof. K.G. Libbrecht).

disc-shaped hail particles grow, which makes them more oblate. The aspect ratio of hail particles decreases from ~1.0 or 0.9 to about 0.6. The hail particles of different shapes, including disc-shaped hailstones, are shown in **Figure 2.4.13**. This figure also compares the shapes and structures of hail and graupel. In strong thunderstorms with maximum updrafts of 35–50 m/s, hailstones may reach several cm in diameter (see **Figure 2.4.14**).

It is accepted to distinguish the dry and wet regimes of hail growth. When the amount of supercooled water is relatively low, the latent heat release caused by the freezing of water droplets at the surface of a growing hail particle is not high enough, thus all liquid water freezes and the surface of the growing particle remains dry. In case the amount of supercooled water collected by growing hail (or graupel) particle is high, the latent heat release on the particle surface prevents complete freezing of the total amount of water collected by falling particles. Consequently, the surface of a growing hailstone is wet even at low environmental temperatures. This mechanism is called the wet growth of hail. Sometimes, the transition from dry to wet growth is referred to as transition from graupel to hail. Before falling to the surface, hail particles often experience several oscillations within clouds (recirculation). In this case, the dry growth and wet growth may interchange several times. The traces of these transitions are seen in cross-sections of hailstones as specific rings indicating the films of frozen water (**Figure 2.4.15**). These rings can also arise when a hail particle falls through the intermittent zones of high and low liquid water content. The density of hailstones with such a multilayer structure changes across the hailstone from ~0.7 g cm^{-3} to 0.9 g cm^{-3}.

The measurements of hail size distributions are quite rare. Sánchez et al. (2009) analyzed databases obtained from several hailpad networks set up in southern France, Spain, and Argentina. The size of hail falling to the surface was evaluated according to the size of the dents left by hailstones on hailpads. In the 8 networks used for the study, slightly more than 13 million dents

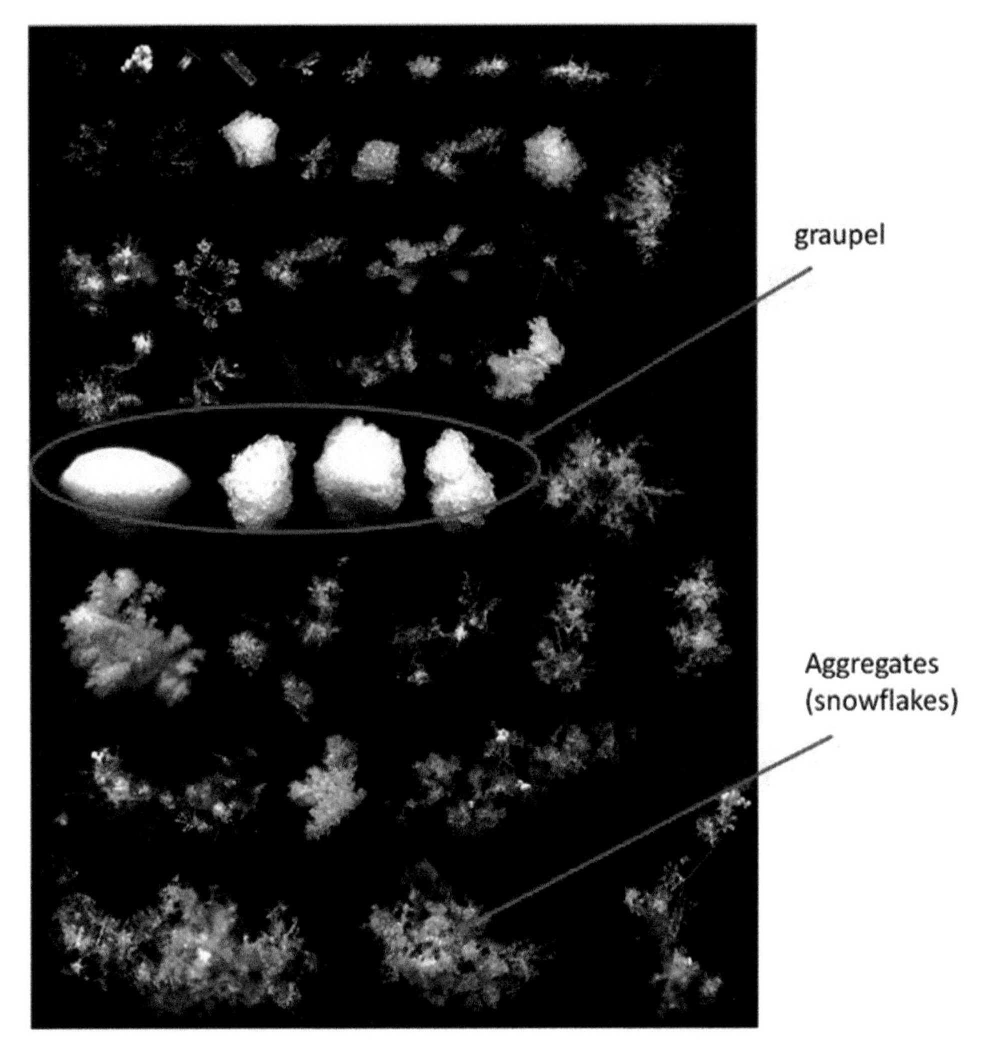

Figure 2.4.11 A collection of snowflakes and graupel photographed automatically as they fell in Alta, Utah, by the new Multi-Angle Snowflake Camera developed at the University of Utah (courtesy of Prof. Tim Garrett, University of Utah).

were analyzed, corresponding to 1,328 hail days. Hail particles with diameters exceeding 5 mm were analyzed. The hail size distributions obtained for cases with big hail are shown in **Figure 2.4.16**. The obtained size distributions were approximated by exponential distributions. Results show that the characteristic parameters of the fitting distributions are different in each area, even in areas that lie geographically close to each other.

One can see the presence of piece-wise distributions with the slope change for the hail size of 15–20 mm in diameter. A possible explanation of the breakpoint could be related to the observation that about 60% of the hailstones larger than 2 cm are elliptical in the concerned areas. At the same time, the fitting was performed under the assumption that the hail particles were spherical. Another explanation is the formation of bimodal size distributions of hail (see Section 7.4). It is noteworthy that size distributions measured on the ground may differ significantly from those in clouds due to the melting and shedding of water from the hail surface.

2.4.5 The Vertical Microphysical Cloud Structure

The vertical pattern of cloud particles and their contents in deep mixed-phase clouds in the Tropics during the monsoon season are shown in **Figure 2.4.17**. These measurements were performed by Takahashi (2006) using videosonde. A video camera recorded images of

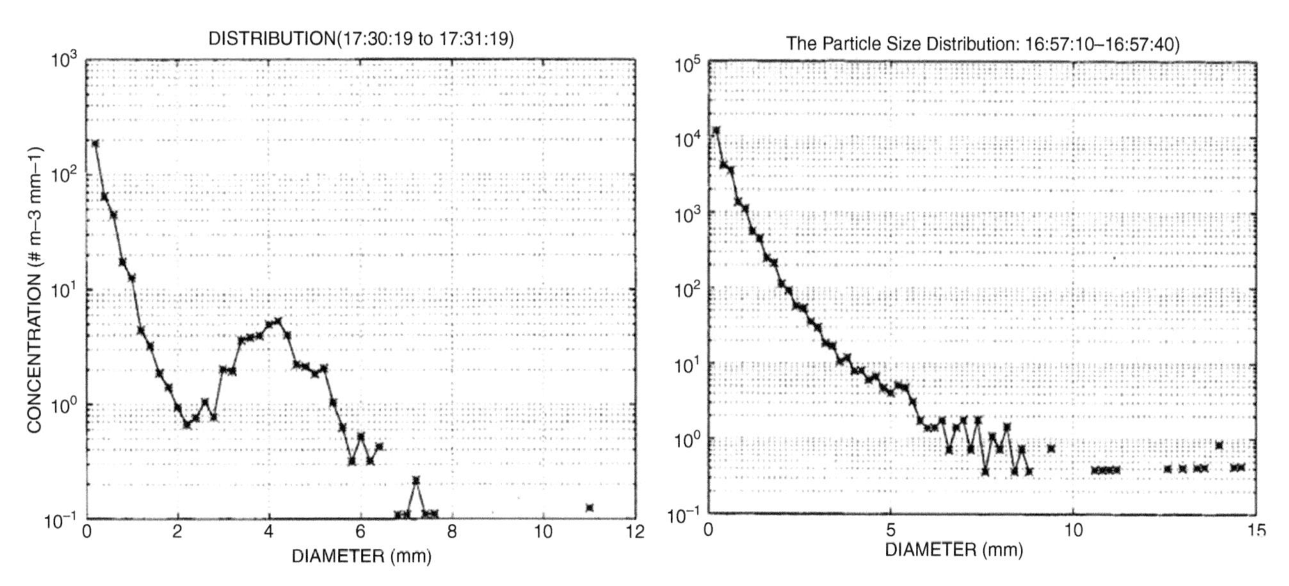

Figure 2.4.12 Particle (mostly graupel) size distributions during a storm on June 20, 1995, measured at the 4-km level by HVPS (from El-Magd et al., 2000; courtesy of © IEEE. Reprinted with permission).

Figure 2.4.13 The images of graupel (left) and hail particles of different shapes (from http://globe-views.com/dcim/dreams/hail/hail-01.jpg).

particles larger than 0.5 mm in diameter. As a result, small crystals are not included in the ice crystal concentration, so this concentration is underestimated. To calculate the mass contents using the measured concentrations, it was assumed that the density is 0.9 g/m^3 for frozen drops/hail, 0.3 g/m^3 for graupel, and 0.1 g/m^3 for ice crystals and snowflakes. One can see that frozen drops form just above the freezing level, where the local maximum of rain water takes place. The largest raindrops form by the melting of frozen drops. The maximum rain water content is reached at ~2 km height. Below this level, the rain water content, as well as the

Figure 2.4.14 A photo of a large hail stone (from http://globe-views.com/dcim/dreams/hail/hail-01.jpg; courtesy of Prof. Tim Garrett, University of Utah).

Figure 2.4.15 Cross-section of a hailstone that experienced recirculation within a cloud (from Mineral Wells, Texas, in May 2013; image credit: posted to The Weather Channel Facebook page by David Adams).

size of rain drops, decreases due to particle breakup and partial evaporation. Above ~5 km, graupel becomes the dominating hydrometeor. Graupel form either by collisions between frozen drops and small supercooled droplets (which were not distinguished by the video camera), or by the riming of ice crystals at heights of about 8–9 km. The cloud tops reach heights of ~14 km. The cloud anvil consists of ice crystals.

The cloud microphysical structure plotted in Figure 2.4.17 is typical of deep maritime clouds during the monsoon period. The cloud microstructure of continental deep convective clouds is quite different. This can be seen by comparison of the radar reflectivity fields in different clouds (e.g., Ryzhkov et al., 2011). The radar reflectivity from deep maritime clouds does not exceed 40–45 dBZ around the freezing level and decreases to 10–20 dBZ at 9–10-km levels (Willoughby et al., 1985). In continental hailstorms with strong updrafts and high supercooled water content, large graupel and hail exist at the heights of ~9–9.5 km, so the radar reflectivity may exceed 55 dBZ at these levels. Formation of big hailstones in such clouds leads to reflectivity values that exceed 65 and even 70 dBZ at the freezing level and below it (Ryzhkov et al., 2011). The melting of graupel and large hailstones leads to an increase in the maximum diameter of raindrops up to 1 cm.

To summarize, ice particle size distributions are quite different, and often have rather complicated shapes depending on the cloud type (convective zones of cloud anvils, stratiform areas of mesoscale systems, mixed-phase stratiform clouds) as well as on the stage of cloud evolution. There is a significant variability of the amount of ice crystals even in clouds of the same

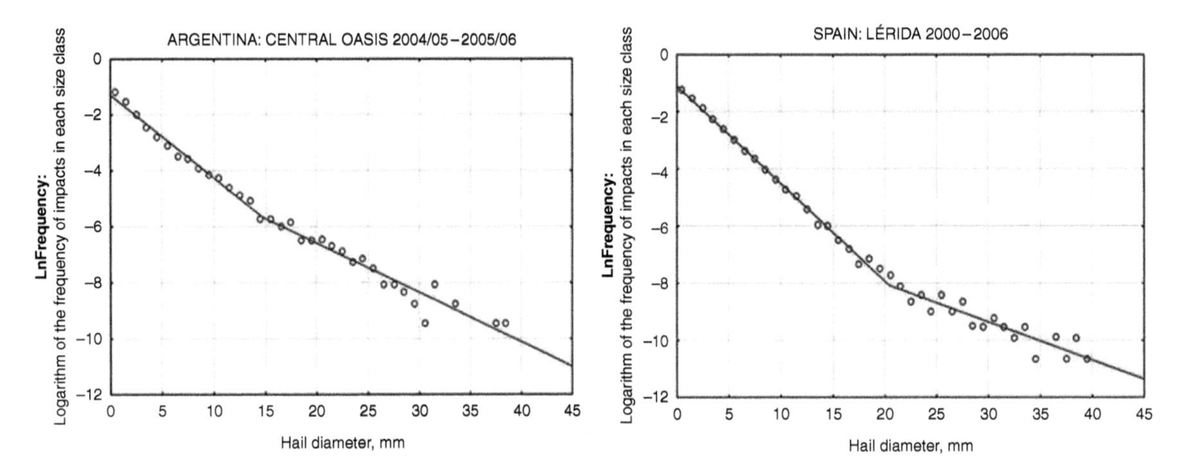

Figure 2.4.16 Piece-wise function for the Central Oasis hailpad network (Argentina) (left) and for the Lérida hailpad network (Spain) (right). Straight lines correspond to exponential distributions (from Sánchez et al., 2009; courtesy of © Elsevier).

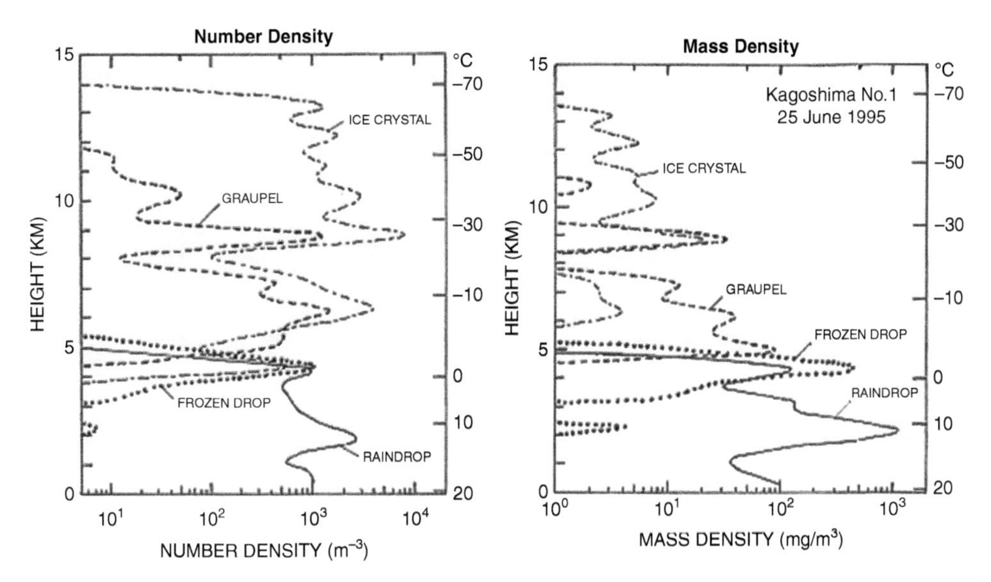

Figure 2.4.17 Number densities and mass densities of hydrometeors measured during the videosonde flight in Kagoshima (Southern Japan). Raindrops (solid line), frozen drops (dotted line), graupel (dashed line), and ice crystals (dash-dotted line) (from Takahashi, 2006; courtesy of © John Wiley & Sons, Inc.).

type. A high variability of snow, graupel, and hail PSD is also observed. Accordingly, approximations of the measured PSD used by different authors are different. For instance, PSD of ice crystals of sizes 50–100 μm were represented by exponential functions (Sekhon and Srivastava, 1970), bimodal functions (Mitchell et al., 1996; Yuter et al., 2006; Lawson et al., 2006), by the sum of an exponential and the Gamma distributions (Field et al., 2005; Heymsfield et al., 2013), and the lognormal function (Tian et al., 2010). Since concentration of ice particles is typically much less than that of liquid droplets, obtaining statistically significant PSDs requires averaging the measurement results over larger areas than those in case of liquid droplets. At the same time, the rates of microphysical processes are determined by local PSDs, which can differ significantly from the averaged ones. Uncertainties related to determination of PSDs of ice particles are discussed in Section 7.2. The wide variety of ice types, shapes, and PSDs presents a serious challenge in modeling mixed-phase clouds. Nevertheless, significant recent progress has been achieved in this field, as well (Chapters 6 and 7).

References

Andreae, M.O., D. Rosenfeld, P. Artaxo, A.A. Costa, G.P. Frank, K.M. Longlo, and M.A.F. Silva-Dias, 2004: Smoking rain clouds over the Amazon. *Science*, **303**, 1337–1342.

Arnott, W.P., Y.Y. Dong, and J. Hallett, 1994: Role of small ice crystals in radiative properties of cirrus: A case study, FIRE II, November 22, 1991. *J. Geophys. Res.*, **99**, 1371–1381.

Bailey, M.P., and J. Hallett, 2009: A comprehensive habit diagram for atmospheric ice crystals: Confirmation from the laboratory, AIRS 2, and other field studies. *J. Atmos. Sci.*, **66**, 2888–2899.

Baker, B., Q. Mo, R. Lawson, D. O'Connor, and A. Korolev, 2009: Drop size distributions and the lack of small drops in RICO rain shafts. *J. Appl. Meteorol. Clim.*, **48**, 616–623.

Bates, T.S., D.J. Coffman, D.S. Covert, and P.K. Quinn, 2002: Regional marine boundary layer aerosol size distributions in the Indian, Atlantic, and Pacific Oceans: A comparison of INDOEX measurements with ACE-1, ACE-2 and Aerosols99. *J. Geophys. Res.*, **107**, 8026.

Berry, E.X., and R.J. Reinhardt, 1974: An analysis of cloud drop growth by collection: Part 1. Double distributions. *J. Atmos. Sci.*, **31**, 1814–1824.

Bleck, R., 1970: A fast approximative method for integrating the stochastic coalescence equation. *J. Geophys. Res.*, **75**, 5165–5171.

Bott, A., 2000: A numerical model of the cloud-topped planetary boundary-layer: Influence of the physicochemical properties of aerosol particles on the effective radius of stratiform clouds. *Atmos. Res.*, **53**, 15–27.

Brandes, E. A., G. Zhang, and J. Vivekanandan, 2002: Experiments in rainfall estimation with a polarimetric radar in a subtropical environment. *J. Appl. Meteorol.*, **41**, 674–685.

Corrigan, C.E., G.C. Roberts, M.V. Ramana, D. Kim, and V. Ramanathan, 2008: Capturing vertical profiles of aerosols and black carbon over the Indian Ocean using autonomous unmanned aerial vehicles. *Atmos. Chem. Phys.*, **8**, 737–747.

Danielsen, E.F., R. Bleck, and D.A. Morris, 1972: Hail growth in a cumulus model. *J. Atmos. Sci.*, **29**, 135–155.

El-Magd, A., V. Chandrasekhar, V.N. Bringi, and W. Strapp, 2000: Multiparameter Radar and *in situ* aircraft observation of graupel and hail. *IEEE Trans. on Geosciences and Remote Sensing*, **38**, 570–577.

Emanuel, K.A., 1994: *Atmospheric Convection*. Oxford University Press p. 580.

Enukashvily, I.M., 1980: A numerical method for integrating the kinetic equation of coalescence and breakup of cloud droplets. *J. Atmos. Sci.*, **61**, 2521–2534.

Fan, J., D. Rosenfeld, Y. Yang, C. Zhao, L.R. Leung, and Z. Li, 2015: Substantial contribution of anthropogenic air pollution to catastrophic floods in Southwest China. *Geophys. Res. Lett.*, **42** (14), 6066–6075.

Feingold, G., B. Stevens, W.R. Cotton, and R.L. Walko, 1994: An explicit cloud microphysical/LES model designed to simulate the Twomey effect. *Atmos. Res.*, **33**, 207–233.

Ferrier, B.S., 1994: A double-moment multiple-phase four-class bulk ice scheme. Part I: Description. *J. Atmos. Sci.*, **51**, 249–280.

Field, P.R., A. J. Heymsfield, and A.B. Bansemer, 2007: Snow size distribution parameterization for midlatitude and tropical ice clouds. *J. Atmos. Sci.*, **64**, 4346–4365.

Field, P.R., R.J. Hogan, P.R.A. Brown, A.J. Illingworth, T.W. Choularton, and R.J. Cotton, 2005: Parameterization of ice particle size distribution for mid-latitude stratiform cloud. *Q. J. Royal Meteorol. Soc.*, **131**, 1997–2017.

Flossmann, A.I., W.D. Hall, and H.R. Pruppacher, 1985: A theoretical study of the wet removal of atmospheric pollutants. Part 1: The redistribution of aerosol particles captured through nucleation and impaction scavenging by growing cloud drops. *J. Atmos. Sci.*, **42**, 583–606.

Flossmann, A.I., and H.R. Pruppacher, 1988: A theoretical study of the wet removal of atmospheric pollutants. Part III: The uptake, redistribution, and deposition of $(NH_4)_2SO_4$ particles by a convective cloud using a two-dimensional cloud dynamics model. *J. Atmos. Sci.*, **45**, 1857–1871.

Freud, E., D. Rosenfeld, M.O. Andreae, A.A. Costa, and P. Artaxo, 2008: Robust relations between CCN and the vertical evolution of cloud drop size distribution in deep convective clouds. *Atmos. Chem. Phys.*, **8**, 1661–1675.

Gerber, H., 1996: Microphysics of marine stratocumulus clouds with two drizzle modes. *J. Atmos. Sci.*, **53**, 1649–1662.

Ghan, S.J., H. Abdul-Razzak, A. Nenes, Y. Ming, X. Liu, M. Ovchinnikov, B. Shipway, N. Meskhidze, J. Xu, and X. Shi, 2011: Droplet nucleation: Physically based parameterizations and comparative evaluation. *J. Adv. Model. Earth Syst.*, **3**, M10001, doi:10.1029/2011MS000074.

Hall, W.D., 1980: A detailed microphysical model within a two-dimensional dynamic framework: Model description and preliminary results. *J. Atmos. Sci.*, **37**, 2486–2507.

Hallett, J., and S.C., Mossop, 1974: Production of secondary ice crystals during the riming process. *Nature*, **249**, 26–28.

Heymsfield, A., C. Schmitt, and A. Bansemer, 2013: Ice cloud particle size distributions and pressure-dependent terminal velocities from in situ observations at temperatures from 0 to −86 C. *J. Atmos. Sci.*, **70**, 4123–4154.

Heymsfield, A.J., A. Bansemer, G. Heymsfield, and A. Fierro, 2009: Microphysics of maritime tropical convective updrafts at temperatures from −20 to −60C. *J. Atmos. Sci.*, **66**, 3530–3562.

Heymsfield, A.J., A. Bansemer, and C.H. Twohy, 2007: Refinements to ice particle mass dimensional and terminal velocity relationships for ice clouds. Part I: Temperature dependence. *J. Atmos. Sci.*, **64**, 1047–1067.

Heymsfield, A.J. and Platt, C.M.R., 1984: A parameterization of the particle size spectrum of ice clouds in terms of the ambient temperature and the ice water content. *J. Atmos. Sci.*, **41**, 846–855.

Heymsfield, G.M., L. Tian, A.J. Heymsfield, L. Li, and S. Guimond, 2010: Characteristics of deep tropical and subtropical convection from nadir-viewing high-altitude airborne Doppler radar. *J. Atmos. Sci.*, **67**, 285–308.

Hobbs, P., 1993: *Aerosol-cloud-climate interactions*, edited by Peter Hobbs, Academic Press, p. 237.

Hobbs, P.V., D.A. Bowdle, and L.F. Radke, 1985: Particles in the lower troposphere over the High Plains of the United States. 1: Size distributions, elemental compositions and morphologies. *J. Clim. Appl. Meteorol.*, **24**, 1344–1356.

Ito, T., 1982: On the size distribution of submicron aerosols in the Antarctic atmosphere. *Antarctic Record*, **76**, 1–19.

Jaenicke, R., 1988: *Properties of atmospheric aerosols*, in *Meteorology: Properties of the Air, vol. V/4b*, edited by G. Fischer, Springer-Verlag, New York, pp. 405–428.

1993: Chapter 1: Tropospheric Aerosols. *Int. Geophysics*, **54**, 1–31 in book Aerosol–Cloud–Climate Interactions. Edited by Peter V.Hobbs. Elsevier.

Jaenicke, R., Dreiling V., Lehmann E., Koutsenoguii P. K., and Stingl J., 1992: Condensation nuclei at the German Antarctic Station "Georg von Neumayer." *Tellus*, **44B**, 311–317.

Jaenicke, R., and L. Schutz, 1982: Arctic Aerosols in Surface Air. *J. Hungarian Meteorol. Service*, **86**, 235–241.

Jorgensen, D.P., and M.A. LeMone, 1989: Vertical velocity characteristics of oceanic convection. *J. Atmos. Sci.*, **46**, 621–640.

Jorgensen, D.P., E.J. Zipser, and. M.A. LeMone, 1985: Vertical motions in intense hurricanes. *J. Atmos. Sci.*, **42**, 839–856.

Junge, C.E., 1955: The size distribution and aging of natural aerosol as determined from electrical and optical data on the atmosphere. *J. Meteorol.*, **12**, 13–25.

1969: Comments on "Concentration and size distribution measurements of atmospheric aerosols and a test of the theory of self-preserving size distributions." *J. Atmos. Sci.*, **26**, 603–608.

Khain, A.P., V. Arkhipov, M. Pinsky, Y. Feldman, and Ya Ryabov, 2004a: Rain enhancement and fog elimination by seeding with charged droplets. Part I: Theory and numerical simulations. *J. Appl. Met.*, **43**, 1513–1529.

Khain, A.P, A. Pokrovsky, M. Pinsky, A. Seifert, and V. Phillips, 2004b: Effects of atmospheric aerosols on deep convective clouds as seen from simulations using a spectral microphysics mixed-phase cumulus cloud model Part 1: Model description. *J. Atmos. Sci.*, **61**, 2963–2982.

Khain, A.P., D. Rosenfeld, A. Pokrovsky, U. Blahak, and A. Ryzhkov, 2011: The role of CCN in precipitation and hail in a mid-latitude storm as seen in simulations using a spectral (bin) microphysics model in a 2D dynamic frame. *Atmos. Res.*, **99**, (Issue 1), 129–146.

Koch, W., 1996. Solarer Strahlungstransport in Arktischem Cirrus. PhD Thesis. GKSS 96/E/60, 99pp.

Korolev, A.V., 1994: A study of bimodal droplet size distributions in stratiform clouds. *Atmos. Res.*, **32**, 143–170.

Korolev, A.V., G.A. Isaac, and J. Hallett, 1999: Ice particle habits in Arctic clouds. *Geophys. Res. Lett.*, **26**, 1299–1302.

2000: Ice particle habits in stratiform clouds. *Quart. J. Roy. Meteorol. Soc.*, **126**, 2873–2902.

Kostinski, A., and A. Jameson, 1999: Fluctuation properties of precipitation. Part III: On the ubiquity and emergence of the exponential drop size spectra. *J. Atmos. Sci.*, **56**, 111–121.

Lawson, R.P., B. Baker, B. Pilson, and Q. Mo, 2006: In situ observations of the microphysical properties of wave, cirrus, and anvil clouds. Part II: Cirrus clouds. *J. Atmos. Sci.*, **63**, 3186–3203.

Lawson, R.P., R.E. Stewart, and L.J. Angus, 1998: Observations and numerical simulations of the origin and development of very large snowflakes. *J. Atmos. Sci.*, **55**, 3209–3229.

Levin, Z., and W.R. Cotton, 2009: *Aerosol pollution impact on precipitation: A scientific review*. Springer, p. 386.

Levin, Z., E. Ganor, and V. Gladstain, 1996: The effects of desert particles coated with sulfate on rain formation in the Eastern Mediterranean. *J. Appl. Meteorol.*, **35**, 1511–1523.

Lewis, E.R., and S.E. Schwartz, 2004: Sea salt aerosol production. Mechanisms, methods, measurements, and models. American Geophysical Union.

Li, S.M., and J.W. Winchester, 1989: Resolution of ionic components of late winter Arctic aerosols. *Atmospheric Environment*, **23**, 2387–2399.

Liu, P.C. Zhao, Q. Zhang, C. Deng, M. Huang, and X. Tie, 2009: Aircraft study of aerosol vertical distributions over Beijing and their optical properties. *Tellus*, **61B**, 756–767.

Low, T.B., and R. List, 1982: Collision, coalescence and breakup of raindrops, Part II: Parameterization of fragment size distributions. *J. Atmos. Sci.*, **39**, 1607–1618.

Lynn, B.H., A.P. Khain, J.W. Bao, S.A. Michelson, T. Yuan, G. Kelman, D. Rosenfeld, J. Shpund, and N. Benmoshe, 2016: The sensitivity of hurricane Irene to aerosols and ocean coupling: Simulations with WRF spectral bin microphysics. *J. Atmos. Sci.*, **73**, 467–486.

Magaritz, L., M. Pinsky, O. Krasnov, and A. Khain, 2009: Investigation of droplet size distributions and drizzle formation using a new trajectory ensemble model. Part II: Lucky parcels. *J. Atmos. Sci.*, **66**, 781–805.

Magono, C., and C.W. Lee, 1966: Meteorological classification of natural snow crystals, *J. Fac. Sci.* Hokkaido Univ. Ser. **7**, 2, 321–335.

Maki, M., T. Keenan, Y. Sasaki, and K. Nakamura, 2001: Characteristics of the raindrop size distribution in tropical continental squall lines observed in Darwin, Australia. *J. Appl. Meteorol.*, **40**, 1393–1412.

Marshall J.S. and W. Mc K Palmer, 1948: The distribution of raindrops with size. *J. Met.*, **5**, 165–166.

Martin, G.M., D.W. Johnson, and A. Spice, 1994: The measurements and parameterization of effective radius of droplets in warm stratocumulus clouds. *J. Atmos. Sci.*, **51**, 1823–1842.

Martner, B., S. Yumer, A. White, S. Matrosov, D. Kingsmill, and F. Ralph, 2008: Raindrop size distributions and rain characteristics in California coastal rainfall for periods with and without a radar bright band. *J. Hydrometeorology*, **9**, 408–425.

Mason, B.J., 1971: *The physics of clouds (2nd edition)*. Oxford University Press, p. 544.

Mazin, I.P., and S.M. Shmeter, 1983: *Clouds, their Structure and Formation*. Gidrometizdat, p. 279.

Milbrandt, J.A, and R. McTaggart-Cowan, 2010: Sedimentation-induced errors in bulk microphysics schemes. *J. Atmos. Sci.*, **67**, 3931–3948.

Mitchell, D.L., S.K. Chai, Y. Liu, A.J. Heymsfield, and Y. Dong, 1996: Modeling Cirrus Clouds. Part I: Treatment of bimodal size spectra and case study analysis. *J. Atmos. Sci.*, **53**, 2952–2966.

Mitchell, D.L., and A.J. Heymsfield, 2005: Refinements in the treatment of ice particle terminal velocities, highlighting aggregates. *J. Atmos. Sci.*, **62**, 1637–1644.

Noll, K.E., and M.J. Pilat, 1971: Size distribution of atmospheric giant particles. *Atmos. Environment*, **5**, 527–540.

Ochs, H.T., and C.S. Yao, 1978: Moment conserving techniques for warm cloud microphysical computations, Part 1: Numerical techniques. *J. Atmos. Sci.*, **35**, 1947–1958.

Pinsky, M., and A. Khain, 2002: Effects of in-cloud nucleation and turbulence on droplet spectrum formation. *Quart. J. Roy. Meteorol. Soc.*, **128**, 501–533.

Pinsky, M., A. Khain, L. Magaritz, and A. Sterkin, 2008: Simulation of droplet size distributions and drizzle formation using a new trajectory ensemble model of cloud topped boundary layer. Part 1: Model description and first results in non-mixing limit. *J. Atmos. Sci.*, **65**, 2064–2086.

Pinsky, M., I.P. Mazin, A. Korolev, and A. Khain, 2014: Supersaturation and diffusional droplet growth in liquid clouds: Polydisperse spectra *J. Geophys. Res. Atmos.*, **119**, 12872–12887,

Politovich, M.K., 1993: A study of the broadening of droplet size distribution in cumuli. *J. Atmos. Sci.*, **50**, 2230–2244.

Prabha, T., A.P. Khain, B.N. Goswami, G. Pandithurai, R.S. Maheshkumar, and J.R. Kulkarni, 2011: Microphysics of pre-monsoon and monsoon clouds as seen from in-situ measurements during CAIPEEX. *J. Atmos. Sci.*, **68**, 1882–1901.

Pruppacher, H.R., and J.D. Klett, 1997: *Microphysics of clouds and precipitation, 2nd edition*. Oxford: Kluwer Academic Publishers.

Reisin, T., Z. Levin, and S. Tzivion, 1996: Rain production in convective clouds as simulated in an axisymmetric model with detailed microphysics. Part 1: Description of the model. *J. Atmos. Sci.*, **53**, 497–519.

Respondek, P.S., A.I. Flossmann, R.R. Alheit, and H.R. Pruppacher, 1995: A theoretical study of the wet removal of atmospheric pollutants: Part V. The uptake, redistribution, and deposition of $(NH_4)_2SO_4$ by a convective cloud containing ice. *J. Atmos. Sci.*, **52**, 2121–2132.

Rissler, J., A. Vestin, E. Swietlicki, G. Fisch, J. Zhou, P. Artaxo, and M.O. Andreae, 2006: Size distribution and hygroscopic properties of aerosol particles from dry-season biomass burning in Amazonia. *Atmos. Chem. Phys.*, **6**, 471–491.

Rogers, R.R, and M.K. Yau, 1996: *Short Course in Cloud Physics*, Butterworth-Heinemann, p. 304.

Rosenfeld, D., M.O. Andreae, A. Asmi, M. Chin, G. Leeuw, D.P. Donovan, R. Kahn, S. Kinne, N. Kivekäs, M. Kulmala, W. Lau, K.S. Schmidt, T. Suni, T. Wagner, M. Wild, and J. Quaas, 2014: Global observations of aerosol-cloud-precipitation-climate interactions. *Rev. Geophys.*, **52** (4), 750–808.

Rosenfeld, D., and G. Gutman, 1994: Retrieving microphysical properties near the tops of potential rain clouds by multispectral analysis of AVHRR data. *Atmos. Res.*, **34**, 259–283.

Rosenfeld, D., and W.L. Woodley, 2000: Deep convective clouds with sustained highly supercooled liquid water until −37.5°C. *Nature*, **405**, 440–442.

Ryzhkov, A., M. Pinsky, A. Pokrovsky, and A. Khain, 2011: Polarimetric radar observation operator for a cloud model with spectral microphysics. *J. Appl. Met. Clim.*, **50**, 873–894.

Saleeby, S.M., and W.R. Cotton, 2004: A large-droplet mode and prognostic number concentration of cloud droplets in the Colorado State University Regional Atmospheric Modeling System (RAMS). Part I: Module descriptions and supercell test simulations. *J Appl. Meteorol.*, **43**, 182–195.

Sánchez, J.L., B. Gil-Robles, J. Dessens, E. Martin, L. Lopez, J.L. Marcos, C. Berthet, J.T. Fernández, and E. García-Ortega, 2009: Characterization of hailstone size spectra in hailpad networks in France, Spain, and Argentina. *Atmos. Res.* **93**, 641–654.

Seifert, A., and K. Beheng, 2006: A two-moment cloud microphysics parameterization for mixed-phase clouds. Part 1: Model description. *Meteorol. Atmos. Phys.*, **92**, 45–66.

Sekhon, R.S., and R.C. Srivastava, 1970: Snow size spectra and radar reflectivity. *J. Atmos. Sci.*, **27**, 299–307.

Shaw, G.E., 1986: On physical properties of aerosols at Ross Island, Antarctica. *J. Aerosol Sci.*, **17**, 937–945.

Stewart, R.E, 1991: Canadian Atlantic Storms Program: Progress and plans of the meteorological component. *Bull. Amer. Meteorol. Soc.*, **72**, 364–371.

Stewart, R.E., and R.W. Crawford, 1995: Some characteristics of the precipitation formed within winter storms over eastern Newfoundland. *Atmos. Res.*, **36**, 17–37.

Straka, J.M., 2009: *Cloud and Precipitation microphysics. Principles and parameterizations*. Cambridge: Cambridge University Press.

Takahashi, T., 2006: Precipitation mechanisms in East Asian monsoon: Videosonde study. *J. Geophys. Res.*, **111**, D09202.

Thurai, M., M. Szakall, V.N. Bringi, K.V. Beard, S.K. Mitra, and S. Borrmann, 2009: Drop shapes and axis ratio distributions: Comparison between 2D video disdrometer and wind-tunnel measurements. *J. Atmos. Ocean Tech.*, **26**, 1427–1432.

Tian, L., G.M. Heymsfield, A.J. Heymsfield, A. Bansemer, L. Li, C.H. Twohy, and R.C. Srivastava, 2010: A study of cirrus ice particle size distribution using TC4 observations. *J. Atmos. Sci.*, **67**, 195–216.

Twohy, C.H., M.D. Petters, J.R. Snider, B. Stevens, W. Tahnk, M. Wetzel, L. Russell, and F. Burnet, 2005: Evaluation of the aerosol indirect effect in marine stratocumulus clouds: Droplet number, size, liquid water path and radiative impact, *J. Geophys. Res.*, **110**, D08203, doi: 10.1029/2004JD005116.

Tzivion, S., G. Feingold, and Z. Levin, 1987: An efficient numerical solution to the stochastic collection equation. *J. Atmos. Sci.*, **44**, 3139–3149.

1989: The evolution of raindrop spectra. Part 2. Collisional collection/breakup and evaporation in a rainshaft. *J. Atmos. Sci.*, **46**, 3312–3327.

VanZanten, M.C., B. Stevens, G. Vali, and D.H. Lenschow, 2005: Observations in nocturnal marine stratocumulus. *J. Atmos. Sci.*, **62**, 88–106.

Van den Heever, S.C., G.G. Carrió, W.R. Cotton, P.J. Demott, and A.J. Prenni, 2006: Impacts of nucleating aerosol on Florida storms. Part I: Mesoscale simulations. *J. Atmos. Sci.*, **63**, 1752–1775.

Wang, P.K., 1997: Characterization of ice crystals in clouds by simple mathematical expressions based on successive modification of simple shapes. *J. Atmos. Sci.*, **54**, 2035–2041.

Warner, J., 1969: The microstructure of cumulus clouds. Part 1. General features of the droplet spectrum. *J. Atmos. Sci.*, **26**, 1049–1059.

Whitby, K.T., 1973: On the multimodal nature of atmospheric aerosol size distributions. VIII-th I*nt. Conferencel. on nucleation*., Leningrad, U.S.S.R.

Willoughby, H.E., D.P. Jorgensen, R.A. Black, and S.L. Rosenthal, 1985: Project storm-fury: A scientific chronicle, 1962–1983. *Bull Amer. Met. Soc.*, **66**, 505–514.

Yum, S.S., and J.G. Hudson, 2002: Maritime/continental microphysical contrasts in stratus. *Tellus*, Series B, **54**, 61–73.

Yuter, S.E., D. Kingsmill, L.B. Nance, and M. Löffler-Mang, 2006: Observations of precipitation size and fall speed characteristics within coexisting rain and wet snow. *J. Appl. Meteorol. Climatol.*, **45**, 1450–1464.

Zappoli, S., A. Andracchio, S. Fuzzi, M.C. Facchini, A. Gelencser, G. Kiss, Z. Krivacsy, A. Molnar, E. Meszaros, H.C. Hansson, K. Rosman, and Y. Zebuhr, 1999: Inorganic, organic and macromolecular components of fine aerosol in different areas of Europe in relation to their water solubility. *Atmos. Environ.* **33**, 2733–2743.

Zhang, Y., A. Macke, and F. Albers, 1999: Effect of crystal size spectrum and crystal shape on stratiform cirrus radiative forcing. *Atmos. Res.* **52**, 59–75.

3 Basic Equations

The atmosphere is characterized by fields of pressure, wind, temperature, humidity, and many other quantities. To investigate changes in these fields over space and time, it is necessary to formulate the corresponding equations. These equations are based on the three laws of conservation: conservation of mass, conservation of momentum, and conservation of energy. The first law allows us to formulate the continuity equation. The second law allows us to formulate the motion equations, and the third law to formulate equations for heat, moisture, budgets, etc. Let us first consider the main thermodynamic relationships.

3.1 Thermodynamics of Dry and Moist Air

3.1.1 Thermodynamic Characteristics of Dry Air

Air is a mixture of many different gases including water vapor. Each gas component of air (marked by the index *i*) is usually considered as an ideal gas whose partial pressure p_i and density ρ_i obey the state equation:

$$p_i = \frac{\mathrm{R}}{M_i}\rho_i T = R_i \rho_i T, \tag{3.1.1}$$

where M_i is the molecular weight of a particular gas, R is the universal gas constant (8.314 J mole^{-1} K^{-1}), T is temperature (in K), and $R_i = \frac{\mathrm{R}}{M_i}$ is the specific gas constant. According to the Dalton law, the total pressure of the air is the sum of partial pressures: $p = \sum_i p_i$. Since water vapor plays a special role in the atmosphere and especially in clouds, we start our analysis introducing the expressions for dry air and then proceed to the effects of air humidity.

Pressure of dry air p_a can be written as

$$p_a = T \sum_i \rho_i R_i = \rho_a R_a T, \tag{3.1.2}$$

where ρ_a is the density of dry air and R_a is the specific gas constant of dry air. In calm atmosphere, pressure at a certain height level is determined by the weight of the air column above this level. This relationship is expressed by the equation of hydrostatic equilibrium

$$\frac{dp_a}{dz} = -\rho_a g, \tag{3.1.3}$$

where g is gravitational acceleration. Equation (3.1.3) expresses the balance between the pressure and the gravity force. In cases when vertical air accelerations are significant, this balance does not take place. In fact, the difference between the two sides of the equation of hydrostatic equilibrium determines the buoyancy force and air accelerations. However, even in this case Equation (3.1.3) is still a good approximation that can be used, for instance, to evaluate the pressure dependence on height. This is why Equation (3.1.3) is often called the quasi-static equation. It is often used to determine the reference (basic) state in which air pressure and density depend only on height: $p_a = \bar{p}_a(z)$, $\rho_a = \overline{\rho_a}(z)$. Air temperature in the reference state also depends only on height: $T = \overline{T}(z)$.

Combining the quasi-static Equation (3.1.3) with the state Equation (3.1.2) allows calculating the vertical profile of pressure as

$$p_a(z) = p_a(z=0)\exp\left(-\int_0^z \frac{g}{R_a T(z)}dz\right). \tag{3.1.4}$$

This expression is known as the barometric formula. It is often used in cloud models to calculate the vertical profiles of air pressure and density using the vertical temperature profile.

Basic equations for temperature and other thermodynamic quantities can be derived from the first law of thermodynamics (e.g., Laichtman, 1976):

$$\frac{dE_{in}}{dT}dT = dQ - p_a dV, \tag{3.1.5}$$

where E_{in} is the internal energy and Q is heat influx in an air volume V. Equation (3.1.5) reflects the energy conservation law, according to which an influx of heat dQ performs mechanical work $p_a dV$, which is necessary to increase the air volume and increase the temperature, thus increasing the internal energy $\frac{dE_{in}}{dT}dT$. Using Equation (3.1.2), Equation (3.1.5) for dry air can be rewritten in the form:

$$dQ = \left(\frac{dE_{in}}{dT} + R_a\right)dT - Vdp_a \tag{3.1.6}$$

Let us consider two thermodynamic processes: the isochoric process, characterized by constant volume ($dV = 0$), and the isobaric process, characterized by constant pressure ($dp_a = 0$). In case of the isochoric process

$$\frac{dE_{in}}{dT} = \left(\frac{dQ}{dT}\right)_v = c_v. \tag{3.1.7}$$

In case of the isobaric process

$$\left(\frac{dE_{in}}{dT} + R_a\right) = \left(\frac{dQ}{dT}\right)_p = c_p, \tag{3.1.8}$$

where c_v and c_p are specific heat capacity values at constant volume and constant pressure, respectively. Using Equations (3.1.7) and (3.1.8), one can derive

$$c_p = c_v + R_a \tag{3.1.9}$$

Using Equations (3.1.6), (3.1.7), and (3.1.8), the first law of thermodynamics can be written in two equivalent forms describing two physically similar processes:

$$dQ = c_p dT - R_a T \frac{dp_a}{p_a} \tag{3.1.10}$$

and

$$dQ = c_v dT - p_a dV. \tag{3.1.11}$$

Assuming that the heat influx is given, using Expressions (3.1.10) and (3.1.11) and the state Equation (3.1.2), one can write down the relationship between different thermodynamic characteristics of air during the thermodynamic process:

$$\frac{T_2}{T_1} = \left(\frac{p_{a2}}{p_{a1}}\right)^{R_a/c_p} \exp\left(\frac{1}{c_p}\int_{Q_1}^{Q_2} dQ/T\right)$$
$$= \left(\frac{\rho_{a2}}{\rho_{a1}}\right)^{R_a/c_v} \exp\left(\frac{1}{c_p}\int_{Q_1}^{Q_2} dQ/T\right), \tag{3.1.12}$$

where T_1, p_{a1} and ρ_{a1} are temperature, pressure and density at the initial state of the process, respectively, and T_2, p_{a2} and ρ_{a2} are the corresponding current values of the process.

The heat influx variations are often assumed proportional to temperature variations:

$$dQ = c dT. \tag{3.1.13}$$

The process described by Equation (3.1.13) is known as a polytropic process. Substitution of Equation (3.1.13) into Equation (3.1.12) leads to the following expressions:

Table 3.1.1 The main thermodynamic processes in the atmosphere.

Process	Adiabatic	Isochoric	Isobaric	Isothermal
Condition	$dQ = 0$	$dV = 0$	$dp_a = 0$	$dT = 0$
Value of c	0	c_v	c_p	∞

$$\frac{T_2}{T_1} = \left(\frac{p_{a2}}{p_{a1}}\right)^{R_a/(c_p - c)} = \left(\frac{\rho_{a2}}{\rho_{a1}}\right)^{R_a/(c_v - c)}$$
$$= \left(\frac{V_1}{V_2}\right)^{R_a/(c_v - c)} \tag{3.1.14}$$

It follows from Equation (3.1.14) that

$$p_a V^n = const, \tag{3.1.15}$$

where $n = \frac{c_p - c}{c_v - c}$ is the polytrophic exponent. Using different values of c, one can describe different thermodynamic atmospheric processes (**Table 3.1.1**).

The adiabatic processes are of high importance in analysis of atmospheric processes. In adiabatic processes the heat influx is equal to zero, i.e., the air parcel is isolated and not involved in any heat exchange with the surrounding air. Although an adiabatic process is a kind of approximation (because some heat exchange always exists), the processes are considered adiabatic if the value of heat influx dQ is much lower than the values of other terms in Equations (3.1.10) and (3.1.11). In particular, in air parcels ascending comparatively rapidly, the work of pressure needed to increase the parcel volume may substantially exceed the heat energy that the parcel obtains through its lateral boundaries. In the first approximation, such a process can be considered adiabatic. In other sections of this book, we will often illustrate different microphysical processes using models of ascending adiabatic parcels. Assuming $c = 0$ in Equation (3.1.14), we can get the following relationships between thermodynamic quantities of an adiabatic process:

$$\frac{T_2}{T_1} = \left(\frac{p_{a2}}{p_{a1}}\right)^{R_a/c_p} = \left(\frac{\rho_{a2}}{\rho_{a1}}\right)^{R_a/c_v} = \left(\frac{V_1}{V_2}\right)^{R_a/c_v}. \tag{3.1.16}$$

In meteorology, a quantity known as potential temperature θ is widely used:

$$\theta = T\left(\frac{p_{ref}}{p_a}\right)^{R_a/c_p} = T/\Pi, \tag{3.1.17}$$

where p_{ref} = 1,000 hPa and

$$\Pi = \left(\frac{p_a}{p_{ref}}\right)^{R_a/c_p} \quad (3.1.18)$$

is the Exner function (which is typically <1). Sometimes the static energy of dry air h is used instead of potential temperature:

$$h = c_p T + gz. \quad (3.1.19)$$

3.1.2 Changes of Temperature in an Ascending Parcel. The Dry Adiabatic Gradient

To determine temperature changes in an ascending parcel, we use Equation (3.1.10) to get

$$\frac{dT}{dz} = \frac{1}{c_p} dQ - \frac{R_a T}{p_a} \frac{dp_a}{dz}. \quad (3.1.20)$$

Assuming the process to be adiabatic ($dQ = 0$) and quasi-static ($p_a = p_0$, $\rho_a = \rho_0$ and $\frac{dp_0}{dz} = -\rho_0 g = -\frac{p_0 g}{R_a T_0}$), Equation (3.1.20) yields:

$$\frac{dT}{dz} = -\frac{g}{c_p} \frac{T}{T_0} \approx -\frac{g}{c_p} = -\gamma_a, \quad (3.1.21)$$

where $\gamma_a = g/c_p$ is (~9.8°C/km) and T_0 denotes the temperature in the air surrounding the parcel. In Equation (3.1.21), the approximate equality $\frac{T}{T_0} \approx 1$ is used. The dry adiabatic gradient γ_a is one of the most important meteorological constants. The gradient of potential temperature can be expressed via the temperature gradient using Equation (3.1.17):

$$\frac{d\theta}{dz} = \frac{\theta}{T} \frac{dT}{dz} + \frac{\theta}{T_0} \gamma_a \approx \frac{dT}{dz} + \gamma_a. \quad (3.1.22)$$

In Equation (3.1.22), condition $\frac{\theta}{T} \approx \frac{\theta}{T_0} \approx 1$ is used. The error of this approximation increases with height.

It follows from Equations (3.1.20) and (3.1.21) that in case of an adiabatic updraft, the air temperature decreases according to the dry adiabatic gradient $-\frac{dT}{dz} = \gamma_a$, i.e., 9.8°C/km. According to Equation (3.1.22), in a parcel ascending adiabatically, $\frac{d\theta}{dz} = 0$. It means that the potential temperature is the conservative quantity in a dry adiabatic process. From Equation (3.1.22) it follows that

$$\theta(z) \approx T(z) + \gamma_a z. \quad (3.1.23)$$

The approximate expression, Equation (3.1.23), is strictly valid if at $z = 0$ $p = 1{,}000$ mb, as $\theta(z = 0) \approx T(z = 0)$ (see the definition of potential temperature (Equation (3.1.17). In case the pressure at the ground surface differs significantly from 1,000 mb, the relation between T and θ should be found from Equation (3.1.17). Comparison of the expression for the static energy of dry air, Equation (3.1.19), with Equation (3.1.23) shows that

$$\theta \approx h/c_p. \quad (3.1.24)$$

Thus, in an adiabatic process the static energy does not change.

3.1.3 Thermodynamic Characteristics of Moist Air

Humid air is a mixture of dry air and water vapor and can be considered as a two-component system. Water vapor plays a very important role in different atmospheric processes, especially in clouds; therefore, different characteristics are used to describe the amount of vapor and its behavior in the atmosphere. Since water vapor in the atmosphere behaves, to a good approximation, as an ideal gas, its state equation is

$$e = \rho_v R_v T, \quad (3.1.25)$$

where e is water vapor pressure (or partial pressure), ρ_v is the vapor density (also called absolute humidity), $R_v = \frac{R}{M_w} \approx 1.6 R_a$ is the specific gas constant for water vapor (461.5 J kg^{-1} K^{-1}), and M_w is the molecular weight of water. The amount of water vapor in the air (air humidity) is often characterized by two quantities: mixing ratio q and specific humidity q_v. The mixing ratio is defined as the mass of water vapor per unit mass of dry air:

$$q = \frac{\rho_v}{\rho_a} = \frac{R_a}{R_v} \frac{e}{p - e} \approx 0.622 \frac{e}{p} \quad (3.1.26)$$

Specific humidity q_v is defined as the mass of water vapor per unit mass of moist air:

$$q_v = \frac{\rho_v}{\rho} = \frac{q}{1+q} = \frac{R_a}{R_v} \frac{e}{p - (1 - R_a/R_v)e} \approx 0.622 \frac{e}{p - 0.378e} \approx 0.622 \frac{e}{p}. \quad (3.1.27)$$

In Equation (3.1.27), the sum $\rho = \rho_a + \rho_v$ is the density of moist air. Equations (3.1.26) and (3.1.27) show that the values of the mixing ratio and the specific humidity are quite close.

According to the Dalton law, the state equation for moist air can be written as

$$p = p_a + e = \rho_a R_a T + \rho_v R_v T = \rho R_a T \left(1 + q \frac{R_v - R_a}{R_a}\right) = \rho R_a T (1 + 0.6q), \quad (3.1.28)$$

where p is the pressure of the mixture of dry air and water vapor. Equation of state (3.1.28) is usually written in a more convenient form by introducing virtual temperature T_v:

$$T_v = T(1 + 0.6q). \quad (3.1.29)$$

Using the virtual temperature, the state equation for moist air becomes similar to that for dry air:

$$p = \rho R_a T_v. \quad (3.1.30)$$

In the barometric formula (Equation 3.1.4) for moist air, the virtual temperature should be used instead of regular temperature. Since $q << 1$, $T_v \approx T$. However, the difference between T and T_v (that can reach 1–3 K) may be of high importance for correct calculation of buoyancy force in moist air. Since $T_v > T$, the buoyancy of moist air is larger than that of dry air. It is known that thermal stability/instability (Section 3.1.7) depends on the temperature gradient: instability takes place if $-\frac{dT}{dz} > \gamma_a$, and stability takes place if $-\frac{dT}{dz} < \gamma_a$. In moist air, the stability condition should be determined by the gradient of virtual temperature as $\frac{dT_v}{dz} \approx \frac{dT}{dz} + 0.6T\frac{dq}{dz}$. Sometimes in the maritime boundary layer in Tropics $-\frac{dT}{dz} < \gamma_a$, which corresponds to stability of dry layer. However, if the gradient of the mixing ratio is high enough, the gradient of the virtual temperature becomes larger than γ_a, i.e. $-\frac{dT_v}{dz} > \gamma_a$, which indicates the instability of the layer and possible development of convection.

3.1.4 Saturation Vapor Pressure

The concept of saturated and non-saturated air is of key importance in cloud physics. Let us consider a closed and thermally insulated air volume located over a planar water surface. Two processes take place within this volume. The first process is condensation, when a flux of water vapor molecules is directed toward the water surface; molecules collide with the surface and stick to it. The second process is a flux of water vapor molecules from the water surface into the air, which determines the rate of evaporation. If, initially, the air over the water contains a low amount of water molecules, evaporation dominates over condensation and the air humidity increases. The increase in the concentration of water vapor molecules in the air leads to an increase in the condensate flux.

At a certain stage, the condensate flux and evaporation flux become equal. It means that *equilibrium* between the liquid water phase and the gas phase is reached. The air over the water surface is then considered saturated with water vapor. The corresponding water vapor pressure is referred to as the equilibrium, or the *saturation* vapor pressure. In case the water vapor pressure exceeds its equilibrium value, condensation prevails until a new equilibrium is reached. An expression for saturation water vapor pressure e_w can be obtained from the well-known Clausius–Clapeyron equation (Rogers and Yau, 1996):

$$\frac{1}{e_w}\frac{de_w}{dT} = \frac{L_w}{R_v T^2}, \quad (3.1.31)$$

where L_w is the heat required to convert a unit mass of liquid water to vapor, at pressure and temperature being constant. L_w is known as *latent heat of evaporation/condensation*. The latent heat depends on temperature only slightly: at $T_C = 0°C$, $L_w = 2.501 \cdot 10^6$ J/kg. Different approximating expressions proposed for calculation of $e_w(T)$ are presented in **Table 3.1.2**.

Figure 3.1.1 shows the dependence $e_w(T)$ plotted according to a formula presented by Flatau et al., (1992). One can see that the saturation pressure is highly sensitive to temperature. Within the temperature range from −40°C to 40°C, the values of $e_w(T)$ change by more than three orders of magnitude. All the formulas presented in Table 3.1.2 provide quite close magnitudes of $e_w(T)$. However, for precise calculation of drop growth rate in clouds, the values of $e_w(T)$ should be calculated at high accuracy (e.g., of 0.01–0.001%).

Figure 3.1.2 shows the relative difference (in percent) between the values of saturation vapor pressure $e_w(T)$ in respect to the values obtained by Flatau et al. (1992), reported by different authors. The equation developed by Murphy and Koop (2005) appears to be best applicable in cloud models since it is valid for a wide range of temperatures.

Ratio e/e_w is known as the saturation ratio. Expressed in percent, it is called relative humidity (RH). In case water vapor pressure is equal to the saturation value, the saturation ratio is equal to unity, i.e., the relative humidity is equal to 100%. An important quantity in cloud physics is *supersaturation* with respect to water, i.e., deviation of the saturation ratio from the value of unity, or deviation of the relative humidity from 100%.

$$S_w = \frac{e}{e_w} - 1 \quad (3.1.32)$$

Similar to the relative humidity, supersaturation is also measured in percentage. Positive supersaturation corresponds to RH > 100%, while negative supersaturation corresponds to RH < 100%. Figure 3.1.2 shows that at low and high temperatures, the formula presented by Rogers and Yau (1996) deviates significantly from other formulas, which may lead to some errors in evaluation of supersaturation at these temperatures in case this formula is used. In case supersaturation is negative ($S_w < 1$), a quantity called saturation deficit is also used, being equal to $-S_w = (100 - RH)\%$.

Table 3.1.2 Expressions for $e_w(T)$.

$e_w = \sum_{k=0}^{6} a_k T_C^k$ $a_0 = 6.1117675 \times 10^2$, $a_1 = 4.43986062 \times 10^1$, $a_2 = 1.43053301$, $a_3 = 2.65027242 \times 10^{-2}$, $a_4 = 3.02246994 \times 10^{-4}$, $a_5 = 2.03886313 \times 10^{-6}$, $a_6 = 6.38780966 \times 10^{-9}$ Temperature range: $-50°C \le T_C \le 50°C$ (Flatau et al., 1992)
$e_w = a \exp\left(-\frac{b}{T_C + 273.15}\right)$ $a = 2.53 \times 10^{11}, b = 5.42 \times 10^3$ Temperature range: $-30°C \le T_C \le 30°C$ (Rogers and Yau, 1996)
$e_w = 10^2(a_0 + T_C(a_1 + T_C(a_2 + T_C(a_3 + T_C(a_4 + T_C(a_5 + a_6 T_C))))))$ $a_0 = 6.107799961$, $a_1 = 4.436518521 \times 10^{-1}$, $a_2 = 1.428945805 \times 10^{-2}$, $a_3 = 2.650648471 \times 10^{-4}$, $a_4 = 3.031240396 \times 10^{-6}$, $a_5 = 2.034080948 \times 10^{-8}$, $a_6 = 6.136820929 \times 10^{-11}$ Temperature range: $-50°C \le T_C \le 50°C$ (Pruppacher and Klett, 1997)
$e_w = \frac{6.1078 \times 10^2}{(a_0 + T_C(a_1 + T_C(a_2 + T_C(a_3 + T_C(a_4 + T_C(a_5 + T_C(a_6 + T_C(a_7 + T_C(a_8 + a_9 T_C)))))))))^8}$ $a_0 = 0.99999683, a_1 = -0.90826951 \times 10^{-2}, a_2 = 0.78736169 \times 10^{-4}$, $a_3 = -0.61117958 \times 10^{-6}, a_4 = 0.43884187 \times 10^{-8}, a_5 = -0.29883885 \times 10^{-10}$, $a_6 = 0.21874425 \times 10^{-12}, a_7 = -0.17892321 \times 10^{-14}, a_8 = 0.11112018 \times 10^{-16}$, $a_9 = -0.30994571 \times 10^{-19}$ Temperature range: $-50°C \le T_C \le 100°C$ (Wobus approximation)
$e_w = \exp\left[a_0 - \frac{a_1}{T} - a_2 \ln(T) + a_3 T + \tanh(a_4(T - a_5)) \times \left(a_6 - \frac{a_7}{T} - a_8 \ln(T) + a_9 T\right)\right]$ $a_0 = 5.4842763 \times 10^1, a_1 = 6.76322 \times 10^3, a_2 = 4.21, a_3 = 3.67 \times 10^{-4}, a_4 = 4.15 \times 10^{-2}$, $a_5 = 2.188 \times 10^2, a_6 = 5.3878 \times 10^1, a_7 = 1.33122 \times 10^3, a_8 = 9.44523, a_9 = 1{,}4025 \times 10^{-2}$ Temperature range: $123\ K \le T \le 332\ K$ (Murphy and Koop, 2005)

S_w and $e_w(T)$ in Equation (3.1.32) and Table 3.1.2 are determined for a large volume of air over a planar water surface (i.e., the surface curvature is neglected). In cloud physics, it is important to determine supersaturation over small spherical droplets with a very high surface curvature that cannot be neglected. This problem is considered in Section 5.1.

3.1.5 Conservative Thermodynamic Quantities of Moist Air

Due to the presence of three phases of water (water vapor, liquid water, and ice) in the atmosphere, the first law of thermodynamics can be written as (e.g., Cotton and Anthes, 1987):

$$c_p d \ln T - R_a d \ln p + \frac{L_w}{T} dq - \frac{L_m}{T} dq_{iw} = 0 \quad (3.1.33)$$

In Equation (3.1.33), L_m is the latent heat of liquid–ice mixed phase, called *latent heat of melting*/freezing (or *latent heat of fusion*) and q_{iw} is the mixing ratio of ice particles. In Equation (3.1.33), all the terms except those related to condensation and freezing are omitted. The signs on the left-hand side of Equation (3.1.33) are chosen in order to show that condensation ($dq < 0$) and freezing ($dq_{iw} > 0$) are accompanied by heating. From Equation (3.1.17) it follows that

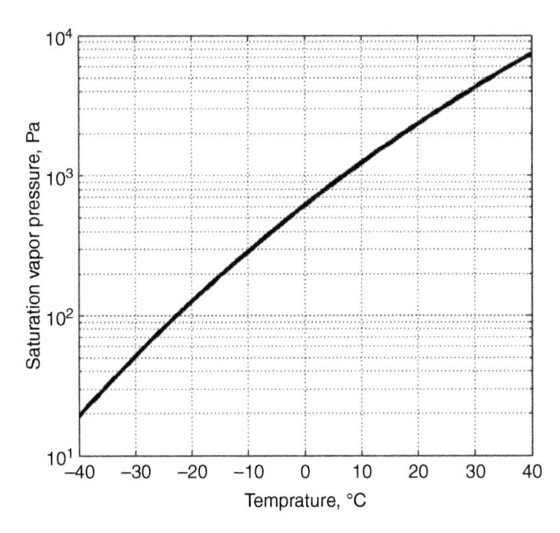

Figure 3.1.1 Dependence of saturation vapor pressure on temperature $e_w(T_C)$, calculated according to Flatau et al. (1992).

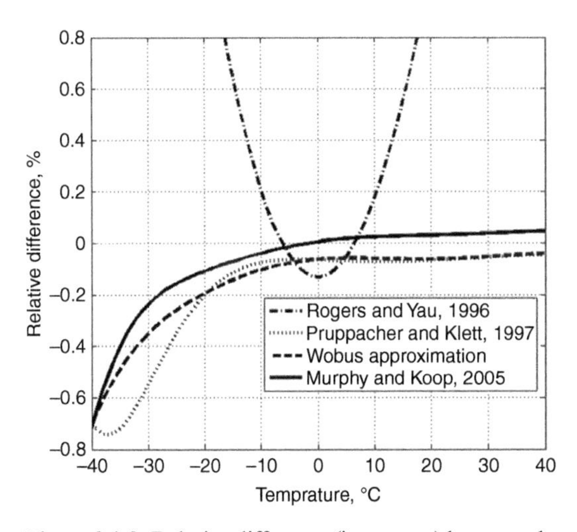

Figure 3.1.2 Relative difference (in percent) between the values of saturation vapor pressure $e_w(T)$ with respect to values calculated by the formula suggested by Flatau et al. (1992) (see also http://cires.colorado.edu/~voemel/vp.html).

$$d\ln\theta = d\ln T - \frac{R_a}{c_p} d\ln p. \tag{3.1.34}$$

Combining Equations (3.1.33) and (3.1.34) we obtain

$$d\ln\theta = -\frac{L_w}{c_p T} dq - \frac{L_m}{c_p T} dq_{iw}. \tag{3.1.35}$$

Equation (3.1.35) is obtained neglecting the heat stored in a condensed water substance.

The conservation of total water can be expressed as

$$dq + dq_l + dq_{iw} = 0, \tag{3.1.36}$$

where q_l is the liquid water mixing ratio. The total specific humidity q_t, i.e., the sum of specific humidity and liquid water mixing ratio $q_t = q_l + q_v$, is often used as invariant (conservative value) in case of the absence of the ice phase. Substituting Equation (3.1.36) alongside with the well-known expression for latent heating (Pruppacher and Klett, 1997)

$$L_m = L_i - L_w. \tag{3.1.37}$$

where L_i, is the latent heat of transition from water vapor to ice (ice sublimation), into Equation (3.1.35) yields

$$d\ln\theta = \frac{L_w}{c_p T} dq_l + \frac{L_i}{c_p T} dq_{iw}. \tag{3.1.38}$$

Equation (3.1.38) shows that the potential temperature, which is conservative in dry adiabatic processes, changes due to phase transitions. It is often desirable to obtain a thermodynamic variable that is conservative under liquid-to-ice transformations. Therefore, using Equation (3.1.38), one can define a conservative variable of *ice–liquid water potential temperature* θ_{il}:

$$d\ln\theta_{il} = d\ln\theta - \frac{L_w}{c_p T} dq_l - \frac{L_i}{c_p T} dq_{iw}. \tag{3.1.39}$$

From Equation (3.1.39), it follows that θ_{il} is a conservative value if no sedimentation of droplets or ice particles is assumed. θ_{il} can also change due to radiative heating and diffusion. If no ice particles arise within the air parcel, the liquid water potential temperature θ_l is a conservative (or invariant) value. The liquid water potential temperature is defined as

$$d\ln\theta_l = d\ln\theta - \frac{L_w}{c_p T} dq_l = 0. \tag{3.1.40}$$

Due to their conservative properties, temperatures θ_{il} and θ_l do not experience any leaps within cloud edges.

Another thermodynamic variable widely used in cloud modeling is the equivalent potential temperature θ_e, defined as

$$d\ln\theta_e = d\ln\theta + \frac{L_w}{c_p T} dq = 0. \tag{3.1.41}$$

The expression for the equivalent potential temperature can be generalized for an ice-containing system (Cotton and Anthes, 1987):

$$d\ln\theta_{e\_iw} = d\ln\theta + \frac{L_w}{c_p T} dq - \frac{L_m}{c_p T} dq_{iw} = 0. \tag{3.1.42}$$

The equivalent potential temperatures θ_e and θ_{e_iw} (its analog in the presence of ice) are conservative in a pseudoadiabatic process, i.e., sedimentation of condensate is neglected. These variables are useful in diagnostic studies as tracers of an air parcel motion. Similar to θ_{il}, θ_{e_iw} is conservative with respect to phase transitions.

The values L_w, L_i, and L_m depend relatively slightly on temperature. (Corresponding approximating dependences and figures can be found in Section 6.1.) Assuming that L_w, L_i, and L_m are totally independent of temperature, Equations (3.1.39), (3.1.40), (3.1.41), and (3.1.42) can be integrated to get analytical expressions for the corresponding temperatures:

$$\theta_{il} = \theta \exp\left(-\frac{L_w q_l}{c_p T} - \frac{L_i q_{iw}}{c_p T}\right) \tag{3.1.43}$$

$$\theta_l = \theta \exp\left(-\frac{L_w q_l}{c_p T}\right) \tag{3.1.44}$$

$$\theta_{e\_iw} = \theta \exp\left(\frac{L_w q}{c_p T} - \frac{L_m q_{iw}}{c_p T}\right) \tag{3.1.45}$$

$$\theta_e = \theta \exp\left(\frac{L_w q}{c_p T}\right). \tag{3.1.46}$$

Equations (3.1.43), (3.1.44), (3.1.45), and (3.1.46) can be further simplified using the expansion of exponents into the Taylor series. For instance, θ_{il} can be written as

$$\theta_{il} \approx \theta\left(1 - \frac{L_w q_l}{c_p T} - \frac{L_i q_{iw}}{c_p T}\right). \tag{3.1.47}$$

The assumption concerning independence of L_w, L_i, and L_m of temperature used while deriving Equations (3.1.43), (3.1.44), (3.1.45), and (3.1.46) leads to significant errors at low temperatures (in the upper troposphere). The corresponding corrections were introduced by Tripoli and Cotton (1981). In particular, these authors improved the accuracy of diagnosing θ_{il} by keeping the temperature constant (equal to 253 K) when in fact it falls below 253 K.

In parallel to quantities determined in Equations (3.1.44), (3.1.45), and (3.1.46), the static energy of moist air (which is an analog of the equivalent potential temperature) is often used:

$$h_l = c_p T + gz + L_w q. \tag{3.1.48}$$

3.1.6 Moist Adiabatic Gradient

In Section 3.1.2, the dry adiabatic gradient γ_a was introduced, which shows the rate of decrease/increase of temperature in an ascending/descending dry adiabatic air parcel. It is of interest to know the rate of temperature changes in saturated air parcels where phase transitions take place. The rate of temperature decreases with height in a saturated air parcel (where only condensation/evaporation takes place) is known as the *moist adiabatic gradient* (or the moist adiabatic lapse rate), γ_{ma}. Assuming saturation mixing ratio ($q = q_S$) from Equation (3.1.41), one can get $0 = \frac{1}{\theta}\frac{d\theta}{dz} + \frac{L_w}{c_p T}\frac{dq_s}{dz} \approx \frac{1}{\theta}\left(\frac{dT}{dz} + \gamma_a\right) + \frac{L_w}{c_p T}\frac{dq_s}{dT}\frac{dT}{dz}$. Since approximately $\theta \approx T$, this expression can be rewritten as $\frac{dT}{dz} = -\gamma_a - \frac{L_w}{c_p}\frac{dq_s}{dT}\frac{dT}{dz}$. Using expression $q_s = 0.622\frac{e_w}{p}$ and the Clausius–Clapeyron Equation (3.1.31), the moist adiabatic lapse rate expression $\gamma_{ma} = -\frac{dT}{dz}$ can be written as

$$\gamma_{ma} = \gamma_a \frac{1 + \dfrac{L_w q_s}{R_v T}}{1 + \dfrac{L_w^2 q_s}{c_p R_v T^2}}. \tag{3.1.49}$$

The moist adiabatic lapse rate characteristic value depends on temperature and pressure; obviously, $\gamma_{ma} < \gamma_a$. The physical cause of the inequality is clear: ascending of a saturated air parcel is accompanied by condensation and latent heat release. As a result, saturated air in updrafts is warmer than the surrounding atmosphere. γ_{ma} reaches its maximum near the condensation level where q_s is maximum, and decreases with increasing height. At high levels, where temperature is low, $q_s \to 0$ and $\gamma_{ma} \to \gamma_a$. The typical value of γ_{ma} that can be used for different estimations is about 7.5°C/km. In a similar way, one can introduce the moist adiabatic lapse rate taking into account liquid–ice transformations.

3.1.7 Convective Instability

Convective instability/stability of an air mass refers to its ability to resist vertical motion caused by the buoyancy force. In a stable atmosphere, vertical movement is hindered, so small vertical disturbances are damped out and disappear. In an unstable atmosphere, vertical air movements tend to become larger, resulting in convective activity that can lead to formation of convective cells in the boundary layer, or cumulus clouds and severe weather phenomena such as thunderstorms.

Let us consider a calm atmospheric layer with a vertical temperature gradient (lapse rate) $\frac{d\overline{T}}{dz} = -\gamma$, where the bar denotes mean (environmental) values. Suppose that there is a non-saturated air parcel with temperature T, and at a certain height level, $z = z_0$, the temperatures are equal, i.e., $T(z_0) = \overline{T}(z_0)$. Pressure in the parcel is assumed equal to that in the environment, which means that the air density in the parcel is equal to

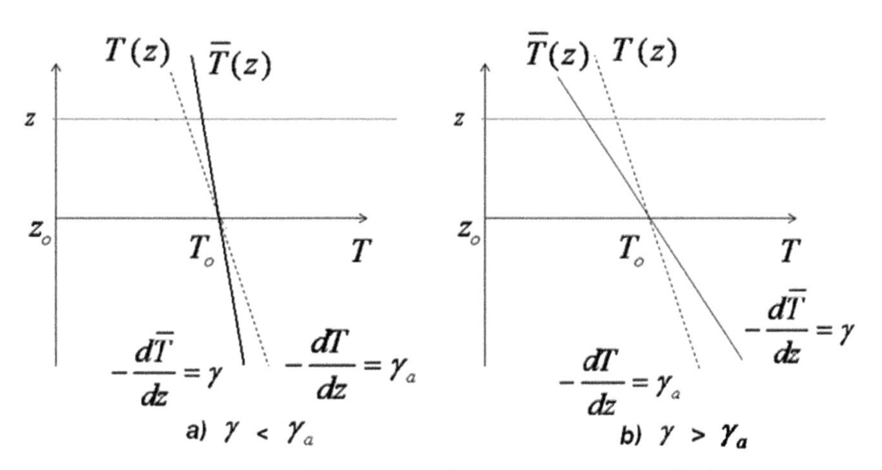

Figure 3.1.3 Temperature profiles corresponding to emergence of oscillations (a) and convective instability (b), during a parcel movement around the equilibrium point $z = z_0$.

that in the environment. The buoyancy force is equal to $B(z) = \frac{g}{\bar{T}}\left(T(z) - \bar{T}(z)\right)$. At $z = z_0$, the buoyancy force $B(z_0) = 0$, i.e., the parcel, is in equilibrium (**Figure 3.1.3**).

Suppose that due to a random perturbation the parcel shifts slightly from the point of equilibrium. Since the parcel is non-saturated, its vertical shift is accompanied by a temperature change according to the dry adiabatic gradient. The buoyancy force at the distance of $z - z_0$ from the equilibrium level is $B(z - z_0) = \frac{g}{\bar{T}}[T(z - z_0) - \bar{T}(z - z_0)] = -\frac{g}{\bar{T}}(\gamma_a - \gamma)(z - z_0)$. Accordingly, the motion equation of an air parcel moving under buoyancy force is

$$\frac{d^2z}{dt^2} = -\frac{g}{\bar{T}}(\gamma_a - \gamma)(z - z_0). \tag{3.1.50}$$

In case $\gamma < \gamma_a$ (Figure 3.1.3a), the buoyancy force is directed toward the equilibrium point. Indeed, in this case an ascending parcel becomes colder than the surrounding air, while a descending parcel becomes warmer than the surrounding. This situation indicates the existence of convective stability, since the buoyancy force tends to return the parcel to the equilibrium point, producing oscillations. Equation (3.1.50) describes oscillations at N_B frequency:

$$N_B = \sqrt{\frac{g}{\bar{T}}(\gamma_a - \gamma)} = \sqrt{\frac{g}{\bar{\theta}}\frac{d\bar{\theta}}{dz}}. \tag{3.1.51}$$

Equation (3.1.51) is obtained using the relationship between temperature and potential temperature (Equation 3.1.23). Frequency N_B is called convective frequency, or the *Brunt–Väisälä frequency*. The oscillations at convective frequency play an important role in the atmosphere because they account for gravity waves arising in the dry atmosphere under condition $\gamma < \gamma_a$. If the friction force is included into Equation (3.1.50), this equation describes oscillations around the equilibrium point, having a decreasing amplitude.

In case $\gamma > \gamma_a$ (Figure 3.1.3b), the buoyancy force is positive and directed away from the equilibrium point, so the parcel continues ascending at velocity increasing with time. Thus, in case $\gamma > \gamma_a$, the equilibrium point is unstable and the atmosphere is convectively unstable. Due to this instability, convective plums and thermals arise in the atmospheric boundary layer.

In the atmosphere, the condition $\gamma > \gamma_a$ usually takes place near the surface in the lowest tens or hundreds of meters. However, we see that vertical updrafts in convective clouds are strong above the boundary layer, so convective clouds can reach heights exceeding 10–12 km. These updrafts form as a result of moist convective instability when the temperature in ascending saturated air volume changes according to the moist adiabatic gradient γ_{ma}. The condition of convective instability in this case is $\gamma > \gamma_{ma}$. This condition is quite widespread in the atmosphere, allowing formation of deep convective clouds.

The case when $\gamma_a > \gamma > \gamma_{ma}$ – typical of atmosphere in the Tropics – is of special interest. Since air in clouds ascends along the moist adiabat, the condition $\gamma > \gamma_{ma}$ means instability of the atmosphere with respect to updrafts. As regards downdrafts, which are assumed to take place along the dry adiabat, the atmosphere is stable. The situation when $\gamma_a > \gamma > \gamma_{ma}$ is called *conditional convective instability*.

3.1.8 Tephigrams, Skew-T Diagrams and the Available Convective Potential Energy

Vertical profiles of temperature and humidity are different in different meteorological conditions. To evaluate convective instability and the probability of hazardous storms, it is necessary to evaluate the potential work of the buoyancy force in the atmosphere. To simplify the weather analysis and the forecast, tephigrams and Skew-T Log-P diagrams (also called Skew-T diagrams) are commonly used in meteorological centers. Tephigrams and Skew-T diagrams are widely used for plotting results of the actual observations or radiosonde soundings, which enable us to obtain the vertical temperature and the *dew point* T_d profiles all through the atmosphere above a certain point on the ground. The dew point T_d is the temperature at which the water vapor within a volume of humid air condenses into liquid water. The dew point is associated with the relative humidity. High relative humidity indicates that the dew point is close to the current air temperature. Relative humidity of 100% indicates that the dew point is equal to the current temperature and that the air is saturated. When the dew point remains constant and temperature increases, relative humidity decreases.

An example of a tephigram is presented in **Figure 3.1.4**. The vertical axis and corresponding solid straight horizontal lines denote the potential temperature (measured in kelvins, K). It means that horizontal solid straight lines show dry adiabats. Air temperature (in °C) is plotted along the low boundary of the tephigram. Correspondingly, the straight vertical lines are isotherms. Dashed lines tilted counterclockwise denote saturating mixing ratio q_S in g kg^{-1}. The air pressure in mb is plotted on a logarithmic scale by parallel lines titled, or skewed, 45 degrees clockwise off of the customary vertical direction. These lines are isobars. The main curves plotted in Figure 3.1.4 are the isotherm, dry adiabats, and moist adiabats. Curves (a), (b), and (c) denote possible vertical profiles of the temperature in the atmosphere. Curve (a) shows a decrease in the potential temperature with decreasing pressure, i.e., with increasing height. This curve corresponds to the condition of absolute instability $\gamma > \gamma_a$. Such temperature profiles usually take place within the surface layer and atmospheric BL. Profile (b) corresponds to the conditional convective instability that often takes place in the atmosphere: $\gamma_a > \gamma > \gamma_{ma}$. Under such temperature profiles, moist convection with formation of clouds can exist. The temperature profile (c) corresponds to the stable condition$\gamma < \gamma_{ma}$.

We now consider several examples of when to use tephigrams.

First example. From the practical point of view, it is important to determine the mixing ratio if the temperature and the dew point are given, or to determine the dew point if the temperature and the mixing ratio are given. The way to do this is illustrated in **Figure 3.1.5** (left). Let us assume that at the pressure level of 900 mb, the temperature of air is equal to 21°C (blue point) and the mixing ratio is equal to 5 g kg^{-1}. To find the dew point, it is necessary to move along the same pressure level of 900 mb (blue arrow) toward the temperature at which the given mixing ratio of 5 g kg^{-1} becomes saturating (red point). The temperature corresponding to this point is the dew point $T_d = 1$°C. The task of finding the mixing ratio under the given temperature and dew point can be solved analogously.

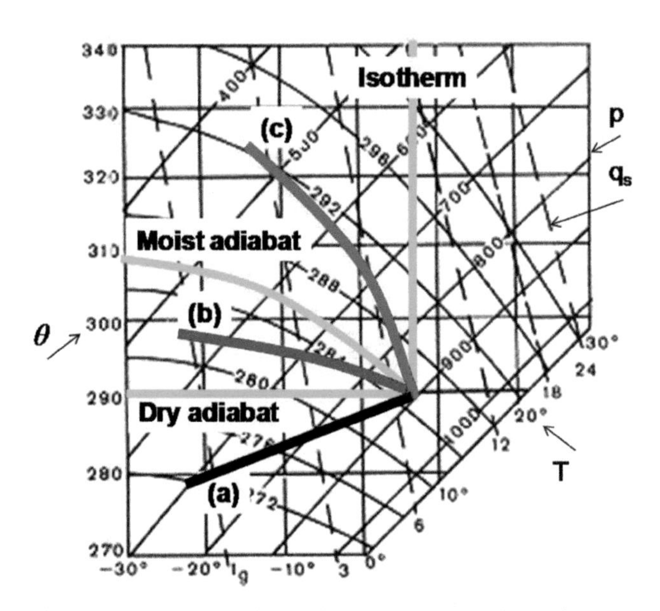

Figure 3.1.4 Example of a tephigram. The curves plotted in green are isotherms, dry adiabats, and moist adiabats. Curves (a), (b), and (c) denote possible profiles of temperature (in °C) in the atmosphere. Vertical axis and horizontal straight lines denote potential temperature (in K); horizontal axis (and straight vertical lines) denote temperature (in °C). Tilted straight lines are isobars, i.e., lines of constant pressure. Dashed lines tilted counterclockwise and the tilted axis show saturating mixing ratios (in g/kg).

Second example. Another practical task is to find the lifting condensation level (LCL), i.e., the level at which an adiabatic parcel with given initial temperature and mixing ratio becomes saturated during its ascent along dry adiabat. Let us consider the previous example, when the temperature at the pressure level of 900 mb is equal to 21°C, and the mixing ratio is equal to 5 g kg^{-1}. In case the air volume ascends adiabatically, its temperature changes according to the dry adiabat (blue arrow). The LCL is reached at the point where the value 5 g kg^{-1} of the mixing ratio becomes saturating. This point is marked by the yellow triangle in Figure 3.1.5 (right). The pressure corresponding to this point is about 690 mb.

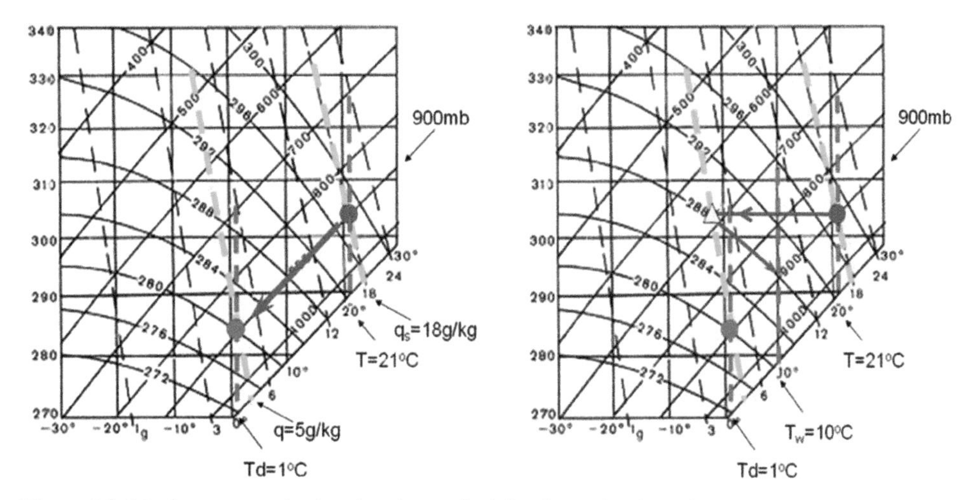

Figure 3.1.5 Left: an example showing the method for determination of the dew point temperature using the temperature and the mixing ratio. Right: an example showing the method for determination of LCL and the wet-bulb temperature.

Third example. In order to measure the air humidity, meteorologists often use the wet-bulb temperature T_w. It is the temperature one feels when one's skin is wet and is exposed to moving air. The adiabatic wet-bulb temperature is the temperature a volume of air would have if cooled adiabatically to the saturation level (i.e., moving along the dry adiabat) and then compressed adiabatically to the original pressure in a moist-adiabatic process, i.e., moving along the moist adiabat. The method for determination of the wet-bulb temperature is shown in Figure 3.1.5 (right) by two blue arrows. In the example considered, $T_w = 10°C$. The task is solved in two steps. First, the LCL points are found (horizontal blue arrow). Second, we move along the moist adiabat down to the initial pressure of 900 mb (second blue arrow).

Fourth example. Tephigrams are often used to analyze meteorological situations and predict formation and intensity of clouds. **Figure 3.1.6** illustrates two different meteorological situations. Sounding plotted in Figure 3.1.6 (left) indicates no-rain weather with the existence of two cloud layers, which can be identified by the high dew point whose values become equal to the air temperature or only slightly lower than the air temperature. The lower cloud layer is located within the pressure layer from 920 to 800 mb. This layer contains boundary layer cumulus clouds located below the stable isothermal layer. The statement concerning the convective nature of these clouds follows from the fact that the boundary layer is well mixed with a nearly adiabatic temperature gradient and with a nearly constant mixing ratio. The upper layer from 600 to 550 mb contains altostratus. Below the layer of altostratus clouds, the mixing ratio increases with height. Such an increase can be explained only by the horizontal advection of moist air at the upper levels. The appearance of such altostratus clouds suggests significant changes in the weather conditions as a strong convective system producing these altostratus is approaching.

Figure 3.1.6 (right) shows atmospheric sounding data under unstable conditions typical of a severe weather system (severe storm). The dew point coincides with the air temperature within the deep layer, indicating a strong deep convection.

The rate of atmosphere instability is often measured as an amount of work that can produce buoyancy force during an air parcel's ascent from the surface. This parameter is known as Convective Availible Potential Energy (CAPE). As shown in Figure 3.1.6 (right), the parcel ascends from 1,000 mb level. Below LCL, the parcel is non-saturated, so its ascent takes place along the dry adiabat. At about 960 mb, the relative humidity reaches 100% and the parcel begins ascending along the moist adiabat. The adiabatic parcel rising along moist adiabat remains warmer than the surrounding air up to the so-called level of free convection (LFC). The zone of free convection where an ascending parcel is warmer than its surrounding is shown by the shaded area. CAPE represents the work of the buoyancy force within this layer:

$$CAPE = g \int_{z_{lcl}}^{z_{fc}} \frac{T - \overline{T}}{\overline{T}} dz, \tag{3.1.52}$$

where z_{lcl} and z_{fc} are the heights of the LCL and the LFC (neutral buoyancy), respectively. CAPE is usually

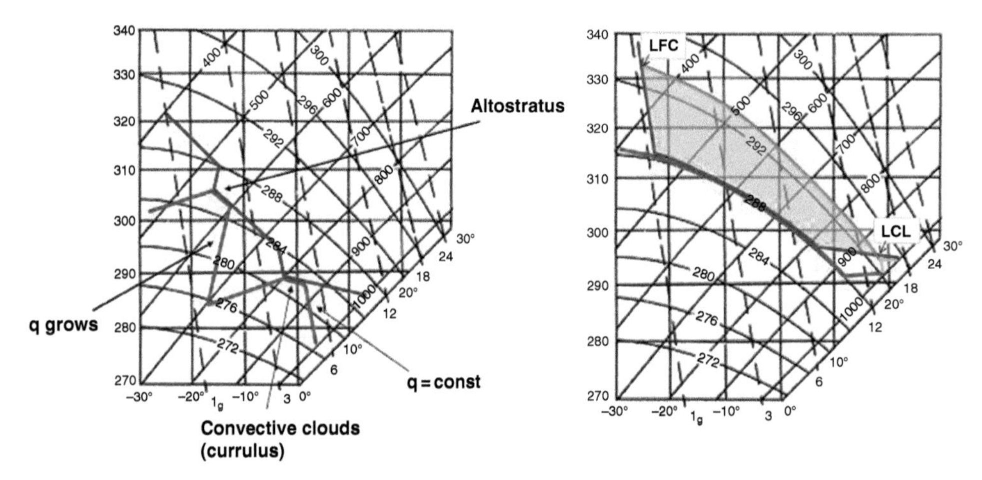

Figure 3.1.6 Left: an example of synoptic situations characterized by existence of two cloud layers: convective clouds at the top of the boundary layer and altostratus near the 600 mb level. Right: sounding data typical of a severe storm. The shaded area denotes high CAPE.

expressed in J kg^{-1} or, equivalently, in m^2 s^{-2}. In fact, CAPE is sometimes referred to as the positive buoyant energy (PBE). To calculate CAPE more accurately, virtual temperatures of a parcel $T_v(z)$ and its environment $\overline{T}_v(z)$ should be used. Neglecting the virtual temperature correction may result in substantial relative errors in calculations. The typical value of CAPE are ~1,000 J kg^{-1}. In some thunderstorms and tornados, values of CAPE exceeding 5,000 J kg^{-1} were measured. Extreme CAPE values may exceed 8,000 J kg^{-1}, e.g., in the deadly F5 tornadoes that hit Plainfield, Illinois, on August 28, 1990 and Jarrell, Texas, on May 27, 1997.

CAPE is the measure of the convective energy available to parcels. In case the work of the buoyancy force contributes to the kinetic energy of the parcel, the vertical velocity maximum of a buoyant parcel can be evaluated using the formula

$$w_{\max} = 2\sqrt{CAPE}. \tag{3.1.53}$$

In an ideal case, such velocity should be reached at the LFC. In real atmosphere, the amount of energy available for convection is rarely realized. There are three reasons for this. First, real parcels are of a finite size and their accelerations are balanced by pressure gradient forces that arise from the need to push the air ahead of them out of the way. Second, there are viscous forces decreasing the parcel velocity. Third, the ascending parcel mixes with surrounding air, which decreases the parcel buoyancy. Therefore, the maximum velocity in real clouds is about twice as low than follows from Equation (3.1.53), and reaches well below the LFC. The LFC is close to the cloud top level.

Similar thermodynamic analysis can be performed using a Skew-T diagram, shown in **Figure 3.1.7**. Along the left vertical axis, the air pressure is plotted on a logarithmic scale with the highest pressure (1,050 mb) at the bottom (which is sea level) and 100 mb at the top. On the right vertical axis, the heights of standard atmosphere (in feet) corresponding to the pressure are presented. These values can serve as an estimation of the height under the measured pressure. Along the horizontal axis, the increments of temperature in °C are given. The straight slanting lines across the figure from lower left to upper right are the lines of constant temperature. These lines are titled or skewed 45° clockwise off the customary vertical direction. The curved and solid green lines and the dashed purple lines represent dry and moist adiabats, respectively, as shown in Figure 3.1.7. The dashed, straight, parallel purple lines, running steeply from lower left to upper right, indicate the saturation mixing ratio q_S.

As a result of the elongation of the pressure axis in the vertical direction, soundings plotted on Skew-T diagram show the vertical profiles of the corresponding values. Figure 3.1.7 shows an example of an atmospheric sounding under unstable conditions of a severe weather system. The atmospheric soundings on the Skew-T diagram consist of the air temperature profile (thick, solid red line) and the dew-point temperature profile (thick, solid blue line), each plotted against the pressure and the temperature. The thin red line shows the temperature of an adiabatic parcel ascending from the surface. The hatched area between LCL and LFC is proportional to CAPE. The value of CAPE is high in this example, which is typical of thunderstorms.

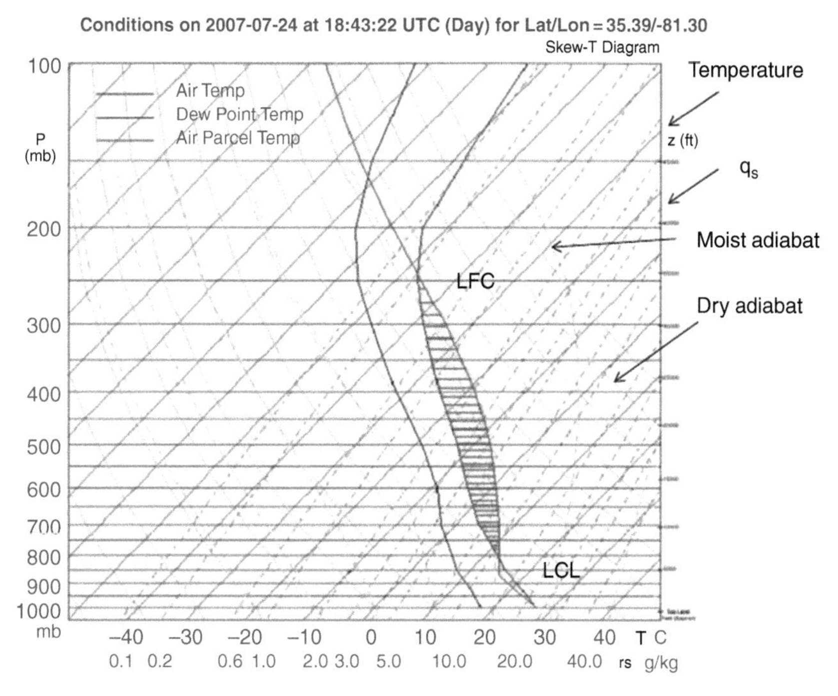

Figure 3.1.7 Example of Skew-T diagram plotted for unstable atmospheric conditions (thunderstorm at 2:43 PM, on July 24, 2007, in Northern Georgia just east of the Appalachian Mountains) (from http://airsnrt.jpl.nasa.gov/SkewT_info.html; courtesy of the author, meteorologist J. Haby).

3.2 Budget and Motion Equations

In this section the equations for conservation of mass, heat, and momentum are formulated, and some properties of these equations are discussed. Although these equations are from first principles, they cannot be directly used in cloud modeling, but represent the basis for deriving averaged equations (Section 3.4) used in cloud models and other atmospheric models.

3.2.1 The Continuity Equation

Let us consider a small virtual box of volume $V = \Delta x \Delta y \Delta z$ placed inside an air flow (**Figure 3.2.1**). The mass of air within this box is ρV. Since mass in the atmosphere has neither source nor sink, the change of the mass within volume V over time is determined by the difference between influxes and outfluxes passing through the sides of the box: $\frac{\rho V|_{t+\Delta t} - \rho V|_t}{\Delta t} = \left(\rho u|_x - \rho u|_{x+\Delta x}\right)\Delta y \Delta z + \left(\rho v|_y - \rho v|_{y+\Delta y}\right)\Delta x \Delta z + \left(\rho w|_z - \rho w|_{z+\Delta z}\right)\Delta x \Delta y$, where u, v, and w are the velocity components along the x, y, and z axes, respectively. Dividing this expression by volume V and assuming that $\Delta t \to 0$, $\Delta x \to 0$, $\Delta y \to 0$, and $\Delta z \to 0$, we obtain

$$\frac{\partial \rho}{\partial t} + \frac{\partial \rho u_i}{\partial x_i} = 0. \tag{3.2.1}$$

Equation (3.2.1) is written in the tensor form (the indexes $i = 1,2,3$ correspond to directions x, y, z), assuming summation with respect to repeating indexes. Equation (3.2.1) is known as the *continuity equation*. In cases when temporal and spatial changes of the air density ρ are small enough to be neglected (as, e.g., within the atmospheric boundary layer, as well as in narrow stratiform and small cumuli clouds), Equation (3.2.1) can be rewritten in a simpler form valid for an incompressible fluid:

$$\frac{\partial u_i}{\partial x_i} = 0 \tag{3.2.2}$$

3.2.2 Budget Equations for Conservative and Nonconservative Values

Let us formulate a budget equation for some specific scalar quantity a per air mass unit. For instance, a can be a mixing ratio (kg kg^{-1}). In this case, the amount of a within volume V is $\rho a V$. The fluxes of a through the sides of the box are equal to the fluxes of the air mass multiplied by a. There is an important difference between the budget of quantity a and the air mass, since within volume V a source of a may exist. Let us denote this source as C. As a result, the equation for conservation of a is

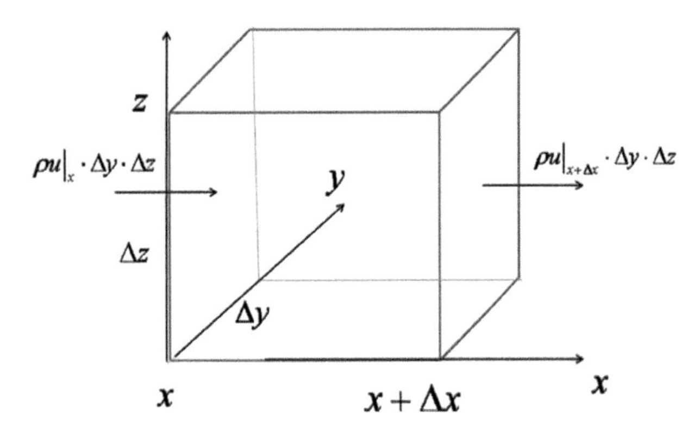

Figure 3.2.1 Applying the concept of mass conservation within a certain air volume for derivation of the continuity equation.

$$\frac{\partial \rho a}{\partial t} + \frac{\partial \rho a u_i}{\partial x_i} = C. \tag{3.2.3}$$

Equation (3.2.3) is written in a so-called divergent form, which is convenient for solving this equation using finite-difference methods (Section 4.2). Using the continuity Equation (3.2.1), Equation (3.2.3) can be rewritten as

$$\frac{\partial a}{\partial t} + u_i \frac{\partial a}{\partial x_i} = \frac{C}{\rho}, \tag{3.2.4}$$

where $\frac{C}{\rho}$ is the power of the source per unit of air mass. The type of source C depends on the nature of the quantities and, in particular, on conservative properties of those quantities. For instance, the potential temperature is conservative in the dry adiabatic process, but is nonconservative in the moist adiabatic process. Therefore, Equation (3.2.4), written for the potential temperature, should contain the source component expressing the heating rate by condensation/evaporation of drops and deposition/sublimation of ice. In cases when a is the equivalent potential temperature that does not change during condensation/evaporation processes, source C should include the rate of heating caused only by formation of ice. In case the ice–liquid water potential temperature is used as a variable in Equation (3.2.4), source C should include only the radiative heating/cooling expressions. If a quantity is conservative with respect to all the processes considered, the source C is equal to zero; hence,

$$\frac{da}{dt} = \frac{\partial a}{\partial t} + u_i \frac{\partial a}{\partial x_i} = 0. \tag{3.2.5}$$

Equation (3.2.4) is derived under the assumption that changes of quantity a are caused by fluxes through the lateral sides of an air volume, occurring due to a directed nonzero mean air velocity (i.e., due to advection of a through the lateral sides). However, the fluxes of a may be of another nature, when the mean velocity of molecules is equal to zero. Such fluxes are known as *diffusion fluxes* and arise due to the spatial gradients of a. For instance, heat flux is directed from a warmer body to a cooler one. Diffusion fluxes are proportional to gradients of quantity a:

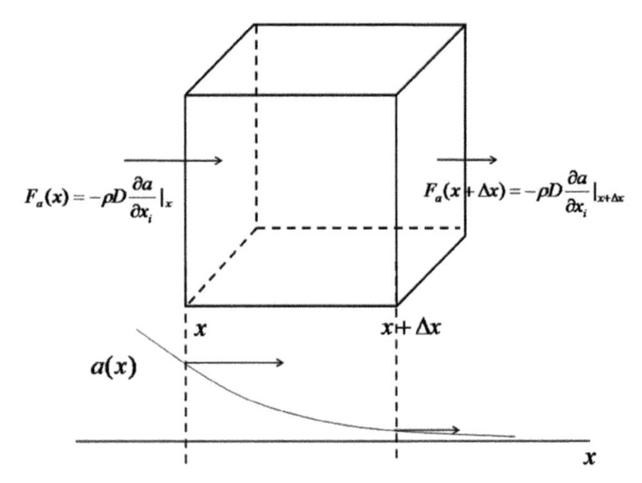

Figure 3.2.2 Illustration of a change of quantity a caused by diffusion fluxes through the lateral sides of an air volume. In the case illustrated, the influx is higher than the outflux, hence the magnitude of a increases due to the divergence of the diffusion fluxes.

$$F_{ai} = -\rho D \frac{\partial a}{\partial x_i}. \tag{3.2.6}$$

Here, D is the *coefficient of molecular diffusion* of a (diffusivity of a). The units of D are $m^2 s^{-1}$. In case a is temperature, D is the *coefficient of molecular heat conductivity*. Changes of quantity a caused by diffusion fluxes are illustrated in **Figure 3.2.2.**

Using considerations similar to those applied while deriving Equation (3.2.3), changes of a caused by the divergence of diffusion fluxes can be expressed as

$$\frac{\partial \rho a}{\partial t} = \frac{\partial}{\partial x_i} \rho D \frac{\partial a}{\partial x_i}. \tag{3.2.7}$$

In case $\rho = const$ and $D = const$, Equation (3.2.7) reduces to the classical equation of diffusion:

$$\frac{\partial a}{\partial t} = D \frac{\partial^2 a}{\partial x_i^2} \tag{3.2.8}$$

Combining Equations (3.2.3) and (3.2.7), we obtain

$$\frac{\partial \rho a}{\partial t} + \frac{\partial \rho a u_i}{\partial x_i} = C + \frac{\partial}{\partial x_i} \rho D \frac{\partial a}{\partial x_i} \tag{3.2.9}$$

Equation (3.2.9) is known as the *advection-diffusion equation*. Equations for different quantities (e.g., potential temperature, mixing ratio, liquid water content, ice content, etc.) differ in formulation of source/sink C, which should describe contributions of corresponding microphysical and radiative processes.

3.2.3 Equations for Microphysical Variables (Kinetic Equations)

Equation (3.2.9) can be applied to thermodynamic variables such as temperature, water vapor, etc. In cloud microphysics, the budget equations should be written for condensates, i.e., for liquid and ice particles. These equations should take into account settling of particles, as well as express the multiple transitions between different species related to the processes of melting, freezing, riming (collision of ice with supercooled water), etc.

There are two main approaches to formulation of these equations. The first approach is used in most bulk microphysical parameterization schemes. As was discussed in Section 2.1, in bulk parameterizations the shape of size distributions is prescribed a priori. This allows writing equations for integral quantities such as CWC, RWC, IWC, etc. (e.g., Straka, 2009). The general form of equations for microphysical species is the following:

$$\frac{\partial \rho q_j}{\partial t} + \frac{\partial \rho q_j (u_i - \delta_{i,3}\overline{V}_{gj})}{\partial x_i} = \rho \sum_k \left[\frac{\delta q_j}{\delta t}\right]_k + \frac{\partial}{\partial x_i} \rho D \frac{\partial q_j}{\partial x_i}, \tag{3.2.10}$$

where q_j is the mixing ratio of j-th species, $\overline{V}_{gj}$ is the mass-averaged sedimentation velocity, $\left[\frac{\delta q_j}{\delta t}\right]_k$ is the rate of q_j due to k-th microphysical process, and $\delta_{i,3}$ is the Kronecker symbol (see Appendix A), which in our case means that only the vertical component of the sedimentation velocity is nonzero. The last term on the right-hand side of Equation (3.2.10) expresses the changes of the variable due to the divergence of molecular diffusion fluxes.

We now present the equations for cloud water mixing ratio q_c and rain water mixing ratio q_r in a warm cloud (containing no ice). For the sake of simplicity, we excluded the ice component from the equations. In bulk-parameterization schemes, liquid water is separated into cloud droplets and raindrops. Such separation is reasonable because drop size distribution indicates the minimum near drop radius ~35 μm (see Figure 2.3.3), which can be considered the boundary between these two modes. These equations are written as

$$\frac{\partial \rho q_c}{\partial t} + \frac{\partial \rho q_c u_i}{\partial x_i} = \rho \left\{ \left[\frac{\delta q_c}{\delta t}\right]_{nucl} + \left[\frac{\delta q_c}{\delta t}\right]_{c/e} - \left[\frac{\delta q_c}{\delta t}\right]_{aut} - \left[\frac{\delta q_c}{\delta t}\right]_{acr} \right\} + DIFF_{q_c} \tag{3.2.11}$$

$$\frac{\partial \rho q_r}{\partial t} + \frac{\partial \rho q_r (u_i - \delta_{i,3}\overline{V}_{g\_m})}{\partial x_i} = \rho \left\{ \left[\frac{\delta q_r}{\delta t}\right]_{c/e} + \left[\frac{\delta q_c}{\delta t}\right]_{aut} + \left[\frac{\delta q_c}{\delta t}\right]_{acr} \right\} + DIFF_{q_r}, \tag{3.2.12}$$

where $\left[\frac{\delta q_c}{\delta t}\right]_{nucl}$ is the rate of droplet formation by nucleation, $\left[\frac{\delta q_c}{\delta t}\right]_{c/e}$ is the rate of growth/decrease of CWC due to condensation/evaporation, $\left[\frac{\delta q_c}{\delta t}\right]_{aut}$ is the rate of autoconversion – i.e., the rate of raindrop formation by collisions between cloud droplets – and $\left[\frac{\delta q_c}{\delta t}\right]_{acr}$ is the rate of accretion (rain drop-cloud droplet collisions). It is noteworthy that some of these terms have opposite signs in Equations (3.2.11) and (3.2.12). For instance, a decrease in the mass of cloud droplets due to autoconversion is accompanied by a corresponding increase in the mass of raindrops. The terms $DIFF_{q_l}$ and $DIFF_{q_r}$ are diffusion terms of the corresponding values describing the mixing process. The expression of these rates is an important task of microphysical parameterizations (analyzed in Chapter 5).

Equation (3.2.11) does not contain the term for droplet sedimentation, as it is typically assumed in bulk schemes that sedimentation of cloud droplets can be neglected (the fall velocity of cloud droplets with radii lower than ~40 μm does not exceed 16 cm s^{-1}). At the same time, sedimentation of raindrops is taken into account. According to the definition, the mass-averaged sedimentation velocity is

$$\overline{V}_{g\_m} = \frac{1}{\rho q_r} \int_{m_{\min}}^{\infty} V_g(m) m f(m) dm, \tag{3.2.13}$$

where $f(m)$ is the size distribution function (PSD) defined in Equation (2.1.2). In two-moment bulk-parameterization microphysical schemes, the budget equations similar to Equations (3.2.11) and (3.2.12), are also written for number concentrations. In the equation for raindrop concentration, the settling velocity averaged over number density is used:

$$\overline{V}_g = \frac{1}{N_r} \int_{m_{\min}}^{\infty} V_g(m) f(m) dm, \tag{3.2.14}$$

where N_r is the concentration of raindrops. Since large fast-falling drops account for most of the total drop mass, while the number concentration is determined by drops of lower mass falling slowly (i.e., $\overline{V}_{g\_m} > \overline{V}_g$), raindrop mass falls at a higher velocity than raindrop concentration does. Although this phenomenon has a clear physical meaning, there are some technical difficulties impeding its representation in two-moment bulk-parameterization schemes.

The second approach to formulation of the budget equations is used in spectral bin microphysics models (e.g., Khain et al., 2004), including the system of kinetic equations for size distribution functions. As was discussed in Section 2.1, PSD are defined on mass grids containing from several tens to several hundreds of bins (categories). The shape of SD is calculated in the course of integration of these equations. Each equation describes the mass (and concentration) budget for each bin (category). Hence, the number of the kinetic equations is equal or proportional to the number of bins (depending on the methods used to solve the equations). The general form of a budget equation for j-th hydrometeor type for the m-th bin is

$$\frac{\partial \rho f_{j,m}}{\partial t}+\frac{\partial \rho f_{j,m}\left(u_i-\delta_{i,3}V_{g,j,m}\right)}{\partial x_i}=\rho\sum_k\left[\frac{\delta f_{j,m}}{\delta t}\right]_k+DIFF_{j,m}, \tag{3.2.15}$$

where $\left[\frac{\delta f_{j,m}}{\delta t}\right]_k$ are the rates of changes in the size distribution of j-th hydrometeor type and $DIFF_{j,m}$ is the diffusion term. The number of hydrometeor types j is different in different models. In the mixed-phase HUCM, there are seven hydrometeor types: water drops, three types of ice crystals (columnar, plate-like, and dendrites), aggregates (snow), graupel, and hail. The model also contains an equation for the SD function for aerosols playing the role of cloud condensational nuclei, as well as equations for some other parameters such as liquid water mass within melting hydrometeors, etc. As a result, the number of equations of type (3.2.15) can be quite large, in fact, significantly larger than that used in bulk-parameterization schemes. This makes SBM cloud models computationally time consuming.

The approaches to describing microphysical processes used in bulk-parameterization schemes (Equations 3.2.11 and 3.2.12) are quite different from those used in SBM (Equation 3.2.15). In SBM, each microphysical process is described by equations mostly based on the first principles (Chapters 5 and 6), i.e., expressing certain physical laws formulated by the theory. The rates of microphysical processes depend on particle size involved in the processes. In the bulk parameterization schemes, the rates of microphysical processes are averaged over the entire particle size range. This decreases the sensitivity of predicted rates to particle size.

The accuracy of the bin-microphysics models is limited largely by the lack of knowledge of particular physical mechanisms. For instance, collision rates between nonspherical ice crystals are not well established, which accounts for the fact that simulations of aggregate formation by collisions of ice crystals are prone to errors although the stochastic collision equation is applied (Chapter 5). As soon as the sufficient knowledge of particular processes is acquired (either theoretically or as a result of new observations), the accuracy of SBM can be improved without any changes in the structure of the basic Equation (3.2.15). Bulk-parameterization schemes assume the main microphysical cloud structure as given *a priori*. Besides, in these schemes microphysical processes are often described by simplified semi-empirical relationships, which, in principle, should mimic solutions of complicated equations based on microphysical laws. In practice, however, solutions of semi-empirical expressions do not mimic solutions of complicated, highly nonlinear basic microphysical equations used in SBM. As a result, SBM cloud models usually simulate evolution of clouds more accurately than models based on bulk-parameterization schemes (see Section 7.2)

3.2.4 Motion Equations

Motion equations represent the second law of Newton. Being written in the tensor form for the velocity components u_i (where $u_1 = u, u_2 = v, u_3 = w$), the equations of motion have the following form (Schlichting, 1960):

$$\rho\frac{du_i}{dt}=\rho\left(\frac{\partial u_i}{\partial t}+u_j\frac{\partial u_i}{\partial x_j}\right)=X_i-\frac{\partial p}{\partial x_i}+\frac{\partial \tau_{ij}}{\partial x_j}, \tag{3.2.16}$$

where X_i are the components of the volume forces, the second term on the right-hand side expresses the gradient of pressure (the pressure force), and the third term on the right represents the friction force. τ_{ij} are the components of a stress tensor representing the surface forces per unit of surface, as shown in **Figure 3.2.3.**

The first index of this tensor shows the direction perpendicular to the stress component, while the second index denotes the direction parallel to the stress component. The stress tensor is symmetric with respect to its main diagonal, i.e., $\tau_{ij} = \tau_{ji}$ (Schlichting, 1960). According to Stokes's law of friction, the stress tensor can be expressed in the following form:

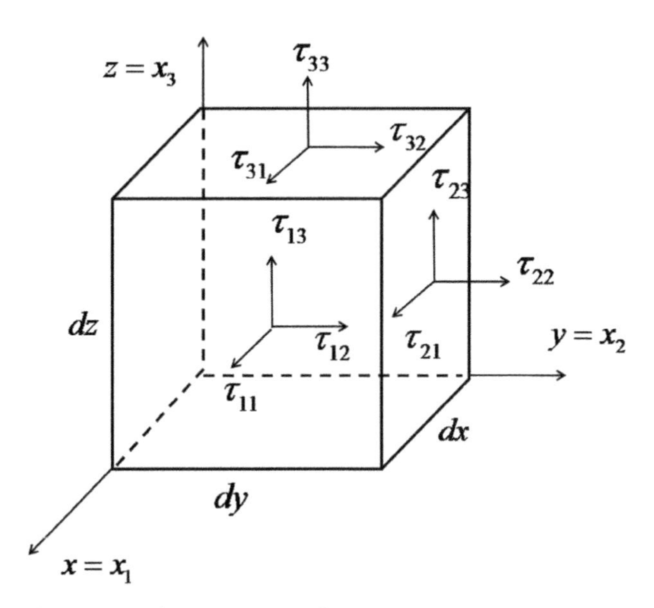

Figure 3.2.3 The components of stress tensor τ_{ij}.

$$\tau_{ij}=\begin{pmatrix} 2\mu\frac{\partial u}{\partial x} & \mu\left(\frac{\partial u}{\partial y}+\frac{\partial v}{\partial x}\right) & \mu\left(\frac{\partial u}{\partial z}+\frac{\partial w}{\partial x}\right) \\ \mu\left(\frac{\partial u}{\partial y}+\frac{\partial v}{\partial x}\right) & 2\mu\frac{\partial v}{\partial y} & \mu\left(\frac{\partial w}{\partial y}+\frac{\partial v}{\partial z}\right) \\ \mu\left(\frac{\partial u}{\partial z}+\frac{\partial w}{\partial x}\right) & \mu\left(\frac{\partial w}{\partial y}+\frac{\partial v}{\partial z}\right) & 2\mu\frac{\partial w}{\partial z} \end{pmatrix}, \tag{3.2.17}$$

where $\mu=\rho v$ is the coefficient of dynamic viscosity and v is the coefficient of molecular kinematic viscosity. The dynamic viscosity μ is almost pressure independent and strongly decreases with increasing temperature. The kinematic viscosity v depends on the density, which exponentially decreases with height. Substituting Equation (3.2.17) into Equation (3.2.16), we get the motion equations for air velocity components. These equations are known as Navier–Stokes equations:

$$\begin{aligned} \rho\frac{du}{dt} &= X-\frac{\partial p}{\partial x}+\frac{\partial}{\partial x}\left(2\mu\frac{\partial u}{\partial x}\right)+\frac{\partial}{\partial y}\left[\mu\left(\frac{\partial u}{\partial y}+\frac{\partial v}{\partial x}\right)\right] \\ &\quad+\frac{\partial}{\partial z}\left[\mu\left(\frac{\partial w}{\partial x}+\frac{\partial u}{\partial z}\right)\right] \\ \rho\frac{dv}{dt} &= Y-\frac{\partial p}{\partial y}+\frac{\partial}{\partial y}\left(2\mu\frac{\partial v}{\partial y}\right)+\frac{\partial}{\partial x}\left[\mu\left(\frac{\partial u}{\partial y}+\frac{\partial v}{\partial x}\right)\right] \\ &\quad+\frac{\partial}{\partial z}\left[\mu\left(\frac{\partial v}{\partial z}+\frac{\partial w}{\partial y}\right)\right] \\ \rho\frac{dw}{dt} &= Z-\frac{\partial p}{\partial z}+\frac{\partial}{\partial z}\left(2\mu\frac{\partial w}{\partial z}\right)+\frac{\partial}{\partial x}\left[\mu\left(\frac{\partial w}{\partial x}+\frac{\partial u}{\partial z}\right)\right] \\ &\quad+\frac{\partial}{\partial y}\left[\mu\left(\frac{\partial v}{\partial z}+\frac{\partial w}{\partial y}\right)\right]. \end{aligned} \tag{3.2.18}$$

In case $\mu=const$, using the continuity equation in the form Equation (3.2.2) (which is applicable for stratiform clouds in the boundary layer and low convective clouds), Equation (3.2.18) can be written in a simpler form that is often used in many hydrodynamic applications. In the tensor form, these equations look like:

$$\rho\frac{du_i}{dt}=\rho\left(\frac{\partial u_i}{\partial t}+u_j\frac{\partial u_i}{\partial x_j}\right)=X_i-\frac{\partial p}{\partial x_i}+\mu\frac{\partial^2 u_i}{\partial x_j^2} \tag{3.2.19}$$

The pressure gradient in the atmosphere is caused by differences in the air temperature and is the main source of atmospheric motion. Due to the work of the pressure force, the potential energy of the atmosphere transforms into kinetic energy.

Two volume forces in the atmosphere are typically considered: the Coriolis force caused by Earth's rotation and the gravity force. Substituting components of these forces into Equation (3.2.19), we obtain well-known motion equations:

$$\begin{aligned} \frac{du}{dt} &= -\frac{1}{\rho}\frac{\partial p}{\partial x}+fv+v\nabla^2 u \\ \frac{dv}{dt} &= -\frac{1}{\rho}\frac{\partial p}{\partial y}-fu+v\nabla^2 v, \\ \frac{dw}{dt} &= -\frac{1}{\rho}\frac{\partial p}{\partial z}-g+v\nabla^2 w \end{aligned} \tag{3.2.20}$$

where $f=2\Omega\sin\varphi$ is the Coriolis parameter, Ω is the angular velocity of Earth's rotation, ϕ is the latitude, and $\nabla^2(u_i)=\frac{\partial^2 u_i}{\partial x_j^2}$.

It is convenient to rewrite Equation (3.2.20) introducing deviations of pressure, density, and temperature from the reference state in which the atmosphere is calm, obeying the quasi-static equation (see Section 3.1), denoting the values in the reference state as $p_0(z)$, $\rho_0(z)$, $T_0(z)$, $\theta_0(z)$, etc. For the reference state one can write

$$-\frac{1}{\rho_0}\frac{\partial p_0}{\partial z}=g,\ p_0=\rho_0 RT_0. \tag{3.2.21}$$

The deviations from the reference state are defined as

$$p''=p-p_0(z);\rho''=\rho-\rho_0(z);T''=T-T_0(z). \tag{3.2.22}$$

These deviations obey the inequalities

$$|p''|\ll p_0(z),|\rho''|\ll\rho_0(z)\text{ and }|T''|\ll T_0(z). \tag{3.2.23}$$

In fact, the differences between the temperature in clouds and the surrounding temperature are of a few degrees, which is much lower than the basic temperature value of order of 300 K. Similar evaluations can be done for other inequalities.

Using Equations (3.2.22) and (3.2.23), the terms in the third Equation (3.2.20) can be approximately written as $-\frac{1}{\rho}\frac{\partial p}{\partial z}-g=-\frac{1}{\rho_0+\rho''}\frac{\partial(p+p'')}{\partial z}-g\approx-\frac{1}{\rho_0}\frac{\partial p''}{\partial z}-g\frac{\rho''}{\rho_0}$. In this expression, only the linear terms with respect to the deviations remain. The nonlinear term containing multiplications p'' and ρ'' is neglected because it is much smaller than the remaining terms. The term $-g\frac{\rho''}{\rho_0}$ represents the buoyancy force. It is positive when $\rho''<0$, i.e., when the local air density is lower than that in the environment. The expression for buoyancy force is often written in the form containing fluctuations of temperature. To rewrite the expression for the buoyancy force, the *Boussinesq approximation* (Equation 3.2.24) is used:

$$\frac{\rho''}{\rho_0}\approx-\frac{T''}{T_0}\approx-\frac{\theta''}{\theta_0} \tag{3.2.24}$$

This expression comes from the gas law, which can be written in the differential form as $\frac{dp}{p}=\frac{d\rho}{\rho}+\frac{dT}{T}$. Since perturbation of pressure dp at a certain height is actually determined by the fluctuation of the air column weight above this level (caused by various smaller scale fluctuations of density and temperature, both positive and negative), dp represents some kind of integral characteristic, so $\frac{dp}{p}$ is typically much less than each term on the right-hand side of this equality. This leads to Equation (3.2.24). The accuracy of Equation (3.2.24) decreases with height. Within the validity range of the Boussinesq approximation, the buoyancy force per unit of mass can be written as $B=g\frac{\theta''}{\theta_0}$, showing that this force is positive if the air volume is warmer than the environmental air. Since the reference pressure depends on z only, the equation system (Equation 3.2.20) can be rewritten as

$$\begin{aligned}\frac{du}{dt}&=-\frac{1}{\rho_0}\frac{\partial p''}{\partial x}+fv+\nu\nabla^2u\\ \frac{dv}{dt}&=-\frac{1}{\rho_0}\frac{\partial p''}{\partial y}-fu+\nu\nabla^2v\\ \frac{dw}{dt}&=-\frac{1}{\rho_0}\frac{\partial p''}{\partial z}+g\frac{\theta''}{\theta_0}+\nu\nabla^2w.\end{aligned} \tag{3.2.25}$$

For future considerations, it is useful to write down the motion equations (Equation 3.2.25) in the divergent form. Combining Equation (3.2.25) with the continuity equation $\frac{\partial\rho_0}{\partial t}+\frac{\partial\rho_0u_i}{\partial x_i}=0$, we obtain

$$\frac{\partial\rho_0u_i}{\partial t}+\frac{\partial\rho_0u_iu_j}{\partial x_j}=-\frac{\partial p''}{\partial x_i}+\delta_{i3}\rho_0g\frac{\theta''}{\theta_0}+\varepsilon_{ijk}\rho_0f_ku_j+\mu\nabla^2u_i, \tag{3.2.26}$$

where ε_{ijk} is the antisymmetric Levi–Civita tensor (see Appendix A) determining the sign of the Coriolis force terms.

For clouds, the buoyancy force equation should be rewritten to take into account the effects of the lower density of moist air (see Equation 3.1.29), as well as the weight of the condensate:

$$B=g\left(\frac{\theta''}{\theta_0}+0.61q-q_l-q_{iw}\right). \tag{3.2.27}$$

The weight of condensate is called *mass loading*. It decreases the air buoyancy in updrafts and fosters formation of downdrafts in deep convective clouds.

3.2.5 Similarity Concept and the Reynolds Number

It is important to establish conditions under which two flows can be considered dynamically similar. This problem arises, for instance, when results obtained in a laboratory should be applied to real atmosphere, where scales of flow or even a type of liquid can be different (e.g., air instead of water). In the stationary state, the motion equation for the neutral atmosphere (assuming there is no buoyancy force and neglecting the Coriolis force) can be written as

$$u_j\frac{\partial u_i}{\partial x_j}=-\frac{1}{\rho_0}\frac{\partial p''}{\partial x_i}+\nu\frac{\partial^2u_i}{\partial x_j^2} \tag{3.2.28}$$

The term on the left-hand side is usually referred to as the inertial force. It is clear that a necessary condition for dynamical similarity of two flows is the identity of the motion equation (Equation 3.2.28) written in the nondimensional form. One possible way to obtain a nondimensional equation is to multiply the whole equation by factor $\frac{\rho_0l}{U^2}$, where U is the characteristic velocity and l is the characteristic spatial scale of the flow. Denoting nondimensional values with an asterisk, we can write

$$u^*=\frac{u}{U},\ x^*=\frac{x}{l},\ p^*=p\frac{1}{\rho U^2},\ \frac{\partial}{\partial x^*}=l\frac{\partial}{\partial x}. \tag{3.2.29}$$

Substituting these expressions into Equation (3.2.28), we obtain

$$u_j^*\frac{\partial u_i^*}{\partial x_j^*}=-\frac{\partial p^{*\prime\prime}}{\partial x_i^*}+\frac{1}{Re}\frac{\partial^2u_i^*}{\partial x_j^{*2}}. \tag{3.2.30}$$

Equation (3.2.30) contains one nondimensional parameter only known as the Reynolds number:

$$Re=\frac{Ul}{\nu}. \tag{3.2.31}$$

Equation (3.2.30) shows that dynamic equations for different flows written in the nondimensional form differ only by the value of the Reynolds number. It follows from Equation (3.2.30) that all flows with the same Reynolds number are similar. This fact plays a crucial role in dynamics of fluids and in atmospheric physics, in particular. The value Re^{-1} has a physical meaning of nondimensional viscosity. It is noteworthy that in Equation (3.2.30), as $Re \to \infty$, the viscosity terms disappear. Thus, flows with a high Reynolds number are approximately inviscid streams. The Re value shows the ratio between the inertial force and the friction force. It is important to stress that in order to be fully similar, the flows should also be geometrically similar and have similar boundary conditions. The Reynolds number is not only a measure of the similarity of two flows, but also indicates the level of hydrodynamic stability/instability of the flows that determines the flow structure.

In more sophisticated cases when both the effects of buoyancy and the Coriolis force are taken into account, other nondimensional numbers appear in the dynamical equations representing more complicated conditions of similarity.

3.3 Turbulence

Motion equations are not directly applicable to the atmospheric conditions since atmospheric motions are turbulent. It means that even slight random changes of initial or boundary conditions lead to dramatic changes in velocity, temperature, and other hydrodynamic and thermodynamic fields. As a result, evolution of these fields occurs randomly and the instant values of velocity and other quantities cannot be predicted. The major feature of a turbulent motion is characterized by random irregular changes in time and space while the environmental conditions remain almost unchanged. Samples of turbulent fluctuations of temperature and velocity obtained by means of a balloon experiment at a particular point (time evolution) and along the horizontal balloon trajectory (spatial evolution) are shown in **Figures 3.3.1** and **3.3.2**, respectively.

Although in turbulent flows the initial and boundary conditions determine future velocity values in principle, these values depend also on random unpredictable fluctuations of these conditions. Therefore, integration of the corresponding differential equations describing instantaneous realizations of turbulent fields is practically pointless. There are models of a special kind known as the Direct Numerical Simulation (DNS), which calculate turbulent motions using the Navier–Stokes equations. However, these models, usually applied for investigation of turbulence, are currently able to describe motions within air volumes with linear scales smaller than a few tens of centimeters only. Each simulation represents only a single sample, so a repeated solving of thermodynamic equations (with some random fluctuations of the initial and boundary conditions) will yield different results, which can be used for investigation of probability distributions of different quantities or averaged characteristics of these random quantity fields. In a sense, dealing with turbulent fields resembles the situation in the kinetic gas theory. A precise description of the motion of each molecule is practically impossible (and in most cases is not necessary). The gas theory predicts only the averaged statistical characteristics of the majority of molecules.

Turbulence plays an important role in practically all atmospheric phenomena. It acts like friction for motions of scales higher than several hundred meters. Turbulence also determines fluxes of heat, moisture, and momentum in BL, as well as effects such as turbulent diffusion. Turbulence leads to formation of well-known vertical wind profiles in the BL (e.g., Garratt, 1992). Turbulence plays an important role in cloud microphysics and dynamics, being responsible for fluxes of momentum, heat, and humidity in clouds. It determines mixing between clouds and surrounding air, as well as mixing between different volumes inside clouds. Such way turbulence strongly influences microphysical processes in clouds. In this section a brief description of parameters and properties of turbulent flows is presented. More information on turbulence can be found both in a number of earlier fundamental monographers (Monin and Yaglom, 1971, 1975; Frisch, 1995; Pope, 2000) and recent works (Tsinober, 2009; Wyngaard, 2010).

3.3.1 Conditions for Emergence of Turbulence

Formation of turbulence is a complex and still not-well-understood process. Here, we present only the qualitative considerations (Laichtman, 1976) that allow us to understand factors favoring or hindering the development of turbulence.

Assume that a small turbulent vortex has been formed within a homogeneous flow. The characteristic linear scale of the vortex is l, the characteristic velocity related to the vortex is u_l. The characteristic time scale (lifetime) of the vortex is $\tau_l = l/u_l$. In order to create such a vortex, the environment flow (or larger vortex) should perform the work per unit of time of $R_{1l} \sim u_l^2/\tau_l = u_l^3/l$. There are several sinks and sources of energy of the vortex. First, its kinetic energy is lost due to friction. The friction force can be approximated

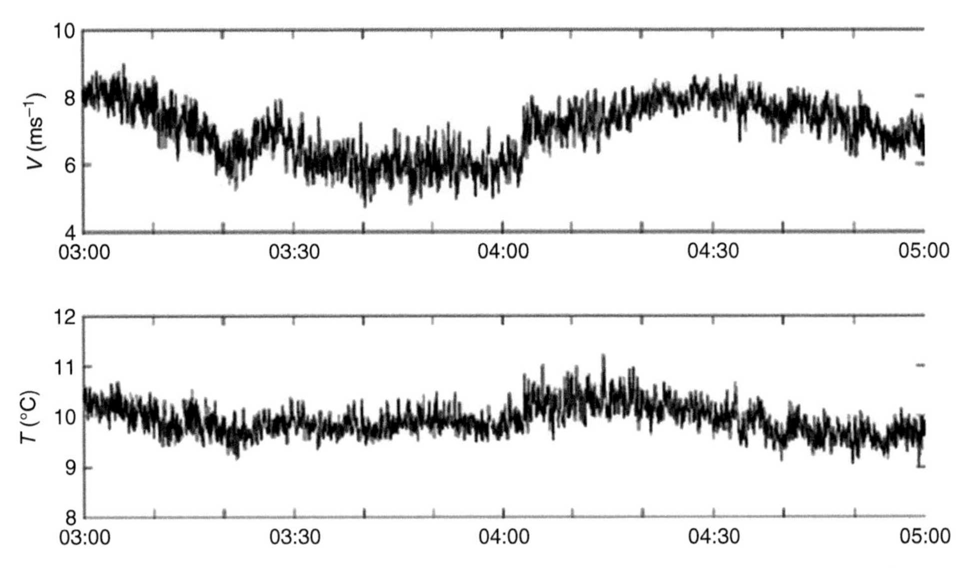

Figure 3.3.1 Samples of wind speed and air temperature records obtained using Tethered Lifting System (TLS) in the night-time atmosphere within the lowest few hundred meters (from Muschinski et al., 2004; courtesy of © Cambridge University Press).

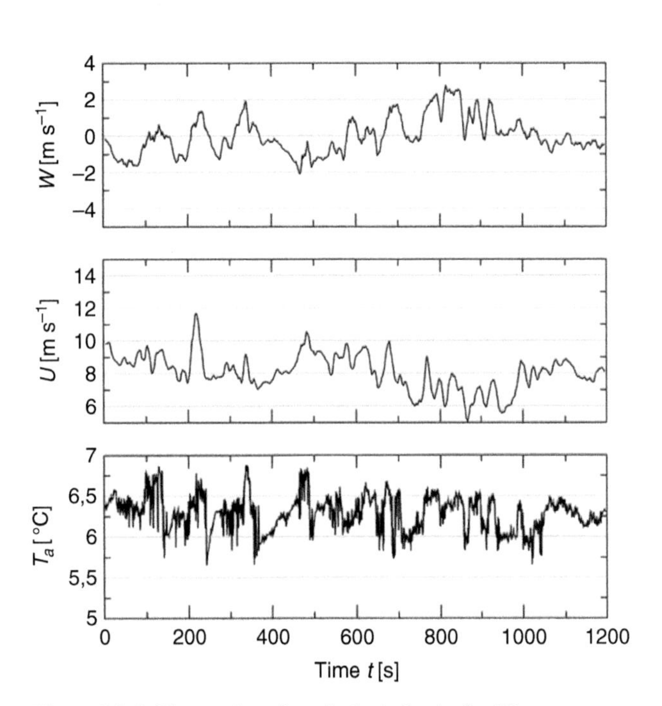

Figure 3.3.2 Time series of vertical wind velocity W, horizontal wind velocity U, and temperature T_a, as measured with ACTOS (Airborne Cloud Turbulence Observation System) at the height of about 1,540 m AGL on May 16, 2004, during the INSPECTRO campaign (from Siebert et al., 2006; courtesy of © American Meteorological Society. Used with permission).

as $\nu u_l/l^2$ (see Equation 3.2.25). The work against the friction force per unit of time is $R_{2l} \sim \nu u_l^2/l^2$. Second, the existence of turbulence also depends on the atmospheric stability. Assuming that the potential temperature of a turbulent vortex deviates from that of the environment by θ'', the buoyancy force can be written as $\frac{g}{\theta_0}\theta''$ (Section 3.2). The work against the buoyancy force per unit of time is $R_{3l} = \frac{g}{\theta_0}\theta'' u_l$. The sign of this work (and the role of the buoyancy force in formation of turbulence) depends on the sign of the buoyancy force. Under stable conditions, the buoyancy hinders the motion of a turbulent vortex; the unstable conditions favor the turbulence. These three factors (friction, buoyancy, and atmospheric stability) determine the three conditions needed for turbulence to emerge.

a) **First condition:** a vortex develops if the kinetic energy that it obtains from a larger scale flow exceeds the work against the friction force:

$$\frac{R_{1l}}{R_{2l}} = \frac{u_l l}{\nu} = Re_l > 1 \tag{3.3.1}$$

It follows from Equation (3.3.1) that the development of turbulence is impossible if the Reynolds number of the vortex is less than one. The flow becomes unstable when the inertial forces arising within it (i.e., the forces corresponding to the nonlinear advection term in Navier–Stokes (Equation 3.2.19) are much larger than the viscous forces described by the viscous term of the equation (Monin and Yaglom, 1971). The nondimensional ratio of these terms is determined by Reynolds number $Re = \frac{lU}{\nu}$, defined in Equation (3.2.31), where l is the characteristic spatial scale of the flow, U is the characteristic velocity of the flow, and ν is kinematic viscosity. As was mentioned in Section 3.2.5, the value Re^{-1} can be interpreted as nondimensional viscosity. Larger Re numbers lead to smaller relative friction between different parts of the air and to smaller attenuation of random perturbations in the flow. Therefore, Re is a measure of hydrodynamic instability of the flow. Analysis of hydrodynamic instability (Monin and Yaglom, 1971) shows that turbulence arises when $Re > Re_{cr}$, where Re_{cr} is the critical value determined by the geometry of the air flow.

Transition of a laminar flow to a turbulent one, with increasing Re in a channel within a flow behind a small obstacle, is illustrated in **Figure 3.3.3.** One can see that at $Re = 126$ the fluctuations decrease along the flow, i.e., the flow is fully laminar. At $Re = 378$ the flow becomes unstable. At $Re = 945$ the flow pattern becomes irregular and turbulent.

The typical values of Re_{cr} for different flows (in pipes, over plates, etc.) are $(1-3) \times 10^3$ (Schlichting, 1960). At the same time, in the atmospheric BL of 1 km depth at the typical wind speed of 10 $\mathrm{m\,s^{-1}}$, the typical value of Re is 10^9. Evaluations of Re in different types of clouds also give values of 10^8–10^9 that are many orders of magnitude higher than the known critical Re_{cr}. These evaluations indicate that motions in the atmosphere are practically always turbulent. These very large values of Re also mean that turbulence in the atmosphere is fully developed and intermittent

b) **Second condition:** takes into account buoyancy and environment stratification: turbulence develops if the work against the buoyancy force (under stable atmospheric condition) is lower than the energy that the turbulent vortex obtains from the environmental larger scale flow:

$$\frac{R_{3l}}{R_{1l}} = \frac{g}{\theta_0}\frac{l\theta''}{u_l^2} = Ri_l < 1, \tag{3.3.2}$$

where Ri_l is the Richardson number of vortex with scale l. More detailed considerations show that there exists critical Richardson number $Ri_{cr} \approx 0.25$, so that turbulence exists if $Ri < Ri_{cr}$. Recent studies (Zilitinkevich et al., 2009) showed that turbulence can also form under conditions $Ri > 0.25$ because of the formation of gravity waves.

Conditions (a) and (b) stress the role of dynamical factors in development of turbulence. However, turbulence can be induced by thermal factors, as well, when the background wind speed is low while the atmosphere is thermally unstable. For thermally induced turbulence to emerge, the condition $\frac{R_{3l}}{R_{2l}} > 1$ should take place. Turbulence is actually generated due to temperature gradients. Turbulent diffusion and viscosity smooth these gradients and hinders development of turbulence.

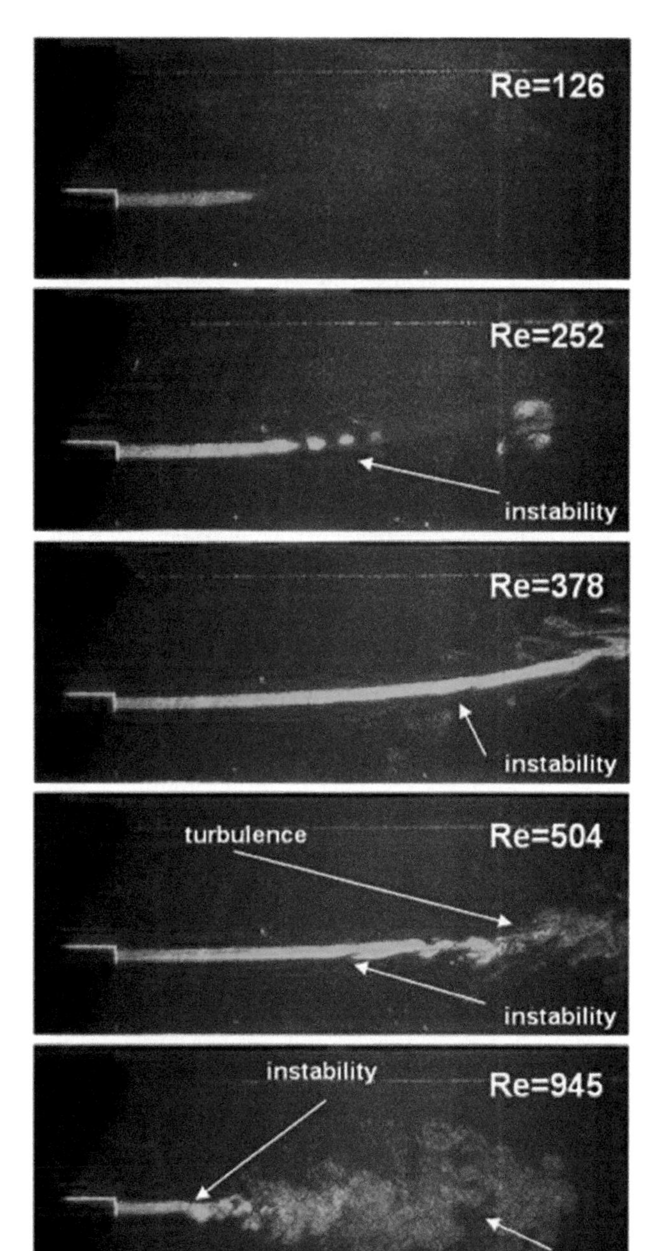

Figure 3.3.3 Laminar to turbulent transition of a submerged jet flow (from http://en.wikiversity.org/wiki/Fluid_Mechanics_for_MAP_Chapter_1._Introduction; courtesy of O. Ertunc, Institute of Fluid Mechanics, Germany).

Taking into account these two factors, one can get the condition of instability of an atmospheric layer (Emanuel, 1994):

$$Ra > Ra_{cr}, \tag{3.3.3}$$

where $Ra = -h^4\left(\frac{g}{\bar{\theta}}\frac{d\bar{\theta}}{dz}\right)\frac{1}{\nu\kappa}$ is the Rayleigh number, h is the depth of the layer where the instability arises, and κ is the thermodiffusion coefficient. The value of Ra_{cr} is about 10^3. At values of Ra, slightly larger than Ra_{cr}, cells with a linear size of the order of h arise in the layer. With further increasing Ra, motions within a wider wavelength range become unstable, so at $Ra \sim 10^3 Ra_{cr}$ velocity fluctuations with random phases grow within a very wide wavelength range, which indicates the emergence of a thermally induced turbulent flow. **Figure 3.3.4** shows a schematic zone of turbulence within the Re – Ri space.

3.3.2 Energy Cascade and Characteristic Turbulent Scales

The simplest scenario of energy transitions in a developed homogeneous and isotropic turbulence is shown in **Figure 3.3.5**. Due to its instability, a non-turbulent flow produces vortexes with scales L, which can be assigned to the largest turbulent vortices. L, known as the external turbulent scale, is typically ten to twenty times smaller than the scale of the host background flow. Since Re are high, the vortices of scale L are also unstable and break down into smaller vortices. Although the resulting smaller vortexes are characterized by lower Re and are less unstable, the values of Re are still high enough, and the smaller vortexes in turn break down into smaller ones. The vortexes are divided and redivided, forming a turbulent cascade. At each stage Re decreases, i.e., the role of friction increases. This division continues until Re becomes close to one. The size of the vortices with $Re = 1$ is known as *internal turbulent scale*, or Kolmogorov turbulent microscale μ_k. So, the ratio in Equation (3.3.1) becomes equal to one if $l = \mu_k$. The kinetic energy of the flow in this process transfers from larger vortex to smaller ones and finally, when the flow becomes laminar, transfers to heat. Most of the vortexes, with the exception of the largest ones, become homogeneous and isotropic.

According to the Kolmogorov theory (1941), the kinetic energy constantly arrives from the mean air flow to the largest turbulent vortices of the turbulent cascade, and then transfers toward smaller vortices. It is assumed that there is no external source of energy feeding the turbulent motions within the cascade. It means that the energy flux from larger scales to smaller ones is constant and does not depend on the scale itself. This kind of turbulence is known as quasi-stationary turbulence. The flux of the turbulent kinetic energy from the largest scales to smaller ones and eventually to heat is known as the mean *turbulent dissipation rate* ε. As a result, the dissipation rate ε shows the rate of the decrease in the kinetic energy of turbulence, since it transfers to heat due to friction.

As the mean ε in the stationary state is constant over the spectrum of turbulent motions, the dissipation rate shows the energy flux that turbulence obtains from non-turbulent sources. In this sense, the dissipation rate

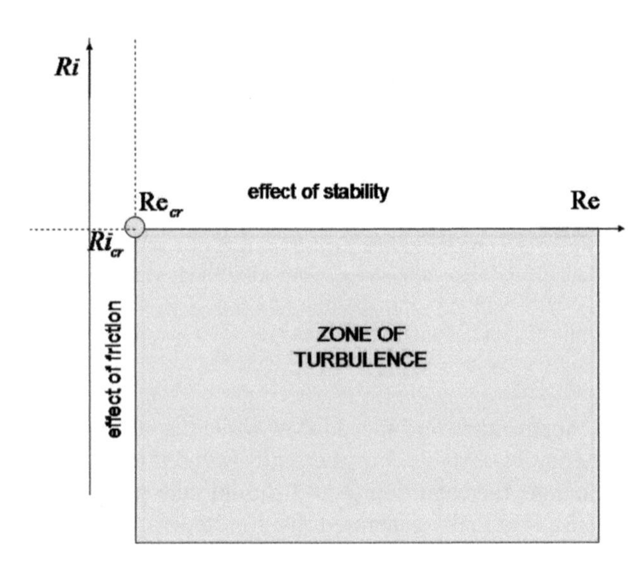

Figure 3.3.4 Schematic illustration of the zone of turbulence within the Re – Ri space (from Laichtman, ed., 1976).

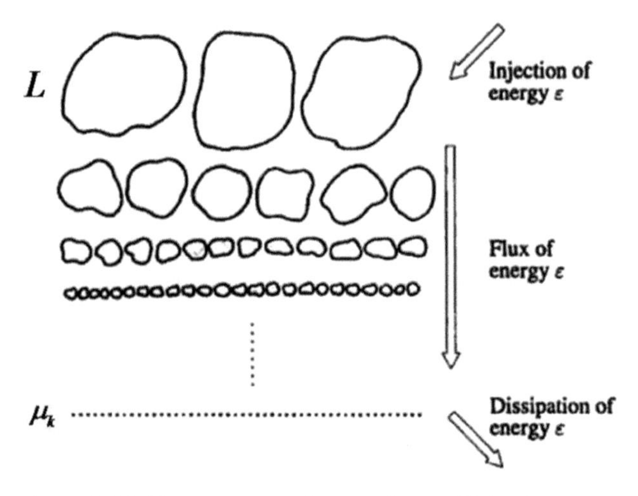

Figure 3.3.5 The energy cascade according to Richardson (from Frisch, 1995, with changes; courtesy of © Cambridge University Press).

Table 3.3.1 Expressions for turbulent parameters within viscous and inertial subranges.

Parameter	Inertial Subrange	Dissipation Subrange
Spatial scale	l	$\mu_k = \varepsilon^{-1/4}\nu^{3/4}$
Wave number	$k(l) = 2\pi/l$	$k_k = 2\pi\varepsilon^{1/4}\nu^{-3/4}$
Time scale	$\tau(l) = \varepsilon^{-1/3}l^{2/3}$	$\tau_k = \varepsilon^{-1/2}\nu^{1/2}$
Frequency	$f(l) = \varepsilon^{1/3}l^{-2/3}$	$f_k = \varepsilon^{1/2}\nu^{-1/2}$
Turbulent velocity	$u(l) = (\varepsilon l)^{1/3}$	$u_k = (\varepsilon\nu)^{1/4}$
Turbulent shear	$S(l) = \varepsilon^{1/3}l^{-2/3}$	$S_k = \varepsilon^{1/2}\nu^{-1/2}$
Reynolds number	$Re(l) = \varepsilon^{1/3}\nu^{-1}l^{4/3}$	$Re = 1$
Lagrangian acceleration		$A_k = \varepsilon^{3/2}\nu^{-1/4}$
Kinetic energy per unit mass	$E(l) = (\varepsilon l)^{2/3}$	$E_k = (\varepsilon\nu)^{1/2}$
Turbulent coefficient (4/3 Richardson law)	$K(l) = C_1\varepsilon^{1/3}l^{4/3}$, $C_1 = 0.2$,	$K_k = \nu$
Spectrum of kinetic energy (−5/3 Kolmogorov law)	$F(k) = \frac{dE}{dk} = C\varepsilon^{2/3}k^{-5/3}$, $C = 1.52$	

ε characterizes the intensity of turbulence. The mean dissipation rate ε is universal and characterizes vortexes of any size. The size range between external and internal scales is referred to as the inertial subrange. The inertial subrange in developed turbulence covers scales of several orders of magnitude, from about 1 cm to a few hundred meters. The dissipation of kinetic energy within this subrange is negligible in comparison to the turbulent kinetic energy itself. The role of viscosity in the inertial range is negligible because the values of *Re* are large enough. In contrast, vortexes of sizes of Kolmogorov microscale μ_k have small kinetic energy, thus viscosity plays a dominating role. Accordingly, this size range in the turbulent cascade is called the *dissipation* (or *viscous*) subrange. The distribution of turbulent kinetic energy (TKE), turbulent characteristic scales, and subranges is schematically shown in **Figure 3.3.6.**

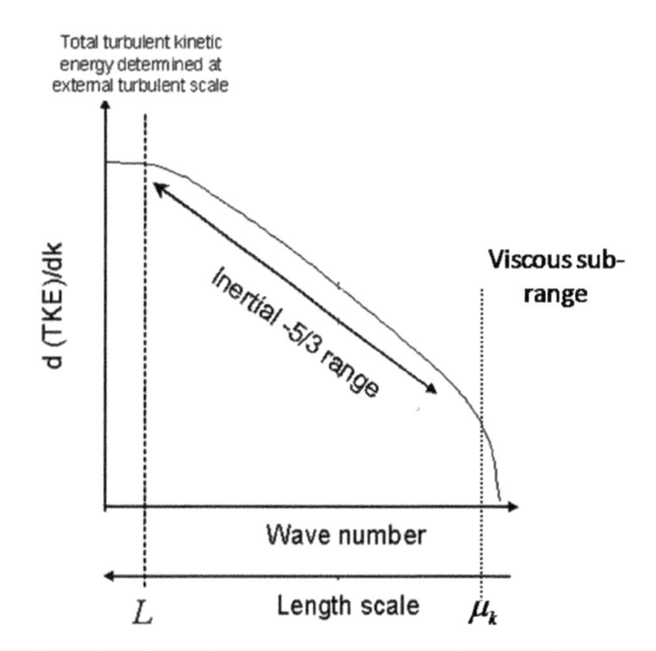

Figure 3.3.6 Turbulent spectrum, turbulent scales, and subranges according to the Kolmogorov (1941) theory.

Parameters ε and ν fully characterize the turbulent cascade. Using the physical concept of the turbulent energy cascade and the dimensional theory, one can estimate many turbulent parameters within both the inertial and the viscous subranges, including distribution of kinetic energy between turbulent vortexes in the cascade. Such estimations are widely used both in turbulent theory, as well as in its applications in cloud physics and modeling, and are presented in **Table 3.3.1.**

Although the formulas presented in Table 3.3.1 are defined with the accuracy of constant factors, they reflect the main concepts of the turbulent theory and are widely applied in cloud physics. The concepts of the Richardson energy cascade and the Kolmogorov theory are used in all modern theories of turbulence. However, turbulence in the real atmosphere and especially in clouds somewhat differs from Kolmogorov's 41 turbulence. Turbulence structure can deviate from homogeneous and isotropic due to buoyancy, existence of underlying surface, vertical stratification of temperature (preferential vertical direction), etc. Atmospheric convection and wind shears make turbulence anisotropic in horizontal directions, as well. The work of buoyancy forces may inject additional energy into the inertial subrange and make the energy cascade deviate from Kolmogorov's (Wyngaard, 2010). Some properties of atmospheric turbulence at very high *Re* are discussed in the following sections.

Table 3.3.2 Moments used to describe turbulence.

Quantity	Definition	Physical Meaning
Mean value	$\bar{a}$	Center of mass of random variable a
Variation	$\sigma_a^2 = \overline{(a-\bar{a})^2}$	
STD	$\sigma_a = \sqrt{\overline{(a-\bar{a})^2}}$	Width of PDF
Skewness	$Sk = \dfrac{\overline{(a-\bar{a})^3}}{\sigma_a^{3/2}}$	Asymmetry of PDF. For symmetric PDFs $Sk = 0$
Flatness (kurtosis)	$F = \dfrac{\overline{(a-\bar{a})^4}}{\sigma_a^4}$	Deviation of PDF from Gaussian one. For Gaussian PDF $F = 3$. The large value of F means the existence of an elongated tail of the PDF.
Covariance	$\overline{(a'-\bar{a}')(a''-\bar{a}'')}$	
Correlation coefficient	$R = \dfrac{\overline{(a'-\bar{a}')(a''-\bar{a}'')}}{\sigma_{a'}\sigma_{a''}}$	Linear statistical relationship between random values a' and a'' . The conditions $R = 1$ and $R = -1$ mean linear functional positive and negative relationships, respectively.
Turbulent flux	$\overline{a'\mathbf{u}'}$	Turbulent flux of quantity a' across unit square that is perpendicular to turbulent velocity $\mathbf{u}'$.

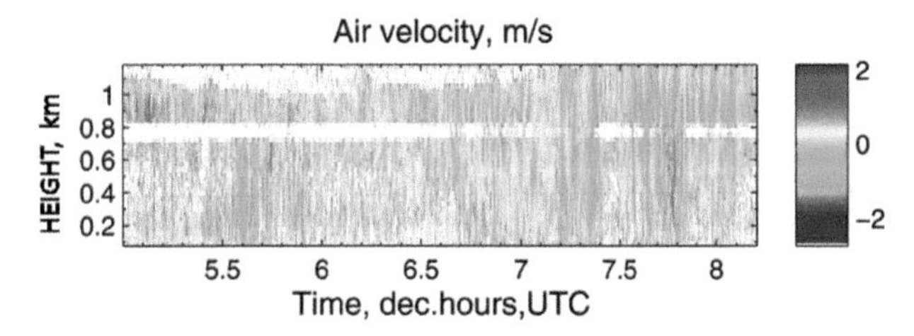

Figure 3.3.7 Field of the retrieved vertical air velocity obtained from the Cabauw Radar observation data (The Netherlands). The color bar corresponds to the velocity range of ± 2 m s^{-1} (from Pinsky et al., 2010b; courtesy of © American Meteorological Society. Used with permission).

3.3.3 Statistical Description of Turbulence

As was mentioned, turbulent motion is of stochastic nature. In clouds, both aerodynamic quantities (velocity, turbulent shears, acceleration, etc.) and thermodynamic quantities (temperature and humidity) can be considered as random values, random processes, and random fields. Examples of realization of such random processes are shown in Figures 3.3.1 and 3.3.2. An example of a random field of vertical velocity in cloud-topped BL is shown in **Figure 3.3.7**.

Turbulent quantities are described by 1D, 2D, and 3D probability distribution functions (PDF) and their moments. The appropriate averaging can be done and interpreted as time averaging, spatial averaging, or averaging with respect to a great number of realizations. A particular interpretation of the averaged values depends on conditions of measurement or modeling, the physics of the processes studied, and other factors. The k-th statistical moment of a 1D PDF of random variable a is defined as

$$\overline{a^k} = \int_{-\infty}^{\infty} a^k W(a)da, \tag{3.3.4}$$

where $W(a)$ is PDF defined by the condition that a random variable a falls within the interval of $a_1 \le a < a_2$ with probability equal to $P = \int_{a_1}^{a_2} W(a)da$. The 1D PDF obey the normalization condition

$$\int_{-\infty}^{\infty} W(a)da = 1. \tag{3.3.5}$$

Similarly, the product moments of 2D PDF are defined as

$$\overline{a_1^m a_2^n} = \int_{-\infty}^{\infty} a_1^m a_2^n W(a_1, a_2)da_1 da_2, \tag{3.3.6}$$

where $W(a_1, a_2)$ is a 2D PDF.

The most applicable moments used in cloud physics as the characteristic values of turbulent quantities, as well as their physical meaning, are presented in **Table 3.3.2**.

An example of moments is turbulent dissipation rate ε, introduced previously. It is related to the random turbulent shear tensor (or *velocity gradient tensor*) $\frac{\partial u_i}{\partial x_j}$ as follows

$$\varepsilon = \frac{\nu}{2}\overline{\left(\frac{\partial u_i}{\partial x_j} + \frac{\partial u_j}{\partial x_i}\right)^2}. \tag{3.3.7}$$

In order to characterize space and time variations of turbulent values (considered as realizations of a random process), correlation functions are widely used. A two-point correlation function determines the linear dependence between random variables at two different points in space or time. The correlation function is usually a function of spatial or temporal distance between the points. Correlation functions include information about amplitudes and characteristic time, as well as about spatial scales of fluctuations. The correlation function $B(t_1, t_2)$ is a second moment of a 2D distribution of a random process $a(t)$ at two time instances, t_1 and t_2:

$$B(t_1, t_2) = \overline{[a(t_1) - \overline{a}][a(t_2) - \overline{a}]}. \tag{3.3.8}$$

In case the fluctuations are statistically stationary, the correlation function depends on the difference of times $\tau = t_2 - t_1$ only and is an even function: $B(t_1, t_2) = B(t_2 - t_1) = B(\tau) = B(-\tau)$. The correlation function has a maximum at $\tau = 0$, which is equal to the variation of the random process $B(0) = \max\{B(\tau)\} = \sigma_a^2$. The correlation between turbulent fluctuations typically decreases with the increase of τ. It is possible to choose the value of τ_{cor} in such a way that for $t_2 - t_1 > \tau_{cor}$ turbulent fluctuations can be considered non-correlated. This value of τ_{cor} is often referred to as the correlation radius.

A typical correlation function in the turbulent theory is the Lagrangian correlation function of turbulent velocities measured or calculated along an air parcel trajectory:

$$B^{(L)}(\tau) = \overline{u(t)u(t+\tau)}. \tag{3.3.9}$$

As a measure of the correlation radius, the integral time scale known as Lagrangian time scale T_L is often used. The integral time scale T_L is defined as

$$T_L = \frac{1}{B^{(L)}(0)}\int_0^\infty B^{(L)}(\tau)d\tau. \tag{3.3.10}$$

In case the correlation function is approximated by an exponential function, T_L is the time scale at which the correlation function decreases $e = 2.7$ times.

Along with the correlation function $B(\tau)$, its Fourier transform $F(\omega)$, called the spectral density function, is also widely used to describe stationary turbulent fluctuations. These two functions are related by the pair of Fourier transforms

$$F(\omega) = \frac{1}{2\pi}\int_{-\infty}^{\infty} B(\tau)\exp(-i\omega\tau)d\tau \tag{3.3.11}$$

$$B(\tau) = \int_{-\infty}^{\infty} F(\omega)\exp(i\omega\tau)d\omega. \tag{3.3.12}$$

The following energetic relationship is noteworthy:

$$\sigma_a^2 = B(0) = \int_{-\infty}^{\infty} F(\omega)d\omega. \tag{3.3.13}$$

Since spectral density has an energy sense, this function is non-negative $F(\omega) \geq 0$.

Along with correlation functions, a structure function $D(t_1, t_2)$ is often used to analyze turbulence. The structure function is defined as follows:

$$D(\tau) = D(-\tau) = D(t_2 - t_1) = \overline{[a(t_1 + \tau) - a(t_1)]^2} \tag{3.3.14}$$

This function increases with increasing τ, and reaches a plateau at large τ where turbulent fluctuations become non-correlated. The advantage of the utilization of the structure functions for turbulence analysis is that they filter out non-turbulent low-frequency variations and, therefore, reflect only the local properties of turbulence. The relationship between $B(\tau)$ and $D(\tau)$ is described by the following expressions:

$$D(\tau) = 2[B(0) - B(\tau)]; \quad B(\tau) = \frac{1}{2}[D(\infty) - D(\tau)] \tag{3.3.15}$$

Measurements in atmosphere may be performed at unmovable ground-based stations or moving platforms (airplanes, helicopters). In the first case, one measures the time variations of quantity; in the second case, the spatial variations. The question arises of how time and spatial variations of different quantities are related. To relate spatial and time variations, a hypothesis of a frozen fluctuation is usually used. Let us consider the situation when an atmospheric object is embedded into a large-scale flow having mean velocity $\overline{u}$ **(Figure 3.3.8).** There are many examples of such situations in the atmosphere. For instance, a cumulus cloud can be transported downwind by background wind. Cyclones and hurricanes also can be considered as coherent structures moving within a larger-scale stirring current. Turbulent vortices also can be examples of such objects embedded into a larger-scale flow.

Let us assume that the object is characterized at certain time instances by a known function of

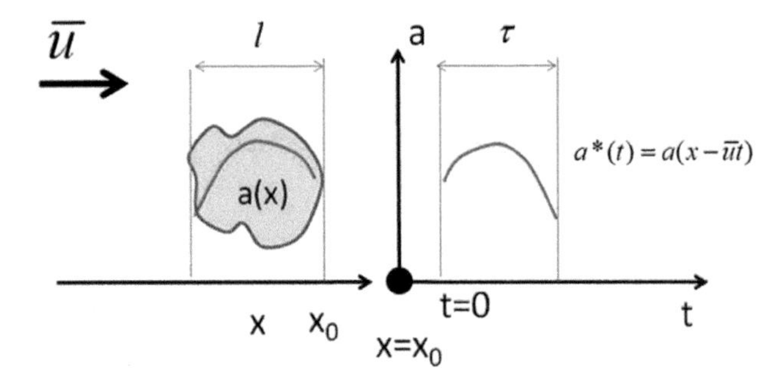

Figure 3.3.8 Illustration of an object frozen within a background flow. l is the characteristic spatial scale of the object; τ is the time period during which the object is located over the point x_0. Red line on the left panel shows dependence of quantity $a(x)$ within the object. Red line on the right panel shows the time changes of the signal from the object when it moves through point x_0.

coordinate $a(x)$. This function can be, for instance, temperature or humidity changing along x. The spatial scale of the object is equal to l. This object moves with the flow velocity over fixed point $x = x_0$ (say, over a meteorological station), where measurements of the values a are performed. Let us determine the time $\tau = \frac{l}{\overline{u}}$ needed for an object to pass point $x = x_0$. The spatial averaging of quantity a over the distance l is equivalent to time averaging of this quantity over the time interval τ if the object does not change during the period τ. In this case, one can say that the object is completely frozen within the flow. If the object is frozen, the value of $a^*(t)$ to be measured in point $x = x_0$ relates to spatial changes of function $a(x)$ as $a^*(t) = a(x - \overline{u}t)$. It is clear, however, that all objects evolve with the time. Therefore, the hypothesis of frozenness of an object is valid if the characteristic time scale of the object changes (lifetime of object) τ_l is significantly longer than the time period τ, i.e., $\tau_l >> \tau = \frac{l}{\overline{u}}$.

Now, consider as an object a turbulent vortex with characteristic spatial scales l, moving within mean flow with velocity $\overline{u}$. The characteristic lifetime of such vortex is $\tau_l = \frac{l}{u'}$, where u' is the characteristic velocity related to the turbulent vortex. In case the vortex belongs to inertial turbulent subrange, its lifetime can be estimated as $\tau_l = \varepsilon^{-1/3} l^{2/3}$. Applying the condition that fluctuation is frozen, one has $\frac{l}{u'} \gg \frac{l}{\overline{u}}$, i.e., the turbulent fluctuation can be considered as frozen if $u' \ll \overline{u}$. If the vortex is within the inertial subrange, the latter inequality is equivalent to the condition $(\varepsilon l)^{1/3} \ll \overline{u}$. This is the condition of the validity of the *Taylor hypothesis* or *frozen turbulence.* It should be noted that real turbulence is close to frozen since turbulent velocities as a rule are several times smaller than the velocity of the mean flow. In case the hypothesis of frozen turbulence is valid, the time measured correlation and the structure functions, as well as the spectral density function, are related to the corresponding space functions as

$$B(x) = B(\overline{u}\tau), \; D(x) = D(\overline{u}\tau), \; F(k) = F(\omega/\overline{u}). \tag{3.3.16}$$

Spatial random fluctuations of scalar fields, such as temperature $T(\mathbf{x})$ or humidity $\rho_v(\mathbf{x})$, are characterized by the spatial correlation, the spatial structure functions, and the spectral density function, similar to those characterizing temporal fluctuations. In this case, instead of arguments t and τ, vectors $\mathbf{x}$ and $\mathbf{r}$ are used. The following spatial functions are defined and used when analyzing the scalar fields:

- The correlation function of the vector argument:

$$B(\mathbf{r}) = B(-\mathbf{r}) = \overline{[a(\mathbf{x}+\mathbf{r}) - \overline{a}][a(\mathbf{x}) - \overline{a}]}. \tag{3.3.17}$$

- The structure function of the vector argument:

$$D(\mathbf{r}) = D(-\mathbf{r}) = \overline{[a(\mathbf{x}+\mathbf{r}) - a(\mathbf{x})]^2}. \tag{3.3.18}$$

- 3D spatial spectrum. The spatial spectrum and the correlation function are related by the 3D Fourier transform:

$$F(\mathbf{k}) = \frac{1}{8\pi^3} \int_{-\infty}^{\infty} B(\mathbf{r}) \exp(-i\mathbf{kr}) d^3\mathbf{r} \tag{3.3.19}$$

$$B(\mathbf{r}) = \int_{-\infty}^{\infty} F(\mathbf{k}) \exp(i\mathbf{kr}) d^3\mathbf{k}, \tag{3.3.20}$$

where $\mathbf{k}$ is a wave number vector. The physical meaning of these functions is similar to the physical meaning of Equations (3.3.11), (3.3.12), and (3.3.13). In case the turbulence is homogeneous and isotropic, the Equations (3.3.17), (3.3.18), and (3.3.19) depend on the modulus of vectors $r = |\mathbf{r}|$ or $k = |\mathbf{k}|$.

Spatial turbulent structure of vector fields, such as velocity field $\mathbf{u}(\mathbf{x})$ or field of Lagrangian acceleration $\mathbf{A}(\mathbf{x})$, are characterized by correlation, structure, and spectral density tensors, having nine components (see Appendix A). In case of homogeneous and isotropic turbulence the tensors depend on two scalar functions only. For example, the structure tensor can be written in the following general form:

$$D_{jk}(\mathbf{r}) = \overline{[v_j(\mathbf{x}+\mathbf{r}) - v_j(\mathbf{x})][v_k(\mathbf{x}+\mathbf{r}) - v_k(\mathbf{x})]} = [D_{ll}(r) - D_{nn}(r)]\frac{r_j r_k}{r^2} + D_{nn}(r)\delta_{jk}, \quad \text{j,k} = 1,2,3 \tag{3.3.21}$$

Table 3.3.3 Turbulent quantities in the viscous and inertial subranges, according to the Kolmogorov-41 theory.

Quantity	Inertial Sub-range	Dissipation Sub-range
Longitudinal structure function	$D_{ll}(r) = C\varepsilon^{2/3}r^{2/3}$ $C = 2$	$D_{ll}(r) = \frac{1}{15}\frac{\varepsilon}{\nu}r^2$
Transverse structure function	$D_{nn}(r) = \frac{4}{3}C\varepsilon^{2/3}r^{2/3}$	$D_{nn}(r) = \frac{2}{15}\frac{\varepsilon}{\nu}r^2$
Third order structure function (4/5 Kolmogorov law)	$D_{lll}(r) = \frac{4}{5}\varepsilon r$	$D_{lll}(r) \sim r^3$
Spectrum of kinetic energy (−5/3 Kolmogorov law)	$F(k) = C_1\varepsilon^{2/3}k^{-5/3}, C_1 = 1.52$	
Taylor micro-scale and Re_λ	$\lambda = \sqrt{\langle u'^2\rangle \frac{15\nu}{\varepsilon}}$ $\quad Re_\lambda = \langle u'^2\rangle\sqrt{\frac{15}{\varepsilon\nu}}$	

Figure 3.3.9 Illustration of the definition of the structure functions and the correlation functions: longitudinal (the direction of velocities is parallel to the line connecting the two points) and transverse (the direction of velocities is perpendicular to the line connecting the two points).

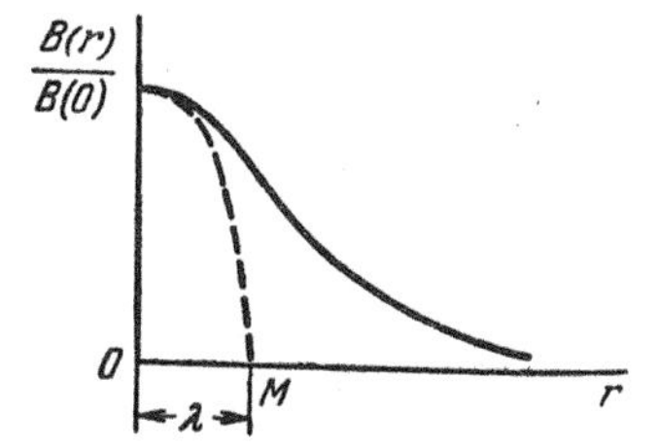

Figure 3.3.10 Definition of Taylor microscale (from Monin and Yaglom, 1975; courtesy of © MIT Press).

where δ_{jk} is unit tensor. Scalar functions $D_{ll}(r)$ and $D_{nn}(r)$ are called longitudinal and transverse structure functions, respectively. The definitions of these functions, as well as the definition of the corresponding correlation functions $B_{ll}(r)$ and $B_{nn}(r)$, are illustrated in **Figure 3.3.9.** According to the definitions, longitudinal and transverse structure functions and correlation functions are expressed as

$$D_{ll}(r) = \overline{[u_l(x+r) - u_l(x)]^2}; D_{nn}(r) = \overline{[u_n(x+r) - u_n(x)]^2}$$
$$B_{ll}(r) = \overline{u_l(x+r)u_l(x)}; B_{nn}(r) = \overline{u_n(x+r)u_n(x)} \tag{3.3.22}$$

The spatial correlation functions allow one to define the two main characteristic spatial scales, namely: the integral turbulent scale L (which can be interpreted as the radius of correlation) and the Taylor micro-scale λ. The integral scale, which is a measure of external turbulent scale, is defined as

$$L = \frac{1}{B_{ll}(0)}\int_0^\infty B_{ll}(r)dr \tag{3.3.23}$$

The Taylor microscale characterizes turbulence properties both in the viscous subrange and the inertial subrange. According to the definition

$$\lambda = \sqrt{-\frac{2B_{nn}(0)}{B''_{nn}(0)}}. \tag{3.3.24}$$

The definition of the Taylor microscale is illustrated in **Figure 3.3.10.** Using the Taylor microscale, one can define Taylor microscale Reynolds number Re_λ, whichis more suitable for descriptions of turbulent flows than the ordinary Re. The definition of Re_λ is

$$Re_\lambda = \frac{\sqrt{\langle u'^2\rangle}\lambda}{\nu}, \tag{3.3.25}$$

where $\sqrt{\langle u'^2\rangle}$ is RMS of turbulent velocity component. The Taylor microscale Reynolds number Re_λ is approximately related to Re of the largest turbulent vortices as (Pope, 2000)

$$Re_\lambda = \sqrt{\frac{20}{3}Re}. \tag{3.3.26}$$

The Kolmogorov 41 theory allows estimating different turbulent quantities characterizing the structure of homogeneous and isotropic turbulence (**Table 3.3.3).**

The formulas introduced in Tables 3.3.1 and 3.3.3 were obtained under the assumption of locally homogeneous and isotropic turbulence characterized by the Richardson cascade of kinetic energy (see Figures 3.3.5 and 3.3.6). In particular, it means that the dissipation rate ε is constant and should be interpreted as averaged over some air volume. To apply these formulas in cloud models, grid spacing should not exceed the external turbulent scale. At the same time, this grid spacing should be large enough to exclude strong centimeter-scale fluctuations of ε caused by small-scale intermittency.

3.3.4 Small-Scale Intermittency

Under very high Reynolds numbers (which is typical of the atmosphere and specifically for the clouds) turbulence becomes intermittent. Small-scale intermittency determines the interesting spatial structure of a turbulent flow. The elementary turbulent vortexes do not fill the total volume but concentrate in narrow elongated zones (filaments) randomly distributed over the air volume, forming fractal structures. In such a structure, every vortex should be characterized not by one spatial scale but by two or three scales. Strictly speaking, turbulence has a multi-fractal structure (Frisch, 1995). An example of an intermittent turbulent structure is shown in **Figure 3.3.11**. The intermittent structure of turbulence reflects the fact that the cascade of turbulent energy is more complicated than that described in Section 3.3.2. The dissipation rate experiences strong cm-scale random fluctuations in space, therefore, the Richardson energetic cascade can be interpreted as spatially averaged. However, these fluctuations do not significantly affect the statistical parameters of velocity fluctuations within the inertial subrange.

The analysis of intermittent fluctuations was first performed by Kolmogorov (1962). According to this study, fluctuations of ε obey the lognormal PDF (illustrated in **Figure 3.3.12**). The PDF has an elongated tail reflecting the existence of zones where the local dissipation rate is much larger than the mean dissipation rate in accordance with Figure 3.3.11. The log-normality of PDF is related to the deviation of PDF of the turbulent shear values from normal ones.

Turbulence becomes more intermittent with increasing Re_λ. In particular, changes occur in the properties of the turbulent shears $S = \frac{\partial u}{\partial x}$, which reach their maximum at the scales of ~1 cm corresponding to the transition from the inertial turbulence subrange to the viscous one, and the Lagrangian acceleration $A = \frac{du}{dt}$, which is actually related to the fourth order moment of turbulent velocity (**Figure 3.3.13**). The absolute values of skewness and flatness of the shear increase with Re_λ

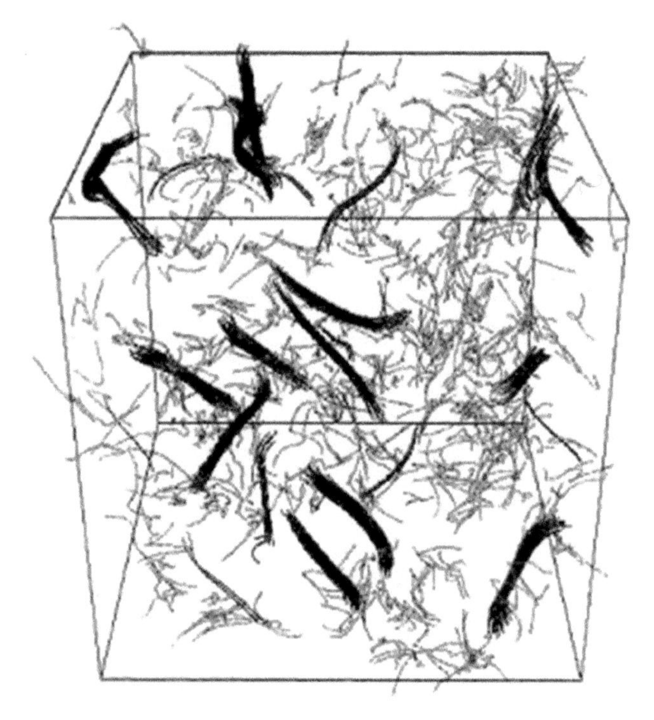

Figure 3.3.11 A 3D perspective view of vortex lines in a homogeneous turbulent flow with $Re_\lambda = 77$ obtained by DNS (from She et al., 1991; courtesy of © The Royal Society).

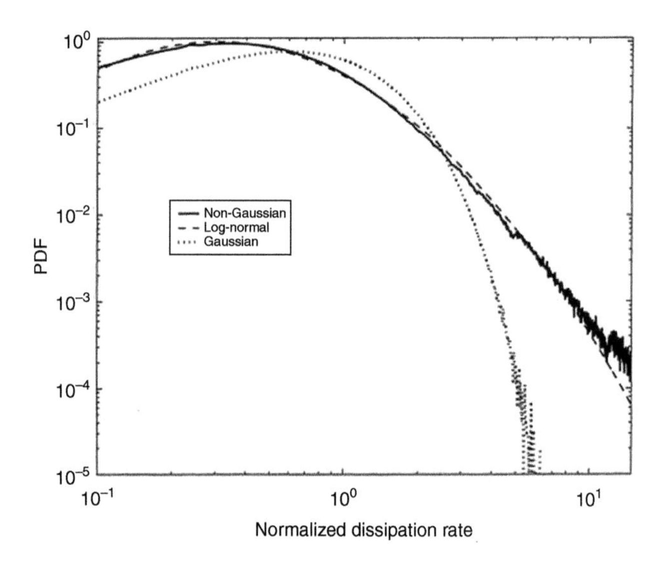

Figure 3.3.12 PDF of local dissipation rate and its approximation by log-normal function. The dotted line corresponds to normal (Gaussian) PDF of turbulent shears (from Pinsky et al., 2004; courtesy of © Elsevier).

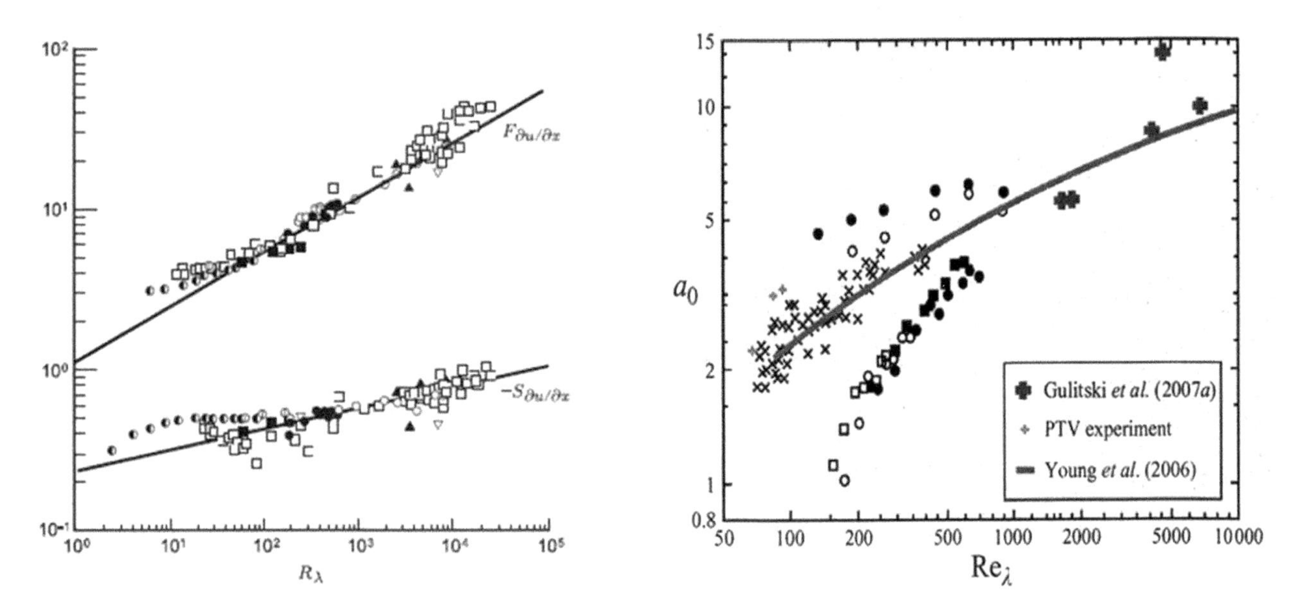

Figure 3.3.13 Left: dependences of the mean absolute value of skewness and flatness of turbulent shears on Re_λ (from Sreenivasan and Antonia, 1997; reproduced with permission of Annual Review). Right: normalized Lagrangian acceleration variance as a function of Re_λ (from Gylfason et al., 2004) with added experimental data from field experiment (Gulitski et al., 2007) and from PTV experiments (Lüthi et al., 2005) (from Tsinober, 2009; with permission of the author).

(Figure 3.3.13, left). This increase leads to an increase in Lagrangian acceleration variation (Figure 3.3.13, right). Since these quantities influence droplet motion and collisions on microscales (Sections 5.5, 5.8), the dependences shown in Figure 3.3.13 can be used in cloud models for parameterization of turbulent effects on drop collisions. Dependences shown in Figure 3.3.13 are approximated by the following formulas in parameterization suggested by Hill (2002):

$$Sk = \frac{\overline{S^3}}{\overline{S^2}^{3/2}} = -0.27Re_\lambda^{0.11}, \quad F = \frac{\overline{S^4}}{\overline{S^2}^2} = 1.33Re_\lambda^{0.32} \tag{3.3.27}$$

$$\begin{aligned}\overline{A^2} &= \frac{1}{3}\varepsilon^{3/2}\nu^{-1/2}\left(2F^{0.79} + 0.3|Sk|\right) \\ &= \frac{1}{3}\varepsilon^{3/2}\nu^{-1/2}\left(2.5Re_\lambda^{0.25} + 0.08Re_\lambda^{0.11}\right)\end{aligned} \tag{3.3.28}$$

These formulas were used in modeling of the collision process in clouds (Pinsky et al., 2008; Benmoshe et al., 2012). Therefore, there is a significant difference in the statistics of the turbulent shears and especially of the Lagrangian accelerations between intermittent and non-intermittent turbulent flows.

3.3.5 Turbulence in Clouds

Clouds are known as zones of enhanced turbulence, although their intensity and other turbulent parameters depend on cloud type and change substantially within clouds. The turbulence intensity characterized by the mean turbulent dissipation rate ε in cumulus clouds can be by two to three orders of magnitude higher than that in stratiform clouds. In stratiform and stratocumulus clouds, the relative role of turbulent motion is larger, and turbulent motions with velocity fluctuations up to 1–1.5 m/s determine the cloud dynamics. At the same time, in convective clouds the turbulent velocities, which can reach 2–4 m/s, are several times weaker than the velocities caused by thermal (larger-scale) convection. The typical evaluated parameters of turbulence in clouds of different types are presented in **Table 3.3.4**.

Although observations of turbulence in cumulus clouds started as long as thirty to forty years ago, the detailed measurements of turbulent parameters in strong convective clouds are scarce. Observations show that turbulence in these clouds corresponds, to a great extent, to the Kolmogorov theory. The −5/3 slope of turbulent kinetic energy spectrum is clearly seen in these measurements (**Figure 3.3.14**)

Table 3.3.4 Turbulent parameters and time/spatial scales of turbulent fluctuations for clouds of different type (from Pinsky et al., 2006; American Meteorological Society©; used with permission).

Turbulent parameters:	Stratiform clouds	Cumulus clouds	Cb	
Dissipation rate, ε	$\varepsilon = 10^{-3}\ \mathrm{m^2\ s^{-3}}$	$\varepsilon = 2 \times 10^{-2}\ \mathrm{m^2\ s^{-3}}$	$\varepsilon = 10^{-1}\ \mathrm{m^2\ s^{-3}}$	Shmeter (1987); Mazin et al. (1989);
Taylor microscale Reynolds number*	$Re_\lambda = 5 \times 10^3$	$Re_\lambda = 2 \times 10^4$	$Re_\lambda = 2 \times 10^4$	Weil et al. (1993); Panchev (1971); Pinsky and Khain (2003)
Kolmogorov scales				Monin and Yaglom (1975)
$l_k = \nu^{3/4}\varepsilon^{1/4}$	1.4 mm	0.65 mm	0.44 mm	
$\tau_k = (\nu/\varepsilon)^{1/2}$	0.124 s	0.028 s	0.0124 s	
Turbulent shear				
$l_s \approx 3l_k$	4.2 mm	2.0 mm	1.3 mm	
τ_S	several τ_K	several τ_K	several τ_K	
Lagrangian acceleration				Hill (2002)
$l_A = 6.8 l_k \times Re_\lambda^{-0.033}$	7.1 mm	3.2 mm	2.1 mm	
τ_A	several τ_K	several τ_K	several τ_K	
Droplet concentration fluctuations, l_c	Several l_K	Several l_K	Several l_K	Pinsky and Khain (2003); Pinsky et al. (1999)

* Estimations of Re_λ were performed using the values of typical velocity fluctuations and the typical dissipation rates in clouds of different type presented in Mazin et al. (1989) and Shmeter (1987).

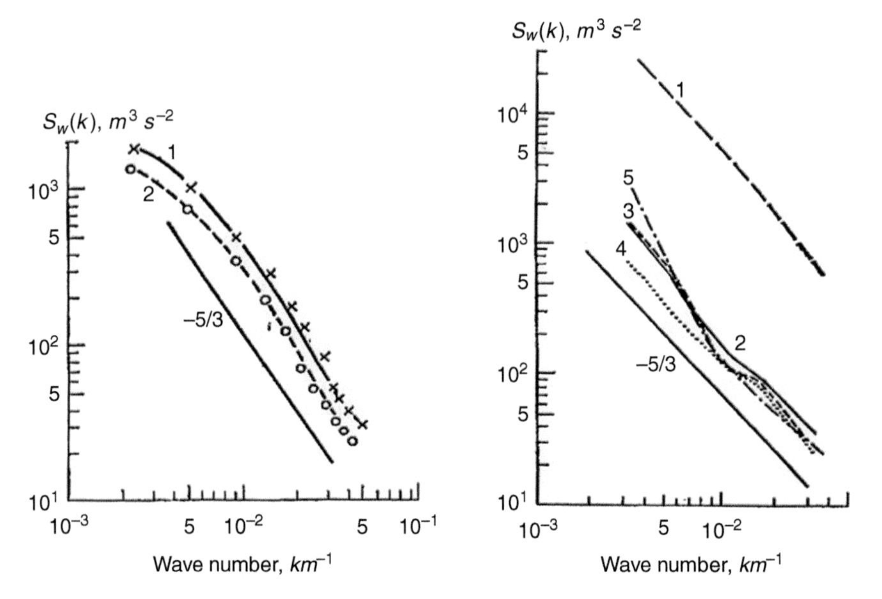

Figure 3.3.14 Spectrum of turbulent kinetic energy in convective clouds (left: 1 – upper cloud part; 2 – lower cloud part) (from Mazin et al., 1989); and in cumulonimbus clouds (right: 1 – central part; 2 – near top; 3 – above cloud; 4 – lateral; near bottom) (from Shmeter, 1987).

A theoretical study of turbulence in zones of wet convection was performed by Kabanov and Mazin (1970), who investigated turbulent fluctuations of temperature and other thermodynamic parameters caused by phase transition (the condensation/evaporation processes). They came to the conclusion of a possible existence of three subranges in inertial turbulent range, namely, a subrange corresponding to the moist adiabatic temperature gradient, a subrange corresponding to the dry adiabatic temperature gradient, and an intermediate zone between the two subranges. In principal, the three subranges should be reflected by the shape of the structure functions and energy spectra. However, the scarcity of accurate turbulence measurements in clouds does not currently allow us to clearly distinguish these subranges.

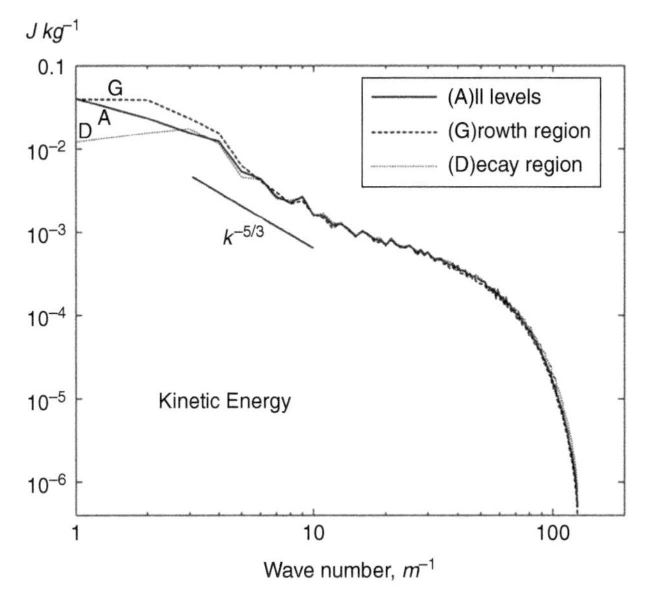

Figure 3.3.15 Kinetic energy spectra: total average all-level (solid line), growth region (dashed line), and decay region (dotted line) (from Spyksma and Bartello, 2008; courtesy of © American Meteorological Society. Used with permission).

Recently, high-resolution large-eddy simulation (LES) models of convective plumes (e.g., Grabowski and Clark, 1991; Spyksma and Bartello, 2008) enabled investigating turbulence in dry and moist convection and obtaining important information about their turbulent nature. Grabowski and Clark (1991) evaluated the external turbulent scale L and found its relationship with the cloud size L_{cl} to be $L = L_{cl}/10$. This evaluation agrees with the theoretical estimations by Elperin et al. (2002). Spyksma and Bartello (2008) studied turbulence in dry and moist convection, and found their kinetic energy spectrum to deviate from the $-5/3$ Kolmogorov law. This spectrum is illustrated in **Figure 3.3.15**.

Turbulent structure in strong convective clouds was simulated in a model by Benmoshe et al. (2012). The turbulent parameters were calculated using the $k - \varepsilon$ theory with 1.5 order closure (see Section 3.5). The fields of turbulent quantities (**Figure 3.3.16**) demonstrate a strong spatial variability characterized by

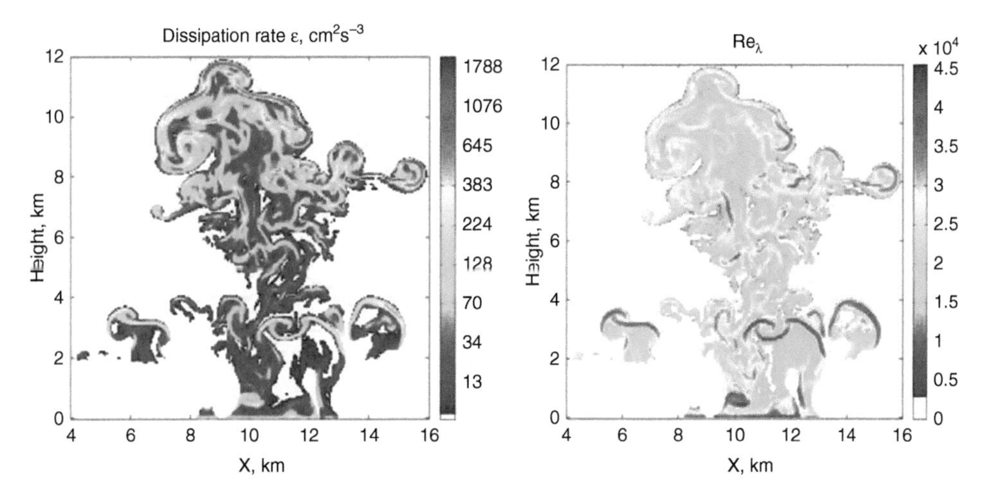

Figure 3.3.16 The fields of turbulent dissipation rate (left) and of Re_λ (right) obtained in a simulation of strong convective cloud by HUCM (from Benmoshe et al., 2012; courtesy © John Wiley & Sons, Inc.).

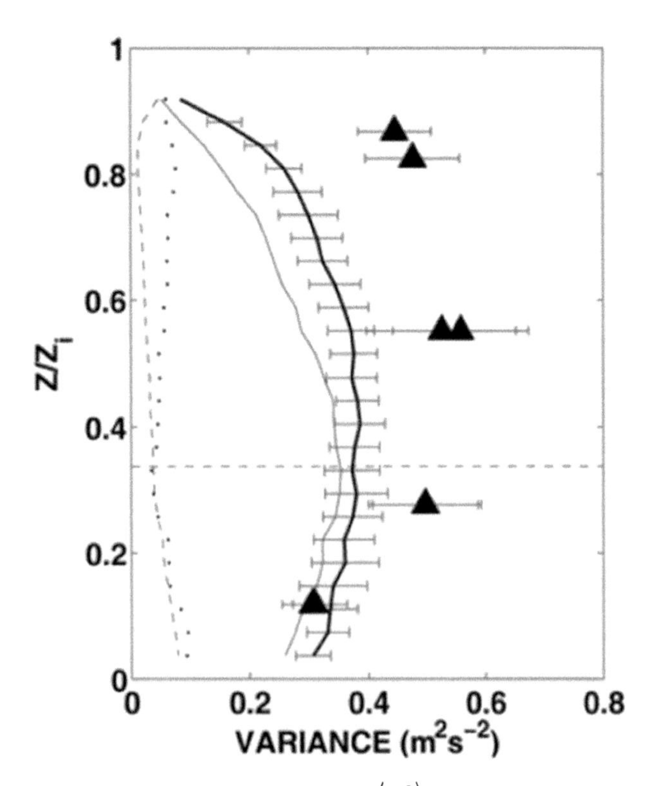

Figure 3.3.17 Solid lines: profiles of $\langle w'^2 \rangle$ found by fitting the observed structure functions of the nadir Doppler velocity to structure functions modeled over wavelengths of <50 m (thick line) and of <1,000 m (thin line). Dashed and dotted lines are the respective white-noise contributions obtained from the fits. *In situ* measurements of variance are indicated by triangles. The horizontal dashed line is cloud base (from Lothon et al., 2005; courtesy of © John Wiley & Sons, Inc.).

elongated zones of enhanced turbulent intensity. The turbulent dissipation rate in these zones reaches 0.18 m^2 s^{-3}, and Re_λ can be as large as 4.5×10^4. Spatial variability of turbulent parameters with characteristic scales of several tens to several hundreds of meters is called, sometimes, large-scale intermittency. The zones of enhanced turbulence were observed both in stratocumulus clouds and in convective clouds, in the vicinity of cloud top and cloud edges. A strong large-scale inhomogeneity of turbulent dissipation rate leads to differential influence of turbulence on the microphysical processes in different cloud areas and, eventually, to inhomogeneity in optical characteristics, rain production rates, and other important cloud properties.

Measurements of turbulent parameters in stratiform, stratocumulus, and small cumulus clouds were carried out using both Doppler radars (Kollias and Albrecht, 2000; Kollias et al., 2001; Lothon et al., 2005; Pinsky et al., 2010b) and moving platforms such as a balloon (Siebert et al., 2006), helicopter (Siebert et al., 2010), or aircraft (Strunin, 2013). The measurements show that turbulent dissipation rate in these clouds is of the order 1–50 cm^2 s^{-3}, and Re_λ is of the order 10^4. Siebert et al. (2006) observed in clouds of these types the existence of narrow zones with enhanced turbulence with $\varepsilon \sim 10^3$ cm^2 s^{-3}. These zones can be identified as zones of embedded convection.

An important turbulent characteristic of cloud-capped BL is vertical profile of the vertical turbulent velocity variance$\langle w'^2 \rangle (z)$. An example of such a profile measured by an airborne Doppler radar is shown in **Figure 3.3.17**. It is typical of $\langle w'^2 \rangle$ to reach the maximum values in the middle or the upper part of BL. The velocity variation decreases toward the surface and toward the upper bound due to temperature inversion existing near the upper bound.

An example of vertical profiles of different turbulent quantities derived from measurements by the Transportable Atmospheric Radar (TARA) located in Cabauw, Netherlands, and the measured transverse structure functions are shown in **Figures 3.3.18** and **3.3.19,** respectively. The slope of the structure function deviates from 2/3, indicating that the turbulence at low levels in the vicinity of the surface is not isotropic. The periodical changes of the lateral structure function at scales of several hundred meters (Figure 3.3.19, left) indicate the existence of large eddies of the corresponding scales in cloud-capped BL. This situation is typical of atmospheric BL.

Precise measurements with resolution of 10–15 cm were carried out in small cumulus and stratocumulus clouds by Siebert et al. (2006, 2010). These measurements allow estimating intermittent properties of cloud turbulence. **Figures 3.3.20** and **3.3.21** show the PDF of the turbulent dissipation rate ε and the increments of air velocity, measured at lags of 15 cm length, respectively. The first PDF demonstrates a good agreement with Kolmogorov's 62 theory, according to which ε should obey the log-normal distribution. The second PDF, having a sharp peak and an elongated tail, is characterized by a large value of velocity increment flatness, indicating strong deviations from the normal distribution. The intermittent properties of cloud turbulence are used in Chapter 5 to evaluate the effects of turbulence on drop collisions.

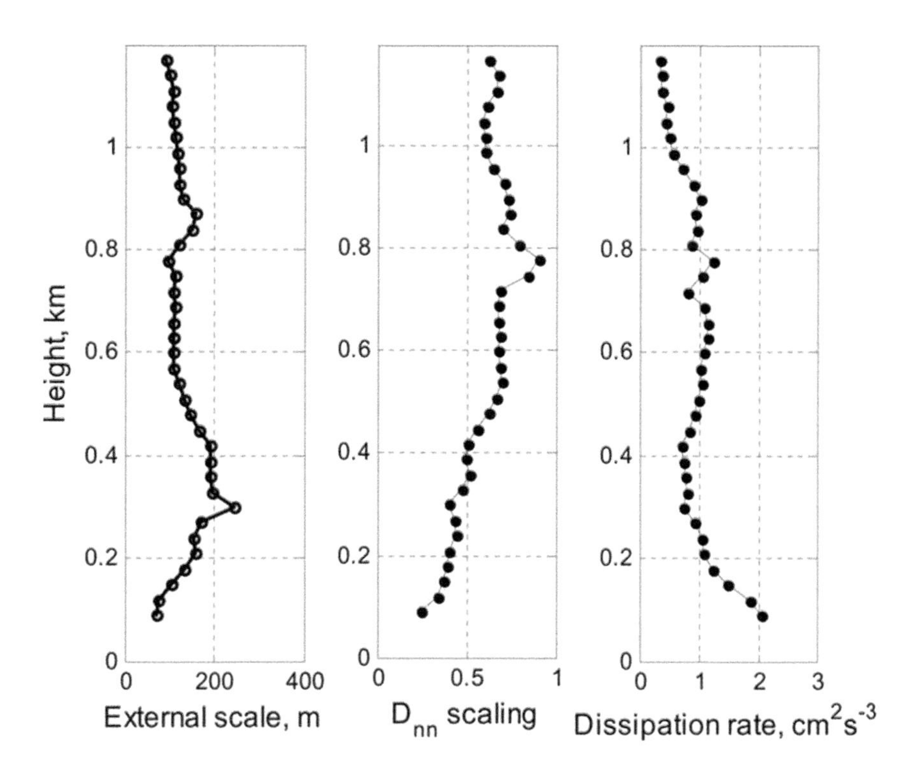

Figure 3.3.18 Vertical profiles of the turbulent integral scale (left), scaling of structure function (middle), and turbulent dissipation rate (right). The profiles were obtained from the data shown in Figure 3.3.7 (from Pinsky et al., 2010b; courtesy of © American Meteorological Society. Used with permission).

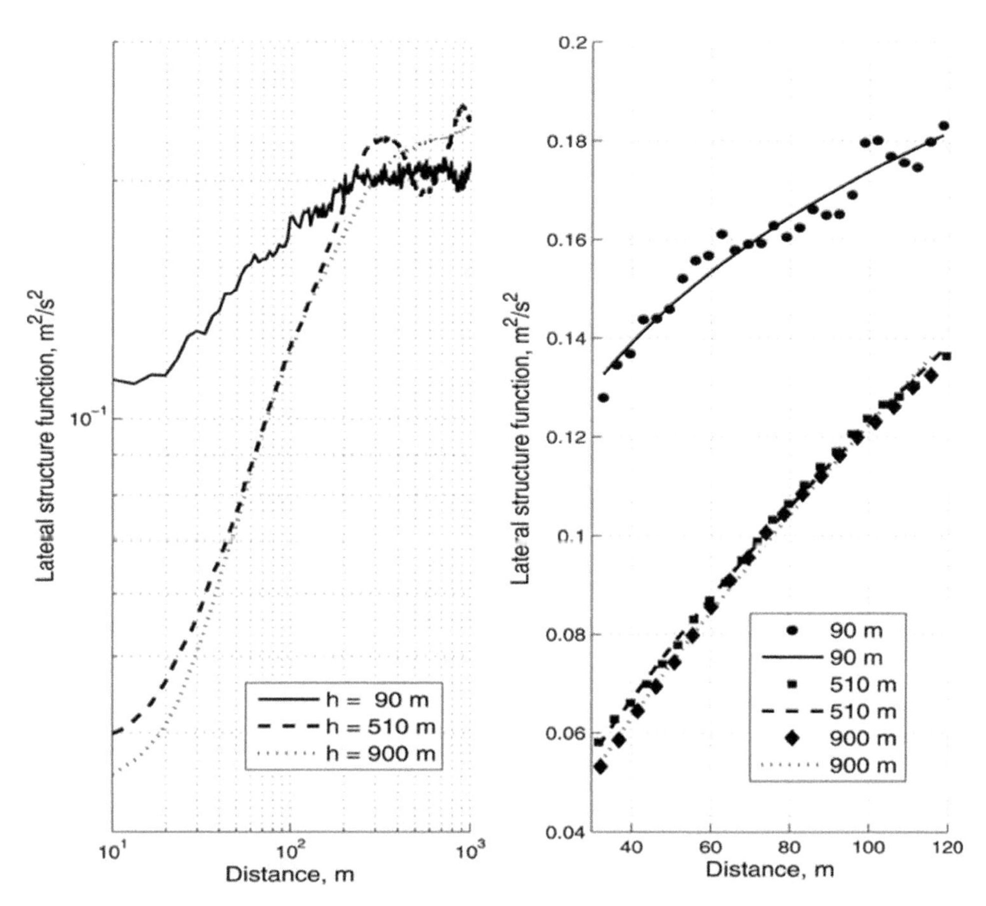

Figure 3.3.19 The structure functions at different height levels in the log–log coordinates (left) and the linear coordinates (right). The lines in the right panel show the approximation of the structure functions by the power function. The functions were obtained from the data shown in Figure 3.3.7 (from Pinsky et al., 2010b; courtesy of © American Meteorological Society. Used with permission).

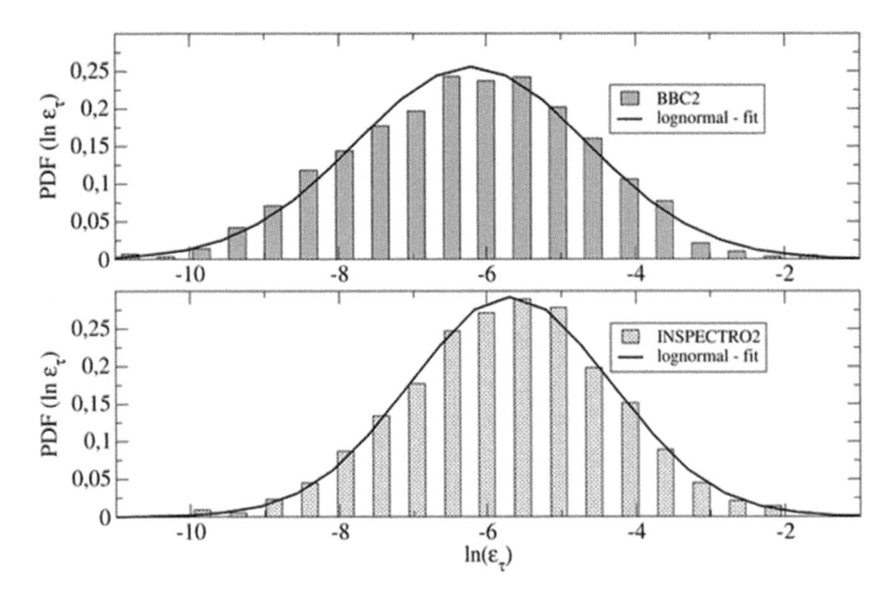

Figure 3.3.20 PDF of the natural logarithm of local energy dissipation rate $\mathrm{Ln}(\varepsilon_r)$. A Gauss fit is included for reference (from Siebert et al., 2006; courtesy of © American Meteorological Society. Used with permission).

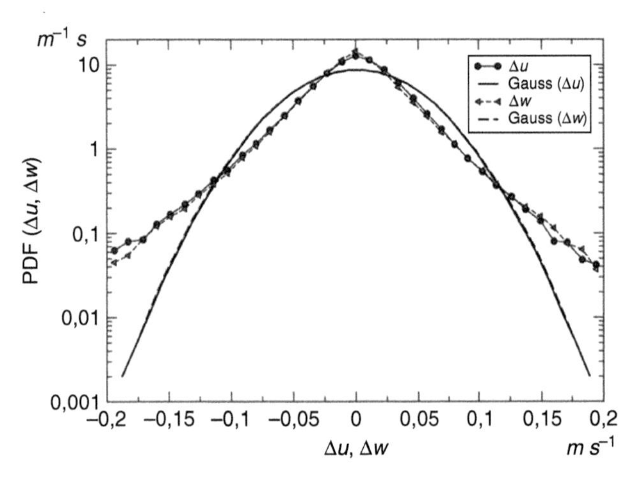

Figure 3.3.21 Semilogarithmic plot of the PDFs of the velocity increments Δu and Δw based on the BBC2 data. A Gaussian fit is included for reference (from Siebert et al., 2006; courtesy of © American Meteorological Society. Used with permission).

3.4 Scales of Atmospheric Motions and Equation Averaging

3.4.1 Scales of Atmospheric Motions and the Necessity of Equation Averaging

Atmospheric motions have a wide range of spatial scales (from about 1 mm to several thousand kilometers) and of time scales (from fractions of a second to weeks, months, etc.) There is a close relationship between spatial and time scales of different atmospheric phenomena, as shown in **Figure 3.4.1**. Large-scale phenomena have, as a rule, higher energy and longer characteristic time (lifetime). Typically, the characteristic time of a certain phenomena τ can be evaluated as

$$\tau \approx \frac{L}{U}, \tag{3.4.1}$$

where L is the characteristic spatial scale and U is the characteristic velocity related to the phenomena.

Figure 3.4.2 shows schematically the spectrum of wind velocity variations in the atmospheric boundary layer. The solid line denotes the spectrum measured in the surface atmospheric layer (the lowest layer of ~50 m depth) (Van der Hoven, 1957). The dashed line denotes the spectra of convective motions measured at the 300 m-high meteorological tower (Obninsk, Russia) at heights above 100 m (Bizova et al., 1989). The area below the curve within a certain range of frequencies is equal to the kinetic energy of the atmospheric motions within this frequency range. There are two distinct peaks of the kinetic energy spectrum in the surface layer: the first peak corresponds to time scales of a few days and to spatial scales of ~1,000 km. This peak is caused by synoptic motions such as atmospheric cyclones and anticyclones. The other peak is located within the high-frequency part of the spectrum with characteristic time scales of several tenths of a second and a spatial scale of ~100 m. This high-frequency peak is caused by small-scale turbulent motions. There is an additional small peak in the spectrum corresponding to the time period of 24 h. This peak reflects daily changes of meteorological fields.

Multiple measurements and observations in the BL (BL is the layer of 1–2 km height above the surface layer)

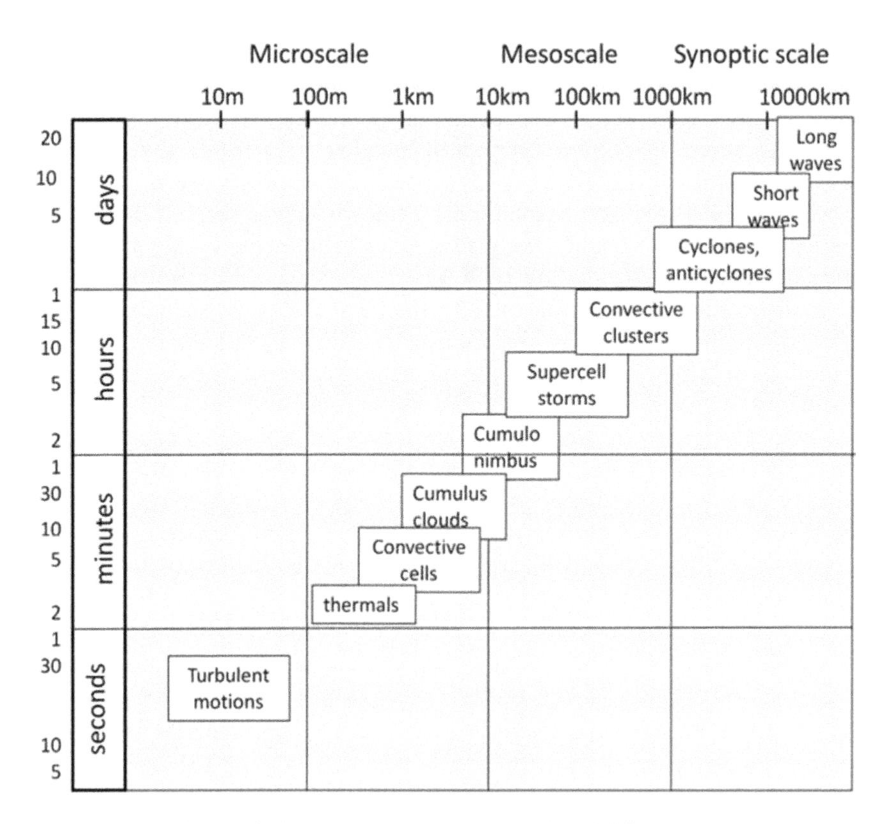

Figure 3.4.1 Relationship between spatial and time scales of different atmospheric processes.

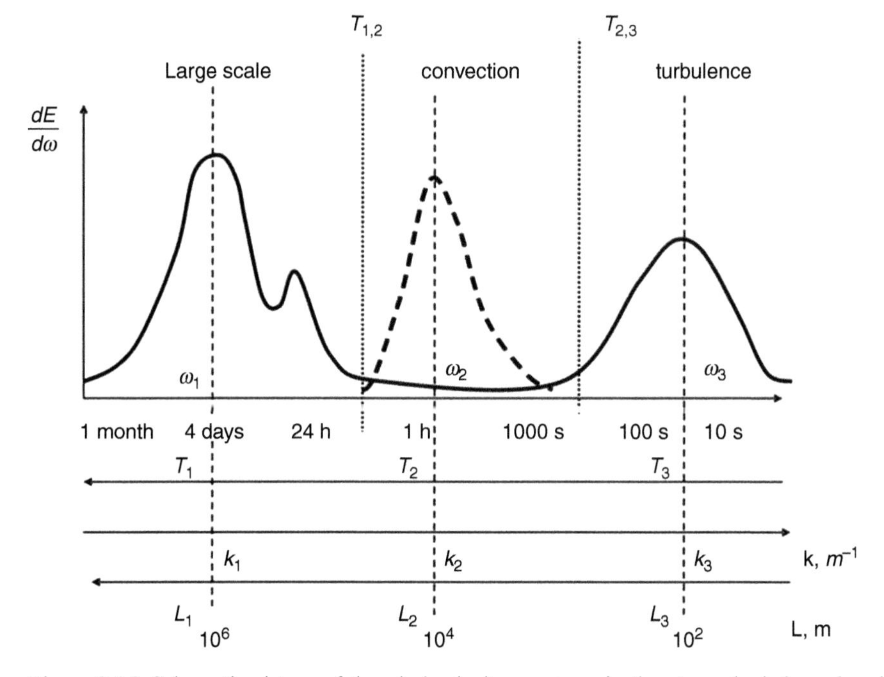

Figure 3.4.2 Schematic picture of the wind velocity spectrum in the atmospheric boundary layer. The solid line denotes the Van der Hoven spectrum measured in 1957 in the surface layer of the atmosphere. The dashed line denotes the spectrum measured at the 300 m-high meteorological tower (Obninsk, Russia) at heights above 100 m.

indicate the existence of a very important peak at time scales from about tens of minutes to a few hours and at spatial scales from ~1 to ~10 km. This peak is caused by convective motions including clouds (Section 1.2). Convective cells in BL are located above the surface layer where small-scale turbulence dominates (e.g., Bizova et al., 1989). The roots of convective clouds are also located above the surface layer. This is the reason why surface observations performed by Van der Hoven do not reveal the convection-related peak within the energy spectrum.

As discussed in Section 3.3, turbulent motions experience stochastic changes over space and time, so a turbulent flow is usually characterized by averaged characteristics. However, the necessity of averaging arises not only due to the stochastic nature of turbulent flows. All numerical atmospheric models available today are characterized by their resolution, i.e., the minimum spatial scale resolvable by the model. Most cloud models use the finite-difference approach to solve the model equations (Chapter 4). In this approach, the entire computational area is covered by a finite-difference grid where the distance between the neighboring grid points is known as grid spacing length. The smaller the grid spacing length, the higher the model resolution is. A grid cannot explicitly resolve motions with scales smaller than a double grid spacing length (the reasons are discussed in Section 4.2). In terms of the theory of finite-difference methods, at these scales also known as subgrid scales, finite-difference schemes lose the approximation. However, motions of subgrid scales affect the representation of larger scales on the finite-difference grid. Let us suppose, by way of example, that a motion represents a sum of two harmonics (two signals) with different wavelengths (**Figure 3.4.3**). The first

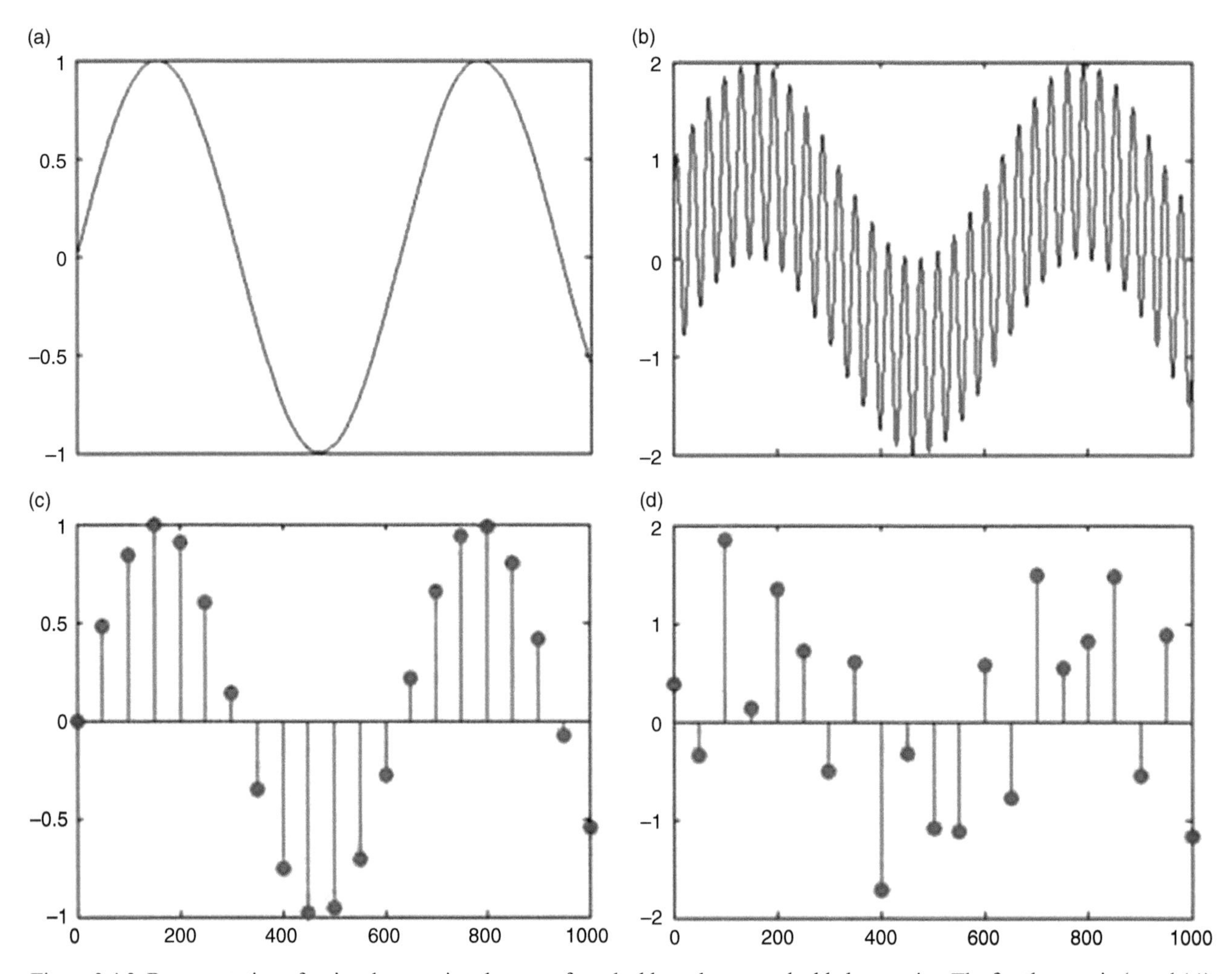

Figure 3.4.3 Representation of a signal expressing the sum of resolvable and non-resolvable harmonics. The first harmonic (panel **(a)**) has a wavelength significantly larger than the double grid spacing. It is well represented on the finite-difference grid (panel **(b)**). The second harmonic has a non-resolvable wavelength. The sum of these two signals is shown in panel **(c)**. Panel **(d)** shows that the harmonic with a non-resolvable wavelength substantially distorts the representation of both the signals and the large wavelength harmonic.

signal has a wavelength that is significantly larger than the double grid spacing (Figure 3.4.3a). This signal is well represented on the finite-difference grid (Figure 3.4.3b). The second harmonic has a wavelength less than the double grid spacing, i.e., falls into the non-resolvable range of scales. The sum of these harmonics is shown in Figure 3.4.3c. Panel (d) of Figure 3.4.3 shows how this sum is represented on the finite-difference grid. One can see that the harmonic with the short wavelength substantially distorts the representation of both the total signal and the large wavelength signal. Such distortion is related to so-called aliasing error, when small non-resolved scales are erroneously represented on the finite-difference grid at large scales (Section 4.2).

In climatic and weather forecast models, the grid spacing length exceeds several kilometers or even several tens of kilometers. In such models, both turbulent and convective motions represent non-resolvable subgrid motions. In addition, these models often consider convective motions as stochastic ones, because the location, the time of initiation, and the intensity of each convective cell or cloud can be considered random. The cumulative effect of space- and time-averaged clouds is of importance for these models. In high-resolution cloud models, the grid spacing length ranges from several tens to several hundred meters. Such models resolve both large-scale motions and convective motions. Only turbulent motions, either as a whole or partially, remain subgrid. In many cases, the effects of non-resolvable motions on large-scale resolvable motions may be quite significant, but the methods for describing these effects depend on the averaging, related to the model grid spacing.

Thus, the necessity of proper representation of motions of resolvable scales on model grids requires a preliminary time- or space-averaging, which would eliminate subgrid scale fluctuations and the related aliasing errors.

3.4.2 Averaging Operators; Selecting Averaging Intervals

As was discussed previously, turbulent motions are stochastic, and only their statistical characteristics (i.e., averaged values) can be predicted. Reynolds (1895) proposed representing any quantity a in a turbulent flow as a sum of the mean values and fluctuations, i.e., deviations from these mean values:

$$a = \overline{a} + a', \tag{3.4.2}$$

where the overbar denotes averaging. According to the definition of averaging:

$$\overline{a'} = 0. \tag{3.4.3}$$

In Equations (3.4.2) and (3.4.3), a, $\overline{a}$, and a' are the instantaneous, averaged, and perturbation values, respectively. Equation (3.4.2), though quite simple, causes numerous problems related to separation of the fields of different atmospheric quantities into the averaged fields and perturbation fields. To solve these problems, it is first necessary to formulate the meaning of averaging as regards turbulent flows. As the instantaneous fields of a turbulent flow are random, turbulent flows with identical external conditions are realized with different probabilities. In contrast, in laminar flows equal external conditions yield identical realizations. Thus, averaging of different variables in a turbulent flow should be the ensemble averaging over a great number of realizations. However, practically it is almost impossible to use ensemble averaging. Instead, time and spatial averaging are used. In case the grid spacing is larger than the scales of turbulent motions, instantaneous values represent a combination of convective and turbulent motions.

The time-averaging operator is defined as:

$$\overline{a}_t(\mathbf{x}, t) = \frac{1}{T} \int_{-T/2}^{T/2} a(\mathbf{x}, t + \tau) d\tau, \tag{3.4.4}$$

where T is the time-averaging interval. Time averaging of measurements is performed at a particular spatial point with coordinates $\mathbf{x}$. The procedure of time averaging (also known as moving averaging) is illustrated in **Figure 3.4.4**. The time variations include variations of different time scales: from low-frequency synoptic variations to high-frequency turbulent fluctuations.

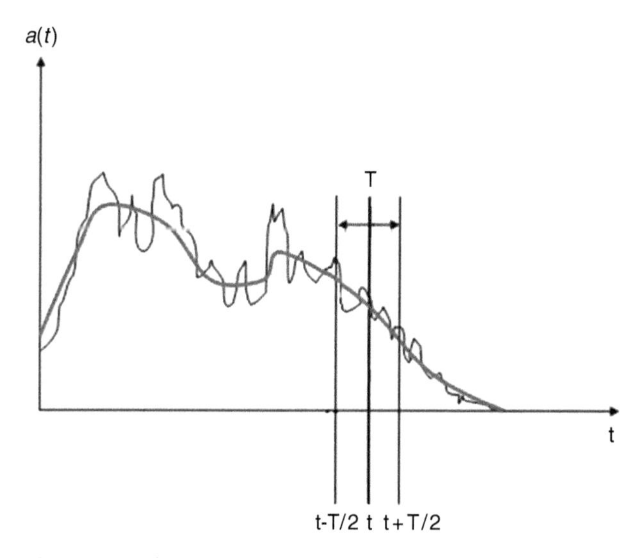

Figure 3.4.4 Illustration of the time-averaging procedure. The red line denotes the averaged value.

Averaging time interval T should be significantly longer than the period of high-frequency fluctuations. At the same time, T cannot be too long, since value $\overline{a}$ should not vary significantly within this interval. It means that T should be significantly shorter than the characteristic time scale of the larger scale flow. Actually, the moving averaging operator performs a low-frequency filtration that results in elimination of high-frequency motions.

The operator of spatial averaging along a chosen direction is defined as

$$\overline{a}(\mathbf{x}, t) = \frac{1}{l} \int_{-l/2}^{l/2} a\left(\mathbf{x} + \frac{\mathbf{x}}{|\mathbf{x}|} l', t\right) dl', \qquad (3.4.5a)$$

where l is the spatial-averaging interval. Spatial averaging is performed at a certain time using simultaneous measurements at different coordinate points. For instance, spatial averaging is used in analysis of synoptic maps. The requirements for choosing the spatial-averaging interval are similar to those for the time-averaging interval: l should be significantly larger than the characteristic spatial scale of fluctuations. At the same time, l should be significantly smaller than the spatial scale of an average large-scale field. In 3D cases, the averaging is performed in 3D space:

$$\overline{a_V}(\mathbf{x}, t) = \frac{1}{V} \iiint_V a(\mathbf{x} + \mathbf{x}', t) d\mathbf{x}' \qquad (3.4.5b)$$

In order to relate spatial scales to time scales, the concept of frozen flows is usually used. This concept is an extension of the hypothesis of frozen turbulence to a wider range of atmospheric motions (Section 3.3, Figure 3.3.8). According to this hypothesis, perturbations of smaller scales that are transported by motions of a larger scale do not undergo any significant internal changes during this transport. In this case, the spatial scale of an atmospheric phenomenon L can be evaluated as $L \approx \overline{u}T$, where T is the time period during which the phenomenon (object) passes over the measuring device. If the object is frozen, this time period should be much smaller than the characteristic time scale τ of the phenomenon. Although the hypothesis of frozen flows is not always strictly valid, it is widely used in investigations of atmospheric processes for transformation of time scales into spatial scales, e.g., Figure 3.4.1.

The intervals of both time averaging and spatial averaging should be chosen larger than the corresponding characteristic scales of fluctuations, but significantly smaller than the corresponding scales of an average flow. To perform such averaging, scales of fluctuations and of average fields should be well separated. Thereupon, two questions should be answered: Are there natural scales of atmospheric processes that allow separation of average (larger scale) fields and perturbation fields? and How can these natural scales be determined?

Since this book is dedicated to the description of cloud processes, it is quite important to understand whether it is possible to separate convective scales from turbulent scales by averaging. The answer is positive, as will be illustrated using a simple example. Let us characterize each scale of motions by one frequency corresponding to the peaks in the spectra (these peaks are shown in Figure 3.4.2). In this way, any atmospheric quantity can be approximately represented by a sum of three harmonics:

$$A(t) = a_1 \sin \omega_1 t + a_2 \sin \omega_2 t + a_3 \sin \omega_3 t, \qquad (3.4.6)$$

where the first, second, and third peaks correspond to synoptic large scale motions, convective scale motions, and turbulent motions, respectively. The characteristic values of frequencies ($\omega = 2\pi/T$), which can be evaluated using the values of T presented in Figure 3.4.2, are: $\omega_1 \approx 1.7 \cdot 10^{-5}\ \mathrm{s}^{-1}$; $\omega_3 \approx 10^{-1}\ \mathrm{s}^{-1}$; $\omega_2 \approx 1.5 \cdot 10^{-3}\ \mathrm{s}^{-1}$. Applying the operator of time averaging (Equation 3.4.4), one obtains

$$\overline{A}(t) = a_1 \sin \omega_1 t \frac{\sin \frac{\omega_1 T}{2}}{\frac{\omega_1 T}{2}} + a_2 \sin \omega_2 t \frac{\sin \frac{\omega_2 T}{2}}{\frac{\omega_2 T}{2}} + a_3 \sin \omega_3 t \frac{\sin \frac{\omega_3 T}{2}}{\frac{\omega_3 T}{2}} \qquad (3.4.7)$$

In Equation (3.4.7), the coefficients $\frac{\sin \frac{\omega_i T}{2}}{\frac{\omega_i T}{2}}$ indicate the effects of averaging on the amplitudes of the corresponding i-th harmonic.

As was discussed (see Figure 3.4.2), there are two possibilities for separating atmospheric quantities into average values and perturbation values: a) when a large scale flow is considered as an average one while convective and turbulent motions represent fluctuations and b) when a large scale flow includes convective motions so fluctuations are represented by small-scale turbulence only. In the first case, in order to split the process (Equation 3.4.6) into large-scale (average) processes and fluctuations (convection + turbulence), the following constraints on the coefficients in Equation (3.4.7) should be imposed: $\frac{\sin\frac{\omega_1 T}{2}}{\frac{\omega_1 T}{2}} \approx 1$ and $\frac{\sin\frac{\omega_2 T}{2}}{\frac{\omega_2 T}{2}} << 1$. It means that the following inequalities should be valid: $\omega_1 T << 2\pi$ and $\omega_2 T >> 2\pi$. The first condition follows from a well-known limit: $\lim_{x\to 0}\frac{\sin x}{x} = 1$. To obey both inequalities, the time-averaging interval T_{12}, suitable to separate the first harmonic from the second harmonic (the third harmonic will then be separated even better), can be chosen as the geometric average between $\frac{2\pi}{\omega_1}$ and $\frac{2\pi}{\omega_2}$, or

$$T_{12} = \frac{2\pi}{\sqrt{\omega_1\omega_2}}. \tag{3.4.8}$$

Substituting the characteristic frequencies for synoptic and convective motions, we obtain the optimum value of the time-averaging interval $T_{12} \approx 4 \cdot 10^4$ s.

Finite-difference grids in cloud models must have a resolution high enough to resolve convective motions. Typical grid spacing used in cloud models varies from several tens of meters to about one kilometer. In such models, convective fields and larger-scale fields are the average fields, and only turbulent motions represent fluctuations. The time-averaging interval T_{23}, suitable to separate the first harmonic from the second, can be chosen by a similar condition:

$$T_{23} = \frac{2\pi}{\sqrt{\omega_2\omega_3}}. \tag{3.4.9}$$

Substituting the characteristic frequencies for convective and turbulent motions into Equation (3.4.9), the optimum value of the time-averaging interval T_{23} can be evaluated as ~500 s. The averaging intervals T_{12} and T_{23} are shown in Figure 3.4.2.

It is clear that averaging cannot separate scales perfectly. The errors caused by averaging can be evaluated by the values of coefficients $\frac{\sin\frac{\omega_i T}{2}}{\frac{\omega_i T}{2}}$. For instance, coefficients $\frac{\sin\frac{\omega_2 T_{23}}{2}}{\frac{\omega_2 T_{23}}{2}} \approx 0.98$ and $\frac{\sin\frac{\omega_3 T_{23}}{2}}{\frac{\omega_3 T_{23}}{2}} \approx 0.2$, in case T_{23} is chosen. The values of the two coefficients give an idea about the two types of errors caused by averaging: first, averaging distorts the amplitude of large-scale processes, and second, the averaged values include the effect of smaller-scale motions, which in the ideal case should be fully eliminated.

The spatial averaging is similar to the time averaging considered previously. The optimum averaging lengths L_{12} and L_{23} can be written as follows:

$$L_{12} = \frac{2\pi}{\sqrt{k_1 k_2}}; L_{23} = \frac{2\pi}{\sqrt{k_2 k_3}}, \tag{3.4.10}$$

where $k_1 = \frac{2\pi}{L_1}$ and $k_3 = \frac{2\pi}{L_3}$ are the wave numbers corresponding to the peaks of the energetic spectrum. Substituting the characteristic values of the wave numbers of different motions (Figure 3.4.2) into Equation (3.4.10), one can evaluate the spatial averaging lengths as $L_{12} \approx 20 - 30$ km and $L_{23} \approx 500$ m.

In addition to these natural averaging scales related to the energy spectrum of atmospheric motions, numerical models introduce their own characteristic scales determined by their grid spacing. As is shown in Section 4.3.1, the models are unable to resolve motions with scales equal to or lower than two grid spacing lengths. At the same time, the models explicitly describe processes with scales larger than two grid-spacing lengths. It means that the choice of the grid spacing determines the separation between resolved (averaged or large scale) scales and non-resolvable (subgrid) scales. In some cases, a model resolution is consistent with the natural separation between scales of atmospheric motions, but often it is not so. For instance, large scale models with resolution of several tens of kilometers separate the large-scale synoptic motions (average values) and convective-scale motions that are subgrid processes in such models. At the same time, the grid spacing length in the many mesoscale models intended for description of convective processes is of 1–3 km. It means that convective motions of larger scales belong to resolvable scales, while smaller scale convective motions remain irresolvable and are assigned to fluctuations. The utilization of model resolution that is not consistent with the natural separation scales can lead to errors in reproduction of atmospheric motions of both resolvable and subgrid scales. The numerical errors are maximum at scales of double–triple grid spacing lengths (Chapter 4). Thus, utilization of grid spacing lengths of the order of 1 km may introduce significant errors in representation of convective motions with characteristic scales of 2–6 km. Similarly, using a resolution of a few tens of meters does not allow us to assign all turbulent motions to subgrid fluctuations because some turbulent scales are explicitly resolved by the model.

Assuming the optimum averaging intervals are determined and averaged values, and fluctuations are well separated, one can formulate the following rules for averaging:

$$\overline{a_1 + a_2} = \overline{a}_1 + \overline{a}_2$$
$$\overline{\overline{a}} = \overline{a}$$
$$\overline{\frac{\partial a}{\partial x_i}} = \frac{\partial \overline{a}}{\partial x_i}$$
$$\overline{\overline{a} \cdot a_1} = \overline{\overline{a}(\overline{a}_1 + a_1')} = \overline{\overline{a} \cdot \overline{a}_1} + \overline{\overline{a} \cdot a_1'} = \overline{a} \cdot \overline{a}_1$$
$$\overline{a_1 \cdot a_2} = \overline{(\overline{a}_1 + a_1')(\overline{a}_2 + a_2')} = \overline{a}_1 \cdot \overline{a}_2 + \overline{a_1' \cdot a_2'} \ . \tag{3.4.11}$$

All these rules can be derived from the definition of the time- and spatial-averaging operators (Equations 3.4.4 and 3.4.5), as well as from the statistical definitions considered in Section 3.3.

The values of terms of type $\overline{a_1' a_2'}$ (correlations) are often not equal to zero. These terms play a very important role in atmospheric physics. For example, if one assumes $a_1' = w'$ and $a_2' = \theta'$, $\overline{a_1' a_2'} = \overline{w'\theta'}$ is the heat flux through unit of square in the vertical direction due to fluctuations of temperature and vertical velocity. In case of ascending air volumes that are warmer than the surrounding air ($\theta' > 0$), i.e., $w' > 0$, the flux $\overline{w'\theta'} > 0$, and the heat flux is directed upward. The buoyancy force is equal to $\frac{g}{\theta_0}\theta'$. Therefore, $\frac{g}{\theta_0}\overline{w'\theta'}$ is the average work of the buoyancy force caused by temperature fluctuations per unit of time. The terms of type $\overline{a_1' a_2'}$ describe surface fluxes of heat, moisture, and momentum at the land and the ocean surfaces.

3.4.3 Averaging of Budget and Dynamic Equations

It is convenient to perform the averaging if the equations are written in the divergent form. Averaging of the budget Equation (3.2.9) (with fluctuations of density being neglected) using the averaging rules (Equation 3.4.11) yields

$$\frac{\partial \rho \overline{a}}{\partial t} + \frac{\partial \rho \overline{u}_i \overline{a}}{\partial x_i} = \overline{C} - \frac{\partial \rho \overline{u_i' a'}}{\partial x_i} + \frac{\partial}{\partial x_i} D \frac{\partial \rho \overline{a}}{\partial x_i} \tag{3.4.12}$$

One can see that on the right-hand side the averaged equation contains new terms expressing the divergence of fluxes caused by fluctuations. If the averaging scale is chosen to separate convective motions from turbulent motions – which is true for most cloud models – these fluxes are caused by turbulence diffusion. If the averaging scale is between large scales and convective scales, these fluxes are caused by both convective and turbulent fluctuations. The most important term is $\frac{\partial \rho \overline{w'a'}}{\partial z}$, representing the divergence of vertical fluxes $\overline{w'a'}$. In the surface layer, $\overline{w'a'}$ represents the surface flux of a.

Averaging of motion Equation (3.2.26) written in the tensor form leads to the following averaged equation:

$$\frac{\partial \rho_0 \overline{u}_i}{\partial t} + \frac{\partial \rho_0 \overline{u}_i \overline{u}_j}{\partial x_j} = -\frac{\partial \overline{p''}}{\partial x_i} - \frac{\partial \rho_0 \overline{u_i' u_j'}}{\partial x_j} + \varepsilon_{ijk} \rho_0 f_k \overline{u}_j + \delta_{i3} \rho_0 \frac{g}{\theta_0} \overline{\theta''} + \mu \frac{\partial^2 \overline{u}_i}{\partial x_j^2} . \tag{3.4.13}$$

The terms $\tau_{ij} = -\rho_0 \overline{u_i' u_j'}$ are known as the Reynolds stresses. The form of the terms containing the Reynolds stresses is similar to that of the components of stress tensors introduced in Section 3.2. However, the Reynolds stresses are of a different physical nature since they are caused by fluctuations; in particular, by turbulent fluctuations. Within the surface layer, τ_{ij} represent the momentum flux or the surface wind stress, directed from the atmosphere down onto the surface.

The mean products $\overline{u_i' a'}$ in Equation (3.4.12) and $\overline{u_i' u_j'}$ in Equation (3.4.13) are known also as moments of the second order (Section 3.3). Since these moments are unknown, the averaged equation system is not closed. Thus, the problem of the equation closure arises; therefore, the unknown moments of the second order should be expressed as a function of the averaged values. We will show how to perform closure of the first order as well as the closure of so-called 1.5 order.

There are other approaches to closure of the equation system that imply deriving equations for the second-order moments, which contain moments of the third order on the right-hand side. In case the third-order moments are written as functions of the second-order moments, the closure is referred to as closure of the second order. In some models, the closure of the third order is applied, according to which moments of the fourth order are expressed as functions of third-order moments. Since the number of moments of higher orders is significantly larger than that of lower orders, using closure schemes of high order requires more and more equations, which makes the equation system very complicated. Besides, the relationships between moments of different order are not well known. Therefore, although closures of high order (usually of the second order) are used in some atmospheric models, the closure of 1.5 order is the most widespread approach.

3.4.4 Closure of Budget Equations; K-Theory for Conservative and Nonconservative Values

Representation of Turbulent Fluxes $\overline{u_i'a'}$

Let us consider a case when fluctuations represent turbulence only, i.e., averaging using length scale of L_{23} was applied. In most atmospheric models, and cloud models in particular, the expressions for the second-order moments are based on the Prandtl's mixing-length concept. For the sake of simplicity, we consider the averaged equation for variable a containing the divergence of diffusion fluxes along the vertical axis z:

$$\frac{d\bar{a}}{dt} = -\frac{\partial \overline{a'w'}}{\partial z} \tag{3.4.14}$$

To find the expressions for turbulent fluxes $\overline{a'w'}$, we follow the approach proposed by Prandtl (1925) with modifications suggested by Leichtman (1976). The basic assumptions of the Prandtl approach are the following:

- When a turbulent vortex arises, its thermodynamic quantities (temperature, humidity, velocity, etc.) are equal to the corresponding mean values in its surrounding;
- While the vortex is moving over the distance l, known as the mixing length (e.g., Stull, 1988), no mixing with the environment takes place;
- Having moved by the mixing length l, the vortex immediately mixes with its environment. The turbulent fluctuation of the corresponding quantities is defined as the difference between the values of these quantities in the vortex and in the vortex environment.

These assumptions allow us to derive expressions for turbulent fluctuations and then for turbulent fluxes. Although the Prandtl's concept may appear quite simple, it is, in fact, the basis of several important theories, among them the theory of turbulent jets, of the surface layer, of the boundary layer (Garratt, 1992), and a number of other theories. Moreover, the assumptions form the basis for parameterization of turbulent diffusion in all cloud models and other numerical atmospheric models, with the exception of a few models using high-order closure. Many studies formulated requirements to a proper application of the Prandtl concept (e.g., Corrsin, 1974; Stull, 1988; Garratt, 1992). One of the requirements states that the mixing length should be small enough to consider the changes of the background fields at the mixing-length distances as linear changes. This allows us to neglect the curvatures of the flow on the mixing-length spatial scales and to assume that turbulent fluxes are proportional to the gradients of the background (averaged) values.

Let us consider, in order of increasing complexity, several examples where the Prandtl mixing-length concept is applied to calculate turbulent fluxes.

Turbulent Diffusion Fluxes of Conservative Values

This case is relevant, for instance, for the potential temperature in a dry adiabatic process (**Figure 3.4.5**). The turbulent vortex forms at level $z_1 = z_2 - l$ with the averaged quantity $\bar{a}(z_1)$ (point 1) and moves up, reaching level $z = z_2$, where it mixes with the environment. Since a is a conservative value, it does not change during the vortex ascent, and the fluctuation a' at level $z = z_2$ can be written as

$$a'(z_2)_+ = \bar{a}(z_2 - l) - \bar{a}(z_2) = -l\frac{\partial \bar{a}}{\partial z} + \cdots, \tag{3.4.15}$$

where the second-order term and the higher-order terms are omitted. Turbulent fluctuation at level $z = z_2$ (point 2), caused by the vortex descent from level $z_3 = z_2 + l$, is

$$a'(z_2)_- = -a'(z_2)_+ \tag{3.4.16}$$

Since the values are conservative, fluctuations in the updrafts and in the downdraughts are of the same absolute value, but have opposite signs. Turbulent fluxes in the updrafts and downdrafts are equal. In Equations (3.4.15) and (3.4.16), the signs + and – denote the values in the updrafts and the downdrafts, respectively. Since in a homogeneous turbulence $\overline{w'} = 0$, half of the area is covered by updrafts and half of the area is covered by downdrafts, thus, the turbulent fluxes at level $z = z_2$ can be expressed as

$$\overline{w'a'} = \frac{1}{2}\left[\left(\overline{w'a'}\right)_+ + \left(\overline{w'a'}\right)_-\right]. \tag{3.4.17}$$

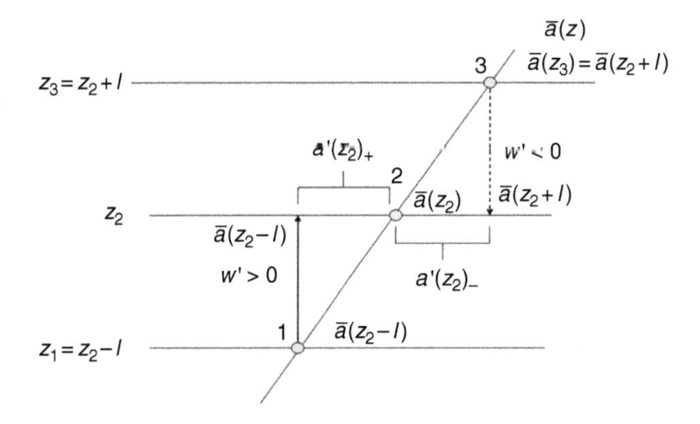

Figure 3.4.5 Formation of turbulent fluctuations at level $z = z_2$ due to turbulent updrafts from level $z_1 = z_2 - l$ and downdrafts from level $z_3 = z_2 + l$.

Using Equation (3.4.15), the turbuSlent flux formed by vortices arising at level $z_1 = z_2 - l$ can be expressed for level z_2 as $\left(\overline{w'a'}\right)_+ = -\overline{lw'}\frac{\partial \bar{a}}{\partial z}$. The same expression describes the turbulent flux transported to level z_2 by vortices located initially at level $z_3 = z_2 + l$ (point 3). Hence, the total turbulent flux is

$$\overline{w'a'} = -\overline{lw'}\frac{\partial \bar{a}}{\partial z} = -K\frac{\partial \bar{a}}{\partial z}. \tag{3.4.18}$$

The value of l is positively correlated with w', which allows us to introduce $K = \overline{lw'} > 0$, where K is the turbulent coefficient (Table 3.3). Equation (3.4.18) represents the K-theory, according to which the turbulent fluxes are proportional to the gradients of the mean quantities.

Turbulent Fluxes of Nonconservative Values

Let us consider a general case when value a in turbulent updrafts changes differently than in downdrafts (**Figure 3.4.6**). Such a situation arises, for instance, when the updrafts occur along the moist adiabat, while the downdrafts occur along the dry adiabat. In this case, fluctuations of value a at level z_2, caused by turbulent updrafts and downdrafts, are different. Correspondingly, the turbulent fluxes in the updrafts and the downdrafts are different, as well.

The turbulent diffusion flux caused by the ascending vortices at level $z = z_2$ is

$$\begin{aligned}\overline{(a'w')}_+ &= -\overline{w'l}\left(\frac{\partial \bar{a}}{\partial z} - \left[\frac{\delta \bar{a}}{\delta z}\right]_+\right)\\ &= -K\left(\frac{\partial \bar{a}}{\partial z} - \left[\frac{\delta \bar{a}}{\delta z}\right]_+\right)\end{aligned} \tag{3.4.19}$$

and the turbulent flux caused by the descending vortices at level z_2 is

$$\begin{aligned}\overline{(a'w')}_- &= -\overline{w'l}\left(\frac{\partial \bar{a}}{\partial z} - \left[\frac{\delta \bar{a}}{\delta z}\right]_-\right)\\ &= -K\left(\frac{\partial \bar{a}}{\partial z} - \left[\frac{\delta \bar{a}}{\delta z}\right]_-\right)\end{aligned} \tag{3.4.20}$$

The total flux at level z_2 is equal to

$$\begin{aligned}\overline{(a'w')}(z_2) &= -\overline{w'l}\left(\frac{\partial \bar{a}}{\partial z} - \frac{1}{2}\left\{\left[\frac{\delta \bar{a}}{\delta z}\right]_+ + \left[\frac{\delta \bar{a}}{\delta z}\right]_-\right\}\right)\\ &= -K\left(\frac{\partial \bar{a}}{\partial z} - \frac{1}{2}\left\{\left[\frac{\delta \bar{a}}{\delta z}\right]_+ + \left[\frac{\delta \bar{a}}{\delta z}\right]_-\right\}\right),\end{aligned} \tag{3.4.21}$$

where $\left[\frac{\delta \bar{a}}{\delta z}\right]_\pm$ are changes of value a due to nonconservativity of a. Equations (3.4.19), (3.4.20), and (3.4.21) reflect the fact that the fluxes in the updrafts and downdrafts are different, thus the procedure of turbulent-diffusion calculation should take into account the direction of turbulent-velocity fluctuations.

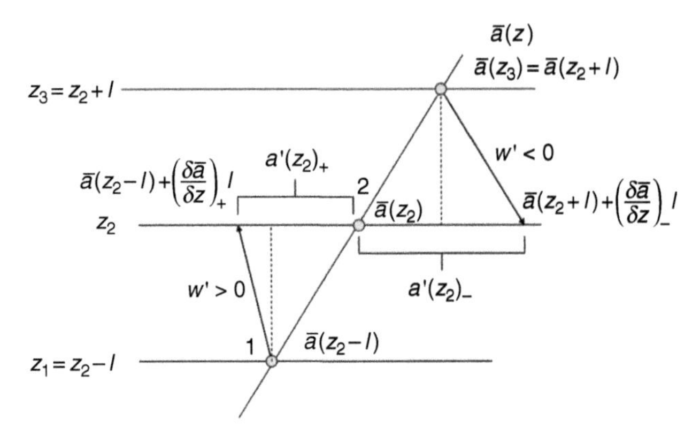

Figure 3.4.6 Scheme illustrating determination of turbulent fluxes and influxes in a nonsymmetric case. Fluctuation $a'(z_2)_+$ is formed at level z_2 by turbulent vortices coming from level z_1. Fluctuation $a'(z_2)_-$ is formed at level z_2 by turbulent vortices coming from level z_3 (from Pinsky et al., 2010a; courtesy of © John Wiley & Sons, Inc.).

In the case of a, changes in turbulent updrafts and downdrafts are similar (for instance, when a is the temperature in both the ascending and descending air volumes that changes according to the dry or the moist adiabat), $\left[\frac{\delta \bar{a}}{\delta z}\right]_+ = \left[\frac{\delta \bar{a}}{\delta z}\right]_-$. In this case, Equation (3.4.21) can be rewritten as

$$\begin{aligned}\overline{(a'w')}(z_2) &= -\overline{w'l}\left(\frac{\partial \bar{a}}{\partial z} - \left[\frac{\delta \bar{a}}{\delta z}\right]\right)\\ &= -K\left(\frac{\partial \bar{a}}{\partial z} - \left[\frac{\delta \bar{a}}{\delta z}\right]\right)\end{aligned} \tag{3.4.22}$$

For instance, if a is the temperature T, as $\frac{\delta \bar{a}}{\delta z} = -\gamma_a$, the temperature turbulent flux can be written as

$$\overline{w'T'} = -K\left(\frac{\partial T}{\partial z} + \gamma_a\right) \approx -K\frac{\partial \theta}{\partial z}, \tag{3.4.23}$$

where θ is the potential temperature. As a result, the averaged equation determining the evolution of temperature can be written as

$$\frac{dT}{dt} = \overline{Q}_{nc} + \frac{\partial}{\partial z}\left(K\frac{\partial \theta}{\partial z}\right), \tag{3.4.24}$$

where $\overline{Q}_{nc}$ is an averaged nonconservative source of T (for instance, latent heat release). In case K is known, expressions (3.4.19), (3.4.20), (3.4.21), (3.4.22), and (3.4.23) represent the closure of the first order.

It should be stressed that turbulent coefficients can be different for different variables. In case a diffusion momentum flux is considered, K is called turbulent viscosity and is often denoted as K_m. In Equation (3.4.23), K is the coefficient of turbulent thermal conductivity, often denoted as K_θ. The ratio $Pr = K_m/K_\theta$ is known as the turbulent Prandtl number in analogy to the laminar Prandtl number ν/κ. While the laminar Prandtl number for air is 0.71, the turbulent Prandtl number varies depending on temperature stratification within the range of 1–3 (Stull, 1988).

3.4.5 Closure of the Motion Equations; Parameterization of the Reynolds Stresses

It is customary to express turbulent wind stresses in the form similar to that used to describe stresses in a laminar flow. Namely, for $\tau_{ij} = -\rho\overline{u_i'u_j'}$ the expressions are similar to Equation (3.2.17):

$$\rho\overline{u_i'u_j'} = -\tau_{ij} = -\rho K_m\left(\frac{\partial u_i}{\partial x_j} + \frac{\partial u_j}{\partial x_i}\right) \tag{3.4.25}$$

There are, however, three very important differences between the representations of turbulent fluxes and laminar fluxes. First, since the turbulent viscosity coefficient K_m is by five to six orders of magnitude higher than molecular viscosity ν, turbulent momentum fluxes are also by several orders of magnitude higher than laminar momentum fluxes. Accordingly, it is possible to neglect the molecular viscosity terms in the motion equations, since the viscosity values are much lower compared to turbulent values everywhere except the lowest laminar layer of ~1 mm depth near the surface. Second, both turbulent coefficients of heat exchange and turbulent viscosity depend on the wind field and the rate of stability/instability of the atmosphere, i.e., on time and spatial coordinates. Third, turbulent viscosity coefficient K_m (as well as turbulent coefficients of heat exchange) are scale dependent and increase with an increasing mixing length according to the Richardson 4/3 law (Table 3.3.1). At the same time, molecular viscosity or heat conductivity do not depend on motion scales. Parameterization of turbulent viscosity coefficient K_m used in numerical models, including cloud models, is presented in Section 3.5.

3.5 Dynamic, Thermodynamic, and Kinetic Equations in Cloud Models

Atmospheric models including cloud models imply solving the system of averaged equations describing thermodynamics, dynamics, and kinetics of atmospheric processes. Summarizing the derivations presented in Section 3.4, we can write down the averaged equations for several thermodynamic quantities, as well as the motion equations. In all of these equations, terms representing molecular diffusion are usually neglected.

3.5.1 Thermodynamic Equations

The equation for equivalent potential temperature $\overline{\theta_e}$ for warm processes has the form

$$\frac{\partial\overline{\theta_e}}{\partial t} + \overline{u_i}\frac{\partial\overline{\theta_e}}{\partial x_i} = \overline{C}_{rad} + \frac{\partial}{\partial x_i}\left(K_\theta\frac{\partial}{\partial x_i}\overline{\theta_e}\right), \tag{3.5.1}$$

where $\overline{C}_{rad}$ is the source/sink existing due to the radiation processes. In cases where ice processes are taken into account, the source term should include also the latent heat caused by water ice transitions.

The equation for the potential temperature can be written as

$$\frac{\partial\overline{\theta}}{\partial t} + \overline{u_i}\frac{\partial\overline{\theta}}{\partial x_i} = \overline{C} + \frac{\partial}{\partial x}\left(K_\theta\frac{\partial}{\partial x}\overline{\theta}\right) + \frac{\partial}{\partial y}\left(K_\theta\frac{\partial}{\partial y}\overline{\theta}\right) + \frac{\partial}{\partial z}\left(K_\theta\left[\frac{\partial}{\partial z}\overline{\theta} - \frac{\delta\overline{\theta}}{\delta z}\right]\right), \tag{3.5.2a}$$

where $\overline{C}$ is the source/sink existing due to condensation/evaporation, sublimation and deposition of ice, as well as to radiation processes. This expression takes into account the fact that the potential temperature is nonconservative in the moist adiabatic process. The last term on the right-hand side reflects the fact that the potential temperature should change during turbulent diffusion. However, in most cloud models the nonconservativity of quantities is ignored in turbulence terms, so the equation for the potential temperature is usually written as

$$\frac{\partial\overline{\theta}}{\partial t} + \overline{u_i}\frac{\partial\overline{\theta}}{\partial x_i} = \overline{C} + \frac{\partial}{\partial x_i}\left(K_\theta\frac{\partial}{\partial x_i}\overline{\theta}\right) \tag{3.5.2b}$$

The equation for the mixing ratio $\overline{q}$ is

$$\frac{\partial\overline{q}}{\partial t} + \overline{u_i}\frac{\partial\overline{q}}{\partial x_i} = \overline{Q_q} + \frac{\partial}{\partial x}\left(K_\theta\frac{\partial}{\partial x}\overline{q}\right) + \frac{\partial}{\partial y}\left(K_\theta\frac{\partial}{\partial y}\overline{q}\right) + \frac{\partial}{\partial z}\left(K_\theta\left[\frac{\partial}{\partial z}\overline{q} - \frac{\delta\overline{q}}{\delta z}\right]\right), \tag{3.5.3a}$$

where $\overline{Q_q}$ is the source/sink existing due to condensation/evaporation as well as deposition/sublimation. Equation (3.5.3a) takes into account the fact that the mixing ratio is a nonconservative value in moist adiabatic processes, changing during turbulent mixing. This fact is often not taken into account. The equation for the mixing ratio is written as

$$\frac{\partial \overline{q}}{\partial t}+\overline{u}_i\frac{\partial \overline{q}}{\partial x_i}=\overline{Q}_q+\frac{\partial}{\partial x_i}\left(K_\theta\frac{\partial}{\partial x_i}\overline{q}\right). \tag{3.5.3b}$$

In order to use a quantity that is conservative with respect to the condensation/evaporation process, the equation for the total water mixing ratio $\overline{q}_t$ ($\overline{q}_t=\overline{q}+\overline{q}_l$) is usually applied:

$$\frac{\partial \overline{q}_t}{\partial t}+\overline{u}_i\frac{\partial \overline{q}_t}{\partial x_i}=\overline{Q}_{qt}+\frac{\partial}{\partial x_i}\left(K_\theta\frac{\partial}{\partial x_i}\overline{q}_t\right), \tag{3.5.4}$$

where $\overline{Q}_{qt}$ is the source/sink existing due to some non-adiabatic processes such as sedimentation of liquid water. All of these equations are derived under the assumption that the coefficients of turbulent heat exchange, turbulent mixing ratio exchange, and liquid water exchange are equal.

In bulk-parameterization cloud models, the microphysical equations should be written for integral values such as mixing ratios of liquid water and of ice water. As an example, we present below two equations written for liquid water mixing ratio $\overline{q}_l$ and ice water mixing ratio $\overline{q}_{iw}$. These equations are typically written in the form

$$\underbrace{\frac{\partial \rho_0\overline{q}_l}{\partial t}+\frac{\partial \rho_0(\overline{u}_i-\overline{V}_{g\_m}\delta_{i3})\overline{q}_l}{\partial x_i}}_{\text{Advection+sedimentation}}=\underbrace{\rho_0\left[\frac{\delta \overline{q}_l}{\delta t}\right]_{c/e}+\rho_0\left[\frac{\delta \overline{q}_l}{\delta t}\right]_{f/m}}_{\text{Microphysics}}+\cdots+\underbrace{\frac{\partial}{\partial x_i}\left(K_\theta\frac{\partial}{\partial x_i}\rho_0\overline{q}_l\right)}_{\text{Turbulent diffusion}} \tag{3.5.5}$$

$$\underbrace{\frac{\partial \rho_0\overline{q}_{iw}}{\partial t}+\frac{\partial \rho_0(\overline{u}_i-\overline{V}_{g\_iw}\delta_{i3})\overline{q}_{iw}}{\partial x_i}}_{\text{Advection+sedimentation}}=\underbrace{\rho_0\left[\frac{\delta \overline{q}_{iw}}{\delta t}\right]_{d/s}+\rho_0\left[\frac{\delta \overline{q}_{iw}}{\delta t}\right]_{f/m}}_{\text{Microphysics}}+\cdots+\underbrace{\frac{\partial}{\partial x_i}\left(K_\theta\frac{\partial}{\partial x_i}\rho_0\overline{q}_{iw}\right)}_{\text{Turbulent diffusion}}. \tag{3.5.6}$$

In Equations (3.5.5) and (3.5.6), $\overline{V}_{g\_m}$ and $\overline{V}_{g\_iw}$ are mass-weighted mean fall velocities of drops and ice hydrometeors, respectively. The first two terms on the right-hand sides of these equations describe the rates of drop condensation/evaporation $\left[\frac{\delta \overline{q}_l}{\delta t}\right]_{c/e}$, ice deposition/sublimation $\left[\frac{\delta \overline{q}_{iw}}{\delta t}\right]_{d/s}$, and two terms corresponding to freezing/melting processes $\left[\frac{\delta \overline{q}_l}{\delta t}\right]_{f/m}$ and $\left[\frac{\delta \overline{q}_{iw}}{\delta t}\right]_{f/m}$ (Chapter 6). The turbulent diffusion terms are written in these equations neglecting the nonconservative nature of these values with respect to phase transitions. In two-moment bulk-parameterization schemes, equations similar to Equations (3.5.5) and (3.5.6) are written also for number concentrations (in units kg^{-1}) using the concentration-averaged fall velocities, as was described in Section 3.2. In cloud models equations similar to Equations (3.5.5) and (3.5.6) are written for mixing ratios of different species, such as cloud droplets, raindrops, ice crystals, aggregates, graupel, hail.

3.5.2 The Averaged Kinetic Equations for Size Distributions

In spectral bin microphysics, the equations for size (or mass) distributions are solved instead of equations for integral quantities such as mass content and number concentrations. The definitions of size distribution functions are given in Section 2.1. In this approach, the entire range of particle sizes is divided into several tens of categories (bins). Each bin contains a certain mass and concentration of particles, so the sum of masses over all the bins is equal to the total mass content of a particular hydrometeor mass.

The averaged kinetic equations for size distributions used in the SBM models are often written in the form

$$\frac{\partial \overline{f}_{nk}}{\partial t}+\frac{\partial(\overline{u}_i-\delta_{i3}V_{g,nk})\overline{f}_{nk}}{\partial x_i}=\left[\frac{\delta \overline{f}_{nk}}{\delta t}\right]_{nucl}+\left[\frac{\delta \overline{f}_{nk}}{\delta t}\right]_{c/e}+\left[\frac{\delta \overline{f}_{nk}}{\delta t}\right]_{d/s}+\left[\frac{\delta \overline{f}_{nk}}{\delta t}\right]_{f/m}+\left[\frac{\delta \overline{f}_{nk}}{\delta t}\right]_{col}\cdots+\frac{\partial}{\partial x_i}\left(K_\theta\frac{\partial}{\partial x_i}\overline{f}_{nk}\right), \tag{3.5.7}$$

where index n denotes the type of hydrometeor (drops, ice crystals, graupel, hail, etc.) and k is the number of mass bin (category) determining the mass of particles. Particles belonging to a certain bin have their own fall velocity, depending on the type of the hydrometeor (including the characteristics of particle shape) and the air density. In Equation (3.5.7), the diffusion term is written neglecting the nonconservative nature of size distributions. A more accurate description of the turbulent diffusion terms taking into account the nonconservativity of size distributions is presented in Section 5.10.

3.5.3 Dynamic Equations

Averaging Equation (3.2.26) leads to the following motion equations written in the tensor form:

$$\frac{\partial \rho_0\overline{u}_i}{\partial t}+\frac{\partial \rho_0\overline{u}_i\overline{u}_j}{\partial x_j}=-\frac{\partial \overline{p''}}{\partial x_i}+\varepsilon_{ijk}\rho_0 f_k\overline{u}_j+\delta_{i3}\rho_0\frac{g}{\theta_0}\overline{\theta''}-\frac{\partial \rho_0\overline{u_i'u_j'}}{\partial x_j}+\mu\nabla^2\overline{u}_i. \tag{3.5.8}$$

Using Equation (3.4.25) and neglecting molecular viscosity term Equation (3.5.8) can be written in the form

$$\frac{\partial \rho_0 \overline{u_i}}{\partial t} + \frac{\partial \rho_0 \overline{u_i u_j}}{\partial x_j} = -\frac{\partial \overline{p''}}{\partial x_i} + \varepsilon_{ijk}\rho_0 f_k \overline{u}_j + \delta_{i3}\rho_0 \frac{g}{\theta_0}\overline{\theta''} + \frac{\partial}{\partial x_j}\rho_0 K_m \left(\frac{\partial \overline{u_i}}{\partial x_j} + \frac{\partial \overline{u_j}}{\partial x_i}\right). \tag{3.5.9}$$

It is noteworthy that in Equation (3.5.8) and Equation (3.5.9), $\overline{p''}$ and $\overline{\theta''}$ are not equal to zero because the values p'' and θ'' are not deviations from averaged values, but deviations from the values in the reference state. The quantities p'' and θ'' contain all the scales and play the role of non-averaged variables.

3.5.4 Equations for Deviations and Second Moments

In some atmospheric models, the equation system is separated into two coupled equation subsystems for averaged values and deviations (Khain and Ingel, 1988, 1995; Ginis et al., 2004; Khairoutdinov et al., 2005; Majda, 2007). In these studies, the averaged values represent a mesoscale flow, while deviations represent convective-scale motions. The equations for deviations are also used to derive equations for second moments, necessary for constructing closure schemes of higher order (Mellor and Yamada, 1982). Therefore, it is useful to formulate the equations for deviation and for moments of the second order to be used for deriving the equation for turbulent kinetic energy.

First, using the initial equation for velocity (Equation 3.2.26) and the averaged Equation (3.5.8), one can write the equations for velocity deviations from the averaged velocity by subtracting the averaged equation from the non-averaged. The equations for i-th and j-th components of velocity deviations are as follows (the terms related to the Coriolis force are neglected since spatial scales of turbulent motions are small):

$$\frac{\partial \rho_0 u'_i}{\partial t} + \frac{\partial \rho_0 u'_i \overline{u_\alpha}}{\partial x_\alpha} + \frac{\partial \rho_0 \overline{u_i} u'_\alpha}{\partial x_\alpha} + \frac{\partial \rho_0 u'_i u'_\alpha}{\partial x_\alpha} = -\frac{\partial p'}{\partial x_i} + \frac{\partial \rho_0 \overline{u'_i u'_\alpha}}{\partial x_\alpha} + \delta_{i3}\rho_0 g \frac{\theta'}{\theta_0} + \mu \frac{\partial^2 u'_i}{\partial x_\alpha^2} \tag{3.5.10}$$

$$\frac{\partial \rho_0 u'_j}{\partial t} + \frac{\partial \rho_0 u'_j \overline{u_\alpha}}{\partial x_\alpha} + \frac{\partial \rho_0 \overline{u_j} u'_\alpha}{\partial x_\alpha} + \frac{\partial \rho_0 u'_j u'_\alpha}{\partial x_\alpha} = -\frac{\partial p'}{\partial x_j} + \frac{\partial \rho_0 \overline{u'_j u'_\alpha}}{\partial x_\alpha} + \delta_{j3}\rho_0 g \frac{\theta'}{\theta_0} + \mu \frac{\partial^2 u'_j}{\partial x_\alpha^2}. \tag{3.5.11}$$

In these equations, $\theta' = \theta'' - \overline{\theta''}$, $p' = p'' - \overline{p''}$, and $\mu = \rho_0 \nu$. Multiplying the Equations (3.5.10) and (3.5.11) by u'_j and u'_i, respectively, then summing up the equations, applying the continuity equation for averaged values and fluctuations $\frac{\partial \rho_0 \overline{u}_\alpha}{\partial x_\alpha} = 0$; $\frac{\partial \rho_0 u'_\alpha}{\partial x_\alpha} = 0$, and averaging the equation thus obtained, we get

$$\frac{\partial \overline{u'_i u'_j}}{\partial t} + \overline{u}_\alpha \frac{\partial \overline{u'_i u'_j}}{\partial x_\alpha} = -\left(\overline{u'_j u'_\alpha}\frac{\partial \overline{u_i}}{\partial x_\alpha} + \overline{u'_i u'_\alpha}\frac{\partial \overline{u_j}}{\partial x_\alpha}\right) - \frac{1}{\rho_0}\left(\overline{u'_i \frac{\partial p'}{\partial x_j}} + \overline{u'_j \frac{\partial p'}{\partial x_i}}\right) - \frac{\partial \overline{u'_i u'_j u'_\alpha}}{\partial x_\alpha} + \frac{g}{\theta_0}\left(\overline{u'_i \theta'}\delta_{j3} + \overline{u'_j \theta'}\delta_{i3}\right) + \nu\left(\overline{u'_j \frac{\partial^2 u'_i}{\partial x_\alpha^2}} + \overline{u'_i \frac{\partial^2 u'_j}{\partial x_\alpha^2}}\right). \tag{3.5.12}$$

Equation (3.5.12) is the equation for the Reynolds (wind) stresses or moments of the second order. This equation is used in many meteorological applications and, in particular, in higher-order closure of the equation system for averaged quantities. The second term on the left-hand side represents advection of $\overline{u'_i u'_j}$. The first term on the right-hand side is the production term showing that the Reynolds stresses arise due to wind shears of the mean (averaged) flow. The second term on the right-hand side describes the work of pressure fluctuations. The third term includes the moment of the third order. The fourth term, $B' = \frac{g}{\theta_0}\left(\overline{u'_i \theta'}\delta_{j3} + \overline{u'_j \theta'}\delta_{i3}\right)$, represents the work of the buoyancy force per unit of air mass per unit of time. The fifth term describes the dissipation due to molecular friction. Equations similar to Equation (3.5.12) can be written for other second-order moments such as $\overline{u'_i \theta'}$, $\overline{u'_i q'}$, $\overline{\theta'^2}$, etc.

3.5.5 Equation for Turbulent Kinetic Energy

Equations (3.5.1–3.5.9) are still not closed because the expressions for turbulent coefficients are not determined. It is usual practice to determine these coefficients using the equation for the kinetic energy of turbulence. This equation is used in cloud models with grid spacing from several tens to a few hundred meters. With such model resolution, turbulence is a subgrid process. The turbulent kinetic energy per unit mass of air is defined as

$$E = \frac{1}{2}\overline{u_i'^2} = \frac{\overline{u'^2 + v'^2 + w'^2}}{2}. \tag{3.5.13}$$

This quantity characterizes the intensity of turbulence and plays a very important role in investigations of atmospheric processes, especially in the boundary layer and in clouds, where turbulence is intense.

The equation for the turbulent kinetic energy is derived from Equation (3.5.12) assuming $i = j$ and dividing each term by factor of 2:

$$\frac{\partial E}{\partial t} + \overline{u}_\alpha \frac{\partial E}{\partial x_\alpha} = -\overline{u_i' u_\alpha'} \frac{\partial \overline{u_i}}{\partial x_\alpha} - \frac{1}{\rho_0} \overline{u_i' \frac{\partial p'}{\partial x_i}} - \frac{1}{2} \frac{\partial \overline{u_i'^2 u_\alpha'}}{\partial x_\alpha} + \frac{g}{\theta_0} \overline{w' \theta'} - \varepsilon, \quad (3.5.14)$$

where ε is the mean turbulent dissipation rate indicating the transformation of the turbulent kinetic energy to heat. This term appears from the equality

$$\varepsilon = -\nu \overline{u_i' \frac{\partial^2 u_i'}{\partial x_\alpha^2}} \approx \nu \overline{\left(\frac{\partial u_i'}{\partial x_\alpha} \right)^2}. \quad (3.5.15)$$

The meaning of the mean turbulent dissipation rate ε was discussed in detail in Section 3.3.

Taking into account the continuity equation, the expression for pressure term can be rewritten as

$$\frac{1}{\rho_0} \overline{u_i' \frac{\partial p'}{\partial x_i}} \approx \frac{1}{\rho_0} \frac{\partial \overline{u_i' p'}}{\partial x_i}. \quad (3.5.16)$$

Integrating the Equation (3.5.16) over any air volume shows that the turbulent kinetic energy within this volume changes due to the work of pressure fluctuations at the volume boundaries. Within the volume, the pressure only redistributes the energy between different directions. These terms are important in zones of inhomogeneous and non-isotropic turbulence. Assuming that turbulence is homogeneous and isotropic, the pressure term in the equation for turbulent kinetic energy is often neglected.

The first term on the right-hand side of Equation (3.5.14) is so-called shear production term $P = \overline{u_i' u_\alpha'} \frac{\partial \overline{u_i}}{\partial x_\alpha}$, showing that the turbulent kinetic energy is produced dynamically by shears of the mean flow. Using Equation (3.4.25) and denoting

$$\tau_{ij} = -\rho K_m \left(\frac{\partial \overline{u_i}}{\partial x_j} + \frac{\partial \overline{u_j}}{\partial x_i} \right) = -\rho K_m D_{ij}, \quad (3.5.17)$$

the shear production term can be written as

$$P = -\overline{u_i' u_\alpha'} \frac{\partial \overline{u_i}}{\partial x_\alpha} = K_m \sum_i \sum_\alpha D_{i\alpha} = K_m D. \quad (3.5.18)$$

Applying K-theory, the work of buoyancy force per unit of time (buoyancy term) in Equation (3.5.14) is typically written as

$$B' = \frac{g}{\theta_0} \overline{w' \theta'} = -K_\theta \frac{g}{\theta_0} \frac{\partial \overline{\theta}}{\partial z} = -\alpha K_m \frac{g}{\theta_0} \frac{\partial \overline{\theta}}{\partial z} \approx -\alpha K_m \frac{g}{\theta_0} (\gamma_a - \gamma) \quad (3.5.19)$$

In Equation (3.5.19)

$$\alpha = Pr^{-1} = \frac{K_\theta}{K_m}, \quad (3.5.20)$$

where Pr is the turbulent Prandtl number, γ_a is the dry adiabatic gradient, and $\gamma = -\frac{d\overline{T}}{dz}$. Equation (3.5.19) is valid only in an undersaturated atmosphere where the potential temperature can be considered a conservative value (Section 3.1). This expression means that $B' > 0$ if $\frac{\partial \overline{\theta}}{\partial z} < 0$, which agrees with the instability condition $\gamma > \gamma_a$. Under subsaturation conditions, gravity waves arise when $\frac{\partial \overline{\theta}}{\partial z} > 0$ or $\gamma < \gamma_a$. If the effects of moisture are taken into account, $\overline{\theta}$ and $\overline{T}$ should be replaced by the corresponding virtual values $\overline{\theta}_v$ and $\overline{T}_v$, respectively.

In clouds, the instability condition is $\gamma > \gamma_{ma}$, where γ_{ma} is the wet adiabatic gradient (Section 3.1). Using the approach discussed in Section 3.4, the work of the buoyancy force within clouds per unit of time can be written as

$$B' = \frac{g}{\theta_0} \overline{w' \theta'} = -K_\theta \frac{g}{\theta_0} \left(\frac{\partial \overline{\theta}}{\partial z} - \frac{\delta \overline{\theta}}{\delta z} \right) = -\alpha K_m \frac{g}{\theta_0} \left(\frac{\partial \overline{\theta}}{\partial z} - \frac{\delta \overline{\theta}}{\delta z} \right) \approx -\alpha K_m \frac{g}{\theta_0} (\gamma_{ma} - \gamma), \quad (3.5.21)$$

where term $\frac{\delta \overline{\theta}}{\delta z}$ shows the changes of the potential temperature during turbulent motions within saturated air. In cases where the virtual temperature corrections and weight of liquid water are taken into account, the expression for the buoyancy term becomes more complicated (Lalas and Einaudi, 1974; Durran and Klemp, 1982):

$$B' = -\alpha K_m \left[\frac{g}{\theta_0} (\gamma_{ma} - \gamma) \left(1 + \frac{L_w q_S}{RT} \right) - \frac{g}{1 + (q + q_l)} \frac{d(q + q_l)}{dz} \right], \quad (3.5.22)$$

where q, q_s, and q_l are the water vapor mixing ratio, the saturated mixing ratio, and the liquid water mixing ratio, respectively. As was shown by Durran and Klemp (1982), the difference between Equations (3.5.22) and (3.5.21) is quite small. In the presence of ice, the effect of ice on the buoyancy should be taken into account.

The moment of the third order $\frac{1}{2} \frac{\partial \overline{u_i'^2 u_\alpha'}}{\partial x_\alpha}$ in Equation (3.5.14) is usually parameterized as

$$\frac{1}{2}\frac{\partial \overline{u_i'^2 u_\alpha'}}{\partial x_\alpha} = -\frac{\partial}{\partial x_\alpha} K_E \frac{\partial E}{\partial x_\alpha}. \tag{3.5.23}$$

Usually the turbulence coefficient K_E is assumed equal to K_m. Substituting Equation (3.5.23) into Equation (3.5.14) and neglecting the pressure term, we obtain

$$\frac{\partial E}{\partial t} + \overline{u_\alpha}\frac{\partial E}{\partial x_\alpha} = K_m D + \frac{\partial}{\partial x_\alpha} K_m \frac{\partial E}{\partial x_\alpha} + B' - \varepsilon, \tag{3.5.24}$$

where, in case of unsaturated air, the work of the buoyancy force should be written as in Equation (3.5.19); while for the inside-cloud case it should be as in Equation (3.5.21) or Equation (3.5.22).

To close Equation (3.5.24), one needs to express the turbulence coefficients and the dissipation rate via the turbulence kinetic energy E. It is widely accepted to assume that these quantities are determined by E and mixing length l. Applying arguments for units (Section 3.3), we obtain

$$\varepsilon = CE^{3/2} l^{-1/2} \tag{3.5.25}$$

$$K_m = C_k l E^{1/2}. \tag{3.5.26}$$

Using expressions (3.5.25) and (3.5.26) is the simplest and most widespread approach. There are more complicated approaches that use prognostic equations for dissipation rate (Pope, 2000).

In numerical models, mixing length is often related to the model grid spacing. For instance, in the Weather Research Forecasting model (WRF) (Skamarock et al., 2005) the mixing length is calculated as

$$l = \begin{cases} \min\left[(\Delta x \cdot \Delta y \cdot \Delta x)^{1/3}, 0.76\dfrac{E^{1/2}}{N_B}\right], & \text{if } N_B^2 > 0 \\ (\Delta x \cdot \Delta y \cdot \Delta x)^{1/3}, & \text{if } N_B^2 \le 0 \end{cases} \tag{3.5.27}$$

and

$$C = 1.9C_k + \frac{(0.93 - 1.9C_k)}{(\Delta x \cdot \Delta y \cdot \Delta x)^{1/3}} l, \tag{3.5.28}$$

where $C_k = 2$ is an empirical coefficient. In Equation (3.5.27), N_B is the Brunt–Väisälä frequency, which in WRF is written as

$$N_B^2 = \frac{g}{\theta_0}\left(\frac{\partial T_v}{\partial z} + \gamma_a\right). \tag{3.5.29}$$

According to Equation (3.5.29), the simplified expression for the Brunt–Väisälä frequency in clouds is (Fraser et al., 1973)

$$N_B^2 = \frac{g}{\theta_0}\left(\frac{\partial T}{\partial z} + \gamma_{ma}\right) = \frac{g}{\theta_0}(\gamma_{ma} - \gamma). \tag{3.5.30}$$

A more exact expression for the Brunt–Väisälä frequency under saturated conditions follows from Equation (3.5.22):

$$N_B^2 = \frac{g}{\theta_0}(\gamma_{ma} - \gamma)\left(1 + \frac{L_w q_s}{RT}\right) - \frac{g}{1 + (q + q_l)}\frac{d(q + q_l)}{dz} \tag{3.5.31}$$

Equation system (3.5.24), (3.5.25), (3.5.26), together with expressions (3.5.29), (3.5.30), and (3.5.31) enables us to solve the equations with respect to the turbulence coefficient both for undersaturated and saturated air. Using the velocity components, the dissipation rate, and the turbulence coefficients known at a time instance t, one can solve Equation (3.5.24) numerically and find the value of the turbulent kinetic energy at the next time step $t + \Delta t$. Knowing $E^{t+\Delta t}$, one can calculate $K_m^{t+\Delta t}$ from Equation (3.5.26) at the next time step $t + \Delta t$.

Calculation of the turbulence coefficients closes the thermodynamic and motion equations. This approach is known as 1.5-order closure. This terminology is related to the fact that this method uses both the equation for the turbulence kinetic energy (which is a moment of the second order) and the K-theory which represents the first-order closure since it directly relates the second moments with the gradients of averaged values. Thus, such parameterization uses some properties of both the first-order closure and the second-order closure.

This approach is time consuming. For this reason, the equation for turbulent kinetic energy is often simplified. Assuming the turbulence to be stationary and isotropic, Equation (3.5.24) can be written in a simplified form as

$$K_m D + B' - \varepsilon = 0, \tag{3.5.32}$$

where D is determined in Equation (3.5.17). This expression indicates a simple balance between the generation of turbulence and its transformation to heat due to friction. From Equations (3.5.25) and (3.5.26) we obtain $\varepsilon \sim K_m^4 l^{-4}$. Substituting this expression into (3.5.32) yields

$$K_m = C_1 l^2 \sqrt{D - \frac{\alpha}{\overline{\theta}} g(\gamma_a - \gamma)} \tag{3.5.33}$$

in dry air and

$$K_m = C_1 l^2 \sqrt{D - \frac{\alpha}{\bar{\theta}} g(\gamma_{ma} - \gamma)} \qquad (3.5.34)$$

in saturated air; coefficient C_1 is a constant. Taking into account the uncertainty related to the definition of the mixing length, this constant is usually assumed equal to one. Expressions (3.5.33–3.5.34) allow calculation of turbulence coefficients without solving the equation for turbulent kinetic energy. They have a simple physical meaning: in case turbulence production is higher than the work against the buoyancy force (if atmospheric conditions are stable), turbulence is generated and $K_m > 0$. Otherwise, turbulence cannot develop, being suppressed by the negative buoyancy. Formally, it means that the expression under the square root becomes negative. In these cases, most atmospheric models assume the turbulence coefficient to be of the minimum value $K_{m\_\min}$ of order of 1 $m^2 s^{-1}$ (e.g., Ginis et al., 2004).

In unstable conditions when $\gamma > \gamma_a$ in undersaturated air and $\gamma > \gamma_{ma}$ in clouds, the buoyancy force contributes to turbulence generation. In case the turbulence is caused mainly by air buoyancy, the regime is called *free convection*, and the corresponding turbulence is known as thermally driven turbulence. In case the production term dominates, turbulence is driven by dynamic factor.

There are several specific cases requiring different expressions of the production term. In a horizontally homogeneous atmospheric boundary layer, only the derivatives with respect to vertical coordinates remain. Besides, the mean vertical velocity is equal to zero. In this case, $D_{11} = D_{12} = D_{22} = D_{33} = D_{12} = 0$ and the production term is

$$P = K_m D = K_m \left[\left(\frac{\partial \bar{u}}{\partial z} \right)^2 + \left(\frac{\partial \bar{v}}{\partial z} \right)^2 \right]. \qquad (3.5.35)$$

When the equation system for averaged values is derived for the boundary layer, the concept of the BL horizontal homogeneity is often used. It means that the averaging of the equations is performed over spatial scales exceeding the convective scale. Accordingly, Equation (3.5.35) does not take into account the effects of convective motions in BL (for instance, convective cells, large eddies, etc.). These effects are determined by the air velocity derivatives in both the vertical and the horizontal directions. To simulate 2D roll vortices, 2D models are applicable (e.g., Khain and Ingel, 1995; Ginis et al., 2004). In 2D models (in the x-z plane), $D_{11} = 2\frac{\partial \bar{u}}{\partial x}$, $D_{33} = 2\frac{\partial \bar{w}}{\partial z}$, $D_{13} = \frac{\partial \bar{u}}{\partial z} + \frac{\partial \bar{w}}{\partial x}$, $D_{12} = D_{22} = D_{23} = 0$, and

$$D = 2\left(\frac{\partial \bar{u}}{\partial x} \right)^2 + 2\left(\frac{\partial \bar{w}}{\partial z} \right)^2 + \left(\frac{\partial \bar{u}}{\partial z} + \frac{\partial \bar{w}}{\partial x} \right)^2. \qquad (3.5.36)$$

It is worthwhile to stress that implementing 1D geometry or 2D geometry does not mean that Equations (3.5.35) and (3.5.36) describe 1D or 2D turbulence, respectively. Both equations describe the properties of 3D turbulence. The values of K_m calculated using Equations (3.5.35) and (3.5.36) parameterize the properties of 3D turbulence, but at different levels of accuracy and sophistication.

3.6 Similarity of Averaged and Non-Averaged Equations

As was discussed in Section 3.3, motions with the Reynolds numbers exceeding the critical values (of about 1,000) are turbulent. Since molecular viscosity of air is very low, the Reynolds numbers characterizing atmospheric motions exceed the critical values by many orders of magnitude. It means that atmospheric motions are always turbulent. In a turbulent flow, air motions are stochastic and represent a composition of motions of different scales that have random amplitudes and phases. At the same time, there exist multiple clearly visible and quite distinguishable atmospheric structures with scales much larger than turbulent ones (e.g., cyclones, clouds, convective cells, etc.), which can be considered as regular coherent structures within turbulent atmosphere. Coherent structures remain clearly distinguished for time periods that are by several orders of magnitude longer than the lifetime of turbulent vortices of the maximum size. It means that these regular structures survive although small-scale motions within and around them try to destroy them. These structures resemble those observed in laminar flows at the *Re* slightly exceeding the critical values.

Figure 3.6.1 shows so-called Karman vortex streets observed in laboratory (a) and in the atmosphere where they arise downstream of an island (b). In the laboratory, the Karman vortex streets were observed within the Reynolds numbers range of 60–5,000 (Schlichting, 1960), i.e., at subcritical and slightly supercritical *Re* values. Another example of similarity between flows observed in the laboratory and in nature is presented in **Figure 3.6.2**. The upper panel shows meanders observed in the laboratory at a slightly supercritical $Re = 4{,}000$; the bottom panel shows meanders arising in the sea behind a ship at $Re = 10^7$. One can see similarity of the flows whose Reynolds numbers differ by several orders of magnitude. In the examples

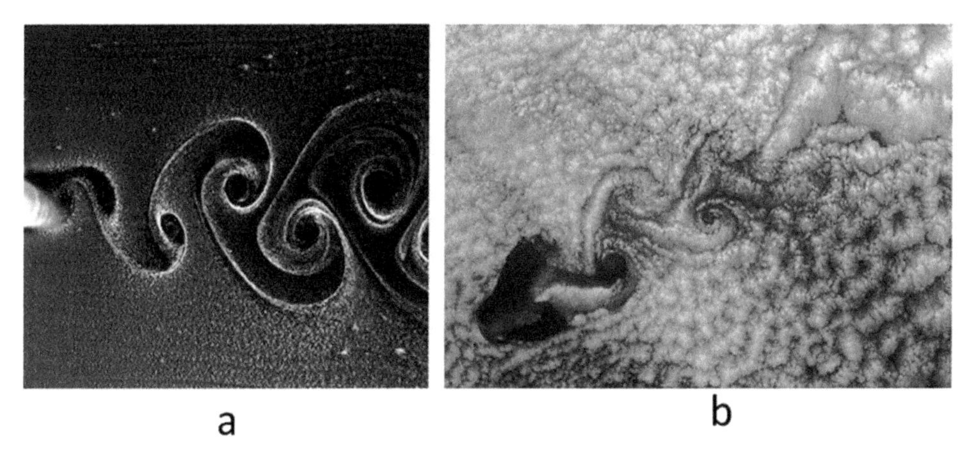

Figure 3.6.1 (a) The Karman vortex street observed in laboratory behind a 6.35 mm diameter circular cylinder in water at a Reynolds number of 168 (photos by Sanjay Kumar and George Laughlin, Department of Engineering, the University of Texas at Brownsville; permission of authors), and (b) satellite image of the cloud pattern downstream of Guadulupe Island off the coast of Baja California from June 11, 2000; (http://www.jpl.nasa.gov/spaceimages/details.php?id=PIA02635; image credit: NASA/GSFC/JPL, MISR Team).

presented the values of the Reynolds numbers were evaluated using the expression

$$Re = \frac{UL}{\nu}, \tag{3.6.1}$$

where U, L, and ν are characteristic velocity, characteristic spatial scale, and molecular viscosity, respectively.

The structures shown in Figures 3.6.1 and 3.6.2 form due to dynamical factors. Likewise, one can observe similarity of structures forming due to thermal mechanisms. For instance, **Figure 3.6.3** shows convective structures (rolls and 3D cells) obtained in a narrow layer of liquid placed between two horizontal plates heated from below (a) and similar structures arising in the atmospheric boundary layer (b).

According to laboratory experiments and the Rayleigh theory (1916), convective structures of this type form when the difference between temperatures of the horizontal boundaries exceeds a certain critical value. Convection arising in such a layer is called the Rayleigh-Benard convection (Emanuel, 1994). The critical temperature gradient written in a nondimensional form is known as the Rayleigh number:

$$Ra = \frac{g}{\theta}\frac{\Delta\theta h^3}{\nu\kappa}, \tag{3.6.2}$$

where $\Delta\theta = \theta_{low} - \theta_{up}$ is the difference in temperatures between the lower and upper boundaries (in the atmosphere this is the difference in potential temperatures), h is the depth of the convective layer, and ν and κ are the coefficients of molecular kinematic viscosity and heat conductivity, respectively. The critical values of Ra_{cr} are about 1,000 and depend on boundary conditions. In the atmospheric boundary layer, Ra determined by means of Equation (3.6.2) exceeds the critical value by several orders of magnitude. This indicates that atmosphere is fully turbulent due to both dynamic and thermodynamic factors. At the same time, phenomena such as convective cells, roll vortices, and cloud streets, resembling cells arising in a laminar flow at slightly supercritical Ra, are quite frequent in the atmospheric boundary layer.

These facts raise two questions: Why do similar structures form in flows that differ dramatically in their scales? and Why do coherent or organized structures form in turbulent flows whose Reynolds and Rayleigh numbers are much higher than critical values? The answer is the following. Phenomena with scales exceeding turbulent scales are described by averaged equations in a form similar to that of the initial non-averaged equations. Amid the similarities, however, there is one important difference: the coefficients of molecular viscosity and molecular heat conductivity in non-averaged equations are replaced by turbulent viscosity and turbulent heat conductivity coefficients (compare the basic equations in Section 3.2 and the averaged equations in Section 3.5). It means that friction forces and heat exchange in the form of heat diffusion fluxes are now determined not by molecular diffusion, but by a much stronger turbulent diffusion. Therefore, instabilities of averaged motions are determined by turbulent Reynolds and Rayleigh numbers,

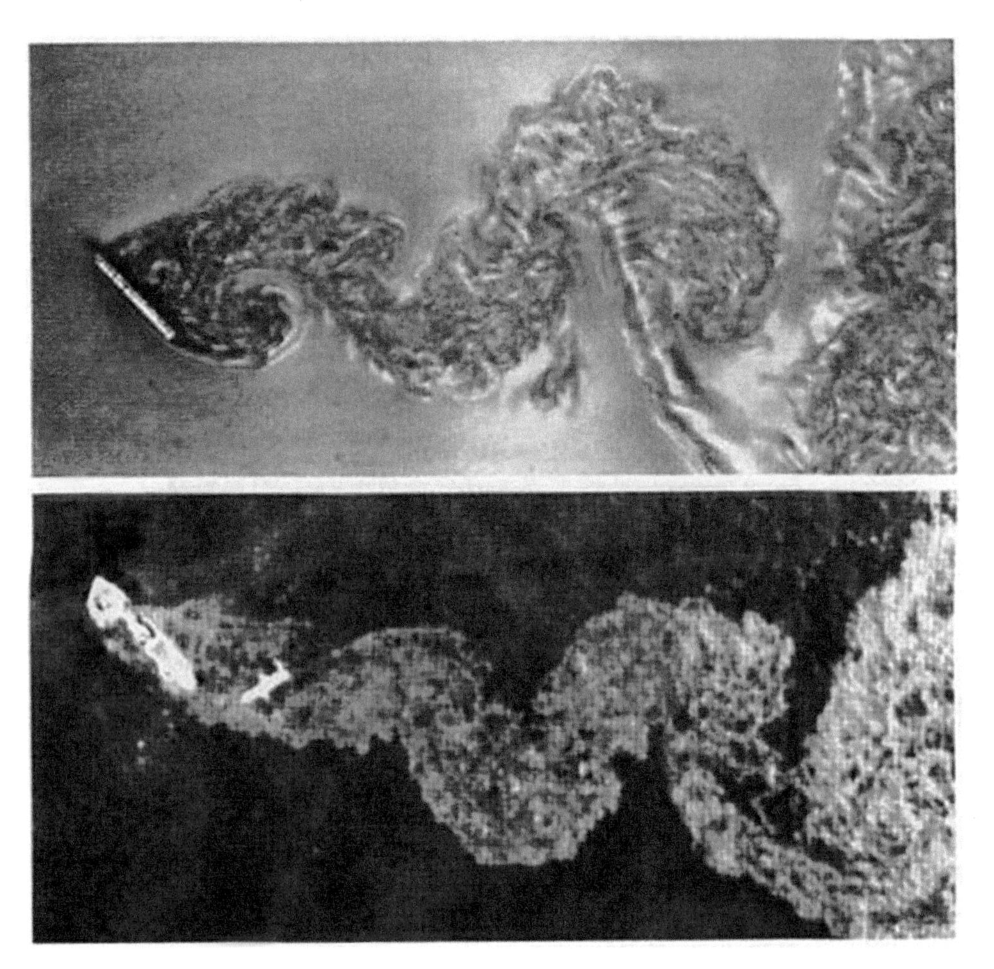

Figure 3.6.2 Top: laboratory flow of aluminum flakes suspended in water past an inclined flat plate. The plate is several centimeters long and the Reynolds number is 4,300. Bottom: Tanker Argo Merchant aground on Nantucket shoals in 1976. The ship is inclined by about 45° in respect to the mean flow, and the leaking oil shows a wake pattern remarkably similar to the one in the top photo. The Reynolds number is ~10^7 (permission of Prof. Cantwell).

and the coefficients of molecular viscosity and heat conductivity in the corresponding equations are replaced by their turbulent analogs K_m and K_θ:

$$Re_t = \frac{UL}{K_m}; \quad Ra_t = \frac{g}{\theta}\frac{\Delta\theta h^3}{K_m K_\theta}. \tag{3.6.3}$$

It should be noted that the critical values of turbulent Reynolds and Rayleigh numbers, defined in Equation (3.6.3), are the same as the critical values of laminar Reynolds and Rayleigh numbers defined in Equations (3.6.1) and (3.6.2). Since K_m and K_θ are by five to six orders of magnitude larger than the corresponding molecular values, the values of Re_t and Ra_t in the atmosphere are quite close to the critical values at which regular atmospheric structures can exist. The substantial difference between ν and κ on the one hand and K_m and K_θ on the other hand is not only in the magnitudes of these values. While the molecular values depend on temperature and type of liquid (e.g., water, air), K_m and K_θ depend on intensity of turbulence determined by the gradients of the background wind and the buoyancy force. Ivanov and Khain (1976) performed numerical calculations of Ra_t values in the atmospheric BL using a numerical BL model. Initially, at t = 0, the air was assumed motionless and the values of the turbulent coefficients were chosen very small, so the initial Rayleigh numbers exceeded the critical values by several orders of magnitude. Formation of air motions with substantial wind gradients led to a rapid

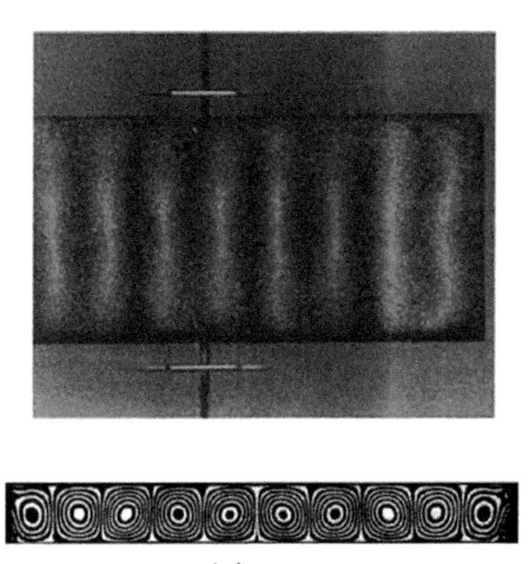

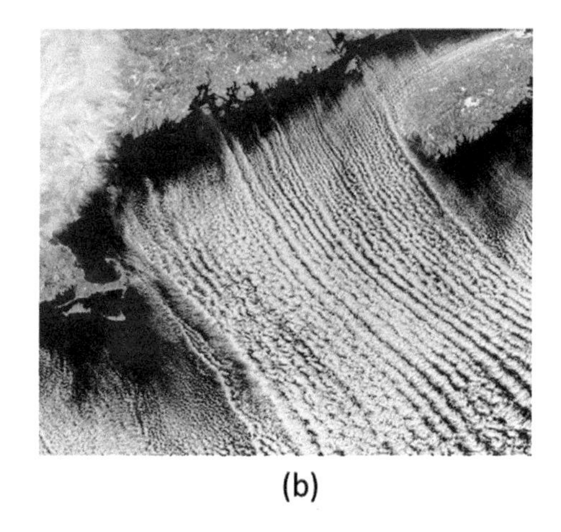

Figure 3.6.3 (a) Laboratory observed the Rayleigh-Benard flow pattern of rolls arising in silicone oil under uniform heating from below. Upper panel: a plan view of roll cells. Lower panel: a side view of convective instability patterns (from Drazin, 2002; courtesy of © Cambridge University Press). (b) Banded clouds seen in MODIS imagery during several days in January and February over the Gulf of Maine, Stack Overflow Team, license CC-BY-SA.

increase of turbulent coefficients K_m and K_θ according to Equation (3.5.33). As a result, the turbulent Rayleigh number rapidly decreased, so when the stationary state was reached, the Rayleigh number exceeded the critical value by a factor as low as three to ten. This feedback is negative since intensifying convection increases turbulence-induced friction, which prevents further intensification of convection. As a result, convective cells formed in these simulations resembled those observed at slightly subcritical Rayleigh numbers.

The analysis of parcel instability presented in Section 3.1 indicates formation of convective motions and, in particular, convective clouds when the temperature gradient exceeds certain critical values ($\gamma > \gamma_a$ in undersaturated air or $\gamma > \gamma_{ma}$ in saturated air). This analysis was performed for comparatively large motion scales, so the temperature gradients should be considered at its averaged values. The difference between the conditions of parcel instability ($\gamma > \gamma_a$) and the Rayleigh–Benard instability ($Ra_t > Ra_{cr}$) is that in the latter case the effects of turbulent viscosity and heat conductivity are taken into account. It is easy to show that the condition $Ra_t > Ra_{cr}$ is almost similar to the condition $\gamma > \gamma_a$ in cases when the atmospheric layers are 1 km deep. In case the depth of a layer is less than 100 m, the condition $Ra_t = Ra_{cr}$ corresponds to values of γ that significantly exceed γ_a.

In the examples presented, we considered atmospheric structures that can be described by averaged equations obtained using the spatial averaging scale L_{23} that separates convective scales from turbulent scales (Section 3.4). With respect to convective motions, turbulent motions are random (chaotic) and their net effect can be expressed by the equations for turbulent friction and turbulent heat conductivity by means of the turbulent coefficients K_m and K_θ. This analogy can also be extended to larger scales. For instance, if averaging of the equations is performed using the averaging scale L_{12}, which separates synoptic and convective scale processes (Section 3.4), convective scales (e.g., clouds) represent random fluctuations with respect to larger scales such as a cyclone.

In mesoscale models with grid spacing exceeding ~10 km, convection is a subgrid process. In this case, the moments of the second order are determined largely by convective motions. In the mesoscale and large-scale atmospheric models, it is accepted to parameterize the moments of second order using expressions similar to those in the K-theory (Section 3.5). In particular, Equation (3.5.33) is used for calculation of effective viscosity and effective heat conductivity. The mixing length is calculated using formula (3.5.27). The coefficient of effective viscosity calculated in this way is not related to turbulence, but rather reflects the aggregated friction or mixing effect of subgrid scales on motions of synoptic scale.

This analogy can be extrapolated toward global climatic structures having spatial scales of several thousands of kilometers and time scales from about a month to a few months. The presence of such structures can be revealed by averaging atmospheric motions in the east–west direction over the globe or over a significant fraction of the globe. Examples of global climatic structures are the Hadley cells in the tropics and the Ferrel cells in

Figure 3.6.4 NOAA GOES-8 satellite image for larger view of parade of hurricanes and tropical storms taken at 10:45 a.m. EDT on August 24, 1995, NOAA News Online (story 2217) (courtesy of © NOAA).

mid-latitudes. With respect to global scale motions, synoptic scale motions such as waves, cyclones, and anticyclones play the role of turbulent vortices providing thermodynamic mixing with respect to motions of the global scale. **Figure 3.6.4** shows formation of several tropical cyclones in the tropical Atlantic in the zones of enhanced vorticity of easterly waves. Using the analogy between a turbulent vortex in a background flow and a tropical cyclone (TC) in a global scale flow, TCs can be characterized by the same parameters as a turbulent vortex. For instance, TCs move by thousands of kilometers during their lifetime, so their mixing length l is of several thousand kilometers. The corresponding viscosity coefficient related to TCs can be evaluated as $K_{eff} \sim \overline{u'l} \sim 10^7 - 10^8$ m^2 s^{-1}, where u' is the mean speed of TC (about 10 m/s). The values of K_{eff} are by several orders of magnitude larger than those of the turbulent viscosity coefficients.

Interestingly, the energy spectrum of atmospheric pressure plotted within a wide range of scales confirms that it is correct to represent mutual interaction between smaller and larger scales in which divergence of the movement on smaller scales creates the effect of the effective friction and the heat conductivity (**Figure 3.6.5**). One can see that the spectrum is close to the $-5/3$ Kolmogorov law (Section 3.3) within ranges of scales far exceeding the turbulent scales.

These examples show that the analogy of averaged equations with the laminar flow equations has a deep physical meaning. The divergence of the second moments actually describes the friction force produced by smaller-scale (subgrid) motions and affects larger-scale background motions. The larger the spatial scale of averaging, the wider the range of fluctuation determines the effect of friction. This allows us to explain formation of regular (coherent) structures in the atmosphere at different scales, characterized by the corresponding instability parameters whose values are close to the critical ones.

A close interaction and energy exchange between motions of different scales often takes place due to instability mechanisms when larger scales become unstable with respect to smaller ones (e.g., when the Reynolds and the Rayleigh numbers calculated with respect to the corresponding scales exceed their critical values). Since there are many types of instability, the critical values of the Reynolds and the Rayleigh numbers are only part of the parameters determining stability/instability of flows.

Atmospheric instabilities are classified into two major types. Fluctuations often gain the energy for growth from the energy of the larger-scale flow. These fluctuations may be caused by *barotropic instability* if they grow at the expense of the kinetic energy of the

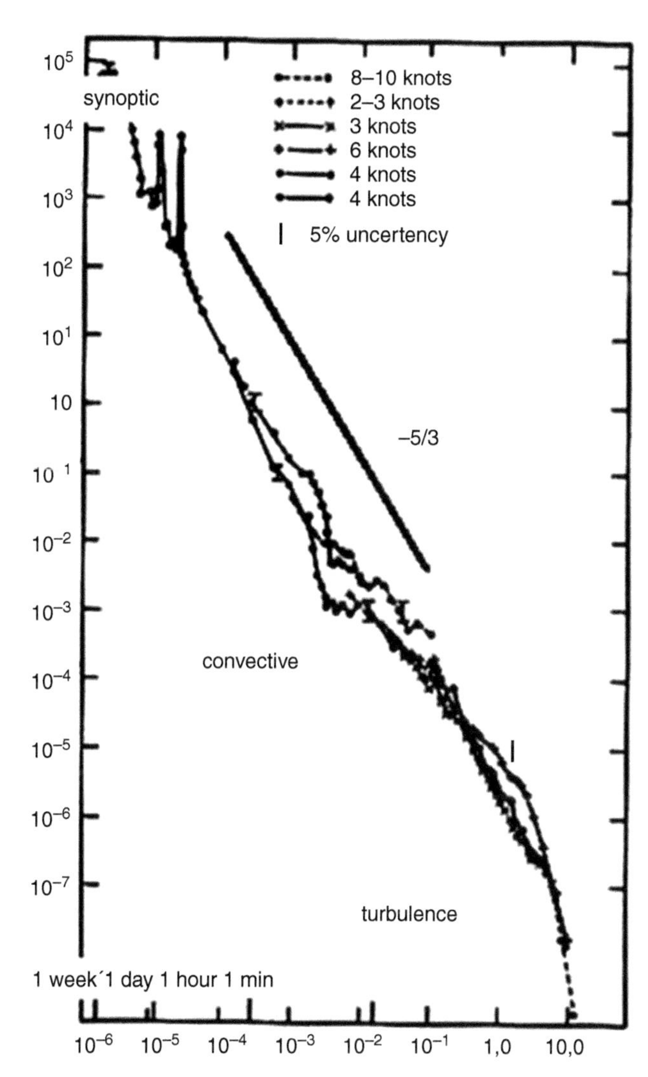

Figure 3.6.5 The energy spectrum of atmospheric pressure fluctuations plotted within a wide range of scales. One can see that the $-5/3$ law works within a range much wider than the inertial turbulent range. The deviations from the $-5/3$ law at convective scales suggest that at these scales the atmosphere obtains its kinetic energy from phase transitions in clouds (from Gossard, 1960; courtesy of © John Wiley & Sons, Inc.).

background large-scale flow, or by *baroclinic instability* if they borrow the potential energy of this flow. Baroclinic instability is typically related to the vertical profiles of thermodynamic values (e.g., temperature), as well as to the horizontal temperature gradients leading to the corresponding pressure gradients. The flow instability when the temperature gradient exceeds the dry adiabatic one is an example of baroclinic instability. Barotropic instability is often caused by wind shears in the horizontal direction, by a certain horizontal profile of absolute vorticity, etc. Hydrodynamic instability arising when *Re* exceeds the critical value is an example of batrotropic instability. Very often instability is caused by both factors. Typically, instability of a flow of a certain scale leads to formation and intensification of motions of smaller scales, which in turn can be unstable with respect to perturbations of even smaller scales, etc.

Barotropic and baroclining instabilities of global-scale motions trigger formation of waves with wavelength of several thousand kilometers (e.g., the Rossby waves in mid-latitude or easterly waves in the tropics) (e.g., Drazin, 2002). The amplitude of easterly waves grows due to both barotropic and baroclinic instability, which in turn leads to formation of zones of enhanced vorticity, giving rise to hurricanes and typhoons with scales of several hundred kilometers. Analyzing motions generated by models of global atmospheric circulation, one can see the formation of vortices (cyclones) that resemble formation of turbulent vortices in smaller-scale streams. Strong gradients of temperature and wind-speed gradients at atmospheric fronts can lead to the formation of more tropical cyclones. Flows of cyclone scales may also be unstable and produce motions of smaller scales. Flows with convective scales of a few kilometers can also be unstable with respect to fluctuations of smaller scales. Motions with scales below several hundred meters that get their energy from larger (convective) scales become highly complicated and fully stochastic. The properties of these motions, called small-scale turbulence, are discussed in Section 3.3.

Motions within a wide range of scales, from cyclones to convective cells, obtain their energy not only from larger-scale motions, but also from convective scales due to latent heat release in clouds, the work of the buoyancy force, etc. It is interesting that interaction of different scales in the atmosphere sometimes leads to a situation where motions of larger scale get their energy from motions of smaller scale. For instance, latent heat release in clouds is the main energy source for tropical cyclones, leading to so-called self-organization.

References

Benmoshe, N., M. Pinsky, A. Pokrovsky, and A. Khain, 2012: Turbulent effects on the microphysics and initiation of warm rain in deep convective clouds: 2-D simulations by a spectral mixed-phase microphysics cloud model. *J. Geophys. Res*, **117**, D06220.

Bizova, N.L., V.N. Ivanov, and E.K. Garger, 1989: *Turbulence in the Boundary Layer of the Atmosphere*. Gidrometeoizdat, p. 263.

Cantwell, B.J. 1981: Organized motion in turbulent flow. *Ann. Rev. Fluid Mech.*, **13**, 457–515.

Corrsin, S., 1974: Limitations of gradient transport in random walks and turbulence. *Adv. In Geophys.*, **18A**, Academic Press.

Cotton, W., and R. Anthes, 1987: *Storm and Cloud Dynamics*. Academic Press, p. 882.

Drazin, A.P., 2002: *Introduction to Hydrodynamic Stability*. Cambridge University Press, p. 263.

Durran, D.R., and J.B. Klemp, 1982: On the effects of moisture on the Brunt-Vaisala frequency. *J. Atmos. Sci.*, **39**, 2152–2158.

Elperin, T., N. Kleeorin, and I. Rogachevskii, 2002: Formation of large scale semiorginized structures in turbulent convection. *Phys. Rev. E,* **66**, 066305, 5–10.

Emanuel, K.A., 1994: *Atmospheric Convection*. Oxford University Press, p. 580.

Flatau, P.J., R.L. Walko, and W.R. Cotton, 1992: Polynomial fits to saturation vapor pressure. *J. Appl. Meteorol.*, **31**, 1507–1513.

Fraser, A.B., R.C. Easter, and P.V. Hobbs, 1973: A theoretical study of the flow of air and fallout of solid precipitation over mountainous terrain. Part 1. Air flow model. *J. Atmos. Sci.,* **30**, 801–812.

Frisch, U., 1995: *Turbulence*. Cambridge University Press, p. 296.

Garratt, J.R., 1992: *The Atmospheric Boundary Layer*. Cambridge University press, p. 316.

Ginis, I., A. Khain, and E. Morosovsky, 2004: Effects of large eddies on the structure of the marine boundary layer under strong wind conditions. *J. Atmos. Sci.*, **61**, 3049–3063.

Gossard, E.E, 1960: Spectra of Atmospheric Scales. *J. Geophys. Res.*, **65**, 3339–3351.

Gossard, E.E., and W.H. Hooke, 1975: *Waves in Atmosphere*. Elsevier, p. 532.

Grabowski, W.W., and T.L. Clark, 1991: Cloud-environment interface instability: Rising thermal calculations in two spatial dimensions. *J. Atmos. Sci.*, **48**, 527–546.

Gulitski, G., M. Kholmyansky, W. Kinzlebach, B. Luthi, A. Tsinober, and S. Yorish, 2007: Velocity and temperature derivatives in high Reynolds number turbulent flows in the atmospheric surface layer. Part. 1. Facilities, methods and some general results. *J. Fluid Mech.*, **589**, 57–81.

Gylfason, A., and Z. Warhaft, 2004: On higher-order passive scalar structure functions in grid turbulence. *Phys. Fluids*, **16**, 4012–4019.

Hill, R.J., 2002: Scaling of acceleration in locally isotropic turbulence. *J. Fluid Mech.,* **452**, 361–370.

Ivanov, V.N., and A.P. Khain, 1976: On characteristic values of Rayleigh numbers during the development of cellular convection in turbulent atmosphere. *Atmospheric and Oceanic Physics*, **12**, 23–28.

Kabanov, A.S., and I.P. Mazin, 1970: The influence of phase transitions on the turbulence in clouds. *Trudy CAO*, **98**, 113–121.

Khain, A.P., and L.Kh. Ingel, 1988: A numerical model of the atmospheric boundary layer above the ocean in the presence of convection. *Atmos. Ocean. Phys.*, **24**, 24–32.

1995: Numerical modeling of interaction of a nonstationary divergent flow with convective processes in the boundary layer over the ocean. *Atmos. Ocean. Phys.*, **31**, 496–506.

Khain, A.P., A. Pokrovsky, M. Pinsky, A. Seifert, and V. Phillips, 2004: Effects of atmospheric aerosols on deep convective clouds as seen from simulations using a spectral microphysics mixed-phase cumulus cloud model. Part 1: Model description. *J. Atmos. Sci.*, **61**, 2963–2982.

Khairoutdinov, M., D. Randall, and C. DeMott, 2005: Simulations of the atmospheric general circulation using a cloud-resolving model as a superparameterization of physical processes. *J. Atmos. Sci.*, **62,** 2136–2154.

Kollias, P., and B.A. Albrecht, 2000: The turbulence structure in a continental stratocumulus cloud from millimeter wavelength radar observations. *J. Atmos. Sci.*, **57,** 2417–2434.

Kollias, P., B.A. Albrecht, R. Lhermitte, and A. Savtchenko, 2001: Radar observations of updrafts, downdrafts, and turbulence in fair-weather cumuli. *J. Atmos. Sci.*, **58,** 1750–1766.

Kolmogorov, A.N., 1941: The local structure of turbulence in incompressible viscous fluid for very large Reynolds numbers. *Dokl. Akad. Nauk SSSR*, **30,** 301–304.

1962: A refinement of previous hypotheses concerning the local structure of turbulence in a viscous incompressible fluid at high Reynolds number. *J. Fluid Mech.*, **13,** 82–85.

Laichtman, D.L. (editor), 1976: *Dynamic Meteorology*. Gidrometizdat, Leningrad, p. 607.

Lalas, D.P., and F. Einaudi, 1974: On the correct use of wet adiabatic lapse rate in stability criteria of a saturated atmosphere. *J. Appl. Meteorol.*, **13**, 318–324.

Lothon, M., D.H. Lenschow, D. Leon, and G. Vali, 2005: Turbulence measurements in marine stratocumulus with airborne Doppler radar. *Q. J. Royal Meteorol. Soc.* **131**, 2063–2080.

Lüthi, B., A. Tsinober, and W. Kinzelbach, 2005: Lagrangian measurement of vorticity dynamics in turbulent flow. *J. Fluid Mech.*, **528**, 87–118.

Majda, A.J., 2007: Multiscale models with moisture and systematic strategies for superparameterization. *J. Atmos. Sci.*, **64**, 2726–2734.

Mazin, I.P., A.Kh. Khrgian, and I.M. Imyanitov, 1989: *Handbook of Clouds and Cloudy Atmosphere.* Gidrometeoizdat.

Mellor, G.L., and T. Yamada, 1982: Development of a turbulence closure model for geophysical fluid problems. *Rev. Geophys. Space Phys.*, **20**, 851–875.

Monin, A.S., and A.M. Yaglom, 971: Statistical Fluid Mechanics, Volume **1**: Mechanics of Turbulence. MIT Press.

Monin, A.S., and A.M. Yaglom, 1975: *Statistical Fluid Mechanics: Mechanics of Turbulence*, vol. **2**, MIT Press.

Murphy, D.M., and T. Koop, 2005: Review of the vapour pressures of ice and supercooled water for atmospheric applications. *Q. J. Royal Meteorol. Soc.*, **131**, 1539–1565.

Muschinski, A., R.G. Frehlich, and B.B Balsley, 2004: Small-scale and large-scale intermittency in the nocturnal boundary layer and the residual layer. *J. Fluid Mech.*, **515**, 319–351.

Panchev, S., 1971: Random Fluctuations in Turbulence. Pergamon Press, p. 444.

Pinsky, M., and A. Khain, 2003: Fine structure of cloud droplet concentration as seen from the Fast-FSSP measurements. Part II: Results of in situ observations. *J. Appl. Meteorol.*, **42,** 65–73.

Pinsky, M., A. Khain, B. Grits, and M. Shapiro, 2006: Collisions of small drops in a turbulent flow. Part III: Relative droplet fluxes and swept volumes. *J. Atmos. Sci.*, **63**, 2123–2139.

Pinsky, M., A. Khain, and H. Krugliak, 2008: Collisions of cloud droplets in a turbulent flow. Part 5: Application of detailed tables of turbulent collision rate enhancement to simulation of droplet spectra evolution. *J. Atmos. Sci.*, **65,** 357–374.

Pinsky, M., A. Khain, and L. Magaritz, 2010a: Representing turbulent mixing of non-conservative values in Eulerian and Lagrangian cloud models. *Quart. J. Roy. Meteorol. Soc.*, **136**, 1228–1242.

Pinsky, M., O. Krasnov, H.W.J. Russchenberg, and A. Khain, 2010b: An investigation of turbulent structure of cloud-capped mixed layer by means of a Doppler radar. *J. Appl. Met. Clim.*, **49**, 1170–1190.

Pinsky, M., M. Shapiro, A. Khain, and H. Wirzberger, 2004: A statistical model of strains in homogeneous and isotropic turbulence. *Physica D.*, **191**, 297–313.

Pope, S.B., 2000: *Turbulent Flows*. Cambridge University Press, p. 771.

Prandtl, L., 1925: Bericht über Untersuchungen zur ausgebildeten Turbulenz. *Z. Angew. Math, Meth.*, **5**, 136–139.

Pruppacher, H.R., and J.D. Klett, 1997: Microphysics of Clouds and Precipitation, 2nd edition. London: Kluwer Academic Publishers.

Rayliegh, L., 1916: On convection currents in a horizontal layer of fluid, when the higher temperature is on the underside. *Phil. Mag.*, **32**, 529.

Reynolds, O., 1895: On the dynamical theory of turbulent incompressible viscous fluids and the determination of the criterion. *Phil Trans. Royal. Soc.*, **186**, 123–161.

Rogers, R.R., and M.K. Yau, 1996: Short Course in Cloud Physics. Butterworth-Heinemann, p. 304.

Schlichting, H., 1960: Boundary Layer Theory. McGraw-Hill Book Company, INC., p. 647.

She, Z., E. Jackson, and S.A. Orszag, 1991: Structure and Dynamics of Homogeneous Turbulence: Models and Simulations. *Proc. R. Soc. Lond. A*. **434**, 101–124.

Shmeter, S.M., 1987: Thermodynamics and Physics of Convective Clouds. Gidrometizdat, p. 288.

Siebert, H., K. Lehmann, and M. Windisch, 2006: Observations of small-scale turbulence and energy dissipation rates in the cloudy boundary layer. *J. Atmos. Sci.*, **63**, 1451–1466.

Siebert, H., R.A. Shaw, and Z. Warhaft, 2010: Statistics of small-scale velocity fluctuations and internal intermittency in marine stratocumulus clouds. *J. Atmos. Sci.*, **67**, 262–273.

Skamarock, W.C., J.B. Klemp, J. Dudhia, D.O. Gill, D.M. Barker, W. Wang, and J.G. Powers, 2005: A description of the Advanced Research WRF Version 2. NCAR Tech Notes-468+STR.

Spyksma, K., and P. Bartello, 2008: Small-scale moist turbulence in numerically generated convective clouds. *J. Atmos. Sci.*, **65**, 1967–1978.

Sreenivasan, K.R., and R.A. Antonia, 1997: The phenomenology of small-scale turbulence. *Annu. Rev. Fluid Mech.*, ***29***, 435–472.

Straka, J.M. 2009: *Cloud and Precipitation Microphysics. Principles and parameterizations*. Cambridge University Press, p. 392.

Strunin, M.A., 2013: *Turbulence in a Cloudy Atmosphere (Clouds and Cloud Environs)*. An empirical model of turbulence in a cloudy atmosphere. Fizmatkniga, p. 191.

Stull, R.B., 1988: *An Introduction to Boundary Layer Meteorology*. London: Kluwer Academic Publishers.

Tripoli, G.J., and W.R. Cotton, 1981: The use of ice-liquid water potential temperature as a thermodynamic variable in deep atmospheric models. *Mon. Wea. Rev.*, **109**, 1094–1102.

Tsinober, A., 2009: An Informal Conceptual Introduction to Turbulence, 2nd edition. Springer, p. 464.

Van der Hoven, I., 1957: Power spectrum of horizontal wind speed in the frequency range from 0.0007 to 900 cycles per hour. *J. Meteorol.*, **14**, 160–164.

Weil, J.C., R.P. Lawson, and A.R. Rodi, 1993: Relative dispersion of ice crystals in seeded cumuli. *J. Appl. Meteorol.*, **32,** 1055–1073.

Wyngaard, J.C., 2010: Turbulence in the Atmosphere. Cambridge University Press, p. 393.

Zilitinkevich, S.S., T. Elperin, N. Kleeorin, V. L'vov, and I. Rogachevskii, 2009: Energy- and flux-budget turbulence closure model for stably stratified flows. Part II: The Role of internal gravity waves. *Boundary-Layer Meteorol.*, **133**, 139. DOI 10.1007/s10546-009-9424-0.

4 Numerical Methods Used in Cloud Models

Numerical methods applied in atmospheric models in general and in cloud models in particular are described in several books (Marchuk, 1974, 1980; Pielke, 2013; Mesinger and Arakawa, 1976; Durran, 2010). In this chapter we present only a brief overview of the main concepts and approaches. The pivot element of a numerical method is a system of averaged equations that describes several basic processes, among them advection of various quantities by a background wind, diffusion, gravity-inertial waves caused by the interaction of fields of pressure, and velocity in the presence of the Coriolis force, etc. (Sections 4.4 and 4.5). With the help of the splitting-up method (Marchuk, 1974, 1980; Temam, 1977), these processes can be treated successively at each time step, so the solution obtained for one process is used as the initial condition for solving the equations describing the next process. This allows for performing numerical treatment of each particular physical process separately.

4.1 Finite-Difference Approximation and Representation of Derivatives

Thermodynamic and dynamic equations formulated in Sections 3.4 and 3.5 are nonlinear differential equations that cannot be solved by analytical methods. Therefore, numerical methods are applied, among them the finite-difference method widely used in most cloud models. According to this method, the entire computational area is covered with a mesh containing numerous grid points. The distance between any two neighboring grid points is called grid spacing length or *grid spacing*. Grid spacings in x, y, and z directions are defined as $\Delta x_i = x_{i+1} - x_i$, $\Delta y_j = y_{j+1} - y_j$ and $\Delta z_k = z_{k+1} - z_k$, respectively. Here, the indices i, j, and k denote the number of a grid point along the corresponding direction. Typically, grid spacing in the horizontal direction is uniform, although some models, for instance, Goddard Cloud Ensemble Model-GCE, use nonuniform grid spacing (Tao et al., 2003). In case of a uniform grid spacing, the horizontal coordinates of a particular grid point are determined as $x_i = \sum_i \Delta x_i = i\Delta x$ and $y_j = \sum_j \Delta y_j = j\Delta y$, where $i = 0, \ldots N$ and $j = 0, \ldots M$. The grid spacing in the vertical direction can be either uniform, e.g., in HUCM (Khain et al., 2004), or nonuniform, e.g., in GCE (Tao et al., 2003). In the nonuniform versions, the grid spacing is small near the surface and increases over height. This approach is applied to obtain a more accurate description of the dynamic processes near the underlying surface where the gradients of meteorological quantities reach their maximum. The vertical coordinate of the k-level is calculated as $z_k = \sum_k \Delta z_k$.

The process of time integration of differential equations is discretized and performed by successively using time steps Δt, which can be either uniform or variable. Different physical processes require different time steps. The ordinal number of a time step (or a time instance) is denoted as τ. Assuming that time steps are uniform, the model time can be calculated as $t_\tau = \tau\Delta t$. According to the finite-difference method, any continuous function $a(x, y, z, t)$ is represented by its discrete analog, defined at grid points i, j, and k and at time t_τ as $a(x_i, y_j, z_k, t_\tau) = a^\tau_{i,j,k}$. Thus, solution of model equations is reduced to determination of values of $a^\tau_{i,j,k}$. **Figure 4.1.1** shows the simplest 1D spatial grid containing $N + 1$ grid points. The values of $a_i^{\tau=0}$ $(0 \le i \le N)$ at time instance $\tau = 0$ are given as the initial conditions. The time integration is reduced to successive calculation of $a_i^{\tau=1}$ at the first time instance, followed by calculation of $a_i^{\tau=2}$, etc., until $a_i^{\tau=T}$.

The dynamic and thermodynamic equations (Sections 3.4 and 3.5) contain the first-order derivative with respect to time and coordinates, as well as the second-order derivatives with respect to coordinates. The differences δ between values of $a^\tau_{i,j,k}$ at different grid points are called finite spatial differences and are used for approximation of the derivatives. The finite spatial differences can be either centered or uncentered, the latter subdivided into forward and backward types. In 1D geometry, depending on the location of grid points with respect to the point where the derivative is to be calculated, the finite differences can be either centered if $\delta a_i = a_{i+1} - a_{i-1}$, or uncentered ($\delta a_i = a_{i+1} - a_i$ is the forward difference and $\delta a_i = a_i - a_{i-1}$ is the backward difference). Centered differences can be used in the form

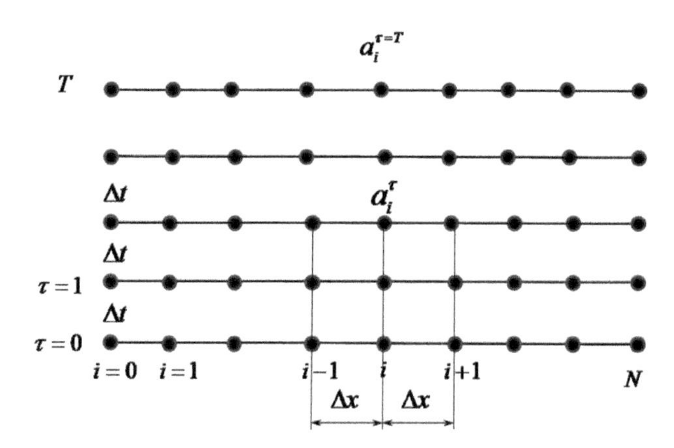

Figure 4.1.1 1D spatial finite–difference grid containing $N+1$ grid points. The values of $a_i^{\tau=0}$ at time instance $\tau = 0$ are given as the initial conditions (the bottom horizontal line). The time integration is reduced to successive calculation of $a_i^{\tau=1}$ at the first time instance (horizontal line the second from below), and is followed by calculation of $a_i^{\tau=2}$, etc., until $a_i^{\tau=T}$(the upper horizontal line).

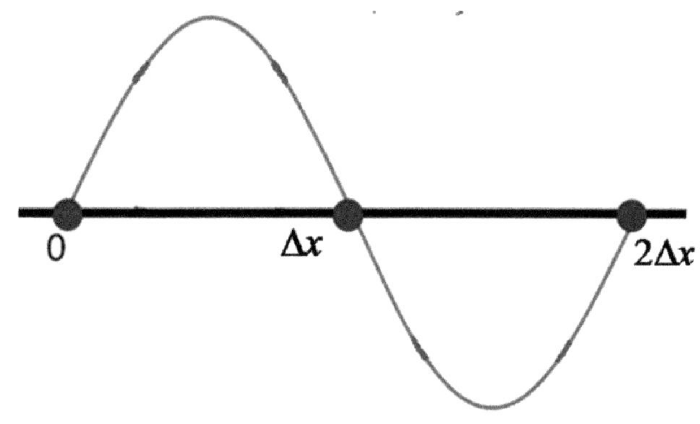

Figure 4.1.2 Loss of approximation consistency at wavelength equal to $2\Delta x$.

of $\delta a_{i+1/2} = a_{i+1} - a_i$. The derivative $\left(\frac{\partial a}{\partial x}\right)_i$ at i-th grid point can be approximated on the mesh by a simple replacement of the derivative for an appropriate *finite difference quotient*, for instance:

$$\left(\frac{\partial a}{\partial x}\right)_i \cong \frac{a_{i+1} - a_i}{\Delta x}, \text{ or } \left(\frac{\partial a}{\partial x}\right)_i \cong \frac{a_i - a_{i-1}}{\Delta x}, \text{ or }$$
$$\left(\frac{\partial a}{\partial x}\right)_i \cong \frac{a_{i+1} - a_{i-1}}{2\Delta x}. \tag{4.1.1}$$

The finite differences presented in Equation (4.1.1) are *forward, backward*, and *centered spatial differences for* approximation to the first derivative, respectively. The finite-difference quotients represent only three of many possible approximations to the first derivative at point i. There are also other expressions for approximation of spatial derivatives.

Finite-difference approximations of the second spatial derivative can be written, for instance, as

$$\left(\frac{\partial^2 a}{\partial x^2}\right)_i \cong \frac{1}{\Delta x}\left(\frac{a_{i+1} - a_i}{\Delta x} - \frac{a_i - a_{i-1}}{\Delta x}\right)$$
$$= \frac{a_{i+1} - 2a_i + a_{i-1}}{(\Delta x)^2}. \tag{4.1.2}$$

Similarly to the finite difference representation of spatial derivatives, time derivatives can be approximated as

$$\left(\frac{\partial a}{\partial t}\right)^\tau \cong \frac{a^{\tau+1} - a^\tau}{\Delta t}, \text{ or } \left(\frac{\partial a}{\partial t}\right)^\tau \cong \frac{a^{\tau+1} - a^{\tau-1}}{2\Delta t}. \tag{4.1.3}$$

4.1.1 Resolvable and Non-Resolvable Scales and the Consistency of Approximation

Utilization of finite differences instead of continuous derivatives inevitably leads to errors in calculations. Therefore, the first requirement to finite-difference schemes is the consistency of approximation. This means that an approximation of a derivative should approach the derivative itself when the grid spacing tends to zero. In view of this requirement, let us consider the finite difference quotients.

Assuming that function $a(x)$ is defined within interval L, this function can be represented by the Fourier series as

$$a(x) = \frac{a_0}{2} + \sum_{n=1}^{\infty}\left(a_n \cos 2\pi n\frac{x}{L} + b_n \sin 2\pi n\frac{x}{L}\right). \tag{4.1.4}$$

Let us assume then that interval L is divided by $N+1$ grid points (N is the number of space increments Δx). Since the function is determined at grid points, only $N+1$ values of function $a(x_i) = a_i$ are known. These $N+1$ values can be used to determine the first $N+1$ different coefficients in this Fourier series, i.e., the values of a_0 and of coefficients a_n and b_n for $n = 1,2,\ldots.N/2$. The shortest wave length of the harmonic with number $n = N/2$ resolved by the grid is

$$L/n = 2L/N = 2\Delta x. \tag{4.1.5}$$

Equation (4.1.5) is of principal importance since no finite difference model is able to represent processes with scales less than $2\Delta x$. At scales below $2\Delta x$, a numerical scheme loses the consistency of its approximation. This loss of consistency is illustrated in **Figure 4.1.2**. Let us assume that function $a(x)$ represents a sinusoidal harmonic of wavelength equal to $2\Delta x$. One can see that at neighboring grid points the values of function $a(x_i) = 0$. In this case, the approximation of the derivative of $a(x)$ at point $x = \Delta x$ is $\frac{a_{i+1}-a_{i-1}}{2\Delta x} = 0$, i.e., the finite difference quotient does not approach the derivative when $\Delta x \to 0$.

4.1.2 The Order of Approximation Accuracy

At scales larger than $2\Delta x$, approximation is consistent. However, approximations (Equation 4.1.1) lead to errors. To evaluate the errors introduced by finite-difference approximations, one can use the Taylor series:

$$a_{i+1} = a_i + \left(\frac{\partial a}{\partial x}\right)_i \Delta x + \left(\frac{\partial^2 a}{\partial x^2}\right)_i \frac{\Delta x^2}{2} + \left(\frac{\partial^3 a}{\partial x^3}\right)_i \frac{\Delta x^3}{6} + O(\Delta x^4) \tag{4.1.6a}$$

$$a_{i-1} = a_i - \left(\frac{\partial a}{\partial x}\right)_i \Delta x + \left(\frac{\partial^2 a}{\partial x^2}\right)_i \frac{\Delta x^2}{2} - \left(\frac{\partial^3 a}{\partial x^3}\right)_i \frac{\Delta x^3}{6} + O(\Delta x^4). \tag{4.1.6b}$$

Using Equations (4.1.6a) and (4.1.6b), one can evaluate the error ε of representation of derivatives (known as the truncation errors). For instance, ε for uncentered difference is

$$\varepsilon = \frac{a_{i+1} - a_i}{\Delta x} - \left(\frac{\partial a}{\partial x}\right)_i = \left(\frac{\partial^2 a}{\partial x^2}\right)_i \frac{\Delta x}{2} + \cdots. \tag{4.1.7}$$

The truncation error gives a measure of accuracy at which the difference quotient approximates the derivative at small values of Δx. This error is represented in the Taylor series with respect to Δx. The truncation error is determined by the term that contains Δx in the lowest power. This minimum power is known as the order of approximation accuracy. Equation (4.1.7) shows that uncentered finite differences approximate the derivatives with the first-order accuracy of approximation, i.e., $\varepsilon = O(\Delta x)$. For the derivative to be consistent, the approximation must be at least of the first order of accuracy.

The first-order approximation is a comparatively crude one. The second-order accuracy of approximation of the spatial derivative can be reached using the centered difference by applying Equation (4.1.6):

$$\varepsilon = \frac{a_{i+1} - a_{i-1}}{2\Delta x} - \left(\frac{\partial a}{\partial x}\right)_i = \left(\frac{\partial^3 a}{\partial x^3}\right)_i \frac{\Delta x^2}{3} + \cdots, \text{ i.e. } \varepsilon = O(\Delta x^2). \tag{4.1.8}$$

One can see that the centered difference approximates the first derivative with second-order accurate approximation, i.e., with accuracy higher than that obtained using uncentered differences. Since we assumed Δx to be small, $O(\Delta x^2) \ll O(\Delta x)$.

The fourth-order accurate approximation provides a significant improvement in the accuracy of approximation of derivatives. By applying the Taylor series to the difference

$$\frac{a_{i+2} - a_{i-2}}{4\Delta x} = \left(\frac{\partial a}{\partial x}\right)_i + \frac{2}{3}\left(\frac{\partial^3 a}{\partial x^3}\right)_i \Delta x^2 + O(\Delta x^4) \tag{4.1.9}$$

alongside Equation (4.1.8), we obtain the expression

$$\left(\frac{\partial a}{\partial x}\right)_i = \frac{4}{3}\frac{a_{i+1} - a_{i-1}}{2\Delta x} - \frac{1}{3}\frac{a_{i+2} - a_{i-2}}{4\Delta x} + O(\Delta x^4), \tag{4.1.10}$$

which represents the fourth-order accurate approximation of the first derivative.

Theoretically, higher-order approximation accuracy should increase the overall accuracy of calculations, but it is not always the case in practice. The reason is that higher-order approximation accuracy requires more grid points in the finite-difference expression. For instance, to reach the fourth-order accuracy in approximation of the first derivative Equation (4.1.10) at the i-th point, it is necessary to calculate the values of the function at points $i-2$, $i-1$, $i+1$, and $i+2$. By way of example, we may assume that the i-th grid point is located near the cloud edge and the grid spacing is 300 m. In this case, the derivative of any variable at the cloud edge will be affected by the values of this variable at distances of 600 m outside the cloud and 600 m inside the cloud. This may decrease the accuracy of reproduction of the derivative, taking into account that in the vicinity of cloud boundaries the values of gradients are high. Therefore, high-order approximations may improve calculation results in cases where the simulated field is comparatively smooth and the grid spacing is small.

Likewise, it is possible to evaluate the order of approximation to the second derivative. It is easy to show that Equation (4.1.2) represents the second order of approximation accuracy for the second derivative.

4.1.3 Schemes of Time Integration

In a general form, the differential equations with respect to time variable can be written as

$$\frac{\partial a}{\partial t} = F(\mathbf{x}, a, t), \tag{4.1.11}$$

where $F(\mathbf{x}, a, t)$ describes all of the terms on the right-hand side of the averaged equations presented in

Chapter 3. In the finite-difference form, these equations can be approximated in different ways. Numerous studies are dedicated to description and analysis of time integration schemes (e.g., Durran, 2010). Here we mention only some of the schemes applied in different cloud models.

Two-level schemes. These schemes relate the values of the dependent variable at two successive time instances, τ and $\tau+1$. Only a two-level scheme can be used to integrate the equation over the first-time step when a single initial condition at $t=0$ is given. The general form of a two-level scheme is

$$\frac{a^{\tau+1}-a^{\tau}}{\Delta t}=\alpha F^{\tau}+(1-\alpha)F^{\tau+1}, \tag{4.1.12}$$

where $1 \geq \alpha \geq 0$ is the weighting coefficient. In case $\alpha=1$, scheme (4.1.12) is called explicit, while at $\alpha \neq 1$ it is implicit. The advantage of explicit schemes is the possibility to calculate the variables at the next time step at each grid point successively.

a) An explicit scheme is the *Euler scheme* (for sake of convenience, indices standing for the spatial coordinates are omitted), which corresponds to $\alpha=1$:

$$a^{\tau+1}=a^{\tau}+\Delta t F^{\tau}, \tag{4.1.13}$$

where $F^{\tau}=F(a^{\tau},\tau\Delta t)$. This scheme is of first order of approximation accuracy $O(\Delta t)$.

b) The simplest two-level implicit scheme is a *backward scheme* (Mesinger and Arakawa, 1976), corresponding to $\alpha=0$:

$$a^{\tau+1}=a^{\tau}+\Delta t F^{\tau+1}, \tag{4.1.14}$$

where $F^{\tau+1}=F\left(a^{\tau+1},(\tau+1)\Delta t\right)$. This scheme is also of first order of approximation. For an ordinary differential equation it is usually simple to calculate $a_{i,j,k}^{\tau+1}$. But in case of partial differential equations, the values of $a_{i,j,k}^{\tau+1}$ at all grid points should be calculated simultaneously by solving a system of algebraic equations, including one equation for each grid point of the computation area and obeying the corresponding boundary conditions.

c) The *trapezoidal scheme* is an implicit scheme and can be obtained from Equation (4.1.12) at $\alpha=1/2$:

$$a^{\tau+1}=a^{\tau}+\frac{\Delta t}{2}\left(F^{\tau}+F^{\tau+1}\right). \tag{4.1.15}$$

This scheme is of second order of approximation accuracy $O\left(\Delta t^2\right)$.

Iterative schemes are widely used to improve the accuracy. The two schemes presented here are constructed in the same way as Equations (4.1.14) and (4.1.15), except that a two-step iterative procedure is performed to make them explicit.

a) The *Matsuno scheme* (also known as the *Euler-backward scheme*). In this scheme, the first step is made using the Euler scheme. The value of a^* obtained for the time instance $\tau+1$ is then used to determine the approximate (intermediate) value $F^* \equiv F(a^*,(\tau+1)\Delta t)$. This approximate value is then used to make a backward second step described by Equation (4.1.14):

$$a^*=a^{\tau}+\Delta t F^{\tau};\quad a^{\tau+1}=a^{\tau}+\Delta t F^*. \tag{4.1.16}$$

This first-order-of-accuracy scheme is explicit. This scheme has been used, for instance, for time integration in HUCM (Khain and Sednev, 1996) and in the convective model of the boundary layer (Ginis et al., 2004).

b) The *Houn scheme*. The first step of this scheme is made using the Euler scheme. The second step is designed on the basis of the trapezoidal scheme:

$$a^*=a^{\tau}+\Delta t F^{\tau};\ a^{\tau+1}=a^{\tau}+\frac{\Delta t}{2}(F^{\tau}+F^*). \tag{4.1.17}$$

The Houn scheme is of second-order-of-approximation accuracy with respect to time while actually remaining explicit.

c) The *Runge–Kutta* iterative schemes have recently been widely used. The two-step second-order Runge–Kutta scheme (RK2) has the form of

$$a^*=a^{\tau}+\frac{\Delta t}{2}F^{\tau};a^{\tau+1}=a^{\tau}+\Delta t F^*. \tag{4.1.18a}$$

The third-order Runge–Kutta scheme (RK3) consists of three steps:

$$a^*=a^{\tau}+\frac{\Delta t}{3}F^{\tau};a^{**}=a^{\tau}+\frac{\Delta t}{2}F^*;a^{\tau+1}=a^{\tau}+\Delta t F^{**}. \tag{4.1.18b}$$

The Runge–Kutta schemes are used for time integration in the Weather Research Forecasting (WRF) model (Skamarock et al., 2005a, b). The Runge–Kutta schemes including four iteration steps (RK4) provide a very high accuracy. A RK4 was used by Pinsky et al. (2007) to calculate the relative motion of two interacting droplets in a turbulent flow.

Three-level schemes. Three-level schemes that are also widely applied in cloud modeling use variable values obtained at three time instances to calculate values of variables at time instance $\tau+1$:

$$\frac{a^{\tau+1}-a^{\tau-1}}{2\Delta t}=\alpha F^{\tau}+(1-\alpha)F^{\tau+1}, \tag{4.1.19}$$

where a is the weighting coefficient indicating the level of implicitly. Now we present two examples of a three-level scheme.

a) The *leapfrog scheme*. The simplest way to evaluate the integral $\int F(\mathbf{x}, a, t)dt$ following from Equation (4.1.11) is to assume that F is a constant value equal to that at the middle of the time instance $2\Delta t$. This scheme is known as centered in respect to time:

$$a^{\tau+1} = a^{\tau-1} + 2\Delta t F^{\tau}. \tag{4.1.20}$$

This scheme has second order of accuracy of approximation over time – $O(\Delta t^2)$ – and is widely used in atmospheric models, e.g., in the mesoscale model MM5 (the Pennsylvania State University/National Center for Atmospheric Research). To start model integration using this scheme, the first time step is performed by the Euler scheme.

b) The *Adams–Bashforth scheme* is based on approximating function F by the value obtained at the center of time instance Δt by means of a linear extrapolation using the values $F^{\tau-1}$ and F^{τ}. This scheme has a form:

$$F^{\tau+1} = F^{\tau} + \Delta t\left(\frac{3}{2}F^{\tau} - \frac{1}{2}F^{\tau-1}\right). \tag{4.1.21}$$

This scheme is of second-order-approximation accuracy.

In addition to the examples described in this section, there are many other ways of constructing numerical schemes for solving Equation (4.1.11) (see review by Durran, 2010). **Table 4.1.1** summarizes the schemes mentioned.

4.2 Equation of Advection. Stability and Errors of Numerical Schemes

The algebraic equation obtained when derivatives in a differential equation are replaced by the appropriate finite-difference approximations is called a finite-difference approximation of a differential equation, or a finite-difference scheme. Let us consider the linear 1D advection equation along x direction:

$$\frac{\partial a}{\partial t} = -u\frac{\partial a}{\partial x}. \tag{4.2.1}$$

This equation describes advection of quantity a at velocity u, which, for sake of simplicity, is assumed constant ($u = const$). Advection of momentum and of different thermodynamic quantities is one of the most important physical processes simulated in atmospheric models in general and in cloud models in particular. By way of example, we consider a finite-difference approximation of Equation (4.2.1) in the form

$$\frac{a_i^{\tau+1} - a_i^{\tau}}{\Delta t} = \begin{cases} -u\dfrac{\left(a_i^{\tau} - a_{i-1}^{\tau}\right)}{\Delta x}, & \text{if } u > 0 \quad (4.2.2a) \\ -u\dfrac{\left(a_{i+1}^{\tau} - a_i^{\tau}\right)}{\Delta x}, & \text{if } u \le 0\cdot \quad (4.2.2b) \end{cases}$$

The scheme (4.2.2) is known as the forward-in-time and upstream advection scheme. The word "upstream" indicates that the space derivative is approximated by the finite differences in the upwind direction from a grid point. This scheme is of first order of accuracy both in space and in time, i.e., the approximation error can be written as $\varepsilon = O(\Delta x) + O(\Delta t)$ or just $\varepsilon = O(\Delta x, \Delta t)$.

Table 4.1.1 Properties of different time integration schemes.

Scheme Name	Type	Order of Accuracy	Equation
	Two-Level schemes		
a) Euler scheme	Explicit	1st	4.1.13
b) Backward scheme	Implicit	1st	4.1.14
c) Trapezoidal scheme	Implicit	2nd	4.1.15
	Iterative schemes		
a) Matsuno scheme	Explicit	1st	4.1.16
b) Houn scheme	Explicit	2nd	4.1.17
c) Runge–Kutta iterative schemes	Explicit	2nd to higher order depending on amount of iterations	4.1.18 a,b
	Three-level schemes		
a) Leapfrog scheme	Explicit	2nd	4.1.20
b) Adams–Bashforth scheme	Explicit	2nd	4.1.21

Equation (4.2.2) is, obviously, only one of many possible finite-difference schemes for the differential Equation (4.2.1).

4.2.1 Stability Analysis

The error ε of a solution decreases with a decrease in grid spacing. To provide the condition $\varepsilon \to 0$ if $\Delta x \to 0$, the scheme should have approximation accuracy of not less than first order. However, the consistency of the scheme does not guarantee that the solution of a particular numerical scheme converges to that of differential equation. The numerical solution converges to the differential equation solution if the scheme is stable. For instance, if the solution of the original differential equations does not tend to infinity, the stability of the scheme requires the numerical solution also be limited. Stability of a scheme is a property of great practical importance. There are consistent schemes of high order of accuracy that still yield solutions diverging from the true one. Thus, one should know the stability conditions of a scheme to be used.

There are several methods for analyzing the stability of finite-difference schemes (Mesinger and Arakawa, 1976). We illustrate the stability analysis using a widespread method known as the von Neumann method. It is used in analysis of linearized versions of corresponding differential equations. A solution of a linear equation can be expressed in the form of a Fourier series (Equation 4.1.4), where each harmonic component is also a solution. Thus, this method tests the stability of a single harmonic solution. Stability of all admissible harmonics will then be a necessary condition for stability of the scheme as a whole.

We are looking for a solution of Equation (4.2.2) in the form of a plane wave:

$$a_i^\tau = a(i\Delta x, \tau\Delta t) = A(k,\omega)\exp\left\{\hat{i}\,(ki\Delta x + \omega\tau\Delta t)\right\}, \tag{4.2.3}$$

where $\hat{i} = \sqrt{-1}$, k is the wave number, and $\omega = \omega_r + \hat{i}\omega_{im}$ is the complex angular frequency with the corresponding real part ω_r and imaginary part ω_{im}. It should be noted that only the real part of the equation has a physical meaning. The expressions for values of variable a at the next time step $a_i^{\tau+1}$ and at the grid point $i-1$ a_{i-1}^τ at time τ follow from Equation (4.2.3):

$$a_i^{\tau+1} = a[i\Delta x, (\tau+1)\Delta t] = a_i^\tau \exp\left\{\hat{i}\,(\omega\Delta t)\right\} = \lambda_0 a_i^\tau \tag{4.2.4}$$

$$a_{i-1}^\tau = a[(i-1)k\Delta x, \tau\Delta t] = a_i^\tau \exp\left(-\hat{i}\,k\Delta x\right), \tag{4.2.5}$$

where

$$\lambda_0 = \exp\left\{\hat{i}\,(\omega_r\Delta t)\right\}\exp(-\omega_{im}\Delta t) = \lambda\exp\left\{\hat{i}\,(\omega_r\Delta t)\right\} \tag{4.2.6}$$

and

$$\lambda = \exp(-\omega_{im}\Delta t). \tag{4.2.7}$$

It follows from Equation (4.2.4) that

$$\left|a_i^{\tau+1}\right| = |\lambda_0|\left|a_i^\tau\right|. \tag{4.2.8}$$

The value $|\lambda_0|$ represents an *amplification factor* by which the amplitude of the numerical solution changes during one time step. During τ time steps, the amplitude of the solution changes by factor $|\lambda_0|^\tau$. In case $|\lambda_0| > 1$, the amplitude increases exponentially with time. If a solution of a differential equation is time-limited, the condition $|\lambda_0| > 1$ indicates instability of the numerical scheme leading to rapid divergence between the exact solution and the numerical solutions. In order to prevent the tendency of numerical solutions to unlimited growth, the condition of the scheme stability

$$|\lambda_0| = |\lambda| \le 1 \tag{4.2.9}$$

should be obeyed. According to the analytical solution of Equation (4.2.1) – $x - ut = const$ representing moving of the initial condition $a_i^{\tau=0}$ along x-direction at velocity u (the corresponding straight line $x = ut + const$ is known as a characteristic of the equation) – the amplitude of the solution should not change, so the exact solution obeys condition $|\lambda_0| = 1$ (the neutral scheme).

Let us consider the case $u > 0$ in Equation (4.2.2a), which is similar to condition $u < 0$ in Equation (4.2.2b). Substituting Equations (4.2.3), (4.2.4), (4.2.5), and (4.2.6) for Equation (4.2.2a) yields

$$\lambda\exp\left\{\hat{i}(\omega_r\Delta t)\right\} - 1 = -\frac{u\Delta t}{\Delta x}\left[1 - \exp\left(-\hat{i}k\Delta x\right)\right]. \tag{4.2.10}$$

The equations for the real part and the imaginary part of Equation (4.2.10) are

$$\begin{aligned}\lambda\cos\omega_r\Delta t &= 1 - C_0(1 - \cos k\Delta x)\\ \lambda\sin\omega_r\Delta t &= -C_0\sin k\Delta x,\end{aligned} \tag{4.2.11}$$

where

$$C_0 = \frac{u\Delta t}{\Delta x} \tag{4.2.12}$$

is the *Courant number*, which plays a crucial role in the theory of finite-difference schemes determining their stability. Squaring and summering Equation (4.2.11) yields

$$|\lambda_0|^2 = |\lambda|^2 = 1 + 2C_0(\cos k\Delta x - 1)(1 - C_0), \quad (4.2.13)$$

i.e.,

$$\lambda_0 = \pm\sqrt{1 + 2C_0(\cos k\Delta x - 1)(1 - C_0)}. \quad (4.2.14)$$

Analysis of Equation (4.2.14) shows that $|\lambda_0| \leq 1$ if $C_0 \leq 1$. Thus, the scheme is stable if $C_0 \leq 1$ and becomes unstable if $C_0 > 1$. The numerical schemes that are stable only when C_0 is smaller than a critical value C_{0_cr} are known as *conditionally stable*. For such schemes, the stability condition is

$$C_0 = \frac{u\Delta t}{\Delta x} < C_{0\_cr}. \quad (4.2.15)$$

Equation (4.2.15) is known as the Courant–Friedrichs–Lewy condition (CFL condition). The value of C_{0_cr} depends on a particular equation or equation system to be solved and does not depend either on Δt or on Δx. In many cases (for instance, for the scheme considered here), $C_{0\_cr} = 1$.

Unequality (4.2.15) should be taken into account while choosing Δt under given Δx and u. In this unequality, u has to be evaluated as a maximum velocity that can be described by the model equations. For instance, in spite of the fact that the typical wind speed in the atmosphere is 10–20 ms^{-1}, the value Δt has to be chosen, taking into account that the speed of sound waves is ~330 ms^{-1}. In case the equation system that is used describes acoustic waves, the CFL condition represents a very limiting constraint on the time step Δt. It should be noted that the CFL condition shows that the improvement of model resolution (a decrease in the grid spacing Δx) requires utilization of a smaller time step to keep the scheme stable, which makes the high-resolution computations very time consuming. To increase a time step during the integration of the model equations, different approaches are used. For instance, neglecting the term $\frac{\partial \rho}{\partial t}$ in the continuity equation excludes the sound waves from the solutions of model equations. For this reason, the continuity equation in the form $\frac{\partial \rho u_i}{\partial x_i} = 0$ is often used for the description of convective processes (e.g., Khairoutdinov and Randall, 2003; Khain et al., 2004).

The CFL conditions are typically applied to explicit schemes. Implicit schemes are often stable under any conditions, i.e., they are absolutely stable. However, utilization of implicit schemes is computationally less convenient and more complicated than that of explicit schemes. The way of treating sound waves in modern cloud-resolved models is described in Section 4.4.

In case 2D equation of advection is used, the stability condition is expressed as proposed by Mesinger and Arakawa (1976):

$$\sqrt{2}\sqrt{u^2 + v^2}\frac{\Delta t}{\Delta x} \leq 1, \quad (4.2.16)$$

where u and v are wind components in the corresponding directions.

4.2.2 Amplitude Errors; Numerical Viscosity

As mentioned, condition $|\lambda_0| = 1$ corresponds to the case when the solution does not decay and does not amplify (in agreement with the solution of the differential Equation 4.2.1). Thus, in case $|\lambda_0| = 1$ there is no amplitude error. If $|\lambda_0| < 1$, the value of $|\lambda_0|$ is the measure of the amplitude error of the numerical scheme. Since condition $|\lambda_0| < 1$ corresponds to a decrease in the amplitude, it indicates the existence of numerical viscosity in the scheme. **Figure 4.2.1** shows the dependence of $|\lambda|^2$ in Equation (4.2.13) on C_0 for different wavelengths $L = \frac{2\pi}{k}$. One can see that the rate of amplitude attenuation decreases with a decrease in the model grid spacing, i.e., the amplitude of a harmonic with wavelength of $20\Delta x$ does not experience any significant attenuation. The attenuation rate increases with decreasing wavelengths (i.e., the scheme damps the short waves stronger). The largest errors occur for waves with the shortest wavelength of $2\Delta x$.

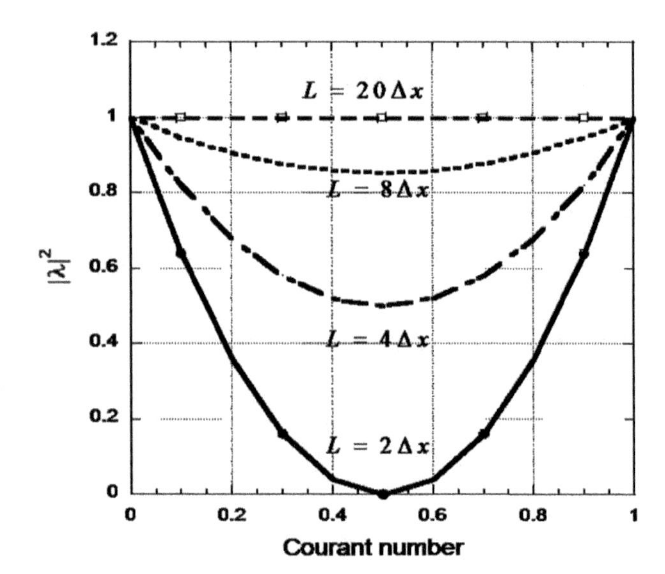

Figure 4.2.1 Dependencies of $|\lambda|^2$ in Equation (4.2.13) on C_0 for different wavelengths, illustrating the amplitude error.

Figure 4.2.1 also indicates that the amplitude error of the finite difference schemes depends on the wavelength. The amplitude errors for different schemes are presented in **Table 4.2.1**.

The concept of numerical viscosity can be demonstrated using the same example of the finite difference representation of advection Equation (4.2.2). Applying the truncated Taylor series approximation to values $a_i^{\tau+1}$ and a_{i-1}^{τ} we get $a_i^{\tau+1} \cong a_i^{\tau} + \left(\frac{\partial a}{\partial x}\right)_i \Delta t + \left(\frac{\partial^2 a}{\partial t^2}\right)_i \frac{\Delta t^2}{2}$ and $a_{i-1}^{\tau} \cong a_i^{\tau} - \left(\frac{\partial a}{\partial x}\right)_i \Delta x + \left(\frac{\partial^2 a}{\partial x^2}\right)_i \frac{\Delta x^2}{2}$. Substitution of these expressions into Equation (4.2.2a) (for $u > 0$) leads to the following equation for value a at point i and at time instance τ:

$$\frac{\partial a}{\partial t} + u\frac{\partial a}{\partial x} = \frac{1}{2}u\frac{\partial^2 a}{\partial x^2}\Delta x - \frac{1}{2}\frac{\partial^2 a}{\partial t^2}\Delta t \qquad (4.2.17)$$

Equation (4.2.17) shows that due to numerical reasons (utilization of finite time steps and finite grid spacing) the equation to be solved contains additional terms on the right-hand side, which results from using forward upstream differencing. If $\Delta x \to 0$ and $\Delta t \to 0$, Equation (4.2.17) reduces to the proper differential Equation (4.2.1). Using the original expression for the advection (4.2.1), one can obtain expressions for the computational diffusion terms. Indeed, from Equation (4.2.1) it follows that $\frac{\partial^2 a}{\partial t^2} = -u\frac{\partial}{\partial t}\frac{\partial a}{\partial x} = -u\frac{\partial}{\partial x}\frac{\partial a}{\partial t} = u^2\frac{\partial^2 a}{\partial x^2}$. Substituting this expression into Equation (4.2.17), we get the following equation containing the term of the same form as the diffusion term:

$$\frac{\partial a}{\partial t} + u\frac{\partial a}{\partial x} = K_{num}\frac{\partial^2 a}{\partial x^2}, \qquad (4.2.18)$$

where K_{num} is referred to as the coefficient of numerical viscosity (or the coefficient of numerical diffusion)

$$K_{num} = \frac{1}{2}u\Delta x\left(1 - \frac{u\Delta t}{\Delta x}\right). \qquad (4.2.19)$$

One can see that $K_{num} = 0$ when $C_0 = 1$.

In practice, due to many reasons that may arise in numerical modeling, the velocities may increase during numerical simulations to unrealistically high values. The common approach preventing this velocity growth is to decrease the time step without changing the grid spacing. Equation (4.2.19) shows that decreasing Δt increases computational viscosity, which explains the desirable decrease in velocity. Evaluations show that the numerical viscosity of the scheme (4.2.2) is significant. Indeed, if $u = 10\,\mathrm{ms}^{-1}$, Δx is 500 m, and $C_0 = 0.5$ $K_{num} = 1250\,\mathrm{m}^2\mathrm{s}^{-1}$, which is more than by order of magnitude larger than the values of the typical physical turbulent coefficients.

Most of the iteration schemes considered in Section 4.1 also have numerical viscosity. For instance, the scheme suggested by Matsuno (Equation 4.1.16) for the advection equation can be written as

$$\frac{a_i^* - a_i^{\tau}}{\Delta t} = -u\frac{a_{i+1}^{\tau} - a_{i-1}^{\tau}}{2\Delta x} \qquad (4.2.20)$$

$$\frac{a_i^{\tau+1} - a_i^{\tau}}{\Delta t} = -u\frac{a_{i+1}^* - a_{i-1}^*}{2\Delta x}. \qquad (4.2.21)$$

Substitution of values a_{i+1}^* and a_{i-1}^* from Equation (4.2.20) written for $i+1$ and $i-1$ into Equation (4.2.21) leads to the equation:

$$\frac{a_i^{\tau+1} - a_i^{\tau}}{\Delta t} = -u\frac{a_{i+1}^{\tau} - a_{i-1}^{\tau}}{2\Delta x} + u^2\Delta t\frac{a_{i+2}^{\tau} - 2a_i^{\tau} + a_{i-2}^{\tau}}{(2\Delta x)^2} \qquad (4.2.22)$$

Without the last right-hand-side term, Equation (4.2.22) represents the Euler scheme, which is unstable (Pielke, 2013). The second right-hand term tends to $u^2\Delta t\frac{\partial^2 a}{\partial x^2}$ when $\Delta x \to 0$. Thus, this finite difference approximation contains the diffusion term with the effective coefficient of computational viscosity $K_{num} = u^2\Delta t$. Assuming as we did in the previous example that the Courant number is equal to 1/2, we have $K_{num} = \frac{u\Delta x}{2}$, which is even higher than it follows from Equation (4.2.19). The viscosity term makes the scheme conditionally stable. The damping effect of scheme (4.2.22) occurs for a wave of length $4\Delta x$. At the same time, there is no attenuation of the shortest wave of wavelength $2\Delta x$. Even if a damping effect is desirable when solving the advection equation, this particular dependence on wavelength is not desirable. This scheme is, therefore, not suitable for accurate description of the advection process. In principle, computational viscosity is not a desirable property in numerical schemes. It is desirable that the physical coefficients of turbulent viscosity should be higher than the coefficients of computational viscosity. However, in many cases, utilization of numerical schemes with the numerical viscosity helps to suppress undesirable small-scale fluctuations.

Different numerical schemes have different computational viscosity. There are neutral stability schemes with $|\lambda| = 1$ that are free from numerical viscosity. One such scheme is the leapfrog scheme (e.g., Pielke, 2013):

$$\frac{a_i^{\tau+1} - a_i^{\tau-1}}{2\Delta t} = -u\frac{a_{i+1}^{\tau} - a_{i-1}^{\tau}}{2\Delta x}. \qquad (4.2.23)$$

Table 4.2.1 Values of amplitude errors and phase errors per a time step as a function of wavelength for different computational approximations of the advection equation (from Pielke, 2013, with changes; courtesy of Elsevier).

Scheme		Wavelength	0.001	0.01	0.1	0.2	0.3	0.4	0.5	0.6	0.7	0.8	0.9	1.0	1.1
I. Forward-in-time linear interpolation upstream	λ	$2\Delta x$	0.998	0.980	0.800	0.600	0.400	0.200	0.000	0.200	0.400	0.600	0.800	1.000	$\lvert\lambda\rvert > 1$
		$4\Delta x$	0.999	0.990	0.906	0.825	0.762	0.721	0.707	0.721	0.762	0.825	0.906	1.000	
		$10\Delta x$	1.000	0.998	0.983	0.969	0.959	0.953	0.951	0.953	0.959	0.969	0.983	1.000	
		$20\Delta x$	1.000	1.000	0.996	0.992	0.990	0.988	0.988	0.988	0.990	0.992	0.996	1.000	
	$\tilde{c}_\phi/U$	$2\Delta x$	0.000	0.000	0.000	0.000	0.000	0.000	1.000	1.667	1.429	1.250	1.111	1.000	
		$4\Delta x$	0.637	0.643	0.704	0.780	0.859	0.936	1.000	1.043	1.060	1.055	1.033	1.000	
		$10\Delta x$	0.936	0.937	0.953	0.968	0.981	0.992	1.000	1.005	1.008	1.008	1.005	1.000	
		$20\Delta x$	0.984	0.984	0.988	0.992	0.995	0.998	1.000	1.001	1.002	1.002	1.001	1.000	
II. Centered-in-time, centered-in-space (leapfrog)	λ	$2\Delta x$	1.000	1.000	1.000	1.000	1.000	1.000	1.000	1.000	1.000	1.000	1.000	1.000	$\lvert\lambda\rvert > 1$
		$4\Delta x$	1.000	1.000	1.000	1.000	1.000	1.000	1.000	1.000	1.000	1.000	1.000	1.000	
		$10\Delta x$	1.000	1.000	1.000	1.000	1.000	1.000	1.000	1.000	1.000	1.000	1.000	1.000	
		$20\Delta x$	1.000	1.000	1.000	1.000	1.000	1.000	1.000	1.000	1.000	1.000	1.000	1.000	
	$\tilde{c}_\phi/U_{\text{physical mode}}$	$2\Delta x$	0.000	0.000	0.000	0.000	0.000	0.000	0.000	0.000	0.000	0.000	0.000	0.000	
		$4\Delta x$	0.637	0.637	0.638	0.641	0.647	0.655	0.667	0.683	0.705	0.738	0.792	1.000	
		$10\Delta x$	0.935	0.935	0.936	0.938	0.940	0.944	0.950	0.956	0.964	0.974	0.986	1.000	
		$20\Delta x$	0.984	0.984	0.984	0.984	0.985	0.986	0.988	0.989	0.991	0.994	0.997	1.000	

4.2.3 Phase Errors of Finite–Difference Schemes: Numerical Dispersion

In addition to amplitude errors, numerical schemes have phase errors. For instance, from Equation (4.2.11) it follows that

$$\tan \omega_r \Delta t = -\frac{C_0 \sin k\Delta x}{1 + C_0(\cos k\Delta x - 1)}. \quad (4.2.24)$$

Since the phase speed C_{ph} of a wave of frequency ω_r and wave number k is defined as $C_{ph} = -\frac{\omega_r}{k}$, the expression for numerical phase speed that can be derived from Equation (4.2.24) is

$$C_{ph} = \frac{1}{k\Delta t} \text{atan} \frac{C_0 \sin k\Delta x}{1 + C_0(\cos k\Delta x - 1)}. \quad (4.2.25)$$

Since the physical phase speed for advection (Equation 4.2.1) is equal to u, the ratio C_{ph}/u characterizing the relative error in the calculation of phase speed is

$$\frac{C_{ph}}{u} = \frac{1}{ku\Delta t} \text{atan} \frac{C_0 \sin k\Delta x}{1 + C_0(\cos k\Delta x - 1)}. \quad (4.2.26)$$

The absence of the phase error corresponds to the condition $\frac{C_{ph}}{u} = 1$. Equation (4.2.26) shows that no phase error corresponds to the case $C_0 = 1$.

Equation (4.2.26) shows that errors in phase speed values depend on the wavelength. Table 4.2.1 shows relative amplitude errors for different wavelengths and phase errors C_{ph}/u for various combinations of C_0 and k. One can see that wavelength of $2\Delta x$ generally provides the poorest representation of phase speed. When the Courant number falls within the interval $0.5 < C_0 < 1.0$, waves travel faster than in the true solution, whereas they are slower than in the true solution when $0 < C_0 < 0.5$. The amplitude errors and phase errors of the leapfrog scheme (Equation 4.2.22) are also presented in Table 4.2.1. Although the leapfrog scheme preserves amplitudes exactly as long as the Courant number obeys the condition $|C_0| \leq 1.0$ (Pielke, 2013), the accuracy of phase-speed calculation decreases markedly for shorter wavelengths. Both schemes described in Table 4.2.1 are dispersive, since when waves of different wavelengths are linearly superimposed they travel at different speeds even if the advection velocity u is constant. Retention of these dispersive shorter waves in the solution can cause computational problems due to nonlinear instability, as discussed in Section 4.2.4. The example of a leapfrog scheme shows that an exact representation of the amplitude does not guarantee successful simulations since a fictitious dispersion of waves of different lengths can generate errors.

Utilization of schemes that accurately represent the phase speed for all wavelengths is quite important for an accurate representation of advection in weather forecast models, propagation of cyclones, squall lines propagation, etc. Use of advection terms written with the fourth order of approximation decreases errors in the phase speed.

4.2.4 Nonlinear Instability and Aliasing Errors

Let us consider the nonlinear advection equation

$$\frac{\partial a}{\partial t} = -u(x,t)\frac{\partial a}{\partial x}, \quad (4.2.27)$$

which is a natural generalization of the linear Equation (4.2.1). In order to understand the effect related to the nonlinear term in Equation (4.2.27), we consider an example where a and u are represented by harmonics with wave numbers k_1 and k_2, respectively:

$$a = A_0 \sin k_1 x \\ u = U_0 \cos k_2 x. \quad (4.2.28)$$

Using Equation (4.2.28), the nonlinear term in Equation (4.2.27) can be written as

$$u(x,t)\frac{\partial a}{\partial x} = U_0 A_0 k_1 \cos k_1 x \cos k_2 x \\ = \frac{U_0 A_0}{2} k_1 [\cos(k_1 + k_2)x + \cos(k_1 - k_2)x]. \quad (4.2.29)$$

As a result of the nonlinear interaction, two waves with wave numbers $k_1 + k_2$ and $k_1 - k_2$ arise. The wave with a larger wave number is shorter in length than the minimal wavelength of the interacting waves. For instance, interaction of waves with wavelength $\lambda_1 = 2\Delta x$ and $\lambda_2 = 4\Delta x$ (the corresponding wave numbers are $k_1 = \frac{2\pi}{2\Delta x}$ and $k_2 = \frac{2\pi}{4\Delta x}$) leads to formation of waves with wavelengths of $\frac{\lambda_1 \lambda_2}{\lambda_1 + \lambda_2} = 1.33\Delta x$ and $\frac{\lambda_1 \lambda_2}{\lambda_2 - \lambda_1} = 4\Delta x$. While the latter wave is resolvable, the $1.33\Delta x$-length wave cannot be resolved on the finite-difference grid. However, it does not disappear from the system, but is fictionally seen as a wave with wavelength of $4\Delta x$ (**Figure 4.2.2**). Waves that erroneously appear in this process are said to have aliased, or folded to longer wavelength. As seen from Equation (4.2.29), in order to generate a wave with wavelength less than $2\Delta x$ (i.e., when aliasing occurs), one of the interacting waves should have a wavelength less then $4\Delta x$. **Table 4.2.2** shows the results of interaction of motions of different wavelengths that should be produced theoretically and the way they will be represented on the finite-difference grid due to aliasing.

Table 4.2.2 Examples of wave–wave interactions that produce aliased waves (from Pielke, 2013; courtesy of Elsevier).

Interactive wavelengths	Should produce	Will produce due to aliasing
$2\Delta x$ and $2\Delta x$	$1\Delta x$	*Add a constant to the entire model*
$2\Delta x$ and $4\Delta x$	$1.33\Delta x$	$4\Delta x$
$2\Delta x$ and $6\Delta x$	$1.5\Delta x$	$3\Delta x$
$2\Delta x$ and $8\Delta x$	$1.6\Delta x$	$8\Delta x$
$2\Delta x$ and $10\Delta x$	$1.67\Delta x$	$5\Delta x$

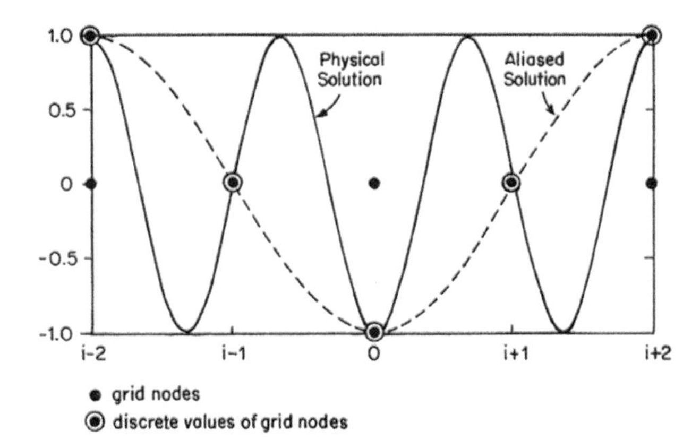

Figure 4.2.2 Schematic illustration of how a wave with a wavelength of $1.33\Delta x$, caused by the nonlinear interaction of waves of $2\Delta x$ and $4\Delta x$ in length, is seen as a computational $4\Delta x$ wave in the computational grid (from Pielke, 2013; courtesy of © Elsevier).

In reality, formation of the smallest wavelength is controlled by molecular dissipation. In numerical models, waves with wavelengths shorter than $2\Delta x$ are erroneously seen as larger-scale waves that interact with other waves transferring the energy toward smaller scales. Because the natural influx of energy to smaller and smaller wavelengths is interrupted by the finite-difference grids, the energy is continuously and erroneously accumulated at the smallest resolvable scales. If dissipation of this energy is improperly represented, the increase in magnitude without bound leads to a *nonlinear instability*. Thus, linearly stable schemes may be nonlinearly unstable.

There are several ways to eliminate a nonlinear instability. As mentioned, short waves are inadequately represented on a computational grid and cause significant amplitude and phase errors, and thus should be removed. As shown in Chapter 3, the averaged equations contain diffusion terms where turbulent viscosity coefficients and turbulent conductivity coefficients are used instead of their molecular analogs from non-averaged equations. These terms parameterize the effect of small-scale turbulence in the models, when the sink of kinetic energy at small scales caused by friction is mimicked by the attenuation of short waves' amplitudes. This approach is attractive since it is based on fundamental physical concepts. However, the 1.5 closure approach (Sections 3.4 and 3.5) is relatively well justified when subgrid scales are represented only by small-scale turbulence. In many mesoscale and global models, the averaging is performed over scales larger (sometimes significantly) than turbulent scales. Nevertheless, the moments of the second order are parameterized using the expressions similar to those used in the K-theory. The effective viscosity coefficient is calculated using expressions similar to Equations (3.5.33) and (3.5.34):

$$K_m = C_1 l^2 \sqrt{D - \alpha \frac{g}{\theta_0} \frac{\partial \bar{\theta}}{\partial z}}, \tag{4.2.30}$$

where the mixing length is determined by the grid spacing, like in expressions $l = \sqrt{\Delta x \Delta y}$ in 2D models or $l = (\Delta x \Delta y \Delta z)^{1/3}$ in 3D models. In these cases, K_m loses the meaning of the coefficient of turbulent viscosity, but describes the effective friction induced by small scales with respect to larger ones (Section 3.6). It is clear that there are significant uncertainties as regards the values of the effective viscosity coefficient. Therefore, these values should be chosen to control nonlinear aliasing, and not to represent the actual physical processes. The magnitude of coefficient C_1 is tuned to get the desirable rate of damping the short wavelength motions.

There are other approaches to preventing nonlinear instability, such as using the computational viscosity or a spatial smoother (filter) that removes the shortest waves, but leaves the longer ones relatively unaffected. We will describe several physically grounded and efficient approaches to construct finite-difference schemes free of nonlinear instability. First, it is important that a significant flux of energy to small scales that can lead to nonlinear instability is a result of errors of particular finite-difference approximations of the advective terms. The following example illustrates this point. For instance, the advective term $u\frac{\partial u}{\partial x}$ is obviously equal to $\frac{\partial u^2/2}{\partial x}$. From this expression, one can see that the integral with respect to x is equal to the difference of the values of $u^2/2$ taken at the right and left boundaries. It means that Equation (4.2.1) should not lead to an infinitely large increase in energy. If the amplitude of small wavelength motions is low from the very beginning, it would remain small if the erroneous generation of such a wave is eliminated. A finite-difference approximation of this term in form $\frac{1}{2}\frac{u_{i+1}^2-u_i^2}{\Delta x}$ has similar properties: being summated with respect to i, it leads to the difference of $(u_N^2 - u_0^2)/2$, i.e., the values taken at the right and left boundaries (indices 0 and N denote the left and right boundaries, respectively). In contrast, term $u\frac{\partial u}{\partial x}$ approximated in the form $u_i\frac{u_{i+1}-u_i}{\Delta x}$ does not obey this condition, but contains products of velocities taken at different points, which leads to erroneous production of short wavelength fluctuations and may generate nonlinear instability. This example shows that while the first approximation is free of nonlinear instability, the second approximation may lead to such instability due to the erroneous energy flux to small scales.

In order to prevent erroneous energy flux to small scales and, accordingly, the infinitely large energy increase, finite-difference schemes should obey the same conservation laws as the initial differential equations. For this purpose, it is preferable to use the differential equations written in the divergent form to construct finite-difference approximations.

The budget Equation (3.2.3) is given in Section 3.2 in the form $\frac{\partial \rho a}{\partial t} + \frac{\partial \rho a u_i}{\partial x_i} = Q_a$, where Q_a is the source of a that is not related to advection. Integration of this equation over the computational volume yields

$$\frac{\partial}{\partial t}\int_V (\rho a)dxdydz = \rho au|_{\Gamma_{x0}} - \rho au|_{\Gamma_{xf}} + \rho au|_{\Gamma_{y0}} - \rho au|_{\Gamma_{yf}} + \rho au|_{\Gamma_{z0}} - \rho au|_{\Gamma_{zf}} + \int_V Q_a dxdydz, \quad (4.2.31)$$

where Γ_{x0}, Γ_{xf}, etc. denote the boundaries of the computational area. Equation (4.2.31) shows that the amount of quantity a within the computational volume changes only due to fluxes via the boundaries and due to the internal sources. In case the fluxes at the boundaries are equal to zero, the advective terms cannot change the amount of quantity a within the computational area. Quantity a can be, for instance, kinetic energy per unit of volume, heat content, vorticity, etc. Equation (4.2.31) means that advection terms should not lead to an increase in the kinetic energy within the computational area due to nonlinear interactions. Thus, finite-difference schemes obeying the conservation laws are free of nonlinear instability.

A simple way to construct finite-difference schemes free of nonlinear instability is illustrated in **Figure 4.2.3**,

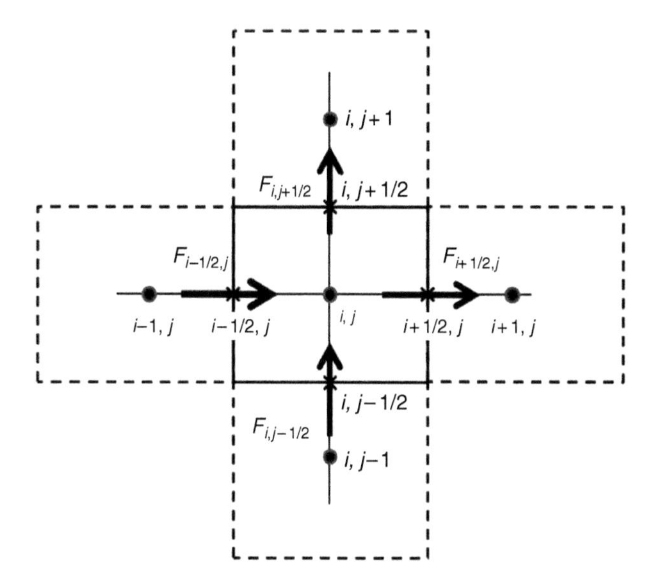

Figure 4.2.3 An example of a box of a finite-difference grid that can be used for approximation of equations written in the divergent form. Arrows indicate fluxes through the boundaries of the *i, j-th* box.

where a 2D cell (box) of the finite-difference mesh is shown.

Let us introduce fluxes of quantity a through the boundaries of the box around point (i,j) (Figure 4.2.3). Using the fluxes, the corresponding finite-difference scheme approximating Equation (3.2.3) in case $Q_a = 0$ is

$$\frac{\partial \rho a}{\partial t} + \left(\frac{\partial \rho au}{\partial x} + \frac{\partial \rho av}{\partial y}\right)_{i,j} = 0 \rightarrow \frac{(\rho a)_{i,j}^{\tau+1} - (\rho a)_{i,j}^{\tau}}{\Delta t} + \frac{F_{i+1/2,j} - F_{i-1/2,j}}{\Delta x} + \frac{F_{i,j+1/2} - F_{i,j-1/2}}{\Delta y} = 0, \quad (4.2.32)$$

where the fluxes through the box boundaries can be calculated, for instance, as

$$F_{i+1/2,j} = \frac{1}{2}\left[(\rho au)_{i,j} + (\rho au)_{i+1,j}\right];$$
$$F_{i-1/2,j} = \frac{1}{2}\left[(\rho au)_{i,j} + (\rho au)_{i-1,j}\right];$$
$$F_{i,j+1/2} = \frac{1}{2}\left[(\rho av)_{i,j} + (\rho av)_{i,j+1}\right];$$
$$F_{i,j-1/2} = \frac{1}{2}\left[(\rho av)_{i,j} + (\rho av)_{i,j-1}\right]. \quad (4.2.33)$$

Summation of expressions similar to Equation (4.2.33) over all the grid points leads to a mutual compensation of the fluxes between adjacent boxes (see **Figure 4.2.4**), so the changes of quantity a within the computational

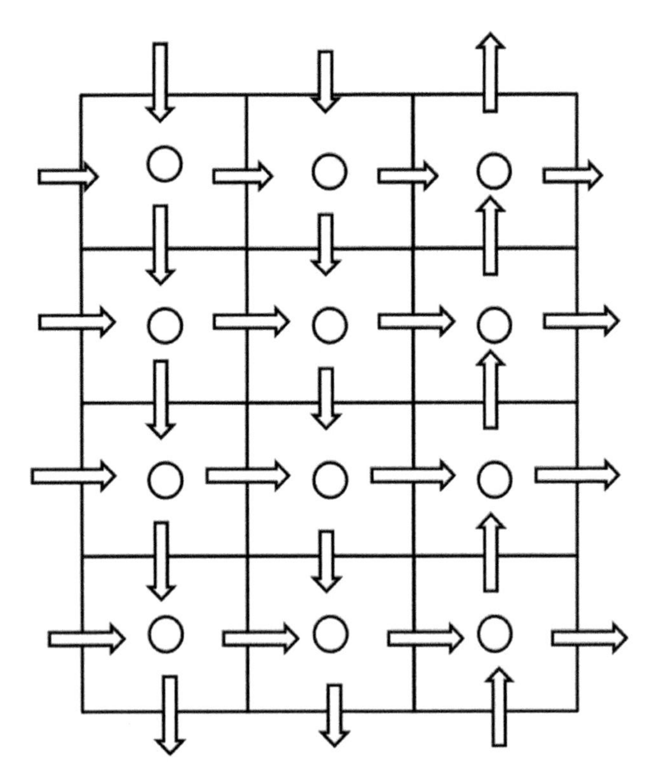

Figure 4.2.4 Illustration of the box method. Circles denote the grid points. Arrows denote fluxes through the boundaries of the grid boxes. The time changes of any variable a within any air volume (or over the computational area) are determined by fluxes of this variable through the outer boundaries of the volume. The internal fluxes between neighboring boxes are eliminated by integration (summation) over the volume.

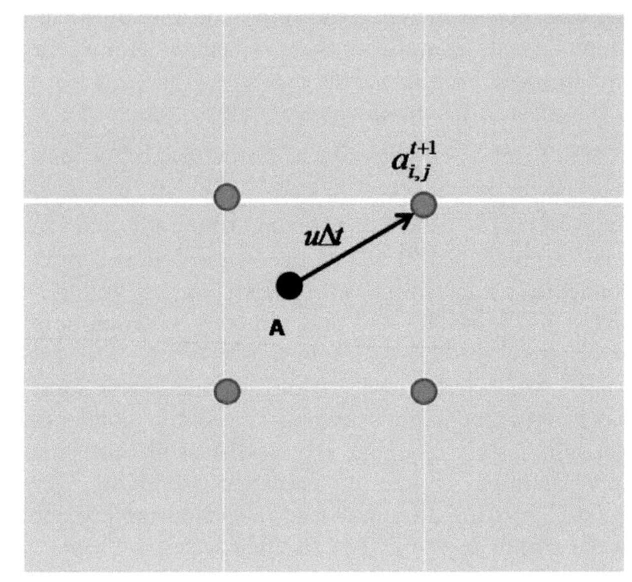

Figure 4.2.5 Illustration of the Lagrangian approach to representation of advection in numerical models.

area are determined by fluxes through the outer boundary of the computational area only. It means that scheme in Equations (4.2.32) and (4.2.33) obeys the same conservation laws as the initial differential equation, and thus is free of nonlinear instability. The approximation of equations written in the divergent form with the help of the finite differences using the fluxes between adjacent boxes is known as the *box method*. This method was first proposed by Bryan (1966) and since then has been widely used in many atmospheric and cloud models. The differences between the different numerical schemes are largely reduced to utilization of different ways of calculation of fluxes through the boundaries of the grid boxes (see, e.g., Bott, 1989).

There are other methods to prevent formation of nonlinear instability. A special form of finite-difference representation of advection terms was proposed by Arakawa (1966). The method is used in many cloud models (for instance, in HUCM). Although Arakawa used a non-divergent form of the advective terms, his scheme conserves the kinetic energy, the mean value of vorticity, and the mean square of vorticity in the same way the original differential equation does. The finite difference approximations have a quite complicated form and are not presented here.

Another approach to prevent formation of nonlinear instability is utilization of the Lagrangian approach to representation of advective terms (Leith, 1965). This approach is illustrated in **Figure 4.2.5**. The value of quantity $a_{i,j}^{t+1}$ calculated at grid point (i,j) is equal to the value of quantity a_A^t at point A located in the upstream direction from point (i,j) at the distance $u\Delta t$. The value a_A^t is calculated by interpolation of the corresponding values at the nearest grid points (marked by blue circles).

4.2.5 Positive Definite Schemes

Many meteorological values and functions such as humidity, size distributions of hydrometeors, masses, concentrations, etc., are positive by definition. At the same time, any explicit scheme $(\rho a)_{i,j}^{\tau+1} = (\rho a)_{i,j}^{\tau} + \Delta t \cdot RHS$ with a negative right-hand side ($RHS < 0$) leads to appearance of negative values if $|\Delta t \cdot RHS| > (\rho a)_{i,j}^{\tau}$. Strictly speaking, the latter condition imposes limitations on the time step of explicit schemes. However, such time steps may be too small and the calculations can become time consuming. To avoid the problem and to increase the computational time step (even at the expense of the accuracy of calculations) positive definite schemes are widely used. Previous attempts to construct positive-definite advection schemes involved filling algorithms, where the solution obtained after each integration step was corrected by filling in any negative values. To conserve the total mass of the advected species, negative values cannot simply be set to zero; the compensating mass must be removed from regions where the values are positive. There is a variety of filling algorithms designed for this purpose. Some of them attempt to fill local regions with negative values from the adjacent regions with positive values. According to Durran (2010), "This may be a physically satisfying way to remove dispersive undershoots, but it requires a great deal of logical testing that cannot be performed efficiently."

In other approaches, the compensating mass is removed from the entire field by multiplying the value at every grid point by the ratio of the total original mass to the total nonnegative mass. Multiplicative compensation is computationally efficient, but it preferentially damps the values in the regions of highest tracer concentration. Different filling algorithms are analyzed by Rood (1987). Although empirical testing has shown the efficiency of filling algorithms, the theoretical basis for these schemes is largely undeveloped. Practical utilization of such algorithms may lead to serious errors.

The box method is often used as a basis for developing positive-definite schemes that do not generate negative values in the course of integration. The method is reduced to a correction of the internal fluxes (Smolarkiewicz and Grell, 1990; Bott, 1989; Durran, 2010). For instance, in the approach described by Durran (2010), the outgoing flux should be chosen to keep $(\rho a)_{i,j}^{\tau+1}$ positive. Accordingly, the actual flux $F_{i+1/2}^{\tau}$ (for sake of simplicity, here we consider a 1D problem) is replaced with the corrected value $C_{i+1/2}F_{i+1/2}^{\tau}$, in which the correction factor $C_{i+1/2}$ is defined as

$$C_{i+1/2} = \min\left[1, \left(\rho a_i^{\tau}\frac{\Delta x}{\Delta t}\right)\Big/F_{i+1/2}^{\tau}\right]. \qquad (4.2.34)$$

This procedure can be extended for the general case where the fluxes may have an arbitrary sign. Suppose P_i is the net flux out of the grid volume i:

$$P_i = \max\left(0, F_{i+1/2}\right) - \min\left(0, F_{i-1/2}\right) + \varepsilon, \qquad (4.2.35)$$

where ε is a small value added to ensure that $P_i \neq 0$. Suppose also Q_i is the maximum outward flux that can be supported without forcing $(\rho a)_i^{\tau+1}$ to negative values:

$$Q_i = \rho a_i^{\tau}\frac{\Delta x}{\Delta t}. \qquad (4.2.36)$$

Now we can determine the ratio by which the fluxes $F_{i+1/2}^{\tau}$ and $F_{i-1/2}^{\tau}$ should be reduced to ensure that a negative value is not created at the grid point i:

$$R_i = \min(1, Q_i/P_i). \qquad (4.2.37)$$

Finally, one can choose the actual limiter for the corrected flux $C_{i+1/2}F_{i+1/2}^{\tau}$ so that negative values are not created within the volume:

$$C_{i+1/2} = \min\begin{cases} R_i, & \text{if} \quad F_{i+1/2} \geq 0 \\ R_{i+1}, & \text{if} \quad F_{i+1/2} < 0. \end{cases} \qquad (4.2.38)$$

Clearly, it is possible to further generalize this procedure (e.g., Bott, 1989; Skamarock, 2006; Durran, 2010). However, using expressions such as Equation (4.2.37) distorts the solution of the advection equation. Indeed, instead of physically grounded application of smaller time steps, Equation (4.2.37) artificially decreases internal fluxes, which is equivalent to utilization of a lower speed. Therefore, such schemes can be used only in situations when negative values are unlikely to appear.

4.3 Equations of Friction and Diffusion

4.3.1 Friction Equation

An important element of numerical models and cloud models in particular, is the equation of friction:

$$\frac{da}{dt} = -K_f a, \qquad (4.3.1)$$

where K_f is the friction coefficient. This equation describes, for instance, a decrease in supersaturation during condensation droplet growth (see Equation 5.2.13) and a decrease in concentration of large raindrops because of spontaneous breakup (first term in Equation 5.5.13). The friction terms are typically added to averaged equations near the upper boundary of the computational area to damp erroneous reflection of waves from the upper boundary (so-called Rayleigh damping). The analytical solution of this equation is

$$a = a_0 \exp\left(-K_f t\right). \qquad (4.3.2)$$

Let us consider the finite difference scheme corresponding to Equation (4.3.1):

$$\frac{a_i^{\tau+1} - a_i^{\tau}}{\Delta t} = -K_f a_i^{\tau}. \qquad (4.3.3)$$

Looking for the solution of Equation (4.3.3) in the form of plane wave (Equation 4.2.3) we have $\exp(i\omega\Delta t) - 1 = -K_f\Delta t$ or

$$\lambda \cos \omega_r \Delta t = -K_f \Delta t + 1$$
$$\lambda \sin \omega_r \Delta t = 0. \tag{4.3.4}$$

From Equation (4.3.4) it follows that $|\lambda|^2 = (1 - K_f \Delta t)^2$ and the stability condition is

$$0 \le K_f \Delta t \le 2. \tag{4.3.5}$$

However, the time step should be chosen using the condition

$$0 < K_f \Delta t \le 1. \tag{4.3.6}$$

Indeed, from Equation (4.3.3) it follows that $a_i^{\tau+1} = (1 - K_f \Delta t) a_i^{\tau}$, so that when $K_f \Delta t = 1$ $a_i^{\tau+1} = 0$. At $K_f \Delta t > 1$ the sign of the solution changes at each time step (**Figure 4.3.1**). One can see that a numerical solution obeying the stability conditions may deviate dramatically from the analytical one. To describe the analytical solution properly (red line in Figure 4.3.1), time step should be chosen at least by the condition $K_f \Delta t < \frac{1}{3}$, i.e., shorter than the characteristic time of the exponentially decreasing solution. These conclusions concerning the choice of the time step are important for solving many equations of cloud microphysics.

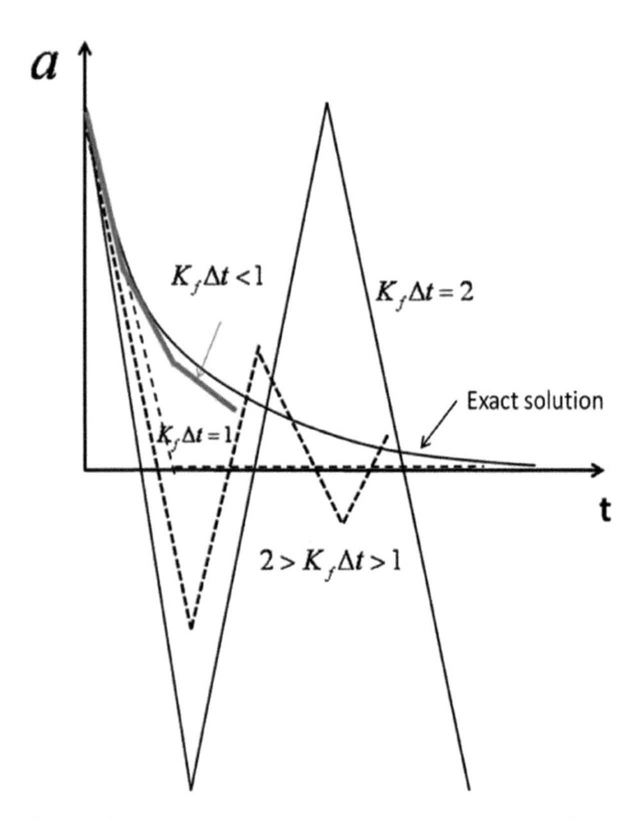

Figure 4.3.1 Analytical (exact) and numerical solutions of the equation of friction at different time steps.

For instance, condensation accompanied by a decrease in supersaturation physically cannot result in formation of subsaturation. Similarly, evaporation of droplets accompanied by an increase in the relative humidity cannot result in formation of supersaturation. The wrong choice of the time step can lead to results that are so physically unrealistic that they may lead to artificial droplet nucleation during evaporation, etc.

It is widely assumed that an increase in the friction coefficient leads to a decrease in the calculated value. It was shown that an increase in the friction coefficient without a change of the time step may lead to instability. To avoid the erroneous change in the sign of a numerical solution, the implicit scheme can be applied:

$$\frac{a_i^{\tau+1} - a_i^{\tau}}{\Delta t} = -K_f a_i^{\tau+1}. \tag{4.3.7}$$

It follows from Equation (4.3.7) that

$$a_i^{\tau+1} = \frac{a_i^{\tau}}{1 + \Delta t K_f}. \tag{4.3.8}$$

One can see from Equation (4.3.8) that the implicit scheme is stable and the solution does not change its sign. Despite the fact that the solution of Equation (4.3.8) does not change its sign at any Δt, the value of the time step should be used small to ensure the condition $\Delta t K_f << 1$ (at least $\Delta t K_f < 1/3$) to provide necessary closeness to the exact solution.

4.3.2 Criteria of Stability of the Diffusion Equation

The basic averaged equations (Section 3.5) include terms describing process of diffusion. The turbulent diffusion plays an important role near the surface andcloud boundaries as well as in clouds themselves. Again, it is widely assumed that an increase in the diffusion coefficient (the coefficient of turbulent viscosity) causes fluctuation damping, leading to more smoothed solutions.

Let us consider 1D equation of diffusion in the form

$$\frac{\partial a}{\partial t} = K \frac{\partial^2 a}{\partial z^2} \tag{4.3.9}$$

Let us analyze the linear stability of the finite difference scheme corresponding to Equation (4.3.9):

$$\frac{a_i^{\tau+1} - a_i^{\tau}}{\Delta t} = K \left(a_{i+1}^{\tau} - 2a_i^{\tau} + a_{i-1}^{\tau} \right). \tag{4.3.10}$$

Assuming there is no damping of the solution in z-direction, the exact solution of Equation (4.3.9) can be written as (Pielke, 2013)

$$a = a_0 \exp\left\{\widehat{i}(kz + \omega t)\right\} = a_0 \exp(-\omega_{im} t) \exp\left\{\widehat{i}(kz + \omega_r t)\right\}. \quad (4.3.11)$$

Substituting this expression into Equation (4.3.9) yields $\widehat{i}\,\omega_r - \omega_{im} = -Kk^2$. It follows from this expression that $\omega_r = 0$, so that the exact solution can be in the form

$$a = a_0 \exp(-Kk^2 t) \exp\left(\widehat{i}\,kz\right). \quad (4.3.12)$$

Substitution of Equation (4.3.12) for Equation (4.3.10) shows that the amplitude factor $\lambda = \exp\left(-Kk^2\Delta t\right)$ can be represented as follows:

$$\lambda = 1 + \gamma_F\left[\exp\left(\widehat{i}\,k\Delta z\right) - 2 + \exp\left(-\widehat{i}\,k\Delta z\right)\right] = 1 + 2\gamma_F(\cos k\Delta z - 1) \quad (4.3.13)$$

In Equation (4.3.13), $\gamma_F = K\frac{\Delta t}{\Delta z^2}$ is the so-called Fourier number. The Equation (4.3.13) shows that the minimum value of $\lambda = 1 - 4\gamma_F$ reached when $k\Delta z = \pi$, i.e., wavelength, is equal to $2\Delta z$. Accordingly, the stability criteria (requiring $|\lambda| < 1$) is $\gamma_F \leq \frac{1}{2}$. However, to avoid changes of the sign of solution at each time step, λ should be positive. Assuming $\lambda > 0$, the stability criteria reduces to $\gamma_F \leq \frac{1}{4}$. It means that the time step should be chosen as

$$\Delta t \leq \frac{\Delta z^2}{4K}. \quad (4.3.14)$$

In case $\Delta z = 10\,\text{m}$, $K = 20\,\text{m}^2\text{s}^{-1}$, the time step should be less than $\Delta t \leq 1\,\text{s}$. Typically, the addition of the diffusion (viscosity) term into equations of advection increases the stability of the scheme. However, when diffusion process is dominating, it requires its own limitation on the time step (Equation 4.3.14). Utilization of implicit schemes allows increasing the possible time step (e.g., Pielke, 2013).

4.4 Gravity and Inertia Gravity Waves

In previous sections we described application of the finite-difference schemes to solving a single differential equation. Below, we consider the method used for solving equation systems. First of all, we should analyze one of the most important processes in the atmosphere in general and in clouds in particular, namely, the mutual adaptation of the pressure field and the wind field accompanied by formation of gravity waves. The mutual adaptation of those two fields takes place within a short time and requires utilization of small time steps in order to solve the corresponding finite-difference equations describing these processes.

We will consider the specific features of the finite-difference schemes applied for description of these processes using the simplest case of shallow water equations, in which the gradient of the depth h of the atmospheric layer plays the role of the pressure gradient. The equations for motion and continuity can be written as

$$\frac{\partial u}{\partial t} = -g\frac{\partial h}{\partial x};\ \frac{\partial v}{\partial t} = -g\frac{\partial h}{\partial y};\ \frac{\partial h}{\partial t} = -H\left(\frac{\partial u}{\partial x} + \frac{\partial v}{\partial y}\right), \quad (4.4.1)$$

where H is the averaged depth of the liquid (or air) layer and g is the gravity acceleration. The solution of this equation system can be written in the wave form

$$\begin{aligned} u &= \mathrm{Re}\left[u_0 \exp\left\{\widehat{i}(kx + ly + \omega t)\right\}\right] \\ v &= \mathrm{Re}\left[v_0 \exp\left\{\widehat{i}(kx + ly + \omega t)\right\}\right] \\ h &= \mathrm{Re}\left[h_0 \exp\left\{\widehat{i}(kx + ly + \omega t)\right\}\right]. \end{aligned} \quad (4.4.2)$$

Simple transformation of Equation (4.4.1) leads to the following hyperbolic equation for a layer of depth h:

$$\frac{\partial^2 h}{\partial t^2} = gH\left(\frac{\partial^2 h}{\partial x^2} + \frac{\partial^2 h}{\partial y^2}\right). \quad (4.4.3)$$

Substitution of solution (4.4.2) into Equation (4.4.3) gives the dispersion relationship:

$$\omega^2 = gH\left(k^2 + l^2\right). \quad (4.4.4)$$

Equation (4.4.4) shows that the phase speed of the gravitational waves C_{ph} can be written as

$$C_{ph} = \frac{\omega}{\sqrt{k^2 + l^2}} = \sqrt{gH}. \quad (4.4.5)$$

The phase speed can be quite high. For instance, if H is equal to 10 km, C_{ph} exceeds 300 m/s. Equation (4.4.5) describes the phase speed of so-called external gravity waves, which does not depend on the wave number, so there is no wave dispersion. Another type of gravity waves arising within the atmosphere due to the difference in densities of adjacent air layers (so-called internal-gravity waves) is characterized by a much lower phase speed.

4.4.1 Stability of Finite Difference Schemes for a Gravity Wave Equation

In a simplified 1D case, when the wind velocity is directed along the x-axis,

$$\frac{\partial u}{\partial t} = -g\frac{\partial h}{\partial x};\ \frac{\partial h}{\partial t} = -H\frac{\partial u}{\partial x}. \quad (4.4.6)$$

The explicit finite difference scheme, approximating the Equation system (4.4.6), is expressed as

$$\frac{u_i^{\tau+1}-u_i^{\tau}}{\Delta t}=-g\frac{h_{i+1}^{\tau}-h_{i-1}^{\tau}}{2\Delta x}$$
$$\frac{h_i^{\tau+1}-h_i^{\tau}}{\Delta t}=-H\frac{u_{i+1}^{\tau}-u_{i-1}^{\tau}}{2\Delta x} \quad (4.4.7)$$

Substituting the solution (4.4.2) (assuming the wave number $l=0$) into Equations (4.4.7) and denoting, as in Section 4.2, $\lambda_0=\lambda\exp\left\{\hat{i}(\omega_r\Delta t)\right\}$, where $|\lambda_0|$ is the amplitude of the amplification factor and $\lambda=\exp(-\omega_{im}\Delta t)$ (the frequency is a complex value: $\omega=\omega_r+\hat{i}\omega_{im}$), we obtain the following equation system with respect to variables u and h:

$$\begin{cases}(\lambda_0-1)u+\dfrac{g\Delta t}{2\Delta x}\left(\exp\left(\hat{i}k\Delta x\right)-\exp\left(-\hat{i}k\Delta x\right)\right)h=0\\ \dfrac{H\Delta t}{2\Delta x}\left(\exp\left(\hat{i}k\Delta x\right)-\exp\left(-\hat{i}k\Delta x\right)\right)u+(\lambda_0-1)h=0\cdot\end{cases} \quad (4.4.8)$$

Equation (4.4.8) has a nontrivial solution if its determinant is equal to zero. Taking into account the equality $\left(e^{\hat{i}k\Delta x}-e^{-\hat{i}k\Delta x}\right)=2\hat{i}\sin k\Delta x$, this determinant can be written as

$$\begin{vmatrix}(\lambda_0-1) & \hat{i}\dfrac{g\Delta t}{\Delta x}\sin k\Delta x\\ \hat{i}\dfrac{H\Delta t}{\Delta x}\sin k\Delta x & (\lambda_0-1)\end{vmatrix}=0. \quad (4.4.9)$$

The solution of quadratic Equation (4.4.9) is

$$\lambda_0=1\pm\hat{i}\sqrt{gH}\frac{\Delta t}{\Delta x}\sin k\Delta x. \quad (4.4.10)$$

Equation (4.4.10) is equivalent to two equations, one for the real part and the second for the imaginary part. Taking into account the relationship between parameters λ_0 and λ yields

$$\lambda\cos\omega_r\Delta t=1$$
$$\lambda\sin\omega_r\Delta t=\pm\sqrt{gH}\left(\frac{\Delta t}{\Delta x}\right)\sin k\Delta x. \quad (4.4.11)$$

It follows from Equation (4.4.11) that

$$|\lambda|^2=1+gH\left(\frac{\Delta t}{\Delta x}\right)^2\sin^2 k\Delta x>1. \quad (4.4.12)$$

Equation (4.4.12) shows that $|\lambda|^2>1$ is always obeyed, i.e., scheme (4.4.7) is unstable.

A typical approach to designing stable schemes is using semi-implicit schemes, where the values calculated by the explicit equation at the first step are used in the second equation, which is implicit. Let us show that the following finite difference scheme

$$\frac{u_i^{\tau+1}-u_i^{\tau}}{\Delta t}=-g\frac{h_{i+1}^{\tau}-h_{i-1}^{\tau}}{2\Delta x}$$
$$\frac{h_i^{\tau+1}-h_i^{\tau}}{\Delta t}=-H\frac{u_{i+1}^{\tau+1}-u_{i-1}^{\tau+1}}{2\Delta x} \quad (4.4.13)$$

is conditionally stable. Performing transformations similar to those used for Equation (4.4.7) we obtain the following determinant:

$$\begin{vmatrix}(\lambda_0-1) & \hat{i}\dfrac{g\Delta t}{\Delta x}\sin k\Delta x\\ \hat{i}\dfrac{H\Delta t}{\Delta x}\lambda_0\sin k\Delta x & (\lambda_0-1)\end{vmatrix}=0. \quad (4.4.14)$$

Solution of Equation (4.4.14) leads to expression

$$\lambda_0=\frac{2-\gamma^2\pm\gamma\sqrt{\gamma^2-4}}{2}, \quad (4.4.15)$$

where

$$\gamma^2=\frac{gH\Delta t^2}{\Delta x^2}\sin^2 k\Delta x. \quad (4.4.16)$$

Equation (4.4.15) demonstrates the existence of two possible solutions: a) in case $\gamma^2\le 4$ the equation yields two complex roots and b) in case $\gamma^2>4$ the same equation yields two real roots. In the first case, the real part and the imaginary part of solution (4.4.15) are

$$\lambda\cos\omega_r\Delta t=\left(2-\gamma^2\right)/2$$
$$\lambda\sin\omega_r\Delta t=\pm\frac{\gamma}{2}\sqrt{4-\gamma^2}. \quad (4.4.17)$$

It follows from Equation (4.4.17) that

$$|\lambda|^2=\frac{1}{4}\left(4-4\gamma^2+\gamma^4+4\gamma^2-\gamma^4\right)=1. \quad (4.4.18)$$

This condition means that in case $\gamma^2\le 4$, the scheme (4.4.13) is neutrally stable. This condition can be reduced to the following stability condition valid for any wavelength:

$$\frac{\sqrt{gH}\Delta t}{\Delta x}\le 2. \quad (4.4.19)$$

Unequality (4.4.19) is the CFL condition (Section 4.2), where the phase speed of a gravity wave $C_{ph}=\sqrt{gH}$ is used instead of the wind speed. It can be shown that when $\gamma^2>4$, the scheme (4.4.13) is unstable. The approach applied for constructing the numerical scheme (Equation 4.4.13) is used in many numerical models, for instance, in the Weather Research and

Table 4.4.1 Values of amplitude and phase errors per time step as a function of wavelength and parameter $\frac{\sqrt{gH}\Delta t}{\Delta x}$ for the centered-in-space, implicit, forward-in-time approximation to linearized 2D equations (from Pielke, 2013; courtesy of Elsevier).

		$\sqrt{gH}\Delta t/\Delta x$													
	Wavelength	0.001	0.01	0.1	0.2	0.3	0.4	0.5	0.6	0.7	0.8	0.9	1.0	1.5	2.0
λ	$2\Delta x$	1.0	1.0	1.0	1.0	1.0	1.0	1.0	1.0	1.0	1.0	1.0	1.0	1.0	1.0
	$4\Delta x$	1.0	1.0	1.0	1.0	1.0	1.0	1.0	1.0	1.0	1.0	1.0	1.0	1.0	1.0
	$10\Delta x$	1.0	1.0	1.0	1.0	1.0	1.0	1.0	1.0	1.0	1.0	1.0	1.0	1.0	1.0
	$20\Delta x$	1.0	1.0	1.0	1.0	1.0	1.0	1.0	1.0	1.0	1.0	1.0	1.0	1.0	1.0
$\frac{C_{ph}}{\sqrt{gH}}$	$2\Delta x$	0.0	0.0	0.0	0.0	0.0	0.0	0.0	0.0	0.0	0.0	0.0	0.0	0.0	0.0
	$4\Delta x$	0.637	0.637	0.637	0.638	0.639	0.641	0.643	0.647	0.650	0.655	0.660	0.667	0.613	0.0
	$10\Delta x$	0.935	0.935	0.936	0.936	0.937	0.938	0.939	0.940	0.942	0.944	0.947	0.950	0.969	1.0
	$20\Delta x$	0.984	0.984	0.0984	0.984	0.984	0.984	0.0985	0.985	0.986	0.986	0.987	0.988	0.993	1.0

Forecasting model (WRF) (Skamarock et al., 2005a, b). Unequality (4.4.19) requires using very small time steps because the phase speed can be as large as ~300 m/s.

From Equation (4.4.17), one can derive the following expression for phase errors:

$$\frac{C_{ph}}{\sqrt{gH}} = \pm \frac{1}{k\Delta t\sqrt{gH}} \operatorname{asin}\left(\frac{\gamma\sqrt{4-\gamma^2}}{2}\right). \quad (4.4.20)$$

Equation (4.4.20) shows that the numerical phase speed obtained depends on the wave length, i.e., the numerical scheme (Equation 4.4.13) produces wave dispersion. One can see from Equation (4.4.20) that the phase speed is equal to zero in two cases: when $\gamma = 0$ (corresponding to the case when $\sin k\Delta x = 0$ and the wavelength is $2\Delta x$) and when $\gamma = 2$. If the grid spacing $\Delta x \to 0$ accompanied by a simultaneous decrease of Δt, $C_{ph} \to \sqrt{gH}$, which means a decrease in the numerical dispersion and fewer phase errors. Dependencies of the values of amplitude and phase errors on wavelength and parameter $\frac{\sqrt{gH}\Delta t}{\Delta x}$ for the centered-in-space implicit, forward-in-time schemes are presented in **Table 4.4.1**.

In case the Coriolis force is taken into account (which means that the scales of the simulated phenomena exceed several tens of kilometers), the following equations can be used instead of Equation system (4.4.1):

$$\frac{\partial u}{\partial t} = -g\frac{\partial h}{\partial x} + fv; \; \frac{\partial v}{\partial t} = -g\frac{\partial h}{\partial y} - fu; \; \frac{\partial h}{\partial t} = -H\left(\frac{\partial u}{\partial x} + \frac{\partial v}{\partial y}\right), \quad (4.4.21)$$

where f is the Coriolis parameter (Section 3.2). These equations include two types of motions: low frequency quasi-geostrophic motions and high-frequency inertia gravity waves. The inertia-gravity waves are permanently generated in the atmosphere and rapidly leave the areas of their "birth." As a result, a quasi-geostrophic adjustment takes place. Quasi-geostrophic motions are slow and change mostly due to advection. Stability of the schemes containing the Coriolis terms can be achieved if the velocity components in those terms are used at time instance τ (i.e., fv^τ, fu^τ), but in the continuity equation the velocity gradients are calculated at time instance $\tau + 1$, i.e., $h^{\tau+1}$ is calculated from the equation $\frac{\partial h}{\partial t} = -H\left(\frac{\partial u^{\tau+1}}{\partial x} + \frac{\partial v^{\tau+1}}{\partial y}\right)$ (e.g., Pielke, 2013).

4.4.2 Types of Finite-Difference Grids

Staggered grids. When different variables of an equation system are calculated, it becomes possible to construct different types of finite-difference grids. This differs from an earlier assumption that all the variables are calculated at the same grid points, as shown in **Figure 4.4.1a**. As far as the Equation systems (4.4.7) or (4.4.13) are concerned, the variables underlined in Figure 4.4.1a depend only on other underlined variables at the neighboring grid points. The same statement holds for the variables that are not underlined. Thus, since the grid in Figure 4.4.1a contains two elementary subgrids, a solution on one of these subgrids is completely decoupled from that on the other. In this case, it is better to use the grid shown in Figure 4.4.1b, where different variables are calculated at different grid points. Grids of this type shown in Figure 4.4.1b are called staggered grids. The computational time required to solve equations on a staggered grid is by a factor of

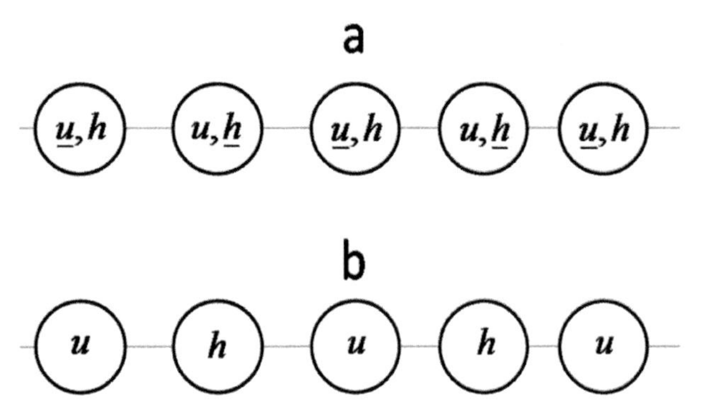

Figure 4.4.1 Schemes of 1D non-staggered (a) and staggered (b) grids.

two lower as compared to a non-staggered grid, while the accuracy of approximation (truncation errors) remains the same.

There are several possibilities for constructing staggered grids. Arakawa (1972) analyzed the properties of five grid versions with different distribution of variables in space (**Figure 4.4.2**). The five versions are known as Arakawa schemes A, B, C, D, and E. In Figure 4.4.2, d denotes the shortest distance between the grid points at which the same quantities are calculated.

Since finite-difference approximation of equations containing many terms can be quite complicated, it is accepted to use a compact way to write these

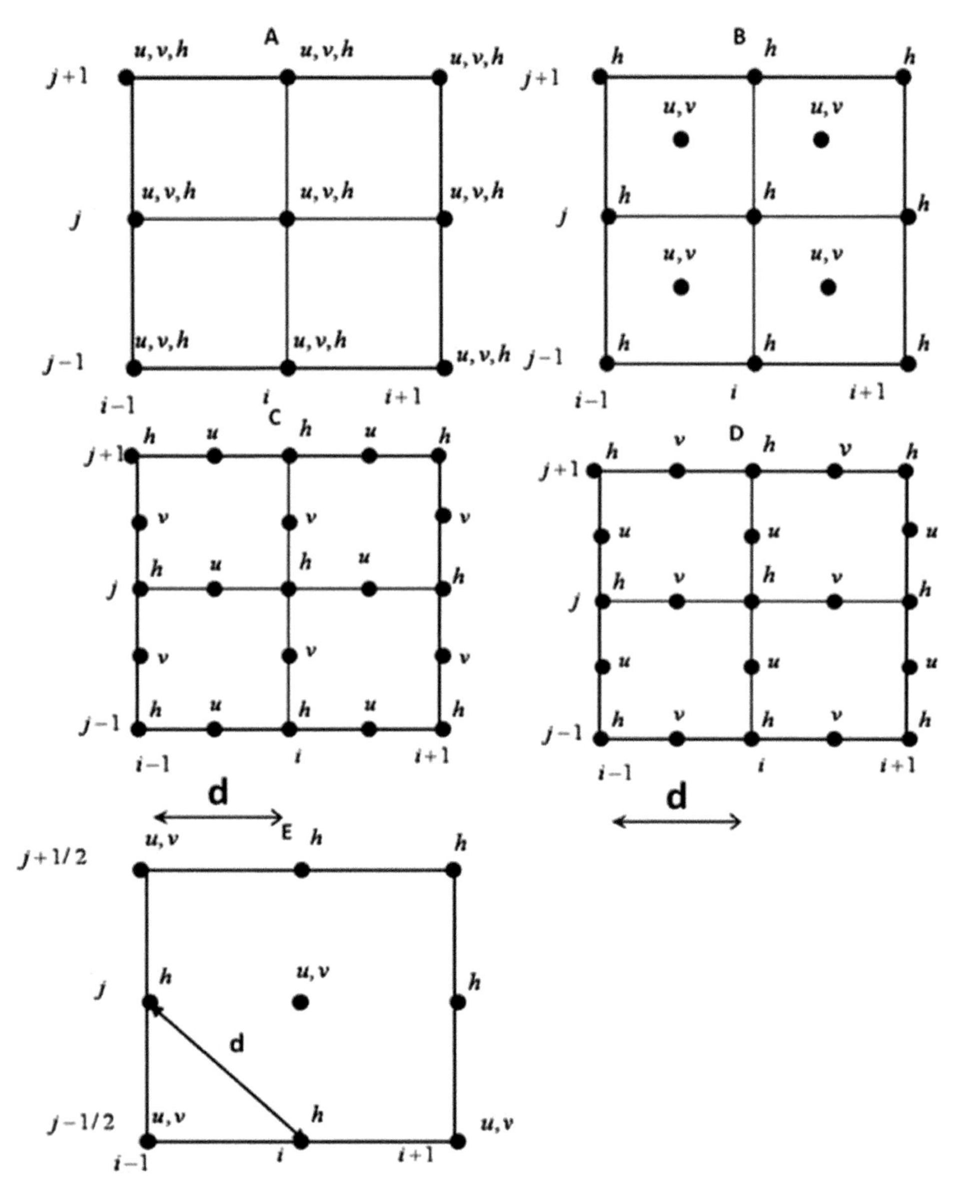

Figure 4.4.2 Five types of grids used in atmospheric numerical models (from Mesinger and Arakawa, 1976; courtesy of © World Meteorological Organization. Used with permission).

approximations. The spatial difference operator can be written as

$$(\delta_x a)_{i,j} = \frac{1}{d}\left(a_{i+1/2,j} - a_{i-1/2,j}\right). \quad (4.4.22)$$

This definition is applicable to all grids. The averaging operator for the same two grid points can also be expressed as

$$(\overline{a}^x)_{i,j} = \frac{1}{2}\left(a_{i+1/2,j} + a_{i-1/2,j}\right). \quad (4.4.23)$$

The quantities $\left(\delta_y a\right)_{i,j}$ and $(\overline{a}^y)_{i,j}$ are expressed similarly, but with respect to the y-axis. Finally, there is the averaging operator for both x- and y-directions:

$$(\overline{a}^{xy})_{i,j} = \overline{(\overline{a}^x)_{i,j}}^y = \frac{1}{4}\left(a_{i+1/2,j} + a_{i-1/2,j} + a_{i,j+1/2} + a_{i,j-1/2}\right). \quad (4.4.24)$$

The simplest finite difference approximations of the spatial derivatives and the Coriolis terms of Equation (4.4.21) are

Grid A: $\frac{\partial u}{\partial t} = g\overline{(\delta_x h)}^x + fv, \frac{\partial v}{\partial t} = g\overline{(\delta_y h)}^y - fu,$
$\frac{\partial h}{\partial t} = -H\left(\overline{\delta_x u}^x + \overline{\delta_y v}^y\right);$

Grid B: $\frac{\partial u}{\partial t} = g\overline{(\delta_x h)}^y + fv, \frac{\partial v}{\partial t} = g\overline{(\delta_y h)}^x - fu,$
$\frac{\partial h}{\partial t} = -H\left(\overline{\delta_x u}^y + \overline{\delta_y v}^x\right);$

Grid C: $\frac{\partial u}{\partial t} = g\delta_x h + f\overline{v}^{xy}, \frac{\partial v}{\partial t} = g\delta_y h - f\overline{u}^{xy},$
$\frac{\partial h}{\partial t} = -H\left(\delta_x u + \delta_y v\right);$

Grid D: $\frac{\partial u}{\partial t} = g\overline{(\delta_x h)}^{xy} + f\overline{v}^{xy}, \frac{\partial v}{\partial t} = g\overline{(\delta_y h)}^{xy} - f\overline{u}^{xy},$
$\frac{\partial h}{\partial t} = -H\left(\overline{\delta_x u}^{xy} + \overline{\delta_y v}^{xy}\right);$

Grid E: $\frac{\partial u}{\partial t} = g\delta_x h + fv, \frac{\partial v}{\partial t} = g\delta_y h - fu,$
$\frac{\partial h}{\partial t} = -H\left(\delta_x u + \delta_y v\right).$

It should be taken into account that in Equations (4.4.22), (4.2.23), and (4.4.24), the indices i and j denote the grid point at which the time derivatives are written. Thus, indices $i \pm 1/2$ and $j \pm 1/2$ denote locations shifted from this grid point by $d/2$ in the corresponding directions. Analysis done by Arakawa shows that grid C is the best one as regards the description of pressure-velocity adjustment and geostrophic adaptation. It describes the phase velocities and group velocities more accurately than grids of the other types. Currently, grid C is most widely used, for instance, in the WRF model.

4.4.3 Distribution of Variables Over Grid Points and the Vertical Structure of Finite-Difference Grids

In non-staggered grids, all the variables are calculated at all grid points. In staggered grids, the thermodynamical variables such as density, temperature, humidity, and microphysical variables are calculated at the same points where the pressure or the geopotential are calculated (in Figure 4.4.2 the depth of the layer h is calculated at these points). It allows us to accurately utilize (i.e., without any interpolation between values at neighboring grid points) the gas law equation.

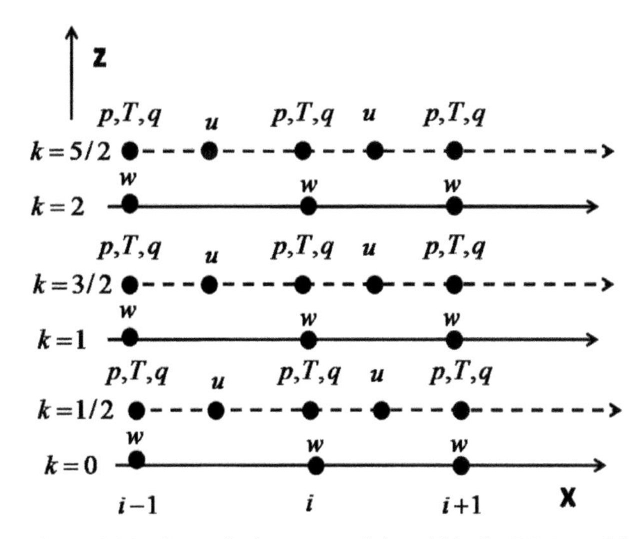

Figure 4.4.3 The vertical structure of the grid in the WRF model (from Skamarock et al., 2005 (courtesy of © UCAR 2005. Used with permission).

The vertical velocity is usually calculated at the levels located between the levels where the horizontal velocity is calculated. For grid C, the vertical velocity is calculated in the same columns as the pressure. The vertical structure in the x–z plane used in grid C is shown in **Figure 4.4.3**. The grid structure in the x–y cross-section is similar to that in the x–z plane, except that the velocity component v is used instead of u. Such a structure provides a number of advantages. First, the vertical velocity is typically known at the surface. In case of a plane surface, the boundary conditions for vertical velocity are easily determined (e.g., $w = 0$). Second, the mutual location of grid points where the velocity components are calculated allows application of the simplest and accurate utilization of the continuity equation. Third, the values of the thermodynamic variables at the surface are determined using the corresponding boundary conditions. In case of sea surface, its temperature is typically a known function of time. Sometimes, instead of prescribed values at the surface, the surface fluxes are used as boundary conditions. In the latter case, the fluxes and the surface values can also be determined using the similarity theory of the surface layer (Stull, 1988).

4.5 Numerical Schemes for Non-Hydrostatic Models

4.5.1 Anelastic Models and Fully Compressible Models

Numerical cloud models as well as advanced mesoscale models are non-hydrostatic. This means that they use the full equation for the vertical velocity component (Equation 3.2.20) instead of the hydrostatic equation (Equation 3.2.21). Equation (3.2.20) contains the vertical velocity acceleration, which means that solutions of the equation system also contain the acoustic modes. Although sound waves are not meteorologically significant, their rapid propagation imposes severe limitations on the time steps used in models.

Two types of non-hydrostatic models have been developed that treat the acoustic modes differently. These types are known as *anelastic* and *elastic*. The earlier non-hydrostatic models filtered sound waves using equations based on the anelastic approximation introduced by Ogura and Philips (1962). There are two main simplifications in the anelastic models: a) the continuity equation is written neglecting the time derivative of the density and b) the air density fluctuations are taken into account only in the buoyancy force term, while in other terms of the motion equations the density is assumed to be dependent on height only. It is known that omitting the time derivative term in the continuity equation removes acoustic modes from the system. The corresponding continuity equation

$$\frac{\partial \rho_0 u_i}{\partial x_i} = 0 \tag{4.5.1}$$

is known as continuity equation for deep convection, where $\rho_0(z)$ is the reference air density (Equation 3.2.21).

Many cloud models (e.g., HUCM, SAM) are anelastic. Scale analysis shows that Equation (4.5.1) is possible under the assumption that fluctuations of potential temperature are much smaller than the reference potential temperature $\theta' \ll \theta_0$. In most cases, this simplification is accurate enough. However, in cases when gravity waves in the stratosphere or deep convection in the upper troposphere are modeled, the accuracy of the simplification decreases. There are several different anelastic approximations that are valid for a wider range of temperatures and heights than those based on the approximation suggested by Ogura and Philips (Durran, 1989).

Non-hydrostatic models of the second type use the full continuity equation containing the time derivative of air density. In this type of elastic models, air density varies over time and space. The first numerical integrations of fully compressible equations were performed by Hill (1974) using an explicit scheme with time steps limited by the acoustic modes. An explicit integration of a fully compressible equations system is computationally time consuming. Two approaches are used to reduce the cost of simulations. Tapp and White (1976) integrated a compressible system with a semi-implicit scheme where the terms responsible for the acoustic modes are integrated implicitly. Klemp and Wihelmson (1978) and Skamarock and Klemp (1992) introduced schemes in which these terms are integrated using explicit schemes with a time step smaller than that used for integration of the other terms. This approach is more economical as calculation of the acoustic mode terms is inexpensive compared with that of nonlinear terms describing advection, microphysics, radiation, etc. This approach is used now in advanced mesoscale models such as MM5, RAMS, COSMO, and WRF.

4.5.2 Computational Aspects of Explicit Treating of Acoustic Modes

In many models, the gradient of pressure in the motion equations is replaced by the gradient of the Exner function Π (Section 3.1):

$$\Pi = \left(\frac{p}{p_0}\right)^{R_a/c_p}. \tag{4.5.2}$$

It simplifies the numerical algorithm allowing using potential temperature instead of density. Representing pressure and the Exner function as sums of reference (hydrostatic values) and deviations, $p(x,y,z) = p_0(z) + p''(x,y,z)$ and $\Pi(x,y,z) = \Pi_0(z) + \pi''(x,y,z)$, one can obtain from Equation (4.5.2):

$$\frac{1}{\rho}\frac{\partial p''}{\partial x_i} = c_p \theta_v \frac{\partial \pi''}{\partial x_i}, \tag{4.5.3}$$

where θ_v is the virtual potential temperature. Using Equation (4.5.3), averaged motion equations can be written (symbol standing for averaging is omitted):

$$\frac{\partial u_i}{\partial t} + u_j \frac{\partial u_i}{\partial x_j} = -c_p \theta_v \frac{\partial \pi''}{\partial x_i} - \varepsilon_{ijk} f_k u_j + \delta_{i3} B + D_u, \tag{4.5.4}$$

where B is the buoyancy force and D_u is the viscosity term.

In anelastic models, the equation for the pressure (or equation for π'') is a diagnostic one, i.e., it does not contain the time derivative. This equation can be obtained from the motion equations and using continuity Equation (4.5.1). The elliptic equation for π'' has the form

$$\nabla \cdot \rho_0 \nabla \pi'' = RHS, \tag{4.5.5}$$

where RHS denotes the right-hand side of the equations and ρ_0 is the background density, which does not vary in the horizontal direction but does vary vertically. There are two methods for solving elliptic equations such as Equation (4.5.5): an iteration method and direct method (Clark, 1977; Durran, 2010). In most direct methods, the solution is suggested in the form of the Fourier series where the coefficients are calculated using Fast Fourier Transform (FFT). There are two disadvantages of the approach based on solving the diagnostic equation for pressure. First, it is computationally time consuming, especially in 3D cases. Second, it is difficult to formulate the boundary conditions for function π''. These conditions should be derived from the motion equations themselves (so-called Neumann boundary conditions).

In case a fully compressible equation system is to be solved, the continuity equation used is

$$\frac{\partial \rho}{\partial t} + \frac{\partial \rho u_j}{\partial x_j} = 0. \tag{4.5.6}$$

Thus, the equation for π'' serves as a prognostic one, i.e., it contains the time derivative. Using definition of the Exner function alongside the continuity Equation (4.5.6), one can derive the equation for fluctuations of π'' as suggested by Klemp and Wilhelmson (1978):

$$\frac{\partial \pi''}{\partial t} + u_j \frac{\partial \pi''}{\partial x_j} = -\frac{c^2}{c_v \rho_0 \theta_v^2} \frac{\partial \rho_0 \theta_v u_j}{\partial x_j} + \frac{R\Pi_0}{c_p} \frac{\partial u_j}{\partial x_j} + \frac{c^2}{c_v \theta_v^2} \frac{d\theta_v}{dt} + D_\pi, \tag{4.5.7}$$

where $c = \sqrt{\frac{c_p}{c_v} R_a \Pi_0 \theta_v}$ is the sound velocity and D_π is diffusion term, and index 0 denotes values in the reference state. Equations (4.5.4) and (4.5.7) contain both high-frequency and low-frequency terms: values of terms representing acoustic and external gravity waves change fast; at the same time, the values of the advection terms, the Coriolis term, and the diffusion term change relatively slowly. This fact allows transforming Equations (4.5.4) and (4.5.7) by putting terms representing rapidly changing values on the left-hand side and the terms representing slowly changing ones on the right-hand side:

$$\frac{\partial u_i}{\partial t} + c_p \bar{\theta}_v \frac{\partial \pi''}{\partial x_i} = -u_j \frac{\partial u_i}{\partial x_j} - \varepsilon_{ijk} f_k u_j + \delta_{i3} B + D_u = R_u \tag{4.5.8}$$

$$\frac{\partial \pi''}{\partial t} + \frac{c^2}{c_v \rho_0 \theta_v^2} \frac{\partial \rho_0 \theta_v u_j}{\partial x_j} = -u_j \frac{\partial \pi''}{\partial x_j} + \frac{R\Pi_0}{c_p} \frac{\partial u_j}{\partial x_j} + \frac{c^2}{c_v \theta_v^2} \frac{d\theta_v}{dt} + D_\pi = R_\pi \tag{4.5.9}$$

The terms R_u and R_π on the right-hand sides of Equations (4.5.8) and (4.5.9) change slowly. According to the evaluations by Klemp and Wilhelmson (1978), term R_π is not important and can be set equal to zero without any significant effect on the solution (this simplification is known as quasi-compressible approximation). At the same time, the terms on the left-hand side of these equations are responsible for fast adaptation of the velocity field and of the pressure field due to high-frequency gravity and sound waves. Using the splitting-up method developed by Marchuk (1974), one can use different time steps, namely, Δt for slow terms and $\Delta \tau$ for rapid terms ($\Delta t >> \Delta \tau$).

Let us consider several examples of applications of the method. The scheme proposed by Klemp and Wilhelmson (1978) was applied in the MM5 mesoscale model used in numerous studies including simulation of clouds, cloud ensembles, and supercell storms (e.g., Lynn et al., 2005a,b; Lynn and Khain, 2007). The sequence of calculations proposed by Klemp and Wilhelmson is illustrated in **Figure 4.5.1**. In Equations (4.5.8) and (4.5.9), the right-hand-side terms R_u and R_π are calculated using a leapfrog scheme with a time step $2\Delta t$. A forward-backward scheme with a time step $\Delta \tau$ was applied for time integration of the terms representing the high-frequency acoustic modes.

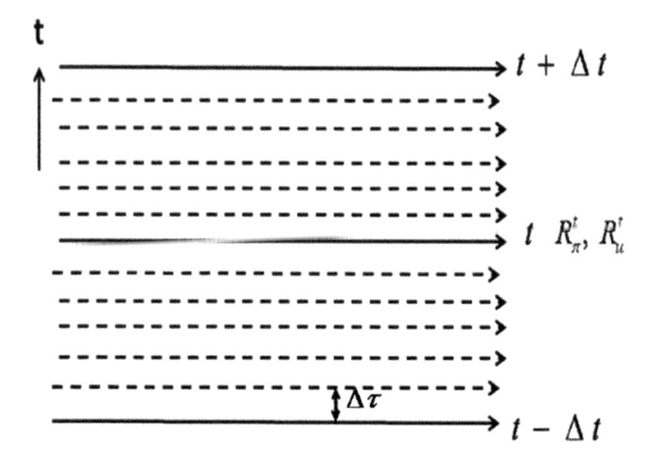

Figure 4.5.1 A scheme illustrating the sequence of time integration proposed by Klemp and Wilhelmson (1978) for treatment of sound waves and gravity waves.

Let us assume that at the beginning of the larger time step Δt used for integration of low frequency terms the values R_u^t and R_π^t on the right-hand side are known. Then the high-frequency terms are integrated using the time step $\Delta\tau$ over $n = \frac{2\Delta t}{\Delta\tau}$ small steps:

$$\begin{aligned}
&\frac{u_i^{\tau+1}-u_i^\tau}{\Delta\tau}+c_p\overline{\theta}_v\delta_x\pi^\tau=R_u^t\\
&\frac{v_j^{\tau+1}-v_j^\tau}{\Delta\tau}+c_p\overline{\theta}_v\delta_y\pi^\tau=R_v^t\\
&\frac{\widetilde{w}^{\tau+1}-\widetilde{w}^\tau}{\Delta\tau}+c_p\rho_0\theta_v^2\overline{\delta_z\pi}^\tau=\rho_0\overline{\theta}_vR_w^t\\
&\frac{\pi^{\tau+1}-\pi^\tau}{\Delta\tau}+\frac{c^2}{c_p\theta_v^2}\left(\delta_xu^{\tau+1}+\delta_yv^{\tau+1}\right)+\frac{c^2}{c_p\rho_0\theta_v^2}\overline{\delta_z\widetilde{w}}^\tau=0.
\end{aligned} \tag{4.5.10}$$

Equations (4.5.10) are approximations of the differential Equations (4.5.8) and (4.5.9) written using the Equations (4.4.22), (4.4.23), and (4.4.24). In Equation (4.5.10) $\tilde{w} = \rho_0\overline{\theta}_v w$. To provide numerical stability, the time integration expressed by Equations (4.5.10) is performed using the approach described by Equation (4.4.13). After reaching the time instance $t + \Delta t$, new values of R_u^{t+1} and R_π^{t+1} are calculated and new time step is performed.

Other examples of splitting the terms up into the low-frequency and high-frequency modes can be found in Wicker and Skamarock (1998) and Wicker and Skamarock (2002). Unlike Klemp and Wilhelmson (1978), these authors use Runge–Kutta (RK) schemes: Wicker and Skamarock (1998) used RK2 and Wicker and Skamarock (2002) used RK3 schemes (see Equations 4.1.18a, 4.1.18b) to perform integration of the low-frequency mode. The RK3 scheme is currently applied in the WRF model, which is widely used for simulation of many meteorological phenomena, from individual clouds to synoptic systems. In particular, it was applied for simulation of tropical cyclones with different bulk parameterizations of convection, as well as with spectral bin microphysics (e.g., Khain et al., 2010, 2016). In case the RK2 scheme is used, the right-hand sides of the corresponding equations are recalculated for time instances of $\Delta t/2$ and Δt. In the RK3 scheme, the right-hand sides are recalculated for time instances of $\Delta t/3$, $\Delta t/2$, and Δt. Each of these intervals is divided into the corresponding small time steps of $\Delta\tau$ in order to integrate high-frequency terms (see Skamarock et al., 2008 for more details). Wicker and Skamarock (2002) showed a significant advantage of RK3 over RK2 regarding the scheme stability. In a 1D case, the RK3 scheme is stable for maximum Courant numbers equal to 1.08 and 1.61 for advection schemes with approximation accuracy of 6-th and 3-rd order, respectively. In a 3D case, the maximum Courant number should be taken $\sqrt{3}$ times smaller than in the 1D case. Regarding the integration of high-frequency terms, small time steps should be chosen from the condition $\Delta\tau < 2\frac{\Delta x}{c}$, where $c \approx 300\,\mathrm{m/s}$ is the sound speed. The number of small time steps in WRF with a RK3 scheme is determined from the condition $n = \max\left(2\cdot\left[300\frac{\Delta t}{\Delta x}+1\right],4\right)$ (Skamarock et al., 2008).

4.5.3 Boundary Conditions and Nesting Grid Systems

Boundary conditions play a very important role in numerical models, in particular in cloud modeling. At the underlying surface the momentum, heat, and moisture fluxes are either prescribed or calculated. In the latter case, the Monin–Obukhov similarity theory of the surface layer is typically used.

The lateral boundaries inevitably introduce errors in calculations, because the real atmosphere does not contain such boundaries. When the described phenomena (for instance, clouds) approach the lateral boundaries, it can cause erroneous fluctuations near the boundaries, which may affect the solution. The easiest way to avoid these problems is to use periodic boundary conditions in the horizontal direction. These conditions are used in models of startiform clouds (LES models), in models of convective cells in the boundary layer, in direct numerical simulations of turbulent processes, etc. The scheme of application of periodic (cyclic) boundary conditions is illustrated in **Figure 4.5.2**. In case cyclic boundary conditions are used, the derivatives at the lateral boundaries can be approximated, for instance, as follows:

$$\left(\frac{\partial a}{\partial x}\right)_N \sim \frac{a_2 - a_{N-1}}{2\Delta x};\ \left(\frac{\partial^2 a}{\partial x^2}\right)_N \sim \frac{a_2 - 2a_N + a_{N-1}}{(\Delta x)^2}. \tag{4.5.11}$$

Figure 4.5.2 A scheme of application of cyclic boundary conditions.

Application of cyclic boundary conditions has its drawbacks. Even in cases when the problem to be solved allows utilization of such conditions, the choice of the horizontal size of the computational area determines the spatial periodicity of the solution. For instance, utilization of cyclic conditions may affect the size of convective cells, their aspect ratio, and other parameters.

When noncyclic boundary conditions are used, simulated clouds (or other phenomena) should remain at a significant distance from the lateral boundaries. In some cases (e.g., simulation of a convective cell in the boundary layer) so-called symmetric boundary conditions can be used (Skamarock et al., 2008). In this case, velocity components normal to the lateral boundaries are equal to zero. The field of normal velocity is antisymmetric with respect to the boundary, i.e., $u(x_b - x) = -u(x_b + x)$, where x_b is the coordinate of the boundary. Other variables are assumed to be symmetric with respect to the boundary, i.e., $a(x_b - x) = a(x_b + x)$. As an example, we can mention the model of the boundary layer (Pinsky et al., 2008), where cyclic boundary conditions are used at the lateral boundaries, and the antisymmetric boundary condition is used at the low boundary to design a turbulent-like velocity field.

The conditions of zero horizontal derivatives of the horizontal velocity component as well as various thermodynamic values are often used at the lateral boundaries. Any gradients of temperature and velocity at the lateral boundaries cause the emergence of erroneous vertical velocity near the boundaries, formation of erroneous clouds, etc. The lateral boundaries can reflect the waves generated within the computational area backward, causing appearance and growth of computational noise. This noise can be eliminated by using the adequate filters, diffusion terms, etc.

It is of high importance that the boundary conditions allow a free outward passage of gravity waves and advected phenomena. In several studies (e.g., Orlanski, 1976; Falkovich et al., 1995) the radiation boundary conditions eliminating the wave reflection back to the computational area were proposed. One of the versions of the radiation boundary condition is illustrated in **Figure 4.5.3**. This method assumes that near a lateral boundary (e.g., the right boundary), the simulated phenomenon moves as a linear propagating wave at some phase velocity C_{ph}. In this case, the changes of variable a (which can also be a velocity component normal to the boundary) near the boundary are described by the following equation:

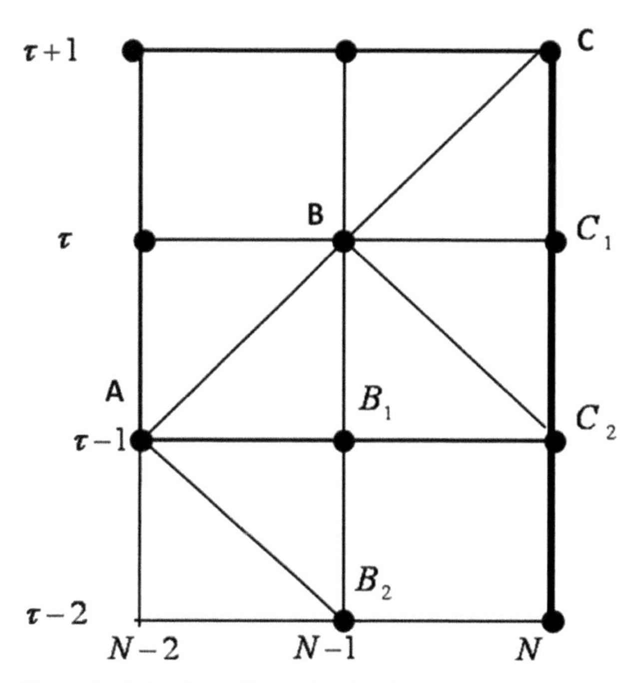

Figure 4.5.3 A scheme illustrating the algorithm of radiation condition at the right lateral boundary. Column N is the right lateral boundary (thick black line).

$$\frac{\partial a}{\partial t} = -C_{ph}\frac{\partial a}{\partial x}. \tag{4.5.12}$$

Let us consider the finite difference grid near the right boundary as shown in Figure 4.5.3. The problem is to calculate the value $a_N^{\tau+1}$ at point C. The algorithm is based on the assumption that the phase speed does not change at small distances near the boundary. Accordingly, the calculation algorithm consists of two substeps (Falkovich et al., 1995), namely: a) determination of the phase speed C_{ph} from triangle ABB_2:

$$C_{ph} = -\left(\frac{\partial a}{\partial t}\right)\Big/\left(\frac{\partial a}{\partial x}\right) = \frac{\Delta x}{2\Delta t}\left(a_{N-1}^{\tau} - a_{N-1}^{\tau-2}\right)\Big/\left[\frac{1}{2}\left(a_{N-1}^{\tau} + a_{N-1}^{\tau-2}\right) - a_{N-2}^{\tau-1}\right] \tag{4.5.13}$$

and

b) determination of $a_N^{\tau+1}$ from triangle BCC_2:

$$a_N^{\tau+1} = a_N^{\tau-1} - 2\frac{\Delta t}{\Delta x}C_{ph}\left[\frac{1}{2}\left(a_N^{\tau+1} + a_N^{\tau-1}\right) - a_{N-1}^{\tau}\right]. \tag{4.5.14}$$

The algorithm described by Equations (4.5.13) and (4.5.14) is close to that used by Orlanski (1976).

There are other algorithms of the radiation boundary conditions (e.g., Klemp and Lilly, 1978; Klemp and Wilhelmson, 1978). In these algorithms, the horizontal velocities perpendicular to the boundaries are determined from Equation (4.5.12), where the phase velocity is determined as $C_{ph} = \min(u - c_b, 0)$ at the western boundary ($x = 0$) and as $C_{ph} = \max(u + c_b, 0)$ at the eastern boundary. The phase speed of a gravity wave c_b is assumed to be a given constant. The radiative conditions at the north and south boundaries are defined in a similar way, but the y-component of the wind speed is used. Spatial derivatives in the direction perpendicular to the boundaries are calculated by the difference between the values at the boundary and the values at one grid point away from the boundary. This approach makes it unnecessary to determine the phase speed by means of Equation (4.5.13). A more detailed description of radiative conditions can be found in Klemp and Lilly (1978) and Klemp and Wilhelmson (1978). Clark (1977) analyzed the advantages and disadvantages of these two versions of radiative boundary conditions. He found some advantages of the two-step approach in a particular case, when C_{ph} is determined at an interior point next to the boundary.

The upper boundary of the computational area can also reflect waves ascending from below, and thus significantly affect the solution. There were several attempts to formulate the upper boundary condition that could prevent such reflection. In fully compressible elastic models, an acoustic radiation condition is often used (Tripoli and Cotton, 1982). There are also more complicated expressions for radiative conditions at the upper boundary.

Typically, a simpler approach to damping the reflected waves is used. In this approach, in the vicinity of the upper boundary the friction terms are added (so-called Rayleigh damping), which nudge the solution to the prescribed values (taken, for example, from the sounding data). The equation with the additional term in the right-hand side has the form

$$\frac{\partial a}{\partial t} = \cdots\cdots + (a_0 - a)/\tau, \qquad (4.5.15),$$

where a_0 is the value to which the solution should be nudged, and τ is the characteristic time of the nudging. With other terms neglected, the Rayleigh damping term leads to an exponential decrease of the difference $a_0 - a$, i.e., $a \to a_0$. The friction terms are added at heights where they cannot affect the parameters of interest. For instance, in convective cloud models the Rayleigh damping can be applied at heights above 14–15 km where clouds do not exist. Typically, the value of τ is decreased toward the upper boundary accelerating the nudging and damping non-desirable perturbations.

Many current mesoscale models applied for simulation of clouds, storms, etc. use nesting grid systems, when one or several grids with higher resolution are nested into a grid with lower resolution. Grids with different resolution within the same model are needed due to the nature of the atmospheric phenomena. For instance, to resolve the central zone of a TC, one needs small grid spacing that provides a better description of physical processes and high gradients in this zone. The TC zone can have 500 x 500 km size with a grid spacing of a few km or even 1 km. At the same time, TC during their evolution move by several thousands of kilometers within a steering current that does not contain strong gradients and does not require such fine grid spacing. **Figure 4.5.4** shows the Arakawa–C grid staggering for a

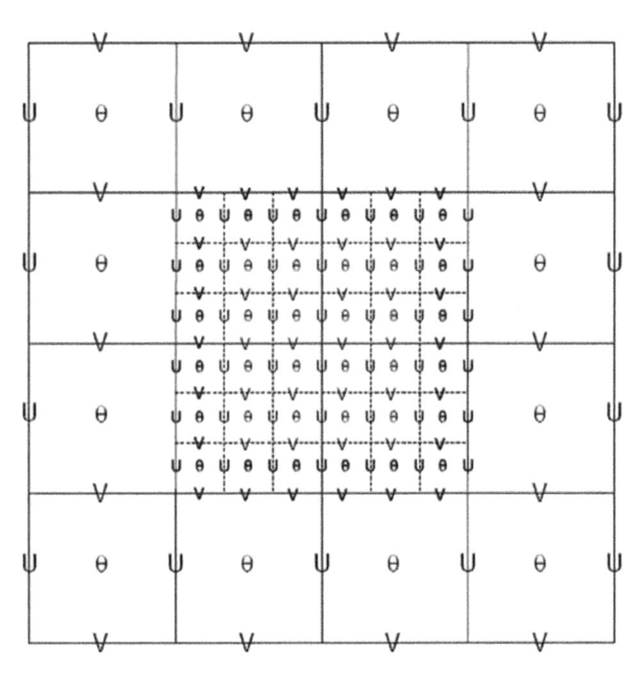

Figure 4.5.4 Arakawa–C grid staggering for a portion of a parent domain and an imbedded nest domain with a 3:1 grid spacing ratio. The solid lines denote coarse grid cell boundaries, and the dashed lines are the boundaries for each fine grid cell. The horizontal components of velocity (U and V) are defined along the cell face, and the thermodynamic variables (#) are defined at the center of the grid cell (each square). The bold-typeface variables along the interface between the coarse grid and the fine grid define the locations where the specified lateral boundaries for the nest are in effect (from Skamarock et al., 2008; courtesy of © UCAR 2008. Published with permission).

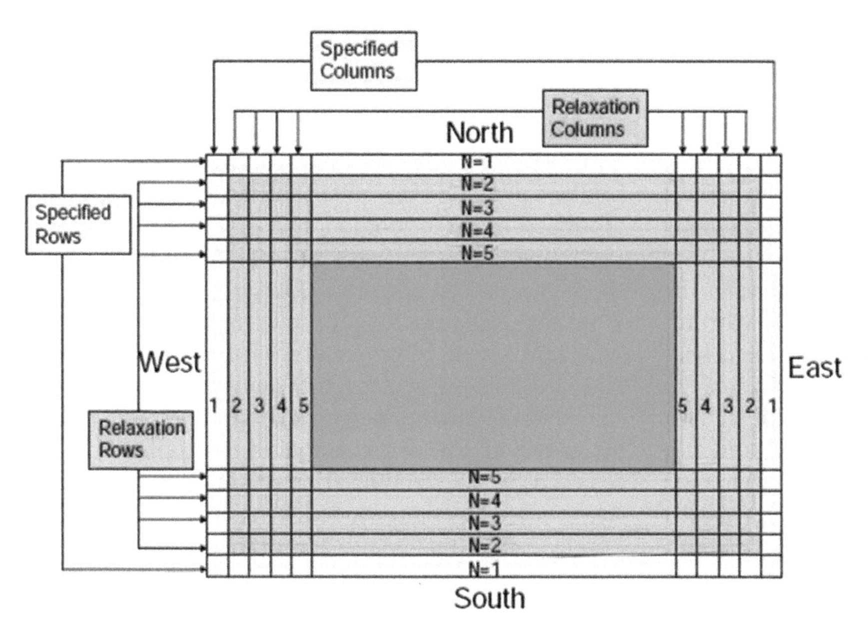

Figure 4.5.5 Specified zones and relaxation zones for an internal grid with a single specified row and column. The relaxation zones consist of four rows and columns. These are the typical values used for a specified lateral boundary condition in a real-data case (from Skamarock et al., 2008; courtesy of © UCAR 2008. Published with permission).

portion of the parent domain and an imbedded nest domain with a 3:1 grid size ratio used in a WRF model (Skamarock et al., 2008). Since the scales resolved by the imbedded grid are not resolved by the outer grid, the boundaries between the grids are sources of different perturbations affecting the solution within the imbedded grid. Several methods for matching the inner and outer grids were proposed. Falkovich et al. (1995) used the radiation condition to avoid formation of erroneous fluctuations near the boundaries of the inner imbedded grid. According to this study, at the interfaces between nesting grids the radiative conditions should be applied to the difference between the solutions obtained for fine grids and coarse grids.

At present, the method using relaxation zones between grids is widely used. **Figure 4.5.5** shows the scheme of matching parent grid with imbedded grids. Specified zones belong to the external grid. In these zones the interpolation of values from the external to internal grid is performed. The relaxation zone contains four rows and four columns. Within the relaxation zone, perturbations are damped using a special smoother. The strength of the smoother depends on the ordinal number of a particular relaxation rows, and columns and is higher for the outer rows and the outer columns. More details can be found in Skamarock et al. (2008).

4.6 Comments Concerning Application of Numerical Schemes in Cloud Models

This chapter gives a brief overview of numerical methods used in cloud and mesoscale models. Many existing numerical methods, such as the finite elements method and spectral methods based on representation of variables in the form of the Fourier series that are widely used in LES models, are beyond the scope of this book. The same is true about methods for solving 3D elliptic equations. The description of these methods can be found in other works (e.g., Durran, 2010).

The purpose of this overview is to present the main methods used in numerical models, largely in cloud models, and to relate these methods to requirements consistent with the physical nature of simulated phenomena. We have shown that different numerical schemes may yield different solutions for the same equations, which in turn may lead to different results when simulating the same phenomena. Correct choice of time steps is of crucial importance for getting physically grounded results. To save computation time, modelers tend to use time steps of permissible maximum, which enables us to keep the schemes stable (e.g., time steps obeying the CFL criteria). However, physical processes simulated by models impose their own requirements on the choice of time steps. In Section 4.5, we considered the methods using different time steps to integrate both the low-frequency and high-frequency terms. In these methods, most physical processes (e.g., microphysical processes in clouds) are assigned to the low-frequency terms having large characteristic time scales. At the same time, each microphysical process has its own characteristic time. For instance, processes of diffusion growth/evaporation have a characteristic time scale known as phase relaxation time (Section 5.2), which is often as low as a few seconds. To describe the simulated processes accurately (which is important because these processes are responsible for latent heat release affecting the dynamics and microphysics of the simulated phenomena), time steps should be chosen smaller than the phase relaxation time. Various microphysical processes such as collisions leading to changes in droplet size distributions and precipitation formation may also have quite short characteristic time steps. Application of large time steps in explicit schemes may lead to the erroneous appearance of negative values of positive-defined values (e.g., the example of the friction equation in Section 4.3).

Utilization of stable explicit schemes or positive-defined schemes may avoid appearance of negative values, yet does not solve the problem of proper physical treatment of the corresponding physical processes. As a result, inconsistency may arise during time integration of equations describing different processes that have different characteristic time scales.

Correct choice of the model grid resolution is also of high importance. In current mesoscale CRM and cloud models, the grid spacing of 2–3 km is widely used. The use of the model grid spacing of 2–3 km instead of ~10–20 km in mesoscale models has become possible due to increased computer power. This improvement leads first of all to a better representation of synoptic (large-scale) processes. At the same time, most convective phenomena (such as individual convective clouds) cannot be adequately resolved in such models, because approximation is lost at scales lower than a double grid spacing, and the maximum errors of numerical schemes are at spatial scales of 2–4 grid spacing – i.e., at scales of 4–6 km; exactly within the convective spatial range. Currently, a kind of a hybrid approach is used in such models, i.e., explicit calculation of clouds is complemented by the description of subgrid cloudiness.

In order to improve the representation of convective processes, grid spacing lower than a few hundred meters should be used. Such grid spacing would agree with the natural separation of atmospheric motions into convective and turbulent motions (Section 3.4). Then, subgrid motions could be considered as small-scale turbulence. In this case, utilization of the K-theory (which is now widely used for any grid spacing) would be physically grounded. We believe that with further growth of computer power, it will be possible to reduce errors induced when time steps and grid spacing are too large. However, even at present much more accurate results can be obtained due to utilization of physically grounded model resolution and time steps in research studies which do not have as strict time limitations as those imposed on operative (forecast) models.

In large-scale models, accurate high-order advection schemes provide better results, which encourage scientists to use such schemes. Note that these schemes use a significant number of neighboring grid points. In cloud models used for investigation of microphysical processes in clouds, the utilization of high-order advection schemes is sometimes undesirable, because the neighboring grid points used in these schemes may turn out to be in different parts of simulated cloud or even outside of the cloud. In these cases, the utilization of lower-order approximation schemes can be recommended. Of course, if grid spacing is low enough, the utilization of the schemes of higher order of approximation is justified.

One of the main problems hindering the wide application of cloud models with spectral bin microphysics in the operation practice is that these models are computationally expensive. Bin schemes typically have 30–40 bins per hydrometeor type and one or two prognostic variables per bin. State-of-the-art mixed-phase bin schemes contain 5–10 hydrometeor types: liquid drops, ice crystals, snowflakes, graupel, hail, aerosols, etc. This means that several hundred scalar quantities must be transported. More than 70% of the computer cost of these models is the transport of these quantities in physical space. Using the fact that at each time step these quantities are transported within the same wind field, Morrison (2016) proposed some modification of advection procedure allowing significant decrease in the computer time.

References

Arakawa, A., 1966: Computational design for long-term numerical integration of the equations of fluid motion: Two-dimensional incompressible flow. Part 1. *J. Comput. Phys.* **1**, #1.

1972: Design of the UCLA general circulation model. Numerical simulation of weather and climate. Dept. of Meteorology, Univ. of California, Los Angeles, Tech Rept. 7, p. 116.

Bott, A., 1989: A positive definite advection scheme obtained by nonlinear renormalization of the advective fluxes. *Mon. Wea. Rev.*, **117**, 1006–1015.

Bryan, K., 1966: A scheme for numerical integration of the equations of motion on an irregular grid free of nonlinear instability. *Mon. Wea. Rev.*, **94**, 1.

Clark, T.L., 1977: A small scale dynamic model using a terrain following coordinate transformation. *J. Comput. Phys.*, **24**, 186–215.

Durran, D.R., 1989: Improving the anelastic approximation. *J. Atmos. Sci.*, **46**, 1453–1461.

2010: *Numerical methods for fluid dynamics with applications to Geophysics*. Springer, p. 515.

Falkovich, A.I., A.P. Khain, and I. Ginis, 1995: Evolution and motion of binary tropical cyclones as revealed by experiments with a coupled atmosphere–ocean movable nested grid model. *Mon. Wea. Rev.*, 123, 1345–1363.

Ginis, I., A. Khain, and E. Morosovsky, 2004: Effects of large eddies on the structure of the marine boundary layer under strong wind conditions. *J. Atmos. Sci.*, **61**, 3049–3063.

Hill, G.E., 1974: Factors controlling the size and spacing of cumulus clouds as revealed by numerical experiments. *J. Atmos. Sci.*, **30**, 1672–1690.

Khain, A.P., B. Lynn, and J. Dudhia, 2010: Aerosol effects on intensity of landfalling hurricanes as seen from simulations with WRF model with spectral bin microphysics. *J. Atmos. Sci.*, **67**, 365–384.

Khain, A., B. Lynn, and J. Shpund, 2016: High Resolution WRF Simulations of Hurricane Irene: Sensitivity to Aerosols and Choice of Microphysical Schemes. *Atmos. Res.*, 167, 129–145.

Khain, A., A. Pokrovsky, M. Pinsky, A. Seifert, and V. Phillips, 2004: Effects of atmospheric aerosols on deep convective clouds as seen from simulations using a spectral microphysics mixed-phase cumulus cloud model. Part 1: Model description. *J. Atmos. Sci.*, **61**, 2963–2982.

Khain, A.P., and I. Sednev, 1996: Simulation of precipitation formation in the Eastern Mediterranean coastal zone using a spectral microphysics cloud ensemble model. *Atmos. Res.*, **43**, 77–110.

Khairoutdinov, M.F., and D.A. Randall, 2003: Cloud-resolving modeling of the ARM summer 1997 IOP: Model formulation, results, uncertainties and sensitivities. *J. Atmos. Sci.*, **60**, 607–625.

Klemp, J.B., and D.K. Lilly, 1978: Numerical simulation of hydrostatic mountain waves. *J. Atmos. Sci.*, **35**, 78–107.

Klemp, J.B., and R. Wilhelmson, 1978: The simulation of three-dimensional convective storm dynamics. *J. Atmos. Sci.*, **35**, 1070–1096.

Leith, G., 1965: Lagrangian advection in an atmospheric model. WMO-IUGG Symposium on Research and development Aspects of long-range forecasting. Boulder, Colo. WMO Tech. Note **66**, 168–176.

Lynn, B., and A. Khain, 2007: Utilization of spectral bin microphysics and bulk parameterization schemes to simulate the cloud structure and precipitation in a mesoscale rain event. *J. Geophys. Res.* **112**, D22205.

Lynn, B., A. Khain, J. Dudhia, D. Rosenfeld, A. Pokrovsky, and A. Seifert, 2005a: Spectral (bin) microphysics coupled with a mesoscale model (MM5). Part 1. Model description and first results. *Mon. Wea. Rev.*, **133**, 44–58.

2005b: Spectral (bin) microphysics coupled with a mesoscale model (MM5). Part 2: Simulation of a CaPe rain event with squall line. *Mon. Wea. Rev.*, **133**, 59–71.

Marchuk, G.I., 1974: *Numerical methods in weather prediction*. Academic Press, p. 277.

1980: *Methods of computational mathematics* (in Russian). Nauka Press, p. 536.

Mesinger, F., and A. Arakawa, 1976: Numerical methods used in atmospheric models. **1**, GARP Publication Series, 17. WMO.

Morrison, H., A.A. Jensen, J.Y. Harrington, and J.A. Milbrandt, 2016: Advection of coupled hydrometeor quantities in bulk cloud microphysics schemes. *Mon. Wea.Rev.*, **144**, 2809–2829.

Ogura, Y., and N.A. Phillips, 1962: Scale analysis of deep and shallow convection in the atmosphere. *J. Atmos. Sci.*, **19**, 173–179.

Orlanski, I., 1976: A simple boundary condition for unbounded hyperbolic flows. *J. Comput. Physics*, **21**, 251–269.

Pielke, R.A., 2013: *Mesoscale meteorological modeling*, Third Edition. Academic Press, p. 722.

Pinsky, M., A. Khain, L. Magaritz, and A. Sterkin, 2008: Simulation of droplet size distributions and drizzle formation using a new trajectory ensemble model of cloud topped boundary layer. Part 1: Model description and first results in non-mixing limit. *J. Atmos. Sci.*, **65**, 2064–2086.

Pinsky, M.B., A.P. Khain, and M. Shapiro, 2007: Collisions of cloud droplets in a turbulent flow. Part 4: Droplet hydrodynamic interaction. *J. Atmos. Sci.*, **64**, 2462–2482.

Rood, R.B., 1987: Numerical advection algorithms and their role in atmospheric transport and chemistry models. *Rev. Geophys.*, **25**, 71–100.

Skamarock, W.C., 2006: Positive-definite and monotonic limiters for unrestricted-time-step transport scheme. *Mon. Wea. Rev.*, **134**, 2241–2250.

Skamarock, W.C., and J.B. Klemp, 1992: The Stability of Time-Split Numerical Methods for the Hydrostatic and the Nonhydrostatic Elastic Equations. *Mon. Wea. Rev.*, **120**, 2109–2127.

Skamarock, W.C., J.B. Klemp, J. Dudhia, D.O. Gill, D.M. Barker, W. Wang, and J.G, 2005a: A Description of the Advanced Research WRF Version 2. Powers, NCAR Technical Note.

Skamarock, W.C., J.B. Klemp, J. Dudhia, D.O. Gill, D.M. Barker, W. Wang, and J.G. Powers, 2005b: A Description of the Advanced Research WRF Version 2, NCAR, Boulder, Colorado.

Skamarock, W.C., J.B. Klemp, J. Dudhia, D.O. Gill, D.M. Barker, M.G. Duda, X-Y. Huang, W. Wang, and J.G.

Powers, 2008: A Description of the advanced research WRF version 3. NCAR Technical Note. Boulder, Colorado.

Smolarkiewicz, P.K., and G.A. Grell, 1990: A class of monotone interpolation schemes. *J. Comp. Phys.*, **101**, 431–440.

Stull, R.B., 1988: *An Introduction to Boundary Layer Meteorology*. Kluwer Academic Publishers, p. 666.

Tao, W.-K, J. Simpson, D. Baker, S. Braun, M. Chou, B. Ferrier, D. Johnson, A. Khain, S. Lang, B. Lynn, C. Shie, D. Starr, C-h. Sui, Y. Wang, and P. Wetzel, 2003: Microphysics, radiation and surface processes in the Goddard Cumulus Ensemble (GCE) model. *Meteorol. Atmos. Phys.*, **82**, 97–137.

Tapp, M.C., and P.W. White, 1976: A non-hydrostatic mesoscale model. *Quart. J. Royal Meteorol. Soc.*, **102**, 277–296.

Temam, R., 1977: *Theory and Numerical Analysis of the Navier-Stokes Equations*. North-Holland, p. 465.

Tripoli, G.I., and Cotton, W.R., 1982: The Colorado State University three-dimensional cloud/mesoscale model-1982. Part 1: General theoretical framework and sensitivity experiments. *J. de Rech. Atmos.*, **16**, 185–220.

Wicker, L.J., and W.C. Skamarock, 1998: A time splitting scheme for the elastic equations incorporating second-order Runge-Kutta time differencing. *Mon. Wea. Rev.*, **126**, 1992–1999.

2002: Time-splitting methods for elastic models using forward time schemes. *Mon. Wea. Rev.*, **130**, 2088–2097.

5 Warm Microphysical Processes

Microphysical processes that are not related to ice formation are often referred to as warm microphysical processes. It does not mean that these processes take place at positive temperatures only. Drops of a particular kind (supercooled drops) can exist at temperatures as cold as −38°C; nevertheless their diffusional growth and collisions are considered as warm microphysical processes. The major warm microphysical processes and terms of kinetic equations describing their rates are listed in **Table 5.1.1**.

Warm microphysics includes a wide range of microphysical processes taking place during cloud formation and evolution, and afterward during rain formation. The typical stages are generation of small droplets growing on cloud condensational nuclei (nucleation), their diffusional growth up to the radii of about 22–25 μm, and finally, collisions between droplets, targeting formation and growth of raindrops that later settle down to the surface. Rain forming this way is called *warm rain*. The relative velocity between falling raindrops and the surrounding air intensifies the heat exchange and the

Table 5.1.1 Rates of major warm microphysical processes.

Warm Process	Term in a Kinetic Equation
Nucleation	$\frac{\partial f(m(r_{Ncr}),t)}{\partial m} = \begin{cases} -\frac{\partial f_{CCN}(r_{Ncr})}{\partial r_{Ncr}} & S_w > S_{w\_\max} \\ 0 & S_w < S_{w\_\max} \end{cases}$
Condensation/evaporation with a ventilation effect	$-F_v \frac{\partial}{\partial m}\left[\frac{dm}{dt} f(m)\right]$
Sedimentation	$-\frac{\partial}{\partial z}\left[V_g(m) f(m)\right]$
Spontaneous breakup	$\int_0^{\infty} f(m')P_r(m')P(m,m')dm' - f(m)P_r(m)$
Collisions and coalescence	$\frac{1}{2}\int_0^{m} f(m')f(m-m')K(m-m',m')dm' - f(m)\int_0^{\infty} f(m')K(m,m')dm'$
Collisional breakup	$\frac{1}{2}\int_0^{\infty} f(m'')dm''\int_0^{\infty} f(m'')B(m,m'')P(m;m'',m')dm' - f(m)\int_0^{\infty}\frac{f(m'')B(m,m'')}{m+m''}dm''\int_0^{\infty} m'P(m';m,m'')dm'$
Conservative mixing	$\frac{\partial}{\partial x_i} K_{ij} \frac{\partial}{\partial x_j} f(m)$
Non-conservative mixing	$\left(\frac{\partial}{\partial x_i} + G_i \frac{\partial}{\partial m}\right) K_{ij} \left(\frac{\partial}{\partial x_j} + G_j \frac{\partial}{\partial m}\right) f(m)$

mass exchange between the drops and the surrounding air. This leads to an increase in the rate of drop size change that can be either condensational (at positive supersaturation) or evaporational (at negative supersaturation, i.e., at saturation deficit). This effect is known as the ventilation effect. As a result of collisional growth, raindrops reach sizes at which their surface tension cannot keep the drop shape any longer. Raindrops gain asymmetric disc-like shapes and can break up spontaneously, or due to collisions with other particles. During their fall below cloud base, raindrops evaporate partially or totally, depending on their size and the relative environmental humidity. The rates of condensation/evaporation depend on the environmental thermodynamic conditions that are affected by turbulent mixing between the neighboring air parcels. Some quantities such as the equivalent potential temperature and the total water mixing ratio (including water vapor) do not change during a rapid adiabatic mixing. Nonconservative values (referred to sometimes as invariant) such as supersaturation, size distribution (SD) functions, temperature, etc. change during an adiabatic mixing.

5.1 Droplet Nucleation

5.1.1 Equation for Condensation Growth of Drops

Equilibrium between the two phases (in our case, between water vapor and liquid water) exists if the following equalities for both phases are obeyed (Pruppacher and Klett, 1997):

a) temperature equality: $T_1 = T_2$;
b) pressures equality: $p_1 = p_2 = e$;
c) equality of chemical potentials: $\mu_1 = \mu_2$.

The pressure of saturation water vapor upon drops with small radii is not equal to that upon the planar surface. Two factors determine this difference: the curvature of a drop surface and the presence of a dissolved substance (salt molecules, ions) in drops. During condensation growth or evaporation of drops, the surface area changes, thus the surface energy changes accordingly.

We can formulate the expression for the equilibrium (saturation) pressure upon spherical drops. Since the saturation water vapor pressure upon a drop depends on its radius, we denote it as $e_w(r)$. If the pressure of the environmental water vapor is larger than $e_w(r)$, the number of water vapor molecules that condensate is larger than the number of evaporating molecules, and drops grow. In case of a spherical drop, the surface tension creates an additional pressure inside the drop, thus the condition b) is to be replaced by

$$p_2 - p_1 = \frac{2\sigma_w}{r}, \tag{5.1.1}$$

where $p_1 = e_w(r)$ is the equilibrium vapor pressure around the drop, which now depends on the drop radius r, and σ_w is the surface tension equal to the work needed to increase the surface by a unit of square. The surface tension slightly depends on the temperature (Figure 6.1.4). Applying Equation (5.1.1) under the assumption of the equality of the chemical potentials yields (Pruppacher and Klett, 1997)

$$e_w(r) = e_w(\infty)\exp\left(\frac{2\sigma_w}{rR_v\rho_w T}\right) = e_w(\infty)\exp\left(\frac{A}{r}\right), \tag{5.1.2}$$

where $e_w(\infty)$ is the saturation water vapor pressure over a planar water surface, as it was determined in Section 3.1.4 (Table 3.1.2), and ρ_w is the water density. When $r \to \infty$, $e_w(r) \to e_w(\infty)$. In Equation (5.1.2)

$$A = \frac{2\sigma_w}{\rho_w R_v T} \tag{5.1.3}$$

The $A(T_C)$ dependence (T_C is temperature in °C) is shown in **Figure 5.1.1**. At $T_C = 0$°C, $A \approx 1.2 \times 10^{-3}$ µm.

Equation (5.1.2) represents the Kelvin's law and shows that saturation water pressure upon curved drops is higher than that upon a flat surface, i.e., $e_w(r) > e_w(\infty)$. This result has a clear physical meaning: in order to increase the surface area, condensed molecules at the surface have to do additional work against the surface tension forces. This effect is schematically illustrated in **Figure 5.1.2**. Water molecules on the surface of a drop have more energy than those on a flat surface and, therefore, are more likely to evaporate.

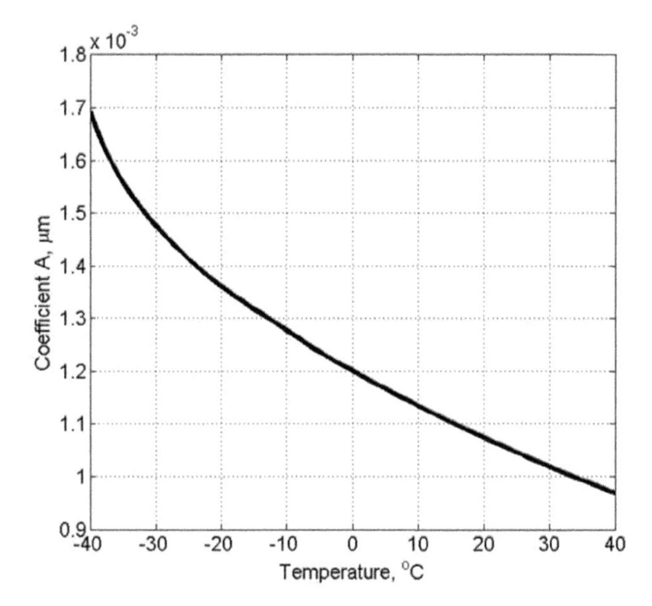

Figure 5.1.1 Dependence of the A coefficient on the temperature.

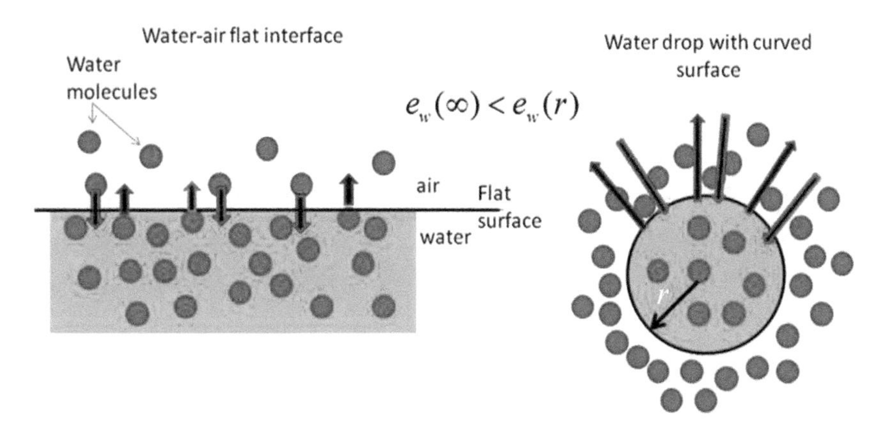

Figure 5.1.2 Schematic illustration of the curvature effect: the saturated water vapor pressure upon a droplet is larger than that over a flat surface.

Thus, evaporation rate from a curved surface is higher than that from a flat surface. In Figure 5.1.2, the increase in the evaporation flux is shown by longer blue arrows. To keep the equilibrium, the water vapor pressure in the environment should also be larger to increase the rate of molecular condensation and to compensate for the increased rate of evaporation (the fluxes marked with longer red arrows). For this reason, the saturation water vapor pressure $e_w(r)$ upon a drop should be larger than that upon a flat surface $e_w(\infty)$.

There is another important factor affecting the value of saturation over drops. As was discussed in Section 2.3, haze particles and drops are weak solutions. A nonvolatile dissolved matter tends to lower the equilibrium pressure of a liquid. The effect may be simplistically described in the following way: when a solute is added to a liquid, some of the liquid molecules are replaced by the solute molecules (Rogers and Yau, 1996). If the vapor pressure of the solute is less than that of the solvent, the net vapor pressure is reduced proportionally to the amount of the solute present. This effect can drastically lower the equilibrium vapor pressure upon a drop. As a result, a solution drop can be in equilibrium with the environment at much lower supersaturations than a pure water drop of the same size.

This relationship between the saturation pressures is expressed by Raoult's Law (Rogers and Yau, 1996):

$$\frac{e'_w(r)}{e_w(\infty)} = \frac{N}{N+n} \approx 1 - \frac{n}{N}, \tag{5.1.4}$$

where $e'_w(r)$ is the equilibrium vapor pressure upon a solution consisting of N water molecules and n molecules of the solute. The derivation of Raoult's Law is based on the multicomponent system in which the phase equilibrium is reached in case there is an equality of the chemical potentials of each component (Mazin and Shmeter, 1983; Pruppacher and Klett, 1997). Raoult's law is valid for weak perfect solutions. For solutions in which the dissolved molecules are dissociated, Equation (5.1.4) is modified by multiplying n by the factor $i > 1$ characterizing the degree of ionic dissociation. This factor is known as the Van't Hoff factor, whose values are not well known. Typically, the value $i = 2$ is used. The value $i \sim 2$ appears to be a reasonable approximation to be used in calculations in the absence of more precise information.

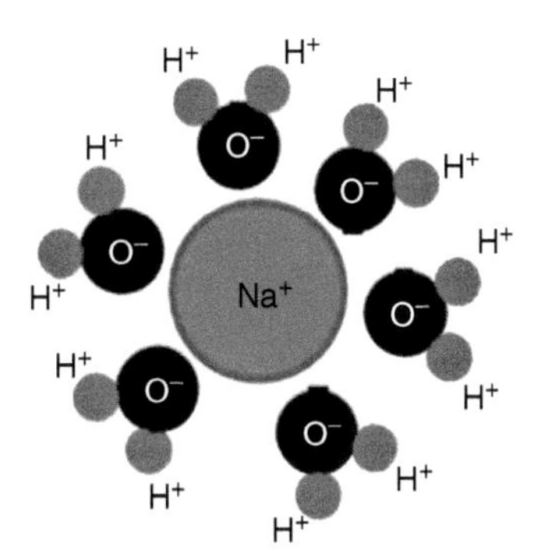

Figure 5.1.3 Ions in water solution create germs of water molecules.

The physical mechanism making it necessary to modify Raoult's Law in case of dissociated ions is illustrated in **Figure 5.1.3**. Ions tend to attract molecules of water, creating germs. Water molecules belonging to such germs have lower probability for evaporation. Thus, ions lead to a decrease in the effective amount of free water molecules able to evaporate. As a result, the rate of evaporation over the solution decreases.

Using the modified Raoult's Law, Pruppacher and Klett (1997) obtained the following expression for saturated pressure upon a salty drop:

$$e'_w(r) = e_w(\infty)\exp\left(-\frac{Br_N^3}{r^3 - r_N^3}\right) \quad (5.1.5)$$

In Equation (5.1.5)

$$B = \frac{\nu_N \Phi_s \delta_s M_w \rho_N}{M_N \rho_w}, \quad (5.1.6)$$

where δ_s is the soluble fraction of the aerosol particle (AP); Φ_s is the molecular osmotic coefficient characterizing a deviation from perfect solutions; ν_N is the total number of ions produced by one molecule of salt in the solution; ρ_N is the density of a dry salty AP; r_N is the radius of the dry AP; and M_w and M_N are molecular masses of water and salt, respectively. The value $i = \nu_N \Phi_s \approx 2$ is typically used in cloud modeling. More accurate analysis of the magnitude of factor i can be found in Lewis (2008).

Table 5.1.2 shows the values of ν_N and ρ_N, as well as the molecular weight of salt M_N and parameter B standing for different chemical matters. Since water can dissolve only a limited amount of salt (depending on the temperature), there is the maximum value of water salinity s_N known as the solubility of aerosol water solution (Table 5.1.2). Equation (5.1.6) allows the evaluation of the minimum relative humidity (humidity of deliquesce RH_{del}) at which condensation of water molecules starts:

$$RH_{del} = \exp\left(-\frac{\nu_N M_w s_N}{M_N}\right) \quad (5.1.7)$$

If the environmental $RH < RH_{del}$, aerosol particles remain dry and start growing only at $RH > RH_{del}$. Table 5.1.2 shows the values of RH_{del} calculated by the formula presented in Equation 5.1.7, in comparison with the values given by Pruppacher and Klett (1997) that were obtained using more complicated approaches. A good agreement between the values of RH_{del} is obvious.

The combination of Equation (5.1.2) describing the curvature effect and Equation (5.1.5) describing the salinity effect leads to the expression for saturation (equilibrium) pressure on a curved salty drop of radius r (Pruppacher and Klett, 1997):

$$\frac{e_w(r)}{e_w(\infty)} = \exp\left(\frac{A}{r} - \frac{Br_N^3}{r^3 - r_N^3}\right) \quad (5.1.8)$$

Taking into account that in most cases the following inequality is valid:

$$\frac{A}{r} - \frac{Br_N^3}{r^3 - r_N^3} \ll 1, \quad (5.1.9)$$

expression (5.1.8) can be rewritten as

$$\frac{e_w(r)}{e_w(\infty)} - 1 = \frac{A}{r} - \frac{Br_N^3}{r^3 - r_N^3} \quad (5.1.10)$$

Denoting equilibrium supersaturation as $S_{w\_eq} = \frac{e_w(r)}{e_w(\infty)} - 1$, we get

$$S_{w\_eq} = \frac{A}{r} - \frac{Br_N^3}{r^3 - r_N^3} \quad (5.1.11)$$

The first right-hand term of Equation (5.1.11) is known as the curvature term, while the second term is known as the chemical term. The value of S_{w_eq} is determined only by the size and the salinity of drops. Equation (5.1.11) is valid when $r \gg \frac{\sqrt{2}A}{3|S_{w\_eq}|^{1/2}}$ (this inequality follows from Equations (5.1.9) and (5.1.10), for instance, for $S_{w\_eq} = 0.1\%$ $r \gg 0.017$ μm. Expression (5.1.11) determines the equilibrium value of supersaturation over a salty drop of radius r. If $r \to \infty$ and $\frac{r_N}{r} \to 0$,

Table 5.1.2 Chemical parameters affecting condensation.

	ν_N	M_N, kg ml^{-1}	s_N	ρ_N, kg m^{-3}	RH_{del}, % (P&K)	RH_{del}, % (Eq. 5.1.7)	B
NaCl	2	0.058	0.3	2,170	75.28	80.1	1.34
KCl	2	0.074	0.3	1,990	84.26	84.8	0.962
Li_2CO_3	2	0.073	0.1	2,100		92.8	1.02
MgO	0	0.02	0.0	3,580		100	0
$(NH_4)_2S$	3	0.132	0.7	1,770	79.97	75.1	0.72
$CaCl_2$	2	0.111	0.7	2,160		78.6	0.70
Na_2SO_4	2	0.142	0.6	2,680	82	84.6	0.68
NH_4NO	2	0.080	1.1	1,725	61.83	58.8	0.78
NH_4Cl	2	0.053	0.5	1,527	77.1	71.4	1.03

the equilibrium supersaturation $S_{w\_eq} \to 0$, i.e., to the equilibrium value over the flat surface of pure water (Section 3.1).

If S_w in the environment exceeds S_{w_eq}, the equilibrium is broken and the drop begins growing. The growth of a drop of radius r is described by the equation of condensation growth. This equation is widely known as the equation for diffusional growth, because growth of drops takes place due to diffusion of water vapor molecules on the drop surface. A detailed derivation of the equation for diffusional growth is presented by Pruppacher and Klett (1997). The most general form of this equation is

$$r\frac{dr}{dt} = \frac{1}{F}\left[1 + S_w - \exp\left(\frac{A}{r} - \frac{Br_N^3}{r^3 - r_N^3}\right)\right], \quad (5.1.12)$$

where

$$F = \frac{\rho_w R_v T}{e_w D} + \frac{\rho_w L_w^2}{R_v T^2 k_a}, \quad (5.1.13)$$

where D is the coefficient of molecular diffusion of water vapor and k_a is the air heat conductivity. Both quantities depend on temperature (Pruppacher and Klett, 1997) (Section 6.1). At the temperature of 273.15 K and the pressure of 1.01325×10^5 Pa, $D = 2.36 \times 10^{-5}$ m^2 s^{-1} and $k_a = 0.0245$ W K^{-1} m^{-1}. Therefore, coefficient F in Equation (5.1.13) depends on temperature (**Figure 5.1.4**).

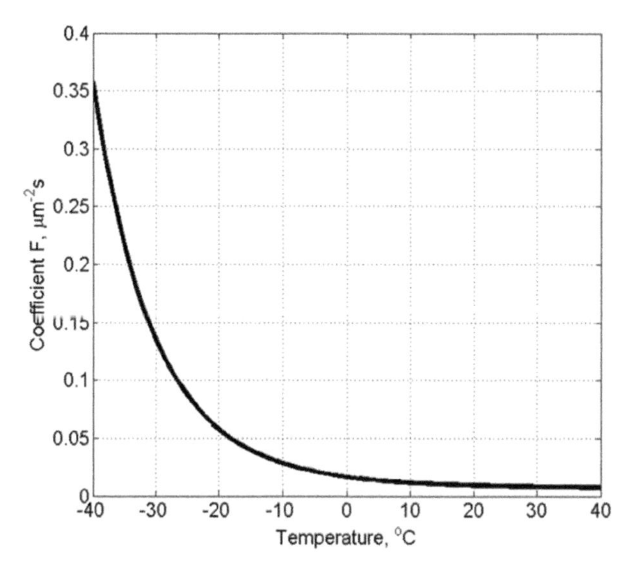

Figure 5.1.4 Dependence of coefficient F (Equation 5.1.13) on temperature.

Using the condition (5.1.9), Equation (5.1.12) can be rewritten in a more convenient and widely used form:

$$r\frac{dr}{dt} = \frac{1}{F}(S_w - S_{w\_eq}) = \frac{1}{F}\left(S_w - \frac{A}{r} + \frac{Br_N^3}{r^3 - r_N^3}\right) \quad (5.1.14)$$

Equation (5.1.14) describes the process of diffusional growth/evaporation of atmospheric particles that are water solutions of soluble aerosols.

We used the term "drops" for all atmospheric liquid particles. In cloud modeling, it is generally accepted to separate liquid particles into haze particles and actual drops. Accordingly, the process of diffusional growth is separated into three stages: a) growth of AP to their equilibrium size – particles at this stage are assigned to wet aerosols or haze; b) nucleation, i.e., formation of drops from haze; and c) diffusion growth of drops. It is important to note that all the three stages are described by Equation (5.1.14). In most cloud models, separating drop growth into three stages is described using some parameterizations. These parameterizations are required due to the inability of models to provide a direct solution of Equation (5.1.14) for small AP. The direct calculation of AP growth requires very small time steps (below 0.01 s), which makes the computations extremely time consuming. In most models, the process of haze particle growth is not considered at all, assuming the process of drop formation starts with droplet nucleation. There are only a few models (e.g., Pinsky et al., 2008b; Magaritz et al., 2009), where AP growth is calculated directly using time steps of 0.001–0.01 s.

5.1.2 Nucleation of Cloud Droplets and the Köhler Curves

The problem of nucleation may be formulated as follows: "How readily can chance collisions and aggregation of water molecules lead to formation of an embryonic droplet that will be stable and continue to exist under given environment temperature and humidity?" (Rogers and Yau, 1996). There are two types of droplet nucleation: *homogeneous* nucleation and *heterogeneous* nucleation (Pruppacher and Klett, 1997). Homogeneous nucleation is formation of pure water droplets from vapor. These droplets grow from small clusters of water molecules that emerge as a result of random fluctuations in the concentration of the molecules. If the clusters are large enough (when their radius exceeds a certain critical value r_{cr}), they continue growing with time. Otherwise, the clusters disappear. Clusters of molecules with radii exceeding the critical value are actually condensational nuclei.

The critical size of such a cluster is determined by the Kelvin formula:

$$r_{cr} = \frac{2\sigma_w}{R_v \rho_w T \ln\left[e_w(r_{cr})/e_w(\infty)\right]} \tag{5.1.15}$$

To form droplets from water vapor by homogeneous nucleation, supersaturation of several hundred percent is required. Since such supersaturation values are not observed in the atmosphere, homogeneous nucleation does not take place in clouds and is not considered in this book.

Heterogeneous nucleation is formation of droplets on AP that takes place at comparatively low supersaturation. It is the main mechanism of droplet formation in the atmosphere. First, let us consider the case when AP are fully soluble, i.e., $\delta_s = 1$. As follows from Equation (5.1.11), for each soluble AP of dry radius r_N there exists an equilibrium supersaturation S_{w_eq} at which a wet particle with radius r does not grow or evaporate; in this case, $\frac{dr}{dt} = 0$. Particles grow if $S_w > S_{w\_eq}(r)$ and evaporate if $S_w < S_{w\_eq}(r)$. Dependencies $S_{w\_eq}(r)$ (e.g., Equation 5.1.11)) known as the Köhler curves (Rogers and Yau, 1996; Pruppacher and Klett, 1997), are determined by the radius or the mass of dry particles (**Figure 5.1.5**).

The Köhler curves are widely used for analysis of droplet nucleation (e.g., Mazin and Shmeter, 1983; Rogers and Yau, 1996; Pruppacher and Klett, 1997). We illustrate different possible patterns of AP behavior by a contraposition of the Köhler curves and the environmental supersaturations (the latter denoted in **Figure 5.1.6** by the horizontal straight line).

Let us consider three cases that differ in the values of radius r_N of solid AP.

a) The horizontal straight line $S_w = const$ crosses the Köhler curve at two points a and b (Figure 5.1.6). The corresponding radii of the particles are r_a and r_b. Points a and b are the points of equilibrium. It is important to determine whether these points are stable or unstable. At point a, $r = r_a$. For particles with $r < r_a$, supersaturation is larger than the equilibrioum value, i.e., $S_w > S_{w\_eq}$, thus the particles grow tending to $r = r_a$. For particles with $r_a < r < r_b$, $S_w < S_{w\_eq}$, and these particles partially evaporate until the value of $r = r_a$ is reached. Thus, the point where $r = r_a$ is a stable point of equilibrium. This fact is illustrated in Figure 5.1.6 by the arrows directed toward the point. Let us now

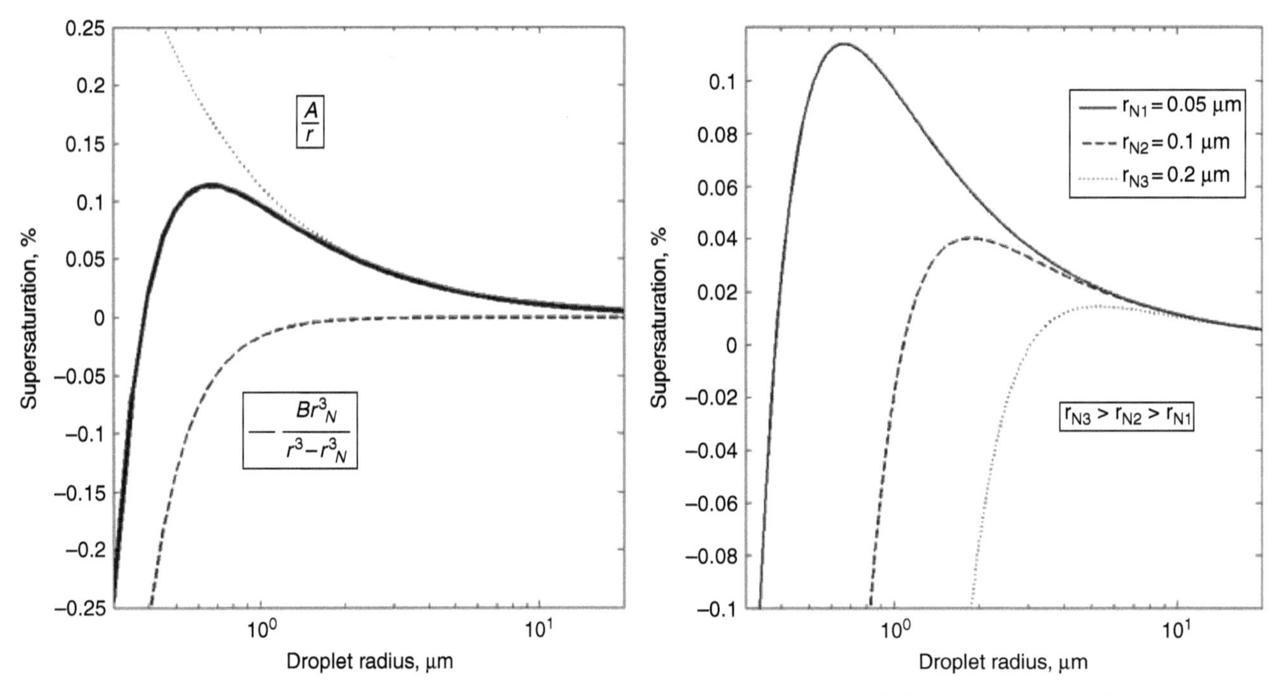

Figure 5.1.5 Illustration of the Köhler curves. Left panel presents the curvature term (dotted line), the chemical term (dashed line), and the Köhler curve (thick solid line) plotted for $r_N = 0.05$ μm. The Köhler curve determining the dependence $S_{w\_eq}(r)$ is the difference between these terms. Right panel shows the Köhler curves plotted for AP of three different radii. Each curve corresponds to a certain value of r_N. The Köhler curves shift downward and to the right with increasing r_N, thus in the figure $r_{N1} < r_{N2} < r_{N3}$.

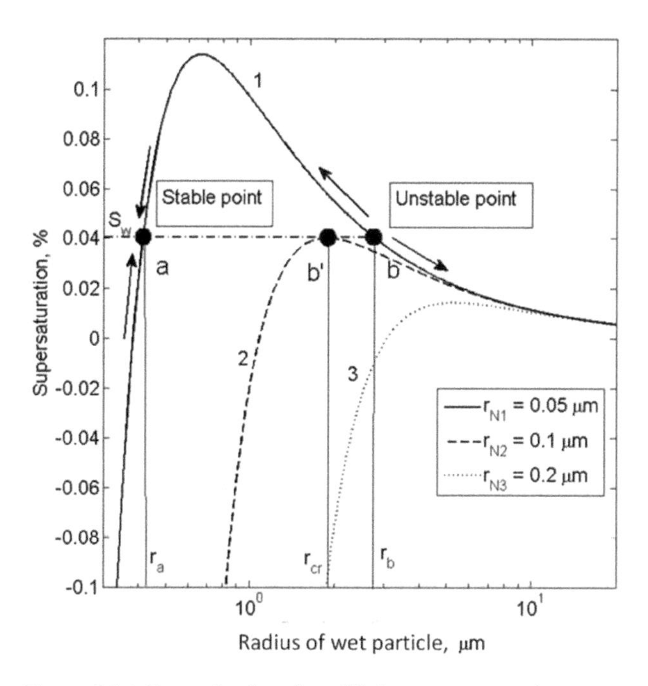

Figure 5.1.6 Dependencies of equilibrium supersaturation S_{w_eq} on the radius r of a wet particle (the Köhler curves). Horizontal line denotes environmental supersaturations S_w.

analyze the particle behavior at point b. In case $r < r_b$, $S_w < S_{w\_eq}$ and the particle evaporates until it reaches the stable radius of $r = r_a$. If $r > r_b$, $S_w > S_{w\_eq}$ and the particles grow unlimitedly. Hence, the point where $r = r_b$ is a point of unstable equilibrium. This conclusion is illustrated by the arrows directed away from point b (Figure 5.1.6).

b) The line $S_w = const$ touches the Köhler curve at a single point b' (Figure 5.1.6, curve 2). In this case $S_w > S_{w\_eq}$ for all the particles of radius r except at one point b' that corresponds to $r = r_{cr}$. Point b' is a point of unstable equilibrium. Supersaturation corresponding to the maximum of the curve is referred to as the critical supersaturation S_{w_cr}. If the radius of a wet particle r exceeds r_{cr}, droplet nucleation takes place. There is no spike in the particle size during droplet nucleation. Droplet nucleation is just a continuous transition from $r < r_{cr}$ to $r \geq r_{cr}$. The dry radius of an aerosol particle r_N corresponding to curve 2 can be referred to as the critical radius of a dry aerosol particle ($r_N = r_{Ncr}$) under external supersaturation equal to S_{w_cr}. At this supersaturation value, aerosols with dry radii of $r_N > r_{Ncr}$ become droplets with radii of $r > r_{cr}$. This transformation of AP into droplets is known as AP *activation*. Thus, the condition $S_w > S_{w\_cr}$ for certain AP with radius r_N means that $r_N > r_{Ncr}$, i.e., these wet AP (haze) become activated under the environmental supersaturation S_w.

c) If $S_w > S_{w\_eq}$ for all wet AP, these AP grow within the whole range of radii r like cloud droplets (Figure 5.1.6, curve 3).

Analysis of stability of the equilibrium points on the Köhler curves can also be performed using a standard mathematical approach. The problem is reduced to analysis of stability of the equilibrium points of the equation $r\frac{dr}{dt} = f(r, S_{w\_eq})$, with the equation for equilibrium points $f(r, S_{w\_eq}) = 0$. This analysis shows that the equilibrium points are stable if $\frac{dS_{w\_eq}}{dr} > 0$, and are unstable if $\frac{dS_{w\_eq}}{dr} \leq 0$.

It is important to present the expressions for the critical values r_{cr} and r_{Ncr}, since these expressions are widely applied in cloud physics. As can be seen from Figure 5.1.6, r_{cr} corresponds to the maximum of the Köhler curves. Hence, r_{cr} and r_{Ncr} can be obtained from Equation (5.1.11) by applying the condition $\frac{dS_{w\_eq}}{dr} = 0$. Typically, the radius of a wet particle is much larger than that of a dry AP, i.e., $r_N^3 \ll r^3$. Neglecting r_N^3 in the denominator of Equation (5.1.11), the critical values of the wet radius of a nucleated droplet r_{cr} and the critical dry radius of an activated AP particle r_{Ncr} can be expressed as

$$r_{cr} = \frac{2A}{3}\frac{1}{S_{w\_cr}} \tag{5.1.16}$$

and

$$r_{Ncr} = \frac{A}{3}\left(\frac{4}{BS_{w\_cr}^2}\right)^{1/3} \tag{5.1.17}$$

From Equations (5.1.16) and (5.1.17), we obtain the relationship between r_{cr} and r_{Ncr}:

$$r_{cr} = r_{Ncr}^{3/2}\left(\frac{3B}{A}\right)^{1/2} \tag{5.1.18}$$

In cloud models, it is necessary to determine the sizes of dry AP and haze particles whose activation/nucleation occurs at external supersaturation S_w. S_w is calculated at each time step and at each grid point. The problem is reduced to calculation of r_{cr} and r_{Ncr} using Equations (5.1.16) and (5.1.17) and assuming $S_w = S_{w\_cr}$. One can see that at low S_w only large AP are activated. With increasing S_w, the values of r_{Ncr} decrease, i.e., AP within a wider size range are activated. **Figure 5.1.7** shows the dependencies of the critical radii of dry and wet particles on the critical

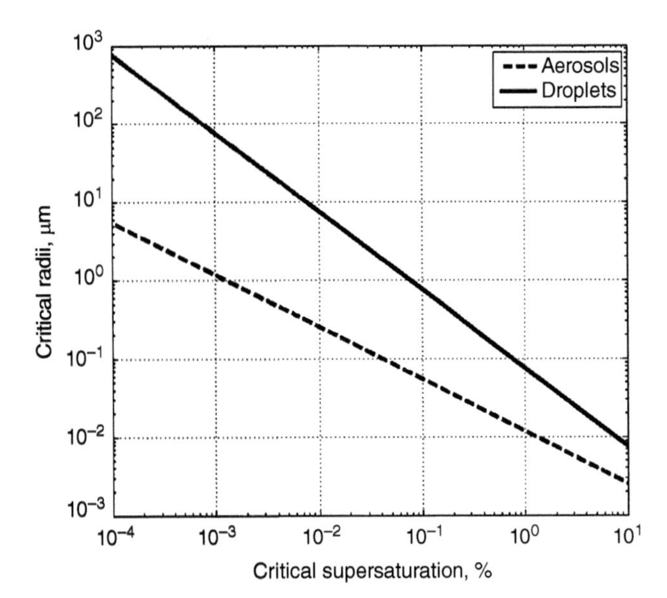

Figure 5.1.7 Dependencies of the critical radii on the critical supersaturation values. Solid line denotes the critical radii of nucleated droplets; the dashed line denotes the critical radii of dry soluble AP.

supersaturation values. One can see that the critical radii of giant AP with $r_N > 5$ μm exceed 800 μm. According to the Köhler theory, the critical supersaturation value for large soluble AP is very low, therefore, for these particles usually $S_w > S_{w\_cr}$.

The Köhler theory is a theory of equilibrium and thus does not contain information about the time required for particles to reach their equilibrium size. To reach their critical radius, giant AP should be located under supersaturation conditions for a time period exceeding cloud lifetime. The maximum radius that cloud droplets can reach as a result of diffusion growth does not exceed ~25 μm. Thus, wet particles growing on giant AP by diffusion may never reach their critical radius and will remain haze particles. Hence, haze particles in clouds can exist under two completely different conditions, namely, when these particles are small and $S_w < S_{w\_cr}$, or when they are giant AP that do not grow enough to reach their critical radius. Therefore, particles of, e.g., 20 μm in radius, can turn either into droplets if they grow on small AP, or into haze if they grow on giant AP.

Figure 5.1.7 shows that high values of supersaturation are necessary to activate very small AP (the Aitken mode – Section 2.2). Such supersaturations (of several percent) can be reached far above cloud base (Section 5.2).

The process of AP nucleation and droplet growth is illustrated in **Figure 5.1.8**. The left panel in Figure 5.1.8 shows the time dependence of the altitude of a Largangian parcel (one of about 2000) in a Lagrangian–Eulerian stratocumulus cloud model, LEM (Magaritz et al., 2010). The parcel ascends during the time period 20–45 min. Panels (a–c) show PSD and the relationship between dry aerosol dry radius and the radius of wet particles. At t = 0, the parcel contains only wet non-activated AP, since RH = 99%. At t = 30 min, AP with radii of $r_N > 0.3$ μm are activated, turning into droplets. AP with $r_N < 0.3$ μm remain nonactivated. One can see that while droplets continue growing during the parcel ascent, nonactivated AP are in equilibrium at $S_w = 0.17\%$ (measured in the parcel at t = 30 min). As a result, a gap arises between the SD of droplets and the SD of nonactivated wet AP (haze). The relationship between r_N and r for nonactivated AP obeys the Köhler theory. At t = 45 min, the gap increases. During the parcel ascent (t = 30–45 min), an additional portion of AP is activated, creating a new second mode of DSD. This small peak is denoted as "in-cloud nucleation" in Figure 5.1.8.

5.1.3 The Role of Chemical Composition of Aerosols in Droplet Nucleation

Figure 5.1.5 shows the Köhler curves plotted for fully soluble AP when the soluble fraction $\delta_s = 1$. The value of the soluble fraction is the parameter of AP chemical composition that has the strongest impact on droplet nucleation. This fact is illustrated in **Figure 5.1.9,** showing the Köhler curves for AP of the same radius of 0.05 μm but with different values of their soluble fractions. The critical sizes of AP and the critical supersaturation values depend largely on the soluble fraction (or the size of a soluble particle). A decrease in the soluble fraction from 1 to 0.01 leads to an increase in the critical supersaturation from 0.13 to 1%. Accordingly, the critical radius of wet AP r_{cr} decreases from 0.65 μm to 0.08 μm.

The limiting case is when AP is fully insoluble. In this regard, a question arises: can insoluble particles grow by condensation, e.g., can droplets form out of sand particles? If particles are insoluble, the diffusion growth equation is reduced to the following:

$$r\frac{dr}{dt} = \frac{1}{F}\left(S_w - \frac{A}{r}\right) \tag{5.1.19}$$

The particles start growing if $r > \frac{A}{S_w}$. For $S_w \sim 1\%$, $r > 2$ μm, which means that insoluble AP can also grow by condensation. However, to do so they should be large enough at quite high supersaturation values. The growth rate is equal to zero if $S_w = \frac{A}{r}$. When

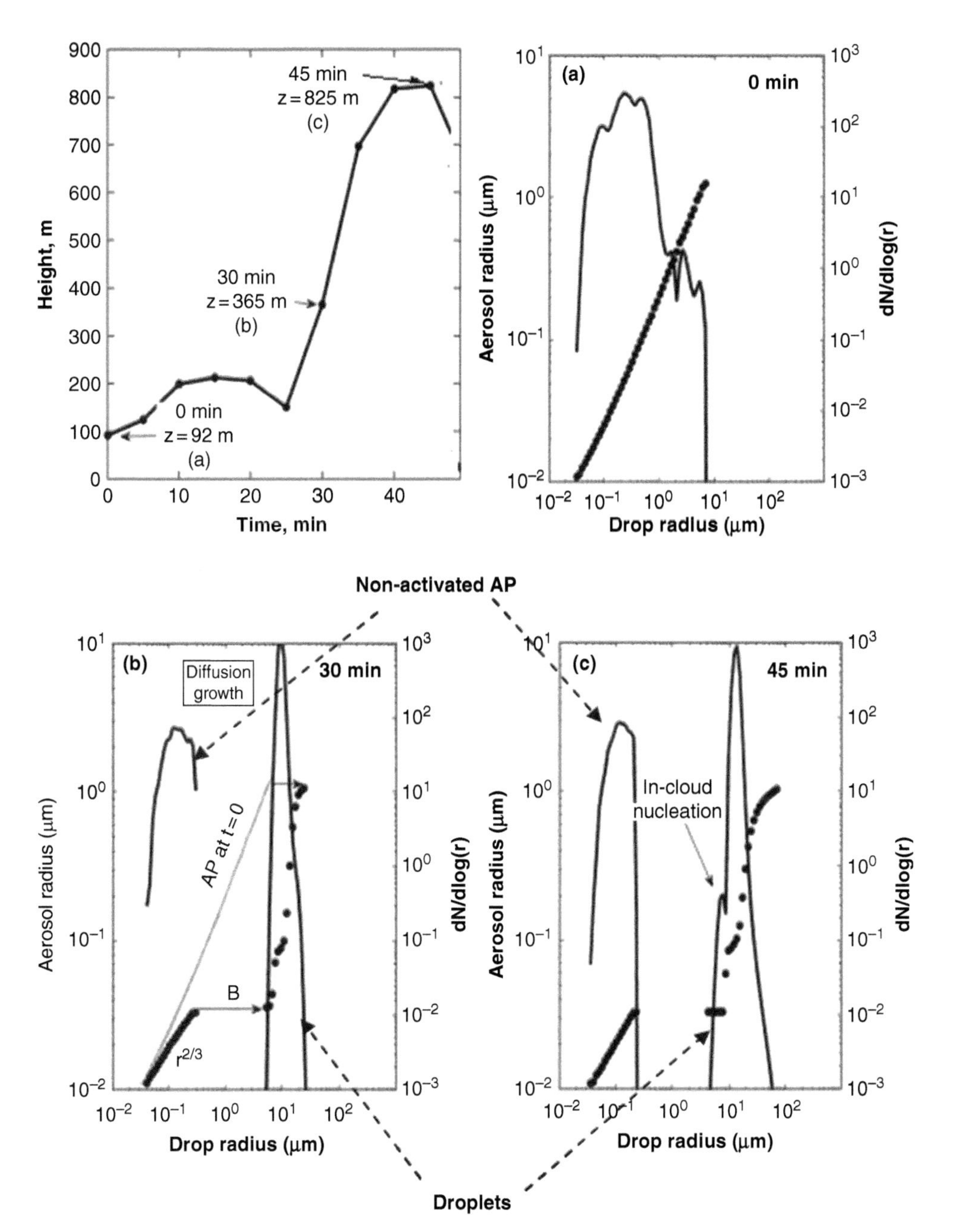

Figure 5.1.8 Left panel: time dependence of an air parcel elevation. Panels (a–c): SD functions (thin solid lines) and the relationship between dry aerosol radius and the radius of wet particles (thick solid line) (from Magaritz et al., 2010; courtesy of Elsevier).

supersaturation is lower than $\frac{A}{r}$, wet AP lose water due to evaporation.

Nucleation is only slightly sensitive to the chemical formulas of AP. The impact of the chemical formulas is determined by the constant B in the chemical term of Equation (5.1.11). For different salts, B varies relatively insignificantly, in most cases not more than by factor of two (Table 5.1.2). For instance, the ratio $B_{NaCl}/B_{Na_2SO_4} = 1.97$. These comparatively low ratios indicate a small effect of chemical composition of AP on nucleation. Natural mineral AP typically contain mixtures of different salts. In this case, it is convenient to reduce all the various chemical compositions to a single equivalent composition, e.g., to NaCl. To accomplish this procedure without affecting the process of droplet growth, one needs to change the size of AP in

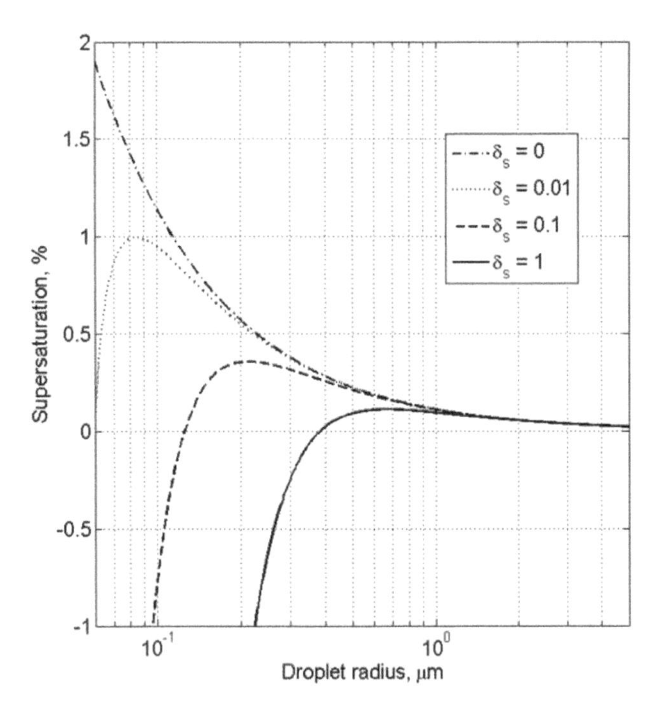

Figure 5.1.9 The Köhler curves plotted for AP with radius of 0.05 μm having different soluble fractions.

order to keep the value Br_N^3 unchanged. It means that the new equivalent radius of AP expressed in terms of the NaCl equivalent can be calculated as follows:

$$r_{N_{NaCl}} = r_N \left(\frac{B}{B_{NaCl}} \right)^{1/3} \tag{5.1.20}$$

From Equation (5.1.20) it follows that for AP consisting of Na_2SO_4 and having a radius $r_{N_{Na_2SO_4}}$ one can introduce an equivalent radius of AP consisting of NaCl $r_{N_{NaCl}}$ to get the same value of the chemical term. Due to the low value of $B_{NaCl}/B_{Na_2SO_4}$ ratio $r_{N_{NaCl}}$ is only by factor of $1.97^{1/3} = 1.25$ smaller than $r_{N_{Na_2SO_4}}$. If, for instance, r_{Ncr} for NaCl is 0.01 μm, r_{Ncr} will be 0.0125 μm for Na_2SO_4. Taking into account that the range of AP sizes is wide enough (from 0.001 μm to ~5 μm), such changes in r_{Ncr} lead to comparatively small changes in the number of activated AP. So, the main difference between the amounts of droplets nucleated on AP of different chemical compositions is related not to the chemical composition itself, but to the significant difference in the SDs of these AP and in the values of their soluble fraction (e.g., Phillips et al., 2008; Levin and Cotton, 2009). Reducing AP of different chemical compositions to single equivalent composition simplifies treatment of nucleation in cloud models, allowing the use of only one AP type.

Parameters responsible for the values of the curvature and the chemical term are determined with uncertainties that can exceed 10% As was shown by Wex et al. (2008), reducing droplet surface tension by only 10% can cause a change in the activated fraction (i.e., in cloud droplet number concentration) at least by 10–20%. Since the presence of organic molecules reduces surface tension, a sufficient concentration of organic molecules in AP may be an important factor that determines AP activation. Investigating the role of various chemical reactions in formation of AP of different chemical compositions is one of the main objectives of atmospheric chemistry. There are multiple studies on the issue (see references in Levin and Cotton, 2009).

5.1.4 Cloud Condensational Nuclei and the Relationship Between CCN Concentration and Supersaturation

First, some important definitions should be introduced. APs that potentially can become droplets are referred to as CN. To become droplets, the wet radius of AP must be greater than their critical radius. Dry APs that give rise to droplets at certain supersaturation S_w are referred to as CCN. The CCN concentration depends, therefore, on S_w. If supersaturation becomes very large ($S_w \to \infty$), CCN concentration values tend to CN concentration values. It is widely accepted to characterize CCN concentration by its value at $S_w = 1\%$.

Along with this definition of CCN, Pruppacher and Klett (1997) suggest classifying CN into *nonactivated* CCN and *activated* CCN. Dry aerosol nuclei with radii of $r_N > r_{Ncr}$ are referred to as activated CCN. Other particles are considered nonactivated CCN. Concentration of CN is the sum of concentrations of activated and nonactivated CCN.

The ratio of N_{CCN}/N_{CN} is known as the *CCN efficiency* (Levin and Cotton, 2009). Dependence of the CCN efficiency on supersaturation is the *activation spectrum*. An example of an activation spectrum measured for different types of clouds in Amazonia is presented in **Figure 5.1.10**. Figure 5.1.10 presents information on the SD of CCN, as well as the fractions of large and small CN. For instance, if N_{CCN}/N_{CN} is equal to 0.6 (i.e., 60%) at $S_w \sim 1.2\%$, the CN size spectrum contains comparatively small CN, thus the remaining 40% of CN (nonactivated CCN) can be activated at higher S_w. According to Pruppacher and Klett (1997), the ratio N_{CCN}/N_{CN}, measured at 1% supersaturation varies for maritime clouds from 0.2 to 0.6 with a median value of 0.5. Values of the CCN efficiency for maritime atmosphere measured in

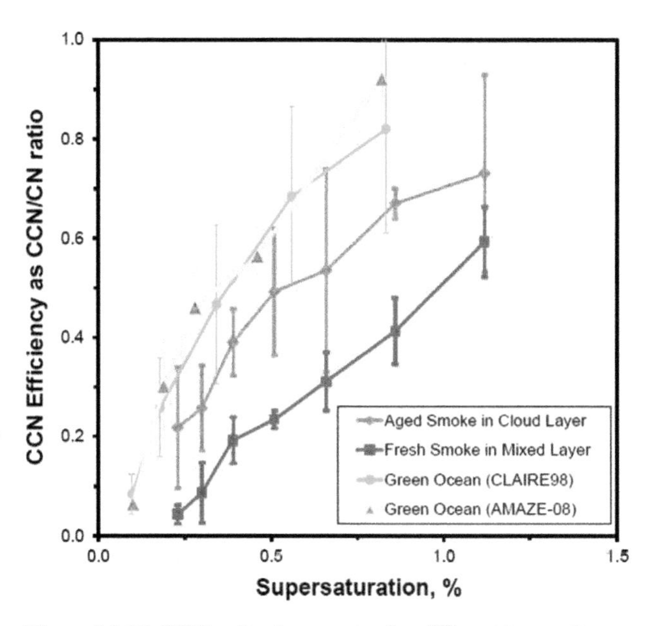

Figure 5.1.10 CCN activation spectra for different types of aerosol particles in Amazonia. Green Ocean measurements were performed in biogenic conditions during the wet season. Fresh smoke and aged smoke measurements were performed during heavy emissions from biomass burning (courtesy of Prof. M. Andreae).

different regions are shown in **Table 5.1.3**. One can see that in the maritime atmosphere over the sea the CCN efficiency is often below 0.5, indicating the existence of a significant amount of small CN (nonactivated CCN).

Much lower values of CCN efficiency were found for continental aerosols (Pruppacher and Klett, 1997). Median worldwide dependencies $N_{CCN} = f(S_w)$ are presented in **Figure 5.1.11**. $N_{CCN}(S_w)$ dependencies vary from location to location, from one season to another, and even from day to night. Examples of the $N_{CCN}(S_w)$ dependencies measured by different researchers over oceans are presented in **Figure 5.1.12**.

The concentration of activated CCN is often represented by the dependence $N_{CCN}(S_w)$, which can be expressed by a semiempirical formula known as Twomey's formula (Twomey, 1959; Pruppacher and Klett, 1997):

$$N_{CCN} = N_0 S_w^k, \tag{5.1.21}$$

where N_{CCN} is the concentration of activated CCN at supersaturation S_w (in %), and N_0 and k are measured parameters. Parameter k is known as the slope parameter (in log–log scale). The dependencies in Equation (5.1.21) were determined both under laboratory conditions and in situ measurements using a particle

Table 5.1.3 CCN efficiency at 1% supersaturation in maritime air (from Hegg and Hobbs, 1992).

Location	N_{CCN}/N_{CN}	Observer
Pacific Ocean	0.6	Hoppel et al. (1973)
Arctic	0.5	Saxena and Rathore (1984)
Cape Grim, Australia	0.4–0.6	Gras (1989)
Cape Grim, Australia	0.4	Gras (1989)
Northeast Pacific Ocean	0.2	Hegg et al. (1991a)
Northeast Pacific Ocean	0.2	Hegg et al. (1991a)
Northeast Pacific Ocean	0.3	Hegg et al. (1991a)
East Pacific Ocean	0.6	Hudson and Frisbie (1991)

measurement system (PMS) and other CCN probes (e.g., FSSP-300, PCASP-100X, ASASP-100X probes). In these studies, the temperature of an air probe could be decreased, thus increasing the supersaturation within a range determined by a CCN counter used (see Kim and Boatman, 1990; Strapp et al., 1992). The values of k vary from 0.3 to 1.3 within a wide range of supersaturation values in different zones over the ocean (**Table 5.1.4** and Figure 5.1.10), as well as within different air masses at the same geographical location (Hudson and Li, 1995). According to Pruppacher and Klett (1997), the averaged value of k for all maritime clouds is close to 0.9, which indicates the existence of a significant amount of small CCN in the maritime atmosphere. According to Levin and Cotton (2009), the typical value of k is close to 0.3 within the 0.1%–8% supersaturation range. A decrease in the slope parameter means a decrease in the fraction of small CCN. The data obtained by Levin and Cotton (2009) correspond to very remote maritime areas. **Tables 5.1.4 and 5.1.5** show the magnitudes of N_0 and k for CCN, measured at 1% supersaturation in maritime air masses and in continental cloud masses, respectively.

The slope parameter cannot be constant within the entire range of supersaturation values. The dependence (5.1.21) must be modified both for the largest particles corresponding to very low supersaturations and for the smallest CCN corresponding to very high supersaturations. As stressed by Cooper et al. (1997), a total CCN mass tends to infinity in case $k < 2$ for very low supersaturation values (very large CCN). Measurements carried out by Alofs and Liu (1981) in continental air at Rolla (Missouri) suggest that a change in the slope

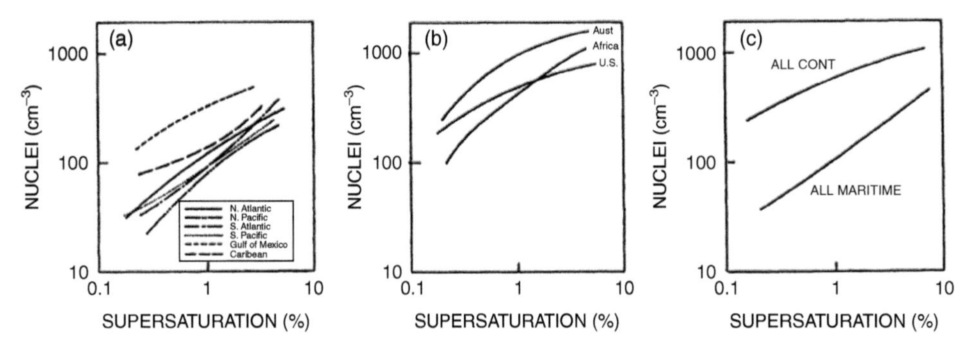

Figure 5.1.11 Median world-wide concentration of CCN as a function of supersaturation required for activation: (a) in the air over oceans, (b) in the air over continents and (c) all the observations (from Twomey and Wojciechowski, 1969; American Meteorological Society©; used with permission).

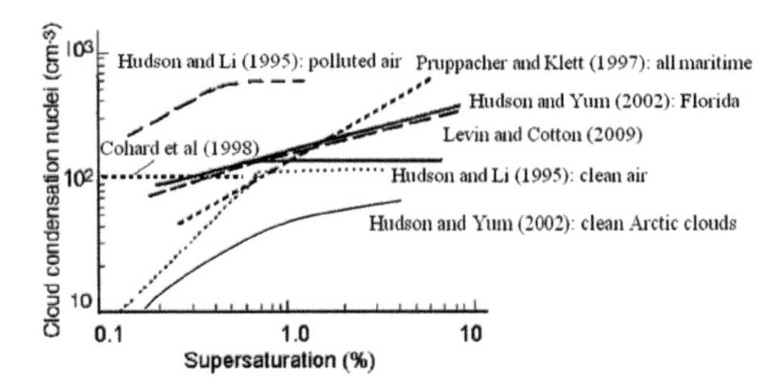

Figure 5.1.12 Dependencies of concentration of activated CCN on supersaturation over the sea, reported by different researchers. The dependencies presented by Pruppacher and Klett (1997) for "all maritime" cases is close to the steep-slope case (k = 0.9).

occurs at supersaturation of about 0.05%, and the slope parameter k is about 3.85 for supersaturations lower than 0.05% (**Figure 5.1.13**).

Formula (5.1.21) with $k = const$ cannot be applied for very high supersaturations values, since if $S_w \to \infty$, $N_{CCN} \to \infty$. To obey the condition of a limited number of CN, the slope parameter should decrease, tending to zero at $S_w \to \infty$. The data obtained by Pruppacher and Klett (1997) show that k value can be considered constant within a wide range of supersaturations up to ~8%. Clark and Kapustin (2002) also reported the existence of very small aerosols (<15 nm in diameter) within the remote maritime boundary layer. Observations of bimodal DSD spectra show that high concentrations of ice crystals and large optical depths of deep tropical cloud anvils are consistent with the existence of small CCN that can be activated under S_w exceeding several percent (Khain et al., 2012). Small CCN (belonging to the Aitken mode) can be of continental nature. They may form under the impact of various factors such as fossil fuel combustion or slightly hygroscopic dust, like Saharan dust, which was often found in convective storms near the Eastern African coast, as well as in storms and hurricanes reaching the American coast, or as a result of various chemical reactions over the sea (Covert et al., 1992; Hobbs, 1993; Pruppacher and Klett, 1997; Clark and Kapustin, 2002). The presence of significant concentrations of CCN with diameters ranging from 0.001 μm to 0.06 μm in marine atmosphere was reported by Jaenicke (1993) (Section 2.2).

Some observational studies (e.g., Hudson, 1984; Hudson and Frisbie, 1991; Hudson and Li, 1995; Hudson and Yum, 1997, 2002) and laboratory studies (Jiusto and Lala, 1981) showed a substantial deviation of the activation spectra from that calculated by the Twomey formula, especially in clean polar air. **Figure 5.1.14** shows dependencies $N(S_w)$ measured by Hudson and Li (1995) during the Atlantic Stratocumulus Transition Experiment (ASTEX) in the mid-Atlantic atmospheric boundary layer. According to these results, no new CCN can be activated at supersaturation exceeding some threshold of about 0.6%. In other studies, the threshold supersaturation values are even lower (Cohard et al., 1998; Emde and Wacker, 1993). Such observations point at lack of small dry CCN with dry diameters below 0.05–0.12 μm in the sampled air.

Due to the reasons given, the Twomey formula for CCN activation spectra is often modified. For instance, Cohard et al. (1998) and Shipway and Abel (2010) express the activation spectra in the form

$$N_{CCN} = N_0 S_w^k \times {}_2F_1(S_w, k, \mu, \beta), \qquad (5.1.22)$$

where hypergeometric function ${}_2F_1(S_w, k, \mu, \beta)$ plays the role of a limiting factor preventing nucleation at high supersaturation. Parameters μ and β allow the adjustment of the asymptotic behavior of $N_{CCN}(S_w)$

Table 5.1.4 Values of the empirical constants N_0 and k for CCN at 1% supersaturation in a maritime air mass (from Hegg and Hobbs, 1992).

N_o (cm^{-3})	k	Location	Observer
125	0.3	Australian Coast	Twomey (1959a)
53–105	0.5–0.6	Hawaii	Jiusto (1967)
100	0.5	North & South Atlantic Ocean, North & South Pacific Ocean, and Caribbean	Twomey and Wojciechowski (1969)
190	0.8	North & South Pacific Ocean	Dinger et al. (1970)
250	1.4	North Atlantic Ocean (1,500 ft)	Dinger et al. (1970)
250	1.3	North & South Pacific Ocean	Hoppel et al. (1973)
145–370	0.4–0.9	North Atlantic Ocean	Saxena and Fukuta (1976)
100–1,000	–	Arctic	Saxena and Rathore (1984)
140	0.4	Cape Grim, Australia	Gras (1990)
250	0.5	North Atlantic Ocean	Hoppel et al. (1990)
25–128	0.4–0.6	North Pacific Ocean	Hudson and Frisbie (1991)
27–111	1.0	North Pacific Ocean	Hegg et al. (1991a)
400	0.3	Polluted North Pacific Ocean	Covert (1992)
100	0.4	Equatorial Pacific Ocean	Covert (1992)
290	0.7	High Planes, Montana	Hobbs et al. (1978)

Table 5.1.5 Values of the empirical constants N_0 and k for CCN at 1% supersaturation in a continental air mass (from Pruppacher and Klett, 1997; courtesy of Springer).

N_0 (cm^{-3})	k	Location	Observer
600	0.5	Australia, Africa, USA	Twomey and Wojciechowski (1969)
2,000	0.4	Australia	Twomey (1959a)
3,500	0.9	Buffalo, N.Y.	Kocmond (1965)
300–4,000	0.9	Alps	Stein et al. (1985)
3,000–5,000	0.8	Texas	Hobbs et al. (1985)
2,000	0.9	High Planes, Montana	Hobbs et al. (1978)

for high and low supersaturation to particular observations. Shipway and Abel (2010) parameterized parameters k, μ, and β as functions of the underlying aerosol physics chemistry. The solid line in Figure 5.1.14 corresponds to the nonlinear fit of Equation (5.1.22), while the dashed line corresponds to the linear fit (in log–log scale) of a power law. The adjusted coefficients for polluted air and clean air are given in **Table 5.1.6**. In view of the drawbacks of the parameterization (5.1.21), Khvorostyanov and Curry (2008) proposed generalizing this formula in the form

$$N_{CCN} = N_0(S_w)S_w^{k(S_w)}, \qquad (5.1.23)$$

where parameters N_0 and k are functions of supersaturation. The authors calculated the differential activation spectrum of the CCN assuming their log-normal SD.

A simpler approach enabling us to avoid the problem of infinitely large CCNs at $S_w \to \infty$ is used in the two-moment bulk parameterization scheme WDM6 (e.g., Hong et al., 2010), where the number of activated CCN is expressed as

$$N_{CCN} = N_{CN}(S_w/S_{\max})^k, \qquad (5.1.24)$$

where N_{CN} is the total number of particles equal to the sum of concentrations of activated and nonactivated CCN. $S_{\max}$ represents the supersaturation value needed to activate the total particle count N_{CN}. Formula (5.1.24) takes into account the decrease in the amount of activated CCN occurring at increasing supersaturation, thus $N_{CCN} \to N_{CN}$ when $S_w \to S_{\max}$ and $\frac{dN_{CCN}}{dS_w} \to 0$ when $S_w \to S_{\max}$. Utilization of Equation (5.1.24) does not abolish the necessity of decreasing the slope parameter with increasing S_w. Otherwise, CCN concentration may be overestimated.

Equation (5.1.21) allows the calculating of $\frac{dN_{CCN}}{dS_w}$ which is called the *differential activation spectrum* of CCN:

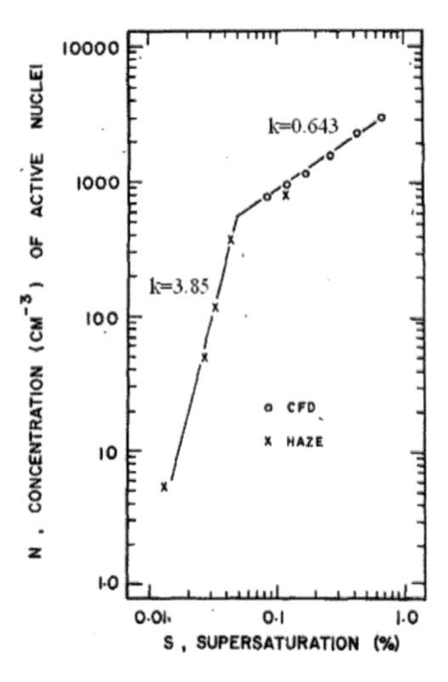

Figure 5.1.13 Activation spectrum of CCN (from Alofs and Liu, 1981; American Meteorological Society©; used with permission).

$$\frac{dN_{CCN}}{dS_w} = kN_0 S_w^{k-1} \tag{5.1.25}$$

To take into account the deviations of the slope parameter both at low and high supersaturation, Cohard et al. (1998) proposed a more general equation for the differential activation spectrum:

$$\frac{dN_{CCN}}{dS_w} = kN_0 S_w^{k-1}\left(1 + \eta S_w^2\right)^{-\mu}, \tag{5.1.26}$$

where η and μ are additional parameters that modify the asymptotic behavior for high and low supersaturation. Shipway and Abel (2010) generalized Equation (5.1.26) as

$$\frac{dN_{CCN}}{dS_w} = kN_0 S_w^{k-1}\left(1 + \eta S_w^{v}\right)^{-\mu}, \tag{5.1.27}$$

where the fifth parameter v replaces the quadratic exponent in the denominator of Equation (5.1.26). Expressions (5.1.24) and (5.1.25) are used for analytic calculation of the number of nucleated droplets at cloud base.

Comparison of activation spectra presented in Figures 5.1.11–5.1.14 indicate significant differences in supersaturation ranges within which the Twomey formula is valid. The differences in the activation spectra at high supersaturation values indicate dramatic differences in the amount of small CCN in the size CCN distribution. These differences in concentration of smallest CCN can be attributed to the variability of

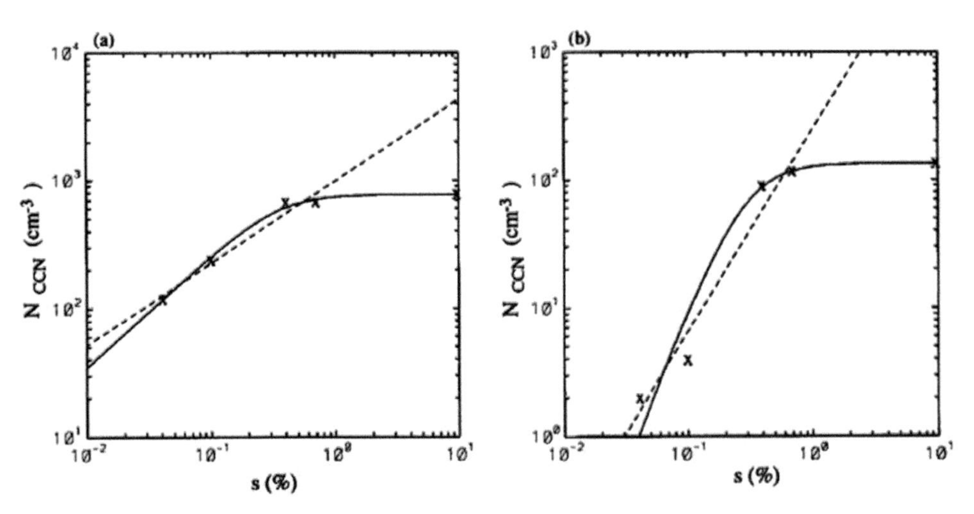

Figure 5.1.14 Representative CCN activation spectra in log–log scale as measured by Hudson and Li (1995) for polluted air (a) and for clean air (b). Crosses mark the observed data, solid line denotes the adjusted expression (5.1.22), and the dashed line corresponds to the best fit by the Twomey power law. The experimental data point at $S_w = 10\%$ corresponds to the total CCN amount (from Cohard et al., 1998; American Meteorological Society©; used with permission).

Table 5.1.6 Values of the adjusted parameters for activation spectra described by the Twomey power law and by Equation (5.1.22) for polluted air and clean air, corresponding to measurements presented in Figure 5.1.14 (from Cohard et al., 1998; American Meteorological Society©; used with permission).

	$N_{CCN} = N_0 S^k$		$N_{CCN} = N_0 S^k F(\mu, \frac{k}{2}, \frac{k}{2}+1; -\beta S^2)$			
	N_0 (cm^{-3})	k	N_0 (cm^{-3})	k	β	μ
Polluted air	986	0.64	1,865	0.86	6.80	1.50
Clean air	243	1.57	1,104	2.07	6.84	1.90

CCN over height and space and, possibly, to different characteristics of CCN counters used in the observations. Since knowledge on CCN activation spectra is of high importance in cloud modeling, accurate measurements are required to determine the behavior of the slope parameter both at high and very low supersaturations.

5.1.5 Calculation of CCN Distributions Using the $N_{CCN}(S_w)$ Dependencies

Knowing CCN SD is necessary to calculate droplet concentration in models with spectral bin microphysics. However, SDs are rarely available, since in most cases – in particular, in situ observations – information about CCN is obtained from measurements of CCN concentration at several values of supersaturation. In such cases, the $N_{CCN}(S_w)$ dependence is often represented in the form of Equation (5.1.21). The empirical formula (5.1.21) (as well as any empirical formula for $N_{CCN}(S_w)$) determines, within the range of its validity, the SD of CCN and nucleated droplets (Mazin and Shmeter, 1983; Emde and Wacker, 1993; Khain et al., 2000). Indeed, each value of $r_{Ncr}(S_w)$ corresponds to a certain value of supersaturation S_w. Using Equation (5.1.21), one can write the following expression for aerosol SD $\frac{dN_{CCN}}{dr_{Ncr}}$:

$$\frac{dN_{CCN}}{dr_{Ncr}} = N_0 k S_w^{k-1} \frac{dS_w}{dr_{Ncr}} \qquad (5.1.28)$$

Using Equation (5.1.17), the SDs of CCN can be rewritten as follows:

$$\frac{dN_{CCN}}{d\ln r_{Ncr}} = \frac{3}{2} N_0 k S_w^k \qquad (5.1.29)$$

For values of CCNs and of supersaturation falling outside the range of observations (i.e., for the smallest and the largest CCN), the CCN SD should be taken from direct measurements or from additional parameterizations.

5.2 Condensational (Diffusional) Drop Growth

Condensational (or *diffusional*) growth of drops, i.e., drop growth by diffusion of water vapor molecules upon drop surface is accompanied by latent heat release and is the main energy source of cloud updrafts in warm clouds. Diffusional growth of drops is one of the slowest microphysical processes. Drops growing by diffusion reach 10–20 μm in radius, depending on cloud type and drop concentration. The time needed to reach these sizes ranges from several minutes to several tens of minutes. During this time, air parcels containing growing drops may reach heights of several kilometers. The relative rate of diffusional growth decreases with increasing drop size. Diffusional growth does not cause the emergence of rain drops that are a result of drop collisions.

Growing drops absorb water vapor, which leads to a decrease in supersaturation S_w. Supersaturation in clouds is typically less than 1% and can drop rapidly to a value close to zero during a few seconds. During this time, the mass of drops changes insignificantly. In this sense, the supersaturation field adjusts to the water drops population. For drops to grow by diffusion, a continuous source of supersaturation is required. The phenomenon of the rapid adjustment of supersaturation is highly important when diffusional growth is simulated in cloud models. Therefore, the equations for diffusional growth and supersaturation should be solved together (in coupling).

5.2.1 Equation for Supersaturation and Water Budget

Using equations for the mixing ratio of water vapor, temperature and the Clayperon–Clausius equation in quasi-hydrostatic approximation, one can obtain the equation for supersaturation over water (Squires, 1952; Pruppacher and Klett, 1997; Korolev and Mazin, 2003) (see Appendix 5.2.9):

$$\frac{dS_w}{dt} = (1 + S_w)\left(A_1 w - A_2 \frac{dq_w}{dt}\right) \quad (5.2.1)$$

Since supersaturation in clouds usually obeys the condition $|S_w| \ll 1$, the supersaturation equation often can be written in the form

$$\frac{dS_w}{dt} = A_1 w - A_2 \frac{dq_w}{dt} \quad (5.2.2)$$

In Equations (5.2.1) and (5.2.2), the values of A_1 and A_2 are calculated as

$$A_1 = \frac{g}{T}\left(\frac{L_w}{c_p R_v T} - \frac{1}{R_a}\right); \quad A_2 = \frac{1}{\rho_v} + \frac{L_w^2}{R_v c_p T^2 \rho_a}, \quad (5.2.3)$$

where ρ_v and q_w are water vapor density (absolute humidity) and LWC, respectively, g is the acceleration of gravity, and w is the vertical velocity of the cloud parcel. These coefficients slightly depend on temperature (**Figure 5.2.1**).

The first term on the right-hand side of Equation (5.2.2) describes an increase in supersaturation due to adiabatic cooling of air during the parcel's ascent; the second right-hand-side term describes the supersaturation sink caused by condensation of water vapor on drops. Equations (5.2.1) and (5.2.2) are applied to air volumes where the supersaturation is assumed to be uniform for all the drops.

Let us consider a well-mixed adiabatic cloud volume ascending at velocity $\frac{dz}{dt} = w(z)$ from the lifting condensation level. Integration of Equation (5.2.2) with respect to time leads to the following equation describing the water balance in an ascending adiabatic parcel:

$$S_w = A_1 z - A_2 q_w + C, \quad (5.2.4)$$

where the integration constant $C = A_2 q_0 + S_0$ is determined by the initial conditions at $z = 0$. The diagnostic equation (5.2.4) shows that the sum $S_w + A_2 q_w$ is a linear function of z in an isolated cloud parcel. In the process termed a moist adiabatic process in cloud physics, the moisture excess over saturation immediately drops to zero (i.e., $S_w = 0$) and q_w becomes a linear function of height. The value of q_w is known as the adiabatic liquid water content:

$$q_{ad} = q_0 + \frac{A_1}{A_2} z \quad (5.2.5)$$

The profile of q_{ad} is only a certain approximation of the real LWC profile even in an adiabatic parcel, because S_w is not equal to zero. A simple diagnostic equation (5.2.4) allows evaluating a possible deviation of LWC from the adiabatic value by comparing the values of the terms in this water-balance equation. As an example, **Table 5.2.1** compares the values of the terms in Equation (5.2.4) at different heights at drop concentration of $N = 100\ \mathrm{cm}^{-3}$ and updraft velocity of 5 m/s (these conditions are suitable for deep convective clouds over the ocean). The evaluations were performed under the condition $C = 0$. One can see that value S_w is of the same order or even larger than $A_2 q_w$ near cloud base, i.e., at the level of supersaturation maximum. With increasing height, q_w increases and S_w decreases, so at 180 m above the lifting condensation level $S_w \ll A_2 q_w$. The condition $S_w \ll A_2 q_w$ explains why fluctuations of supersaturation do not lead to significant changes in the vertical profile of $q_w(z)$ and why this profile is close to the adiabatic one.

5.2.2 Equations of Diffusional Drop Growth and Supersaturation

Generally, diffusional growth of drops is described by Equation (5.1.14). However, for drops with radii

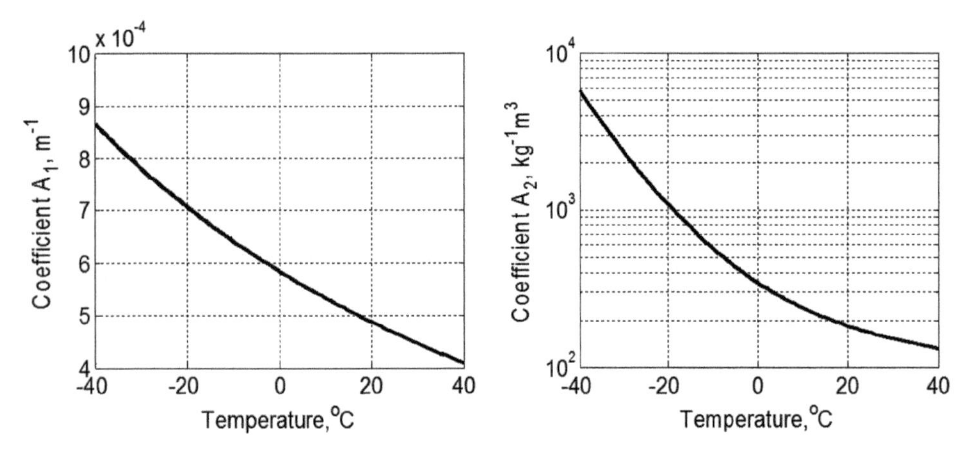

Figure 5.2.1 Temperature dependencies of coefficients A_1 and A_2 ($p = 1{,}013$ mb).

Table 5.2.1 Evaluation of magnitude of terms in Equation (5.2.4).

Height Levels	Terms in Equation (5.2.4).		
	q_w, g m^{-3}	$A_2 \cdot q_w$, %	S_w, %
Maximum supersaturation	2.56×10^{-2}	1.4	1.75
180 m above cloud base	2.92×10^{-1}	8.1	0.75

exceeding a few μm (except when drops grow on giant CCN and have high salinity), the sum of the curvature term and the chemical terms is much lower than S_w (i.e., $S_{w\_eq} \ll S_w$), therefore these terms are usually omitted. As a result, in most cloud models the equation of diffusional drop growth $r\frac{dr}{dt} = \frac{1}{F}\left(S_w - S_{w\_eq}\right)$ is replaced by a simpler equation:

$$r\frac{dr}{dt} = \frac{1}{F}S_w \tag{5.2.6}$$

Some authors use the equation for condensational growth in the form $\frac{d(r+r^*)^2}{dt} = aS_w$, where parameter a slightly depends on temperature and radius r^* characterizes the transfer from the diffusion condensation regime to the free-molecular regime in the vicinity of the drop surface (e.g., Devis, 2006; Mazin and Merkulovich, 2008). Since evaluations show that r^* is as small as several hundredth of μm, this parameter is usually omitted. In this book we use the equation of condensational growth in the form (5.2.6).

Solution of Equation (5.2.6) can be written as

$$r^2(t) = r_0^2 + \frac{2}{F}\int_0^t S_w(t')dt' = r_0^2 + Q(t), \tag{5.2.7}$$

where $r_0 = r(t = 0)$ and

$$Q(t) = \frac{2}{F}\int_0^t S_w dt' \tag{5.2.8}$$

According to Equation (5.2.6), any supersaturation obeying $S_w > 0$ leads to a drop growth. Equation (5.2.6) shows that the rate of drop radius growth is inversely proportional to the drop radius itself. It means that radii of small drops grow faster than those of large drops. Equations (5.2.6–5.2.8) also show that the change in the square of the drop radius is determined by the integral of supersaturation with respect to time. In case $S_w = const$ during the time period t, the solution of Equation (5.2.6) can be written as

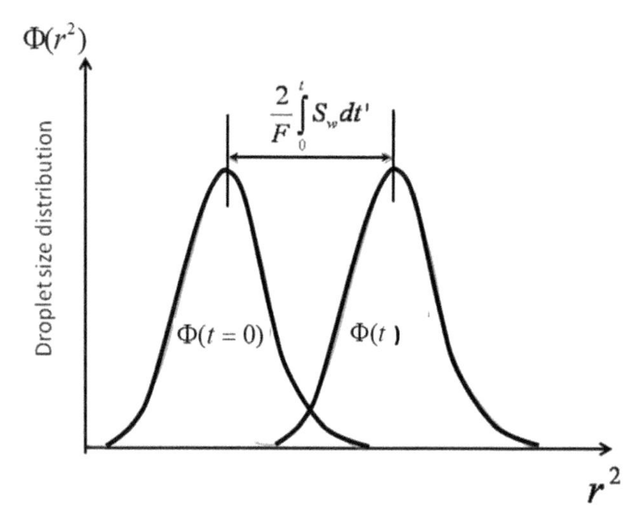

Figure 5.2.2 Changes of function $\Phi(r^2)$ due to diffusional growth/evaporation.

$$r^2(t) = r^2(t = 0) + \frac{2}{F}S_w t \tag{5.2.9}$$

Let us introduce the drop SD $\Phi(r^2)$ with normalization

$$N = \int_0^\infty \Phi(r^2)d(r^2), \tag{5.2.10}$$

where N is the drop concentration. The changes of $\Phi(r^2)$ due to diffusional growth/evaporation are reduced to a simple shift of $\Phi(r^2)$ to the right/left by the value of Q (see **Figure 5.2.2**). This fact is used in some cloud models to calculate the changes in DSD caused by diffusional growth (e.g., Khvorostyanov et al., 1989; Khain et al., 1993).

Equation (5.2.6) can be rewritten in the form

$$\frac{dm}{dt} = \frac{4\pi\rho_w r}{F}S_w, \tag{5.2.11}$$

where m is the drop mass. This equation shows that the mass of larger drops increases faster than that of small ones. Equations (5.2.6) and (5.2.11) for diffusional growth are written for drops that do not move relative to the air. For drops with radii below ~40 μm, the effects of drop sedimentation on drop growth can be neglected. For drops of larger sizes, it is necessary to take into account the fact that diffusional fluxes of water vapor molecules between falling drops and the surrounding air increase with increasing relative velocity. These additional fluxes can be taken into account by multiplying the drop growth/evaporation rate by a factor known as the ventilation coefficient (see Section 5.5).

Equation (5.2.6) should be solved together with Equation (5.2.1) or with Equation (5.2.2) for supersaturation. The rate of LWC changes caused by condensation/evaporation of cloud drops can be written as

$$\frac{dq_w}{dt} = \frac{d}{dt}\left(\frac{4}{3}\pi\rho_w \int_{r_{\min}}^{r_{\max}} r^3 f_r(r)dr\right) = 4\pi\rho_w \int_{r_{\min}}^{r_{\max}} r^2 f_r(r)\frac{dr}{dt}dr = \frac{4\pi\rho_w \bar{r} N}{F} S_w, \quad (5.2.12)$$

where $\bar{r}$ is the mean radius (Section 2.1). Using Equations (5.2.2) and (5.2.12), the equation for supersaturation can be written as a linear equation:

$$\frac{dS_w}{dt} = A_1 w - \frac{1}{\tau_{pr}} S_w, \quad (5.2.13)$$

where τ_{pr} is called the *phase relaxation time* (following Mazin, 1968). This parameter is calculated as (Korolev and Mazin, 2003)

$$\tau_{pr} = \left(\frac{4\pi\rho_w A_2 \bar{r} N}{F} - A_1 w\right)^{-1} \approx (4\pi D\bar{r}N - A_1 w)^{-1}, \quad (5.2.14a)$$

where D is the diffusivity of water vapor. In cases when $4\pi D\bar{r}N \gg A_1 w$, the phase relaxation time can be written in the form

$$\tau_{pr} = \left(\frac{4\pi\rho_w A_2 \bar{r} N}{F}\right)^{-1} \approx (4\pi D\bar{r}N)^{-1} \quad (5.2.14b)$$

The quantity $N\bar{r}$ is known as the *integral length* (Politovich and Cooper, 1988).

Within the ranges of time or height where the mean radius, the drop concentration and the vertical velocity can be assumed constant, Equation (5.2.13) has the following solution:

$$S_w(t) = S_0 \exp(-t/\tau_{pr}) + S_{qs}(1 - \exp(-t/\tau_{pr})), \quad (5.2.15)$$

where $S_0 = S_w(t = 0)$ and

$$S_{qs} = A_1 \tau_{pr} w \approx \frac{A_1 w}{4\pi D\bar{r}N} \quad (5.2.16a)$$

is the *quasi-steady supersaturation* (Squires, 1952; Rogers, 1975). As seen from Equation (5.2.15), the phase relaxation time is the characteristic time during which S_w approaches its quasi-steady value. It should be stressed that Equations (5.2.15) and (5.2.16a) are derived under the condition $\tau_{pr} = const$. It is clear, however, that in the natural conditions the phase relaxation time changes with height in the course of diffusional growth. In case the vertical velocity and the mean radius change with time relatively slowly and τ_{pr} is small, supersaturation adjusts to its quasi-steady value, which changes with time:

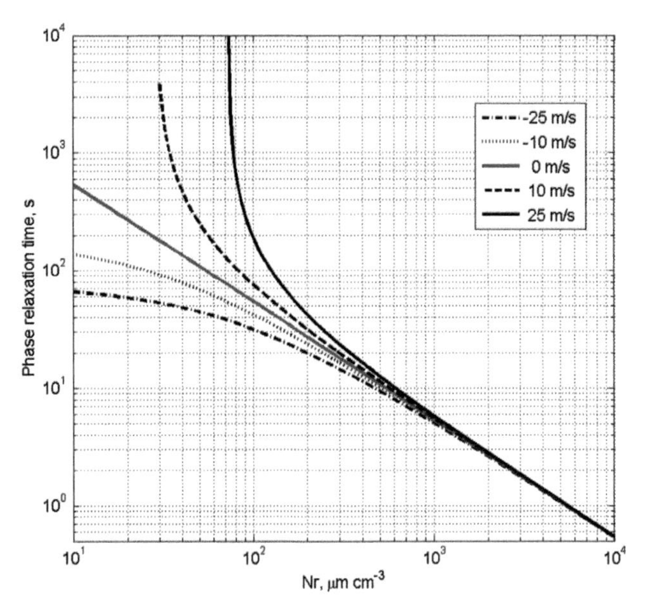

Figure 5.2.3 Phase relaxation time vs $N\bar{r}$ in liquid clouds at different vertical velocities. $T_C = 10°C$ and $p = 1,013$ mb. A similar figure can be found in Korolev and Mazin (2003).

$$S_{qs}(t) = A_1 \tau_{pr}(t) w(t) \approx \frac{A_1(t) w(t)}{4\pi DN\bar{r}(t)} \quad (5.2.16b)$$

This solution corresponds to the case when an increase in the relative humidity due to adiabatic cooling of the air is approximately balanced by the decrease caused by condensation of water vapor on cloud droplets. Equation (5.2.16b) allows evaluating supersaturation from in situ observations.

Figure 5.2.3 shows the relationship between the phase relaxation time and the integral length $N\bar{r}$ at different vertical velocities and $T_C = 10°C$, $p = 1,013$ mb. Values of $N\bar{r}$ vary within a wide range. In a typical liquid cloud, the droplet concentration is a few hundred per cm^3, and the mean droplet radius is about 10 μm. For such conditions $N\bar{r}$ is of the order of a few thousands μm · cm^{-3}, and τ_{pr} is below 5–10 s, which indicates a quite rapid adaptation of supersaturation to its quasi-steady value S_{qs}. In maritime clouds developing in clean air, the droplet concentration can be as low as 50 cm^{-3} and the mean radius is of the order of 15 μm. In this case, as well as at significant vertical velocity, the relaxation time can be as long as 20 s. A similar figure can be found in Korolev and Mazin (2003).

Figure 5.2.4 shows S_{qs} calculated from Equation (5.2.16) for different values of $N\bar{r}$ and w. For vertical

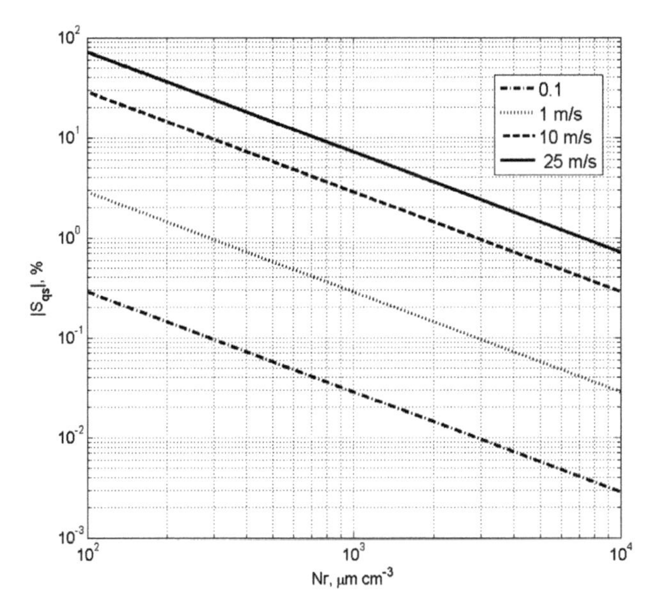

Figure 5.2.4 Quasi-steady supersaturation S_{qs} vs. $N\bar{r}$ in liquid clouds at different vertical velocities. $T_C = 10°C$ and $p = 1{,}013$ mb.

velocities within the range of $-1 < w < 1$ m/s that are typical of stratocumulus clouds (e.g., Martin et al., 1994), the quasi-steady supersaturation varies within the range of $-0.5\% < S_{qs} < 0.5\%$. In convective clouds where the vertical velocities vary within a wider range, w can reach 20–25 m/s. In this case S_{qs} can reach 5% even if $N\bar{r}$ values are comparatively large. When high w are accompanied by low $N\bar{r}$ as in deep maritime clouds developing in clean air, S_{qs} can sometimes exceed 10%. A similar figure can be found in Korolev and Mazin (2003).

As shown by Korolev and Mazin (2003), the assumption $N\bar{r} = const$ can be applied during the time interval considered, the mean drop radius does not change significantly, i.e., if

$$\overline{r^2}(t=0) \gg \left| \frac{2}{F} \int_0^t S_w dt' \right| \qquad (5.2.17)$$

Condition (5.2.17) is valid when the time period t is short or when LWC is significant. In contrast, this condition is invalid for small values of LWC. It means, for instance, that S_{qs} cannot be used for estimation of supersaturation near cloud base.

5.2.3 Vertical Profiles of Microphysical Characteristics in Adiabatic Parcels

The vertical profiles of supersaturation and vertical profiles of cloud-droplet radius are analyzed here, mainly following the study by Pinsky et al. (2013) with a simplifying assumption that DSDs are monodisperse. Monodisperse spectra can be a reasonable approximation to the real ones in case of narrow DSDs (e.g., with width equal or below 1–2 μm).

Vertical Profile of Supersaturation

For an ensemble of monodisperse droplets with concentration N and radii r, LWC can be written as

$$q_w = \frac{4}{3}\pi\rho_w N r^3 \qquad (5.2.18)$$

Substituting Equation (5.2.18) into Equation (5.2.11) yields

$$\frac{dq_w}{dt} = BN^{2/3} S q_w^{1/3}, \qquad (5.2.19)$$

where $B = \frac{3}{F}\left(\frac{4\pi\rho_w}{3}\right)^{2/3}$. Excluding q_w from Eqs. (5.2.4) and (5.2.19) leads to the following differential equation for supersaturation S_w:

$$\frac{dS_w}{dt} = A_1 w(t) - B(A_2 N)^{2/3}(A_1 z(t) + C - S_w)^{1/3} S_w \qquad (5.2.20a)$$

Equation (5.2.20a) can be rewritten for the independent variable z as

$$\frac{dS_w}{dz} = A_1 - \frac{1}{w(z)} B(A_2 N)^{2/3}(A_1 z + C - S_w)^{1/3} S_w \qquad (5.2.20b)$$

Equations (5.2.20a and 5.2.20b) are non-linear differential equations with variable coefficients. The supersaturation equation in the form of Equation (5.2.20b) is a closed differential equation with just one dependent variable. Other forms of closed equations for S_w were considered by Sedunov (1974) and Korolev and Mazin (2003). However, these representations of the closed supersaturation equations have integral-differential form, which makes them essentially more complex and more difficult for analysis, in comparison to Equation (5.2.20b). When the vertical updraft does not increase strongly with height, S_w decreases with height (above the supersaturation maximum near cloud base). At heights large enough, $S_w \ll A_1 z + C$, and the supersaturation (in brackets on the right-hand side of Equation (5.2.20)) can be neglected. In this case, Equation (5.2.20b) can be written in a linearized form:

$$\frac{dS_w}{dz} = A_1 - B(A_2 N)^{2/3} \frac{(A_1 z + C)^{1/3}}{w(z)} S_w \qquad (5.2.21)$$

Analysis of Equation (5.2.21) shows that when $S \ll A_1 z + C$, the value of the left-hand term becomes

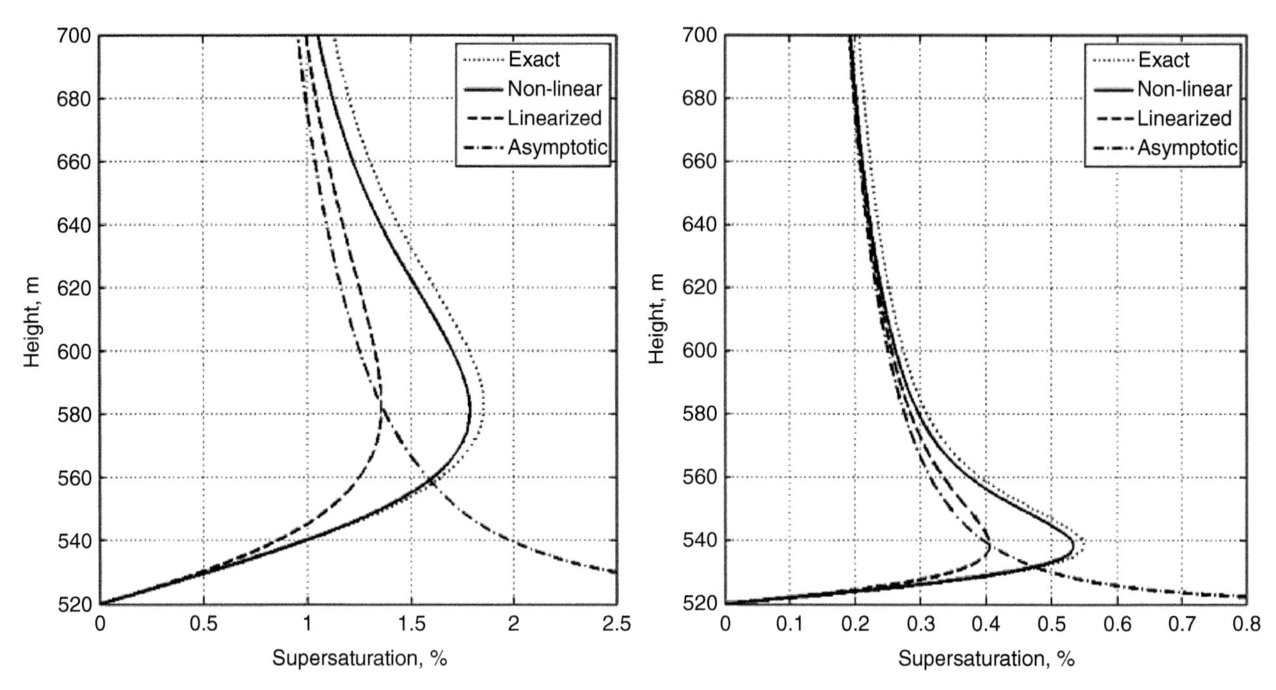

Figure 5.2.5 Vertical profiles of supersaturation in a cloud volume ascending at $w = 5$ m/s (left) and $w = 1$ m/s (right) at $N = 100$ cm^{-3}, calculated using different expressions: the exact solution (dotted line); nonlinear Equation (5.2.20b) (solid line); linearized Equation (5.2.21) (dashed line) and asymptotic approximation by Equation (5.2.22) (dash-dotted line). The initial conditions at $z = 520$ m are $T = 10°$C, $p = 950$ mb and $S_0 = 0$.

much smaller than values of either of the two terms on the right-hand side. In this case, neglecting the term $\frac{dS_w}{dz}$ in Equation (5.2.21) and assuming $C = 0$, it yields

$$S_{as} = \frac{A_1}{A_2^{2/3} B} w(z) N^{-2/3} z^{-1/3} \tag{5.2.22}$$

Equation (5.2.22) represents an asymptotic supersaturation forming in an adiabatic cloud parcel and can be written in a form similar to Equation (5.2.16b) for quasi-steady supersaturation. The only difference is that the mean droplet radius r in Equation (5.2.16b) is replaced by its adiabatic analog r_{ad}, corresponding to the adiabatic LWC q_{ad} in Equation (5.2.5). Equation (5.2.22) shows that supersaturation decreases with increasing height as $z^{-1/3}$.

Figure 5.2.5 shows the supersaturation profiles calculated from Equations (5.2.20b), (5.2.21), and (5.2.22), and the profile deduced from the numerical integration of the full system of equations describing a collective growth of droplets in an adiabatic parcel (the exact solution). Figure 5.2.5 shows a well-known maximum of supersaturation, i.e., ~20–60 m above cloud base, and a rapid decrease in S_w with height above this maximum. Equation (5.2.20b) is integrated assuming that coefficients A_1, A_2 and B remain constant, while in the numerical model (exact solution) the dependences of A_1, A_2 and B on T and on p were taken into account. As seen from Figure 5.2.5, Equation (5.2.20b) accurately describes the changes of supersaturation and agrees quite well with the numerically modeled supersaturation. Under particular conditions used in calculations, the difference between the modeled supersaturation and the one calculated using Equation (5.2.20b) does not exceed a few percent. It should be noted that neglecting the dependences of coefficients A_1, A_2, and B on T and on p gives quite accurate solutions for S_w within the vertical scale of the order of a few hundred meters. However, for displacements beyond one kilometer these dependences should be accounted for. The linearized Equation (5.2.21) approximates the exact solution well for $z > 30$ m at $w = 1$ m/s and $z > 150$ m at $w = 5$ m/s above the level of the supersaturation maximum. When these altitudes are converted into the time required for the parcel to reach them, it turns out that this time remains approximately the same (i.e., 30 s in this specific case). Equation (5.2.21) also leads to formation of a local supersaturation maximum near cloud base, but this maximum is lower than that obtained from the numerical model. In addition, Figure 5.2.5 shows that with altitude S_{as} (Equation 5.2.22) asymptotically approaches $S_w(z)$, but S_{as} does not reproduce the maximum and monotonically decreases with increasing height, being quite close to the exact solution.

Comparison of Equations (5.2.20a) and (5.2.13) shows that the value $\left[B(A_2N)^{2/3}(A_1z + C - S_w)^{1/3}\right]^{-1}$ has a meaning of phase relaxation time depending on height z. For the initial condition $C = 0$ one can write the following expression for profile of τ_{pr}:

$$\tau_{pr} = \left[B(A_2N)^{2/3}(A_1z - S_w)^{1/3}\right]^{-1} \quad (5.2.23)$$

Equation (5.2.23) coincides exactly with the standard form of τ_{pr} from Equation (5.2.14b) for the case of monodisperse droplets, i.e., when the mean radius of droplets is equal to their mean cubic radius. **Figure 5.2.6** shows changes of τ_{pr} versus the height of cloud parcels with different droplet concentrations ascending through a cloud base at $w = 1$ m/s, calculated by Equation (5.2.23). One can see that τ_{pr} is maximal in the vicinity of the cloud base supersaturation maximum and rapidly decreases with height. At the distance exceeding about 40 m above cloud base and at low droplet concentration, τ_{pr} is less than a few seconds. At higher droplet concentrations τ_{pr} is lower than 1 s. At these distances, supersaturation adapts to its quasi-steady value quite rapidly. According to Equation (5.2.23), τ_{pr} asymptotically decreases with increasing height as $\tau_{pr} \sim z^{-1/3}$.

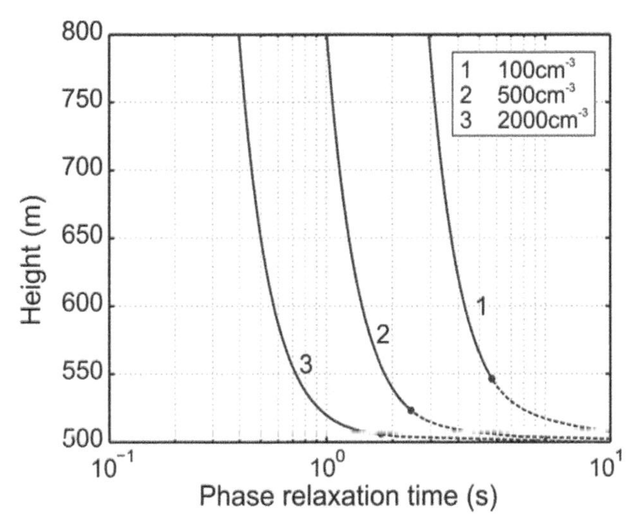

Figure 5.2.6 Changes of $\tau_{pr}(z)$ calculated from Equation (5.2.23) for cloud parcels ascending at $w = 1$ m/s and having different droplet concentrations. The dashed segment of the curves indicates the regions where quasi-steady approximation of supersaturation cannot be applied. The solid segment of the curves follows $z^{-1/3}$ law with a good approximation (from Pinsky et al., 2013; American Meteorological Society©; used with permission).

Maximum of Supersaturation Near Cloud Base

An important feature of Equation (5.2.20b) is that it allows estimating the supersaturation maximum S_{max} and the height z_{max} corresponding to the level where S_{max} is reached. For the sake of simplicity, assume $C = A_1q_0 + S_0 = 0$. We also introduce two new variables, namely, a nondimensional altitude $h = A_1z$ and a nondimensional parameter $R = \frac{BN^{2/3}A_2^{2/3}}{wA_1} = \frac{3}{FA_1w}\left(\frac{4\pi\rho_w NA_2}{3}\right)^{2/3}$. Using the new variables, Equation (5.2.20b) can be rewritten in the nondimensional form

$$\frac{dS_w}{dh} = 1 - R(h - S_w)^{1/3}S_w, \quad (5.2.24)$$

with the initial condition $S_w|_{h=0} = 0$ at cloud base. Analysis of Equation (5.2.24) suggests that the solution $S_w(h)$ depends on the sole parameter R. The condition $\frac{dS_w}{dh}\Big|_{h=h_{max}} = 0$ in Equation (5.2.24) yields an expression relating S_{max} and h_{max}:

$$1 - R(h_{max} - S_{max})^{1/3}S_{max} = 0 \quad (5.2.25)$$

A solution of Equation (5.2.25) can be written in the form

$$S_{max}(R) = C_1R^{-3/4}; h_{max}(R) = C_2R^{-3/4};$$
$$z_{max}(R) = \frac{C_2}{A_1}R^{-3/4}, \quad (5.2.26)$$

where coefficients $C_1 = 1.058$ and $C_2 = 1.904$ were obtained from numerical calculations. It should be noted that coefficients C_1 and C_2 are linked by Equation (5.2.25) and are not related to any physical variables (i.e., T, p, N, w, etc). Substituting R into Equation (5.2.25) and using the balance Equation (5.2.4) enables deriving the dependences of S_{max}, z_{max} and q_{max} on droplet concentration and vertical velocity:

$$S_{max} = C_1(FA_1/3)^{3/4}\left(\frac{3}{4\pi\rho_w A_2}\right)^{1/2} w^{3/4}N^{-1/2} \quad (5.2.27)$$

$$z_{max} = \frac{C_2}{A_1}(FA_1/3)^{3/4}\left(\frac{3}{4\pi\rho_w A_2^3}\right)^{1/2} w^{3/4}N^{-1/2} \quad (5.2.28)$$

$$q_{max} = (C_2 - C_1)(FA_1/3)^{3/4}\left(\frac{3}{4\pi\rho_w A_2^3}\right)^{1/2} w^{3/4}N^{-1/2} \quad (5.2.29)$$

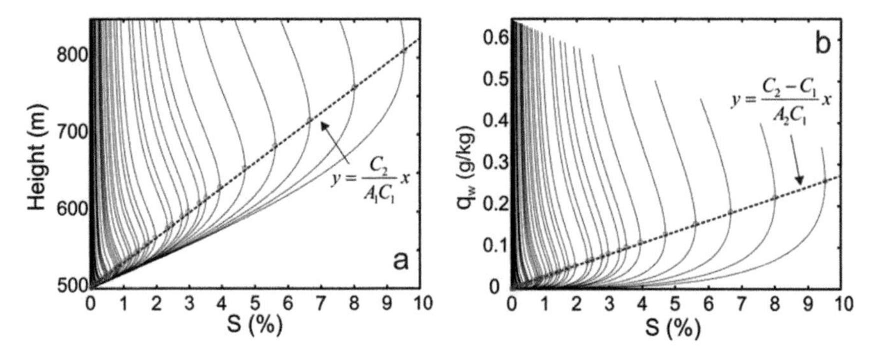

Figure 5.2.7 Modeled dependences (a) $S(z)$ and (b) $S(q)$ calculated for various combinations of vertical velocities of ascending parcels and droplet number concentrations from the respective ranges $0.01 \leq w \leq 10$ m/s and $10 \leq N \leq 1{,}000$ cm^{-3}. The small circles indicate the locations of S_{max}. The calculations were performed for the following initial conditions: $T_0 = 10°$C, $p_0 = 950$ mb, $S_0 = 0$, and $r_0 =$ 0.01 μm (from Pinsky et al., 2013; American Meteorological Society©; used with permission).

Equations (5.2.27–5.2.29) show that all three variables S_{max}, z_{max} and q_{max} are proportional to $w^{3/4}N^{-1/2}$.

Equations (5.2.27–5.2.29) suggest an important conclusion regarding the universal nature of the following ratios:

$$\frac{S_{max}}{A_1 z_{max}} = \frac{C_1}{C_2} = 0.556; \frac{S_{max}}{A_2 q_{max}} = \frac{C_1}{C_2 - C_1} = 1.25 \tag{5.2.30}$$

In other words, S_{max}, z_{max} and q_{max} are linearly related to one another. This finding was verified with the help of a numerical simulation of droplet growth in an ascending adiabatic parcel. The diagrams in **Figure 5.2.7** show that the modeled relationships between S_{max}, z_{max}, and q_{max} follow Equation (5.2.30) with high accuracy for a wide range of vertical velocities w and droplet number concentrations N, which occur in the tropospheric liquid clouds. An interesting feature of the initial stage of the cloud formation, which follows from Equation (5.2.30) is that at the level of supersaturation maximum, 45% of potentially condensed water exist in liquid phase, whereas the remaining 55% exist in the form of supersaturated vapor.

Note that non-activated CCN and cloud droplets affect the value of supersaturation maximum. At cloud bases where supersaturation is equal to zero, dry soluble aerosols with radii exceeding 0.1 μm have the equilibrium radius of a few microns (Kogan, 1991; Segal et al., 2003). In this case, the simplified equation for diffusional growth (5.2.6) can also be applied to these particles, at least in the first approximation.

Drop Growth by Diffusion

Excluding S_w from Equations (5.2.19) and (5.2.4), one can get a closed equation for q_w:

$$\frac{dq_w}{dz} = -\frac{BN^{2/3}}{w(z)}\left(A_2 q_w^{4/3} - (A_1 z + C) q_w^{1/3}\right) \tag{5.2.31}$$

At heights large enough, the values of the two terms on the right-hand side of Equation (5.2.31) become significantly larger than $\frac{dq_w}{dz}$. The balance between these large values leads to a linear dependence of q_w on height. Equations (5.2.18) and (5.2.31) yield differential equations for the changes of r:

$$w(z)\frac{dr}{dz} = -\frac{4\pi A_2 \rho_w N}{3\rho_a F} r^2 + \frac{(A_1 z + C)}{Fr} \tag{5.2.32}$$

At heights large enough that r only slowly changes with height, solutions of Equation (5.2.32) can be approximated by the adiabatic dependence:

$$r(z) = \left[\frac{3(A_1 z + C)}{4\pi A_2 \rho_w N}\right]^{1/3} = \left(\frac{3q_{ad}(z)}{4\pi \rho_w N}\right)^{1/3} \tag{5.2.33}$$

One can see that $r \sim N^{-1/3}$ in this case.

Figure 5.2.8 shows the changes of water content and droplet sizes computed using Equations (5.2.31) and (5.2.32) for the two different vertical velocities during of ascending/descending of a parcel. As seen from Figure 5.2.8, integration of Equations (5.2.31) and (5.2.32) provides a very good agreement with q_w and r, respectively, calculated by the numerical model. Figure 5.2.8a also shows the changes of the mixing ratio of the supersaturated or subsaturated fraction of water vapor q_{vs}. One can see that the deviation of r from its adiabatic value is the largest near cloud base where the supersaturation in updraft is maximal and decreases

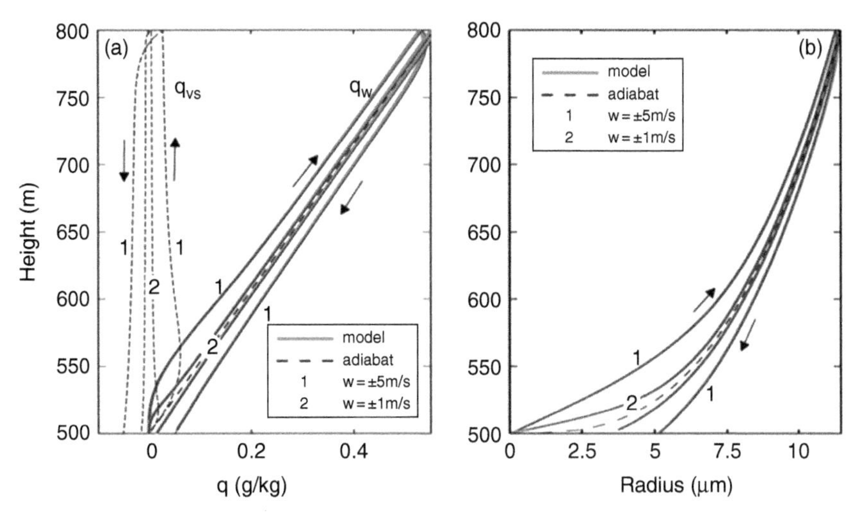

Figure 5.2.8 Vertical changes of liquid water mixing ratio calculated using Equation (5.2.31) (solid lines) and supersaturated vapor mixing ratio (dashed lines) (a) and drop radius calculated by Equation (5.2.32) (b). The calculations were performed for vertical velocities $w = \pm 1$ m/s and $w = \pm 5$ m/s. The arrows show the direction of the air parcel movement (ascent vs. descent). The initial conditions are $T_0 = 10°$C, $p_0 = 950$ mb, and $S_0 = 0$, $r_0 = 0.01$ μm (from Pinsky et al., 2013; American Meteorological Society©; used with permission).

with increasing altitude. It follows from Figure 5.2.8 that the deviation of q_w and r from q_{ad} and r_{ad}, respectively, increases with the increasing vertical velocities.

Equation of balance (5.2.4) shows that the process of diffusional growth/evaporation is not reversible due to different values of supersaturations in updrafts and downdrafts. Figure 5.2.8 illustrates this nonreversibility. While supersaturation is positive in updrafts, liquid water content q_w and droplet radii are smaller than in downdrafts where supersaturation is negative. This effect is another indication of the fact that supersaturation tends to adjust to the field of droplets: while lower q_w in updrafts allows S_w to positive, larger q_w in downdrafts lead to negative supersaturation.

Universal Height Dependences in Adiabatic Parcel

Introducing new variables $S^* = S_w R^{3/4}$, $h^* = hR^{3/4}$, and $Q^* = A_2 q_w R^{3/4}$ and substituting them into Equations (5.2.24) and (5.2.31) yields universal equations for S_w and q_w:

$$\frac{dS^*}{dh^*} = 1 - \left(h^* - S^*\right)^{1/3} S^* \qquad (5.2.34)$$

$$\frac{dQ^*}{dh^*} = \left(h^* - Q^*\right) Q^{*1/3} \qquad (5.2.35)$$

The solutions $S^*\left(h^*\right)$ and $Q^*\left(h^*\right)$ are universal and are valid for any values of droplet concentration, vertical velocity, temperature, and pressure. The maximum of universal supersaturation $S^*_{\max} = C_1$ and its altitude $h^*_{\max} = C_2$ are constant. In order to obtain the actual $S_w(z)$ and $q_w(z)$ for specific w, N, T, and p, the normalized solutions $S^*\left(h^*\right)$ and $Q^*\left(h^*\right)$ should be scaled using $S_{\max}$, $q_{\max}$ and $h_{\max}$ from Equations (5.2.27–5.2.29). For the new variables, the normalized adiabatic water content is $Q^*_{ad} = h^*$. Equations (5.2.34) and (5.2.35) are dependent, and either of them can be derived from the other using the mass balance equation (5.2.4) written in the normalized form as $S^* + Q^* - h^* = 0$. Such universal dependencies can be effectively used for parameterization of supersaturation and CCN nucleation in different cloud and large-scale models.

Nonconstant Vertical Velocity

Previous analysis of supersaturation and droplet radius evolution was presented for the simple case of constant vertical velocity in updrafts and downdrafts. Equations (5.2.20b) and (5.2.32) are also valid in a more general case of arbitrary profiles of vertical velocity. Next, we present several examples of supersaturation evolution and droplet growth in parcels with a variable vertical velocity. **Figure 5.2.9** shows the vertical profiles of radii and supersaturation in cases when vertical velocity increases with height at different rates. One can see that even when the vertical velocity increases with height at a comparatively low rate, supersaturation starts increasing at several hundred meters above the

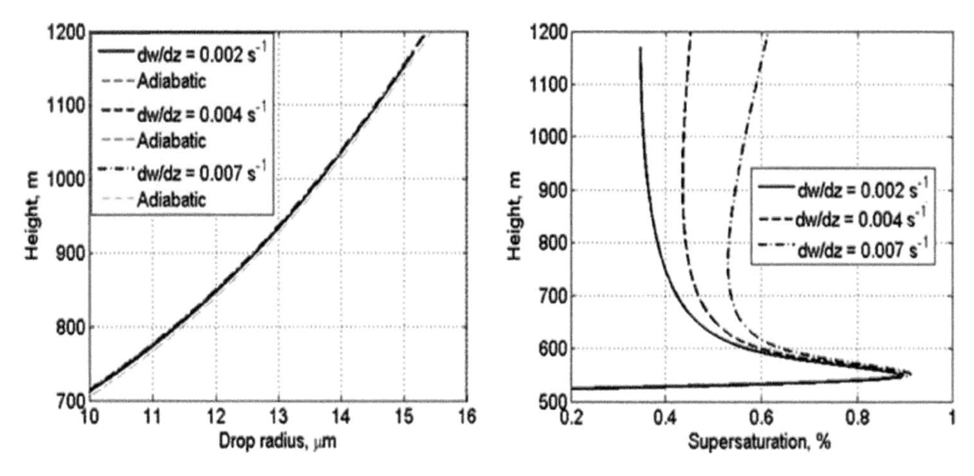

Figure 5.2.9 Vertical profiles of drop radius (left) and supersaturation (right) in cases when vertical velocity increases with height at different rates. The vertical velocity at cloud base is assumed equal to 3 m/s. Drop concentration $N = 100\ \mathrm{cm}^{-3}$.

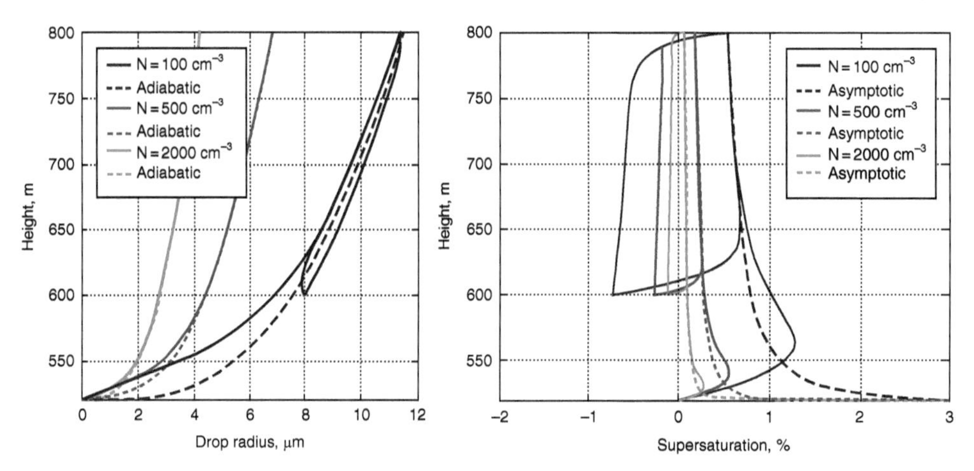

Figure 5.2.10 Vertical profiles of drop radius (left) and supersaturation (right) in case of periodic vertical oscillations of a parcel within the layer of 600–800 m. The asymptotic profiles of supersaturation calculated using Equation (5.2.22) are shown by dashed lines.

supersaturation maximum level. This increase may lead to supersaturation that exceeds its maximum near cloud base. This in turn triggers in-cloud nucleation of droplets (Section 5.11). Supersaturation typically remains smaller than $A_2 q_w$, therefore the droplet radii remain quite close to their adiabatic values (Figure 5.2.9, left).

It is known that oscillations of a cloud parcel in the vertical direction is a typical phenomenon in stratocumulus clouds (e.g., Korolev, 1995). **Figure 5.2.10** shows the vertical profiles of droplet radii and supersaturation when a parcel experiences several oscillations in the vertical direction within the layer of 600–800 m. One can see that the vertical profiles of radii, as well as the vertical profiles of supersaturation obtained during several successive oscillations, are coincident. Although the droplet radii are different in the updraft branch and the downdraft branch of the parcel motion, oscillations take place along the same height profiles of radii and do not cause any continuous shift of the radii loop with time. Figure 5.2.10 shows the existence of a stable cycle of radius oscillations located around the adiabatic curve. This can be explained by Equation (5.2.32). The adiabatic curve represents a stable curve, since in the adiabatic case ($S_w = 0$) the right-hand side of Equation (5.2.32) is equal to zero, and since the derivative $\frac{\partial}{\partial r}\frac{dr}{dt} < 0$ is negative for any height z. Therefore, the adiabatic curve attracts the profiles of radii. At the same time, a parcel moves up and down making the radius spin around the adiabatic curve. The width of this stable cycle is proportional to the updraft/downdraft velocity, namely, the larger the velocity the wider the cycle is. The stable velocity oscillations cycle generates a limit cycle of supersaturation oscillations around the line $S_w = 0$.

The limit cycle reflects the equivalence of the increase of droplet mass by condensation during the parcel ascent and the mass loss by evaporation during the decent. During the first oscillation, the mean mass of water droplets increases because droplets in downdraft are larger than in updraft (Figures 5.2.8, right panel, and 5.2.10, left panel) so that supersaturation maximum during the second cycle decreases, which leads to a decrease in the water mass generation during the second ascent (Figure 5.2.10, right panel). Due to low phase relaxation time, the equilibrium between mass generation in the ascent and its loss in the descent, the branches of the parcel's trajectory become equal and the limit cycle is established. Averaging over a set of uniformly oscillating parcels should not lead to any broadening of the DSD.

5.2.4 Characteristics of Droplet Size Distribution During Diffusional Growth

Evaluation of the DSD Width

In situ-measured width of DSD is determined by two factors: a) formation of local wide DSD in each small cloudy volume and b) averaging of different individual DSD along the aircraft traverse. It is a typical practice to analyze DSD obtained over 100 m-length intervals. Modern high-frequency probes allow averaging of DSD over a 10 m aircraft traverse. Averaged DSD should be wider than local DSD, as averaging may lead to a wide DSD even if local DSD are narrow but centered at different radii. Droplet spectrum dispersion (also known as the variation coefficient) measured as the ratio of DSD width σ_r to the mean radius $\bar{r}$ (Section 2.1) in clouds varies from ~0.1 to 0.6. No significant changes in droplet spectrum dispersion with height have been found (Politovich, 1993; Martin et al., 1994; Prabha et al., 2011, see also Figure 2.3.5).

In this subsection, we evaluate the width of DSD that can be formed by spatial averaging of individual narrow DSD of different cloudy parcels. DSD in each parcel is assumed to be monodisperse with the droplet concentration N constant and equal in all the parcels. From Equations (5.2.4) and (5.2.18), the difference in droplet radii cubes at a certain height in two adiabatic parcels ascending at the same initial conditions (the same constant C) but at different velocities can be expressed as

$$\Delta(r^3) = \frac{3}{4\pi\rho_w A_2}\frac{|\Delta S|}{N} \tag{5.2.36}$$

We introduce the ratio $\frac{\Delta r}{r}$ as the measure of droplet spectrum dispersion in the analyzed case. Using Equation (5.2.36) yields for narrow spectra:

$$\frac{\Delta r}{r} = \frac{1}{3}\frac{|\Delta S|}{A_2 q_w} \tag{5.2.37}$$

Equation (5.2.37) allows for evaluating the difference between supersaturation values in two air parcels, needed to obtain certain dispersion under conditions typical of different clouds. Evaluations presented in **Table. 5.2.2** are performed for $\frac{\Delta r}{r} = 0.2$. As seen from Table 5.2.2, the values of $|\Delta S|$ required to get the dispersion of 0.2 are much higher than the values of supersaturation in clouds. These evaluations lead to the following important conclusion: if adiabatically ascending parcels have similar initial conditions, condensation does not lead to a wide average DSD.

The maximum difference between supersaturation in parcels at the same level and the maximumal difference between droplet sizes can be obtained between ascending ($S_w > 0$) and descending ($S_w < 0$) parcels. Accordingly, the maximum difference in the droplet

Table 5.2.2 Difference in supersaturation needed to obtain spectrum dispersion equal to 0.2. The parameters used in the table are chosen according to the references presented (from Pinsky et al., 2013; American Meteorological Society©; used with permission).

Cloud Type	N, cm^{-3}	q_w, g m^{-3}	r, μm	$\|\Delta S\|$, %
Deep maritime Cu (Andreae et al., 2004)	100	2.0	16.8	28.9
Maritime stratocumulus (e.g., Martin et al., 1994; Stevens et al., 2003, 2005; Magaritz et al., 2009)	100	0.5	10.6	7.2
Weak stratocumulus (Stevens et al., 2003, 2005)	100	0.2	7.8	2.9
Deep continental Cu (Andreae et al., 2004; Prabha et al., 2011)	1,000	3	8.9	43.2

size takes place between ascending and descending parcels. The maximum difference in radii takes place near cloud base where the difference in supersaturation in updrafts and downdrafts is maximum and the LWC is low (see Figure 5.2.8). Within this zone, the difference in radii can reach a few μm. The difference in droplet radii increases with increasing vertical velocity and decreases with increasing droplet concentration, as follows from Equation (5.2.36). At higher levels, the deviation from the adiabatic value rapidly decreases to ~0.1–0.15 μm at droplet concentration equal to 100 cm^{-3}.

Evolution of DSD During Diffusional Growth

Evolution of polydisperse DSD $f_r(r,t)$ due to diffusional growth was investigated in analytical studies (Pinsky et al., 2014) as well as in studies using a bin-microphysics models (e.g., Pinsky and Khain, 2002; Khain et al., 2008; Benmoshe et al., 2012). Using Equation (5.2.7), the time evolution of the initial DSD $f_r(r,t) = f_{0r}(r_0)\frac{dr_0}{dr}$ in an ascending adiabatic parcel can be written in the form (Pinsky et al., 2014)

$$f_r(r,t) = \begin{cases} \dfrac{r}{\sqrt{r^2 - Q(t)}} f_{0r}\left(\sqrt{r^2 - Q(t)}\right), & r \geq \sqrt{Q(t)} \\ 0, & r < \sqrt{Q(t)} \end{cases}, \quad (5.2.38)$$

where $Q(t)$ (Equation 5.2.8) is proportional to the integral of supersaturation. From Equation (5.2.38), the k-th moment of DSD can be calculated as

$$\overline{r^k}(t) = \frac{1}{N}\int_{\sqrt{Q(t)}}^{\infty} r^k f_r(r^2 - Q(t))\,dr = \frac{1}{N}\int_0^{\infty} \left(r^2 + Q(t)\right)^{k/2} f_{0r}(r)\,dr \quad (5.2.39)$$

Pinsky et al. (2014) obtained the following close differential equation for $Q(z)$:

$$\frac{dQ(z)}{dz} = \frac{2}{Fw(z)}\left(A_1 z - \frac{4}{3}\pi\rho_w A_2 \int_0^{\infty} \left(r_0^2 + Q(z)\right)^{3/2} f_{0r}(r_0)\,dr_0 + C\right) \quad (5.2.40)$$

The initial condition for this equation is: at $z = 0$ $Q(z)|_{z=0} = 0$ and $\left.\frac{dQ(z)}{dz}\right|_{z=0} = \frac{2S_0}{Fw(0)}$. The vertical profile of supersaturation can be calculated using Equation (5.2.40) as

$$S_w(z) = \frac{Fw(z)}{2}\frac{dQ(z)}{dz} = A_1 z - \frac{4}{3}\pi\rho_w A_2 \int_0^{\infty} \left(r_0^2 + Q(z)\right)^{3/2} f_{0r}(r_0)\,dr_0 + C \quad (5.2.41)$$

In the ascending parcel, function $Q(z)$ increases (as DSD shifts toward larger droplets) and condition $Q(z) \gg \overline{r_0^2}$ becomes valid. In this case, Equation (5.2.40) can be simplified to its asymptotic form:

$$\frac{dQ(z)}{dz} = \frac{2}{Fw(z)}\left(A_1 z - \frac{4}{3}\pi\rho_w N A_2 Q^{3/2}(z) + C\right) \quad (5.2.42)$$

The asymptotic behavior of $Q(z)$ in Equation (5.2.42) is universal and does not depend on the shape of the initial DSD. In case supersaturation tends to zero with increasing height, $\frac{dQ(z)}{dz}$ decreases as well, resulting in the adiabatic profile of $Q(z)$:

$$Q_{ad}(z) = \left(\frac{3(A_1 z + C)}{4\pi\rho_w N A_2}\right)^{2/3} \approx (\chi z)^{2/3}, \quad (5.2.43)$$

where $\chi = \frac{3A_1}{4\pi\rho_w N A_2}$. Equations (5.2.39) and (5.2.43) lead to the asymptotic profiles of DSD moments, the corresponding formulas are presented in **Table 5.2.3**. The asymptotic equations in Table 5.2.3 indicate that all the moments tend to their adiabatic values with increasing height, so $\overline{r^k}(z) \sim z^{k/3}$. At the same time, central moments $M_k(z) \sim z^{-k/3}$ as well as the variation coefficient $\delta_r(z) \sim z^{-2/3}$ decrease to zero with increasing height, indicating DSD narrowing. The asymmetry coefficient and the kurtosis of DSD tend to constant values that depend on the parameters of the initial DSD at cloud base.

Figure 5.2.11 illustrates evolution of DSDs forming for different aerosol conditions in the course of parcel updraft. The dry aerosols, on which the DSD form, are described by 3-mode lognormal distributions (Equation 2.2.2) with parameters presented in Table 2.2.2. The set of parameters covers practically all possible aerosol conditions in clouds. These distributions together with distributions of wet aerosols serving as initial DSD at cloud base are shown in the Figure 2.2.3. Figure 5.2.11 shows that diffusional growth changes the DSD shape and shifts DSD toward larger droplet radii in updrafts. The distributions are truncated from the left by radius equal to $\sqrt{Q(z)}$, which increases with height. The main difference between DSDs in the maritime and continental cases is the values of this increasing truncated radius. This radius reaches 13 μm at 480 m above cloud base in

Table 5.2.3 Asymptotic formulas for moments and moment functions (from Pinsky et al., 2014; courtesy of John Wiley & Sons, Inc. ©).

Quantity	Definition	Asymptotic Formula
Moments	$\overline{r^k}(z) = \frac{1}{N}\int_0^\infty r^k f(r,z)dr$	$\overline{r^k}(z) = (\chi z)^{k/3}$
Central moments	$M_k = \frac{1}{N}\int_0^\infty (r - \bar{r})^k f(r,z)dr$	$M_k \sim z^{-k/3}$
Mean radius	$\bar{r}(z) = \frac{1}{N}\int_0^\infty rf(r,z)dr$	$\bar{r}(z) = (\chi z)^{1/3}$
Mean volume radius	$r_v(z) = \left[\overline{r^k}(z)\right]^{1/3}$	$r_v(z) = (\chi z)^{1/3}$
Effective radius	$r_{eff}(z) = \frac{\overline{r^3}(z)}{\overline{r^2}(z)}$	$r_{eff}(z) = (\chi z)^{1/3}$
Variation	$\sigma_r^2(z) = M_2$	$\sigma_r^2(z) = \frac{1}{4}\left[\overline{r_0^4} - \left(\overline{r_0^2}\right)^2\right](\chi z)^{-2/3}$
Dispersion (variation coefficient)	$\delta_r(z) = \frac{\sigma_r(z)}{\bar{r}(z)}$	$\delta_r(z) = \frac{1}{4}\sqrt{\overline{r_0^4} - \left(\overline{r_0^2}\right)^2}(\chi z)^{-2/3}$
Asymmetry coefficient	$\xi(z) = \frac{M_3}{M_2^{3/2}}$	$\xi = \frac{\overline{r_0^6} - \overline{3r_0^4}\,\overline{r_0^2} + \overline{2r_0^2}^3}{\left(\overline{r_0^4} - \overline{r_0^2}^2\right)^{3/2}}$
Kurtosis	$\eta(z) = \frac{M_4}{M_2^2}$	$\eta = \frac{\overline{r_0^8} - 4\overline{r_0^6}\,\overline{r_0^2} + 6\overline{r_0^4}\,\overline{r_0^2}^2 - 3\overline{r_0^2}^4}{\left(\overline{r_0^4} - \overline{r_0^2}^2\right)^2}$

marine clouds but only 3 μm in urban clouds. Another characteristic feature of DSD shapes is their fast narrowing with height, so the right tail of DSD can be seen only if DSD are plotted on the logarithmic scale. It is unlikely that such kinds of shapes can be well approximated by the Gamma distribution usually used for parameterization of DSD in different bulk-parameterization cloud models.

Profiles of supersaturation obtained for polydisperse DSDs turn out to be similar to monodisperse DSDs. This similarity is not an accidental coincidence but rather a reflection of the general laws of diffusional growth in clouds. Equation (5.2.20b) can be considered as monodisperse analog of Equation (5.2.41). A comparison of the solutions of these two equations under the same initial conditions allows for understanding the difference between the condensation processes in the monodisperse versus polydisperse cases. Note that the solutions of both equations tend to the same asymptotic profile of supersaturation $S_w(z) \sim z^{-1/3}$ (Equation (5.2.22)).

Figure 5.2.12 compares profiles of supersaturation calculating using Equations (5.2.20b) and (5.2.41) for different aerosol types. The mean radius of haze particles $\bar{r}_0$ at cloud base for polydisperse DSD was chosen as the initial parameter for solving Equation (5.2.20b). Figure 5.2.12 shows that for each polydisperse DSD (solid lines) there exists an equivalent monodisperse DSD with the initial radius equal to the mean radius of polydisperse DSD at cloud base $\bar{r}_0$. The difference between the two profiles is very small, despite the fact that supersaturation is a very sensitive quantity characterizing the condensation process. An excellent agreement is obtained for a very wide range of aerosol conditions, from low-concentration marine aerosols to extremely polluted air over urban zones. This fact also indicates that there is excellent agreement between the LWC profiles for all DSD types due to the tight connection between $S_w(z)$ and $q_w(z)$ imposed by the balance Equation (5.2.4). Figure 5.2.12 also presents supersaturation profiles calculated using the parcel

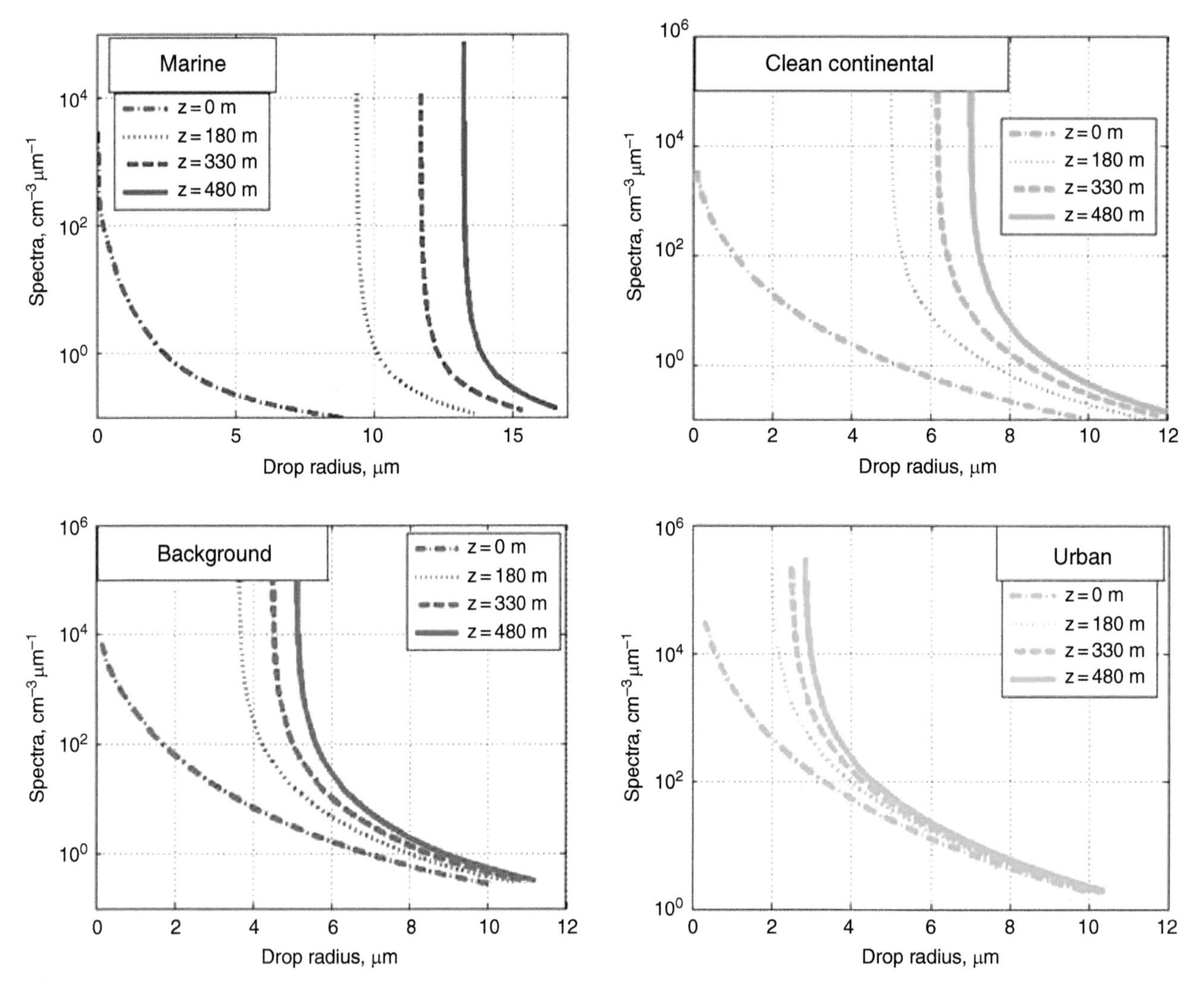

Figure 5.2.11 Drop SDs developing in adiabatic ascending parcel at different heights for aerosols of different types. The distributions of wet aerosols shown in Figure 2.2.3 are used as the initial DSD. The updraft velocities are: 3.21 m/s for Marine, 2.9 m/s for Clean continental, 3.032 m/s for Background, and 5.87 m/s for Urban (from Pinsky et al., 2014; courtesy of John Wiley & Sons, Inc. ©).

model in which the full equation of diffusion growth was used. One can see a good agreement of the theoretical and numerically determined supersaturation profiles for marine, clean continental, and background aerosols. However, in case of urban aerosols the agreement is not that good. In this case, droplets do not reach sufficient sizes (the effective radius reaches only 2.5 μm (see Figure 5.2.11), so the simplified equation of diffusion growth (5.2.6) is not applicable.

The coincidence of vertical profiles of supersaturation under different conditions indicates the universality of the laws of condensational growth. One consequence of this universality is the possibility to use the theory of supersaturation maximum near cloud base (Section 5.2.3) with small changes for polydisperse DSD (Pinsky et al., 2014). This finding opens ample opportunities for applying the theory in investigation and modeling of the condensation process near cloud base.

5.2.5 Stochastic Condensation in the Adiabatic Case

The equation of diffusional growth predicts narrowing DSD in the space of radii (Figure 5.2.11). Since the width of observed DSD increases with height (e.g., Prabha et al., 2011, 2012), several mechanisms were proposed to explain this phenomenon. In a number of studies (e.g., Belyaev, 1961; Mazin, 1965; Mazin and Smirnov, 1969; Barlett and Jonas, 1972; Manton, 1979;

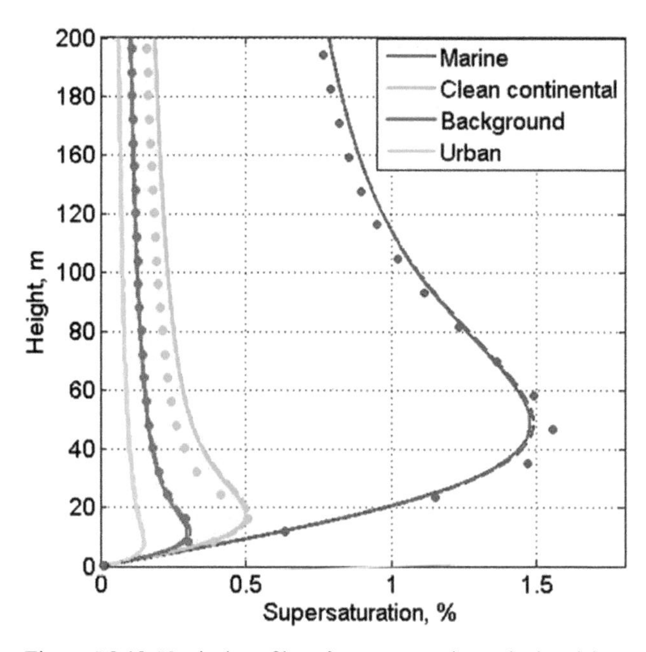

Figure 5.2.12 Vertical profiles of supersaturation calculated for the polydisperse case (solid lines) and for monodisperse case using Equation (5.2.20b) (dashed lines) for different aerosol types. Parameters of the initial DSDs and the updraft velocities are the same as in Figure 5.2.11. The mean radius of haze particles at cloud base was chosen as the initial radius of monodisperse DSD. The results of parcel model are shown by circles (from Pinsky et al., 2014; courtesy of John Wiley & Sons, Inc. ©).

overview by Mazin and Merkulovich, 2008), the theory of stochastic condensation was proposed that attributes the DSD broadening to random fluctuations of supersaturation in a turbulent flow. The idea behind the theory is that different droplets in a turbulent flow experience different supersaturation history and grow to different sizes. Fluctuations of supersaturation were attributed either to fluctuations of temperature, or to fluctuations of the vertical velocity or were prescribed a priori. Indeed, formulation of the supersaturation equation over individual droplets is a complicated problem that has not been finally solved despite several attempts (e.g., Sedunov, 1965; Srivastava, 1989; Cooper, 1989).

The typical approach is to consider DSD evolution in the radii space. Regular condensation described by Equation (5.2.6) is regarded as the process of advection of DSD within the radii space (Section 5.4). The effect of supersaturation fluctuations is represented by a diffusion of DSD within this space and leads to some DSD broadening. For instance, assuming normally distributed fluctuations of supersaturation with a priori-prescribed parameters, Jeffery et al. (2007) got the Fokker–Plank-type equation for DSD evolution:

$$\frac{\partial f_r}{\partial t} = -\frac{\overline{S}_w}{F}\frac{\partial}{\partial r}\left(\frac{1}{r}f_r\right) + D_r\frac{\partial}{\partial r}\left(\frac{1}{r}\frac{\partial}{\partial r}\left(\frac{1}{r}f_r\right)\right), \tag{5.2.44}$$

where $\overline{S}_w$ is a mean supersaturation and D_r has a meaning of a diffusion coefficient in the radii space. This equation is written for the Lagrangian adiabatic parcel. The second term on the right-hand side of Equation (5.2.44) is responsible for DSD broadening at large D_r. A priori normal distributions of supersaturation fluctuations were also assumed by Belyaev (1961) and Mazin (1965), who obtained substantial DSD broadening.

In a number of studies, diffusional droplet growth in turbulent flows was investigated using DNS in which turbulence is explicitly simulated. Theoretically, DNS simulations are quite close to the original concept of stochastic condensation. This concept looks for the reasons of DSD broadening in differences between supersaturation histories experienced by individual droplets. An important difference between DNS and real clouds is that DNS have no large-scale updrafts and downdrafts determining regular condensation. Paoli and Shariff (2009) reported a significant DSD broadening in DNS as a result of diffusional growth. In these DNS, droplet growth was forced by random low-frequency fluctuations of supersaturation prescribed a priori. No negative feedback of supersaturation caused by water vapor sink on droplets (which is the basic mechanism of the changes of supersaturation in real clouds) was taken into account in this study. The DSD broadening in their study can be interpreted in such a way that some "lucky" droplets spent more time in zones of high supersaturation, while some "unlucky" droplets spent more time in zones of low supersaturation.

Lanotte et al. (2009) simulated diffusional growth of droplets in DNS with the grid spacing of about 1 mm. Evolution of two initially monodisperse DSDs with droplet radii of 13 μm and 5 μm was simulated taking into account the effects of water vapor sink on droplets. It was found that DSD broadening increases with increasing turbulence intensity. However, the maximum DSD width obtained in these simulations was ~0.02 μm, i.e., negligibly small. The negligible effect of turbulence on diffusional droplet growth was also reported in DNS performed by Vaillancourt et al. (2002). If supersaturation fluctuations are assumed proportional to fluctuations of the vertical velocity, the situation resembles that of regular condensation and DSD narrowing is expected (Mazin and Merkolovich, 2008).

Mazin (1967) showed that when turbulent velocity fluctuates at high frequency, supersaturation does not reach its quasi-steady value and, consequently, is not proportional to the vertical velocity. Mazin supposed that a disruption of this proportionality may result in DSD broadening. The strongest effect of turbulence was expected at turbulent time scales equal to the phase relaxation time. At these time scales the disproportion between fluctuations of S_w and fluctuations of w is significant and supersaturation fluctuations are comparatively strong (Mazin and Merkulovich, 2008).

Equation (5.2.4) does not indicate any specific preferable wavelength or frequency of velocity fluctuations affecting droplet size in adiabatic parcels. This equation shows that droplet size is fully determined by the initial conditions and the value of supersaturation. Thus, in order to understand the possible effects of turbulent fluctuations of velocity on droplet size one should determine whether turbulent fluctuations can create significant fluctuations of supersaturation comparable with the value A_1z. Results presented in Table 5.2.2 suggest a negative answer to this question: turbulent velocity fluctuations of scales corresponding to the phase relaxation time are typically significantly weaker than velocity fluctuations of larger turbulent scales and than velocities in convective updrafts and downdrafts. Response of supersaturation to high-frequency turbulent velocity fluctuations monotonically weakens with increasing frequency. This conclusion is supported by simulations of DSD evolution by means of a spectral (bin) microphysics parcel model (Pinsky and Khain, 2002) (**Figure 5.2.13**). In these simulations fluctuations of vertical velocity were superimposed on the mean

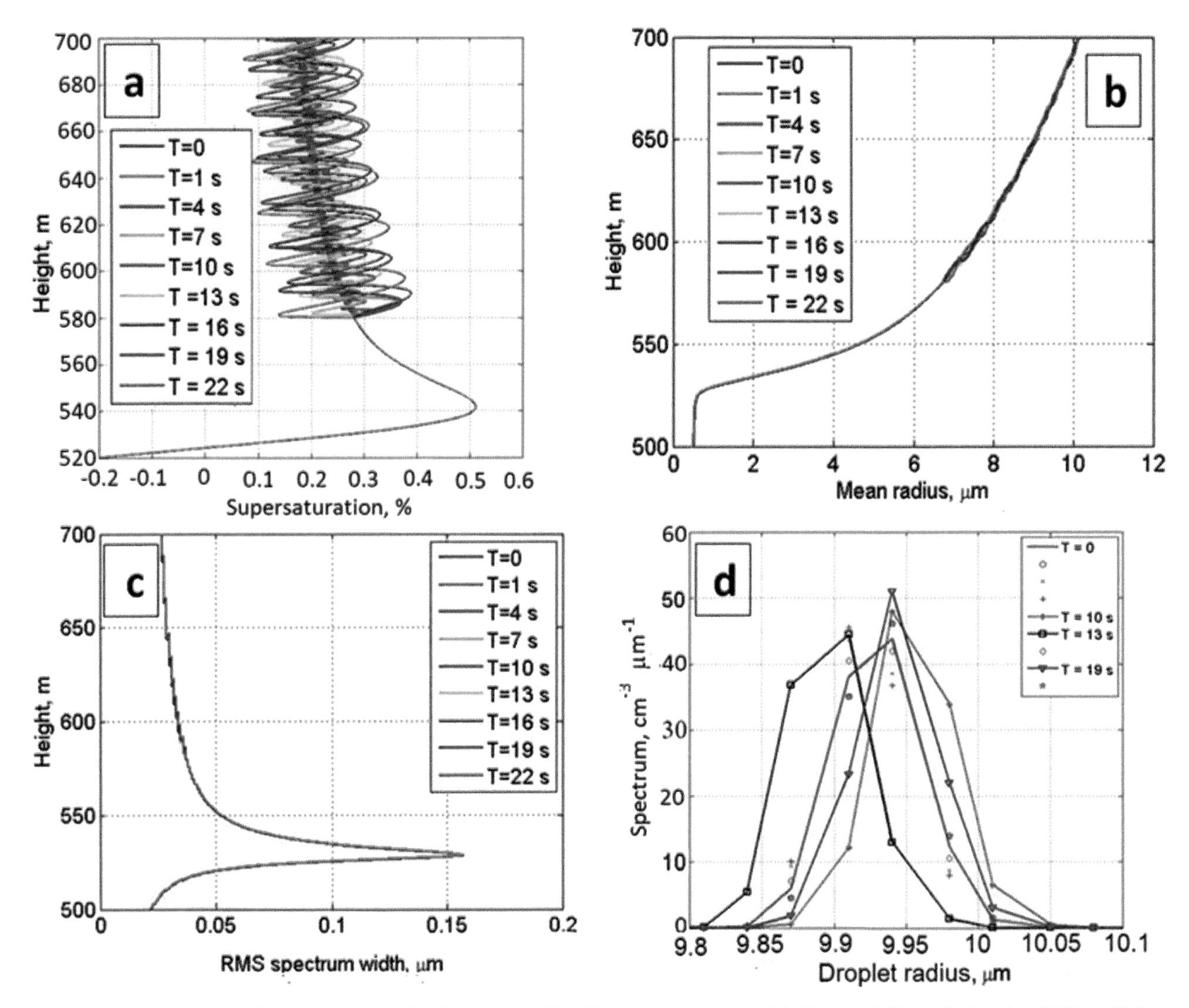

Figure 5.2.13 Vertical profiles of supersturation (a), mean radius (b), drop spectrum width (c), and DSD at the height of 700 m (d) in a parcel ascent simulations with superimposed fluctuations of vertical velocity. The periods of fluctuations are specified in the icons.

vertical velocity of 1 m/s. The amplitude of fluctuations was assumed equal to 90% of the mean velocity (Figure 5.2.13a). Even in presence of fluctuations of such significant amplitude, the maximum of DSD was shifted by ~0.05 μm only (Figure 5.2.13d). Thus, high frequency fluctuations of supersaturation do not lead to DSD broadening, regardless of the wavelength (or frequency) of turbulent velocity fluctuations. This result is obtained despite the fact that fluctuations of supersaturation are not correlated with high-frequency fluctuations of vertical velocity. Therefore, despite the contradictory results obtained in different studies, it appears that diffusional growth alone is not the major mechanism of or maybe even has no effect whatsoever on formation of the observed wide DSDs.

5.2.6 Effects of the Curvature Terms and the Chemistry Terms on DSD Behavior

Equation (5.2.6) predicts droplet growth if $S_w > 0$ and evaporation if $S_w < 0$. Further, it predicts DSD narrowing ($\frac{\partial}{\partial r}\frac{dr}{dt} < 0$) if $S_w > 0$, and DSD broadening ($\frac{\partial}{\partial r}\frac{dr}{dt} > 0$) if $S_w < 0$. The situation becomes more complicated since if droplets are small enough, the curvature terms and the chemical terms become important (Korolev, 1995). Let us express a divergence of droplet radius change rate as $\psi_1(r) = \frac{\partial}{\partial r}\frac{dr}{dt}$, and the divergence of mass change rate as $\psi_2(r) = \frac{\partial}{\partial m}\frac{dm}{dt}$. These functions characterize local DSD broadening or narrowing. Equation (5.1.14) yields

$$\psi_1(r) = \frac{1}{Fr^2}\left(-S_w + \frac{2A}{r} - \frac{4Br_N^3}{r^3}\right) \tag{5.2.45}$$

$$\psi_2(r) = \frac{1}{Fr^5}\left(S_w r^3 - 2Br_N^3\right) \tag{5.2.46}$$

From Equations (5.2.45) and (5.2.46) one can obtain the values of supersaturation at which the local narrowing (or broadening) is equal to zero:

$$S_1^* = \frac{2A}{r} - \frac{4Br_N^3}{r^3} \tag{5.2.47}$$

$$S_2^* = \frac{2Br_N^3}{r^3} \tag{5.2.48}$$

Functions $\psi(r)$ at supersaturation $S_w = 0.05\%$ and $r_N = 0.1$ μm, as well as functions $S^*(r)$ at $r_N = 0.1$ μm, are presented in **Figure 5.2.14**. If droplets are large enough ($r > r_3$) $\psi_1(r) < 0$, while $\psi_2(r) > 0$. This means that DSD narrows within the radii space, but broadens within the mass space. This regime is typical of diffusional growth corresponding to simplified Equation (5.2.6). At $r_2 < r < r_3$, both $\psi_1(r) > 0$ and $\psi_2(r) > 0$, which indicates a broadening of DSD within both the radii space and the mass space. At $r_1 < r < r_2$ $\psi_1(r) < 0$, but $\psi_2(r) > 0$, which indicates a local narrowing of DSD within the radii space and broadening within the mass space. At $r < r_1$, $\psi_1(r) < 0$ and $\psi_2(r) < 0$, which indicates a narrowing of DSD within both the radii space and the mass space. Therefore, within the small-droplet radii range there exist some zones of local narrowing and local broadening of DSD. Korolev (1995) showed that taking into account the curvature term and the chemical term may lead to a partial evaporation of some small droplets (with radius around 1.5 μm) and a growth of droplets of other sizes. These effects determine the asymmetry in diffusional growth and evaporation of droplets. This asymmetry together with some other mechanisms may lead to DSD broadening in clouds (Section 5.11).

5.2.7 Factors Affecting Supersaturation in Clouds

In previous sections, the equation for supersaturation was presented and analyzed for an idealized case of an adiabatic cloud volume where droplet concentration was assumed constant. This equation takes into account only two factors, namely, an increase in supersaturation caused by adiabatic cooling during a parcel updraft and water vapor sink on cloud droplets that grow by diffusion. There are several other factors affecting supersaturation, which are typically taken into account in cloud models, where supersaturation is determined by different microphysical and dynamical processes. Using Equations (3.5.2b) and (3.5.3b) for temperature and mixing ratio (Section 3.5) together with the Clayperon–Clausius equation, the following equation for supersaturation can be obtained (see also Clark, 1973):

$$\frac{\partial S_w}{\partial t} = (S_w+1)A_1 w - A_2(S_w+1)\frac{dq_w}{dt} - \frac{1}{\overline{\rho}}\frac{\partial}{\partial x_i}\overline{\rho}u_i S_w + \frac{1}{\overline{\rho}q_S}\left[\frac{\partial}{\partial x_i}\left(\overline{\rho}K_e\frac{\partial q}{\partial x_i}\right) - \frac{qL_w}{R_v\overline{T}}\frac{\partial}{\partial x_i}\left(\overline{\rho}K_e\frac{\partial \theta_v}{\partial x_i}\right)\right], \tag{5.2.49}$$

where A_1 and A_2 are given in Equation (5.2.3) and $q_S = 0.622\frac{e_w}{p}$ is the specific humidity of saturation. The first term on the right-hand side of Equation (5.2.49) describes the increase in supersaturation caused by adiabatic cooling of air in updrafts. The second term is the sink of water vapor onto drops during diffusional

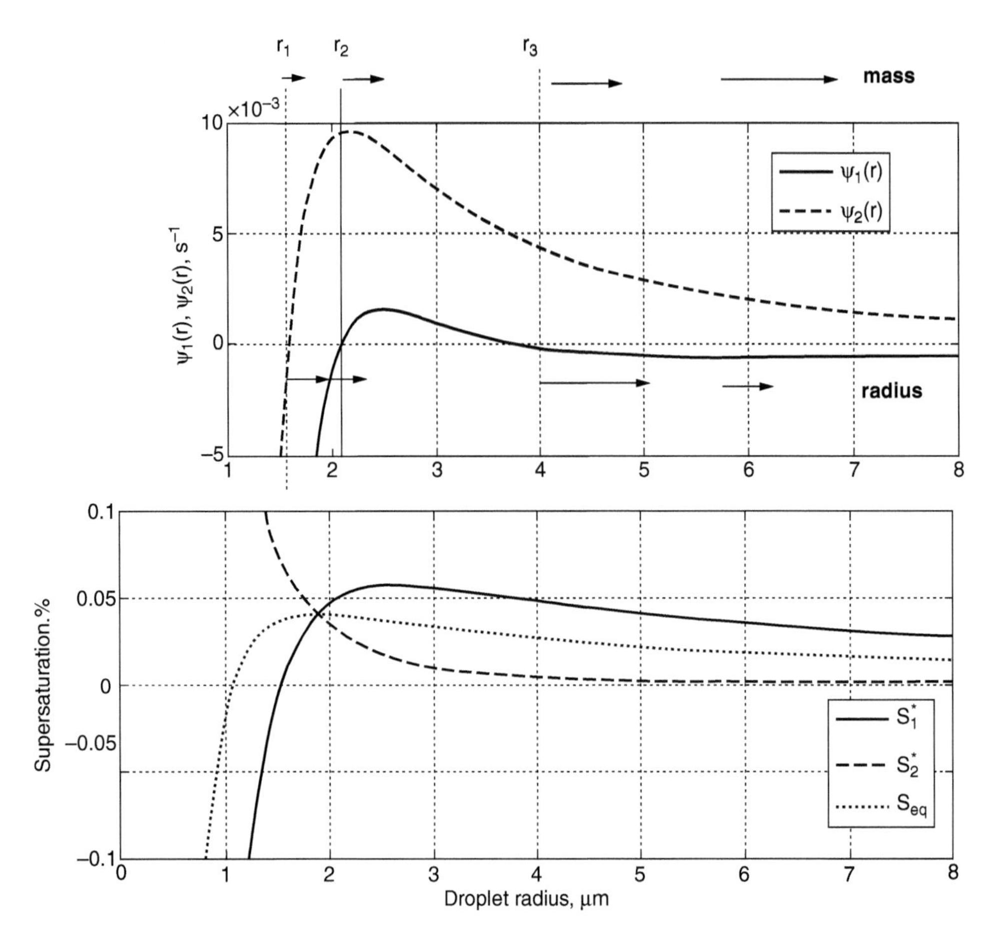

Figure 5.2.14 Dependencies of functions $\psi_1(r)$ and $\psi_2(r)$ on drop radius at supersaturation $S_w = 0.05\%$ and $r_N = 0.1$ μm (upper panel) and dependencies of functions $S_1^*(r)$ and $S_2^*(r)$ on drop radius at $r_N = 0.1$ μm (lower panel). Arrows above the upper panel denote the rate of drop mass change, while arrows located at the bottom denote the rate of radii change.

growth. The third term describes the change of supersaturation caused by air advection. The last right-hand term describes the change of supersaturation due to turbulent fluxes of humidity and temperature. K_e is the turbulent coefficient of heat exchange, which is assumed here to be equal to the coefficient of moisture exchange. Equation (5.2.49) is written neglecting the fact that the potential temperature and mixing ratio are not conservative quantities in a moist adiabatic process (Section 3.5).

The equation for supersaturation should be solved together with the kinetic equation for DSD (Equation 3.5.7), which describes all the processes leading to changes of drop concentration such as drop nucleation, drop collisions, sedimentation, etc. In particular, the following processes must be taken into account in cloud models while calculating supersaturation:

- drop nucleation at cloud boundaries in the course of lateral entrainment of aerosols;
- in-cloud droplet nucleation in accelerating cloud updrafts;
- decrease in concentration due to drop collisions;
- change in LWC due to rainfall;
- diffusional growth of ice particles;
- drop-ice collisions at low temperatures (ice riming) in mixed-phase clouds;
- mixing between cloud volumes having different LWC and temperatures, as well as mixing of cloudy air with drop-free air;
- effects of radiative cooling/heating.

These microphysical processes are closely related to cloud dynamics and affect the vertical velocity, which in turn affects supersaturation. There are two kinds of processes affecting supersaturation via the vertical velocity:

- processes accompanied by latent heat release;
- processes related to the influence of hydrometeors' weight on the buoyancy force (loading)

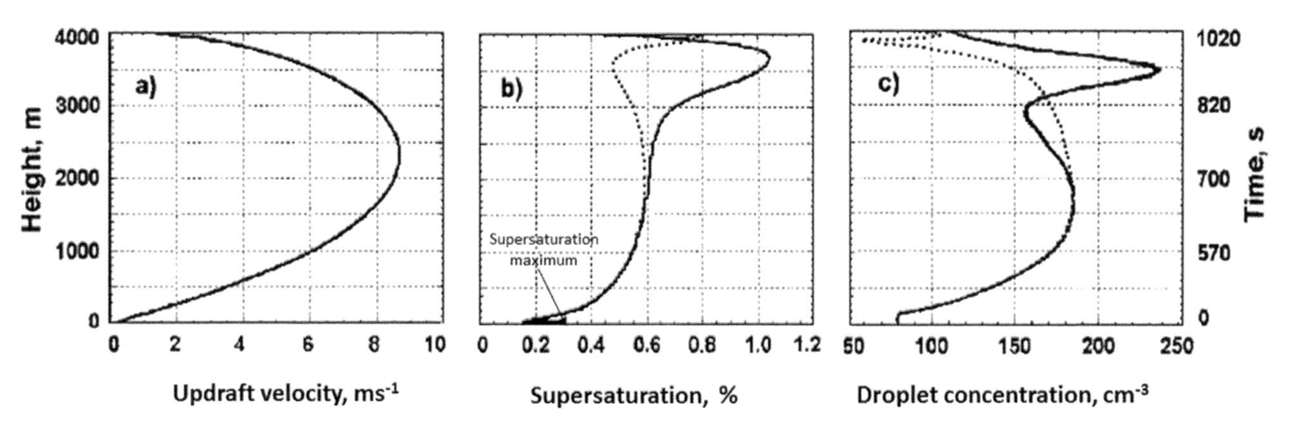

Figure 5.2.15 Vertical profiles above cloud base: (a) vertical velocity, (b) supersaturation, and (c) drop concentration in a maritime-type cloud with updraft speed of $w_0 = 25$ cm/s at cloud base in turbulent (solid lines) and nonturbulent (dotted lines) simulations. The local maximum of supersaturation near cloud base is shown by an arrow (from Pinsky and Khain, 2002; courtesy of John Wiley & Sons, Inc. ©).

(Equation (3.2.27)). Rainfall leads to an unloading that increases buoyancy and w. In turn, changes in supersaturation can lead to changes in droplet concentration, size, etc.

Most of these processes will be considered further in this book. Here, we illustrate the effects of the vertical velocity and drop collisions on supersaturation. **Figure 5.2.15** shows the vertical profiles of vertical velocity, supersaturation, and drop concentration above the lifting condensational level in an ascending cloudy volume as simulated using the Largangian parcel model (Pinsky and Khain, 2002). The time step of 0.001 s was used to simulate the growth of small aerosol particles. The model also describes the process of drop fall out of the parcel. The supersaturation reaches its local maximum (of about 0.4%) at cloud base, then decreases down to 0.2% at ~100 m and finally increases with height due to the increase in w. The increase in supersaturation leads to activation of CCN ascending from cloud base and to in-cloud nucleation of droplets. As a result, drop concentration increases. At distances exceeding 2,500 m above cloud base, w decreases since in the simulation a stable temperature lapse rate is used. In spite of the decrease in w, supersaturation significantly increases due to the decrease in drop concentration caused by intense collisions and raindrop fallout. This increase in supersaturation is especially pronounced in cases when collision enhancement caused by turbulence is taken into account (solid lines in Figure 5.2.15). The appearance of a new supersaturation maximum in the zone of intense collisions results in new in-cloud nucleation at 3,500 m above cloud base (Section 5.11). Comparing the supersaturation profile in Figure 5.2.15 with that in Figure 5.2.5 illustrates the complexity of cloud processes responsible for supersaturation formation.

5.2.8 Examples of Supersaturation Fields Obtained in *in situ* Measurements and Cloud Models

Since supersaturation in clouds is not measurable, values of S_{qs} are used to estimate S_w. **Figure 5.2.16** shows the vertical profiles of S_{qs} in comparatively small cumulus clouds (Politovich and Cooper, 1988) calculated from in situ measurements performed during the Cooperative Convective Precipitation Experiment (CCOPE, 1981). One can see that the values of S_{qs} at the cloud base are high and apparently significantly exceed the real supersaturation in clouds. Above cloud base S_{qs} varies from ~0.1% to 0.5%. High values of S_{qs} were observed in cloud regions diluted by air from entrainment by about 80%. **Figure 5.2.17** shows the values of S_{qs} calculated for deep convective clouds using in situ measurement performed during Cloud Aerosol Interaction and Precipitation Enhancement Experiment (CAIPEEX, 2009) over the peninsular Indian region in the area of Hyderabad (17.45 °N, 78.46 °E) and Bengaluru (12.97 °N, 77.63 °E) in June, 2009: June 16 (pre-monsoon period), June 21 (transition period), and June 22 (monsoon onset). The cloud top was at about 7.5–8 km in height. Measurements were performed up to heights of about 7 km. In addition to S_{qs}, Figure 5.2.17 also shows the values of w (in color scale) and drop number concentrations scaled by means of symbol. There is a considerable variation in S_{qs} along the aircraft traverses. While in most cases $S_{qs} < 1 - 2\%$, at certain heights the peaks of S_{qs} reach 5–10%.

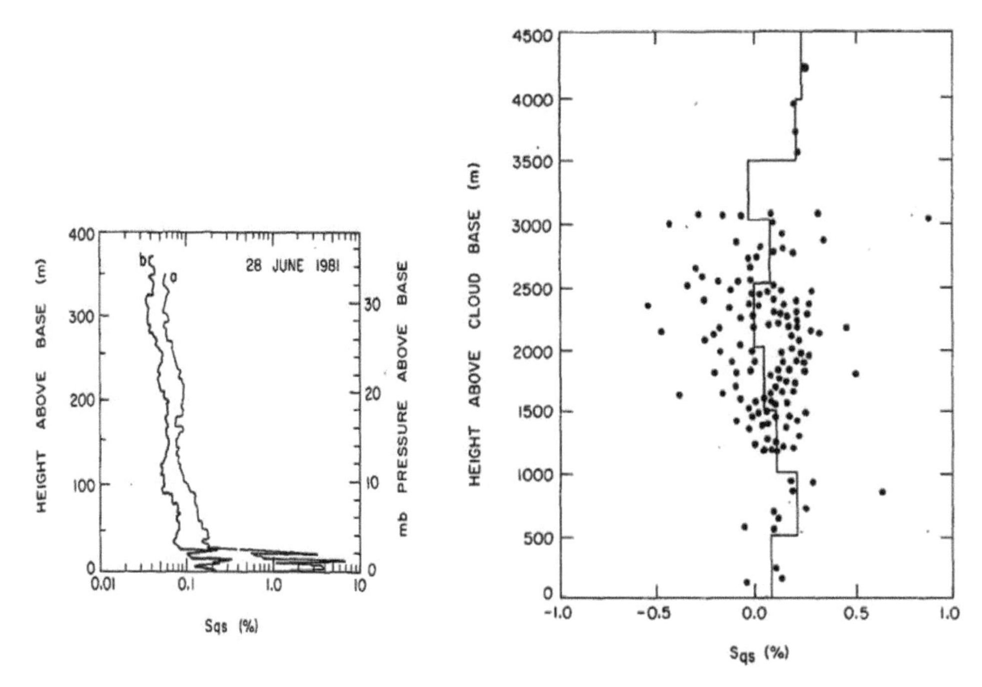

Figure 5.2.16 Left: quasi-steady supersaturation plotted against the height above cloud base for two ascents of a small cumulus through the updraft core, sampled June 28, 1981: 1903:41–1903:57 (a) and 1905:52–1906:07 UTC (b). Vertical velocities used in calculations were 3.4 ms^{-1} (a) and 2.2 ms^{-1} (b). Right: The quasi-steady supersaturation plotted against the height above cloud base. Each point represents a value averaged over a region of ~200–2,700 m. The data were taken for thirteen clouds sampled over eight days during CCOPE. The mean S_{qs} for each 500 m-high vertical interval is indicated by a solid line (from Politovich and Cooper, 1988; American Meteorological Society©; used with permission).

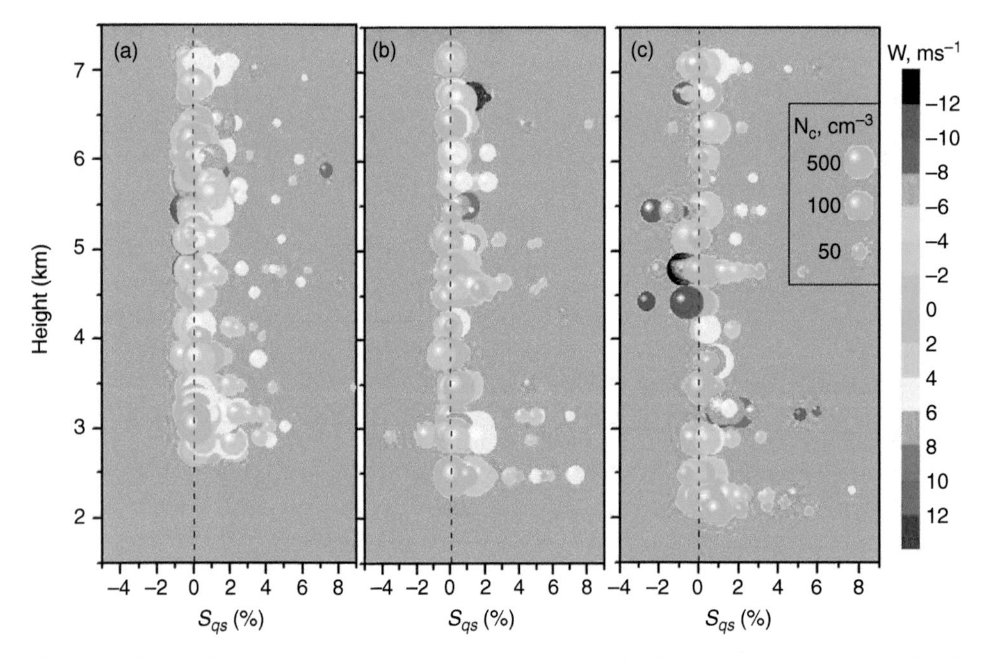

Figure 5.2.17 Quasi-steady supersaturation in various types of clouds: pre-monsoon (a), transition (b), and monsoon (c), measured *in situ* in CAIPEEX, 2009. The vertical velocity is color mapped; the drop concentration is scaled by means of symbol sizes (from Prabha, personal communication, 2011; courtesy of the author).

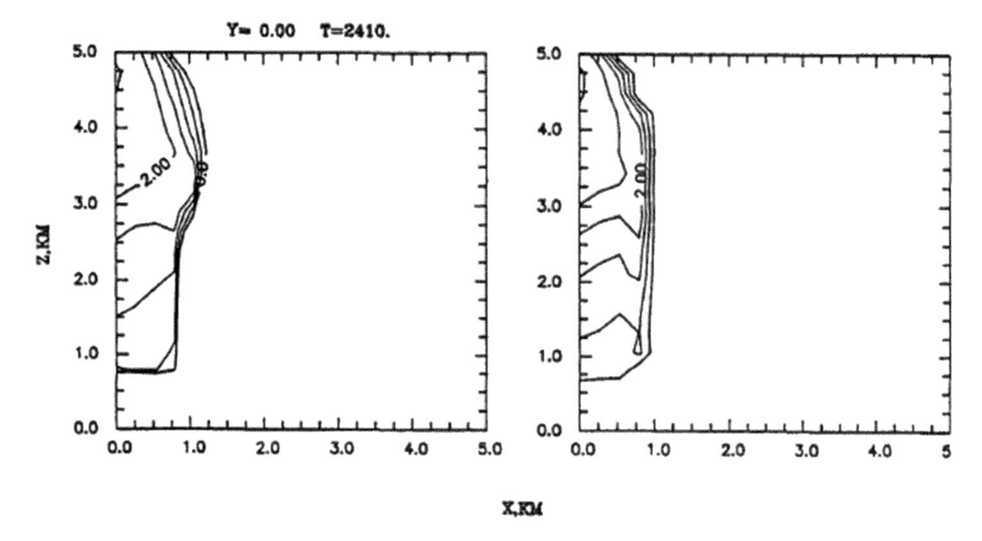

Figure 5.2.18 Left: supersaturation values calculated in a warm cumulus cloud as simulated by the spectral bin microphysics cloud model. Right: values of quasi-steady supersaturation. Only positive values are shown. Intervals between the contours are 0.5% (from Kogan, 1991; American Meteorological Society©; used with permission).

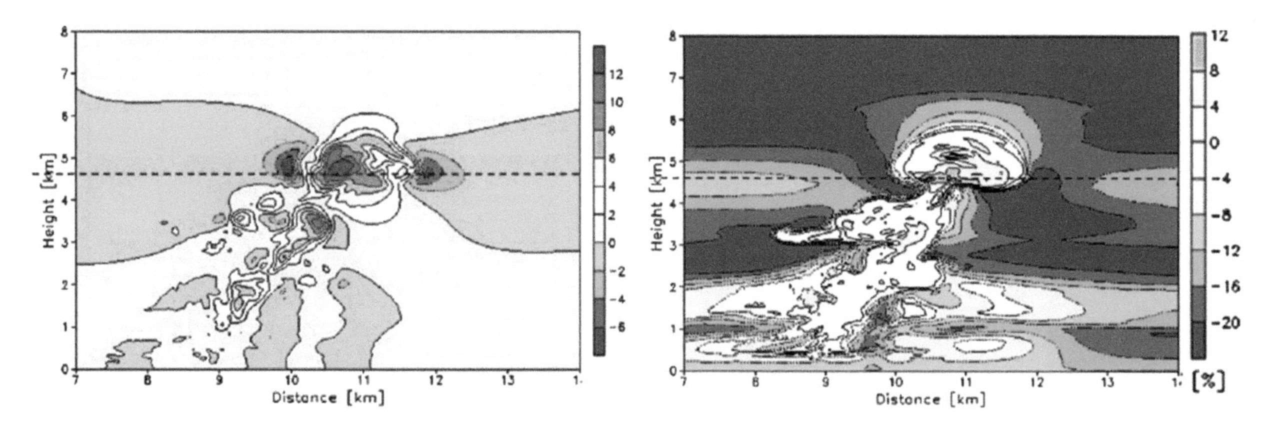

Figure 5.2.19 Fields of vertical velocity (left) and supersaturation (right) in a developing maritime cloud simulated using HUCM with 50-m resolution (from Khain et al., 2012; American Meteorological Society©; used with permission).

These high values can lead to activation of the smallest CCN within the CCN spectra (Hobbs, 1993; Pruppacher and Klett, 1997). As was mentioned, simulated S_{qs} near cloud base may significantly exceed the real supersaturation.

The values of supersaturation are calculated in numerical cloud models. **Figure 5.2.18** shows fields of saturation and of quasi-steady supersaturation in comparatively small cumulus clouds simulated by means of a cloud model with warm spectral bin microphysics (Kogan, 1991). The maximum values of supersaturation of about 2% are reached in the upper part of clouds. The values of S_{qs} in the simulation exceed the values of actual supersaturation. The relative difference between S_w and S_{qs} is maximumal in the lowest part of the cloud.

Figure 5.2.19 shows the vertical velocity field and the supersaturation field in a developing maritime cloud simulated using HUCM with 50-m resolution. A high correlation between both is obvious. In the zone of increasing vertical velocity (just above warm rain formation level at heights of 4–5 km), the supersaturation reaches 4–8%, which is in agreement with the estimations made for deep monsoon clouds (Figure 5.2.17). Figures 5.2.17 and 5.2.19 show that within certain layers, supersaturation values can reach several percent both in continental clouds and especially in maritime clouds (where the drop concentration is about 100 cm^{-3}). Such supersaturation values are high enough to activate very small aerosol particles. The maximum supersaturation is located several kilometers above cloud base (~3 km above cloud base in the presented example). This result indicates that in convective clouds nucleation of the smallest aerosols and formation of new cloud droplets can take place several kilometers above cloud base (similar to Figure 5.2.15).

Figures 5.2.20a-d show fields of S_w and S_{qs} in a maritime stratiform cloud at non-drizzling stages (a, b) and drizzling stages (c, d), calculated by means of the LEM of stratiform cloud (Pinsky et al., 2008b; Magaritz et al., 2009). The meteorological and aerosol conditions were close to those observed during the research flight RF07 in the field experiment DYCOMS II. One can see that supersaturation does not exceed 0.5%. One can see a high correlation between the supersaturation field and the vertical velocity field, as well as between the fields of S_w and S_{qs}. The maximum values

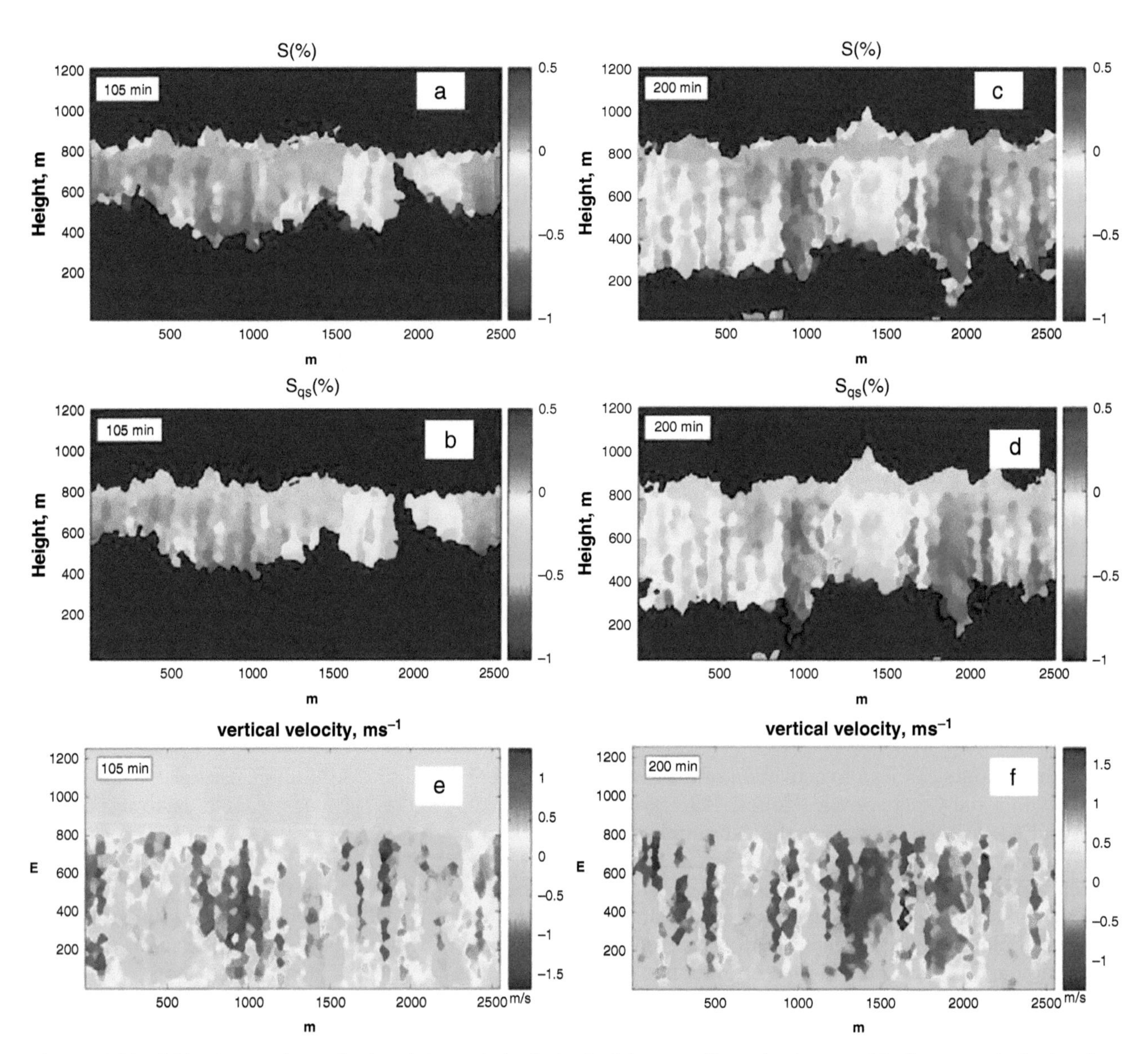

Figure 5.2.20 Fields of S_w (a, c), S_{qs} (b, d) and the vertical velocity (e, f) in a stratiform cloud at non-drizzling stages (a, b, e) and drizzling stages (c, d, f), calculated by means of the LEM of stratiform cloud (Pinsky et al., 2008; Magaritz et al., 2009; American Meteorological Society© used with permission).

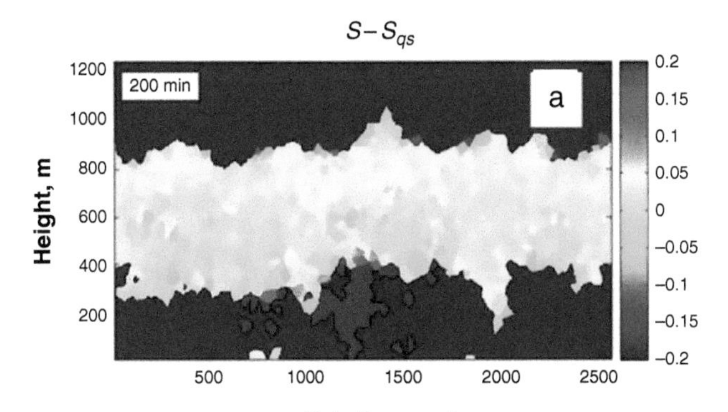

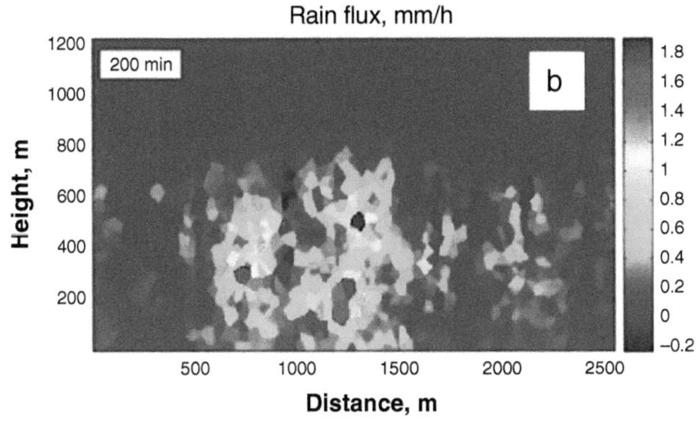

Figure 5.2.21 The fields of difference $S_w - S_{qs}$ (in %) (a), and the rain flux (b) at t = 200 min. The fields of supersaturations themselves are shown in Figure 5.2.20.

of supersaturation are reached near cloud base. At the same time, there are areas inside the cloud where supersaturation is higher than at cloud base. Those are zones of enhanced vertical velocity, where nucleation of new droplets (in-cloud nucleation) takes place. In downdrafts, superstuartion is usually negative, which can lead to a partial or even total evaporation of drops.

Figure 5.2.21 (together with **Figure 5.2.20**) shows that difference $S_w - S_{qs}$ is typically significantly smaller than the values of supersaturation themselves, which indicates that supersaturation is quite close to its quasi-steady value. As shown in Section 2.3 (see also Pinsky et al., 2008b, Magaritz et al., 2009), DSD within a stratiform cloud varies significantly with time and space. The bi modal DSD can arise at distances of several tens of meters above cloud base. DSD widths in cloud updrafts differ from those in downdrafts. The fact that S_w is close to S_{qs} indicates that drop relaxation time is quite short, so S_w rapidly approaches the S_{qs} in spite of effects of in-cloud nucleation, diffusional growth/evaporation, collisions tending to increase the difference. The values of S_w and S_{qs} can become close even when the shapes of DSD are different. Thus, the closeness of S_w and S_{qs} does not mean similarity of DSD. In zones of drizzle fall below cloud base (Figure 5.2.21b), S_w is higher than its quasi-steady value S_{qs}. The reason is that in these zones the phase relaxation time is high, and S_w cannot adjust to S_{qs} which rapidly decreases due to drop evaporation in downdrafts.

5.2.9 Appendix. Supersaturation in a Vertically Moving Parcel (by Korolev and Mazin, 2003)

In the following illustration, supersaturation is defined as

$$S_w = \frac{e - e_w}{e_w} \tag{A1}$$

where e is the water vapor pressure and e_w is the saturated water vapor pressure. The rate of supersaturation change in a vertically moving adiabatic parcel can be found by differentiating Equation (A1):

$$\frac{dS_w}{dt} = \frac{1}{e_w}\frac{de}{dt} - \frac{e}{e_w^2}\frac{de_w}{dt} \tag{A2}$$

To find $\frac{de}{dt}$ the water vapor pressure is expressed as

$$e = qp\frac{R_v}{R_a} \tag{A3}$$

Here, p is the pressure of dry air and q is the mixing ratio of water vapor (i.e., the mass of water vapor per mass of dry air). Differentiating Equation (A3) yields

$$\frac{de}{dt} = \frac{R_v}{R_a}\left(p\frac{dq}{dt} + q\frac{dp}{dt}\right) \tag{A4}$$

Using the Clausius–Clapeyron equation $\frac{de_w}{dT} = \frac{L_w e_w}{R_v T^2}$, the changes of the saturated vapor pressure can be presented as

$$\frac{de_w}{dt} = \frac{de_w}{dT}\frac{dT}{dt} = \frac{L_w e_w}{R_v T^2}\frac{dT}{dt} \tag{A5}$$

The term $\frac{dT}{dt}$ can be found from the energy conservation equation for an adiabatic parcel:

$$c_p dT - R_a T\frac{dp}{p} - L_w dq_l - L_i dq_{iw} = 0 \tag{A6}$$

Differentiating Equation (A6) and substituting $\frac{dT}{dt}$ in Equation (A5) yields

$$\frac{de_w}{dt} = \frac{R_a L_w e_w}{p c_p R_v T}\frac{dp}{dt} + \frac{L_w^2 e_w}{c_p R_v T^2}\frac{dq_l}{dt} + \frac{L_i L_w e_w}{c_p R_v T^2}\frac{dq_{iw}}{dt} \tag{A7}$$

Substituting Equations (A4) and (A7) in Equation (A2) results in

$$\frac{dS_w}{dt} = \frac{R_v}{e_w R_a}\left(p\frac{dq}{dt} + q\frac{dp}{dt}\right) - \frac{e}{e_w^2}\left(\frac{R_a L_w e_w}{p c_p R_v T}\frac{dp}{dt} + \frac{L_w^2 e_w}{c_p R_v T^2}\frac{dq_l}{dt} + \frac{L_i L_w e_w}{c_p R_v T^2}\frac{dq_{iw}}{dt}\right) \quad \text{(A8)}$$

Using the equation for quasi-hydrostatic approximation:

$$\frac{dp}{dt} = -\frac{gp}{R_a T}w, \quad \text{(A9)}$$

the equation for conservation of total water mass:

$$\frac{dq}{dt} + \frac{dq_l}{dt} + \frac{dq_{iw}}{dt} = 0, \quad \text{(A10)}$$

and Equations (A1) and (A3) to derive $\frac{dp}{dt}, \frac{dq}{dt}, \frac{e}{e_w}$, and p, respectively, and substituting into Equation (A8) yields

$$\frac{1}{S_w+1}\frac{dS_w}{dt} = \left(\frac{gL_w}{c_p R_v T^2} - \frac{g}{R_a T}\right)w - \left(\frac{1}{q} + \frac{L_w^2}{c_p R_v T^2}\right)\frac{dq_l}{dt} - \left(\frac{1}{q} + \frac{L_i L_w}{c_p R_v T^2}\right)\frac{dq_{iw}}{dt} \quad \text{(A11)}$$

For warm cloud, when $q_{iw} = 0$ one obtains

$$\frac{1}{S_w+1}\frac{dS_w}{dt} = \frac{g}{T}\left(\frac{L_w}{c_p R_v T} - \frac{1}{R_a}\right)w - \left(\frac{1}{\rho_v} + \frac{L_w^2}{R_v c_p T^2 \rho_a}\right)\frac{dq_w}{dt} \quad \text{(A12)}$$

where ρ_v is absolute humidity and q_w is LWC. This expression coincides with Equation (5.2.1).

In the derivations presented above, approximation $p_a \approx p$ was used, where p_a and p are the pressures of dry and moist air, respectively. The quasi-hydrostatic approximation Equation (A9) requires w be lower than about 15 ms^{-1}.

5.3 Parameterization of Droplet Nucleation in Cloud Models

As shown in Section 5.1, cloud droplets arise on AP when the size of these particles exceeds the critical value. Only a few cloud models (e.g., Pinsky et al., 2008b; Magaritz et al., 2009; Magaritz-Ronen et al., 2016a) explicitly describe haze particle growth and their conversion to droplets. The reason is that such a description requires extremely small time steps of the order of 0.01 s. Therefore, most models treat only dry AP and parameterize conversion of dry particles into droplets, i.e., the process of droplet nucleation. The parameterization allows taking into account the effects of aerosols on cloud microphysics and dynamics. Treating droplet nucleation within a cloud body and that at the cloud base only requires different approaches.

5.3.1 The General Methods of Parameterization

There are several algorithms for parameterization of droplet nucleation that are applied depending on the degree of a model's sophistication. Models based on bulk-parameterization of cloud microphysics often use the simplest approach to parameterization of droplet nucleation. In this approach, the aerosol particles' budget is not calculated. Potentially available amount of CCN is assumed to be infinitely large, and the amount of nucleated droplets is determined by the Twomey's dependence (5.1.21), including parameters that are assumed to be time-independent. The number of new droplets nucleated at any given time step is determined as (e.g., Reisner et al., 1998)

$$\Delta N = \begin{cases} N_0 S_w^k - N, & \text{if} \quad N_0 S_w^k > N \\ 0, & \text{if} \quad N_0 S_w^k \le N \end{cases}, \quad (5.3.1)$$

where N is the droplet concentration at a particular grid point just prior to applying nucleation procedure described by Equation (5.3.1). Expression (5.3.1) is based on the concept of an ascending adiabatic parcel, in which droplets ascend together with the air parcel and do not settle. In this case, the amount of droplets at a particular height is equal to the total amount of CCN activated previously along the parcel track. Since in bulk models DSD are often presented as Gamma distributions (Section 2.1), the newly nucleated droplets have the same SD as the other droplets. Parameters determining the DSD shape after nucleation are recalculated using the new values of droplet concentration.

It is clear that the parameterization represented by Equation (5.3.1) is quite simplified, since the droplet concentration is not a conservative value, but changes due to collisions, settling, and mixing. Thus, droplet concentration at a certain level only weakly reflects the history of an air volume ascending from the cloud base. From the physical point of view, the number of newly nucleated droplets should not depend on the amount of droplets within the air volume under consideration. Grabowski et al. (2011) used an expression similar to Equation (5.3.1) with one important difference: instead of droplet concentration, the concentration of activated CCN is advected. In this case, activated CCN are considered a passive scalar whose role is to keep the history of droplet nucleation in the ascending air. Using activated CCN instead of droplets has an advantage over Equation (5.3.1), because in this way concentration of activated CCN is not affected by drop collisions and sedimentation, but is determined by nucleation and advection only. However, concentration of activated CCN can change due to mixing, and then the value at a grid point does not fully reflect the amount of CCN actually activated along the air parcel track.

In more sophisticated approaches, such as those used in models with spectral bin microphysics, CCN are described using the SD function $f_{CCN}(\mathbf{x}, r_N)$ depending on the size and spatial coordinates of CCN (e.g., Respondek et al., 1995; Khain et al., 2004, 2008). The initial size spectrum $f_{CCN}(\mathbf{x}, r_N)$ of nonactivated CCN is assumed known from measurements (including, supposedly, some assumption about concentration of giant CCN, GCCN). The time evolution of $f_{CCN}(\mathbf{x}, r_N)$ is described by equation:

$$\frac{\partial f_{CCN}}{\partial t} = -div(\mathbf{u}f_{CCN}) - \left[\frac{\delta f_{CCN}}{\delta t}\right]_{nucl} + \left[\frac{\delta f_{CCN}}{\delta t}\right]_{evap}, \tag{5.3.2}$$

where the first term on the right-hand side describes the advection of nonactivated CCN SD, term $\left[\frac{\delta f_{CCN}}{\delta t}\right]_{nucl}$ describes the rate of the $f_{CCN}(\mathbf{x}, r_N)$ decrease due to nucleation of droplets, and term $\left[\frac{\delta f_{CCN}}{\delta t}\right]_{evap}$ reflects the CCN return to the atmosphere due to complete evaporation of drops.

The nucleation scheme is presented in **Figure 5.3.1**. At each time step and at each spatial grid point, the supersaturation S_w is calculated (methods of calculating supersaturation in numerical models are described in Section 5.4). Typically, time steps in cloud models are large enough, so the changes of S_w during a single time step should be taken into account. Modelers face the problem of at which time instance the procedure of droplet nucleation should be applied. Supersaturation increases in the course of the air-volume ascent and decreases due condensation. If the dynamically caused increase in S_w exceeds its decrease caused by condensation, the S_w value at the end of the microphysical time step will be larger than that at the beginning of the time step. In this case, it appears reasonable to use the S_w calculated to the end of the microphysical time step to perform the procedure of droplet nucleation. When supersaturation decreases during a time step of diffusional growth, using the S_w value calculated at the end

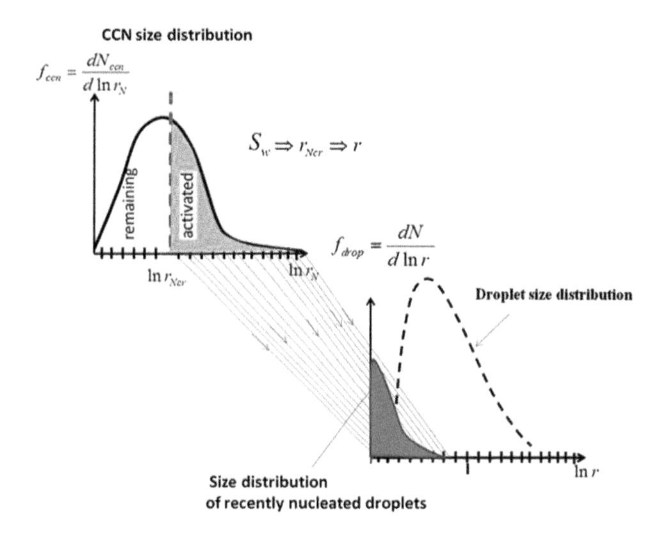

Figure 5.3.1 A scheme of drop nucleation in spectral bin microphysics models. SD of CCN and of droplets are represented in the corresponding logarithmic mass grid or size grid. Using supersaturation S_w, the critical value of the CCN radius r_{Ncr} is calculated. CCN with $r_N \geq r_{Ncr}$ are activated and converted to drops. The corresponding bins of the aerosol mass grid become empty. The radii of newly nucleated droplets are typically below several μm. Distribution of nucleated CCN is shown by the blue area. Droplets grow by diffusion and form the entire DSD containing cloud droplets with radii up to ~20 μm (dashed line).

of the time step may lead to an underestimation of the nucleation rate. At the same time, using the S_w value obtained right after an advection sub-step may lead to an overestimation of the nucleation rate. Khain et al. (2004, 2008) performed nucleation at the middle and at the end of each time step, which allows taking into account both scenarios of S_w behavior during different time steps. As soon as S_w is determined, it is used to calculate r_{Ncr} according to Equation (5.1.17). All CCN from SD $f_{CCN}(\mathbf{x}, r_N)$ with radii exceeding r_{Ncr} are activated and assigned to droplets. Thus, the corresponding bins in the CCN SD become empty (Figure 5.3.1).

The next step of stimulating nucleation is calculation of the droplet sizes (or haze sizes in case of GCCN) using the size of dry activated CCN. Formula (5.1.16) can be applied when CCN are small and their equilibrium size r_{cr} is also small and can be rapidly achieved. The formula can be applied, for instance, for $r_N \approx 0.05-0.1$ μm corresponding to the critical haze sizes below one μm. For CCN of a larger size, the equilibrium radius of a wet particle r_{cr} rapidly increases with an increase of dry CCN size. For instance, for $r_N = 0.48$ μm, the equilibrium size is 19 μm (Figure 5.1.7). To achieve these values, a significant amount of time is required, and thus the Köhler

Table 5.3.1 Relationship between size of dry AP and the size of wet particles at cloud base (from Kogan, 1991; American Meteorological Society©; used with permission).

r_N	k	r_0	S_c
0.0076	—	0.022	2.081
0.0120	—	0.044	1.042
0.0191	—	0.085	0.523
0.0296	—	0.180	0.262
0.0545	—	0.340	0.131
0.130	8.3	1.08	<0.03
0.220	8.2	1.8	—
0.430	7.0	3.0	—
0.63	6.3	4.0	—
0.82	6.1	5.0	—
1.03	6.1	6.3	—
1.30	6.1	8.0	—
1.66	6.1	10.1	—
2.10	6.0	12.7	—
2.75	5.8	16.0	—
3.5	5.8	20.2	—
4.4	5.8	25.4	—
5.5	5.8	32.0	—
7.6	5.3	40.3	—

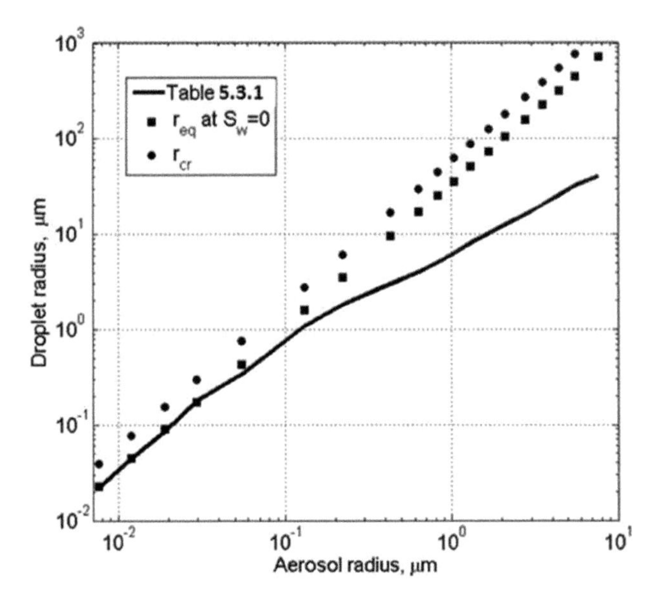

Figure 5.3.2 Dependence of a nucleated drop radius on a dry aerosol radius in numerical cloud models (solid line). The circles denote the critical radii r_{cr} of wet particles according to the Köhler curves. The diamonds denote the equilibrium radii at RH = 100%.

equilibrium theory cannot be applied just because the wet radius does not reach its equilibrium value. To calculate the size of wet particles at cloud base, Lagrangian parcel models (e.g., Ivanova et al., 1977) were used. The size of wet AP depends on the dry AP radius and on the residential time of particles affected by the vertical velocity below cloud base. However, due to many uncertainties arising in estimation of the residential time, the effects of the vertical velocity are often neglected. Using results of the study by Ivanova et al. (1977), the radius of wet CCN at cloud base for dry CCN radius $r_N > 0.03$ μm can be calculated as

$$r = Kr_N, \tag{5.3.3}$$

where factor K varies from five to eight **(Table 5.3.1** and **Figure 5.3.2).** For GCCN, Mechem and Kogan (2008) proposed an expression:

$$K = aw^{-0.12}r_N^{0.214}, \tag{5.3.4}$$

where $a = 5.8$, w is the vertical velocity, and r_N is dry GCCN radius (μm). Mechem and Kogan (2008) used Equation (5.3.4) for parameterization of the effect of GCCN in stratocumulus clouds. For vertical velocity of 0.6 ms^{-1} typical of stratocumulus clouds, and GCCN radii ranging from 1 μm to 10 μm, factor K ranges from 6.2 to 3.8. The dependence of a nucleated droplet radius on a dry aerosol radius in numerical cloud models is shown in Figure 5.3.2 by solid line. The dependence is plotted using the data presented in Table 5.3.1. The circles denote the critical radii r_{cr} of wet particles according to the Köhler curves. The diamonds denote the equilibrium radii at RH = 100%. One can see that the parameterization (5.3.3) does not allow for nucleated droplets to be larger than ~10 μm (if the dry radii are lower than 2 μm). Application of the Köhler relationship would yield particles exceeding 100 μm at cloud base, which is unrealistic.

Because the radius of CCN is typically lower than one μm, the radii of newly nucleated droplets are typically smaller than several μm. Thus, nucleation increases the concentration of the smallest droplets in the DSD (Figure 5.3.1). Later droplets grow by diffusion and form the entire DSD containing cloud droplets with radii up to ~20 μm. So, in SBM models, formation of DSD due to condensation of water vapor includes two steps: nucleation and diffusion growth. This approach differs from those usually used in bulk schemes, where the distribution of newly nucleated droplets is assumed to be the same as all cloud droplets.

5.3.2 Calculation of Droplet Concentration Near Cloud Base

The method of calculating droplet nucleation discussed and illustrated in Figure 5.3.1 suggests that supersaturation can be explicitly represented, that is, well approximated on the model finite-difference grid. This is correct if the spatial scales of supersaturation variation are substantially larger than the doubled grid spacing. As shown in Section 5.2, supersaturation above cloud base rapidly increases to its maximum S_{max} located at the distance of 10–50 m above cloud base, and then again rapidly decreases above this maximum (Figure 5.2.5). The vertical grid spacing of most high-resolution cloud models (and certainly of large-scale models and global circulation models) is too coarse to resolve this supersaturation maximum. At the same time, droplet concentration that affects cloud microphysics and dynamics is determined mainly by the value of S_{max}. Therefore, the problem arises of parameterizing droplet nucleation near cloud base in cloud models and other atmospheric models.

Parameterizations based on analytical considerations were proposed in several studies (Ghan et al., 1993, 1995, 2011; Bedos et al., 1996; Cohard et al., 1998; Abdul-Razzak et al., 1998; Abdul-Razzak and Ghan, 2000; Fountoukis and Nenes, 2005; Shipway and Abel, 2010). There is a common idea underlying these studies. In updrafts below the supersaturation maximum, CCN of different sizes are activated at different levels. The largest CCN are activated within the lower several meters above cloud base, while the smallest CCN are activated at the level of the supersaturation maximum. The growth of droplets nucleated at different levels below the supersaturation maximum decreases supersaturation. The purpose of the parameterizations is first to calculate the local supersaturation maximum and then the correct amount of activated CCN (i.e., droplets nucleated near cloud base).

Combining the supersaturation equation (5.2.2) with the diffusional growth equation (5.2.6) and taking into account that $q_w = \int_{m_{min}}^{m_{max}} mf(m)dm$ and expression $\frac{dN}{dt} = \frac{\partial N_{CCN}}{\partial S_w}\frac{dS_w}{dt}$ (where $N = \int_{m_{min}}^{m_{max}} f(m)dm$ is droplet concentration), the equation for supersaturation near cloud base can be approximately written as (Pruppacher and Klett, 1997; Shipway and Abel, 2010):

$$\frac{dS_w}{dt} = A_1 w - A_3 S_w \int_0^{S_w} \left[\int_{\tau(S'_w)}^{t} S_w dt\right]^{1/2} \frac{\partial N_{CCN}}{\partial S'_w} dS'_w, \tag{5.3.5}$$

where A_1 is defined in Equation (5.2.3), A_3 is the coefficient slightly depending on temperature:

$$A_3 = 2\pi\frac{\rho_w}{\rho_a}(2G(T,p))^{3/2}\left(\frac{p}{0.622e_w} + \frac{0.622L_w}{c_p R_a T^2}\right) \tag{5.3.6}$$

and

$$G(T,p) = \frac{1}{\rho_w}\left(\frac{R_v T}{e_w D_v} + \frac{L_w}{k_a T}\left(\frac{L_w}{R_v T} - 1\right)\right)^{-1} \tag{5.3.7}$$

In Equation (5.3.5), $\frac{\partial N_{CCN}}{\partial S'_w} dS'_w$ is the number of CCN activated when supersaturation increases by dS'_w, and $\tau(S'_w)$ is the time instance at which these CCN are activated. The inner integral in Equation (5.3.5) represents the net supersaturation affecting nucleated droplets. Actually, Equation (5.3.5) is the supersaturation equation, in which the change in LWC is the result of continuous nucleation and subsequent condensational droplet growth.

Twomey (1959b) showed that the inner integral in Equation (5.3.5) can be written as

$$\int_{\tau(S'_w)}^{t} S_w dt \approx \frac{1}{2A_1 w}\left(S_w^2 - S'^2_w\right) \tag{5.3.8}$$

Setting the left-hand side of Equation (5.3.5) to zero (since at the supersaturation maximum $\frac{dS_w}{dt} = 0$) and applying Equation (5.3.8), Shipway and Abel (2010) got the following equation for the maximum value of supersaturation S_{max} in the form:

$$\frac{\sqrt{2}(A_1 w)^{3/2}}{A_3} - S_{max} \int_0^{S_{max}} \left.\frac{\partial N_{CCN}}{\partial S_w}\right|_{S_w = S_{max}} \left(S_{max}^2 - S'^2_w\right)^{1/2} dS'_w \tag{5.3.9}$$

The left side of Equation (5.3.9) is assumed to be known from the previous model calculations. Several authors (Ghan et al., 1993, 1995, 2011; Bedos et al., 1996; Abdul-Razzak et al., 1998; Nenes and Seinfeld, 2003) proposed another form of the transcendental equation for S_{max}, different from Equation (5.3.9). These equations enable one to determine the minimum

CCN size to be activated at the point of maximum supersaturation.

Khvorostyanov and Curry (2006) suggested a further development of the approach based on utilization of Equation (5.3.5). SD of aerosols in the study was approximated by a sum of log-normal modes. However, applying the integral-differential equation (5.3.5) together with complicated expressions for CCN SD represents a quite complicated mathematical problem.

Pinsky et al. (2012) proposed a simple method for calculating the supersaturation maximum and droplet concentration at cloud base, which is suitable for various formulations of aerosol SD. The method is based on using Equation (5.2.27) for the supersaturation maximum. This equation was obtained using simplified assumptions concerning monodisperse DSD, while omitting the curvature term and the chemical terms in the equation of diffusional growth. When the activation spectra are given in the form of the Twomey formula (Section 5.1), the droplet concentration equal to the maximum concentration of activated CCN is determined as

$$N = N_0 S_{\max}^k, \qquad (5.3.10)$$

Substitution of Equation (5.3.10) in Equation (5.2.27) leads to the following relationship between the maximum supersaturation and parameters of activation spectrum:

$$S_{\max} = C_3^{\frac{2}{2+k}} N_0^{\frac{1}{2+k}} w^{\frac{3}{4+2k}}, \qquad (5.3.11)$$

where $C_3 = C_1 (F A_1/3)^{3/4} \left(\frac{3\rho_a}{4\pi\rho_w A_2}\right)^{1/2}$ is the coefficient depending on temperature and $C_1 = 1.058$. Using Equations (5.3.10) and (5.3.11), droplet concentration can be calculated as

$$N = C_3^{\frac{2k}{2+k}} N_0^{\frac{2}{2+k}} w^{\frac{3k}{4+2k}} \qquad (5.3.12)$$

Figure 5.3.3 shows dependencies of droplet concentration on vertical velocity for polluted air (left) and clean air (right), calculated using activation spectra in the forms (5.1.21) and (5.1.22), at values of parameters given in Table 5.1.6. One can see that the activation spectrum given by Equation (5.1.22) which describes the CCN spectrum that does not contain small CCN, leads to a lower droplet concentration at high vertical velocities, as compared to the concentration predicted by Twomey's formula (5.1.21).

CCN SD sometimes has a complicated form. The distribution is often represented using a log-normal distribution (Ghan et al., 2011) or as a sum of three log-normal modes (Section 2.2) with parameters varying within wide ranges (Respondek et al., 1995; Segal and Khain, 2006). In spectral bin microphysics models, CCN distribution is represented on a mass grid (Section 2.1). The shape of CCN SD changes with time as a result of cloud-aerosol interaction accompanied by washout of

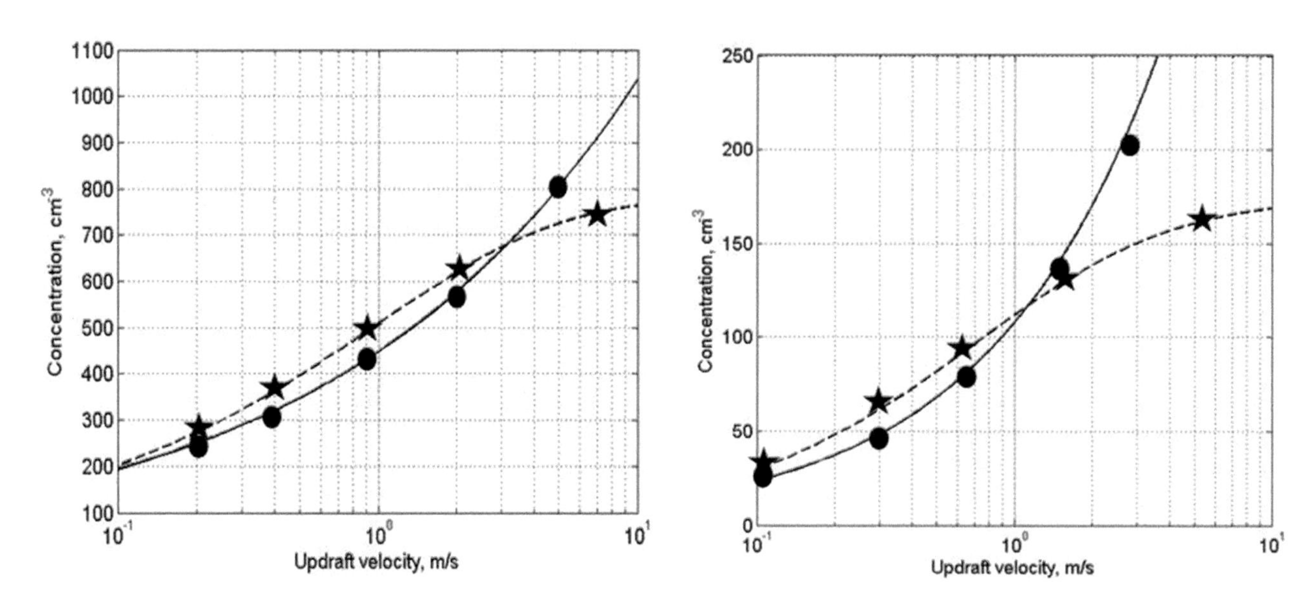

Figure 5.3.3 Dependencies of droplet concentration on vertical velocity for polluted air (left) and clean air (right) at $T = 283$ K and $p = 800$ hPa. Symbols show the results of parameterization performed by Cohard et al. (1998). Solid and dashed curves are plotted using results of the approach proposed by Pinsky et al. (2012) for activation spectra given in Table 5.1.6 (from Pinsky et al., 2012; courtesy of John Wiley & Sons, Inc. ©).

CCN, mixing, penetration of air from higher levels that has lower CCN concentration, nucleation of some fraction of CCN, etc. So, during the development of a cloud system, parameters of CCN activation spectrum change with time. Hence, using a priori prescribed CCN nucleation spectra (e.g., in the form of Equation (5.1.22)) becomes questionable, and it is preferable to use CCN SD directly. Since the droplet concentration to be nucleated is equal to the concentration of CCN activated at $S = S_{max}$, it can be calculated as

$$N = \int_{r_{Ncr}(S_{max})}^{\infty} f(r_N)dr_N, \quad (5.3.13)$$

where $f(r_N)$ is a given SD of dry aerosol particles and r_{Ncr} is the critical radius of CCN particles activated at S_{max}. r_{Ncr} is related to S_{max} via Equation (5.1.17) (Section 5.1). From Equations (5.2.27) and (5.3.13), one obtains the closed transcendental equation for S_{max}:

$$S_{max}\left[\int_{r_{Ncr}(S_{max})}^{\infty} f(r_N)dr_N\right]^{1/2} = C_3 w^{3/4} \quad (5.3.14)$$

Equation (5.3.14) is the equation with respect to S_{max} that can be easily solved numerically. The right-hand side of Equation (5.3.14) is assumed to be known, while its left-hand side is a monotonically increasing function of S_{max} that simplifies the finding S_{max} numerically. As soon as S_{max} is found, droplet concentration N is calculated using the equation that follows from Equations (5.3.13) and (5.3.14):

$$N = C_3^2 w^{3/2} S_{max}^{-2} \quad (5.3.15)$$

Pinsky et al. (2012) performed calculations of the supersaturation maximum and droplet concentration for cases when the CCN distribution was represented by three-mode log-normal distribution using the values of parameters of different modes, presented in Table 2.2.2. **Figure 5.3.4** compares the dependencies of the supersaturation maximum on the vertical velocity, obtained using different parameterization schemes including the approach presented by Pinsky et al. (2012) (denoted as PKMK).

Figure 5.3.5 shows the dependencies of the fraction of activated CCN of the entire aerosol concentration on the vertical velocity. All the parameterizations in Figures 5.3.4 and 5.3.5 indicate an increase of the supersaturation maximum and in the number fraction of activated aerosols with increasing updraft velocity. There is, however, a significant dispersion of the results of different parameterizations. This dispersion increases with increasing value of the updraft velocity. The values of the supersaturation maximum as well as of the number of activated aerosols predicted by various methods differ by a factor of 2–3. The method proposed by Pinsky et al. (2012) produces results quite similar to those simulated by an accurate parcel model.

The approaches discussed are based on the assumption that particles begin growing and affecting supersaturation only after their activation and conversion to droplets. There are more accurate approaches (mostly in numerical modeling using high-accuracy parcel models), in which the absorption of water vapor on haze particles as well as the effects of particle curvature and salinity are taken into account (e.g., Boucher and Lohmann, 1995; Segal and Khain, 2006; Kivekäs et al., 2008). Segal and Khain (2006) calculated look-up tables presenting concentration of droplets at cloud base as a function of concentration and width of the CN size spectrum for different vertical velocities at cloud base. A detailed parcel model by Pinsky and Khain (2002) and Segal et al. (2003) was used in calculations. The main feature of the model is an accurate description of diffusional growth of wet aerosols and droplets. The model is briefly discussed in Section 5.11. To describe the DSD of particles (nonactivated aerosols and drops), 2,000 mass bins were used within the radii range from 0.01 μm to 2,000 μm. The time step of 0.005 s was used to calculate diffusion growth of drops and aerosol particles.

Segal and Khain (2006) represented the SD of CN by a three-mode log-normal distribution. In the simulations, parameters of aerosol SD were chosen as $R_1 = 0.006$ μm $= const$, while R_2 varied from 0.02 μm to 0.04 μm. Similar parameters of the CN SD were used by Hobbs et al., 1985; Respondek et al., 1995 and Yin et al., 2000). **Tables 5.3.2 a–h** show the values of droplet concentration at different values of aerosol concentration and different vertical velocities, and can be used as look-up tables for parameterization of droplet nucleation near cloud base.

Pinsky et al. (2012) compared the values of droplet concentration calculated using Equations (5.3.140 and (5.3.15) with those from look-up **Tables 5.3.2a–h** (exact solutions) and obtained an excellent agreement. This fact sheds light on the major physical mechanisms determining CCN activation and the role of CCN SD. Notably, for a number of reasons, larger CCN contribute more to formation of S_{max} as compared to smaller ones. First, the Köhler theory predicts the equilibrium radii of haze particles to be proportional to $r_N^{3/2}$, and

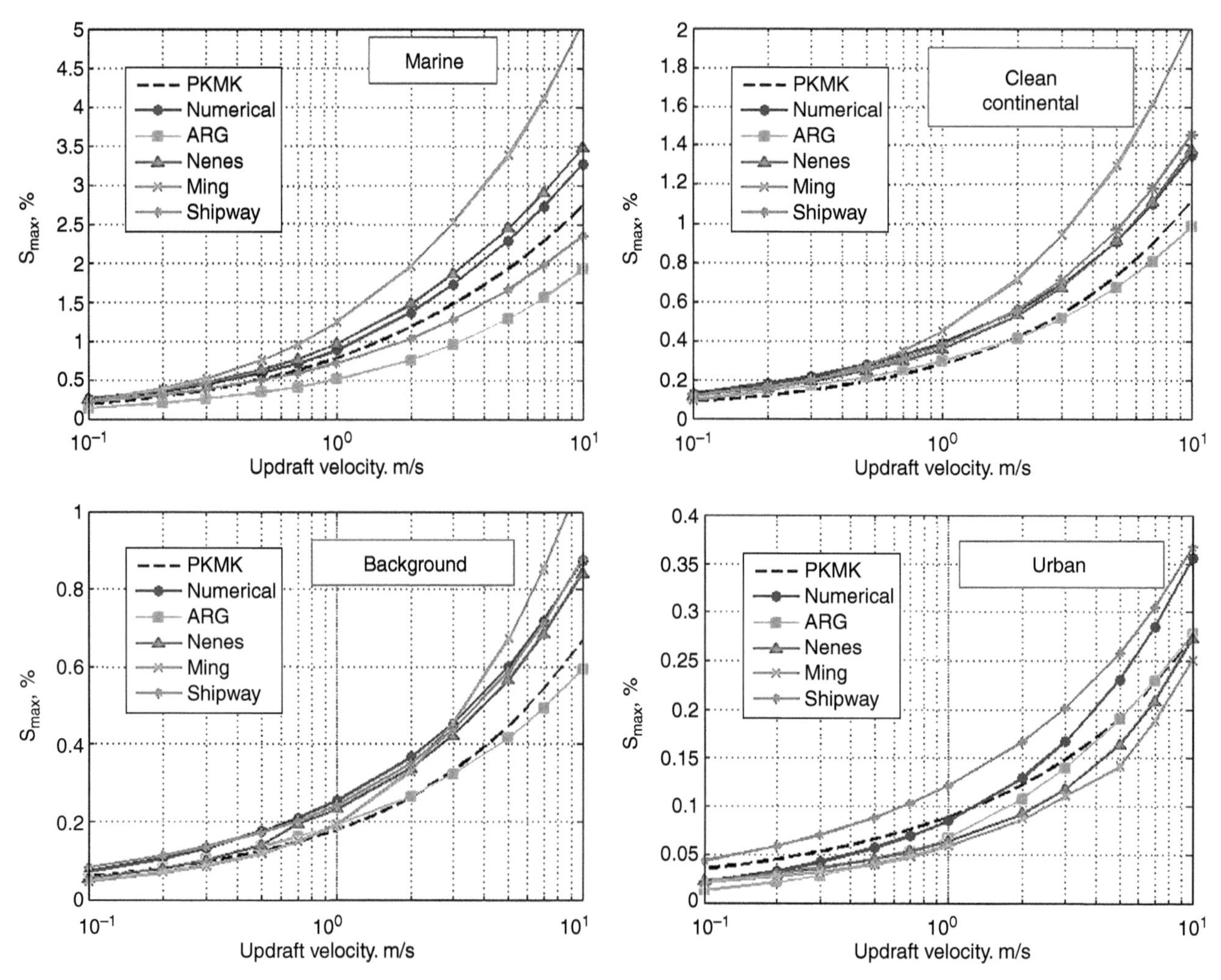

Figure 5.3.4 Dependencies of S_{max} on the vertical velocity obtained for aerosol distributions given in Table 2.2.2. The figure is plotted for different parameterization schemes: ARG is the Abdul-Razzak and Ghan (2000) modal parameterization; "Nenes" is the Fountoukis and Nenes (2005) scheme; "Ming" is the Ming et al. (2006) scheme; "Shipway" is the Shipway and Abel (2010) scheme. The curves corresponding to the approach presented in Pinsky et al. (2012) are denoted as "PKMK" (from Ghan et al., 2011 and Pinsky et al., 2012; courtesy of John Wiley & Sons, Inc. ©).

therefore larger CCN absorb larger amounts of water than the smaller ones. The second and a more important reason is that the larger CCN are activated at smaller distances above cloud base (~1 m) than the smaller ones, and continue rapidly growing as droplets, whereas nonactivated CCNs (haze particles) grow slowly, remaining in equilibrium with the ambient water vapor. The difference in the masses of the activated droplets and of haze near cloud base can reach several orders of magnitude. This is the outcome of the competition between large and small CCN: droplets forming on large CCN absorb water vapor and decrease supersaturation, preventing nucleation of the smallest CCN in the CCN SD. This competition tends to narrow the DSD at the level of the supersaturation maximum, i.e., results are close to those in case monodisperse CCN SD is assumed.

The similarities between the supersaturation profiles calculated for monodisperse and polydisperse CCN SDs (see also Section 5.2) can be accounted for by the fact that the supersaturation profile is determined by equivalent CCN of comparatively large size. The equivalent size of a monodisperse CCN has the same effect on supersaturation and droplet concentration as a polydisperse CCN spectrum. Pinsky et al. (2012) assume that the equivalent CCN radius is close to the mean volume radius of activated CCN. Estimations show that for CCN distributions with the mean radius

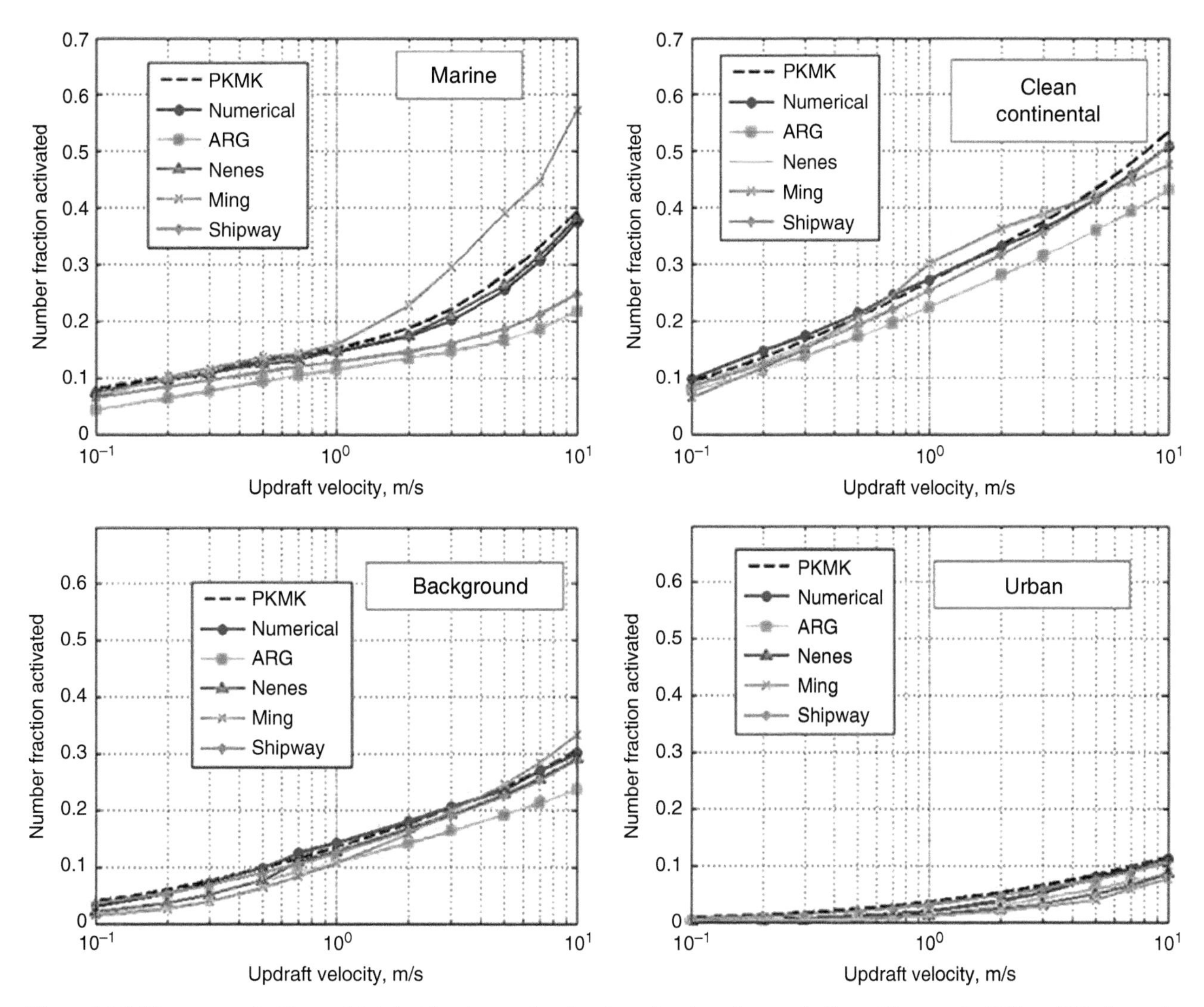

Figure 5.3.5 The same as in Figure 5.3.4, but for the number fraction of activated aerosols (from Ghan et al., 2011 and Pinsky et al., 2012; courtesy of John Wiley & Sons, Inc.©).

of the second mode (which contribution to the droplet concentration is dominating) of $R_2 = 0.03$ µm and $\log\sigma \sim 0.3$, the equivalent radius is about of 0.2 µm.

The approaches described in this section are useful for precise determination of concentration of droplets nucleated at cloud base. As an illustration, **Figure 5.3.6** shows two fields of droplet concentration in a deep polluted cumulus cloud simulated using HUCM at the grid spacing of 50 m. The left panel describes the simulation in which droplet nucleation was performed according to the approach illustrated in Figure 5.3.1. In the right panel, droplet concentration near cloud base was calculated using the method proposed by Pinsky et al. (2012), according to which Equation (5.3.14) was solved and then Equation (5.3.15) was used. One can see that even at high model resolution, calculation of supersaturation maximum may significantly increase droplet concentration. Therefore, special parameterization of droplet nucleation near cloud base is a necessary component of a cloud model aimed at accurate descriptions of cloud microphysics. Of course, it is especially important to use parameterization of droplet nucleation near cloud base in large-scale models with spatial resolutions cruder than those in cloud models.

In conclusion, the methods described in this subsection were developed to determine droplet concentration near cloud base only. They cannot be applied to calculating a possible nucleation of droplets above the cloud base supersaturation maximum. As was shown in

Tables 5.3.2a-h Droplet concentration at cloud base for different values of CCN SD parameters and different vertical velocities at cloud base (from Segal and Khain, 2006; courtesy of John Wiley & Sons, Inc.©).

Table 5.3.2a R_2 = 0.02 μm, w = 0.5 ms^{-1}

	N_N, cm^{-3}					
log(σ)	50	100	200	400	800	1,600
0.1	42.2	70.2	112.2	173.1	263.7	397.5
0.2	35.5	60.1	100	163.9	264.5	418.4
0.3	32.6	56.3	96.7	163.9	272	438.5
0.4	30.9	54.4	94.6	162.4	271.9	433.5
0.5	29.4	51.9	89.9	150.6	236.5	364.4

Table 5.3.2b R_2 = 0.02 μm, w = 1.0 ms^{-1}

	N_N, cm^{-3}					
log(σ)	100	200	400	800	1,600	3,200
0.1	91.5	158.7	264.4	423.1	672.5	1,018.2
0.2	77.1	133	224.9	376.5	615.7	975.6
0.3	70	122.5	212	362.1	605.3	964.8
0.4	65.8	116.4	204	350.6	584.4	914.3
0.5	62.3	110.1	191.3	320.6	501.3	760.4

Table 5.3.2c R_2 = 0.02 μm, w = 2.5 ms^{-1}

	N_N, cm^{-3}					
log(σ)	200	400	800	1,600	3,200	6,400
0.1	201	373.1	664.7	1,132.8	1,876.8	2,973.7
0.2	176.2	314	546.9	941.4	1,579.2	2,542.2
0.3	158.9	283.4	498.9	865.9	1,462.6	2,355.8
0.4	148	264.6	468.3	813.3	1,371.3	2,137.2
0.5	138.8	246.9	432.9	737.8	1,176.7	1,733

Table 5.3.2d R_2 = 0.02 μm, w = 2.5 ms^{-1}

	N_N, cm^{-3}				
log(σ)	400	800	1,600	3,200	6,400
0.1	412.2	788	1,453.1	2,585.1	4,382.5
0.2	372.6	675.2	1,202.8	2,098	3,556.9
0.3	337.6	606.7	1,078.5	1,889	3,206.9
0.4	312.7	562.2	1,000.3	1,741.1	2,910.1
0.5	291.6	521	916.1	1,551.1	2,444.6

Tables 5.3.2a-h (cont.)

Table 5.3.2e $R_2 = 0.04$ μm $w = 0.5$ ms^{-1}

	N_N, cm^{-3}					
log(σ)	50	100	200	400	800	1,600
0.1	50.6	100.3	196.5	374.7	677.3	1,138.9
0.2	48.4	91.9	170.6	306.9	529.2	862.4
0.3	44.4	82.5	150.3	266.4	448	740.7
0.4	40.9	75	134.7	231.9	382.1	657.6
0.5	34.7	59.3	93.5	156.8	301.9	603.8

Table 5.3.2f $R_2 = 0.04$ μm $w = 1.0$ ms^{-1}

	N_N, cm^{-3}					
log(σ)	100	200	400	800	1,600	3,200
0.1	101.7	201.8	398.8	773.7	1,420.8	2,411.8
0.2	98.9	189.7	356.2	649.5	1,117.9	1,805.2
0.3	91.8	171.5	314.9	559	932.8	1,501.6
0.4	84.7	155.8	280.5	481.9	779	1,321.6
0.5	72.1	124.4	198.4	319.1	603.8	1,207.6

Table 5.3.2g $R_2 = 0.04$ μm $w = 2.5$ ms^{-1}

	N_N, cm^{-1}					
log(σ)	200	400	800	1,600	3,200	6,400
0.1	205.7	406.9	807.6	1,597.5	3,072.2	53,93.9
0.2	203.6	396.0	760.4	1,422.1	2,517.4	4,062.8
0.3	193.8	367.3	684.0	1,238.3	2,087.3	3,287.1
0.4	180.8	335.7	611.2	1,066.3	1,713.4	2,780.3
0.5	155.5	273.7	455.2	702.2	1,230.7	2,453.7

Table 5.3.2h $R_2 = 0.04$ μm $w = 5.0$ ms^{-1}

	N_N, cm^{-3}				
log(σ)	400	800	1,600	3,200	6,400
0.1	414.6	818.3	1,622.2	3,216.8	6,243.9
0.2	412.5	805.3	1,557.4	2,940.4	5,210.1
0.3	396.7	755.5	1,414.3	2,565.3	4,288.1
0.4	371.9	692.9	1,262.0	2,188.3	3,461.2
0.5	319.4	561.7	953.9	1,493.9	2,461.7

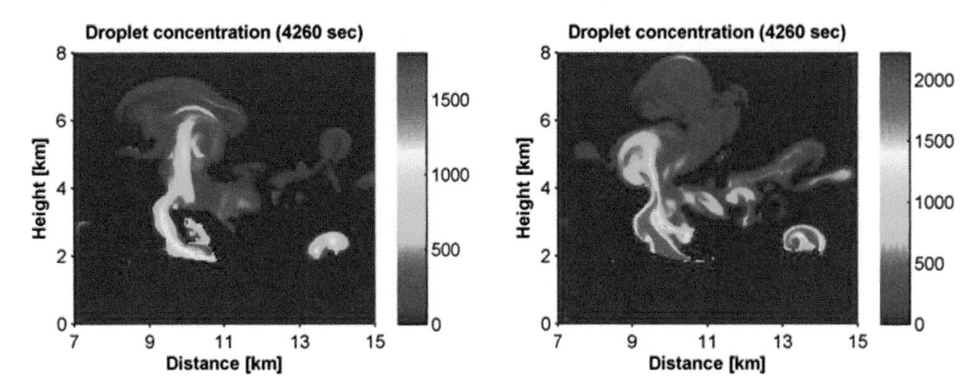

Figure 5.3.6 Fields of droplet concentration in deep very polluted cumulus cloud. Left: droplet concentration is calculated using the approach illustrated in Figure 5.3.1. Right: droplet concentration near cloud base is calculated using the method proposed by Pinsky et al. (2012).

Section 5.2, supersaturation inside clouds can sometimes be significantly higher than that at cloud base, which leads to in-cloud nucleation affecting droplet spectra. To describe droplet nucleation above the cloud base supersaturation maximum the approach illustrated in Figure 5.3.1 should be used.

5.3.3 Droplet Nucleation in Cases of Very High CCN Concentration

The last problem to be discussed in this section concerns the source of water for nucleated droplets. This problem is of special importance for modeling of convective clouds developing in extremely polluted air (like pyro-clouds developing in zones of forest fire), in which droplet concentration can exceed several thousand per cm^3. In such cases, the parameterization procedure described in Section 5.3.1 leads to the appearance of a significant amount of droplets whose mass may not be negligible. When calculating the exact conservation of water vapor mass, the mass of newly forming droplets should be subtracted from the water vapor mixing ratio, which in cases of very high CCN concentration may lead to an artificially large decrease in supersaturation.

To avoid this decrease in supersaturation, the procedure of the CCN nucleation may be performed beginning from the largest bin, and supersaturation can be recalculated after nucleation of CCN in each bin, the value of r_{Ncr} thus increasing after each recalculation. As a result, the condition $r_N \geq r_{Ncr}$ will be reached at larger values of r_N. The number of nucleated droplets thus will be lower than that determined by the initial supersaturation value, and an artificial decrease in supersaturation will be avoided. However, this successive nucleation bin by bin may lead to some underestimation of the number of droplets nucleated at a particular grid point, as it gives too much advantage to the largest CCN to absorb water vapor. It is necessary to take into account that the event of nucleation occurs instantaneously when a wet particle exceeds its critical size and becomes a droplet whose mass does not significantly increase during this conversion. To obtain an accurate budget of water vapor at very high AP concentration, models dealing with droplet nucleation parameterization should take into account that haze particles get their water within a layer of significant depth, including the layer from cloud base to the level of supersaturation maximum, as well as within some layer below the cloud base.

5.4 Calculation of Diffusional Growth and of Supersaturation in Numerical Cloud Models

5.4.1 Representation of Diffusional Growth by Mass Fluxes

Let us write a balance equation for a drop population. The diffusion growth/evaporation can be considered movement of DSD within the drop-size space at the rate equal to $\frac{dr}{dt}$ (**Figure 5.4.1**).

Suppose there is a size category (bin) having boundaries r and $r + \Delta r$. Since dr/dt shows the velocity of the radius change, the influx of drop concentration into the bin through boundary r is $\left(f_r \frac{dr}{dt}\right)_r$ and the outflux of drop concentration through the boundary $r + \Delta r$ is $\left(f_r \frac{dr}{dt}\right)_{r+\Delta r}$. As a result, the kinetic equation describing change of DSD due to diffusional growth within this bin can be written in a differential form for $\Delta r \to 0$ as (Rogers and Yau, 1996)

$$\frac{\partial f_r}{\partial t} = -\frac{\partial}{\partial r}\left(f_r \frac{dr}{dt}\right) \tag{5.4.1}$$

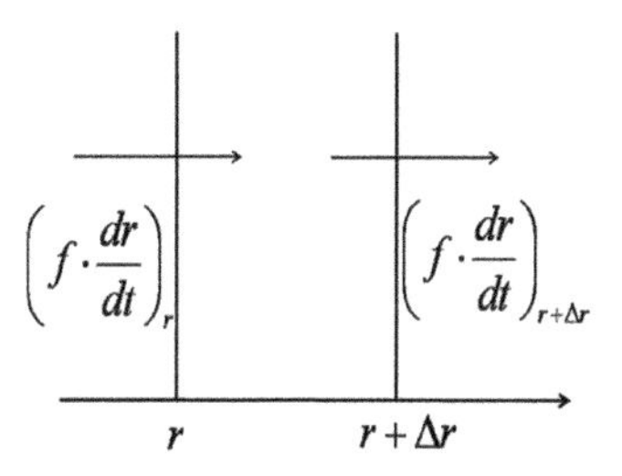

Figure 5.4.1 A scheme of droplet concentration fluxes within the droplet size space.

In case drops are comparatively large (larger than about 1–2 μm), the rate of drop growth $\frac{dr}{dt}$ is determined from Equation (5.2.6). If $S_w = const$, the solution of Equation (5.4.1) is

$$f_r(r,t) = \frac{r}{\sqrt{r^2 - \frac{2}{F}S_w t}} f_{r0}\left(\sqrt{r^2 - \frac{2}{F}S_w t}\right) \tag{5.4.2}$$

The assumption $S_w = const$ can be applied at time steps $\Delta t < \tau_{pr}$, where τ_{pr} is the phase relaxation time (Section 5.2). As shown in Figure 5.2.6, τ_{pr} decreases with height and is less than 1 s in intermediate and continental clouds. It means that supersaturation substantially changes over a time period shorter than 1 s. This imposes strong limitation on the time step in case assumption $S_w = const$ is used in calculations of DSD evolution.

5.4.2 Calculation of DSD Evolution Caused by Condensational Growth/Evaporation in Cloud Models

There are several approaches to calculate changes of DSD caused by condensational growth/evaporation in numerical models. In many bulk parameterization schemes, saturation adjustment hypothesis is used, according to which supersaturation formed at the advection sub-step due to adiabatic cooling in updrafts vanishes at the diffusional growth sub-step, so the final relative humidity RH = 100%. The final temperature and humidity are determined using the iteration method in which temperature increases and humidity decreases, so the final temperature corresponds to RH = 100%. The increase in LWC is equal to the difference between absolute humidity before and after the diffusional growth sub-step. The parameters of the Gamma distribution used to describe DSD (Section 2.1) are recalculated to obey the normalization condition. In case a two-moment bulk scheme is used, implementation of the nucleation procedure and saturation adjustment allows recalculating two integral parameters, namely, drop concentration and drop mass. Using these two new values, it is possible to recalculate the two parameters of the Gamma distribution. Besides, knowing the total mass and concentration values allows evaluating the averaged drop size.

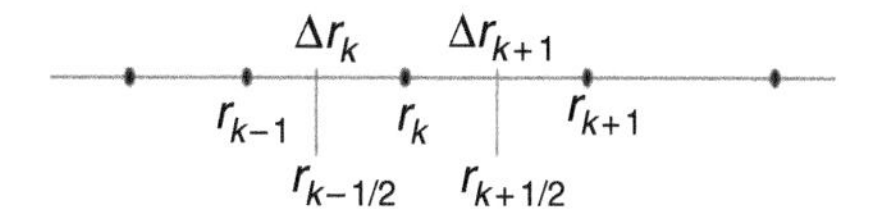

Figure 5.4.2 Schematic picture of a grid used to calculate droplet SD functions. This scheme illustrates the finite difference scheme (5.4.3).

Spectral bin microphysics applies more sophisticated approaches. The first approach is based on the direct solving of Equation (5.4.1) for condensation/evaporation, assuming that no nucleation takes place during the diffusion growth/evaporation sub-step. Equation (5.4.1) should be solved on the mass grid where SD is defined. For instance, Clark (1973) uses the following finite difference scheme:

$$f_k^{t+\Delta t} = f_k^t - \frac{\Delta t}{\Delta r_k}\left[f_{k+1/2}^t\left(\frac{dr}{dt}\right)_{k+1/2} - f_{k-1/2}^t\left(\frac{dr}{dt}\right)_{k-1/2}\right], \tag{5.4.3}$$

where k denotes the number of the mass bin, t is the current time, f_k^t is the DSD in the k-th bin before the diffusional growth sub-step, $f_k^{t+\Delta t}$ is the DSD in the k-th bin after the diffusional growth sub-step, Δt is the time step used to solve the equation, and $\Delta r_k = r_k - r_{k-1}$ is the difference between the radii of neighboring bins. Indexes $k + 1/2$ and $k - 1/2$ denote the bin boundaries (**Figure 5.4.2**). The values of DSD at the boundaries of the k-th bin can be determined by linear interpolation:

$$f_{k+1/2} = \frac{f_k \Delta r_{k+1} + f_{k+1}\Delta r_k}{\Delta r_{k+1} + \Delta r_k} \tag{5.4.4}$$

The values of $\frac{dr}{dt}$ at intermediate points are determined using the equation for diffusional growth. Numerical schemes (like those in Equation (5.4.3)) allow utilization of any form of the diffusional growth equation, including its full form, which includes the curvature and the chemical terms. However, in most models changes of DSD caused by diffusional growth are calculated using a simpler form of the equation (5.2.6) because drop radius typically exceeds a few μm. Formation of smaller drops is treated at the stage of droplet nucleation. High variation of DSD from one bin to another may lead to serious numerical problems when using Equation (5.4.3).

The second approach includes two sub-steps (e.g., Kogan, 1991; Khain and Sednev, 1996; Khain et al., 2004). To solve Equation (5.4.1) at the first sub-step, the value of $Q = \int_t^{t+\Delta t} S_w dt'$ is determined. The method for calculation of Q is described next. It is clear that as a result of the first sub-step the values of drop mass do not coincide with the mass of drops corresponding to the bins of the regular mass grid. Therefore, the second sub-step requires interpolation of obtained DSD onto a regular grid. This procedure is known as remapping. Let m_i be the mass of drop belonging to the i-th bin of a regular mass grid. Suppose that during one time step of diffusion growth, mass m_i grows up to $m_{i,new}$ (according to Equation (5.2.11)). As a result, a new non-regular sequence of drop masses arises (**Figure 5.4.3a**). Concentration of drops with mass of $m_{i,new}$ is equal to concentration of drops with mass m_i, because drop concentration does not change during condensation growth.

A widely used remapping procedure was proposed by Kovetz and Olund (1969) and is schematically illustrated in Figure 5.4.3b. The Kovetz–Olund scheme splits mass $m_{i,new}$ and concentration of hydrometeors (drops) $N_{i,new}$ between neighboring bins of a regular grid. This procedure conserves the concentration and the mass (i.e., the two moments of DSD), which can be described as

$$\begin{cases} N_{i,new} = N_i + N_{i+1} \\ m_{i,new} N_{i,new} = m_i N_i + m_{i+1} N_{i+1} \end{cases}, \tag{5.4.5}$$

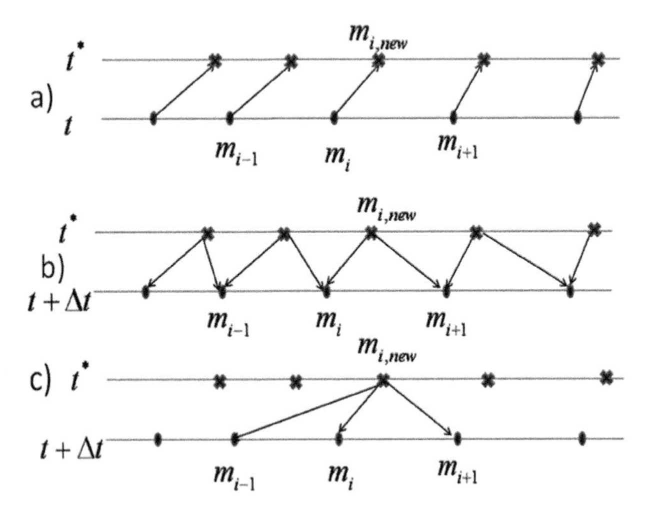

Figure 5.4.3 The schemes of remapping. Panel a) illustrates the step of diffusion growth and formation of droplets on a nonregular grid. Panel b) illustrates the remapping using the scheme proposed by Kovetz and Olund (1969). Panel c) illustrates the remapping used by Khain et al. (2008).

where N_i and N_{i+1} are concentration values added into the i-th and the i + 1-th bins. The solution of Equation (5.4.5) is

$$\begin{cases} N_i = N_{i,new} \dfrac{(m_{i+1} - m_{i,new})}{(m_{i+1} - m_i)} \\ N_{i+1} = N_{i,new} \dfrac{(m_{i,new} - m_i)}{(m_{i+1} - m_i)} \end{cases} \tag{5.4.6}$$

The scheme (5.4.6) of remapping has a drawback, since it artificially accelerates formation of large drops in DSD. Indeed, as seen from Figure 5.4.3b, each remapping step leads to appearance of drops with masses $m_{i+1} > m_{i,new}$. So during diffusion growth the maximum drop size is shifted to the right by one bin. In case of drop evaporation, the minimum drop size is shifted to the left at each remapping step. As a result, remapping leads to an artificial DSD broadening caused exclusively by numerical factors, which is akin to numerical viscosity of a numerical scheme describing advection (Section 4.2). It is clear the concentration of drops in the largest bins may be very low. Nevertheless, the artificial DSD broadening is undesirable because it leads to earlier formation of cloud droplets with radii about 20–21 μm that are able to trigger intense collisions and raindrop formation at unrealistically low distances above cloud base.

Remapping procedure (5.4.6) conserves the zero moment and the third moments of DSD $f_r(r)$, but overstates the higher moments of DSD. Decrease in DSD broadening can be reached by increasing the number of bins. Several versions of remapping schemes were proposed aimed at decreasing the artificial DSD broadening. Liu et al. (1997) developed a method implying a variational optimization technique and conserving the four moments of drop spectra. However, this method was found applicable mostly for DSD containing only small droplets and was used in simulations of stratocumulus clouds.

It is clear that a perfect remapping scheme must conserve all the DSD moments. The Kovetz–Olund scheme overestimates the higher moments, which leads to an increase of the large-size tail of DSD. To exclude this phenomenon it is necessary to conserve the higher DSD moments. Khain et al. (2008) proposed a remapping procedure that conserves the zero moment (the total concentration), the third moment (the total mass), and the sixth moment (the radar reflectivity). The scheme of this remapping is presented in Figure 5.4.3c. The three equations expressing the conservation laws were solved for each mass bin of a non-regular grid:

$$\begin{cases} N_{i,new} = & N_{i-1} + N_i + N_{i+1} \\ M_{i,new} = m_{i,new} N_{i,new} = & m_{i-1} N_{i-1} + m_i N_i + m_{i+1} N_{i+1}, \\ Z_{i,new} = m_{i,new}^2 N_{i,new} = & m_{i-1}^2 N_{i-1} + m_i^2 N_i + m_{i+1}^2 N_{i+1} \end{cases} \tag{5.4.7}$$

where $m_{i-1} < m_i < m_{i,new} < m_{i+1}$. One can solve Equation (5.4.7) analytically. For example the concentration in i+1-th bin after remapping is calculated as

$$N_{i+1} = N_{i,new} \frac{(m_{i,new} - m_i)(m_{i,new} - m_{i-1})}{(m_{i+1} - m_{i-1})(m_{i+1} - m_i)} \tag{5.4.8}$$

Comparison of Equations (5.4.8) and (5.4.6) shows that concentration in i+1-th bin calculated using Equation (5.4.8) is always less than that calculated using Equation (5.4.6). It means the remapping procedure by Khain et al. (2008) leads to weaker DSD broadening then the Kovetz and Olund procedure. If the final concentration in a ceratin bin turns out to be negative as a result of application of Equation (5.4.8), Equation (5.4.6) is used.

Figure 5.4.4 compares the DSDs calculated in the simulations of green ocean (clean air), smoky (polluted air), and pyro clouds (arising above forest fires where concentration of CCN is very high) observed in the Amazon region (Andreae et al., 2004). The simulations used the Kovetz–Olund scheme (1969) and the Khain et al. (2008) scheme. The DSDs calculated using the latter scheme are narrower and broaden with height slower than those obtained via the Kovetz–Olund scheme. As shown next, the Khain et al. (2008) scheme allows reproduction of DSD measured in situ. Remapping performed by the scheme proposed by Khain et al. (2008) decreases the numerically induced DSD broadening but does not completely eliminate it. Therefore, the number of remapping steps in cloud models should be reduced while solving the diffusional growth equation. There are several approaches to perform this reduction. For instance, Kogan, (1991) and Khain et al. (2004, 2008) perform the remapping with time increments several times larger than the time sub-steps used for solving the diffusional growth equation.

In Lagrangian cloud models where DSD are calculated in air parcels, movable mass grids are used (Feingold et al., 1996; Pinsky and Khain, 2002; Pinsky et al., 2008b). The bins of mass grid in such models shift according to the equation of diffusion growth/evaporation. Figure 5.4.3a shows such a shift of a mass grid during a diffusion growth time sub-steps. Pinsky et al. (2008b) use time sub-step of 0.01 s (or lower) to solve the diffusional growth equation. Treatment of collisions, drop sedimentation, and mixing require utilization of a regular mass grid, on which these processes are simulated with time steps of 1 s, i.e., remapping is performed after 100 sub-steps of diffusion growth. There are also attempts to create some "hybrids" of the Lagrangian and the Eulerian approaches to solve the equation of diffusional growth (e.g., Cooper et al., 1997).

In the method developed by Khvorostyanov et al. (1989) and later used by Khain et al. (1993), the SD function $\Phi(r^2)$ is defined on a r^2 grid. Diffusion growth/evaporation on this grid is reduced to a shift of SD by the value of $Q = \frac{2}{F}\int_t^{t+\Delta t} S_w dt'$. The procedure of diffusion growth/evaporation used by Khvorostyanov et al. (1989) is illustrated in **Figure 5.4.5**. The values of $\Phi(r_i^2)$ at points r_i^2 at $t + \Delta t$ are equal to $\Phi(r_i^2 - Q)$ at points $r_{0,i}^2 = r_i^2 - Q$ at time instance t. The values of DSD at

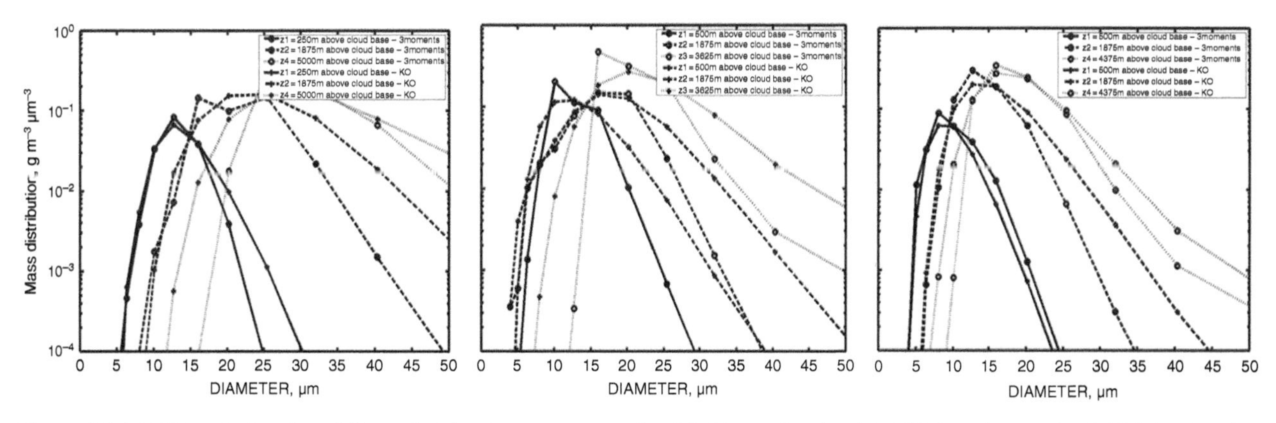

Figure 5.4.4 DSDs calculated at different heights in green-ocean cloud (left), smoky cloud (middle), and pyro cloud (right) using the Kovetz–Olund remapping scheme (1969) and the remapping scheme proposed by Khain et al. (2008) (from Khain et al., 2008; American Meteorological Society©; used with permission).

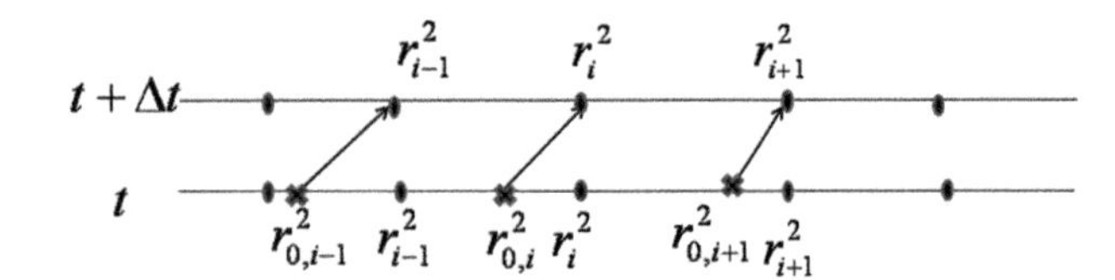

Figure 5.4.5 The scheme of diffusional growth applied at the square radius grid according to the method proposed by Khvorostyanov et al. (1989).

points $r_{0,i}^2$ are determined by interpolation of the values of $\Phi(r_i^2)$ at the neighboring points of a regular grid at time instance t. This scheme does not lead to any artificial DSD broadening. However, the interpolation does not exactly conserve the mass and the concentrations of drops. There are more serious shortcomings of this method. First, it requires very small time steps because supersaturation is assumed constant during a single time step. Second, it requires too many bins in the grid where $\Phi(r_i^2)$ is defined to cover the whole range of drops. For these reasons, this method was used only in calculating diffusional growth of cloud droplets.

Another approach for calculation of drop diffusional growth was proposed by Tzivion et al. (1989) (see also Reisin et al., 1996) and is in agreement with the description of DSD by the method of moments. The method of moments describes drop distribution using moments of DSD calculated for each size category (Section 2.1). Therefore, diffusion growth/evaporation is described in this approach by changes of DSD moments in each category. The equation used by Tzivion et al. (1989) to describe the evolution of DSD during diffusional growth is similar to Equation (5.4.1), but written for fluxes of mass:

$$\frac{\partial f(m)}{\partial t} = -\frac{\partial}{\partial m}\left(f(m)\frac{dm}{dt}\right) \tag{5.4.9}$$

The rate of the drop mass growth is written as

$$\frac{dm}{dt} = C(p,T)\Delta S_w m^{1/3}, \tag{5.4.10}$$

where ΔS_w is the specific humidity surplus and $C(p,T)$ is the known function of pressure p and temperature T. Using the analytical solution of Equations (5.4.9) and (5.4.10), the equation for mass in the k-th bin (M_k is the first moment of DSD $f(m)$) at time instance $t+\Delta t$ is calculated as

$$M_k(t+\Delta t) = \int_{z_k}^{z_{k+1}}\left(m^{2/3}+\frac{2}{3}\tau\right)^{3/2} f(m)dm, \tag{5.4.11}$$

where $\tau = C\int_t^{t+\Delta t}\Delta S_w(t)dt$. Obviously, drop concentration in the bin N_k (the zero moment of DSD $f(m)$) is conserved in this solution. The method also takes into account that the boundaries of the k-th category $[z_k;\ z_{k+1}]$ shift during diffusion growth/evaporation. A hypothesis about the relation between the moment of 1/3 order, the zero moment, and the first moment is also used in this method to recalculate the moments of DSD in the categories. The derivation formulas and a detailed description of the method can be found in Tzivion et al. (1989).

5.4.3 Calculation of Supersaturation in Cloud Models

Knowing supersaturation values is necessary for solving the equation of diffusional growth/evaporation and for calculating the increase/decrease in LWC caused by condensation/evaporation. An accurate reproducing of the supersaturation field is a key element of any cloud model. The equation system for microphysical and thermodynamical quantities to be solved in bin-microphysics cloud models is the following (the equations contain only the terms describing latent heat release, advection and turbulent diffusion, i.e., the main processes affecting supersaturation):

$$\frac{\partial \theta}{\partial t} = \frac{L_w}{\Pi c_p}C - \frac{1}{\rho}\nabla\cdot\rho\mathbf{u}\theta + DIFF_\theta \tag{5.4.12}$$

$$\frac{\partial q_v}{\partial t} = -C - \frac{1}{\rho}\nabla\cdot\rho\mathbf{u}q_v + DIFF_{q_v} \tag{5.4.13}$$

$$\frac{\partial f(m)}{\partial t} = -\frac{1}{\rho}\nabla\cdot\rho(\mathbf{u}-\mathbf{V_g})f(m) + DIFF_f - \frac{\partial}{\partial m}\left(f(m)\frac{dm}{dt}\right) \tag{5.4.14}$$

$$\frac{1}{r}\frac{dm}{dt} = \frac{4\pi\rho_w}{F}S_w(\theta,q_v), \tag{5.4.15}$$

where C is the condensation rate, $DIFF$ terms describe turbulent diffusion, $\Pi = \left(\frac{p}{p_0}\right)^{R/c_p} = \frac{T}{\theta}$ is the Exner function (Section 3.1), p is the ambient pressure profile, and $p_0 = 1{,}000$ hPa. The dependence $S_w(\theta,q_v)$ is discussed in Section 3.1.4. The rate of condensation C is calculated as

$$C = \frac{1}{\rho}\int_0^\infty f(m)\frac{dm}{dt}dm, \tag{5.4.16}$$

where $\frac{dm}{dt}$ is determined by the diffusion growth equation (5.4.15). $\mathbf{V_g}$ is the fall velocity of drop with mass m, relative to the air. In the models with bulk parameterization of microphysical processes the equation for SD function (5.4.14) is replaced by the equation for liquid water mixing ratio:

$$\frac{\partial q_l}{\partial t} = C - \frac{1}{\rho}\nabla\cdot\rho\left(\mathbf{u} - \overline{\mathbf{V}}_{\mathbf{g}}\right)q_l + DIFF_l, \qquad (5.4.17)$$

where $\overline{\mathbf{V}}_{\mathbf{g}}$ is the mass-averaged fall velocity of raindrops. Note that bulk schemes typically neglect the sedimentation of cloud droplets. So, settling of rain drops only is taken into account.

Bulk- and spectral bin microphysical models often use the alternating-direction implicit method (ADI) to solve the equation system (5.4.12–5.4.16). ADI is also known as the splitting-up method (i.e., treating physical processes successively) (Temam, 1977; Marchuk, 1974, 1980). In ADI, advection and turbulent diffusion are treated at the first sub-step, i.e., the following equations are solved:

$$\frac{\partial \theta}{\partial t} = -\frac{1}{\rho}\nabla\cdot\rho\mathbf{u}\theta + DIFF_\theta \qquad (5.4.18)$$

$$\frac{\partial q_v}{\partial t} = -\frac{1}{\rho}\nabla\cdot\rho\mathbf{u}q_v + DIFF_{q_v} \qquad (5.4.19)$$

$$\frac{\partial f}{\partial t} = -\frac{1}{\rho}\nabla\cdot\rho\left(\mathbf{u} - \mathbf{V_g}\right)f + DIFF_f$$
(in bin microphysics) (5.4.20)

$$\frac{\partial q_l}{\partial t} = -\frac{1}{\rho}\nabla\cdot\rho\left(\mathbf{u} - \mathbf{V_g}\right)q_l + DIFF_l$$
(in bulk-parameterization schemes) (5.4.21)

To solve these equations, various finite-difference methods are applied (Chapter 4). As a result of this sub-step, the values of θ^*, T^*, q_v^*, f^* (or q_l^*) after advection, sedimentation, and turbulent diffusion are determined. These values are used as the initial conditions to perform the second sub-step, at which the effects of condensation/evaporation on drop size, temperature, and humidity are calculated. At the second sub-step, the following equation system is solved:

$$\frac{\partial \theta}{\partial t} = \frac{L_w}{\Pi c_p}C \qquad (5.4.22)$$

$$\frac{\partial q_v}{\partial t} = -C \qquad (5.4.23)$$

$$\frac{\partial q_l}{\partial t} = C \qquad (5.4.24)$$

$$\frac{\partial f}{\partial t} = -\frac{\partial}{\partial m}\left(f\frac{dm}{dt}\right) \qquad (5.4.25)$$

These equations are solved together with the diffusional growth equation (5.4.15).

The order of solving an equation of the system (5.4.22–5.4.25) is the following. In case short time steps $\Delta t_{c/e}$ are used ($\Delta t_{c/e} \ll \tau_{pr}$), Equations (5.4.25) and (5.4.15) are solved under supersaturation S_w^* calculated using T^* and q_v^*. As a result, a new DSD value $f\left(t + \Delta t_{c/e}\right)$ is calculated. Using this new DSD value, the change in q_l is calculated as

$$\Delta q_l = \frac{1}{\rho}\int_0^\infty \left[f\left(t + \Delta t_{c/e}\right) - f^*\right]m dm. \qquad (5.4.26)$$

The rate of condensation/evaporation is calculated by Equations (5.4.22–5.4.24) as

$$C = \frac{\Delta q_l}{\Delta t_{c/e}}. \qquad (5.4.27)$$

New values of water vapor mixing ratio and temperature are calculated as

$$q_v\left(t + \Delta t_{c/e}\right) = q_v^* - \Delta q_l \qquad (5.4.28)$$

$$\theta\left(t + \Delta t_{c/e}\right) = \theta^* + \frac{L_w}{c_p\Pi}\Delta q_l \qquad (5.4.29)$$

Values of supersaturation S_w^* calculated after the advection sub-step and turbulent diffusion sub-step take into account different sources of supersaturation including cooling in adiabatic updraft described by term $A_1 w$ in Equation (5.2.49).

Since S_w^* decreases rapidly in the course of diffusional growth of drops, the assumption that S_w^* is constant during the second sub-step $\Delta t_{c/e}$ may overestimate the drop growth rate. Thus, it is desirable to take into account the changes of supersaturation during diffusional growth/evaporation. Due to the changes in diffusional growth, the supersaturation changes according to Equation (5.2.13) $\frac{dS_w}{dt} = -\frac{S_w}{\tau_{pr}}$. The analytic solution of this equation is an exponential function:

$$S_w(t) = S_w^* \exp\left(-t/\tau_{pr}\right), \qquad (5.4.30)$$

which tends to zero at $t \to \infty$. To solve the diffusional growth equation during the time step $\Delta t_{c/e}$, it is necessary to calculate the integral $Q = \frac{2}{F}\int_0^{\Delta t_{c/e}} S_w dt'$, where F

is calculated using Equation (5.1.13). Using Equation (5.4.30) and assuming that the relaxation time does not change significantly during the time step $\Delta t_{c/e}$, one can write

$$Q = \frac{2}{F}\int_0^{\Delta t_{c/e}} S_w dt' = \frac{2}{F}\tau_{pr} S_w^* \left[1 - \exp\left(-\frac{\Delta t_{c/e}}{\tau_{pr}}\right)\right] \tag{5.4.31}$$

Using the value of Q, the DSD, temperature, and humidity are recalculated as was described earlier in this section.

Although formally Equation (5.4.31) allows using any large time step $\Delta t_{c/e}$, $\Delta t_{c/e}$ should be chosen smaller than τ_{pr}. Usually, the solution of Equations (5.4.22–5.4.25) requires smaller time steps than the dynamic time step Δt_{dyn} applied to solve the equation system (5.4.18–5.4.21). Kogan (1991), Khain and Sednev (1996), and Khain et al. (2004) use $\Delta t_{c/e}$ chosen within the range from $\Delta t_{dyn}/10$ to $\Delta t_{dyn}/2$. The decrease in $\Delta t_{c/e}$ leading to a better representation of diffusional growth leads to an increase in the number of remappings to be performed for recalculation of DSD, which in turn may lead to a numerical DSD broadening. To avoid this problem, the procedure for calculating Q is successively repeated n times (where $n = \Delta t_{dyn}/\Delta t_{c/e}$), so by the end of dynamical step $Q_\Sigma = \sum_{i=1}^{n} \frac{2}{F}\int_{(i-1)\Delta t_{c/e}}^{i\Delta t_{c/e}} S_w dt'$ is determined. The final change of drop mass is calculated using Q_Σ, so remapping is performed only once at time instance $\Delta t_{dyn} = n\Delta t_{c/e}$.

A disadvantage of this method is utilization of supersaturation S_w^* value calculated after the dynamical sub-step. In fact, each model designer has to decide at which time instance the procedure of droplet nucleation should be included. Without taking into account the supersaturation decrease caused by condensational droplet growth, S_w^* calculated at the end of the dynamical sub-step may reach unrealistically large values. These values then decrease exponentially due to diffusional growth. In case the droplet nucleation procedure is used just after the advection sub-step, it may lead to nucleation of unrealistically high amounts of droplets. This problem is especially serious in simulations of clouds in highly polluted air. For instance, in zones of biomass burning the AP concentration may exceed 10^4 cm^{-3}. Launching the nucleation procedure just after the advection may lead to nucleation of all these AP. At the same time, supersaturation substantially decreases by the end of the diffusion growth sub-step. Launching the nucleation procedure at this time instance may underestimate the amount of activated AP. In real clouds, the processes of advection and diffusional drop growth take place simultaneously, so the value of supersaturation remains lower than S_w^*. To decrease S_w leaps during a single time step, caused by the successive treatment of dynamical and microphysical processes, Khain et al. (2004) divide the change in supersaturation $\Delta S_{w,dyn}$ caused integrally by all the processes except condensation/evaporation between number of microphysical time sub-steps $N_{sub} = \Delta t/\Delta t_{c/e}$ (where $\Delta t_{c/e}$ is the time sub-step used to solve the diffision growth equation) and add the value $\Delta S_{w,dyn}/N_{sub}$ to the supersaturation value at each sub-step.

In a more general approach, Khain et al. (2008) solve the following equation for supersaturation at each microphysical time step:

$$\frac{dS_w}{dt} = -\frac{S_w}{\tau_{pr}} + \frac{\Delta S_{w,dyn}}{\Delta t_{dry}} \tag{5.4.32}$$

The term $\frac{\Delta S_{w,dyn}}{\Delta t_{dyn}}$ may be called the *dynamic tendency* of supersaturation. Actually, Equation (5.4.32) is an analog of Equation (5.2.49). Equation (5.4.32) is solved using sub-steps $\Delta t_{c/e}$. The initial condition used for solving Equation (5.4.32) at the fist sub-step is the supersaturation value before advection. Solution of Equation (5.4.32) is

$$S_w(t) = S_{w0}\exp(-t/\tau_{pr}) + \tau_{pr}\frac{\Delta S_{w,dyn}}{\Delta t_{dyn}}(1 - \exp(-t/\tau_{pr})) \tag{5.4.33}$$

In order to recalculate τ_{pr} at each i-th microphysical step, the value $Q_i = \frac{2}{F}\int_0^{\Delta t_{c/e}} S_w dt'$ is calculated as

$$Q_i = \frac{2}{F}\left\{\left(\tau_{pr}S_{w0} - \tau_{pr}^2\frac{\Delta S_{dyn}}{\Delta t_{c/e}}\right)\left[1 - \exp\left(\frac{\Delta t_{c/e}}{\tau_{pr}}\right)\right] + \tau_{pr}\frac{\Delta S_{w,dyn}}{\Delta t_{dyn}}\Delta t_{c/e}\right\} \tag{5.4.34}$$

Summation of Q_i allows calculating the value $Q(t + \Delta t_{dyn})$ the end of the dynamic time step. Then the new radius and the new drop mass belonging to each bin is calculated using expression $r^2(t + \Delta t_{dyn}) = r^2(t) + Q(t + \Delta t_{dyn})$ following from Equation (5.2.7). Accordingly, the new DSD is calculated using remapping on a regular mass grid. Using the new DSD, the changes in temperature and humidity caused by condensation/evaporation at each model spatial grid point are calculated. Using these new temperatures and humidity values, the supersaturation at the end of the dynamic step is calculated.

Figure 5.4.6 illustrates the algorithm of calculation of supersaturation at a grid point located 200 m above cloud base within an updraft of a deep maritime convective cloud simulated using HUCM. The model grid spacing is 50 m × 50 m. The value $N_{sub} = 10$, so $\Delta t_{c/e} = \Delta t_{dyn}/10$. The straight solid line denotes the dynamic tendency of supersaturation in this grid point. The dashed curve shows a semi-analytic solution of the supersaturation equation. This curve goes below the straight solid line because the condensation calculated during Δt_{dyn} decreases supersaturation. The black circle denotes the value of supersaturation after time step Δt_{dyn}. It can be seen that supersaturation after advection (in updrafts) is quite high. Utilization of the dynamic tendency of supersaturation together with the water vapor sink in the same Equation (5.4.32) does not allow supersaturation to reach unrealistically high values, in contrast to methods treating advection separately.

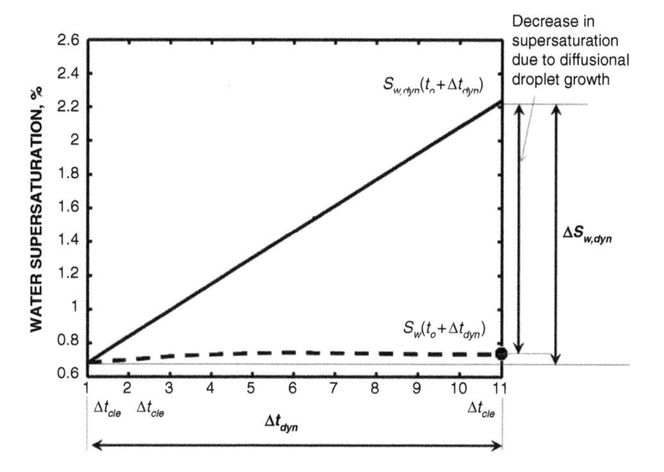

Figure 5.4.6 The scheme illustrating the algorithm for calculating supersaturation at each microphysical sub-step $\Delta t_{c/e}$. The algorithm takes into account all the factors affecting droplet growth.

Figure 5.4.7 illustrates time changes of supersaturation S_w during several successive time steps in a simulated pyro cloud developing in extremely polluted air (Khain et al., 2008). The left panel shows the changes in supersaturation obtained with the scheme that treats advection and diffusion growth separately. The periodic peaks of supersaturation seen in the left panel are the result of the advection sub-step; a decrease in supersaturation takes place due to diffusional growth of drops. The right panel shows supersaturation obtained according to the algorithm illustrated in Figure 5.4.6. The curves in the right panel of Figure 5.4.7 correspond to the dashed curve in Figure 5.4.6. One can see that when both factors are treated simultaneously, supersaturation does not reach unrealistically high values and its sensitivity to time step duration is much lower. The amplitudes of the supersaturation peaks depend on the dynamic time step. A decrease in the dynamic time step leads to a decrease in the amplitude of the supersaturation peaks after advection. In this case, the choice of time instance most appropriate for performing nucleation becomes less critical. For instance, in different versions of HUCM, droplet nucleation is performed with nearly similar results either once at the end of the dynamic time step or several times during the dynamic time step.

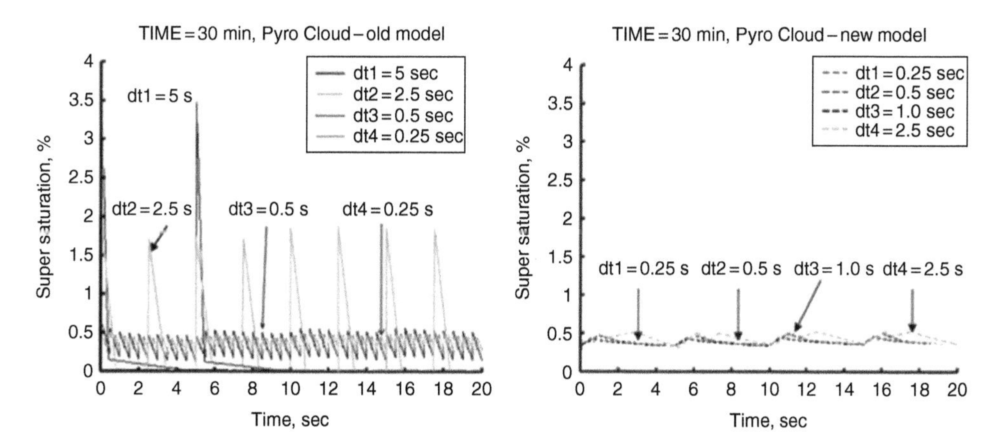

Figure 5.4.7 Time changes of supersaturation S_w during several successive dynamic time steps at a grid point near the base of a simulated pyro cloud: using a scheme that treats supersaturation changes at the advection sub-step and the diffusion growth sub-step separately (left); using a scheme that treats supersaturation changes due to advection and diffusional growth of droplets simultaneously (right) (from Khain et al., 2008; American Meteorological Society©; used with permission).

There are other methods for calculating supersaturation and drop diffusional growth. For instance, Clark (1973) and Hall (1980) used the equation system (5.4.12–5.4.15) integrated with a dynamic time step Δt_{dyn}. At the beginning of each time step, at time instance t, the condensation rate values C are known. During the time step Δt_{dyn} the condensation rate values can change significantly, which requires a special iteration procedure to adjust the condensation rate. In order to use larger microphysical time steps, Clark (1973) applies a semi-analytical solution of the supersaturation equation (5.2.15) at each microphysical time sub-step $\Delta t_{c/e}$. During $\Delta t_{c/e}$, the phase relaxation time and the vertical velocity are assumed constant, and S_0 is the supersaturation value at the beginning of the time step. This solution is used to calculate S_w value averaged over time interval $\Delta t_{c/e}$. This averaged supersaturation value is then used to solve the diffusional growth equation, which allows us to utilize $\Delta t_{c/e}$ of the same order of magnitude as Δt_{dyn}. However, an accurate calculation of supersaturation changes requires time steps shorter than the phase relaxation time. Otherwise, it can lead to the erroneous appearance of negative supersaturation during condensational growth or of positive supersaturation during drop evaporation (which resembles the numerical solution of the friction equation, Section 4.3). Although the implicit scheme theoretically allows using large time steps, the increasing time step over the phase relaxation time may lead to a decrease in the averaged value of supersaturation, affecting the rate of diffusional drop growth.

There are other challenges related to calculating supersaturation. Since the supersaturation equation (5.2.49) is the consequence of Equations (5.4.12) and (5.4.13), it should yield the same supersaturation value as Equations (5.4.12) and (5.4.13). However, time steps often used to solve temperature and humidity equations (5.4.12) and (5.4.13) are larger than the phase relaxation time. In addition, small errors in calculation of temperature and humidity may lead to significant errors in the values of supersaturation. Hence, the value of supersaturation calculated by Equations (5.4.12) and (5.4.13) may not be the same as that obtained by Equation (5.2.49). Such inconsistency may remain even after the adjustment of condensation rate applied by Clark (1973) and Hall (1980). Grabowski and Morrisson (2008) argue that this inconsistency is the reason of spurious cloud-edge supersaturation fluctuations near the upper boundaries of stratocumulus clouds reported by Klaassen and Clark (1985) and Stevens et al. (1996). To eliminate this inconsistency, Grabowski and Morrisson (2008) proposed to use the solution of the supersaturation equation to correct temperature and humidity values obtained by the corresponding equations. The authors use the absolute supersaturation value $\delta = q - q_S(p, T)$, where $q_S(p, T)$ is the water-vapor mixing ratio at saturation. A simplified equation for absolute supersaturation presented in this study is

$$\frac{d\delta}{dt} = wg\left(\rho\frac{\partial q_S}{\partial p} + \frac{1}{c_p}\frac{\partial q_S}{\partial T}\right) - \frac{\delta}{\tau_{pr}}, \tag{5.4.35}$$

The first term on the right-hand side represents forcing caused by the vertical velocity (all other explicit forces in the equations for temperature and for moisture are set at zero here), whereas the last right-hand term represents the tendency of the absolute supersaturation due to condensational growth of cloud droplets. The solution of Equation (5.4.35) represents a physically reliable prediction of supersaturation. In contrast, predicting supersaturation calculated using temperature and humidity obtained by solving Equations (5.4.12) and (5.4.13) leads to errors. Therefore, the values of temperature and moisture calculated by Equations (5.4.12) and (5.4.13) should be adjusted to produce a supersaturation value that matches solution of Equation (5.4.35). The adjustment procedure is described by the following equations:

$$q_v = q_v^* - \varepsilon; \quad q_l = q_l^* + \varepsilon; \quad \theta = \theta^* + \frac{L}{\Pi c_p}\varepsilon, \tag{5.4.36}$$

where the variables with an asterisk represent solutions of Equations (5.4.12) and (5.4.13) and the variables without an asterisk represent adjusted solutions; ε is the factor correcting the mixing ratio, cloud-water value, and temperature. While $\varepsilon > 0$ corresponds to an additional condensation, $\varepsilon < 0$ corresponds to an additional evaporation needed to match the temperature field and moisture field to the predicted supersaturation value. The value of ε should be chosen in such a way as to get the true value of the absolute supersaturation δ_{true} calculated by Equation (5.4.35):

$$\delta_{true} = q_v^* - \varepsilon - q_s\left(p, T^* + \frac{L}{c_p}\varepsilon\right) \tag{5.4.37}$$

Since values of ε required for the adjustment described by Equation (5.4.37) are typically small, it is sufficient to use the linearized version of Equation (5.4.37). In this case, ε can be derived analytically as

$$\varepsilon = \frac{q_v^* - q_s(p, T^*)\delta_{true}}{\Gamma(p, T^*)}, \tag{5.4.38}$$

where $\Gamma(p, T) = 1 + \frac{L_w}{c_p}\frac{dq_S(p,T)}{dT}$ is the psychrometric correction.

In Largangian cloud models (Pinsky and Khain, 2002; Pinsky et al., 2008b; Magaritz et al., 2009), advection and diffusional growth of droplets are calculated using a very small uniform time step that varies from $\Delta t_{dyn} = \Delta t_{c/e} = 0.001$–$0.01$ s. At each time step, the supersaturation equation in the form (5.2.1) is solved. The obtained supersaturation value is used to solve the diffusion growth equation in its full form, including the curvature terms and the chemistry terms. Since $\Delta t_{c/e}$ is very small, the superstaturation is assumed constant during $\Delta t_{c/e}$. Utilizing such small time steps allows calculating supersaturation with high accuracy.

5.5 Drop Dynamics

Drop motion has a major impact on collisions between drops as well as on raindrop formation. Drop sedimentation forms liquid water fluxes directed downward and thus determines the water balance in different parts of a cloud. Fluxes of large drops form rain both inside cloud and near surface. To a first approximation, drops move relative to the surrounding air at a constant velocity V_g called settling or terminal velocity, which depends on drop mass. However, many important processes in clouds require taking into account the inertia effects of drops; this leads to formation of additional drop velocity relative to the air.

5.5.1 Equation of Drop Motion

The two main forces that influence drop motion are the gravity force $\mathbf{F_g}$ and the drag force $\mathbf{F_d}$. Accordingly, the drop-motion equation is usually written in the form

$$m\frac{d\mathbf{V}}{dt} = \mathbf{F_g} + \mathbf{F_d}, \qquad (5.5.1)$$

where m is the drop mass. The gravity force with a small compensation by the Archimedes force is written as $\mathbf{F_g} = \frac{4}{3}\pi(\rho_w - \rho)r^3\mathbf{g}$, where $\mathbf{g}$ is the gravity acceleration, r is the equivalent drop radius (Section 2.3), and ρ_w and ρ are densities of the water and the environmental air, respectively. The drag force depends on drop velocity relative to the air and is written as $\mathbf{F_d} = -\frac{1}{2}\rho SC_d \mid \mathbf{V} - \mathbf{u} \mid (\mathbf{V} - \mathbf{u})$, where $\mathbf{V} - \mathbf{u}$ is the relative drop velocity and S is the area of cross section of the equivalent drop, i.e., of a spherical drop having the same cross section as a real drop. C_d is the drag coefficient, which is a function of the drop Reynolds number Re:

$$Re = \frac{2r\rho}{\mu} \mid \mathbf{V} - \mathbf{u} \mid , \qquad (5.5.2)$$

where μ is air dynamic viscosity. The drag coefficient is $C_d = \frac{24}{Re}[1 + f(Re)]$ (Pruppacher and Klett, 1997). The Reynolds number characterizes airflow regime around a drop. For cloud droplets of radii smaller than 20 μm, the surrounding flow is laminar, $Re \ll 1$, and the function $f(Re)$ is close to zero. The largest drops, $Re \gg 1$, and the surrounding flow can be turbulent. While falling, large drops create air vortices behind them, which leads to a rapid increase in $f(Re)$ and in the drag coefficient. Other forces are usually not taken into account when analyzing drop motion in cloud (such as the added mass force caused by acceleration of the surrounding air, or the Basset history force caused by the lagging boundary layer development leading to changes in the drop relative velocity). The error related to the omission of the added mass force is of the order of $\rho/\rho_w = 10^{-3}$. The Basset history force can be important in local zones where accelerations in turbulent flow are strong: $\frac{du}{dt} \gg g$. Such accelerations, however, are rare in real clouds.

Equations (5.5.1) and (5.5.2) allow rewriting the drop motion equation in the form

$$\frac{d\mathbf{V}}{dt} = -\frac{1}{\tau_d}(\mathbf{V} - \mathbf{u}) + \left(1 - \frac{\rho}{\rho_w}\right)\mathbf{g}, \qquad (5.5.3)$$

where τ_d is the *drop relaxation time* characterizing drop inertia:

$$\tau_d = \frac{2\rho_w r^2}{9\mu}[1 + f(Re)] \qquad (5.5.4)$$

The drop relaxation time increases with increasing drop size. For instance, $\tau_d = 10^{-3}$ s for small 3–μm radius cloud droplets and $\tau_d = 1$ s for large rain drops.

5.5.2 Sedimentation of Drops in Calm Air

Drops falling in calm air reach their terminal fall velocity $V_g = const$ during the time period of order of drop relaxation time τ_d (P.K. Wang and Pruppacher, 1977). This velocity is an important parameter since it is used in numerous cloud models for calculation of drop collisions, rain fluxes, and other cloud characteristics. The terminal velocity depends on a) drop size, b) on the degree of drop deformation that increases with increasing drop size (Section 2.3), and c) on the density and the temperature of the environmental air. The terminal velocity can be found from Equations (5.5.3) and (5.5.4) assuming the balance between gravity and drag forces under conditions $\frac{d\mathbf{V}}{dt} = 0$ and $\mathbf{u} = 0$:

$$V_g = \tau_d g\left(1 - \frac{\rho}{\rho_w}\right) = \frac{2\rho_w r^2 g}{9\mu}[1 + f(Re)]\left(1 - \frac{\rho}{\rho_w}\right) \qquad (5.5.5)$$

For cloud droplets with $r < 20\ \mu\text{m}$ (called also the Stokes droplets) the following approximation formulas are often used to calculate the parameters of droplet sedimentation:

$$\tau_d = \frac{2\rho_w r^2}{9\mu};\ V_g = \frac{2\rho_w r^2 g}{9\mu};\quad Re = \frac{4\rho\rho_w r^3 g}{9\mu^2}. \tag{5.5.6}$$

As follows from Equation (5.5.6), the relaxation time and the terminal velocity of such droplets are proportional to the square of droplet radius, while the Reynolds number is proportional to the cube of droplet radius.

Many approximations of terminal-velocity dependence on drop radius have been proposed (e.g., Pruppacher and Klett, 1997; Khain and Pinsky, 1995). The most precise approximation that takes into account drop deformation, dependence on the air temperature and density, as well as other factors, is proposed by Beard (1976). The dependencies $V_g(r)$, $\tau_d(r)$ and $Re(r)$ calculated using the expressions introduced by Beard (1976) are shown in **Figures 5.5.1(left), 5.5.1(right), 5.5.2**, and **5.5.3**, respectively.

In many cloud models, dependence of the terminal velocity on drop radius and on the thermodynamic factors can be written as a product of two dependencies, namely, on drop radius and on air density. For instance, Khain and Sednev (1995) approximate the dependence of the terminal velocity on drop radius by a piecewise function:

$$V_g = \begin{cases} 1.19 \times 10^6 r^2 & r < 40\ \mu\text{m} \\ 8.00 \times 10^3 r & 40\ \mu\text{m} \le r < 600\ \mu\text{m} \\ 1.78 \times 10^3 r^{1/2} & 600\ \mu\text{m} \le r < 2500\ \mu\text{m} \\ 890 & r \ge 2500\ \mu\text{m} \end{cases}, \tag{5.5.7}$$

where the velocity is in cm/s and the drop radius is in cm. Figure 5.5.1 demonstrates the difference between the precise formula (Beard, 1976) and the approximation (Equation 5.5.7). In bulk microphysics models, the terminal velocity of cloud droplets is usually neglected. The dependence of raindrops terminal velocity on their radius is approximated by the power law $V_g = \alpha r^\beta$ (Straka, 2009), or by a more complicate formula $V_g = \alpha r^\beta \exp(-\gamma r)$ (Ferrier, 1994). Figure 5.5.1 shows dependencies of the terminal velocity on the raindrop radius used in different bulk models. The dependence of V_g on the thermodynamic parameters is usually approximated by the factor $\left(\frac{\rho_0}{\rho}\right)^{1/2}$ describing the effects of the air-density decrease with height. The terminal velocity can increase with height by 50% or more due to the air-density decrease and the corresponding drag-force decrease.

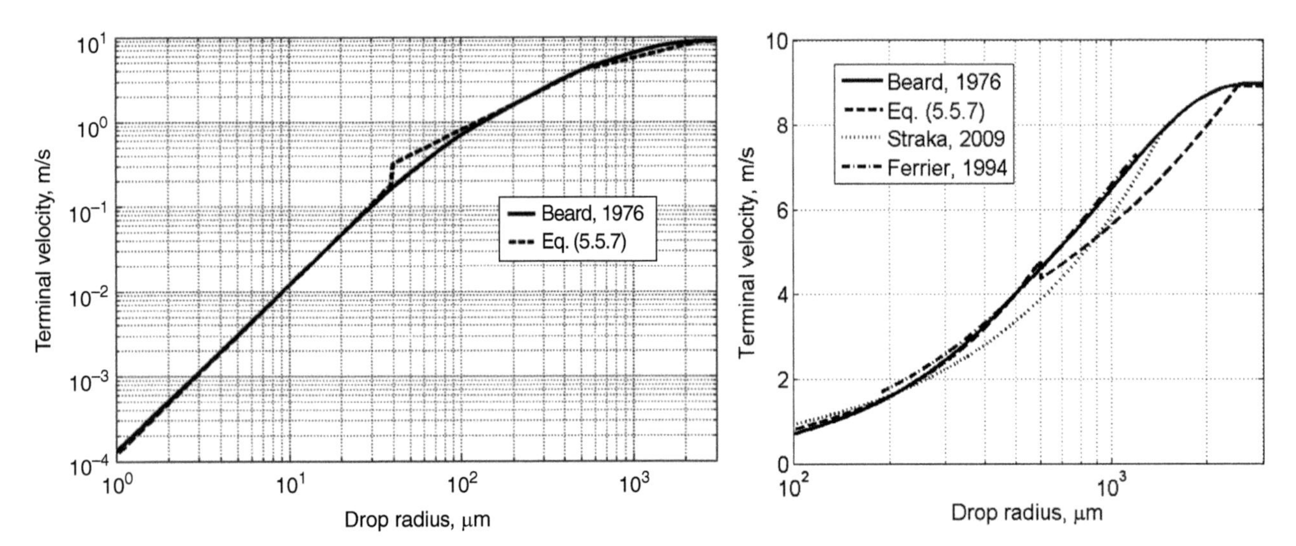

Figure 5.5.1 Dependencies of the terminal velocity on drop radius according to Beard (1976) at $T_C = 20°\text{C}$ and $\rho = 1.2\ \text{kg/m}^3$ (solid line) and using different approximations (dashed and dotted lines). The left panel presents the dependence within a wide range of drop radii in log–log coordinates; the right panel is related to raindrops and plotted in log–linear coordinates.

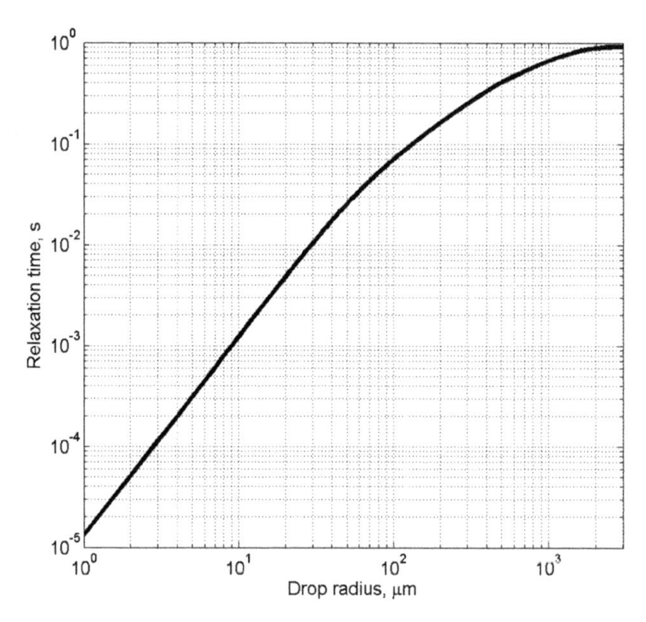

Figure 5.5.2 Dependence of the drop relaxation time on the equivalent drop radius, calculated according to Beard (1976) at $T_C = 20°C$ and $\rho = 1.2\ kg/m^3$.

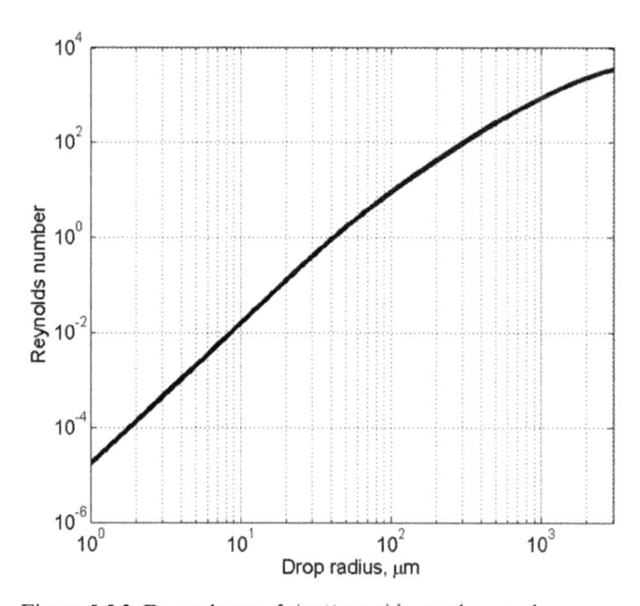

Figure 5.5.3 Dependence of the Reynolds number on the equivalent drop radius, calculated according to Beard (1976) at $T_C = 20°C$ and $\rho = 1.2\ kg/m^3$.

5.5.3 Representation of Drop Sedimentation and Rain Flux in Cloud Models

Drop sedimentation is one of the most important microphysical processes determining precipitation and affecting collisions and vertical distribution of drops of different sizes (so-called *size sorting*). Treatment of drop sedimentation in numerical cloud models depends on the microphysical schemes used. In SBM models, sedimentation of drops belonging to each bin is calculated separately, by means of two alternative approaches. In the first approach, drop sedimentation is calculated together with advection at the same time sub-step (e.g., Khain and Sednev, 1996). In the second approach, drop advection is calculated first and the sedimentation equation is solved at the next sub-step. The equation describing sedimentation has the form

$$\frac{\partial \rho f_k^*}{\partial t} = \frac{\partial V_{g,k} \rho f_k^*}{\partial z}, \tag{5.5.8}$$

where f_k^* is the value of the SD function for the k-th bin obtained after the first sub-step (advection). Actually, f_k^* is liquid water mixing ratio (in units kg/kg) of the k-th bin. Equation (5.5.8) has the form of the 1D advection equation and can be solved by methods described in Section 4.2. At the top of the computational area, the condition $f_k^* = 0$ is usually assumed, while the condition $\frac{\partial \rho f_k^*}{\partial z} = 0$ is often used at the surface (the bottom of the computational area). Application of the CFL stability criteria (Section 4.2) may require utilization of small time steps if large raindrops sediment is treated. The rain rate at the surface (in units of $kg\ m^{-2}\ s^{-1}$) is calculated as $\sum_k \rho V_{g,k} f_k^*$. Equation (5.5.8) is typically solved by the explicit upstream difference scheme (Equation 4.2.2) or by a box–Lagrangian scheme with properties close to those of the upstream difference scheme (Kato, 2005).

The process of drop sedimentation in bin models is illustrated in **Figure 5.5.4**. Assume, for example, that at some time instance there exists a DSD at height level h and drops belonging to this DSD begin falling with their terminal velocity. During one time step Δt, drops belonging to the k-th bin fall by distance $V_{g,k}\Delta t$. Larger drops fall covering longer distances. As a result, at each height level drops have different sizes: larger drops located at lower levels, while smaller drops remain at high levels. This way, the vertical profile of drop concentration forms (Figure 5.5.4). This phenomenon of stretching of DSD in the vertical direction according to drop sizes is called size sorting. In addition, due to viscosity of the finite difference scheme, drop concentration profile becomes broader in the vertical direction. Since for sake of simplicity the fall velocities of drops belonging to the same mass bin are assumed identical, a limited number of bins may form an intermittent vertical structure of concentration of drops with some spatial intervals between drops belonging to

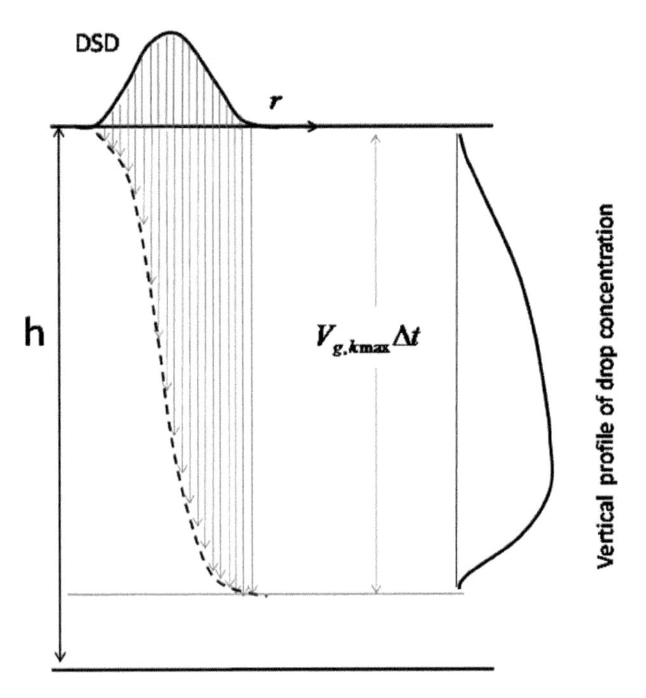

Figure 5.5.4 Illustration of drop sedimentation process leading to size sorting in bin schemes. Initially, drops having a given DSD are located at height h. During one time step Δt droplets belonging to the k-th bin fall by distance $V_{g,k}\Delta t$. Larger drops fall faster and thus cover longer distances (dashed curve). The largest drops belonging to the $k = k_{\max}$ bin cover the maximum distance $V_{g,k\max}\,\Delta t$. After several time steps, drops of different sizes are located at different height levels, forming the vertical profile of drop concentration. This profile describing size sorting is shown by the solid curve.

neighboring bins. To provide continuous drop concentration distribution in the vertical, the droplet fall velocity should be, in an ideal case, distributed within each bin. The utilization of viscous numerical schemes eliminates this inhomogeneity, the spectral bin models with a few tens of bins and comparatively high vertical model resolution reproduce sedimentation and size sorting quite accurately.

In bulk-parameterization models, sedimentation of cloud droplets is usually neglected. Sedimentation of raindrops having SD $f(m)$ is described by equations for sedimentation of the moments of $f(m)$:

$$\frac{\partial M^i}{\partial t} = \frac{\partial \overline{V}_g M^i}{\partial z}, \tag{5.5.9}$$

where $N = M^0 = \int_{m_{\min}}^{m_{\max}} f(m)dm$ is the zero moment representing the total drop concentration and $M^1 = \int_{m_{\min}}^{m_{\max}} mf(m)dm$ is the first moment representing the total drop mass content. For M^0, $\overline{V}_g$ is concentration-weighted terminal drop velocity (Equation 3.2.14). For M^1, $\overline{V}_g = \overline{V}_{g\_m}$ is mass-weighted terminal drop velocity (see Equation 3.2.13).

In a one-moment bulk scheme, sedimentation is performed for the first moment, so that only the mass-weighted velocity is used. This approach assumes that all raindrops fall at the same averaged velocity. The rain rate at the surface is calculated as $\rho \overline{V}_{g\_m} q_l^*$. In one-moment bulk schemes, the concentration of particles is fully determined by the mass content, so the concentration settles as fast as drop mass, which does not agree with the mechanism of size sorting illustrated in Figure 5.5.4.

In two-moment bulk-parameterization schemes, drop sedimentation is described using two equations, one for drop concentration and the other for drop mass. Both the mass-weighted and concentration-weighted velocities are used. The concentration-averaged sedimentation velocity $\overline{V}_g$ is less than the mass-averaged velocity $\overline{V}_{g\_m}$. This difference reflects the fact that smaller drops fall slower than larger ones. Sometimes, the difference between these sedimentation velocities leads to a numerical problem, when in some cloud zones only the mass exists at negligible concentration, while in other cloud regions only drop concentration exists, while mass is negligible. To avoid these problems, some limiting constrains are added into the schemes. For instance, an important parameter of two-moment bulk-parameterization schemes is the mean volume radius of drops. It is determined by the ratio of the total drop mass to the number concentration. If the value of this radius falls outside the range of reasonable values, the value of the radius is artificially changed to shift it into the range. In more detail, the problem of size sorting and the ways to improve representation of particle sedimentation in bulk-parameterization schemes is considered in Section 6.4, where sedimentation of hydrometeors in mixed-phase clouds is discussed.

Implementation of the averaged raindrop velocity may substantially affect both the amount and the time evolution of calculated precipitation. The problem arising in reproduction of duration and intensity of precipitation when mass-averaged raindrop fall velocities are used, is schematically illustrated in **Figure 5.5.5.** Assume for example that the air vertical velocity in a cloud increases (the stage of cloud development) and then decreases (the stage of cloud decay). Since the largest precipitating drops fall with velocity of 10 m/s, they are able to fall down even within a comparatively strong updraft. As a result, both in real clouds and in bin models, precipitation at the surface takes place when the updraft is relatively high. The smallest drops fall slowly and form a zone of weak precipitation when the vertical velocity in the cloud becomes negligible.

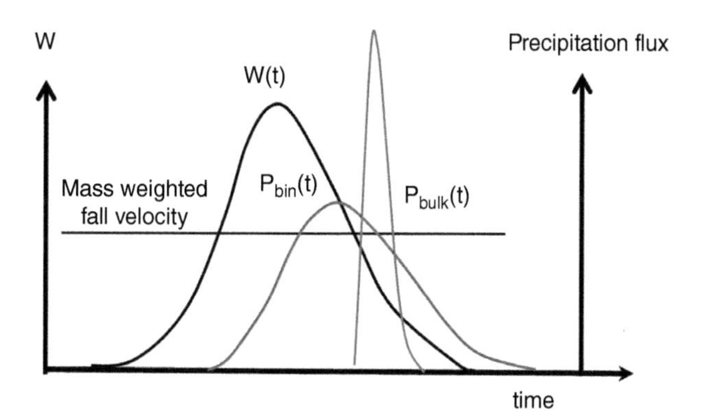

Figure 5.5.5 A schematic representation of time dependencies of the vertical velocity (black line) and of precipitation fluxes in a bin scheme, $P_{bin}(t)$ (red), and in a bulk scheme, $P_{bulk}(t)$ (blue). The mass-weighted fall velocity is shown by the horizontal straight line. Left vertical axis denotes the vertical velocity; the right vertical axis denotes the precipitation fluxes. Precipitation in the bin scheme starts earlier and ends later than in the bulk scheme. The maximum precipitation rate is higher in the bulk scheme.

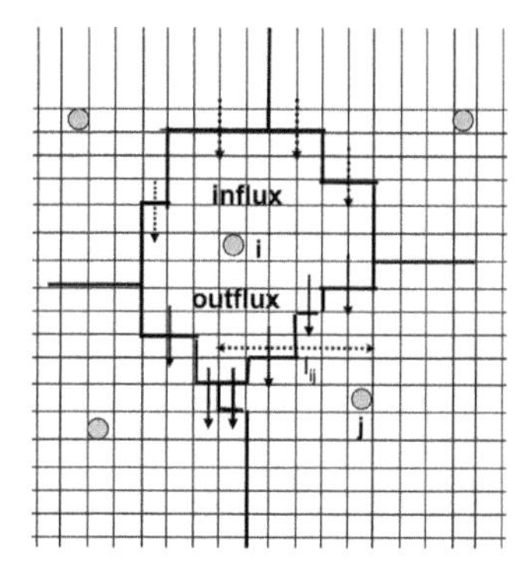

Figure 5.5.6 Illustration of the sedimentation procedure used by Pinsky et al. (2008). The circles denote parcels' centers. Arrows denote drop fluxes through the interface (from Pinsky et al., 2008; American Meteorological Society©; used with permission).

At the same time, precipitation in bulk-parameterization schemes starts only when the updraft velocity becomes lower than the mass-weighted fall velocity of precipitating hydrometeors. Since all rain drops fall at the same velocity, the precipitation rate is intensive, but the duration of heavy precipitation is short. Although Figure 5.5.5 represents only a qualitative conceptual scheme, simulations with different bulk models indicate that these schemes indeed tend to overestimate the intensity of strong precipitations (convective rain) and underestimate the rate of low stratiform precipitation, as well as the area covered by stratiform (weak) precipitation. Spatial distribution of convective and stratiform zones is an important characteristic of precipitating systems since it affects both precipitation amounts and radiative properties of clouds. The examples of storm simulations obtained using SBM and bulk schemes are presented in Section 7.2.

In advanced bulk schemes (e.g., RAMS, Walko et al., 1995), a bin-emulating procedure for drop sedimentation is used. The DSD given in the form of Gamma distribution and obtained after advection is separated into several tens of bins, and sedimentation is performed using an equation similar to Equation (5.5.8). It is assumed that the drop SDs after drop sedimentation can still be approximated by the Gamma distribution or the exponential distribution. After the sedimentation procedure is completed, the parameters of these distributions are recalculated using the total mass and the total concentrations at model grid points. The advantages and disadvantages of bin-emulating procedures are discussed in more detail in Section 6.4.

In some models, the processes of diffusion growth and collisions are simulated within the Lagrangian framework, but the sedimentation is performed within the Eulerian coordinate framework. Pinsky et al. (2008b) extended the flux method used for description of advection and sedimentation in the Eulerian models with regular finite difference grids (Chapter 4) to irregular grid formed by the centers of the parcels, located randomly within the modeled volume. The procedure is illustrated in **Figure 5.5.6**. The entire computational zone is covered with a regular mesh with grid spacing of 1 m to 5 m (depending on the model version). The interface between the parcels is designed in such a way that the corresponding Eulerian cell belongs to the parcel with the nearest center. Such combination of Lagrangian and Eulerian approaches is known as the hybrid method.

5.5.4 The Ventilation Effect

The movement of drops relative to the air increases the flux of vapor molecules from the supersaturated air upon a drop-in case of supersaturation, as well as the flux of water molecules from a drop into the air in case of undersaturation as compared with the case of unmovable drop. In cloud physics this effect is called the *ventilation effect*. Numerical simulations of ventilation effect were performed by Woo and Hamielec (1971). They found that the ventilation effect is strongest, i.e., evaporation/condensation is the highest on the upstream side of the falling drop. The increase in the evaporation/condensation rate over the entire drop surface is described by the ventilation coefficient $F_v \geq 1$, which is the ratio of the mass growth rate of a falling drop to that of motionless drop:

$$F_v = \frac{dm/dt}{(dm/dt)_{|\mathbf{V}-\mathbf{u}|=0}} \tag{5.5.10}$$

The ventilation coefficient is used in the kinetic equation of diffusion growth/evaporation (Table 5.1.1). According to Pruppacher and Klett (1997), the ventilation coefficient depends on two dimensionless characteristic numbers. The first is the Reynolds number (Re) determined in Equation (5.5.2). The second is the Schmidt number (Sc), which is the ratio of the kinematic viscosity to the molecular diffusion coefficient:

$$Sc = \frac{\mu}{\rho D}, \tag{5.5.11}$$

where D is the diffusivity of water vapor in the air. Basing on the theory of the boundary layer arising around the falling drop, the ventilation coefficient can be considered as function of $Z = Sc^{1/3} \times Re^{1/2}$ (Pruppacher and Klett, 1997). The parameterization formula for the ventilation coefficient is

$$F_v = \begin{cases} 1.00 + 0.108Z^2 & Z < 1.4 \\ 0.78 + 0.308Z & 1.4 \le Z \le 51.4 \end{cases} \tag{5.5.12}$$

The first expression in Equation (5.5.12) is valid for small drops with radii below about 60 μm. The second expression is valid for drops with radii ranged from 60 μm to 2,500 μm. The dependencies $F_v(Z)$ is presented in **Figure 5.5.7**. The dashed line is the dependence proposed by Pruppacher and Klett (1997). Results of laboratory experiments are shown in Figure 5.5.7 by symbols. Pruppacher and Klett (1997) attributed deviations of results of laboratory experiments performed by Kinzer and Gunn (1951) from the dashed line by the limited accuracy of their experiments. In general, the results of laboratory measurements and parameterization (5.5.12) are close. The deviations of the results of calculations and the laboratory experiments can be attributed to the deformation of large drops which was not taken into account in the simulations.

Using dependencies of fall velocity on drop radius shown in Figure 5.5.1, one can represent dependence of ventilation coefficient on drop radius $F_v(r)$. Such dependence calculated for $T_C = 20°C$ is shown in **Figure 5.5.8.** It is clear from Figure 5.5.8 that the ventilation effect perceptibly affects the process of condensation/evaporation for raindrops with radii exceeding

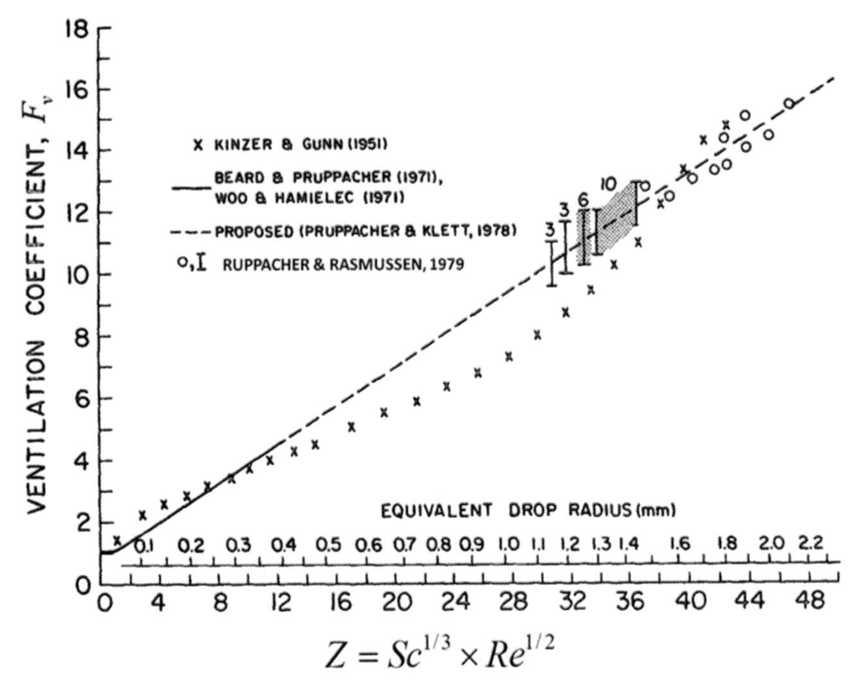

Figure 5.5.7 The dependencies of ventilation coefficient $F_v(Z)$. The dependencies $F_v(Z)$ calculated for large Re are extrapolated for moderate values of Re. The results of laboratory experiments are denoted by symbols. Drop size range in laboratory experiments by Pruppacher and Rasmussen (1979) is $1,100\ \mu m \le r_0 \le 2,500\ \mu m$ ($29.7 \le Z \le 51.4$). Drop size range in laboratory experiments by Beard and Pruppacher (1971) is $60\ \mu m \le r_0 \le 400\ \mu m$ ($1.4 \le Z \le 12.3$) (from Pruppacher and Rasmussen, 1979; American Meteorological Society©; used with permission).

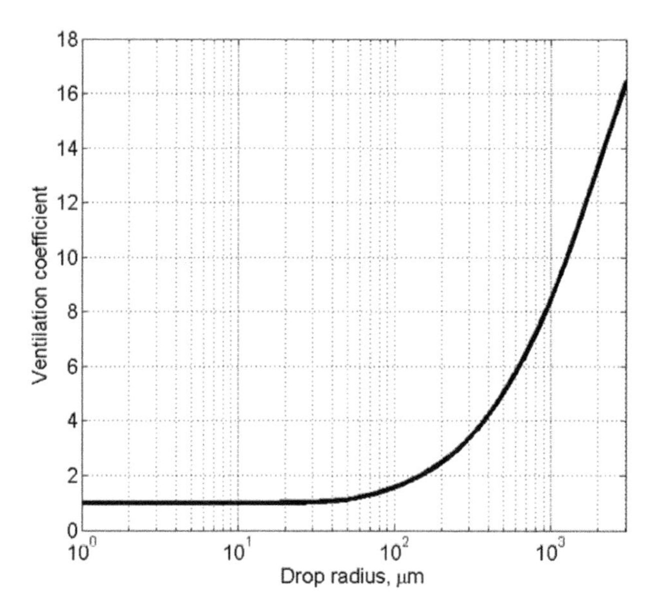

Figure 5.5.8 Dependence of the ventilation coefficient on the equivalent drop radius, calculated using the Beard (1976) approximation of Re at $T_C = 20°\mathrm{C}$ and $\rho = 1.2\ \mathrm{kg/m^3}$.

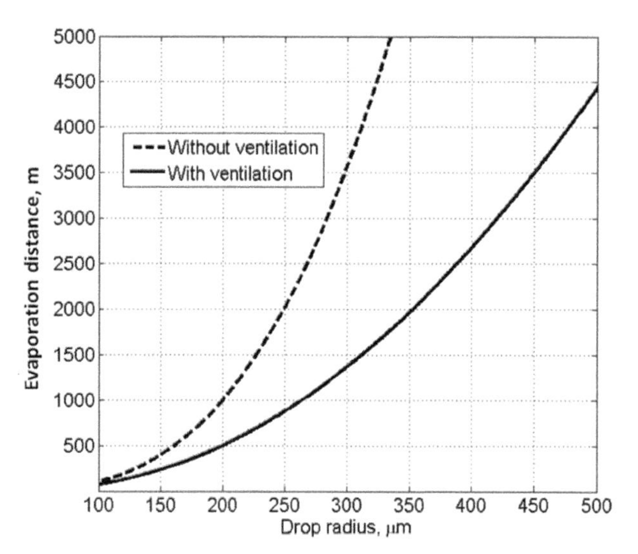

Figure 5.5.9 Dependencies of the total evaporation distance on the initial raindrop radius ($RH = 80\%$).

100 μm, and thus can be important for evaluation of rain characteristics.

Figure 5.5.9 compares the dependence of the total evaporation distance on the initial radii of raindrops, with and without the ventilation effect taken into account. In calculations the formula for fall velocity of Beard (1976) (Figure 5.5.1) was used. The comparison shows that the ventilation effect can decrease the evaporation distance several times. Thus, the ventilation effect should be taken into account to reproduce rain rates correctly.

5.5.5 Drop Oscillations and Spontaneous Breakup

During their fall, rain drops oscillate incessantly changing their shapes. These oscillations were investigated in several laboratory and theoretical studies (e.g., Pruppacher and Klett (1997) and Szakáll et al. (2010)). The amplitude and the frequency of the oscillations were found to be dependent on the equivalent drop radius. **Figure 5.5.10** shows dependencies of amplitude and frequency of drop oscillations on the equivalent drop diameter, obtained in different studies. With drop diameter increasing, the amplitude of the oscillations increases and the frequency decreases. Although these oscillations can somehow influence drop collisions due to differences in the effective cross section of drops, this phenomenon is usually taken into account in cloud models indirectly via a decrease in fall velocity and an increase in the probability of spontaneous breakup.

When rain drops' radii reach 1–2 mm they become flattened (Section 2.3), the amplitude of their oscillations increases, and they become unstable. This hydrodynamic instability leads to a so-called *spontaneous breakup* of large raindrops into several smaller drops (Pruppacher and Klett, 1997). The process of spontaneous breakup observed in a laboratory experiment is illustrated in **Figure 5.5.11.**

The size of the largest drop that still remains stable is determined by the balance of the drag force and the surface tension: when the drag stress exceeds the surface tension stress, the drop breaks up. The maximal stable drop diameters reported by different authors are about 6–8 mm. Drops of smaller size do not break spontaneously. Spontaneous breakup was investigated in several laboratory studies (e.g., Komabayasi et al., 1964; Kamra et al., 1991). They found maximum raindrop diameter to be about 8.2 mm. Super-large raindrops with diameters exceeding 8 mm were detected in winter clouds during *in situ* measurements (Szakáll et al., 2010). Such drops can efficiently form as a result of melting of large ice particles such as snowflakes and hail. The existence of super-large raindrops with maximum dimensions of at least 8.8 mm and possibly up to 1 cm was also reported by Hobbs and Rangno (2004). In turbulent flow, the maximum diameter of super-large raindrops may decrease to about 5 mm.

Spontaneous breakup is an important mechanism of formation of specific shapes of rain SD, for instance, the exponential (or Marshall–Palmer) RSD (Section 2.3). The changes of DSD in the course of spontaneous

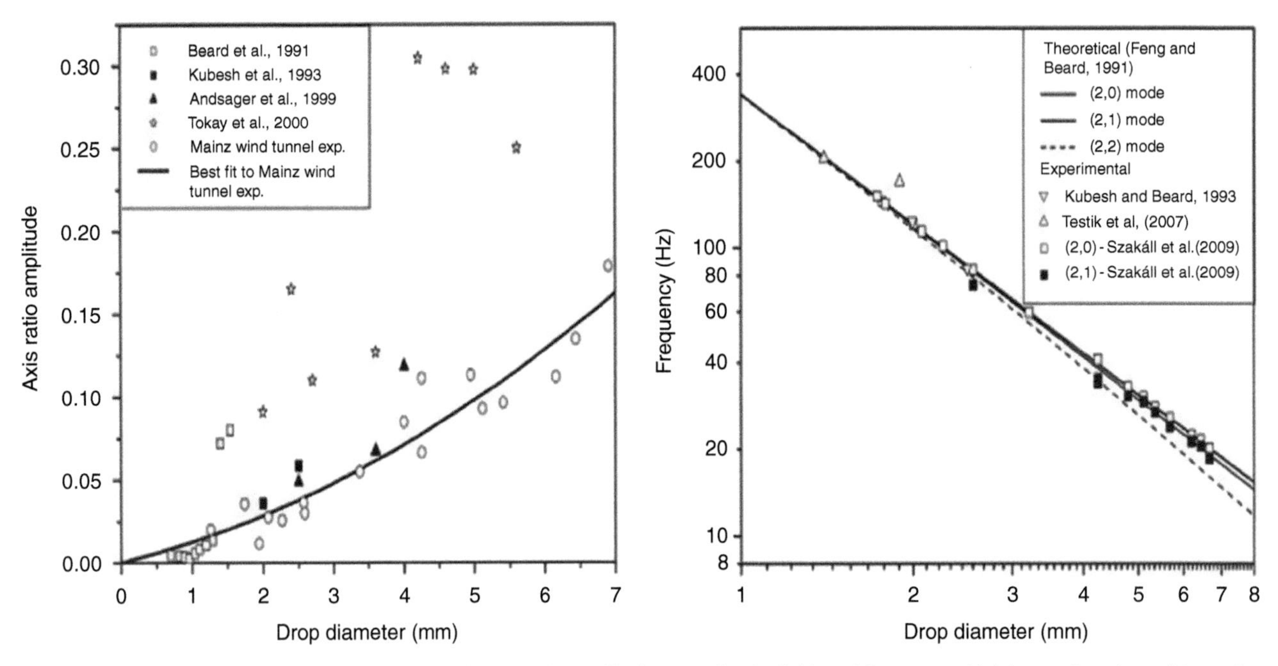

Figure 5.5.10 Theoretical and experimental findings on the oscillation amplitude (left) and frequency (right) as a function of drop size (from Szakáll et al., 2010; with permission from Elsevier).

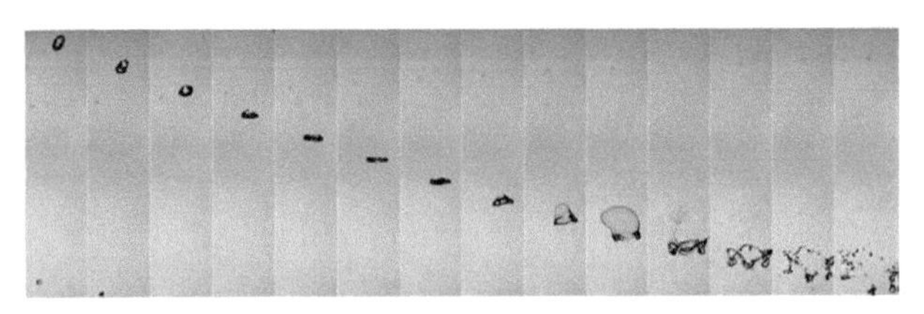

Figure 5.5.11 Series of events in the fragmentation of a 6 mm diameter water drop falling in an ascending air stream. The time interval between the images is t = 4.7 ms. One can see that drop breakup leads to formation of many smaller drops of different size (from Villermaux and Bossa, 2009; courtesy of Macmillan Publishers Ltd.).

breakup are determined by two factors: the probability of a spontaneous drop breakup and the subsequent mass distribution (MD) of drop fragments. In cloud models with spectral bin microphysics, a spontaneous breakup is described by the following kinetic equation (Pruppacher and Klett, 1997):

$$\left.\frac{\partial f(m)}{\partial t}\right|_{sp\_breakup} = -f(m)P_r(m) + \int_m^\infty f(m')P_r(m')Q(m,m')dm', \quad (5.5.13)$$

where $P_r(m)$ is the probability of a spontaneous breakup of a raindrop of mass m, and $Q(m,m')$ is the SD function for the drop fragments formed as a result of breakup, which can be interpreted as a conditional distribution showing the probability of formation of drops with mass m as a result of a spontaneous breakup of a drop with mass m'. The first term on the right-hand side of Equation (5.5.13) describes the rate of decrease of $f(m)$ due to the breakup of drops with mass m. The second term on the right-hand side shows the increase of $f(m)$ due to the breakup of other drops with masses $m' > m$.

Based on the laboratory data obtained by Komabayasi et al. (1964), Srivastava (1971) proposed the following expression for the probability of breakup (in s^{-1}):

$$P_r(m) = 2.94 \times 10^{-7} \exp(3.4r), \quad (5.5.14)$$

where r is the drop radius in mm, $P_r(m)$ is in s^{-1}. According to Equation (5.5.14), the probability of a spontaneous breakup rapidly increases with increasing drop size. Laboratory experiments performed later by Kamra et al. (1991) showed a somehow smaller probability of a spontaneous breakup of the largest raindrops. **Table 5.5.1** shows averaged life time (averaged time before breakup) and the probability of breaking obtained in laboratory measurements of spontaneous

Table 5.5.1 Average lifetime and the probability of spontaneous drop breakup (from Kamra et al., 1991; courtesy of John Wiley & Sons, Inc. ©).

Drop diameter, mm	6.6	7.1	7.6	8.0	8.2
Averaged lifetime, s	48.8	34.6	25.9	18.7	14.4
Probability of breakup $\times 10^{-2}$, s^{-1}	2.05	2.89	4.01	5.34	6.90

breakup of raindrops suspended over a vertical wind tunnel (Kamra et al., 1991). The size dependence of raindrops' lifetime s^{-1} was approximated by Kamra et al. (1991) as $\tau = 6.34 \times 10^3 \exp(-1.466r)$, where r is drop radius in mm. Since $P_r = \tau^{-1}$, the expression for the probability of a spontaneous breakup of drops within the diameter range 6.6 mm to 8.2 mm is

$$P_r = 0.155 \times 10^{-3} e^{1.466r} \tag{5.5.15}$$

Considering mass conservation laws, the sum of mass fragments resulting from a breakup of a single raindrop should be equal to the mass of this raindrop prior to the breakup:

$$m' = \int_0^{m'} mQ(m, m')dm \tag{5.5.16}$$

Here the units of the SD of drop fragments $Q(m, m')$ are g^{-1}. The number of drop fragments increases with the size of the initial drops and can exceed several tens and even a few hundred. In many numerical studies (e.g., Hall, 1980; Flossmann and Pruppacher, 1988; Kogan, 1991) the maximum drop radius of 2.58 mm is assumed. A drop of the maximum size breaks up due to hydrodynamic instability, producing fragments with MD shown in **Table 5.5.2**. One can see from the table that the SD of the drop fragments is rather wide, with the maximum fragment size of about five times lower than the drop size before the breakup.

Using the laboratory data obtained by Komabayasi et al. (1964), Srivastava (1971) proposed the following expression for $Q(m, m')$:

$$Q(m,m') = 145.37 \cdot m^{-1}\left(\frac{r}{r'}\right)\exp\left(-7\frac{r}{r'}\right) \tag{5.5.17}$$

In Equation (5.5.17) r' is the radius of a drop of mass m' before breakup. This expression obeys the condition (5.5.16). Later laboratory experiments performed by Kamra et al. (1991) suggest that the SD of drop fragments can be approximated by the exponential distribution $N_{frag}(D) \sim \exp(-\lambda_{sb}D)$, where drop diameters D are in mm and $\lambda_{sb} = 0.453\ \mathrm{mm}^{-1}$. Since the distribution of drop fragments should obey the condition (5.5.16), their exponential distribution can be written as

Table 5.5.2 MD of fragments of a 2,580 μm-radius drop (in percentage of the original mass) after a breakup caused by hydrodynamic instability (from Hall, 1980 courtesy of John Wiley & Sons, Inc. ©).

Drop Radius (μm)	%	Drop Radius (μm)	%	Drop Radius (μm)	%
143.7	0	406.4	5.5	1,149.0	4.0
161.3	0.2	456.1	7.5	1,290.0	2.5
181.0	0.3	512.0	9.0	1,625.0	1.5
203.2	1.5	574.7	10.0	1,825.0	1.0
228.1	1.0	645.1	12.0	2,048.0	0.5
256.0	1.5	724.1	10.0	2,299.0	0.3
287.4	2.0	812.7	9.0	2,580.0	0.2
322.5	2.5	912.3	7.5	2,600.0	0
362.0	4.0	1,024.0	5.5		

$$N_{frag}(D) = N_{sb}\exp(-\lambda_{sb}D), \tag{5.5.18}$$

where the value N_{sb} can be written as

$$N_{sb} = \lambda_{sb}^4 D'^3 [6 - \exp(-\lambda_{sb}D') \{\lambda_{sb}^3 D'^3 + 3\lambda_{sb}^2 D'^2 + 6\lambda_{sb}D' + 6\}]^{-1}, \tag{5.5.19}$$

where D' is the diameter of the initial raindrop supposed to breakup. Despite the differences in the breakup probabilities of largest raindrops, reported by Komabayasi et al. (1964) and Kamra et al. (1991), the effects of a spontaneous breakup on the PSD, computed employing the bin-mirophysics model, turned out to be quite similar.

Figure 5.5.12 compares fields of radar reflectivity from liquid drops in a Cb, obtained in simulations with HUCM with and without a spontaneous breakup using the formulas by Kamra et al. (1991). One can see a significant decrease in the area of high reflectivity from raindrops in the simulation with a spontaneous breakup. These reflectivity values are realistic, while in the case without a spontaneous breakup they are obviously overestimated.

Another mechanism of raindrop breakup is interdrop collisions (a collisional breakup), which is more efficient than the spontaneous breakup in case of frequent collisions (see Section 5.6).

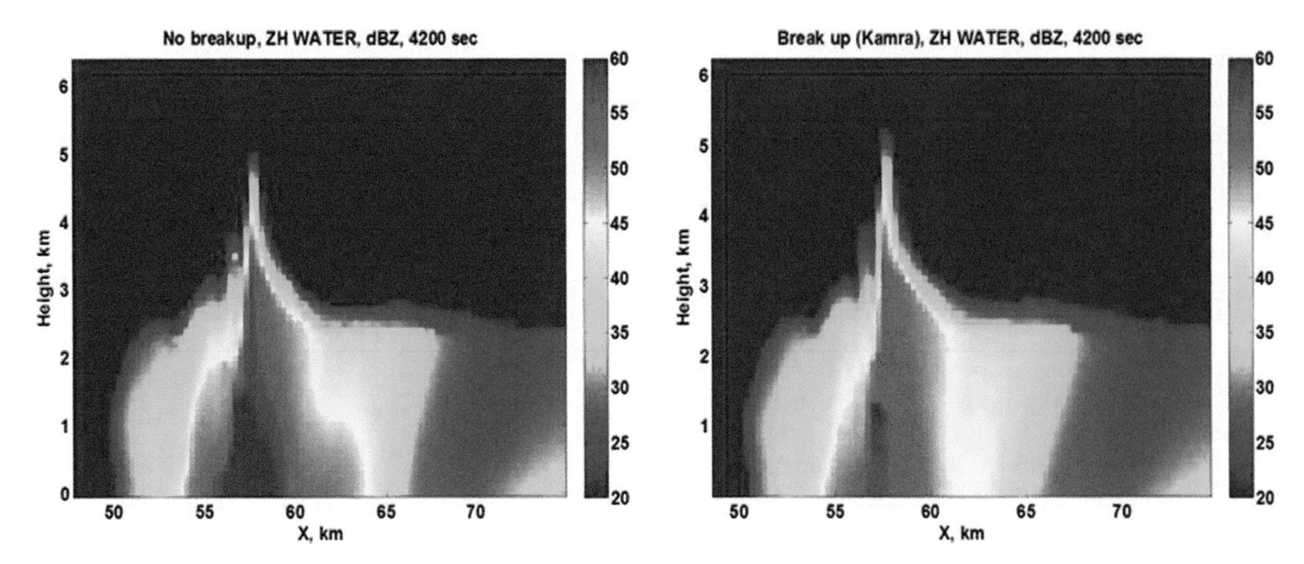

Figure 5.5.12 Fields of radar reflectivity from liquid drops in a Cb simulated by HUCM without a spontaneous breakup (left) and with a spontaneous breakup taken into account according to a study by Kamra et al. (1991) (right).

5.5.6 Drop Motion in Nonuniform and Turbulent Flows

Investigations of drop motion in different nonuniform and turbulent air flows were carried out in many studies (e.g., Manton, 1977; Maxey, 1987; Fung, 1993; Khain and Pinsky, 1995; Pinsky and Khain, 1996, 1997a, 1997b). The inertia of drops moving in a nonuniform flow leads to deviations of drop velocity from the air velocity. In contrast to drop motion in a uniform flow, the velocity of drops falling in a nonuniform flow does not tend to its terminal value with time, and thus eventually deviates from the terminal velocity. The mechanism of this deviation in a linear shear flow and in a curved flow is shown in **Figure 5.5.13**.

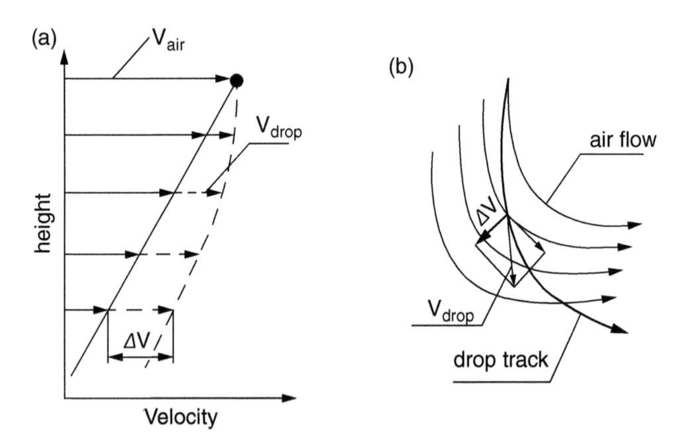

Figure 5.5.13 Formation of drop velocity deviation from the surrounding air velocity within a linearly sheared air flow (a) and a curved flow (b). In (a), the drop velocity deviations are determined by the tendency to conserve the drop momentum in the horizontal direction. In (b), the drop velocity deviations are induced by the centrifugal forces (from Pinsky and Khain, 1997b; courtesy of John Wiley & Sons, Inc. ©).

For low-inertia cloud droplets moving in a nonuniform flow, the relative velocity tends to the quasi-steady value that depends on flow shears and drop mass (Khain and Pinsky, 1995). For heavier drops, the motion is more complicated. For example, due to the centrifugal forces arising in the air vortexes, inertial drops follow an untwisted spiral trajectory and as a result may leave the vortex area (Fung, 1993). This effect was called the sling effect (Falkovich and Pumir, (2007). This type of drop motion is illustrated in **Figure 5.5.14**.

The motion of drops in a turbulent flow is stochastic. Since the air velocities are turbulent, the velocity gradients and the air accelerations randomly change in space and time (Section 3.3), thus the drag force that influences the drop motion changes randomly, as well. It leads to highly complicated stochastic drop trajectories, to increasing frequency of inter-drop collisions and to a nonuniform spatial distribution of drops. The effect of drop inertia manifests itself in the stochasticity of the field of the relative-to-air drop velocity. Using Equation (5.5.3), the equation for the turbulence-induced drop velocity deviation along the drop track $\mathbf{V}' = \mathbf{V} - \mathbf{u} - \mathbf{V}_g$ can be written in the tensor form (Khain et al., 2007):

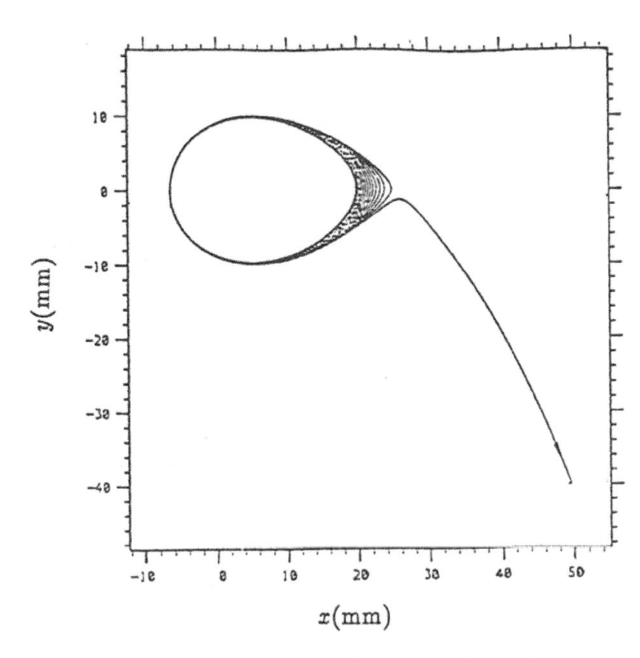

Figure 5.5.14 Typical dense particle (drop) trajectory in a Rankine vortex (from Fung, 1993; courtesy of John Wiley & Sons, Inc.©).

$$\frac{dV_i'(x_i,t)}{dt} = -V_j'\left[\frac{1}{\tau_d}\delta_{ij} + S_{ij}(x_i,t)\right] - \left[A_i(x_i,t) + V_g S_{i3}(x_i,t)\right], \quad (5.5.20)$$

where $S_{ij} = \frac{\partial u_i}{\partial x_j}$ is the air velocity shear tensor and $A_i = \frac{\partial u_i}{\partial t} + u_j \frac{\partial u_i}{\partial x_j}$ is the Lagrangian acceleration of the air in the vicinity of the drop. Equation (5.5.20) demonstrates that statistical characteristics of drop-velocity deviations depend on drop inertia and on two turbulent fields: the field of turbulent velocity shears and the Lagrangian acceleration field (Section 3.3). For small cloud droplets at $\tau_d S_{ij}(x_i,t) < 1$, the relative velocity tends to its local quasi-steady values:

$$V_i' = -\tau_d\left[A_i(x_i,t) + V_g S_{i3}(x_i,t)\right] \quad (5.5.21a)$$

In this case, the droplet velocity can be written as

$$V_i = u_i + V_g\delta_{i3} + V_i'. \quad (5.5.21b)$$

The physical meaning of Equation (5.5.12) is that droplet flux adjusts to the turbulent air flow with some velocity shift.

Motion of large cloud drops and small raindrops in a turbulent flow is more complicated than the motion of smaller droplets. It follows from Equation (5.5.21) that larger cloud drops can be locally accelerated by the drag

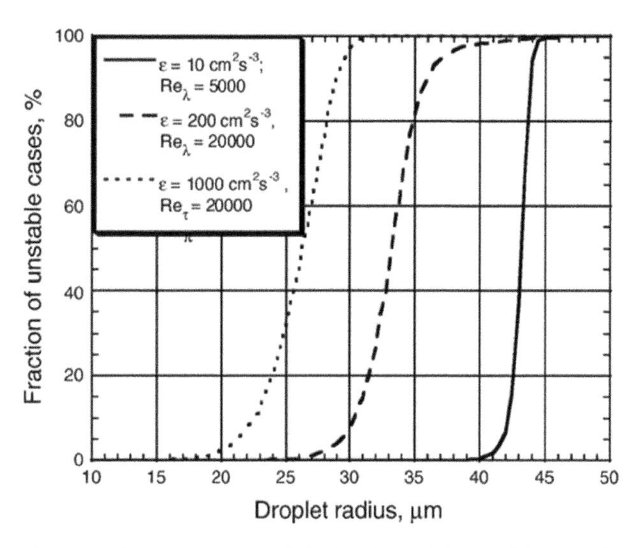

Figure 5.5.15 Dependence of the fraction of drops with unstable (non-stationary) trajectory on drop size under turbulent conditions typical of stratiform, cumulus, and cumulo-nimbus clouds. The definitions of ε and Re_λ are given in Section 3.3 (from Khain et al., 2007; courtesy of Elsevier).

force, leading to strong deviations from the air velocity (Pinsky et al., 2006; Khain et al., 2007). These drops do not adjust to the turbulent flow, and their movement becomes unstable. In order to evaluate the fraction of such drops in the total drop amount, multiple turbulent shear realizations were generated using the model developed by Pinsky et al. (2004). The dependencies of the fraction of unstable drops on drop radius under turbulent conditions typical of stratiform, cumulus, and cumulo–nimbus clouds are presented in **Figure 5.5.15**. The figure shows that drops with radii exceeding 45 µm do not adjust to a turbulent air flow in all types of clouds, and apparently experience strong velocity fluctuations. The analysis of motion of these drops requires application of Equation (5.5.20) in its full form. This strong impact of turbulent vortices leading to complicated drop-motion tracks can be expected for raindrops with radii smaller than about 250 µm. The motion of larger raindrops is simpler since these heavy drops do not respond to turbulent vortices and fall at their terminal velocity, exceeding the velocity of the largest turbulent vortices. For example, the terminal velocity of a 250 µm-radius drop is about 2 m/s (Figure 5.5.1) which is typically larger than the turbulent velocities in stratiform and stratocumulus clouds being of order of 0.5–1 m/s. Therefore, turbulence influences raindrop motion only slightly, although the turbulence-induced relative velocity is maximal for these drops.

5.5.7 Drop Clustering

As was mentioned, nonuniformity of air flow leads to inhomogeneous drop concentration. In a turbulent flow, drop concentration changes randomly in space and time. This phenomenon is called preferential concentration or drop clustering. There are multiple studies investigating drop clustering (see reviews by Vaillancourt and Yau, 2000; Shaw, 2003; Khain et al., 2007). This strong interest is motivated by a possible impact of drop clustering on diffusional growth and especially on collision processes in clouds. Raindrop clustering can influence rain characteristics (Jamerson and Kostinski, 2000) and collision-induced breakup (McFarquhar, 2004a). It creates random supersaturation fluctuations (Equations (5.2.13) and (5.2.14)), which may disrupt the balance between the terms in the supersatration equation (5.2.13) and lead to drop-spectrum broadening. The effects of drop clustering on DSD formation are discussed in Section 5.8.

The mechanism of drop clusters formation is illustrated in **Figure 5.5.16.** Due to the centrifugal forces, drops tend to leave the areas of turbulent vortices and to accumulate in the areas between vortices.

The degree of drop accumulation depends on drop–vortices interaction. A turbulent flow consists of numerous vortices with a wide range of sizes (Section 3.3). The kinetic energy of these vortices within the inertial subrange decreases as a vortex size decreases according to the −5/3 Kolmogorov law. For each drop size, there is a specific vortex size providing the strongest interaction. Small cloud droplets interact intensively with the smallest vortexes of the Kolmogorov micro-scale sizes, despite the lowest intensity of the latter. Larger drops respond more strongly to vortices of larger size. This interaction can be characterized by the spectrum of the relative-to-air drop velocity. Examples of such spectra are shown in **Figure 5.5.17**. The spectra characterize the dependence of inertia-induced, relative to air drop velocity on the vortex size. The maxima of these spectra are located at the wavelengths of 1.5 cm, 8 cm, and 70 cm for drops of 30 μm, 50 μm, and 100-μm radii, respectively. These maxima represent the characteristic sizes of turbulent vortices that have the strongest impact on relative-to-air drop velocity. Large raindrops respond to vortices of several tens of meters in size.

To characterize drop-vortex interaction, the turbulent Stokes number St is used. This number is defined as the ratio of drop relaxation time (Equation 5.5.4) and the Kolmogorov timescale (Table 3.3.1):

$$St = \frac{\tau_d}{\tau_k} \tag{5.5.22}$$

For cloud droplets, the value of the Stokes number can be found from Equation (5.5.6). It is proportional to the square of a droplet radius:

$$St = \frac{2\rho_w \varepsilon^{1/2}}{9\rho \nu^{3/2}} r^2, \tag{5.5.23}$$

where ν is the kinematic air viscosity and ε is the turbulent dissipation rate (Section 3.3). Clustering of drops of the same size for different Stokes numbers is illustrated in **Figure 5.5.18**, showing the positions of drops moving in a turbulent flow. It shows the results of DNS of a drop motion in a turbulent flow for different St numbers. Figure 5.5.18a corresponds to $St = 0$ and demonstrates a uniform spatial distribution of drops.

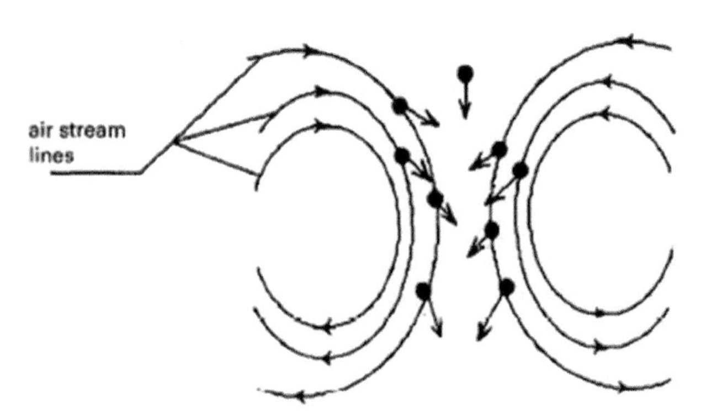

Figure 5.5.16 Schematic illustration of drop clusters formation.

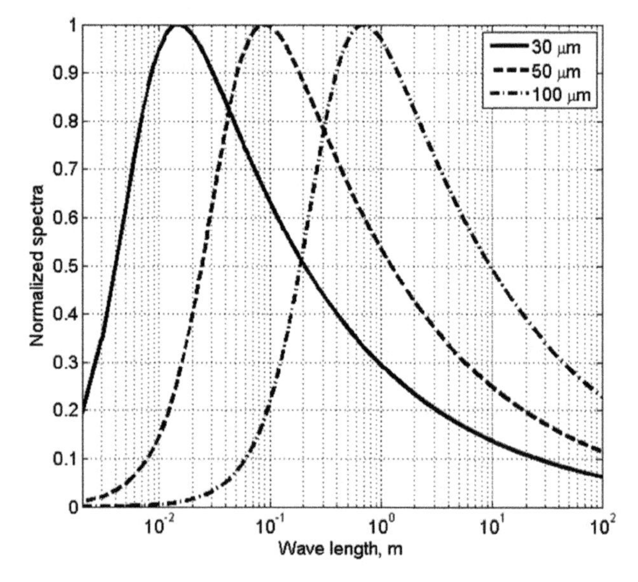

Figure 5.5.17 Spectra of the relative-to-air drop velocity for drops of 30 μm, 50 μm and 100-μm radii. The spectra are normalized over their maximum value.

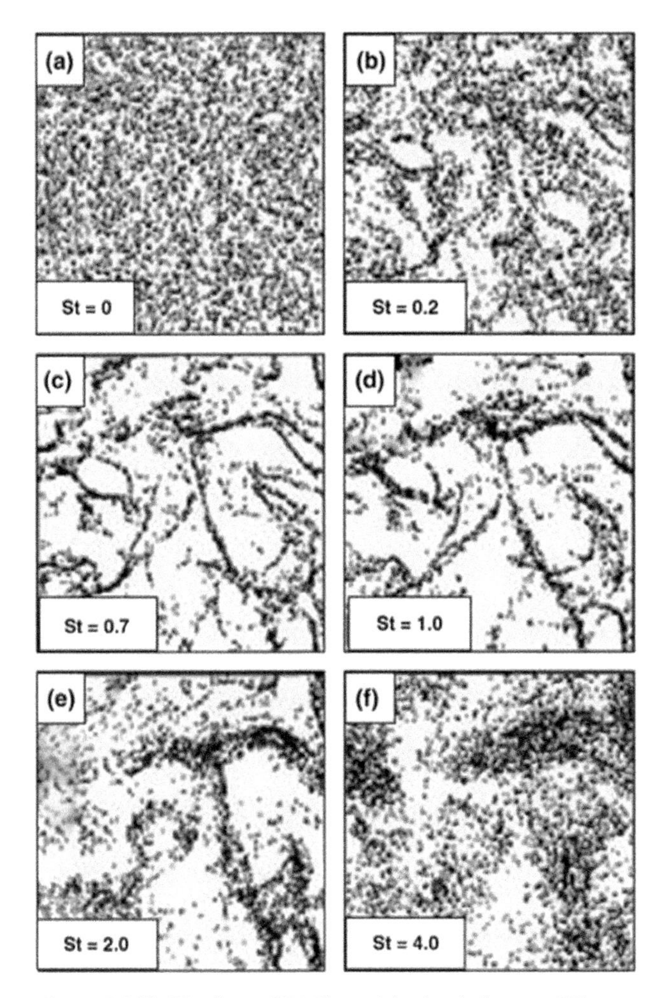

Figure 5.5.18 2D slices of DNS particle simulations at different Stokes numbers. Dots correspond to particle center locations. The particles are of the same size (from Reade and Collins, 2000; reproduced with permission from AIP Publishing LLC).

In this case the probability of finding N drops within volume V obeys the Poisson PDF (Shaw, 2003):

$$p(N) = \frac{(\overline{N}V)^N}{N!}\exp(-\overline{N}V), \tag{5.5.24}$$

where $\overline{N}$ is the mean drop concentration. Drop inertia leads to additional super-Poisson fluctuations of drop locations within a given volume. The most pronounced clustering effect is observed at Stokes numbers close to 1 (Figure 5.5.18d) when the strongest drop-turbulence interaction takes place. However, DNS simulations presented in Figure 5.5.18 do not take into account two important factors affecting clustering. First, there is gravitational settling that decreases the clustering effect since it shortens the time a drop spends within a turbulent vortex. Second, in the DNS simulations

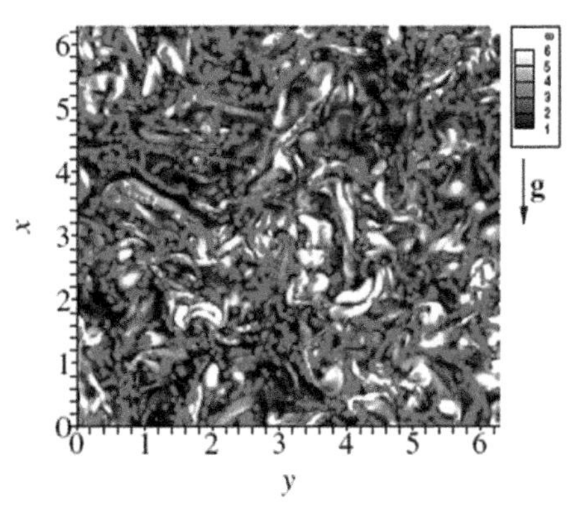

Figure 5.5.19 Preferential concentration of particles with $St = 1$ in the quasi-equilibrium state. Particle locations are shown by red dots. The background is the vorticity contours normalized by the average vorticity. The arrow below the legend represents the direction of gravity (from Jin et al., 2010; reproduced with permission from AIP Publishing LLC).

illustrated above the turbulent flow was frozen, so the life time of each vortex was infinitely large. Both simplifications lead to overestimation of the clustering effect. **Figure 5.5.19** shows the results of more realistic simulations. Here the clustering at $St = 1$ is substantially weaker compared to that shown in Figure 5.5.18d.

The clustering effect for small St particles (cloud droplets) was also studied analytically (Maxey, 1987; Pinsky and Khain, 1997c; Elperin et al., 1996, 2002a; Falkovich and Pumir, 2004). The major outcome of droplet inertia characterized by droplet relaxation time τ_d is that the droplet velocity field becomes divergent within a non-divergent turbulent flow. This divergence can be evaluated as (Maxey, 1987)

$$div(\mathbf{V}) = -\tau_d \frac{\partial u_i}{\partial x_j}\frac{\partial u_j}{\partial x_i} \tag{5.5.25}$$

According to this formula, the sign of the divergence does not depend on droplet size, so formation of clusters of droplets having different sizes takes place in the same zones of a turbulent flow. Droplets tend to leave areas of positive divergence and accumulate in the areas of negative divergence.

Since turbulent shears $\frac{\partial u_i}{\partial x_j}$ are random, the field of $div(\mathbf{V})$ is stochastic. Pinsky et al. (1999a) evaluated the statistical properties of $div(\mathbf{V})$, assuming the turbulent velocity field to be homogeneous and isotropic. The longitudinal structure function of turbulent velocities (Section 3.3) in these evaluations is set by the Batchelor formula $D_{ll}(r) = r^2(1 + Cr^2)^{-2/3}$, which is valid in both the inertial and the viscous turbulent subranges. The field of $div(\mathbf{V})$ was also shown to be homogeneous and

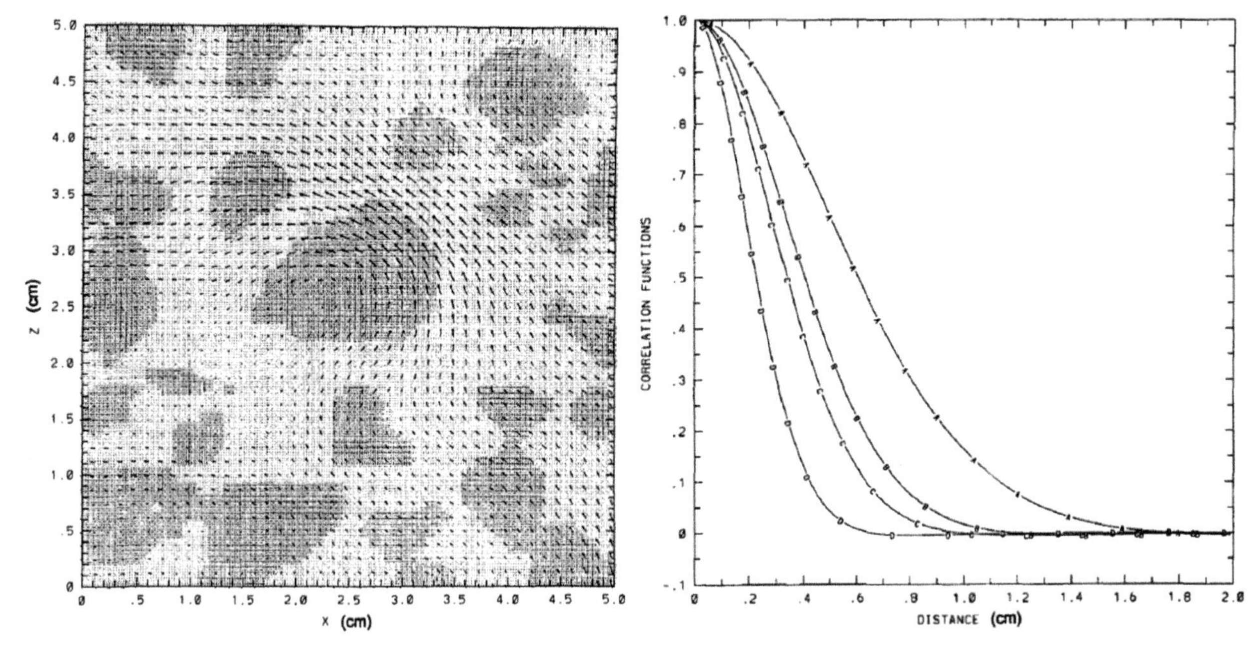

Figure 5.5.20 Left: the areas of positive (dark) and negative (light) drop velocity field divergence. The turbulent velocity field is shown by arrows. Right: the correlation functions of the drop velocity field divergence for different values of dissipation rate: (A) $\varepsilon = 10\ \text{cm}^2\ \text{s}^{-3}$, (B) $\varepsilon = 50\ \text{cm}^2\ \text{s}^{-3}$, (C) $\varepsilon = 100\ \text{cm}^2\ \text{s}^{-3}$, (D) $\varepsilon = 500\ \text{cm}^2\ \text{s}^{-3}$ (from Pinsky et al., 1999; courtesy of John Wiley & Sons, Inc.©).

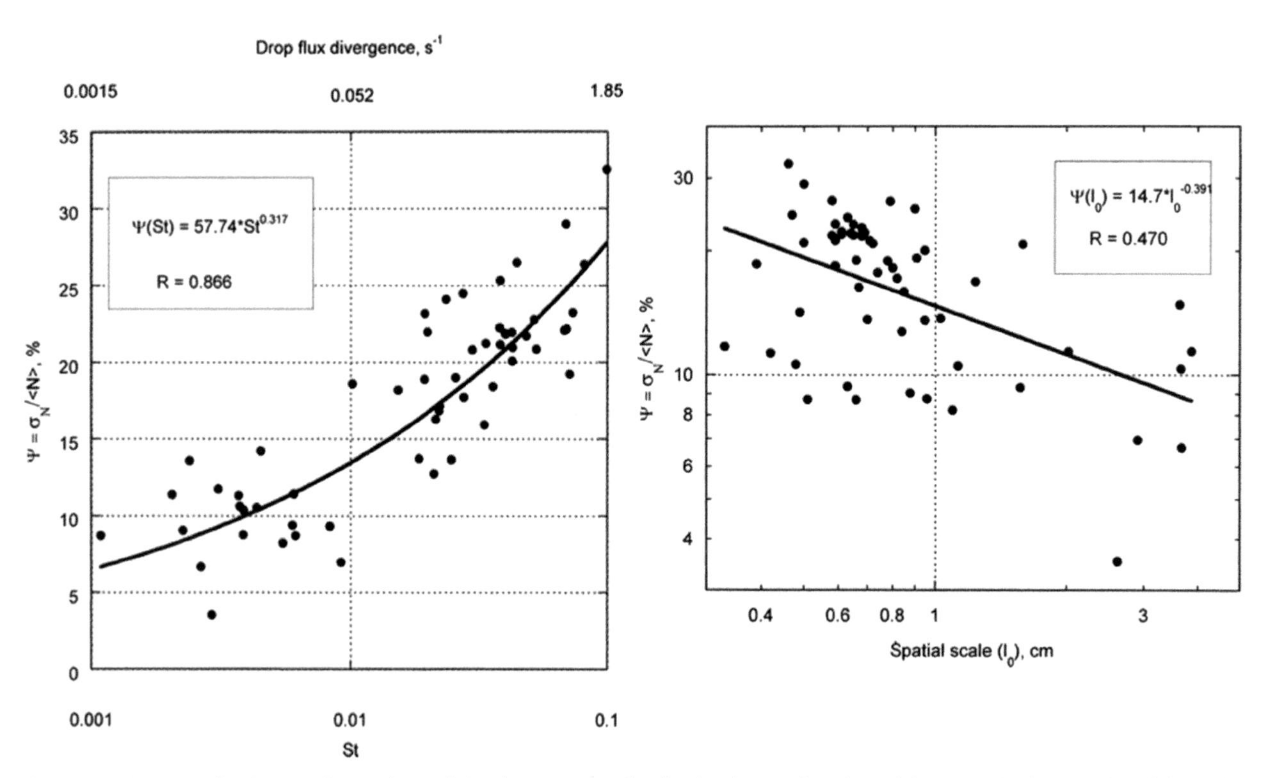

Figure 5.5.21 Normalized RMS fluctuations of droplet concentration in clouds as a function of the mean Stokes number (left) and of spatial scales (right). Points correspond to different traverses. $\langle N \rangle$ and σ_N are the mean and the standard deviation values of droplet concentration, respectively (from Pinsky and Khain, 2003; American Meteorological Society©; used with permission).

isotropic with zero mean value $\overline{div(\mathbf{V})} = 0$. Squaring the left and the right parts of Equation (5.5.25) and using the Batchelor formula to the average resulting equation, Pinsky et al. (1999a) derived the following RMS value of $div(\mathbf{V})$:

$$\left(\overline{[div(\mathbf{V})]^2}\right)^{1/2} = \frac{1}{9}\frac{\varepsilon\rho_w}{\nu^2\rho}r^2 \quad (5.5.26)$$

This value is proportional to the turbulent dissipation rate ε and to the square of droplet radius r^2. The latter dependence indicates an amplification of the clustering process with increasing droplet inertia. Comparison of Equations (5.5.23) and (5.5.26) reveals a close relationship between the Stokes number and the RMS value of $div(\mathbf{V})$. Both equations are applicable only to parameterization of clustering of small cloud droplets (so-called Stokesian particles).

The spatial correlation functions of $div(\mathbf{V})$ were evaluated, and the characteristic spatial scales of $div(\mathbf{V})$ were found to be of order of 1 cm, decreasing with the ε increase. **Figure 5.5.20** demonstrates an example of the droplet velocity field divergence (left) and the spatial correlation functions of $div(\mathbf{V})$ (right). The cm scale is preferable here, since the velocity shears reach their maximum at the scales of ~1 cm, which corresponds to the transition from the inertial turbulence subrange to the viscous one (Section 3.3). At scales below ~1 cm the velocity shears decrease due to viscosity effects. The fluctuations of $div(\mathbf{V})$ lead to cm-scale fluctuations of small droplet concentration, as can be seen in Figures 5.5.18 and 5.5.19.

In real clouds, the centimeter-scale droplet concentration fluctuations caused by the turbulence-inertia effects were found by Pinsky and Khain (2001, 2003) and Kostinski and Shaw (2001) via statistical analysis of in situ data obtained by fast FSSP probe (Brenguier and Chaumat, 2001). **Figure 5.5.21** shows the results obtained by Pinsky and Khain (2003) in statistical analysis of long series of droplet arrival times in ~60 cumulus clouds. The results indicate that droplet clustering does occur in clouds. The amplitude of droplet concentration fluctuations was found to increase with the mean St in agreement with the theoretical predictions. The mean St was calculated by averaging the values of St of individual droplets using Equation (5.5.23). One can see an increase in the clustering rate when the spatial scale of concentration inhomogeneities decreases. The data described allowed introducing the regression dependence between the Stokes number and the magnitude of droplet clustering. This dependence can be written as

$$\frac{\sigma_N}{\overline{N}} = \psi(St) = 57.74 \times St^{0.317} \quad (5.5.27)$$

Dependence (5.5.27) characterizes the average effect of droplet inertia on clustering. This dependence is used in cloud models for parameterization of turbulent effect on droplet collisions (Benmoshe et al., 2012) (Section 5.8).

5.5.8 Turbulence Impact on Drop Sedimentation

The preferential concentration leads to an interesting effect: an increase in the effective settling velocity in a turbulent flow (Wang and Maxey, 1993). **Figure 5.5.22** illustrates the mechanism leading to a preferential sweeping of inertial particles to downdraft zones. On the whole, the tangential velocity of turbulent vortices adds to the terminal velocity of particles, increasing their effective settling. The maximum increase in the effective-settling velocity was found for particles with $St = 0.7$, typical of small drizzle drops. These additional velocities are of order of several cm/s and can hardly influence rain fluxes evaluation in cloud models.

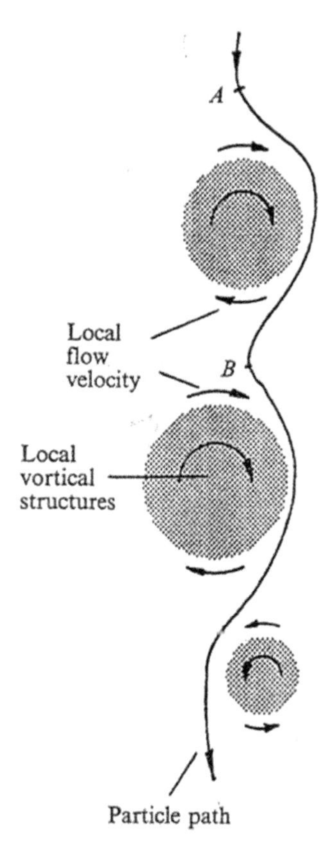

Figure 5.5.22 Schematic illustration of preferential sweeping for a heavy particle interacting with local flow structures. The preferential sweeping is caused by the inertia and the body force of the drop (from Wang and Maxey, 1993; courtesy of Cambridge University Press).

5.6 Gravitational Drop Collisions

Drop collision is the main mechanism of creating large cloud droplets and raindrops (warm rain). In the past, drop collisions in clouds were regarded as collisions in calm air or in laminar flows. In laminar flows, drops collide due to the difference between their terminal fall velocities (Section 5.5), as smaller drops fall more slowly than larger drops that collect them on the way. This kind of collision is known as gravitational coagulation, since the difference in fall velocities is caused by gravitation. Gravitational coagulation is dominating when a particle size (in our case, drop size) exceeds ~1 μm. For smaller particles (aerosols) the coagulation is caused by the Brownian diffusion and other processes. Methods developed for treating gravitational coagulation in calm air are still used in the majority of cloud models. However, during the past decade it was recognized that drop motion and collisions in turbulent media-like clouds differ substantially from those in laminar flows (Section 5.8). Taking the turbulent effects into account usually implies calculating the enhancement factor showing by how many times collisions in turbulent flows are more efficient than in laminar flows. Turbulent effects on drop collisions increase with increasing turbulence intensity. In clouds with low turbulence (stratiform, stratocumulus), especially in clouds or fogs developing in stable atmosphere, the turbulence-induced collision-rate enhancement factor is close to one, so the gravitational collision mechanism is dominating.

5.6.1 Main Concepts Related to Drop Collision Process

Collision is a stochastic process. Cloud drops are separated by distances of about 1–4 mm depending on their concentration. These distances are much larger than the sizes of cloud drops themselves. In clouds where the mean droplet radius is 10 μm, raindrops with radii of 1 mm are formed as a result of about 10^6 collisions. To form one raindrop, droplets should be collected within a relatively large volume of about 10^3–10^4 cm^3 containing a great number of droplets. Considering the volumes of this size allows analyzing collision using the mean values of various parameters such as drop concentration, drop SD and the probability of drop collisions. Relatively large distances between drops justify neglecting triple collisions, so drop collisions can be viewed as successive acts of hydrodynamic interaction between two drops. When two drops collide, three outcomes are possible: a) the colliding drops coalesce forming one drop; b) the colliding drops rebound, keeping their initial sizes; and c) the colliding drops coalesce and then break into several drops of different sizes.

The geometry of cloud drop collisions is shown in **Figure 5.6.1**. The drop collector falls at the terminal fall velocity V_{g1}. Smaller drops fall at fall velocity V_{g2}. During unit of time, the drop collector with radius r_1 may collide with smaller drops of radius r_2 located within a cylindrical air volume $V_{swept} = S_g\left|V_{g1} - V_{g2}\right| = \pi(r_1 + r_2)^2\left|V_{g1} - V_{g2}\right|$, known as the swept volume. Here S_g is the geometric cross section. Assuming the concentration of small drops collected by the drop collector is equal to N_d, the number of smaller drops within swept volume is $N = \pi(r_1 + r_2)^2 (V_{g1} - V_{g2})N_d$. All the drops within the swept volume V_{swept} experience collisions with a drop collector.

In fact, a drop falling in a calm air induces an air flow around itself. Examples of such flows arising around a small drop and a comparatively large drop are presented in **Figure 5.6.2**. A falling drop experiences the impact of two main forces: the gravity force and the drag force (Section 5.5). In case drops are separated by very small distances, an additional drag force arises due to the influence of the velocity field induced by its counterpart. The interaction between drops arising at small separation distance when the influence of the induced fields becomes significant is called *hydrodynamic drop interaction* (HDI). When the distance between drops becomes smaller than about 10 radii of the largest drop in a drop pair, the velocity field induced by the drop collector changes the trajectory of the

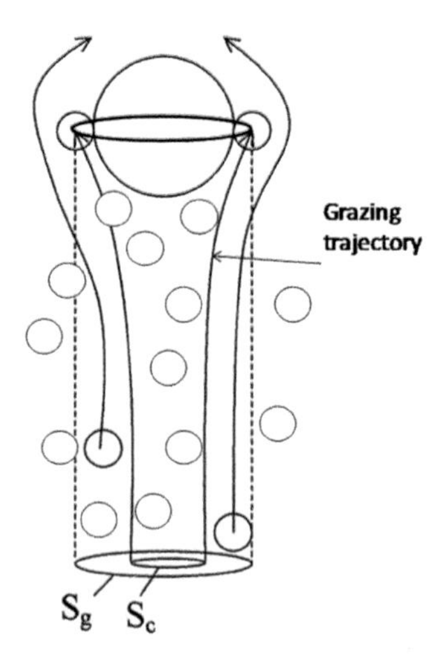

Figure 5.6.1 Geometry of gravitational drop collisions and drop hydrodynamic interaction. Collision cross-section S_c is less than the geometrical cross-section S_g (the collision efficiency $E = S_c/S_g < 1$).

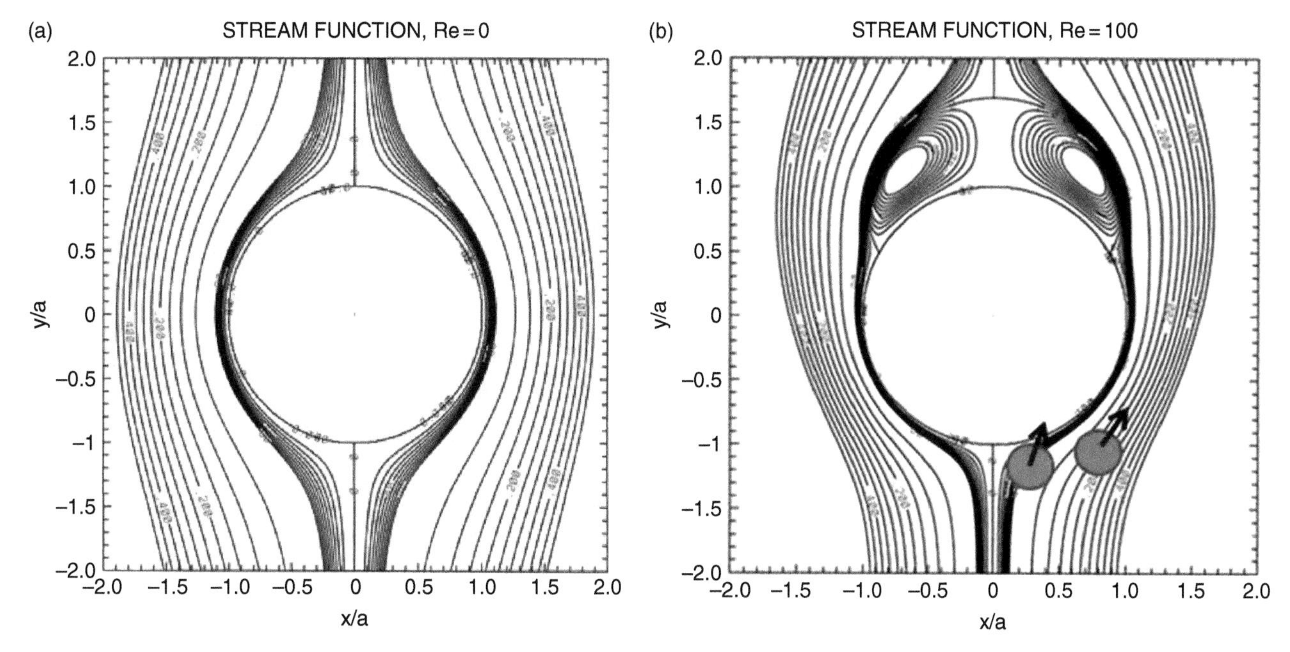

Figure 5.6.2 The stream-function fields around a falling drop for (a) $Re \ll 1$ and (b) $Re = 100$ at $p = 1,000$ mb in the fixed coordinate frame. The Reynolds number of a drop is determined as $Re = \frac{2r}{\nu} V_g$ (Section 5.5). Vortices behind the moving drop with $Re = 100$ are clearly seen. Panel (b) also shows small droplets collected by the drop collector. Due to their inertia, the collected droplets deviate from the air flow induced by the drop collector, which may lead to a collision (from Pinsky et al., 2001; American Meteorological Society©; used with permission).

smaller drop, which tends to avoid the collision. Due to the inertia of the smaller drop, while approaching the collector it deviates from the air flow surrounding the collector (Figure 5.6.2b). This deviation increases with increasing drop inertia (i.e., drop size), so the probability of collisions increases with a drop size and increasing relative velocity between the drops.

Since the inertia of cloud droplets is very low, most of them avoid collisions, moving together with the air flow around the drop collector. If the smaller drop is located in the vicinity of the trajectory of the larger drop center, the collision takes place. If the smaller drop is initially shifted with respect to this trajectory, it deviates from the trajectory while approaching the drop collector. There exists a trajectory (or trajectories) of the smaller drop around the drop collector that passes at the minimum distance between the drop centers equal to the sum of their radii. Such trajectory is called a *grazing* trajectory (Figure 5.6.1). Grazing trajectories form a grazing surface, and collisions take place if the smaller drop is located within the volume inside the grazing surface. At separation distance of about 10 radii of the drop collector, the shape of the volume limited by the grazing surface becomes a cylinder with cross section S_c known as the collision cross section. In case of gravitational collisions, the collision cross section has the same center as the geometric cross section, as shown in Figure 5.6.1.

The fraction of drops experiencing collisions within the swept volume can be represented by the ratio of the collision cross-section S_c to the geometrical cross section S_g (Figure 5.6.1). This ratio is known as the *collision efficiency*:

$$E(r_1, r_2) = S_c/S_g \tag{5.6.1}$$

The collision efficiency depends on sizes of colliding drops. The quantity

$$K_g(r_1, r_2) = \pi(r_1 + r_2)^2 E(r_1, r_2)\left|V_{g1} - V_{g2}\right| \tag{5.6.2}$$

is known as the *gravitational collision kernel.* The product of the collision kernel and the concentration of smaller drops shows the number of collisions between a drop collector and small drops in calm air per unit of time. Statistically, the collision kernel characterizes the probability of collision between two drops with radii r_1 and r_2 per unit of time. Taking into account that collision efficiency is defined here as the ratio of the collision cross section S_c to the geometrical cross sections S_g of the two corresponding cylinders. Equation (5.6.1) defines of the *cylindrical* gravitational collision kernel. More general formulation of gravitational collision kernels, namely a *spherical* formulation of the collision kernel, is introduced in Section 5.8, where collisions of drops in a turbulent flow are considered.

Not all drop collisions lead to drop coalescence. The fraction of collisions that yields coalescence is known as the *coalescence efficiency* $E_{coal}(r_1, r_2)$. In fact, laboratory measurements do not determine the coalescence efficiency but rather the collection efficiency, which is the product of collision efficiency and coalescence (merger) efficiency. Introducing the coalescence efficiency into Equation (5.6.2), we obtain a quantity known as the *collection kernel*:

$$K_{g\_col}(r_1, r_2) = \pi(r_1 + r_2)^2 E(r_1, r_2) E_{coal}(r_1, r_2) \left| V_{g1} - V_{g2} \right| \quad (5.6.3)$$

From Equation (5.6.3) it follows that the gravitational collection kernels for drops of the same sizes are equal to zero. Kernels written in the forms (5.6.2) and (5.6.3) are known as the *hydrodynamical kernels.*

5.6.2 Equations Describing Drop Collisions

In case the drop collector is much larger than the collected drops (for instance, when a raindrop collects small cloud droplets in a process known as *accretion*), the growth of the drop collector mass can be calculated using the *continuous growth equation* (Pruppacher and Klett, 1997):

$$\frac{dm}{dt} = \frac{4\pi\rho_w}{3} \int_0^{r'_{max}} r'^3 f_r(r') K_{g\_col}(r, r') dr', \quad (5.6.4)$$

where m and r are the mass and the equivalent radius of the collector, respectively, ρ_w is water density and $f_r(r')$ is DSD of small drops. Equation (5.6.4) can be applied when the concentration of drop collectors is low and collisions between the collectors can be neglected. In particular, this equation has been used to interpret laboratory experiments in a wind tunnel, where a single drop collector grew by collecting a large number of small drops (Vohl et al., 2007).

Collisions in bulk-parameterization schemes are described using the continuous growth equation. Application of this equation to calculation of DSD means that drop collectors are assumed to be independent and no competition for collecting droplets is considered, because of lack of collisions between collectors concentration of the collectors remains unchanged. The schemes using the continuous growth equation to describe the rain formation cannot simulate formation of correct distribution of raindrops, in particular, the schemes cannot simulate formation of a small number of the largest raindrops.

When the concentration of drop collectors is high and collisions between them are not negligible, it is hardly possible to distinguish between collecting drops and collected drops. In this case, collisions are described using the *stochastic collection equation* (SCE), which is a version of the well-known Smoluchowski equation. The equation can be written as follows (Pruppacher and Klett, 1997):

$$\frac{df(m,t)}{dt} = \int_0^{m/2} f(m') f(m-m') K_{g\_col}(m-m', m') dm' - \int_0^{\infty} f(m) f(m') K_{g\_col}(m, m') dm', \quad (5.6.5)$$

where m and m' are drop masses, $f(m)$ is DSD (Sections 2.1 and 2.3) and $K_{g\_col}(m, m')$ is the collection kernel. The number of collisions per unit of time is proportional to the product of concentrations of drops with masses m and m'. The first integral on the right-hand side of Equation (5.6.5) is known as the *gain integral; it* describes the rate of generation of drops with mass m by coalescence of drops with masses m' and $m - m'$. The reason the gain integral is evaluated within the boundaries from zero to $m/2$ is illustrated in **Figure 5.6.3.** When masses of smaller drops m' increase from zero to $m/2$, masses of their counterparts *m-m'* vary from m to $m/2$. It means that the gain integral takes into account all pair collisions of drops with masses ranging from 0 to m. The gain integral can be also written as $\frac{1}{2}\int_0^m f(m') f(m-m') K(m-m', m') dm'$, where the coefficient 1/2 is inserted to avoid double counting of pair drop collisions. The second term on the right-hand side of Equation (5.6.5) is known as the *loss integral*; it describes the decrease in the number of drops with mass m, caused by coalescence with drops of any other sizes.

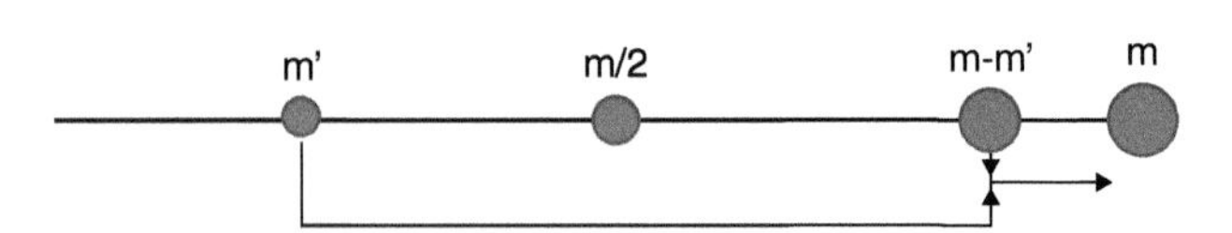

Figure 5.6.3 Illustration of the gain integral calculation in Equation (5.6.5).

Equation (5.6.5) is the averaged equation (over a certain volume or over many realizations). The averaging of the equation means the existence of a great number of drop pairs within the volume where collisions are analyzed. Actually, the term "stochastic collection equation" is not precise, since Equation (5.6.5) is not stochastic, so it is also called the quasi-stochastic collection equation. Equation (5.6.5) is a nonlinear integral-differential equation, for which analytical solutions are known only for several specific forms of the collision kernels. A heuristic justification of the applicability of SCE in cloud physics can be found in the book by Voloshchuk and Sedunov (1977). The main assumptions used in SCE derivation are:

a) SCE takes into account binary (pair) collisions only, while neglecting triple collisions and any higher-order ones. This assumption is valid with high accuracy for clouds, since the volume fraction of drops is of order of $10^{-5} - 10^{-6}$, so

$$\frac{1}{\rho_w}\int_0^\infty mf(m)dm \ll 1 \qquad (5.6.6)$$

Condition (5.6.6) is relevant when the mean size of drops is much smaller than the mean distance between neighboring drops.

b) The number of drops within the analyzed volume is large:

$$\Delta x \Delta y \Delta z \int_0^\infty f(m)dm \gg 1 \qquad (5.6.7a)$$

In practice, this condition is too weak. If DSD is represented on a mass grid by bins within mass ranges of Δm_i, condition (5.6.7a) should be replaced by a stronger condition:

$$\Delta x \Delta y \Delta z \Delta m_i f(m_i) \gg 1 \qquad (5.6.7b)$$

Condition (5.6.7b) is usually fulfilled for cloud droplets and is not always observed for raindrops.

c) Successive collision events are independent, so the collision process is considered as a Poisson flux of events. It is fulfilled if the characteristic time of drop hydrodynamic interactions is much smaller than the characteristic time of DSD temporal changes. This condition can be written as

$$f\left(\frac{df}{dt}\right)^{-1} \gg \tau_{HI} \qquad (5.6.8)$$

Since the typical time of drop hydrodynamic interaction τ_{HI} is of order of 0.1 s and does not exceed 0.5 s, this condition is usually fulfilled.

d) Histories of individual drop formation are independent, so the 2D DSD $f(m, m')$ is equal to the product of 1D distributions:

$$f(m, m') = f(m)f(m') \qquad (5.6.9)$$

It is difficult to check the validity of this condition.

Equation (5.6.5) is used to describe drop collisions in spectral bin microphysics cloud models. Analytical solution of Equation (5.6.5) is known only for specific collision kernels, for instance, for the Golovin kernel K_G:

$$K_G(m_1, m_2) = b(m_1 + m_2) \qquad (5.6.10)$$

where b is constant. Analytical solutions are also known for kernels having the form $K(m_1, m_2) = Cm_1m_2$ and $K(m_1, m_2) = b(m_1 + m_2) + Cm_1m_2$ (Long, 1974). The Golovin kernel is often used to verify different numerical methods used to solve Equation (5.6.5). At the same time, this kernel cannot be applied for description of drop collisions in real clouds. Some approximations of the hydrodynamical collision kernel were proposed by Long (1974):

$$K_L(m_1, m_2) = \begin{cases} k_c(m_1^2 + m_2^2), & r^* \le 50\ \mu\text{m} \\ k_r(m_1 + m_2), & r^* > 50\ \mu\text{m} \end{cases}, \qquad (5.6.11)$$

where $k_c = 9.44 \times 10^9$ cm^3 g^{-2} s^{-1}, $k_r = 5.78 \times 10^3$ cm^3 g^{-1} s^{-1}, and r^* is the radius of the larger drop. The Long kernel is closer to the hydrodynamic kernel and is widely used for verification of different methods used to solve SCE.

5.6.3 Gravitational Collision Efficiencies

Multiple studies deal with calculation of collision efficiencies between drops of different sizes. Most of them were conducted in the 1970's using different numerical methods and laboratory measurements (see the review in Pruppacher and Klett, 1997). Calculation of collision efficiencies involves a complicated hydrodynamic problem of drop interaction. Each drop induces a velocity field that influences the motion of this drop's counterpart. Due to the differences in approaches and the numerical schemes used, there are discrepancies between the values of collision efficiencies obtained by different authors (Pruppacher and Klett, 1997). These discrepancies are especially significant for collisions between small drops, as well as for collisions of small drops with larger drops. Hall (1980) compiled results of different studies and tabulated them to be used in microphysical models by means of interpolation. However, the Hall table does not contain information about collision efficiencies of drop pairs containing the smallest cloud droplets.

The widely used method for calculation of collision efficiency between drops is the method of superposition (Pruppacher and Klett, 1997). According to this method, the trajectories of drops about to collide are calculated under the assumption that they move within the velocity field induced by their counterparts. When the distance between the drop centers becomes equal to the sum of their radii, the collision is assumed to take place. The motion equation for each drop in the drop pair in calm air can be written as

$$\frac{d\mathbf{V}}{dt} = -\frac{1}{\tau_d}(\mathbf{V} - \mathbf{U}^*) + \left(1 - \frac{\rho}{\rho_w}\right)\mathbf{g}, \tag{5.6.12}$$

where $\mathbf{U}^*$ is the air velocity induced by the counterpart at the drop center. Equation (5.6.12) is an extension of the drop motion equation (5.5.3) to the case of interacting drops. When the separation distance between drops is large enough and the hydrodynamic interaction is negligible, each drop falls at the terminal fall velocity V_g induced by gravity (Equation (5.5.5)). This condition is used as the initial condition in calculations of collision efficiency. The accuracy of the superposition method for cloud droplets was justified by Wang et al. (2005a).

Using the superposition method, Pinsky et al. (2001) calculated the collision efficiencies within a wide range of drop radii from 1 to 300 μm. The velocity fields induced by falling drops were calculated by matching two analytical solutions: the Stokes analytical solution for $Re < 1$ (cloud droplets) (Kim and Karilla, 1991) and the solution proposed by Hamielec and Johnson (1962) for $Re > 1$. The matching allows calculating the velocity fields induced by moving drops with the Reynolds numbers up to $Re = 100$, which corresponds to drop radii equal to about 300 μm. The corresponding high-resolution tables of collision efficiency between drops calculated for $p = 1{,}000$ mb are presented in **Appendix B (Tables B1–B3)**.

Collision efficiencies for drops with radii below 40 μm calculated in different studies as a function of the radii ratio are presented in **Figure 5.6.4a**. Collision efficiencies between large drops and small droplets as a function of small droplet radius are presented in Figure 5.5.4b. One can see that collision efficiencies between cloud droplets vary by two orders of magnitude. For pairs containing droplets with radii below 10 μm and 20 μm, the collision efficiency does not exceed 0.01 and 0.1, respectively. Collision efficiency between small rain drops and cloud droplets dramatically increases with increasing cloud droplet radius. For raindrops with radii exceeding 250 μm, the collision efficiency is close to one for all sizes of collected drops except the smallest ones.

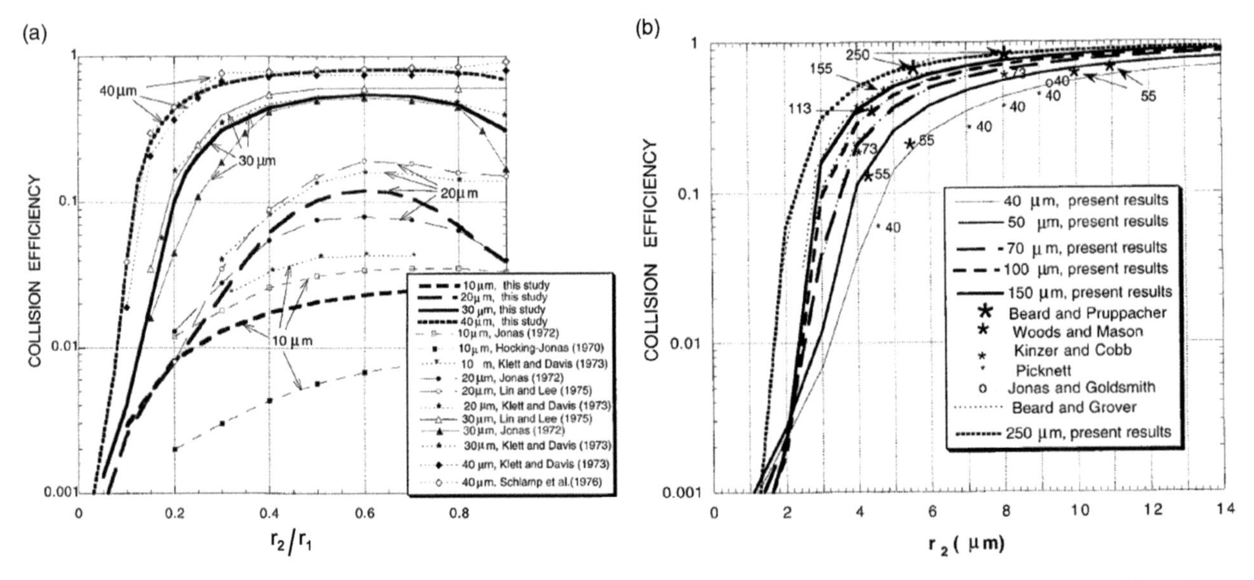

Figure 5.6.4 Collision efficiencies calculated in different studies for a) different radii ratios of colliding cloud droplets with radii below 40 μm and b) collisions between large and small drops (from Pinsky et al., 2001; American Meteorological Society©; used with permission).

Pinsky et al. (2001) calculated the height dependence of collision efficiency and collision kernels. Cumulus clouds can easily reach heights of several kilometers. The air density at 500-mb level is about half of that at the surface (1,000 mb). An increase in terminal drop fall velocities with height leads to increasing differences in terminal drop fall velocities and to an increase in the swept volume. Evaluations show that the increase in the collision kernel caused by increasing swept volume is about 10–25%, depending on the height and on drop sizes. Another effect is an increase in the collision efficiency with height. In fact, the collision efficiency depends not only on the size of colliding drops, but also on the relative velocity between them. The increase in the collision kernel with height is determined, therefore, by the product of increasing swept volume and increasing collision efficiency. **Figure 5.6.5** shows the ratio of collision efficiencies and collision kernels at 750-mb and 500-mb levels to those at 1,000-mb level, as calculated by Pinsky et al. (2001). One can see that the increase of the collision kernel with height is rather significant. It is the strongest for relatively small drop collectors within the radius range of 15 μm to 30 μm, colliding with droplets within the radius range from 5 μm to 10 μm. For these drop pairs, the collision kernel at the 500-mb level is more than twice as large as that at the 1,000-mb level.

5.6.4 Coalescence Efficiency

The current knowledge about the coalescence efficiency E_{coal} comes mostly from a few laboratory experiments (Beard and Ochs, 1984, 1993, 1995; Low and List, 1982b) and numerical studies by Straub et al. (2010). Beard and Ochs (1984, 1993) showed that the coalescence efficiency decreases with increasing drop sizes from unity for the drop collector radius $r_1 < 50$ μm and the collected droplet radii $r_2 < 5$ μm to 0.5 for $r_1 > 500$ μm and $r_2 < 25$ μm. The results of measurements of coalescence efficiency for drops of different sizes are presented in **Table 5.6.1**.

Beard and Ochs (1995) analyzed data from several experiments with relatively small raindrops. The dependence of the coalescence efficiency on two dimensionless quantities was studied: a) the size ratio $q = r_1/r_2$ where r_1 and r_2 are the radii of the smaller and the larger drops and b) the Weber number $We = \rho_w r_1 \Delta V_g^2/\sigma_w$, where ρ_w is the density, σ_w is the surface tension of liquid water, and ΔV_g is the difference of fall velocities of the two drops. To analyze the experimental results, Beard and Ochs (1995) parameterized the coalescence efficiency of small raindrops $E_{coal,BO1}$ as

$$E_{coal,BO1} = 0.767 - 10.14 \frac{2^{3/2}}{6\pi} \frac{q^4(1-q)}{(1+q^2)(1+q^3)} We^{1/2} \tag{5.6.13a}$$

Parameterization (5.6.13a) can be applied to raindrops of radii between 100 and 400 μm. For drops smaller than these sizes, Beard and Ochs (1995) proposed another parameterization formula for coalescence efficiency $E_{coal,BO2}$ expressed as a root of the following cubic equation:

$$a_0 + a_1 E_{coal,BO2} + a_2 E^2_{coal,BO2} + a_3 E^3_{coal,BO2} = \ln(r_1) + \ln\left(\frac{r_2}{200}\right), \tag{5.6.13b}$$

where the radii of drops are in μm and $a_0 = 5.07$, $a_1 = -5.94$, $a_2 = 7.27$, and $a_3 = -5.29$. Seifert et al. (2005) combined these two formulas into a continuous parameterization for small raindrops by taking the maximum value of both formulas:

$$E_{coal,BO} = \max(E_{coal,BO1}, E_{coal,BO2}) \tag{5.6.14}$$

Figure 5.6.6a shows the coalescence efficiency as a function of diameters of the colliding drops. The coalescence efficiency was calculated according to Equation (5.6.14).

For larger drops, Low and List (1982a) proposed a parameterization of E_{coal} as a function of three quantities: the collision kinetic energy CKE, the total surface energy of the two drops S_T, and the surface energy of the equivalent spherical drop with mass equal to the sum of masses of the colliding drops S_C. This parameterization takes into account the phenomenon of collisional drop breakup. The main idea of the parameterization is that breakup leads to an increase in the drop surface. Such increase in the surface requires a certain work against surface tension. Thus, in order for the collisional breakup to take place, the collision kinetic energy should exceed a certain threshold. The collision kinetic energy is defined as

$$CKE = \frac{1}{2}\frac{m_L m_S}{m_L + m_S}\Delta V_g^2 = \frac{\pi}{12}\rho_w \frac{D_L^3 D_S^3}{D_L^3 + D_S^3}\Delta V_g^2, \tag{5.6.15}$$

where m_S, D_S, and m_L, D_L are the masses and the diameters of the smaller drop and the larger drop, respectively. The total surface energy of the two drops is calculated as $S_T = \pi\sigma_w(D_L^2 + D_S^2)$, where σ_w is the surface tension of liquid water. The surface energy of the spherical equivalent of the drop with mass equal to

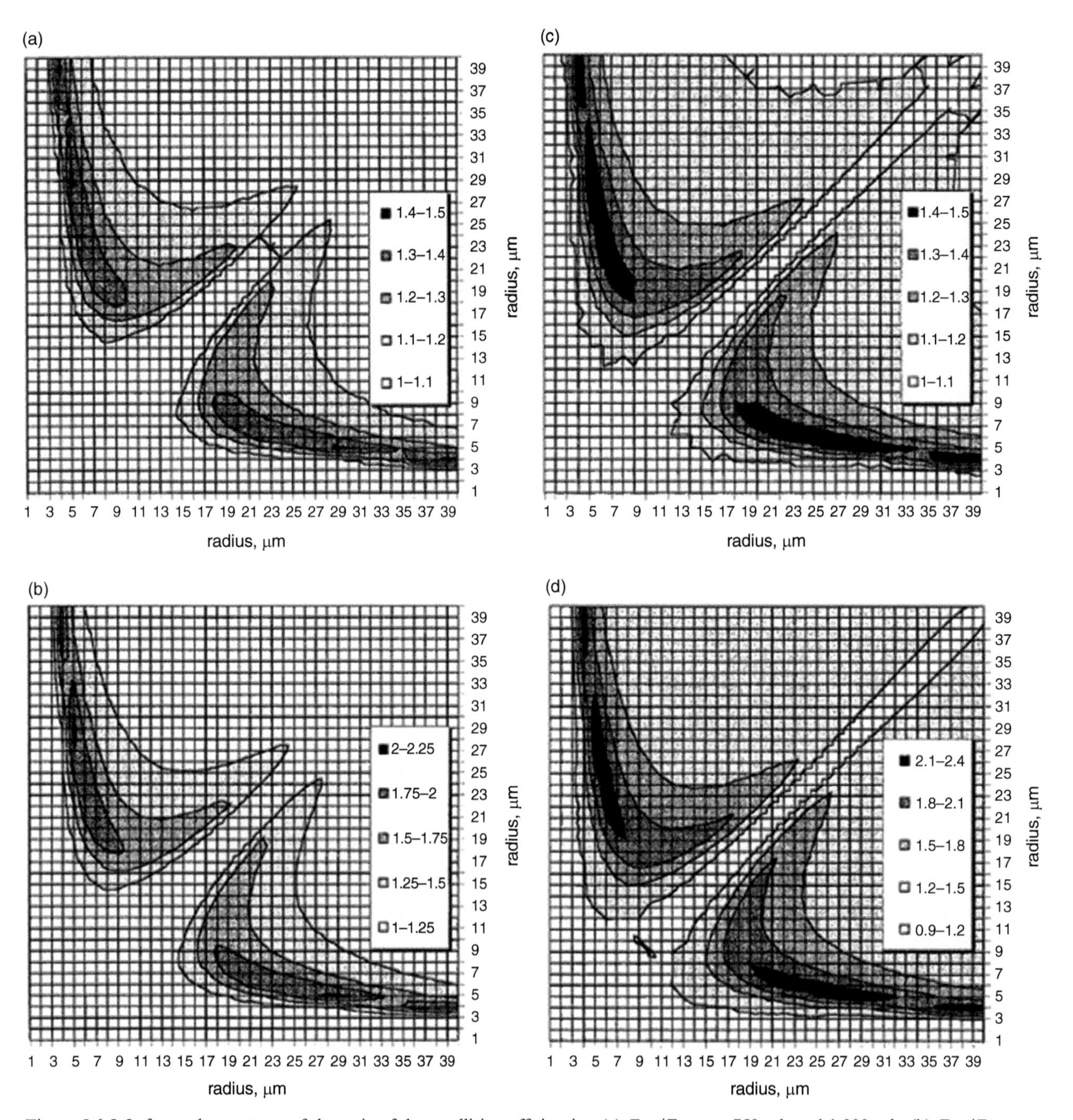

Figure 5.6.5 Left panels: contours of the ratio of drop collision efficiencies: (a) $E_{750}/E_{1,000}$ at 750 mb and 1,000 mb; (b) $E_{500}/E_{1,000}$ at 500-mb and 1,000-mb levels. Right panels: contours of the ratio of drop collision kernels at 750 mb and 1,000 mb, $K_{750}/K_{1,000}$ (c) and at 500 mb and 1,000 mb levels, $K_{500}/K_{1,000}$ (d) (from Pinsky et al., 2001; American Meteorological Society©; used with permission).

the sum of the masses of the colliding drops is $S_C = \pi\sigma_w\left(D_L^3 + D_S^3\right)^{2/3}$. The sum

$$E_T = CKE + S_T - S_C, \tag{5.6.16}$$

is called the total coalescence energy. This energy has to be dissipated through oscillations and deformation to reach a permanent coalescence, otherwise drop breakup takes place. Based on the experiments with

Table 5.6.1 Empirical coalescence efficiencies. Efficiencies in %; drop radii in μm (from Beard and Ochs, 1984; courtesy of John Wiley & Sons, Inc.©).

Cloud Droplet Radius	Collector Drop Radius r_1										
r_2	50	63	79	100	126	158	200	251	316	398	501
31.6	64.8	61.5	58.0	54.4	50.6	50.0	50.0	50.0	50.0	50.0	50.0
25.1	71.6	68.7	65.7	62.4	59.0	55.4	51.6	50.0	50.0	50.0	50.0
20.0	77.5	75.0	72.3	69.5	66.5	63.3	59.9	56.4	52.7	50.0	50.0
15.8	82.6	80.5	78.1	75.7	73.1	70.3	67.4	64.2	60.9	57.4	53.7
12.6	87.2	85.3	83.2	81.1	78.8	76.4	73.8	71.1	68.2	65.1	61.8
10.0	91.4	89.6	87.8	85.8	83.8	81.7	79.4	77.0	74.5	71.8	69.0
7.94	95.1	93.5	91.8	90.1	88.3	86.4	84.4	82.3	80.0	77.7	75.2
6.31	98.5	97.1	95.5	93.9	92.3	90.6	88.8	86.9	84.9	82.8	80.7
5.01	100.0	100.0	98.9	97.5	95.9	94.4	92.7	91.0	89.3	87.4	85.5
3.98	100.0	100.0	100.0	100.0	99.3	97.9	96.4	94.8	93.2	91.5	89.8
3.16	100.0	100.0	100.0	100.0	100.0	100.0	99.7	98.3	96.8	95.2	93.6
2.51	100.0	100.0	100.0	100.0	100.0	100.0	100.0	100.0	100.0	98.7	97.2
2.00	100.0	100.0	100.0	100.0	100.0	100.0	100.0	100.0	100.0	100.0	100.0
1.58	100.0	100.0	100.0	100.0	100.0	100.0	100.0	100.0	100.0	100.0	100.0

six different drop pairs, Low and List (1982a) suggest an empirical relation:

$$E_{coal,LL} = \begin{cases} a\left(1+\dfrac{D_S}{D_L}\right)^{-1/2} \exp\left(-\dfrac{b\sigma_w E_T^2}{S_C}\right), & E_T < 5.0\times 10^{-6}\,\mathrm{J} \\ 0, & E_T \geq 5.0\times 10^{-6}\,\mathrm{J} \end{cases} \quad (5.6.17)$$

with constants $a = 0.778$ and $b = 2.61 \times 10^6\ \mathrm{m^2\ J^{-2}}$. Expression (5.6.17) is valid for drops with diameters $D_S \geq 250\ \mu\mathrm{m}$ and $D_L \geq 500\ \mu\mathrm{m}$. Figure 5.6.6b shows the coalescence efficiency determined by Equation (5.6.17) as a function of the diameters of the colliding drops. The two formulas, i.e., Equation (5.5.14) for small raindrops and Equation (5.6.17) for large raindrops, do not match well for drops of intermediate sizes, as can be seen in Figures 5.6.6a and 5.6.6b. For a smooth transition between both parameterizations, Seifert et al. (2005) used the following interpolation between the formulas to represent the coalescence efficiency for both small and large raindrops:

$$E_{coal} = \begin{cases} E_{coal,BO}, & D_S < 300\,\mu\mathrm{m} \\ E_{coal,BO}\cos^2\theta + E_{coal,LL}\sin^2\theta, & 300\,\mu\mathrm{m} \leq D_S \leq 600\,\mu\mathrm{m}, \\ E_{coal,LL}, & D_S > 600\,\mu\mathrm{m} \end{cases} \quad (5.6.18)$$

where $\theta = \frac{\pi}{2}\frac{D_S - 300\ \mu\mathrm{m}}{300\ \mu\mathrm{m}}$. The coalescence efficiency determined by Equation (5.6.18) as a function of the diameters of the colliding drops is shown in Figure 5.6.6c. The parameterization of coalescence efficiency in the form (5.6.17) was used by Brown (1986), Tzivion et al. (1989), and List and McFarquhar (1990). The parameterization of the coalescence efficiency in the form (5.6.18) is used in the mixed-phase cloud model HUCM (Khain et al., 2008).

Using the results of numerically investigated binary collisions of thirty-two drop pairs, Straub et al. (2010) proposed a new parameterization of the coalescence efficiency as a function of the Weber number We:

$$E_{coal} = \exp(-1.15We), \quad (5.6.19)$$

where the Weber number $We = \rho_w r_1 \Delta V_g / \sigma_w$, where r_1 is the radius of the smaller drop, and ΔV_g is the differential fall speed of the two colliding drops. **Figure 5.6.7** shows the coalescence efficiency E_{coal} as a function of the Weber number. Results obtained by Low and List (1982a) are presented, as well. **Figure 5.6.8a** shows contours of E_{coal} as a function of drop diameters D_L and D_S with numerical values indicated by the color bar. The comparison with E_{coal} proposed by Seifert et al. (2005) (Figure 5.6.8b) indicates significant differences. The parameterization proposed by Straub et al. (2010) provides a more smoothed field of E_{coal} with values

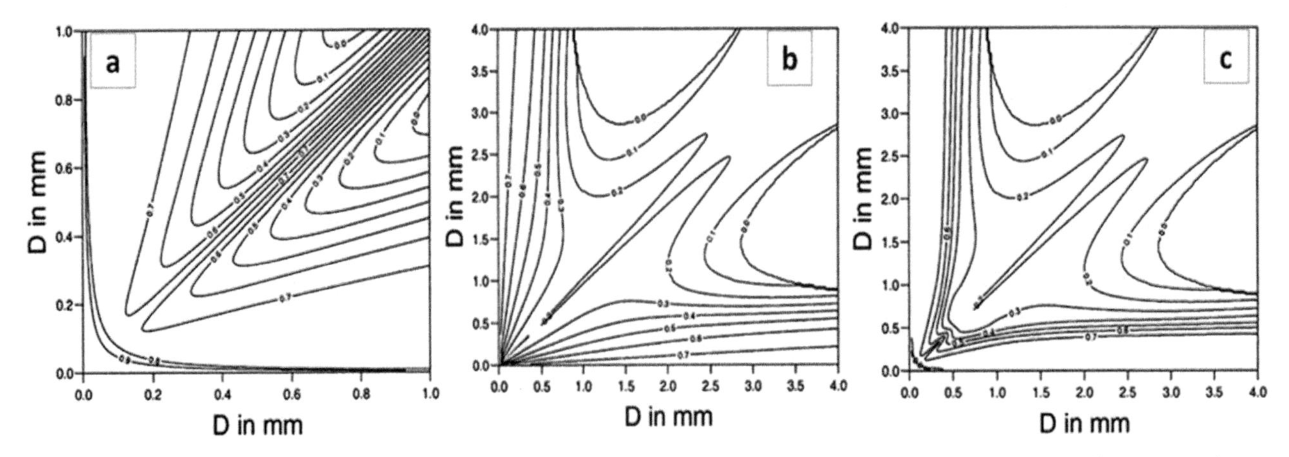

Figure 5.6.6 (a) Coalescence efficiency $E_{coal,BO}$ as a function of the diameters of the colliding drops (in mm) according to Equation (5.6.14) (from Beard and Ochs, 1995); (b) Coalescence efficiency $E_{coal,LL}$ as a function of the diameters of the colliding drops according to Equation (5.6.17) (from Low and List, 1982b); and (c) Coalescence efficiency determined by Equation (5.6.18) as a function of the diameters of the colliding drops (from Seifert et al., 2005; American Meteorological Society©; used with permission).

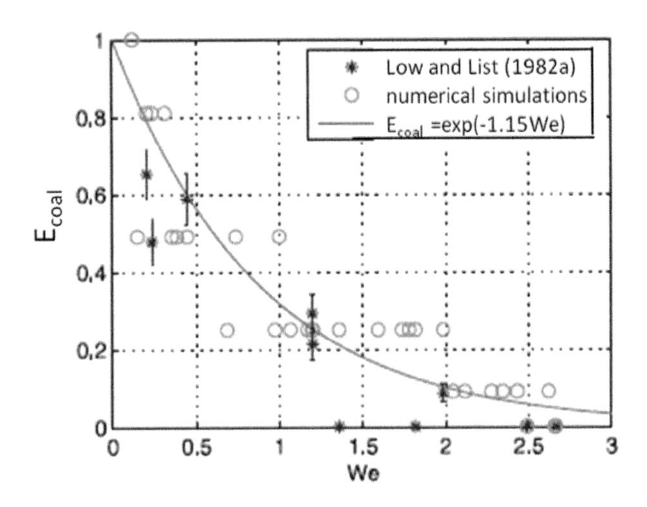

Figure 5.6.7 Coalescence efficiency E_{coal} as a function of the Weber number We (from Straub et al., 2010; American Meteorological Society©; used with permission).

close to one for drop pairs containing drops of very different sizes.

5.6.5 Collisional Breakup of Drops

SCE Equation (5.6.5) predicts a continuous drop growth with decreasing drop concentration. At the same time, the typical diameter of raindrops in warm rain is of 4–5 mm, while maximum diameters of 8–9 mm are very rare (Section 2.3). It is impossible to explain the observed distribution of raindrops without taking into account the breakup of large raindrops. In Section 5.5, a spontaneous breakup was analyzed. In case collisions of raindrops with other drops are not rare events, collisional breakup is more efficient than a spontaneous breakup and determines the maximum radius of raindrops and the DSD shape. Besides, both spontaneous and collisional breakup increase the evaporation rate below cloud base and affect the surface precipitation amount. The equation for stochastic collisional breakup is

$$\left.\frac{df(m,t)}{dt}\right|_{breakup} = \frac{1}{2}\int_0^\infty\int_0^m f(m')f(m'')B(m',m'')R(m,m',m'')dm'dm'' - \int_0^\infty f(m)f(m')B(m,m')dm', \quad (5.6.20)$$

where the breakup collision kernel is defined as

$$B(m',m'') = K_g(1-E_{coal}) = K_{g\_col}(1-E_{coal})/E_{coal}, \quad (5.6.21)$$

In Equation (5.6.20), $R(m,m',m'')$ is the fragment SD that determines the average number of fragments of mass m for a single event collision between two drops of masses m' and m''. The number of fragments within the mass range $[m, m+dm]$ is given as $R(m,m',m'')dm$. So, the first term on the right-hand side of Equation (5.6.20) shows how many drops of mass m are produced by collisions of drops with masses m' and m''. This integral can be called the gain integral for collisional breakup. The second integral shows the rate of a

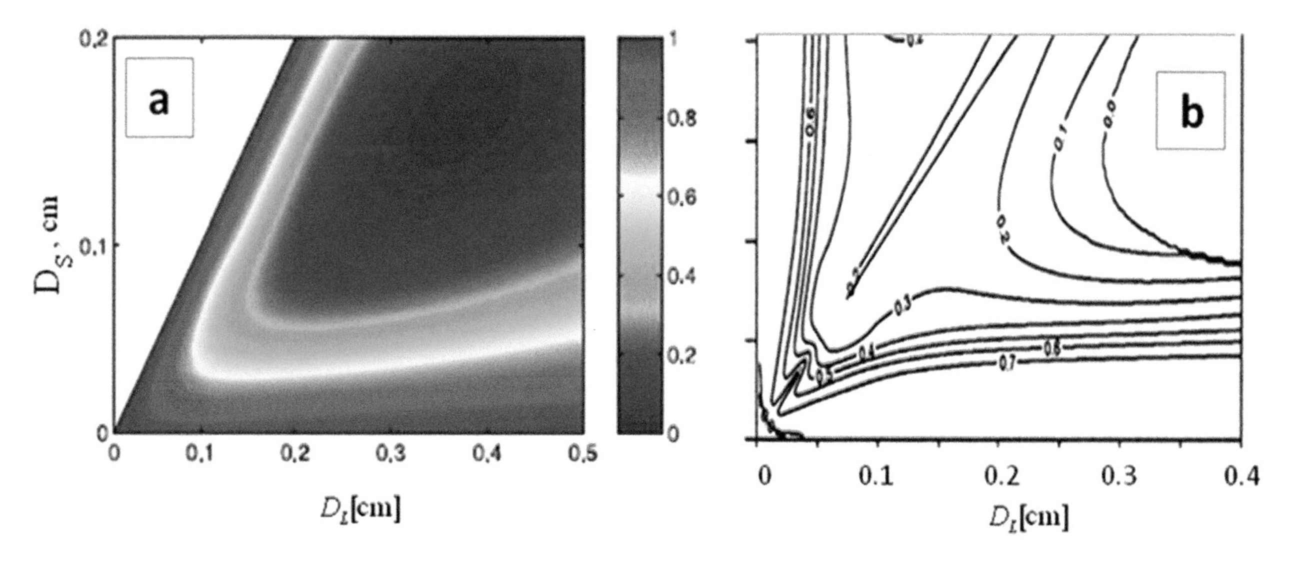

Figure 5.6.8 (a) Contours of E_{coal} as a function of diameters D_L and D_S of the colliding drops. Numerical values are indicated by the color bar (from Straub et al., 2010). (b) Isolines of E_{coal} as a function of the diameters of the colliding drops D_L and D_S (from Seifert et al., 2005; American Meteorological Society©; used with permission).

decrease in the number of drops with mass m caused by collisions with other drops ending up in breakups. This integral can be called the loss integral for collisional breakup.

Combining SCE (Equation 5.6.5) and Equation (5.5.13) for spontaneous breakup and Equation (5.6.20) for collisional breakup, we get the *collection-breakup equation* describing the evolution of DSD:

$$\begin{aligned}\frac{df(m,t)}{dt} &= \frac{1}{2}\int_0^m f(m')f(m-m')K_{g\_col}(m-m',m')dm' \\ &-f(m)\int_0^\infty f(m')K_{g\_col}(m,m')dm' \\ &+\frac{1}{2}\int_0^\infty\int_0^m f(m')f(m'')B(m',m'')R(m,m',m'')dm'dm'' \\ &-f(m)\int_0^\infty f(m')B(m,m')dm' \\ &+\int_0^\infty f(m')P_r(m')Q(m,m')dm' - f(m)P_r(m)\end{aligned} \tag{5.6.22}$$

The two last terms describe the rate of DSD change due to spontaneous breakup (Section 5.5). In cloud models, Equation (5.6.22) is often treated as three separate equations describing drop collisions and coalescence (SCE, Equation (5.6.5)), spontaneous breakup (Equation (5.5.13)), and collisional breakup (Equation (5.6.20)).

To solve Equation (5.6.20) it is necessary to know the SD of fragments $R(m,m',m'')$. Low and List (1982a) discriminate three main modes of collisional breakup for the filament mode, sheet mode, and disc-breakup mode. Each mode has different SD of fragments. A filament breakup occurs as a result of a glancing contact leading to formation of a water neck between the two drops. As the drops separate, the neck breaks into two main drops and several satellite drops of smaller size. A sheet breakup leads to a rotation of the larger drop, while a film or a sheet of water forms around the impact zone. Disintegration of the sheet leads to formation of about eight drops with sizes close to that of the original smaller drop. A disc breakup takes place when the smaller drop hits the larger one close to its center. As a result, an unstable disc forms, which sheds drops. Eventually, the disc disintegrates, producing a large amount of small drops. The breakup occurring in these three modes is illustrated in **Figure 5.6.9**. The filament mode is the dominating breakup mode, especially for small raindrops.

The complicated shape of the fragment SD requires description of all the peaks in the distributions. Low and List (1982b) present a parameterization for the modes using either the Gaussian or the lognormal distributions. The overall fragment SD is represented

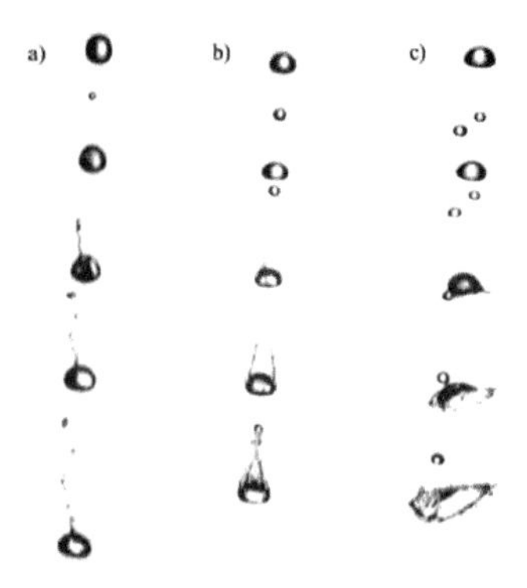

Figure 5.6.9 Sequences of high-speed camera images displaying three different types of breakup: (a) filament, (b) sheet, and (c) disc. The time interval between the frames is 2 ms. As a result of the collisional breakup, one large drop and several smaller satellites form (from Barros et al., 2008; American Meteorological Society©; used with permission).

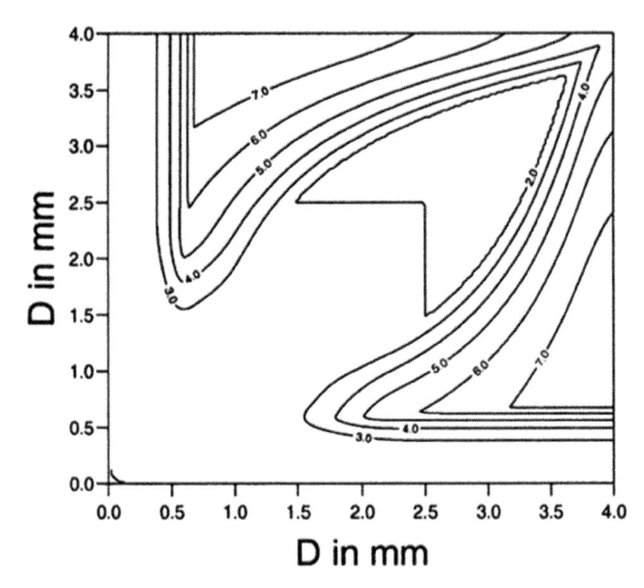

Figure 5.6.10 The total number of fragments in the filament mode. Calculation is based on parameterizations proposed by Low and List (1982b) and Beard and Ochs (1995) (from Seifert et al., 2005; American Meteorological Society©; used with permission).

as a sum of three fragment SDs in the breakup modes: filament (f), sheet (s), and disk (d):

$$R = R_f P_f + R_s P_s + R_d P_d, \tag{5.6.23}$$

where $P_f(D_i)$, $P_s(D_i)$ and $P_d(D_i)$ are defined as the average number of fragments of diameter D_i within the interval $D_i \pm \Delta D_i/2$, formed in the corresponding mode for one collision between two drops of diameter D_L (large drop) and D_S (small drop), i.e.,

$$\begin{aligned} P_f(D_i) &= N_{fi} C_f^{-1} \\ P_s(D_i) &= N_{si} C_s^{-1} \\ P_d(D_i) &= N_{di} C_d^{-1}, \end{aligned} \tag{5.6.24}$$

where N_{fi}, N_{si}, and N_{si} are the numbers of fragments of diameter D_i per size interval ΔD_i for the corresponding breakup mode, and C_f, C_s, and C_d are the total numbers of collisions for the corresponding breakup mode. The contribution of each breakup mode to the overall SD of fragments is determined by fractions R_f, R_s, and R_d of each mode: $R_f = C_f/(C_f + C_s + C_s)$, $R_s = C_s/(C_f + C_s + C_s)$, and $R_d = C_d/(C_f + C_s + C_s)$. Long and List (1982b) parameterize fractions R_f, R_s and R_d of each mode as functions of CKE and surface energy.

The parameterization procedure consists of several steps. First, the total number of drop fragments for each mode is calculated using the laboratory data, and is then parameterized as a function of the diameters of the colliding drops, of the total surface energy S_T and of the collision kinetic energy CKE. Laboratory experiments showed that the number of fragments increases from several (for a drop pair containing comparatively small colliding drops, e.g., a pair of drops with diameters 0.1 cm and 0.3 cm) to several tens of fragments for drop pairs containing large raindrops (e.g., 0.4 cm and 0.18 cm in diameter). Finally, the concentration of drops at each peak of fragment SD is determined for each mode.

Seifert et al. (2005) describe the number of fragments for the fragment mode P_f by matching parameterizations suggested by Beard and Ochs (1995) and Low and List (1982b). **Figure 5.6.10** shows the total number of fragments in the filament mode calculated according to this parameterization.

One can see that maximum numbers of fragments in the filament mode appear in collisions between drop pairs with the smaller drop in the pair of 0.5 mm. Number of fragments in the act of collision can be as large as seven.

The SD of fragments is determined as a sum of the fragments in each breakup mode. For instance, for the filament mode the distribution centered at the larger drop size looks like

$$P_{f1}(D_i) = A(D_L) \exp\left[-\frac{1}{2}\left(\frac{D_i - \mu}{\sigma_{f1}}\right)^2\right], \tag{5.6.25}$$

where σ_{f1} is the standard deviation of P_{f1}, $A(D_L)$ is a constant chosen to represent the peak value of P_{f1}, and μ is the parameter determining the location of the peak. In this particular case, the peak is centered at the diameter of the larger drop, so $\mu = D_L$. Using the experimental data, the coefficient $A(D_L)$ was approximated as $A(D_L) = 50.8D_L^{-0.718}$. Applying the normalization condition of the fragment SD allows us to calculate σ_{f1}. This method was used to determine the distributions of drop fragments within the entire range of the drop fragment sizes. Results reported by Long and List (1982b) are obtained under air pressure of 50 kPa. Later, List et al. (2009) reconsidered their results and expanded their analysis to the atmospheric pressure of 100 kPa. The effect of pressure was found to be not very significant. The new results were in agreement with those obtained by Long and List (1982b) under low and intermediate rain rates, but are more realistic for high rain rates (>50 mm/h). Laboratory experiments (Barros et al., 2008) performed using updated high-speed imaging techniques showed that SDs of the fragments are in agreement with the Low and List results. At the same time, a lower fragment number was reported for collisions of small rain drops of $D_S > 1$ mm with large drops of $D_L > 3$ mm.

A different parameterization of the SD of drop fragments as a function of fragment diameter D was proposed by Straub et al. (2010). This parameterization is based on the results of numerical simulations of collisions of different drop pairs. Obtained SDs mostly do not differ substantially from those found in the Low and List (1982b) parameterization **(Figure 5.6.11).** However, Straub et al. (2010) do not discriminate between different breakup modes. They separate the whole range of fragment sizes into four different diameter ranges. The first two ranges correspond to small satellites, the third range contains drops with diameters of about D_S, and the fourth range contains drops with diameters of about D_L. These four ranges are seen in Figure 5.6.11, where the SDs of fragments as a function of a fragment diameter D for different D_L and D_S are shown. In the first three ranges, functions $P_i(D, D_S, D_L)$ are parameterized by lognormal distributions with parameters depending on CKE (Equation 5.6.15) and We. In the fourth range, the fragment SD is described by the delta function, since the simulations show that the number of fragments within this range is equal to one. A single drop belonging to the fourth range has a diameter only slightly smaller than D_L because satellite drops are much smaller than the remaining large drop. The overall parameterized distribution of the breakup fragments is calculated as the sum of distributions within different diameter ranges:

$$R(D, D_S, D_L) = \sum_{1}^{4} R_i(D, D_S, D_L) \tag{5.6.26}$$

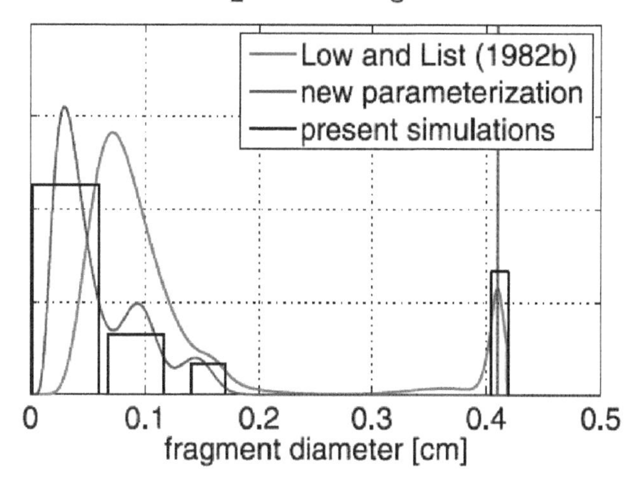

Figure 5.6.11 SD of fragments (cm^{-1}) as a function of fragment diameter (cm) for different D_L and D_S. Black rectangles show the simulation results, blue lines show the results of parameterization proposed by Straub et al. (2010), and red lines show the parameterization proposed by Low and List (1982b) (from Straub et al., 2010; American Meteorological Society©; used with permission).

Figure 5.6.11 shows examples of SD of fragments (cm^{-1}) as a function of fragment diameter (cm) for different drop pairs with drop diameters D_L and D_S. These distributions were obtained in laboratory experiments by Low and List (1982a, 1982b) and in numerical simulations by Straub et al. (2010). One can see that each fragment SD has a bimodal or a trimodal shape. The SD of drop fragments contains a drop with size close to that of the larger drop before the collision, drops with diameters close to that of small drop and several smaller satellites.

Table 5.6.2 summarizes the main results obtained by Straub et al., (2010) in numerical simulations of collisions of all the thirty-two drop pairs. The total number N_{res} of collision outcomes for a specific drop pair k is given by presented in the Table 5.6.2 is determined as

$$N_{res} = N_f(1 - E_{coal}) + E_{coal}, \tag{5.6.27}$$

where N_f is the number of fragments due to breakup. Equation (5.6.27) considers the formation of a single permanently coalesced drop. The table shows that the number of the fragments varies from two to about ten.

Table 5.6.2 The main results of numerical simulations and the parameterization by Straub et al., (2010) of collisional breakup. Notations: N_f is the number of fragments due to breakup for a specific drop pair k and N_{res} is the total number of collision outcomes as derived from the simulations. The last three columns contain values derived from the parameterization (from Straub et al., 2010; American Meteorological Society©; used with permission).

				Simulation			Parameterization		
k	d_L	d_S	d_L/d_S	E_c	$\overline{N}_f$	$\overline{N}_{res}$	E_c	$\overline{N}_f$	$\overline{N}_{res}$
1	0.18	0.0395	4.56	0.49	2.00	1.51	0.60	2.00	1.40
2	0.40	0.0395	10.13	0.81	3.00	1.38	0.76	2.00	1.24
3	0.44	0.0395	11.14	0.81	2.00	1.19	0.79	2.00	1.21
4	0.18	0.0715	2.52	0.25	2.00	1.75	0.25	2.00	1.75
5	0.18	0.10	1.80	0.25	2.00	1.75	0.25	2.00	1.75
6	0.30	0.10	3.00	0.25	3.28	2.71	0.10	3.58	3.31
7	0.36	0.10	3.60	0.25	5.20	4.15	0.12	4.52	4.08
8	0.46	0.10	4.60	0.25	5.84	4.63	0.21	5.02	4.18
9	0.36	0.18	2.00	0.00	4.95	4.95	0.05	5.73	5.51
10	0.46	0.18	2.56	0.00	8.21	8.21	0.06	10.24	9.72
11	0.06	0.035	1.71	1.00	2.00	1.00	0.87	2.00	1.13
12	0.12	0.035	3.43	0.49	2.00	1.51	0.67	2.00	1.33
13	0.12	0.06	2.00	0.25	2.00	1.75	0.45	2.00	1.55
14	0.25	0.0395	6.33	0.49	2.00	1.51	0.64	2.00	1.36
15	0.24	0.09	2.67	0.25	2.00	1.75	0.14	2.35	2.16
16	0.27	0.15	1.80	0.09	2.79	2.63	0.09	2.70	2.56
17	0.32	0.0395	8.10	0.81	2.00	1.19	0.70	2.00	1.30
18	0.41	0.14	2.93	0.09	8.59	7.91	0.06	7.83	7.41
19	0.24	0.06	4.00	0.25	2.00	1.75	0.33	2.00	1.67
20	0.30	0.07	4.29	0.25	2.85	2.39	0.26	2.45	2.07
21	0.36	0.07	5.14	0.49	3.88	2.47	0.32	2.72	2.18
22	0.45	0.07	6.43	0.49	3.00	2.02	0.43	2.81	2.04
23	0.12	0.10	1.20	0.49	2.00	1.51	0.84	2.00	1.16
24	0.41	0.10	4.10	0.25	6.48	5.11	0.16	4.93	4.30
25	0.25	0.12	2.08	0.09	2.53	2.39	0.10	2.56	2.41
26	0.30	0.12	2.50	0.09	3.85	3.59	0.07	3.99	3.79
27	0.36	0.12	3.00	0.09	3.49	3.27	0.07	5.56	5.23
28	0.46	0.12	3.83	0.25	6.16	4.87	0.13	6.60	5.87
29	0.36	0.14	2.57	0.09	4.90	4.55	0.05	6.20	5.95
30	0.18	0.16	1.13	1.00	2.00	1.00	0.87	2.00	1.13
31	0.41	0.16	2.56	0.00	9.72	9.72	0.05	9.05	8.67
32	0.25	0.18	1.39	0.25	2.75	2.31	0.29	2.00	1.71

One can see that the maximum number of drops forms as a result of collisions of drop pairs having zero collection efficiency. Numerical simulations indicate a lower number of the fragments in comparison with the Low and List parameterization. Laboratory experiments performed by Barros et al. (2008) also indicate that the Low and List parameterization overestimates the number of fragments formed by collisions of large drops.

5.7 Methods Used for Solving Stochastic Collection Equation and Stochastic Breakup Equation

5.7.1 Methods Used for Solving SCE

Collisions between drops lead to rain formation. Understanding the processes of raindrop formation and prediction of rain amount are among the major problems in cloud physics. In SBM models, DSD evolution due to collisions is described by SCE (Equation (5.6.5)). This equation is complicated and has no analytical solution for the hydrodynamical kernel, which imposes complicated requirements to numerical methods of SCE solution. The methods for solving SCE can be roughly separated into two main groups: the *point-based* methods and the *spectral-moment* methods.

In point-based methods (L.-P.Wang et al., 2007) the DSD $f(m,t)$ or some transformed form of DSD is defined on a discrete mass grid. The main problem of these methods is to represent the continuous distribution function as precisely as possible using the values of DSD at discrete points, in order to accurately evaluate the Loss and Gain integrals. The point-based methods can be in turn separated into two subgroups. In the first subgroup, the calculations of the Loss and Gain integrals are performed independently at regular grid points, and then the DSD variation rate is determined by the difference between these integrals. All possible drop-pair collisions are considered to form a particular target bin. Changes of DSD in all the target bins are considered. This method was proposed by Berry and Reinhard (1974a).

The basic idea of the methods belonging to the second subgroup is to calculate changes in the Loss and Gain integrals by summating the results of all pair collisions of the source drops corresponding to the source bins. The full set of the source bins is treated while analyzing the consequences of the binary-pair interactions, and the mass resulting from a binary interaction is separated between two adjacent bins. The methods proposed by Kovetz and Olund (1969), Bott (1998), and the linear discrete method used by Simmel et al. (2002) can be assigned to this subgroup. Wang et al. (2007) referred to the methods belonging to this subgroup as the bin-based pair-interaction methods.

Spectral-moment methods were pioneered by Bleck (1970) and Enukashvily (1980), and further developed by Tzivion et al. (1987, 1999, 2001). In these methods, SCE is written for DSD moments within discrete mass intervals or mass categories $m_i \leq m < m_{i+1}$, where m_i and m_{i+1} are considered the left and the right boundaries of the i-th category, respectively. The algorithm used in realization of these methods aims at representation of the category-based Loss and Gain integrals in terms of the spectral moments. The calculation of these integrals is performed independently on a regular grid formed by the interval boundaries, as in the Berry and Reinhardt method. The spectral-moment methods require averaging of SCE within each interval. Wang et al. (2007) proposed a version of combined spectral-moment method with Gauss quadrature integration, in which the ideas of the bin-based pair-interaction methods are used. There are also several methods that do not belong to either group. Next, we consider methods most often used in SBM cloud models for SCE solution.

The Kovetz and Olund Method

In the method proposed by Kovetz and Olund (1969), the DSD defined on the mass grid is determined by drop concentration in each bin as $N(m_i) = N_i$, where m_i is the mass of each drop belonging to the i-th bin. SCE is used in the form that is a discrete analog of Equation (5.6.5):

$$\frac{\partial N_i}{\partial t} = \sum_{n=1}^{i-1} \sum_{k=n+1}^{i} B(n,k,i) K(m_n, m_k) N_n N_k - N_i \sum_{n=1}^{i_{\max}} K(m_n, m_i) N_n \tag{5.7.1}$$

In Equation (5.7.1), the first term on the right-hand side represents the Gain integral and the second term is the discrete analog of the Loss integral. The method is illustrated in **Figure 5.7.1**. The first term describes all possible collisions between drops with masses m_k and m_n. At certain k and n, the resulting drop with mass $m_c = m_k + m_n$ is interposed either between bins with numbers $i-1$, and i (i.e., $m_{i-1} \leq m_c < m_i$) or between bins with numbers i and $i+1$, (i.e., $m_i \leq m_c < m_{i+1}$). In either case, the remapping procedure is used (similar to that discussed in Section 5.4) to divide the total drop mass and drop concentrations at an intermediate point c between the nearest grid points $i-1$ and i of the regular mass grid. The remapping procedure is

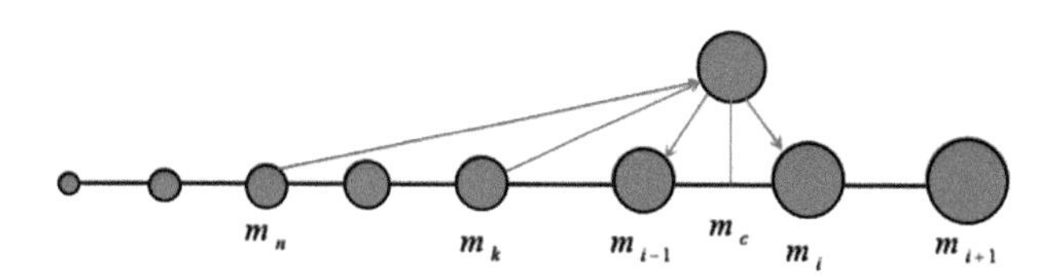

Figure 5.7.1 The collision procedure according to Kovetz and Olund (1969).

described by factor $B(n,k,i)$ in Equation (5.7.1), which is equal to

$$B(n,k,i) = \begin{cases} (m_c - m_{i-1})/(m_i - m_{i-1}), & \text{if } m_{i-1} < m \le m_i \\ (m_{i+1} - m_c)/(m_{i+1} - m_i), & \text{if } m_i < m < m_{i+1} \\ 0, & \text{if } m \le m_{i-1}, m \ge m_{i+1} \end{cases} \tag{5.7.2}$$

The values of the remapping coefficients depend on m_c, i.e., on the indexes of the source bins n and k.

The advantage of the Kovetz and Olund method is that it conserves the drop mass and the drop concentration obtained after collisions. At the same time, each remapping act leads to formation of drops with mass m_{i+1} that is larger than m_c, i.e., leads to an artificial numerical broadening of the drop spectrum. This method can be applied when the mass grid has a high resolution, which decreases the rate of largest drops production. The numerical DSD broadening caused by remapping is especially undesirable in simulations of first raindrop formation.

The Berry and Reinhardt Method

Berry and Reinhardt (1974a) used SCE written for MD function $\tilde{g}(\ln r)$ (see Section 2.1). This function obeys the normalization condition $M = \int_{\ln r_0}^{\ln r_{max}} \tilde{g}(\ln r) d \ln r$, where M is the liquid water content. The function is related to other distribution functions such as $\tilde{g}(\ln r) = rmf_r(r) = 3m^2 f(m)$ (Equations (2.1.15) and (2.1.16)). To write SCE in terms of function $\tilde{g}(\ln r)$, Berry and Reinhardt introduced a new nondimensional variable:

$$i = 1 + \ln (m/m_0)^{1/\ln \alpha}, \tag{5.7.3}$$

where m_0 is the minimum drop mass. This variable enables using a logarithmic scale that is convenient since drop masses change by several orders of magnitude. The minimum value of i is equal to one. Parameter α indicates scaling transformation from m to i. For convenience, Berry and Reinhardt introduced another parameter:

$$J_0 = \frac{3}{\ln \alpha} \tag{5.7.4}$$

The dependencies of drop radius and drop mass on the new variable i can be written as

$$r(i) = r_0 \exp \frac{(i-1)}{J_0}, \; m(i) = m_0 \exp \frac{3(i-1)}{J_0}, \tag{5.7.5}$$

where r_0 is the minimum drop radius corresponding to mass m_0.

Introducing the continuous function $G(i) = \tilde{g}(\ln r)$, Berry and Reinhardt rewrite SCE (5.6.5) in the form

$$\frac{\partial G(i)}{\partial t} = \frac{1}{J_0} \left\{ \int_1^{i_d} \frac{m^2(i)}{m^2(i_c) m(i')} G(i_c) K(i_c, i') G(i') di' - G(i) \int_1^{i_{max}} \frac{K(i_c, i')}{m(i')} G(i') di' \right\} \tag{5.7.6}$$

The upper limit in the Gain integral of SGE in Equation (5.6.5) equal to $m/2$ is replaced in Equation (5.7.6) by the value

$$i_d = i - J_0 \frac{\ln 2}{3} = i - \frac{\ln 2}{\ln \alpha} \tag{5.7.7}$$

According to Equation (5.7.6), collisions of drops with masses $m(i_c)$ and $m(i')$ trigger formation of drops of mass $m(i) = m(i_c) + m(i')$. Taking into account Equations (5.7.3) and (5.7.5), this equality allows expressing i_c in Equation (5.7.6) in terms of variables i and i' as

$$i_c = i + \frac{1}{\ln \alpha} \ln \left[1 - \alpha^{(i'-i)} \right] \tag{5.7.8}$$

The factor J_0^{-1} appears in Equation (5.7.6) since the variable of integration $d \ln r$ is substituted for di, as follows from Equation (5.7.5): $d \ln r = \frac{1}{J_0} di$.

Further, Berry and Reinhard (1974a) implement the discrete logarithmically equidistant mass grid where i becomes an integer value denoting a bin number. The values of function $G(i)$ at the grid point i are denoted as $G_i = \tilde{g}(\ln r_i)$. The differentials in the integrals of Equation (5.7.6) are substituted for finite differences:

$$d \ln r \sim \ln r_{i+1} - \ln r_i = \frac{1}{3} \ln \alpha = \frac{1}{J_0} \tag{5.7.9}$$

After the discretization, the value $\alpha = m_{i+1}/m_i$ represents the ratio of masses of successive bins. The fact that $d \ln r = const$ shows that the mass grid implemented is both logarithmically equidistant with respect to masses and is uniform with respect to i.

The method of evaluating the value i_c (Equation 5.7.8) is illustrated in **Figure 5.7.2**. The scheme is plotted for the case $\alpha = 2$, which means doubling of mass in neighboring bins. The scheme illustrates a situation when point i_c is located far enough from the DSD edges, so the values of G_{i-3}, G_{i-2}, G_{i-1}, G_i, G_{i+1}, and G_{i+2} are already calculated and are not equal to zero. As shown in the figure, formation of drops of mass m_i is determined by collisions of drops of mass $m_{i'}$ and drops of mass $m_c = m_i - m_{i'}$. Since $m_{i'} \le m_i/2$, i.e., $m_{i'} \le m_{i-1}$, mass m_c should be larger than $m_i/2$, so m_c

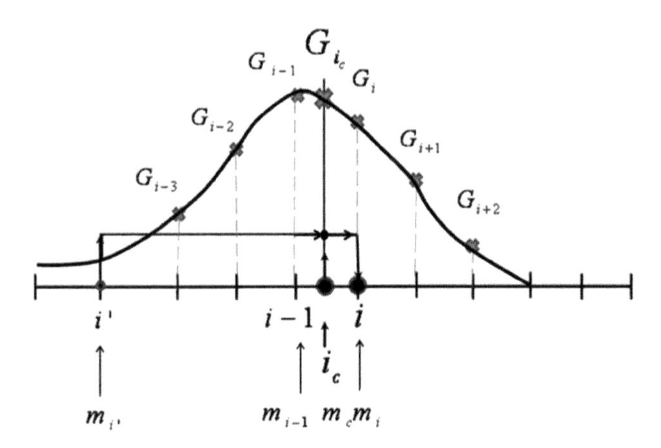

Figure 5.7.2 Scheme illustrating the algorithm proposed by Berry and Reinhardt in case when point i_c is located far enough from the DSD edges, so G_{i-3}, G_{i-2}, G_{i-1}, G_i, G_{i+1}, and G_{i+2} are known and not equal to zero. Solid line denotes continuous DSD. Crosses denote values of DSD at points of the mass grid. Drops of mass m_i belonging to the i-th bin are obtained by coalescence of drops of mass $m_{i'}$, belonging to the i'-th bin, and drops of mass m_c.

is located between m_{i-1} and m_i. Berry and Reinhardt calculated the unknown value of G_{i_c} using a six-point Lagrange interpolation formula applied to $\ln G_{i_c}$:

$$\ln G_{i_c} = \sum_{k=1}^{6} A_k \ln G_{i-4+k}, \tag{5.7.10}$$

where coefficients A_k are polynomials of value $A = i_c - i$.

A particular problem arises while describing formation of the largest drops that did not exist at the beginning of the collision time step. Emergence of new largest drops determines the rate of DSD expansion to larger drop sizes. Assuming $G_{i_{\max}}$ to be a nonzero value at the right edge of the DSD, all G_i with $i > i_{\max}$ are equal to zero, as shown in **Figure 5.7.3**. It means that the G_i values must be calculated when $i_c > i_{\max}$. For the case illustrated in Figure 5.7.3, a simple linear interpolation of logarithms of G_i is used:

$$\ln G_{i_c} = \ln G_{i_{\max}+1}(i_c - i_{\max}) + \ln G_{i_{\max}}(i_{\max} + 1 - i_c) \tag{5.7.11}$$

As soon as the G_{i_c} values are determined, the integrals in Equation (5.7.6) are calculated using the Lagrangian integration coefficients. For most bin points, three-point Lagrangian coefficients are used. However, in cases when i' is close to i (i.e., the sizes of the colliding drops are close), more precise four- and five-point Lagrange integration coefficients are used.

Simulations performed using the Golovin kernel (Equation (5.6.10)) and the Long kernel (Equation

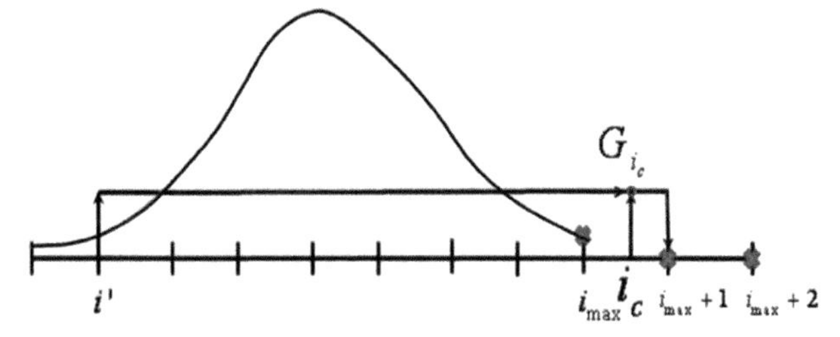

Figure 5.7.3 Schematic illustration of the approach used to calculate the DSD tail propagation in the Berry and Reinhardt method.

5.6.11)) showed that the Berry and Reinhardt scheme is quite accurate. It was used in many cloud models (e.g., Kogan, 1991; Khain and Sednev, 1996). Typically, this scheme does not produce artificial DSD broadening (i.e., it does not overestimate the rate of the DSD tail growth). When the method is implemented using the hydrodynamic kernels, it works well in cases when the number of bins is larger than several tens and the resolution is comparatively high (e.g., when $\alpha = \sqrt{2}$) and when the DSD are quite smoothed. However, the Berry and Reinhardt method has important shortcomings. First, the interpolations (Equation (5.7.10)) do not guarantee mass conservation. As soon as the DSD shape becomes complicated (e.g., bimodal or multimodal) due to nucleation or settling of drops of different sizes, the interpolations can lead to significant errors and even to a numerical instability in the DSD tail region (Carrio and Levi, 1995). Ice processes can lead to a quite complicated DSD shape with sharp variations and even zero values. In these cases, utilization of the Lagrange interpolation formula may lead to highly fluctuating distributions, and the application of simplier (e.g., linear) interpolation increases the stability of the scheme and enables the avoidance of formation of unrealistic DSD values.

The Bott Flux Method

Bott (1998) solved Equation (5.7.6) using a method different from that suggested by Berry and Reinhardt. The Bott procedure is illustrated in **Figure 5.7.4**. Pair collisions of drops with masses m_k and m_n are treated, and the target bins with numbers i and $i+1$ are determined from the condition $m_i \le m_c = m_k + m_n \le m_{i+1}$. The DSD values in bins n and k are calculated as

$$G_n^*(n,k) = G_n - G_n \frac{\overline{K}(m_n, m_k)}{J_0 m_k} G_k \Delta t,$$

$$G_k^*(k,n) = G_k - G_k \frac{\overline{K}(m_k, m_n)}{J_0 m_n} G_n \Delta t \tag{5.7.12}$$

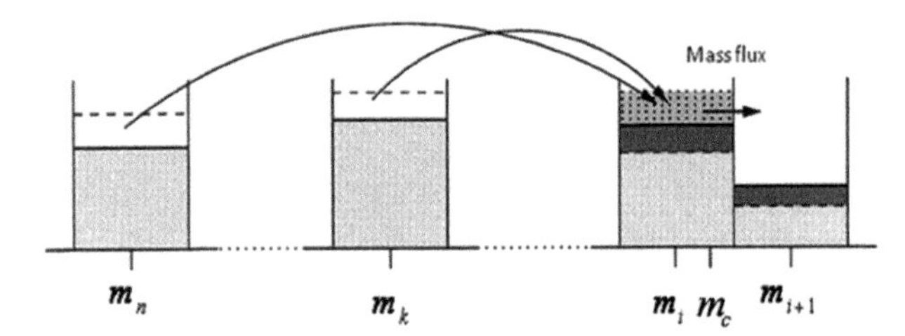

Figure 5.7.4 Schematic illustration of the flux method (from Bott, 1998; American Meteorological Society©; used with permission).

Equation (5.7.12) represent components of the loss integral calculated for collisions of drops belonging to source bins k and n. Bott used the averaged values of collision kernel in his procedure:

$$\overline{K}(m_n, m_k) = \frac{1}{8}[K(m_{n-1}, m_k) + K(m_n, m_{k-1}) + 4K(m_n, m_k) + K(m_{n+1}, m_k) + K(m_n, m_{k+1})] \quad (5.7.13)$$

The collision kernel is symmetrical, so $\overline{K}(m_n, m_k) = \overline{K}(m_k, m_n)$.

Due to collision of drops with masses m_n and m_k, new drops of mass m_c are produced with the total mass equal to

$$G^*_{i_c}(k, n) = \frac{m_c}{m_n m_k} G_k \frac{\overline{K}(m_k, m_n)}{J_0} G_n \Delta t \quad (5.7.14)$$

This total mass is separated between bins i and $i+1$. First, $G^*_{i_c}(k, n)$ is added to grid box i, yielding

$$G^*_i(k, n) = G_i + G^*_{i_c}(k, n) \quad (5.7.15)$$

Then, some fraction of mass is transferred from bin i to bin $i+1$, yielding the following values of DSD in the i-th and i + 1-th bins:

$$G_i^{t+\Delta t}(k, n) = G^*_i - flux_{i+1/2}(k, n),$$
$$G_{i+1}^{t+\Delta t}(k, n) = G_{i+1} + flux_{i+1/2}(k, n) \quad (5.7.16)$$

Equation (5.7.16) represents components of the Gain integral calculated for collisions of drops belonging to source bins k and n. Bott determined the value of $flux_{i+1/2}$ using an analogy with the higher-order advection schemes he suggested earlier and applied in a study (Bott, 1989a, 1998) for solving the dynamical equations. The first-order advection scheme corresponds to the remapping scheme by Kovetz and Olund (1969). As was mentioned, this scheme has a significant numerical viscosity and leads to an artificial drop spectrum broadening. Bott used the second-order approximation that does not introduce a significant artificial drop spectrum broadening. Testing the Bott scheme indicated a close agreement with the analytical solution in cases where the Golovin kernel is used. A good agreement with the Berry and Reinhardt scheme was also demonstrated.

The Bott method has several specific features:

a) In contrast to the Berry and Reinhardt method, the Bott method treats collisions of different drop pairs successively, i.e., the values of SDs obtained after collisions of any drop pair are afterward used to calculate collisions between other drop pairs within the same collision time step.
b) In case of low concentration of large raindrops when the application of SCE is questionable (Section 5.6), the Bott scheme coincides with the continuous growth equation (5.6.4), which is applicable for such conditions.
c) An important advantage of the Bott method is stability of results under different conditions, making the method attractive in cases of complicated DSD shapes caused by different processes such as sedimentation, freezing, riming, etc.
d) The Bott method is much less time consuming as compared to the Berry and Reinhardt method.

A drawback of the Bott method is that in contrast to the Kovetz and Olund method, it conserves mass but does not conserve drop concentration, which may lead to some artificial increase in drop concentration.

Bott (2000) extended his approach to the case when drops contain aerosols inside and the collision process is analyzed using 2D SDs (Section 2.1). This approach is essential for investigations of cloud–aerosol interaction and the effects of aerosols on DSD formation.

When solving SCE, the problem of choosing time step arises. This problem is discussed using the Bott scheme as an example. It is clear that the values of SDs $G^*_n(n, k)$ and $G^*_k(k, n)$ in Equation (5.7.12) should not be negative. This imposes the following condition on a time step:

$$\Delta t \leq \min\left\{\frac{J_0 m_k}{G_k \overline{K}(m_n, m_k)}, \frac{J_0 m_n}{G_n \overline{K}(m_k, m_n)}\right\} \quad (5.7.17)$$

Typically, concentration of rain drops is much less than that of cloud droplets. Since in each collision event two bins lose the same number of drops, application of a time step exceeding the limit determined by Equation (5.7.17) leads to the appearance of negative concentration in the bin. At the same time, time steps in most models are chosen large enough to make computations efficient. In this case, the values of terms such as $G_k \frac{\overline{K(m_k, m_n)}}{J_0 m_n} G_n$ in Equation (5.7.12) should be artificially corrected to prevent formation of negative values of SDs. These corrections are similar to those used while constructing positive defined numerical schemes (Section 4.5). This widely accepted technique actually means that the corresponding collision kernels are artificially decreased, which introduces errors in collisions description. Utilization of the doubled mass grid ($\alpha = 2$) in the Bott method allows us to avoid formation of negative concentrations; in this case, $m_k \le m_c \le m_{k+1}$ (Figure 5.7.4). It means that application of Equation (5.7.12), which can lead to negative values of $G_k^*(k,n)$), and the subsequent use of Equation (5.7.15) lead to a nonnegative final value of drop concentration in the k-th bin.

Solving SCE Using the Method of Moments

In spectral moment methods, SCE equation is formulated for moments of DSD (instead of DSD itself) written for discrete mass categories $m_i \le m < m_{i+1}$, where m_i and m_{i+1} are considered as the left and right boundaries of the categories, respectively. The moments of order I within each category are expressed as (Section 2.1)

$$M_i^I = \int_{m_i}^{m_{i+1}} m^I f(m) dm, \qquad (5.7.18)$$

where $I = 0$ corresponds to the drop concentration in the i-th category, $I = 1$ corresponds to the drop mass in the category, and $I = 2$ corresponds to reflectivity. Equations for M_i^I can be obtained by integration of Equation (5.6.5) (Tzivion et al., 1999). These equations can be written in the form (L.-P. Wang et al., 2007)

$$\frac{\partial M_i^I}{\partial t} = \int_{m_i}^{m_{i+1}} m^I dm \int_{m_0}^{m/2} f(m-m')K(m,m')f(m')dm' - \int_{m_i}^{m_{i+1}} m^I f(m) dm \int_{m_0}^{\infty} K(m,m')f(m')dm', \qquad (5.7.19)$$

where m_0 is the left boundary of the first category. Boundaries of the categories are defined on the logarithmically equidistant grid, i.e., $m_{i+1}/m_i = \alpha > 1$. Bleck (1970) solved Equation (5.7.19) for the first moment only, assuming that DSD is constant in each category. Tzivion et al. (1999) solved Equation (5.7.19) for the first two moments ($I = 0$ and $I = 1$). To close Equation (5.7.19), it is necessary to represent both the Loss and Gain integrals in terms of moments M_k^0 and M_k^1. To perform this closure, various approximations of SDs and collision kernels within each category were used. For instance, Tzivion et al. (1999) approximated DSD within i-th category as

$$m^I f(m) = \frac{m_i^I}{\alpha - 1} f(m_i)\left(\alpha - \frac{m}{m_i}\right) + \frac{m_{i+1}^I}{\alpha - 1} f(m_{i+1})\left(\frac{m}{m_i} - 1\right) \qquad (5.7.20)$$

Using Equations (5.7.18) and (5.7.20), as well as some closure assumption concerning the relationship between three neighboring moments in each category, Tzivion et al. (1999) obtained the following expressions for DSD at the boundaries of the i-th category:

$$f(m_i) = \frac{2M_i^0}{(\alpha-1)^2 m_i}\left[(\alpha-1)\frac{\overline{m}_i}{m_i} - \left(\frac{\overline{m}_i}{m_i}\right)^2 + \alpha\right]$$

$$f(m_{i+1}) = \frac{2M_i^0}{(\alpha-1)^2 m_i}\left[(\alpha-1)\frac{\overline{m}_i}{m_i} - \left(\frac{\overline{m}_i}{m_i}\right)^2 - \alpha\right], \qquad (5.7.21)$$

where $\overline{m}_i = \frac{M_i^1}{M_i^0}$ is the averaged mass of drops in i-th category. Equations (5.7.19–5.7.21) represent the closed equation system with respect to the two first moments in each category. Substitution of Equation (5.7.21) into Equation (5.7.20) and then into Equation (5.7.19) leads to a set of complicated equations for the two moments in each category, which contain sums of double integrals. Integrals related to the Gain integral should be calculated within variable boundaries. Integrands are of the type $m^n m'^l K(m,m')$ $(n,l = 0,1,2,3)$, in which all the values are assumed known, so most double integrals are calculated a priori. This procedure is also applicable if the coalescence kernel depends on time; however, in this case it can be approximated as a function of mass multiplied by a function of time. As a result, the Equation (5.7.19) is reduced to a system of differential equations having the following form:

$$\frac{dM_i^I}{dt} = F_i^I, \qquad (5.7.22)$$

where the right-hand side depends on all the moments under consideration. This equation system can be solved using an explicit scheme.

The numerical realization of the method is illustrated in **Figure 5.7.5** (Wang et al., 2007). For the example category 8, the Loss integral can be calculated in a

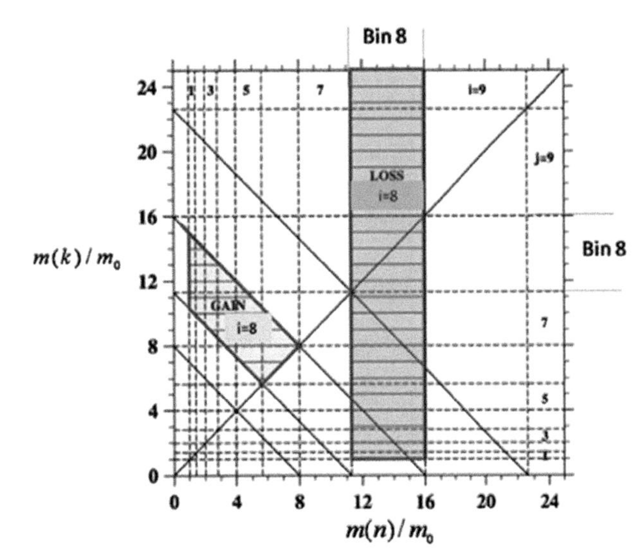

Figure 5.7.5 Illustration of the integration domains in the spectral moment method suggested by Tzivion et al. (1999). Both mass axes are normalized by m_0, and $\alpha = \sqrt{2}$. The numbers on the right and at the top indicate the number of source categories. Lines with a slope of -1 indicate the boundaries of the Gain integrals for the target categories. Vertical lines are the boundaries for the Loss integrals for target categories. Colored areas denote the corresponding domains for category number 8, chosen as an example of a target category. The tilted quadrilateral on the left indicates the domain of integration for the Gain integral, and the vertical rectangular denotes the domain of integration for the Loss integral (from Wang et al., 2007; courtesy of Elsevier).

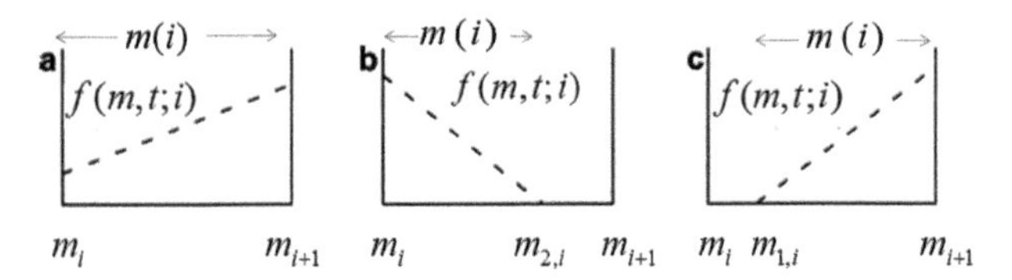

Figure 5.7.6 Three possible scenarios for representation of DSD within i-th category. m_i and m_{i+1} are masses corresponding to the category boundaries. Number SDs may be equal to zero within some fraction of the category, as shown in panels (b) and (c) (from Wang et al., 2007; courtesy of Elsevier).

straightforward manner because the integration domain boundaries coincide with the category boundaries. The integration domain for the Loss integral, where drops belonging to category 8 collide with all other drops, is plotted as the vertical rectangular integration domain (Figure 5.7.5). However, the Gain term for the same target category 8 is calculated over a tilted quadrilateral domain (the area marked by GAIN in Figure 5.7.5). This area is located between the two lines with a slope of -1, which obey conditions $m(n) + m(k) = m_7$ (left boundary) and $m(n) + m(k) = m_8$ (right boundary). The boundaries of the integration zone corresponding to the Gain integral do not coincide with the boundaries of source categories. For example, in Figure 5.7.5 one can see that drops in category 8 can be obtained by collisions of drops belonging to category 8 with drops belonging to categories 1–4, or by collisions of drops belonging to category 7 with drops belonging to categories 1–6. Such integration domain leads to very complicated and lengthy expressions for the Gain integral. Utilization of these complicated expressions requires computation of spectral moments of order higher than those explicitly considered (for instance, second-order moment) and require certain closure assumptions concerning the relationship between the moments.

Another disadvantage of the method is the existence of independent equations for masses and concentrations, which doubles the number of kinetic equations for collisions and makes the method time consuming. Difficulties in application of this method dramatically increase when mixed-phase microphysics is considered.

Wang et al. (2007) proposed a modification of the method of moments, where both the Loss and Gain integrals in Equation (5.7.19) are treated as summations of bin-based pair interactions, similarly to that used by Bott (1998) and Simmel et al. (2002). Wang et al. (2007) expressed number SDs within each i-th category by linear functions, as shown in **Figure 5.7.6**. Using this representation of DSD within the categories, Wang et al. (2007) found expressions for masses corresponding to the category boundaries (Figure 5.7.6, panel (a)) or for masses $m_{2,i}$ and $m_{1,i}$ (Figure 5.7.6, panels (b) and (c)) via zero M_i^0 and first M_i^1 moments defined in Equation (5.7.18). The meaning of these expressions is similar to that of expression (5.7.21) in the Tzivion et al. (1999) method.

The principal difference between the methods proposed by Wang et al. (2007) and by Tzivion et al. (1999) is in the way that the Loss and Gain integrals are calculated. Tzivion et al. (1999) calculate these integrals for a specific target category (as is also the case in the Berry and Reinhardt method (1974a)). Wang et al. (2007) considered collisions of drops belonging to source bins n and k (this process is referred to as cross interaction $n \leftrightarrow k$) leading to formation of larger drops belonging to target categories i and $i + 1$ (as suggested in the Bott scheme). The decrease in mass in the n-th and k-th categories was calculated as a part of the Loss integral for these categories, while the increase in mass within i-th and $i + 1$-th categories was regarded as a component of the Gain integral. For cross interaction $n \leftrightarrow k$, the reductions in drop concentration and drop mass in the source categories are written as

$$\Delta\tilde{M}_n^0 = \Delta\tilde{M}_k^0(n \leftrightarrow k) = -\Delta t \int_{m_{1,k}}^{m_{2,k}} dm(k) \int_{m_{1,n}}^{m_{2,n}} f(m;n)K[m(n),m(k)]f(m;k)dm(n)$$

$$\Delta\tilde{M}_n^1(n \leftrightarrow k) = -\Delta t \int_{m_{1,n}}^{m_{2,n}} dm(n) \int_{m_{1,k}}^{m_{2,k}} f(m;k)m(n)K[m(n),m(k)]f(m;n)dm(k) \quad (5.7.23)$$

$$\Delta\tilde{M}_k^1(n \leftrightarrow k) = -\Delta t \int_{m_{1,n}}^{m_{2,n}} dm(n) \int_{m_{1,k}}^{m_{2,k}} f(m;k)m(k)K[m(n),m(k)]f(m;n)dm(k)$$

The meaning of Equation (5.7.23) is similar to that of Equation (5.7.12) in the Bott method. The changes in concentration and masses within target categories i and $i+1$ due to drop cross-interaction $n \leftrightarrow k$ can be written as

$$\Delta\tilde{M}_i^0(n \leftrightarrow k) = \Delta t \iint_{\Omega_i} f(m;n)K[m(n),m(k)]f(m;k)dm(n)dm(k)$$

$$\Delta\tilde{M}_{i+1}^0(n \leftrightarrow k) = \Delta t \iint_{\Omega_{i+1}} f(m;n)K[m(n),m(k)]f(m;k)dm(n)dm(k)$$

$$\Delta\tilde{M}_i^1(n \leftrightarrow k) = \Delta t \iint_{\Omega_i} f(m;n)[m(n)+m(k)]K[m(n),m(k)]f(m;k)dm(n)dm(k) \quad (5.7.24)$$

$$\Delta\tilde{M}_{i+1}^1(n \leftrightarrow k) = \Delta t \iint_{\Omega_{i+1}} f(m;n)[m(n)+m(k)]K[m(n),m(k)]f(m;k)dm(n)dm(k)$$

where $m(n)$ and $m(k)$ are drop masses within the n-th and k-th categories and $f(m;n)$, and $f(m;k)$ are DSD values within categories n and k, respectively. Equation (5.7.24) represents components of the Gain integral in Equation (5.7.19), i.e., play a role similar to that of Equation (5.7.16) in the Bott method. In Equation (5.7.24), Ω_i and Ω_{i+1} are the sub-domains of integration, separating the cases when collisions of drops belonging to the n-th category with those in the k-th category lead to formation of drops belonging to the i-th category and the i + 1-th category.

Figure 5.7.7 shows a sketch of integration domains for the Gain integrals for the i-th and the i + 1-th categories, respectively. The ranges of integration over $m(n)$ and $m(k)$ vary depending on the location of the line $m(n)+m(k)=m_{i+1}$ cutting through the shaded area. To integrate Equations (5.7.23) and (5.7.24), the collision kernel within each category is calculated using the bilinear interpolation of the kernel values at the category boundaries. The Gauss quadrature formula is used to perform this integration. The procedure is completed with the summation of all the integrals to calculate the net Loss and Gain integrals for each category for $M_i^0(t)$ and $M_i^1(t)$ at time t, and to perform integration of Equation (5.7.24) over time step Δt.

The main characteristics of the methods just described are summarized in **Table 5.7.1.** A detailed comparison of different methods used for solving SCE is presented in Section 5.7.3.

5.7.2 Solving the Stochastic Breakup Equation (SBE)

Typically, the kinetic collection-breakup equation (5.6.22) is solved in cloud models in two steps. At the first step, the SCE equation is solved, as discussed in Section 5.7.1. At the second step, SBE (5.6.20) is solved. It is widely accepted to use the Bleck method to solve SBE (Pruppacher and Klett, 1997; Seifert et al., 2005).

Table 5.7.1 The main characteristics of the methods used to solve SCE.

Reference	Function	Source/ Target	Conservative Value	Artificial Spectrum Broadening	Computation Efficiency
Kovetz and Olund (1969)	$f(m)$	Source	Concentration, mass	Very large	Efficient
Berry and Reinhardt (1974a)	$\tilde{g}(\ln r)$	Target	No conservation	Very small	Intermediate
Bott (1998)	$\tilde{g}(\ln r)$	Source	Mass	Small	Efficient
Bleck (1970)	$M^1(m)$	Target	No conservation		Time consuming
Tzivion et al. (1999)	$M^0(m)$, $M^1(m)$	Target	No conservation	Very small	Highly time consuming
Simmel et al. (2002)	$M^0(m)$, $M^1(m)$	Source	No conservation	Small	Efficient
Wang et al. (2007)	$M^0(m)$, $M^1(m)$	Source	No conservation	Small	Time consuming

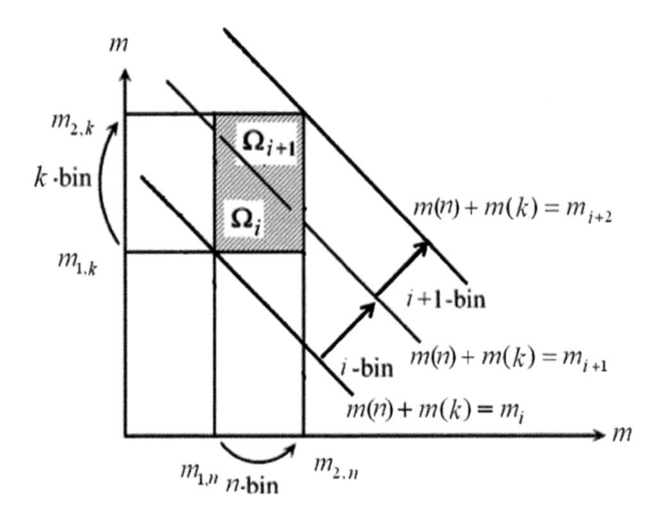

Figure 5.7.7 Sketch showing the domains of integration for the Gain integral of the i-th and the i + 1-th categories, respectively. The ranges of integration over the x and the y axes vary depending on the exact location of the line $m(n) + m(k) = m_{i+1}$ cutting through the shaded area (from Wang et al., 2007, with changes; courtesy of Elsevier).

The Bleck method is a version of the moment approach where mass-averaged variables for each category are used. Bleck defined the averaged DSD value $\bar{f}_k$ within the k-th category as

$$\bar{f}_k = \frac{2}{m_{k+1}^2 - m_k^2} \int\limits_{m_k}^{m_{k+1}} mf(m)dm \cong f(m_k) \quad (5.7.25)$$

A discrete analog of Equation (5.6.20) can be rewritten in the form (Pruppacher and Klett, 1997)

$$\left.\frac{\partial \bar{f}_k}{\partial t}\right|_{breakup} = \frac{2}{m_{k+1}^2 - m_k^2}\left(\sum_1^N \sum_{j=1}^{i} \bar{f}_i \bar{f}_j R_{kij} - \bar{f}_k \sum_{j=1}^{N} \bar{f}_j Q_{kij}\right), \quad (5.7.26)$$

where the breakup coefficients R_{kij} and Q_{kij} (so-called Bleck integrals) are given as

$$R_{kij} = \iiint\limits_{C_{kij}} mB(m', m'')R(m, m', m'')dm'dm''dm$$
$$Q_{kij} = \iiint\limits_{D_{kij}} \frac{m}{m + m''} m'B(m', m'')dm'dm''dm, \quad (5.7.27)$$

where the integration domains are determined as (List and Gillespie, 1976)

$$C_{kij} : m_k \le m \le m_{k+1}; m_i \le m' \le m_{i+1}; m_j'' \le m'' \le m_{j+1}'', \text{for } j < i$$
$$C_{kij} : m_k \le m \le m_{k+1}; m_i \le m' \le m_{i+1}; m_j'' \le m'' \le m, \text{for } j = i$$
$$D_{kij} : m_k \le m \le m_{k+1}; m_i \le m' \le m_N; m_j'' \le m'' \le m_{j+1}'' \quad (5.7.28)$$

Calculation of these multidimensional integrals, which include the fragment SD P presents a complicated problem since P is discontinuous and has high narrow peaks within C_{kij} (Brown, 1983). Integrals (5.7.27) are calculated just once at t = 0 and stored in tables to be used in consequent simulations.

Tzivion et al. (1989) solved SBE using the two-moment method. Solutions obtained with the two-moment scheme and with the Bleck one-moment scheme are in an excellent agreement. Pruppacher and Klett (1997) gave the following interpretation of this fact. During collisional growth, the drop mass is progressively redistributed toward larger drops where the category width increases. As a result, the deviation between the actual average drop mass and the mass corresponding to the category center increases. In contrast, drop breakup redistributes the drop mass toward smaller drops and narrower categories. Consequently, the differences between the actual average mass and the mass of the category center rapidly decrease; hence, errors related to this deviation decrease, as well. It appears that the simpler single-moment method is suitable for treating the stochastic breakup.

5.7.3 Examples of Drop Spectra Evolution Caused by Collisions and Breakups

One of the first analyses of DSD evolution was performed by Berry and Reinhardt (1974a) (see Figure 2.3.4 and Section 2.3). The figure illustrates time evolution of the initial MD function $g(\ln r)$ that consists of two modes: the cloud droplet mode centered at drop radius of 10 μm and the small raindrop mode centered at radius 50 μm. Since function $g(\ln r)$ describes MD, its utilization allows us to see the raindrop mode in spite of the fact that concentration of raindrops is much lower than that of cloud droplets. This example shows that if raindrops exist within the drop spectra, DSD evolution is determined largely by collisions of raindrops with cloud droplets. As a result, the raindrop mass increases and the cloud droplet mass decreases. In the example shown in Figure 2.3.4, it takes only about twenty min for all cloud droplets to be collected by raindrops. While the raindrop mode shifts toward larger sizes, indicating the growth of raindrops by collisions, the cloud droplet mode, until its full dissipation, remains centered at about 10 μm droplet radius.

The numerical solution of SCE depends on the resolution of the mass grid. As was shown by Tzivion et al. (1999) and Wang et al. (2007), most methods indicate convergence of results as the resolution of the mass grid increases. The simulations also show that different methods produce similar results at a very high mass grid resolution when, for instance, $m_{i+1}/m_i = \alpha = 2^{1/16}$. At cruder resolutions, different methods yield close results if the Golovin kernel is used. At the same time, utilization of cruder resolutions while using the Long kernel shows significant dispersion of results toward the time of ~40 min (**Figure 5.7.8**). DSDs obtained using different methods at $\alpha = 2^{1/2}$ are compared to DSD obtained by the Berry and Reinhardt (BRM) method at $\alpha = 2^{1/16}$, which is considered the benchmark solution. Figure 5.7.8 shows that both the BRM and the Wang et al. (BIMGQ) solutions are closer to the benchmark, especially near the peak region. At t = 20 min, no significant differences in DSD produced by different methods are seen. At forty min, the mass density function demonstrates formation of two peaks. BRM provides the best solution not only at the first peak but also at the second peak. The BIMGQ method overestimates the distribution at the first peak, while LFM and LDM underestimate it at the first peak. Consequently, LFM and LDM overestimate the rate of raindrop production, while the scheme BIMGQ underestimates it.

One can see that BRM does not introduce any significant drop spectrum broadening. As was shown by Khain et al. (2000), the Kovetz and Olund method leads to the highest DSD broadening among the schemes discussed. DSD broadening is quite important while considering formation of the largest cloud droplets and first rain drops. After the commencement of intense collisions, the effects of artificial DSD broadening become less significant. One can see that DSDs obtained by different methods at t = 40 min are quite different. For instance, the maximum raindrop radius obtained by BIMGQ is 0.8 mm, while the BRM produces raindrops as large as 1.4 mm (Figure 5.7.8c). However, these differences might not be of crucial importance since raindrop size increases very rapidly and rain drops grow from 0.8 mm to 1.4 mm in radius within only a few minutes.

SCE is often used to investigate the sensitivity of DSD evolution to different physical factors. We will present several examples. As was mentioned in Section 5.6, collision efficiencies and collision kernels increase with increasing height. **Figure 5.7.9** compares the drop spectrum evolutions in simulations using collision kernels calculated for different heights. The height dependence was calculated following Pinsky et al. (2001) (Section 5.6). The figure shows the forty-min evolution of the DSD (with the time increment of 10 s) at levels of 1,000 mb and 500 mb. One can see that the rate of the drop spectrum development significantly increases with height. For instance, toward the twenty-min point raindrops with radii as large as 300 μm form at $p = 500$ mb. At the same time, there are no raindrops at $p = 1,000$ mb. Thus, the effect of collision kernel increase with height is a factor significantly accelerating DSD development and rain formation.

The next example illustrates the effect of large drops that can form on giant CCN (GCCN) on the DSD development. Acceleration of warm rain formation caused by the effects of GCCN was found in several studies (e.g., Beard and Ochs, 1993; Lasher-Trapp et al., 1998; Yin et al., 2000). Under the assumption that raindrop formation takes place on GCCN only, the concentration of GCCN must be of the same order as that of raindrops, i.e., 100–1,000 m^{-3}. We illustrate the effects of GCCN on the evolution of drop spectrum by solving SCE using the Bott method (1998). The mass grid contains 400 mass bins to resolve DSD from 1 μm to 2,000 μm. **Figure 5.7.10** shows the evolution of the DSD initially centered at the radius of 7 μm, the cloud water

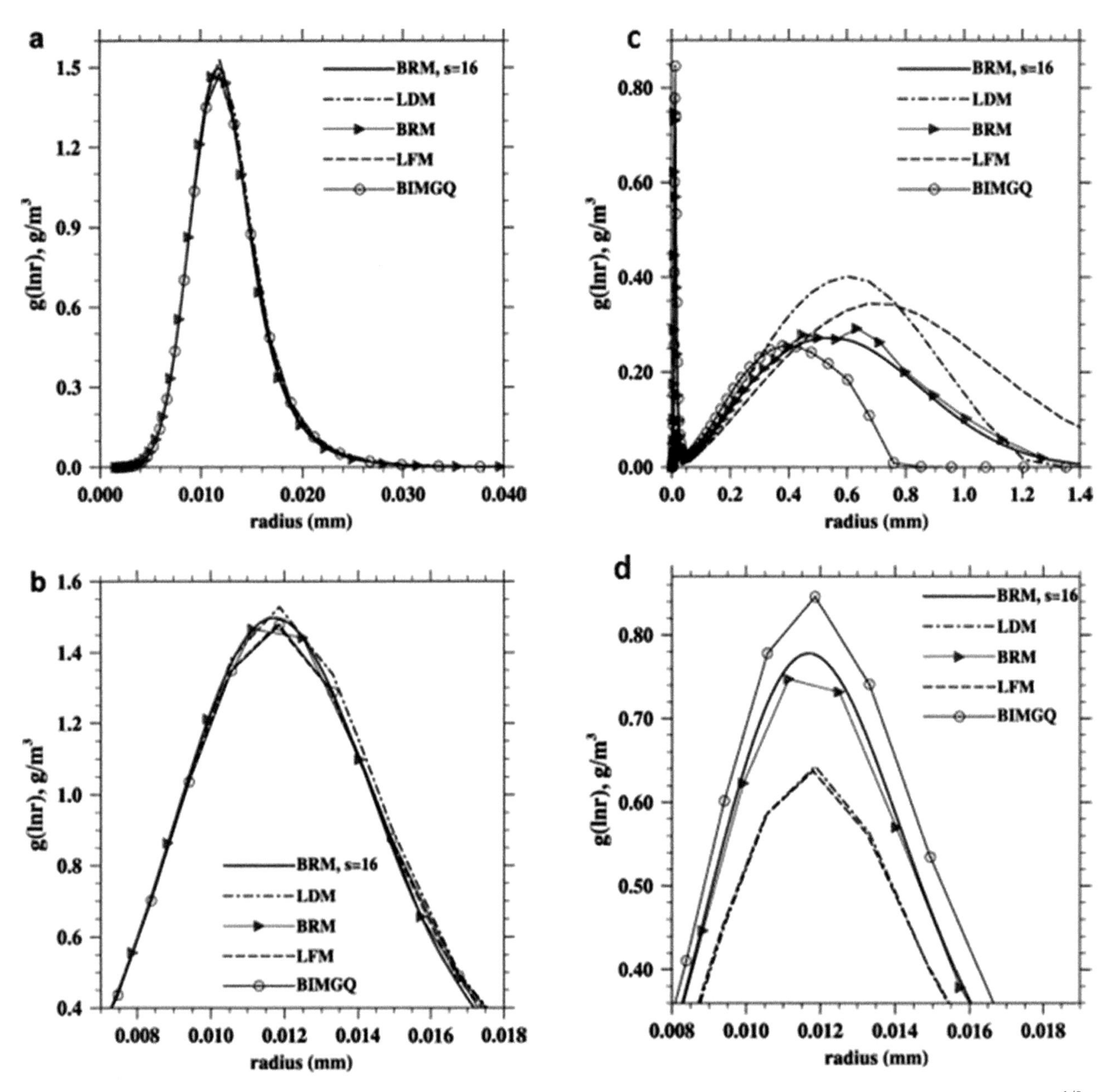

Figure 5.7.8 Panels (a) and (b): Mass density function at t = 20 min calculated for the Long kernel. Numerical solutions at $\alpha = 2^{1/2}$ obtained by different methods are compared with the solution obtained using BRMa at $\alpha = 2^{1/16}$. Panels (c) and (d): the same as in panels (a) and (b), respectively, but for t = 40 min. Panels (b) and (d) show a zoom-up of figures in panels (a) and (c), respectively. Notations: LDM is the Simmel et al. (2002) linear discrete method; LFM is the Bott linear flux method by Bott (1998) and BIMGQ is the bin integral method with the Gauss quadrature (Wang et al., 2007). Solid line denotes BRM with $\alpha = 2^{1/16}$ (from Wang et al., 2007; courtesy of Elsevier).

mixing ratio of 1 g/kg, and the drop concentration of 1,100 cm^{-3}, for two cases (with and without large drops in the initial drop spectrum). The concentration of drops with radii above 30 μm was set equal to 400 m^{-3}. One can see that in the absence of larger drops, no raindrops are formed (panel a). If large drops are added, the rain production does occur, but its rate is comparably low: only s about 8% of the total LWC was transferred to raindrops by or during the time instance of forty min.

Figures 5.7.10c and 5.7.10d present the DSD evolution in similar simulations with the initial DSD centered at droplet radius of 12 μm, which illustrates formation

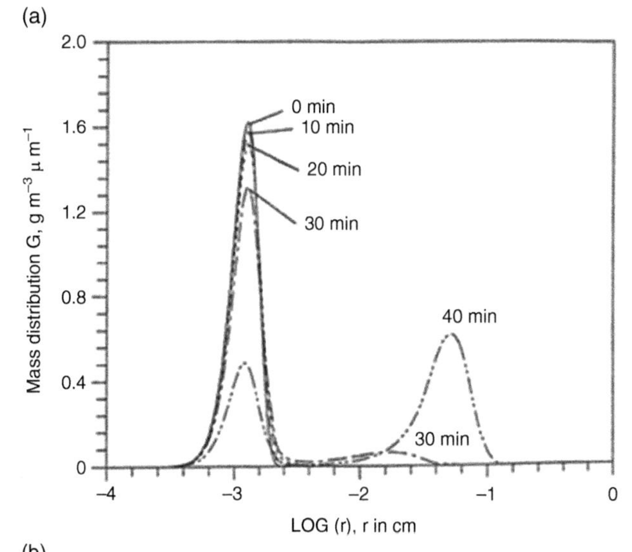

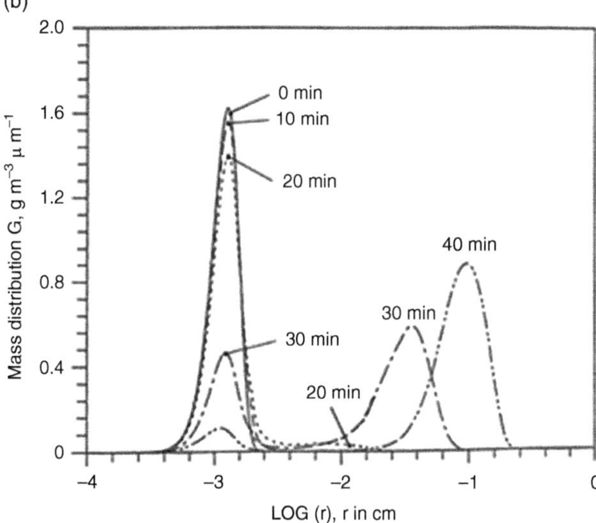

Figure 5.7.9 Drop spectrum evolutions in simulations that mimic drop collisions at different heights at the pressure of 1,000 mb (panel (a)) and of 500 mb (panel (b)) (from Khain et al., 2000; courtesy of Elsevier).

of raindrops in a maritime cloud. Both (c) and (d) simulations indicate nearly identical DSD development, demonstrating the insignificant effect of large drops added to the initial DSD. Thus, these simulations show a comparatively small role of the largest drops associated with GCCN on DSD development, especially in maritime clouds where DSD and raindrops form as a result of droplet collisions.

These conclusions were derived under a comparatively small CWC and low concentration of large drops. An increase in the CWC, e.g., up to 5 g m^{-3}, as in simulations performed by Lasher-Trapp et al. (1998), increases the contribution of the GCCN-associated largest drops to rain formation. At a very narrow drop spectra and high CWC, the effect of the GCCN would be dominating (Cooper et al., 1997), since in gravity-induced drop collisions small droplets of similar size do not collide. The sensitivity of DSD evolution to turbulence intensity is analyzed in Section 5.8.

Examples presented next illustrate the effect of collisional drop breakup on DSD. As was mentioned previously, the solution of SCE predicts a continuous shift of DSD toward larger drop sizes. At the same time, formation of raindrops intensifies the counter process of breakup, which shifts DSD toward smaller sizes. One can expect that there exists an equilibrium raindrop distribution when both processes compensate each other (Pruppacher and Klett, 1997). It is also known that distribution of rain drops is often close to the Marshall–Palmer distribution (Section 2.3) with a rapid decrease in larger drops concentration. Two questions arising in this connection are: to which extent the equilibrium raindrop distribution is close to the Marshall–Palmer distribution and to which extent this kind of distribution is close to observed raindrop distributions. **Figure 5.7.11** shows equilibrium DSD that were obtained by solving the collection-breakup stochastic equation (5.6.22) (spontaneous breakup is not considered) using several breakup parameterizations. The calculations of breakups were performed using the Bleck method. The DSD is represented on a logarithmic grid where the mass doubles at every third category. Initially, the Marshall–Palmer distribution $f_0(D) = N_0 \exp(-\lambda D)$ was assumed, with $N_0 = 8 \times 10^6\ \mathrm{m}^{-4}$, $\lambda = 4.1 \times 10^3 R^{-0.21}\ \mathrm{m}^{-1}$, with diameter D in m and the rain rate $R = 54\ \mathrm{mm\,h}^{-1}$ prescribed according to McFarquhar (2004b). The collection-breakup equation was integrated until the stationary state was achieved, which required DSD evolution of about 2 h.

The Low and List parameterization (1982b) demonstrates a stronger breakup rate, so the maximum raindrop diameter is of 3 mm only. In the Straub et al. parameterization (2010) the maximum drop diameter is 3.6 mm. The shapes of the equilibrium DSD are also different: while the Low and List scheme predicts three peaks occurring at 0.26, 0.80, and 1.95 mm, the Straub et al. (2010) scheme predicts a bimodal DSD. There are also substantial differences in the amount of intermediate raindrop with diameters around 1 mm. The slope of the Straub et al. (2010) curve for drop diameters exceeding about 2 mm is estimated as 22 cm^{-1}, which agrees very well with the slope obtained by Hu and Srivastava (1995) from surface measurements at high rainfall rates. Moreover, this slope also agrees

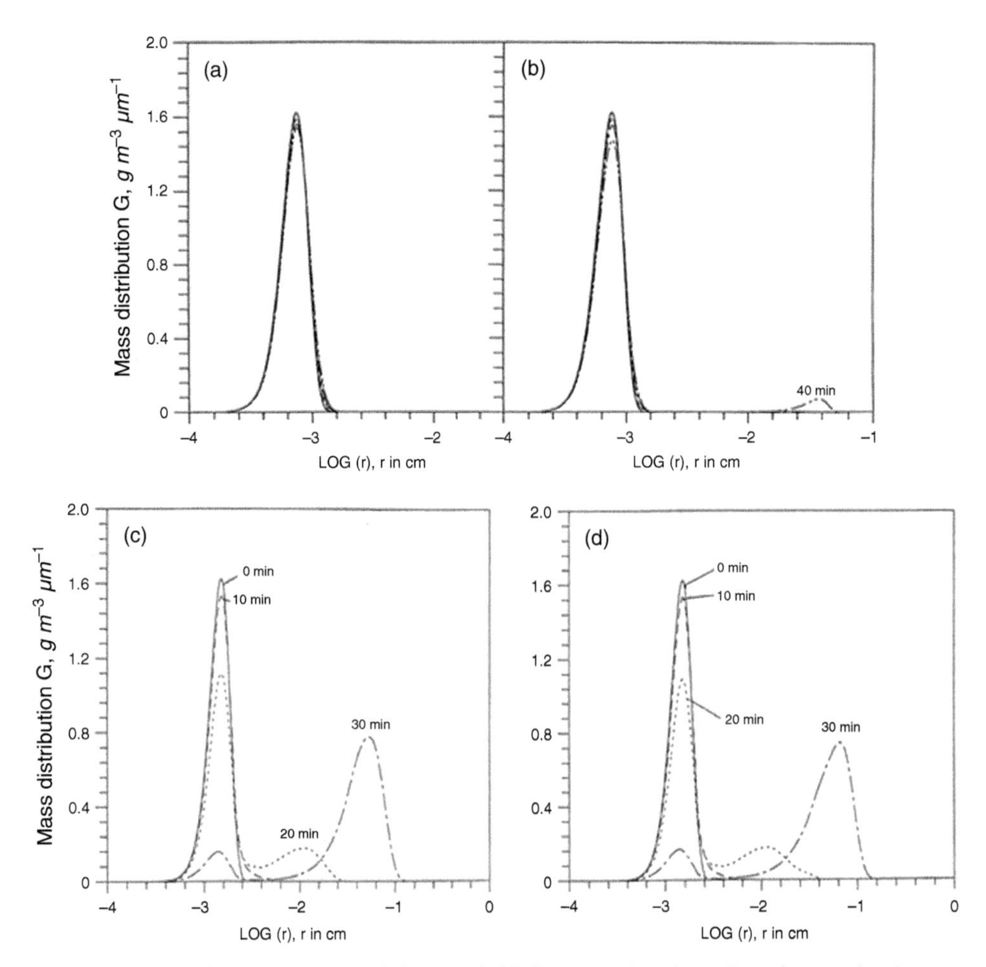

Figure 5.7.10 Panels a, b: evolution of the DSD initially centered at the radius of 7 μm, droplet concentration of 1,100 cm^{-3} and the cloud–water mixing ratio of 1 g kg^{-1} in two cases, namely: a) no large drops in the initial drop spectrum and b) large drops exceeding 30 μm in radius with concentration of 400 m^{-3} are added. Panels c, d are the same as (a) and (b), respectively, but for the DSD initially centered at the radius of 12 μm (from Khain et al., 2000; courtesy of Elsevier).

with multiple measurements performed in Karlsruhe (Germany) at the rain rates higher than about 10 mm h^{-1}. None of the solutions predicts Marshall–Palmer distribution, which in the log-linear coordinates should be represented by a straight line without any modes. In real clouds, the stationary state may not be reached and spatial averaging may lead to DSD closer to the Marshall–Palmer distribution.

Another example illustrates the effect of breakup on surface precipitation. In this example, a maritime convective system was simulated using profiles from the 1,200 UTC sounding on day 261 of the Global Atlantic Tropical Experiment (GATE) (Simpson et al., 1982; Ferrier and Houze, 1989). The weather system during this day was quite typical of maritime conditions and characterized by a rather humid boundary layer and a weak westerly background flow leading to intensive surface precipitation. The cloud microphysics of this convective system is to a large extent dominated by warm rain processes. To investigate the effects of collisional breakup under strong rain conditions, two simulations of the GATE case were performed, with and without breakup taken into account (Seifert et al., 2005). The simulations were performed using the mixed-phase HUCM with spectral bin microphysics (Khain et al., 2004). SCE was solved using the Bott method (1998), while SBE was solved using the Bleck method. The breakup was parameterized as described by Seifert et al. (2005) (Section 5.6). **Figure 5.7.12** compares the surface rain rate in simulations of both kinds (with and without breakup). Breakup decreases the maximum rain rate from 178 mm h^{-1} to 90 mm h^{-1}. In case breakup is taken into account, the rain rate reaches its maximum several minutes later. The

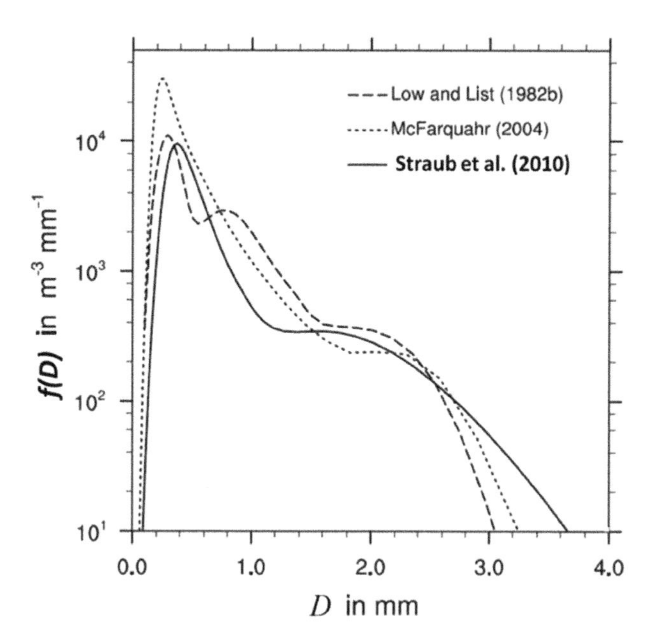

Figure 5.7.11 Equilibrium DSD that were obtained by solving the collision-breakup stochastic equation (5.6.22) (spontaneous breakup not considered) using several breakup parameterizations (from Straub et al., 2010; American Meteorological Society©; used with permission).

decrease in the maximum rain rate can be caused by the decrease in the mean fall velocity and in more intense evaporation. The second conspicuous difference between the results of the two kinds of simulations is breakup-related formation of a strong secondary precipitation event. This can be attributed to stronger evaporation of rain drops, which increases air cooling and leads to a stronger gust front and to a more intensive secondary convection. This example shows that a single microphysical process like breakup is able to affect the evolution of a mesoscale cloud system due to the strong nonlinear coupling between cloud microphysics and cloud dynamics.

5.7.4 Description of Collisions in Bulk-Parameterization Schemes

In bulk-microphysics models, the DSD is separated into two modes related to two different species: cloud droplets and raindrops. This distinction is not an artificial assumption used to simplify parameterization, but is based on the natural separation of the drop spectrum into the cloud mode and rain mode (Figure 2.3.3). It is accepted to distinguish four types (or processes) of collisions: a) *selfcollection* of droplets: collisions between cloud droplets, leading to formation of larger cloud droplets; b) *autoconversion*: collisions of cloud droplets leading to formation of raindrops; c) *accretion*: collisions between raindrops and cloud droplets; and d) *self-collection* of raindrops: collisions between raindrops. The four types of collisions are schematically illustrated in **Figure 5.7.13**.

The first parameterization of autoconversion was proposed by Kessler (1969). According to Kessler, the rate of rain mass production due to droplet collisions is proportional to the mixing ratio of cloud droplets q_c:

$$\frac{\partial q_r}{\partial t} = \begin{cases} k(q_c - q_{cr}), & \text{if } q_c > q_{cr} \\ 0 & \text{otherwise} \end{cases} \tag{5.7.29}$$

In Equation (5.7.29), k and q_{cr} are model parameters to be tuned to simulate rain formation in different clouds. The threshold value q_{cr} is often chosen equal to values from 0.5 to 1 g cm^{-3}, and $k = 10^{-3}$ s^{-1} (Straka, 2009). This formula does not take into account the shape of DSD and is usually used in single-moment bulk-parameterization schemes (e.g., Lin et al., 1983; Reisner et al., 1998). Equation (5.7.29) produces similar amounts of rain water under different DSD that are characterized, however, by the same CWC. As was shown (e.g., Figures 5.7.10a and 5.7.10c) rain production strongly depends on the DSD parameters, even under the same CWC. As is shown in Section 7.3, an increase in concentration of small aerosols typically leads to an increase in CWC, but makes the DSD narrower and causes a decrease in raindrop production rate. Equation (5.7.29), with given k and q_{cr}, does not describe this basic effect of aerosols. Typically, Equation (5.7.29) overestimates the raindrop production rate in continental and polluted clouds. To simulate raindrop production in polluted clouds using Equation (5.7.29), parameter q_{cr} should be increased up to 3–4 g cm^{-3}; to simulate rain formation in clouds developing in clean atmosphere, value q_{cr} should be decreased down to ~0.5 g cm^{-3}. Since aerosol loading changes over space and time, such adaptation of parameters is not practically possible (Beheng and Doms, 1986). Therefore, more advanced model schemes tend to avoid utilization of Equation (5.7.29) to describe autoconversion.

An ideal bulk parameterization of collision should reproduce solution of SCE in terms of different collision modes (autoconversion, selfcollection, etc.). However, the process of DSD evolution is quite complicated, so the currently available parameterizations are rather crude approximations of SCE solution. A theoretical background of bulk parameterization of collisions is presented by Beheng (1994) and Seifert and Beheng (2001). Seifert and Beheng (2001) illustrate their

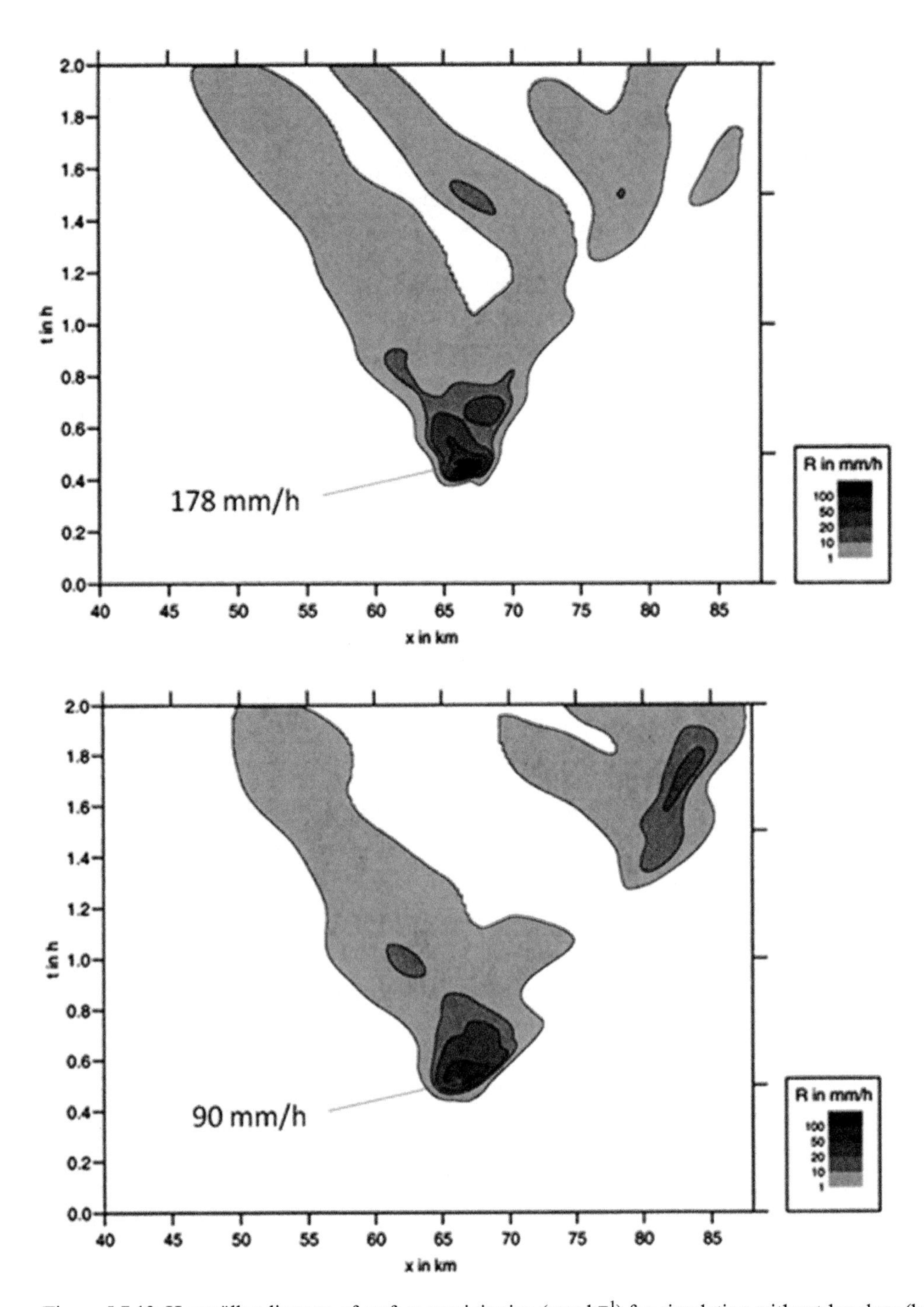

Figure 5.7.12 Hovmöller diagram of surface precipitation (mm h^{-1}) for simulation without breakup (left) and with breakup taken into account (right). Clouds observed in GATE (1974) were simulated (from Seifert et al., 2005; American Meteorological Society©; used with permission).

approach by starting parameterization of collision rates using SCE written for the moments Equation (5.7.19) with the Long collision kernel (Equation 5.6.11). Inserting Equation (5.6.11) into Equation (5.7.19) yields the following expressions for the zero moment (concentration, N_d), the first moment (liquid water mixing ratio, q_l), and the second moment (radar reflectivity, Z):

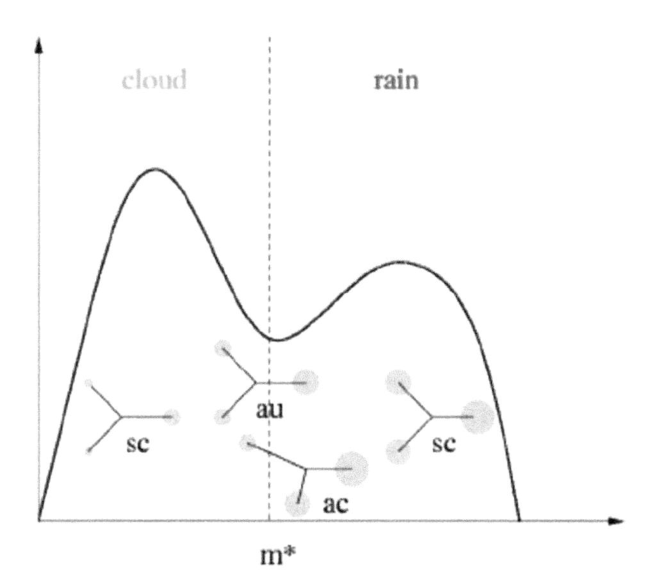

Figure 5.7.13 Four types of collisions: *sc* (selfcollection of cloud droplets, and raindrops), *au* (autoconversion) and *ac* (accretion). m^* is the drop mass separating cloud droplet and rain drop modes (from Khain et al., 2015; Courtesy of John Wiley & Sons, Inc.©).

$$\frac{\partial N_d}{\partial t} = \frac{\partial N_c}{\partial t} + \frac{\partial N_r}{\partial t} = -k_c N_c Z_c - k_r(N_c q_r + N_r q_c) - k_r N_r q_r \tag{5.7.30}$$

$$\frac{\partial q_l}{\partial t} = \frac{\partial q_c}{\partial t} + \frac{\partial q_r}{\partial t} = 0 \tag{5.7.31}$$

$$\frac{\partial Z}{\partial t} = 2k_c q_c M_c^3 + k_r(q_c Z_r + q_r Z_c) + k_r q_r Z_r, \tag{5.7.32}$$

where index "c" denotes cloud droplets, "r" denotes raindrops, and k_c and k_r are parameters in the expressions for the Long collision kernel. The complete moment is the sum of the partial moments for cloud droplets and rain drops:

$$M^k = M_c^k + M_r^k. \tag{5.7.33}$$

The first terms on the right-hand side of Equations (5.7.30–5.7.32) describe the rates of changes of the corresponding values, occurring due to selfcollection and autoconversion of cloud droplets. The second terms describe accretion. The third terms describe changes of the corresponding quantities due to selfcollection of raindrops. Equation (5.7.31) describes mass conservation during collisions.

The equation system (5.7.30–5.7.32) is not closed for two reasons. First, the equation of each moment M^k includes terms with a higher partial moment M^{k+1}. This closure problem is analogous to that in the turbulence theory (Section 3.5) and results from the nonlinearity of the collection kernel. The second closure problem arises due to separation of DSD into the cloud-droplet mode and the rain-drop mode: if the value of M^k is given, an additional assumption is necessary to calculate M_c^k and M_r^k. Both closure problems are circumvented by representing DSD $f(m)$ in the form of a specific mathematical function, which is completely determined by only a few parameters (Section 2.1). To derive an analytical approximation of autoconversion rate, Seifert and Beheng approximate the distribution of cloud droplets by a Gamma distribution function with parameter ν given a priori:

$$f_c(m) = Am^\nu \exp(-Bm), \nu = const \tag{5.7.34}$$

and distribution of raindrops by the exponential function

$$f_{ra}(D) = \alpha \exp(-\beta D), \tag{5.7.35}$$

where D is drop diameter. From a mathematical point of view, the expressions for DSD in each mode are defined within the radii range from zero to infinity. However, according to the main concept of bulk parameterization, the entire range of drop sizes should be separated into two subranges. Seifert and Beheng (2001) use radius $r^* = 50$ µm applied in Equation (5.6.11) in the Long kernel formula as an unmovable boundary value separating droplets and rain drops. With an appropriate choice of parameters of distribution (5.7.34), $f_c(m)$ contains a negligible number of drops with radii $r > r^*$; hence the cloud droplet concentration can be expressed as

$$N_c = \int_0^{m*} f_c(m)dm \approx \int_0^{\infty} f_c(m)dm = \frac{A\Gamma(\nu+1)}{B^{\nu+1}}, \tag{5.7.36}$$

where m^* is the mass of cloud droplets with radius r^* and Γ is the Gamma function (Section 2.1). Using Equations (5.7.34) and (5.7.35) and taking into account that the mean masses of cloud droplets and raindrops are $\overline{m}_c = \frac{q_c}{N_c}$ and $\overline{m}_r = \frac{q_r}{N_r}$, respectively, the expressions for the second partial moments can be written as $Z_c = M_c^2 = \frac{(\nu+2)}{(\nu+1)}\overline{m}_c q_c$ and $Z_r = M_r^2 = 20\,\overline{m}_r q_r$, respectively.

At the stage of the first raindrop formation, the mean mass of raindrops is assumed to be equal to the maximum mass of cloud droplets: $\overline{m}_r \approx m^*$ and $\overline{m}_c \ll m^*$. At this stage, autoconversion is the dominating process, so the terms in Equations (5.7.30–5.7.32) describing accretion and raindrop self-collection can be neglected. As a result, Equations (5.7.30–5.7.32) can be rewritten as

$$\frac{\partial N_c}{\partial t} \approx \left.\frac{\partial N_c}{\partial t}\right|_{au,sc} = -k_c N_c Z_c = -k_c \frac{(\nu+2)}{(\nu+1)} q_c^2 \tag{5.7.37}$$

$$\left.\frac{\partial q_r}{\partial t}\right|_{au} = -\left.\frac{\partial q_c}{\partial t}\right|_{au} = \frac{k_c}{20m^*}\frac{(\nu+2)(\nu+4)}{(\nu+1)^2} q_c^2 \overline{m}_c^2 \tag{5.7.38}$$

$$\left.\frac{\partial Z}{\partial t}\right|_{au,sc} = 2k_c \frac{(\nu+2)(\nu+3)}{(\nu+1)^2} q_c^2 \overline{m}_c^2, \tag{5.7.39}$$

where indices "au" and "sc" denote autoconversion and selfcollection, respectively. Equation (5.7.38) describes the rate of raindrop production and shows that it depends on CWC, on the mean mass of droplets and parameter ν characterizing DSD shape.

Equations (5.7.37–5.7.39) are valid for the initial period of cloud evolution only, when the first raindrops' mass is close to m^*. With further DSD development, the mean mass of raindrops increases and becomes larger than m^*. In addition, Equations (5.7.37–5.7.39) are derived for the Long kernel, which differs from the hydrodynamic collision kernel. In order to take into account the time evolution of autoconversion during a typical rain event, as well as to use a more realistic collision kernel, Seifert and Beheng (2001) introduced "universal functions" $\Phi_{au}(\tau)$ and $\Phi_{ac}(\tau)$ depending on a dimensionless parameter $\tau = 1 - \frac{q_c(t)}{q_l}$. This parameter shows the mass fraction of raindrops within the total LWC. The universal functions were derived by comparison with the exact solution of SCE and are given as $\Phi_{au}(\tau) = 600\tau^{0.698}\left(1 - \tau^{0.68}\right)^3$ and $\Phi_{ac}(\tau) = \left(\frac{\tau}{\tau + 5\cdot 10^{-4}}\right)^4$. These functions are chosen in such a way to reproduce solutions of SCE with the Hall collision kernel (Hall, 1980). In case other kernels are used, the expressions for the universal functions should be reconsidered. The final parameterization formulas for different collision modes are presented in **Table 5.7.2.**

The approach proposed by Seifert and Beheng (2001) is supposedly the most advanced among the available parameterization schemes. At the same time, in most bulk schemes the rates of different collision modes (mainly, the autoconversion rate) are based on parameterizations proposed by Berry and Reinhardt (1974b and 1974c). The parameterizations of autoconversion rate were derived from results of simulations of DSD evolution, where the initial DSD distributions were presented in the form of Gamma distribution, with the mean volume radius varying from 10 μm to 18 μm in different simulations. SCE was solved using the collision efficiencies calculated by Hocking and Jonas (1970) (Figure 5.6.4, Section 5.6.3). All the collisions were assumed to result in coalescence. Berry and Reinhardt (1974b) presented curve fits that show the relationship between the mean mass and the dispersion of the initial DSD at $t = 0$ (DSD1) and the mean mass and concentration of DSD taken at certain time $t = T_2$

Table 5.7.2 Formulas for rates of autoconversion, selfcollection, and accretion for mass contents and number concentrations according to the scheme suggested by Seifert and Beheng (2001).

	Rain Drops Transformation Rates	Cloud Droplet Transformation Rates
Autoconversion, mass contents	$\left.\frac{\partial q_r}{\partial t}\right\|_{au} = \frac{k_c}{20m^*}\frac{(\nu+2)(\nu+4)}{(\nu+1)^2} q_c^2 \overline{m}_c^2 \left[1 + \frac{\Phi_{au}(\tau)}{(1-\tau)^2}\right]$	$\left.\frac{\partial q_c}{\partial t}\right\|_{au} = -\left.\frac{\partial q_r}{\partial t}\right\|_{au}$
Autoconversion, concentrations	$\left.\frac{\partial N_r}{\partial t}\right\|_{au} = -\frac{1}{2}\left.\frac{\partial N_c}{\partial t}\right\|_{au}$	$\left.\frac{\partial N_c}{\partial t}\right\|_{au} = \frac{2}{m^*}\left.\frac{\partial q_c}{\partial t}\right\|_{au}$
Selfcollection of droplets	$\left.\frac{\partial N_r}{\partial t}\right\|_{sc} = -k_r N_r q_r$	$\left.\frac{\partial N_c}{\partial t}\right\|_{sc} = -k_c \frac{\nu+2}{\nu+1} q_c^2 - \left.\frac{\partial q_c}{\partial t}\right\|_{au}$
Accretion, mass contents	$\left.\frac{\partial q_r}{\partial t}\right\|_{ac} = k_r q_c q_r \Phi_{ac}(\tau)$	$\left.\frac{\partial q_c}{\partial t}\right\|_{ac} = -\left.\frac{\partial q_r}{\partial t}\right\|_{ac}$
Accretion, concentration		$\left.\frac{\partial N_c}{\partial t}\right\|_{ac} = \frac{1}{\overline{m}_c}\left.\frac{\partial q_c}{\partial t}\right\|_{ac}$

(DSD2). The DSD2 is the result of accretion and self-collection. Berry and Reinhardt characterize the shape of DSD by three characteristic masses: $m_f = \frac{q_l}{N_d}$, which is the mean drop mass, mass $m_g = \frac{Z}{q_l}$, characterizing the level of DSD elongation toward the right boundary, and mass $m_b = \sqrt{m_f m_g - m_f^2}$, characterizing the DSD width. Berry and Reinhardt define mass m_b as a standard deviation of mass from the mean value m_f, and introduce the characteristic time T_2 during which radius $r_g = \frac{3}{4\pi\rho_w} m_g^{1/3}$ reaches 50 μm. Based on the results of numerical simulations, Berry and Reinhardt obtained an approximation formula to determine time T_2. To avoid misinterpretations, Gilmore and Straka (2008) wrote the expression for T_2 together with units of different values as

$$T_2 = 3.72\{\mathrm{s \cdot kg \cdot m^{-3} \mu m}\} \times \left[\left(10^6\left\{\frac{\mu\mathrm{m}}{\mathrm{m}}\right\} r_b|_{t=0}\{\mathrm{m}\} - 7.5\{\mu\mathrm{m}\}\right) q_c\Big|_{t=0}\{\mathrm{kgm^{-3}}\}\right]^{-1} \tag{5.7.40}$$

In Equation (5.7.40), r_b is the radius corresponding to mass m_b. The approximated amount of cloud–water mass Δq_{c2} converted to rainwater after T_2 is calculated as

$$\Delta q_{c2}\{\mathrm{kgm^{-3}}\} = \left[10^{20}\left\{\frac{\mu\mathrm{m}^4}{\mathrm{m}^4}\right\}\left(r_b|_{t=0}\right)^3\{\mathrm{m}^3\} r_f\{\mathrm{m}\} - 0.4\{\mu\mathrm{m}^4\}\right] \times 2.7 \times 10^{-2}\{\mu\mathrm{m}^{-4}\} q_c|_{t=0}\{\mathrm{kgm^{-3}}\} \tag{5.7.41}$$

Combining Equations (5.7.40) and (5.7.41) allows us to express the average rate of rain-mixing ratio production during time interval T_2 due to autoconversion as

$$\left.\frac{dq_r}{dt}\right|_{au}\{\mathrm{kg\,kg^{-1}\,s^{-1}}\} = \frac{1}{\rho_w}\frac{\Delta q_{c2}}{T_2} \tag{5.7.42}$$

Berry and Reinhardt parameterize the number concentration of raindrops ΔN_{r2} formed during period T_2 as: $\Delta N_{r2}\{\mathrm{m^{-3}}\} = 3.5 \times 10^6\{\mathrm{g^{-1}}\}\Delta q_{c2}\{\mathrm{gm^{-3}}\}$. Verlinde and Cotton (1993) used this expression to parameterize the net rate of raindrop concentration change caused by autoconversion, as well as raindrop selfcollection over T_2 by means of a double moment bulk-parameterization scheme:

$$\frac{dN_r}{dt}\{\mathrm{m^{-3}s^{-1}}\} = \frac{1}{3.5 \times 10^9\{\mathrm{kg}\}}\rho\{\mathrm{kgm^{-3}}\}\frac{dq_r}{dt}\{\mathrm{kgkg^{-1}s^{-1}}\}, \tag{5.7.43}$$

where $\frac{dq_r}{dt}$ is determined by Equation (5.7.42). Analysis shows that Equation (5.7.43) corresponds to production of raindrops with the mean volume diameter of ~82 μm. Equations (5.7.42) and (5.7.43) show that raindrop production is proportional to CWC and increases with an increase in DSD width characterized by parameter r_b.

Despite the fact that Equations (5.7.42) and (5.7.43) are applicable only for the time period T_2 required for the first raindrops to form, most microphysical schemes use these formulas for all time of cloud evolution. Besides, the Berry and Reinhardt parameterization is based on a comparatively low number of simulations with the mean radii of the initial DSD 10 μm $\le r_f \le$ 18 μm and for the same LWC of 1 g m^{-3}. Berry and Reinhardt stressed that extrapolating their results beyond these limits should be made with caution. In addition, as mentioned by Gilmore and Straka (2008), the adequate mass and number concentration rates are difficult to determine using this method, since DSD1 and DSD2 overlap. These complications motivated the development of more than ten parameterization versions following the Berry and Reinhardt–like approach (Gilmore and Straka, 2008). These versions differ by the choice of DSD shape (the distribution functions being monodisperse, generalized Gamma, lognormal, etc.), as well by different values of the mean diameter and parameters characterizing the DSD shape. Some schemes use the mean volume radius as a governing parameter, other schemes use the mean radius, thus arriving at different parameterization formulas. Cohard and Pinty (2000) and Milbrandt and Yau (2005) use the time corresponding to formation of a hump in DSD as a characteristic time of first raindrop formation. These differences account for differences in rates of raindrop formation, sometimes as high as by orders of magnitude, obtained in different versions of the same scheme (Gilmore and Straka, 2008).

Berry and Reinhardt did not propose formulas for changes of cloud droplet concentration caused by self-collection and accretion. To fill this gap, Cohard and Pinty (2000), Milbrandt and Yau (2005), and Carrio and Nicolini (1999) approximated the rate droplet concentration decrease caused by selfcollection and accretion, using the expression derived from solving SCE with the Long self-collection kernel. Gilmore and Straka (2008) wrote this expression with SI units as

$$\frac{dN_c}{dt}\{\mathrm{m^{-3}s^{-1}}\} = -9.44 \times 10^9\{\mathrm{kg^{-2}m^3s^{-1}}\}(\rho q_c)^2\{\mathrm{kg^2m^{-6}}\}a_w, \tag{5.7.44}$$

where a_w is a value depending on the parameters of the generalized Gamma distribution (Section 2.1).

Most bulk schemes (with the exception of Seifert and Beheng, 2001) do not treat parameterization of raindrop production rate at later time instances when raindrop diameter exceeds ~80 μm. Strictly speaking, these schemes can be attributed only to the stage of the first raindrops' formation. Gilmore and Straka (2008) attribute the lack of schemes capable of treating autoconversion at later time instances to the fact that most bulk schemes cannot effectively deal with DSD that are not unimodal but bimodal or multimodal. Likewise, separate modes for rain and drizzle cannot be well represented with a unimodal Gamma distribution or a log-normal distribution. Cohard and Pinty (2000) proposed a simple method that takes into account (at least partially) the changes in rain production over time. Instead of the predominant rain mass $m_b(T_2)$, which corresponds to drop diameter of ~82 μm, this method uses the mean rain mass. As a result, lower concentrations of raindrops are produced.

Investigators paid special attention to parameterization of drizzle formation in stratocumulus clouds. A set of parameterization formulas was developed by Khairoutdinov and Kogan (2000) using the results of large eddy simulations (see also Section 7.4), including several formulas for rate of drizzle formation by autoconversion:

$$\left.\frac{\partial q_r}{\partial t}\right|_{au} = 1{,}350\, q_c^{2.47} N_c^{-1.79} \tag{5.7.45a}$$

$$\left.\frac{\partial q_r}{\partial t}\right|_{au} = 4.1 \times 10^{15} \bar{r}_c^{5.67}, \tag{5.7.45b}$$

where $\bar{r}_c$ is the mean-volume droplet radius in μm. Two parameters of DSD distributions are used in Equation (5.7.45a), while Equation (5.7.45b) incorporates only a single parameter $\bar{r}_c$. Assuming that all newly formed drizzle drops have a radius of $r_{dr0} = 25$ μm, the rate of drizzle concentration production can be written as

$$\left.\frac{\partial N_r}{\partial t}\right|_{au} = \frac{\left.\frac{dq_r}{dt}\right|_{au}}{\left(\frac{4}{3}\pi\rho_w r_{dr0}^3\right)} \tag{5.7.46}$$

Parameterization of the accretion in different bulk schemes by means of the hydrodynamic kernel Equation (5.6.3) is based on utilization of formulas for continuous drop growth (Equation (5.6.4)) (e.g., Walko et al., 1995):

$$\left.\frac{\partial q_r}{\partial t}\right|_{ac} = \frac{\pi}{4\rho}\overline{E}\int_0^\infty\int_0^\infty m_r(D_c+D_r)^2\left|V_{g_r} - V_{g_c}\right| f_c f_r dr_c dr_r, \tag{5.7.47}$$

where DSD values of droplets and raindrops are fixed by Equations (5.7.34) and (5.7.35), and $\overline{E}$ is the averaged collection efficiency between cloud droplets and raindrops. In Equation (5.7.47), integration over DSD of both droplets and raindrops is performed from zero to infinity, i.e., it is assumed that overlapping of these distributions can be neglected. Verlinde et al. (1990) found an analytic solution for Equation (5.7.47) in terms of known functions. To reduce computational efforts during model runtime, a large number of solutions are precomputed and tabulated in 3D lookup tables.

In two-moment schemes, Equation (5.7.47) for drop mass is supplemented by the equation for a change in raindrop concentration (Meyers et al., 1997):

$$\left.\frac{\partial N_r}{\partial t}\right|_{ac} = -\frac{\pi}{4\rho}\overline{E}\int_0^\infty\int_0^\infty (D_c + D_r)^2\left|V_{g_r} - V_{g_c}\right| f_c f_r dr_c dr_r \tag{5.7.48}$$

Like to Equation (5.7.47), multiple solutions of Equation (5.7.48) are computed and tabulated. An attempt to avoid overlapping between DSD for droplets and raindrops while calculating the integrals (5.7.47) and (5.7.48) was performed by Seifert and Beheng (2001), who integrated SD for raindrops beginning with a minimum raindrop radius r^*.

Collisional breakup of raindrops is taken into account by modifying the mean collection efficiency (Meyers et al., 1997):

$$E_c(D_m) = \begin{cases} 1 & \text{for } D_m < D_{cut} \\ 2 - \exp\left[A(D_m - D_{cut})\right] & \text{for } D_m > D_{cut} \end{cases}, \tag{5.7.49}$$

where $A = 2300$, D_m is the mean diameter of raindrops (m) and D_{cut} is the critical value of raindrop diameter. Meyers et al. (1997) used $D_{cut} = 6 \times 10^{-4}$m. In two-moment bulk schemes, breakup of raindrops results in a change of raindrop concentration.

To sum up, many formulas have been proposed to parameterize the rates of changes in raindrop mass and concentrations caused by collisions. Seifert and Beheng (2001) present a certain scientific basis for such parameterizations. Most of the formulas for rain production by autoconversion are applicable to the initial stage of the first raindrop formation only. Even so, the rates predicted by these formulas differ by orders of magnitude (Gilmore and Straka, 2008). These huge differences between the predictions can be attributed to the highly nonlinear nature of SCE, where the rates depend on DSD shape and its change with time, as well as on the mass content. In case DSD contains raindrops (which is a typical situation in clouds), the rate of raindrop formation depends on the shape of a cloud, as well as on the rain drop modes and relationship between the drop masses in the modes. It should be added that natural collision kernels depend on height. As shown in Section 5.8, collision kernels depend also on the intensity of turbulence, i.e., vary over time and space. Possible solutions of SCE (especially with kernels variable in space and time) are so numerous that it is difficult to parameterize all of them using comparatively simple expressions. It is necessary to take into consideration that the parameterization formulas were derived under the assumption that DSD are unimodal and can be represented by the Gamma distributions and exponential distributions. The assumption that the DSD shapes can be represented by these analytical formulas simplifies the microphysical parameterization. However, this representation does not always agree with the shape of real DSD in clouds where distributions of cloud droplets and raindrops can be bimodal or multimodal. As a result, it is difficult to propose universal parameterization formulas suitable for all conditions and in all kinds of clouds. The development of parameterizations based on adjusting solutions of SCE might be an answer to the problem. However, such an adjustment decreases uniformity of parameterization formulas because over a limited number of SCE evolution simulations it is impossible to reproduce the entire variety of actual situations in clouds. The problem of parameterization becomes even more complicated when collisions in mixed-phase clouds are analyzed.

5.8 Turbulent Collisions of Drops and Their Parameterization in Cloud Models

Difference in gravity-induced velocities (Section 5.6) is not a single reason of drop collisions in clouds; another major factor is turbulence in clouds. Investigation of turbulence effects on drop collisions started with the study by Arenberg (1939). A significant progress was achieved in the classical study by Saffman and Turner (1956). Since then, turbulence effects on collisions have attracted many researchers. In earlier studies, the estimations of collision-rate enhancement varied by three orders of magnitude from 1 (no effect) to about 10^3 due to the crude representation of turbulent flow and lack of understanding of mechanisms underlying turbulence effects on drop motion. The reasons for the high discrepancies between the results were analyzed in several overviews and studies (e.g., Pruppacher and Klett, 1997; Pinsky and Khain, 1996, 1997a; Pinsky et al., 2000; Vaillancourt and Yau, 2000). These discrepancies were related largely to different and quite crude representation of turbulent flows. Overviews of more recent results (e.g., Shaw, 2003; Franklin et al., 2007; Khain et al., 2007; Pinsky et al., 2008a) showed a significant progress in the field. Magnitudes of collision-rate enhancement obtained using different methods (Pinsky et al., 2008a; Ayala et al., 2008a,b; Wang and Grabowski, 2009) were quite close, indicating a convergence of the estimations of turbulent collision enhancement factors. These results indicate that in cumulus clouds turbulence can be an important and in some cases the dominating mechanism of drop collisions.

Laboratory measurements performed in the Wind Tunnel at the University of Mainz (Vohl et al., 1999) clearly demonstrated a strong influence of turbulence on collisions between small raindrops with radii of 80–240 μm and small droplets with radii of about 3 μm. An increase of 50–60% in collision rate compared with pure gravity collisions is shown **Figure 5.8.1**. Laboratory experiments conducted by Duru et al. (2007) also demonstrated significant effects of turbulence on collisions of small μm-size particles. Thus, important effects of turbulence on drop collisions were demonstrated in theoretical, numerical, and laboratory studies.

5.8.1 Relative Drop Velocity and the Stochastic Nature of Drop Collisions

Clouds are zones of enhanced turbulence. As was shown in Section 5.5, the inertia forces drops to deviate from the trajectories of the environmental air flows. Drops respond to air velocity fluctuations with some delay that depends on drop mass: the larger the drop mass the larger the delay is. As a result, drop inertia in a turbulent flow leads to formation of turbulence-induced relative velocity between drops, in addition to the gravity-induced relative velocity. If the turbulence-induced relative velocity is of the same order or exceeds the gravity-induced relative velocity, one can expect a significant turbulence effect on drop collisions. Since turbulent air velocities are of stochastic nature, all parameters determining drop collisions in a turbulent flow, such as relative velocity, swept volume, collision kernel, and drop concentration, are random values varying over time and space. The main reason of stochasticity of drop collisions is the random nature of the relative drop velocities in a turbulent flow (Section 5.5.6).

It was shown in Section 5.5 that drop velocity relative to the environment air depends on turbulent shears S_{ij} and turbulent Lagrangian accelerations A_i

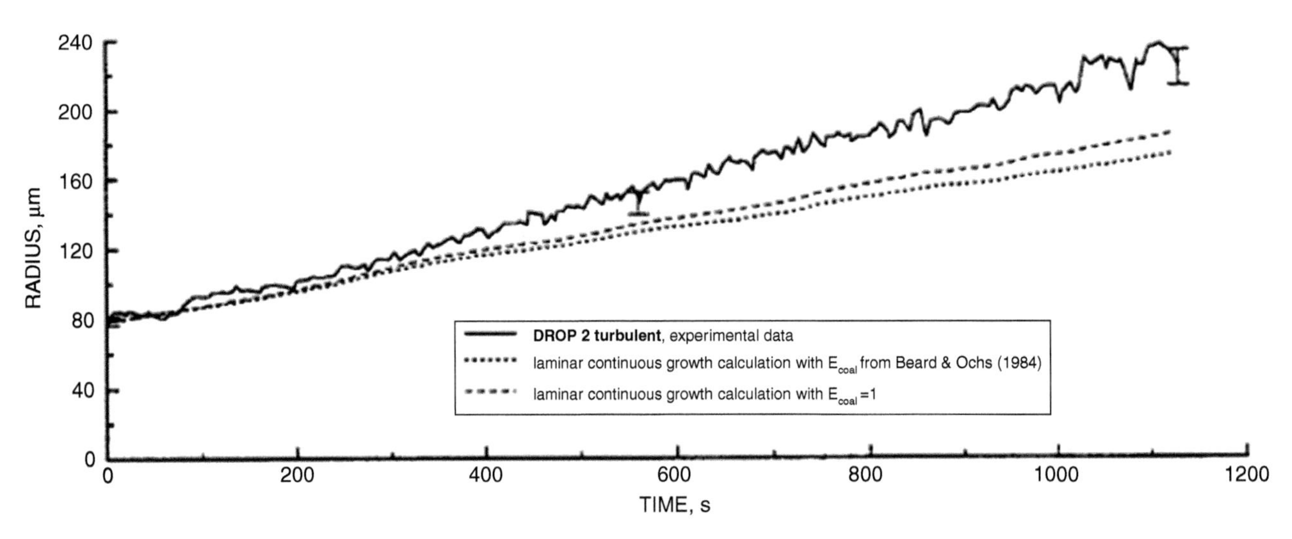

Figure 5.8.1 Comparison of experimental drop growth data in turbulent conditions with those of laminar continuous growth (from Vohl et al., 1999; American Meteorological Society©; used with permission).

(Equation (5.5.20)). For cloud droplets characterized by the Stokes number $St < 1$, one can get simple expressions for the relative velocities and relative trajectories between two colliding droplets within small volumes (referred to as elementary air volumes) with linear scales of the order of Kolmogorov microscale (Pinsky et al., 2006):

$$\frac{d\tilde{x}_i}{dt} = \tilde{V}_i = S_{ij}\tilde{x}_j + \tilde{V}_i' \tag{5.8.1}$$

$$\tilde{V}_i' \approx \left(V_{g2} - V_{g1}\right)\delta_{i3} + \left(\tau_{d2} - \tau_{d1}\right)A_i + O\left(St^2\right) \tag{5.8.2}$$

In Equations (5.8.1) and (5.8.2), $\tilde{x}_i$ is the position vector between the centers of the two droplets, $\tilde{V}_i$ is their relative velocity, V_{g1}, V_{g2} are the terminal velocities, and τ_{d1}, τ_{d2} are the droplet relaxation times (Section 5.5). The relative velocity between droplets depends on two local properties of the turbulent flow, namely, on the Lagrangian acceleration vector A_i and the turbulent shear tensor $S_{ij} = \frac{\partial u_i}{\partial x_j}$, which within each elementary volume can be considered constant as well as the value $\tilde{V}'$. Equations (5.8.1) and (5.8.2) do not take into account the hydrodynamic interaction between droplets. Since shears and accelerations stochastically change from one elementary volume to another, the field of relative velocities between droplets becomes random.

The turbulence-induced relative velocity is mainly the consequence of drop inertia. A simplified scheme illustrating mechanisms of the formation of the turbulence-induced relative velocity is shown in **Figure 5.8.2**.

In the first example (left panel), two drops are located within a turbulent air vortex. Non-inertial drops would rotate together with airflow keeping their distance from the vortex center. In contrast, inertial drops tend to leave the vortex. Centrifugal force affecting the larger (red) inertial drop is higher than that affecting the smaller (blue) drop. As a result, larger drop leaves the vortex faster than the smaller drop and the relative velocity between drops arise in the direction perpendicular to the direction of air flow. This relative velocity forms even in case of no gravitation.

In the second example (right panel) falling drops of different mass are located within a flow with a velocity shear. Initially, the larger (red) drop is situated above the smaller (blue) drop and moves faster within stronger flow. During its fall, the larger drop tends to keep its larger horizontal velocity because of inertia. As a result, the drops collide having different horizontal velocities. This example shows that intertial drops can collide at different angles, and not only in the vertical direction.

Pinsky et al. (2006) analyzed the properties of relative droplet motion for different samples of shears and accelerations. Equation (5.8.1) was integrated in the backward direction (using negative time steps) with the initial positions of droplets on the spherical surface of $(r_1 + r_2)$ radius. The trajectories were calculated for time periods corresponding to the distances equal to ten radii of the larger droplet in a droplet pair (**Figure 5.8.3**). The final points of the backward trajectories form source surfaces. The droplets that start moving from these source surfaces along the forward-in-time trajectories simultaneously collide the target that is a sphere

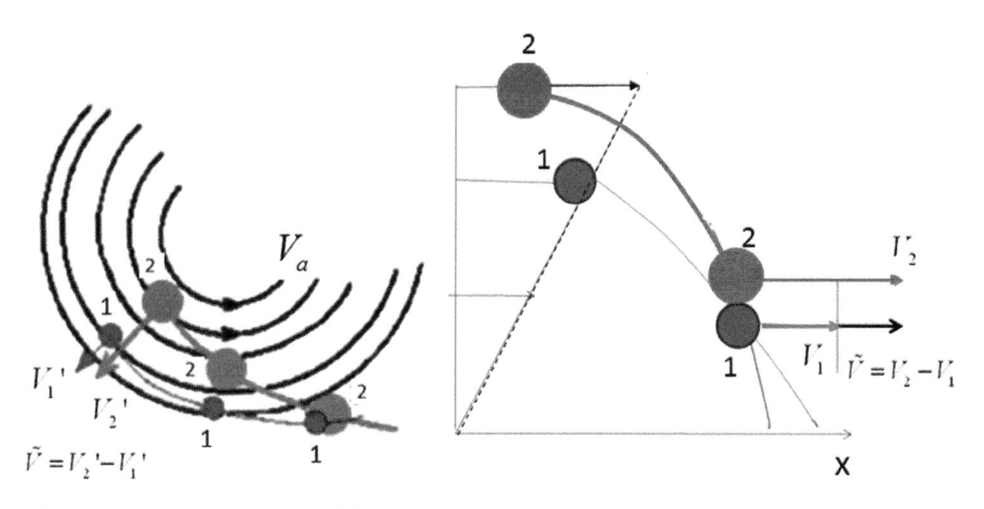

Figure 5.8.2 Formation of the relative drop velocity arising in a curved flow (left) and in the zone of wind shear (right).

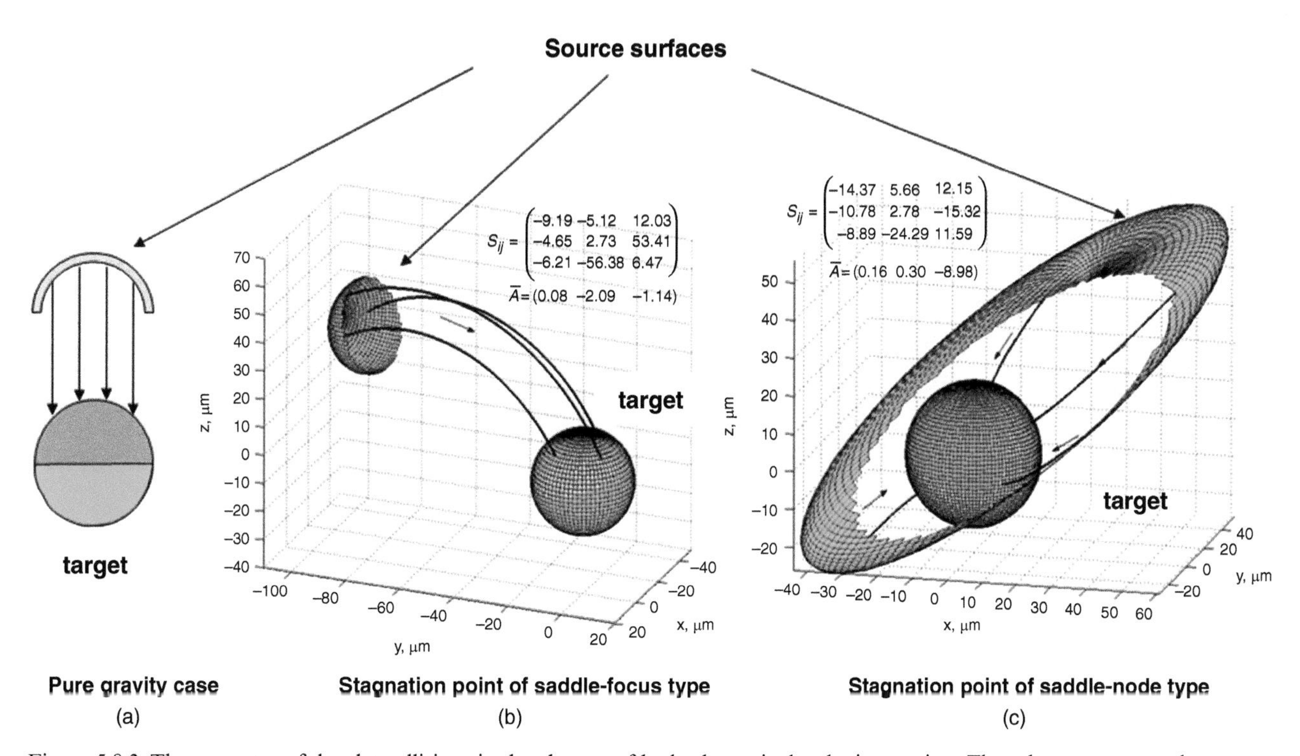

Figure 5.8.3 The geometry of droplet collisions in the absence of hydrodynamic droplet interaction. The spheres represent the target of $r_1 + r_2$ radius, where r_1 and r_2 are the radii of approaching droplets. The droplets starting from the nonspherical source surfaces collide the target at the same time instance. The fraction of the target surfaces, which is crossed by the colliding droplets, is colored yellow. The relative droplet trajectories are shown by solid lines. The components of the velocity gradient and the Lagrangian acceleration are shown in the upper part of the figures. A dramatic difference in the geometry of droplets approaching each other in turbulent flows and in a pure gravity case (panel (a)) is clearly seen (from Pinsky et al., 2007; American Meteorological Society©; used with permission).

of $(r_1 + r_2)$ radius. The shape of the relative trajectory depends on the distance between droplets, on the droplet masses, and on turbulent parameters S_{ij} and A_i. In all cases, there is a stagnation point determined by the following linear system of equations: $S_{ij}\tilde{x}_j + \tilde{V}'_i = 0$. The types of stagnation points correspond either to the saddle-node type trajectories (divergent or convergent trajectories) or to the saddle-focus type (spiral trajectories). The surface in the case of a saddle focus is shown in Figure 5.8.3b. In case of a saddle node, the surface resembles a torus (Figure 5.8.3c). The geometry of droplets mutually approaching in calm air ($S_{ij} = 0$, $A_i = 0$) is illustrated in Figure 5.8.3a. One can see that there exist several geometries of droplets approaching, including the cases when the droplets approach a particular droplet from different directions (Figure 5.8.3). These trajectories indicate a significant difference between the relative droplet motion in turbulent flows and in still air, which must affect hydrodynamic droplet interaction (HDI). Knowing the topology of the relative droplet motion simplifies both understanding of the collision process in a turbulent flow and calculation of collision parameters.

Since the relative velocities and all other parameters of collisions are random, one cannot characterize collisions of a particular drop pair by a single value such as a certain collision kernel, as in the gravitational case (Section 5.6). Instead, we need to introduce the PDF and their moments, in particular, the mean values. PDF of the relative velocities and collision kernels are determined by PDF of turbulent shears and accelerations. **Figure 5.8.4** shows PDF of a normalized longitudinal shear component S_{11} measured in a turbulent flow at $Re_\lambda = 1500$ (Belin et al., 1997) and PDF of the Lagrangian acceleration measured at a high Re_λ turbulent flow obtained by La Porta et al. (2001). These PDF were used in a set of studies by Pinsky et al. (2006, 2007, 2008a) for calculation of drop collision characteristics. One can see that the PDF of the Largangian accelerations and shears are quite wide.

All collision characteristics such as collision kernel and collision efficiency change randomly depending on local turbulent shears and accelerations. The angle of drops approaching each other in a turbulent flow is also random. While in calm air larger drops fall faster and collect smaller drops approaching them in the vertical direction (Figure 5.8.3a), collisions in turbulent flows take place at different relative velocities and different angles, as shown by PDF presented in **Figure 5.8.5**.

The appearance of the turbulence-induced relative velocity between drops gives rise to three mechanisms, by means of which turbulence affects collision rate

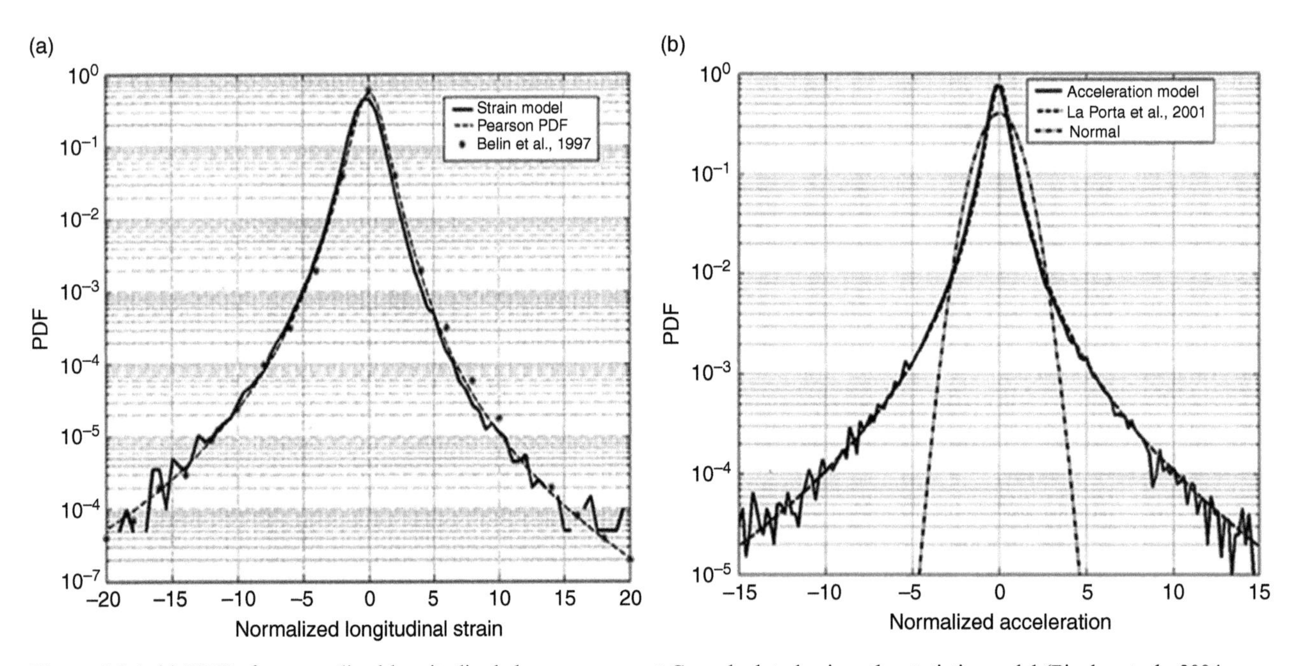

Figure 5.8.4 (a) PDF of a normalized longitudinal shear component S_{11} calculated using: the statistic model (Pinsky et al., 2004; solid line); the Pearson approximating distribution (dashed line) and the distribution measured in a turbulent flow at $Re_\lambda = 1500$ (Belin et al., 1997; asterisks). (b) PDF of a Lagrangian acceleration generated by the statistic model (solid line); the distribution performed by LaPorta et al. (2001; dashed line). The Gaussian PDF with the same variation is shown by the dashed-dotted line (from Pinsky et al., 2006; American Meteorological Society©; used with permission).

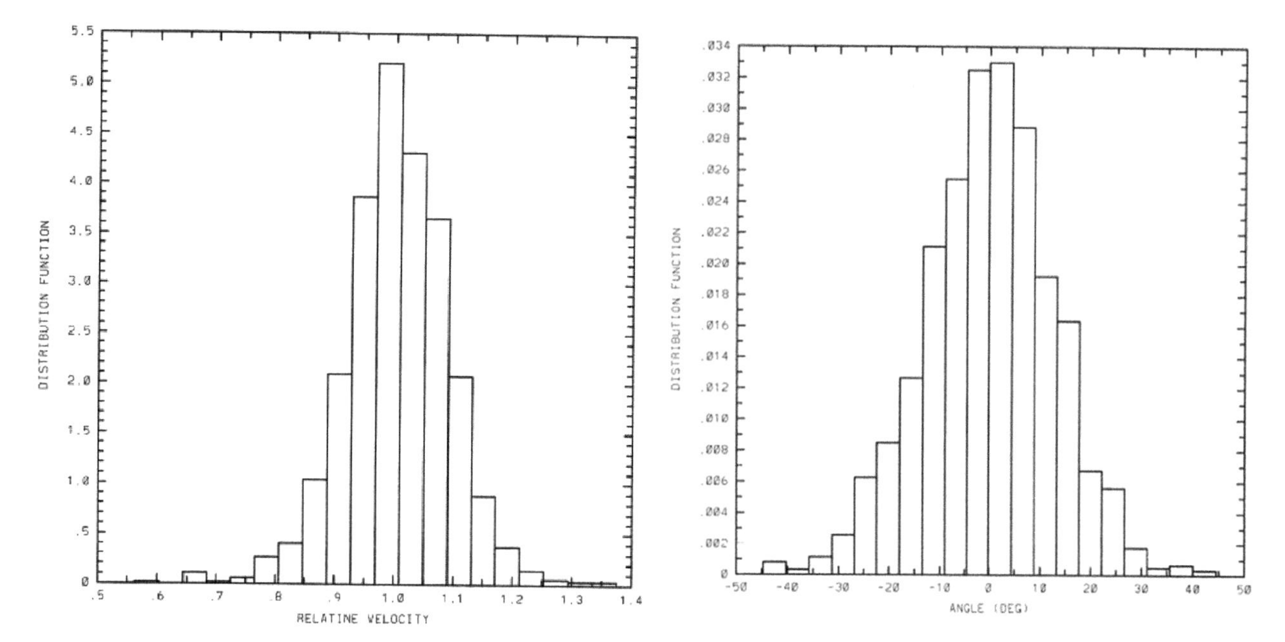

Figure 5.8.5 Left: the distribution of 10-μm and 20-μm-radii inter-drop velocity normalized by the difference in the terminal drop velocities. Right: the distribution of angles at which drops of these sizes approach each other (from Pinsky et al., 2000; courtesy of Elsevier).

between drops. First, turbulence increases the net relative velocity between drops and, consequently, the swept volume (the *turbulent transport effect*). Second, the relative velocity increase and the appearance of nonzero approaching angles leads to an increase in collision efficiency (effect of the hydrodynamic interaction, HDI) Third, turbulence leads to formation of drop concentration inhomogeneity (Section 5.5.7). Collisions within media containing a nonhomogeneous concentration of particles are more frequent than in those with homogeneous concentration of particles. Thus, the third mechanism of collision rate increase is the clustering effect. The three mechanisms are considered in the next sections.

5.8.2 Models of Turbulent Flows Used in Analysis of Drop Collisions

Turbulence effects on drop collisions have been investigated both analytically and numerically by simulation of motion and interaction of drops in turbulent flows. The main tool of numerical approaches is numerical models of three types reproducing fine properties of turbulent flows.

The first type includes *kinematic models* (Kraichnan, 1970; Fung et al., 1992; Pinsky and Khain, 1995, 1996), which generate the turbulent velocity field as a sum of random harmonics of different spatial scales. The velocity field obeys both the turbulent energy laws (e.g., the $-5/3$ Kolmogorov law, Section 3.3) and the correlation laws. A kinematic model was successfully used by Pinsky et al. (1999b) for calculation of realistic collision efficiencies in a turbulent flow. These models usually utilize the Gaussian velocity and do not describe the intermittent properties of turbulence typical of atmospheric flows with Re_λ. Therefore, kinematic models are unable to reproduce fine statistical properties of turbulence.

Models of the second type are DNS models (Franklin et al., 2005, 2007; Ayala et al., 2008b) that are widely used to simulate drop motion and collisions. A significant advantage of DNS models is their ability to directly solve the Navier–Stokes equation (Equation 3.2.19). In most DNS studies, the problem of drop clustering in a turbulent flow is investigated. To use DNS for calculating collision efficiency, a hybrid model has been developed (Ayala et al., 2007; Wang et al., 2008, Wang and Grabowski, 2009). In the hybrid model, drop trajectories are calculated using DNS, and then the hydrodynamic problem of drop interaction is solved using interpolated drop tracks. DNS models have their shortcomings. The values of Re_λ in DNS typically do not exceed 100–200, while Re_λ in clouds can be as high as 5×10^4. In addition, since raindrops respond to vortices with sizes exceeding that of the computational area (Figure 5.5.14), DNS models, with their small computational area, are not able to simulate motion and collisions of drop pairs containing a raindrop.

The third type includes *statistical* Langevin-type *models* that simulate random motion of drops (see Pope, 2000). These models are widely used for simulation of different types of turbulent diffusion, including diffusion of inertial particles. However these models are not suitable to simulate statistics of shears and accelerations required to calculate drop–drop interaction in a turbulent flow.

Pinsky et al. (2004, 2006) developed a statistical model specially designed to describe droplet collisions under conditions typical of atmospheric clouds. According to this model, statistical properties of a turbulent flow characterized by certain values of ε and Re_λ are represented by a set of non-correlated samples of turbulent shears and the Lagrangian accelerations. Each sample can be assigned to a certain elementary volume of the turbulent flow at a certain time. In each elementary volume, the Lagrangian acceleration and the turbulent shears are considered uniform in space and invariable in time. The elementary volumes represent a set of independent samples from the same statistical ensemble. This representation of a turbulent flow at small scales radically simplifies calculation of swept volumes, collision kernels, and collision efficiencies in each elementary volume. The characteristic size of elementary volumes l_{el} was chosen equal to the Kolmogorov microscale μ_k (Section 3.3), which obeys condition $l_{el} = \mu_k < \min(l_A, l_S)$, where l_A and l_S are the characteristic length scales of the Lagrangian acceleration and the turbulent shears, respectively. Evaluation of these scales is presented in Section 3.3.4 for different values of ε and Re_λ, typical of stratiform, cumulus, and cumulonimbus clouds. The characteristic time scales of the elementary volume were assumed to be of order of the Kolmogorov time scale τ_k.

This statistical model by Pinsky et al. (2004, 2006) generates three components of accelerations and nine components of the shear tensor measured in laboratory measurements at high Re_λ up to $2 \cdot 10^{-4}$. The model generator accurately reproduces the shear statistics measured by Antonia et al. (1981), Belin et al. (1997) and Kholmyansky et al. (2001) (Figure 5.8.4). These statistics agrees well with the statistics of atmospheric turbulence. An approximation formula for moments of shear distribution (3.3.27) is used in the model to take into account the dependence of the shears on Re_λ. The statistics of the Lagrangian accelerations are based on the experimental studies by La Porta et al. (2001) and Voth et al. (2002) and on the theoretical analysis performed by Hill (2002). The results obtained in these studies agree well with those obtained in the laboratory experiments for Re_λ as high as about 10^4 (Figure 3.3.13) and are applicable for atmospheric conditions. Pinsky et al. (2004, 2006) used the results obtained by Hill (2002) for the approximation of the Largangian acceleration variations dependence on Re_λ (Equation 3.3.28). Figure 5.8.4 shows that the statistical model by Pinsky et al. (2004, 2006) reproduces PDF of shears and accelerations with high accuracy. Pinsky et al. (2004, 2006) assume the turbulence to be locally isotropic, which determines the correlations between different components of shears. It also determines the zero correlation between shears and accelerations.

Using this statistical model, Pinsky et al. (2006) calculated two long statistical series: the series of the turbulent velocity gradients (nine components of shears), and the series of the Lagrangian accelerations (three components). In each realization of a shear-acceleration pair, the relative droplet trajectories and collision parameters (swept volume, collision efficiency, and collision kernel) were calculated. The statistical model is quite applicable to treat motion and collisions of small cloud droplets. At the same time, the statistical model by Pinsky et al. (2006) has limitations in description of motion or collisions of raindrops whose relaxation time exceeds τ_k.

5.8.3 Drop Fluxes and the Definition of the Spherical Collision Kernel

Since the number of drops in elementary volumes is small (the mean distance between them being about 1 mm), drop concentration should be considered in statistical sense and is defined in terms of probability for a drop to be at a certain point of the flow (e.g., Pinsky and Khain, 2001, 2003). In this case, the frequency of collisions can be characterized by continuous fluxes of drops of one size onto drops of another size. The flux of drops of radius r_2 onto a drop (one) of radius r_1 is equal to the drop flux penetrating a spherical surface of $(r_1 + r_2)$ radius. **Figure 5.8.6** illustrates the drop flux definition. The flux can be calculated using the following formula (Pinsky et al., 2006, 2007):

$$\Phi = \frac{1}{r_1 + r_2} \int_{\Omega+} N_{d1} \tilde{V}_i \tilde{x}_i d\Omega_+, \tag{5.8.3}$$

where N_{d1} is the concentration of r_1-radius drops and Ω_+ is the fraction of the spherical surface with $\tilde{V}_i \tilde{x}_i < 0$; i.e., the relative velocity is directed toward the target volume of radius $(r_1 + r_2)$.

It is more convenient to deal not with the drop flux but rather with the relative velocity flux, which, in the

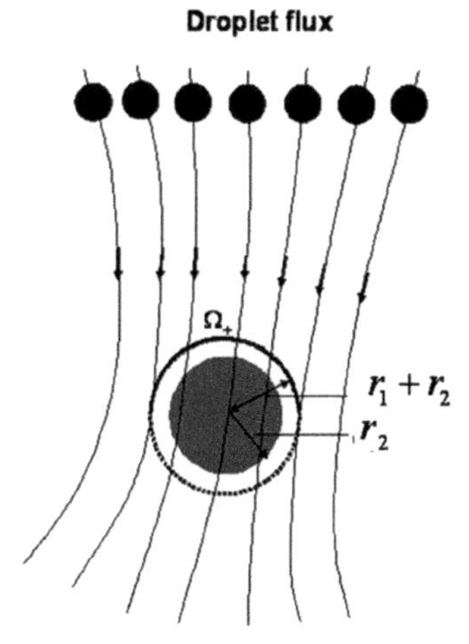

Figure 5.8.6 Illustration of the drop flux definition. Collision rate between drops of radii r_1 and r_2 is determined by the relative drop flux Φ through a fraction of spherical surface Ω_+, where the relative velocity is directed toward the target spherical volume of the radius $(r_1 + r_2)$ (from Pinsky et al., 2006; American Meteorological Society©; used with permission).

equation of stochastic collisions (Equation 5.6.5), represents the collision kernel K as

$$K = \frac{\Phi}{N_1} = \frac{1}{r_1 + r_2} \int_{\Omega+} \tilde{V}_i \tilde{x}_i d\Omega_+ \quad (5.8.4)$$

$$V_{swept} = 2\sqrt{2\pi}(r_1 + r_2)^2 \times \left[\frac{1}{9}(r_1 + r_2)^2 \frac{\varepsilon}{\nu} + \left(1 - \frac{\rho}{\rho_w}\right)^2 (\tau_{d1} - \tau_{d2})^2 \overline{A^2} + \frac{1}{3}\left(1 - \frac{\rho}{\rho_w}\right)^2 (\tau_{d1} - \tau_{d2})^2 g^2\right]^{1/2} \quad (5.8.5a)$$

Expression (5.8.4) represents a collision kernel defined in the spherical geometry. The concept of *spherical collision kernel* was introduced first by Wang et al. (1998, 2000). Thereby, there are two approaches to describing the geometry of drop collisions: the cylindrical geometry (Figure 5.6.1) and the spherical geometry (Equation 5.8.4, Figure 5.8.6). The spherical geometry concept is more complicated since it generally requires calculation of drop fluxes across a spherical surface of $R = r_1 + r_2$ radius. However, in view of the complicated geometry of drops approaching each other in a turbulent flow, it is necessary to use the spherical collision kernel. The spherical geometry describes turbulent collisions more accurately. The concept of the spherical kernel can also be used for description of hydrodynamic drop interaction. In case of purely gravitational collisions, the two representations of collision kernels yield similar results.

In analytical investigations, the collision kernels are calculated by averaging the equations for relative drop velocities (e.g., Saffman and Turner, 1956; Pinsky et al., 1997a,b). In numerical methods, the collision kernels are determined using the trajectories of individual drops in simulated turbulent flows (Figure 5.8.2) with successive averaging.

5.8.4 Turbulence Effects on the Swept Volume

In the absence of HDI ($E_{col} = 1$), the collision kernel is regarded as the geometrical collision kernel, or the swept volume. Several analytical equations for the swept volume were offered. Under the assumption that velocities in a turbulent flow are normally distributed, Saffman and Turner (1956) obtained an expression for the mean swept volume in the cylindrical geometry:

where r_1 and r_2 are the radii of colliding drops, τ_{d1} and τ_{d2} are the relaxation time values for these drops (Equations (5.5.4) and (5.5.6)), $\overline{A^2}$ is the mean square of turbulent acceleration and g is gravity acceleration. In Equation (5.8.5a), the first term in square brackets on the right-hand side reflects the turbulent shear influence, the second term corresponds to the Lagrangian acceleration influence, and the third term corresponds to gravity acceleration influence. A more precise formula for the spherical geometry was derived by Wang et al. (1998):

$$V_{swept} = 2\sqrt{2\pi}(r_1 + r_2)^2 \times \left[\frac{1}{15}(r_1 + r_2)^2 \frac{\varepsilon}{\nu} + \left(1 - \frac{\rho}{\rho_w}\right)^2 (\tau_{d1} - \tau_{d2})^2 \overline{A^2} + \frac{\pi}{8}\left(1 - \frac{\rho}{\rho_w}\right)^2 (\tau_{d1} - \tau_{d2})^2 g^2 + 2\left(1 - \frac{\rho}{\rho_w}\right)^2 \tau_{d1}\tau_{d2}\overline{A^2}\frac{(r_1 + r_2)^2}{\lambda_A^2}\right]^{1/2} \quad (5.8.5b)$$

where λ_A is the Taylor microscale of the Lagrangian acceleration. The last term on the right-hand side of Equation (5.8.5b) takes into account the combined effect of spatial variations of fluid acceleration and particle inertia. Equations (5.8.5a) and (5.8.5b) show that the Lagrangian acceleration works like an additional acceleration directed at a random angle to the gravitational acceleration. Neither acceleration allows collisions between drops of the same size. In contrast, turbulent shears and spatial variations of the Lagrangian acceleration allow such collisions. Wang et al. (1998) showed that utilization of the cylindrical geometry overestimates the value of the mean swept volume by ~20%. Other analytical formulas for the swept volumes were proposed by Pinsky et al. (1997b), Dodin and Elperin (2002) and Ayala et al. (2008b).

Figure 5.8.7 shows PDF of the swept volumes for two droplets of 10-μm and 15-μm in radius in conditions typical of stratiform, moderate cumulus, and deep cumulonimbus, calculated by Pinsky et al. (2006). To reveal the effect of turbulent intermittency on the swept volume distributions, calculations were performed both for the non–Gaussian and the Gaussian PDF of shears and accelerations, the variances of both PDF assumed

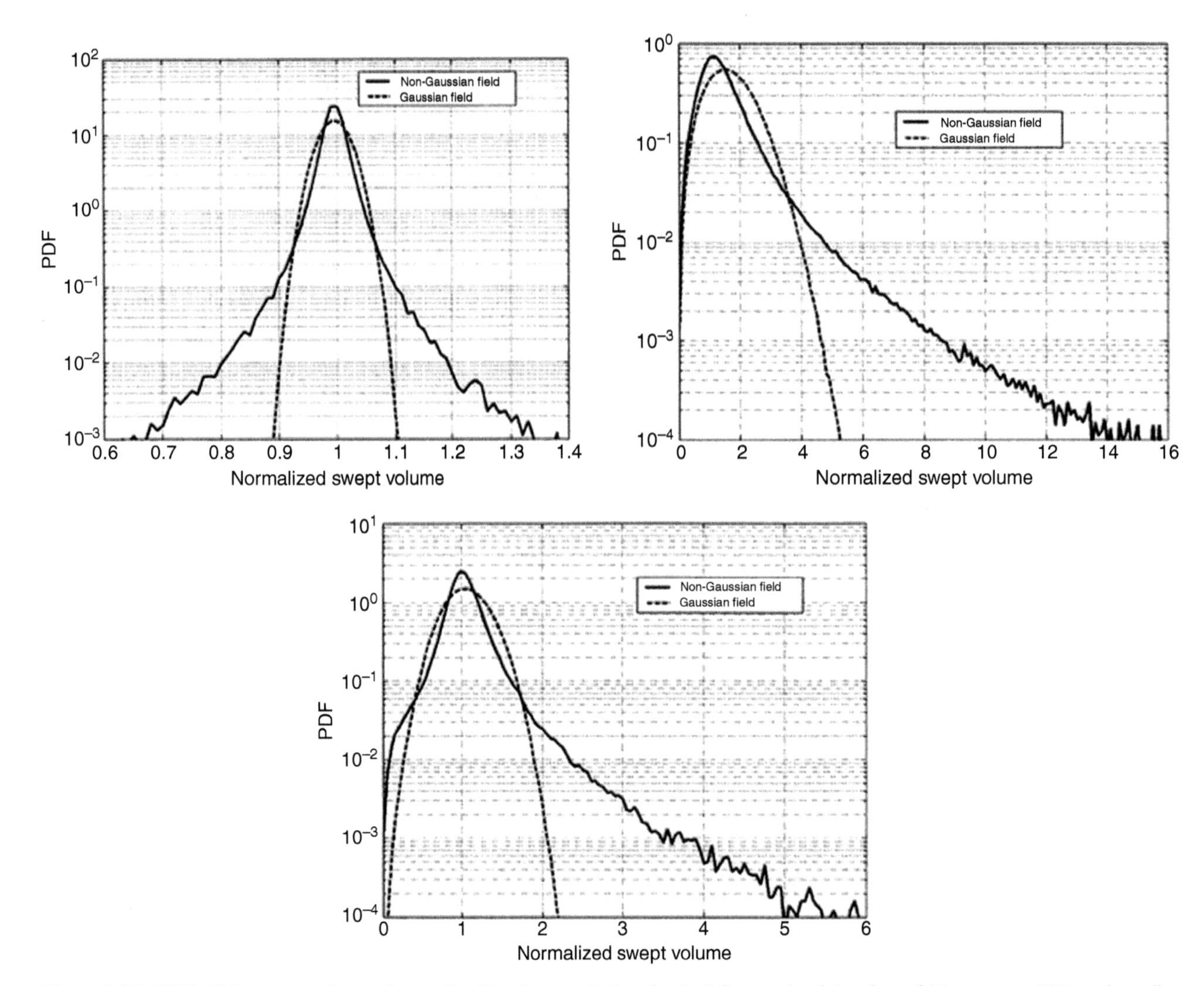

Figure 5.8.7 PDF of the swept volumes (normalized by the gravitational value) for a pair of droplets of 15 μm and of 10 μm in radius for conditions typical of stratiform clouds ($Re_\lambda=5\times10^3$, $\varepsilon=10^{-3}m^2s^{-3}$) (top left), moderate cumulus ($Re_\lambda=2\times10^4$, $\varepsilon=2\times10^{-2}m^2s^{-3}$) (top right), and deep cumulonimbus ($Re_\lambda=2\times10^4$, $\varepsilon=10^{-1}m^2s^{-3}$) (bottom). Calculations were performed using generators of both the non–Gaussian PDF (solid lines) and the Gaussian PDF of shears and accelerations (dashed lines) (from Pinsky et al., 2006; American Meteorological Society©; used with permission).

to be identical. The PDF were calculated using the statistics containing 10^6 acceleration–shear pairs (i.e., 10^6 elementary volumes). Figure 5.8.7 (upper left panel) shows that the PDF of the swept volumes is actually symmetric and centered at the gravitational value both in the non–Gaussian and the Gaussian generators of shear and acceleration. It means that the mean value of the swept volume for turbulent parameters typical of stratiform clouds is in fact equal to the gravitational value. The turbulence intermittency leads to formation of an elongated tail. In both cases the PDF are comparatively narrow; their width being about 10% of the gravitational value. At ε and Re_λ, values typical of developing cumulus clouds (upper right panel), the PDF of the swept volume becomes asymmetric, indicating that turbulence increases the mean swept volumes. The PDF width reaches ~100% of the gravitational value. A further increase in turbulence intensity leads to an increase in the PDF asymmetry; the elongated tail becomes more pronounced (lower panel). The probability of the appearance of very large swept volumes increases with increasing Re_λ and ε. However, the probability corresponding to the PDF tails is very small. Thus, the mean values of the swept volumes are almost not affected by the tails and were found to be actually similar to those obtained for normally distributed accelerations and shears. The maximum difference in the mean values between the non–Gaussian and the Gaussian case was found for turbulent conditions typical of strong cumulus clouds. Additional simulations were carried out that revealed a low sensitivity of PDF of normalized swept volumes to droplet size (with the exception of droplets of almost similar sizes). The results indicate that the swept volumes are determined mainly by flow accelerations and gravity and almost do not depend on the shears. The shears become important when the difference in the colliding droplets radii does not exceed 0.1– 0.2 μm.

Pinsky et al. (2006) calculated the mean values of the swept volumes by averaging series of 10^4 samples for different shear–acceleration pairs. **Figure 5.8.8** shows that the mean swept volume normalized by the gravitational value increases as both ε and Re_λ increase. This dependency is another demonstration of the high importance of using realistic Re_λ values in order to describe collision properties, even when hydrodynamic droplet interactions are not taken into account. However, the increase in the swept volume, even under the extreme conditions of deep cumulus clouds when it may reach ~60%, is quite moderate compared to the gravitational value. As was discussed, the normalized mean swept volume increase shown in Figure 5.8.8

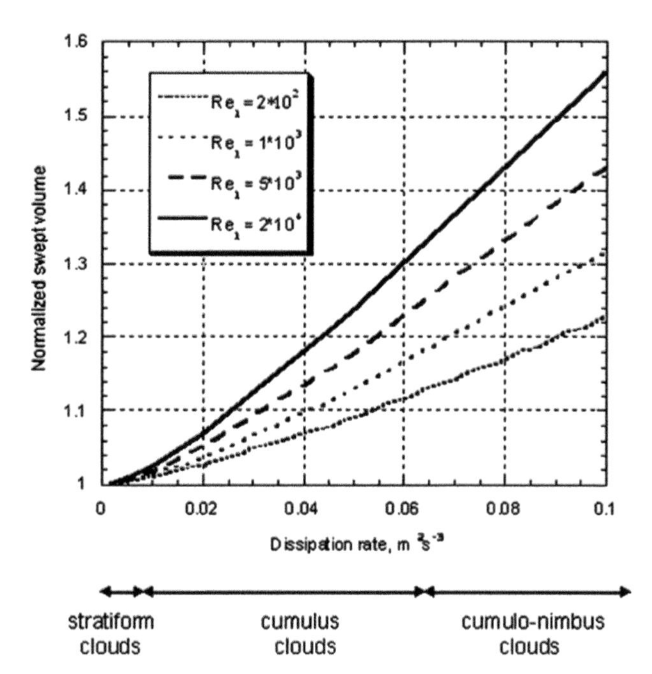

Figure 5.8.8 Dependence of the mean swept volume normalized by the gravity value on turbulent dissipation rate ε for a pair of droplets of 10 μm and of 15 μm in radius at different Re_λ (from Pinsky et al., 2006; American Meteorological Society©; used with permission).

does not depend on the droplet size. Thus, the curves plotted in Figure 5.8.8 are universal and suitable for any droplet pair (with the exception of droplets of almost similar sizes). Comparison of the results presented in Figure 5.8.8 with those obtained using Equations (5.8.5a) and (5.8.5b) indicates a quite good agreement with the difference not exceeding 10–20%. This difference can be attributed to the fact that different formulations of the swept volumes were used: the cylindrical geometry (Saffman and Turner, 1956) and spherical geometry (Wang et al., 1998; Pinsky et al., 2006). Another reason is utilization of the Gaussian distributions assumption in Equations (5.8.5a) and (5.8.5b).

Collisions of drops of equal size are considered theoretically impossible in the pure gravitational case, albeit possible in a turbulent shear flow. Dependencies of the mean swept volume of equal-sized droplets (of 5 μm in radius) on the dissipation rate at $Re_\lambda = 2 \times 10^4$ are shown in **Figure 5.8.9**. One can see a dramatic increase in the swept volume with the increasing dissipation rate. Comparison of the results obtained by Pinsky et al. (2006) with those obtained using formulas developed by Saffman and Turner (1956) and Wang et al. (1998) indicates ~25% difference that can be attributed to the

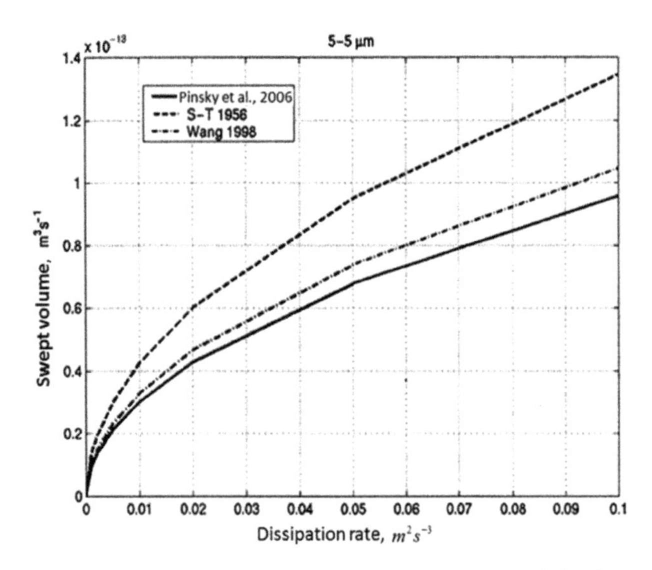

Figure 5.8.9 The dependence of the swept volume on dissipation rate for 5 μm radius droplets according to different authors (from Pinsky et al., 2006; American Meteorological Society©; used with permission).

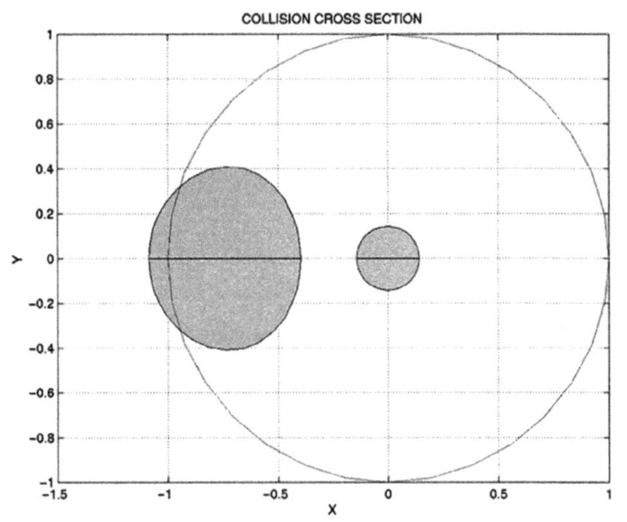

Figure 5.8.10 An example of collision cross sections between two droplets of 10-μm and of 15-μm in radius, approaching each other: along the same direction (small dark circle in the center) and at the angle of 30° between directions of droplet velocities at infinity (large dark area shifted to the left from the center) (from Pinsky and Khain, 2004; American Meteorological Society©; used with permission).

reasons mentioned previously. In spite of the difference in the results, all of the studies mentioned in Figure 5.8.9 indicate that the effect of turbulence intensity on the swept volumes is relatively moderate (not exceeding a few tens of percent).

5.8.5 Hydrodynamic Drop Interaction

The most important mechanism by which turbulence increases the rate of drop collisions in clouds is the turbulence influence on HDI, which increases collision efficiency $E(r_1, r_2)$. Almeida (1976, 1979) was the first to analyze HDI in a turbulent flow (the criticism of his studies can be found in Pruppacher and Klett, 1997). Koziol and Leighton (1996) analyzed the effects of turbulent vortices of scales below the HDI scale, which normally is within the viscous range. The effect of these low-energy vortices on collision efficiency was shown to be negligibly small. Further estimations were made by Pinsky et al. (2000) and Pinsky and Khain (2004), who took into account the effects of turbulent vortices within the inertial subrange and the transition subrange. It was shown that turbulence can increase collision efficiency of cloud drops by several hundred percent.

An increase in the collision efficiency between drops moving in a turbulent flow is accounted for by two main physical factors. The first one is the increase in the absolute value of the relative drop velocity. A similar effect takes place in a pure gravity case with increasing height (Section 5.6). The second factor is the appearance of nonzero angles of drop approaching (Figure 5.8.5). The pressure is maximum ahead of the drop collector and lower at the drop-collector sides. This pressure hinders drop collisions at zero approaching angles (Pinsky et al., 2000), but does prevent counterparts approaching from the sides. Thus, drops can approach each other much more easily at nonzero angles than at zero angles. At nonzero angles, the collision cross sections are no longer concentric circles as in the pure gravity case, but are shifted from the geometric cross-section center (Pinsky et al., 2000). **Figure 5.8.10** shows an example of collision cross sections between droplets of 10-μm and 15-μm radii, calculated for two cases: a) when the approaching droplets move along the same direction and b) when a 30° angle was assumed between the directions of the droplet velocities at infinity. In the second case, the collision cross section shifts from the axis connecting the droplet centers and significantly increases. A significant effect of the drop approaching angles on collision efficiency was also found by Wang et al. (2006a).

The collision efficiency $E(r_1, r_2)$ between any two populations of drops can be generally defined as the ratio of the drop fluxes Φ_{HDI} and Φ_{noHDI} in presence and in absence of hydrodynamic interaction, respectively (Wang et al., 2005; Pinsky et al., 2007, 2008a).

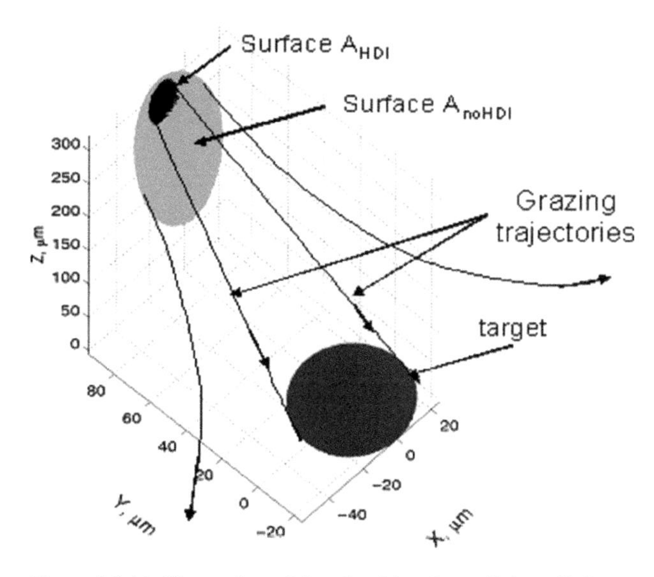

Figure 5.8.11 Illustration of the algorithm for collision efficiency calculation. The black sphere is the $(r_1 + r_2)$ radius target. Drops that start moving from A_{noHDI} surface collide with the target in the absence of HDI. Drops starting from the area A_{HDI} (marked black) collide with the target in the presence of HDI. The figure illustrates the case when the A_{HDI} area is fully located inside the A_{noHDI} area (from Pinsky et al., 2008a; American Meteorological Society©; used with permission).

$$E(r_1, r_2) = \frac{\Phi_{HDI}}{\Phi_{noHDI}} = \frac{K_{HDI}}{K_{noHDI}} \tag{5.8.6}$$

In the pure gravity case, ratio K_{HDI}/K_{noHDI} is reduced to the ratio between the collision area and the geometrical cross-section area.

Pinsky et al. (2007) calculated the collision efficiency as follows. Using Equations (5.8.1) and (5.8.2) for the relative drop velocity, they calculated the drop fluxes in the absence of HDI and in the presence of HDI. The sources of these fluxes are surfaces shown in **Figure 5.8.11** and denoted as A_{noHDI} and A_{HDI}, respectively. These surfaces were separated from the target by the distance of about tenfold of the larger drop radii, at which HDI is negligible. In case HDI was taken into account, the drop trajectories were calculated using a modified superposition method (Pinsky et al., 2007). The ratio of the drop fluxes coming from surfaces A_{HDI} and A_{noHDI} determines the value of the collision efficiency in accordance with definition (Equation (5.8.6)).

Pinsky et al. (2007) calculated series of collision efficiencies for Re_λ and ε typical of atmospheric clouds. To evaluate the role of turbulence intermittency, calculations were performed both for the PDF corresponding to the intermittent turbulence (the non–Gaussian PDF), and for the Gaussian PDF. The variances of the Gaussian and the non–Gaussian PDF were assumed to be identical. **Figure 5.8.12** shows the PDF of the collision efficiency (left) and the collision kernel (right) of droplet pairs of 10 μm and 15 μm in radius. To resolve the tails of the PDF, long series containing 10^5 acceleration–strain pairs were used. One can see that at low turbulence the PDF of the collision efficiency is relatively narrow, with the standard deviation of about 10% of the gravitational value. The PDF obtained using the non–Gaussian and the Gaussian generators are quite close in this case. The collision efficiency corresponding to the PDF maxima is close to the value in the pure gravity case. At higher turbulence intensities, the turbulent flow intermittency leads to formation of elongated PDF tails. At ε and Re_λ, typical of developed cumulus clouds, the PDF of the collision efficiency becomes positively skewed, indicating that turbulence increases the mean collision efficiency. The standard deviation is by tenfold larger than the gravitational collision efficiency. The difference between the PDF shapes becomes pronounced: the standard deviation in the non–Gaussian case is twice as large as that in the Gaussian case.

A further increase in turbulence intensity leads to a more pronounced PDF asymmetry, as the tails become more elongated. Figure 5.8.12 shows that the probability of very large values of collision efficiency increases with an increase in Re_λ and ε. The PDF of the collision kernel for a pair of droplets of 10 μm and 15 μm in radius are wider than the corresponding distributions of the collision efficiency. In case of a highly intense turbulence, the long PDF tail contributes significantly to the mean value of the kernel, which increases by factor of 7.1, whereas the Gaussian PDF is characterized by a smaller 5.65-fold increase. The large magnitudes of the enhancement indicate the dominant effect of turbulence on collisions in highly turbulent clouds as compared to the gravity effect. The increase in the collision kernel at low and moderate turbulent intensities stems mainly from the increase of the collision efficiency. In case of intense turbulence, the swept volume contributes significantly to the increase of the collision kernel.

Figure 5.8.13 shows the dependencies of the average normalized collision kernel for a 10–20-μm radii droplet pair on the dissipation rate ε under different Re_λ. While the increase in the swept volume was found to be 60% at very strong turbulence intensity (Figure 5.8.8), the collision kernel increases by the factor as large as 4.8. Thus, the effect of HDI appears to be the main mechanism by means of which turbulence increases the rate of cloud droplets' collisions. One can see in Figure 5.8.13 that the collision kernel increases both

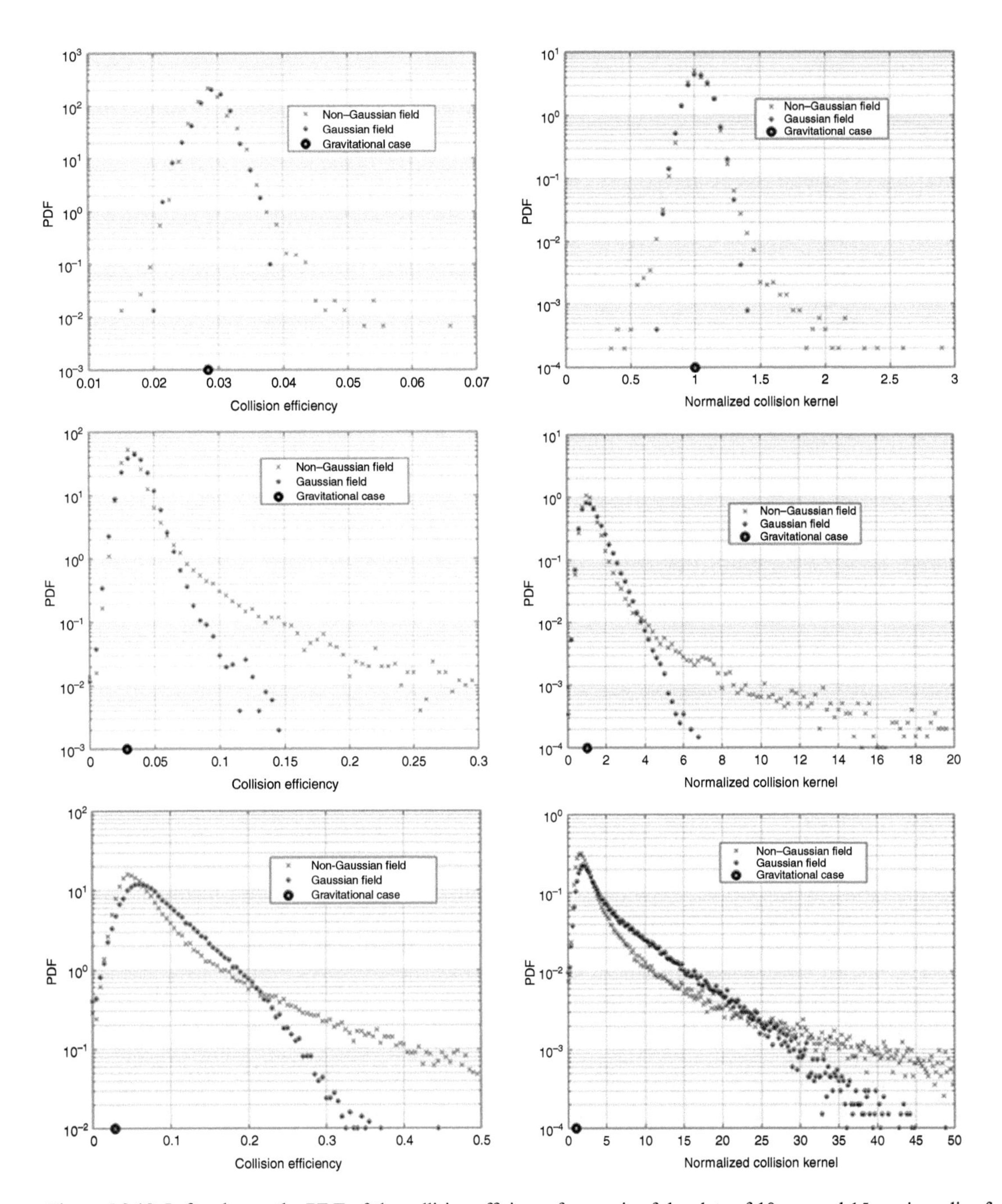

Figure 5.8.12 Left column: the PDF of the collision efficiency for a pair of droplets of 10 μm and 15 μm in radius for three cases corresponding to stratiform clouds (top), moderate cumulus (middle) and deep cumulonimbus (bottom). Right column: the same as in left column, but for the normalized collision kernel. The parameters of turbulence are the same as in Figure 5.8.7 (from Pinsky et al., 2007; ©American Meteorological Society. Used with permission).

with the increase in ε and in Re_λ. Thus, the small-scale intermittency characterized by large Re_λ is of significant importance as regards collision-rate enhancement. The value of the turbulence enhancement factor for a 10–20-μm radii droplet pair is not maximal. The increase in the collision kernel of droplet pairs containing droplets of similar sizes or droplets smaller than 3 μm in radii is more pronounced.

Lookup tables for quantitative evaluation of turbulence-induced enhancement of a collision kernel

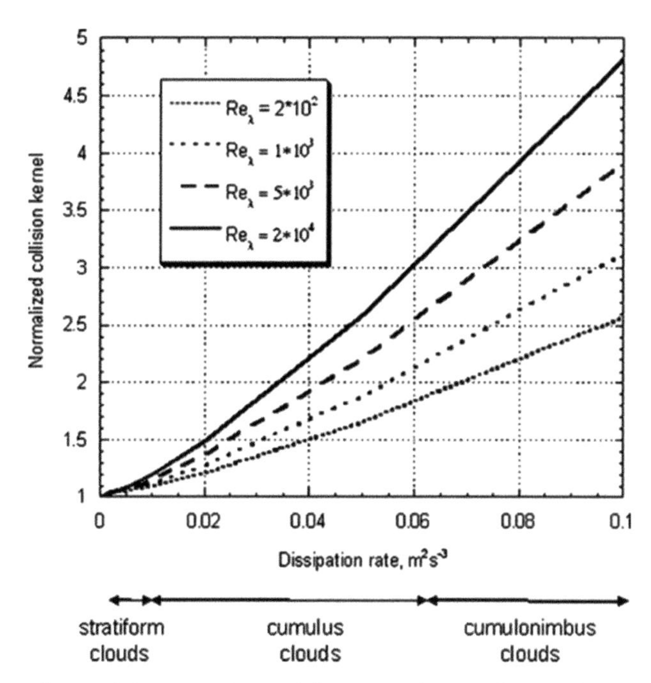

Figure 5.8.13 Dependence of the averaged normalized collision kernel for a pair of droplets of 10 μm and of 20 μm in radius on the dissipation rate ε under different Re_λ (from Pinsky et al., 2007; American Meteorological Society©; used with permission).

for cloud droplets were calculated by Pinsky et al. (2008a). Each value of the collision enhancement factor was obtained by averaging more than 1,500 samples for each droplet pair. Eleven tables corresponding to typical turbulent conditions for clouds of different types were calculated. Three of them are given in **Appendix B (Tables B4–B6)**. The collision enhancement factor depends on droplet radii and turbulent parameters ε and Re_λ. **Figure 5.8.14** shows the values of the mean collision kernels normalized by the gravitational values for stratiform, cumulus, and cumulonimbus clouds, according to **Tables B4–B6**. One can see that the turbulence effect significantly increases with the turbulence intensity. Under the turbulence typical of stratiform clouds, the mean enhancement factor varies from several tens of percent for droplets of different sizes to factor two for droplets of similar sizes. Under strong turbulence typical of cumulonimbus clouds, the collision-kernel enhancement factor exceeds fifteen for droplets of greatly different sizes and exceeds ten for droplets of similar sizes.

Xue et al. (2008) calculated collision efficiency and collision enhancement factors using DNS results extrapolated to large Re_λ. Wang and Grabowski (2009) corrected these evaluations using results obtained from different authors. Their parameterization included the dependence of collision parameters on Re_λ for a wide range of drop sizes. The results are presented in **Figure 5.8.15**. For cloud droplets, the results are quite close to those presented in Figure 5.8.14 and the corresponding **Tables B4–B6** in Appendix B. Comparison of the enhancement factors presented in Figure 5.8.14 with those in Figure 5.8.15 indicates a reasonable agreement, at least by order of magnitude. For instance, the typical values of the collision enhancement factor for conditions characteristic of Cu vary from two to five in both cases.

5.8.6 Effects of Drop Clustering

The third mechanism by means of which turbulence affects drop collisions is related to fluctuations of drop concentration (the clustering effect, or the preferential concentration effect). Turbulent shears create clusters of inertial particles moving in a turbulent flow (as explained in Section 5.4). At the same time, the mean distance between neighboring drops is lower compared to a purely Poisson spatial distribution of drop locations. This decrease in turn leads to an increase in the drop collisions probability. From a statistical point of view, averaging of the DSD product in both the loss integral and the gain integrals in SCE (5.6.5) over concentration fluctuations leads to inequality $\overline{f(m)f(m')} \geq \overline{f(m)} \times \overline{f(m')}$, reflecting the increase in the collision rate.

Most DNS were dedicated to investigation of this effect (see review by Khain et al., 2007). To characterize the effect of drop clustering on drop collisions, Reade and Collins (2000) introduced the radial distribution functions (RDF): $G_{11}(l)$ for monodisperse suspicions and $G_{12}(l)$ for bi-disperse suspicions. These functions reflect the ability of drops to form clusters in a turbulent flow. A physical interpretation of $G(l)$ is the normalized number of particle centers located within a spherical shell at distances between l and $l + dl$ around the central particle. The normalization of the number of particle centers is performed by the expected number of particles, given a uniformly distributed particle field (Reade and Collins, 2000). Although the existence of gravity makes spatial distribution of drop coordinates non-isotropic, these functions, being averaged quantities, are usually considered isotropic in space. The applicability of the isotropic assumption for estimation of drop clustering has not yet been justified.

An example of dependences of $G_{11}(l/\mu_k)$ and $G_{12}(l/\mu_k)$ is shown in the **Figure 5.8.16**. The figure shows that at a large distance between drops the functions are equal to one. This value means that the drop

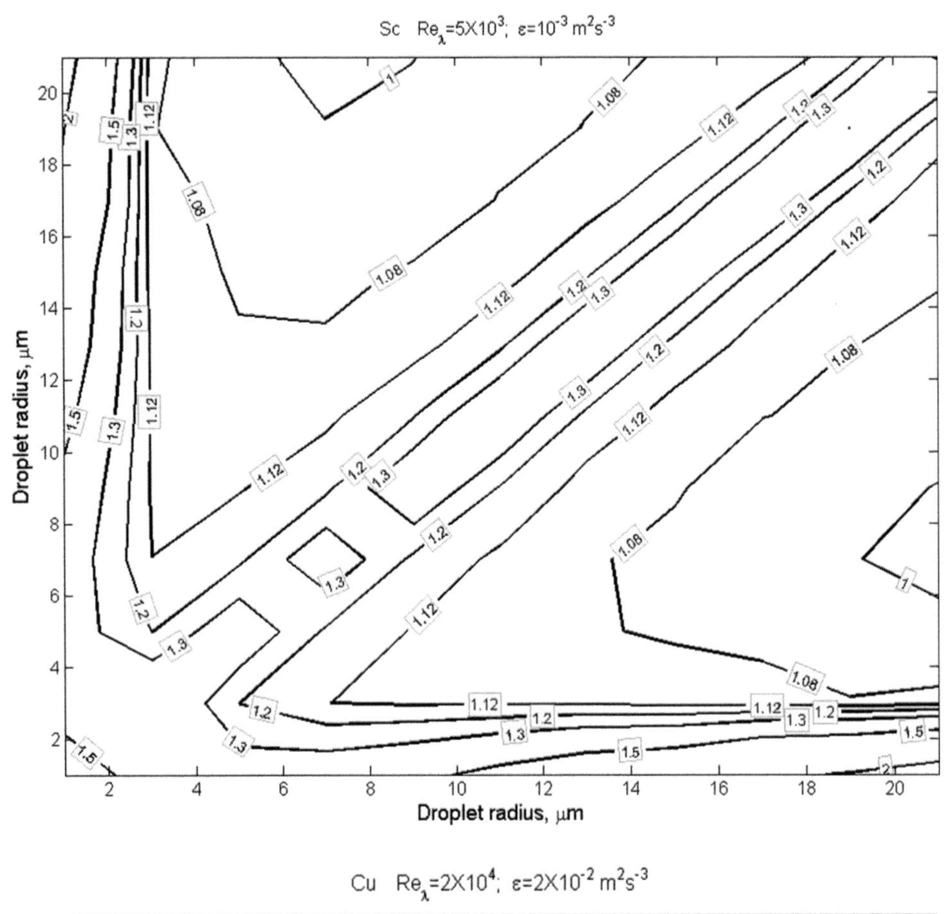

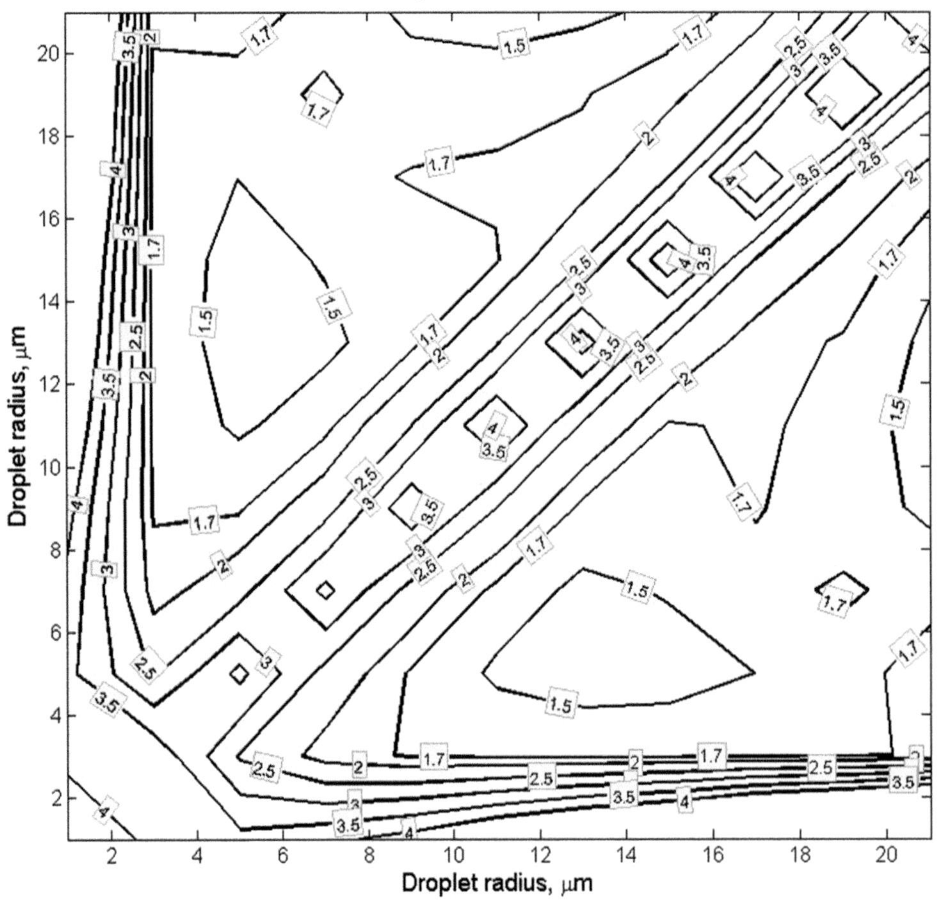

Figure 5.8.14 The mean collision kernel normalized by the gravity values typical of stratiform clouds (top), cumulus clouds (middle), and cumulonimbus clouds (bottom).

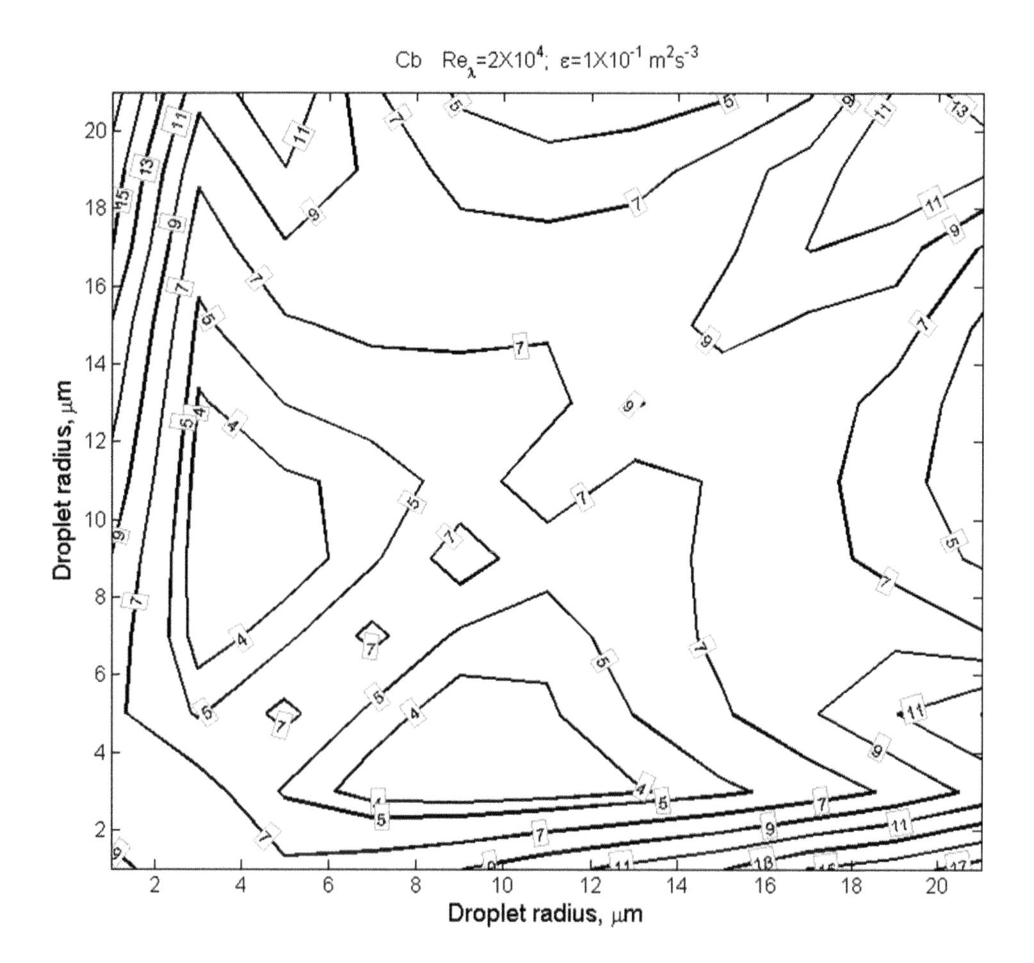

Figure 5.8.14 *(cont.)*

locations obey the Poisson spatial distribution (Equation (5.5.24)) and indicates absence of the clustering effect. At smaller distances between drops approaching each other within the air flow that experiences deformation due to turbulent shears, one sees a monotonic increase of the functions. Function $G_{11}(l)$ demonstrates a strong increase in accordance with the power law, which indicates strong clustering of monodisperse particles. Function $G_{12}(l)$ reaches a plateau at small distances, which indicates a decorrelation of locations of drops of different sizes in relation to one another, caused by different drop inertia. This plateau is determined by the Lagrangian acceleration of a turbulent flow as well as by the gravity acceleration. Both functions depend both on drop sizes and turbulent shears. Functions $G_{11}(l)$ and $G_{12}(l)$ are defined for $l \geq 2r$ and $l \geq r_1 + r_2$, respectively.

The values of functions G_{11} and G_{12} at the point of drop contact (i.e., when $l = r_1 + r_2$) are used as an increasing factor in the stochastic collision equation. The functions at contact points can be interpreted as follows:

$$G_{11}(2r) = \frac{\overline{f^2(r)}}{\overline{f(r)}^2} \geq 1;\; G_{12}(r + r') = \frac{\overline{f(r)f(r')}}{\overline{f(r)} \times \overline{f(r')}} \geq 1, \tag{5.8.7}$$

where r and r' are the radii of drops about to collide, and the overbar denotes space averaging. Inequalities in Equation (5.8.7) show a positive correlation between locations of neighboring drops, leading to an increasing probability of drop collisions in a turbulent flow. The increase factor depends on the drop radii and the parameters of a turbulent flow. The turbulent Stokes numbers of drops (Equations (5.5.22) and (5.5.23)) are often considered the main parameters characterizing clustering.

Figure 5.8.17 shows the dependence of G_{11} on S_t obtained by different authors via theoretical analysis and DNS. The difference between the curves is clearly

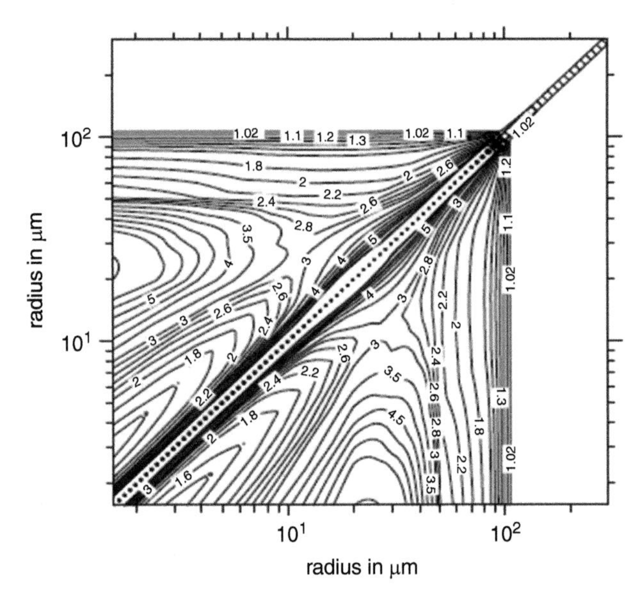

Figure 5.8.15 The ratio of a typical turbulent collection kernel to the Hall kernel. The ratio on the 45° line is undefined due to the zero value of the Hall kernel. The turbulent dissipation rate is 400 $cm^2 s^{-3}$ and the rms velocity is 202 cm/s (from Wang and Grabowski, 2009; courtesy of John Wiley & Sons, Inc.©).

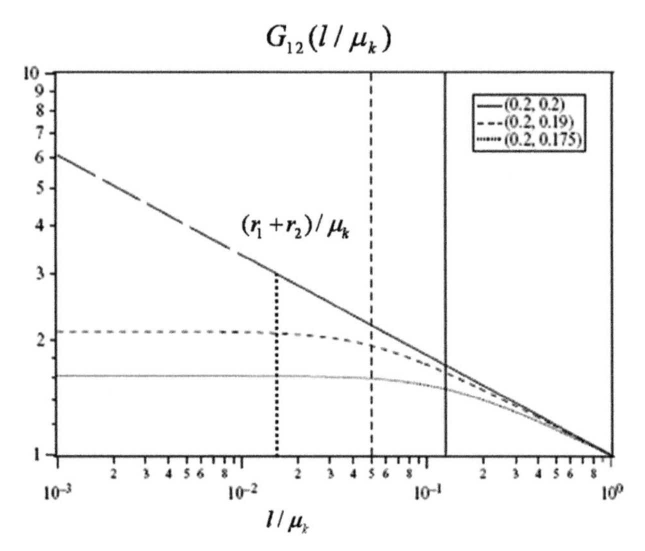

Figure 5.8.16 Radial distribution function for monodisperse mixture (solid line: $St = 0.2$) and bi-disperse mixtures (dashed line: $St_1 = 0.2$, $St_2 = 0.19$; dotted line: $St_1 = 0.2$, $St_2 = 0.175$) (μ_k is the Kolmogorov micro-scale; see Section 3.3.) (from Chun et al., 2005, with changes; courtesy of Cambridge University Press).

small for $St < 0.1$. The differences between the results can be attributed to the differences in the approaches and using small Re_λ values in DNS.

Figure 5.8.18 compares the G_{12} function for bi-disperse suspensions of small drops obtained by Zhou et al. (2001) in DNS (Figure 5.8.18a), and the theoretical results calculated by Chun et al. (2005) for $Re_\lambda = 47$ (Figure 5.8.18b). There is a very good agreement between these results. In the model used by Chun et al. (2005), function G_{12} depends on the Lagrangian acceleration that, in turn, depends on Re_λ. Figure 5.8.18c shows function G_{12} at values of the Lagrangian accelerations measured by La Porta et al. (2001) and Voth et al. (2002) for $Re_\lambda = 10^3$. According to the parameterization scheme proposed by Hill (2002), fluctuations of the Lagrangian acceleration continue increasing with Re_λ (Equation (3.3.28)). G_{12} function calculated using this parameterization for $Re_\lambda = 2 \times 10^4$ is shown in Figure 5.8.18d. The increase of Re_λ leads to a strong decrease in G_{12} for drops of different sizes. This effect is attributed to the Lagrangian acceleration. It is interesting that the Lagrangian acceleration, being the main factor of increase of the relative velocity between drops (i.e., of the collision kernel), tends at the same time to destroy clusters containing drops of different sizes (Chun et al., 2005). The collision enhancement factor is maximum for drops

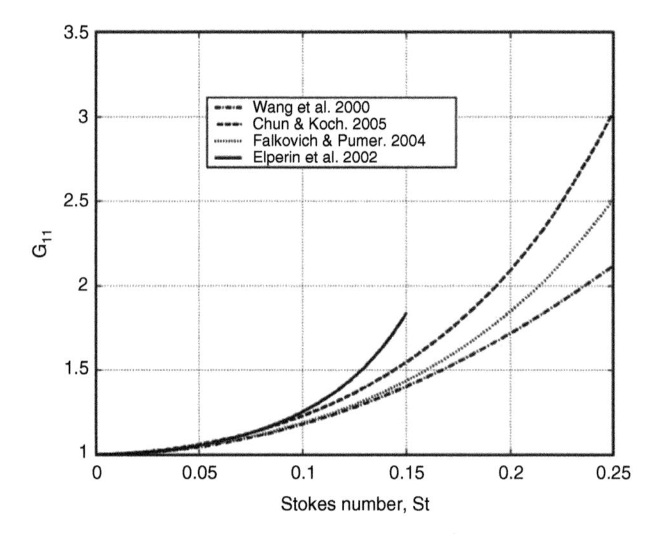

Figure 5.8.17 Dependencies of G_{11} on St obtained by different authors (from Khain et al., 2007; courtesy of Elsevier).

with similar size $St = 0.25$ ($G_{12} \sim 2$). When $St > 0.1$, the difference between drop sizes causes a steep decrease of the G_{12} function because of the Lagrangian acceleration growth.

Wang et al. (2000), Reade and Collins (2000), Zhou et al. (2001), Elperin et al. (2002a), and Falkovich et al. (2002) reported strong drop clustering at $St > 0.3$

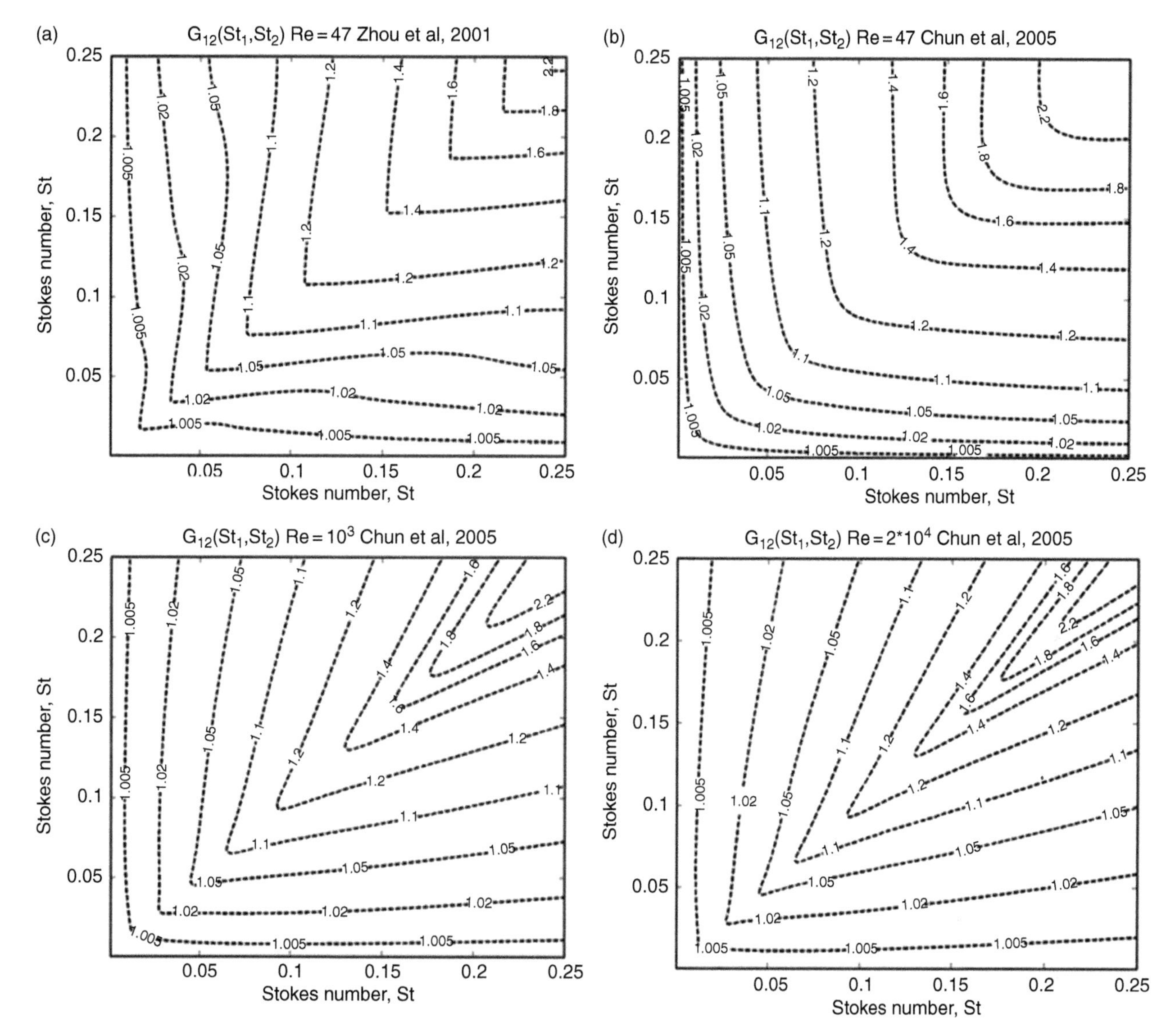

Figure 5.8.18 The G_{12} function for bi-disperse suspensions of small drops. (a) Zhou et al. (2001), $Re_\lambda = 47$; (b) Chun et al. (2005), $Re_\lambda = 47$; (c) Chun et al. (2005), $Re_\lambda = 10^3$; (d) Chun et al. (2005), $Re_\lambda = 2 \times 10^4$. The results were obtained with drop sedimentation neglected (from Khain et al., 2007; courtesy of Elsevier).

(Figure 5.5.18). For $St \sim 0.7 - 1$, functions G_{11} and G_{12} increase by one–two orders of magnitude, with the scale decreasing from the Kolmogorov microscale down to the drop-size scale. Later studies showed that these function values are unrealistically high. Taking into account differential drop sedimentations in DNS yielded lower clustering rates (Grabowski and Vaillancourt, 1999; Franklin et al., 2007). Grits et al. (2006) used the kinematic model of turbulent flow to investigate effects of turbulence on drop clustering. They calculated spatial spectra of drop concentration fluctuations and found the pronounced maximum at ~1 cm. In addition, they found that drop sedimentation decreases the clustering rate, thus the normalized values of RMS amplitudes of drop-concentration fluctuations become close to the experimental results (~25% at $St = 0.07$) (Pinsky and Khain, 2001, 2003).

Figure 5.8.19 shows the RDF at the contact points for collector drops with radii r_2 equal to 10 μm, 20 μm and 30 μm as function of the ratio of the radii of the pairs drops, calculated in DNS by Franklin et al. (2007). In these simulations, the effects of drop

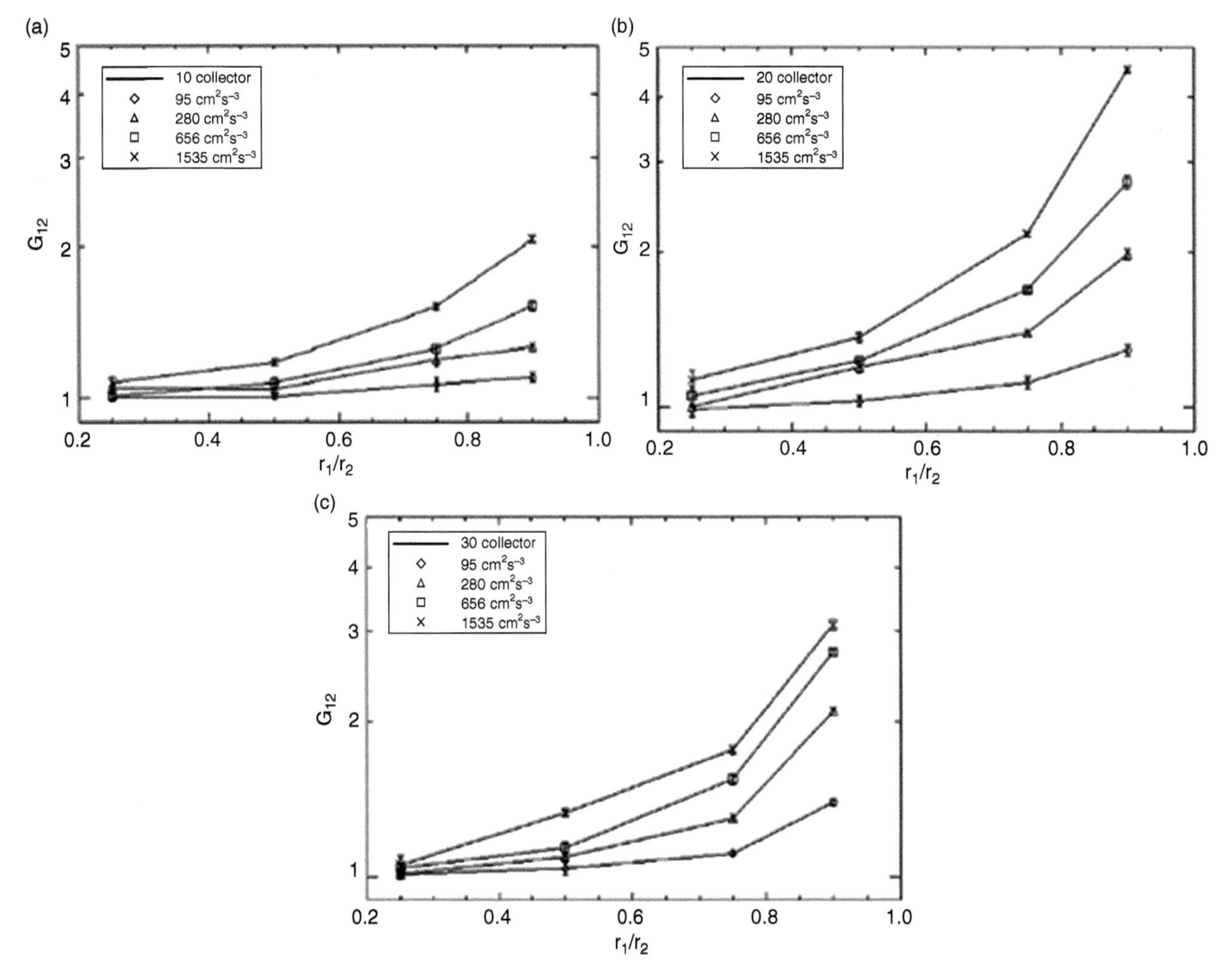

Figure 5.8.19 Normalized radial distribution function for three collector droplets in four flow fields as a function of the radius ratio. Droplet collectors have radii of (a) 10-μm, (b) 20-μm, and (c) 30 μm (from Franklin et al., 2007; American Meteorological Society©; used with permission).

sedimentation were taken into account. One can see that RDF reaches its maximum at the drop collector radius of about 20 μm, and then significantly decreases when the drop collector radius is down to 30 μm. It is reasonable to expect that RDF decreases monotonically as a drop collector radius increases.

Figure 5.8.20 demonstrates the dependence of RDF at drop contact for different turbulent dissipation rates ε. These results were obtained in DNS simulations by Ayala et al. (2008a), who took the gravitation settling of drops into account. Although large values of G_{11} function are typical for monodisperse collisions (the peak in Figure 5.8.20b reaches seventeen), the role of these collisions is not so essential since the width of drop SDs in real clouds usually exceeds several μm (Sections 2.3 and 5.11), i.e., substantially exceeds the width of the peak seen in Figure 5.8.20b.

Figure 5.8.21 shows the dependence of the relative increase in the collision efficiency, in the swept volume, and in the $G_{12}(r_1 + r_2)$ function on the dissipation rate for a pair of droplets of 10 μm and 20 μm in radius, obtained in different studies. The statistical estimations of droplet clustering magnitude performed by Pinsky and Khain (2003) using in situ measurements are also presented. One can see that the increase in the collision efficiency with increasing turbulent intensity is faster than in other quantities. Estimations show that the collision rate increases due to the clustering effect by about 15%, due to the transport effect (swept volume), by about 30–40%, and

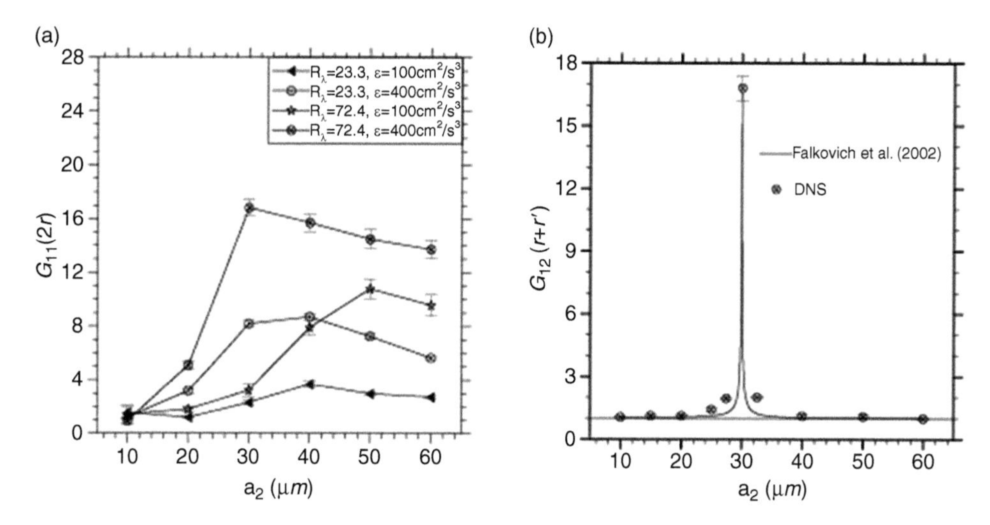

Figure 5.8.20 Radial distribution functions for settling drops in a turbulent flow: (a) monodisperse case and (b) bi-disperse case with $r = 30$ μm, $r + r' = 72.41$μm, and $\varepsilon = 400$ cm^{2}s^{-3} (from Ayala et al., 2008a; courtesy of IOP Publishing & Deutsche Physikalische Gesellschaft©).

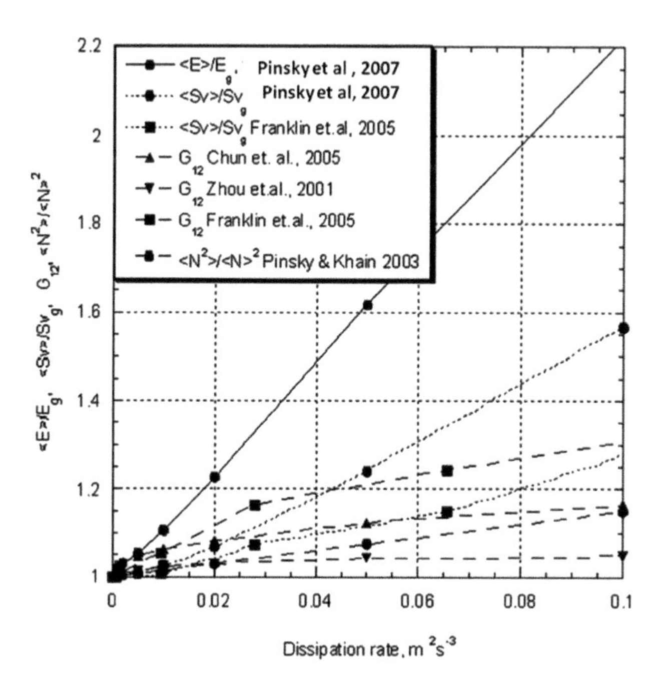

Figure 5.8.21 Dependences of the $G_{12}(r_1 + r_2)$ function, normalized swept volume and normalized collision efficiency on the dissipation rate for pair of droplets of 10 μm and 20 μm in radius (from Pinsky et al., 2007; American Meteorological Society©; used with permission).

by ~ twofold due to HDI. The growth of the collision efficiency is, therefore, the major factor by means of which turbulence increases the collision rate between cloud droplets.

5.8.7 Parameterization of Turbulent Collisions in Bin Microphysics Cloud Models

In numerous theoretical and modeling studies, evolution of DSD is calculated using the SCE with the collection kernel $K_{g\_col}(m_1, m_2) = V_{swept} \cdot E \cdot E_{coal}$, where V_{swept} is the swept volume (Section 5.6, Equation (5.6.5)). The SCE is usually used to describe collisions within air volumes of linear scales ranging from several tens to a few hundred meters. The collision rate is determined by product $f(m_1)f(m_2) \cdot V_{swept} \cdot E \cdot E_{coal}$. In the state-of-the-art numerical cloud models not taking turbulence effects into account, the correlations in the product are neglected, assuming that $\overline{f(m_1)f(m_2)V_{swept}EE_{coal}} = \overline{f}(m_1)\overline{f}(m_2)\overline{V}_{swept}\overline{E}\,\overline{E}_{coal}$. In a turbulent flow, all the terms in the product $f(m_1)f(m_2) \cdot V_{swept} \cdot E \cdot E_{coal}$ depend on the Lagrangian accelerations and the turbulent shears, i.e., they can be statistically dependent. For instance, the clustering effect is related to the positive correlation of fluctuations of drop concentration (Equation (5.8.7)). The increase in the Lagrangian accelerations attenuates the drop clustering effect, but increases the collision efficiency, which is an example of a negative correlation. Pinsky et al. (2007) found a positive correlation between V_{swept} and E under strong turbulence when the Lagrangian accelerations are high. Due to positive and negative correlations between the functions, the product $f(m_1)f(m_2) \cdot V_{swept} \cdot E \cdot E_{coal}$ should be averaged as a whole.

Turbulent collision kernels are random variables whose PDF have elongated tails (Figure 5.8.12). The tails become more pronounced with an increase in turbulence intensity. Employing the mean turbulent kernels (increased by a certain factor in comparison to those in the pure gravity case) actually leads to loss of important information concerning the PDF of the collision kernels. Nevertheless, in cloud models that take turbulence effects on collisions into account, only the mean turbulent enhancement factors are used, which most likely leads to underestimation of the role of turbulence in precipitation formation.

A modified SCE taking into account the turbulence effects is thought to be averaged over an appropriate volume, as in the case of purely gravitational collisions (Section 5.6). Apart from the corresponding enhancement factors introduced, Equation (5.8.8) is similar to Equation (5.6.5) for the purely gravitational case:

$$\frac{d\bar{f}(m,t)}{dt} = \int_0^{m/2} \bar{f}(m')\bar{f}(m-m')P_{clust}P_{kern}K_{g\_col}(m,m')dm' - \int_0^{\infty} \bar{f}(m)\bar{f}(m')P_{clust}P_{kern}K_{g\_col}(m,m')dm' \quad (5.8.8)$$

Since DSD can experience strong fluctuations of different scales in a turbulent medium, we use the overbar in $\bar{f}(m)$ to stress the fact that this function is averaged. In Equation (5.8.8), $P_{kern}(m,m')$ and $P_{clust}(m,m')$ are the factors corresponding to the turbulent increase in collision kernel and to the drop clustering effect, respectively. In cloud models these factors can be calculated using interpolation of **Tables B4–B7** (Appendix B). **Tables B4–B6** take into account the effect of small-scale turbulence intermittency, arising at large Re_λ values conditions, on collision kernels (Pinsky et al., 2008a). Numerous parameterization formulas for P_{clust}, based either on experimental findings or on DNS results, were used in cloud models (Ayala et al., 2008a; Franklin et al., 2007; Pinsky and Khain, 2003). For example, Pinsky et al. (2008a) and Benmoshe et al. (2012) parameterize P_{clust} by a formula based on the relationship between the Stokes number and the amplitude of droplet concentration fluctuations (Equation (5.5.27)) and Figure 5.5.21) as

$$P_{clast}(St_1, St_2) = 1 + 0.333(St_1 St_2)^{0.317}, \quad (5.8.9)$$

where St_1 and St_2 are the Stokes numbers of droplets supposed to collide.

The gravitational collision kernel between drops of the same size is equal to zero, while the turbulent kernel is not. However, K_g decrease to zero for drops of the same size is very sharp, so the zone of very low values of K_g is substantially narrower than the width of the grid bin at which the DSD is determined. Thus, when calculating the gravitational kernel and the collision enhancement factors for drops belonging to the same bins, values averaged over the bin width should be used. For this reason, P_{kern} is defined for drops belonging to the same bin, as well.

5.8.8 Evolution of DSD Within a Turbulent Medium

The role of turbulence in DSD evolution was investigated by solving SCE (Pinsky et al., 2008a; Xue et al., 2008; Wang and Grabowski, 2009). Pinsky et al. (2008a) solved modified SCE (Equation (5.8.8)) for turbulent conditions typical of different types of clouds; the effect of turbulence was parameterized using interpolation of collision enhancement factor (Appendix B, **Tables B4–B6**) and applying Equation (5.8.9) for the clustering effect. Evolution of two DSD typical of Mediterranean clouds (drop concentration $N = 500\ \text{cm}^{-3}$) and clouds of more continental type ($N = 800\ \text{cm}^{-3}$) with a LWC of about 2 g m^{-3} were simulated. The initial DSD were centered at the 9-µm radius. The calculations were performed both for the purely gravitational case and for different dissipation rates ranging from $\varepsilon = 100\ \text{cm}^2\text{s}^{-3}$ to $\varepsilon = 1{,}000\ \text{cm}^2\text{s}^{-3}$. The calculations under turbulent conditions were performed both with and without the effects of drop clustering taken into account. The results of the simulations are shown in **Figure 5.8.22**. One can see that at high drop concentration and low turbulence intensity, the turbulence effect is not very significant (at least, during the time period of 30 min). Turbulence of higher intensity significantly accelerates collisions, indicating formation of raindrops, while no raindrops were formed in the pure gravity case. Therefore, a dramatic effect of turbulence on rain formation is demonstrated. The clustering effect accelerates large raindrop formation by a few minutes. Such acceleration is not very significant in warm rain clouds, but may play an important role in clouds with high vertical velocities where a few minutes' delay can lead to transport of liquid drops to higher levels and their freezing. The results of the calculations depend on the DSD initial shape: the narrower the initial DSD, the more substantial the

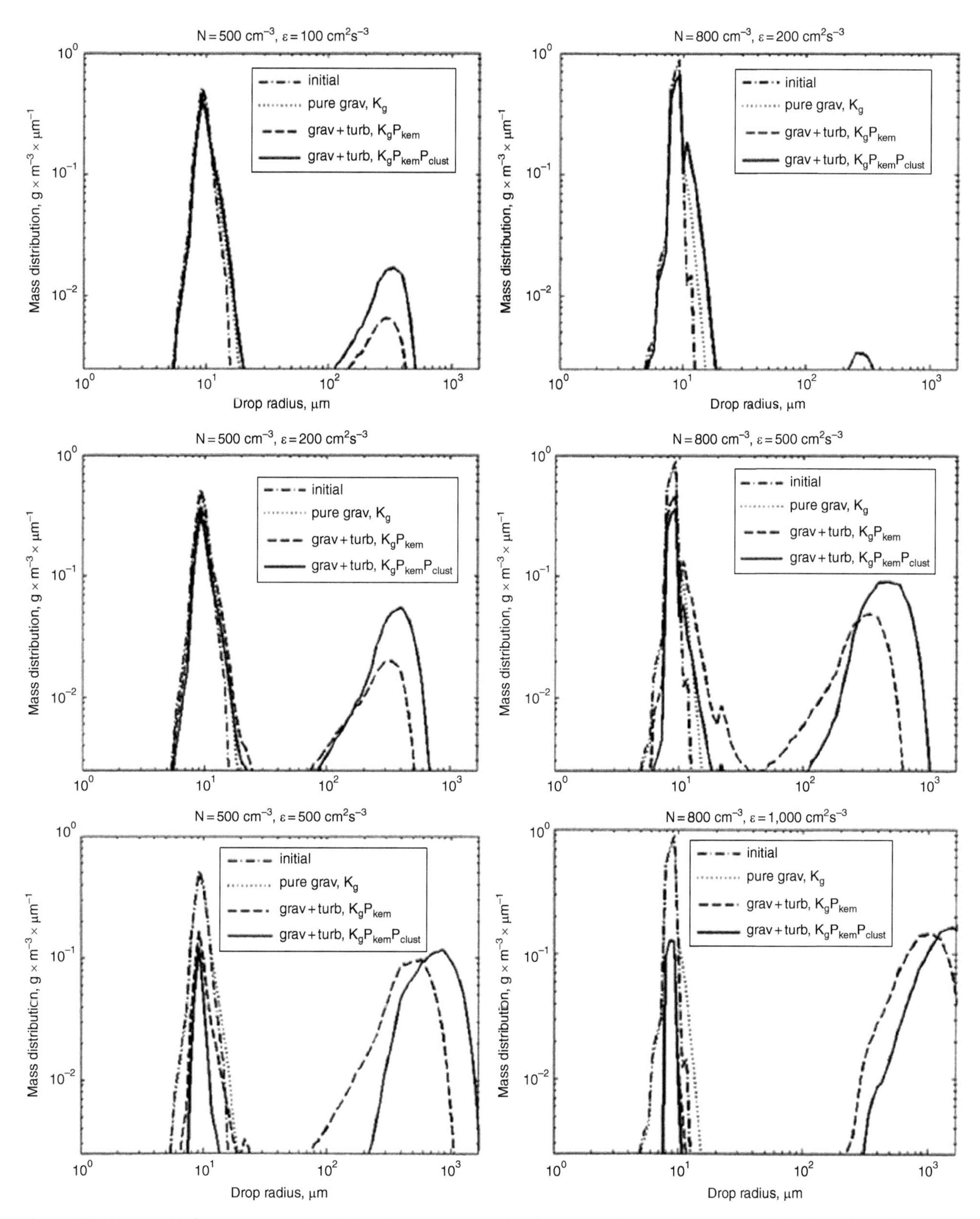

Figure 5.8.22 DSD obtained by solving the stochastic collision equation during t = 30 min. Three levels of turbulence intensity typical of cumulus clouds (top), deep convective clouds (middle), and cumulonimbus clouds (bottom) were used. The DSD obtained in the purely gravitational case (dotted line), in case the collision enhancement is determined only by HDI and the transport effect (dashed line), and when the enhancement factor due to the clustering effect is also taken into account (solid line) (from Pinsky et. al., 2008a; American Meteorological Society©; used with permission).

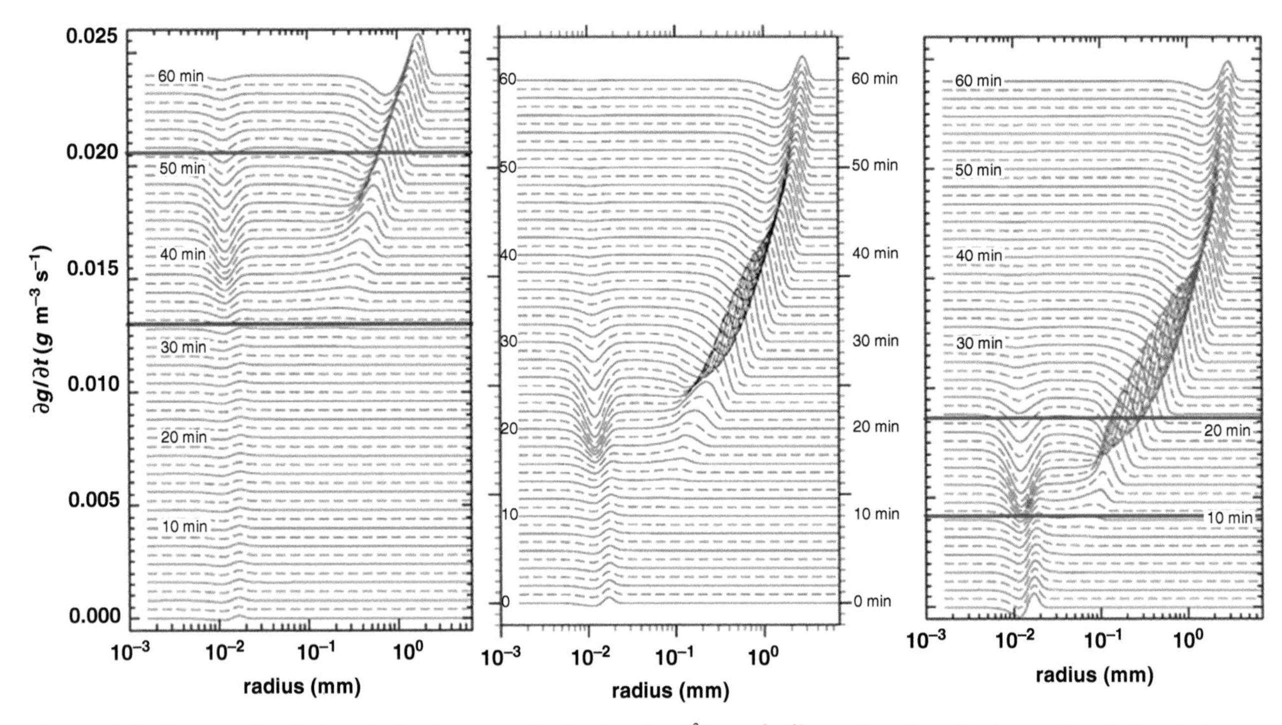

Figure 5.8.23 The rate of variations in the drop mass-density function $\left(\frac{\partial g}{\partial t}, \mathrm{gm}^{-3}\mathrm{s}^{-1}\right)$ as a function of a drop radius. Left: solutions using a purely gravitational kernel; middle: solutions using a geometrical turbulent kernel at $\varepsilon = 300$ cm^2 s^{-3} and the r.m.s. fluctuation velocity of 2.0 m/s; right: solution using a full turbulent collision kernel, taking into account all of the mechanisms of turbulence influence. Each plot contains 61 curves representing the time period from t = 0 to t = 60 min with the increment of 1 min. The two red lines mark the beginning and the end of the accretion phase (from Xue et al., 2008 and Wang and Grabowski, 2009; courtesy of John Wiley & Sons, Inc.©).

turbulence effects. A more detailed analysis of turbulent effects on warm precipitation using HUCM is given in Sections 5.11 and 7.4.

Xue et al. (2008) and Wang and Grabowski (2009) simulated evolution of DSD under turbulent and non-turbulent conditions to study to which extent turbulence affects warm rain development. The turbulent enhancement factor based on DNS and the corresponding parameterization formulas (Ayala et al., 2008a,b; Wang et al., 2008) were used. **Figure 5.8.23** demonstrates the rates of evolution of the MD function in three cases: (a) a purely gravitational collision kernel; 9b) a geometrical collision kernel when the turbulence-induced enhancement of collision efficiency is not taken into account; and (c) a full turbulent collision kernel taking into account all the mechanisms of turbulence influence. The simulations demonstrate a significant turbulence-induced acceleration in raindrop formation: the first raindrops appeared at 33 min, 15 min, and 10 min in cases (a), (b), and (c), respectively.

5.8.9 Parameterization of Turbulence Effects on Drop Collisions in Bulk Microphysics Models

Several studies have been dedicated to parameterization of turbulence effects on drop collisions in bulk cloud models (Franklin, 2008; Seifert et al., 2010). Franklin (2008) developed parameterization of turbulence effect on autoconversion, accretion, and selfcollection (Section 5.7) using the DNS obtained in an earlier study by (Franklin et al., 2007). Since DNS demonstrate a close relationship between turbulence dissipation rate ε and the Reynolds number Re_λ ($Re_\lambda = 21\varepsilon^{0.12}$, where ε has units cm^2 s^{-3}), only a single parameter Re_λ was used for parameterization. The influence of turbulence on the swept volume and the clustering were the only effects investigated in this study. The effects of turbulence on rain formation were investigated via the increase in the rates of autoconversion and of accretion, given in the form $\frac{\partial q_r}{\partial t} = Cq_c^\alpha N_c^\beta$, where q_r is rainwater mixing ratio and q_c and N_c are the cloud water mixing ratio and the

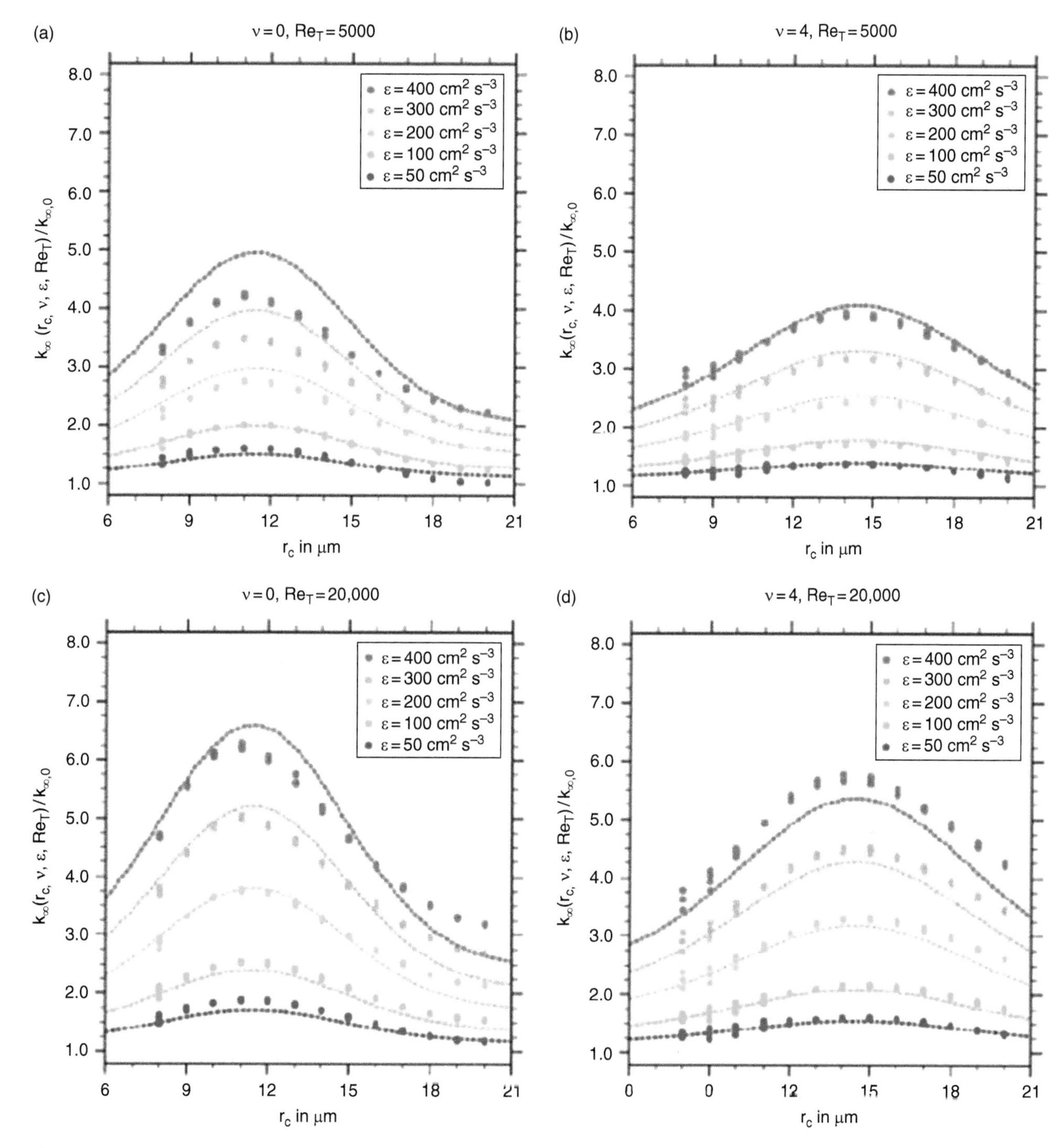

Figure 5.8.24 The enhancement factor of the autoconversion rate k_c/k_{c0} as a function of the mean volume radius r_c of cloud droplets, calculated for various values of dissipation rate ε (from Seifert et al., 2010; courtesy of John Wiley & Sons, Inc.©).

concentration of cloud droplets, respectively. Coefficients C, α and β are regarded as functions of Re_λ. This type of approximation formula corresponds to Equation (5.7.45) for the purely gravitational case. The final fitting equations obtained by Franklin (2008) for autoconversion rate $\frac{\partial q_r}{\partial t}\big|_{au}$, accretion rate $\frac{\partial q_r}{\partial t}\big|_{ac}$, and droplet concentration rate in selfcollection process $\frac{\partial N_c}{\partial t}\big|_{sc}$ are the following:

$$\left.\frac{\partial q_r}{\partial t}\right|_{au} = \left(6.5 \times 10^{1.3} Re_\lambda^{-6.3} + 1.9\right) q_c^{3.4 Re_\lambda^{-0.23}} \; N_c^{-5.3 \; Re_\lambda^{-0.38}}$$
$$\left.\frac{\partial q_r}{\partial t}\right|_{ac} = 5.44 \; \exp\left(-\frac{3.81}{Re_\lambda}\right) q_c q_r$$
$$\left.\frac{\partial N_c}{\partial t}\right|_{sc} = -\left(-2.6 \times 10^5 \times 0.092^{Re_\lambda} + 4.4 \times 10^4\right) q_c^2 \tag{5.8.10}$$

The rates obtainable from Equation (5.8.10) inevitably underestimate the overall effect of turbulence since the turbulence effect on collision efficiency is not taken into account in this parameterization.

Seifert et al. (2010) based their parameterization of the turbulence influence on the rates of autoconversion and accretion using the results obtained by Ayala et al. (2008b) and Wang et al. (2008). They introduced a dependence of coefficient k_c from the autoconversion Equation (5.7.38) on turbulence parameters in the form

$$k_c = k_{c0}\left[1 + \varepsilon Re_\lambda^{1/4} f(v)\right], \tag{5.8.11}$$

where k_{c0} corresponds to the purely gravitational case and $f(v)$ is a function of the shape parameter v in the Gamma distribution (Equation 5.7.34) used for DSD approximation. The values of the enhancement factor of the autoconversion rate, calculated as a function of the mean volume radius r_c of cloud droplets for various values of dissipation rate ε, are shown in **Figure 5.8.24**. The figure demonstrates that the autoconversion enhancement factor induced by turbulence can be as large as 6–7. As for accretion process, Seifert et al. (2010) found only a weak dependence on turbulence that was parameterized by formula

$$k_r = k_{r0}\left(1 + 0.05\varepsilon^{1/4}\right), \tag{5.8.12}$$

where k_{r0} is the coefficient applied in the purely gravitational case (Table 5.7.2). For $\varepsilon = 400\ \text{cm}^2\text{s}^{-3}$, this equation gives a moderate accretion enhancement rate of 18%.

5.8.10 Some Unsolved Problems

In spite of the significant progress in the investigation of turbulence effects on drop collisions, some important problems remain unresolved. Turbulent collision kernels are random variables whose probability density functions have elongated tails. The tails become more pronounced as turbulence intensity increases. The tails indicate, therefore, the existence of a comparatively large number of "lucky" drops experiencing collisions several times more often than other drops, even under spatially uniform concentration. These lucky drops are large enough to trigger enhanced collisions during a time interval shorter than that following from the stochastic collision equation incorporating only the mean values of the collision kernels. The clustering effect also fosters formation of lucky drops. The effect of the stochastic nature of collision kernels on DSD evolution is currently unknown. Although Wang et al. (2006b) proposed some approach to averaging of SCE, the method for proper SCE averaging using simple and accurate enough approaches has not been found yet. In spite of its name, SCE is not actually stochastic, and it would be more appropriate to regard SCE as a quasi-stochastic equation.

Factors enhancing collisions between drop pairs containing large drops still require a justification. Pinsky et al. (2008a) did not consider turbulent effects on drops with radii exceeding 22 μm. Wang and Grabowski (2009) suppose that collision-rate enhancement in drop pairs containing drops with radii equal to or exceeding 100 μm is equal to unity (Figure 5.8.15). At the same time, the results obtained by Vohl et al. (1999) demonstrate a strong influence of turbulence on collisions between small raindrops and small cloud droplets (Figure 5.8.1). These findings indicate that the effects of turbulence on drop collisions are, supposedly, stronger than it follows from these theoretical evaluations. The fact that collision rate increases within a turbulent flow calls for analyzing the problem turbulent effects on drop breakup.

5.9 Turbulent Mixing in Clouds

Mixing between cloud volumes and neighboring air volumes decreases the gradients of thermodynamic variables and affects microphysical cloud structure, DSD shape, and precipitation. The mixing is especially intense near cloud top and cloud edges where the gradients of the thermodynamic quantities are sharp. The crucial role of cloud/air mixing in DSD shaping and precipitation formation has made it the focus of multiple research (e.g., Devenish et al., 2012)

5.9.1 The Concept of Turbulent Mixing and Mixing Operators

Mixing in the atmosphere occurs due to turbulent and molecular diffusion (Monin and Yaglom, 1971; Wyngaard, 2010). Turbulent mixing starts as a mutual diffusion of neighboring air volumes and unfolds in the form of turbulent filaments, elongating and narrowing with time (**Figure 5.9.1**). The filaments change their shape forming fractal-like structures. The number of filaments and their interface area continuously increase

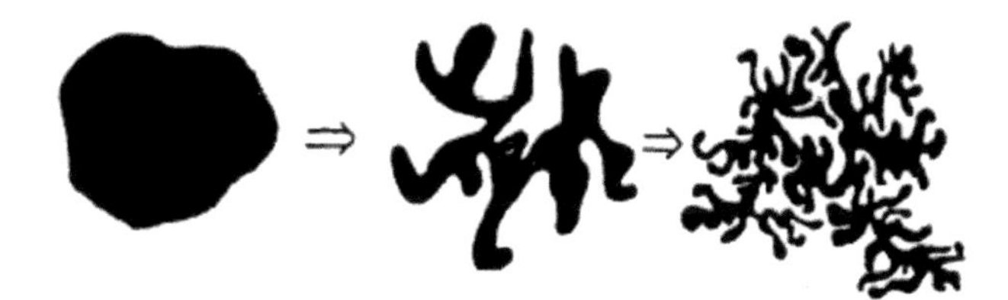

Figure 5.9.1 Formation of filaments of complex shapes during turbulent diffusion (from Monin and Yaglom, 1971; courtesy of MIT Press).

with time, while the width and the distance between filaments decrease down to the Kolmogorov microscale. Amplitudes of fluctuations of different thermodynamic and microphysical quantities decrease with time, creating high-frequency fluctuations of these quantities. Parallel to turbulent diffusion, there is molecular diffusion that is efficient in zones of high gradients at the boundaries of filaments. When the width of filaments reaches several millimeters, molecular diffusion becomes dominating, further increasing the rate of smoothing the neighboring filaments. As a result, both turbulent and molecular mixing decrease the spatial gradients of the thermodynamic and microphysical quantities.

A complete spatial homogenization of an air volume is impossible due to continuous generation of turbulent fluctuations. In clouds, turbulent velocity fluctuations lead to random fluctuations of drop location (the Poisson noise), fluctuations of drop concentration due to the inertial effect (Section 5.5), and creation of supersaturation fluctuations (Section 5.2). These low-energy and high-frequency fluctuations play the role of noise, but their effect is negligibly small. Despite the noise, the volumes can be considered well mixed, or homogenized by the end of mixing process.

Turbulent mixing rate depends on the sizes of air volumes and on the turbulence intensity characterized by dissipation rate ε. The most important parameter characterizing the mixing rate is the characteristic mixing time τ_{mix} of turbulence-caused homogenization of all the parameters within the air volume, e.g., temperature, humidity, and supersaturation. The characteristic mixing time of a volume with linear scale L is estimated as

$$\tau_{mix} = C\left(\frac{L^2}{\varepsilon}\right)^{1/3}, \tag{5.9.1}$$

where C is a constant. After the time τ_{mix}, a volume with a linear scale of L will be mechanically homogenized and all the droplets within it will experience the same supersaturation. The spatial scale L is substantially larger than the Kolmogorov microscale and belongs to the inertial turbulent range (Section 3.3).

Different estimations show that C ranges from 0.1 to 5 (e.g., Monin and Yaglom, 1975; Boffetta and Sokolov, 2002; Giola et al., 2004; Jeffrey and Reisner, 2006; Lehmann et al., 2009). The uncertainty in the value of coefficient C in Equation (5.9.1) leads to uncertainties in evaluating the impact of mixing. The mixing time τ_{mix} is proportional to the characteristic lifetime of a turbulent vortex with linear scale L. Depending on the dissipation rate (different in clouds of different types) and the sizes of turbulent vortices, τ_{mix} varies from a few tenths of a second for centimeter-sized vortices to several minutes for vortices with $L \approx 100$ m. The values of τ_{mix} show that even comparatively large air volumes in clouds lose their initial nature during a period of about 10–15 min, which is substantially less than a lifetime of individual clouds. This fact illustrates the importance of including the mixing process into cloud models.

To describe the fine features of mixing including formation and breakup of filaments, it is necessary to use DNS with grid resolution of about 1 mm. However, the computational area of DNS models is too small to adequately simulate the mixing occurring in real clouds. Instead, in cloud models (as well as in other atmospheric models) turbulent mixing is parameterized using the K-theory, according to which turbulent diffusion fluxes of quantity a are assumed to be proportional to the gradient of the averaged values $\overline{a'u'_i} = -K\frac{\partial \bar{a}}{\partial x_i}$ (Section 3.4). In this expression, a' represents fluctuations related to turbulent vortices and filaments, and u'_i represents the velocity of movement of these fluctuations. The diffusion equation describing turbulent mixing of conservative quantity $\bar{a}$ is written as (Section 3.4)

$$\frac{\partial \bar{a}}{\partial t} = \frac{\partial}{\partial x_i}\left(K\frac{\partial \bar{a}}{\partial x_i}\right) \tag{5.9.2}$$

In fact, the diffusion operator on the right-hand side of Equation (5.9.2) plays the role of a spatially isotropic smoothing filter, averaging the parameters of the neighboring volumes. Although there is no strict physical basis for application of this operator, it is successfully used in all atmospheric and, in particular, cloud models, making the fields of modeled quantities similar to real smooth fields. Since the molecular viscosity is by five to six orders of magnitude lower than the turbulent coefficient K, the effect of molecular diffusion on scales larger than a few millimeters can be neglected. Only few studies investigate mixing at scales as low as the

Kolmogorov scales, where molecular diffusion is important (Baker and Latham, 1979; Krueger et al., 1997, 2006). In case the quantity a is not conservative, the term $K\frac{\partial \bar{a}}{\partial x_i}$ is replaced by $K\left[\frac{\partial \bar{a}}{\partial x_i} - \frac{\delta \bar{a}}{\delta x_i}\right]$, where the term $\frac{\delta \bar{a}}{\delta x_i}$ describes non-conservative changes of quantity $\bar{a}$ at the distance equal to the mixing length (Section 3.4). Accordingly, the mixing operator has the form

$$\frac{\partial \bar{a}}{\partial t} = \frac{\partial}{\partial x_i}\left(K\left[\frac{\partial \bar{a}}{\partial x_i} - \frac{\delta \bar{a}}{\delta x_i}\right]\right) \tag{5.9.3}$$

Typically, the term $\frac{\delta \bar{a}}{\delta x_i}$ is important while analyzing mixing in the vertical direction, where the air density and the temperature vary in a much stronger way than in the horizontal one.

The impact of turbulent mixing on cloud dynamics and microphysics has been an issue for discussion since the 1960s (e.g., Warner, 1969a), but still remains one of the least understood issues in warm cloud microphysics. This fact can be attributed to the high complexity of turbulent mixing that can be accompanied by changes in all the thermodynamic quantities and DSD. The process becomes especially complicated near cloud edges where phase transitions and DSD changes caused by mixing are particularly pronounced.

5.9.2 Mixing at Cloud Boundaries

Entrainment/Detrainment in Clouds

Interaction between clouds and the surrounding air can be conditionally separated into two stages. The first stage is entrainment of drop-free air volumes into a cloudy volume or entrainment of a cloudy volume into drop-free air. There are no phase transitions at this stage. The second stage is mixing between neighboring drop-free volumes and cloudy volumes, accompanied by phase transitions. This stage can lead either to total evaporation of drops and disappearance of cloudy volumes, or to conversion of initially drop-free volumes to cloud ones, which increases the total cloud volume. Entrainment and mixing lead to a decrease in the mean liquid mass content and drop concentration.

Effects of entrainment on cloud properties depend on the initial sizes of the drop-free volumes entrained and thermodynamic parameters of both drop-free and cloudy volumes. In stratocumulus clouds, entrainment of drop-free air takes place largely through the cloud top. According to Devenish et al. (2012), entrainment leads to a decrease in LWC only within a thin layer of a few tens of meters near cloud top of unbroken Sc. Continuous penetration of cloudy volumes into the inversion layer increases the cloud top height.

One can expect that the role of entrainment in small cumulus clouds is more important than in Sc due to relatively larger surface area of Cu. It is important to know whether the entrainment takes place at turbulent scales or at larger convective scales, and whether dry air reaches cloud axis or the dilution is concentrated near lateral cloud boundaries only. Despite the multiple field experiments investigating entrainment, the location of cloud zones where entrainment is most intense is still a subject of debates. Blyth et al. (1988) and Heus and Jonker (2008) assume that entrainment of the environmental air takes place largely through lateral cloud boundaries from a descending shell toward the cloud core where updrafts dominate. According to Gerber et al. (2008), in small cumulus clouds entrainment from the lateral boundaries is more intense than from the cloud top, supposedly because the depth of such clouds is limited by the inversion layer where turbulence is suppressed. Penetration of drop-free air via the lateral cloud boundaries of deep convective clouds was reported by Bera et al. (2016a,b) and Kumar et al. (2016).

The penetration of turbulent scale bulbs from the surrounding through the lateral cloud boundary leads to the local decrease in the LWC near cloud edge. **Figures 5.9.2a** and **5.9.2b** show two examples of such penetration. The figure shows *in situ* measured LWC along the aircraft track as a function of the horizontal distance from the cloud edge toward cloud core. The figure gives an idea on the sizes of air volumes penetrating a small warm cumulus cloud observed in the field experiment RICO (Gerber et al., 2008) (panel (a)), and penetrating a deep convective cloud observed in India (Bera et al., 2016a) (panel (b)).The measurements in RICO Cu were performed using the PVM, allowing measuring microphysical parameters at 10-cm resolution. Measurements in the Indian deep convective cloud were performed using the high frequency (10 Hz) forward scattering spectrometer probe (FSSP), allowing spatial resolution of about 10 m. Zones of depleted LWC with sharp gradients represent penetrated parcels. The sizes of these zones can be interpreted as entrained parcels' lengths. The scales of dry bubbles that initially enter the cloud typically decrease inside the cloud as is shown in panel (b). This decrease in the sizes takes place because the bubbles get fragmented and distorted toward the cloud core. The size of droplet-free bubbles penetrating small Cu is about 10 m. In deep convective clouds the size of entrained drop-free bubbles may exceed 100 m near the cloud edge and then decreases to 10 m and smaller toward the cloud core. The comparatively small scales of volumes

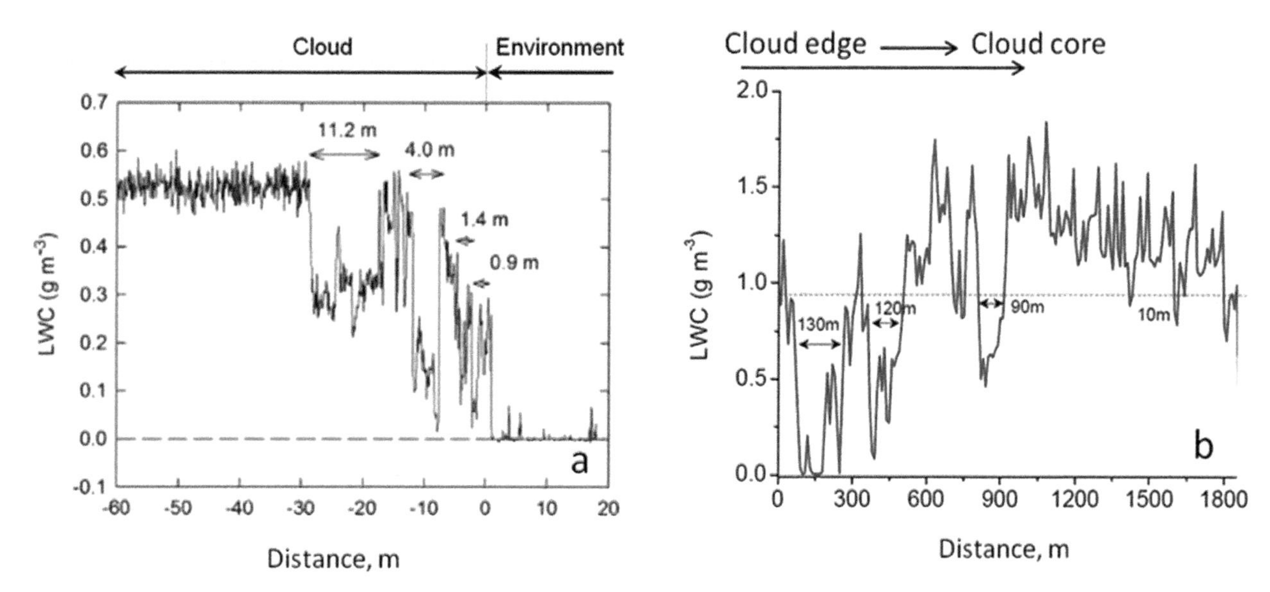

Figure 5.9.2 LWC measured as a function of the horizontal distance from the edge of a convective cloud. (a) small Cu observed in RICO (from Gerber et al., 2008; courtesy of JMS); (b) horizontal pass at 3.1 km altitude in a deep convective cloud measured in a monsoon cloud in India (from Bera et al., 2016a; courtesy of John Wiley & Sons, Inc. ©; used with permission).

penetrating through the lateral cloud boundaries, measured in these studies, indicate the turbulent nature of the entrainment process. The distance at which the bubbles penetrate clouds varies from several tens of meters in small Cu and about 1 km in deep Cb.

Another mechanism of entrainment in deep convective clouds was described by Damiani et al. (2006). Using dual-Doppler radar observations, they found that the cloud structure is characterized by a set of bubbles with well-defined toroidal circulation (vortex rings) with maximum updrafts in the centers. Such vortexes located near cloud edges have scales from several hundred meters to a few km and can be responsible for large-scale entrainment within comparatively large convective clouds. The entrainment at cloud top is determined by the structure of cumulus clouds whose upper part is nonuniform but consists of multiple bubbles (turrets) penetrating the dry air (**Figure 5.9.3**).

The plausible mechanism of the bubble formation is the *cloud-environment interface instability* or *cloud boundary instability*, proposed and justified in simulations performed by Grabowski and Clark (1991, 1993). The authors presented a theoretical analysis of the instability, showing that near the upper boundary of an ascending bubble the wind shear increases, and the shear layer depth decreases. Once the shear layer collapses to several tens of meters, the shear instability rises, leading to formation of bubbles with characteristic sizes of several hundred meters. The instabilities grow and migrate sideways along the interface, increasing their depth perpendicular to the interface. The final size of the bubbles depends on the initial thermal size, on interaction of instabilities agitated at different times and other factors. **Figure 5.9.4** illustrates the formation of the interface instability in a simulation of a rising thermal, obtained by a 3D model with spatial resolution of 6 m.

Figure 5.9.5 shows the vertical velocity field in a deep mixed-phase convective cloud at t = 45 min, obtained in simulations with HUCM at 50-m grid resolution. One can see that clouds simulated at this resolution have a large number of strong updrafts separated by low updrafts and even by downdrafts. The maximum updraft zone is located within the cloud core at z = 8 km, but the structure of the vertical velocity field is quite nonuniform, especially near the cloud top, which means that within the cloud there are both updrafts and downdrafts. The turrets forming near the cloud top have a comparatively low width, which fosters the dilution of the cloudy air through a lateral boundary of each turret. Similar results were obtained in simulations of a maritime squall line by Fierro et al. (2012). Strong downdrafts a few hundred meters below the top of convective clouds at distances of several kilometers above cloud base were reported by Prabha et al. (2011), based on the experimental data obtained in CAIPEEX.

Figure 5.9.3 Photo of cumulus clouds in RICO. The upper part of clouds consists of multiple turrets (courtesy of Bjorn Stevens).

Squires (1958), Telford and Chai (1980), and Telford et al. (1984) assumed that dry-air entrainment in deep convective clouds takes place largely in the the vertical rather than in the horizontal directions. Telford and Chai (1980) and Telford et al. (1984) analyzed the possible effect of such entrainment on microphysical structure of clouds (see Section 5.10).

Two Scenarios of Turbulent Mixing Between Clouds and Surrounding Air

Turbulent mixing at cloud edges and cloud tops accompanied by phase transitions has been the focus of numerous studies, beginning with the pioneering works of Latham and Reed (1977), Baker and Latham (1979), Baker et al. (1980), and Blyth et al. (1980). It is known that in subsaturated air volumes, evaporation of cloud droplets leads to DSD shifts toward smaller droplet sizes. At the same time, Latham and Reed (1977) showed in laboratory experiments that at certain conditions some droplets completely evaporate while others remain unchanged after mixing with subsaturated air. This finding gave rise to the concept of two types of turbulent mixing or two mixing scenarios: *homogeneous* and *inhomogeneous*. Typically, two limiting cases are considered: homogeneous mixing and extremely inhomogeneous mixing. The classical conceptual schemes of homogeneous and extremely inhomogeneous mixing is presented in **Figure 5.9.6**.

According to the classical concept, the process of homogeneous mixing (left branch in Figure 5.9.6) consists of two stages. During the first short stage, the initial gradients of microphysical and thermodynamic variables rapidly decrease to zero. By the end of this stage, the temperature, humidity (hence, supersaturation), and droplet concentration are spatially homogenized and all the droplets within the mixing volume experience an identical saturation deficit. During the relatively long second stage, droplets evaporate and increase the relative humidity in the volume. The evaporation leads to a decrease in the size of all the droplets and to a shift of the entire DSD toward smaller droplet sizes. There are two possible final equilibrium states of homogeneous mixing: either the droplets completely evaporate if the mixed air is too dry or droplets continue evaporating until they saturate the environment. Figure 5.9.6(3) shows the final state in the second case. At a monodisperse DSD, such evaporation does not lead to a decrease in the droplet concentration. At a polydisperse initial DSD, both total and partial evaporation of droplets determine the final DSD. If droplet evaporation is only partial, the droplet concentration remains unchanged.

Extremely inhomogeneous mixing (right branch in Figure 5.9.6) is a relatively slow process in which the spatial gradients of thermodynamic values remain nonzero for the entire period of mixing, so droplets within the mixing volume experience different saturation deficit, which causes different evaporation rates. According to the concept of extremely inhomogeneous mixing, some droplets transported by the turbulent

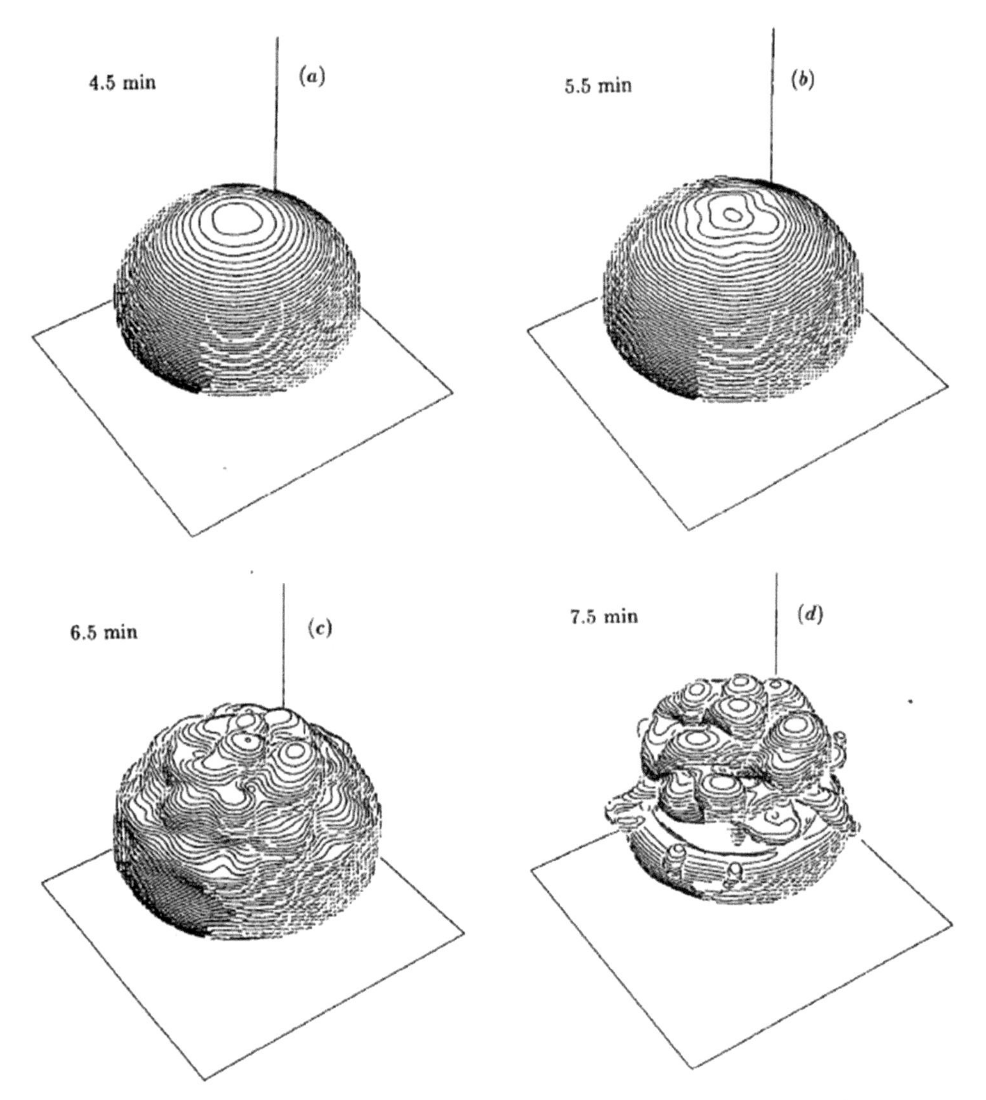

Figure 5.9.4 Three-dimensional perspective of the cloud–water field exceeding 0.01 g kg^{-1} in a numerical simulation of a rising thermal. The square at the bottom shows the extension of the innermost computational domain, which is 0.8 km × 0.8 km (from Grabowski and Clark, 1993; American Meteorological Society©; used with permission).

eddies into the dry environment experience complete evaporation, whereas other droplets remain unchanged. The mixing results, therefore, in the reduced droplet concentration. There is no shift of the DSD in the radii space, but the final DSD is proportional to the initial DSD with the proportionality coefficient below unity. It also means that the change in the droplet mass is proportional to the change in the droplet concentration. The characteristic droplet radii such as the mean volume radius and the effective radius also remain unchanged. Therefore, according to the classical concept, during extremely inhomogeneous mixing the shape of DSD is conserved, while the total droplet concentration decreases (see reviews by Devenish et al., 2012 and Korolev et al., 2016). The mechanism of the extremely inhomogeneous mixing has been a major focus of research since it was assumed to favor formation of large droplets (so-called superadiabatic droplets), which are able to trigger the intense collisions leading to the emergence of first raindrops.

Since mixing is accompanied by droplet evaporation and changes in supersaturation, the evolution of thermodynamic parameters and DSD (including those of the final DSD) depend on the rates of two processes: the turbulent diffusion rate (mixing rate) and the supersaturation change rate. These two rates have the characteristic times: the first is determined by Equation (5.9.1) and the second is droplet relaxation time τ_{pr},

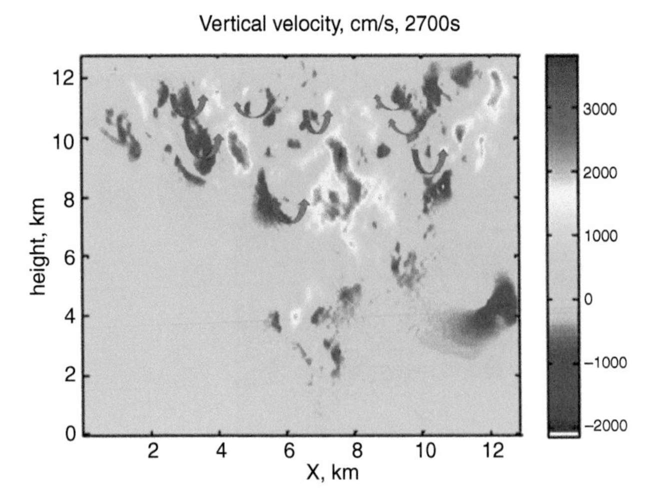

Figure 5.9.5 The vertical velocity field in a deep mixed-phase convective cloud, obtained in simulations using HUCM at 50-m grid spacing. The process of entrainment and mixing (schematically denoted by arrows) takes place within the cloud and is especially pronounced near the cloud top (from Benmoshe et al., 2012; courtesy of John Wiley & Sons, Inc.©; used with permission).

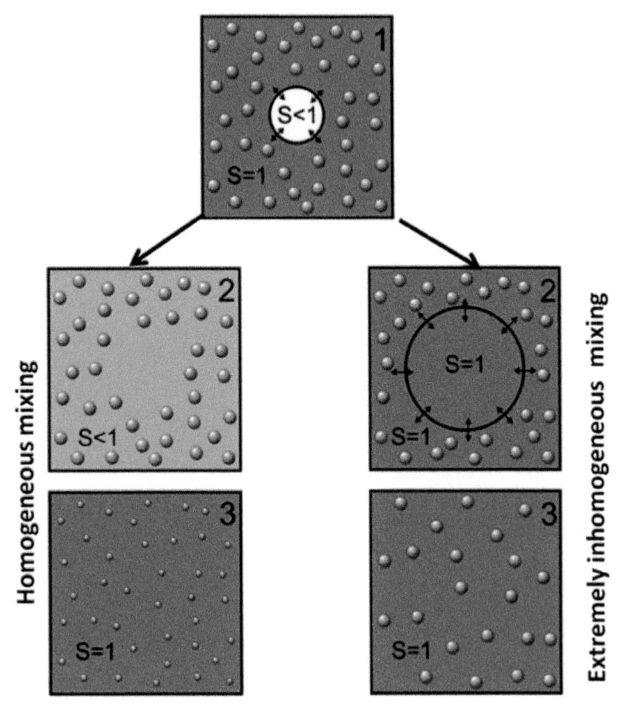

Figure 5.9.6 The classical conceptual schemes of (left) homogeneous mixing and (right) extremely inhomogeneous mixing. (1) initial state; (2) mixing state; (3) final state. The schemes are given for monodisperse droplet SDs (from Korolev et al. (2016), with changes; courtesy of ACP).

determined by Equation (5.2.14). If $\tau_{mix} \ll \tau_{pr}$, the mechanical mixing is much faster than the change in supersaturation and droplet size, so the mixing is considered homogeneous. In case $\tau_{mix} \gg \tau_{pr}$, the mixing is much slower than the adjustment of supersaturation to evaporating droplets and is considered extremely inhomogeneous. The ratio

$$Da = \frac{\tau_{mix}}{\tau_{pr}} = \frac{4\pi D\bar{r}NL^{2/3}}{\varepsilon^{1/3}} \tag{5.9.4}$$

is called the Damköhler number. The value $Da \approx 1$ is usually assumed to be the boundary approximately separating homogeneous and inhomogeneous mixing, $Da \ll 1$ corresponds to homogeneous mixing, while $Da > 1$ corresponds to inhomogeneous mixing. $Da \gg 1$ corresponds to extremely inhomogeneous mixing. Sometimes, the characteristic time scale of total droplet evaporation is used in the definition of the Damköhler number instead of the phase relaxation time (e.g., Lehmann et al., 2009). This time is evaluated from the equation for diffusional growth of a droplet of a typical size.

Mixing Diagrams

Distinguishing between mixing types by analyzing in situ measurements represents significant difficulties. Typically, analysis of mixing is performed using so-called mixing diagrams (Brenguier and Burnet, 1996; Pawlowska et al., 2000; Burnet and Brenguier, 2007; Lehmann et al., 2009; Freud et al., 2011). The original mixing diagrams show dependencies of the normalized cube of the mean volume radius (or the normalized cube of effective radius) on cloud fraction (also called the degree of dilution). The cloud fraction, μ, defined as a ratio of initially cloudy volume and total mixing volume, ranges from zero to unity.

In the classical theory of mixing diagrams, the DSD is assumed to be monodisperse, i.e., all droplets have identical sizes. Mixing diagrams are plotted for final equilibrium stages of two limiting cases, homogeneous and extremely inhomogeneous mixing. Pinsky et al. (2016a) derived analytical expressions for mixing diagrams of homogeneous mixing (see Equation (5.9.15)). An example of a mixing diagram is presented in **Figure 5.9.7**.

Extremely inhomogeneous mixing is depicted in this diagram by a horizontal straight line, indicating that the mean volume radius does not change in the course of mixing despite changes in the cloud fraction. In case of homogeneous mixing and a monodisperse initial DSD,

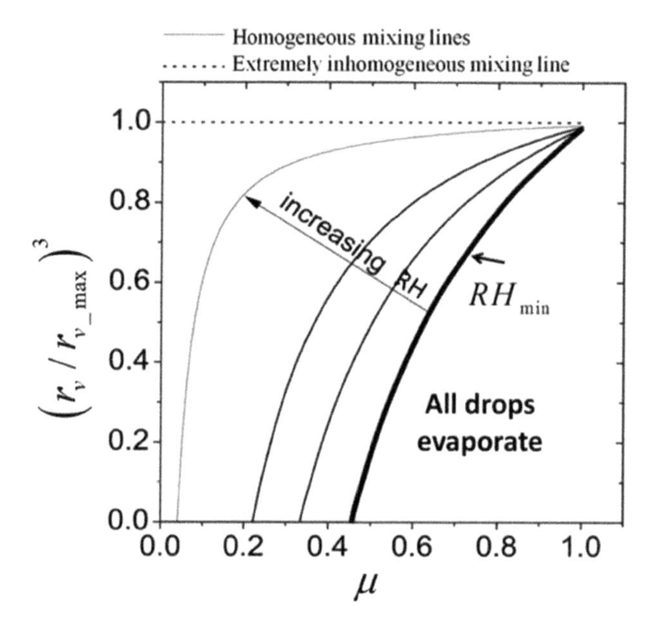

Figure 5.9.7 Example of a mixing diagram. Lines show the dependences of the normalized cube of the mean volume droplet radius on the cloud fraction. Curves are plotted for different relative humidity values in the initially droplet-free volume. The bold curve indicates the limiting case when all droplets evaporate. An increase in *RH* is indicated on the diagram by arrow (from Lehmann et al., 2009, with changes; American Meteorological Society©; used with permission).

the droplet amount in the mixing volume remains unchanged if the RH in the initially droplet-free volume is larger than a minimum value RH_{min}, which determines the right boundary of the diagram (thick solid line in Figure 5.9.7). At $RH < RH_{min}$, all droplets in the final state evaporate. At $RH > RH_{min}$ saturation is reached in the final equilibrium state. The increase in RH in the initially droplet-free volume indicates that the final equilibrium state is reached at a lower rate of partial evaporation of droplets. It means that the final state corresponding to saturation of the mixing volume is reached at larger droplet sizes. So, the increase in RH leads to a shift of the curves on the mixing diagram to the left. At high *RH* (>95%) of the environment, homogeneous mixing and extremely inhomogeneous mixing become undistinguishable in the diagrams. The dependencies corresponding to inhomogeneous mixing for intermediate values of the Damköhler number are not presented in the mixing diagrams.

Since cloud fraction (or degree of dilution) in real clouds is not known, in practice, the parameter μ is replaced by normalized droplet concentration N/N_{max}. Note that the cloud fraction differ from normalized droplet concentration, because normalized concentration depends on the decrease in droplet concentration due to evaporation in course of mixing, while μ is determined by the initial conditions only. Some authors evaluate degree of dilution by the ratio LWC/LWC_{max}. LWC_{max} is often considered equal to adiabatic liquid water content LWC_{ad}.

For analysis of *in situ* observations, scattering diagrams expressing the dependence of normalized values of mean volume, or effective radius on normalized droplet concentration measured along an aircraft traverse, are plotted and depicted on the mixing diagram. Droplet concentration is normalized by the maximum value along the aircraft traverse. Consequently, the values of the effective radii are normalized using the value of effective radius in the cloud volume with maximum droplet concentration.

The type of mixing is evaluated according to closeness of experimental points to the horizontal straight line, corresponding to extreme inhomogeneous mixing (Lu et al., 2014). Scattering diagrams obtained using observational data show high variability (Burnet and Brenguier, 2007; Gerber et al., 2008; Lehmann et al., 2009). In some scattering diagrams, observed points are located far from the straight line, but groups of points are elongated horizontally at different distances from the straight line $(r_v/r_{v\_max})^3 \approx 1$ (Lehmann et al., 2009). In some cases, groups of points correspond to values $(r_v/r_{v\_max})^3 > 1$. An example of a theoretical mixing diagram, superimposed with *in situ* observations in small cumulus clouds when many points of scattering diagrams are concentrated along the straight line corresponding to extremely inhomogeneous mixing, is presented in **Figure 5.9.8**. The significant deviations from this straight line take place largely at low droplet concentrations. However, even this diagram does not allow us to conclude with complete certainty that the mixing is extremely inhomogeneous, since this diagram allows different interpretations. For instance, diagrams of such kind in different studies are interpreted as extremely inhomogeneous (Hill and Choularton, 1986; Bower and Choularton, 1988; Pawlowska et al., 2000; Freud et al., 2011; Freud and Rosenfeld, 2012). At the same time, observations can be interpreted as evidence of homogeneous mixing that takes place at high RH of the surrounding air (Jensen and Baker, 1989; Paluch and Baumgardner, 1989). In some studies the mixing is assigned to be of intermediate type (Lu et al., 2014).

In addition to the uncertainty in the *RH* in the surrounding air, three main reasons impeding the utilization of standard mixing diagrams in analysis of experimental data as regards identification of a mixing

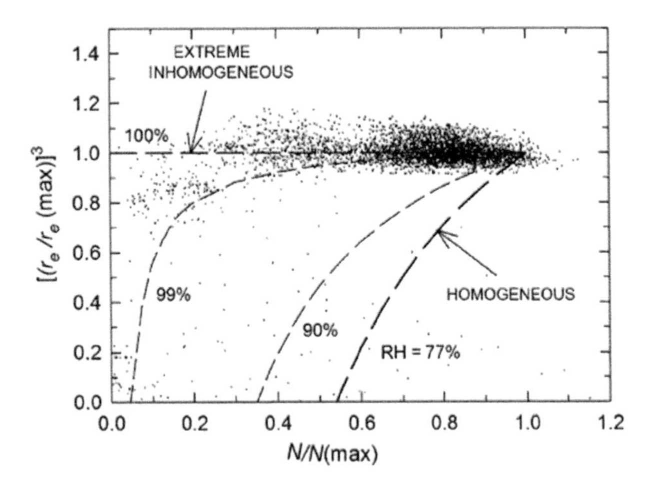

Figure 5.9.8 Normalized values of the effective radius as a function of the dilution ratio N/N_{max} for high resolution *in situ* measurements in the cloud pass # 21 during the RICO field experiment in small maritime clouds. Thick dashed line represents the relative environment humidity RH = 77% (from Gerber et al., 2008; courtesy of MSJ).

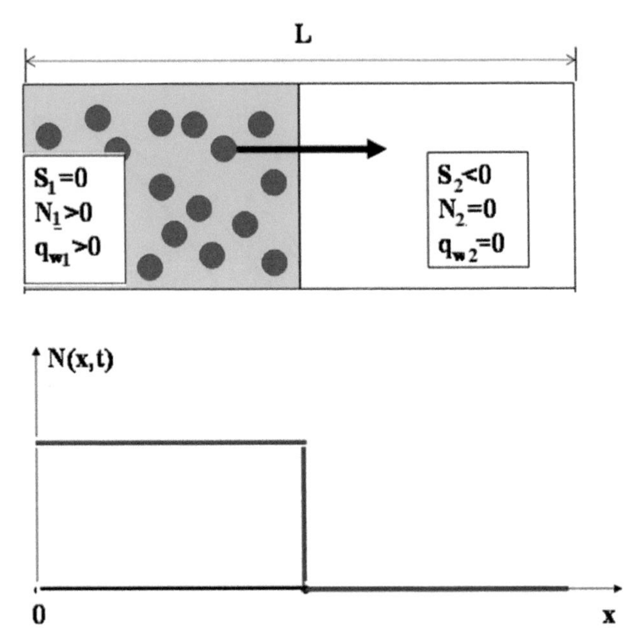

Figure 5.9.9 The geometry of diffusion-evaporation model at the initial state ($t = 0$). The left-side volume is a saturated cloud volume; the right-side volume is a subsaturated air volume from the cloud environment. The bottom panel indicates the initial dependence of dropet concentration on x (from Pinsky et al., 2016b; courtesy of ACP).

type in clouds are: the diagrams are plotted for two limiting cases of homogeneous and extremely inhomogeneous mixing, so they do not reflect the intermediate inhomogeneous mixing regime; the diagrams show the final states of mixing, while observations are likely to show transient regimes; and, finally, the diagrams are designed for monodisperse DSD, while DSDs in clouds are polydisperse. These reasons indicate the necessity to investigate time evolution of the mixing process and, in particular, time evolution of DSD during mixing.

5.9.3 Diffusion-Evaporation Theory of Turbulent Mixing

Equations Describing the Mixing-Evaporation Process
When a cloud volume and the surrounding air mix, there are two processes that determine the changes of their microphysical and thermodynamic variables: turbulent diffusion of temperature, water vapor, and droplets; and droplet evaporation. Pinsky et al. (2016a, 2016b) analyzed a time-dependent mixing process using a simple 1D model. The conceptual scheme of this diffusion-evaporation model is presented in **Figure 5.9.9**, illustrating mixing between two equal volumes: a cloud volume and a droplet-free volume. To simplify the analysis, the linear size of each volume is assumed equal to $L/2$. The scale L is a turbulent scale within the inertial interval, which ranges from several centimeters up to several hundred meters. The total mixing volume consisting of these two volumes is assumed adiabatic. The cloud volume is initially saturated with $RH_1 = 100\%$ i.e., $S_1 = 0$, and initial droplet concentration N_1. Under the simplified assumption that DSD is monodisperse, the initial liquid water content is $q_{w1} = \frac{4\pi\rho_w}{3} N_1 r_0^3$, where r_0 is the initial droplet radius. The droplet-free volume at the initial moment is subsaturated with $RH_2 < 100\%$ (i.e., $S_2 < 0$), $N_2 = 0$, and $q_{w2} = 0$. Therefore, the initial profiles of these quantities along the x-axis are step functions, schematically shown in Figure 5.9.9.

This is the simplest scheme of inhomogeneous mixing where mixing takes place only in the x-direction while the vertical velocity is neglected. The turbulent diffusion is described by a 1D equation with the turbulent coefficient K. The mixing is assumed to be driven by isotropic turbulence within the inertial subrange where Richardson's law is valid. In this case, the turbulent coefficient is evaluated as $K(L) = C\varepsilon^{1/3}L^{4/3}$ (Section 3.3; Table 3.3.1). Therefore, the turbulent mixing takes place at scales much larger than the Kolmogorov microscale, i.e., at scales where the molecular diffusion can be neglected. Since the total volume is adiabatic, the fluxes of different quantities such as droplet concentration, water vapor mixing ratio, etc. through the left and the right boundaries at any time instance are equal to zero.

The basic system of equations describes the processes of diffusion and evaporation occurring simultaneously. The first equation is written for value Γ defined as

$$\Gamma = S_w + A_2 q_w, \tag{5.9.5}$$

where Γ is approximately conservative in a moist adiabatic process, i.e., its value does not change during phase transitions. This quantity can be obtained from the balance equation (5.2.4) relating supersaturation to LWC in an adiabatic parcel $S_w = A_1 z - A_2 q_w + C_1$ (C_1 is a constant determined by the initial conditions), assuming that the parcel is immovable. For simplicity, z is chosen equal to zero. The quantity Γ obeys the diffusion equation

$$\frac{\partial \Gamma(x,t)}{\partial t} = K \frac{\partial^2 \Gamma(x,t)}{\partial x^2} \tag{5.9.6}$$

with the following initial condition at $t = 0$:

$$\Gamma(x,0) = \begin{cases} A_2 q_{w1} & \text{if} \quad 0 \le x < L/2 \\ S_2 & \text{if} \quad L/2 \le x < L \end{cases} \tag{5.9.7}$$

In the course of mixing between the cloud volume and the surrounding air volume, droplets diffuse to subsaturated volume and evaporate. According to the equation of diffusion growth (5.2.6), evaporation leads to a shift of droplet square radii toward smaller values regardless of droplet size (Section 5.2). In this case, it is convenient to write DSD as a function of the droplet radius square $g(\sigma)$, where $\sigma = r^2$. The standard DSD $f_r(r)$ is related to $g(\sigma)$ as $f_r(r) = 2r \cdot g(\sigma)$. The second main equation is the 1D diffusion-evaporation equation for nonconservative function $g(\sigma)$ that can be written in the form suggested by Rogers and Yau (1996) (see also Equation (5.4.14)):

$$\frac{\partial g(\sigma)}{\partial t} = K \frac{\partial^2 g(\sigma)}{\partial x^2} - \frac{\partial}{\partial \sigma}\left(\frac{d\sigma}{dt} g(\sigma)\right) \tag{5.9.8a}$$

Using the equation for droplet evaporation (5.2.6) written in the form $\frac{d\sigma}{dt} = \frac{2S_w}{F}$, Equation (5.9.8a) can be rewritten as

$$\frac{\partial g(x,t,\sigma)}{\partial t} = K \frac{\partial^2 g(x,t,\sigma)}{\partial x^2} - \frac{2S_w(x,t)}{F} \frac{\partial g(x,t,\sigma)}{\partial \sigma} \tag{5.9.8b}$$

Equations (5.9.8a and 5.9.8b) describe the changes of DSD due to spatial diffusion (the first term on right-hand side) and due to evaporation (the second term on right-hand side). Although Equation (5.9.8b) is not closed since supersaturation $S_w(x,t)$ is an unknown function of x and t, the problem can be solved by adding Equations (5.9.5) and (5.9.6).

Using Equation (5.9.8b) one can obtain the equations for LWC, superaturation, DSD moments, and other microphysical quantities. For example, the equation for LWC is written as

$$\begin{aligned}\frac{\partial q_w(x,t)}{\partial t} &= K \frac{\partial^2 q_w(x,t)}{\partial x^2} + \frac{1}{\tau_{pr}(x,t)} \left[\frac{1}{A_2}\Gamma(x,t) - q_w(x,t)\right] \\ &= K \frac{\partial^2 q_w(x,t)}{\partial x^2} + \frac{S_w}{A_2 \tau_{pr}(x,t)},\end{aligned} \tag{5.9.9}$$

where τ_{pr} is the phase relaxation time (Section 5.2). As in Equation (5.9.8), the first term on the right-hand side describes the rate of the spatial diffusion and the second term shows the changes in the microphysical variables due to evaporation.

For analysis, it is convenient to write down the diffusion-evaporation equation and the equations for the microphysical quantities in a nondimensional form. The expressions for normalization of different variables are presented in **Table 5.9.1**. Using these nondimensional parameters, Equation (5.9.7) can be rewritten in a nondimensional form as

$$\frac{\partial \tilde{q}(\tilde{x},\tilde{t})}{\partial \tilde{t}} = \frac{1}{Da} \frac{\partial^2 \tilde{q}(\tilde{x},\tilde{t})}{\partial \tilde{x}^2} + \frac{\tilde{S}_w}{\tilde{\tau}_{pr}(\tilde{x},\tilde{\tau})} \tag{5.9.10}$$

In this equation, the nondimensional parameter Da

$$Da = \frac{\tau_{mix}}{\tau_{1pr}} = \frac{L^2}{K\tau_{1pr}} \tag{5.9.11}$$

is the Damkölher number, where τ_{1pr} is initial droplet relaxation time in the cloudy volume. In Equation (5.9.11), Da is expressed using the turbulence coefficient that proved convenient in different theoretical and numerical studies. Unlike Equation (5.9.4), here Da is defined via the initial (premixing) value of the phase relaxation time.

The conservative nondimensional function $\tilde{\Gamma}(\tilde{x},\tilde{t})$ depends on the two nondimensional parameters defined in Table 5.9.1: the *potential evaporation parameter R* and the Damkölher number Da (Pinsky et al., 2016b), where

$$R = \frac{S_2}{A_2 q_{w1}} \tag{5.9.12}$$

At $t \to \infty$, $\tilde{\Gamma}$ depends solely on the potential evaporation parameter R:

$$\tilde{\Gamma}(\tilde{x},\infty) = \frac{1}{2}(1+R) \tag{5.9.13}$$

Table 5.9.1 Different quantities in a diffusion-evaporation model and their nondimensional forms* (Pinsky et al., 2016b; courtesy of ASP).

Quantity	Symbol	Non-dimensional Form	Range of Normalized Values
Time	t	$\tilde{t} = \frac{t}{\tau_{1pr}}$	$[0\ldots\infty]$
Distance	x	$\tilde{x} = \frac{x}{L}$	$[0\ldots1]$
Square of droplet radius	σ	$\tilde{\sigma} = \frac{\sigma}{r_0^2}$	$[0\ldots1]$
Droplet concentration	N	$\tilde{N} = \frac{N}{N_1}$	$[0\ldots1]$
LWC	q_w	$\tilde{q} = \frac{q_w}{q_{w1}}$	$[0\ldots1]$
Distribution of square of droplet radius	$g(\sigma)$	$\tilde{g}(\tilde{\sigma}) = \frac{r_0^2}{N_1} f(\sigma)$	
Conservative function	Γ	$\tilde{\Gamma} = \frac{\Gamma}{A_2 q_{w1}}$	$[-\infty\ldots1]$
Supersaturation	S	$\tilde{S} = \frac{S}{A_2 q_{w1}}$	$[-\infty\ldots0]$
Relaxation time	τ_{pr}	$\tilde{\tau}_{pr} = \frac{\tau_{pr}}{\tau_{1pr}}$	$[1\ldots\infty]$
The Damkölher number	Da	$Da = \frac{\tau_{mix}}{\tau_{1pr}} = \frac{L^2}{K\tau_{1pr}}$	$[0\ldots\infty]$
Potential evaporation parameter	R	$R = \frac{S_2}{A_2 q_{w1}}$	$[-\infty\ldots0]$

* All the normalized quantities depend on parameters L, N_1, r_0, A_2, S_2, and K.

Hence, the solution of the diffusion-evaporation equation fully depends on the two non-dimensional parameters R and Da.

The potential evaporation parameter R is an important parameter that controls mixing. When $R < 0$ ($S_2 < 0$), droplets can only evaporate in the course of mixing. R is the ratio of amount of liquid water vapour needed to saturate the initially droplet-free volume and the available liquid water that potentially can evaporate. If $R < -1$, $\tilde{\Gamma}(\tilde{x}, \infty) < 0$, which means that the droplet-free volume is too dry and all the droplets eventually evaporate completely. In this case, at the final equilibrium state $RH < 100\%$, i.e., $S(x, \infty) < 0$. If $R > -1$, then $\tilde{\Gamma}(\tilde{x}, \infty) > 0$ and the mixed volume at the final state contains droplets and $RH < 100\%$ (i.e., $S(x, \infty) = 0$). In this case, mixing leads to an increase in the cloud volume. The condition $|R| \ll 1$ corresponds to cases when the relative humidity in the initially droplet-free volume is close to the saturation and/or the liquid water content in the cloud volume is large. In this case, the changes in DSD due to droplet evaporation are negligible. From a mathematical point of view, this condition means that the second term on the right-hand side of Equation (5.9.8a) is very small and the result of mixing is determined solely by the turbulent diffusion independent of the value of Da. Therefore, at $|R| \ll 1$ homogeneous and inhomogeneous mixing are indistinguishable.

The value of Da is a crucial factor since it determines the mixing type. Da emerges in the diffusion-evaporation equation when nondimensionalization is made. This fact clearly shows that the droplet relaxation time evaluated just prior to mixing is the characteristic time scale of the mixing-evaporation process, as we deal with an ensemble of evaporating droplets.

In the extreme case at $Da \to 0$ (homogeneous mixing), the value of the diffusion term at the start of mixing is much larger than that of the evaporation term. It means that within a short period, the total spatial homogenization of all the variables takes place in the mixing volume and all the spatial gradients become close to zero. After this period, the second term becomes dominant. Thus, at very low *Da* the process of mixing consists of two stages: a short stage of inhomogeneous mixing and a longer stage of homogeneous evaporation. Using the equations for mass balance (5.2.4) and the equation for diffusion growth (5.2.6), Pinsky et al. (2016a) formulated a closed system of differential equations for evolution of the normalized LWC and for normalized supersaturation for homogeneous evaporation:

$$\frac{d\tilde{q}}{d\tilde{t}} = \tilde{q}^{1/3}(\gamma - \tilde{q})$$
$$\frac{d\tilde{S}}{d\tilde{t}} = -\left(\gamma - \tilde{S}\right)^{1/3}\tilde{S}, \tag{5.9.14}$$

where $\gamma = 1 + \frac{1-\mu}{\mu}\frac{S_2}{A_2 q_{w1}} = 1 + R\frac{1-\mu}{\mu}$ and μ is cloud fraction. This equation allows an analytical solution. Assuming time derivatives in Equation (5.9.14) equal to zero allows analytical calculation of mixing diagrams for homogeneous mixing:

$$\tilde{q}_{\min}(\mu) = \begin{cases} 1 + R\dfrac{1-\mu}{\mu} & \text{if} \quad \mu > -\dfrac{R}{1-R} \\ 0 & \text{if} \quad \mu \leq -\dfrac{R}{1-R} \end{cases} \tag{5.9.15}$$

Another extreme case is $Da \to \infty$. This condition corresponds to extremely inhomogeneous mixing when the value of the diffusion term is much smaller than that of the evaporation term. The microphysical processes thus take place under significant spatial gradients of *RH*. At the limit of $Da = \infty$, the adjacent volumes do not mix at all, making this case equivalent to the existence of two independent adiabatic volumes. Another interpretation of the limiting case $Da = \infty$ is an infinitely fast droplet evaporation. At intermediate *Da*, mixing is inhomogeneous, so both turbulent diffusion and evaporation form the DSD during the entire period of mixing.

Spatial-Time Dependence of Thermodynamic and Microphysical Quantities

Profiles of normalized supersaturation at different values of *Da* and *R* at different time instances are shown in **Figure 5.9.10**. In all the cases, the final state is

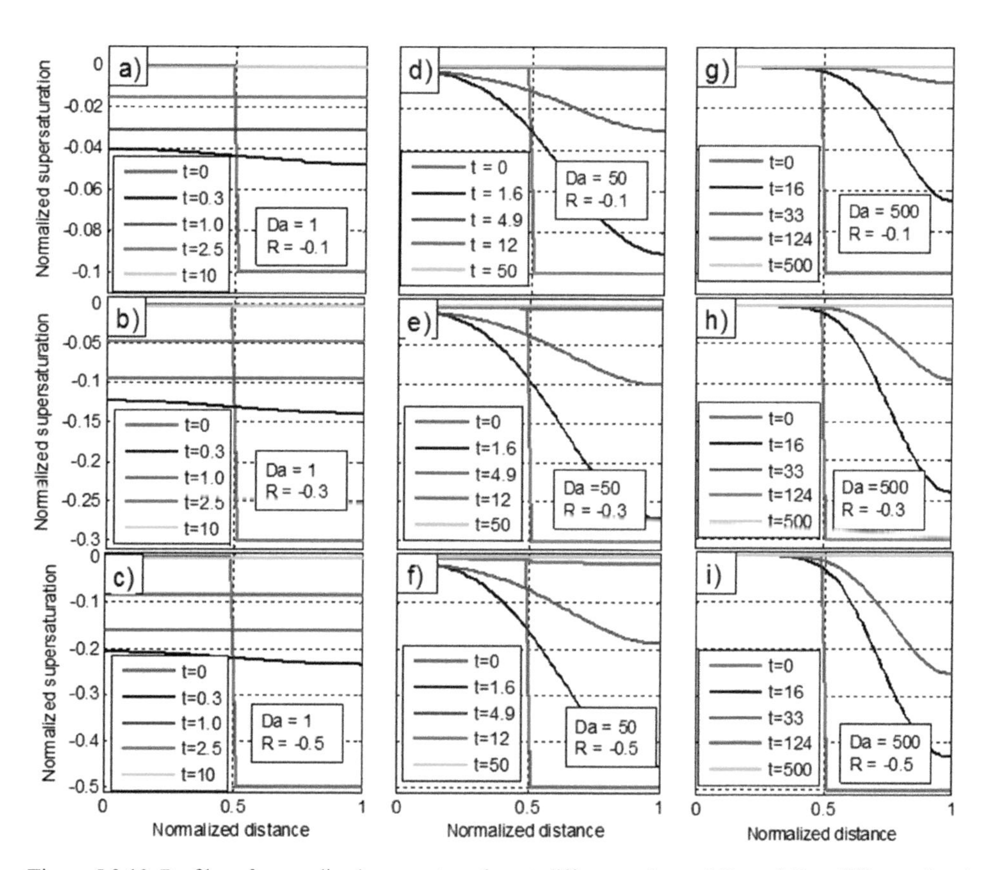

Figure 5.9.10 Profiles of normalized supersaturation at different values of *Da* and *R* at different time instances. Time is normalized by the droplet relaxation time (from Pinsky et al., 2016b; courtesy of ACP).

characterized by establishing the equilibrium supersaturation $\tilde{S} = 0(\text{RH} = 100\%)$. However, the history of reaching this final value is different depending on *Da*. At $Da = 1$, rapid mixing leads to formation of horizontally homogeneous humidity and supersaturation within a few tenths of τ_{pr}. Then, supersaturation within the entire volume grows due to evaporation of droplets that are uniformly distributed within the total volume. At $Da = 500$, the changes in supersaturation take place mostly within the initially droplet-free volume, while the changes of RH in the initially cloud volume are small. This case agrees with the classical concept of extremely inhomogeneous mixing. The supersaturation gradient remains strong for a long time (tens of τ_{pr}) also within the initially droplet-free volume. At $Da = 50$, the situation is intermediate: mixing is intensive enough to eliminate gradient of RH during about $t \approx 12\tau_{pr}$. After this period, the mixing continues according to the homogeneous scenario. Thus, at $Da = 50$ mixing that starts as inhomogeneous switches to the homogeneous type.

Figure 5.9.11 shows the profiles of normalized LWC for different values of *Da* and *R*. For the same values of parameter *R*, the final equilibrium values of LWC are the same, as follows from Equation (5.9.13). The final LWC decreases with decreasing *R*. Decrease of the LWC in the cloud volume is caused largely by diffusion of droplets from the cloud volume to the initially droplet-free volume at any values of *Da*. The evaporation in the cloud volume at $Da = 500$ is slow because $\tilde{S}$ remains high during the mixing (Figure 5.9.10). At $Da = 1$, spatial homogenization takes place during tenths of τ_{pr}. Then, during a comparatively long time of $10\tau_{pr}$, evaporation leads to a decrease in LWC over the entire mixing volume, as it should in a homogeneous mixing scenario. Time needed to reach the equilibrium is different for different values of *Da*. Reaching the final uniform LWC takes one τ_{pr} at $Da = 1$, about $15\tau_{pr}$ at $Da = 50$, and about $100\tau_{pr}$ at $Da = 500$.

Figure 5.9.12 shows the profiles of normalized droplet concentration for different values of *Da* and *R*. In contrast to LWC, the final concentration depends on *Da* and can be different at the same *R*. The difference between the initial and final values of droplet concentration depends on the value of *R*. At $R = -0.1$ (which corresponds to high RH in the initially droplet-free volume), there is no total droplet evaporation, so the final normalized droplet concentration is equal to

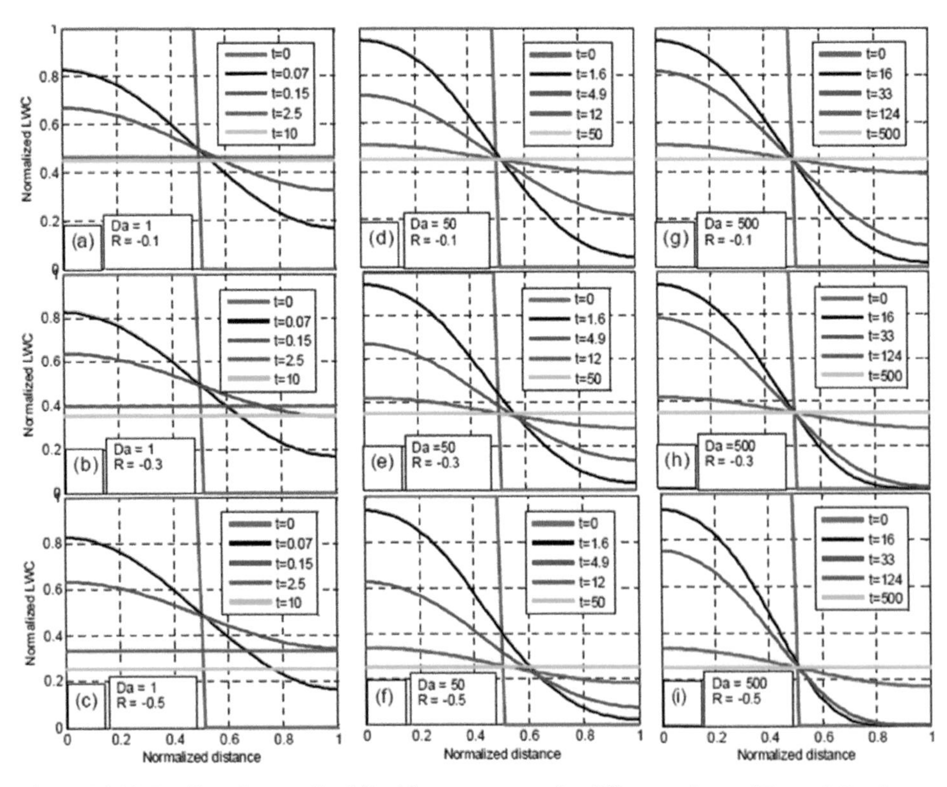

Figure 5.9.11 Profiles of normalized liquid water content for different values of *Da* and *R* values at different time instances. Time is normalized by the droplet relaxation time (from Pinsky et al., 2016b; courtesy of ACP).

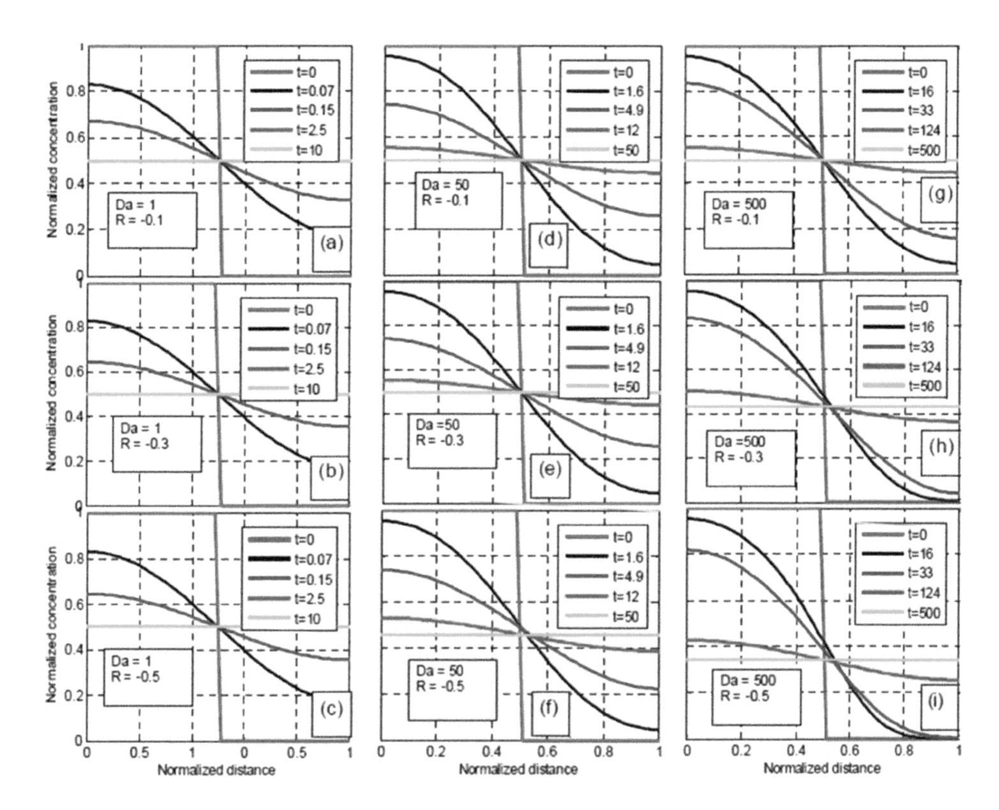

Figure 5.9.12 Profiles of normalized droplet concentration for different values of Da and R at different time instances. Time is normalized by the droplet relaxation time (from Pinsky et al., 2016b; courtesy of ACP).

$\tilde{N} = 1/2$. It means that all droplets in the volume are now uniformly distributed between the initially cloud volume and the initially droplet-free volume. At R lower than about -0.3, some fraction of droplets totally evaporates and final droplet concentration becomes lower than the initial one. The final droplet concentration decreases also with increase in Da.

Formation of DSD in the Course of Mixing

Mixing affects the shape of DSD, and as a result, all the DSD moments. The mean volume and the effective radii of DSD are important parameters that are typically used to determine the mixing type. We present simulation results employing the diffusion-evaporation model where the initial DSD was assumed to be monodisperse. **Figure 5.9.13** presents examples of DSD evolution in an initially cloud volume and in an initially droplet-free volume at $R = -0.5$ and at different Da values. Several typical properties of DSD are clearly seen. At $Da = 1$ (panels a and b), identical DSDs rapidly form in both volumes (black lines at $t = 0.317\tau_{pr}$). Formation of a polydisperse DSDs in case $Da = 1$ takes place during the short inhomogeneous stage when droplets experience different saturation deficit. Therefore, droplets penetrating the initially dry volume first evaporate more intensively than those penetrating this volume later. Further evolution is similar in both volumes and is characterized by a shifting and broadening of the DSD toward smaller droplet sizes, as during homogeneous evaporation. Such a shift is typical of homogeneous mixing when a decrease in the mass content occurs at constant droplet concentration. Droplets diffuse into the initially droplet-free volume and partially evaporate, reaching the minimum size at $\tilde{x} = 1$. The DSD broadening continues during the homogeneous stage in accordance with the diffusion growth equation.

At $Da = 50$ and $Da = 500$, the shapes of DSD substantially differ from those at $Da = 1$. There are two main differences: the peak of the distribution shifts only slightly ($Da = 50$) or does not shift at all ($Da = 500$). At the same time, a long tail of small droplets develops. Since the mixing rate at these Da values is slow, droplets penetrating at longer distances into the initially dry volume decrease more in size. As a result, at moderate and large Da, a polydisperse DSD arises with normalized droplet sizes ranging from zero to one. Formation of a long tail of small droplets in inhomogeneous mixing was found in DNSs by Kumar et al. (2013, 2014), as well as by Krueger et al. (2006), Su et al.

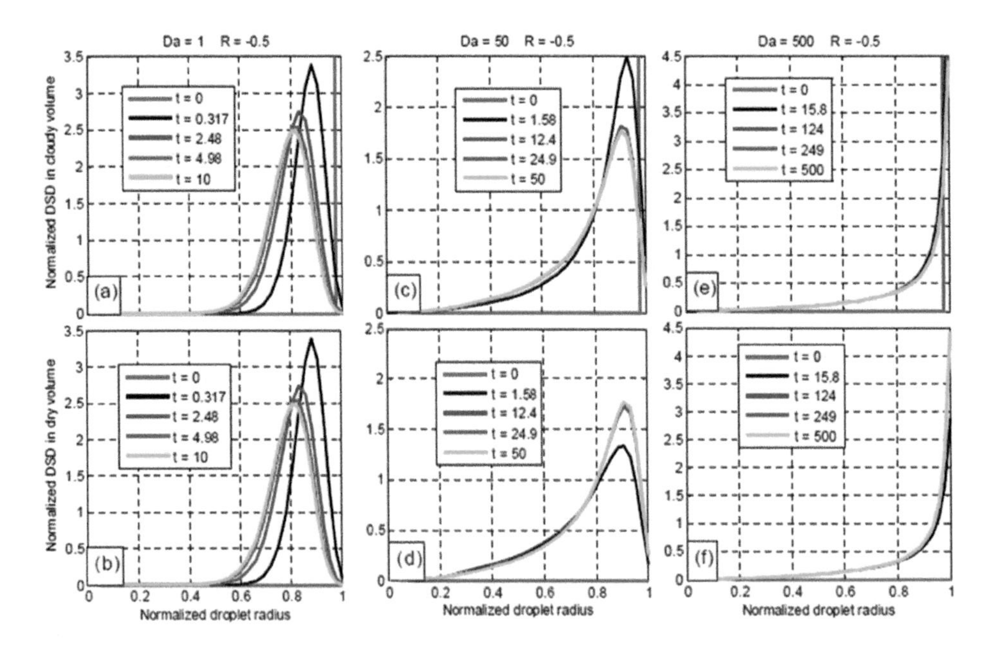

Figure 5.9.13 Examples of DSD evolution in the initially cloud volume ($\tilde{x} = 1/4$) (upper row) and in the initially droplet-free volume ($\tilde{x} = 3/4$) (bottom row) at $R = -0.5$ and different Da values (from Pinsky et al., 2016b; courtesy of ACP).

(1998), and Schlüter (2006) using the explicit-mixing parcel model (see Section 5.10). The differences in DSD shapes determine the differences in DSD moments. In particular, the values of DSD relative dispersion at $R = -0.5$ reaches 0.11 at $Da = 1$ and about 0.2 at $Da \geq 50$. Observed values of DSD dispersion in different clouds typically range from 0.1 to 0.4 (Section 2.3). Therefore, mixing at cloud edges can be an important mechanism of DSD broadening, especially at large Da.

The effective radius is an important DSD characteristic, which, according to the classical concept, remains unchanged during extreme inhomogeneous mixing and, in contrast, decreases in the course of homogeneous mixing. **Figure 5.9.14** shows the spatial dependencies of the effective radius at different time instances and values of Da and R. At $R = -0.1$ (high RH in the initially droplet-free volume), the effective radius is the same at all Da values. Thus, at high *RH* the behaviour of the effective radius does not allow us to determine the mixing type. At a given R, changes in the effective radii become less significant with increasing Da. For instance, if $R = -0.5$, the effective radius at the final state differs from its initial value by 20% at $Da = 1$, and by less than 6% for $Da = 500$. Changes in the effective radius are especially insignificant within initially cloud volumes.

Figure 5.9.14 also shows the evolution of the effective radius in the initially droplet-free volume. One can see that during a few tens of τ_{pr}, it remains small (less than 0.5 in the normalized units). Then r_{eff} rapidly increases from small values toward the initial r_{eff} value in the cloud volume. For instance, at $Da = 500$ and $R = -0.5$ the values of r_{eff} become uniform over the entire mixing volume much faster than the equilibrium state is reached (Figure 5.9.14i). Figures 5.9.13 and 5.9.14 show that uniformity of r_{eff} is reached earlier than the homogeneity of the DSD. This result illustrates the mechanism leading to low variability of r_{eff} in the horizontal direction despite the high variability of LWC (Section 5.9.4).

The results just discussed were obtained for monodisperse initial DSD. Pinsky et al. (2016a) showed that the initial DSD shape substantially affects both the shape and the parameters of the final DSD at $Da = 0$. For initially wide DSD containing small droplets, homogeneous mixing also leads to total evaporation of the smallest droplets and the corresponding increase in the effective radius. This feature is typically attributed to inhomogeneous mixing. These results indicate that at wide DSD it is difficult to determine the mixing type based only on the behavior of the effective radius with increasing dilution.

Quantitative Delimitation Between Mixing Types

A more pronounced difference between mixing types is seen in diagrams presenting relationships between

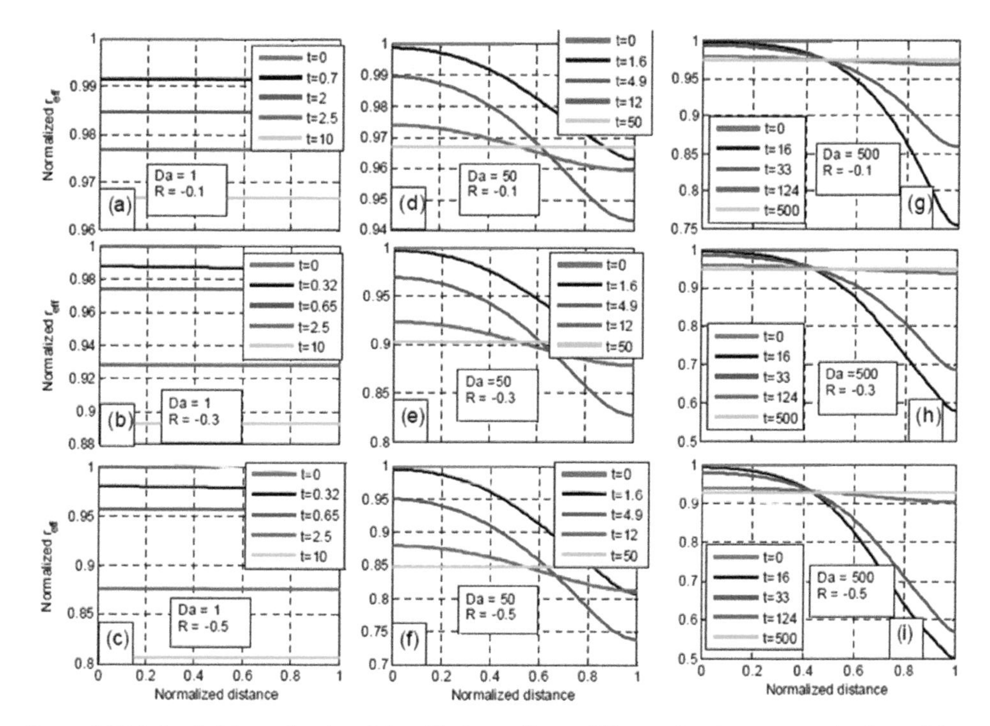

Figure 5.9.14 Spatial dependencies of the effective radius at different time instances and values of *Da* and *R* (from Pinsky et al., 2016b; courtesy of ACP).

different DSD moments. **Figure 5.9.15** shows the relationships between normalized droplet concentration and normalized LWC at $R = -0.5$ and at different values of $\tilde{x}$ and *Da*. Dots forming the relationship line are plotted at the same time increment, so the fractions of curves corresponding to low density of dots along the curves indicate short time periods, and, in contrast, high dot density corresponds to long time periods. Point A corresponds to initially droplet-free volume; point B corresponds to initially cloud volume. Diagonal straight line AB corresponds to the classical definition of extremely inhomogeneous mixing. At $Da = 1$ (panel a), a very fast mixing leads to a rapid decrease in LWC and in droplet concentration in the initially cloud volume, as well as to an increase of these quantities in the initially droplet-free volume. In the end of this short period of inhomogeneous mixing, all the curves corresponding to different values of $\tilde{x}$ coincide at point C (left panel). After this time instance, spatially homogeneous evaporation takes place. During this stage, the droplet mass decreases at unchanged droplet concentration; this agrees with the classical concept of homogeneous mixing. The process comes to the final equilibrium state at point F.

At $Da = 50$ (panel b) and $Da = 500$ (panel c), the three curves coincide only at the final point F. It means that the spatial gradients of the microphysical values exist until the final equilibrium state is reached. This is inhomogeneous mixing. At $Da = 500$, the $\tilde{N} - \tilde{q}$ relationship is closer to linear, in accordance to the classical definition of extremely inhomogeneous mixing. The linearity of $\tilde{N} - \tilde{q}$ relationship means that the mean volume radius does not change in the course of mixing. Thus, the fact that the measured *in situ* $N - q_w$ relationships are close to linear dependences, as well as the low variability of the mean volume radius near cloud boundary may indicate that the mixing is close to extremely inhomogeneous.

Analysis of time-dependent mixing allows us to develop delimitation criteria for differentiation between homogeneous and inhomogeneous mixing. We can distinguish three characteristic time periods: a) the homogenization period T_{mix}, during which the spatial gradients are fully smoothed out; b) the evaporation period T_{ev}, during which the supersaturation reaches zero value; and c) the total mixing period T_{tot}, that ends up when complete equilibrium is reached so $T_{tot} = \max\{T_{ev}, T_{mix}\}$. In case $T_{tot} = T_{mix} \geq T_{ev}$, droplet evaporation takes place under spatial gradients of supersaturation, so mixing is inhomogeneous. In case $T_{tot} = T_{ev} \geq T_{mix}$, mixing between two different volumes starts as inhomogeneous, but at $t > T_{mix}$ turns

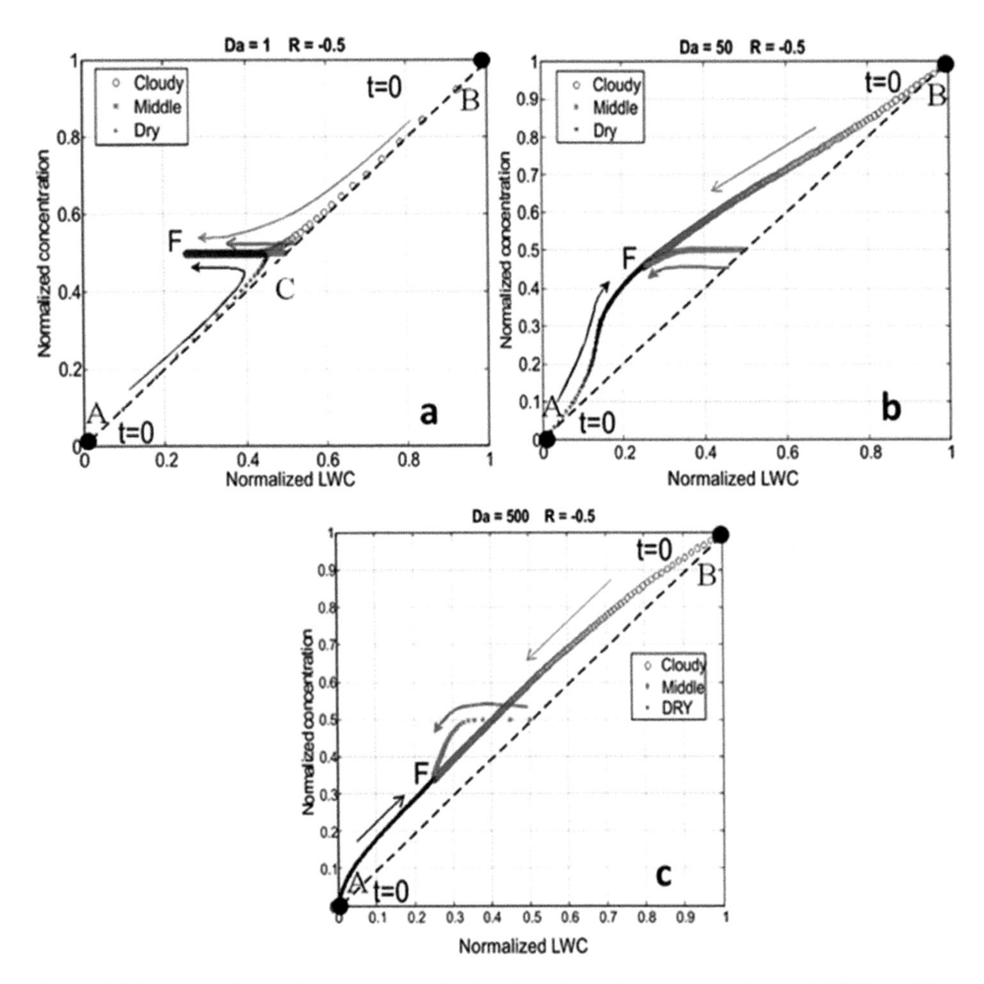

Figure 5.9.15 Dependencies between normalized values of droplet concentration and LWC at different Da and $R = -0.5$. Blue circles correspond to the center of the cloudy volume ($\tilde{x} = 1/4$), red symbols correspond to the interface ($\tilde{x} = 1/2$), and black crosses correspond to $\tilde{x} = 3/4$, which is the center of the initially dry volume. Arrows show the direction of the movement of points A and B with time. Point C denotes the start of homogeneous evaporation. Point F denotes the final stationary state of the system. Dashed line indicates the relationship between $\tilde{N}$ and $\tilde{q}$ for extremely inhomogeneous mixing according to its classical definition (from Pinsky et al., 2016b, with changes; courtesy of ACP).

into homogeneous. Point C in Figure 5.9.15a corresponds to time $t > T_{mix}$.

Pinsky et al. (2016b) proposed a quantitative criterion for delimitation between mixing types, introducing parameter λ as the ratio between evaporated masses at the first (nonhomogeneous) stage and the total evaporated mass:

$$\lambda = \frac{\langle \tilde{q}(t=0) \rangle - \langle \tilde{q}(T_{mix}) \rangle}{\langle \tilde{q}(t=0) \rangle - \langle \tilde{q}(t=T_{tot}) \rangle} \tag{5.9.16}$$

According to the criteria, condition $\lambda < 0.5$ can be associated with homogeneous mixing when most of evaporation takes place during the homogeneous stage described by the track C–F in Figure 5.9.15a. The condition $0.5 \leq \lambda < 1$ corresponds to the intermediate mixing type, when larger loss of droplet mass takes place at the first inhomogeneous stage. Condition $\lambda = 1$ means that entire evaporation takes place under inhomogeneous conditions, which corresponds to inhomogeneous mixing.

Figure 5.9.16 shows the differentiation between different types of mixing on the $Da - R$ plane. Parameter λ indicates that at Da below 4–10 and $R < -0.1$ mixing can be considered homogeneous. Values Da of several tens are a delimiter between intermediate and inhomogeneous mixing. Within the area of inhomogeneous mixing in Figure 5.9.16, one can distinguish a zone of extremely inhomogeneous mixing marked by the dashed line. The closeness of black-blue curve to the diagonal straight line in Figure 5.9.15 is the measure of

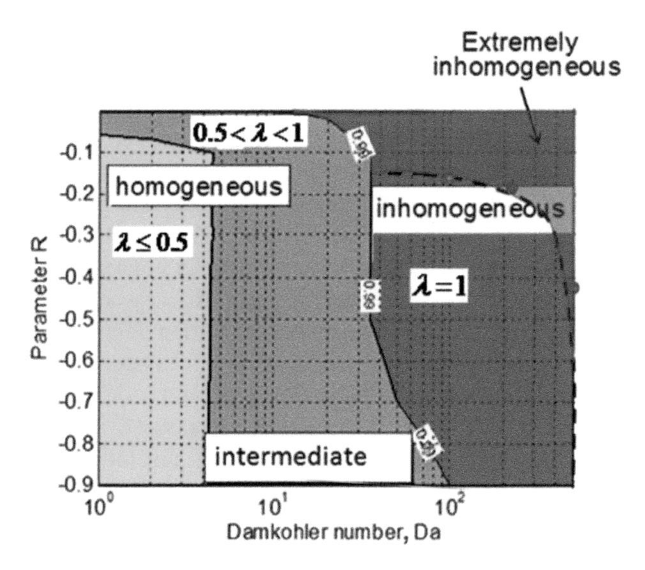

Figure 5.9.16 The boundaries between mixing types on the $Da - R$ plane designed according to criteria (5.9.16). Dashed line showing the boundaries of extremely inhomogeneous mixing corresponds to the deviation of the effective radius (in the equilibrium state) from the initial value by a few percents (from Pinsky et al., 2016b, with changes; courtesy of ACP).

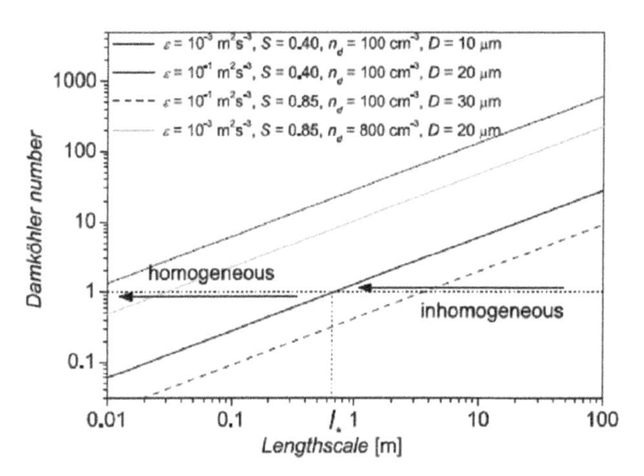

Figure 5.9.17 The Damkölher number as a function of the entrainment length scale, for four exemplary values of energy dissipation, saturation ratio, droplet concentration, and droplet diameter. The transition length scale at $Da = 1$ is indicated for one of the exemplary values (from Lehmann et al., 2009; American Meteorological Society©; used with permission).

the closeness of the mixing to extremely inhomogeneous. The deviation of the equilibrium effective radius from the initial value was also used to separate this zone. The zone where the deviation of the equilibrium effective radius from the initial value is below a few percents was chosen as the boundary of the zone of extremely inhomogeneous mixing. Figure 5.9.16 shows that in addition to regimes of homogeneous and extremely inhomogeneous mixing, reflected in the mixing diagrams (Figures 5.9.7) and (5.9.8), there are regimes of intermediate and inhomogeneous mixing, which are not plotted in the diagrams. This hampers interpretation of a mixing type using the standard mixing diagrams.

5.9.4 Observations of Mixing Phenomena in Clouds

Typical Values of R *and* Da *in Clouds*

There are uncertainties regarding the value of potential evaporation parameter R, which can be attributed to high variability of LWC in volumes near cloud edges, where LWC changes from zero to nearly undiluted values (Gerber et al., 2008; Prabha et al., 2011; Khain et al., 2013), as well as to the variability of the relative humidity in the surrounding air. Penetration of droplets into the environmental air and their total evaporation, as well as diffusion of water vapor with time, lead to an increase in humidity in the non-cloudy air near cloud edges.

Gerber et al. (2008) observed a very high relative humidity of the environment air at cloud edges. These findings were supported by radar observations (Knight and Miller, 1998) showing the existence of the Bragg scattering mantle echo covering the region near the cloud surface and arising due to the refractive index variations related to water vapor fluctuations. The existence of a 100–150 m-deep layer of high RH around cumulus clouds was also reported by Bar-Or et al. (2012). Lehmann et al. (2009) concluded that RH is high in droplet-free volumes neighboring cloudy ones. The results presented in Figures 5.9.11–5.9.13 show that at high Da, some parts of the initially dry volume remain droplet-free for a long time. At the same time, RH in these volumes substantially increases. It means that cloud edges should be surrounded by an air layer of enhanced humidity. It means that in close air surrounding the values of $|R|$ are small.

The reliable values of Da in real clouds are currently not well known due to the uncertainties regarding the value of coefficient C in Equation (5.9.1). According to evaluations made by Lehmann et al. (2009), Da values in clouds of different types range from 0.1 to a few hundred (**Figure 5.9.17**). Estimations of the delimiting value of Da (Figure 5.9.16) show that homogeneous mixing takes place at scales of about 1 m and lower. Since cloud elements, bubbles, thermals, etc. have scales significantly larger than 1 m, mixing in clouds should be considered inhomogeneous, at least at scales resolvable in in situ measurements and in most cloud models. *Da* values in stratocumulus clouds may be similar or even higher than in cumulus clouds, since both τ_{mix} and τ_{pr} in stratocumulus clouds are larger than in *Cu*.

Behavior of DSD and DSD Parameters

Observations (e.g., Paluch and Knight, 1984; Paluch, 1986; Gerber et al., 2008; Bera et al., 2016a, 2016b) indicate a specific feature of DSD in cloud zones affected by the environment air, namely, the similarity of DSD shapes (with the exception of the tail of the smallest droplets) despite very different rates of dilution along the horizontal traverses. As an example, **Figure 5.9.18** shows typical DSD measured during the CCOPE. The CCOPE was conducted within a wide range of cloud conditions, vertical velocities, and turbulence intensities. Figure 5.9.18 shows that in most observations DSD peaks remain unchanged or changed only slightly. DSD shapes are similar, except the zone of small droplets (panel C). When droplet concentration is low, the peak of large droplets sometimes shifts toward smaller sizes, as shown in panels (C and D). This shift typically occurs near cloud edges or in decaying cloud volumes. The similarity of DSD and the negligible variation in the DSD peak droplet radii at different droplet concentrations indicate that the effective radius (as well as mean volume radius) is independent (or dependent only slightly) of droplet concentration. The similarity of DSD also means that LWC and droplet concentration are nearly linearly related. Similar features of DSD behavior and LWC-droplet concentration relationship were found in deep convective clouds measured during CAIPEEX, both in cloud updrafts and downdrafts. **Figure 5.9.19a** shows typical DSD measured in CAIPEEX with an increment of 1 s (spatial resolution is of about 100 m). Substantial variation in droplet concentration is seen, while the droplet size corresponding to the DSD peak remains nearly unchanged. Figure 5.9.19b shows that the coefficient of correlation between LWC and droplet concentration over this traverse exceeds 0.99. The linear proportionality between DSD moments, typical of extremely inhomogeneous mixing, as well as low variability of the effective radius were found also in

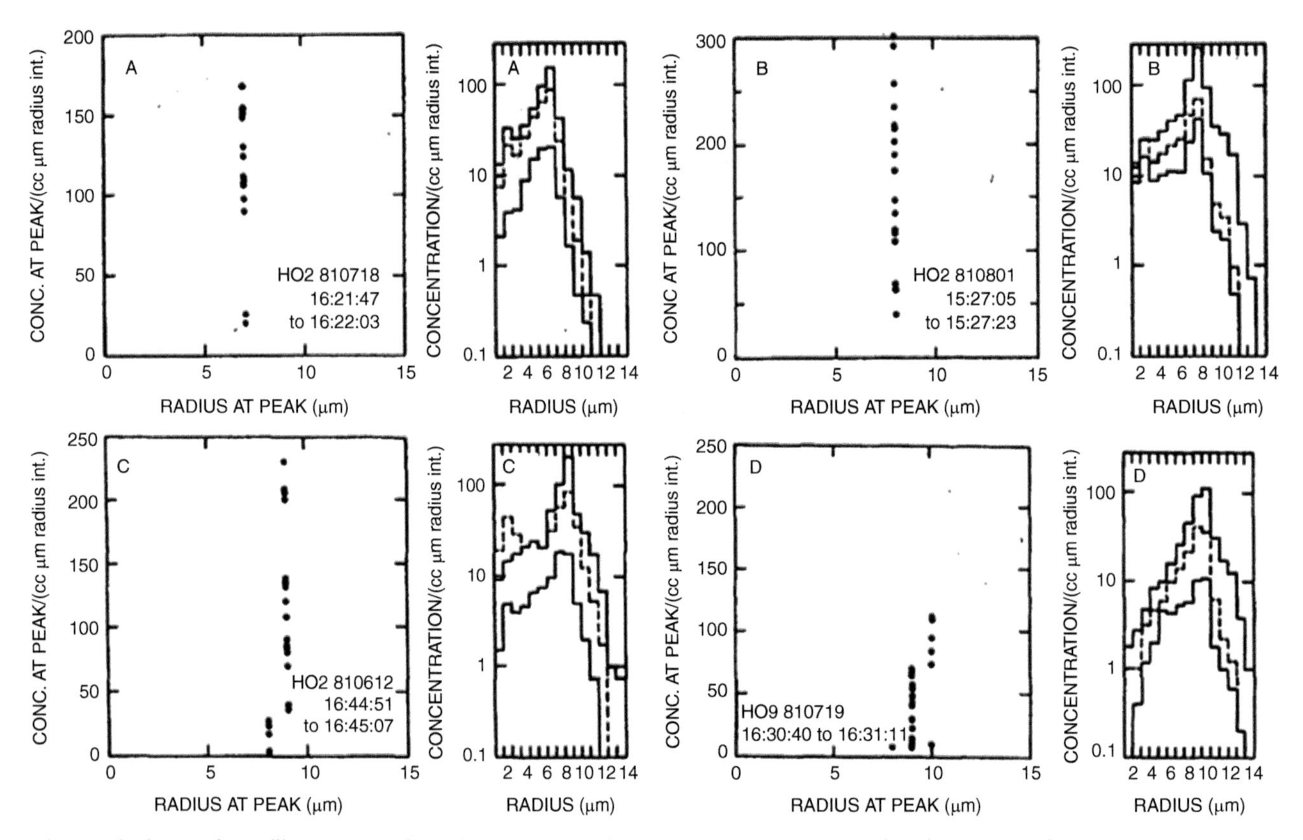

Figure 5.9.18 Droplet radii at DSD peaks and DSD measured along different traverses during the CCOPE (from Paluch, 1984; American Meteorological Society©; used with permission).

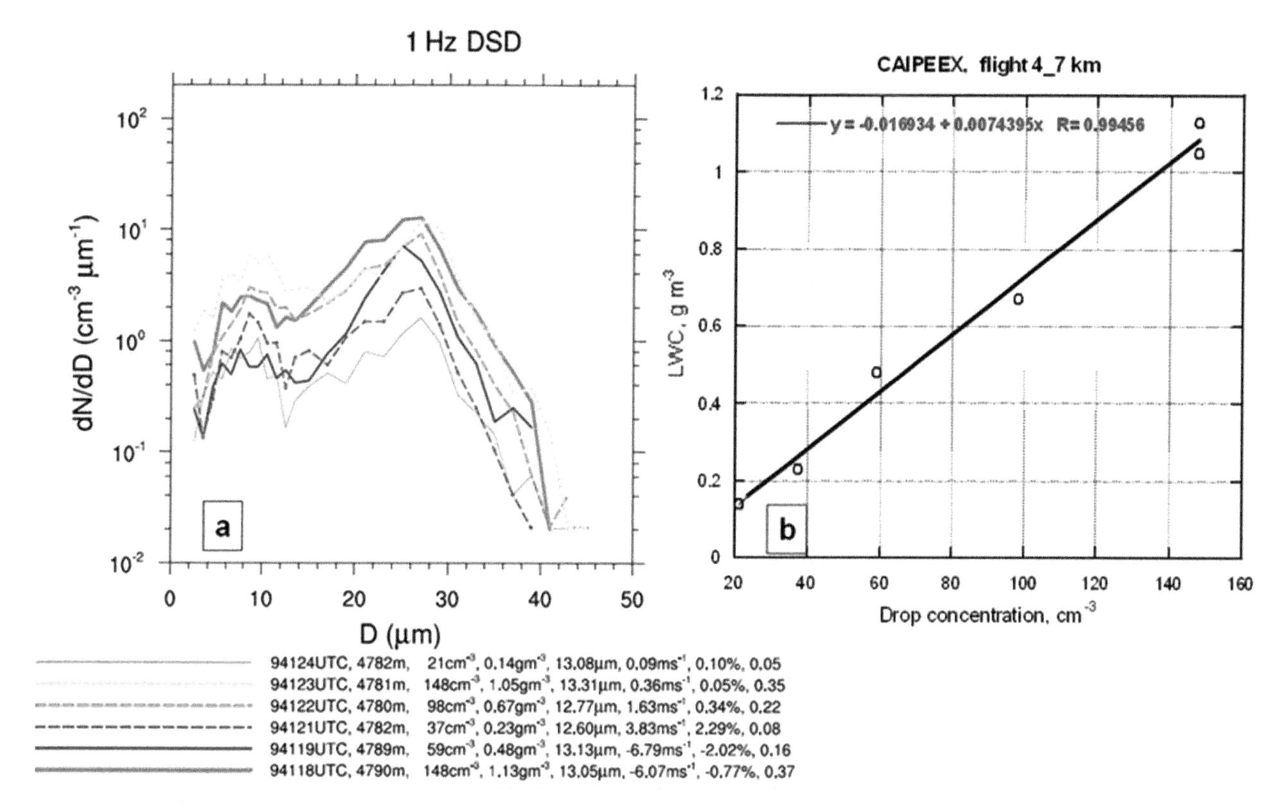

Figure 5.9.19 (a) DSDs along the horizontal traverse in a deep convective monsoon cloud (CAIPEEX, June 22, 2009) and (b) LWC-droplet concentration relationship. Panel (a) also presents the parameters of the flight: UTC, height of flight, droplet concentration, LWC, the effective radius, the vertical velocity, the quasi-steady value of supersaturation, and the AF, determined as LWC/LWC_{ad} (from Prabha et al., 2011, with changes; American Meteorological Society©; used with permission).

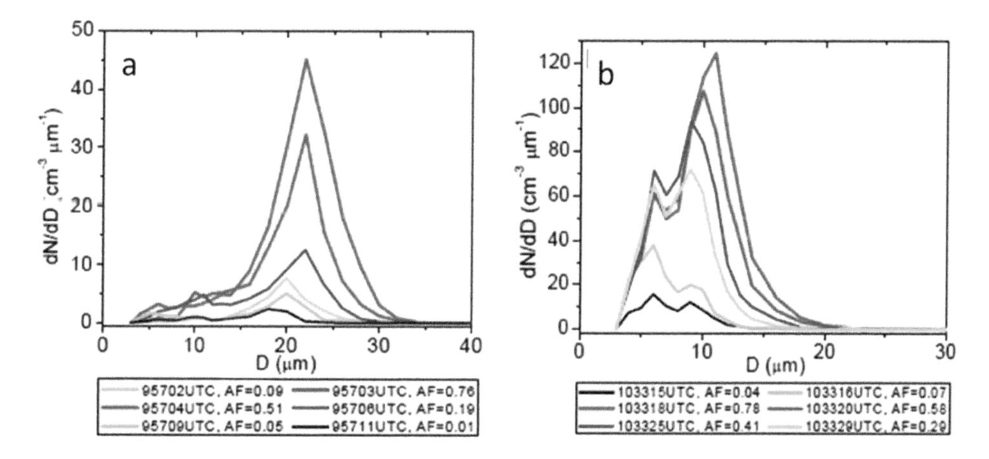

Figure 5.9.20 Droplet SD measured along the aircraft traverses in deep convective clouds at the level of 3.2.km on June, 22 (panel a) and at the level of 3.1 km on June, 16 (panel b) in the field experiment CAIPEX (2009). UTC time (hhmmss) of observations and the corresponding AF of the samples are indicated in the legend below the DSD plots (from Bera et al., 2016a, with changes; courtesy of John Wiley & Sons, Inc.©).

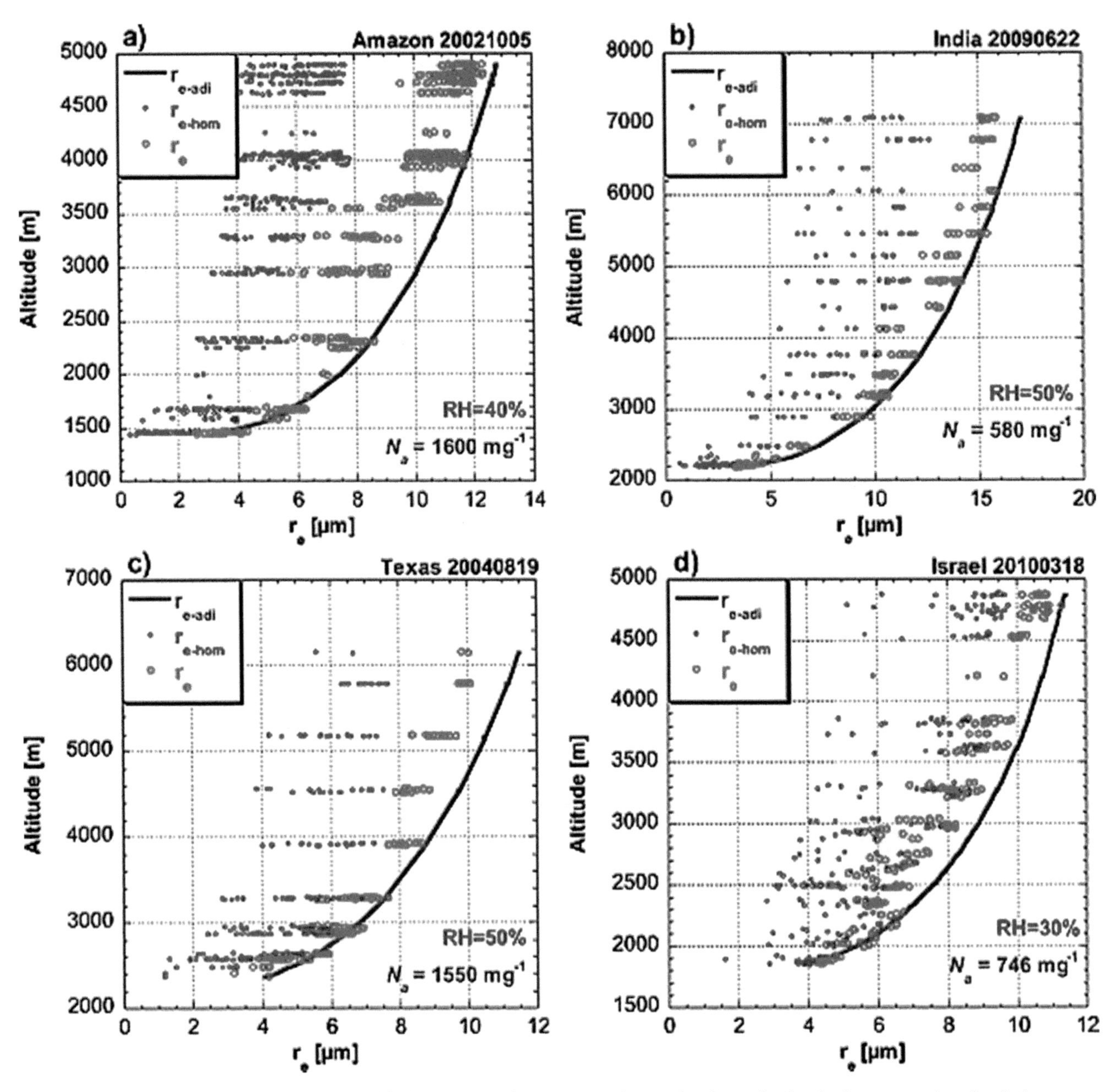

Figure 5.9.21 Aircraft observations of the effective radius values measured near cloud tops in developing convective clouds in different field experiments (red dots). Each data point represents a flight path of ~80–100 m length. Panel (a) shows measurements from SMOCC flight 20021005 over the pristine Amazon. The data shown in panel (b) were obtained near Hyderabad, India, during CAIPEEX-I flight 20090622. Panel (c) displays data from SPECTRA flight 20040819 over West Texas, and (d) data from a winter cloud from flight 20100318 over Israel. Dark gray curves show the profiles of the adiabatic values of the effective radius r_{eff_a}, calculated using a parcel model. Blue circles indicate the expected effective radii in case of fully homogeneous mixing with entrained air that has RH, as denoted in each panel, obtained from the nearest meteorological station or from reanalysis data (from Freud and Rosenfeld, 2012; courtesy of ACP).

in situ 10-m resolution measurements performed using the University of Wyoming King Air aircraft during the COPE-MED project in the Southwestern part of the UK July–August, 2013 (Leon et al., 2016).

Bera et al. (2016a) used the data obtained during CAIPEX in India (2009) to analyze DSD at the lateral edge of deep convective clouds where the adiabatic fraction (AF) increases from zero in droplet-free

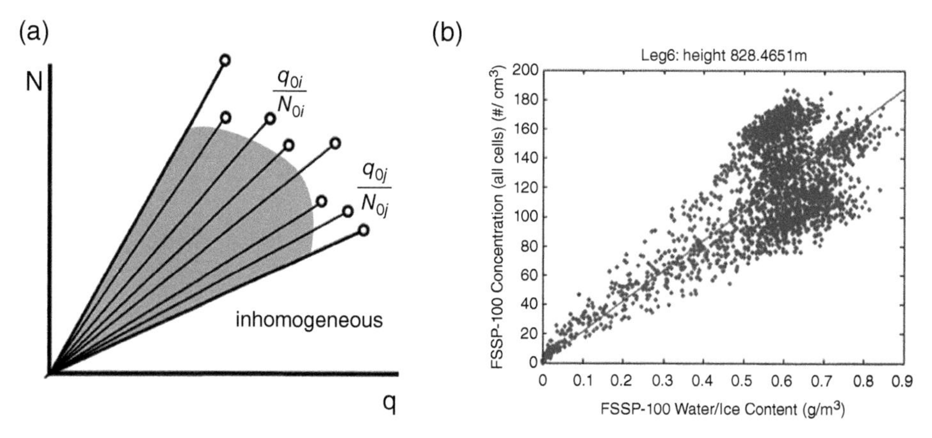

Figure 5.9.22 (a) Conceptual $N - q$ scattering diagrams for extremely inhomogeneous mixing. Circles indicate different initial states of cloud volumes prior to mixing. Straight lines indicate the trajectory of the volume permanently mixing with the environment. (b) Scattering diagram plotted using the data of RF01 of DYCOMS for the upper cloud layer (from Korolev et al., 2016; Courtesy of ACP).

environment toward the cloud core. AF is defined as the ratio of LWC to the maximum value of LWC along the traverse. They found that at small values of AF, DSD are shifted to smaller droplet sizes (**Figure 5.9.20**). With increase in AF values, the peaks of DSD rapidly increase and reach values close to those observed at large AF. In all the cases, pronounced tails of small droplets in the DSD are clearly seen. Bera et al. (2016b) found that the DSD relative dispersion increases toward the cloud edge where AF is small, where it reaches values of 0.3–0.35. They showed that such values of relative dispersion are reached at the transient period in initially droplet-free volumes at intermediate and large *Da*.

Another observed feature related to similarities in DSD is a low variation of the effective radius along horizontal tracks. This feature is typical of both convective and stratocumulus clouds. For instance, Figure 5.9.19b shows that the effective radius changes along the traverse by about 1 μm, i.e., by less than 10% of its mean value of 13 μm. At the same time, the LWC changes along the traverse from 0.14 g m^{-3} to 1.13 g m^{-3}, which corresponds to the change of AF from 0.05 (very diluted) to 0.37. **Figure 5.9.21** shows the values of r_{eff} measured in situ near tops of developing non-precipitating convective clouds during different field experiments. The measurements of droplet size were carried out at 1-Hz frequency within the diameter range from 2 μm to 50 μm. The obtained vertical profiles of r_{eff} indicate that $r_{eff} \sim (z - z_0)^{1/3}$, which is a characteristic of adiabatic parcels (Section 5.2). Measurements of droplet concentration above the supersaturation maximum near cloud base allow calculating the vertical profile of the effective radius that could be expected in an ascending adiabatic parcel. Figure 5.9.21 shows that r_{eff} is close to its adiabatic value. At high enough altitudes, the standard deviation of the effective radius does not exceed 10–15%. Figure 5.9.21 also presents the values of the effective radius that could be expected in case of homogeneous mixing under RH of the environment obtained from the nearest meteorological station or reanalysis data. These estimations show that homogeneous mixing would lead to the values of the effective radius that are substantially lower than the observed values.

Korolev et al. (2016) paid attention to specific features of the scattering diagrams that should be observed in case of extremely inhomogeneous mixing. Several successive mixing events do not change the value of the mean volume radius, i.e., the slope of the $N - q$ dependence does not change. The slope is determined by the initial values of LWC and droplet concentration in a cloud volume that gets involved into mixing with surrounding volumes. It means that in case of extreme inhomogeneous mixing, observed points should be located inside a cone with the apex at point (0.0) (**Figure 5.9.22a**). The shape of the diagram in Figure 5.9.22b plotted using the data of RF01 of DYCOMS for the upper layer of a stratocumulus cloud resembles the cone in Figure 5.9.22a, which may indicate that the mixing at the chosen altitude is inhomogeneous and, likely, close to extremely inhomogeneous.

The results presented in Figures 5.9.18–5.9.22 show that mixing, at least at the measurement frequencies (1 Hz and 10 Hz), i.e., at scales of 10–100 m, is likely to be inhomogeneous, and at scales of 100 m is close to extremely inhomogeneous. This conclusion also agrees with the evaluations of *Da* in clouds (Figure 5.9.17) and the delimitation diagram (Figure 5.9.16). At the same time, the DSD change with time depending on the ratio between cloud volumes and environment air volumes, thermodynamic and microphysical parameters of these volumes as well as on intensity of turbulence.

5.10 Numerical Modeling of Mixing Effects

As was shown in Section 5.9, mixing is one of the main mechanisms of formation of DSD in clouds. Thus, proper treatement of mixing processes in cloud models is one of the important problems of cloud modeling. The description of mixing in cloud models depends on model resolution, dimensionality, and sophistication in description of cloud dynamics and microphysics.

5.10.1 Parcel Models

Nonadiabatic parcel models applied in the theory of turbulent jets and thermals use the simplest approach to parameterization of the effects of entrainment and mixing on cloud dynamics and microphysics (Pruppacher and Klett, 1997). Entrainment and mixing are parameterized by introducing the entrainment parameter λ into the motion equation and into the thermodynamic equation describing ascending plumes. The equation for vertical velocity of an ascending parcel is (Pruppacher and Klett, 1997)

$$\frac{dw}{dt} = g\left(\frac{\theta - \overline{\theta}}{\overline{\theta}(1+\beta)} - q_l\right) - \lambda w^2, \tag{5.10.1}$$

where θ is the potential temperature of an ascending parcel, $\overline{\theta}$ is the potential temperature of the environment air, and q_l is the liquid water mixing ratio. The first term on the right-hand side of Equation (5.10.1) represents the buoyancy force, coefficient $\beta \approx 0.5$ takes into account the effects of pressure decreasing the parcel's acceleration and λ is the entrainment parameter determining the rate of cloudy air dilution by the environmental air. In most applications $\lambda = \frac{0.2}{R}$, where R is the radius of a bubble (of a cloud parcel) or the radius of the cross section of a turbulent jet. One can see that the entrainment term (the second term on the right-hand side) decreases the parcel's updraft velocity.

The equation for the potential temperature in an ascending parcel with the entrainment effects taken into account is written as

$$\frac{d\theta}{dt} = C - \lambda\left(\theta - \overline{\theta}\right)w \tag{5.10.2}$$

where C is the rate of the potential temperature increase due to the latent heat release. Equation (5.10.2) shows that entrainment of dry air tends to adjust the parcel's temperature to some intermediate value between the environment temperature and the adiabatic one. This adjustment decreases the cloud top height. The equation for the potential (or the equivalent potential) temperature is supplemented by the equation for water vapor mixing ratio in the parcel:

$$\frac{dq}{dt} = Q - \lambda(q - \overline{q})w, \tag{5.10.3}$$

where Q is the condensation/evaporation rate. These equations are supplemented by microphysical equations. The thermodynamic equations for temperature and mixing ratio (5.10.2) and (5.10.3) indicate that the parcel is considered uniform, so all the drops within the parcel experience the same supersaturation. There are, however, parcel models in which mixing between penetrated air volumes and cloud air is parameterized according to either homogeneous or inhomogeneous scenario (Section 5.10.4)

The simplicity of parcel model dynamics allows utilization of complicated microphysics. Time steps in parcel models can be as small as 0.001 s, which allows direct simulation of diffusional growth of submicron aerosol particles. An example of a complicated microphysics included into a parcel model is presented in Section 5.11. Due to their simplicity, parcel models based on equations such as (5.10.1) and (5.10.2) are also used as a basis for convective parameterization in most large-scale convection models and climatic models (e.g., Arakawa and Shubert, 1974).

5.10.2 Representation of Mixing in Cloud Models at Different Spatial Scales

In most cloud models that calculate time changes of different microphysical and thermodynamic parameters, turbulent mixing is performed by turbulent vortices or filaments that are mostly of subgrid scale. The first question that arises with regard to application of mixing algorithms in cloud models is: What type of mixing (homogeneous or inhomogeneous) does the model describe? This question pertains to both the Euilerian models, which calculate microphysical variables on finite different grids (e.g., Benmoshe et al., 2012), and to the LEM where these values are calculated within movable air parcels (e.g., Pinsky et al., 2008b; Magaritz et al., 2009; Magaritz-Ronen et al., 2014, 2016a, 2016b). Mixing involves two sub-steps, namely, the resolvable sub-step and the subgrid sub-step. At the resolvable sub-step, divergence of thermodynamic and microphysical quantities of the turbulence flux is calculated at each grid point, and the time changes of these quantities are calculated by solving the turbulence-diffusion equation. When mixing between two neighbouring grid points is calculated, the mixing volume containing both grid points is never uniform, i.e., the spatial gradients of the microphysical variables remain between the neighbouring grid points. So, at resolved scales the mixing is inhomogeneous.

At the subgrid sub-step, the results of mixing within the volume related to a particular grid point (or within a particular Lagrangian parcel) should be determined. Typically, these volumes are assumed uniform, so the modeled subgrid mixing is treated as homogeneous. Therefore, in most numerical studies, mixing is treated as inhomogeneous at resolved scales and as homogeneous at subgrid scales. In some models, however, mixing at subgrid scales is also parameterized as extremely inhomogeneous (or sometimes as inhomogeneous). These particular parameterizations are based on the classical concepts of extremely inhomogeneous mixing according to which a decrease in drop concentration does not change the mean volume radius or the effective drop radius (Section 5.9).

Representation of Mixing on Resolved Scales in Cloud Models

Mixing at resolved scales in the Eulerian models is parameterized using the operators of turbulent diffusion that have the form $\frac{\partial}{\partial x_i}\left(K\frac{\partial \bar{a}}{\partial x_i}\right)$ for conservative quantities and $\frac{\partial}{\partial x_i}\left(K\left[\frac{\partial \bar{a}}{\partial x_i}-\frac{\delta \bar{a}}{\delta x_i}\right]\right)$ for nonconservative quantities (Equation (5.9.3)), where the term $\frac{\delta \bar{a}}{\delta x_i}$ describes nonconservative changes of quantity $\bar{a}$ at the distance equal to the mixing length (Section 5.9). Mixing of nonconservative DSD is of special interest since drop size changes in the course of mixing are accompanied with latent heat release. The latent heat release during mixing is typically not taken into account in cloud models, so DSD mixing is performed by applying the simple mixing operator $\frac{\partial}{\partial x_i}\left(K\frac{\partial f}{\partial x_i}\right)$ (e.g., Takahashi, 1976; Hall, 1980; Khvorostyanov et al., 1989; Feingold et al., 1996; Kogan, 1991; Khain and Sednev, 1996; Khairoutdinov and Kogan, 2000; Yin et al., 2000; Stevens et al., 2005, etc.). Such mixing can be referred to as passive or mechanical.

Pinsky et al. (2010) applied a numerical analog of operator $\frac{\partial}{\partial x_i}\left(K\left[\frac{\partial \bar{a}}{\partial x_i}-\frac{\delta \bar{a}}{\delta x_i}\right]\right)$ to perform DSD mixing between Lagrangian parcels. The proposed approach is schematically illustrated in **Figure 5.10.1**. During ascending/descending of mass in the filaments, the value of a in these filaments changes due to its

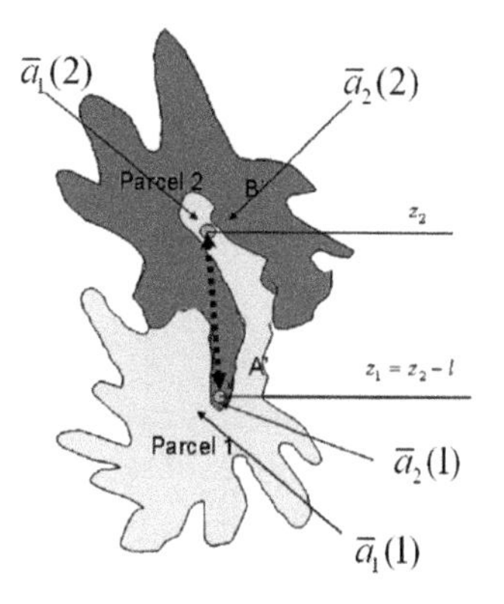

Figure 5.10.1 Scheme illustrating the concept and the procedure of mixing between two Lagrangian air parcels. The initial values of a nonconservative quantity a at the centers of the parcels are denoted as $\bar{a}_1(1)$ and $\bar{a}_2(2)$. The indices denote the number of the parcel and the arguments show the height at which these values are calculated (from Pinsky et al., 2010, with changes; courtesy of John Wiley & Sons, Inc.©; used with permission).

nonconservativity. During an ascent of mass in the filament from the center of parcel one to the parcel two level, a changes from $\bar{a}_1(1)$ to $\bar{a}_1(2)$. During the descent of air in the filament belonging to parcel two, a changes from $\bar{a}_2(2)$ to $\bar{a}_2(1)$. Changes in the value of a over time due to turbulent fluxes between the two volumes are expressed according to Pinsky et al. (2010) as

$$\begin{aligned}\frac{d\bar{a}_1}{dt} &= -\frac{1}{2}\frac{\partial}{\partial l}\left\{K\left(\frac{\bar{a}_1(2)-\bar{a}_2(1)}{l}-\frac{\delta \bar{a}_2}{\delta l}\right)\right\}\\ \frac{d\bar{a}_2}{dt} &= -\frac{1}{2}\frac{\partial}{\partial l}\left\{K\left(\frac{\bar{a}_2(2)-\bar{a}_1(1)}{l}-\frac{\delta \bar{a}_1}{\delta l}\right)\right\}\end{aligned} \tag{5.10.4}$$

where $K(l)$ is determined according to Richardson's four-thirds law (Section 3.3) and l is the distance between the centers of the parcels, associated to the mixing length. The terms $\frac{\delta \bar{a}_1}{\delta l}$ and $\frac{\delta \bar{a}_2}{\delta l}$ take into account the changes of values $\bar{a}$ in the course of mixing due to phase transition. When the procedure was applied to DSD mixing, drop radii changes δr_i are calculated during the ascent and descent of cloudy air in the filaments, using the equation of diffusional drop growth. At this sub-step, the nonconservative nature of DSD is taken into account. This procedure is applied to each of the parcel pairs. As a result, the divergence of turbulent fluxes at the center of each parcel is calculated. During turbulent mixing there is no net transport of air mass in the vertical direction.

The role of nonadiabaticity in formation of DSD is illustrated by a simple example of two air parcels located one above the other. The initial DSD in the parcels are

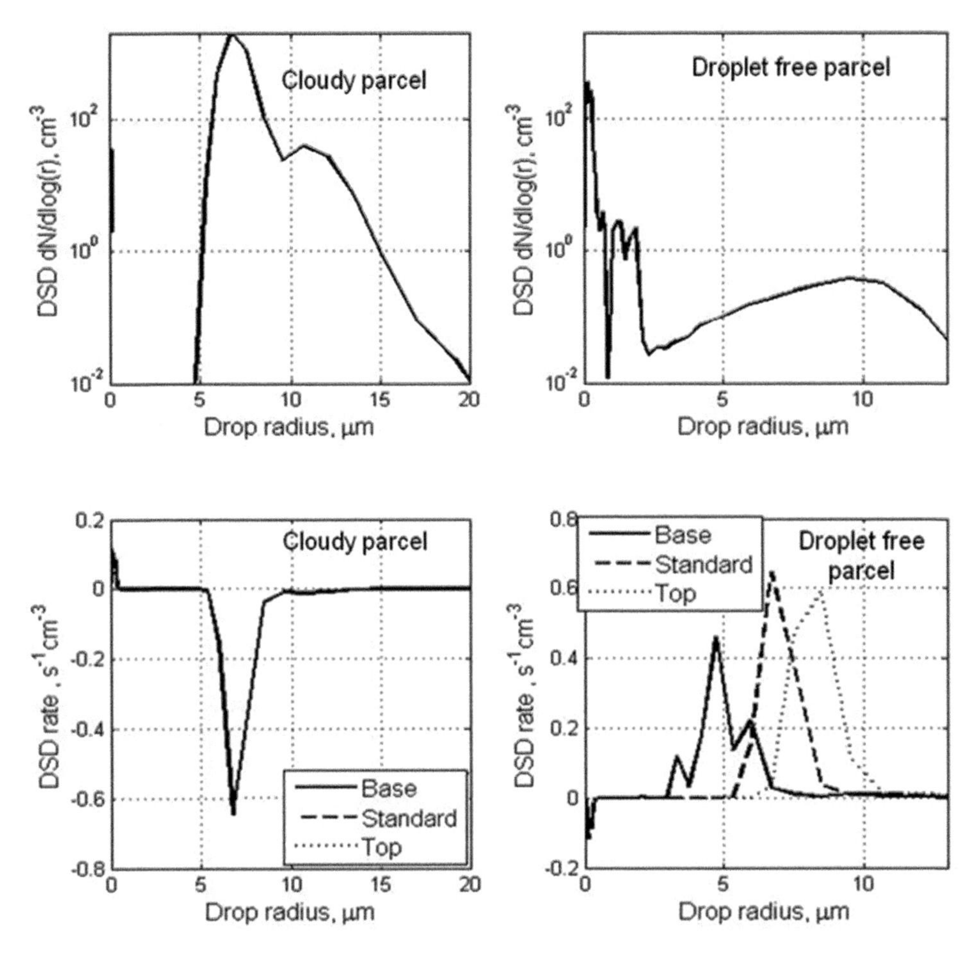

Figure 5.10.2 Upper row: the initial DSD in cloud parcel (left) and the initial DSD in the droplet-free parcel (right). Bottom row: the mixing-induced rates of DSD changes in the cloud parcel (left) and in the initially droplet-free parcel (right) in different simulations (from Pinsky et al., 2010; courtesy of John Wiley & Sons, Inc.©).

shown in **Figure 5.10.2**. The cloud parcel with LWC = 0.31 gm^{-3} was assumed supersaturated, while the second one was a subsaturated droplet-free parcel containing haze particles with LWC = 3×10^{-4} gm^{-3}. The dissipation rate was assumed to be $\varepsilon = 10\ \text{cm}^2\ \text{s}^{-3}$, i.e., typical of stratocumulus clouds. The rates of DSD changes in three simulations are compared: in the simulation referred to as Top, the droplet-free parcel was located above the cloud parcel (like at a cloud top); in the simulation referred to as Base, the cloud parcel was located above the droplet-free parcel (like at cloud base); and in the simulation referred to as Standard, no latent heat release during mixing was taken into account. The rates of the DSD changes in the cloud volumes in all simulations were found to be actually similar. The substantial difference in the DSDs rates between Top and Base is pronounced only in the initially droplet-free volumes. In Top, mixing leads to a growth of larger droplet concentration. The Standard simulation that simulated passive mixing indicates the growth in the concentration of droplets of intermediate size. In Base, the mixing leads to an increase in the concentration of smaller droplets. This difference in the growth rates of droplets of different sizes in Top and Base can be attributed to the following factors. In Top, the droplets in the cloud filaments penetrating the dry air move upward and grow during the ascent. In contrast, in Base the subsaturated parcel obtains droplets that descend in the cloud filaments from above and partially evaporate. The differences in the rates of DSD changes indicate that near cloud-top mixing is more efficient than near the cloud base. In Standard simulation, the results of mixing are insensitive to the relative location of the mixing parcels.

Another approach to representing DSD mixing is used in studies attributed to the nonadiabatic stochastic condensation. To take into account the nonconservativeness of DSD, the operator of turbulent diffusion $K\frac{\partial}{\partial x_i}$ is replaced by the operator $K\frac{\partial}{\partial x_i} + A\frac{\partial}{\partial s}$, where s is the square of the drop surface and A is proportional to the vertical velocity. Several attempts were made to combine the stochastic kinetic equation for DSD with other thermodynamic equations in order to investigate the effects of turbulent fluctuations on the DSD evolution (e.g., Levin and Sedunov, 1966; Sedunov, 1974; Stepanov, 1975; Voloshchuk and Sedunov, 1977; Manton, 1979; Merkulovich and Stepanov, 1977). In all these studies, the supersaturation fluctuations are highly correlated with the vertical velocity fluctuations. Although some authors reported DSD broadening (e.g., Manton, 1979), the width of the obtained DSD was still narrower than that observed in convective clouds. The detailed review of these studies was presented by Mazin and Merkolovich (2008).

Kvorostyanov and Curry (1999a, 1999b) developed a modified theory of stochastic condensation. Using the standard algorithm of separating variables into mean values and fluctuating values and taking into account that DSD depends on coordinates and drop size, they obtained an equation for DSD averaged over an ensemble of realizations:

$$\frac{\partial \bar{f}}{\partial t} + \frac{\partial}{\partial x_i}\left(\bar{u}_i \bar{f}\right) + \frac{\partial}{\partial r}\left(\bar{\dot{r}}\bar{f}\right) = \frac{\partial}{\partial x_i}\overline{u_i' f'} - \frac{\partial}{\partial r}\left(\overline{\dot{r}' f'}\right) \tag{5.10.5}$$

Here $\dot{r}$ is the rate of drop radius increase due to diffusion growth; the mean values are denoted by the overbar; and fluctuating values are denoted by the prime. The averaging can also be interpreted as a grid-volume averaging in an LES model. In this case, the averaged values are of a resolved scale, while the deviations represent subgrid unresolved scales. Equation (5.10.5) takes into account both the advection of DSD and DSD changes caused by the regular diffusion growth/evaporation. The authors derived analytical expressions for terms $\overline{\dot{r}' f'}$ and $\overline{u_i' f'}$, and obtained an equation for DSD evolution describing advection and diffusion growth of drops. The equation is as follows:

$$\frac{\partial \bar{f}}{\partial t} + \frac{\partial}{\partial x_i}\left[u_i - V_g \delta_{i3}\bar{f}\right] + \frac{\partial}{\partial r}\left(\frac{bS}{r}\bar{f}\right) = \left(\frac{\partial}{\partial x_i} + \delta_{i3} G \frac{\partial}{\partial r}\right) K_{ij}\left(\frac{\partial}{\partial x_j} + \delta_{j3} G \frac{\partial}{\partial r}\right)\bar{f} \tag{5.10.6}$$

The term $\frac{\partial}{\partial r}\left(\frac{bS}{r}\bar{f}\right)$ describes advection in the space of drop radii with the effective speed equal to $\frac{bS}{r}$ that is inversely proportional to the drop radius. This advection is faster for smaller drops, so this term alone causes DSD narrowing and describes the changes of DSD due to regular condensation. The term $\frac{\partial}{\partial x_i} K_{ij} \frac{\partial}{\partial x_j}\bar{f}$ represents the regular mixing operator in the space coordinates (K_{ij} is the tensor of turbulent coefficients). The turbulent flux $K_{ij} G \delta_{j3} \frac{\partial}{\partial r}\bar{f}$ represents the DSD flux in the space of drop radii. The product $K_{ij}G$ is the tensor of the effective diffusion coefficients in the space of radii. This diffusion in the radii space tends to smoothing and broadening of the DSD. The resulting DSD is determined by the spatial advection and by diffusion in the radii space. Khvorostyanov and Curry (1999a) hypothesized that at certain conditions, larger drops experience larger supersaturations than smaller ones. These supersaturation fluctuations do not correlate with fluctuations of the vertical velocity. This hypothesis allowed Khvorostyanov and Curry (1999a) to obtain DSD broadening toward larger drop sizes. The authors showed theoretically that representation of DSD evolution by Equation (5.10.6) leads to asymptotical DSD having the shape of the Gamma distribution. They related the parameters of the obtained Gamma distribution, including the relative dispersion, to the values of the thermodynamic parameters (turbulent coefficient, vertical velocity, and vertical gradient of temperature) and the microphysical parameters (the drop concentration and the mean radius). The obtained dependencies were used to determine the DSD shape in the bulk parameterization scheme implemented by Morrison et al. (2005).

Parameterization of Mixing on Subgrid Scales

Grid spacing of cloud-resolving models is typically larger than several hundred meters, while the transition between mixing types takes place at scales of about 1 m. It means that within the air volumes that can be attributed to particular grid points, mixing can also be considered inhomogeneous. There were several attempts to develop parameterization of inhomogeneous mixing on subgrid scales in the Eulerian models of cumulus clouds (e.g., Grabowski, 2006, 2007; Morrison and Grabowski, 2008; Jarecka et al., 2009, 2013, etc.) and in LES models of stratocumulus clouds (Hill et al., 2009). The parameterization proposed by Morrison and Grabowski (2008) mimics the results of mixing according to the simplified mixing concept, i.e., it does not change the effective radius but reduces the droplet concentration. Morrison and Grabowski (2008) separated the process of the droplet concentration change into two steps. At the first step, advection and standard mixing on the resolved scale are calculated using the K-theory operators (Equation (5.9.2)). The droplet concentration obtained after this step, denoted as N_i, served as an initial condition for the second step of the mixing procedure (on the subgrid scale) that could be either homogeneous or inhomogeneous. If the mixing is assumed homogeneous, the final droplet concentration after homogenization N_f is equal to N_i, since homogeneous mixing does not change droplet concentration (according to the classical concept). The parameterization of inhomogeneous mixing is reduced, therefore, to a decrease in droplet concentration. Droplet concentration changes in such a way that the mean volume radius remains unchanged. The final droplet concentration N_f under an arbitrary mixing scenario is calculated as

$$N_f = N_i\left(\frac{q_f}{q_i}\right)^{\alpha}, \qquad (5.10.7)$$

where $0 \leq \alpha \leq 1$, q_i is the initial cloud water mixing ratio (i.e., the value obtained after the first step), and q_f is the final cloud water mixing ratio. For pure homogeneous mixing $\alpha = 0$; for extremely inhomogeneous mixing $\alpha = 1$. In their simulations, Morrison and Grabowski (2008) used $\alpha = 0.5$. Numerical tests show that implementation of droplet concentration correction that mimics the effects of inhomogeneous mixing leads to a decrease in droplet concentration by 10–15%, while other quantities are practically unchanged. A further development of this parameterization was performed by Jarecka et al. (2009), who introduced an equation for change of the entrained filaments size with time, as well as the equation for the cloud volume fraction within the model grid cell. This parameterization was tested in simulations of small cumuli using a high-resolution bulk parameterization cloud model. A certain effect of inhomogeneous mixing parameterization was found for grid spacing of 25m × 25 m. At grid spacings of 50m × 40 m and 100m × 40 m, the effect of inhomogeneous mixing parameterization was not distinguishable from that of homogeneous mixing.

Hill et al. (2009) further developed the Morrison and Grabowski (2008) parameterization in simulation of stratocumulus clouds using an LES model with spectral bin microphysics. They expressed the changes of the temperature and the mixing ratio at each time step as the following sums:

$$\begin{aligned} T^{t+\Delta t} - T^t &= \left[\left(\frac{\Delta T}{\Delta t}\right)_{res} + \left(\frac{\Delta T}{\Delta t}\right)_{sub}\right]\Delta t; \\ q_v^{t+\Delta t} - q_v^t &= \left[\left(\frac{\Delta q_v}{\Delta t}\right)_{res} + \left(\frac{\Delta q_v}{\Delta t}\right)_{sub}\right]\Delta t;, \end{aligned} \qquad (5.10.8)$$

where index "res" denotes changes caused by a resolved processes (e.g., advection) and index "sub" denotes changes caused by subgrid mixing. Accordingly, changes in the supersaturation and in the liquid water mixing ratio were also separated into two steps: changes on the resolved scales and changes on the subgrid scales. The standard model configuration for evaporation was used to calculate evaporation on the resolved scales. The local water mass mixing ratio before the subgrid evaporation q_0 was calculated as $q_0 = \sum_k q_{k0}$, where k is the bin number, q_{k0} is the water mass mixing ratio in the k-th bin, and index "0" stands for "initial." Then the local water mass mixing ratio q_f after subgrid evaporation is calculated using the semi-analytical solution of the supersaturation equation (e.g., Clark, 1973). Evaporation on the subgrid scales is treated as extremely inhomogeneous. The fractional change in the total mass due to the subgrid evaporation was used to correct the mass ratio and the number concentration mixing ratio in each bin. The change in the mass ratio in the k-th bin caused by extremely inhomogeneous mixing on the subgrid scales was calculated as

$$n_{kf} = n_{k0}\left(\frac{q_f}{q_0}\right); q_{kf} = q_{k0}\left(\frac{q_f}{q_0}\right). \qquad (5.10.9)$$

where indexes "0" and "f" denote concentrations and mass mixing ratios before and after the subgrid evaporation. Equation (5.10.9) shows that concentrations and masses in the bins decrease in the same proportion, so the mean volume radius does not change with reducing droplet concentration.

5.10.3 Effects of Mixing on Cloud Microstructure, Simulated by Means of Cloud Models

Stratocumulus Clouds

Figures 5.10.3 and **5.10.4** illustrate the main dynamic effects of the mixing operator (5.9.3) applied in the LEM of a stratocumulus cloud (Magaritz-Ronen et al., 2014, 2016a, 2016b). The model calculates a stratocumulus cloud simulating motion and interaction of about 2,000 adjacent parcels. The results of two simulations are compared: the simulation referred to as Mixing (MI) between parcels, and the simulation named No mixing (NoMI) that does not include inter-parcel mixing.

Figure 5.10.3 shows the LWC and the number concentration fields in the simulations of a non-drizzling stratocumulus cloud observed during flight RF01 of the Second Dynamics and Chemistry of the Marine Stratocumulus field study (DYCOMS II). One can see that turbulent mixing has a profound effect on the geometrical and microphysical structure of the cloud layer, mainly by spatial smoothing and homogenization of the microphysical and thermodynamic fields. In MI, the cloud base and the cloud top are well pronounced.

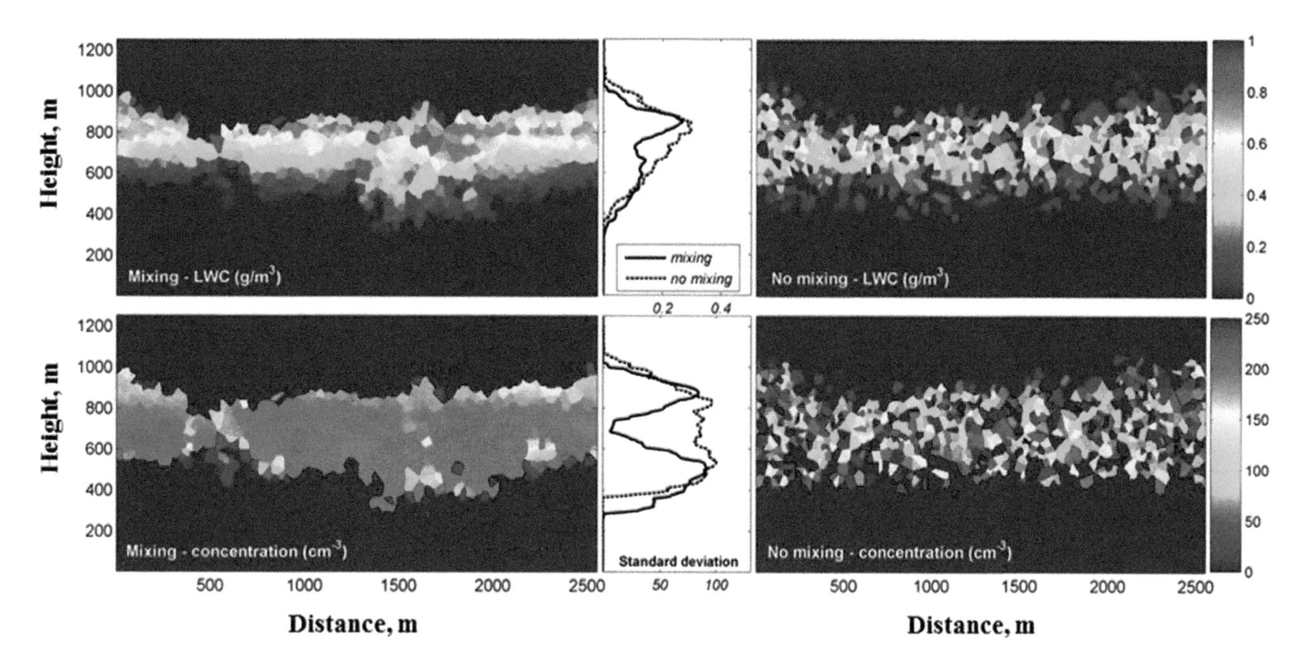

Figure 5.10.3 Fields of LWC (top) and droplet concentration (bottom) in MI (left) and NoMI (right). All fields are plotted at t = 310 min. The profiles of standard deviation are presented, as well. The cloud forms below the inversion layer in which the air is warm and dry (from Magaritz-Ronen et al., 2014; American Meteorological Society©; used with permission).

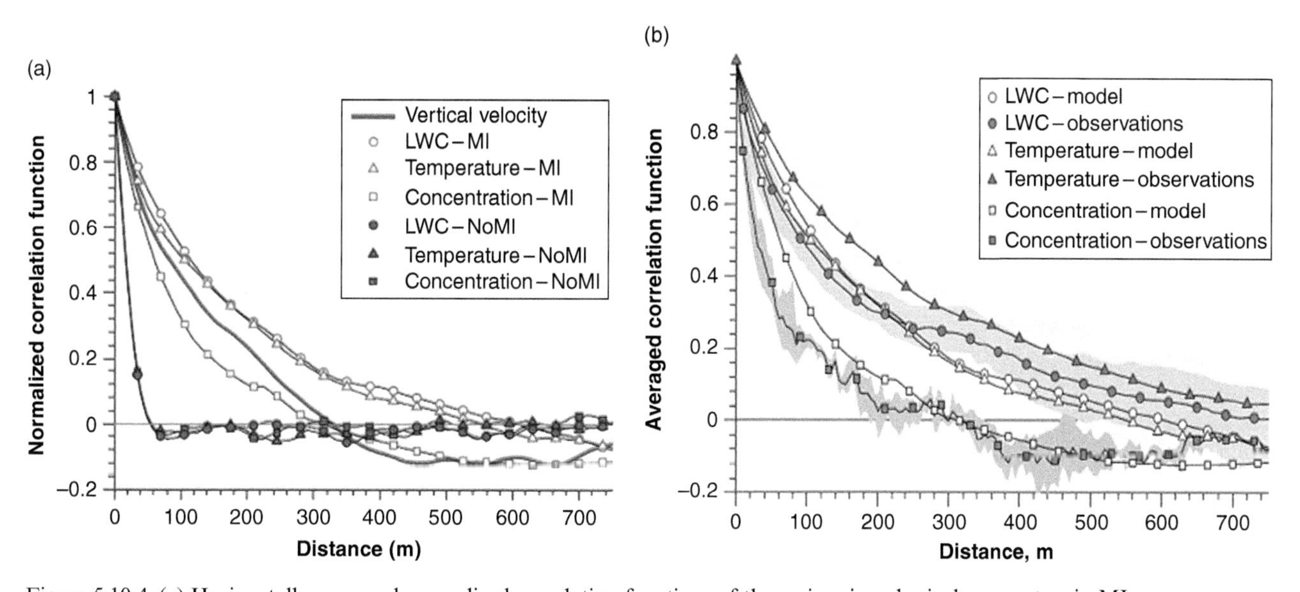

Figure 5.10.4 (a) Horizontally averaged normalized correlation functions of the main microphysical parameters in MI (open markers) and NoMI (filled markers). The correlation function for the vertical velocity (line without symbols) is identical in the two simulations. (b) Horizontally averaged correlation functions for LWC, temperature, and droplet concentration calculated using the data collected during RF01 (filled markers) and in MI (open markers). The shaded area indicates the minimum–maximum values over flight lags. All the correlation functions are calculated at the height of 700 m during the time period of 170–310 min (from Magaritz-Ronen et al., 2014; American Meteorological Society©; used with permission).

The LWC is maximum near the cloud top and decreases toward cloud boundaries. In NoMI, the cloud layer contains many drop-free air volumes and the cloud structure is highly nonuniform. Thus, the cloud boundaries are not easy to recognize. The cloud simulated in MI is characterized by low variability of the drop concentration in the cloud layer. This agrees with the measurements made during flight RF01 and is a typical feature of stratocumulus clouds (Pawlowska et al., 2000). This feature is not seen in NoMI where many parcels with very high or very low drop concentrations can be found. The standard deviation profiles indicate that the variability in LWC in both cases reaches its maximum near the cloud top within the interface area between parcels from the cloud and the inversion layers. Below this maximum, the standard deviation is substantially lower in MI. The areas of reduced LWC and reduced drop concentration at the cloud inversion are largely formed by dry and warm air parcels entrained in the cloud through its top. In MI, such parcels mix with cloudy parcels and become an integral part of the cloud, while in NoMI they form holes in the cloud layer.

An important characteristic of cloud structure is the spatial correlation functions of the thermodynamic and microphysical values (Section 3.3). Figure 5.10.4 presents the normalized horizontal correlation functions of different microphysical and dynamical quantities. For all the parameters, the correlation lengths in MI are considerably larger than those in NoMI. While in MI the correlation length is of the scale of a few hundred meters, i.e., close to the correlation length of the vertical velocity, in NoMI the correlation length is of the order of the linear size of the parcel. The difference in the correlation lengths reflects the difference in the cloud structures simulated with and without mixing, as seen in Figure 5.10.4. Figure 5.10.4b compares the correlation functions of LWC, temperature and drop concentration derived from the model simulation with the correlation functions derived from in situ measurements. In agreement with the results of the simulations, the correlation length of the temperature is the longest, and the correlation length of the drop concentration is the shortest.

Figure 5.10.5 shows the vertical profile of the effective radius PDF in MI. The median profile of r_{eff} in NoMI, where the inter-parcel turbulent mixing was excluded, is shown, as well. The values of the effective radius at the cloud top agree with observations (van Zanten et al., 2005). Figure 5.10.5 shows that turbulent mixing taken into account leads to an increase in the effective radius over all the cloud depth. The mechanisms by which mixing increases r_{eff} are considered in several studies.

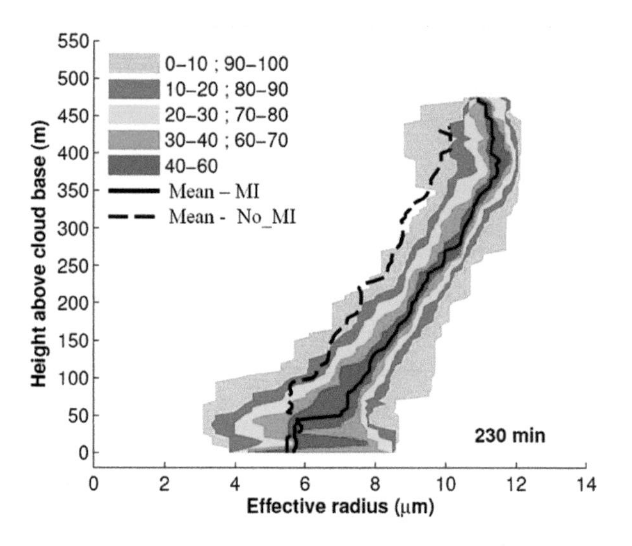

Figure 5.10.5 The vertical profile of PDF of the effective radius. Colors mark every tenth percentile on both sides of the distributions. Black line is the median of the distributions. Dashed line shows the vertical profile of the horizontally averaged effective radius in the NoMI simulation (from Maragritz-Ronen et al., 2016b, with changes; courtesy of John Wiley & Sons, Inc.©).

Kogan (2006) showed that air recirculation accompanied by mixing of droplets in downdrafts and in updrafts leads to an increase in the droplet size. Magaritz et al. (2010) showed that droplet recirculation in the boundary layer accompanied by droplet–droplet collisions within the Sc cloud leads to an increase in the mass of salt (mass of AP) in the largest droplets. These droplets may partially evaporate below cloud base and turn into haze. Being activated, these haze particles lead to formation of larger droplets.

Korolev et al. (2013) proposed a conceptual model according to which mixing of cloud volumes in downdrafts and updrafts near cloud base leads to a substantial DSD broadening toward larger droplet sizes. As was shown in Section 5.2, droplets in downdrafts are larger than droplets in updrafts. Mixing of ascending and descending volumes, especially near cloud base where supersaturation is maximum, leads to the appearance of larger droplets in updrafts. Estimations performed by Korolev et al. (2013) showed that only several repetitive recirculations are needed to get quite large droplets that are able to trigger drizzle formation. The increase r_{eff} in MI as compared with NoMI just above the cloud base (Figure 5.10.5) leads to the increase r_{eff} over the entire cloud layer, which fosters drizzle formation. Examples presented in Figures 5.10.3–5.10.5 indicate the crucial role of turbulent mixing (largely in-cloud mixing) in formation of dynamic and microphysical structure of Sc clouds.

In-cloud mixing takes place at high values of RH, when mixing types are indistinguishable. The influence of mixing types on cloud microphysics become pronounced near the upper cloud boundary, where entrainment and mixing with dry environment takes place and where mixing is accompanied by phase transition. So, the first question concerning the effects of the mixing type is: How does non-conservativity of DSDs affect cloud microphysical structure? **Figure 5.10.6** shows time dependencies of the effective radius, averaged in the horizontal direction and obtained using the LEM simulation of Sc observed during research flight RF07 of DYCOMS II. Figure 5.10.6a shows the results of a simulation in which nonconservativity of DSD was taken into account (Equations (5.9.3) and (5.10.4)) were used. Figure 5.10.6b shows the results obtained by application of the standard mixing operator (Equation (5.9.2)) for DSD mixing. One can see that taking into account nonconservativity of the DSD increases the effective radius in the upper part of the stratocumulus cloud and causes earlier formation of drizzle. These

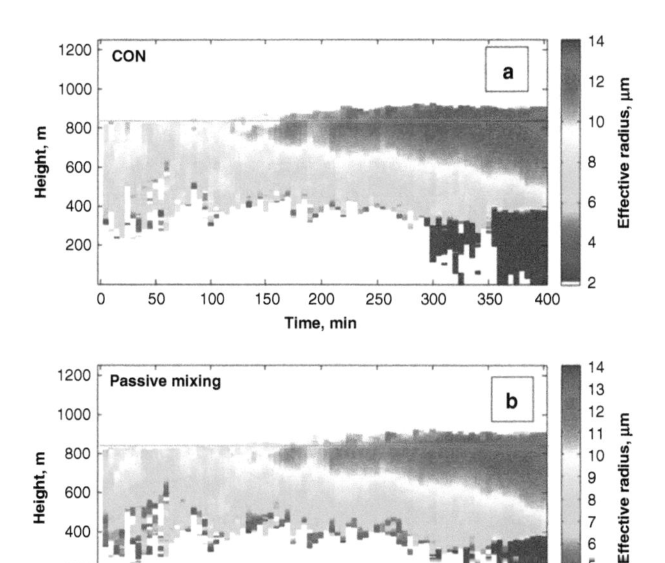

Figure 5.10.6 Time dependencies of the effective radius obtained by means of the LEM in simulation of RF07 and averaged in the horizontal direction. (a) Non-conservativity of DSD is taken into account: CON in panel (a) stands for "control experiment." (b) Results obtained using the standard mixing operator. Drizzle below the cloud is marked brown (from Magaritz-Ronen et al., 2016a; courtesy of ACP).

results are in agreement with the mechanisms illustrated in Figure 5.10.2. However, the difference in the cloud microstructures between these simulations is small, much weaker than the difference between the cloud structures in simulations MI and NoMI.

The effect of extremely inhomogeneous mixing on subgrid scales on the microstructure and dynamics of a Sc was investigated by Hill et al. (2009). The parameterization of inhomogeneous mixing was included according to Equations (5.10.8) and (5.10.9). The authors performed simulations of a non-drizzling stratocumulus cloud at different aerosol concentrations corresponding to CCN concentration of 100 cm^{-3} at 1% of supersaturation (clean atmosphere) and of 1,000 cm^{-3} (polluted atmosphere). Calculations were performed at two model resolutions: a higher resolution ($\Delta z = 10\,m$, $\Delta x = \Delta y = 20\,m$) and a lower resolution ($\Delta z = 20\,m$, $\Delta x = \Delta y = 40\,m$). **Figure 5.10.7** shows the time dependences of domain-averaged liquid water path for the high resolution (a) and the low resolution (b). One can see that the difference caused by application of different mixing assumptions is very small. Regardless the resolution, applying a parameterization of extremely inhomogeneous mixing rather than of homogeneous mixing on subgrid scales led to a reduction of drop concentration by less than 5%, to an increase in the effective radius by a ~1% and to ~1% increase in the cloud optical depth. The changes in aerosol loading and the model resolution affect the results much more strongly than the difference between mixing scenarios.

Hill et al. (2009) explain the insensitivity of Sc evolution to a mixing type on subgrid scales by the fact that the condensation/evaporation rates caused by a resolved dynamics are by two orders of magnitude greater than those caused by subgrid processes. In addition, the low sensitivity of the microphysical structure to different mixing parameterizations on subgrid scales in LES cloud models is related to the fact that the characteristic spatial scales of variation of the microphysical variables (the spatial scales of correlation) are of several hundred meters (Figure 5.10.4), i.e., larger than the grid spacing in LES. In this case, the difference between the neighboring grid points is small. Larger errors can be expected at larger grid spacing. A negligible effect of a mixing type on formation of large droplets in Sc was also reported by Small and Chuang (2008).

Formation of DSDs and Interpretation of Mixing Diagrams

As was discussed is Section 5.9, observations show that the relative variability of the effective radius along

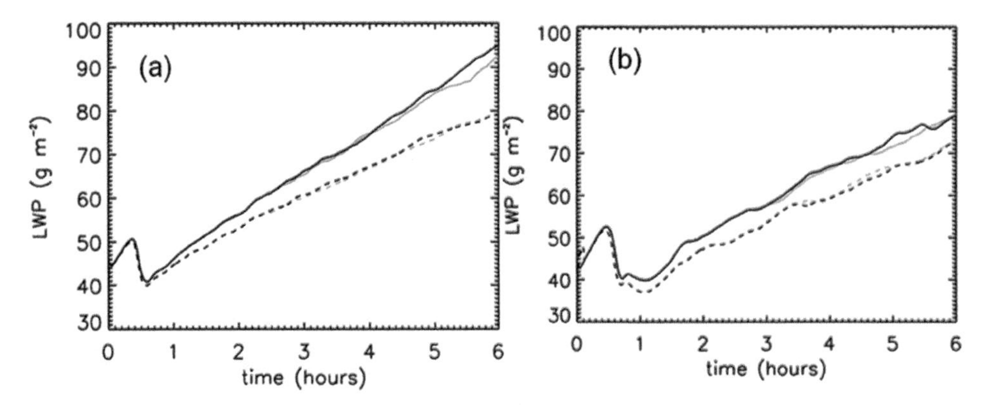

Figure 5.10.7 Time dependences of the domain-averaged liquid water path in simulations at high resolution (a) and low resolution (b). Black lines represent homogeneous mixing; gray lines represent extremely inhomogeneous mixing. Solid lines show the low CCN concentration simulation, whereas dashed lines show the high CCN concentration simulations (from Hill et al., 2009; American Meteorological Society©; used with permission).

traverses near cloud top is low compared to the relative variations in drop concentration or in LWC. Figure 5.10.5 shows that the LEM reproduces these observational findings. Analysis of microphysical characteristics of cloud parcels located near cloud top reveals the existence of three typical scenarios of the impact of mixing on DSD evolution and on the value of r_{eff}. These scenarios are illustrated by analysis of evolution of DSD and of other microphysical parameters in three parcels chosen from the 2,000 parcels in LEM. Parcel 1 (P1) ascending from the cloud base to the cloud top is adiabatic or diluted only slightly. Parcel 2 (P2) moves horizontally near the cloud top after the parcel ascends. During the movement, P2 mixes with the air from the inversion layer. Parcel 3 (P3) is an initially drop-free parcel from the inversion layer. **Figure 5.10.8** illustrates the evolution of DSD in the selected parcels through 170–185 min of simulation (left column) and the trajectories of the parcels on the LWC-r_{eff} diagrams (middle column). Time dependencies of parcels' elevation levels and different microphysical variables are presented in the right column.

P1 ascends and reaches the cloud top at 185 min. During the ascent, the LWC and the effective radius increase due to diffusional growth. The LWC in P1 is larger than in other parcels located at the same height, and slightly exceeds 1 g m^{-3} at the cloud top. In the LWC-r_{eff} diagrams, P1 moves along the outer bottom edge of the diagram, following the dependence expected for an adiabatic parcel with $q_w \sim r_v^3 \approx r_{eff}^3$. One can see several modes in the DSD at t = 170 min. The modes of smaller droplets are formed largely by in-cloud nucleation. Mixing with neighboring cloud parcels does not affect P1 significantly.

Mixing of P2 (middle row) with the inversion air leads to more than a double decrease in LWC, while both the radius of the main mode equal to 10 μm and the effective radius remain nearly constant. These results are in accordance with results of the diffusional-evaporation model at intermediate and high *Da* (Figures 5.9.13 and 5.9.14). The peak in the concentration of smallest droplets and haze particles with radii below 2–3 μm is caused by several factors. Droplets penetrating the initially droplet-free volume decrease in size by partial evaporation. An increase in humidity in the initially droplet-free volumes from inversion leads to growth of haze particles within them. Both processes lead to formation of a large number of particles with radii below 2–3 μm. These particles penetrate back to the cloud parcel. A partial evaporation of droplets in the cloud parcel also contribute to the peak in the concentration of these smallest droplets.

P3 (bottom row, Figure 5.10.8) is initially droplet free. Due to mixing and evaporation of the entrained droplets, the relative humidity in this parcel increases from 50% to saturation. Penetration of droplets from the surrounding cloud parcels and their full or partial evaporation leads to random fluctuations of the effective radius at $t < 120$ min. When the parcel becomes comparatively close to saturation, relatively large droplets do not change their sizes forming DSD with the peak at the 10-μm radius. The resulting DSD has peak around 10 μm and a long tail of smaller droplets. During the period of 170–185 min, the droplet concentration in P3 is lower by an order of magnitude than that in typical cloud parcels. It means that at 185 min the equilibrium is not reached. At the same time, the effective radius is close to that in cloud parcels. The

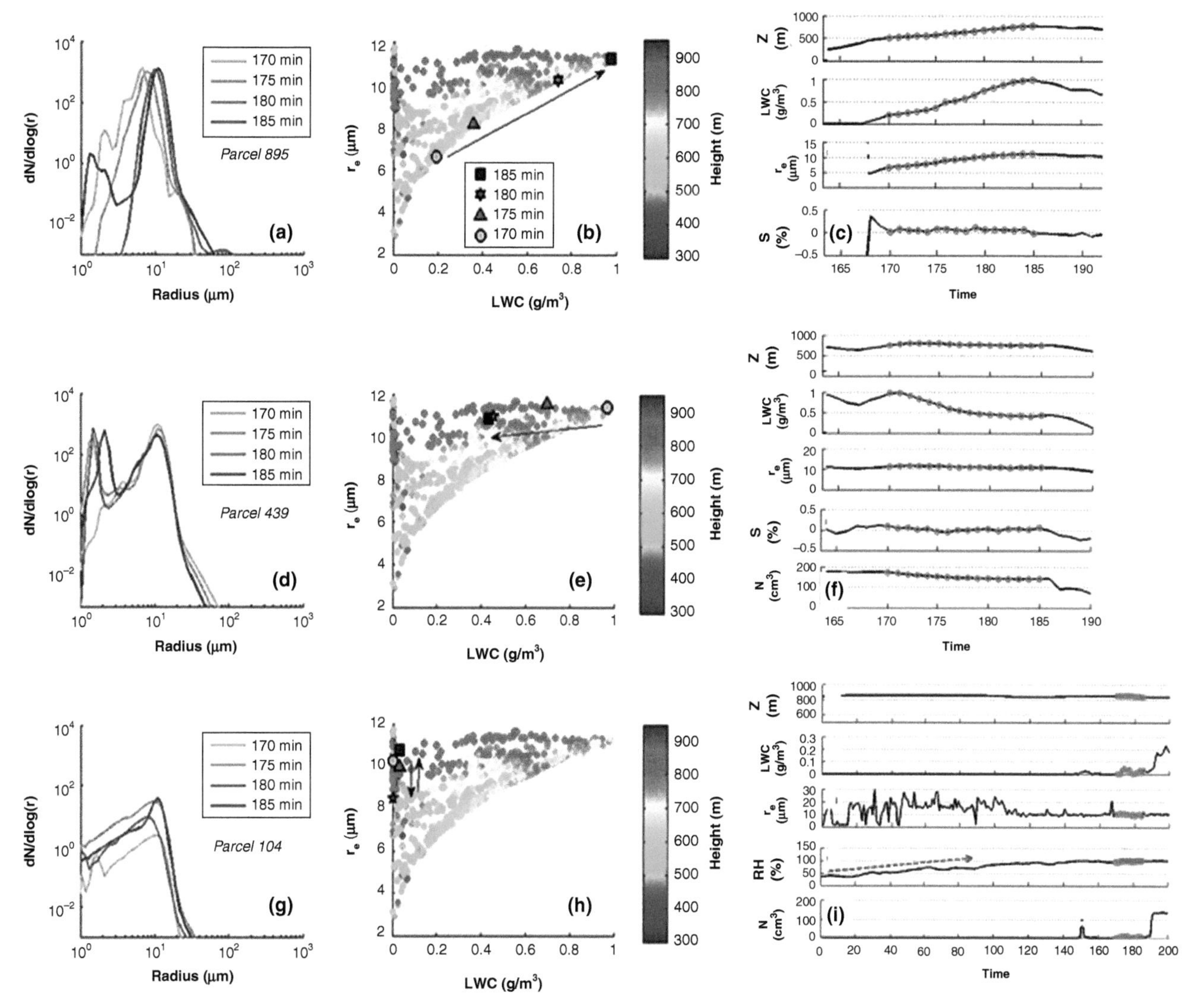

Figure 5.10.8 Time evolution of parameters in three selected parcels: P1 (upper row), P2 (middle row), and P3 (bottom row). Three columns show: DSD (left), LWC-r_{eff} scatter diagrams (middle), and time dependencies of elevation and different microphysical variables in parcels (right). In the scatter diagrams, each point corresponds to a single parcel in the computational area at t = 185 min. The color of dots represents the height of the parcels within the domain. Locations of the chosen parcels on the LWC-r_{eff} diagram are shown by different symbols for each 5-min interval. Right column: the history of the selected parcels is presented with 1-min time resolution. Dashed arrows show an increase in RH due to droplet evaporation (from Magaritz-Ronen et al., 2016b, with changes; courtesy of John Wiley & Sons, Inc.©).

DSD shape at 185 min resembles that in the initially dry volume, as was obtained using the diffusion-evaporation model at early stages of mixing (e.g., Figure 5.9.12d). As can be seen from Figure 5.10.8, the values of the microphysical parameters of P3 become closer to those typical for cloud parcels only at $t > 195$ min.

Figures 5.10.9a and **5.10.9b** present a scattering diagram of $\left(r_{eff}/r_{eff\_\max}\right)^3$ vs. $N/N_{\max}$ plotted for the cloud top layer using the results of the LEM simulation of Sc observed in research flight RF07. Each point in this diagram corresponds to a single parcel among parcels located within the layer 800–900 m, which includes cloud top. The points are plotted each minute during the 10-min time period (180–190 min). The values of droplet concentration are normalized by the maximum value of droplet concentration in the layer during this time period. Such normalization is similar to that often used to analyze in situ data along an aircraft traverse (Gerber et al., 2008). The values of the effective radius are normalized by the value of effective radius in a parcel with the droplet concentration maximum. The

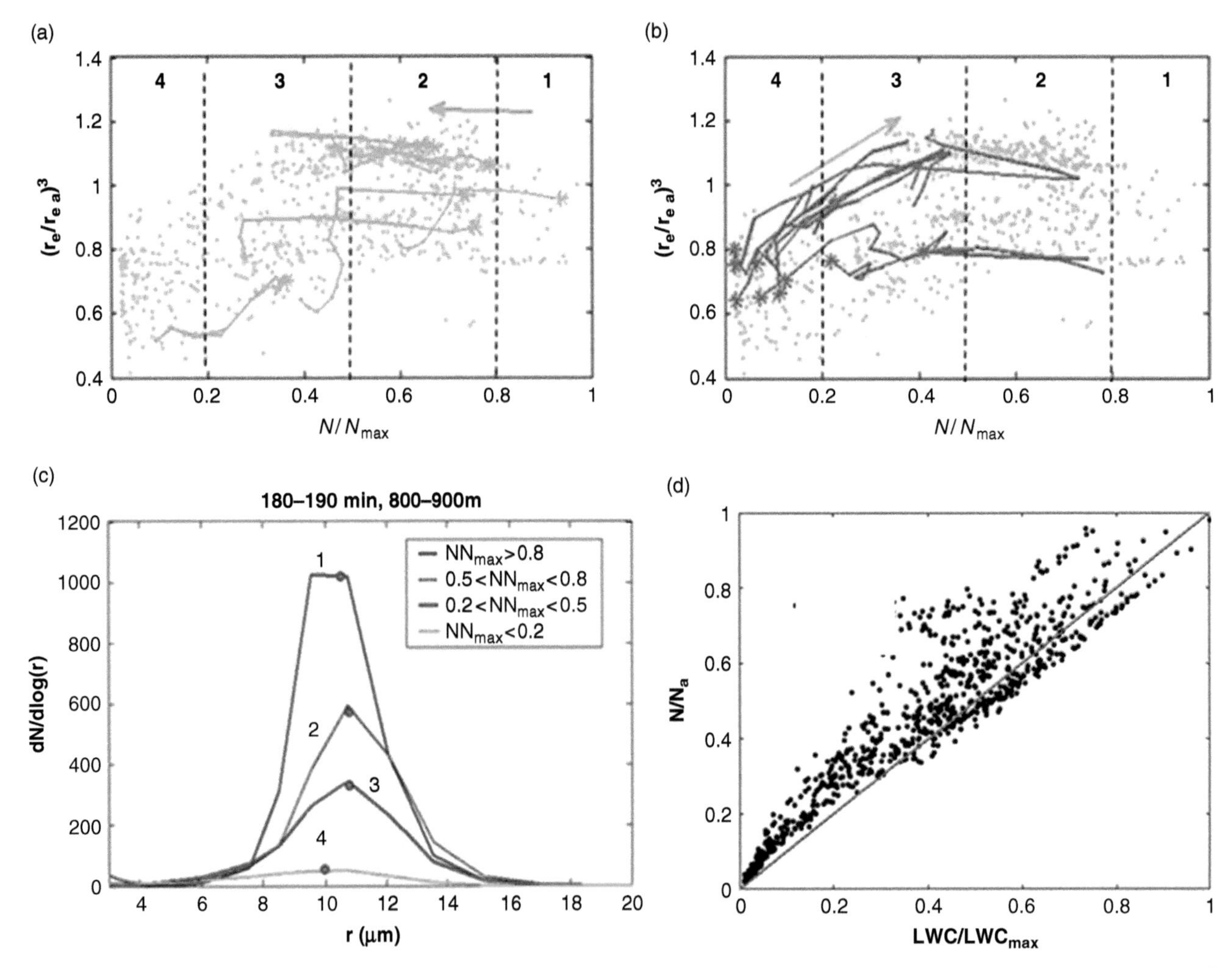

Figure 5.10.9 Upper row: scattering diagram showing the relationship between the cube of normalized effective radius $(r_{eff}/r_{eff\_max})^3$ and the normalized droplet concentration. Each point in the diagram marks a single parcel located between 800 m and 900 m at a certain time instance. Points are inserted into the diagram with time increment of 1 min during the period 180–190 min of simulation. All the points in the diagram are separated by vertical straight lines into four groups according to values of N/N_{max}. The 10-min trajectories are plotted of parcels having initially large N/N_{max} (panel a) and of parcels having initially low N/N_{max} (panel b). Asterisks show the beginning of the trajectories. Arrows show the direction of the parcel movement in course of mixing. Bottom row: (c) DSDs averaged within each group. (d) Normalized droplet concentration vs. normalized LWC scattering diagram plotted using the LEM.

scattering diagram resembles those plotted using in situ measurements in tops of Sc and small Cu. Vertical straight lines divide the total range of N/N_{max} on four subranges. The parcels of Type 1 (cloud parcel close to adiabatic) are located in subrange 1 in Figure 5.10.9a, while parcels of Type 3 (just penetrated cloud from inversion) are located within subrange 4. Parcels in the subranges 2–3 are at different stages of mixing.

Figure 5.10.9c shows DSDs averaged within each subrange of N/N_{max}. One can see that these DSDs are quite similar to those measured in in situ measurements (Figures 5.9.18–5.9.20). Indeed, there is a pronounced similarity of DSD, which indicates nearly linear correlation between the droplet concentration and LWC within a wide range of variation of these quantities.

Points on the scattering diagram do not correspond to final stationary equilibrium state (as assumed in mixing diagrams), but rather correspond to different transient stages of mixing. Accordingly, in the course of mixing the points denoting the parcels move inside the mixing diagram. The trajectories of several parcels in the mixing diagram during 10-min time period are shown in Figures 5.10.9a and 5.10.9b. The initial locations of parcels are shown by asterisks. One can see that most parcels having initially high N/N_{max} (cloudy parcels) move toward lower values of $N/N_{\max}$, while most parcels having initially low $N/N_{\max}$ (initially droplet-free parcels) move toward larger values of $N/N_{\max}$. This is a natural result of mixing between cloudy and initially droplet-free volumes. The effective radius of droplets in the parcels having initially high $N/N_{\max}$ typically does not change during mixing, so the trajectories of these parcels in the mixing diagram are largely horizontal (Figure 5.10.9a). At the same time, mixing of the parcels with initially low $N/N_{\max}$ leads to a sharp increase in the effective radius of droplets in these parcels, which is seen by the slope of their trajectories (Figure 5.10.9b). The values of r_{eff} rapidly become close to those in typical cloud parcels. This evolution of the effective radius in initially cloud parcels and initially droplet-free parcels during their mixing agrees well with the time evolution of the effective radius predicted by the diffusion-evaporation model (Figure 5.9.14) and in the LEM (Figure 5.10.8). A similar increase in r_{eff} with increase $N/N_{\max}$ in volumes with low LWC is seen in observations (Figure 5.9.20). At $N/N_{\max} > 0.3$, r_{eff} does not change by mixing. The values of r_{eff} are close to the values in cloud parcels. Although the values of r_{eff} can vary by 20–25% within the layer of 100-m depth, these variations are much lower than those of $N/N_{\max}$.

Simulations were performed for unchanged background conditions characterized by a certain (low) humidity of inversion air, given dissipation rate and averaged parcel size. Estimations show that such conditions correspond to *Da* from several tens to a few hundred, i.e., to inhomogeneous mixing scenarios. Figure 5.10.9d shows that $N - LWC$ scattering diagram has a form typical for an inhomogeneous mixing regime (see Figure 5.9.22).

The dispersion of the points in the scattering diagrams is due to different initial conditions and different mixing stages of different parcels. The amount of points with low ratios of $N/N_{\max}$ characterizes the rates at which air volumes penetrate the cloud and the rate of their transformation to cloudy volumes by mixing. Analysis of mixing diagrams in the LEM simulations shows that at larger distances from cloud top inside the cloud, the number of parcels with low $N/N_{\max}$ sharply decreases and may disappear completely.

Convective Clouds

The results presented next were obtained by 2D HUCM model and 3D System for Atmospheric Modeling (SAM) model with spectral bin microphysics (Khain et al., 2013). Deep convective clouds with characteristics similar to those observed in the Cloud Aerosol and Precipitation Enhancement Experiment (CAIPEEX-2009) were simulated. The model spatial resolution was 50 m. The results (DSD, the vertical profiles of the effective radius, and the level of rain drop formation) obtained by HUCM and SAM were found to be similar.

Figure 5.10.10 shows the statistics of CWC and the effective radius in a developing convective cloud simulated by SAM by $t = 31$ min when first raindrops arise (the radar reflectivity exceeds 0 dBZ). Histograms representing distributions of these parameters are calculated within a 200-m layer below the ascending cloud top during the cloud development. The histograms are presented as a function of the ascending cloud top height. Similar to the observations, the model-predicted CWC at any given level changes within a wide range from zero to the adiabatic values. At the moment that can be considered the time of the first raindrops formation, the maximum of CWC is close to the adiabatic value. The results of the modeling show that some cloud volumes near the cloud top are close to adiabatic volumes ascending from the lifting condensation level. Despite the high variability of LWC (Figure 5.10.10, left panel), the variability of the effective radius is small, being close to the adiabatic values, which is in agreement with observations. The low variability of the effective radius together with the high variability of LWC indicate that the model simulates inhomogeneous mixing. The closeness of the effective radius to the adiabatic values indicates the dominating role of basic large-scale processes of regular condensation in formation of the microstructure in convective clouds.

5.10.4 Modeling Formation of Superadiabatic Droplets

As was just discussed, mixing accompanied by total droplet evaporation, especially extremely inhomogeneous mixing, leads to a decrease in droplet concentration. In case a dry volume penetrates a cloud through its lateral boundaries, volumes with a decreased droplet concentration can ascend in cloud updrafts. Many researchers hypothesized that a decreased droplet concentration may lead to increasing supersaturation and to formation of large cloud droplets able to trigger intense collisions and raindrop formation. These droplets were called superadiabatic, i.e., their sizes are larger than the maximum size of droplets in parcels adiabatically ascending from cloud base. Since the process of raindrop formation is the key subject in cloud physics, the important role of extremely inhomogeneous mixing in precipitation formation was widely recognized and elaborated in research.

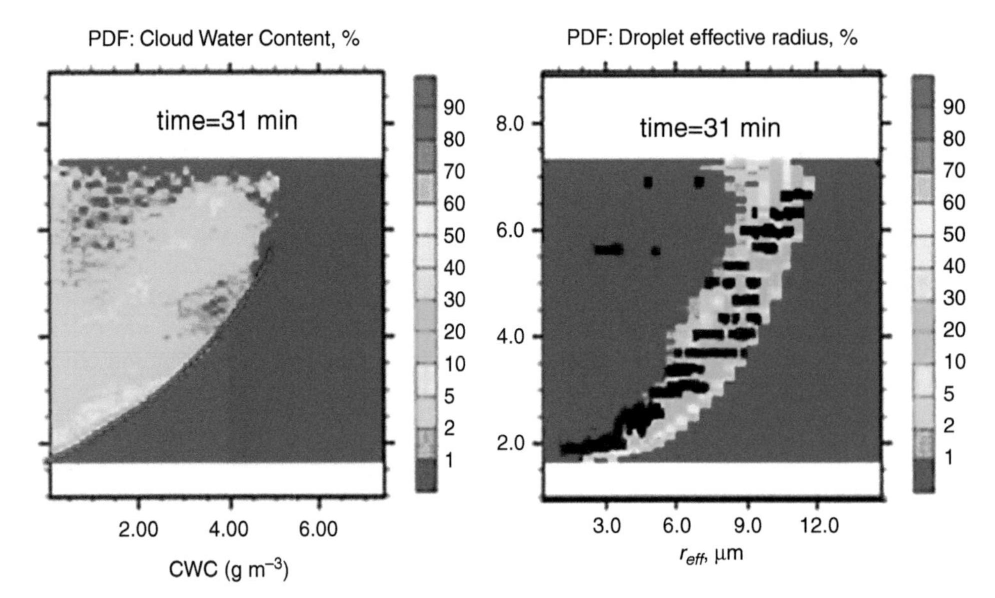

Figure 5.10.10 Statistics of CWC (left) and of the effective radius (right) near the top of a developing convective cloud. The time instance t = 31 min is the time of first raindrops arising (the radar reflectivity exceeds 0 dBZ). Red curve denotes the profile of the adiabatic LWC. The height is measured from the surface. Observations of the effective radius in deep polluted convective clouds are marked with black symbols. The points corresponding to low values of the effective radii at the height of about 6 km do not belong to the deep convective cloud under investigation (from Khain et al., 2013; courtesy of John Wiley & Sons, Inc.©).

The possibility of formation of superadiabatic droplets in the course of inhomogeneous mixing of an ascending volume with droplet-free surrounding was first studied by Baker et al. (1980) and Baker and Latham (1982). Baker et al. (1980) simulated DSD evolution in a cloud parcel ascending at a constant vertical velocity. In the course of the ascent, subsaturated air blobs of 80% relative humidity were entrained together with the environmental CCN into the cloud parcel. In accordance with the concept of extremely inhomogeneous mixing, some fraction of droplets of various sizes was removed from the DSD, while other droplets were left untouched. The evaporation of droplets within the parcel continued until the relative humidity within the entrained blob rose to 100%. Then the blob was mixed instantaneously with the surrounding cloud air. The CCN entrainment rate and CCN activation spectrum were chosen in such a way that all the entrained nuclei were activated during the subsequent ascent to keep the total droplet concentration unchanged. The DSD in the simulations of the parcel ascent accompanied by inhomogeneous mixing were compared with the DSD in the simulations assuming homogeneous mixing where all the droplets in the parcel experienced the same supersaturation changes. Collisions between droplets and droplet sedimentation were neglected.

Figure 5.10.11a shows DSD in a small cumulus cloud simulated by Baker et al. (1980) using an ascending parcel model (a) and DSD calculated by Baker and Latham (1982) using a diffusive mixing model at conditions observed in small cumulus clouds (b). The observed DSD measured by Warner (1969a) is shown, as well. One can see that the DSD in the simulation with extreme inhomogeneous mixing contains larger droplets than the homogeneous mixing simulation and is closer to the observed DSD.

Baker and Latham (1982) applied a diffusion mixing model resembling the theoretical model discussed in Section 5.9.3 to analyze turbulent mixing between a spherical dry volume surrounded by a spherical shell of a cloudy volume. The vertical velocity of the mixing volume was assumed either equal to zero or a constant value. Examples of DSD simulated in the mixing volume using the homogeneous mixing model and the diffusion mixing model are given in Figure 5.10.11b. Simulations with zero and nonzero vertical velocities of the mixing volume show that the results obtained by the diffusion mixing model lie between those obtained using the homogeneous mixing scenario and the extremely inhomogeneous mixing scenario, while being closer to the latter. Baker et al. (1980) and Baker and Latham (1982) found that cloud droplets did not reach sizes large enough to trigger collisions. Thus, despite the fact

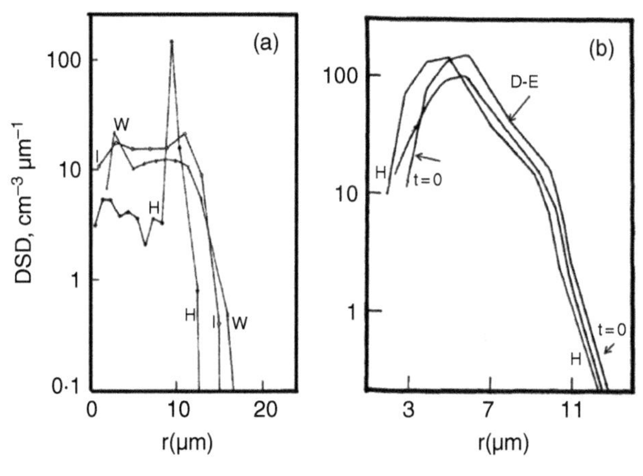

Figure 5.10.11 DSD in a small cumulus cloud simulated by Baker et al. (1980) using an ascending parcel model (a) and DSD calculated by Baker and Latham (1982) using a diffusive mixing model (b). (a) Notations: W: DSD measured by Warner (1969); H: DSD calculated for homogeneous mixing; and I: DSD calculated using the extremely inhomogeneous mixing algorithm. The parcel ascends at the vertical velocity of 1 ms^{-1}, droplet concentration is 200 cm^{-3}, and LWC = 0.4 g m^{-3}. (b) Notations: t = 0 indicates the initial DSD in the cloud volume; H: DSD calculated for homogeneous mixing; and D–E: DSD calculated at t = 72 s. The initial LWC in the cloud volume is equal to 0.5 g m^{-3}. The initial relative humidity in the initially dry volume is 70%. The initial diameters of the dry volume and of the cloud volume are 60 m and 144 m, respectively. The initial temperatures of the initially dry volume and of the cloud volume are 281 K and 279 K, respectively (from Baker et al. (1980), with changes, and Baker and Latham (1982), with changes; courtesy of John Wiley & Sons, Inc.©).

that the extremely inhomogeneous scheme yields droplets larger than in the homogeneous mixing scheme, the role of inhomogeneous mixing in formation of raindrops remains hypothetic.

Another model describing mixing on small spatial scales at a very high resolution was developed by Cooper et al. (2011) and actually combines two models. The first model is a bulk-parameterization atmospheric model (Straka and Anderson, 1993) modified by Carpenter et al. (1998). This basic cloud model calculates background temperature, humidity, and aerosol fields at the grid spacing of 25 m and does not include collisions. The second model is the Lagrangian parcel model with spectral bin microphysics (Cooper et al., 1997). This model was modified to take into account nucleation of the entrained CCN, as well as of CCN ascending from the cloud base. The model is used to simulate formation of DSD due to diffusional and coalescence growth of droplets, as well as due to mixing. The largest droplets exceeding 20 μm in radii are considered raindrop embryos. The modified Lagrangian parcel model was used to calculate the DSD evolution along multiple trajectories within the background velocity field and the thermodynamic field simulated by the basic model.

During their motions along different trajectories, the parcels develop different supersaturation patterns since the initial conditions, the vertical velocities, and the environment conditions are different. As a result, parcels meeting at the same point inside a cloud have different DSD. The DSD in each parcel is calculated taking into account the effect of entrainment of the background air. The study considered mixing as averaging of different DSD formed in different parcels at the same point of the background field, calculated by the basic model. The major premise assumed by Cooper et al. (2011) is that parcels moving along different trajectories and mixed with different environments have different DSD, so combining these DSD can produce a broad DSD-favoring coalescence. However, raindrop formation by the inhomogeneous mixing scenario was not simulated in this study.

Telford and Chai (1980) and Telford et al. (1984) proposed a theory known as the entity-type entrainment mixing theory (ETEM). According to the ETEM, large droplets able to trigger intense collisions form in the course of vertical oscillations of cloud volumes and mixing of these volumes with surrounding air. Cloud plumes rise to the convection level where their buoyancy drops to zero. Mixing of cloudy parcels with dry air leads to evaporation of droplets both inside the cloudy parcels and inside the initially dry air parcels. If evaporation of droplets cools a parcel sufficiently, it descends down and crosses the level where its density matches that of the surrounding air. The buoyancy force causes vertical oscillations of the air parcels. As a new ascent begins, the droplet concentration in the parcels decreases, leading to formation of larger droplets, which shifts the DSD to larger droplet sizes. According to ETEM, the vertical recycling of mixing parcels leads to a situation when different cloudy parcels have different droplet concentrations and ascend from different lifting condensation levels. Therefore, wide DSD and droplets of superadiabatic size form not at the initial stages of cloud evolution, but later on, as the oscillations and mixing with the oscillating parcels proceed. Telford et al. (1984) stressed that recycling of mixing parcels in the vertical direction leads to DSD broadening. Due to the lack of sufficient observational and numerical results related to oscillations of parcels within deep convective clouds, the ETEM should be tested using both *in situ* and radar measurements and high-resolution cloud models allowing representation of vertical parcel oscillations and mixing.

5.10.5 Modeling of Turbulent Mixing Down to Millimeter Scales

Attempts to explicitly describe mixing and entrainment within a wide range of scales down to the Kolmogorov microscale were performed by Krueger et al. (1997), Su et al. (1998), and Schlüter (2006) using the explicit mixing parcel model (EMPM). EMPM links the conventional 1D parcel model with another 1D fine-resolution model describing mixing between a parcel and the entrained blobs. The internal structure of the parcel evolves as a consequence of discrete entrainment events and turbulent mixing, explicitly described following the approach developed by Kerstein (1988) and Krueger (1993). According to this approach, mixing is simulated on a 1D horizontal computational domain of length D that mimics the cross section of a cloud. At each entrainment event, environmental air volume with a linear scale d is inserted into segment D randomly, i.e., at any place within the cloud at the entrainment level. A schematic diagram of the entrainment and random rearrangement events by subgrid-scale diffusion is shown in **Figure 5.10.12**.

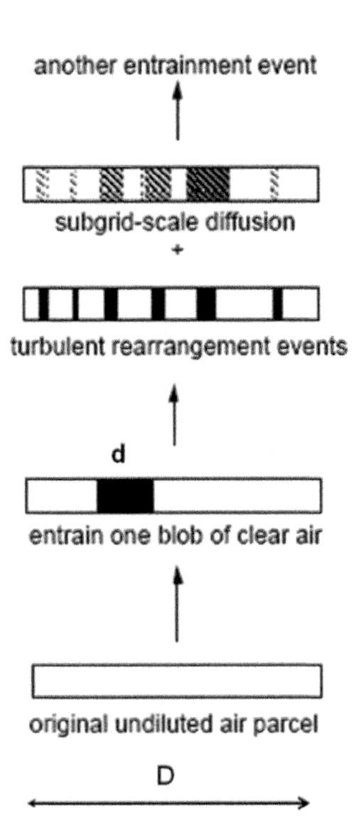

Figure 5.10.12 Schematic diagram of entrainment events and random rearrangement events by subgrid-scale diffusion in EMPM (from Krueger et al., 1997; American Meteorological Society©; used with permission).

Su et al. (1998) performed modeling in which the length of segment D varied within the range of 500 m to 2,000 m. The corresponding sizes d of the entrained blobs were chosen from 50 m to 200 m to obtain the initial entrainment fraction, determined as $f = d/D = 0.1$. Mixing of an entrained blob with cloud surrounding was simulated by spreading this blob along the horizontal direction calculated at high resolution to include all physically relevant length scales down to 1 mm. The molecular diffusion was implemented explicitly. After mixing, individual droplets were adjusted to their local environments created by EMPM.

EMPM was used to predict the bulk properties of the main turrets of the Hawaiian cumulus cloud observed by Raga et al. (1990). In the simulation, undiluted blobs were entrained near the cloud base and ascended while mixing with the background air. Krueger et al. (1997) found that utilization of 100-m entrained blobs provided the best agreement with the observations. Schlüter et al. (2006) analyzed the sensitivity of DSD to the relative humidity of the entrained segment, the turbulent dissipation rate, and the size of the entrained blobs. The same data from the Hawaiian observations were used. The entrainment level was 750 m above the cloud base, the vertical velocity was 2 m/s. The domain size was 20 m (x) × 1 mm (y) × 1 mm (z) with 12,000 grid cells; 2,055 droplets initially existed in the domain and were assigned random locations. After an adiabatic ascent, a single entrainment event occurred. After entrainment, the parcel was held at the same pressure level, and mixing occurred until homogenization. Nucleation of new CCN was not taken into account. It was shown that reducing the moisture in the entrained air or increasing the entrained blob size requires more evaporation and complete evaporation of some droplets to regain saturation and broadens the spectrum toward smaller droplet sizes. An increase in the entrained fraction (i.e., an increase in the entrained blob size) led to a decrease of the maximum droplet size by several μm. A decrease in the humidity of the entrained blobs led to a further decrease of droplet size and of LWC (**Figure 5.10.13**). Multiple entrainment events led to a monotonic decrease of the effective radius. Instant mixing assumes inhomogeneities are gone instantly. The outcoming DSD for that is shown in Figure 5.10.13 (red).

The DSD shape in the EMPM simulations crucially depends on model parameters such as the size and humidity of entrained blobs, the frequency of the

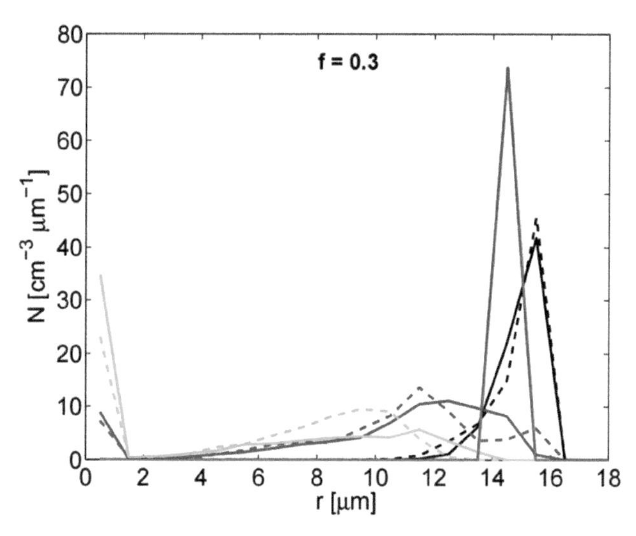

Figure 5.10.13 Droplet spectra for different RH values: 87.4% (black), 43.7% (blue), and 21.9% (green); the entrained fraction $f = 0.3$. Solid and dashed lines show the results of two different realizations of entrainment. The droplet spectra obtained as a result of instant mixing is marked red (from Schlüter, 2006, with changes).

entrainment events, the heights at which the entrainment takes place, etc. The EMPM simulations showed that the mean volume radius decreased with decreasing droplet concentration, and no super-adiabatic droplets were reported to form.

5.10.6 Comments

Section 5.10 presents a review of model studies of entrainment and mixing that demonstrates the important impact of mixing on the dynamical and microphysical structure of clouds. At the same time, despite multiple research efforts and significant advancements, many important questions concerning DSD modification in the course of mixing still remain unclear. This includes the very basic questions, namely: Where does entrainment take place in clouds and how intensive is it? What are the volumes of the entrained blobs and SD of these volumes? Entrainment is not only a turbulent process, it is also related to regular convection in clouds on scales from several hundred meters to several kilometers, when the air is entrained near cloud base and detrained at different levels, largely, in anvils. Hence, more detailed high-frequency measurements and further theoretical efforts, as well as advanced cloud models, are required to quantitatively evaluate the impact of entrainment and mixing on DSD and its evolution.

5.11 Formation of DSD and Raindrops via Warm Processes

In this section we present an overview of different processes discussed in previous sections of this chapter to give a more general picture of warm processes leading to formation of observed DSD and raindrops. The comparative contribution of different microphysical processes leading to rain formation will be illustrated using a model of an ascending parcel, which is simple in dynamics but rich in microphysics as well as other more dynamically complicated models. We explain the mechanism of formation of multimodal DSD often observed in clouds at all height levels. Also, we analyze the mechanism of formation of superadiabatic drops giving rise to raindrop formation. Finally, we discuss the role of atmospheric aerosols in formation of DSD and raindrops.

A practical important problem of remote sensing is determination of the parameters that can indicate the onset of the raindrop formation in convective clouds and of drizzle in stratocumulus clouds. It was found in observations (e.g., Rosenfeld and Gutman, 1994; van Zanten et al., 2005) and numerical studies (e.g., Magaritz et al., 2009; Khain et al., 2013; Magaritz-Ronen et al., 2016a and 2016b) that the magnitude of the effective radius that can be determined by satellites can be used for this purpose. In this section, the physics behind the threshold value of the effective radius is discussed. We also present results concerning the effects of turbulence on the first raindrop formation, obtained in simulations using the HUCM. Detailed descriptions of existing cloud models and results obtained using these models is presented in Chapter 7.

5.11.1 Spectral Bin Microphysics Parcel Model and the Design of Simulations

The equation system of the parcel model includes: the diffusion growth equation applied both for aerosol particles and for water drops (5.1.12), the equations for supersaturation (5.2.1), for the vertical velocity (5.10.1), and for the potential temperature of an ascending parcel (5.10.2). The process of collisional drop growth is described by the stochastic collection equation for drop collisions (5.6.5). The acceleration of a cloud parcel is determined by the sum of the buoyancy force and the friction force. The friction force is assumed to be proportional to the square of the vertical velocity. The proportionality parameter (the entrainment parameter) λ is parameterized as $\lambda = 0.2/R$, where R is the cross-section radius of the parcel (Section 5.10). The term with λ in Equation

(5.10.2) is omitted. No fresh CCN penetration into the cloud parcel from the surroundings is allowed. In this sense, this parcel is an adiabatic one, whose vertical velocity, however, is controlled by both the buoyancy and the friction through parameter λ.

The main feature of the model is the precise description of diffusional growth of aerosols and of drops, as well as of drop collisions. Several specific features of the model enable us to perform high-precision calculations of drop spectrum formation.

a) 2,000 mass bins are used to describe DSD of particles (nonactivated aerosols and drops) within the range of 0.05 μm to 2,000 μm. The grid resolution is 0.001 μm on the left side and gradually decreases down to 8 μm on the right side of the grid. This resolution is high enough to explicitly describe the process of separation of all particles into growing drops and nonactivated haze particles. Droplet nucleation is treated directly without using any parameterization.

b) To describe diffusion growth, a non-regular grid with a variable set of masses is used. The masses related to the corresponding bins are shifted over time according to the equation of diffusion growth. Correspondingly, no remapping is applied, i.e., no artificial spectrum broadening is introduced when drop growth by diffusion is calculated. The time step of 0.001s is used to calculate diffusion growth of drops and aerosol particles.

c) The SCE is solved using the method proposed by Bott (1998) (Section 5.7). Drop growth by collisions is simulated using a time increment of 0.5 s. The minimum size of collected drops was set at 1 μm. Utilization of the time step of 0.5 s ensures the condition $f^{-1}(m_i)\frac{df(m_i)}{dt}\Delta t \ll 1$ (the change of mass in a particular bin during one time step is much less than the drop mass in the bin).

d) The 1-μm-resolution table of drop–drop collision efficiencies calculated by Pinsky et al. (2001) is used (see tables in Appendix B). Such a high resolution is especially important for collisions between small cloud droplets, as well as between small cloud droplets and raindrops. In these cases, collision efficiency crucially changes with the size of collecting droplets. The values of collision efficiency were tested against the laboratory data obtained in a wind-tunnel by Vohl et al. (1999).

e) The increase of collision efficiency with height, found by Pinsky et al. (2001) was taken into account. For this purpose, the collision efficiency tables calculated for the pressure levels of 1,000 mb, 750 mb, and 500 mb were interpolated at the corresponding level of the ascending cloud parcel.

f) The model developed by Pinsky and Khain (2002) was, to the best of our knowledge, the first one to investigate the sensitivity of DSD evolution to turbulence-induced collision-rate enhancement. The turbulence-induced collision enhancement factors were chosen based on the evaluations made by Pinsky et al. (1999b, 2000) for turbulence intensities of 100–200 $cm^2\ s^{-3}$ that are typical of developing convective clouds. The collision kernels were increased by the enhancement factors presented in **Table 5.11.1**. The increase by 10% chosen for other drop pairs including larger drops is mainly due to turbulent effects on the swept volume of drops moving within a turbulent flow. Comparison of the factors presented in Table 5.11.1 and the factors discussed in Section 5.8, obtained in later studies, showed a qualitative agreement. In the model developed by Pinsky et al. (2002), the collision enhancement factor does not depend on time and remains unchanged during the parcel's ascent.

The features of the spectral bin microphysics model described allow us to consider the model as a benchmark model for simulating both diffusion growth and collision growth of drops in an ascending air parcel. Simulations of microphysical structure in ascending parcel were made under conditions typical of the Eastern Mediterranean during the rainy season when tops of cumulus clouds reach 5–7 km height. The DSD evolution by diffusion drop growth is affected by three factors: (a) the vertical velocity at the cloud base; (b) the buoyancy within the cloud layer determining acceleration of the parcel above the cloud base; and (c) the

Table 5.11.1 Turbulence-induced collision rate enhancement factor (from Pinsky and Khain, 2002; courtesy of John Wiley & Sons, Inc.©).

Range of Droplet radii in droplet Pairs	Factor
Smaller than 10 μm	5
10–20 μm	2
20–30 μm	1.5
30–40 μm	1.2
For pairs containing small raindrops ($r > 50$ μm) and the smallest cloud droplets ($r < 3$ μm)	3.0
Other drop pairs	1.1

aerosol particle distribution. Two series of simulations were performed. In the first series, all the parameters remained unvaried, with the exception of the vertical velocity at the cloud base that varied from 5 cm/s to 3 m/s. In the second series the concentration of aerosol particles was varied, other parameters being unchanged. To investigate the possible effects of turbulence, all the experiments in both series were conducted in two versions: the turbulent version, taking the effects of turbulence on drop collision; and the non-turbulent version, where the collision kernels were calculated for non-turbulent conditions.

As was discussed in Section 2.2, concentration of aerosols over continents is typically by order of magnitude higher than that over the oceans According to the CCN concentration, all clouds were separated into three groups: (a) more than 800 cm^{-3} (dirty); (b) within the range of 250 cm^{-3} to 800 cm^{-3} (intermediate); and (c) below 250 cm^{-3} (clean air). The SD of CCN is calculated as the sum of three log-normal modes (Section 2.2), representing small, medium, and large aerosol particles. An example of aerosol particle SD used in the simulations is shown in **Figure 5.11.1**.

As shown in Sections 5.2 and 5.3, supersaturation and, consequently, droplet concentration at cloud base increase with increasing vertical velocity. Therefore, dirty and intermediate clouds were simulated just by varying the vertical velocity at cloud base. In clean clouds, the initial CCN concentration was assumed to be 25% of that shown Figure 5.11.1. No giant or ultra-giant CCN values were allowed.

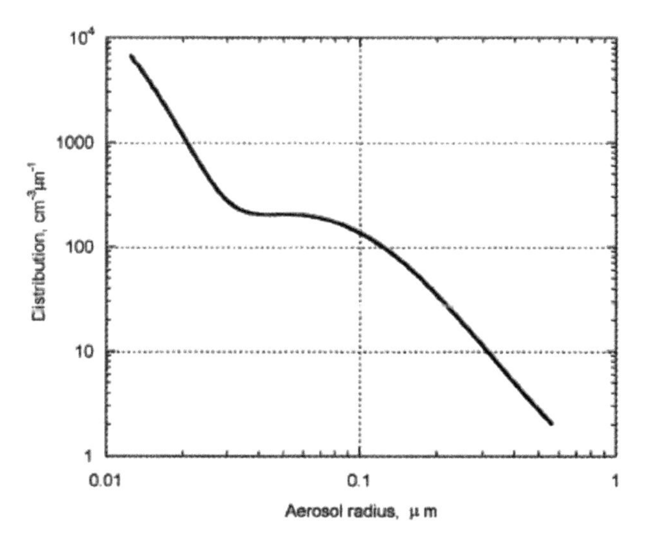

Figure 5.11.1 Aerosol SD used in simulations of dirty and intermediate clouds by means of the parcel model (from Pinsky and Khain, 2002; courtesy of John Wiley & Sons, Inc.©).

5.11.2 In-Cloud Nucleation and Formation of Bimodal DSD

Mechanism of In-Cloud Nucleation

Each DSD measured in situ contains small droplets with radii below ~5 µm at any height level, both near cloud edges and in cloud cores (Andreae et al., 2004; Prabha et al., 2011; Gerber et al., 2008). Sometimes, these small droplets form the second and third DSD modes (**Figures 2.3.6–2.3.10**). The constant droplet concentration observed in small Cu in RICO in the presence of cloudy air dilution is often attributed to the nucleation of small droplets at high levels (Gerber et al., 2008). The most plausible mechanism causing the formation of smallest droplets in DSD is in-cloud nucleation of small CCN. The mechanism of in-cloud nucleation of CCN ascending from cloud base is illustrated by the scheme in **Figure 5.11.2**. At the cloud base, the SD of nonactivated CCN is wide and may contain particles with dry radii ranging from ~0.003 µm to ~2 µm (see Section 2.2). The supersaturation maximum near cloud base usually ranges 0.2–0.8%. To activate the smallest CCN, supersaturation of several percent is required. Therefore, only a fraction of CCN is activated at cloud base, while other CCN rise in the updrafts.

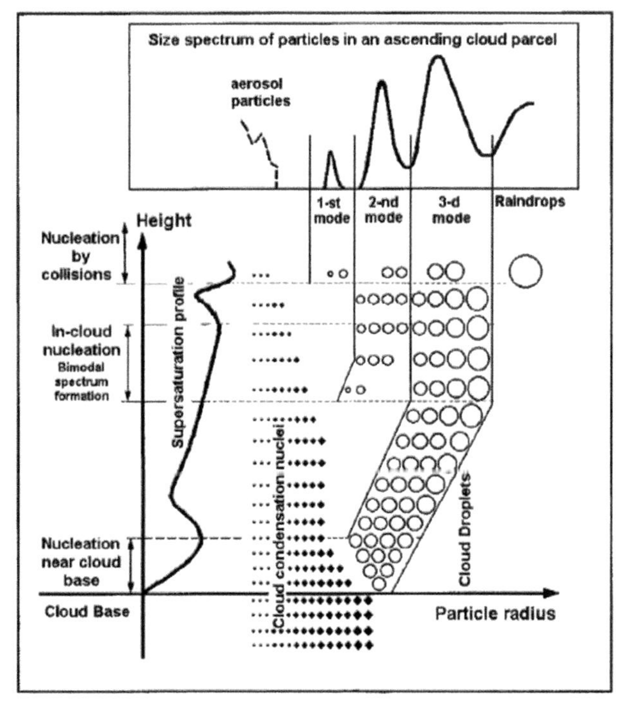

Figure 5.11.2 The conceptual scheme of droplet spectra formation in deep convective clouds (from Pinsky and Khain, 2002; courtesy of John Wiley & Sons, Inc.©).

Droplets nucleated near cloud base trigger the emergence of the first DSD mode, which plays a dominant role in raindrop formation. In case the updraft acceleration is comparatively strong, supersaturation begins growing with increasing height. If supersaturation exceeds the maximum values that were reached during the previous ascent, a new CCN activation (i.e., in-cloud nucleation of droplets) takes place, triggering formation of the second DSD mode. Typically, the second mode forms at heights from a few hundred meters to a few km above cloud base.

In zones of efficient inter-drop collisions, drop concentration decreases and the vertical velocity increases, partially due to the unloading of raindrops and partially due to latent heat release by freezing. As a result, supersaturation may substantially increase up to 10% and even higher, and a new portion of the smallest CCN is activated producing small droplets just above the zone of efficient drop collisions. The scheme in Figure 5.11.2 suggests that an increase in the vertical velocity at cloud base increases the supersaturation maximum near cloud base and decreases the probability of in-cloud nucleation. At the same time, thermal instability of the atmosphere leads to acceleration of updraft above cloud base and increases the probability of in-cloud nucleation. Numerical simulations support these conclusions.

Effects of Cloud-Base Vertical Velocity

Figure 5.11.3 shows the vertical profiles of updraft speed, supersaturation, and the drop concentration for cloud parcels with different values of updraft speed w_0 at the cloud base. As one can see, if a realistic increase in the vertical velocity over height above the cloud base is taken into account, supersaturation in the cloud updraft can exceed the local maximum at the cloud base. It is clear that within the layer where supersaturation increases over height while remaining higher than the local maximum at the cloud base (shaded zone in Figure 5.11.3b), in-cloud nucleation takes place, continuously producing small droplets. These small droplets form the second DSD mode, which leads to formation of a bimodal DSD, corresponding with the DSD broadening. An increase in supersaturation at 3,500 m above the cloud base, seen in Figure 5.11.3b, is caused by a decrease in drop concentration by collisions. This increase can lead to the activation of smallest CCN.

The DSD formed in a cloud parcel moving at low and high speeds at the cloud base are shown in **Figures 5.11.4a** and **5.11.4b**, respectively. One can see that in case of low speed, in-cloud nucleation leads to formation of the bimodal spectra. Droplets nucleated near the cloud base form the right-side maximum in the spectra. Newly nucleated droplets form the left-side maximum. Since the radii of small droplets grow faster than radii of larger droplets, the gap between the modes decreases with height (Figure 5.11.4a) and can disappear, leading to unimodal DSD of a significant width. This process explains the shapes of DSD often measured in natural clouds (e.g., Figures 2.3.10 and 2.3.11).

In high updrafts at the cloud base, the nucleation of all droplets occurs in the vicinity of the cloud base with

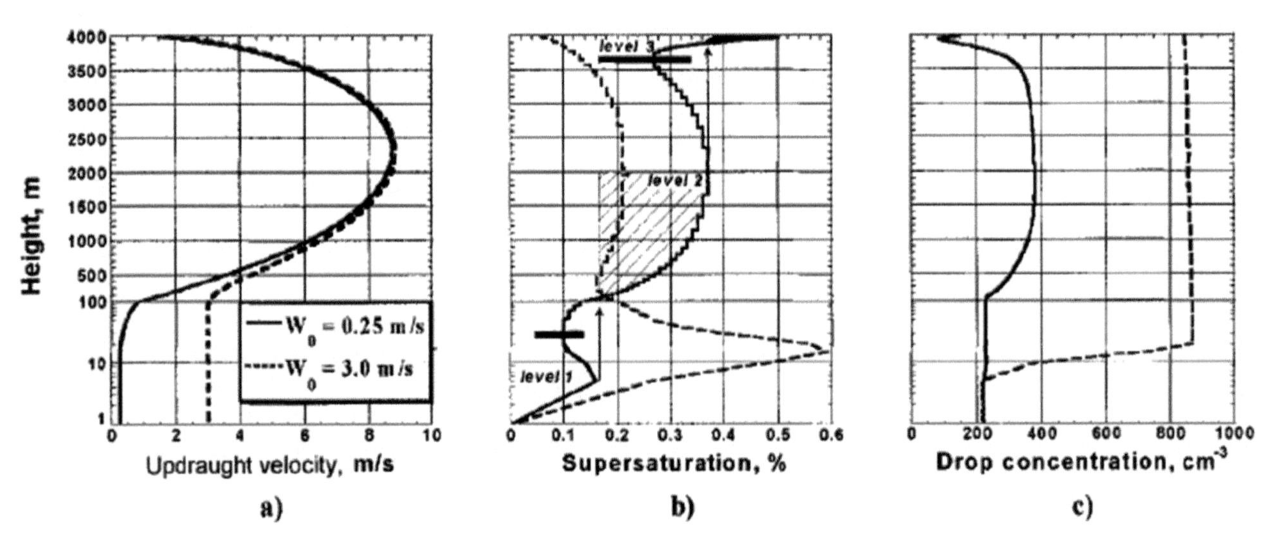

Figure 5.11.3 The vertical profiles of updraft velocity (a), supersaturation (b), and droplet concentration (c). Solid lines correspond to $w_0 = 0.25\,\mathrm{ms}^{-1}$, dashed lines correspond to $w_0 = 3\,\mathrm{ms}^{-1}$. In the shaded zone, supersaturation increases with height, remaining higher than at cloud base (from Pinsky and Khain, 2002; courtesy of John Wiley & Sons, Inc.©).

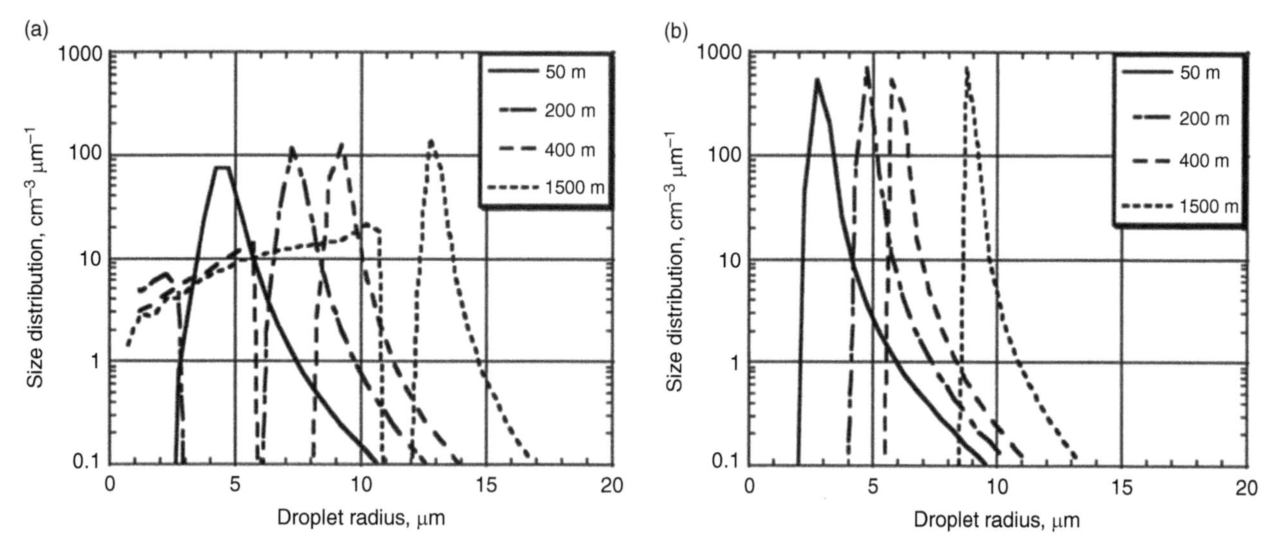

Figure 5.11.4 (a) DSDs formed in a cloud parcel moving at a low vertical velocity (0.25 m/s) at the cloud base. (b) The same as in (a), but for high vertical velocity (3 m/s) at the cloud base (from Khain et al., 2000; courtesy of Elsevier).

no new nucleation of ascending CCN (Figure 5.11.4b). Droplet spectrum in this case is narrow and becomes narrower with increasing height. The droplet concentration and the DSD shape depend, therefore, on the vertical velocity w_0 at the cloud base and on the updraft acceleration. Figures 5.11.3 and 5.11.4 show a significant effect of the vertical velocity at cloud base on further cloud evolution and, in particular, on in-cloud nucleation and formation of bimodal DSD. High vertical velocity at cloud base may prevent in-cloud nucleation aloft. Thus, extending the theory of DSD formation in ascending adiabatic parcels to the case of accelerating updrafts and drop concentration decrease by collisions allows us to explain formation of both wide and narrow local DSD.

As shown in Section 2.3, the significant width of the observed DSD may be a result of spatial averaging of local DSDs along the airplane traverse. The difference between DSD in the parcels may be caused by the inhomogeneity of the vertical velocity at cloud base. According to the measurements performed by Fankhauser et al. (1983), the RMS velocity variation at the base of cumulus clouds is close to the mean velocity value at this level, which indicates the exponential distribution of the vertical velocity. This result agrees well with data reported by Warner (1970) (**Figure 5.11.5**).

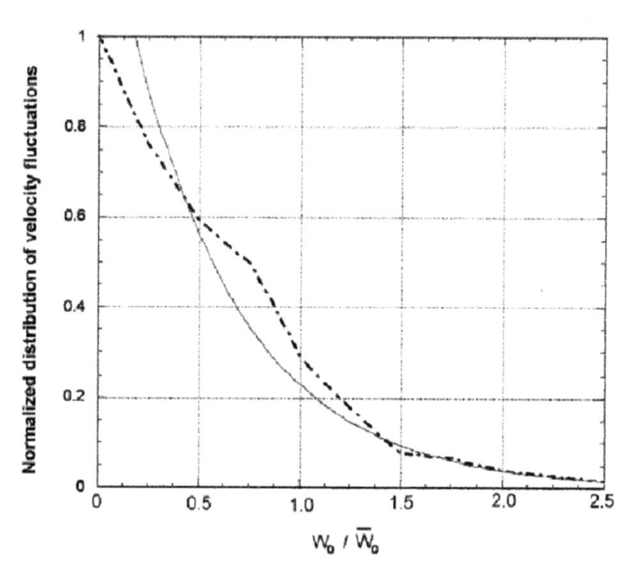

Figure 5.11.5 A normalized distribution of vertical velocity fluctuations of a parcel in the 140 m above the cloud base, calculated using *in situ* data. The approximating exponential distribution (solid line) is also presented (from Pinsky and Khain., 2002; courtesy of John Wiley & Sons, Inc.©).

Figure 5.11.6 shows the vertical profile of the dispersion of DSD averaged over about 500 cloud parcels ascending from cloud base at different velocities distributed according to the exponential law. In-cloud nucleation leads to formation of averaged-size spectra with dispersion of 0.15–0.25, regularly observed in cumulus clouds (Figure 2.3.5). Thus, the observed values of DSD width and dispersion can be reproduced in models taking into account two mechanisms: in-cloud nucleation determining local DSD broadening and the variability of the cloud-base velocity.

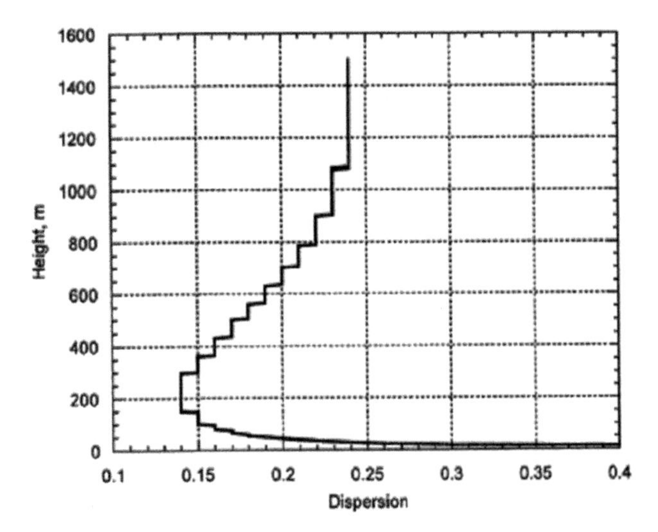

Figure 5.11.6 The vertical profile of the drop size spectrum dispersion averaged over about 500 cloud parcels arising from cloud base at different velocities (Khain et al., 2000; courtesy of Elsevier).

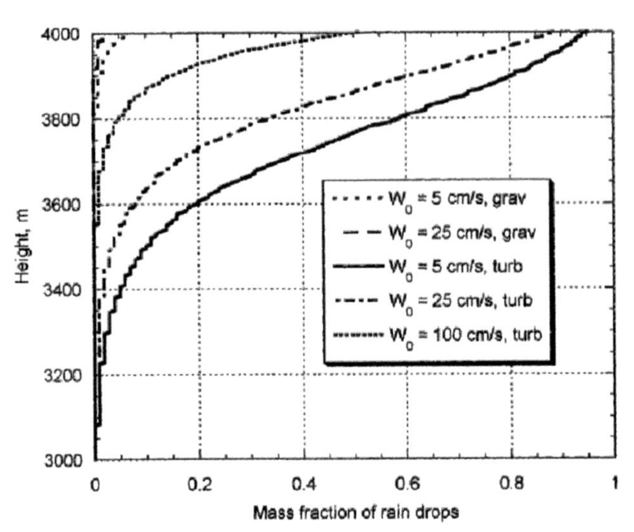

Figure 5.11.7 The vertical profiles of the mass fraction of raindrops with radii exceeding 50 μm, within the total LWC in non-turbulent (grav) and turbulent (turb) simulations (from Pinsky and Khain, 2002; courtesy of John Wiley & Sons, Inc.©).

Inhomogeneity of the vertical velocity at cloud base leads to variability of drop concentration and DSD in different parcels. This inhomogeneity is also a plausible explanation of the existence of small droplets at all height levels, as different vertical velocities determine different levels of in-cloud nucleation.

CCN can penetrate clouds not only through cloud base, but also via lateral boundaries. The vertical velocity of air volumes penetrating just above cloud base is comparatively low and increases with increasing height. The effect of lateral entrainment of CCN near cloud base is similar to that of air volumes ascending through cloud base at a low vertical velocity.

Another mechanism leading to a significant DSD width is turbulent mixing, both homogeneous and inhomogeneous (Section 5.9).

Effects of Aerosols and Turbulence

Figure 5.11.7 shows the mass fraction of raindrops in the total LWC calculated in non-turbulent and turbulent simulations in polluted air. The SD of CCN shown in Figure 5.11.1 and the turbulent enhancement factors shown in Table 5.11.1 were used in these simulations. Intensive raindrop formation takes place in the turbulent simulation at $w_0 \leq 1\,\mathrm{ms}^{-1}$. At the same time, raindrop formation was negligible toward this time in all the non-turbulent simulations. One can see that turbulence-induced increase in the collision kernel lowers by a few hundred meters from the height at which the first raindrops form.

Figure 5.2.15 in Section 5.2 is closely related to the topic of the discussion, showing the vertical profiles of the updraft velocity, supersaturation, and drop concentration in cloud parcel ascending in clean air in turbulent and non-turbulent simulations. In the turbulent run, collisions are more intense, so drop concentration decreases much faster than in the non-turbulent simulation, leading to a sharp increase in supersaturation up to 1% at 3,700 m (Figure 5.2.15b) and to additional in-cloud nucleation (Figure 5.2.15c). As a result, the third DSD mode emerges. Nucleation of new smallest droplets caused by the increase in supersaturation is especially pronounced in maritime clouds developing in clean air. In the absence of turbulence, supersaturation attains magnitudes high enough to trigger in-cloud nucleation at levels ~300 m higher than in the turbulent case.

Figure 5.11.8 shows the evolution of DSDs with height in the non-turbulent and the turbulent simulations. One can see bimodal spectra formation by in-cloud nucleation at the stage of diffusion growth (DSD at 3,300 m) and formation of the third DSD mode near the cloud top. Turbulence accelerates raindrop formation, especially at high CCN concentrations.

Effects of the Cloud Layer Instability

As was mentioned, acceleration of cloud updrafts fosters formation of high supersaturation and is

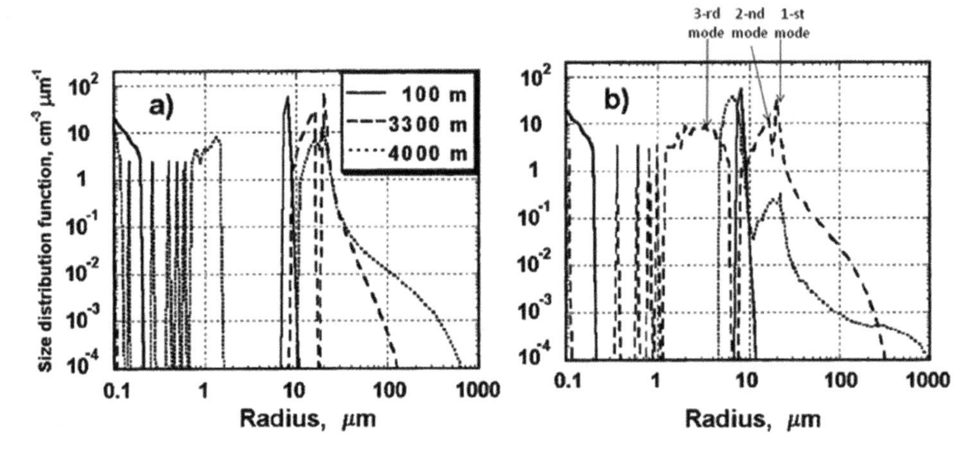

Figure 5.11.8 Drop SDs at three different heights in a maritime cloud parcel in non-turbulent (a) and turbulent (b) simulations. The maxima of the three DSD modes are shown by arrows in panel (b) for 3,300 m height (from Pinsky and Khain, 2002; courtesy of John Wiley & Sons, Inc.©).

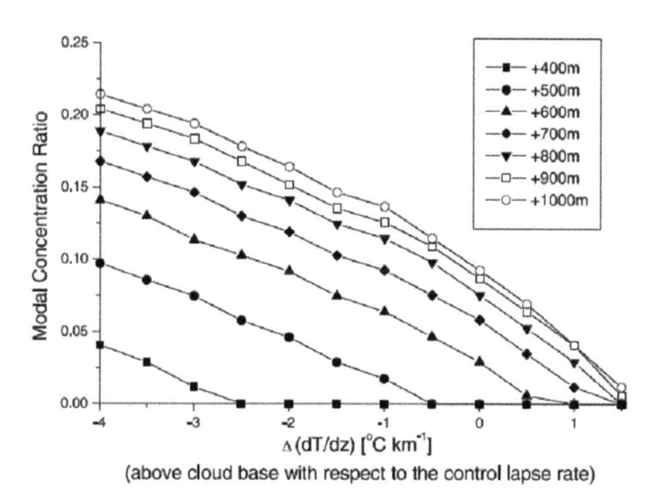

Figure 5.11.9 Modal concentration ratio as a function of the temperature gradient within the cloud layer. The temperature gradient is plotted in terms of its deviation from the moist adiabate. The control lapse rate corresponds to most adiabate. Symbols denote heights above the cloud base (from Segal et al., 2003; courtesy of Elsevier).

favorable for in-cloud nucleation. Segal et al. (2003) investigated the role of atmospheric humidity and the temperature gradient above cloud base in in-cloud nucleation. It was found that an increase in the relative air humidity at the surface from 70 to 90% can decrease the height above cloud base where in-cloud nucleation takes place (from a few km to 250 m). The nature of this effect is lowering the cloud base and causing a corresponding decrease in the vertical velocity at the cloud base level. Results of simulations where the temperature gradient above cloud base varied with respect to the moist adiabatic gradient within the range from −4 K/km (the unstable case) to +1.5 K/km (the stable case) are shown in **Figure 5.11.9**. To characterize the differences in DSD, the modal concentration ratio parameter was defined as the ratio of droplet amount in the second mode (generated by in-cloud nucleation of CCN ascending from the cloud base) to the droplet amount in the first mode. If no bimodal spectrum is generated, the modal concentration ratio is equal to zero. Figure 5.11.9 shows the modal concentration ratio values at different heights above cloud base. One can see that this ratio increases with height and with an increasing instability of the cloud layer. These results agree well with the observed data reported by Warner (1969a).

In-Cloud Nucleation of CCN Penetrating Through Lateral Cloud Boundaries

In simulations just discussed, the existence of small droplets at high levels was attributed to activation of CCN through cloud base. As was shown in several studies (Roesner et al., 1990; Phillips et al., 2005; Fridlind et al., 2004; Yin et al., 2005), penetration of CCN at higher levels during the lateral entrainment can also lead to formation of small droplets. If the size of CCN penetrating clouds at higher levels is large enough, nucleation of these CCN takes place even when the vertical velocities in clouds are relatively small (Phillips et al., 2005). The effect of CCN penetrating clouds at the upper levels on the microphysical cloud structure and DSD was investigated in several studies by means of spectral bin microphysics cloud models (e.g., Khain and Pokrovsky, 2004; Fan et al., 2010). In these studies, the dominating role of CCN in the boundary layer was found. Prabha et al. (2011) found a high correlation between the amount of small droplets in deep convective clouds at distances of ~5 km above cloud base and the concentration of CCN in the boundary layer, which also proves that these CCN are a dominating factor that determines drop concentration and DSD.

5.11.3 The Role of Drop Collisions at the Cloud Development Stage

Drop collisions is the main mechanism of formation of large cloud droplets and raindrops. The collision growth is described by the SCE (Equation 5.6.5); the first term is the Gain integral and the second term is the Loss integral. To analyze the raindrop formation process, Pinsky and Khain (2002) calculated the Loss and Gain difference for each mass bin. If $G-L > 0$, the drop mass contained in the corresponding bin increases, while decreasing at $G–L \leq 0$. In **Figure 5.11.10(a–d)** (upper rows) the difference between Loss and Gain is presented at different height levels as a function of drop radius in a cloud simulated using the parcel model discussed. One can see two negative peaks ($G–L < 0$). A comparison with the drop MD (bottom rows) shows that these peaks coincide with the maxima of the two

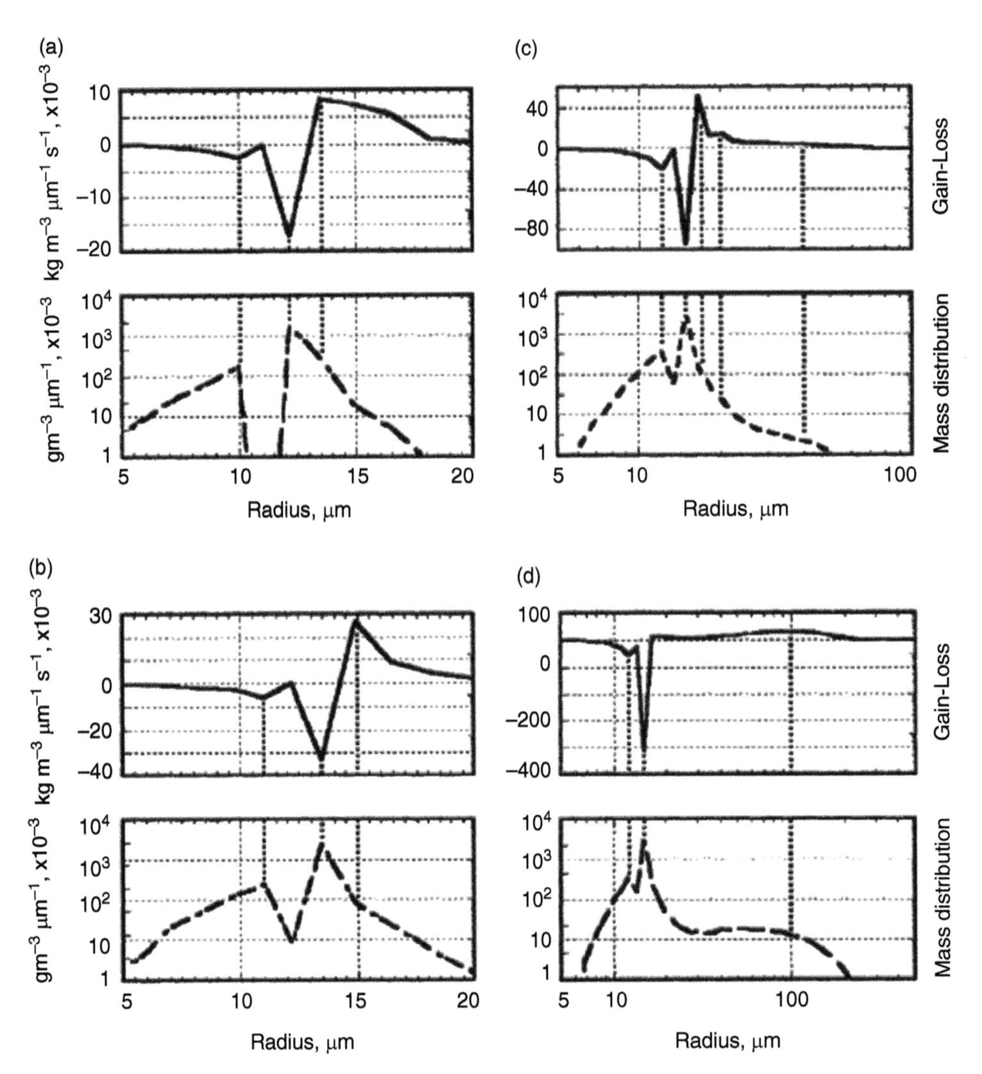

Figure 5.11.10 Loss–Gain (L–G) differences (upper frames) and drop MD (bottom panels) as functions of drop radius for a cloud parcel with $w_0 = 0.25$ ms^{-1} at the cloud base: (a) 1,500 m; (b) 2,000 m; (c) 3,000 m; and (d) 3,500 m above the cloud base. Dotted lines show the positions corresponding to the extreme Loss–Gain values (from Pinsky and Khain, 2002; courtesy of John Wiley & Sons, Inc.©).

MD modes. It indicates that the drop mass rapidly decreases at these radii due to collisions of the small droplets with larger droplets.

Below 3,000 m, there is one positive maximum of G–L > 0, moving from 13 μm at 1,500 m to 16.4 μm at 3,000 m. A comparison with the MD shows that at all the levels this maximum is located at the right slope of the first distribution mode. Thus, the mass of the largest cloud droplets increases at the maximum rate. The Loss and Gain values rapidly increase for large droplets, so at 3,000 m two new positive Gain–Loss peaks arise at drop radii of about 20 μm and 40 μm (Figure 5.11.10c). The first maximum is determined by collisions between cloud droplets. The second Gain–Loss maximum is situated at the right foot of the raindrop mode, rapidly propagating to the right and reaching 100 μm at 3,500 m (Figure 5.11.10d). This maximum forms due to collisions of the largest droplets with all other droplets, especially with those belonging to the first DSD mode.

The drop radius corresponding to the maximum Loss–Gain value can be regarded as the driving radius determining the raindrop formation rate. The emergence of the driving radius and its increase with time are well seen in Figure 5.8.23, where the rates of the evolution of the MD function are shown for a purely gravitational case and a turbulent case. Using the dependence of Loss–Gain on drop radius, one can separate the whole range of drop radii into five zones (**Figure 5.11.11**).

- Analysis of Figure 5.11.11 shows that Zones I and V do not participate in drop collisions.
- Zone II is the source of mass for larger drops (L–G < 0). It is related to the second droplet MD mode (left), caused in our case by in-cloud nucleation. The secondary mode can also be caused by the entrainment-mixing mechanism.
- Zone III is the source of mass for larger drops (L–G < 0). It is related to the primary mode formed by droplets nucleated near the cloud base.
- Zone IV is mass receiver (L–G > 0). The maximum value of Loss Gain profile is denoted by the solid line within Zone IV.

One can see that below 3,000 m the driving radius is located at the right foot of Zone III. Drops of this radius are formed as a result of collisions between cloud droplets. This process of droplet–droplet collisions is known as selfcollection (Section 5.7). At 3,000 m above the cloud base, a bifurcation of the line takes place. The first peak of Loss–Gain attains droplet radius of 18 μm (left), being induced by collisions of cloud droplets. The

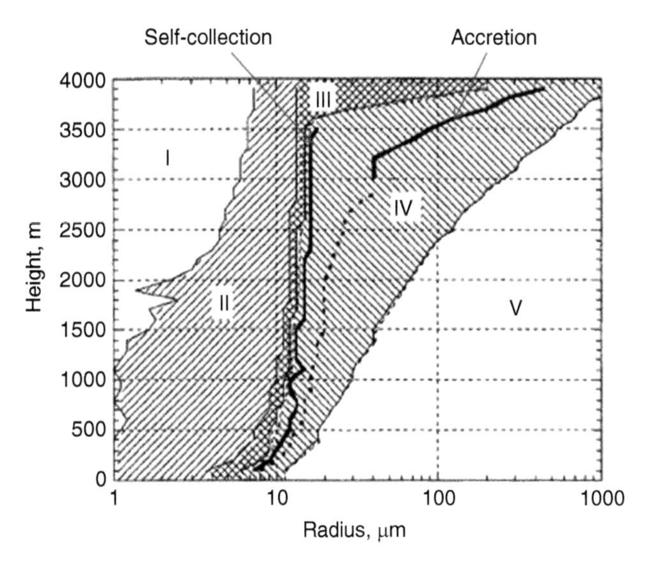

Figure 5.11.11 Zones of different drop mass budgets. Zones II and III have a negative mass budget (L–G < 0). Zone II is related to the secondary DSD mode (left), while Zone III is related to the primary mode formed by droplets nucleated near the cloud base. Zone IV is the area with a positive mass budget (L–G > 0). The maximum value of (L–G) profile is denoted by the solid line within Zone IV. The dotted line corresponds to 5% of the maximum value of (L–G) (from Pinsky and Khain, 2002; courtesy of John Wiley & Sons, Inc.©).

Loss–Gain value corresponding to the peak rapidly decreases with height above 3,000 m due to a decrease in drop concentration. The second peak, rising at 3,000 m at the drop radius of 40 μm (Figures 5.11.10c and 5.11.11) is caused by collisions of the largest cloud droplets with smaller cloud droplets and indicates raindrop formation. The emergence of the drops with driving radius of 40 μm leads to triggering rapid collisions. As a result, the driving radius rapidly increases with height (Figures 5.11.10d and 5.11.11). This sharp acceleration in the collision rate after arriving at the driving radius is caused by a crucial increase of the collision rate between newly formed drops and cloud droplets: a 40 μm radius drop collector collects droplets of the primary mode 100 times more efficiently than a collector of 20 μm radius.

The data presented in Figure 5.11.11 show, therefore, that the process of drop collisions can be separated into three stages: at the first stage, a significant amount of large droplets with radii around 20 μm is produced; at the second stage, collisions of these droplets with smaller ones produce drops with the driving radius of about 40 μm; and at the third stage rapid raindrop formation takes place. These results are in agreement with those obtained by Johnson (1993), who showed

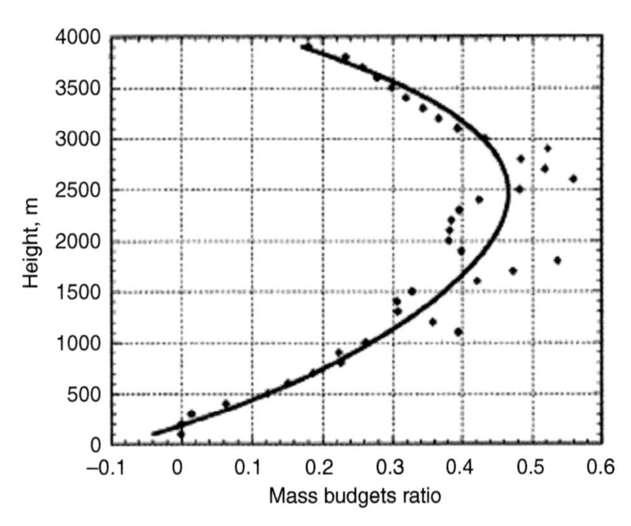

Figure 5.11.12 Ratio of mass budgets (Loss–Gain) integrated over Zones II and III. Points denote results of calculations. The solid curve approximating the results is a fifth-order polynomial curve (from Pinsky and Khain, 2002; courtesy of John Wiley & Sons, Inc.©).

that effective coalescence growth always requires drop embryos larger than about 35 μm in radius.

To evaluate the contribution of DSD modes to formation of the largest droplets and raindrops, Pinsky and Khain (2002) compared the Loss–Gain difference integrated over Zone II an over Zone III. The vertical profile of the ratio of these integrals is presented in **Figure 5.11.12**. One can see that within the layer of 2,000–3,000 m above the cloud base this ratio attains values as high as 0.5. It means that the total contribution of droplets belonging to the secondary DSD maximum (induced by in-cloud nucleation) to the generation of the largest droplets can be as high as 30%. This increase may assume equivalent to a 30% increase in the collision kernel in Zone III, which is known as a highly significant factor in rain formation. This result is closely related to the increase in the collision rate between cloud droplets, caused by the turbulence-inertia mechanism. According to Table 5.11.1, this increase is especially significant for drop pairs containing small droplets of 3–10 μm radius. If the turbulence effect is not taken into account, small droplets cannot be efficiently collected by larger ones.

Comparative Contribution of Diffusion and Collisions to Drop Growth

Cloud evolution is usually separated into two stages: diffusional drop growth and collision drop growth. The second stage is assumed to begin with triggering of intense collisions, i.e., when drops of the driving radius form. Collision growth starts at a certain height above cloud base, while it is believed that below this height DSD forms due to diffusional drop growth. In order to clarify the comparative role of droplet–droplet collisions and diffusional growth, Pinsky and Khain (2002) calculated separately the contributions of diffusion and collisions to the increase in drop size. As a parameter characterizing the evolution of the drop spectrum, the relative rate of the effective drop radius was used, namely: $\frac{1}{r_{eff}}\frac{dr_{eff}}{dt}$.

Figure 5.11.13 shows the changes of r_{eff} growth rates with height (or with time), calculated separately for diffusion and collisions and for three types of clouds: dirty, intermediate, and clean-air clouds. To analyze the effect of turbulence on drop growth, the dependencies are plotted for turbulent and non-turbulent simulations. The relative rate of r_{eff} growth varies within a wide range of several orders of magnitude. Within the lower part of the cloud, the contribution of diffusional growth is many times higher than that of collision growth. While the relative rate of r_{eff} growth caused by diffusion decreases with height, the growth rate caused by collisions increases. Within the zone of intense collisions the role of diffusional growth becomes negligible. The comparison of Figures 5.11.11 and 5.11.13b (data obtained under the same experiment conditions) shows that formation of the second Loss–Gain maximum and triggering rapid collision take place at the height where the contribution of collisions to drop growth starts to exceed the contribution of diffusional growth. Efficient collisions start in less than one minute after the drop growth by collisions becomes dominating. Turbulence increases the rates of collisional growth of r_{eff} by factor a of 20–30 in the dirty clouds, by factor of 5–20 in the intermediate-type cloud, and by a factor of 5 in the clean-air-type cloud. Consequently, turbulence leads to a remarkable acceleration of raindrop formation, especially in dirty clouds. In maritime clouds, precipitation is formed even when turbulence effects on the collision rate are not taken into account.

Figure 5.11.11 shows that at the height of 500 m, collisions already generate droplets with radii of 20 μm that cannot be produced by diffusional growth. Thus, pure diffusional growth takes place only within the lowest 500 m above the cloud base. Toward the beginning of intense collisions (3,000 m above the cloud base), the drop spectrum contains drops as large as 300 μm in radius. This indicates the key role of collisions in large droplet formation. It is noteworthy, however, that Loss–Gain values rapidly decrease to the right from the line of the Loss–Gain maximum

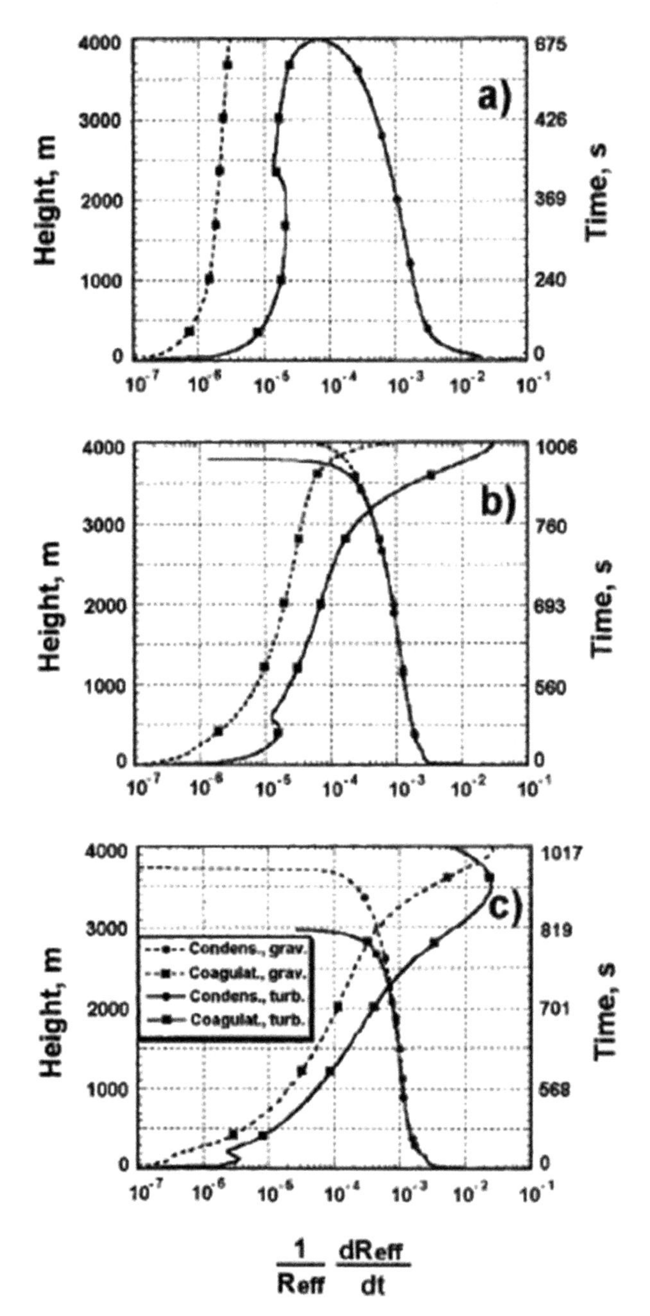

Figure 5.11.13 The rate of the droplet effective radius r_{eff} increase with height (or with time) in cases of (a) dirty, (b) intermediate, and (c) clean-air-type clouds in turb and grav experiments. The rates of increase due to diffusion (condens) and collision (coagulat) are shown separately (from Pinsky and Khain, 2002; courtesy of John Wiley & Sons, Inc.©).

(Figure 5.11.11). The curve marking the Loss–Gain value equal to 5% of the Loss–Gain maximum at the corresponding height level is plotted by the dotted line. The contribution of collisions can be assumed negligible

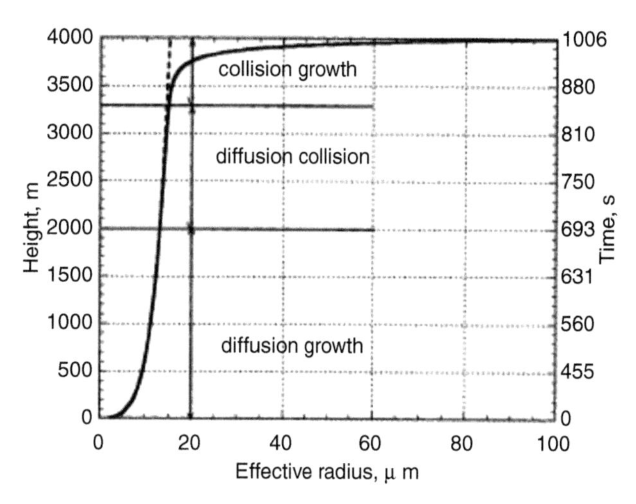

Figure 5.11.14 Time dependencies of the droplet effective radius r_{eff} for the intermediate type cloud in turb (solid line) and grav (dashed line) experiments (from Pinsky and Khain, 2002; courtesy of John Wiley & Sons, Inc.©).

when Loss–Gain is less than 5% of the Loss–Gain maximum, i.e., to the right from this curve. At the same time, droplet collisions in the vicinity of this curve contribute mainly to formation of large drops of the driving radius at 3,000 m (Figures 5.11.10c and 5.11.11) and to triggering intense collisions. It means that the number of largest drops (collectors) should be high enough to produce drops of the driving radius.

As mentioned, diffusion growth cannot produce droplets with sizes exceeding about 20 μm in radii. The 5%-level curve in Figure 5.11.11 indicates that droplets of 20 μm radius arise at about 2,000 m above the cloud base where the contribution of diffusion growth to the total drop growth remains larger by a factor of ten than that of collision growth (Figure 11.13b). Therefore, the stage of cloud evolution preceding intense collisions can be divided into two substages: at the first substage, the contribution of diffusion growth exceeds that of collision growth by more than ten times, hence the droplet growth can be considered to be of pure diffusion nature. At the second substage, as the cloud becomes deeper, collisions produce larger droplets exceeding 20 μm in radius (super-adiabatic droplets). The formation of these droplets is the necessary condition for formation of the drops with driving radius of about 40 μm, leading to triggering intense collisions. This substage can be called the diffusion–collision substage of cloud development when the contribution of collisions is very important for drop spectrum evolution. The separation of the substages of cloud evolution with respect to height/time in case of the intermediate-type cloud is shown in **Figure 5.11.14**.

The turbulence effects significantly decrease the depth of the layer where pure diffusion drop growth takes place, and shorten the duration of the substage where collisions are dominating. Time dependencies of the effective radius r_{eff} in cases of continental-, intermediate-, and maritime-type clouds indicate the following:

a) In the dirty non-precipitating cloud, r_{eff} does not exceed 12 μm, which agrees well with the observations (Rosenfeld and Gutman, 1994; Freud et al., 2008);
b) In the intermediate and clean-air clouds, a rapid increase of r_{eff}, indicating rapid rain formation, starts when r_{eff} attains about 15 μm. The height where r_{eff} exceeds 15 μm is located at about 200 m above the layer where the collision growth rate starts exceeding the diffusion growth rate;
c) The turbulence effects accelerating drop collisions do not significantly change the critical value of r_{eff} corresponding to the commencement of fast rain formation. Turbulence decreases the height and shortens the time period needed for attaining this critical value.

5.11.4 Effects of Turbulence on First Raindrop Formation

Effects of turbulence on collisions and on formation of first raindrops were investigated using HUCM with grid spacing of 50 m × 50 m (Benmoshe et al., 2012). As described in Section 5.8, the gravitational collision kernel in SCE is multiplied by the collision enhancement factor representing the main factors determining turbulence impact on drop collisions (Equation 5.8.8). The collision-enhancement factor at each grid point and at each time step was calculated using the lookup tables (Appendix B). The collision enhancement factor is determined by two parameters: turbulent dissipation rate ε and the Taylor microscale Reynolds number Re_λ. Dissipation rate ε was determined using the values of turbulent kinetic energy E calculated using Equations (3.5.24–3.5.30). $Re_\lambda = \frac{u_{rms}\lambda}{\nu}$ is determined as described in Section 3.3. The RMS velocity fluctuations u_{rms} in a certain direction is determined as

$$u_{rms} = \sqrt{\frac{2}{3}E_{tot}}, \quad (5.11.1)$$

where the total turbulent kinetic energy E_{tot} represents the energy of all the vortices with scales equal to or smaller than the external turbulence scale L. This enegy is calculated as

$$E_{tot} = (\varepsilon L)^{2/3} \quad (5.11.2)$$

The conceptual framework allowing us to determine the external turbulent scale L was presented by Grabowski and Clark (1993), who investigated formation of large turbulent vortices at the upper edge of an ascending thermal. They found a new type of instability called cloud-environment interface instability (Section 5.9). Their concept is based on the balance of factors affecting the gradients of the thermodynamic fields. While a continuous rise of cloud air tends to sharpen these gradients, the turbulent vortices formed as the result of the flow instability tend to decrease them. Assuming that there is a balance between these two tendencies, Grabowski and Clark (1993) evaluated $L \sim R/10$, where R is the characteristic cloud size. A similar ratio between the scale of the convective flow and the turbulent scales is presented by Elperin et al. (2002b), who found the effect of self-organization in a turbulent flow that leads to a scale separation between convective structures and turbulence. The scale of convective structures was evaluated to be ~15 times larger than the maximum scale of turbulent vortices. Therefore, the external turbulent scale was determined by Benmoshe et al. (2012) as $L = L_{cl}/15$, where L_{cl} is the linear cloud size calculated as $L_{cl} = S_{cl}^{1/2}$, and S_{cl} is the cloud area where the total hydrometeor mass content exceeds a certain low threshold value. The value of L increases over time as a cloud develops. When a cloud reaches 10 km height, L reaches several hundred meters. The values of Re_λ obtained using this approach are of the order of $\sim 10^4/4 \times 10^4$, which is typical of the atmosphere. Benmoshe et al. (2012) simulated deep convective clouds under conditions observed during the LBA-SMOCC field experiment. According to the classification used by Andreae et al. (2004), clouds in the region of the experiment were separated into blue-ocean clouds (BO) observed in the vicinity of the coastal line, green-ocean clouds (GO) located over jungles far from the forest fire regions, and extremely polluted smoky clouds (S) formed in the vicinity of the forest fire zones. The CCN concentrations (at 1% supersaturation) were $N_0 = 400$ cm^{-3}, 3,600 cm^{-3}, and 10,000 cm^{-3} for BO clouds, GO clouds and S clouds, respectively (Rissler et al., 2004, 2006; Freud et al., 2008). The slope parameter of the activity spectrum (Equation 5.1.21) k = 0.72 is chosen according to measurements made by Rissler et al. (2006). Aerosol particles gave rise to formation of clouds with the maximum drop concentrations of 500 cm^{-3} in the BO clouds, 700–1,000 cm^{-3} in the GO clouds, and ~2,300 cm^{-3} in the S clouds, being in agreement with in situ measurements. All of the clouds were simulated under the same thermodynamic conditions.

Fields of dissipation rate and the Reynolds number Re_λ in the simulations of a deep convective cloud developed under the thermodynamic conditions observed during the LBA-SMOCC field experiment in the Amazon region are presented in Figure 3.3.16 (Section 3.3). The zones of enhanced turbulence are elongated and cover a relatively low fraction of the cloud volume. This structure reflects the large-scale turbulent intermittency typically observed within clouds (e.g., Mazin et al., 1989). **Figure 5.11.15** shows the fields of CWC, RWC, dissipation rate, and the mean volume radius near the top of a developing convective cloud. The zone of cloud top contains three main turrets (bubbles). Turret A ascends first, followed by turrets B and C. The bubbles have scales of several hundred meters. The maximum intensity of turbulence is reached near cloud top and cloud edges. Such turbulent structure is well known from observations (e.g., Shmeter, 1987; Siebert et al., 2006). The calculations show that the values of the buoyancy term in the equation for TKE (Equation 3.5.24) are typically several times lower than those of the shear production term. It means that TKE in the simulated clouds is generated largely due to transition of energy from convective scales to turbulent scales caused by the wind shears. The buoyancy term that contributes to the inhomogeneity and anisotropy of turbulence plays a comparatively small role. The analysis shows that locations of the buoyancy term maxima and the shear production term do not coincide. The values of the shear production term reach their maximum at bubble edges, mostly bubble tops. The buoyancy term reaches its maximum inside the bubbles.

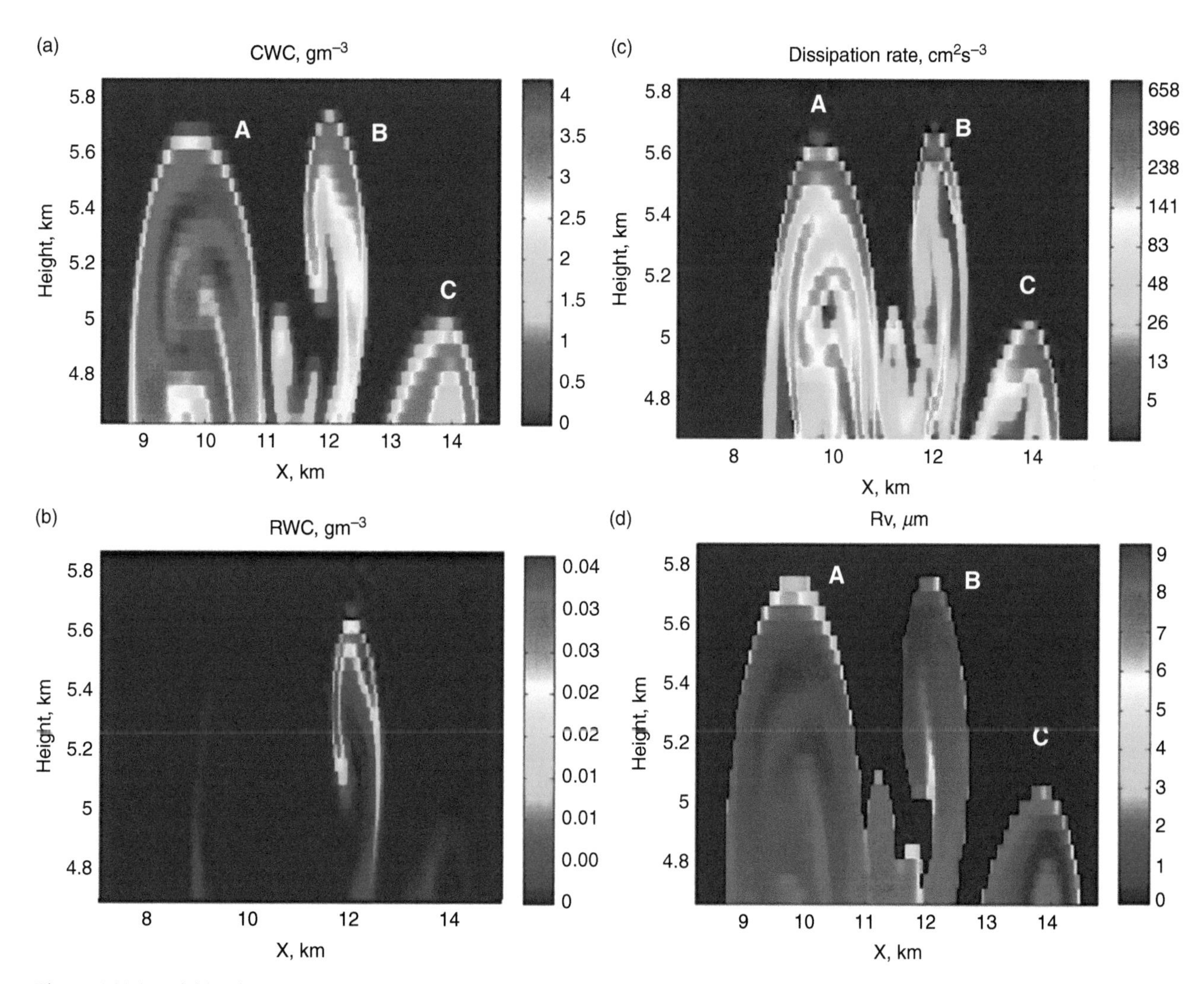

Figure 5.11.15 Fields of CWC, RWC, dissipation rate, and the mean volume radius near the top of a developing convective cloud simulated using HUCM, t = 66 min (from Khain et al., 2013; courtesy of John Wiley & Sons, Inc.©).

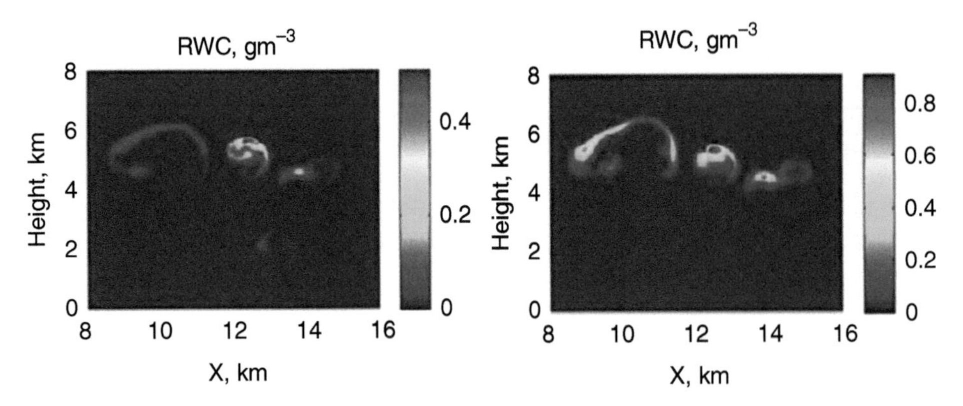

Figure 5.11.16 The RWC fields at t = 68 and 69 min (from Khain et al., 2013; courtesy of John Wiley & Sons, Inc.©).

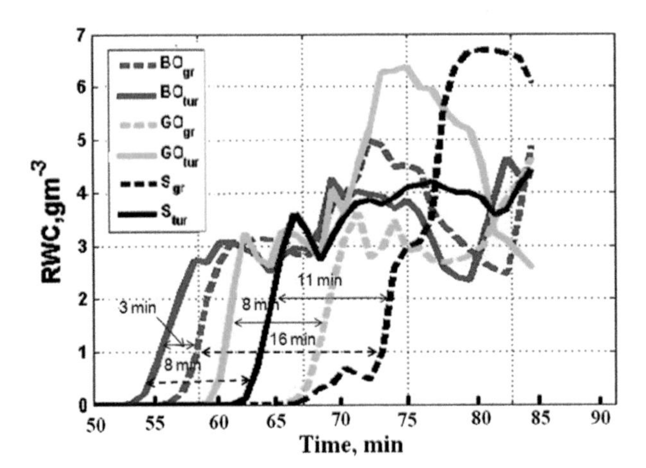

Figure 5.11.17 Time dependences of the maximum raindrop content in different experiments (from Benmoshe et al., 2012; courtesy of John Wiley & Sons, Inc.©).

When the bubbles reach a height of about 5.2 km, the mean volume radius near the upper cloud boundary reaches ~13 μm and effective radius reaches ~14 μm. The first raindrops form in the cores of these turrets where LWC is high (up to ~4 g m^{-3}) and turbulence intensity is maximum. Upon being formed near cloud top, first raindrops spread further along the edges of the bubbles where downdrafts take place (**Figure 5.11.16**).

Figure 5.11.17 shows time dependences of the maximum values of RWC in clouds simulated under conditions observed during the LBA-SMOCC field experiment. As can be expected, in BO the raindrops form earlier and in S later than in GO. The higher the AP concentration, the stronger the effect of turbulence. Turbulence accelerates formation of first raindrops by ~3 min in BO, by ~8 min in GO, and by ~11 min in S. The lowest effect of turbulence on raindrop formation time in BO can be attributed to the fact that drop concentration in BO is low, supersaturation is high, and the efficient diffusion growth shifts drop spectra to larger sizes where even the gravitational coagulation becomes efficient. The effects of CCN and turbulence on formation of first raindrops are opposite: while an increase in aerosol concentration leads to a delay in raindrop formation, turbulence accelerates it.

Results obtained using HUCM qualitatively agree with those obtained by means of the parcel model, showing that the effect of turbulence on acceleration of raindrop formation is most pronounced in polluted clouds.

5.11.5 Relationship Between Drop Concentration and the Level of Raindrop Formation in Cumulus Clouds

Freud and Rosenfeld (2012) found that the height D_p above cloud base where first raindrops form linearly increases with increasing maximum droplet concentration estimated near cloud base. **Figure 5.11.18a** shows the dependence $D_p(N)$ in convective clouds observed at different geographical locations. The raindrop formation height is determined using two threshold values of rain water mixing ratio q_{Pc} of 0.01 g kg^{-1}, and 0.03 g kg^{-1}. Figure 5.11.18b shows the dependence of the height of first raindrop formation on cloud base droplet concentration, obtained in simulations both with the adiabatic parcel model by Pinsky and Khain (2002) described in Section 5.11.1, and with HUCM. In simulations with the parcel model, the height of the first raindrop formation was identified by the values of the mean volume radii r_{v_crit} corresponding to the threshold values of q_{Pc}. In simulations using HUCM, formation of raindrops was identified by the value of radar reflectivity of zero dBz.

The linear dependence of first raindrop formation height (determined as the height where the effective

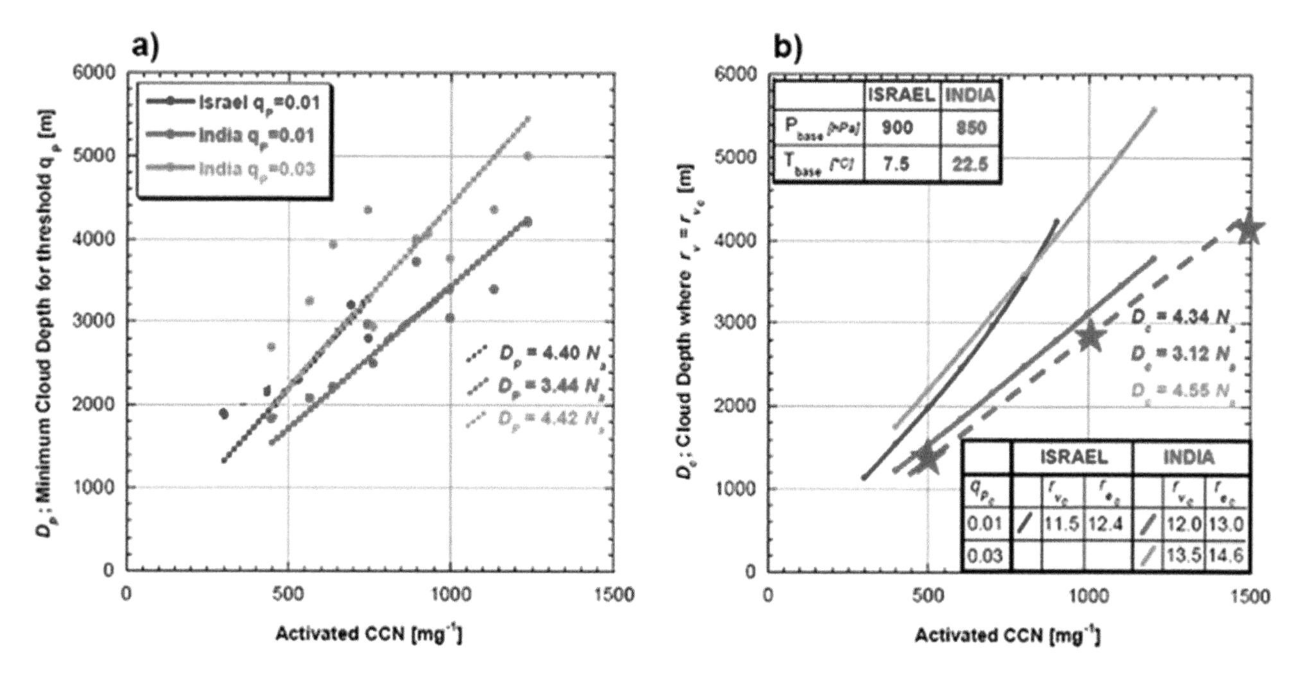

Figure 5.11.18 (a) The relationship between the height of the first raindrop formation and the concentration of activated CCN, i.e., droplet concentration (in mg^{-1}) obtained using in situ measurements. (b) The same relationship obtained using simulations with the cloud parcel model. The values of r_{v_cr} and r_{eff_cr} corresponding to the threshold values of q_{Pc} are presented in the right-bottom-corner table. The equations of the best linear fits are shown. Dashed line with asterisks denotes the results of HUCM simulation, where formation of first raindrops was identified by the value of radar reflectivity of zero dBz (from Freud and Rosenfeld, 2011, with changes, and Khain et al., 2013; courtesy of John Wiley & Sons, Inc.©).

radius reaches its threshold value) on droplet concentration directly follows from the theory of diffusion droplet growth in an ascending adiabatic parcel. Indeed, as follows from Equation (5.2.5), the height of first raindrop formation above cloud base D_p is a linear function of the critical LWC value that is proportional to the droplet concentration at the cloud base: $LWC_{cr} = \frac{A_1}{A_2} D_p$. Taking into account that $LWC_{cr} = \frac{4}{3}\pi r^3_{v\_cr} N \rho_w$, one can get $D_p \sim r^3_{v\_cr} N$. As was shown by Freud and Rosenfeld (2012) and in numerical simulations with HUCM (Section 7.4), in non-precipitating clouds $r_{eff\_cr} \approx r_{v\_cr}$. As a result, we obtain

$$D_p \sim r^3_{eff\_cr} N \tag{5.11.3}$$

The linear relationship (5.11.3) is strictly valid within the range of heights of about ~1 km. Within wider height ranges, there is some deviation from the straight line since the thermodynamic parameters depend on the air temperature and pressure. Simulations performed by a more sophisticated HUCM that takes into account entrainment and mixing with surrounding air also demonstrate this approximately linear dependence of the altitude of first rain formation on the droplet concentration at cloud base. A good agreement between the in situ measurements and the data obtained using the parcel model, as well as the agreement between the results of parcel model simulations and those obtained by means of HUCM, raise an important question: How can the height of the first raindrop formation be reproduced by a dynamically simple parcel model that does not include entrainment and mixing with environment?

The answer to this question can be derived from **Figure 5.11.19**, showing successive LWC fields plotted with a time increment of two min at the growing stage of cloud evolution. All bubbles develop from the same stream rising at the cloud base. At the growth stage, the bubbles rapidly ascend, remaining only slightly diluted. As a result, LWC and drop concentration are close to adiabatic in all undiluted ascending bubbles. Due to the existence of a common source of bubbles at cloud base, the values of the effective radius exceed the critical value at the same height level, and the first raindrops in different bubbles form at approximately the same height (Figure 5.11.16). Thus, the evolution of the microphysical structure in the undiluted bubbles is similar to that in an ascending adiabatic parcel. Different adiabatic bubbles ascend at different velocities and

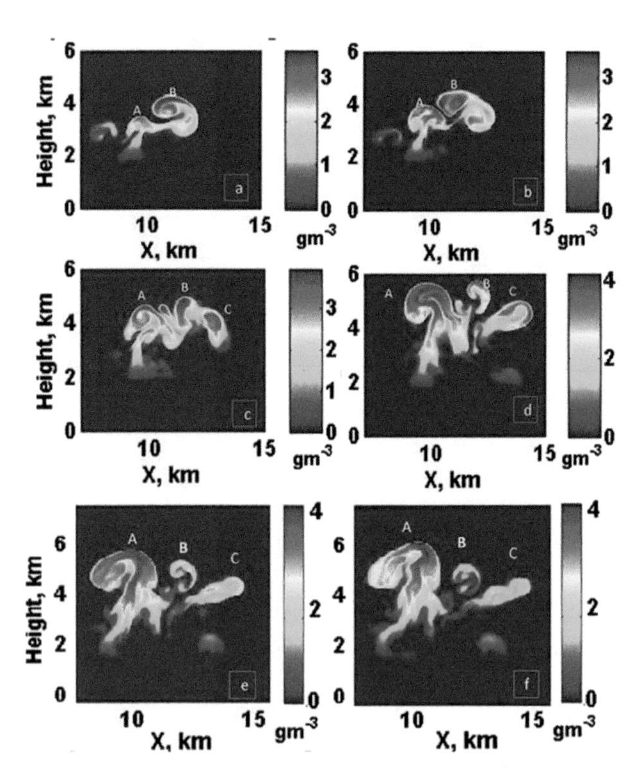

Figure 5.11.19 Fields of LWC plotted with a time increment of 2 min demonstrate the evolution of bubbles A, B, and C. Panel (d) shows the LWC field at t = 66 min. The upper part of this LWC field is shown in Figure 5.11.16 (from Khain et al., 2013; courtesy of John Wiley & Sons, Inc.©).

reach the level of the first raindrop formation at different time instances. Figure 5.11.19 shows that bubble B, which develops first, produces raindrops and decays, while bubble A develops with some delay and produces raindrops later than bubble B. Bubbles starting their ascent from the same cloud base level may lose their roots, as shown in Figure 5.11.19. Nevertheless, the bubbles to a large extent keep their properties, so their LWC remain close to adiabatic values, which means that raindrops form within them almost at the same height levels. The differences in height levels can be easily attributed to possible differences in drop concentrations, in turbulent intensities, etc.

The conclusion concerning formation of the first raindrops in slightly diluted cloud cores does not contradict the finding made by Paluch (1986) regarding the role of mixing in rain formation: "Since such process does not affect the large end of the droplet spectrum, then entrainment and mixing have little immediate effect upon collection efficiencies of hydrometeors growing by accreting the cloud droplets. If mixing does not greatly influence (reduce) the local cloud droplet concentrations and sizes, then the onset of the coalescence process will be relatively uninfluenced by entrainment."

The phenomenon of first raindrops formation in a low number of slightly diluted cloud volumes does not diminish the important role of turbulent mixing, including in-cloud mixing, in formation of the cloud microphysical and thermodynamic structure and precipitation. Intense mixing of cloudy air with entrained air decreases the number of cores and can lead to cloud evaporation, preventing raindrop formation. At the same time, in-cloud mixing can lead to formation of wide local DSD. Mixing affects the rate of raindrop growth during their sedimentation.

The important role of recirculation of largest cloud droplets to their conversion to raindrops in shallow cumulus clouds was analyzed by Naumann and Seifert (2016) in LES simulations.

5.11.6 Formation of DSD in Stratocumulus Clouds

The combined effect of the processes of in-cloud nucleation, diffusion growth/evaporation, mixing, and collisions on DSD formation in a maritime stratocumulus cloud is illustrated next using the results of numerical simulations by means of the LEM (Pinsky et al., 2008b; Magaritz et al., 2009; Magaritz-Ronen, 2016a). In this model, the BL is covered by about 2000 Lagrangian parcels moving within a turbulent-like flow obeying the −5/3 Kolmogorov law. **Figure 5.11.20** shows the DSD evolution in an adiabatic parcel during its motion in case the mixing between neighboring parcels is being neglected. The change of the parcel height over time is shown in the middle panel, indicating that the parcel experiences vertical oscillations. The model takes into account the processes of in-cloud nucleation, drop collisions, and of diffusion growth/evaporation (the latter analyzed in Section 5.2). As follows from the theory of adiabatic parcels (Section 5.2), the diffusion growth/evaporation process is not reversible, so the drops and the DSD width in downdrafts are larger than in updrafts. This difference is, however, not large. In-cloud nucleation in updrafts and drop collisions substantially increase the differences between DSD in updrafts and in downdrafts. As a result of collisions and in-cloud nucleation in the successive vertical oscillations of air volumes lead to DSD broadening.

At t = 0, the parcel contains only nonactivated wet particles. During the ascent (thirty min), some of the largest CCN are activated and start growing as

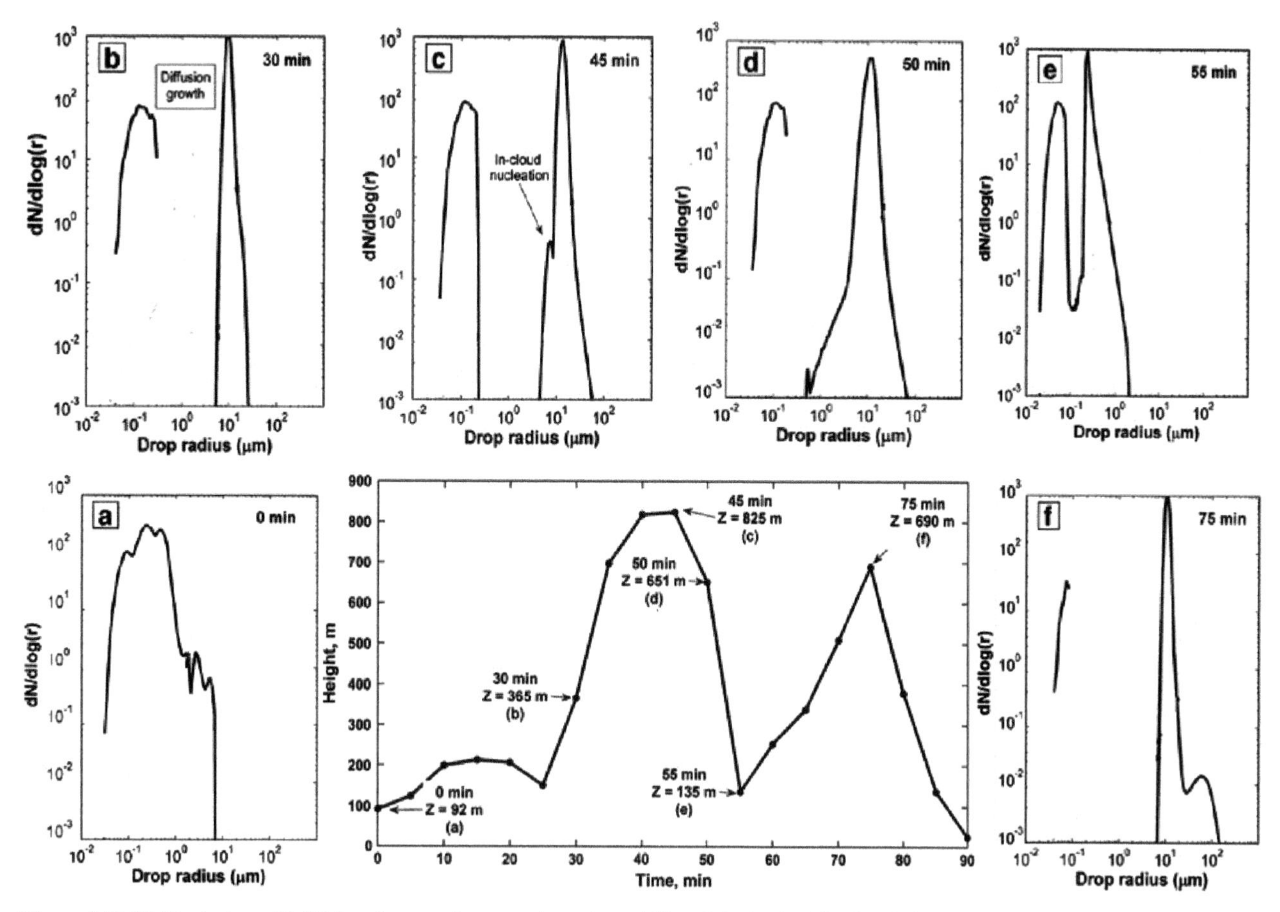

Figure 5.11.20 Evolution of DSD in a Lagrangian parcel moving within a BL obtained in simulations with the Lagrangian model of BL (from Magaritz et al., 2010, with changes; courtesy of Elsevier).

droplets. At the same time, the smallest CCN remain nonactivated and are in equilibrium with the environment. As a result, a gap arises between the nonactivated haze particles and the growing droplets. A rapid ascent leads to in-cloud nucleation and formation of the second mode in about thirty-five min. Droplets belonging to this mode grow fast, and toward forty-five min the gap between the two modes decreases. Collisions during the time period of 35–50 min lead to formation of a small amount of larger drops with the maximum radius of 70 μm. Downdraft occurring between 45 and 50 min is accompanied with subsaturation and leads to a partial evaporation of smaller droplets, and therefore to DSD broadening. Further downdraft leads to evaporation of most droplets.

Drops forming by collisions contain the mass of salt equal to the sum of salt masses in the collided drops. As a result, the spectrum of wet aerosols contains particles larger than in the initial CN distribution. The next ascent of the parcel indicated formation of drizzle drops. This example shows that drop collisions are an important mechanism of formation of superadiabatic drops even in stratocumulus clouds where turbulence intensity is low. Due to in-cloud nucleation, and especially due to DSD broadening during evaporation, DSD dispersion is close to 0.3–0.4, which agrees with the observations (Martin et al., 1994).

Mixing between Lagrangian parcels also leads to DSD broadening, as well as to an increase in the amount of parcels with especially large LWC. **Figure 5.11.21** shows scattering diagrams of DSD widths (standard deviation of DSD) vs. LWC in two simulations of a maritime Sc: with mixing between parcels taken into account: MI (panel a) and NoMI (panel b). The simulations of stratocumulus cloud in RF07 of DYCOMS II were conducted using the LEM (Magaritz-Ronen et al., 2016a). Each dot in the diagrams represents a parcel. Locations of the parcels are

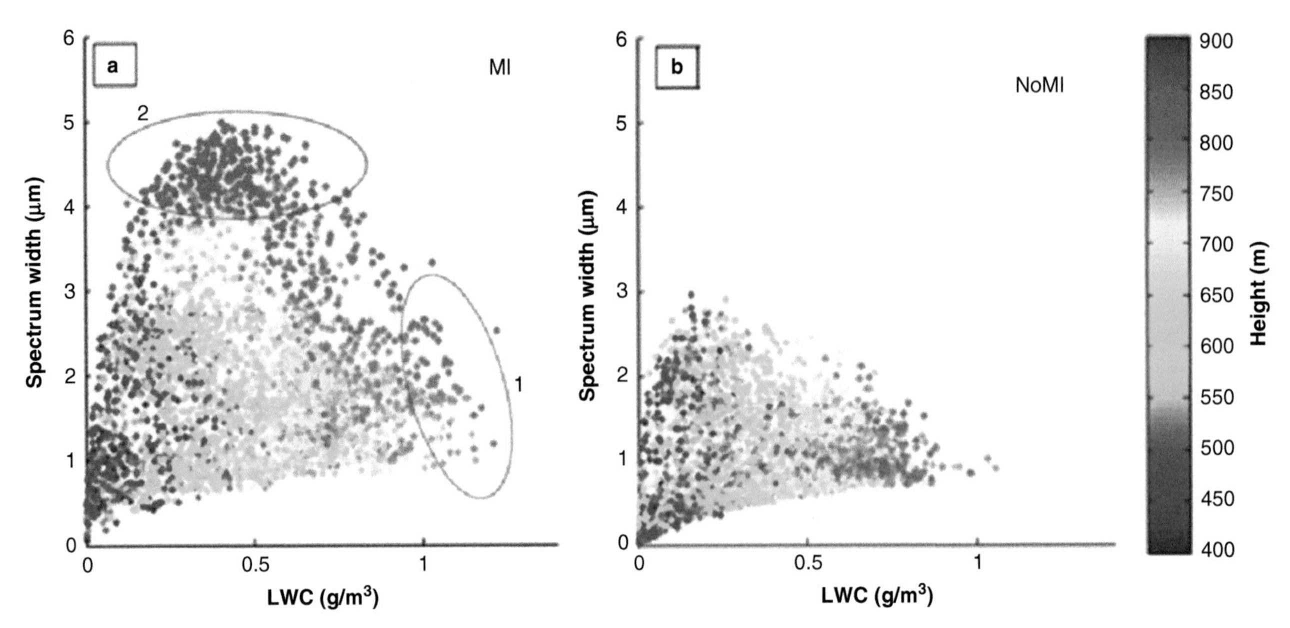

Figure 5.11.21 LWC-spectrum width scatter diagrams for MI (left) and NoMI (right) simulations. Each dot represents a parcel during 195–220 min of simulation. Location of the parcels in the diagrams are plotted with time increment of 5 min. Colors denote the height of the parcels. Ellipse 1 denotes nearly adiabatic parcels with very high LWC. Dots within ellipse 2 denote parcels experiencing mixing with dry parcels penetrating from the inversion layer. The simulations correspond to observations made in the research flight RF07 of the DYCOMS II (from Magaritz-Ronen et al., 2016a; courtesy of ACP).

plotted with time increments of 5 min during time period 195–220 min. Dots within ellipse 1 denote parcels with high LWC close to adiabatic. These parcels reach the cloud top after ascending from the ocean surface and have the DSD width of about 2 μm. Parcels that experience mixing with dry parcels near cloud top are denoted by dots located within ellipse 2. One can see that mixing leads to a decrease in the LWC and to an increase in the DSD width (up to 5 μm) due to formation of small droplets caused by partial droplet evaporation. In NoMI, the LWC maximum does not exceed $1\ \mathrm{g\,m^{-3}}$, and DSD width does not exceed 3 μm. Magaritz et al. (2009) and Magaritz-Ronen et al.(2016a) showed that the DSDs suitable for triggering intense collisions and formation of first drizzle drops (these DSDs contain especially large droplets and large LWC) constitute a comparatively small fraction of parcels (less than 1%) so-called lucky parcels (Section 7.4). Note that most efficient collisions take place in parcels having larger LWC, but not the maximum DSD width.

An interesting mechanism of DSD broadening in the course of vertical oscillation of air parcels in stratocumulus clouds was proposed by Korolev (1995). This broadening is related to the effects of curvature and the chemistry terms in the presence of supersaturation fluctuations occurring during parcel oscillations (see Section 5.2). It was shown that in successive oscillations the drop concentration decreases, which leads to drop spectrum broadening. Korolev (1995) illustrated the effect of the DSD broadening for an isolated adiabatic volume experiencing vertical oscillations with a 50-m amplitude, the vertical velocity of $-0.5 < w < 0.5$ m/s, and the turnover time of ~10 min. Time evolution of the DSD at 40 m and 140 m above the condensation level is shown in **Figure 5.11.22**. The DSD at 140 m level remains narrower than that at 40 m above cloud base up to ~70 min when the second DSD mode arises due to in-cloud nucleation of previously nonactivated CCN. The in-cloud nucleation is caused by supersaturation increasing caused by a decrease in drop concentration within the parcel during the first 70 min.

The DSD broadening during sequent ascent and a descent of parcels in a turbulent-like flow was also reported by Erlick et al. (2005), who showed that turbulent velocity fluctuations lead to penetration of some slowly ascending parcels into strong cloud updrafts, leading to in-cloud nucleation. Another mechanism of DSD broadening, associated with recirculation and mixing of parcels, proposed by Korolev et al. (2013), is described in Section 5.10.

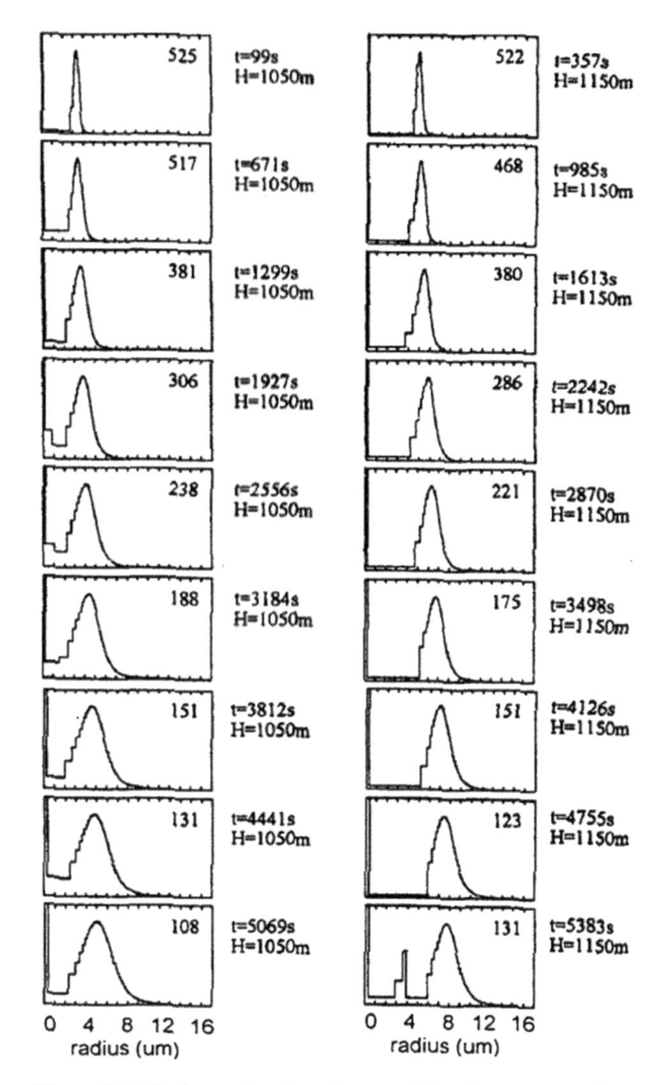

Figure 5.11.22 Successive time changes of droplet spectra at 40 m (left) and 140 m (right) above the condensation level under periodic fluctuations of vertical velocity. Time interval between spectra is approximately 10 min. Numbers in the right corners denote the number concentration (cm^{-3}) for droplets with $r > 0.5$ μm (from Korolev, 1995; American Meteorological Society©; used with permission).

References

Abdul-Razzak, H., and S.J. Ghan, 2000: A parameterization of aerosol activation 2. Multiple aerosol types. *J. Geophys. Res.*, **105** (D5), 6837–6844.

Abdul-Razzak, H., S.J. Ghan, and C. Rivera-Carpio, 1998: A parameterization of aerosol activation. 1. Single aerosol type. *J. Geophys. Res.*, **103** D6, 6123–6131.

Almeida, F.C., 1976: The collisional problem of cloud droplets moving in a turbulent environment-part I: A method of solution. *J. Atmos. Sci.*, **33**, 1571–1578.

1979: The collisional problem of cloud droplets moving in a turbulent environment-part II: Turbulent collision efficiencies. *J. Atmos. Sci.*, **36**, 1564–1576.

Alofs, D.J., and T.-H. Liu, 1981: Atmospheric measurements of CCN in the supersaturation range 0.013–0.681%. *J. Atmos. Sci.*, **38**, 2772–2778.

Andreae, M.O., D. Rosenfeld, P. Artaxo, A.A. Costa. G.P. Frank, K.M. Longlo, and M.A.F. Silva-Dias, 2004: Smoking rain clouds over the Amazon. *Science*, **303**, 1337–1342.

Andsager, K., K.V. Beard, and N.F. Laird, 1999: Laboratory measurements of axis ratios for large raindrops. *J. Atmos. Sci.*, **56** (15), 2673–2683.

Antonia, R.A., A.J. Chambers, and B.R. Satyaprakash, 1981: Reynolds number dependence of high order moments of the streamwise turbulent velocity derivative. *Bound.-Layer Meteor.*, **21**, 159–171.

Arakawa, A., and W.H. Shubert, 1974: Interaction of a cumulus cloud ensemble with the large-scale environment. Part. 1. *J. Atmos. Sci.*, **31**, 674–701.

Arenberg, D., 1939: Turbulence as the major factor in the growth of cloud drops 1939. *Bull. Amer. Meteor. Soc.*, **20**, 444–448.

Ayala, O., W.W. Grabowski, and L.-P. Wang, 2007: A hybrid approach for simulating turbulent collisions of hydrodynamically interacting particles. *J. Comp. Phys.*, **225**, 51–73.

Ayala, O., B. Rosa, and L.-P. Wang, 2008a: Effects of turbulence on the geometric collision rate of sedimenting droplets: Part 2. Theory and parameterization. *New J. Phys.*, **10**, 099802.

Ayala, O., B. Rosa, L.-P. Wang, and W.W. Grabowski, 2008b: Effects of turbulence on the geometric collision rate of sedimenting droplets: Part 1. Results from direct numerical simulation. *New J. Phys.*, **10**, 075015.

Baker, M., R.G. Corbin, and J. Latham, 1980: The influence of entrainment on the evolution of cloud drop spectra: I. A model of inhomogeneous mixing. *Quart. J. Roy. Meteor. Soc.*, **106**, 581–598.

Baker, M., and J. Latham, 1979: The evolution of droplet spectra and the rate of production of embyonic raindrops in small cumulus clouds. *J. Atmos. Sci.*, **36**, 1612–1615.

Baker, M.B., and J. Latham, 1982: A diffusive model of the turbulent mixing of dry and cloudy air. *Q. J. Royal Meteorol. Soc.*, **108**, 871–898.

Barlett, J.T., and P.R. Jonas, 1972: On the dispersion of the sizes of droplets growing by condensation in turbulent clouds. *Quart. J. Roy. Meteor. Soc.*, **98**, 150–164.

Bar-Or, R.Z., I. Koren, O. Altaratz, and E. Fredj, 2012: Radiative properties of 322 humidified aerosols in cloudy environment. *Atmos. Res.*, **118**, 280–294.

Barros, A.P., O.P. Prat, P. Shrestha, and F.Y. Testic, 2008: Revisiting low and list (1982): Evaluation of raindrop collision parameterizations using laboratory observations and modeling. *J. Atmos. Sci.*, **65**, 2983–2993.

Beard, K.V., 1976: Terminal velocity and shape of cloud and precipitation drops aloft. *J. Atmos. Sci.*, **33**, 851–864.

Beard, K.V., R.J. Kubesh, and H.T. Ochs, 1991: Laboratory measurements of small raindrop distortion. Part I: Axis ratios and fall behavior. *J. Atmos. Sci.*, **48** (5), 698–710.

Beard, K.V., and H.T. Ochs, 1984: Collection and coalescence efficiencies for accretion. *J. Geophys. Res.*, **89**, 7165–7169.

Beard, K.V., and H.T. Ochs III, 1993: Warm-rain initiation: An overview of microphysical mechanisms. *J. Appl. Meteorol.*, **33**, 608–625.

1995: Collisions between small precipitation drops. Part II: Formulas for coalescence, temporary coalescence, and satellites. *J. Atmos. Sci.*, **52**, 3977–3996.

Beard, K.V., and H.R. Pruppacher, 1971: A wind tunnel investigation of the rate of evaporation of small water drops falling at terminal velocity in air. *J. Atmos. Sci.*, **28**, 1455–1464.

Bedos, C., K. Suhre, and R. Rosset, 1996: Adaptation of a cloud activation scheme to a spectral-chemical aerosol model. *Atmos. Res.*, **41**, 267–279.

Beheng, K.D., 1994: A parameterization of warm cloud microphysical conversion processes. *Atmos. Res.*, **33**, 193–206.

Beheng, K.D., and G. Doms, 1986: A general formulation of collection rates of clouds and raindrops using the kinetic equation and comparison with parameterizations. *Contib. Atmos. Phys.*, **59**, 66–84.

Belin, F., J. Maurer, P. Tabeling, and H. Willaime, 1997: Velocity gradient distributions in fully developed turbulence: An experimental study. *Phys. Fluids*, **9**, 3843–3850.

Belyaev, V.I., 1961: Drop-size distribution in a cloud during the condensation stages of development. *Akad. Nauk SSSR, Izv, Geophys. Ser.*, 1209–1213.

Benmoshe, N., M. Pinsky, A. Pokrovsky, and A. Khain, 2012: Turbulent effects on microstructure and precipitation of deep convective clouds as seen from simulations with a 2-D spectral microphysics cloud model. *J. Geophys. Res.*, **117**, D06220.

Bera, S., G. Pandithurai, and T.V. Prabha, 2016a: Entrainment and droplet spectral characteristics in convective clouds during transition to monsoon. *Atmos. Sci. Lett.* **17**, 286–293.

Bera, S., T.V. Prabha, and W.W. Grabowski, 2016b: Observations of monsoon convective cloud microphysics over India and role of entrainment-mixing. *J. Geophys. Res.*, **121**, 9767–9788.

Berry, E.X., and R.L. Reinhardt, 1974a: An analysis of cloud droplet growth by collection: Part I. Double distributions. *J. Atmos. Sci.*, **31**, 1814–1824.

1974b: An analysis of cloud drop growth by collection: Part II. Single initial distributions. *J. Atmos. Sci.*, **31**, 1825–1831.

1974c: An analysis of cloud drop growth by collection: Part III. Accretion and selfcollection. *J. Atmos. Sci.*, **31**, 2118–2126.

1974d: An analysis of cloud drop growth by collection: Part IV. A new parameterization. *J. Atmos. Sci.*, **31**, 2127–2135.

Bleck, R., 1970: A fast approximate method for integrating the stochastic coalescence equation. *J. Geophys. Res.*, **75**, 5165–5171.

Blyth, A.M., T.W. Choularton, G. Fullarton, J. Latham, C.S. Mill, M.H. Smith, and I.M. Stromberg, 1980: The influence of entrainment on the evolution of cloud droplet spectra. 2. Field experiments 5 at Great Dun Fell. *Q. J. Royal Meteorol. Soc.*, **106**, 821–840.

Blyth, A.M., W.A. Cooper, and J.B. Jensen, 1988: A study of the source of entrained air in Montana cumuli. *J. Atmos. Sci.*, **45**, 3944–3964.

Boffetta, G., and I.M. Sokolov, 2002: Relative dispersion in fully developed turbulence: The Richardson's law and intermittency corrections. *Phys. Rev. Let.*, **88**, 094501-1–094501-4.

Bott, A., 1989a: A positive definite advection scheme obtained by nonlinear renormalization of the advective fluxes. *Mon. Wea. Rev.*, **117**, 1006–1015.

1989b: Reply. *Mon. Wea. Rev.*, **117**, 2633–2636.

1998: A flux method for the numerical solution of the stochastic collection equation. *J. Atmos. Sci.*, **55**, 2284–2293.

2000: A flux method for the numerical solution of the stochastic collection equation: Extension to two-dimensional particle distributions. *J. Atmos. Sci.*, **57**, 284–294.

Boucher, O., and U. Lohmann, 1995: The sulfate CCN-cloud albedo effect. A sensitivity study with 2 general circulation models. *Tellus, Ser. B Chem. Phys. Meteorol.*, **47**, 281–300.

Bower, K. N., and T. W. Choularton, 1988: The effects of entrainment on the growth of droplets in continental cumulus clouds. *Q. J. Royal Meteorol. Soc.*, 114, 1411–1434.

Brenguier, J.-L., and F. Burnet, 1996: Experimental study of the effect of mixing on droplet spectra. *Proc. 12th Int. Conf. on Clouds and Precipitation*, Zurich, International Commission on Clouds and Precipitation, 67–70.

Brenguier, J.-L., and L. Chaumat, 2001: Droplet spectra broadening in cumulus clouds. Part I: Broadening in adiabatic cores. *J. Atmos. Sci.*, **58**, 628–641.

Brown, P.S., 1983: Some essential details of Bleck's method to the collision-breakup equation. *J. Clim. Appl. Meteorol.*, **22**, 693–697.

Brown, P., 1986: Analysis of the low and list drop-breakup formulation. *J. Climate Appl. Meteorol.*, **25**, 313–321.

Burnet, F., and J.-L. Brenguier, 2007: Observational study of the entrainment-mixing process in warm convective clouds. *J. Atmos. Sci.*, **64**, 1995–2011.

Carpenter, R.L., Droegemeier, K.K. and A.M. Blyth, 1998: Entrainment and detrainment in numerically simulated cumulus congestus clouds. Part I: General results. *J. Atmos. Sci.*, **55**, 3417–3432.

Carrio, G.G., and L. Levi, 1995: On the parameterization of autoconversion. Effects of small-scale turbulent motions. *Atmos. Res.*, **38**, 21–27.

Carrio, G.G., and M. Nicolini, 1999: A double moment warm rain scheme: Description and test within a kinematic framework. *Atmos. Res.*, **52**, 1–16.

Chun, J., and D.L. Koch, 2005: Coagulation of monodisperse aerosol particles by isotropic turbulence. *Phys. Fluid*, **17**, 27102-1–271021-5.

Chun, J., D.L. Koch, Rani, S.L.A. Ahluwalia, and L.R. Collins, 2005: Clustering of aerosol particles in isotropic turbulence. *J. Fluid Mech.*, **536**, 219–251.

Clark, A.D., and V. Kapustin, 2002: A Pacific aerosol survey. Part I: A decade of data on particle production, transport, evolution, and mixing in the troposphere. *J. Atmos. Sci.*, **59**, 363–382.

Clark, T.L., 1973: Numerical modeling of the dynamics and microphysics of warm cumulus convection. *J. Atmos. Sci.*, **30**, 857–878.

Cohard, J.M., and J.P. Pinty, 2000: A comprehensive two-moment warm microphysical bulk model scheme: I: Description and test. *Q. J. Royal Meteorol. Soc.*, **126**, 1815–1842.

Cohard, J.-M., J.-P. Pinty, and C. Bedos, 1998: Extending Twomey's analytical estimate of nucleated cloud droplet concentrations from CCN spectra. *J. Atmos. Sci.*, **55**, 3348–3357.

Cooper, W.A., 1989: Effects of variable droplet growth histories on droplet size distributions. Part 1. Theory. *J. Atmos. Sci.*, **46**, 1301–1311.

Cooper, W.A., R. Bruintjes, and G. Mather, 1997: Calculations pertaining to hygroscopic seeding with flares. *J. Appl. Meteorol.*, **36**, 1449–1469.

Cooper, W.A., S.G. Lasher-Trapp, and A.M. Blyth, 2011: Initiation of coalescence in a cumulus cloud: A beneficial influence of entrainment and mixing. *Atmos. Chem. Phys. Discuss.*, **11**, 10557–10613.

Covert, D.S., V.N. Kapustin, P.K. Quinn, and T.S. Bates, 1992: New particle formation in the marine boundary layer. *J. Geophys. Res.*, **97**(D18), 20581–20589.

Damiani, R., G. Vali, and S. Haimov, 2006: The structure of thermals in cumulus from airborne dual-Doppler radar observations. *J. Atmos. Sci.*, **63**, 1432–1450.

Devenish, B.J., P. Bartello, J.-L. Brenguier, L.R. Collins, W.W. Grabowski, R.H.A. Ijzermans, S.P. Malinovski, M.W. Reeks, J.C. Vassilicos, L.-P. Wang, and Z. Warhaft, 2012: Droplet growth in warm turbulent clouds. *Q. J. Royal Meteorol. Soc.*, **138**, 1401–1429.

Devis, E.J., 2006: A history and state-of-the-art of accommodation coefficients. *Atmos. Res.*, **82**, 561–578.

Dinger, J.E., H.B. Howell, and T.A. Wojciechowski, 1970: On the source of composition of cloud nuclei in subsident air mass over the North Atlantic. *J. Atmos. Sci.*, **27**, 791–797.

Dodin, Z., and T. Elperin, 2002: On the collision rate of particles in turbulent flow with gravity. *Phys. Fluid*, **14**, 2921–2924.

Duru, P., D.L. Koch, and C. Cohen, 2007: Experimental study of turbulence-induced coalescence in aerosols. *Int. J. of Multiph. Flow.*, **33**, 987–1005.

Elperin, T., N. Kleeorin, V.S. L'vov, I. Rogachevskii, and D. Sokoloff, 2002a: Clustering instability of the spatial distribution of inertial particles in turbulent flows. *Phys. Rev.*, **E66**, 36302-1–36302-16.

Elperin, T., N. Kleeorin, I. Rogachevskii, 1996: Self-excitation of fluctuations of inertial particles concentration in turbulent flow. *Phys. Rev. Lett.*, **77**, 5373–5376.

Elperin, T., N. Kleeorin, and I. Rogachevskii, 2002b: Formation of large scale semiorginized structures in turbulent convection. *Phys. Rev. E.*, **66**, 066305, 5–10.

Emde, K., and U. Wacker, 1993: Comments on the relationship between aerosol spectra, equilibrium drop size spectra, and CCN spectra. *Beitr. Phys. Atmosph.*, **66**, 1–2, 157–162.

Enukashvily, I.M., 1980: A numerical method for integrating the kinetic equation of coalescence and breakup of cloud droplets. *J. Atmos. Sci.*, **37**, 2521–2534.

Erlick, C., A. Khain, M. Pinsky, and Y. Segal, 2005: The effect of turbulent velocity fluctuations on drop spectrum broadening in stratiform clouds. *Atmos. Res.*, **75**, 15–45.

Falkovich, G., A. Fouxon, and M.G. Stepanov, 2002: Acceleration of rain initiation by cloud turbulence. *Nature*, **419**, 151–154.

Falkovich, G., and A. Pumir, 2004: Intermittent distribution of heavy particles in a turbulent flow. *Phys. Fluids*, **16**, L47–L50.

Falkovich, G., and A. Pumir, 2007: Sling effect in collisions of water droplets in turbulent clouds. *J. Atmos. Sci.*, **64**, 4497–4505.

Fan, J., J.M. Comstock, and M. Ovchinnikov, 2010: The cloud condensational nuclei and ice nuclei effects on tropical anvil characteristics and water vapor of the tropical tropopause layer. *Environ. Res. Lett.* **5**, doi: 10.1088/1748–9326/5/4/044005.

Fankhauser, J.C., G.M. Barness, C.J. Biter, D.W. Breed, and M.A. LeMone, 1983: Summary of NCAR Technical Note NCAR/TN-207+STR, p. 134. (Available from NCAR, P.O. Box 3000, Boulder, CO 80307.)

Feingold, G., S.M. Kreidenweis, B. Stevens, and W.R. Cotton, 1996: Numerical simulation of stratocumulus processing of cloud condensation nuclei through collision-coalescence. *J. Geophys. Res.*, **101**, 21,391–21,402.

Ferrier, B.S., 1994: A double-moment multiple-phase four-class bulk ice scheme. Part 1: Description. *J. Atmos. Sci.*, **51**, 249–280.

Ferrier, B.S., and R.A. Houze, 1989: One-dimensional time-dependent modeling of GATE cumulonimbus convection. *J. Atmos. Sci.*, **46**, 330–352.

Fierro, A.O., E.J. Zipser, M.A. Lemone, J.M. Straka, and J. (Malkus) Simpson, 2012: Tropical oceanic hot towers: Need they be undilute to transport energy from the boundary layer to the upper troposphere effectively? An answer based on trajectory analysis of a simulation of a TOGA COARE convective system. *J. Atmos. Sci.*, **69**, 195–213.

Flossmann, A.I., and H.R. Pruppacher, 1988: A theoretical study of the wet removal of atmospheric pollutants. Part III: The uptake, redistribution, and deposition of $(NH_4)_2SO_4$ particles by a convective cloud using a two-dimensional cloud dynamics model. *J. Atmos. Sci.*, **45**, 1857–1871.

Fountoukis, C., and A. Nenes, 2005: Continued development of a cloud droplet formation parameterization for global clime models. *J. Geophys. Res.*, **110**, D11212.

Franklin, C.N., 2008: A warm rain microphysics parameterization that includes the effect of turbulence. *J. Atmos. Sci.*, **65**, 1795–1816.

Franklin, C.N., P.A. Vaillancourt, and M.K. Yau, 2007: Statistics and parameterizations of the effect of turbulence on the geometric collision kernel of cloud droplets. *J. Atmos. Sci.*, **64**, 938–954.

Franklin, C.N., P.A. Vaillancourt, M.K. Yau, and P. Bartello, 2005: Collision rates of cloud droplets in turbulent flow. *J. Atmos. Sci.*, **62**, 2451–2466.

Freud, E., and D. Rosenfeld, 2012: Linear relation between convective cloud drop number concentration and depth for rain initiation. *J. Geophys. Res.*, **117**, D02207.

Freud, E., D. Rosenfeld, M.O. Andreae, A.A. Costa, and P. Artaxo, 2008: Robust relations between CCN and the vertical evolution of cloud drop size distribution in deep convective clouds. *Atmos. Chem. Phys.*, **8**, 1661–1675.

Freud, E., D. Rosenfeld, D. Axisa, and J.R. Kulkarni, 2011: Resolving both entrainment-mixing and number of activated CCN in deep convective clouds. *Atmos. Chem. Phys.*, **11**, 12887–12900.

Fridlind, A. et al., 2004: Evidence for the predominance of mid-tropospheric aerosols as subtropical anvil cloud nuclei. *Science*, **304**, 718–722.

Fung, J.C.H., 1993: Gravitational settling of particles and bubbles in homogeneous turbulence. *J. Geoph. Res.*, **98**, 20,287–20,297.

Fung, J.C.H., J.C.R. Hunt, N.A. Malik, and R.J. Perkins, 1992: Kinematic simulation of homogeneous turbulent flows generated by unsteady random Fourier modes. *J. Fluid Mech.*, **236**, 281–317.

Gerber, H., G. Frick, J.B. Jensen, and J.G. Hudson, 2008: Entrainment, mixing, and microphysics in trade-wind cumulus. *J. Meteorol. Soc. Jpn.*, **86A**, 87–106.

Ghan, S.J., C.C. Chuang, R.C. Easter, and J.E. Penner, 1995: A parameterization of cloud droplet nucleation. Pt. 2: Multiple aerosol types. *Atmos. Res.*, **36**, 39–54.

Ghan, S.J., C.C. Chuang, and J.E. Penner, 1993: A parameterization of cloud droplet nucleation. Pt.1: Single aerosol type. *Atmos. Res.*, **30**, 197–221.

Ghan, S.J., A.-R. Hayder, A. Nenes, Y. Ming, L. Xiaohong, M. Ovchinnikov, B. Shipway, N. Meskhidze, J. Xu, and X. Shi, 2011: Droplet nucleation: Physically-based parameterizations and comparative evaluation. *J. Adv. Model. Earth Syst.*, **3**, M10001, p. 33, DOI:10.1029/2011MS000074.

Gilmore, M.S., and J.M. Straka, 2008: The Berry and Reinhardt autoconversion parameterization: A digest. *J. Appl. Meteorol. Climatol.*, **47**, 375–396.

Giola, G., G. Lacorata, E.P. Marques Filho, A. Mazzino, and U. Rizza, 2004: Richardson's law in large-eddy simulations of boundary-layer flows. *Boundary-Layer Met.*, **113**, 187–199.

Grabowski, W.W., 2006: Indirect impact of atmospheric aerosols in idealized simulations of convective–radiative quasi equilibrium. *J. Climate*, **19**, 4664–4682.

2007: Representation of turbulent mixing and buoyancy reversal in bulk cloud models. *J. Atmos. Sci.*, **64**, 3666–3680.

Grabowski, W.W., M. Andrejczuk, and L.-P. Wang, 2011: Droplet growth in a bin warm-rain scheme with Twomey CCN activation. *Atmos. Res.*, **99**, 290–301.

Grabowski, W.W., and T.L. Clark, 1991: Cloud-entrainment interface instability: Rising thermal calculations in two spatial dimensions. *J. Atmos. Sci.*, **48**, 527–546.

1993: Cloud-enviromental interface instability. Part II: Extension to three spatial dimensions. *J. Atmos. Sci.*, **50**, 555–573.

Grabowski, W.W., and H. Morrison, 2008: Toward the mitigation of spurious cloud-edge supersaturation in cloud models. *Mon. Wea. Rev.*, **136**, 1224–1234.

Grabowski, W.W., and P. Vaillancourt, 1999: Comments on "Preferential concentration of cloud droplets by turbulence: effects on the early evolution of cumulus cloud droplet spectra." *J. Atmos. Sci.*, **56**, 1433–1436.

Gras, J.L., 1989: Baseline atmospheric condensation nuclei at Cape grim. *J. Atmos. Chem.*, **11**, 89–106.

1990: Cloud condensation nuclei over the Southern Ocean. *Geophys. Res. Lett.*, **17**, 1565–1567.

Grits, B., M. Pinsky, and A. Khain, 2006: Investigation of small-scale droplet concentration inhomogeneities in a turbulent flow. *Meteorol. Atmos. Phys.*, **92**, 191–204.

Hall, W.D., 1980: A detailed microphysical model within a two-dimensional dynamic framework: Model description and preliminary results. *J. Atmos. Sci.*, **37**, 2486–2507.

Hamielec, A.E., and A.I. Johnson, 1962: Viscous flow around fluid spheres at intermediate Reynolds numbers. *Can. J. Chem. Eng.*, April, 41–45.

Hegg, D.A., and P.V. Hobbs,1992: Cloud condensation nuclei in the marine atmosphere, In N. Fukuta, P.E.

Wagner (Eds.), Nucl. and Atmos. Aerosols. - Proc. 13-th Int. Conf. Nucl. Atmos. Aerosol, A. Deepak Publishing, Hampton, VA. pp. 181–192.

Hegg, D.A., L.F. Radke, and P.V. Hobbs, 1991: Measurements of Aitken nuclei and cloud condensation nuclei in the marine atmosphere and their relation to the DMS-cloud-climate hypothesis. *J. Geophys. Res.*, **96**, 18,727–18,733.

Heus, T., and H. Jonker, 2008: Subsiding shells around shallow cumulus clouds. *J. Atmos. Sci.*, **65**, 1003–1018.

Hill, A.A., G. Feingold, and H. Jiang, 2009: The influence of entrainment and mixing assumption on aerosol–cloud interactions in marine stratocumulus. *J. Atmos. Sci.*, **66**, 1450–1464.

Hill, R.J., 2002: Scaling of acceleration in locally isotropic turbulence. *J. Fluid Mech.*, **452**, 361–370.

Hill, T.A., and T.W. Choularton, 1986: A model of the development of the droplet spectrum in a growing cumulus cloud. *Q. J. Royal Meteorol. Soc.*, **112**, 531–554.

Hobbs, P.V. et al., 1978: *Res. Rept. XIII*, Dept. Atmos. Sci., Univ. Washington, DC.

1993: *Aerosol-cloud-climate interactions*. Academic Press, p. 236.

Hobbs, P.V., D.A. Bowdle, and L.F. Radke, 1985: Particles in the lower troposphere over the High Plains of the United States. 1: Size distributions, elemental compositions and morphologies. *J. Clim. Appl. Meteorol.*, **24**, 1344–1356.

Hobbs, P.V., and A.L. Rangno, 2004: Super-large raindrops. *Geophys. Res. Lett.*, **31**, L13102, doi:10.1029/2004GL020167.

Hocking, L.M., and P.R. Jonas, 1970: The collision efficiency of small drops. *Q. J. Royal Meteorol. Soc.*, **96**, 722–729.

Hong, S.-Y, K.-S. S. Lim, Y.-H. Lee, J.-C. Ha, H.-W. Kim, S.-J. Ham, and J. Dudhia, 2010: Evaluation of the WRF double moment 6-class microphysics scheme for precipitating convection. *Advances in Meteorology*, ID 707253, doi:10.1155/2010/707253.

Hoppel, W.A., J.E. Dinger, and R.E. Ruskin,1973: Vertical profiles of CCN at various geographical locations. *J. Atmos. Sci.*, **30**, 1410–1420.

Hoppel, W.A., J.W. Fitzgerald, G.M. Frick, R.E. Larson, and E.J. Mack, 1990: Aerosol size distributions and optical properties found in the marine boundary layer over the Atlantic Ocean. *J. Geophys. Res.*, **95**, 3659–3686.

Hu, Z., and R.C. Srivastava, 1995: Evolution of raindrop size distribution by coalescence, breakup, and evaporation: Theory and observations. *J. Atmos. Sci.*, **52**, 1761–1783.

Hudson, J.G., 1984: Cloud condensation nuclei measurements within clouds. *J. Climate Appl. Meteorol.*, **23**, 42–51.

Hudson, J.G., and P.R. Frisbie, 1991: Cloud condensation nuclei near marine stratus. *J. Geophys. Res.*, **96**, 20,795–20,808.

Hudson, J.G., and H. Li, 1995: Microphysical contrasts in Atlantic stratus. *J. Atmos. Sci.*, **52**, 3031–3040.

Hudson, J.G., and S.S. Yum, 1997: Droplet spectral broadening in marine stratus. *J. Atmos. Sci.*, **54**, 2642–2654.

2002: Cloud condensation nuclei spectra and polluted and clean clouds over the Indian Ocean. *J. Geophys.Res.*, **107(D19)**, 8022, doi:10.1029/2001JD000829.

Ivanova, E.T., Y.L. Kogan, I.P. Mazin, and M.S. Permyakov, 1977: The ways of parameterization of condensation drop growth in numerical models. *Izv. Atmos. Oceanic Phys.*, **13** (N11), 1193–1201.

Jaenicke, R., 1993: "Tropospheric Aerosols," chapter in book by *Aerosol-Cloud-Climate Interactions*, edited by Peter Hobbs. Academic Press, p. 236.

Jamerson, A.R., and A.B. Kostinski, 2000: Fluctuation properties of precipitation. Part 4: Observations of hyperfine clustering and drop size distribution structures in three-dimensional rain. *J. Atmos. Sci.*, **57**, 373–388.

Jarecka, D., W.W. Grabowski, and H. Pawlowska, 2009: Modeling of subgrid-scale mixing in large-eddy simulation of shallow convection. *J. Atmos. Sci.*, **66**, 2125–2133.

Jarecka, D., H. Pawlowska, W.W. Grabowski, and A.A. Wyszogrodzki, 2013: Modeling microphysical effects of entrainment in clouds observed during EUCAARI-IMPACT field campaign. *Atmos. Chem. Phys. Discuss.*, **13**, 1489–1526, doi:10.5194/acpd-13-1489-2013.

Jeffery, C.A., and J.M. Reisner, 2006: A study of cloud mixing and evolution using PDF methods. Part I: Cloud front propagation and evaporation. *J. Atmos. Sci.*, **63**, 2848–2864.

Jeffery, C.A., J.M. Reisner, and M. Andrejczuc, 2007: Another look at stochastic condensation for subgrid cloud modeling: Adiabatic evolution and effects. *J. Atmos. Sci.*, **64**, 3949–3969.

Jensen, J.B., M. Baker, 1989: A simple model of droplet spectral evolution during turbulent mixing. *J. Atmos. Sci.*, **46**, 2812–2829.

Jin, G., G.-W. He, and L.-P. Wang, 2010: Large-eddy simulation of turbulent collision of heavy particles in isotropic turbulence. *Phys. Fluids*, **22**, 055106.

Jiusto, J.E., 1967: Aerosol and cloud microphysics measurements in Hawaii. *Tellus*, **19**, 359–368.

Jiusto, J.E., and G.G. Lala, 1981: CCN-supersaturation spectra slopes (k). *J. Rech. Atmos.*, **15**, 303–311.

Johnson, D.B., 1993: The onset of effective coalescence growth in convective clouds. *Q. J. Royal Meteorol. Soc.*, **119**, 925–933.

Jonas, P.R., 1972: The collision efficiency of small drops. *Q. J. Royal Meteorol. Soc.*, **98**, 681–683.

Kamra, A.K., R.V. Bhalwankar, and A.B. Sathe, 1991: Spontaneous breakup of charged and uncharged water drops freely suspended in a wind tunnel. *J. Geophys. Res.*, **96** (D9), 17,159–17,168.

Kato, T., 1995: A box–Lagrangian rain-drop scheme. *J. Meteorol. Soc. Jpn.*, **73**, 241–245.

Kerstein, A.R., 1988: Linear eddy modelling of turbulent scalar transport and mixing. *Combust. Sci. Technol.*, **60**, 391–421.

Kessler, E., 1969: On the distribution and continuity of water substance in atmospheric circulations. *Meteorol. Monogr.*, 32.

Khain, A.P., K.D. Beheng, A. Heymsfield, A. Korolev, S.O. Krichak, Z. Levin, M. Pinsky, V. Phillips, T. Prabhakaran, A. Teller, S.C. van den Heever, and J.-I. Yano, 2015: Representation of microphysical processes in cloud resolving models: Spectral (bin) microphysics versus bulk parameterization. *Rev. Geophys.*, **53**, 247–322.

Khain, A.P., N. Benmoshe, and A. Pokrovsky, 2008: Factors determining the impact of aerosols on surface precipitation from clouds: An attempt of classification. *J. Atmos. Sci.*, **65**, 1721–1748.

Khain, A.P., M. Ovtchinnikov, M. Pinsky, A. Pokrovsky, and H. Krugliak, 2000: Notes on the state-of-the-art numerical modeling of cloud microphysics. *Atmos. Res.*, **55**, 159–224.

Khain, A.P., V. Phillips, N. Benmoshe, and A. Pokrovsky, 2012: The role of small soluble aerosols in the microphysics of deep maritime clouds. *J. Atmos. Sci.*, **69**, 2787–2807.

Khain, A.P., and M.B. Pinsky, 1995: Drops' inertia and its contribution to turbulent coalescence in convective clouds: Part 1: Drops' fall in the flow with random horizontal velocity. *J. Atmos. Sci.*, **52**, 196–206.

Khain, A., M. Pinsky, T. Elperin, N. Kleeorin, I. Rogachevskii, and A. Kostinski, 2007: Critical comments to results of investigations of drop collisions in turbulent clouds. *Atmos. Res.*, **86**, 1–20.

Khain, A.P., and A. Pokrovsky, 2004: Effects of atmospheric aerosols on deep convective clouds as seen from simulations using a spectral microphysics mixed-phase cumulus cloud model Part 2: Sensitivity study. *J. Atmos. Sci.*, **61**, 2983–3001.

Khain, A., A. Pokrovsky, M. Pinsky, A. Seifert, and V. Phillips, 2004: Effects of atmospheric aerosols on deep convective clouds as seen from simulations using a spectral microphysics mixed-phase cumulus cloud model Part 1: Model description. *J. Atmos. Sci.*, **61**, 2963–2982.

Khain, A., T.V. Prabha, N. Benmoshe, G. Pandithurai, and M. Ovchinnikov, 2013: The mechanism of first raindrops formation in deep convective clouds. *J. Geophys. Res. Atmos.*, **118**, 9123–9140, doi:10.1002/jgrd.50641.

Khain, A.P., D. Rosenfeld, and I.L. Sednev, 1993: Coastal effects in the Eastern Mediterranean as seen from experiments using a cloud ensemble model with a detailed description of warm and ice microphysical processes. *Atmos. Res.*, **30**, 295–319.

Khain, A.P., and I.L. Sednev, 1995: Simulation of hydrometeor size spectra evolution by water-water, ice water and ice-ice interection. *Atmos. Res.*, **36**, 107–138.

Khain, A.P., and I. Sednev, 1996: Simulation of precipitation formation in the Eastern Mediterranean coastal zone using a spectral microphysics cloud ensemble model. *Atmos. Res.*, **43**, 77–110.

Khairoutdinov, M., and Y. Kogan, 2000: A new cloud physics parameterization in a large-eddy simulation model of marine stratocumulus. *Mon. Wea. Rev.*, **128**, 229–243.

Kholmyansky, M., A. Tsinober, and S. Yorich, 2001: Velocity derivatives in the atmospheric surface layer at $Re_\lambda = 10^4$. *Phys. Fluids*, **13**, 311–314.

Khvorostyanov, V.I., and J.A. Curry, 1999a: Toward the theory of stochastic condensation in clouds. Part 1: A general kinetic equation. *J. Atmos. Sci.*, **56**, 3985–3996.

1999b: Toward the theory of stochastic condensation in clouds. Part 2: Analytical solutions of the gamma-distribution type. *J. Atmos. Sci.*, **56**, 3997–4013.

2006: Aerosol size spectra and CCN activity spectra: Reconciling the lognormal, algebraic, and power laws. *J. Geophys. Res.*, **111**, D12202.

2008: Kinetics of cloud drop formation and its parametrization for cloud and climatemodels. *J. Atmos. Sci.*, **65**, 2784–2802.

Khvorostyanov, V.I., A.P. Khain, and E.L. Kogteva, 1989: A two-dimensional non stationary microphysical model of a three-phase convective cloud and evaluation of the effects of seeding by a crystallizing agent. *Soviet Meteorology and Hydrology*, 5, 33–45.

Kim, S., and S.J. Karrila, 1991: *Microhydrodynamics Principles and Selected Applications*. Butterworth-Heinmann, p. 507.

Kim, Y.J., and J.F. Boatman, 1990: Size calibration corrections for the Active Scattering Aerosol Spectrometer Probe (ASASP-100X). *Aerosol Sci. Technol.*, **12**, 665–672.

Kinzer, G.D., and R. Gann, 1951: The evaporation temperature and thermal relaxation time of freely falling water drops. *J. Meteorol.*, **8**, 71–83.

Kivekäs, N., V.-M. Kerminen, T. Anttila, H. Korhonen, H. Lihavainen, M. Komppula, and M. Kulmala, 2008: Parameterization of cloud droplet activation using a simplified treatment of the aerosol number size distribution. *J. Geophys. Res.*, **113**, D15207.

Klaassen, G.P., and T.L. Clark, 1985: Dynamics of the cloud environment interface and entrainment in small cumuli: Two dimensional simulations in the absence of ambient shear. *J. Atmos. Sci.*, **42**, 2621–2642.

Klett, J.D., and M.H. Davis, 1973: Theoretical collision efficiencies of cloud droplets at small Reynolds numbers. *J. Atmos. Sci.*, **30**, 107–117.

Knight, C.A., and L.J. Miller, 1998: Early radar echoes from small, warm cumulus: Bragg and hydrometeor scattering. *J. Atmos. Sci.*, **55**, 2974–2992.

Kocmond, W.C., 1965: *Res. Rept.* RM-1788-p9, p. 36, Cornell Aeronaut. Lab. Buffalo, NY.

Kogan, Y., 1991: The simulation of a convective cloud in a 3-D model with explicit microphysics. Part 1: Model description and sensitivity experiments. *J. Atmos. Sci.* **48**, 1160–1189.

Kogan, Y.L. 2006: Large-eddy simulation of air parcels in stratocumulus clouds: Time scales and spatial variability. *J. Atmos. Sci.*, **63**, 952–967.

Komabayasi, M., T. Gond, and K. Isono, 1964: Lifetime of water drops before breaking and size distribution of fragment droplets. *J. Meteorol. Soc. Jpn.*, **42**, 330–340.

Korolev, A.V., 1995: The influence of supersaturation fluctuations on droplet size spectra formation. *J. Atmos. Sci.*, **52**, 3620–3634.

Korolev, A., A. Khain, M. Pinsky, and J. French, 2016: Theoretical study of mixing in liquid clouds – Part 1: Classical concept. *Atmos. Chem. Phys.*, **16**, 9235–9254.

Korolev, A.V., and I.P. Mazin, 2003: Supersaturation of water vapor in clouds. *J. Atmos. Sci.*, **60**, 2957–2974.

Korolev, A., M. Pinsky, and A. Khain, 2013: A new mechanism of droplet size distribution broadening during diffusional growth. *J. Atmos. Sci.*, **70**, 2051–2071.

Kostinski, A.B., and R.A. Shaw, 2001: Scale-dependent droplet clustering in turbulent clouds. *J. Fluid Mech.*, **434**, 389–398.

2005: Fluctuations and luck in droplet growth by coalescence. *Bull. Am. Meteorol. Soc.*, **86**, 235–244.

Kovetz, A., and B. Olund, 1969: The effect of coalescence and condensation on rain formation in a cloud of finite vertical extent. *J. Atmos. Sci.*, **26**, 1060–1065.

Koziol, A.S., H.G. Leighton, 1996: The effect of turbulence on the collision rates of small cloud drops. *J. Atmos. Sci.*, **53**, 1910–1920.

Kraichnan, R.H., 1970: Diffusion by a random velocity field. *Phys. Fluid*, **13**, 22–31.

Krueger, S.K., 1993: Linear eddy modeling of entrainment and mixing in stratus clouds. *J. Atmos. Sci.*, **50**, 3078–3090.

Krueger, S., C.-W. Su, and P. McMurtry, 1997: Modeling entrainment and fine-scale mixing in cumulus clouds. *J. Atmos. Sci.*, **54**, 2697–2712.

Krueger, S.K., P.J. Lehr, and C.W. Su, 2006. How entrainment and mixing scenarios affect droplet spectra in cumulus clouds. *12th Conference on Cloud Physics, and 12th Conference on Atmospheric Radiation, Madison WI, USA, July*, 10–14.

Kubesh, R.J., and K.V. Beard, 1993: Laboratory measurements of spontaneous oscillations for moderate-size raindrops. *J. Atmos. Sci.*, **50**(8), 1089–1098.

Kumar, B., S. Bera, T. Prabhakaran, and W.W. Grabowski, 2016: Cloud-edge mixing: Direct numerical simulation and observations in Indian monsoon clouds. Submitted to *JAMES*.

Kumar, B., J. Schumacher, and R.A. Shaw, 2013: Cloud microphysical effects of turbulent mixing and entrainment. *Theor. Comput. Fluid Dyn.*, **27**, 361–376.

2014: Lagrangian mixing dynamics at the cloudy–clear air interface. *J. Atmos. Sci.*, **71**, 2564–2580.

La Porta, A., G.A. Voth, A.M. Crawford, J. Alexander, and E. Bodenschatz, 2001: Fluid particle accelerations in fully developed turbulence. *Nature*, **409**, 1017–1019.

Lanotte, A.S., A. Seminara, and F. Toschi, 2009: Cloud droplet growth by condensation in homogeneous isotropic turbulence. *J. Atmos. Sci.*, **66**, 1685–1697.

Lasher-Trapp, S.G., Knight, C., Straka, J.M., 1998: Ultragiant aerosol growth by collection within a warm continental cumulus. AMS Conference on Cloud Physics, August 17–21, Everett, WA, pp. 494–497.

Latham, J., and R.L. Reed, 1977: Laboratory studies of effects of mixing on evolution of cloud droplet spectra, *Q. J. Royal Meteorol. Soc.*, **103**, 297–306.

Lehmann, K., H. Siebert, and R.A. Shaw, 2009: Homogeneous and inhomogeneous mixing in cumulus clouds: Dependence on local turbulence structure. *J. Atmos. Sci.*, **66**, 3641–3659.

Leon, D.C., J.R. French, S. Lasher-Trapp, A.M. Blyth, V. Abel, S. Ballard, L.J. Bennett, K. Bower, B. Brooks, P. Brown, T. Choularton, P. Clark, C. Collier, J. Crosier, Z. Cui, D. Dufton, C. Eagle, M.J. Flynn, M. Gallagher, K. Hanley, Y. Huang, M. Kitchen, A. Korolev, H. Lean, Z. Liu, J. Marsham, D. Moser, J. Nicol, E.G. Norton, D. Plummer, J. Price, H. Ricketts, N. Roberts, P.D. Rosenberg, J.W. Taylor, P.I. Williams, and G. Young, 2016: The Convective Precipitation Experiment (COPE): Investigating the origins of heavy precipitation in the southwestern UK. *Bull. Amer. Meteor. Soc.*, **97**, 1003–1020.

Levin, L.M., and Y.S. Sedunov, 1966: Stochastic condensation of drops and kinetics of cloud spectrum formation. *J. Rech. Atmos.*, **2**, 425–432.

Levin, Z., and W.R. Cotton (Eds.), 2009: *Aerosol Pollution Impact on Precipitation: A Scientific Review*. Springer, p. 386.

Lewis, E.R., 2008: An examination of Kohler theory resulting in an accurate expression for the equilibrium radius ratio of a hygroscopic aerosol particle valid up to and including relative humidity 100%. *J. Geoph. Res.*, **113**, D03205.

Lin, C.L., and S.C. Lee, 1975: Collision efficiency of water drops in the atmosphere. *J. Atmos. Sci.*, **32**, 1412–1418.

Lin, Y.-L., R.D. Farley, and H.D. Orville, 1983: Bulk parameterization of the snow field in a cloud model. *J. Climate Appl. Meteorol.*, **22**, 1065–1092.

List, R., and J.R. Gillespie, 1976: Evolution of raindrop spectra with collision-induced breakup. *J. Atmos. Sci.*, **33**, 2007–2013.

List, R., and G.M. McFarquhar, 1990: The role of breakup and coalescence in the three-peak equilibrium distribution of raindrops. *J. Atmos. Sci.*, **47**, 2274–2292.

List, R., R. Nissen, and C. Fung, 2009: Effects of pressure on collision, coalescence, and breakup of raindrops. Part II: Parameterization and spectra evolution at 50 and 100 kPa. *J. Atmos. Sci.*, **66**, 2204–2215.

Liu, Q., Y. Kogan, D.K. Lilly, and M.P. Khairoutdinov, 1997: Variational optimization method for calculation of cloud drop growth in Eulerian drop-size framework. *J. Atmos. Sci.*, **54**, 2493–2504.

Long, A., 1974: Solutions to the droplet collection equation for polynomial kernels. *J. Atmos. Sci.*, **31**, 1040–1052.

Low, T.B., and R. List, 1982a: Collision, coalescence, and breakup of raindrops. Part I: Experimentally established coalescence efficiencies and fragment size distributions in breakup. *J. Atmos. Sci.*, **39**, 1591–1606.

1982b: Collision, coalescence, and breakup of raindrops. Part II: Parameterization of fragment size distributions. *J. Atmos. Sci.*, **39**, 1607–1618.

Lu, C., Y. Liu, S. Niu, and S. Endo, 2014: Scale dependence of entrainment-mixing mechanisms in cumulus clouds. *J. Geophys. Res. Atmos.*, **119**, 13,877–13,890.

Magaritz, L., M. Pinsky, and A. Khain, 2010: Effects of stratocumulus clouds on aerosols in the maritime boundary layer. *Atmos. Res.*, **97**, 498–512.

Magaritz, L., M. Pinsky, O. Krasnov, and A. Khain, 2009: Investigation of droplet size distributions and drizzle formation using a new trajectory ensemble model. Part 2: Lucky parcels in non-mixing limit. *J. Atmos. Sci.*, **66**, 781–805.

Magaritz-Ronen, L., M. Pinsky, and A. Khain, 2014: Effects of turbulent mixing on the structure and macroscopic properties of stratocumulus clouds demonstrated by a Lagrangian trajectory model. *J. Atmos. Sci.*, **71**, 1843–1862.

2016a: Drizzle formation in stratocumulus clouds: Effects of turbulent mixing. *Atmos. Chem. Phys.*, **16**, 1849–1862.

2016b: About the horizontal variability of effective radius in stratocumulus clouds. *J. Geophys. Res.*, **121**(16), 9640–9660.

Manton, M.J., 1977: The equation of motion for a small aerosol in a continuum. *PAGEOPH*, **115**, 547–559.

1979: On the broadening of a droplet distribution by turbulence near cloud base. *Q. J. Royal Meteorol. Soc.*, **105**, 899–914.

Marchuk, G.I., 1974: *Numerical Methods in Weather Prediction.* Akademic Press, p. 277.

1980: *Methods of Computational Mathematics* (in Russian). Nauka Press, p. 536.

Martin, G.M., D.W. Johnson, and A. Spice, 1994: The measurements and parameterization of effective radius of droplets in warm stratocumulus clouds. *J. Atmos. Sci.*, **51**, 1823–1842.

Maxey, M.R., 1987: The gravitational settling of aerosol particles in homogeneous turbulence and random flow fields. *J. Fluid Mech.*, **174**, 441–465.

Mazin, I.P., 1965: To the theory of formation of size spectra of particles in clouds and precipitation (in Russian). *Proc. of Central Aerologic Observatory*, **64**, 57–70.

1967: A relationship between fluctuations of supersaturation in clouds and fluctuations of temperature and of vertical flows (in Russian). *Proc. of Central Aerologic Observatory*, **79**, 3–8.

1968: Effect of Phase Transition on Formation of Temperature and Humidity Stratification in Clouds. *Proc. Int. Conf. on Cloud Physics.* Toronto, Ontario, Canada, Amer. Meteor. Soc., 132–137.

Mazin, I.P., A.K. Khrgian, and I.M. Imyanitov, 1989: *Handbook of Clouds and Cloudy atmosphere.* Gidrometeoizdat, p. 647.

Mazin, I.P., and V.M. Merkulovich, 2008: Stochastic condensation and its possible role in liquid cloud microstructure formation (Review). In *Some Problems of Cloud Physics, Collected papers*, Memorial Issue dedicated to Prof. S.M. Shmeter, Moscow, National Geophysical Committee, Russian Academy of Science, 263–295.

Mazin, I.P., and S.M. Shmeter, 1983: *Clouds, Their Structure and Formation.* Gidrometeoizdat, p. 279.

Mazin, I.P., and V.I. Smirnov, 1969: On the theory of cloud drop size spectrum formation by stochastic condensation. *Proceed. CAO*, **89**, 92–94.

McFarquhar, G.M., 2004a: The effect of raindrop clustering on collision-induced break-up of raindrops. *Q. J. Royal Meteorol. Soc.*, **130**, 2169–2190.

2004b: A new representation of collision induced breakup of raindrops and its implications for the shape of raindrop size distributions. *J. Atmos. Sci.*, **61**, 777–794.

Mechem, D. B., and Y.L. Kogan, 2008: A bulk parameterization of giant CCN. *J. Atmos. Sci.*, **65**, 2458–2466.

Merkulovich, V.M., and A.S. Stepanov, 1977: Hygroscopicity effects and surface tension forces during condensational growth of cloud droplet in the presence of turbulence. *Izv. Akad. Sci. USSR, Atmos. Oceanic Phys.*, **13**, 163–171.

Meyers, M.P., R.L Walko, J.Y. Harrington, and W.R. Cotton, 1997: New RAMS cloud microphysics parameterization Part 1: The single-moment scheme. *Atmos. Res.*, **45**, 3–39.

Milbrandt, J.A., and M.K. Yau, 2005: A multimoment bulk microphysics parameterization. Part II: A proposed three-moment closure and scheme description. *J. Atmos. Sci.*, **62**, 3065–3081.

Ming, Y., V. Ramaswamy, L.J. Donner, and V.T.J. Phillips, 2006: A new parameterization of cloud droplet activation applicable to general circulation models. *J. Atmos. Sci.*, **63**(4), 1348–1356, doi:10.1175/JAS3686.1.

Monin, A.S., and A.M. Yaglom, 1971: *Statistical Fluid Mechanics: Mechanics of Turbulence*, vol. **1**. MIT Press, p. 769.

1975: *Statistical Fluid Mechanics: Mechanics of Turbulence*, vol. **2**. MIT Press, p. 874.

Morrison, H., J.A. Curry, and V.I. Khvorostyanov, 2005: A new double-moment microphysics parameterization for application in cloud and climate models. Part I: Description. *J. Atmos. Sci.*, **62**, 1665–1677.

Morrison, H., and W.W. Grabowski, 2008: Modeling supersaturation and subgrid-scale mixing with two-moment bulk warm microphysics. *J. Atmos. Sci.*, **65**, 792–812.

Naumann, A.K., and A. Seifert, 2016: Recirculation and growth of raindrops in simulated shallow cumulus. XVII International Conference on Clouds and Precipitation, Manchester, July 25–29, 2016, ICCP 2016 Conference Guide, p. 34.

Nenes, A., and J.H. Seinfeld, 2003: Parameterization of cloud droplet formation in global climate models, *J. Geophys. Res.*, **108**(D14), 4415, doi: 10.1029/2002JD002911.

Paluch, I.R., 1986: Mixing and the cloud droplet size spectrum: Generalizations from the CCOPE data. *J. Atmos. Sci.*, **43**, 1984–1993.

Paluch, I.R., and D. Baumgardner, 1989: Entrainment and fine-scale mixing in a continental convective cloud. *J. Atmos. Sci.*, **46**, 261–273.

Paluch, I.R., and C.A. Knight, 1984: Mixing and the evolution of cloud droplet size spectra in a vigorous continental cumulus. *J. Atmos. Sci.*, **41**, 1801–1815.

Paoli, R., and K. Shariff, 2009: Turbulent condensation of droplets: Direct simulation and a stochastic model. *J. Atmos. Sci.*, **66**, 723–740.

Pawlowska, H., J.L. Brenguier, and F. Burnet, 2000: Microphysical properties of stratocumulus clouds. *Atmos. Res.*, **55**, 15–33.

Phillips, V.T.J., P.J. DeMott, and C. Andronache, 2008a: An empirical parameterization of heterogeneous ice nucleation for multiple chemical species of aerosols. *J. Atmos. Sci.*, **65**, 2757–2783.

Phillips, V.T.J., S.C. Sherwood, C. Andronache, A. Bansemer, W.C. Conant, P.J. DeMott, R.C. Flagan, A. Heymsfield, H. Jonsson, M. Poellot, T.A. Rissman, J.H. Seinfeld, T. Vanreken, V. Varutbangkul, and J.C. Wilson, 2005: Anvil glaciation in a deep cumulus updraft over Florida simulated with an Explicit Microphysics Model. I: The impact of various nucleation processes. *Q. J. Royal Meteorol. Soc.*, **131**, 2019–2046.

Pinsky, M.B., and A.P. Khain, 1995: A model of homogeneous isotropic turbulence flow and its application for simulation of cloud drop tracks. *Geophys. Astrophys. Fluid Dyn.* **81**, 33–55.

1996: Simulations of drops' fall in a homogeneous isotropic turbulence flow. *Atmos. Res.*, **40**, 223–259.

1997a: Turbulence effects on the collision kernel. Part 1: Formation of velocity deviations of drops falling within a turbulent three-dimensional flow. *Q. J. Royal Meteorol. Soc.*, **123**, 1517–1542.

Pinsky, M.B., A.P. Khain, 1997b: Turbulence effects on the collision kernel. Part 2: Increase of swept volume of colliding drops. *Q. J. Royal Meteorol. Soc.*, **123**, 1543–1560.

Pinsky, M., and A.P. Khain, 1997c: Formation of inhomogeneity in drop concentration induced by the inertia of drops falling in a turbulent flow, and the influence of the inhomogeneity on the drop-spectrum broadening quart. *J. Royal Meteorol. Soc.*, **123**, 165–186.

Pinsky, M.B., and A. Khain, 2001: Fine structure of cloud drop concentration as seen from the Fast-FSSP measurements. Part 1: Method of analysis and preliminary results. *J. Appl. Meteorol.*, **40**, 1515–1537.

Pinsky, M., and A.P. Khain, 2002: Effects of in-cloud nucleation and turbulence on droplet spectrum formation in cumulus clouds. *Q. J. Royol Meteorol. Soc.*, **128**, 501–533.

2003: Fine structure of cloud droplet concentration as seen from the Fast-FSSP measurements. Part 2: Results of in-situ observations. *J. Appl. Meteorol.*, **42**, 65–73.

2004: Collisions of small drops in a turbulent flow. Part 2: Effects of flow accelerations. *J. Atmos. Sci.*, **61**, 1926–1939.

Pinsky, M.B., A.P. Khain, B. Grits, and M. Shapiro, 2006: Collisions of cloud droplets in a turbulent flow. Part 3: Relative droplet fluxes and swept volumes. *J. Atmos. Sci.*, **63**, 2123–2139.

Pinsky, M., A. Khain, H. Krugliak, 2008a: Collisions of cloud droplets in a turbulent flow. Part 5: Application of detailed tables of turbulent collision rate enhancement to simulation of droplet spectra evolution. *J. Atmos. Sci.*, **65**, 357–374.

Pinsky, M.B., A.P. Khain, and Z. Levin, 1999a: The role of the inertia of cloud drops in the evolution of the drop size spectra during drop growth by diffusion. *Q. J. Royal Meteorol. Soc.*, **125**, 553–581.

Pinsky, M., A. Khain, and L. Magaritz, 2010: Representing turbulent mixing of non-conservative values in Eulerian and Lagrangian cloud models. *Q. J. Royal. Meteorol. Soc.*, **136**, 1228–1242.

Pinsky, M.B., A.P. Khain, and M. Shapiro, 1999b: Collisions of small drops in a turbulent flow. Part I: Collision efficiency. Problem formulation and preliminary results. *J. Atmos. Sci.*, **56**, 2585–2600.

2000: Stochastic effects on cloud droplet hydrodynamic interaction in a turbulent flow. *Atmos. Res.*, **53**, 131–169.

Pinsky, M., A. Khain, I. Mazin, and A. Korolev, 2012: Analytical estimation of droplet concentration at cloud base. *J. Geophys. Res.*, **117**, D18211, doi:10.1029/2012JD017753.

Pinsky, M., A.P. Khain, and M. Shapiro, 2001: Collision efficiency of drops in a wide range of Reynolds numbers: Effects of pressure on spectrum evolution. *J. Atmos. Sci.*, **58**, 742–764.

Pinsky, M., A. Khain, and M. Shapiro, 2007: Collisions of cloud droplets in a turbulent flow. Part 4. Droplet hydrodynamic interaction. *J. Atmos. Sci.*, **64**, 2462–2482.

Pinsky, M., A. Khain, A. Korolev and L. Magaritz-Ronen, 2016a: Theoretical investigation of mixing in warm clouds – Part 2: Homogeneous mixing, *Atmos. Chem. Phys.*, **16**, 9255–9272.

Pinsky, M., A. Khain, and A. Korolev, 2016b: Theoretical analysis of mixing in liquid clouds – Part 3: Inhomogeneous mixing, *Atmos. Chem. Phys.*, **16**, 9273–9297.

Pinsky, M., L. Magaritz, A. Khain, O. Krasnov, and A. Sterkin, 2008b: Investigation of droplet size distributions and drizzle formation using a new trajectory ensemble model. Part 1: Model description and first results in a non-mixing limit. *J. Atmos. Sci.*, **65**, 2064–2086.

Pinsky, M., I.P. Mazin, A. Korolev, and A. Khain, 2013: Supersaturation and diffusional droplet growth in liquid clouds. *J. Atmos. Sci.*, **70**, 2778–2793.

2014: Supersaturation and diffusional droplet growth in liquid clouds: Polydisperse spectra. *J. Geophys. Res. Atmospheres*, **119**, 12,872–12,887.

Pinsky, M., M. Shapiro, A. Khain, and H. Wirzberger, 2004: A statistical model of strains in homogeneous and isotropic turbulence. *Physica D.*, **191**, 297–313.

Politovich, M.K., 1993: A study of the broadening of droplet size distribution in cumuli. *J. Atmos. Sci.*, **50**, 2230–2244.

Politovich, M.K., and W.A. Cooper, 1988: Variability of the supersaturation in cumulus clouds. *J. Atmos. Sci.*, **45**, 1651–1664.

Pope, S.B., 2000: *Turbulent Flows*. Cambridge University Press, p. 771.

Prabha, T., A. Khain, B.N. Goswami, G. Pandithurai, R.S. Maheshkumar, and J.R. Kulkarni, 2011: Microphysics of pre-monsoon and monsoon clouds as seen from in-situ measurements during CAIPEEX. *J. Atmos. Sci.*, **68**, 1882–1901.

Prabha, V.T., S. Patade, G. Pandithurai, A. Khain, D. Axisa, P. PradeepKumar, R.S. Maheshkumar, J.R. Kulkarni, and B.N. Goswami, 2012: Spectral width of premonsoon and monsoon clouds over Indo-Gangetic valley during CAIPEEX. *J. Geop. Res.*, **117**, D20205, doi:10.1029/2011JD016837.

Pruppacher, H.R., and J.D. Klett, 1978: *Microphysics of Clouds and Precipitation*. Springer, p. 714.

1997: *Microphysics of Clouds and Precipitation*, second edition. Kluwer Academic Publishers, p. 914.

Pruppacher, H.R., and R. Rasmussen, 1979: A wind tunnel investigation of the rate of evaporation of large water drops falling at terminal velocity in air. *J. Atmos. Sci.*, **36**, 1255–1260.

Raga, G.B., J.B. Jensen, and M.B. Baker, 1990: Characteristics of cumulus band clouds of the coast of Hawaii. *J. Atmos. Sci.*, **47**, 338–355.

Reade, W., and L.R. Collins, 2000: Effect of preferential concentration on turbulent collision rates. *Phys. Fluids*, **12**, 2530–2540.

Reisin, T., Z. Levin and S. Tzivion, 1996: Rain production in convective clouds as simulated in an axisymmetric model with detailed microphysics. Part 1: Description of the model. *J. Atmos. Sci.*, **53**, 497–519.

Reisner, J., R.M. Rassmussen, and R.T. Bruintjes, 1998: Explicit forecasting of supercooled liquid water in winter storms using the MM5 mesoscale model. *Q. J. Royal Meteorol. Soc.*, **124**, 1071–1107.

Respondek, P.S., A.I. Flassman, R.R. Alheit, and H.R. Pruppacher, 1995: A theoretical study of the wet removal of atmospheric pollutants. Part V: The uptake, redistribution, and deposition of $(NH_4)_2SO_4$ by a convective cloud containing ice. *J. Atmos. Sci.*, **52**, 2121–2132.

Rissler, J., E. Swietlicki, J. Zhou, G. Roberts, M.O. Andreae, L.V. Gatti, and P. Artaxo, 2004: Physical properties of the sub-micrometer aerosol over the Amazon rain forest during the wet-to-dry season transition-comparison of modelled and meaqsured CCN concentrations. *Atmos. Chem. Phys. Discuss.*, 4, 3159–3225.

Rissler, J., A. Vestin, E. Swietlicki, G. Fisch, J. Zhou, P. Artaxo, and M.O. Andreae, 2006: Size distribution and hygroscopic properties of aerosol particles from dry-season biomass burning in Amazonia. *Atmos. Chem. Phys.*, **6**, 471–491, doi:10.5194/acp-6-471-2006.

Roesner, S., A.I. Flossmann, and H.R. Pruppacher, 1990: The effect on the evolution of the drop spectrum in clouds of the preconditioning of air by successive convective elements. *Q. J. Royal Meteorol. Soc.*, **116**, 1389–1403.

Rogers, R.R., 1975: An elementary parcel model with explicit condensation and supersaturation. *Atmosphere*, **13**, 192–204.

Rogers, R.R, and M.K. Yau, 1996: *Short Course in Cloud Physics*. Butterworth-Heinemann, p. 304.

Rosenfeld, D., and G. Gutman, 1994: Retrieving microphysical properties near the tops of potential rain clouds by multispectral analysis of AVHRR data. *Atmos. Res.*, **34**, 259–283.

Saffman, P.G., J.S. Turner, 1956: On the collision of drops in turbulent clouds. *J. Fluid Mech.*, **1**, 16.

Saxena, V.K., and N. Fukuta, 1976: *Preprints Cloud Phys. Conf.*, Boulder, p. 26, Am. Meteor. Soc., Boston.

Saxena, V.K., and R.S. Rathore, 1984: *Preprints 11th Int. Conf. on Atmos. Aerosol, Condensation and Ice Nuclei*, p. 292, Hungarian Meteor. Soc., Budapest.

Schlamp, R.J., S.N. Grover, H.R. Pruppacher, and A.E. Hamielec, 1976: A numerical investigation of the effect of electric charges and vertical external electric fields and the collision efficiency of cloud drops. *J. Atmos. Sci.*, **33**, 1747–1755.

Schlüter, M.H., 2006: The effects of entrainment and mixing process on the droplet size distribution in cumuli. A thesis submitted to the faculty of The University of Utah in partial fulfillment of the requirements for the degree of Master of Science, Department of Meteorology, The University of Utah, p. 92 (MS Study).

Schlüter, M.H., S.K. Krueger, and C.-W. Su, 2006: The effects of entrainment and mixing on the droplet size distributions in cumuli, 12th Conference on Cloud Physics. Madison, WI, p 234.

Sedunov, Y.S., 1965: The fine structure of the clouds and its role in formation of the cloud droplet spectra. *Izv. Acad. Sci. USSR, Atmos. Oceanic. Phys.*, **1**, 722–731.

1974: *Physics of Drop Formation in the Atmosphere*. Wiley, p. 234.

Segal, Y., and A. Khain, 2006: Dependence of droplet concentration on aerosol conditions in different cloud types: Application to droplet concentration parameterization of aerosol conditions. *J. Geophys. Res.*, **111**, D15204.

Segal, Y., M. Pinsky, A. Khain, and C. Erlick, 2003: Theromodynamic factors influencing the bimodal spectra formation in cumulus clouds. *Atmos. Res.*, **66**, 43–64.

Seifert, A., and K.D. Beheng, 2001: A double-moment parameterization for simulating autoconversion, accretion and selfcollection. *Atmos. Res.*, **59–60**, 265–281.

Seifert, A., A. Khain, U. Blahak, and K.D. Beheng, 2005: Possible effects of collisional breakup on mixed-phase deep convection simulated by a spectral (bin) cloud model. *J. Atmos. Sci.*, **62**, 1917–1931.

Seifert, A., L. Nuijens, and B. Stevens, 2010: Turbulence effects on warm-rain autoconversion in precipitating shallow convection. *Q. J. Royal Meteorol. Soc.*, **136**, 1753–1762.

Shaw, R.A., 2003: Particle-turbulence interactions in atmospheric clouds. *Annu. Rev. Fluid Mech.*, **35**, 183–227.

Shipway, B.J., and S.J. Abel, 2010: Analytical estimation of cloud droplet nucleation based on an underlying aerosol population. *Atmos. Res.*, **96**, 344–355.

Shmeter, S.M., 1987: *Thermodynamics and Physics of convective clouds*. Gidrometizdat, p. 288.

Siebert, H., K. Lehmann, and M. Wendisch, 2006: Observations of small-scale turbulence and energy dissipation rates in the cloudy boundary layer. *J. Atmos. Sci.*, **63**, 1451–1466.

Simmel, M., T. Trautmann, and G. Tetzlaff, 2002: Numerical solution of the stochastic collection equation – comparison of the linear discrete method with other methods. *Atmos. Res.*, **61**, 135–148.

Simpson, J., G.V. Helvoirt, and M. McCumber, 1982: Three-dimensional simulations of cumulus congestus clouds on GATE day 261. *J. Atmos. Sci.*, **39**, 126–145.

Small, J.D., and P.Y. Chuang, 2008: New observations of precipitation initiation in warm cumulus clouds. *J. Atmos. Sci.*, **65**, 2972–2982.

Squires, P., 1952: The growth of cloud drops by condensation. *Aust. J. Sci. Res.*, **5**, 66–86.

1958: Penetrative downdraughts in cumuli. *Tellus*, **10**, 381–389.

Srivastava, R.C., 1971: Size distribution of raindrops generated by their breakup and coalescence. *J. Atmos. Sci.*, **28**, 410–415.

1989: Growth of cloud drops by condensation: A criticism of currently accepted theory and a new approach. *J. Atmos. Sci.*, **46**, 869–887.

Stein, D., H.W. Georgii, and V. Kramm, 1985: *Meteor. Rundschau*, **38**, 15.

Stepanov, A.S., 1975: Condensational growth of cloud droplets in a turbulized atmosphere. *Izv. Akad. Sci. USSR, Atmos. Oceanic Phys.*, **11**, 27–42.

Stevens, B., et al., 2003: On entrainment rates in nocturnal maritime stratocumulus. *Q. J. Royal Meteorol. Soc.*, **129**, 3469–3492.

Stevens, B.G., et al., 2005: Evaluation of large-eddy simulations via observations of nocturnal marine stratocumulus. *Mon. Wea. Rev.*, **133**, 1443–1455.

Stevens, B., R.L. Walko, W.R. Cotton, and G. Feingold, 1996: The spurious production of cloud-edge supersaturations by Eulerian models. *Mon. Wea. Rev.*, **124**, 1034–1041.

Straka, J.M., and Anderson, J.R., 1993: Numerical simulations of microburst-producing storms – Some results from storms observed during COHMEX. *J. Atmos. Sci.*, **50**, 1329–1348.

Straka, J.M., 2009: *Cloud and Precipitation Microphysics. Principles and Parameterizations*. Cambridge University Press, p. 392.

Strapp, J.W., W.R. Leaitch, and P.S.K. Liu, 1992: Hydrated and dried aerosol-size-distribution measurements from the particle measuring system FSSP-300 probe and the deiced PCASP-100X probe. *J. Atmos. Oceanic Technol.*, **9**, 548–555.

Straub, W., K.D. Beheng, A. Seifert, J. Schlottke, and B. Weigand, 2010: Numerical investigation of collision-induced breakup of raindrops. Part II: Parameterizations of coalescence efficiencies and fragment size distributions. *J. Atmos. Sci.*, **67**, 576–588.

Su, C.-W., S.K. Krueger, P.A. McMurtry, and P.H. Austin, 1998: Linear eddy modeling of droplet spectral evolution during entrainment and mixing in cumulus clouds. *Atmos. Res.*, **47–48**, 41–58.

Szakáll, M., K. Diehl, S.K. Mitra, S. Borrmann, 2009: A wind tunnel study on the shape, oscillation, and internal circulation of large raindrops with sizes between 2.5 and 7.5 mm. *J. Atmos. Sci.*, **66** (3), 755–765.

Szakáll, M., S.K. Mitra, K. Diehl, and S. Borrmann, 2010: Shapes and oscillations of falling raindrops. *Atmos. Res.*, **97**, 416–425.

Takahashi, T., 1976: Hail in axisymmetric cloud model. *J. Atmos. Sci.*, 33, 1579–1601.

Telford, J.W., and S.K. Chai, 1980: A new aspect of condensation theory. *Pageoph*, **118**, 720–742.

Telford, J.W., T.S. Keck, and S.K. Chai, 1984: Entrainment at cloud tops and the droplet spectra. *J. Atmos. Sci.*, **41**, 3170–3179.

Temam, R., 1977: *Theory and Numerical Analysis of the Navier_Stokes Equations*. North-Holland, p. 465.

Testik, F.Y., and A.P. Barros, 2007: Toward elucidating the microstructure of warm rainfall: A survey. *Rev. Geophys.*, **45**, doi:10.1029/2005RG000182.

Tokay, A., R. Chamberlain, and M. Schoenhuber, 2000: Laboratory and field measurements of raindrop oscillations. *Phys. Chem. Earth (B)*, **25**, 867–870.

Twomey, S.A., 1959a: The nuclei of natural cloud formation. part I: The chemical diffusion method and its application to atmospheric nuclei. *Geofis. Pure Appl.*, **43**, 227–242.

Twomey, S., 1959b: The nuclei of natural cloud formation II. The supersaturation in natural clouds and the variation of cloud droplet concentration. *Pure Appl. Geophys.*, **43**, 243–249.

Twomey, S., and T.A. Wojciechowski, 1969: Observations of the geographical variation of cloud nuclei. *J. Atmos. Sci.*, **26**, 684–688.

Tzivion, S., G. Feingold, and Z. Levin, 1987: An efficient numerical solution to the stochastic collection equation. *J. Atmos. Sci.*, **44**, 3139–3149.

1989: The evolution of raindrop spectra. Part 2. Collisional collection/breakup and evaporation in a rainshaft. *J. Atmos. Sci.*, **46**, 3312–3327.

Tzivion, S., T.G. Reisin, and Z. Levin, 1999: A numerical solution of the kinetic collection equation using high spectral grid solution: A proposed reference. *J. Comput. Phys.*, **148**, 527–544.

2001: A new formulation of the spectral multi-moment method for calculating the kinetic collection equation: More accuracy with few bins. *J. Comput. Phys.*, **171**, 418.

Vaillancourt, P.A., M.K. Yau, 2000: Review of particle-turbulence interactions and consequences for Cloud Physics. *Bull. Am. Meteorol. Soc.*, **81**, 285–298.

Vaillancourt, P.A., M.K. Yau, P. Bartello, and W.W. Grabowski, 2002: Microscopic approach to cloud droplet growth by condensation. Part 2: Turbulence, clustering, and condensational growth. *J. Atmos. Sci.*, **59**, 3421–3435.

Van Zanten, M.C., B. Stevens, G. Vali, and D.H. Lenschow, 2005: Observations of drizzle in nocturnal marine stratocumulus. *J. Atmos. Sci.*, **62**, 88–106.

Verlinde, J., and W.R. Cotton, 1993: Fitting microphysical observations of nonsteady convective clouds to a numerical model: An application of the adjoint technique of data assimilation to a kinematic model. *Mon. Wea. Rev.*, **121**, 2776–2793.

Verlinde, J., P.J. Flatau, and W.R. Cotton, 1990: Analytic solution to the collection growth equation: Comparison with an approximate methods and utilization in microphysics parameterization schemes. *J. Atmos. Sci.*, **47**, 2871–2880.

Villermaux, E., and B. Bossa, 2009: Single-drop fragmentation determines size distribution of raindrops. *Nature Physics*, **5**, 697–702.

Vohl, O., S.K. Mitra, S. Wurzler, K. Diehl, and H.R. Pruppacher, 2007: Collision efficiencies empirically determined from laboratory investigations of collisional growth of small raindrops in a laminar flow field. *Atmos. Res.*, **85**, 120–125.

Vohl, O., S.K. Mitra, S.C. Wurzler, and H.R. Pruppacher, 1999: A wind tunnel study on the effects of turbulence on the growth of cloud drops by collision and coalescence. *J. Atmos. Sci.*, **56** (24), 4088–4099.

Voloshchuk, V.M., and Y.S. Sedunov, 1977: A kinetic equation for the evolution of the droplet spectra in a turbulent medium at the condensation stage of cloud development. *Sov. Meteorol. Hydrol.*, **3**, 3–14.

Voth, G.A., A. La Porta, A.M. Crawford, J. Alexander, and E. Bodenschatz, 2002: Measurements of particle accelerations in fully developed turbulence. *J. Fluid Mech.*, **469**, 121–160.

Walko, R.L., W.R. Cotton, M.P. Meyers, and J.Y. Harrington, 1995: New RAMS cloud microphysics parameterization. Part 1: The single-moment scheme. *Atmos. Res.*, **38**, 29–62.

Wang, L.-P., O. Ayala, and W.W. Grabowski, 2005a: Improved formulations of the superposition method. *J. Atmos. Sci.*, **62**, 1255–1266.

Wang, L.-P., O. Ayala, S. Kasprzak, and W. Grabowski, 2005b: Theoretical formulation of collision rate and collision efficiency of hydrodynamically interacting cloud droplets in turbulent atmosphere. *J. Atmos. Sci.*, **62**, 2433–2450.

Wang, L.-P., O. Ayala, B. Rosa, and W. Grabowski, 2008: Turbulent collision efficiency of heavy particles relevant to cloud droplets. *New J. Phys.* **10**, 075013.

Wang, L.-P., C.N. Franklin, O. Ayala, and W. Grabowski, 2006a: Probability distributions of angle of approach and relative velocity for colliding droplets in a turbulent flow. *J. Atmos. Sci.*, **63**, 881–900.

Wang, L.-P., and W. Grabowski, 2009: The role of air turbulence in warm rain initiation. *Atmos. Sci. Let.*, **10**, 1–8.

Wang, L.-P., and M.R. Maxey, 1993: Settling velocity and concentration distribution of heavy particles in homogeneous isotropic turbulence. *J. Fluid Mech.*, **256**, 27–68.

Wang, L.-P., A.S. Wexler, and Y. Zhou, 1998: Statistical mechanical descriptions of turbulent coagulation. *Phys. Fluid*, **10**, 2647–2651.

2000: Statistical mechanical description and modeling of turbulent collision of inertial particles. *J. Fluid Mech.*, **415**, 117–153.

Wang, L.-P., Y. Xue, O. Ayala, and W.W. Grabowski, 2006b: Effects of stochastic coalescence and air turbulence on the size distribution of cloud droplets. *Atmos. Res.*, **82**, 416–432.

Wang, L.-P., Y. Xue, and W.W. Grabowski, 2007: A bin integral method for solving the kinetic collection equation. *J. Comp. Phys.*, **226**, 59–88.

Wang, P.K., and H.R. Pruppacher, 1977: Acceleration to terminal velocity of cloud and raindrops. *J. Appl. Meteorol.*, **16**, 275–280.

Warner, J., 1969a: The microstructure of cumulus cloud. Pt. I, General features of the droplet spectrum. *J. Atmos. Sci.*, **26**, 1049–1059.

1969b: The microstructure of cumulus cloud. Part 2. The effect of droplet size distribution of the cloud nucleus spectrum and updraft velocity. *J. Atmos. Sci.*, **26**, 1272–1282.

1970: The microstructure of cumulus cloud. Part 3. The nature of the updraft. *J. Atmos. Sci.*, **27**, 682–688.

Wex, H., F. Stratmann, D. Topping, and G. McFiggans, 2008: The Kelvin versus the Raoult Term in the Köhler Equation. *J. Atmos. Sci.*, **65**, 4004–4016.

Woo, S.E., and A.E. Hamielec, 1971: A numerical method of determining the rate of evaporation of small water drops falling at terminal velocity in air. *J. Atmos. Sci.*, **28**, 1448–1454.

Wyngaard, J.C., 2010: *Turbulence in the Atmosphere*. Cambridge University Press, p. 393.

Xue, Y., L.-P. Wang, and W. Grabowski, 2008: Growth of cloud droplets by turbulent collision–coalescence. *J. Atmos. Sci.*, **65**, 331–356.

Yin, Y., Z. Levin, T. Reisin, and S. Tzivion, 2000: The effects of giant cloud condensational nuclei on the development of precipitation in convective clouds: A numerical study. *Atmos. Res.*, **53**, 91–116.

Yin, Y., K.S. Carslaw, and G. Feingold, 2005: Vertical transport and processiong of aerosols in a mixed-phase convective cloud and the feedback on cloud development. *Q. J. Royal Meteorol. Soc.*, **131**, 221–245.

Zhou, Y., A. Wexler, and L. Wang, 2001: Modeling turbulent collision of bidisperse inertial particles. *J. Fluid Mech.*, **433**, 77–104.

6 Microphysical Processes in Ice and Mixed-Phase Clouds

Clouds containing liquid drops and ice particles belong to the mixed-phase cloud type. Most deep convective clouds, even in the tropics, are mixed-phase clouds. To a large extent, precipitation from deep convective clouds is related to ice processes. Precipitating ice particles can melt below the melting level or fall to the surface in the form of snow or graupel/hail. Severe weather phenomena such as hailstorms, tornados, icing, and *freezing rain* leading to surface icing are closely related to the ice microphysics. Anvils of deep convective clouds, Arctic stratocumulus clouds and high-altitude cirrus clouds consist mainly of ice particles that determine to a large extent Earth's radiation budget. While Chapter 5 describes warm cloud microphysics, this chapter deals with cold cloud microphysical processes, i.e., those related to formation and transformation of ice.

There are three major features of mixed-phase and ice microphysics. The first feature is that cold microphysics deals with transitions between three phases (gas, liquid, and solid), making its modeling much more complicated than that of warm microphysics. Cold processes involve both drops and ice particles of different shapes and densities (Section 2.4). In some cloud zones, e.g., the melting zone or the zone of hail wet growth, ice particles contain a liquid water fraction. The amount of liquid water in ice particles is determined by complicated thermodynamic processes of freezing/melting. Liquid water exists in melting ice particles at positive temperatures as well as in freezing raindrops at freezing temperatures. Intense collisions of hail or graupel with supercooled droplets can trigger wet growth of hail when the release of the latent heat of freezing at the hail particle surface prevent total liquid water freezing, so at low temperatures hail particles get covered with a film of liquid water.

In warm clouds, supersaturation over water plays the key role in DSD formation (Chapter 5). In modeling of mixed-phase clouds, it is necessary to deal with supersaturations with respect to both water and ice. As mentioned in Section 5.1, supersaturation over a liquid drop depends on the drop's curvature. Ice crystals have complicated shapes and supersaturation is different over different parts of the crystals, leading to faster growth of ice "fingers."

The second feature is related to the lack of theoretical understanding of many ice processes (e.g., there is no reliable theory of collisions of nonspherical ice particles, especially in turbulent flows). Some processes are described by theories that contain multiple uncertain parameters (e.g., theory of ice nucleation, of drop freezing, etc.). As a result, when investigators analyze or model cold microphysical processes, they often have to use empirical and semiempirical descriptions. In comparison to warm microphysical processes, uncertainties in description of cold processes are much larger.

The third feature concerns in situ airborne measurements in mixed-phase clouds or high-level ice clouds that are dangerous to perform due to high vertical velocities, possible airplane icing, etc. According to the current regulations, research airplanes should not penetrate zones of intense convection, which means that most observations of ice microphysics are performed in stratiform clouds or near cloud tops of growing cumulus clouds. As a result, the volume of observed data within convective mixed-phased clouds is limited. In addition, it is much more difficult to perform measurements of size distributions and other parameters of cloud ice than of water drops in warm clouds. Currently available probes are error prone, therefore the ice contents measured along the same aircraft traverses by different probes may differ by several times. Measurements of ice particle concentrations that became widely available in the mid-1970s revealed the existence of multiple small ice crystals within a wide range of temperatures at both subsaturated and supersaturated conditions. These concentrations may significantly exceed concentrations of IN that trigger ice formation in clouds. Attempts to explain these high concentrations were made in many theoretical studies and led to various hypotheses on ice multiplication (e.g., Chisnell and Latham, 1976). However, it has been found recently that some measurements of ice crystal concentration were erroneous due to probe-induced particle shattering (e.g., Korolev and Isaac, 2005). Despite the fact that new ice probes also indicate the crucial role of ice multiplication, these finding required corrections of the corresponding experimental data concerning ice particle concentrations obtained using the old probes.

The purpose of this chapter is to present the current state of knowledge on cold microphysical processes in view of recent findings and to describe representation of these processes in cloud models.

6.1 Main Ice-Related Processes and Ice Particles Description

6.1.1 Overview of Ice-Related Processes

The list of the main cold microphysical processes is presented in **Table 6.1.1**. Ice particles form under negative temperatures (Celsius) above the level of the 0°C isotherm known as the freezing level. The freezing level in the tropics is located at the altitude of 4–4.5 km. At midlatitudes, the altitude of the freezing level depends on the latitude, and is around 2 km during the summer period. During cold periods, the surface temperature can be below zero. When liquid drops ascend above the freezing level, they do not freeze immediately, but ascend without being frozen, sometimes to high levels. Liquid drops existing at freezing temperatures are in the supercooled or metastable state. Small droplets can reach altitudes of ~10 km where the temperature is as low as −35°C and remain unfrozen. Drop freezing at negative temperatures indicates the transfer from the metastable state to the stable one.

Drop freezing can be *homogeneous* or *heterogeneous*. Homogeneous freezing is of a stochastic nature. Random motion of water molecules within a drop can lead to formation of local zones with a molecular lattice similar to that of ice. At low temperatures, these zones become ice embryos forming so-called ice germs. If the size of the germ exceeds its critical value that depends on temperature and some other factors, the germ begins growing leading to the drop freezing. The probability of homogeneous freezing dramatically increases as the temperature falls below −35°C. Homogeneous freezing is the main source of small ice crystals in anvils of deep convective clouds.

Mechanisms of ice formation are usually separated into primary and secondary production. Primary ice production includes nucleation mechanisms (modes), which require the existence IN. IN are insoluble aerosols having specific properties. Schematically, primary ice formation mechanisms (or nucleation modes) are

Table 6.1.1 Main ice-related microphysical processes.

Processes	Short Description
Primary ice formation, homogeneous nucleation, inhomogeneous nucleation, freezing of haze	Formation of ice crystals by condensation of water molecules on ice nuclei or formation of ice crystals on insoluble particles (ice nuclei (IN)) within haze particles.
Secondary ice formation	Formation of new ice crystals (ice fragments) by collisions between ice particles or between ice particles and water drops. Formation of ice fragments during drop freezing.
Deposition/sublimation	Condensation/evaporation growth of ice particles
Sedimentation	Falling of ice particles
Spontaneous breakup, collision breakup	Breakup of ice particles due to random fluctuations of internal stresses or due to collisions with other particles
Collisions and coalescence	Collisions and coalescence between ice particles
Accretion, riming,	Formation of rimed particles via collection of liquid droplets by ice particles. If the collector are snowflakes of low density, riming leads to an increase in bulk density. If the collector is of high density, riming may decrease the ice density forming particles of higher porosity. Riming determines transformation of one type of particles into another, for instance, transformation of aggregates into graupel.
Freezing/Melting	Phase transition of liquid drops to ice/ phase transition of ice to liquid drops
Shedding	Shedding of the water film (or fraction of the film) from the surface of hail particles leading to formation of several liquid drops
Conservative and nonconservative mixing	Mixing of cloud volumes or cloud and environment volumes that is accompanied by phase transitions related to ice

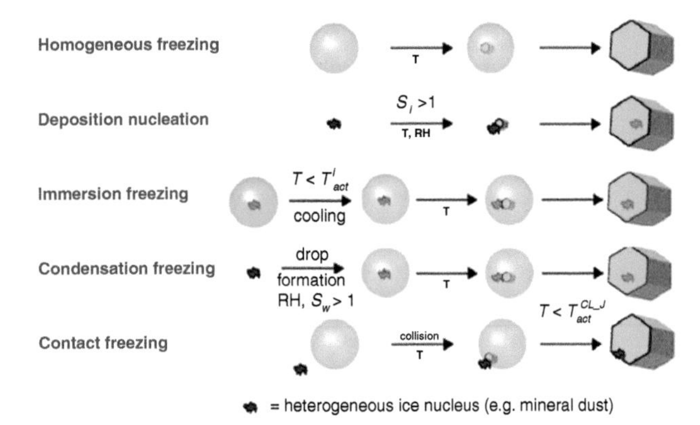

Figure 6.1.1 Schematic representation of ice nucleation modes (primary ice formation) in atmospheric clouds (from Vali, 1999, with changes; courtesy of G. Vali).

shown in **Figure 6.1.1** (see also Table 6.1.1). Heterogeneous freezing, or *immersion* freezing, takes place when an IN initially exists within a drop or a haze particle, and nucleation takes place when the temperature falls below some threshold value that depends on many factors, mainly on the IN size and temperature. Not all aerosols containing an insoluble fraction can act as IN, but only those whose insoluble fraction is large enough. Thus, IN concentration is significantly lower than that of atmospheric aerosols. Immersion freezing is also referred to as nucleation freezing because it implies activation of immersed IN. The main specific feature of immersion freezing is that it produces ice particles at temperatures warmer than those required for homogeneous freezing. In the presence of IN, the germ forms on IN sites, and the energy barrier to overcome for nucleation to take place is much lower than in homogeneous nucleation. The decreasing value of the energy barrier depends on properties of the water–ice interface (so-called wetting parameter). The closer the crystal lattice of an insoluble particle is to that of ice, the higher the efficiency of the particle as IN. Immersion freezing of drops and haze is only one of many mechanisms of ice production. There are several mechanisms of primary ice crystal formation on IN located in the air.

Deposition nucleation takes place during condensation of water vapor molecules onto the surface of IN. It is difficult to distinguish deposition nucleation from condensational freezing, so these modes are often considered in parameterization schemes as a single process.

Condensation freezing takes place when water vapor condenses on the surface of a particle that serves as IN, when the particle is already covered with some water film. This mode is efficient when there is supersaturation with respect to water. Freezing of haze particles, condensational freezing, and immersion freezing, which are thermodynamically undistinguishable, belong to the *deliquescent-heterogeneous* freezing (DHF) mode (Khvorostyanov and Curry, 2005a).

Contact freezing takes place when drops contact IN due to collisions. Drop freezing by contact nucleation can take place at temperatures substantially higher than those in the deposition-nucleation mode.

All mechanisms of primary ice formation require low temperatures and supersaturation with respect to ice.

In addition to primary ice formation, there exist mechanisms of *secondary ice formation*, also known as *ice multiplication*. The most known quantitatively investigated mechanism of ice multiplication (the H-M mechanism) was found by Hallet and Mossop (1974) in laboratory experiments. According to this mechanism, collisions between graupel and water drops with diameters exceeding 24 μm lead to the ejection of ice splinters.

Another type of secondary ice production is fragmentation during freezing of drops while they are spinning or tumbling. Such fragmentation was reported in several studies (Gagin, 1972; Pruppacher and Schlamp, 1975; Kolomeychuk et al., 1975). According to laboratory measurements in a wind tunnel, Pruppacher and Schlamp (1975) showed "explosion" of drops due to their freezing when about 40% of investigated drops produced one splinter.

Upon being formed, ice particles (largely ice crystals) grow by diffusion (or condensational) growth. This process is known as *deposition* of water vapor molecules on the surface of an ice particle. Depending on temperature, deposition growth leads to formation of ice crystals of various types and shapes (Section 2.4). *Sublimation* is the process opposite to deposition and is described by the same equation. Sublimation is strong in anvils of deep convective clouds, leading to an increase in water vapor at the upper levels, but preventing formation of precipitating particles. Thus, formation of large anvils often decreases precipitation efficiency of deep convective clouds.

Collisions between ice crystals accompanied by coalescence result in formation of aggregates or snowflakes. Collisions between ice crystals or aggregates with water drops, as well as collisions of frozen drops with small water droplets lead to formation of rimed particles of higher density. In case their bulk density exceeds about 0.2 g/cm^3, aggregates transform into graupel. The rate of riming determines the porosity of particles. Under some conditions, e.g., when the mass content of supercooled water drops is high and riming is very intense, the density of falling graupel or of frozen drops becomes close to the density of pure ice (0.91 g/cm^3). Such particles are called hail; their diameter D often exceeds 1 cm. Often, the condition $D > 1$ cm is included into hail particles definition. The rate of collisions between ice particles as well as precipitation rates depend on characteristics of ice particle motion, and specifically, on the sedimentation velocity. In case of intense collisions of big hail particles with liquid drops, the release of latent heat of freezing does not allow collected water to freeze totally on the hail surface. As a result, a falling hail stone is coated with liquid water even at very low environment temperatures. The collision growth of liquid-coated hail particles is known as the *wet growth* of hail.

Larger particles or particles of larger density are cold precipitation particles. Depending on the environment temperature and the fall velocity, these particles can either melt below the freezing level or reach the surface. Their melting process is quite complicated. Depending on an ice particle's structure, the melted water can either get sucked into the particle (graupel) or remain on the particle's surface (hail). If the water film depth reaches its threshold value, the process of *shedding* takes place. The shedding triggers formation of one or several rain drops. Melting of snowflakes significantly increases the radar reflectivity of such particles (by ~10 dBZ), forming a horizontal layer of high reflectivity called the *bright band* located just below the freezing level. If snowflakes are large enough and have an asymmetric structure, while the relative humidity is comparatively low (around 70% and lower), freely falling snowflakes undergo breakup and melt.

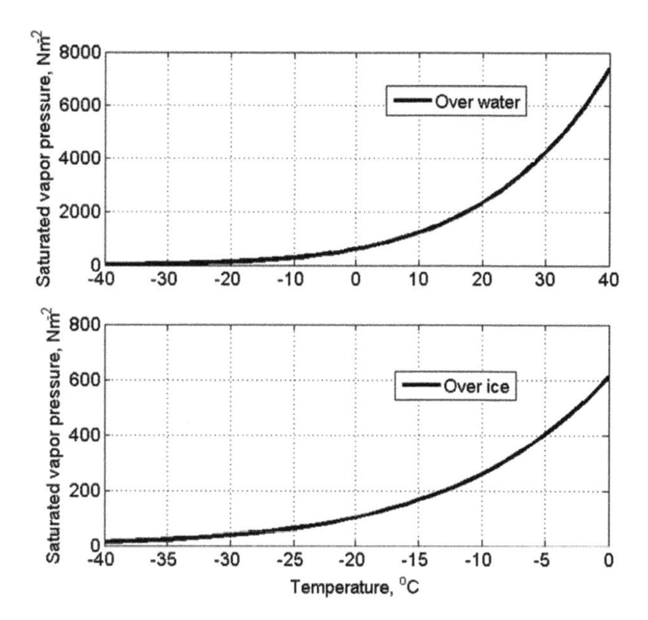

Figure 6.1.2 Temperature dependence of saturated vapor pressure over water and over ice.

6.1.2 Dependences of the Thermodynamic Parameters on Temperature

Similarly to saturation over plane water surface, there exists saturation of water vapor over plane ice surface, which corresponds to an equilibrium state when the flux of evaporating molecules from the ice surface is equal to the flux of water vapor molecules toward the ice surface. Naturally, that kind of equilibrium can be reached only at freezing temperatures. Evaporation of water molecules from a solid ice surface is much weaker than from a surface of liquid water, due to a stronger attraction between molecules in a solid body. It means that a lower flux of water vapor molecules from the environment air to the ice surface is needed to compensate the sublimation of molecules from the ice surface in the equilibrium state. Therefore, the saturation water vapor pressure over ice is lower than that over liquid water, and supersaturation with respect to ice is higher, sometimes significantly, than that over the water surface. The difference in saturation pressures is very important in the microphysical processes in ice clouds and mixed-phase clouds. Under certain conditions, water vapor pressure corresponds to subsaturation with respect to water, but to supersaturation with respect to ice. In this case, drops evaporate and the released water vapor contributes to ice growth. As a result, water mass is transferred from liquid drops to ice particles. This process, known as the *Wegener-Bergeron-Findeizen* (WBF) process, plays an important role in cloud glaciation.

Saturation water vapor pressure over water and over ice strongly depends on the temperature of the environment air. The dependences shown in **Figure 6.1.2** were calculated using the approximation formulas proposed by Flatau et al. (1992) (**Table 6.1.2**). These formulas are widely used in cloud models.

It is known that freezing of water and melting of ice lead to release/absorption of latent heat characterized by heat of fusion L_m. The latent heat of ice sublimation/deposition L_i can be calculated as a sum of the latent heat of condensation/evaporation L_w (Section 3.1) and the latent heat of fusion $L_i = L_m + L_w$ (Pruppacher and Klett, 1997). The cold microphysics operates with the thermodynamic parameters such as the surface tension at water–air interface σ_w, heat conductivity of air k_a, liquid water k_w, and ice k_i; and surface tension at ice–air σ_i and water–ice σ_{iw} interfaces. These parameters depend on temperature. The temperature dependencies obtained in laboratory measurements are often expressed using the polynomial approximation (Table 6.1.2). The formulas in Table 6.1.2 are used in cloud models to describe the microphysical processes. Some temperature dependences presented in Table 6.1.2 are shown in **Figures 6.1.3** and **6.1.4**.

Table 6.1.2 Temperature dependencies of different thermodynamic parameters

Parameters, Units	Approximated Polynomial
Latent heat release of freezing/melting, J/kg	$L_m = \begin{cases} \sum_{k=0}^{6} a_k T_C^k & T_C \le 0°\text{C} \\ 3.3354848 \times 10^5 & T_C > 0°\text{C} \end{cases}$ $a_0 = 3.3354848 \times 10^5$, $a_1 = 2.0390724 \times 10^3$, $a_2 = -3.3637268 \times 10^1$, $a_3 = -3.470628$, $a_4 = -1.93668992 \times 10^{-1}$, $a_5 = -4.7957008 \times 10^{-3}$, $a_6 = -4.646332 \times 10^{-5}$
Latent heat of condensation/ evaporation, J/kg	$L_w = \sum_{k=0}^{3} a_k T_C^k$ $a_0 = 2.50079 \times 10^6$, $a_1 = -2.36418 \times 10^3$, $a_2 = 1.58927$, $a_3 = -6.143419998 \times 10^{-2}$
Latent heat of sublimation/deposition, J/kg	$L_i = L_m + L_w$
Surface tension of water–air interface, N m^{-1}	$\sigma_w = \begin{cases} \sum_{k=0}^{6} a_k T_C^k & T_C \le 5.5°\text{C} \\ 7.566165 \times 10^{-2} - 1.55 \times 10^{-4} T_C & T_C > 5.5°\text{C} \end{cases}$ $a_0 = 7.593 \times 10^{-2}$, $a_1 = 1.15 \times 10^{-4}$, $a_2 = 6.818 \times 10^{-5}$, $a_3 = 6.511 \times 10^{-6}$, $a_4 = 2.933 \times 10^{-7}$, $a_5 = 6.283 \times 10^{-9}$, $a_6 = 5.285 \times 10^{-11}$
Surface tension of ice–air interface, N m^{-1}	$\sigma_i = \begin{cases} \sum_{k=0}^{3} a_k T_C^k & T_C \le -36°\text{C} \\ 2.8 \times 10^{-2} + 2.5 \times 10^{-4} T_C & -36° < T_C \le 0°\text{C} \mid \\ 2.8 \times 10^{-2} & T_C > 0°\text{C} \end{cases}$ $a_0 = 1.89081 \times 10^{-1}$, $a_1 = 1.31625 \times 10^{-2}$, $a_2 = 3.469 \times 10^{-4}$, $a_3 = 3.125 \times 10^{-6}$
Density of liquid water, kg/m^3	$\rho_w = \begin{cases} \sum_{k=0}^{6} a_k T_C^k & T_C \le 0°\text{C} \\ (1 + b_6 T_C)^{-1} \sum_{k=0}^{5} b_k T_C^k & T_C > 0°\text{C} \end{cases}$ $a_0 = 9.9986 \times 10^2$, $a_1 = 6.69 \times 10^{-2}$, $a_2 = -8.486 \times 10^{-3}$, $a_3 = 1.518 \times 10^{-4}$, $a_4 = -6.9984 \times 10^{-6}$, $a_5 = -3.6449 \times 10^{-7}$, $a_6 = -7.497 \times 10^{-9}$ $b_0 = 9.998396 \times 10^2$, $b_1 = 1.8224944 \times 10^1$, $b_2 = -7.92221 \times 10^{-3}$, $b_3 = -5.544846 \times 10^{-5}$, $b_4 = 1.497562 \times 10^{-7}$, $b_5 = -3.932952 \times 10^{-10}$, $b_6 = 1.8159725 \times 10^{-2}$
Density of pure ice, kg/m^3	$\rho_i = \sum_{k=0}^{2} a_k T_C^k$ $a_0 = 9.167 \times 10^2$, $a_1 = -1.75 \times 10^{-1}$, $a_2 = -5.0 \times 10^{-4}$
Specific heat of water, J kg^{-1} K^{-1}	$c_w = \begin{cases} \sum_{k=0}^{4} a_k T_C^k & T_C \le 0°\text{C} \\ \sum_{k=0}^{4} b_k T_C^k & T_C > 0°\text{C} \end{cases}$ $a_0 = 4.187924592 \times 10^3$, $a_1 = -1.13185568 \times 10^1$, $a_2 = -9.721524 \times 10^{-2}$, $a_3 = 1.83167152 \times 10^{-2}$, $a_4 = 1.13537024 \times 10^{-3}$ $b_0 = 4.214961077 \times 10^3$, $b_1 = -3.6346408$, $b_2 = 1.2982952 \times 10^{-1}$, $b_3 = -2.225888 \times 10^{-3}$, $b_4 = 1.58992 \times 10^{-5}$

Table 6.1.2 (cont.)

Parameters, Units	Approximated Polynomial
Specific heat of ice, J kg^{-1} K^{-1}	$c_i = \sum_{k=0}^{1} a_k T_C^k$ $a_0 = 2.1046 \times 10^3$, $a_1 = 7.322$
Saturation pressure of water vapor over plane water surface, Nm^{-2}	$e_w = \sum_{k=0}^{6} a_k T_C^k$ $a_0 = 6.1117675 \times 10^2$, $a_1 = 4.43986062 \times 10^1$, $a_2 = 1.43053301$, $a_3 = 2.65027242 \times 10^{-2}$, $a_4 = 3.02246994 \times 10^{-4}$, $a_5 = 2.03886313 \times 10^{-6}$, $a_6 = 6.38780966 \times 10^{-9}$
Saturation pressure of water vapor over plane ice surface, Nm^{-2}	$e_i = \sum_{k=0}^{6} a_k T_C^k \quad T_C < 0°\mathrm{C}$ $a_0 = 6.10952665 \times 10^2$, $a_1 = 5.01948366 \times 10^1$, $a_2 = 1.86288989$, $a_3 = 4.03488906 \times 10^{-2}$, $a_4 = 5.39797852 \times 10^{-4}$, $a_5 - 4.20713632 \times 10^{-6}$, $a_6 = 1.47271071 \times 10^{-8}$
Heat conductivity of air, WK^{-1} m^{-1}	$k_a = \sum_{k=0}^{1} a_k T_C^k$ $a_0 = 2.380696 \times 10^{-2}$, $a_1 = 7.1128 \times 10^{-5}$
Heat conductivity of ice, WK^{-1} m^{-1}	$k_i = \sum_{k=0}^{2} a_k T_C^k$ $a_0 = 2.2484816$, $a_1 = -6.19232 \times 10^{-3}$, $a_2 = 1.17152 \times 10^{-4}$

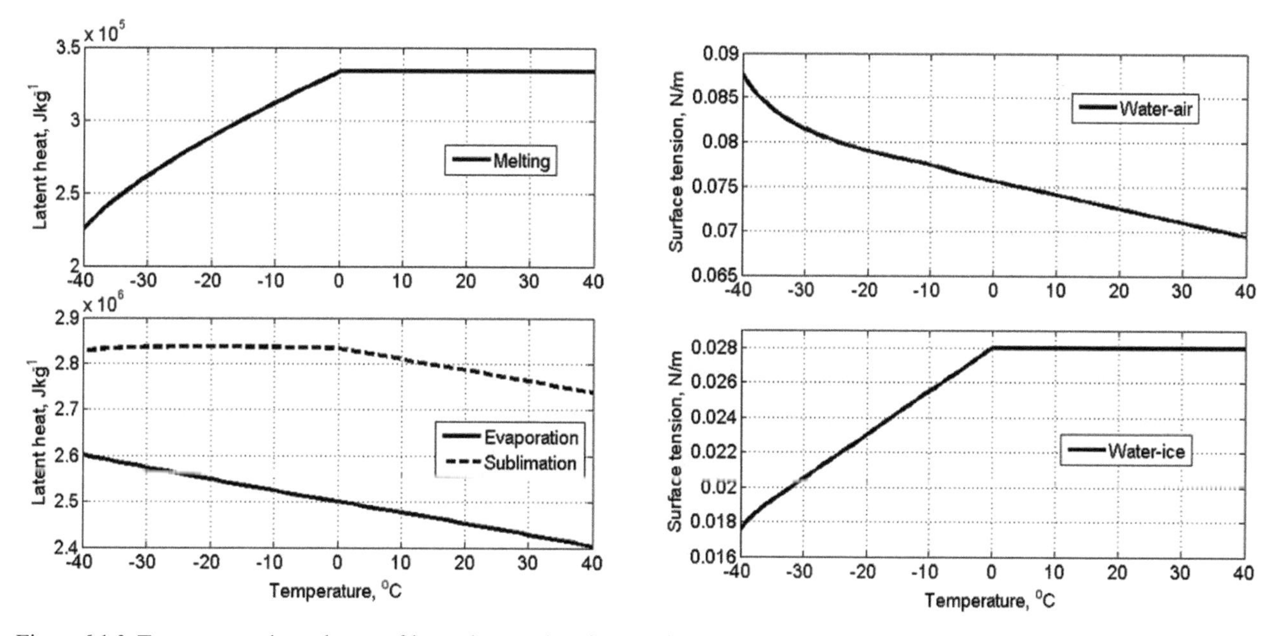

Figure 6.1.3 Temperature dependences of latent heat and surface tension.

6.1.3 Mean Empiric Dependencies Characterizing Ice Particles

As was discussed in Section 2.4, ice particles are characterized by three basic parameters: mass, bulk density, and shape. These parameters determine different properties of ice particles such as fall velocity, efficiency of collisions, reflectivity, etc. The values of these parameters should be known to describe condensational

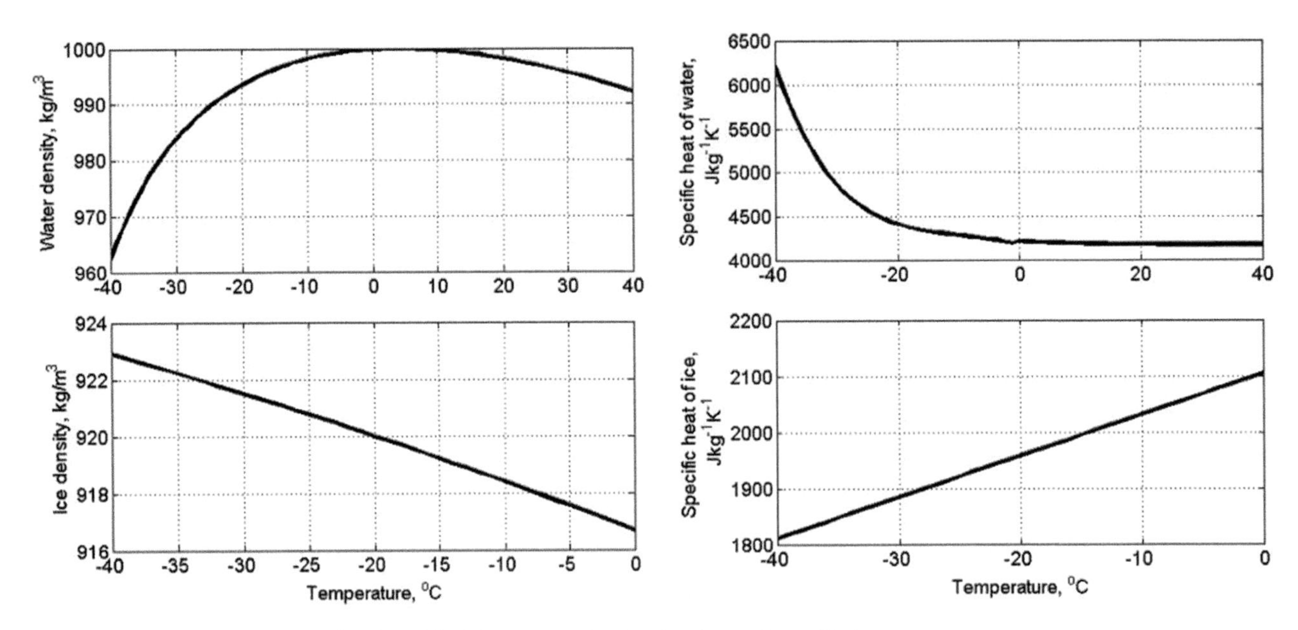

Figure 6.1.4 Temperature dependencies of liquid water density, ice density, and specific heat.

growth and transformation of ice particles, as well as to estimate the collision kernels in cloud models. These parameters are also needed to calculate cloud radiative properties and albedos, as well as to solve different problems of radar/lidar/passive remote-sensing retrievals (including space-borne ones).

There are about seventy regular shapes of ice crystals (Pruppacher and Klett, 1997). Besides, many ice crystals have complicated non-regular shapes. In numerical modeling of processes related to ice particles, the shape of real ice crystals is usually approximated by simple axisymmetric shapes such as sphere, cylinder, disc, or ellipsoid, as shown in Table 2.1.1 (Section 2.1). Ice crystals growing by deposition redistribute their mass in a predictable manner, so the structure of many crystals can be characterized by certain dimensional relations. The shape of plate-like crystals, dendrites, and other branch-like ice crystals is approximated by discs of diameter d and thickness h. The shape of columnar crystals' needles is approximated by cylinders with diameter d and length L. The h/d and L/d ratios, known as aspect ratios, are less than unity for oblate particles and larger than unity for prolate ones.

Diffusional growth of crystals leads to higher aspect ratios for increasing particle size (**Figure 6.1.5**; see also Section 2.1). Diameter-depth and diameter–length relationships are usually approximated by the power laws. Such relationships for some crystal types (according to classifications suggested by Magono and Lee, 1966) observed in different geographic regions are presented by Pruppacher and Klett (1997). Examples of the

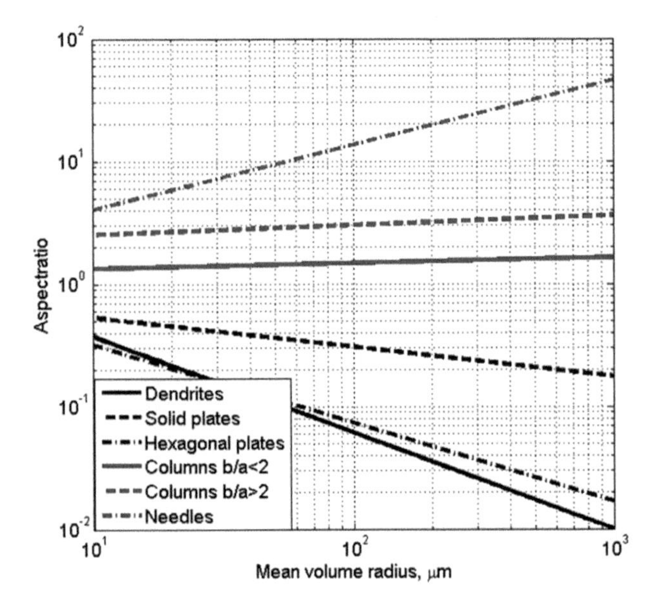

Figure 6.1.5 Dependencies of the aspect ratio on the equivalent crystal radius.

relationships for major types of ice crystals are shown in **Table 6.1.3**. Relationships presented in Table 6.1.3 and in Figure 6.1.5 allow calculation of areas of ice crystal cross sections. The values of the cross sections are needed for simulation of their collisions, reflectivity, and other properties. The dimensional relationships depend on temperature, riming rate, and other parameters. Therefore, the values of parameters characterizing these relationships vary within wide ranges.

Table 6.1.3 Dimensional relationships for various snow crystal types; h, d, and L in cm (from Pruppacher and Klett, 1997; courtesy of Springer).

Crystal Type	Relationship	Validity Range
Hexagonal plates (Pla)	$h(d) = 1.41\times10^{-2}\ d^{0.474}$	10–3,000 μm
Broad branches (Plb)	$h(d) = 1.05\times10^{-2}\ d^{0.423}$	10–2,000 μm
Dendrites (Plc-r, Pld)	$h(d) = 9.96\times10^{-3}\ d^{0.415}$	10–1,500 μm
Solid thick plate (Clg)	$h(d) = 0.138\times10^{-3}\ d^{0.778}$	10–1,000 μm
Solid columns (Cle) $L/d < 2$	$d(L) = 0.578\times L^{0.958}$	10–1,000 μm
Solid columns (Cle) $L/d > 2$	$d(L) = 0.260\times L^{0.927}$	10–1,000 μm
Hollow columns (Clf)	$d(L) = 0.422\times L^{0.892}$	50–1,000 μm
Needles (Nle)	$d(L) = 3.527\times10^{-2}\ L^{0.437}$	10–1,000 μm

Table 6.1.4 Density–size relationships for ice crystals (density in g cm^{-3}, sizes in mm) (from Heymsfield, 1972; © American Meteorological Society; used with permission).

Crystal Type	Relationship	Validity Range
Hexagonal plates (Pla)	$\rho_c = 0.9$	$d \geq 700$ μm
Dendrites	$\rho_c = 0.588\times d^{-0.377}$	$d \geq 300$ μm
Columns, cold region	$\rho_c = 0.65\times L^{-0.0915}$	$L \geq 28$ μm
Columns, warm region	$\rho_c = 0.848\times L^{-0.014}$	$L \geq 14$ μm
Bullets	$\rho_c = 0.78\times L^{-0.0038}$	$L \geq 100$ μm

Another important parameter of ice particles is bulk density, which together with a particle's dimensional relationships allows calculating its mass. Density of ice crystals is typically lower than that of bulk ice due to the existence of air capillary spaces in ice crystals, as well as the tendency of ice crystals to grow in a skeletal fashion. As a result, bulk density typically decreases with increasing size of ice crystals. The relationship between the bulk densities and crystal size can also be described by the power laws (Pruppacher and Klett, 1997). Examples of such dependencies for the main crystal types are presented in **Table 6.1.4**. Density of aggregates strongly depends on the degree of riming, e.g, the density of graupel varies from 0.2 g cm^{-3} to about 0.8 g cm^{-3}; the density of hail stones is close to 0.91 g cm^{-3}.

Along with density–dimension relationships, the mass-dimension relationships are widely used. The mass-diameter (or mass-length) relationships were measured by different authors (e.g., Locatelli and Hobbs, 1974; Brown and Francis, 1995; Mitchell et al., 1990; Mitchell, 1991, 1996; Heymsfield, 1972, 1975, 2007; Heymsfield et al., 2007a, 2007b, 2010). These relationships can also be described by the power law

$$m = \alpha D^{\beta}, \qquad (6.1.1)$$

where D is the maximum dimension of a particle. Some of those relationships for ice crystals are given in **Table 6.1.5** and for aggregates in **Table 6.1.6**. The maximum particle dimension for oblate crystals is diameter d and for prolate crystals length L. It is important to note that utilization of these formulas outside the ranges of their validity may lead to unrealistic results.

In convective clouds, the crystal habit is often not known due to the lack of appropriate measurements. Besides, even when measurements are available, the measured habits are often mixed and complex, thus the dominating ice crystal type cannot be established. Cloud models often deal with a single type of ice crystal,

Table 6.1.5 Mass–diameter (or mass–length) relationships of ice crystals (mass in g, sizes in cm) (from Heymsfield and Kajikawa, 1987; © American Meteorological Society; used with permission).

Crystal Type	Relationship	Validity Range
Hexagonal plates (Pla)	$m = 3.76\times10^{-2}d^{3.31}$	300–1,500 μm
Broad branches (Plb)	$m = 6.34\times10^{-3}\ d^{2.83}$	400–1,600 μm
Dendrites (Plc-r, Pld)	$m = 9.61\times10^{-4}\ d^{2.59}$	400–2,400 μm
Dendrites (Ple)	$m = 6.12\times10^{-4}\ d^{2.29}$	600–5,300 μm
Needles (Nle), m in mg, L in mm	$m = 0.012\times L^{1.8}$	200–1,500 μm
Solid columns, m in mg, L in mm	$m = 0.064\times L^{2.6}$	200–600 μm
Hollow columns (Clf), m in mg, L in mm	$m = 0.037\times L^{1.8}$	400–1,400 μm

Table 6.1.6 Mass–maximum dimension relationship of snow aggregates (mass in mg, maximum dimension D in mm) (from Locatelly and Hobbs, 1974; courtesy of John Wiley & Sons, Inc.).

Aggregate Type	Relationship	Validity Range
Dendrites	$m = 0.037\times D^{1.4}$	2,000–12,000 μm
Plates, side planes, bullets, and columns	$m = 0.037\times D^{1.9}$	1,000–3,000 μm
Side planes	$m = 0.04\times D^{1.4}$	500–4,000 μm

assuming that it should represent ice crystals and aggregates of all shapes. This simplification requires utilization of "effective" ice particles with an averaged mass–dimension relationship. The mass–maximum dimension relationships were proposed for the effective cloud ice that combines ice crystals of different types as well as aggregates. A well-known example of such averaged mass–maximum dimension relationships is the one introduced by Locatelli and Hobbs (1974) and Brown and Francis (1995), with parameter values in Equation (6.1.1) being $\alpha = 0.0029$ and $\beta = 1.9$.

Heymsfield et al. (2010) analyzed observation data obtained from six field campaigns under different temperature conditions, and proposed the mass–maximum dimension relationships. They found that these relationships change depending on temperature, cloud types, and different cloud zones (convective/stratiform). Therefore, they included additional classification parameters such as temperature and cloud type. The parameters of these relationships are $\begin{cases} \alpha = 5.74\cdot10^{-3}, \beta = 2.1 \text{ for } T_c \leq -25°\text{C and } D > 0.0073 \text{ cm} \\ \alpha = 3.59\cdot10^{-3}, \beta = 2.1 \text{ for } T_c \leq -25°\text{C and } D > 0.0043 \text{ cm} \end{cases}$ in stratiform regions, where $\alpha = 6.3\times10^{-3}, \beta = 2.1$ at $D > 0.0081$ cm outside of convection zone, and $\alpha = 0.0011$, $\beta = 2.1$ at $D > 0.0151$ cm in the immediate vicinity of deep convection. For a composite data set, Heymsfield et al. (2010) obtained $\alpha = 0.00528$ and $\beta = 2.1$ for $D > 0.0067$ cm.

6.1.4 Description of Ice Particles in Cloud Models

Models with Spectral Bin Microphysics

In models with SBM size distributions or mass distributions of ice particles are defined on the mass grid. The early SBM mixed-phase cloud models assumed ice particles to be spheres (Khvorostyanov et al., 1989; Hall, 1980). In these models, only one SD of ice particles was used to describe their variety in clouds. Small ice particles were assigned to ice crystals; larger particles were assigned to graupel. Ovchinnikov and Kogan (2000) also used one SD for all kinds of ice particles. The smallest fifteen bins of the mass-doubling grid were considered ice crystals while the largest thirteen categories were considered graupel. Plate-like crystals were chosen to represent ice crystals. Their mass–dimension relationship was assumed as shown in Tables 6.1.5 and 6.1.6. Graupel particles were assumed to be spherical with the bulk density of 0.4 g cm^{-3}.

More complicated SBM models contain several SD to describe different types of ice particles. Typically, these classes are crystals, snow (aggregates), and graupel (e.g., Reisin et al., 1996; Lynn and Khain, 2007; Khain et al., 2010). To describe the variety of ice crystals, Reisin et al. (1996) used columns and Khain et al. (2010) used plate-like crystals. Rasmussen et al. (2002) used four types of ice in their model: pristine ice, rimed ice, aggregates, and graupel. The pristine ice crystals were assumed to be shaped as thin hexagonal plates. To calculate diffusional growth of these crystals, their shape was approximated by an oblate spheroid with two axes. The shape of rimed ice crystals was also approximated by an oblate spheroid. It was assumed that the axis ratio of a crystal increases linearly with crystal mass, while the bulk density decreases linearly with the mass.

HUCM (Khain and Sednev, 1995; Khain et al., 2004, 2008, 2011; Benmoshe et al., 2012) treats evolution and conversion of six types of ice particles: dendrites, plate-like, columnar ice crystals, aggregates (snowflakes), graupel, and hail. Each type of ice particles is described by a size distribution defined on mass grids containing either 33 or 43 bins (depending on the model version used). Ice crystals are described using the parameters presented in Table 6.1.3 and Figure 6.1.5. The dependences of mass on the main geometrical parameters are calculated using the dimensional relationships and density–dimension relationships. To calculate collision rates, graupel and hail are assumed to be spheres with densities of 0.4 g cm^{-3} and 0.91 g cm^{-3}, respectively. The bulk density of unrimed snowflakes decreases from about 0.3 g cm^{-3} for 100-μm radii to 0.09 g cm^{-3} for radii exceeding 0.25 cm (Litvinov, 1974). The dependences of bulk densities of model particles on their "melted" radius (i.e., radius of a drop to be formed after total melting of an ice particle) are given in **Figure 6.1.6**.

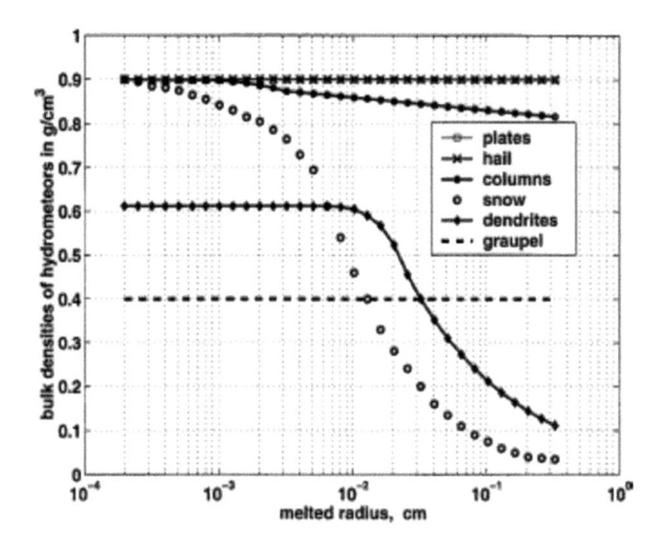

Figure 6.1.6 Dependencies of bulk densities of various ice hydrometeors on their melted radii as used in HUCM. Densities of plates and hail are equal (from Khain et al., 2004; © American Meteorological Society; used with permission).

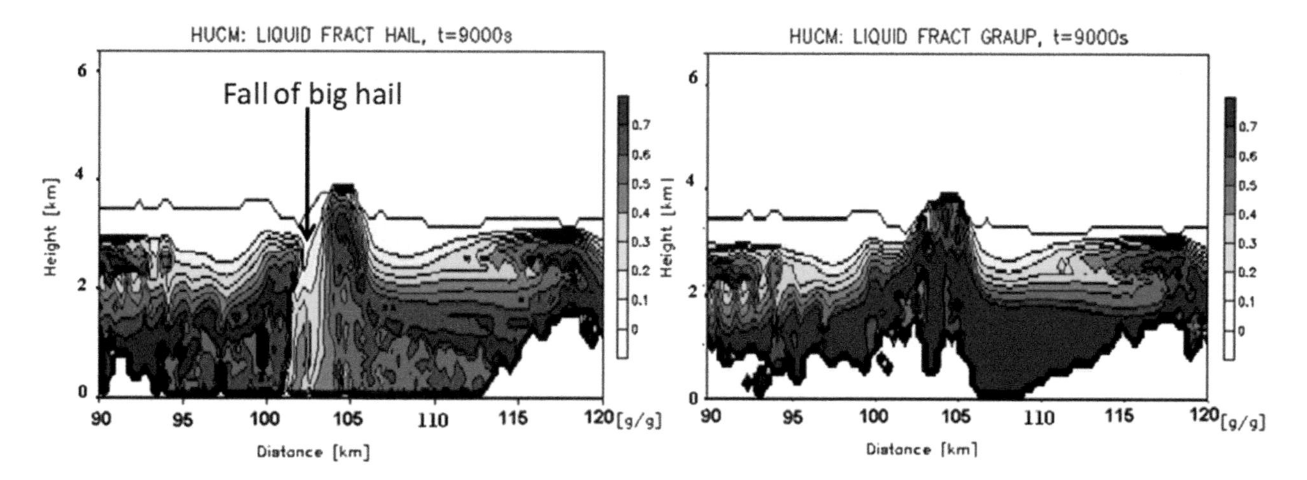

Figure 6.1.7 Fields of averaged liquid water fractions of hail (left) and graupel (right) in a severe hail storm simulated by HUCM (from Ryzhkov et al., 2011; © American Meteorological Society; used with permission).

Knowing the mass and type of an ice particle is often not enough to describe cold microphysical processes. Melting and freezing are time-dependent and may last for several minutes, especially for large particles. Detailed descriptions of melting and freezing procedures require introducing the liquid water fraction (LWF) as an additional characteristic of ice particles (actually, mixed-phase cloud particles). LWF is defined as the ratio of the liquid water mass in a particle to its total mass. In HUCM, three types of ice types of hydrometeors may contain LWF: snow, graupel, and hail. As a result, three additional distribution functions for the liquid water mass within aggregates (snow), graupel, and hail were implemented (Phillips et al., 2007, 2014, 2015). **Figure 6.1.7** shows the LWF fields in melting hail and graupel, calculated in simulations of a severe hailstorm in Oklahoma. The implementation of the LWF parameter required recalculation of the fall velocity and collision kernels of melting/freezing particles in each bin.

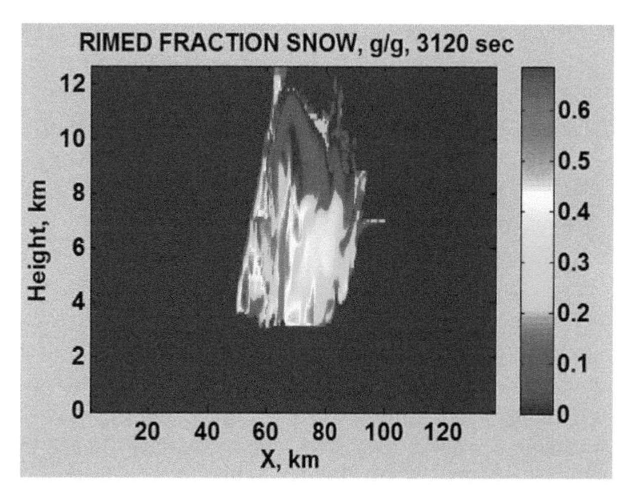

Figure 6.1.8 Spectrum-averaged rimed fraction in snow in a developing hail storm simulated by HUCM.

Density of ice particles is another property that changes during microphysical processes. Riming of snow leads to increasing snow density and fall velocities. Heavily rimed snowflakes are transformed into graupel. To take these effects into account, Khain et al. (2011, 2012), Benmoshe et al. (2012), and Ryzhov et al. (2011) included a time-dependent description of aggregate riming. The mass of rimed ice within aggregates is described by a separate mass distribution function. **Figure 6.1.8** shows a spectrum-averaged rimed fraction in snow in a developing hailstorm simulated by HUCM. One can see that the rimed fraction is high near the column of x = 65 km, where CWC and riming rates are maximum. A significant mass of snow is transformed to graupel in this atmospheric column. Accumulation of the rimed fraction, in turn, increases the particle density. As soon as its density exceeds $0.2\ \text{g cm}^{-3}$, an aggregate is converted to graupel.

The recalculation of density of aggregates in the course of riming is also applied by Morrison and Grabowski (2010) who used plate-like crystals to represent the shapes of nonspherical non-rimed ice crystals using Equation (6.1.1) with $\alpha = 0.00142$ and $\beta = 2.02$ (Mitchell, 1996). Grupel particles were assumed close to spherical with a constant density (Equation (6.1.1) $\alpha = 0.049$, $\beta = 2.8$). In the course of riming, the values of the mass–dimension relationship parameters describing aggregates tend to those of graupel, which imitates transformation of heavy rimed aggregates to graupel.

Models with Bulk Microphysics

In bulk-parameterization schemes, the mass–dimensional relationship for ice particles is often expressed in the form of Equation (6.1.1). In many cases, particles are assumed to be spheres with the effective diameter D (e.g., Lin et. al., 1983). In this case, $\beta = 3$ in Equation (6.1.1). The modern bulk schemes typically consider three types of ice particles: ice crystals, aggregates, and graupel (e.g., Reisner et al., 1998; Li et al., 2008). Reisner et al. (1998) used the "effective mass-size relationship" proposed by Locatelli and Hobbs (1974). In the two-moment bulk scheme developed by Li et al. (2008), the mass–dimension relationships of ice crystals and aggregates are described, using Equation (6.1.1) with $\beta = 2$ and $\alpha = 0.001$ g cm^{-2} for ice crystals and $\alpha = 0.003$ g cm^{-2} for snow (aggregates).

Most bulk-parameterization schemes deal with three types of hydrometeors: pristine ice, aggregates (snow), and graupel (Section 7.2). Only a few bulk-parameterization schemes include hail as a separate hydrometeor type in addition to graupel (Carrió et al., 2007; Noppel, 2010; Loftus et al., 2014). The most detailed bulk-parameterization scheme, apparently, applied in the Regional Atmospheric Modeling System (RAMS) (Walko et al., 1995; Carrió et al., 2007) includes pristine ice, snow, aggregates, graupel, and hail. Snow is defined as large pristine ice crystals with a diameter exceeding 100 μm that grow by vapor deposition and riming, while aggregates are defined as ice particles formed by collision and coalescence of pristine ice, snow, or other aggregates. Pristine ice, snow, and aggregates are assumed to be completely frozen, while graupel and hail are mixed-phase particles that may contain only ice or ice–liquid mixture. Ice particles are assumed to be spheres of different densities.

Morrison and Milbrant (2015) proposed a bulk parameterization in which all ice hydrometeors are described using four physical properties applied to a single ice type: ice–water mixing ratio, rime mass, rime volume, and number concentration. This approach differs from traditional bulk microphysical schemes in which ice-phase hydrometeors are separated into various predefined types (e.g., cloud ice, snow, and graupel) with prescribed characteristics. The corresponding hydrometeor type in the approach proposed by Morrison and Milbrant (2015) can be implicitly diagnosed according to the set of these parameters.

Summarizing, ice processes are far more complicated than warm processes. This is the reason why various simplifications are used in the treatment of ice in cloud models. The differences in the treatment begin with the determination of ice particle characteristics such as type and shape. Even these differences may yield different microphysical structures of mixed-phase clouds in cloud models.

6.2 Nucleation of Ice Particles

6.2.1 The Concept of Homogeneous Nucleation

At certain conditions, formation of ice germs due to random molecular motions can take place in the air (homogeneous deposition) or in liquid (homogeneous freezing). Under atmospheric conditions, homogeneous deposition is not possible since the required supersaturation over ice cannot be reached in the atmosphere. Thus, homogeneous nucleation is determined by homogeneous freezing. According to observations (Phillips et al., 2005; Heymsfield et al., 2009; Rosenfeld and Woodley, 2000) and numerical calculations (e.g., Khain et al., 2012), homogeneous nucleation is an important source of high concentration of small ice crystals in anvils of deep convective clouds and in cirrus clouds. Concentrations of ice crystals, which form by the homogeneous nucleation, can reach huge values from one cm^{-3} to several hundred per cm^{3}, depending on the concentration of supercooled drops, which in turn depends on aerosol concentration. Different theories of homogeneous nucleation in clouds were presented by Pruppacher and Klett (1997). Next, we present the basic concept of homogeneous nucleation.

At negative temperatures, water molecules can exist in the vapor state or the liquid state (supercooled water) or in the solid state (ice). The vapor state and the liquid state are metastable, while the solid state is absolutely stable. Each state can be characterized by its free energy F. The higher the stability, the lower the value of free energy is. The lowest free-energy levels are typical of solid ice, as is schematically shown in **Figure 6.2.1**. Under negative temperatures, supercooled drops tend to become ice particles having lower free energy. However, for this transition an energy should exist to overcome the activation energy barrier to build the water–ice interface (Figure 6.2.1).

Transition of the liquid phase to solid ice (first-order phase transition) begins with a spontaneous creation of ice embryos inside drops and subsequent formation of ice germs consisting of several tens of water molecules and having a structure similar to that of ice. To generate a spherical embryo of radius r, the following free-energy cost is required (Abraham, 1974):

$$\Delta F = -\Delta F_1 + \Delta F_2 = -\frac{4}{3}\pi r^3 E + 4\pi r^2 \sigma_{iw}, \quad (6.2.1)$$

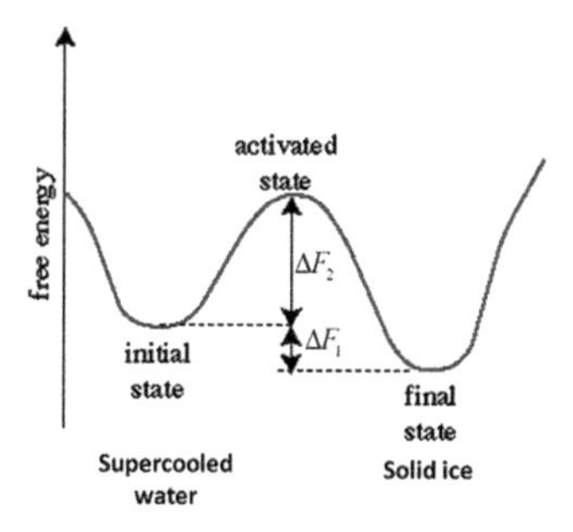

Figure 6.2.1 Illustration of change in the free energy level in the course of homogeneous freezing.

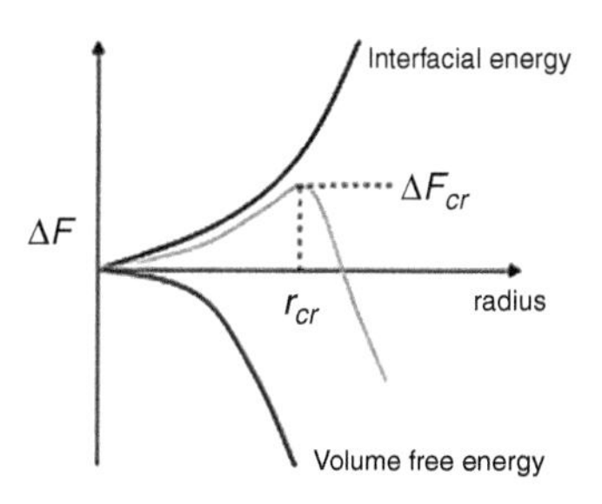

Figure 6.2.2 Dependence of the terms in Equation (6.2.1) on an embryo radius.

where E is the specific energy released during formation of an ice germ of a unit volume and σ_{iw} is the surface tension of water–ice interface. The first term $(-\Delta F_1)$ is proportional to the embryo volume and represents the difference in free energies between ice and supercooled water. The second term (ΔF_2) is proportional to the embryo surface area and represents the energy needed to form a curved interface between liquid water and ice (Figure 6.2.1). As shown in **Figure 6.2.2**, the dependence $\Delta F(r)$ demonstrates the maximum ΔF_{cr} corresponding to the critical radius r_{cr}. The value $\Delta F_{cr} = \Delta F(r_{cr})$ determines the minimum energy needed to transfer water from the supercooled liquid state to the solid state, is calculated from condition $\frac{d\Delta F}{dr} = 0$. The critical point is unstable because $\frac{d^2\Delta F(r)}{d^2r} < 0$ at the point of $r = r_{cr}$:

$$r_{cr} = \frac{2\sigma_{iw}}{E} \tag{6.2.2}$$

$$\Delta F_{cr} = \frac{4}{3}\pi\sigma_{iw}r_{cr}^2 = \frac{16\pi\sigma_{iw}^3}{3E^2} \tag{6.2.3}$$

It is usually assumed that an ice germ and the environmental liquid are in thermodynamic equilibrium, which means, in particular, that they have the same temperatures. In case of pure water E, r_{cr} and ΔF_{cr} can be written as follows (Pruppaher and Klett, 1997; Khvorostyanov and Sassen, 1998):

$$E = \rho_i\overline{L}_m \ln\left(\frac{T_0}{T}\right) \tag{6.2.4}$$

$$r_{cr} = \frac{2\sigma_{iw}}{\rho_i\overline{L}_m \ln\left(\frac{T_0}{T}\right)} \tag{6.2.5}$$

$$\Delta F_{cr} = \frac{16\pi\sigma_{iw}^3}{3\left[\rho_i\overline{L}_m \ln\left(\frac{T_0}{T}\right)\right]^2}, \tag{6.2.6}$$

where T is the drop temperature, $T_0 = 273.15\,\mathrm{K}$ is the temperature at the triple point where all the three phases coexist in an equilibrium, and $\overline{L}_m(T) = \frac{1}{\ln(T_0/T)}\int_T^{T_0}[L_m(T)/T]dT$ is the effective latent heat of freezing obtained by averaging over the temperature range from T_0 to T. According to Khvorostyanov and Sassen (1998), $\overline{L}_m$ can be represented as a function of temperature as

$$\overline{L}_m(T_C) = 79.7 + 0.708T_C - 2.5\times10^{-3}T_C^2\,[\mathrm{cal{\cdot}g}], \tag{6.2.7}$$

where T_C is the temperature in °C.

According to the thermodynamic laws, the probability of forming a germ of the critical radius is proportional to $\exp\left(-\frac{\Delta F_{cr}}{kT}\right)$, where k is the Boltzmann constant. The embryo size fluctuates due to the stochastic movement of the water molecules, leading to random attachment and detachment of molecules from the embryo. As a rule, this process leads to dissipation of embryos of radii $r < r_{cr}$ and to growth of embryos of radii $r > r_{cr}$. The embryos continue to grow and finally fill a drop, forming an ice particle. Since embryo size fluctuations are random, there is always a probability for some embryos to overcome the energy barrier and to form a stable ice germ of radius $r > r_{cr}$. The values of the critical radius and the critical energy depend primarily on drop supercooling rate, i.e., on the temperature difference $\Delta T = T_0 - T$. The higher the supercooling, the smaller the critical radius and the less energy is needed to form the germ with $r > r_{cr}$. Note that supercooling is related to temperature T_C measured in °C as $\Delta T = -T_C$.

An important characteristic of ice formation in clouds is the nucleation rate J, which is the mean rate of producing germs of supercritical radius $r > r_{cr}$ per unit volume of liquid water and per unit time. The nucleation rate is determined by the product of the mean number of germs of the critical size r_{cr} and the rate at which one extra molecule is added to the critical germ due to diffusion of supercooled water molecules. Since the steady–state size distribution of germs is given by the Boltzmann distribution, the mean number of critical germs per unit volume is given by the formula

$$n = n_0 \exp\left(-\frac{\Delta F_{cr}}{kT}\right), \qquad (6.2.8)$$

where n_0 is the concentration of water molecules in supercooled water. The diffusion flux of water molecules across the ice–water interface can be expressed as $A\frac{kT}{h}\exp\left(-\frac{\Delta F_{act}}{kT}\right)$, where $A = 5.85 \times 10^{12}$ cm^{-2} is the number of molecules in contact with an ice area (in cm^2), T is the drop temperature that can be assumed with good accuracy to be equal to the environmental temperature, ΔF_{act} is the activation energy, and h is the Planck constant. The activation energy value can be defined as the minimum energy required to trigger the nucleation process. Khvorostyanov and Sassen (1998) approximated the activation energy as a linear function of temperature:

$$\Delta F_{act}(T_C) = 0.694 \times 10^{-12}[1 + 0.027(T_C + 30)] \qquad (6.2.9)$$

In Equation (6.2.9), ΔF_{act} is measured in ergs. The final equation for the nucleation rate formulated by Pruppacher and Klett (1997) is

$$J = \frac{2A\rho_w kT}{\rho_i h}\left(\frac{\sigma_{iw}}{kT}\right)^{1/2} \exp\left(-\frac{\Delta F_{act} + \Delta F_{cr}}{kT}\right) \qquad (6.2.10)$$

Equation (6.2.10) shows a very high sensitivity of nucleation rates to temperature, which allows us to introduce a certain threshold value of temperature $T_{th} \approx -35°C$ required for the onset of homogeneous nucleation.

The nucleation rate is often associated with the rate of crystal formation, assuming that only one nucleation event per drop is needed to form an ice crystal. Assuming that crystal formation is a stochastic sequence of the Poisson events, the probability that the new crystal will not be formed decreases with increasing time and drop volume. During time interval Δt, the probability of a new crystal not forming is equal to $\exp\left(-\frac{4}{3}\pi r^3 J\Delta t\right)$, where $\frac{4}{3}\pi r^3$ is the drop volume. Correspondingly, the probability of a new crystal forming is equal to $1 - \exp\left(-\frac{4}{3}\pi r^3 J\Delta t\right)$. The concentration of crystals reached during this time interval is equal to

$$\Delta N_i(r) = N_d\left[1 - \exp\left(-\frac{4}{3}\pi r^3 J\Delta t\right)\right], \qquad (6.2.11)$$

where $N_d(r)$ is the concentration of drops of radius r. It is usually assumed that the bulk radius of a crystal is equal to the drop radius. If DSD is multidisperse, the total concentration of ice crystals ΔN_{i_tot}, which is equal to the number of drops freezing homogeneously within time interval Δt, is calculated as

$$\Delta N_{i\_tot} = \int_0^\infty \left[1 - \exp\left(-\frac{4}{3}\pi r^3 J\Delta t\right)\right] f_r(r)dr, \qquad (6.2.12)$$

where $f_r(r)$ is DSD (Section 2.1). **Figure 6.2.3** shows the calculations by Jeffery and Austin (1997) and Pruppacher and Klett (1997) on the dependencies of the crystal nucleation rate on temperature. The nucleation rates measured in laboratory experiments are presented, as well. One can see that the results of theoretical studies agree well with the measurements for temperature values warmer that $-37°C$. A dramatic increase in the nucleation rate occurs as the temperature decreases. For instance, a 1°C temperature decrease leads to the increase in the nucleation rate by two to three orders of magnitude. The nucleation rate is negligible at $-30°C$. Some unfrozen drops can exist at temperatures as low as $-45°C$–$-50°C$. However, the rates predicted for temperatures lower than $-40°C$ are so high that liquid drops freeze rapidly regardless of their size, producing ice crystals.

The effects of drop salinity on homogeneous nucleation of ice are of special importance regarding formation of cirrus clouds, where ice starts forming at subsaturation with respect to water. The effect of solutes on homogeneous nucleation of ice is twofold. First, the freezing temperature is lower in the presence of solutes, and second, solutes impact formation of germs. **Table 6.2.1** gives values of the nucleation rate within the temperature range from $-40°C$ to $-50°C$ at 95% relative humidity for four mass categories of pure ammonium sulfate CCN. One can see that the presence of salt in drops dramatically decreases the rate of homogeneous freezing.

Khvorostyanov and Sassen (1998) extended the classical theory of homogeneous freezing to the case when nonactivated droplets are haze particles, i.e., weak solutions. Assuming an equilibrium at the air–aqueous solution surface, the expression for the critical radius of a germ is

$$r_{cr} = \frac{2\sigma_{iw}}{\rho_i \overline{L}_m \ln\left[\left(\frac{T_0}{T}\right)(S_{w\_eq} + 1)^p\right] - \frac{2\sigma_w}{r_d}}, \qquad (6.2.13)$$

Table 6.2.1 Freezing nucleation rates for ideal (pure water) droplets (3rd column) and nonideal (salty) droplets (4th column) of different masses (from DeMott et al., 1994; © American Meteorological Society; used with permission).

T(°C)	m_s, g	J_{ls}, cm^{-3} s^{-1}	J_{ls0}, T^*
−40.0	1.0×10^{-16}	5.3	6.0×10^{-32}
	1.0×10^{-15}	3.3×10^{4}	8.3×10^{-12}
	1.0×10^{-14}	6.4×10^{5}	2.8×10^{-6}
	1.0×10^{-13}	2.1×10^{6}	3.4×10^{-4}
−45.0	1.0×10^{-16}	1.7×10^{8}	1.1×10^{-9}
	1.0×10^{-15}	5.7×10^{10}	4.2×10^{2}
	1.0×10^{-14}	4.3×10^{11}	6.3×10^{5}
	1.0×10^{-13}	9.5×10^{11}	8.8×10^{6}
−50.0	1.0×10^{-16}	1.4×10^{13}	2.4×10^{3}
	1.0×10^{-15}	8.7×10^{14}	2.2×10^{10}
	1.0×10^{-14}	3.5×10^{15}	2.0×10^{12}
	1.0×10^{-13}	6.2×10^{15}	1.2×10^{13}

where S_{w_eq} is the equilibrium supersaturation over a soluble haze particle, expressed as $S_{w\_eq} \approx \frac{A}{r_d} - \frac{Br_N^3}{r_d^3 - r_N^3}$, where r_N is the radius of a dry aerosol particle (Equation 5.1.11) and $p(T) = \frac{RT}{M_w \overline{L}_m}$ is a dimensionless parameter. This parameter varies from 0.3 to 0.6 within the temperature range −15°C to −40°C. Equation (6.2.13) determines the critical radius of an ice germ accounting for both the chemical effects and the curvature effects. Using Equations (6.2.3) and (6.2.13), the expression for the critical energy can be written in the form

$$\Delta F_{cr} = \frac{4}{3}\pi\sigma_{iw} r_{cr}^2 = \frac{16\pi}{3} \frac{\sigma_{iw}^3}{\left(\rho_i \overline{L}_m \ln\left[\left(\frac{T_0}{T}\right)\left(S_{w\_eq} + 1\right)^p\right] - \frac{2\sigma_w}{r_d}\right)^2} \quad (6.2.14)$$

Equations (6.2.13) and (6.2.14) are derived under assumed equilibrium between a haze particle and its surrounding, so S_{w_eq} for particles is equal to the environmental supersaturation S_w. In this case, formulas (6.2.13) and (6.2.14) are applicable to haze particles smaller than the critical size needed for droplet nucleation (Section 5.1).

When the salinity of a droplet is taken into account, the critical germ size, the critical energy, and the

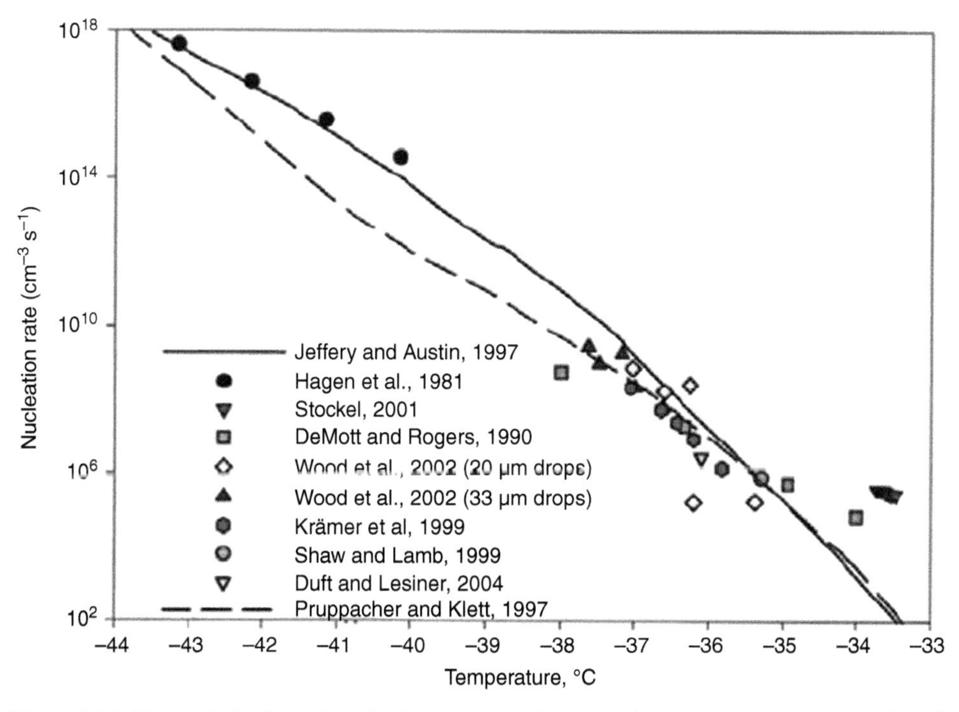

Figure 6.2.3 Theoretical values of nucleation rates as a function of temperature (solid and dashed lines). Symbols show results of different experiments for suspended or freely falling droplets (from Cantrell and Heymsfield, 2005; © American Meteorological Society; used with permission).

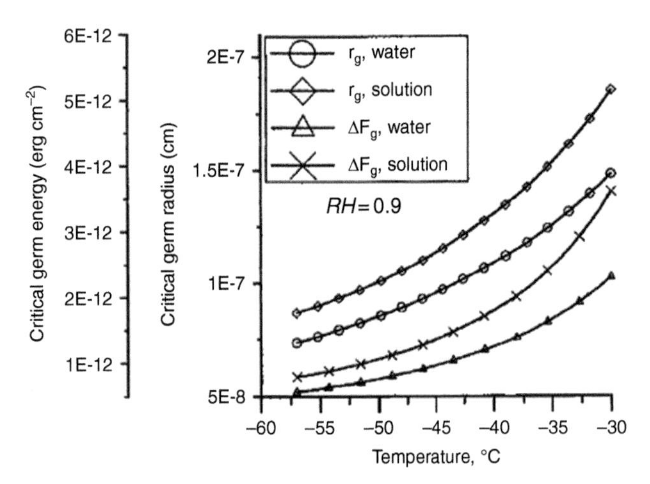

Figure 6.2.4 Dependence of the critical germ radius and the critical energy on temperature calculated at RH equal to 0.9 for pure water, as well as for saline droplets with the curvature effect and the salinity effect taken into account (from Khvorostyanov and Sassen, 1998; courtesy of © John Wiley & Sons, Inc.).

nucleation rate become dependent on the droplet size. Increasing S_w fosters the nucleation. According to Equations (6.2.13) and (6.2.14), the critical radius and the energetic barrier depend on two nonindependent thermodynamic parameters: the temperature and the supersaturation. Since $r_{cr} > 0$, Equation (6.2.13) imposes the following condition on supersaturation for homogeneous nucleation of droplets with radius r_d:

$$S_w + 1 > \left(\frac{T}{T_0}\right)^{1/p} \exp\left(\frac{2\sigma_w}{p\rho_i \overline{L}_m r_d}\right) \tag{6.2.15}$$

The existence of threshold values of S_w were found by Sassen and Dodd (1989) who used a parcel model, and by Heymsfield and Milosevich (1995) on the basis of aircraft data. **Figure 6.2.4** shows the dependence of the critical germ radius and the critical energy on temperature, calculated at $S_w = -10\%$ (i.e., at the relative humidity equal to 0.9) for pure water droplet, as well as for saline droplets when the curvature effect and the solution effect are taken into account. As expected, the critical germ size decreases with decreasing temperature, which means a higher probability of freezing. The presence of salt in haze particles increases the energetic barrier, decreasing the probability of freezing.

6.2.2 Parameterization of Homogeneous Nucleation of Ice in Cloud Models

The key parameter to be parameterized is the nucleation rate. In many models, the nucleation rate is assumed dependent solely on temperature (e.g., DeMott et al., 1994). In more sophisticated approaches, the homogeneous freezing rate depends both on temperature and on drop size (Khvorostyanov and Sassen, 1998). Heymsfield and Milosevich (1993) and DeMott et al. (1994) approximated $J(T_C)$ for pure water by a polynomial:

$$\log_{10} J = -606.3952 - 52.6611T_C - 1.7439T_C^2 - 0.0265T_C^3 - 1.536 \times 10^{-4}T_C^4 \tag{6.2.16}$$

where J is measured in $\text{cm}^{-3}\,\text{s}^{-1}$. This parameterization is valid at $T_C \geq -50°\text{C}$.

The effect of salinity on freezing was parameterized by Sassen and Dodd (1988), who formulated an effective freezing temperature T^* based on observations performed by Rassmussen (1982):

$$T^* = T + \lambda \cdot \Delta T_m, \tag{6.2.17}$$

where ΔT_m is the so-called melting point depression, or the bulk freezing point depression, and $\lambda = 1.7$. The value of T^* can be interpreted as the effective temperature of a pure water drop. This value is used for calculating the freezing rate, with drop salinity taken into account. To calculate ΔT_m, Heymsfield and Sabin (1989) proposed the following formula (derived for ammonium sulfate $(NH_4)_2SO_4$):

$$\Delta T_m = 0.102453 + 3.48484M, \tag{6.2.18}$$

where $M = \frac{10^3 m_N}{M_s\left(\frac{4\pi}{3}\rho_s r_d^3 - m_N\right)}$ is the molality of a drop of radius r_d, containing a salt mass m_N, ρ_s is the density of the solute, and M_s is the gram molecular weight of the solute. To take the salinity into account, the drop temperature in Equation (6.2.16) should be replaced by T^*, calculated according to Equations (6.2.17) and (6.2.18).

In some SBM models, the method for ice crystal nucleation proposed by DeMott et al. (1994) is used. The procedure includes calculation of the nucleation rate using Equation (6.2.16) and Equations (6.2.17) and (6.2.18). Then Equation (6.2.11) is used to calculate the number of ice crystals Δn_i produced by homogeneous freezing of drops for each mass bin.

The parameterization of homogeneous freezing proposed by Khvorostyanov and Sassen (1998) includes several steps. First, the critical germ radius is calculated using Equation (6.2.13). In this expression, $\overline{L}_m(T)$ is calculated using Equation (6.2.7) and the equilibrium supersaturation over water S_{w_eq} is calculated using Equation (5.1.11). Then the critical energy ΔF_{cr} and the activation energy $\Delta F_{act}(T_C)$ are calculated using

Equations (6.2.14) and (6.2.9), respectively. Finally, the activation rate is determined using Equation (6.2.10). This parameterization is applicable to describe homogeneous nucleation both for pure water drops and solutes. Equation (6.2.13), used by Khvorostyanov and Sassen, takes into account that the critical size of germ increases with salinity, which reflects the fact that solutions freeze at lower temperatures than pure water. This freezing was parameterized by DeMott et al. (1994) by means of Equation (6.2.17).

Equation (6.2.17) can be derived within the theory proposed by Khvorostyanov and Sassen (1998) in which parameter $\lambda(T)$ is equal to $\lambda(T) = \overline{L}_m(T_0)/\overline{L}_m(T)$. The value $\lambda(T) \approx 1.7$ used in Equation (6.2.17) is strictly valid for ammonium sulfate at $T = 37°C$. Therefore, it seems that there is a close connection between parameterizations by DeMott et al. (1994) and Khvorostyanov and Sassen (1998).

In bulk schemes, Equation (6.2.12) is used to calculate the total number of activated ice crystals. The DSD in Equation (6.2.12) is expressed by the Gamma distribution, as described in Section 2.1. Calculations are simpler if $\frac{4}{3}\pi r^3 J \Delta t \ll 1$ and J does not depend on drop size. In this case, integration of Equation (6.2.12) yields the following expression:

$$\Delta N_{i\_tot} = NJ\frac{4\pi r_v^3}{3}\Delta t = F_{fr}N, \qquad (6.2.19)$$

where N is drop concentration, r_v is the mean volume radius, and F_{fr} is the fraction of drops that freeze during one time step. Equation (6.2.19) was used by Milbrant and Yau (2005b).

Mesoscale and large-scale models often implement parameterizations of homogeneous nucleation. A typical parametrization was developed by Liu and Penner (2005) for a parcel model. The parcel model calculated the homogeneous freezing rate by means of Equation (6.2.16). The effective freezing temperature was calculated using Equation (6.2.17). The parameterization was derived by fitting the results of a large set of parcel model simulations covering different conditions in the upper troposphere. As a result, a simple expression for concentration of homogeneously nucleated ice crystals was obtained as a function of vertical velocity, temperature, relative humidity, and concentration of sulfate aerosols. The critical value of the relative humidity (in percent) for homogeneous ice formation was fitted as a function of the temperature at which freezing commences, and the updraft velocity w (in m s^{-1}) by expression $RH_{cr} = AT_C^2 + BT_C + C$, where $A = 6 \times 10^{-4}\ln(w) + 6.6 \times 10^{-3}$; $B = 6 \times 10^{-2}\ln(w) + 1.052$ and $C = 1.68\ln(w) + 129.35$.

6.2.3 The Concept of Heterogeneous Nucleation

Homogeneous nucleation of ice is a factor significant mainly in the upper troposphere where the temperatures are below –35°C. In the course of *heterogeneous nucleation*, ice germs form on IN. Heterogeneous nucleation requires less energy than homogeneous nucleation and occurs at warmer temperatures ($T_C < -5°C$). The presence of IN decreases the surface energy and the free energy barrier needed for nucleation. The surface energy decreases as a result of *wetting* and related contact angles. The concept of wettability and contact angles is illustrated in **Figure 6.2.5**. Figure 6.2.5 shows the simplest geometry of an ice germ G forming a spherical cap shape on an insoluble planar substrate S. The substrate plays the role of IN. The contact angle θ_{is} characterizes the relationship between the surface tensions on substrate–liquid interface σ_{SL}, substrate–germ interface σ_{SG}, and germ–liquid interface σ_{GL}. The wettability is characterized by the contact parameter known also as the wettability parameter m_{is} (Vali, 1999):

$$m_{is} \equiv \cos\theta_{is} = \frac{\sigma_{SL} - \sigma_{SG}}{\sigma_{GL}} \qquad (6.2.20)$$

Small angle θ_{is} means high wettability parameter $m_{is} \approx 1$; in this case $\sigma_{SG} = 0$ and $\sigma_{SL} = \sigma_{GL}$. The other limit case $\theta_{is} = \pi (m_{is} = -1)$ corresponds to the ideal non-wettable solid surface. In this case, the germ is a sphere sitting on top of the solid. The interfacial energy of the germ with respect to the substrate is lower than that with respect to the parent liquid phase (L). The larger this difference, the more favorable ice nuclei substrate is for nucleation.

The energy barrier for inhomogeneous nucleation is lower than that for homogeneous nucleation by the geometric factor $f(m_{is}) \leq 1$ (Pruppacher and Klett, 1997)

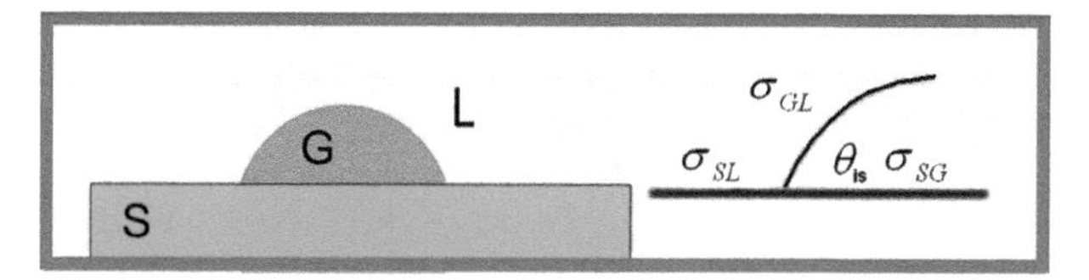

Figure 6.2.5 Illustration of the concept of wettability and contact angles in case of a planar substrate. Notations: G denotes the germ, S is an insoluble planar substrate, L denotes liquid; σ_{SL}, σ_{SG}, and σ_{GL} are the surface tensions on substrate–liquid, substrate–germ and germ–liquid interfaces, respectively. θ_{is} is the ice–substrate contact angle (from Vali, 1999).

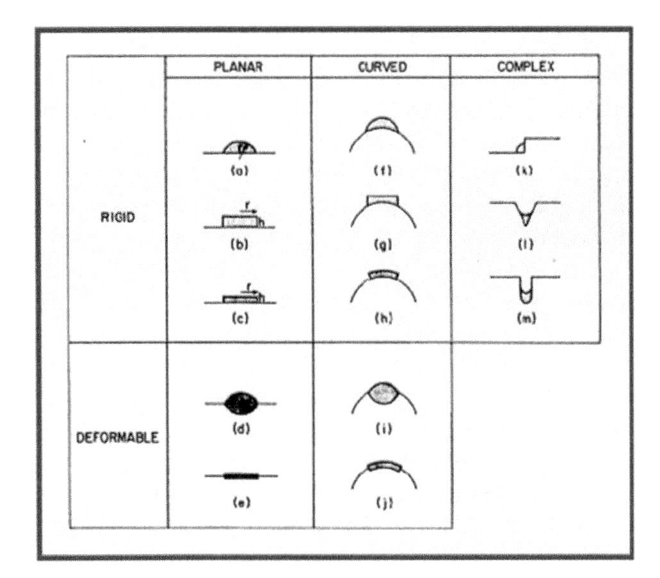

Figure 6.2.6 Some simple shapes of ice germs forming on rigid and deformable substrates (from Knight, 1979; courtesy of Elsevier).

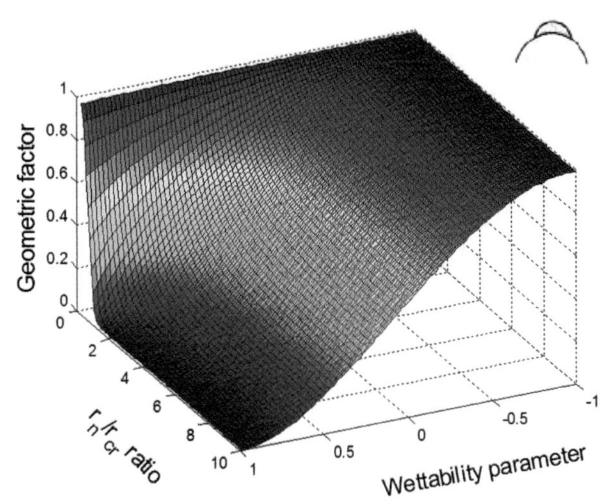

Figure 6.2.7 Dependence of the geometric factor on the wettability parameter and on the r_n/r_{cr} ratio. A shape of an ice germ located on a solid deformable substrate is shown in the right upper corner.

$$\Delta F_{cr\_het} = f(m_{is})\Delta F_{cr\_\mathrm{hom}} \tag{6.2.21}$$

Even a modest value of $f(m_{is}) = 0.1$ may increase the nucleation temperature by a few tens of degrees. In a general case, solid substrates have different shapes, and nucleation takes place on a curved substrate. Some possible simple shapes of ice germs forming on solid substrates are shown in **Figure 6.2.6**. In case of a non-planar substrate, the geometric factor is a function of two parameters: the wettability parameter m_{is} and the ratio of the equivalent radius of the insoluble fraction r_n to the critical radius of the germ $x = r_n/r_{cr}$. If the geometry is similar to one shown in the upper row of Figure 6.2.6, the expression for the geometric factor can be written as (Fletcher, 1962)

$$f(m_{is}, x) = \frac{1}{2}\Big\{1 + [(1 - m_{is}x)/y]^3 + x^3(2 - 3\psi + \psi^3) + 3m_{is}x^2(\psi - 1)\Big\} \tag{6.2.22}$$

For a planar substrate, $x = r_n/r_{cr} \to \infty$ (as in Figure 6.2.5) and the geometric factor is

$$f(m_{is}) = (2 + m_{is})(1 - m_{is})^2/4 \tag{6.2.23}$$

Therefore, at high values of x the geometric factor depends solely on the wettability parameter.

The dependence of the geometric factor on wettability parameter and on the $x = r_n/r_{cr}$ ratio is shown in **Figure 6.2.7**. The geometric factor $f(m_{is}, x)$ is calculated for a spherical substrate according to Equation (6.2.22). One can see that the geometric factor decreases as the size of IN and the wettability parameter increase. Equation (6.2.22) indicates a strong dependence of the geometric factor on the ratio r_n/r_{cr}, rapidly decreasing with decreasing r_n at $r_n/r_{cr} \leq 10$. Thus, inhomogeneous nucleation is efficient only over comparatively large IN. When $r_n/r_{cr} \sim 10$ and the wettability parameter is close to unity, the energetic barrier tends to zero and the probability of nucleation becomes close to one. The high sensitivity of the energetic barrier to the values of r_n/r_{cr} and m_{is} creates a significant uncertainty in the quantitative evaluation of the nucleation rate, since the values of contact angles and their distributions are usually not known.

There are other factors significantly affecting the rate of inhomogeneous nucleation, related to surface properties such as structure of dislocations of the substrates where a germ forms. The effect of the type of IN surface is most pronounced for deposition nucleation where the location of an ice germ can be directly observed, as illustrated in **Figure 6.2.8**. The photos in Figure 6.2.8 show that the sites of ice crystals formation depend on the nonuniform structure of the substrates. However, there is no theory that could describe the relationship between such sites and ice germs. Fletcher (1969) characterized the effect of "active sites" by the relative area α so that an insoluble substrate of radius r_n would have a surface area αr_n^2 with $m_{is} = 1$, while the rest of the surface has $m_{is} < 1$.

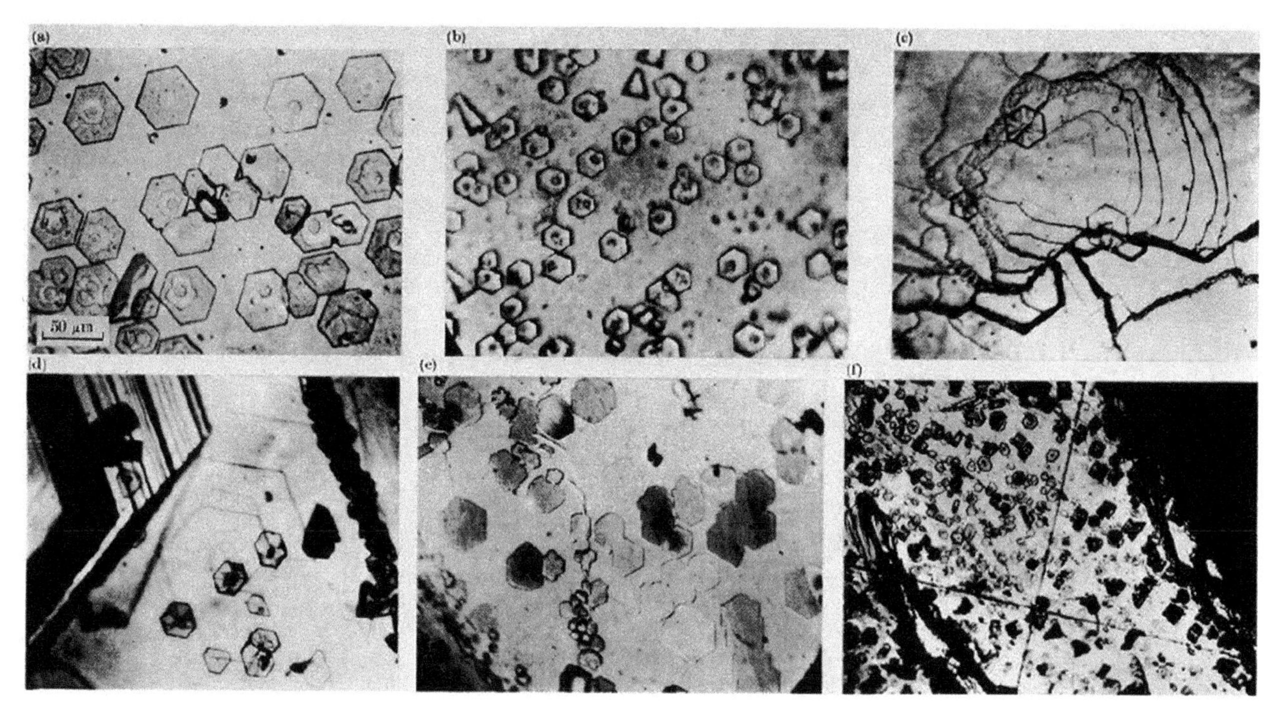

Figure 6.2.8 Microphotographs of ice crystals nucleated from vapor on various inorganic substrates. One can see highly inhomogeneous spatial distribution of growing ice crystals over the IN surface (from Vali, 1999).

6.2.4 Ice Nucleation by Haze Immersion Freezing and by Deposition

Several heterogeneous ice nucleation parameterizations have been suggested based upon theoretical considerations. These parameterizations include analytical fittings applied to a parcel model simulation and various approximations of the basic equations of crystal growth (e.g., Sassen and Benson, 2000; Lin et al., 2002; Gierens, 2003; Karcher and Lohmann, 2003; Khvorostyanov and Curry, 2004, 2005a; Liu and Penner, 2005; Barahona and Nenes, 2008, 2009). Khvorostyanov and Curry (2004, 2005a) derived equations for the critical radius and the critical energy that included both temperature and supersaturation dependencies, generalizing the previous expressions derived using the theory of homogeneous ice nucleation (Section 6.2.1). The authors consider immersion freezing of haze and deposition nucleation as components of a single process called deliquescent–heterogeneous freezing. This approach takes into account the fact that aerosols typically contain both soluble and insoluble fractions. The soluble fraction allows aerosols to be CCN, which triggers both haze formation and droplet formation. The insoluble fraction being immersed into a liquid environment can play the role of IN under certain temperatures and supersaturations.

Under the assumption that haze particles are in equilibrium with the surrounding, Khvorostyanov and Curry (2004) derived the following expression for the critical radius of a germ:

$$r_{cr}(T, S_w) = \frac{2\sigma_{iw}}{\bar{L}_m(T)\rho_i \ln\left[\left(\frac{T_0}{T}\right)(S_w+1)^p\right] - C_\varepsilon \varepsilon^2 - \frac{2\sigma_w}{r_d}} \tag{6.2.24}$$

The difference between Equation (6.2.24) and Equation (6.2.13) is the term $C_\varepsilon \varepsilon^2$, where ε is the elastic strain produced in the ice germ by the insoluble substrate and the constant $C_\varepsilon = 1.7 \times 10^{11}$ $[\text{erg/cm}^3]$. The term $C_\varepsilon \varepsilon^2$ characterizes the misfit between the substrate lattice of IN and that of ice. If that crystallical lattice of the insoluble substrate has a hexagonal structure with parameters close to that of ice, the elastic strain ε produced in the ice embryo by the insoluble IN ranges from ~1% to ~5%. An increase in the misfit leads to increasing in r_{cr}.

Equations (6.2.3) and (6.2.21) together with the correction for the effects of active sites (Section 6.2.3), lead to the following equation for the critical energy required for germ formation (Pruppacher and Klett, 1997):

$$\Delta F_{cr}(T, S_w) = \frac{4}{3}\pi\sigma_{iw} r_{cr}^2 f(m_{is}, x) - a r_n^2 \sigma_{iw}(1 - m_{is}) \quad (6.2.25)$$

Substitution of Equation (6.2.24) into Equation (6.2.25) leads to the following expression for the critical energy:

$$\Delta F_{cr} = \frac{16\pi\sigma_{iw}^3 f(m_{is}, x)}{3\left(\rho_i \bar{L}_m \ln\left[\left(\frac{T_0}{T}\right)(S_w + 1)^p\right] - C_\varepsilon \varepsilon^2 - \frac{2\sigma_w}{r_d}\right)^2} - a r_n^2 \sigma_{iw}(1 - m_{is}) \quad (6.2.26)$$

The geometric factor $f(m_{is}, x) \le 1$ in Equations (6.2.25) and (6.2.26) is the main factor lowering critical energy and making inhomogeneous nucleation much more efficient compared with homogeneous nucleation.

To calculate the nucleation rate per particle, J_{het} Khvorostyanov and Curry (2004) used the expression given by Pruppacher and Klett (1997):

$$J_{het} = C_{het} \exp\left(-\frac{\Delta F_{act}}{kT} - \frac{\Delta F_{cr}}{kT}\right), \quad (6.2.27)$$

where $C_{het} = \frac{kT}{h} C_1 4\pi r_n^2$, $C_1 \sim 10^{15}$ cm^{-2}, and ΔF_{act} is determined using Equation (6.2.9). The units of J_{het} in Equation (6.2.27) are s^{-1} particle^{-1}. Equations (6.2.24–6.2.27) represent the basis of the heterogeneous nucleation scheme.

Figure 6.2.9a shows the dependence of the critical germ radius on the saturation ratio and supercooling $\Delta T = T_0 - T$. One can see that within a reasonable range of parameters, r_{cr} ranges from 3×10^{-3} to 10^{-2} μm. At low saturation ratios and low supercooling, the nucleation is forbidden that corresponds to the negative value of the denominator in Equation (6.2.24). Figures 6.2.9b and 6.2.9c show the values of the critical energy and the nucleation rate per particle as a function of the saturation ratio and supercooling. Figure 6.2.9b indicates very sharp gradients of the nucleation rate, varying by 10–15 orders of magnitude over the temperature range of 5°C or the relative humidity range of 5%.

Figure 6.2.9 is plotted for a particular r_n within the size spectrum of IN. For other aerosol sizes, the curves will be shifted relative to those shown in Figure 6.2.9, so high gradients of the nucleation rate will occur at other values of supercooling and saturation. Since in real conditions r_n varies within a wide range, the ice nucleation takes place within wider ranges of supercooling and supersaturation. Therefore, the gradients of the nucleation rates calculated for the wide aerosol spectra will be smoother than those in Figure 6.2.9. To calculate the probability of nucleation and the number of nucleated ice crystals in case of polydisperse aerosols, one needs to apply a separate nucleation procedure for each particle size within the size spectrum.

As follows from Equation (6.2.24) and Figure 6.2.9, the critical radii of germs are small, so most aerosols containing an insoluble fraction can serve as IN. It remains an open question why the ice crystal concentration is by factor of $10^3 - 10^4$ less than a typical aerosol concentration. Analysis of ice crystal residuals

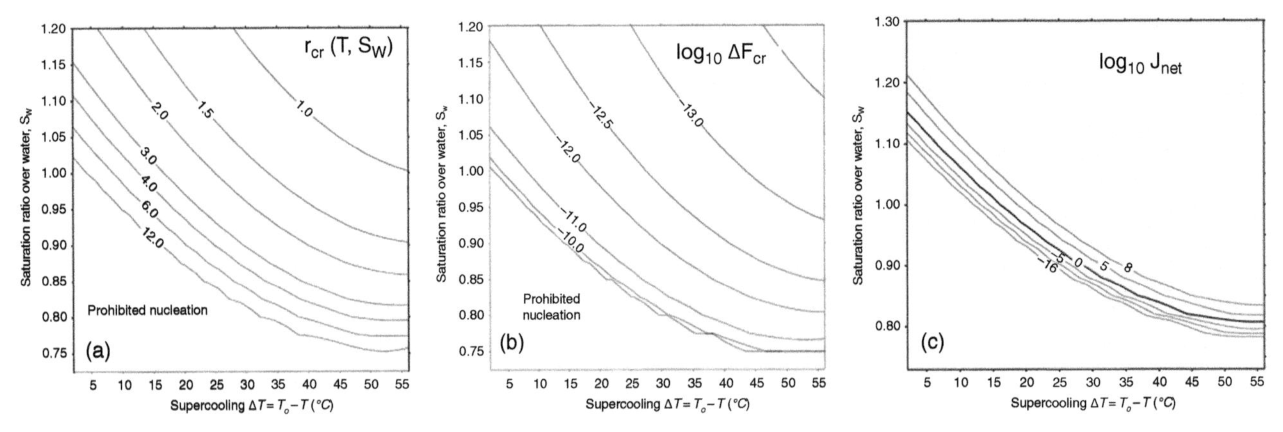

Figure 6.2.9 Dependences of the critical germ radius (in 10^{-7} cm) (a), logarithm of critical energy ΔF_{cr} (erg) (b) and logarithm of nucleation rate in case of immersion haze freezing J_{het} (s^{-1} particle^{-1}) (c) on the saturation ratio and supercooling $\Delta T = T_0 - T$. The wettability parameter $m_{is} = 0.5$ and the radius of the insoluble fraction $r_n = 0.46$ μm. Bold solid line in the right panel corresponds to nucleation rate of 1 s^{-1} (from Khvorostyanov and Curry, 2004; American Meteorological Society©; used with permission).

shows that most ice crystals form on aerosol particles having an insoluble fraction of radii ranging from 0.1 μm to about 15 μm and with a modal radius between 0.5 and 5 μm (Pruppacher and Klett, 1997). Equation (6.2.22) and Figure 6.2.7 show that the value of the critical energy (the energy barrier) dramatically decreases as the size of the insoluble part increases up to $r_n/r_{cr} \sim 10$. Apparently, in reality only aerosol particles of $r_n/r_{cr} > 10$ produce ice crystals. It is clear that the number of activated ice crystals depends on the size distribution of insoluble fractions. Khvorostyanov and Curry (2004) used distributions of the insoluble fraction in the lognormal form with parameters close to that of aerosols, namely, the modal radius was assumed 0.02 μm and the dispersion 2.5 μm. The study shows that the maximum contribution to the ice particle concentration comes from aerosols with the insoluble radii from 0.1 μm to ~0.8 μm. The elastic strain parameter can further increase the size of IN required to produce the observed nucleation rate.

Since atmospheric aerosols are a mixture of different species, their ability to serve as IN depends on specific properties of the material and size distributions of the species. **Figure 6.2.10** shows size distributions of dust, black carbon, and primary-biological aerosols (PBAP) measured in several flights in the Ice in Clouds Experiment – Layer clouds (ICE-L), 2007 field campaign. One can see that the distributions among the flights vary significantly. However, nearly all the values show that the concentration of species with good nucleation properties (e.g., dust, soot) is much smaller than the total aerosol concentration. At the same time, the elongated tail of the aerosol size distributions is formed largely by particles belonging to these species. Due to these reasons, only a tiny fraction ($10^{-6}-10^{-3}$) of aerosols that serve as CCN can be nucleated as ice crystals. To some extent, this fact is reflected in parameterization proposed by DeMott et al. (2010) (see Equation (6.2.32)). Phillips et al. (2013) showed that dust and black carbon are the main contributors to IN in clouds.

Further development of the classical theory of ice nucleation was presented in Marcolli et al. (2007), Broadley et al. (2012), Barahona (2012), and other authors. The main focus of these studies is investigation of the effects of distribution of active sites on the IN surface on the nucleation rate. Marcolli et al. (2007) introduced distribution of contact angles and found a good agreement between theoretical predictions and laboratory results when heterogeneous nucleation rate was averaged over a given distribution of contact angles.

Deposition Nucleation

Deposition nucleation is a process of direct formation of ice crystals on non-soluble particles present in supercooled water vapor. This nucleation mode can be important at subsaturation over water, i.e., when $S_w < 0$. The process of ice germ formation in the deposition mode is similar to that in the freezing mode. Therefore, most equations from this section and the previous sections can be used to describe the deposition mode parameterization after some changes introduced by Khvorostyanov and Curry (2004). For example, ice–water surface tension σ_{iw} in Equations (6.2.24–6.2.26) can be replaced by ice–air surface tension σ_i. Also, substantially different assumptions concerning the prefactor in the nucleation rate equation and the contact angles can be used (Hoose and Mohler, 2012). In general, a comparison of the deposition mode and the freezing mode shows that ice germ formation by freezing is much more efficient from energy balance considerations than freezing by the deposition mode.

6.2.5 Parameterization of Freezing Nucleation and Deposition Nucleation in Numerical Models

Empirical and Semiempirical Relationships

As was shown, the ice nucleation rate strongly depends on many factors such as the wettability, the misfit between the substrate lattice of IN and that of ice, size distribution of insoluble fractions in aerosols, etc. The values of these parameters are not well known. Moreover, atmospheric aerosols are mixtures of different species with different IN properties. From a practical point of view, description of ice production in cloud models requires simple empirical parameterization formulas. Various kinds of such formulas of ice nucleation have been proposed. According to theoretic considerations, ice nucleation rate is a function of temperature and supersaturation, which actually are not independent parameters. At the same time, many parameterization schemes proposed relate the concentration of ice crystals to either temperature or supersaturation. Dependences of concentration of active IN (or ice crystals) on temperature, supersaturation and on properties of IN surfaces are referred to next as the IN activation spectrum. The concept of activation spectrum is similar to that of the CCN activation spectrum describing the dependence of the concentration of activated CCN on supersaturation (Section 5.3).

Temperature Dependencies

The first quantitative description of an average spectrum of ice-forming nuclei in the atmosphere was

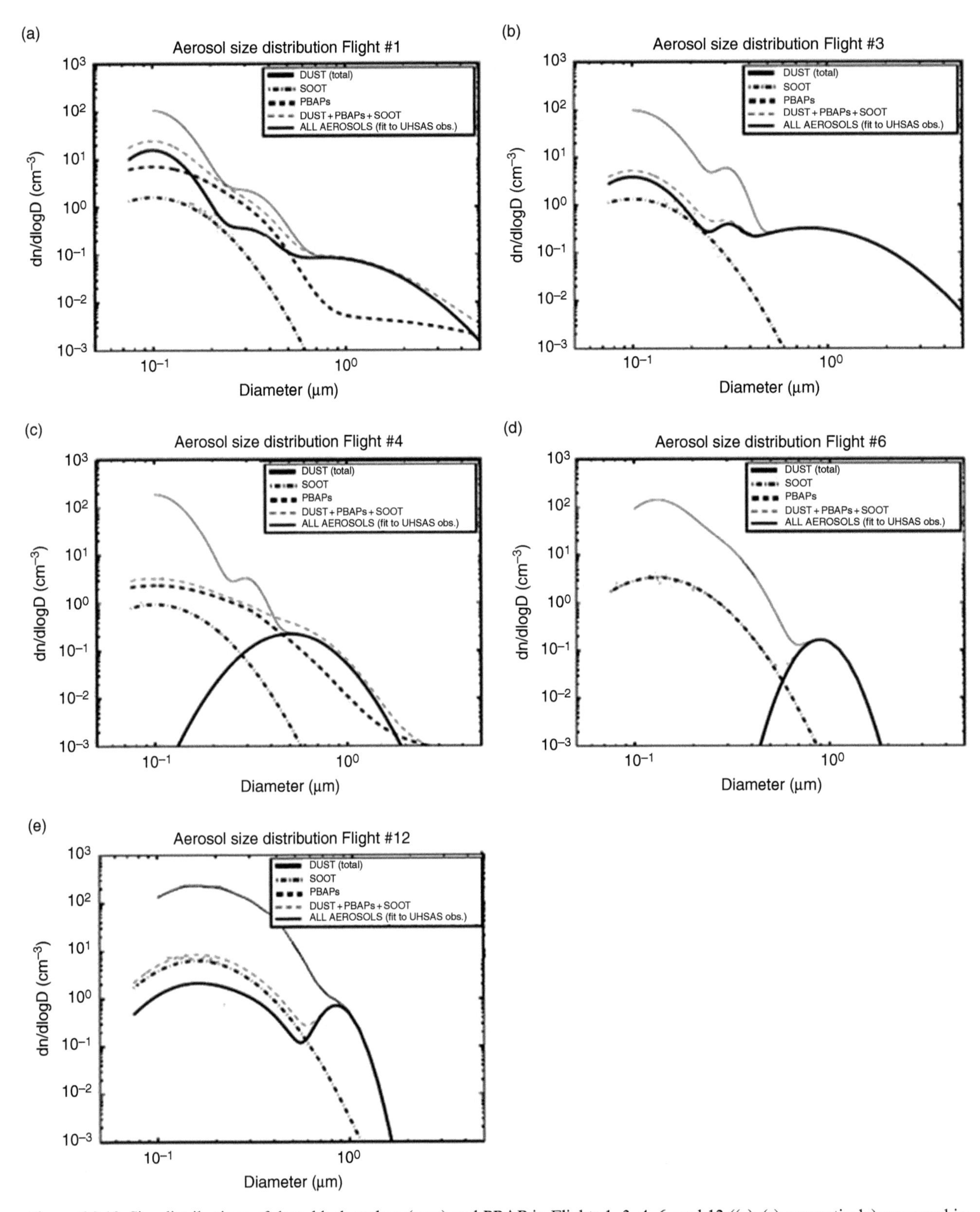

Figure 6.2.10 Size distributions of dust, black carbon (soot) and PBAP in Flights 1, 3, 4, 6, and 12 ((a)–(e), respectively) measured in the ICE-L, 2007 field campaign, averaged for clear air near the sampled wave-cloud. Aerosol concentrations were measured by the Ultra-High Sensitivity Aerosol Spectrometer (UHSAS; Particle Metrics, Inc., Boulder, CO) and photometer SP2 (from Phillips et al., 2013; © American Meteorological Society; used with permission).

probably performed by Fletcher (1962). He combined data from a dozen measurement sets obtained by various instruments to arrive to the exponential expression

$$N_{IN} = N_{i0} \exp(-b_i T_C), \quad (6.2.28)$$

where N_{IN} is the number of nuclei that can be activated and form ice crystals at temperatures exceeding T_C. Fletcher did not distinguish between different nucleation modes, but used measurements conducted within temperature range from $-15°C$ to $-30°C$. The values of parameters in Equation (6.2.28) vary within a wide range. For example, the values of parameter b_i vary between 0.4 and 0.8. The values of N_{i0} vary by several orders of magnitude over different data sets, with the mean value of 10^{-2} m^{-3}. Equation (6.2.28) does not meet the obvious asymptotic condition that at $T_C = 0°C$ there should be no IN nucleation, thus being a rather crude parameterization of the combined effects of different processes. A similar parameterization proposed by Cooper (1986) also describes IN concentration as a function of temperature only.

Supersaturation Dependencies
Huffman (1973) proposed the following power law dependence of concentration of activated IN N_{IN} on the supersaturation with respect to ice, S_i:

$$N_{IN}(S_i) = C_{iH}(S_i)^{b_H}. \quad (6.2.29)$$

According to Huffman, $3 < b_H < 8$, while the dispersion of C_{iH} is also large.

Many model designers use parameterizations based on the formula proposed by Meyers et al. (1992), who used data from two sets of continuous flow diffusion-chamber (CFDC) IN concentration measurements performed by Rogers (1982) and Al-Naimi and Saunders (1985a, 1985b). The Meyers formula is the following:

$$N_{IN} = N_{IN0} \exp(a_i + b_i S_i), \quad (6.2.30)$$

where $N_{IN0} = 10^3$ m^{-3}, $a_i = -0.639$ and $b_i = 12.96$. The formula was derived for the temperature range from $T_C = -7°C$ to $T_C = -20°C$ and for the supersaturation range of $2\% < S_i < 25\%$. This equation predicts formation of about 4×10^3 m^{-3} pristine ice crystals due to deposition and condensation freezing at $-15°C$ at water saturation condition. Nucleation is usually prevented at temperatures higher than $-5°C$. Similar to Equation (6.2.28), Equation (6.2.30) does not obey the asymptotic condition that at $S_i \rightarrow 0$ the number of IN should tend to zero. Equation (6.2.30) is often used in numerical models for a range of conditions wider than it was originally intended for. The errors induced by this extrapolation can be highly significant; hence, some limiting conditions imposed either on the supersaturation range or on the number of nucleated crystals are often used in numerical models (Khain et al., 2004).

Dependencies on Temperature and Supersaturation
Fletcher (1962) developed Equation (6.2.28) based on measurements performed for water saturation conditions. Cotton et al. (1986) modified this equation to include the dependence on supersaturation with respect to ice, proposed by Huffman and Vali (1973), and arrived at the following formula:

$$N_{IN} = aN_{i0} \exp(-b_i T_C)(S_i \cdot 100)^{4.5}, \quad (6.2.31)$$

where $a = 3.5$.

Taking into Account Aerosol Concentration
Equations (6.2.28)–(6.2.31) relate ice crystals or IN concentrations solely to the ambient temperature or supersaturation over ice. Although these parameterizations are based on particular observations, they present averaged dependencies that ignore the temporal–spatial variability of IN in numerical models. IN number concentrations at the same temperature but at different locations and time periods have been observed to vary by more than three orders of magnitude. This high variability is seen in **Figure 6.2.11**, where combined data from nine field studies obtained over fourteen years at a variety of locations are presented (DeMott et al., 2010). One can see that at the same temperature, IN concentrations vary from a few L^{-1} to ~500 L^{-1}. The temperature dependencies of nucleated IN concentrations corresponding to three previously proposed parameterizations do not agree with the observations, particularly at temperatures warmer than -25 °C, even though all of these relationships are observation-based. Such a high dispersion shows that important factors are not taken into account in the parameterization formulas used in the studies.

As was discussed, the main contribution to ice nucleation comes from aerosols containing an insoluble fraction with radii exceeding ~0.1–0.8 μm. It means a correlation can be assumed between ice production and concentration of large aerosols. Indeed, DeMott et al. (2010) showed that a significant correlation exists between the observed IN concentrations and the number concentrations of aerosol particles exceeding 0.5 μm in diameter. Based on this correlation, the authors proposed a parameterization of immersion nucleation and of condensation-freezing nucleation as a simple power law function of temperature:

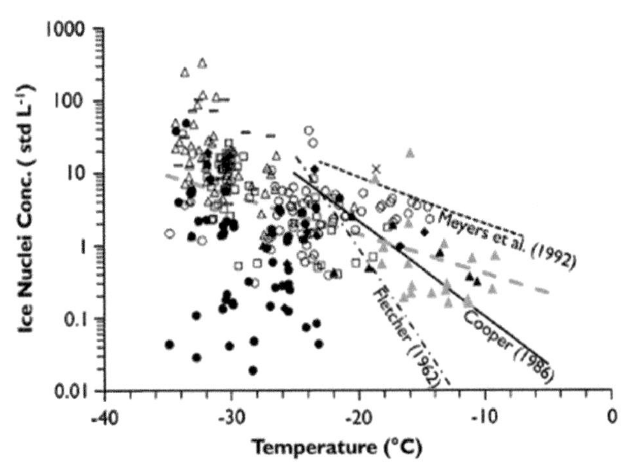

Figure 6.2.11 IN number concentration of active IN at water saturation values or above vs. temperature, measured in different field experiments. The field experiments are marked as: WISP-94 (gray triangles), Alliance Icing Research Study-2 (x), AMAZE-08 (squares), Cloud Layer Experiment-10/ Canadian Cloudsat/CALIPSO Validation Project (open circles), Ice in Clouds Experiment–Layer Clouds (solid circles), Ice Nuclei SPECTroscopy-1 (–), Ice Nuclei SPECTroscopy-2 (diamonds), Mixed-Phase Arctic Cloud Experiment (black triangles), and Pacific Dust Experiment (open triangles). Widely used parameterizations are labeled and plotted over the experimental measurement range on which they were based. The dashed gray line is a T-dependent fit to all the data [$N = 0.117\exp(-0.125\cdot T_C)$; $R^2 = 0.2$] (from DeMott et al., 2010; courtesy of PNAS).

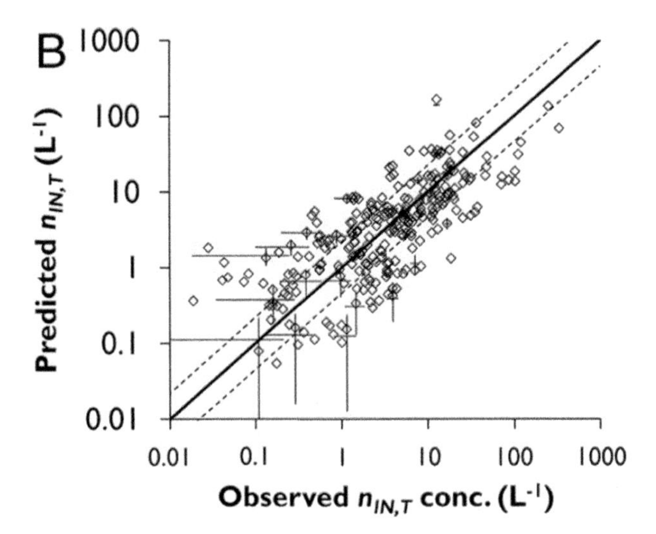

Figure 6.2.12 Relationship between concentrations of activated IN predicted by means of Equation (6.2.32) and the observed concentration of IN with diameters exceeding 0.5 μm (from DeMott et al., 2010; courtesy of PNAS).

$$N_{IN} = a_D(-T_C)^{b_D} N_{a,05}^{-c_D T_C + d_D}, \qquad (6.2.32)$$

where T_C is the temperature in Celsius, N_{IN} is the IN concentration that can be activated and form ice crystals (in L^{-1}) at temperature T_C, $a_D = 0.0000594$, $b_D = 3.33$, $c_D = 0.0264$, $d_D = 0.0033$, and $N_{a,05}$ is the concentration of aerosol particles with diameters exceeding 0.5 μm (in cm^{-3}). Utilization of the dependence of IN concentration on the concentration of large aerosols reduces the spread of potential errors in predicting IN concentrations at a given temperature from a factor of ~10^3 to ~10, as demonstrated in **Figure 6.2.12**.

Taking into Account IN Type and the Area of Active Sites

Phillips et al. (2008) developed an empirical parameterization of ice nucleation. They considered three types of aerosols: dust and metallic compounds, black carbon, and insoluble organics. The aerosols of each type have insoluble fraction that depends on aerosol type and size. The concentration $N_{IN,x}$ of active IN of the x-th kind is determined as

$$N_{IN,x} = \int_{\log(0.1\,\mu m)}^{\infty} (1 - \exp[-\mu_x(D_x, S_i, T_C)]) \frac{dN_x}{d\log D_x} d\log D_x \qquad (6.2.33)$$

where μ_x is the average amount of activated IN per amount of aerosols of diameter D_x. The value μ_x is determined as

$$\mu_x = \pi D_x^2 H_x(S_i, T_C)\xi(T_C)\frac{\alpha_x N_{IN}}{\Omega_x} \qquad (6.2.34)$$

In Equation (6.2.34), N_{IN} is the total number of IN, determined using the modified Equation (6.2.30), α_x is the fraction of aerosols of x-type with respect to the total aerosol concentration and Ω_x is the surface area of all aerosols of x-type with diameters $0.1\ \mu m \le D_x \le 1\ \mu m$. These aerosols cause the observed ice nucleation. H_x is a factor reducing IN activity at low S_i and warm temperatures. The value of H_x ranges from zero to one. At water saturation $H_x = 1$, $\xi(T_C)$ is a function equal to zero for $T_C > -2°C$, and equal to one for $T_C < -5°C$. Within the range from $-2°C$ to $-5°C$ $\xi(T_C)$ is linearly interpolated. At $\mu_x \ll 1$, $N_{IN,x}$ can be written as $N_{IN,x} \approx H_x(S_i,T)\xi(T_C)\alpha_x N_{IN}\frac{\Omega_{xt}}{\Omega_x}$, where Ω_{xt} is the total surface area of all aerosols of x-type with dry diameters exceeding 0.1 μm per unit mass of air (the surface area mixing ratio). Equations (6.2.33) and (6.2.34) express the fundamental concept according to which the number concentration of active IN related to aerosol particles of x-type is approximately proportional to their surface area. As was discussed, ice germs arise at the surface of IN at active sites that have a certain probability of occurrence per unit area of the surface of a given IN material. Such sites are determined by the crystallographic features of the surface. Phillips et al. (2008) provided empirical values for all of the parameters needed to calculate IN concentration. The total number of IN was calculated as a sum of all $N_{IN,x}$. According to the elemental composition analysis of IN residual material from crystals made in a set of field experiments (NASA FIRE_ACE, SUCCESS, NASA CRYSTAL-FACE, INSPECT-1), about 60% of IN were dust and metallic, about 30% carbonaceous, and about 10% could be attributed to bacteria. Later, parameterization (6.2.33) was modified by Phillips et al. (2013) to take into account different temperatures for the onset of freezing of different species. The fourth group of IN was introduced, namely soluble organic aerosols.

The procedure of ice nucleation proposed by Phillips et al. (2008, 2013) included both nucleation freezing of haze and of liquid drops containing IN. The procedure was implemented in HUCM. Since HUCM has bin microphysics, the number and the area of IN within each bin was calculated. If an aerosol serving as CCN was activated to a droplet, this droplet contained IN with mass corresponding to the insoluble fraction of activated CCN. Collisions between droplets led to formation of raindrops containing many IN and having larger total surface. It determined an increase in the probability of raindrop freezing with increasing raindrop size. It meant that parameterization of Phillips et al. (2008) described both primary nucleation of ice crystals and immersion drop freezing.

Parameterizations Based on Theoretical Considerations
Upon performing a set of simulations, Curry and Khvorostyanov (2012) proposed a parameterization of concentration of active IN as a function of two variables, namely T_C and the updraft velocity w:

$$N_{IN} = C_g(-T_C)^{c_T} w^{c_w}, \tag{6.2.35}$$

where N_{IN} is in L^{-1} and $c_w = 1.41$. There are two sets of other constants: $C_g = 0.4 \times 10^{-8}$, $c_T = 8.0$, for $T_C > -15°C$, $C_g = 0.535$, and $c_T = 1.05$ for $T_C < -15°C$. According to Curry and and Khvorostyanov (2012), Equation (6.2.35) can be used as a simple parameterization in cloud models and Global Circulation Models.

Comparison between Parameterizations and Observations
Due to multiple reasons, comparison of different parameterizations of ice nucleation is quite a sophisticated problem. First, the variability of ice crystal concentration observed in different and even in similar field experiments is quite high. Second, the measurements themselves are not perfect and, consequently, the results obtained with different instruments can differ significantly (DeMott et al., 2011). Third, in both real clouds and cloud models, crystal concentration forms as a result of many processes, and it is not simple to isolate the effects of nucleation *per se* due to condensational and haze immersion freezing. The relative contribution of different ice production mechanisms is different in clouds of different types. In stratiform clouds with low vertical velocities, the concentration of ice crystals may be lower than concentration of IN (Mamouri and Ansmann, 2015). At the same time, in convective zones concentration of ice crystals can substantially exceed that of IN (Hobbs, 1969). Fourth, the diagnostic parameterization formulas are treated in different ways in different models, which can lead to discrepancies in results even if the same parameterization formulas are used as a basis. Finally, models include many parameters that affect ice concentration, e.g., the vertical velocity, drop concentration and DSD. Therefore, the same parameterization schemes being implemented into different models may lead to different results.

Comparison of ice crystal concentrations obtained in different parameterization schemes with those measured in several field experiments is shown in **Figure 6.2.13**. The Diehl and Wurzler (2004) parameterization (DW04) used a semiempirical equation for drop freezing in the immersion mode, based on the laboratory experiments of seven different IN types: soot particles from kerosene; three mineral particle types (kaolinite, montmorillonite, and illite); and three biological particles (pollen, leaf litter, and bacteria). Freezing of a drop containing specific insoluble particles is described using the median freezing temperature of the observed monodisperse drop population. PDA08 is an empiric parameterization developed by Phillips et al. (2008) described previously (Equation 6.2.33). In Figure 6.2.13, the results obtained using the parcel model developed by Khvorostyanov and Curry (2005a) and modified by Curry and Khvorostyanov (2012) are presented, as well. These simulations were performed using vertical velocities of 50 cm/s (CK10-1) and 2 cm/s (CK10-2), as well as the wettability parameter $m_{is} = 0.52$. The results of the parameterization performed by Cooper (1986) and analyzed by

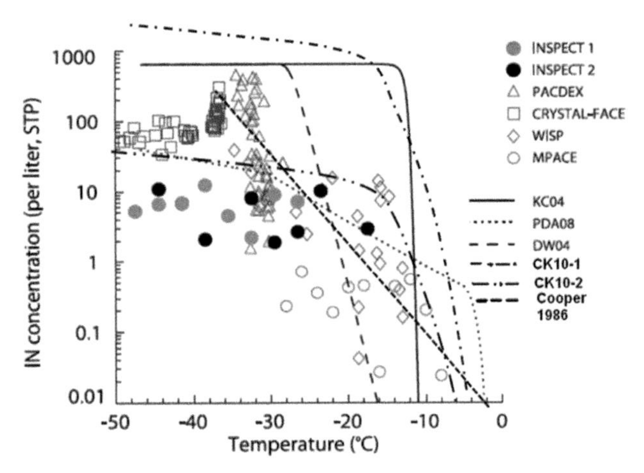

Figure 6.2.13 Ice nuclei number concentration (at standard temperature and pressure) measured with the CFDC from several field campaigns in the free troposphere. The filled symbols represent measurements from INSPECT I (gray: DeMott et al., 2003a) and INSPECT II (black: Richardson et al., 2007). Selected 1-min average concentrations over 3 days during PACDEX (Stith et al., 2009) were obtained over the North Pacific, in and out of intense Asian dust layers. CRYSTAL-FACE measurements were performed within the Saharan Aerosol Layer near Florida (DeMott et al., 2003b; Prenni et al., 2007a). WISP data from the winter–spring transition period in Colorado are adapted from the study by Mohler et al. (2007). MPACE measurements are obtained in Arctic Fall (Prenni et al., 2007b). The curves represent ice crystal concentrations predicted by means of the parcel model by Eidhammer et al. (2009) with an updraft of 500 cm/s and the background aerosol size distribution. The solid curve shows predictions made with the KC04 scheme (the dashed curve with the [DW04] parameterization and the dotted curve with the PDA08 parameterization) (from Eidhammer et al., 2009 and Curry and Khvorostyanov, 2012; courtesy of ACP).

Khvorostyanov and Curry (2005a) are presented, as well. One can see that ice crystal concentrations within the same field experiment may vary by two orders of magnitude. The empiric parameterization suggested by Phillips et al. (2008) gives the best agreement with the observations. Observational results obtained in some of the experiments were used to tune the empirical scheme. The significant difference between results obtained by Eidhammer et al. (2009) and by Curry and Khvorostyanov (2012) applying the Khvorostyanov and Curry's approach (KC04 vs. CK10-1 and CK10-2) can be attributed not only to different vertical velocities used, but to the fact that different models and, possibly, different governing parameters determining the process of ice nucleation were used in these studies.

Further experimental and theoretical studies are required to improve the description of immersion freezing nucleation in cloud models. One way of applying theoretical results is to choose, empirically or theoretically, the values of governing parameters. The first steps in this direction were made by Eidhammeret et al. (2009) and Fan et al. (2012).

6.2.6 Thermodynamic Constraints on Heterogeneous Ice Nucleation Schemes

A theoretical approach allows finding constraints on thermodynamic variables that should be taken into account while considering empirical relationships. For instance, the condition that the critical radius should be positive enables us to derive the threshold supersaturation values with respect to water (see Equation 6.2.15) and the threshold temperature values required for ice nucleation (Curry and Khvorostyanov, 2010). The relationship between the threshold values of temperature and of supersaturation divides the $e/e_w - T_C$ domain into zones where nucleation can take place and where nucleation is prohibited. **Figure 6.2.14** presents saturation ratio–temperature ($e/e_w - T_C$) diagrams within the temperature ranges of $-30°C < T_C < 0°C$ and $0.7 < e/e_w < 1$. Here the threshold difference $\frac{e-e_{th}}{e_w}$ is superimposed. The deep-blue hatched line denotes the boundary $e = e_w$. Below the hatched line (blue field), $r_{cr} > 0$, and the states corresponding to these areas are thermodynamically allowed. The states above the deep-blue hatched line (white field) correspond to the negative values of r_{cr}. Ice germs cannot be nucleated above this line. Figure 6.2.14 shows that the allowed domain is located in the triangle below the temperature range of $-8°C$–$12°C$, and at the water saturation ratio above 0.8. This area covers only about ⅛ of the entire considered domain at the plane $T_C - e/e_w$. The rest ⅞ of this domain are allowed for ice nucleation according to the schemes used by Meyers et al. (1992) and DeMott (2010), but are thermodynamically prohibited because the ice germ radii are negative here. The boundaries of the allowed domain depend on the size r_n of the insoluble fraction of aerosol particles. When r_n increases above the 0.05 μm typical of the fine mode to the 1 μm typical of the coarser mode, the domain allowed for ice nucleation shifts to higher temperatures, in this example to about $-5°C$.

6.2.7 Diagnostic Formulas in Numerical Models

Equations (6.2.28)–(6.2.35) are diagnostic equations and do not explicitly contain IN activation rates, making them difficult to apply in numerical models that should calculate tendencies of variables. The

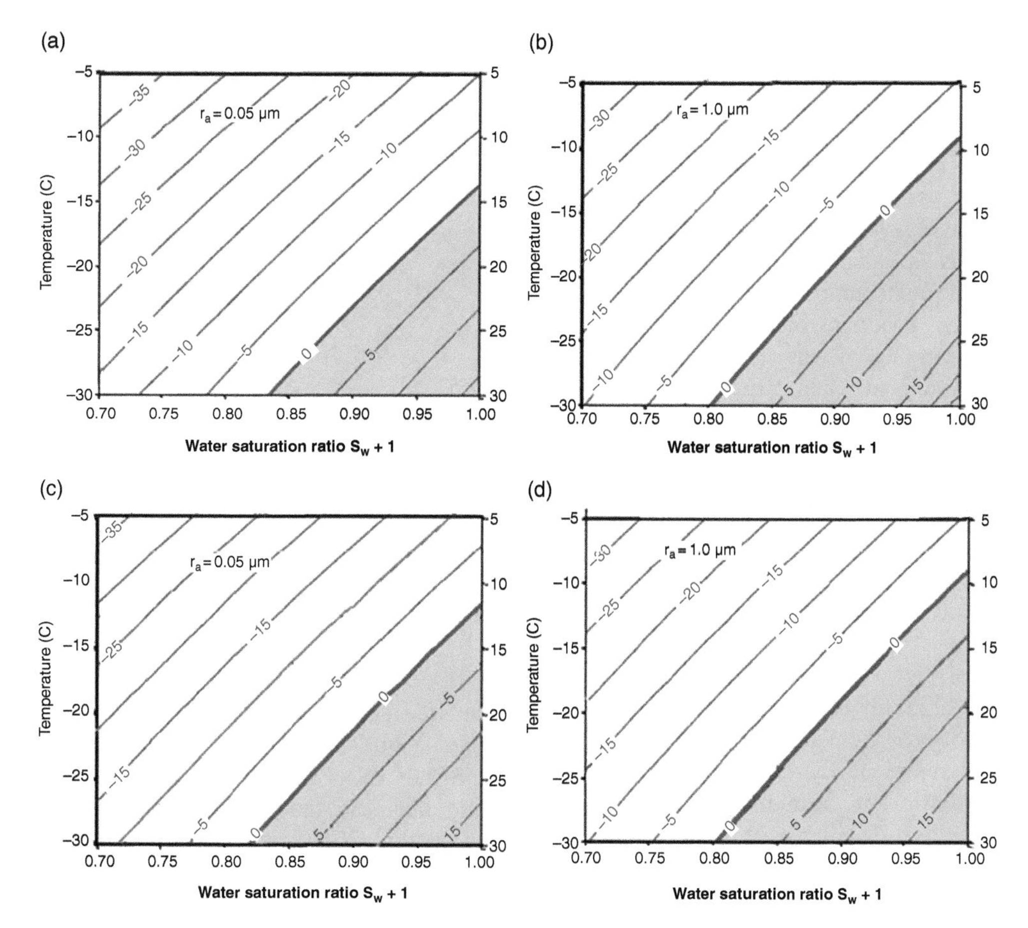

Figure 6.2.14 $T_C - e/e_w$ diagrams with superimposed threshold difference $RH - RH_{\text{th}}$ (blue lines). The $RH = RH_{\text{th}}$ line is hatched and marked deep blue. The entire area of the squares is allowed for parameterizations implemented by Meyers et al. (1992) (a, b) and by DeMott et al. (2010) (c, d). At the same time, only in zones marked blue the critical radius has positive values (from Curry and Khvorostyanov, 2010, with changes; courtesy of ACP).

concentration of IN produced in a numerical model by these equations depends on the way the formulas are applied. In many models, these formulas are used directly, i.e., it is assumed that a certain air parcel contains an infinite source of potential IN which continue to get activated under unchanging conditions (e.g., at the same temperature). Equation (6.2.28) used in this way predicts an unrealistically large IN concentration (for example, 65 cm^{-3} at $T_C = -30°$C). Hence, in this approach the total number of nucleated IN becomes dependent on the length of the model time step. Consequently, one has to limit the number of the IN artificially by some maximum value corresponding, for instance, to the temperature of $-27°$C (e.g., Reisner et al., 1998).

In most models, application of diagnostic formulas such as Equations (6.2.28–6.2.35) is based on the concept of an ascending adiabatic parcel. The concentration of N_{IN} predicted by these formulas is usually considered to be the maximum possible value of ice crystal concentration at given S_i or T_C, like in activation spectrum of CCN (e.g., Khain and Sednev, 1996; Reisin et al., 1996; Reisner et al., 1998; Walko et al., 1995). If ice crystal concentration N_c at a certain grid point exceeds the value predicted by these formulas, next nucleation is prevented. Thus, according to this approach, the number of nucleated ice crystals during one-time step is determined as

$$\Delta N_c = \begin{cases} N_{IN} - N_c, & \text{if} \quad N_{IN} > N_c \\ 0, & \text{if} \quad N_{IN} \le N_c \end{cases} \tag{6.2.36}$$

This enables us to adjust the ice crystal concentration to realistic values, even if the parameterization formulas lead to unrealistically large concentration values. Thus, ice crystal concentrations are actually nudged to the observed values regardless of the parameterization formulas used. At the same time, according to Equation (6.2.36), the number of activated ice crystals at a particular grid point becomes dependent on the concentration of ice crystals at this point. This dependence has a physical meaning only within the conceptual framework of an adiabatic parcel, where the history of IN within Lagrangian parcels is calculated. In real cloud collisions, mixing with the environment, sedimentation, and other microphysical processes change the crystal concentration in the ascending volume.

According to another approach, two quantities are transported: concentration of IN and concentration of already-activated IN (Philips et al, 2008). The concentration of already activated IN is assumed to be a passive scalar that is not affected by collisions and phase transformations. As a result, in each grid point of a computational area nucleation of ice crystals stops if concentration of already activated IN exceeds N_{IN}.

From the physical point of view, activation of IN should depend solely on the thermodynamical parameters and on own properties of a particular IN. Equations (6.2.28–6.2.35) actually represent the activation spectrum related to the distribution of the number of active IN with respect to supersaturation S_i or with respect to temperature. Since these parameters determine the critical germ size, these formulas can be used to calculate distributions of insoluble parts of aerosols, similarly to how Equation (5.1.28) allows us to calculate CCN size distribution (Section 5.1). For instance, it follows from Equation (6.2.30) that

$$\frac{dN_{IN}}{dr_N} = \frac{dN_{IN}}{dS_i}\frac{dS_i}{dr_N} = b_i N_{IN}\left(\frac{dr_N}{dS_i}\right)^{-1} \quad (6.2.37)$$

Since most models do not contain size distributions of IN, a semi-Lagrangian approach can be used as proposed by Khain et al. (2000), according to which the number of the newly activated ice crystals at each time step at a certain grid point dN_c, can be written as

$$dN_c = \begin{cases} b_i N_{IN} dS_i, & \text{if } dS_i > 0 \\ 0, & \text{if } dS_i \leq 0 \end{cases}, \quad (6.2.38)$$

where $dS_i = \left(\frac{\partial S_i}{\partial t} + u\frac{\partial S_i}{\partial x} + v\frac{\partial S_i}{\partial y} + w\frac{\partial S_i}{\partial z}\right)dt$ can be interpreted as an increase in supersaturation in a cloud parcel during its movement from point $(t-dt, x-dx, y-dy, z-dz)$ to a regular grid point (t, x, y, z) (**Figure 6.2.15**). The semi-Lagrangian approach expressed by Equation (6.2.38) and used in HUCM (Khain et al., 2004; 2011) appears to be a reasonable way to describe the sink of IN due to nucleation. However, this approach does not record the history of IN evolution.

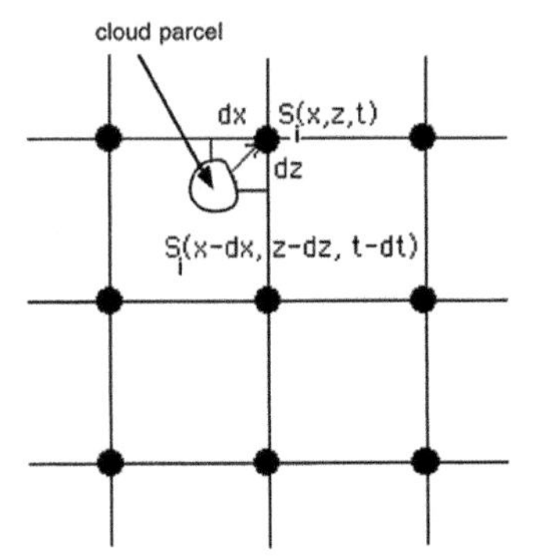

Figure 6.2.15 Scheme of an air parcel track during one time step from point $(x-dx, z-dz)$ to the grid point (x, z). During the motion, supersaturation within the air parcel changes by dS_i (from Khain et al., 2000; courtesy of Elsevier).

In principle, the process of ice crystal nucleation should be treated in models similar to that in case of CCN nucleation in warm microphysics, namely, by implementation of a budget and distributions of IN. As soon as the SD of IN is given, the sink of IN by activation should decrease the number of available IN. Carrio and Cotton (2010) interpreted parameter N_{IN0} in Equation (6.2.30) as the maximum concentration of IN available for activation, and the value $F_m = \exp(a_i + b_i S_i)$ as a fraction of available IN that is activated. F_m is assumed to be equal to one for simulations where supersaturation over ice exceeds 40%. In this approach, N_{IN0} is a forecast variable and is advected, diffused, and sinks due to ice activation.

6.2.8 Nucleation by Drop Immersion Freezing

Drop immersion freezing that takes place under supercooled conditions consists of two stages. The first adiabatic stage is nucleation, i.e., formation of a germ on an immersed IN. The second stage is actually the freezing stage which is a thermodynamic process when the latent heat release of freezing is balanced by the heat fluxes from the freezing particle to the surrounding air. While the first stage is very rapid, the second stage takes much longer. The time of full freezing of a drop depends on the environment temperature and on the drop size. Here we consider only the nucleation stage. The second stage is discussed in Section 6.7.

Drops are not in the equilibrium with the surrounding. This is one of the reasons hindering the development of a theory of the drop immersion freezing. Most descriptions of immersion freezing are based on empirical considerations and laboratory measurements. It is usually assumed that immersion IN are distributed homogeneously throughout liquid cloud water. Thus, larger drops contain a larger amount of insoluble immersion nucleus. Accordingly, the probability for large drops to freeze is higher than the probability of smaller drops. Indeed, it was discovered long ago that the mean freezing temperature of drops increases with drop volume and perhaps with the cooling rate (e.g., Bigg, 1953; Pitter and Pruppacher, 1973).

There are two main hypotheses used as the basis for two types of parameterizations. The stochastic hypothesis treats immersion freezing in a way analogous to homogeneous freezing. According to this hypothesis, freezing results from a random formation of a critical size embryo, while the presence of foreign particles increases the probability of the nucleation without disturbing its stochastic nature. According to the stochastic hypothesis, not all drops with the same properties freeze at the same time, but the fraction of frozen particles increases with time. In the corresponding parameterizations, the freezing fraction depends on time. Many models use the immersion freezing parameterizations based on the stochastic hypothesis formulated by Bigg (1953), according to which the probability of freezing is assumed proportional to the drop mass and the freezing rate. In this case, drop concentration decreases because of freezing as

$$\frac{1}{N_d}\frac{\partial N_d}{\partial t} = -a_b m \exp\left(-b_b T_C\right), \tag{6.2.39}$$

where N_d is the concentration of drops of mass m, $a_b = 10^{-4}\ \mathrm{s}^{-1}\,\mathrm{g}^{-1}$ and $b_b = 0.66\ \left[{}^{\circ}\mathrm{C}^{-1}\right]$. Equation (6.2.39) was used in different models such as MM5 (e.g., Reisner et al., 1998); HUCM (Khain and Sednev, 1996) as well as in the cloud model developed at Tel Aviv University (Reisin et al., 1996). From Equation (6.2.39) one can evaluate the time needed to freeze half of the existing drops with mass m:

$$t_m = \left[\ln 2/(a_b m)\right] \exp\left(b_b T_C\right) \tag{6.2.40}$$

The values of the time parameter t_m (the typical values are presented in **Table 6.2.2**) rapidly decreases with height above the freezing level. One can see that the characteristic freezing time rapidly increases with decreasing drop size and a decrease in the temperature. These evaluations show that rain drops freeze at relatively warm temperatures, while small cloud droplets can remain unfrozen for a comparably long time at very low temperatures. Freezing of small droplets ascending at the temperatures of homogeneous freezing leads to the production of a large amount of small ice crystals (Rosenfeld and Woodley, 2000). Although at temperatures warmer than $T_C = -10^{\circ}\mathrm{C}$ even large drops have a low probability of freezing, such freezing events may be important because they provide instantaneous riming centers and could trigger the rime-splintering multiplication process. Equation (6.2.40) was deduced after laboratory experiments with distilled water that contained numerous solid particles that are relatively small and uniform in composition. Formula (6.2.40) provides a stronger temperature dependence than that measured for cloud and rain water.

Table 6.2.2 Time during which half of the droplets of different sizes freeze at different temperatures. The time was estimated using Equation (6.2.40).

m, g	r, μm	t_m, s ($T_C = -20^{\circ}$C)	t_m, s ($T_C = -30^{\circ}$C)
4×10^{-3}	1,000	3.2	4.4×10^{-3}
4×10^{-6}	100	3.2×10^{3}	4.4
4×10^{-9}	10	3.2×10^{6}	4.4×10^{3}

The parameterization formula (6.2.39) does not take into account the difference between the freezing characteristics of soluble and insoluble materials inside drops. At the same time, such characteristics are quite important. Median freezing temperatures for different insoluble particles obtained in laboratory experiments are shown in **Table 6.2.3**. A possible increase of the freezing temperatures due to the content of insoluble particles acting as ice nuclei was implemented into the model developed by Diehl and Wurzler (2004) based on laboratory investigations. Following Wurzler and Bott (2000), two-component insoluble particles were considered (kaolinite and montmorillonite) to study the freezing point depression change as a function of salt concentration in the drop. The parameterization expression proposed by Diehl and Wurzler (2004) was similar to Equation (6.2.39), but contained coefficient a_b depending on the type of IN. As a result, the probability of freezing for drops containing soot particles was by one to four orders of magnitude lower than that for drops containing mineral particles and by 8–9 orders of magnitude lower than probability of freezing for drops containing biological particles. Numerical simulations showed that Equation (6.2.39) describes the drop immersion freezing well for a "mean" insoluble particle. This approach allowed them to describe the drop freezing rate for defined aerosol particle distributions as a function of the fractions of different insoluble components within the total aerosol particles. Note that

Table 6.2.3 Median freezing temperatures for drops of various sizes containing different insoluble particles (from Diehl and Wurzler, 2004; © American Meteorological Society; used with permission).

Drop Radius	≈ 50 μm	≈ 250 μm	≈ 350 μm
Pure water drops (Mason, 1957)	−35°C	−34°C	−33°C
Soot particles			
From acetylene	for $a < 15$ μm: <−34°C (DeMott, 1990)		
From kerosene			−28°C (Diehl and Mitra, 1998)
Mineral particles			
Kaolinite	−32.5°C (Hoffer, 1961)		−23°C (Pitter and Pruppacher, 1973)
Montmorillonite	−24.0°C (Hoffer, 1961)		−19°C (Pitter and Pruppacher, 1973)
Illite	−23.5°C (Hoffer, 1961)		
Biological particles			
Pollen		−14°C (Diehl et al., 2002)	
Leaf litter			−9°C (Diehl et al., 2001)
Bacteria		−7°C (Levin and Yankofski, 1983)	

some approaches used to parameterize heterogeneous nucleation (e.g., Phillips et al., 2013) also described the first stage of drop freezing.

The alternative to the stochastic hypothesis is the singular hypothesis that assumes that the drop freezing temperature is determined by nucleus properties (Vali, 1994). Unlike the stochastic hypothesis, the number of frozen drops in the singular hypothesis does not change with time at a given supercooling value. All drops with similar properties freeze immediately as the necessary conditions for freezing are satisfied. It is assumed that immersion nuclei are distributed homogeneously throughout liquid cloud water and that their activity increases with decreasing temperature. A parameterization based on the singular hypothesis is also consistent with two experimentally determined tendencies: more drops of a given size freeze at lower temperatures and larger drops are more likely to freeze at a given temperature. Parameterizations based on the singular hypothesis are described by diagnostic relationships that do not contain the time parameter. Ovchinnikov and Kogan (2000) and Khain et al. (2004) used a temperature dependence of immersion nuclei suggested by Vali (1975):

$$N_{im} = N_{im0}(-0.1T_C)^{\gamma}, \qquad (6.2.41)$$

where N_{im} is a number of active immersion nuclei per unit volume of liquid water, $N_{im0} = 10^7$ m^{-3} and $\gamma = 4.4$ for cumuliform clouds. Equation (6.2.41) reveals a much weaker temperature sensitivity of active immersion IN than Equation (6.2.39). For instance, the number of active INs within the temperature range from −20°C to −30°C increases by factor 10^3 when Equation (6.2.39) is used and only by a factor of about seven if Equation (6.2.41) is used.

It is likely that the actual drop freezing is better represented by a combination of the stochastic and the singular approaches than by either one of them alone (Pruppacher and Klett, 1997). However, current models use either the stochastic or the singular approach. Vali (1994) proposed a model for a time-dependent freezing rate that may serve as a basis for an improved parameterization of immersion freezing in cloud models. The main limitations of this parameterization can be accounted for from insufficient knowledge about the content of freezing nucleus within the cloud drops. However, this content is difficult to obtain from either theory or direct sampling.

As Equation (6.2.41) is a diagnostic equation, in order to calculate the number of activated immersion IN per unit volume of a cloud parcel Khain et al. (2004) wrote it in the differential form as $dN_{im} = -10^{-\gamma}N_{im0}(-T_C)^{\gamma-1}dT_C$. Here, dT_C was calculated as a total derivative written similarly to Equation (6.2.38). Assuming that activation of only one immersion IN within a drop is enough to freeze this drop, Khain et al. (2004) distributed the number of activated IN in the bins of the DSD proportional to drop mass in the bins. The change of DSD in the k-th mass category by immersion freezing $(\delta f_k)_{fr}$ is expressed as

$$(\delta f_k)_{fr} = -\frac{4}{3}\pi r_k^3 f_k dN_{im} \qquad (6.2.42)$$

While haze freezing leads to formation of crystals whose type depends on temperature, drop freezing can lead to formation of frozen particles of different sizes depending on the size of the drop. In HUCM, frozen drops with radii smaller than 100 μm are assigned to plate crystals (with density of 0.9 g cm^{-3}). Larger drops are assigned either to hail (Khain et al., 2004) or to freezing drops (Phillips et al., 2015).

6.2.9 Contact Nucleation

Another mechanism of ice particles formation is contact nucleation. **Figure 6.2.16** shows a possible mutual location of a drop and an insoluble particle, whose interaction can lead to drop freezing. Figure 6.2.16a illustrates formation of crystals or frozen drops by immersion freezing. Figures 6.2.16b and 6.2.16c show two possible modes of contact nucleation. Usually, contact of an insoluble particle with drop from outside is considered (Figure 6.2.16b). The presence of active contact IN in the air does not automatically lead to the formation of ice particles. Freezing of a supercooled water drop takes place only upon collision with a contact IN.

The evaluation of contact nuclei concentration in the atmosphere is a great challenge. The nucleation rate within a contact mode crucially depends upon a nucleus size that is not known and is difficult to measure, which creates additional uncertainty.

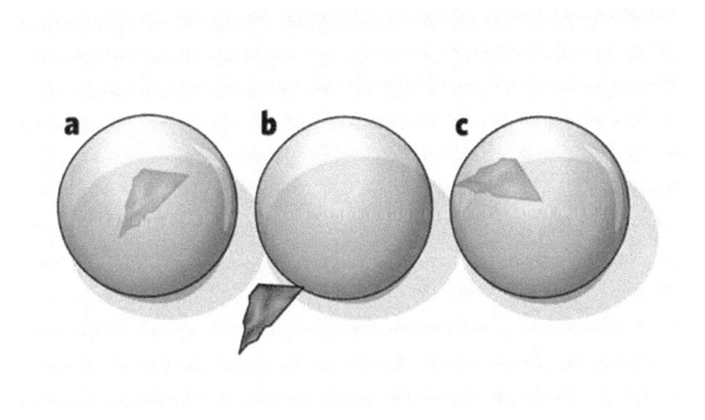

Figure 6.2.16 Mutual locations of an insoluble aerosol and a solute drop leading to different modes of inhomogeneous nucleation (from Sastry, 2005; courtesy of Nature Publishing Group).

Assuming different sizes of contact IN and, therefore, different efficiencies of various capture mechanisms, one may deduce the values of ice nucleus concentrations that vary by orders of magnitude while being based on the same data set. For example, using Blanchard's (1957) data and assuming the contact nucleus size of 0.3–0.5 μm, Young (1974) evaluated the concentration of nuclei active at −4°C to be of 10^5–10^6 m^{-3}. Deshler and Vali (1992) reevaluated Blanchard's experiment assuming the nucleus radius of 0.05 μm and found that the concentration should be ~2×10^4 m^{-3} for the same temperature. Beard (1992) came to a similar estimation for an even smaller nucleus radius of 0.01 μm. He also showed that if the assumed IN size is of 2–5 μm, only 10^2 m^{-3} of such giant nuclei is needed to explain the observed freezing rate. Assuming the initial mass of nucleated ice crystals to be 10^{-9} g, Cotton et al. (1986) approximated the concentration of active contact ice nuclei as

$$N_{cn} = \begin{cases} N_{cn0}(270.16 - T), & \text{if } T < 270.16\ \text{K} \\ 0, & \text{if } T \geq 270.16\ \text{K} \end{cases} \qquad (6.2.43)$$

where T is the air temperature (in K) and $N_{cn0} = 2 \times 10^{-1}$ cm^{-3}. Meyers et al. (1992) analyzed available data and offered a parameterization by fitting its results to the measurements performed by Vali (1976), Cooper (1980), and Deshler (1982):

$$N_{cn} = N_{cn\_m0}(a_{cn} - b_{cn}T_C), \qquad (6.2.44)$$

where $N_{cn\_m0} = 10^3$ m^{-3}, $a_{cn} = -2.8$, and $b_{cn} = 0.2629$ $(°\text{C})^{-1}$. At $T_C = -11°\text{C}$, the dependency (6.2.44) yields the concentration of contact IN of 10^3 m^{-3}.

After determination of amount of the available contact IN, it is necessary to evaluate the number of drops that can be frozen by collection of contact IN. As regards ice formation, such collection leads to ice particle formation. At the same time, this collection leads to aerosol scavenging from the atmosphere and affect the amount of atmospheric aerosols and the chemistry of precipitation (Pruppacher and Klett, 1997). The decrease in the concentration of aerosols is known as *impact scavenging*. Depending on the size of the aerosol particles r_N, the collision mechanisms include: inertia ($r_N > 0.1$ μm) and diffusion ($r_N < 0.1$ μm) as well as thermophoresis and diffusiophoresis (0.01 μm $< r_N <$ 0.1 μm) (Vohl et al., 2001). In the presence of charges, the electrical forces also contribute to collisions (Slinn and Hales, 1971).

Nucleation via Hydrodynamic Collisions

When aerosol particles are large enough (> 0.1 μm), the inertia effects are not negligible and the process of hydrodynamic collisions dominates. Collisions of this type are considered in detail in Sections 5.6 and 6.5. The probability of hydrodynamic collision is described by the value of the collision kernel (Equation 5.6.2). In the expression for the kernel the sedimentation velocity of an aerosol can be neglected. The changes of aerosol SD $f_N(r_N)$ due to inertia-induced collisions with drops can be calculated as (Pruppacher and Klett, 1997)

$$\left(\frac{df_N}{dt}\right)_{in} = -f_N \cdot \int_0^\infty K_g(r_N, r) f_r dr, \qquad (6.2.45)$$

where $f_r(r)$ is DSD and $K_g(r_N, r)$ is the collision kernel. The efficiencies of collisions between drops and aerosol particles depend on the sizes of collecting particles and are typically as small as 10^{-3}–10^{-2}. To calculate the rate of hydrodynamic collisions, large aerosols (such as dust) with sizes exceeding 1 μm are treated as spherical droplets of the same mass. As a result, the efficiencies of collisions between drops and large aerosols are assumed to be equal to those between drop and droplets of aerosol size (e.g., Flossmann and Pruppacher, 1988; Alheit et al., 1990).

Brownian Diffusion

For aerosol particles smaller than 0.1 μm, the Brownian diffusion mechanism is important. Small aerosol particles move randomly and form the diffusion particle flux from the environment where the concentration of aerosol particles is equal to N_{cn} toward the drop surface, where the aerosol concentration is assumed equal to zero. The diffusion flux of aerosols on an unmovable drop of radius r can be written as (Pruppacher and Klett, 1997)

$$F^{BR} = 4\pi D_c r N_{cn}, \qquad (6.2.46)$$

where D_c is the diffusivity of the nuclei in the air, given by following the equation

$$D_c = \frac{kT(1 + Kn)}{6\pi r_N \mu}, \qquad (6.2.47)$$

where k is the Boltzman's constant, μ is the dynamic viscosity of air, and the Knudsen number Kn is defined as the ratio of the mean free path of air molecules λ to the particle radius r_N, that is, $Kn = \lambda / r_N$. The mean free path is a function of temperature, pressure, and dynamic viscosity of the air (Beard, 1976) and is around 0.1 μm at 500 mb.

When the fall velocity of a drop is large enough, the diffusion aerosol flux on the drop increases due to circulation of aerosols within the velocity field induced by the falling drop. This increase in the diffusion flux can be taken into account by introducing the ventilation coefficient F_v defined in Equation (5.5.10). As a result, the aerosol flux caused by Brownian diffusion onto a drop is calculated as

$$F^{BR} = 4\pi D_c r F_v N_{cn} \qquad (6.2.48)$$

The changes in the size distribution of aerosols induced by the Brownian diffusion are

$$\left(\frac{df_N}{dt}\right)_{BR} = -4\pi D_c f_N \cdot \int_0^\infty F_v(r) r f_r dr \qquad (6.2.49)$$

Comparing Equation (6.2.49) with Equation (6.2.45), one can express the Brownian diffusion-induced collision kernel as $K_{BR} = 4\pi r D_c F_v(r)$. According to evaluations made by Pruppacher and Klett (1997), the Brownian diffusion mechanism is able to decrease the concentration of Aitken aerosols in clouds by factor of two during about one hour.

Thermophoresis

The effect of thermophoresis is the motion of small particles relative to the air due to the existence of the temperature gradients. The gas molecules located at the "warm" side of a particle have a larger momentum than those located at the "cold" side. Thereby, a force arises that pushes the particle toward the zone of colder gas. As a result, the particle located within some gas with the temperature gradient begins moving along the temperature gradient direction. The strength of this motion depends on the temperature gradient and on the Knudsen number Kn. The velocity of a particle moving under the thermophoretic force can be written as (Pruppacher and Klett, 1997)

$$V_{TF} = -\frac{Bk_a}{p} \nabla T \qquad (6.2.50)$$

Where p is air pressure, ∇T is temperature gradient at the drop surface, and $B = \frac{0.4[1+1.45\,Kn+0.4\,Kn\exp(-1/Kn)](k_a+2.5\,Kn \cdot k_n)}{(1+3Kn)(2k_a+5\,Kn \cdot k_n+k_n)}$, and k_a and k_n are thermal conductivities of the air and the aerosol particle, respectively. Ovchinnikov and Kogan (2000) assumed that contact nuclei are clay aerosol particles with the thermal conductivity k_n of 0.25 W m^{-1} K^{-1}. Using Equation (6.2.50) and knowing the temperature difference between the environment and the drop surface one can express the flux of small aerosol particles on the surface of a drop of radius r as

$$F^{TF} = -\frac{4\pi r k_a B}{p} N_{cn}(T - T_{sur}), \qquad (6.2.51a)$$

where $T - T_{sur}$ is the temperature difference between the environment and drop surface. Flux F^{TF} is directed outside from the surface of a growing drop and toward the drop that is evaporated. Using Equation (6.2.51a), the collision kernel related to thermophoresis can be written as $K_{TF} = \frac{4\pi r k_a B}{p}|T - T_{sur}|$.

Expression (6.2.51a) resembles the expression for the diffusional heat flux on a drop surface during diffusion drop growth: in both cases, the heat flux is proportional to $T - T_{sur}$. Expressing the temperature difference using the drop growth rate $\frac{dm}{dt}$, one can finally get the following equation for the flux F^{TF} (Ovchinnikov and Kogan, 2000):

$$F^{TF} = \frac{L_w B}{p}\frac{dm}{dt}(1 + 0.3Re^{0.5}Pr^{0.33})N_{cn}, \qquad (6.2.51b)$$

where Pr is the Prandtl number and the term in brackets is analog of ventilation coefficient (see Section 5.5).

Diffusiophoresis

Diffusiophoresis is a quite complicated process in which averaged particle motion is caused by gradients in concentration of water vapor molecules (i.e., water vapor density) in the boundary layer around a drop. The gradient of water vapor density appears in the expression for drop mass growth by diffusion. Thus, the flux of aerosol particles on a drop surface can also be expressed using the rate of the drop radius growth $\frac{dr}{dt}$ (Ovchinnikov and Kogan, 2000):

$$F^{DF} = 4\pi r D_v F_v \left(\frac{M_a}{M_w}\right)^{0.5}\left[q_v - q_S \exp\frac{L_w^2 \rho_w}{R_v k_a T^2} r \frac{dr}{dt}\right] N_{cn}, \qquad (6.2.52)$$

where q_v is specific humidity, q_S is water saturation specific humidity, ρ_w is the liquid water density, and M_a and M_w are the molecular masses for air and water, respectively. Flux F^{DF} can be both positive and negative. In case of drop evaporation, $\frac{dr}{dt} < 0$ and $F^{DF} > 0$. It means that for evaporating drops thermophoretic and diffusiophoretic fluxes are directed oppositely.

Typically, the effects of the diffusion as well as thermophoretic and diffusiophoretic fluxes are considered additive. Carstens and Martin (1982) refined the analysis of the relative importance of the phoretic processes presented by Young (1974). They studied the in-cloud scavenging of submicron particles ($0.05 < r < 1$ μm) caused by thermophoresis, diffusiophoresis, and the Brownian diffusion and found that neither the diffusive nor the phoretic effects are additive. Typically, the rate of collisions caused by thermophoresis and by diffusiophoresis is higher than that caused by the Brownian diffusion.

As mentioned, there is a significant uncertainty in the evaluations of contact nuclei concentration. According to Equation (6.2.30) and Equation (6.2.44), concentrations of immersion nuclei and contact nuclei are of the same order of magnitude. However, since the efficiency of collisions is very low, the concentration of ice particles generated by the contact freezing can be evaluated by a factor of about 10^2–10^3 lower than that generated by the deposition-condensation freezing (Ovtchinnikov, 1998). A similar conclusion about the dominating role of deposition and immersion freezing nucleation on ice crystal formation was made by Cotton et al. (1986). In addition, the concentration of ice crystals obtained in parameterizations of deposition and immersion freezing nucleation is usually compared to the observed ice concentration. It means that this mechanism is considered dominating. At the same time, more investigations are required to reveal the comparative role of contact nucleation in ice crystal formation. It is quite possible that contact nucleation is important at temperatures within the range of −5°C–−10°C, where other mechanisms are not efficient. Note that while deposition–condensation freezing leads to formation of ice crystals, contact freezing can lead to emergence of both ice crystals and large frozen drops. Thus, the effect of contact IN depends not only on concentration, size, and other parameters of contact IN, but also on the size distribution of unfrozen drops.

Despite the lack of reliable data on nuclei concentration and sizes, some of the cloud models are capable of adequate simulation of aerosol–water drop collisions (Respondek et al., 1995; Ovtchinnikov and Kogan, 2000). However, all of these models neglect the aerosol-drop collisions that can be caused by an electrical force. At the same time, Tinsley et al. (2001) and Tinsley (2004) showed that for layered clouds with temperatures within the range of −5°C–−20°C, collision efficiency caused by electric charges is of the same order as that of gravitational collision efficiency. Under these conditions, contact nucleation becomes as efficient as deposition or immersion nucleation. The role of charge in the interaction between particles in clouds might be quite significant, and further efforts are required to evaluate this role quantitatively.

Contact Nucleation Inside Out
Sastry (2005), Shaw et al., (2005), and Durant and Shaw (2005) investigated evaporation freezing by contact nucleation inside-out (Figure 6.2.16c). In these studies, it was shown that ice formation within a liquid water drop by heterogeneous nucleation occurs at higher temperatures if the IN is in contact with the drop surface than if it is fully immersed within the drop volume. The authors presented laboratory evidence for enhanced ice nucleation during drop evaporation. The increased rate of drop freezing in downdrafts during drop evaporation was found earlier in laboratory experiments by Johnson and Hallett (1968). Sastry (2005), Shaw et al. (2005), and Durant and Shaw (2005) hypothesize that "evaporation freezing" in atmospheric clouds where nucleation inside-out is efficient is a plausible explanation for observed high ice concentrations associated with cloud dilution and drop evaporation.

6.3 Phase Transformations in Mixed-Phase Clouds due to Diffusion Growth and Deposition

Pure ice clouds and mixed-phase clouds containing both liquid water drops and ice particles exist in the atmosphere under negative temperatures. In ice clouds, condensation of water vapor on ice particles (deposition) as well as evaporation of ice particles (sublimation) take place. In mixed-phase clouds, where the transition of water molecules between the three phases (water vapor–supercooled water–ice) occurs, deposition/sublimation processes are accompanied by condensation/evaporation of water drops. Existence of the three phases makes cold microphysics more complicated than warm microphysics. In cold microphysical processes the net effect of liquid–ice coexistence is determined by the relationship between two supersaturations: supersaturation over liquid water $S_w = \frac{e}{e_w} - 1$, defined in Equation (3.1.32), and over ice S_i.

6.3.1 Supersaturation and Water Balance in Mixed-Phase and Ice Clouds

Supersaturation with respect to ice is defined as

$$S_i = \frac{e}{e_i} - 1, \tag{6.3.1}$$

where e is vapor pressure and $e_i(T)$ is saturation vapor pressure over ice. Since, due to the existence of crystal lattice in solid ice, evaporation of water vapor molecules from the ice surface is weaker than that over the water surface, lower water vapor pressure is required to balance the evaporation of molecules from ice surface. It means that saturation water vapor pressure over ice is less than that over water, i.e., $e_i(T) \leq e_w(T)$. A good approximation of $e_i(T)$ dependence by polynomials was presented by Flatau et al. (1992) (Table 6.1.2 and Figure 6.1.3). The dependences $e_w(T)$ and $e_i(T)$ as well as the difference between them $e_w(T) - e_i(T)$ are shown in **Figure 6.3.1**.

Figure 6.3.1 demonstrates the existence of three zones characterized by different signs of supersaturations. There is a zone where water vapor pressure is larger than each of the saturated pressures $e > e_w > e_i$. In this zone, both drops and ice particles grow due to condensation. In another zone where $e < e_i < e_w$, both drops and ice particles evaporate. The third zone, where $e_i < e < e_w$ (colored blue in Figure 6.3.1) is the one where the WBF is active (Wegener, 1911; Bergeron, 1935; Findeisen, 1938). By the WBF, drops evaporate, providing ice particles with water vapor enabling the latter to grow, which may lead to a complete evaporation of

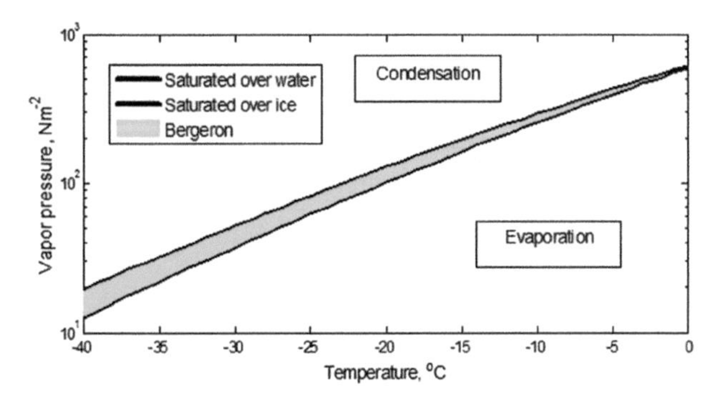

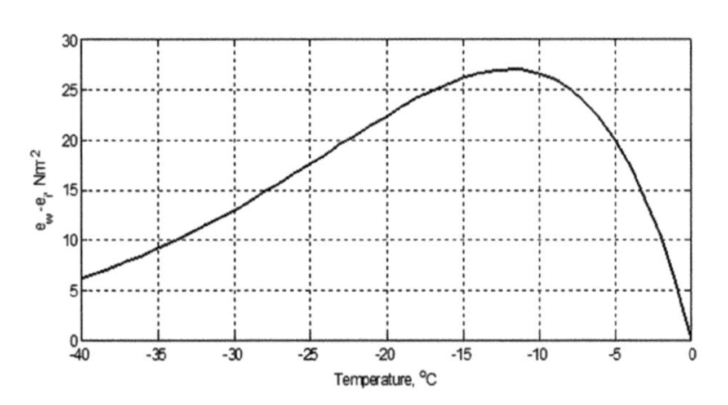

Figure 6.3.1 Dependencies of the saturation water vapor pressure on temperature over water (upper solid line) and over ice (lower solid line) (upper panel) and the difference between the saturation pressures over water and over ice (lower panel). The approximation suggested by Flatau et al (1992) was used. The colored area shows the zone where the Wegener–Bergeron–Findeisen process (WBF) mechanism is active.

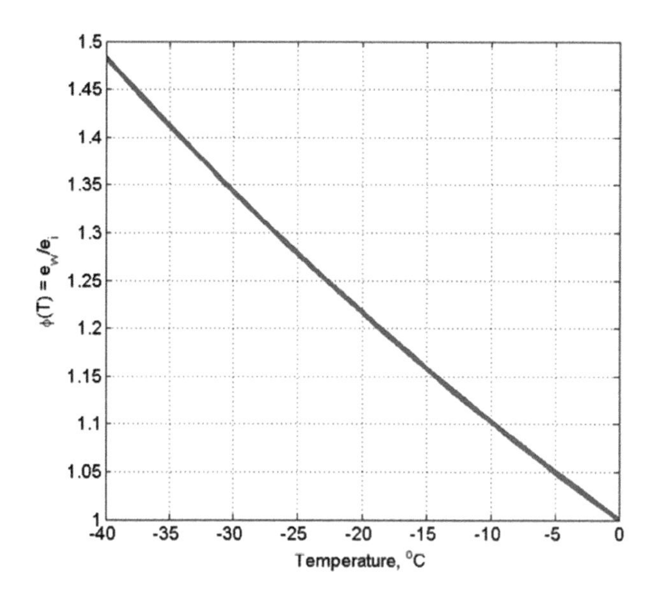

Figure 6.3.2 Dependence of the $\varphi(T_C) = \frac{e_w}{e_i}$ function on temperature. Calculations were performed using the approximating polynomials proposed by Flatau et al. (1992).

the liquid drops. The lower panel of Figure 6.3.1 shows that the maximum value of the $e_w(T) - e_i(T)$ difference is reached at $T_C = -12°\text{C}$, which is the maximum efficiency of the WBF. Since $e_i(T) \leq e_w(T)$, there is an inverse relationship between the supersaturations over ice and water, i.e., $S_i \geq S_w$. The exact equality is reached at $T_C = 0°\text{C}$. As shown by Korolev and Mazin (2003), S_i and S_w are linearly related, which follows directly from the definition of the supersaturations:

$$S_i = \frac{e_w}{e_i}(S_w + 1) - 1 = \varphi(T)S_w + (\varphi(T) - 1), \tag{6.3.2}$$

where $\varphi(T) = \frac{e_w}{e_i} \geq 1$. The dependence of function φ on temperature is shown in **Figure 6.3.2**. This function is equal to unity at $T_C = 0°\text{C}$, i.e., at this temperature $S_i = S_w$. Therefore, knowing the $\varphi(T)$ and one of the supersaturation values (either over water or over ice), one can calculate the second supersaturation value.

Korolev and Mazin (2003) derived and investigated the equation for supersaturation S_w in a moving adiabatic parcel in the presence of ice particles:

$$\frac{1}{S_w + 1}\frac{dS_w}{dt} = A_1 w - A_2\frac{dq_w}{dt} - A_3\frac{dq_i}{dt}, \tag{6.3.3}$$

where q_w is the LWC, q_i is the IWC, w is the vertical velocity, and A_1, A_2, and A_3 are coefficients slightly depending on temperature. The first term on the right-hand side of Equation (6.3.3) determines generation of supersaturation due to the transport of water vapor upward where the air temperature is lower. The second and third terms on the right-hand side describe a decrease in the supersaturation due to absorption of water vapor on drops and ice particles, respectively. Equation (6.3.3) is similar to Equation (5.2.1) for moving adiabatic parcel. Equation (6.3.3) contains, however, an additional third term on the right-hand side. The expressions for the coefficients are

$$A_1 = \frac{g}{T}\left(\frac{L_w}{c_p R_v T} - \frac{1}{R_a}\right); A_2 = \frac{1}{\rho_v} + \frac{L_w^2}{R_v c_p T^2 \rho_a};$$
$$A_3 = \frac{1}{\rho_v} + \frac{L_w L_i}{R_v c_p T^2 \rho_a} \tag{6.3.4}$$

The dependences of A_1, A_2, and A_3 on temperature are presented in **Figure 6.3.4**. Within any temperature interval of 10°C, which corresponds to the height difference of ~1 km, changes of these coefficients do not exceed ~20%. One can also see that $A_3(T_C) \approx A_2(T_C)$.

Since usually $|S_w| \ll 1$ in clouds, Equation (6.3.3) can be rewritten in a simpler form:

$$\frac{dS_w}{dt} = A_1 w - A_2\frac{dq_w}{dt} - A_3\frac{dq_i}{dt} \tag{6.3.5}$$

Let us consider a well-mixed adiabatic cloud volume ascending at the velocity $\frac{dz}{dt} = w(z)$ at negative temperature. Neglecting the dependence of the coefficients A_1, A_2, and A_3 on temperature, i.e., assuming that the coefficients are unchanged with height, Equation (6.3.5) can be integrated to get the mass balance equation for an adiabatic parcel:

$$S_w = A_1 z - A_2 q_w - A_3 q_i + C, \tag{6.3.6}$$

where $C = S_{w0} + A_2 q_{w0} + A_3 q_{i0}$ is a constant determined by the initial conditions at level $z = 0$. Level $z = 0$ can be chosen, for example, at the cloud base. In this case, the initial supersaturation over ice is equal to zero ($S_i = S_{i0} = 0$) and the initial values of LWC and IWC are equal to zero, $q_{i0} = q_{w0} = 0$. Accordingly, at the cloud base $S_{w0} = C < 0$. In the course of the parcel's motion, Equation (6.3.6) determines the water balance when redistribution of water mass between water vapor, q_w and q_i takes place. In case $q_i = 0$, Equation (6.3.6) reduces to the balance equation (5.2.4) for a warm adiabatic parcel (Section 5.2).

The liquid water adiabatic profile $q_{w\_ad}(z)$ and the ice adiabatic profile $q_{i\_ad}(z)$ can be derived from Equation (6.3.6) assuming $S_w = 0$, $q_i = 0$, or $S_w = 0$ and $q_w = 0$:

$$q_{w\_ad} = \frac{A_1}{A_2} z + c_1 \quad (6.3.7)$$

$$q_{i\_ad} = \frac{A_1}{A_3} z + c_2, \quad (6.3.8)$$

where c_1 and c_2 are constants. Since $A_2 \approx A_3$, the slopes of the ice adiabat and the liquid adiabat are nearly the same; the latter is shifted upward with respect to the ice adiabat.

6.3.2 Equation for Deposition/Sublimation of Ice Particles

The equation for supersaturation should be solved together with the equations describing the growth of ice particles and drops. The growth of ice particle of mass m_i under supersaturation over ice $S_i > 0$ is described by the equation for depositional growth (Pruppacher and Klett, 1997):

$$\frac{dm_i}{dt} = F_v \frac{4\pi \rho_i C_i S_i}{F_i}, \quad (6.3.9)$$

where coefficient F_i, slightly depending on temperature, is calculated as

$$F_i = \frac{\rho_i R_v T}{e_i D} + \frac{\rho_i L_i^2}{R_v T^2 k_a} \quad (6.3.10)$$

and C_i is the electrical capacitance of an ice particle; F_v is ventilation coefficient that describes an enhancement of the rates of deposition/sublimation of falling particles. The ventilation coefficient for ice particles differs from that for liquid drops. For small ice particles $F_v = 1$. The expression for ventilation coefficients for ice particles is given in Section 6.4.5.

Equation (6.3.9) has a form similar to that of the equation for diffusional growth/evaporation of drops (5.2.11). This similarity reflects the similarity of the processes of water vapor molecules diffusion onto ice particles and drops. There are, however, important differences between Equation (6.3.9) and Equation (5.2.11). First, a specific feature of ice particles is that the rate of ice particle growth by deposition depends on the particle's shape. **Table 6.3.1** presents expressions for the capacitance of several simplified particle shapes. The capacitance of a spherical particle is equal to its radius. Table 6.3.1 also presents the capacitance normalized on the mean volume radius r_v (i.e., the radius of the sphere with volume equal to the volume of non-spherical particle), $\frac{C_i}{r_v} = f(\gamma_i)$ as a function of the aspect ratio of a particle $\gamma_i = \frac{h}{d}$, which is less than unity for oblates and larger than unity for prolate shapes. The values of the mean volume radius are shown in Table 2.1.1.

Figure 6.3.3 presents the dependence of the normalized capacitance on the aspect ratio of a spheroid particle. The minimum value of the normalized capacitance is reached for spherical particles, so the capacitance grows as the aspect ratio deviates from unity. In clouds, the shape of crystals deviates from a

Table 6.3.1 Expressions for capacitance of ice crystals of several simplified particle shapes $C_i(d, h)$ and the normalized capacitance $\frac{C_i}{r_v} = f(\gamma_i)$ (from Pinsky et al., 2014; © American Meteorological Society; used with permission).

Shape	Sphere	Thin Circular Disk	Oblate Spheroid	Prolate Spheroid
	2r	h, d	h, d	h, d
Capacitance $C_i(d, h)$	r	$\frac{d}{\pi}\left(1 + \frac{\pi - 2}{2}\sqrt{\frac{h}{d}}\right)$	$\frac{d\sqrt{1 - \frac{h^2}{d^2}}}{2\text{a}\sin\sqrt{1 - \frac{h^2}{d^2}}}$	$\frac{\frac{h\sqrt{1-\frac{d^2}{h^2}}}{1+\sqrt{1-\frac{d^2}{h^2}}}}{1 - \sqrt{1 - \frac{d^2}{h^2}}}$
$\frac{C_i}{r_v} = f(\gamma_i)$	1	$\frac{\left(1 + \frac{\pi-2}{2}\sqrt{\gamma_i}\right)}{\pi\left(\frac{3}{16}\gamma_i\right)^{1/3}}$	$\frac{\sqrt{1 - \gamma_i^2}}{\gamma_i^{1/3}\text{a}\sin\sqrt{1 - \gamma_i^2}}$	$\frac{2\sqrt{\gamma_i^2 - 1}}{\gamma_i^{1/3}\ln\frac{\gamma_i + \sqrt{\gamma_i^2 - 1}}{\gamma_i - \sqrt{\gamma_i^2 - 1}}}$

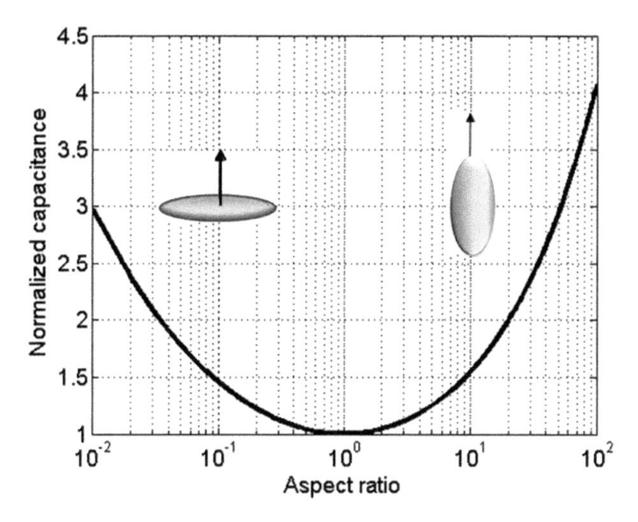

Figure 6.3.3 Dependence of the normalized capacitance $\frac{C_i}{r_v} = f(\gamma_i)$ on the aspect ratio of a spheroid particle γ_i (from Pinsky et al., 2014; American Meteorological Society ©; used with permission).

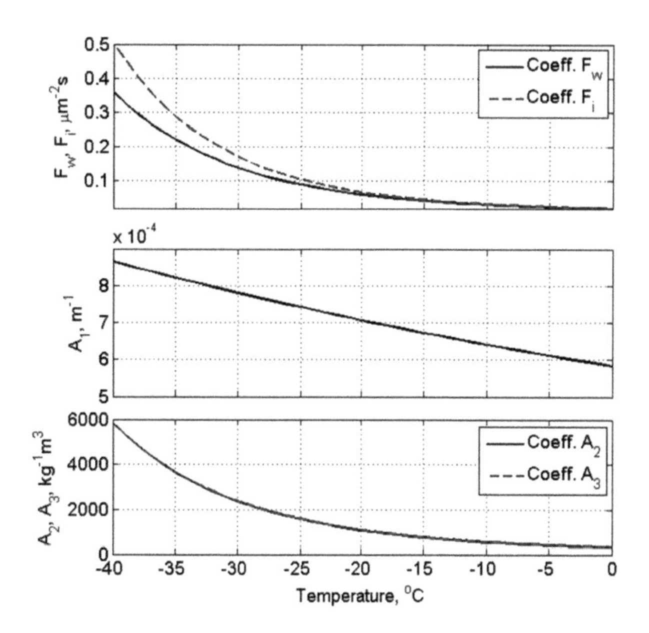

Figure 6.3.4 Dependence of coefficients F, F_i, A_1, A_2, and A_3, on temperature (from Pinsky et al., 2014; American Meteorological Society ©; used with permission).

spherical shape as their mass increases (Figure 6.1.5). For plates, the value C_i/r increases to about three as the aspect ratio decreases. For columnar crystals, the ratio C_i/r increases with increasing aspect ratio (and with increasing particle mass to about 4 for aspect ratio of 100. Thus, a widely used approximation of ice crystals by effective spheres is suitable for comparatively small particles whose aspect ratio is close to unity. For large ice crystals, diffusion growth should be calculated taking into account the particle's shape.

The influence of the aspect ratio of ice crystals on condensation/evaporation in stratocumulus mixed-phase clouds was investigated by Sulia and Harrington (2011) by means of a parcel model. They showed that taking into account the aspect ratio of crystals can lead to a dramatic change in the balance between water vapor, liquid water, and ice. In particular, additional condensation of vapor on ice particles due to their non-sphericity changes the conditions of the liquid phase emergence in clouds, so instead of a mixed phase cloud, a pure ice cloud may form.

The second specific feature of deposition in contrast to diffusion growth of drops is that supersaturation over ice is higher than that over water and can reach tens of percents. The large capacitance of nonspherical ice particles as well as the large values of supersaturation over ice may trigger ice crystal growth up to large sizes, sometimes up to several thousands of microns. At the same time, the maximum droplet radius that can be reached in diffusion growth is about 20 μm. Thus, in contrast to diffusion growth of droplets, ice deposition may lead to formation of precipitating particles.

In many theoretical studies and cloud models, the shape of ice crystals is characterized by the effective sphere of radius r_i. In case the effective sphere has the same volume as an ice crystal, i.e., $r_i = r_v$, the equation for depositional growth is written in the form similar to that for diffusional drop growth:

$$r_i \frac{dr_i}{dt} = f(\gamma_i) \frac{S_i}{F_i} \tag{6.3.11}$$

Equation (6.3.11) is identical to Equation (6.3.9) and also analogous to Equation (5.2.6), i.e., $r\frac{dr}{dt} = \frac{S_w}{F}$ for diffusion growth of water drops. The effect of the non-sphericity of an ice particle in Equation (6.3.11) is taken into account by introducing the factor $f(\gamma_i) = \frac{C_i}{r_v}$. For spherical particles, $f(\gamma_i) = 1$. Dependences of coefficients F_i and $F = F_w$ as well as of A_1, A_2, and A_3 on temperature are presented in **Figure 6.3.4**. Coefficients F_i and F are approximately equal within the temperature range of −20°C–0°C.

6.3.3 Phase Relaxation Time and Quasi-Steady Supersaturation

An important characteristic of the condensation/ deposition process is phase relaxation time τ_p characterizing the rate at which supersaturation adapts to its quasi-steady value due to water vapor absorption on cloud particles. Substitution of Equations (6.3.11) and (5.2.6) into Equation (6.3.5) leads to the following form of the supersaturation equation (Korolev and Mazin, 2003, Pinsky et al., 2015):

$$\frac{dS_w}{dt} = \left[A_1 w - \frac{4\pi(\varphi-1)A_3\rho_i}{F_i}N_i\bar{r}_i\right] - 4\pi\left[\frac{A_2\rho_w}{F}N\bar{r} + \frac{\varphi A_3\rho_i}{F_i}N_i\bar{r}_i\right]S_w, \quad (6.3.12)$$

where $\bar{r}$ and $\bar{r}_i$ are the mean radii of liquid drops and ice particles, respectively. Under the assumption that the coefficients in this equation are constant, Equation (6.3.12) has a solution of the exponential type. The characteristic time of the exponent (the phase relaxation time τ_p) and the quasi-stationary solution when $t \to \infty$, S_{w_qs}, can be represented as proposed by Korolev and Mazin (2003) and Pinsky et al. (2015):

$$\tau_p = \left\{4\pi\left[\frac{A_2\rho_w}{F}N\bar{r} + \frac{\varphi A_3\rho_i}{F_i}N_i\bar{r}_i\right]\right\}^{-1} = \frac{\tau_p^w\tau_p^i}{\tau_p^w + \tau_p^i} \quad (6.3.13)$$

$$S_{w\_qs} = \frac{\left[A_1 w - \frac{4\pi(\varphi-1)A_3\rho_i}{F_i}N_i\bar{r}_i\right]}{4\pi\left[\frac{A_2\rho_w}{F}N\bar{r} + \frac{\varphi A_3\rho_i}{F_i}N_i\bar{r}_i\right]} = \frac{\left(A_1 w\tau_p^i - \frac{\varphi-1}{\varphi}\right)\tau_p^w}{\tau_p^w + \tau_p^i}, \quad (6.3.14)$$

where $\tau_p^w = \left[\frac{4\pi A_2\rho_w}{F}N\bar{r}\right]^{-1}$ and $\tau_p^i = \left[\frac{4\pi\varphi A_3\rho_i}{F_i}N_i\bar{r}_i\right]^{-1}$ are phase relaxation times for pure liquid clouds and pure ice clouds, respectively. It follows from Equation (6.3.13) that the phase relaxation time is determined by the smallest characteristic time from τ_p^w and τ_p^i. Usually, in real mixed-phase clouds τ_p^w is by several orders of magnitude smaller than τ_p^i, so $\tau_p \approx \tau_p^w$ and $S_{w\_qs} \approx A_1 w\tau_p^w$. The equalities mean that liquid microphysics controls supersaturation in mixed-phase clouds.

Equation (6.3.14) can be rewritten in the form allowing a better illustration of changes in S_{w_qs} during an increase in the ice amount:

$$S_{w\_qs} = \frac{S_{w\_qs}^0 - \frac{\varphi-1}{\varphi}\delta}{1+\delta}, \quad (6.3.15)$$

where $S_{w\_qs}^0$ is the quasi-steady supersaturation in the absence of ice and $\delta = \frac{\tau_p^w}{\tau_p^i}$. **Figure 6.3.5** illustrates the dependence of $S_{w\_qs}(\delta)$ at different temperatures. One can see that when $\delta = 0$, quasi-steady supersaturation is equal to that in a pure liquid cloud. With an increase in δ, i.e., with an increase in the ice mass, S_{w_qs} decreases tending to $-\frac{\varphi-1}{\varphi}$ at $\delta \to \infty$ (an entirely ice cloud). Thus, in clouds containing a significant ice mass (large δ), the quasi-steady supersaturation deviates significantly from that in liquid clouds.

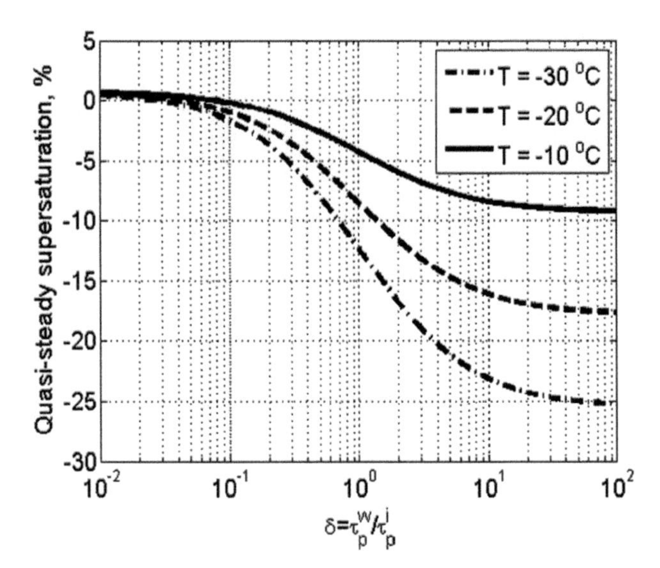

Figure 6.3.5 Dependence of the quasi-steady supersaturation over water on the ratio $\delta = \frac{\tau_p^w}{\tau_p^i}$ (from Pinsky et al., 2015; courtesy of © John Wiley & Sons, Inc.).

6.3.4 Phase Transitions in Ascending Adiabatic Parcels

In clouds with a warm cloud base, the lower part consists of liquid water that becomes supercooled when air volumes cross the freezing level. Depending on the concentration and the chemical composition of ice nuclei and, possibly, on some other factors, different microphysical cloud structures can form at the same negative temperatures. For instance, Rosenfeld et al. (2013, 2014) observed raining pure liquid clouds at temperatures as low as −21°C, while typically clouds at this temperature are glaciated. Typically, mixed-phase clouds exist at temperatures below about −5°C. In deep convective clouds, the mixed-phase zones often remain down to the level of homogeneous freezing, i.e., down to −38°C. In other cases, supercooled water disappears and the cloud glaciates. Observations by Heymsfield et al. (2009) and numerical results obtained by Khain et al. (2012) demonstrated examples of deep convective clouds where supercooled water disappears at heights around 6 km, then rises again above this level and exists up to the homogeneous freezing level. In some stratocumulus clouds, the lower part is glaciated, while the upper cloud layer contains liquid water despite the colder temperatures at cloud top. According to some observations, stratocumulus and frontal clouds consist of three layers, while the lower and the upper levels are fully glaciated and the middle layer is a mixed-phase one. Some clouds glaciate rapidly within less than one hour; in the other cases, a mixed-phase cloud remains stable for days. It is highly important to know the conditions determining the occurrence of these regimes and transitions between them in order to provide proper treatment of precipitation and radiation processes in cloud models.

Phase transitions in clouds at negative temperature take place not only because of diffusion growth/evaporation and deposition/sublimation, but also due to water–ice collisions (riming), freezing, melting, ice nucleation, etc. At the same time, in some cases diffusion growth/evaporation and deposition/sublimation may play an important or even dominating role, for instance, in stratocumulus clouds where collisions are not efficient. It is of theoretical and practical interest to understand to what extent diffusion growth/evaporation and deposition/sublimation are responsible for formation of different microphysical structures in clouds. These processes were investigated in several studies (e.g., Korolev and Mazin, 2003; Korolev and Isaac, 2003; Korolev and Field, 2008). The results described in this section are obtained by Pinsky et al. (2015), who analyzed the ascent of adiabatic mixed-phase cloud parcels.

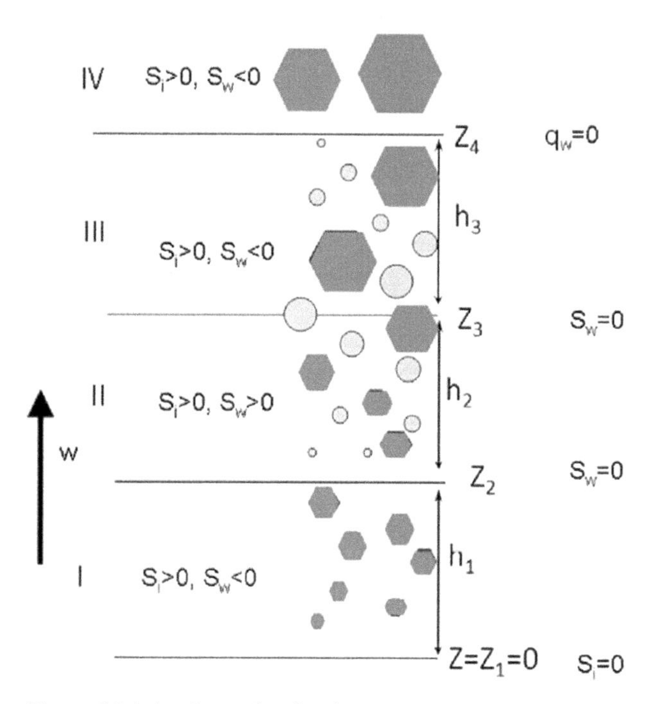

Figure 6.3.6 A scheme showing four zones corresponding to different microphysical regimes experienced by an ascending adiabatic parcel. The regimes are classified according to the signs of the supersaturation values over water and ice (from Pinsky et al., 2015; courtesy of © John Wiley & Sons, Inc.).

Main Microphysical Regimes

A scheme showing different microphysical regimes and their transformation in an adiabatic parcel ascending at negative temperature and containing initially only water vapor can be illustrated by the following scheme (**Figure 6.3.6**). Zones in Figure 6.3.6 are classified according to the signs of supersaturations over water and ice.

Zone I is the lowest layer where $S_w < 0$, $S_i > 0$, and $q_w = 0$. In this zone, ice particles start growing and the cloud parcel contains ice only. We assume that at the beginning of the ascent at $z = Z_1 = 0$, $S_i = 0$, and the ice particles are negligibly small. Since $q_w = 0$, the balance equation (6.3.6) for Zone I can be rewritten as $S_w = A_1 z - A_3 q_i + C$. The lower boundary of Zone I corresponds to conditions $S_{i0} = 0$ and $q_{i0} = 0$, while its upper boundary corresponds to condition $S_w(Z_2) = 0$. Under conditions at which supersaturation over liquid water remains negative, a pure ice cloud forms that belongs entirely to Zone I. Details of ice nucleation process are neglected here.

Zone II is characterized by positive supersaturation values both over water and over ice, i.e., $S_w > 0$, $S_i > 0$; $q_w > 0$, $q_i > 0$. Within Zone II, the mass balance is described by Equation (6.3.6), while droplets and ice particles grow simultaneously. As will be shown, Zone II forms if the updraft velocity exceeds the critical value w^*. Otherwise, the cloud remains a purely ice one, i.e., belongs to Zone I.

Zone III is characterized by supersaturation over ice and subsaturation over liquid water: $S_w < 0$, $S_i > 0$, $q_w > 0$, $q_i > 0$. As in Zone II, the mass balance is described by Equation (6.3.6). Liquid droplets and ice particles ascend in updrafts from Zone II to Zone III. It is known that in warm clouds, after reaching its maximum at the cloud base supersaturation S_w approaches zero. In mixed-phase parcels, water vapor is absorbed by both droplets and ice particles, thus decreasing S_w. The rate of water vapor deposition on ice particles increases with height due to the ice mass growth with height in an ascending adiabatic parcel. As a result, in mixed-phase parcels supersaturation over liquid water at certain heights becomes negative, while supersaturation over ice remains positive. As was shown by Korolev and Mazin (2003) and Korolev (2007b), this situation occurs when $u^0 < w < u^*$, where $u^* \sim N_i\bar{r}_i$ and $u^0 \sim N_w\bar{r}_w$. In Zone II, both droplets and ice particles grow simultaneously in an ascending parcel, and the value u^* grows accordingly. As soon as growing ice particles reach sizes at which the condition $w < u^*$ becomes valid, the parcel enters Zone III, where the liquid droplets start evaporating, whereas the ice particles keep growing, so the WBF mechanism is active in this zone. The WBF mechanism eventually leads to a complete evaporation of droplets and to glaciation of the ascending cloud parcel.

Zone IV represents a glaciated parcel with $q_w = 0$, $q_i > 0$, $S_w < 0$ and $S_i > 0$. At a constant vertical velocity, $S_i \to 0$ when $z \to \infty$. The balance equation is the same as that for Zone I: $S_w = A_1 z - A_3 q_i + C_4$. However, for Zone IV the value of constant C_4 differs and is determined by the values of q_i and S_w at the lower boundary of Zone IV.

Balance equation (6.3.6) and equations for diffusion growth (6.3.11) and (5.2.6) allow analytical analysis of the condensation-deposition processes. Such analysis was carried out by Pinsky et al. (2015) under the following simplifying assumptions: a) size distributions of droplets and ice particles are assumed monodisperse, ice particles assumed spherical; b) the air parcel is assumed to ascend at a constant velocity; and c) coefficients in the equations of diffusional growth are assumed temperature-independent. These assumptions simplify the analytical analysis and allow one to obtain important qualitative (and in many cases quantitative) results regarding the conditions of transformation of microphysical regimes shown in Figure 6.3.6.

Closed Equations for Supersaturation Evolution in Zone I

The equations for supersaturation evolution in Zone I can be obtained by excluding q_i from Equations (6.3.6) and (6.3.11). The equations thus obtained are

$$\frac{dS_i}{dz} = A_1\varphi - \frac{B_i(\varphi A_3 N_i)^{2/3}}{w(z)}(A_1\varphi z - S_i)^{1/3} S_i \tag{6.3.16a}$$

$$\frac{dS_w}{dz} = A_1 - \frac{B_i(\varphi A_3 N_i)^{2/3}}{w(z)}\left(A_1 z + \frac{1-\varphi}{\varphi} - S_w\right)^{1/3}(\varphi S_w + \varphi - 1), \tag{6.3.16b}$$

where coefficient $B_i = \frac{3}{F_i}\left(\frac{4}{3}\pi\rho_i\right)^{2/3}$ is considered a constant. The first terms on the right-hand side of Equations (6.3.16a and 6.3.16b) express the tendency of the supersaturation to increase due to transport of water vapor upward to colder temperatures. The second terms on the right-hand are responsible for the decrease in the supersaturation values due to the condensation of water vapor on ice crystals. The equations are analogous to Equation (5.2.20b) obtained for the supersaturation evolution in a warm adiabatic parcel.

Figure 6.3.7a shows possible profiles of supersaturation over liquid water in Zone I, calculated at $T = -20°C$ by means of Equation (6.3.16b). At comparatively high updrafts and low ice concentrations, S_w changes with height linearly and becomes positive. At low vertical velocities or large ice particle concentrations, S_w reaches its maximum and then decreases with increasing height, remaining negative, so droplets do not form. The boundary case when the S_w maximum is equal to zero is shown in Figure 6.3.7a as well. Figure 6.3.7a demonstrates three different possibilities, namely $S_{w\_\max} < 0$, $S_{w\_\max} = 0$, and $S_{w\_\max} > 0$. At $S_{w\_\max} > 0$, droplets arise during the parcel's ascendance, whereas $S_{w\_\max} = 0$ is a boundary condition that separates the regime with the liquid water from the regime, when no liquid water arises.

Figure 6.3.7b presents the results obtained numerically using a more general equation (6.3.3), with the temperature dependencies of the coefficients taken into account and the temperature decreasing linearly by 6°C/km beginning with the initial level. The middle blue dashed curve was calculated using the same vertical velocity of 10 cm/s, but the ice particle concentration was set equal to 10.74 L^{-1} to get $S_{w\_\max} = 0$. One can see that taking into account the temperature dependencies produces similar profiles of supersaturation as those obtained by Equation (6.3.16b), but a small correction of ice particle concentration of about 10% (from 9.62 L^{-1} to 10.74 L^{-1}) was necessary to get $S_{w\_\max} = 0$ at the same height in both cases. Therefore, we conclude that the approximated equations (6.3.16a and 6.3.16b) provide realistic results and can be used in analysis.

Conditions for Formation of a Mixed-Phase Zone

Equations (6.3.16a) and (6.3.16b) allow calculating the maximum of supersaturation in the same way as it is done in Section 5.2 for warm clouds. The obtained maximum value can then be used to derive conditions for appearance of liquid phase. Introducing a nondimensional height over cloud base $h = A_1\varphi z$ and a nondimensional parameter $R = \frac{B_i N_i^{2/3} A_3^{2/3}}{w A_1 \varphi^{1/3}} = \frac{3}{F_i A_1 w \varphi^{1/3}}\left(\frac{4}{3}\pi\rho_i N_i A_3\right)^{2/3}$, one can estimate the maximum of supersaturation over ice as

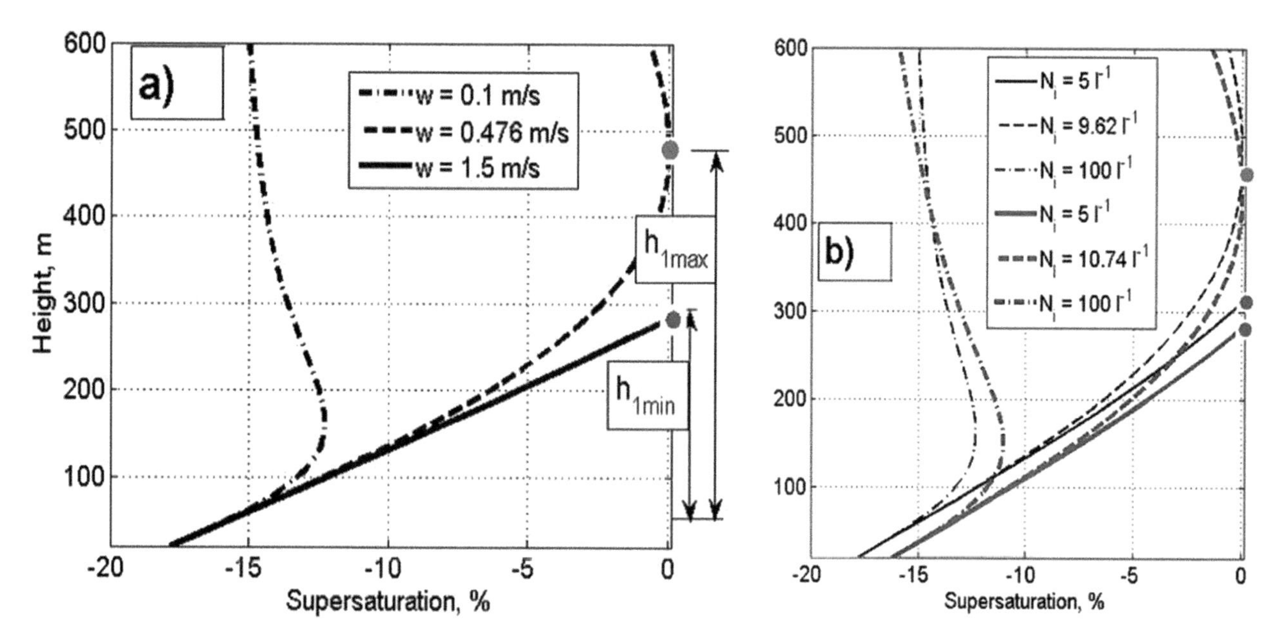

Figure 6.3.7 **(a)** The vertical profiles of supersaturation over liquid water in Zone I calculated using Equation (6.3.18b) under $T = -20°C$, $p = 952$ mb and $N_i = 100\ L^{-1}$ for different vertical velocities. **(b)** Blue lines: the vertical profiles of supersaturation over liquid water in Zone I calculated using Equation (6.3.16b) under $T = -20°C$, $p = 952$ mb, and $w = 10$ cm/s for different ice particle concentrations. Black lines: the profiles calculated by means of a parcel model using Equation (6.3.3), with a linear decrease in temperature according to formula $T = -18 - 0.006(z - 520)°C$ at $p_0 = 952$ mb. The heights where the supersaturation crosses zero value and become positive are marked with blue circles. The heights where supersaturation maximum is equal to zero are marked with red circles (from Pinsky et al., 2015; courtesy of © John Wiley & Sons, Inc.).

$$S_{i\_\max}(R) = C_1 R^{-3/4}, \tag{6.3.17}$$

where $C_1 = 1.058$ (Section 5.2). Since S_i and S_w are linearly related by Equation (6.3.2), the maximum of S_w is reached at the same level as the maximum of S_i. The necessary condition of the liquid phase arising is $S_{w\_\max} = \frac{S_{i\_\max}}{\varphi} - \frac{\varphi-1}{\varphi} > 0$. This condition is equivalent to the condition $C_1 R^{-3/4} > \varphi - 1$. Substituting Equation (6.3.17) into this inequality, one gets the condition for the vertical velocity value determining the appearance of a mixed-phase zone:

$$w > w^* = \frac{3}{F_i A_1 \varphi^{1/3}} \left(\frac{\varphi - 1}{C_1}\right)^{4/3} \left(\frac{4}{3}\pi\rho_i A_3\right)^{2/3} N_i^{2/3} \tag{6.3.18}$$

Figure 6.3.8 shows the dependence of w^* on temperature at different concentrations of ice particles. One can see that the threshold value of the vertical velocity is of order of few cm/s. Comparatively small values of this threshold are accounted for by small concentration and low mass of ice crystals, which, therefore, do not absorb any significant amount of water vapor. The small threshold values of w^* indicate high probability of

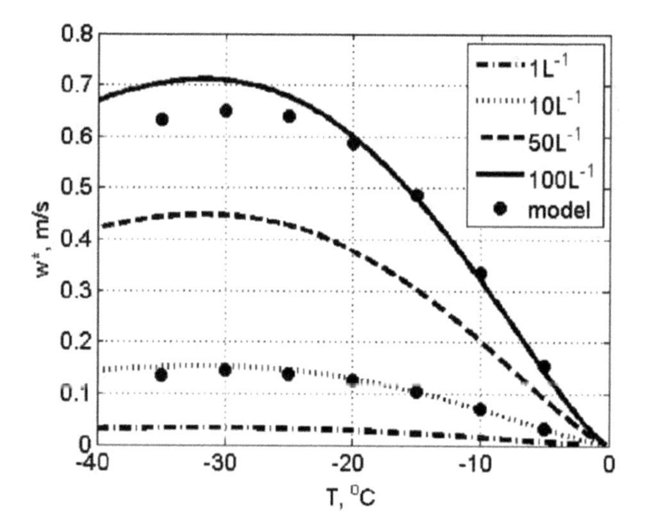

Figure 6.3.8 The temperature dependence of w^* at different concentrations of ice particles; $p = 932$ mb. The circles show the results obtained by means of the parcel model. The circles are located along the corresponding analytical dependencies (from Pinsky et al., 2015; courtesy of © John Wiley & Sons, Inc.).

formation of mixed-phase clouds in cloud updrafts. The value of the threshold increases with ice concentration, because higher ice concentration increases absorption of water vapor on ice crystals and requires larger w to raise supersaturation over water. The threshold value of the vertical velocity reaches its maximum at temperatures around $-15°C$, which is related to the maximum of the difference between the saturation pressure over water and over ice at these temperatures, as shown in Figure 6.3.1.

The Depth of the Pure Ice Layer (Zone I)
When the ascent velocity exceeds its critical value, i.e., $w > w^*$, liquid droplets get activated and Zone II forms. The depth of Zone I reaches its maximum h_{1_max} when $w = w^*$, i.e., when the supersaturation maximum on the curve $S_w(z)$ is equal to zero ($S_{w\_max} = 0$) (Figure 6.3.7). An increase of the vertical velocity w results in a decrease of Zone I depth until it reaches its minimum value h_{1_min}, corresponding to very high vertical velocities at which the supersaturation increases linearly with height. Liquid water cannot form at heights lower than h_{1_min} above the cloud base. Thus, the thickness of the pure ice layer h_1 falls within the range of $h_{1\_min} \leq h_1 \leq h_{1\_max}$.

An approximate equation for the minimum depth of Zone I can be obtained from condition $S_w(z) = 0$ and Equation (6.3.16b) as

$$h_{1\_min}(T) = \frac{1}{A_1(T)} \frac{\varphi(T) - 1}{\varphi(T)} \tag{6.3.19}$$

The maximum depth of Zone I h_{1_max} can be evaluated using the theory developed by Pinsky et al. (2015). The equation for h_{1_max} has the following form:

$$h_{1\_max} = \frac{C_2}{A_1\varphi}[R(w^*)]^{-3/4} = \frac{C_2}{C_1 A_1(T)} \frac{\varphi(T) - 1}{\varphi(T)} = \frac{C_2}{C_1} h_{1\_min} = 1.8 h_{1\_min}, \tag{6.3.20}$$

where $C_2 = 1.904$. The temperature dependencies $h_{1\_min}(T)$ and $h_{1\_max}(T)$ are presented in **Figure 6.3.9**. The depths increase as the temperature decreases. The height boundaries of Zone I ($h_{1\_min}(T)$ and $h_{1\_max}(T)$) increase with decreasing temperature. This dependence is related to the fact that the distance between the level where $S_i = 0$ (the lower boundary of Zone I) and $S_w = 0$ (the upper boundary of Zone I) increases with decreasing temperature. At 0°C, the boundaries coincide, and the height of Zone I is equal to zero. There is good agreement between the analytical calculations and numerical simulations, which verifies the conclusion

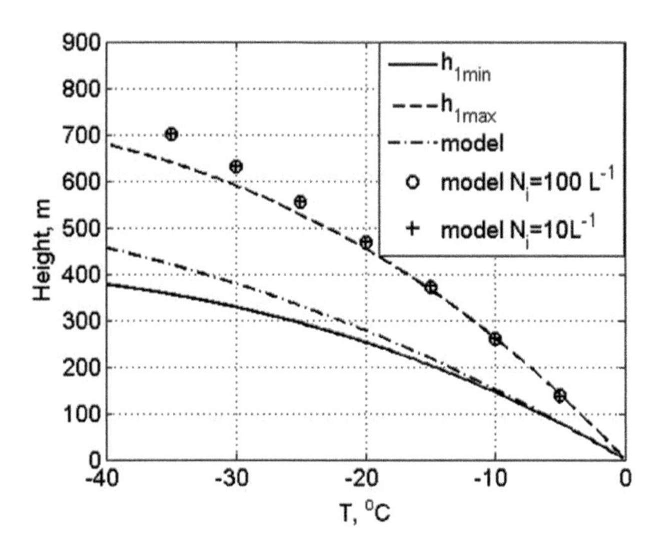

Figure 6.3.9 The temperature dependencies of maximum and minimum depths of Zone I. Symbols indicate the results obtained by means of the parcel model (from Pinsky et al., 2015; courtesy of © John Wiley & Sons, Inc.).

that $h_{1\_min}(T)$ and $h_{1\_max}(T)$ do not depend on ice particle concentration. The ratio $\frac{h_{1\_max}(T)}{h_{1\_min}(T)}$ in the parcel model is about 1.7, which is close to the theoretical value of 1.8. Equations (6.3.16) and (6.3.11) also allow us to evaluate the ice particles' radii and the IWC at the top of Zone I. Calculations show that the characteristic values of ice particle radii are of order of several tenths of μm, while the characteristic values of the IWC are 10^{-2}–10^{-1} g/m³.

Mixed-Phase Zones II and III
Zones II and III are mixed-phase zones where liquid droplets and ice crystals coexist within an ascending parcel. In Zone III, crystal deposition is accompanied by droplet evaporation and the WBF mechanism is active. Pinsky et al. (2015) obtained the following approximate equations for LWC $q_w(z)$ and IWC $q_i(z)$:

$$q_w(z) = \frac{A_1}{A_2} z - \frac{A_3}{A_2}\left\{\left[q_{i0}^{2/3} + Gz\right]^{3/2} - q_{i0}\right\} \tag{6.3.21}$$

$$q_i(z) = \left(q_{i0}^{2/3} + Gz\right)^{3/2}, \tag{6.3.22}$$

where $G = \frac{2}{F_i}\left(\frac{4}{3}\pi\rho_i\right)^{2/3}(\varphi - 1)N_i^{2/3}w^{-1}$. Equations (6.3.21) and (6.3.22) were obtained under the assumption that in a mixed-phase zone, S_w is close enough to zero and can be neglected in the balance equation (6.3.6). Actually, it means that the term $S_w(t)$ on the left-hand side of Equation (6.3.6) is typically much smaller than the

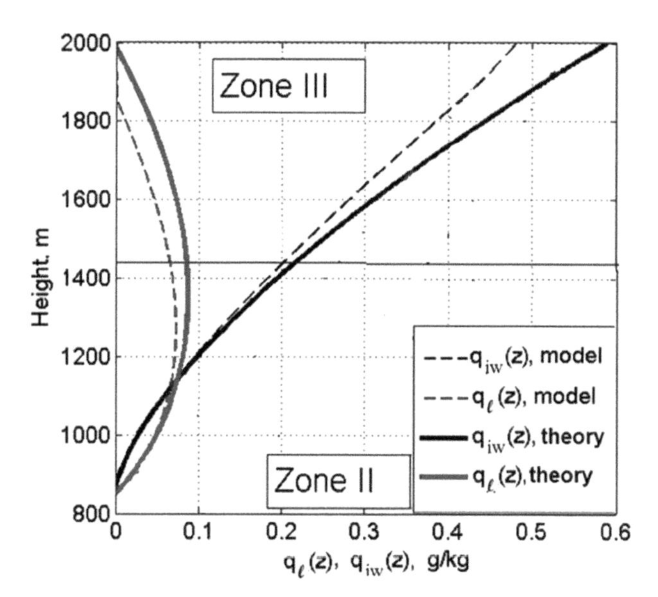

Figure 6.3.10 The profiles of liquid–water mixing ratio $q_l(z)$ and ice–water mixing ratio $q_{iw}(z)$ in the mixed-phase Zones II and III, calculated using Equations (6.3.21) and (6.3.22) and a parcel model. The values of parameters of calculations: $T_C = -20°C$, $p = 887$ mb, $w = 0.1$ m/s, $N_i = 7.5\ L^{-1}$, and $q_{i0} = 0$ g/kg. The horizontal line indicates the boundary between Zones II and III (from Pinsky et al., 2015; courtesy of © John Wiley & Sons, Inc.).

terms on the right-hand side of Equation (6.3.6). Indeed, simple evaluations of the characteristic values of the terms in Equation (6.3.6) are $A_1 z \sim 0.1$ (if the vertical shift of the parcel is 100 m), $A_2 q_w \sim 0.4$ (at $q_w = 0.2\ \mathrm{gm^{-3}}$). At the same time, the typical value of supersaturation over water is $S_w \leq 10^{-2}$.

The height profiles of liquid–water and ice–water mixing ratios $q_l(z)$ and $q_{iw}(z)$, respectively, obtained by Equations (6.3.21) and (6.3.22), as well as the profiles obtained by means of the parcel model, are shown in **Figure 6.3.10.** The value of $q_{iw}(z)$ grows continuously with height. The LWC profile demonstrates the maximum at which supersaturation over water is equal to zero. This maximum divides the mixed-phase zone into Zone II, where $S_w > 0$ and $S_i > 0$, and Zone III, where $S_w < 0$ and $S_i > 0$. In Zone III the WBF mechanism is active. It is interesting that both obtained approximate profiles $q_l(z)$ and $q_{iw}(z)$ do not depend on droplet concentration. The value of $q_{iw}(z)$ is independent of the droplet concentration due to condition $S_w \ll S_i$. The value of $q_l(z)$ is independent of the droplet concentration as follows from the balance equation (6.3.6), where S_w is much smaller than the other terms that do not depend on the droplet concentration.

Equations (6.3.21) and (6.3.22) allow evaluation of major characteristic of LWC and IWC in mixed-phase Zones II and III. The boundary between Zones II and III (i.e., the depth h_2 of Zone II) can be evaluated from condition $\frac{dq_w}{dz} = 0$, which is equivalent to condition $S_w = 0$. Differentiating Equation (6.3.21) with respect to z and equalizing the result to zero leads to an equation for Zone II thickness and to an equation for the LWC maximum:

$$h_2 = \frac{1}{G}\left[\frac{4}{9}\left(\frac{A_1}{A_3 G}\right)^2 - q_{i0}^{2/3}\right] \tag{6.3.23}$$

$$q_{w\_\max} = \frac{4}{27}\frac{A_1^3}{A_2 A_3^2 G^2} - \frac{A_1}{A_2 G} q_{i0}^{2/3} + \frac{A_3}{A_2} q_{i0} \tag{6.3.24}$$

Figure 6.3.11 shows dependences of the thickness of Zone II and the maximum value of LWC reached in this zone on the temperature at different ice particle concentrations. It is noteworthy that, according to Equations (6.3.23) and (6.3.24), both quantities do not depend on the droplet concentration N_w. Figure 6.3.11 also shows that Zone II depth sharply decreases with increasing ice particle concentration. As follows from Equations (6.3.23) and (6.3.24), Zone II depth is approximately proportional to N_i^{-2}. At temperatures $-5°C < T_C < 0°C$, supersaturation over ice is close to that over water, and the growth of ice mass is slow. The presence of ice particles cannot prevent the growth of LWC, and the values of LWC maximum and depth h_2 become large. At $T_C < -20°C$, supersaturation over ice is substantially higher than that over water, so ice particles grow efficiently. If the concentration of ice particles is high enough, supersaturation over water reaches zero at very low h_2. It means that at low temperatures and high concentrations of ice particles, liquid droplets may not get activated, or they may exist within a shallow layer. At low ice particle concentrations, Zone II can be extended in the vertical direction up to the level of homogeneous freezing.

Another parameter that can be evaluated using Equation (6.3.21) is Zone III thickness h_3. One can evaluate first the total depth of the mixed-phase cloud zone as $h_{23} = h_2 + h_3$. The depth h_{23} is evaluated from Equation (6.3.21) by setting $q_w(z)$ to zero at the upper boundary. The obtained quadratic equation is

$$a h_{23}^2 + b h_{23} + c = 0, \tag{6.3.25}$$

with the coefficients calculated as $a = A_3^2 G^3$, $b = 3A_3^2 G q_{i0}^{2/3} - A_1^2$ and $c = A_3 q_{i0}\left(3A_3 G q_{i0}^{1/3} - 2A_1\right)$.

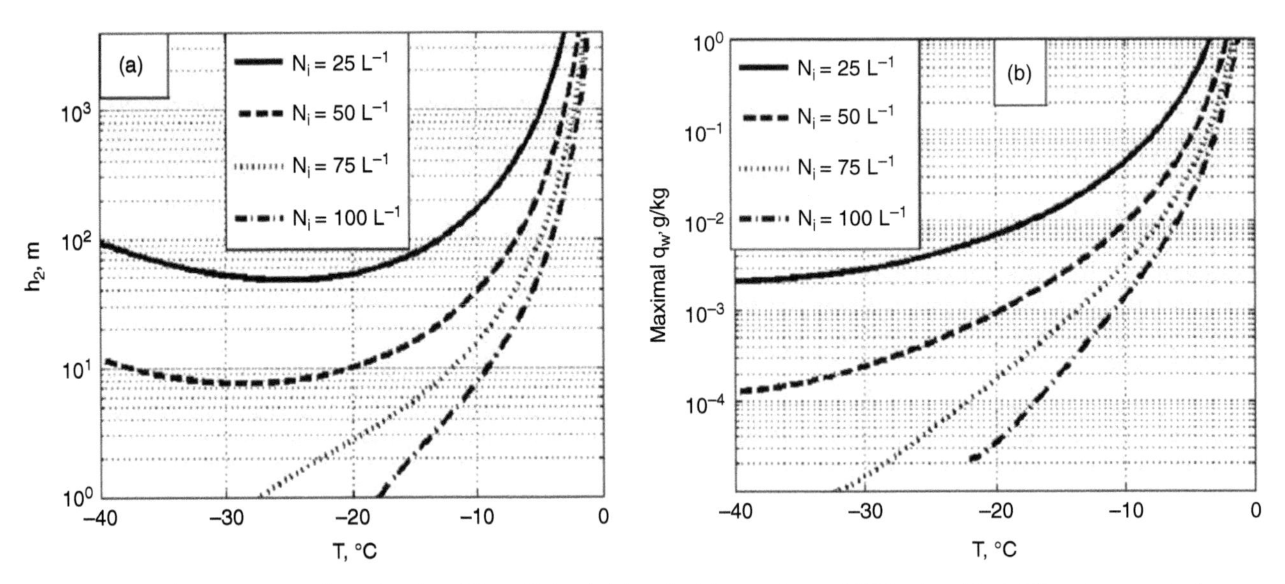

Figure 6.3.11 The temperature dependencies of Zone II depth (a) and of the maximum LWC value (b) at different concentrations of ice particles N_i. The updraft velocity is equal to 0.1m/s and $p = 1{,}000$ mb (from Pinsky et al., 2015; courtesy of © John Wiley & Sons, Inc.).

Correspondingly, the depth of Zone III where the WBF mechanism is active is $h_3 = h_{23} - h_2$. If q_{i0} is negligibly small, the total thickness of the mixed-phased zone is calculated as

$$h_{23} \approx \left(\frac{A_1}{A_3}\right)^2 G^{-3} \tag{6.3.26}$$

Correspondingly, Zone III depth is equal to

$$h_3 = h_{23} - h_2 \approx \frac{5}{9}\left(\frac{A_1}{A_3}\right)^2 G^{-3} = 1.25h_2 \tag{6.3.27}$$

Figure 6.3.12 shows the dependencies of the depth of the entire mixed-phase zone ($h_{23} = h_2 + h_3$) on the updraft velocity and the dependencies of h_{23} on temperature at different ice particle concentrations. The results shown in Figure 6.3.12a. indicate a high sensitivity of the mixed-phase zone thickness to the vertical velocity. From Equation (6.3.27) it follows that the depth is proportional to w^3. The thickness ranges from zero at $w = 10\,\text{cm/s}$ to a few kilometers at $w = 0.4{-}0.5\,\text{m/s}$. At higher vertical velocities, the thickness of the mixed-phase zone becomes unrealistically large. This means that in cumulus clouds with high vertical updrafts, the WBF mechanism is active only near the cloud top (cloud anvils) when the updraft velocity tends toward zero or is not active at all.

The high sensitivity of the mixed-phase zone thickness to ice particle concentration and temperature is demonstrated in Figure 6.3.12b. The increase in the ice concentration from $25\,\text{L}^{-1}$ to $100\,\text{L}^{-1}$ decreases the thickness of the mixed-phase zone by more than an order of magnitude. These results are important since they allow estimation of the ice crystal concentration based on measurements of the thickness of the mixed-phase zone and the temperature. Figure 6.3.12 also shows good agreement between the analytical results and the numerical simulations at $T_C = -20°\text{C}$. However, this agreement diminishes at temperatures $T_C > -10°\text{C}$ because the depth h_{23} increases and magnitude of the error introduced by the assumption of $\varphi(T) = const$ strongly increases. At $T_C > -10°\text{C}$, the depth h_{23} is large due to the low values of supersaturation over ice. At $T_C < -25°\text{C}$, the depth h_{23} increases with the decrease in temperature caused by a decrease in the available water vapor mixing ratio and the corresponding weakening of the ice particle growth rate. The latter leads to a decrease in the efficiency of the WBF mechanism. Thus, the minimum of h_{23} occurs at intermediate temperatures around $T_C \approx -25°\text{C}$.

In Zone IV, ice particles continue growing, and supersaturation with respect to water rapidly decreases since the parcel does not contain liquid water. The microphysical equations determining the growth of ice particles in Zone IV are similar to those in Zone I. The difference is in the initial conditions at the bottom of Zone I and of Zone IV.

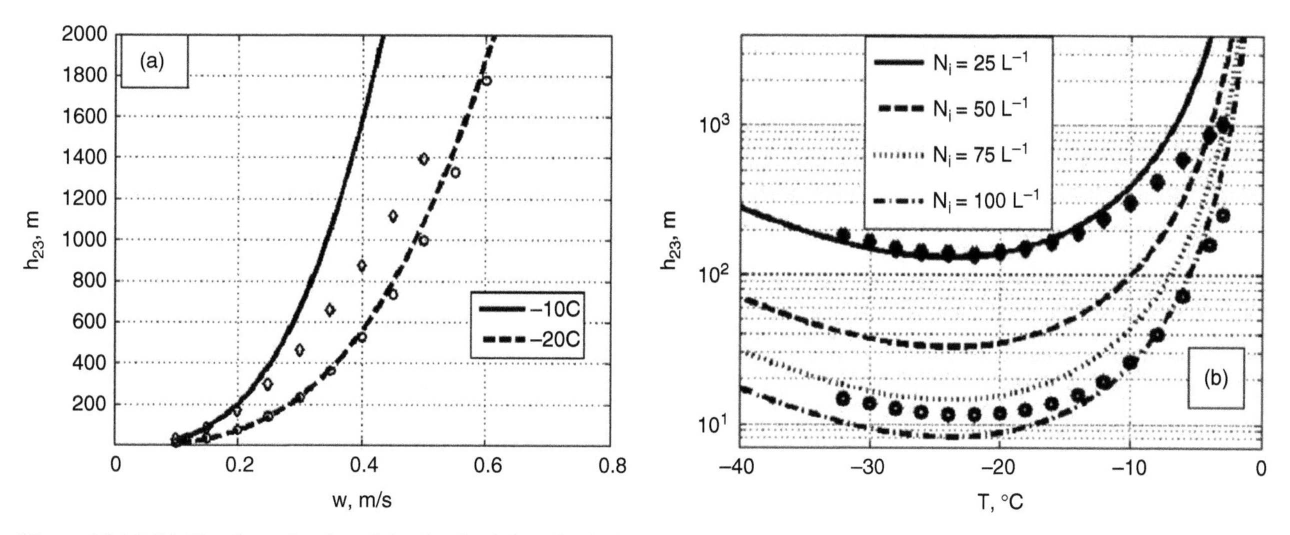

Figure 6.3.12 (a) The dependencies of the depth of the mixed-phase zone (h_{23}) on vertical velocity at two values of air temperature, $N_i = 100\ \mathrm{L}^{-1}$. (b) The temperature dependencies of the depth of the mixed-phase zone at different concentrations of ice particles. The updraft velocity is equal to 0.1 m/s; $p = 1{,}000$ mb. Symbols indicate results obtained by means of the parcel model (from Pinsky et al., 2015; courtesy of © John Wiley & Sons, Inc.).

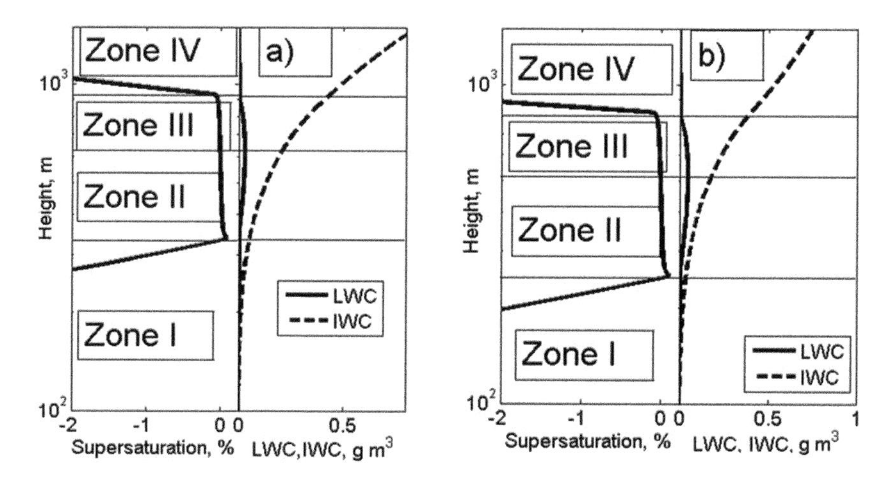

Figure 6.3.13 (a) An example of vertical stratification of zones with different regimes of ice–liquid water transformation. The profile of the supersaturation over liquid water and the profile of the LWC are obtained in an adiabatic parcel ascending at $w = 0.1$ m/s, $N_w = 100\ \mathrm{cm}^{-3}$, $N_i = 10\ \mathrm{L}^{-1}$ and $T_C = -20°\mathrm{C}$. (b) The same as Figure 6.3.13a, but at linearly decreasing temperatures $T_C = -15 - 0.006z$°C (from Pinsky et al., 2015; courtesy of © John Wiley & Sons, Inc.).

The results presented in this section are summarized in **Figure 6.3.13a**, which shows the profiles of supersaturation and LWC in an air parcel ascending at the velocity $w = 0.1$ m/s. Formation of the liquid phase is determined by condition $w > w^*$. In Zones II and III containing the liquid phase, the supersaturation S_w is close to zero due to the low phase relaxation time in a mixed-phase cloud volume. Droplet condensation in Zone II and droplet evaporation in Zone III keep S_w close to zero. Disappearance of liquid droplets leads to strong changes in the supersaturation. In Zone IV, the IWC continues growing.

For comparison, Figure 6.3.13b shows the profiles of supersaturation and the LWC calculated when the temperature decreases linearly with gradient of 6°C/km, so the mean temperature within the layer is approximately equal to −20°C, as in Figure 6.3.13a. The profiles in Figures 6.3.13a and 6.3.13b look qualitatively similar.

On the whole, Figure 6.3.13 illustrates the validity of the conceptual scheme shown in Figure 6.3.6.

As was shown, all four zones in Figure 6.3.13 exist at a certain number of conditions. Formation of one or another zone depends on the vertical velocity and the ice particle concentration. For instance, at low vertical velocities near cloud base the supersaturation over water remains negative and the liquid phase does not arise. If the vertical velocity is of several meters per second, which is typical of convective clouds, the supersaturation over water remains positive up to the upper cloud boundary, and Zone III does not form. In this case the WBF mechanism is not active.

6.3.5 Phase Transitions in Oscillating Adiabatic Parcels

As was shown in Section 6.3.4, cloud dynamics, especially the vertical velocity, dramatically affect the microstructure of mixed-phase clouds and the ice–liquid coexistence. The air in stratocumulus clouds oscillates in large eddies with characteristic vertical scales of the BL depth or of cloud depth. The role of air oscillations in warm clouds is discussed in Section 5.2. The impact of air oscillations on the microphysical structure of stratiform and stratocumulus clouds at negative temperatures was investigated in a number of studies (e.g., Korolev and Mazin, 2003; Korolev and Isaac, 2003; Korolev and Field, 2008; Sulia and Harrington, 2011). Several physical reasons determine the importance of parcel oscillations in the vertical direction. In addition to phase transformations in updrafts that were analyzed in Section 6.3.4, the existence of downdrafts makes the picture of these transformations more complicated. First, the air in downdrafts can be subsaturated with respect to water and supersaturated with respect to ice. It means that the WBF mechanism can dominate in downdrafts leading to ice growth at the expense of evaporating droplets and thus to cloud glaciation. In stronger downdrafts, subsaturation takes place both over ice and over water, so droplets evaporate and ice does not form. Second, ice particles and droplets may evaporate in downdrafts only partially, so that successive parcel ascents may start at new initial conditions, which in turn affects the phase transitions. These changes may accumulate, leading to a gradual disappearance of the originally existing water droplets.

After a certain transition period, periodic oscillations of a parcel lead to formation of limit cycles of the microphysical parameters, i.e., to periodical changes of supersaturation, LWC and IWC. The duration of the transition period as well as the parameters of the limit cycle depend on both the initial values of LWC and of IWC and on numerous dynamic and thermodynamic characteristics such as updraft/downdraft velocity, temperature, the mean height, and the amplitude of fluctuations. Cloudy parcels that reach the limit cycle may be pure ice or mixed phase. Two examples of possible limit cycles are shown in **Figures 6.3.14** and **6.3.15.** Figure 6.3.14 shows supersaturation fluctuations during parcel oscillations at a low vertical velocity of 0.2 m/s. One can see that supersaturation over water is

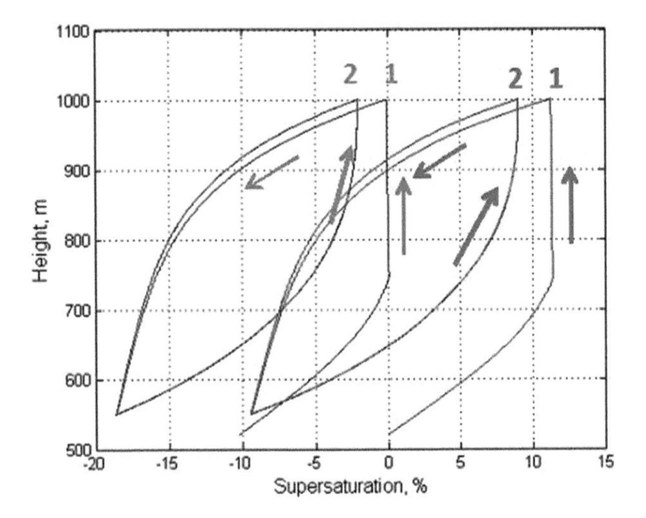

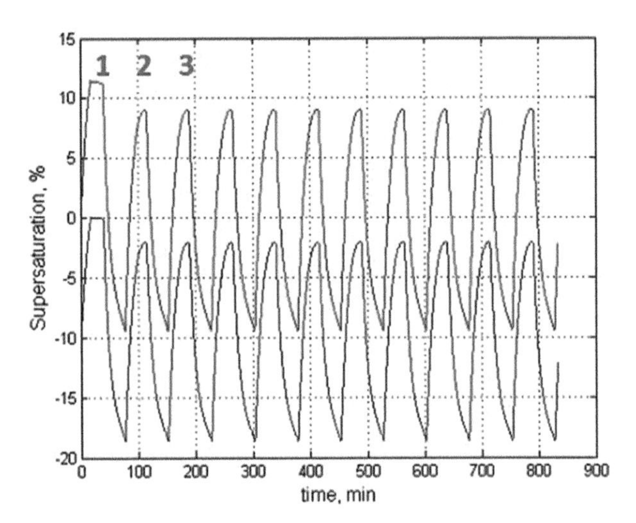

Figure 6.3.14 Height fluctuations (left) and time fluctuations (right) of supersaturations over water (blue) and over ice (red). Arrows show the direction of the parcel's movement (ascending/descending). Numbers denote the number of an oscillation period. The oscillations are assumed within the layer of 550–1,000 m, the vertical velocity in updrafts and downdrafts is $w = \pm 0.2$ m/s, droplet concentration is $N_w = 200$ cm^{-3}, ice particle concentration is $N_i = 100$ L^{-1}, and $T_C = -11$°C.

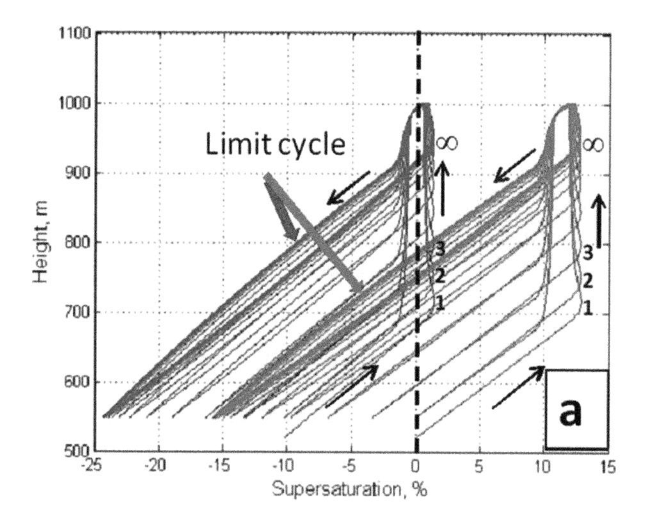

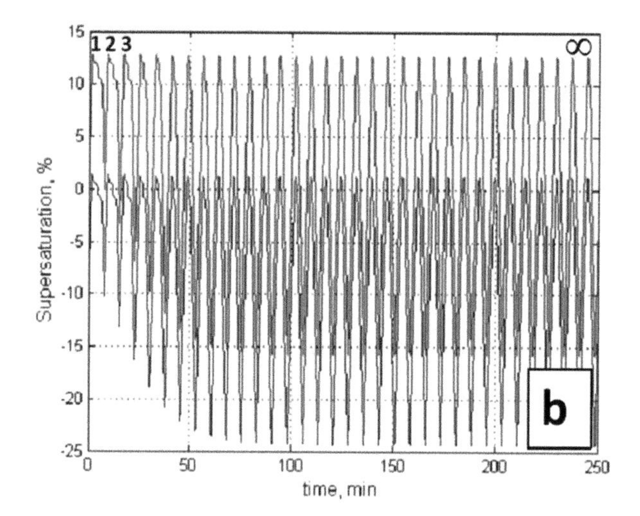

Figure 6.3.15 The same as in the Figure 6.3.14, but for the vertical velocity $w = \pm 2$ m/s.

reached only during the first updraft. Ice growing during the first oscillation absorbs water vapor, so the supersaturation during the second updraft is not reached and water droplets do not form. This oscillating parcel has a pure ice limiting cycle. An example of a limit cycle within which liquid droplets permanently form in the ascending branch of the parcel trajectory is shown in Figure 6.3.15. The difference from the regime illustrated in Figure 6.3.14 is that the vertical velocity was chosen equal to 2 m/s. One can see that the supersaturation over water reaches its maximum value during the first ascent. During several consequent updrafts, the supersaturation maximum decreases and is reached at higher levels. Nevertheless, toward the end of the transition period the maximum supersaturation over water exceeds zero and liquid water forms during each parcel ascent.

As in warm microphysics, the limit cycle is established when the change in the masses of water and ice in the ascent segment of the parcel track is compensated by the corresponding changes in the descend segment. In comparison to a warm parcel, the limit cycle in the mixed-phase case requires more parcel oscillations to emerge. This is supposedly related to a larger relaxation time within the segments of parcel tracks where the parcel contains only ice. During several first oscillations, the IWC increases, causing a decrease in supersaturation in the updraft segments of the parcel track. As a result, generation of ice (or water) in the updrafts decreases, thus establishing the balance between mass generation and mass loss, i.e., triggering the limit cycle. The maximum sizes of ice particles corresponding to the limit cycle are of about a few hundred μm, the maximum droplet radius is about 10 μm. The characteristic feature of the limit cycle illustrated in Figure 6.3.15 is formation of liquid water in the updrafts and its dissipation in the downdrafts.

Conditions for liquid water formation in the parcel updrafts are directly related to parcel oscillations. Korolev and Field (2008) found two necessary conditions for formation of a mixed-phase layer inside a pure ice stratiform cloud. The first condition is satisfied when the amplitude of oscillations exceeds a certain minimum ΔZ. ΔZ depends on the temperature and its characteristic value is of order of several hundred meters. The critical value of ΔZ^* is related to the value h_{1_min} that was defined in Section 6.3.4 (Equation (6.3.19) and Figure 6.3.9). The second condition is satisfied when the updraft/downdraft parcel velocity exceeds a certain critical value u^* that depends on several factors and has the characteristic value of order of 0.1 m s^{-1}. The critical values of u^* are related to the critical velocity value w^* discussed in Section 6.3.4 (Equation (6.3.18) and Figure 6.3.8). The existence of these two conditions is illustrated in **Figure 6.3.16**, presenting the height dependencies of the total water content (IWC+LWC) in the course of parcel oscillations.

The liquid adiabat and the ice adiabat, shown in Figures 6.3.16, are close to straight lines and have similar slopes (Equations (6.3.7) and (6.3.8)). The center of the limit cycle is located on the ice adiabat. If in the course of oscillations, the trajectory reaches the liquid adiabat (i.e., the supersaturation over water becomes positive), the liquid phase forms and the layer becomes

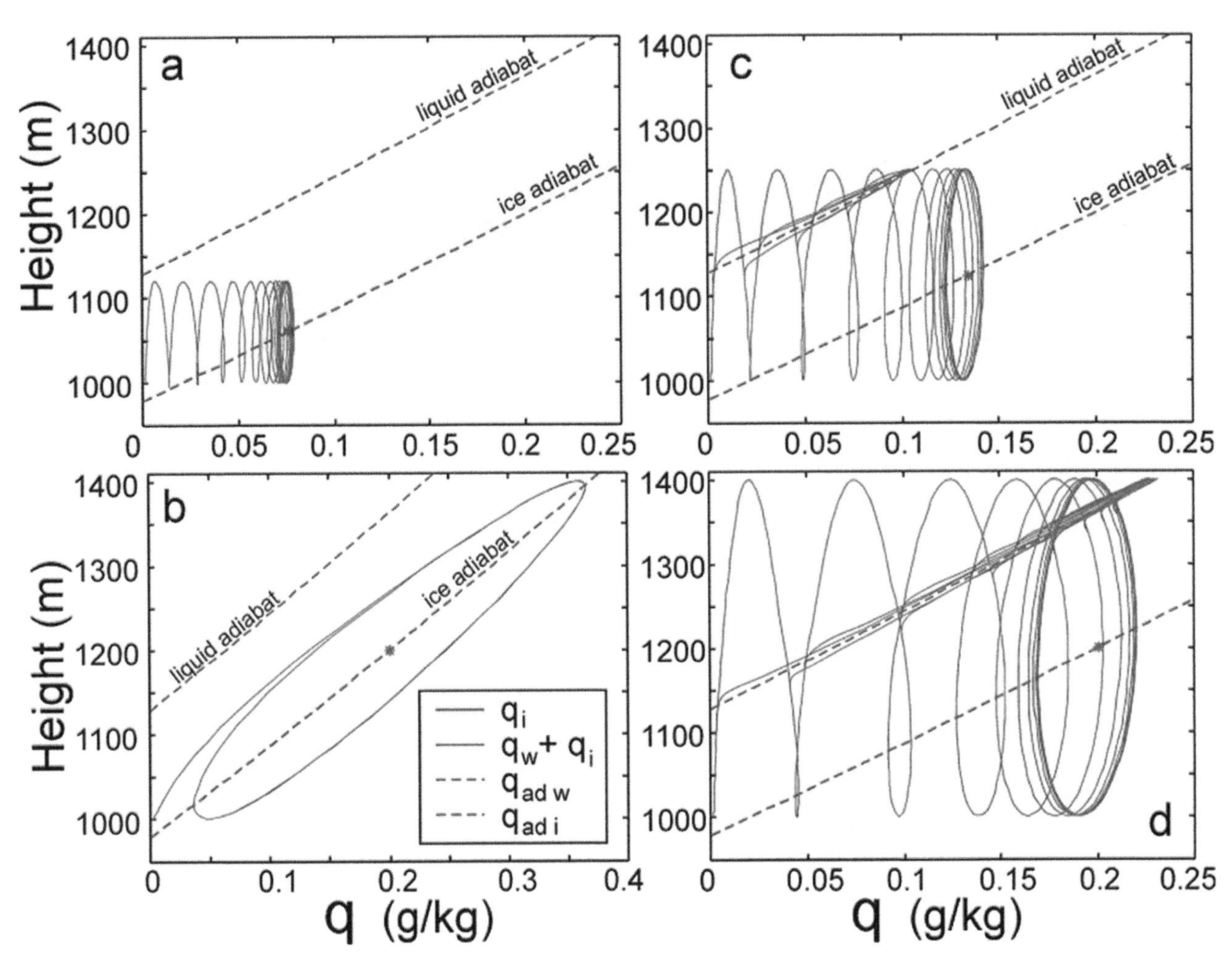

Figure 6.3.16 Numerical modeling of liquid water activation in an ice cloud during vertical harmonic oscillations. For all of the four cases, $N_i = 50\ \mathrm{L}^{-1}$, $r_{i0} = 20\ \mu\mathrm{m}$, $S_{i0} = 0.01$, $T_C = -10°\mathrm{C}$, and $\Delta Z = 153$ m. (a) $\Delta Z < \Delta Z^*$ and $u > u^*$; (b) $\Delta Z > \Delta Z^*$ and $u < u^*$; (c) $\Delta Z < \Delta Z^*$ and $u > u^*$; (d) $\Delta Z > \Delta Z^*$ and $u > u^*$. The red asterisk indicates the expected center of the limit cycle for the pure ice case (from Korolev and Field, 2008; © American Meteorological Society; used with permission).

mixed phase. Figure 6.3.16a illustrates the case when the amplitude of the oscillations is below the critical value, the limit cycle does not cross the liquid adiabat, and supersaturation over water does not occur. In this case, the liquid phase does not form because the first condition for its emergence is violated. Figure 6.3.16b corresponds to the case when the limit cycle does not cross the liquid adiabat because the oscillation's velocity is below the critical value. In this case, the liquid phase does not form because the second condition for its emergence is violated. Figure 6.3.16c shows the case when the liquid phase does arise but disappears after several oscillations. The limit cycle does not contain liquid water. This case is illustrated in Figure 6.3.14 in more detail. Figure 6.3.16d demonstrates the case when both conditions are satisfied and the liquid phase periodically forms in the cloud layer. This case regime is illustrated in Figure 6.3.15 in more detail.

Analysis of the phase transformations caused by diffusion growth/evaporation of droplets and deposition/sublimation of ice indicates that these processes play a highly important role in formation of various regimes and microphysical structures in clouds at negative temperatures, especially in stratiform and stratocumulus clouds where these processes are dominating. They lead to formation of ice crystals with sizes of several hundred of μm that are large enough to create precipitation, as well as to collect droplets and produce graupel.

6.3.6 Glaciation Time

Glaciation time (GT) is typically determined as the period during which the liquid water mass in a cloud (or in a cloud parcel) disappears and turns into ice. In cases when glaciation takes place due to only diffusion growth/evaporation and deposition/sublimation, cloud glaciation is determined by droplet evaporation. At certain conditions, the WBF mechanism can accelerate droplet evaporation and thus significantly affect GT. Korolev and Mazin (2003) and Korolev and Isaac (2003) evaluated GT t_{gl} in an unmovable ($w = 0$) adiabatic parcel. They assumed that the final mass of ice formed as a result of glaciation is equal to the sum of the initial water droplet mass and ice particle mass. Assuming also that during glaciation the supersaturation over water is equal to zero, the following formula was obtained:

$$t_{gl} = \frac{F_i}{2(\varphi - 1)\left(\frac{4}{3}\pi\rho_i N_i\right)^{2/3}}\left((q_{i0} + q_{w0})^{2/3} - q_{i0}^{2/3}\right), \tag{6.3.28}$$

where q_{w0} and q_{i0} are the initial LWC and IWC, respectively. **Figure 6.3.17** shows the dependences of GT on temperature at different concentrations of ice particles. One can see that GT decreases strongly as ice particle concentration increases. Under concentrations typical of stratiform clouds, the GT varies from several tens of minutes to several hours. The GT minimum at $T_C = -15°C$ corresponds to the maximum of the saturation water pressures' difference $e_w(T_C) - e_i(T_C)$ seen in Figure 6.3.1 in the zone where the WBF mechanism is the most pronounced.

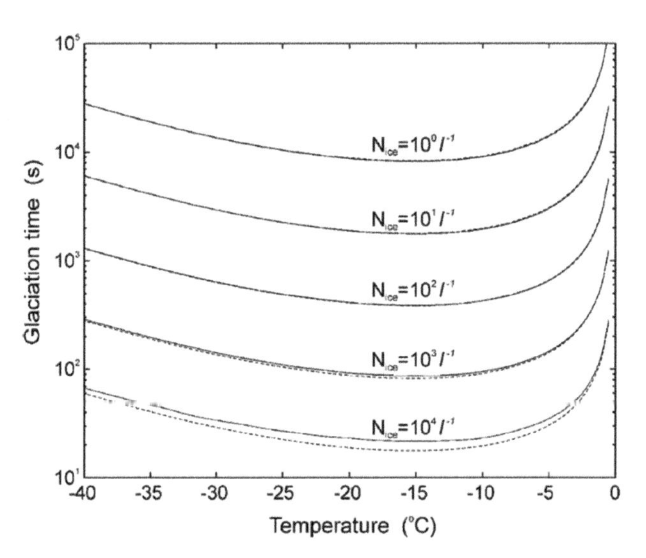

Figure 6.3.17 Dependence of glaciation time on temperature, calculated both using a numerical model (solid lines) and using Equation (6.3.28) (dashed lines). Calculations were performed at zero vertical velocity and the initial liquid-water content of 0.1 g/kg for various number concentrations of ice particles (from Korolev and Isaac, 2003; courtesy of © John Wiley & Sons, Inc.).

Using a parcel model, Korolev and Isaac (2003) showed that the GT strongly depends on the vertical velocity and is distinctly nonsymmetric with respect to updrafts and downdrafts. Similar results were obtained analytically by Pinsky et al. (2014), who used the balance equation (6.3.6) and the equations for diffusional growth, (6.3.11) and (5.2.6), to derive two equations for GT in a moving adiabatic parcel. In the first, equation (6.3.29a), the displacement of parcel Δz is considered an independent variable that is positive for an upward displacement and negative for a downward displacement. Evaluating the GT for a given displacement value is necessary, for instance, when the ascent or descent velocities are not known *a priory*, so only the initial and the final parcel's locations are known. In this case, the knowledge of specific changes of velocities is not necessary. The formula for the GT in this case is

$$t_{gl} = \frac{3}{2(\varphi - 1)}\left\{\frac{1}{B_i N_i^{2/3}}\left[\left(q_{i0} + \frac{A_2}{A_3}q_{w0} + \frac{A_1}{A_3}\Delta z\right)^{2/3} - q_{i0}^{2/3}\right] + \frac{\varphi}{B_w N_w^{2/3}} q_{w0}^{2/3}\right\}, \tag{6.3.29a}$$

where $B_w = \frac{3}{F_w}\left(\frac{4\pi\rho_w}{3}\right)^{2/3}$ and $B_i = \frac{3}{F_i}\left(\frac{4\pi\rho_i}{3}\right)^{2/3}$. Formula (6.3.29a) allows us, for instance, to calculate the GT of a parcel that first shifts upward (or downward) by Δz and then becomes motionless. Obviously, Equation (6.3.29a) is valid if the time of upward/downward displacement is less than the glaciating time. Dependences of the GT on displacement Δz at different concentration of ice particles shown in **Figure 6.3.18** are strongly asymmetric with respect to the zero displacement. As a downward displacement is accompanied by an additional evaporation of cloud droplets, the GT in this case is smaller than for an upward displacement. In contrast, an upward displacement is accompanied by an additional condensation of water vapor on cloud droplets and by increase of GT.

Expression (6.3.29a) is derived for ice crystals approximated by effective spheres. This approximation is suitable for comparatively small particles with the aspect ratio values close to unity. However, for large ice crystals, diffusion growth should be calculated taking into account the actual crystal shape. In case of nonspherical particles, Equation (6.3.29a) is replaced by the following expression:

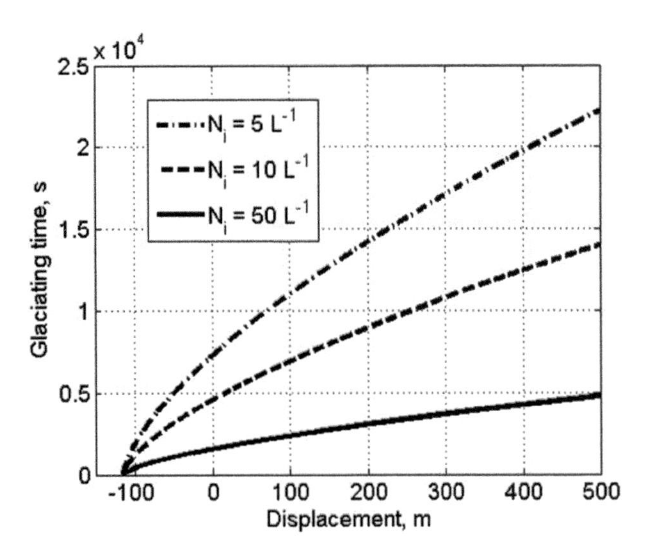

Figure 6.3.18 Dependencies of the glaciation time on a displacement value at different ice particle concentrations, $T_C = -20°\mathrm{C}$, $q_{w0} = 0.2\ \mathrm{g\,m^{-3}}$ and $q_{i0} = 0$ (from Pinsky et al., 2014; © American Meteorological Society; used with permission).

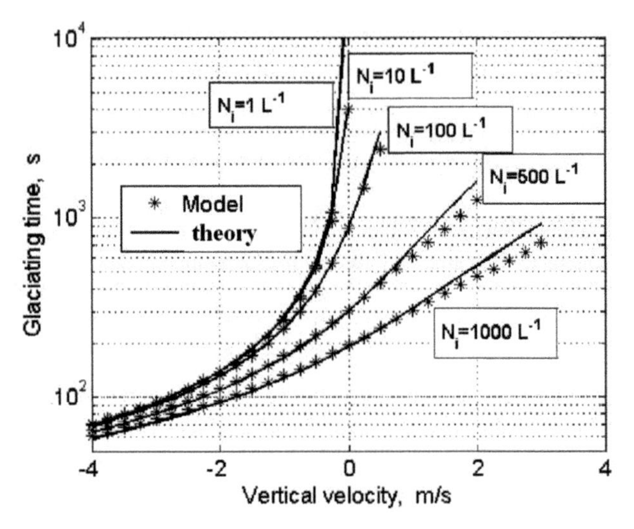

Figure 6.3.19 Dependencies of glaciation time on vertical velocity at different number concentrations of ice particles (solid lines). Parameter values used in the calculations are as follows: $T_C = -15°\mathrm{C}$, $q_{w0} = 0.2\ \mathrm{g/kg}$ and $z_0 = 1{,}000\ \mathrm{m}$. The calculations were terminated when the parcel reached the level of homogeneous droplet freezing. Results obtained using the parcel model are denoted by asterisks (from Pinsky et al., 2014; © American Meteorological Society; used with permission).

$$t_{gl} = \frac{3}{2(\varphi - 1)}\left\{\frac{1}{f(\gamma_i)B_i N_i^{2/3}}\left[\left(q_{i0} + \frac{A_2}{A_3}q_{w0} + \frac{A_1}{A_3}\Delta z\right)^{2/3} - q_{i0}^{2/3}\right] + \frac{\varphi}{B_w N_w^{2/3}} q_{w0}^{2/3}\right\} \quad (6.3.29b)$$

This equation shows that the GT is inversely proportional to the normalized capacitance, so in case the resulting ice particles are nonspherical GT is lower. Since $B_i \sim \rho_i^{-1/3}$, a decrease in the ice particle density leads to a decrease in the GT.

The second equation (6.3.30) for GT glaciation time was derived by Pinsky et el. (2014) for an adiabatic parcel moving at constant vertical velocity w. Displacement Δz was equalized to wt_{gl} in Equation (6.3.29). Introducing a nondimensional time $\tilde{t}$ obeying the following equation

$$t_{gl} = \frac{3}{2(\varphi - 1)G_i N_i^{2/3}}\tilde{t} - \frac{3}{2(\varphi - 1)}\left(\frac{q_{i0}^{2/3}}{G_i N_i^{2/3}} - \frac{\varphi q_{w0}^{2/3}}{G_w N_w^{2/3}}\right) \quad (6.3.30)$$

one can obtain a cubic equation with respect to $\tilde{t}$:

$$\tilde{t}^3 - \gamma^2\tilde{t}^2 - 2q_0\gamma\tilde{t} - q_0^2 = 0, \quad (6.3.31)$$

where the two variables γ and q_0 are defined as

$$\gamma = \frac{3A_1}{2(\varphi - 1)B_i N_i^{2/3} A_3} w \quad (6.3.32)$$

$$q_0 = q_{i0} + \frac{A_2}{A_3}q_{w0} - \left(\frac{3q_{i0}^{2/3}}{2(\varphi - 1)B_i N_i^{2/3}} - \frac{3\varphi q_{w0}^{2/3}}{2(\varphi - 1)B_w N_w^{2/3}}\right)\frac{A_1}{A_3}w \quad (6.3.33)$$

The full solution of Equation (6.3.31) is presented by Pinsky et al. (2014). In a particular case when the initial LWC and IWC are equal to zero (i.e., $q_{w0} = q_{i0} = 0$), the GT is proportional to the square of the vertical velocity:

$$t_{gl} = \left[\frac{3}{2(\varphi - 1)B_i N_i^{2/3}}\right]^3 \left(\frac{A_1}{A_3}\right)^2 w^2 \quad (6.3.34)$$

Figure 6.3.19 shows the dependence of the GT on the vertical velocity at different number concentrations of ice particles as follows from the full solution of Equation (6.3.31). One can see that the GT in updrafts rapidly increases with increasing vertical velocity and decreasing concentration of ice particles. According to Equation (6.3.34), the GT, being inversely proportional to the squared concentration of ice crystals, decreases rapidly with increasing ice particle concentration. Figure 6.3.19 shows that at very high ice particle concentration of $1{,}000\ \mathrm{L^{-1}}$, glaciation may take place during a few tens of minutes even at high updraft velocities of a few meters per second. At vertical velocities exceeding 0.5 m/s and ice particle concentrations of $N_i < 100\,\mathrm{L}^{-1}$, the glaciation time becomes larger than the characteristic lifetime of clouds, indicating that in strong updrafts the WBF mechanism is not efficient. Therefore, under these conditions glaciation in real clouds may be caused by other mechanisms such as droplet freezing and/or riming.

At vertical velocities of $0 < w < 0.25\,\mathrm{m\,s}^{-1}$, which are typical of stratiform clouds, glaciation may occur within a layer of a few hundred meters deep. Analysis shows that evaporation of droplets in adiabatic downdrafts accompanied by a temperature increase plays the major role in droplet disappearance. Thus, the WBF mechanism, i.e., transferring mass of water droplets to ice particles due to the difference in saturation vapor pressures over water and over ice, is effective only at very low vertical velocities. GT calculated using Equation (6.3.31) shows a rather good agreement with those obtained using the parcel model. The maximum deviation of the analytical solution from the parcel model solution takes place at high vertical velocities and low ice concentrations.

Examples of ascending and oscillating parcels discussed here clearly show an important role of condensation/evaporation and deposition/sublimation in formation of the microstructure of mixed-phase clouds. These examples also show that the dynamics and the microphysical processes have a significant impact on the phase composition of clouds.

6.3.7 Calculation of Supersaturation and Diffusional Growth in Mixed-Phase Clouds

Spectral Bin Microphysics Models

Accurate calculation of diffusion growth in mixed-phase clouds is a mandatory requirement to advanced microphysical models. It is necessary to take into account that particle growth is accompanied by instantaneous change in supersaturation, so the model algorithm should include solution of the equations for diffusional growth and supersaturation over water and ice. Algorithms for calculation of supersaturations in SBM models of mixed-phase clouds were proposed by Reisin et al. (1996), Khain and Sednev (1996), Khain et al. (2004), and later modified by Khain et al. (2008). At the stage of diffusion growth/evaporation and deposition/sublimation at $T_C < 0°\mathrm{C}$, the equations to be solved for the mixing ratio of water vapor q and for the potential temperature θ are

$$\left(\frac{\partial q}{\partial t}\right)_{c/e} = -(\varepsilon_1 + \varepsilon_2), \quad \left(\frac{\partial \theta}{\partial t}\right)_{c/e} = \frac{1}{c_p \Pi}(L_w \varepsilon_1 + L_i \varepsilon_2), \tag{6.3.35}$$

where

$$\varepsilon_1 = \frac{1}{\rho}\int_0^\infty f(m)\frac{dm}{dt}dm, \quad \varepsilon_2 = \sum_{i=1}^{I}\frac{1}{\rho}\int_0^\infty f_i(m_i)\frac{dm_i}{dt}dm_i \tag{6.3.36}$$

and $\Pi = T/\theta$ is the Exner function (see Section 3.1). In Equation (6.3.36), ε_1 and ε_2 are the rates of condensation and ice deposition, respectively. In Equation (6.3.36), $f(m)$ and $f_i(m_i)$ are distribution functions of drops and ice particles, respectively. Summation in calculation of ε_2 with respect to index i means taking into account different types of solid hydrometeors. For instance, the HUCM deals with six types of ice particles: planar type, columnar type, and bench type (dendrites) ice crystals, aggregates (snow), graupel, and hail. Therefore, in this model $I = 6$. The terms $\frac{dm}{dt}$ and $\frac{dm_i}{dt}$ are the growth rates of drop mass and ice particle mass, respectively, determined by the equations for diffusional growth of drops (5.2.11) and deposition of water vapor on ice particles of a particular type (6.3.9). A significant advantage of the spectral bin microphysical approach is that the electrical capacitance of nonspherical ice crystals (which is different for different particle masses and types) is taken into account. In HUCM, the capacitance depends on the type of ice crystals as well as on the mass of crystals of a particular type, allowing us to take into account the changes of the crystal shape during depositional growth. As soon as the mass of ice crystal changes, the geometrical parameters also change according to the lookup tables presented in Section 6.1. The capacitance changes accordingly.

Replacing the terms $\frac{dm}{dt}$ and $\frac{dm_i}{dt}$ in Equation (6.3.36) for their expressions from Equations (5.2.11) and (6.3.9), and using the Clausius–Clapeyron relationship, one can derive the following equations for supersaturations over water and over ice (Khain and Sednev, 1996):

$$\frac{dS_w}{dt} = -P_1 S_w - P_2 S_i, \quad \frac{dS_i}{dt} = -R_1 S_w - R_2 S_i, \tag{6.3.37}$$

where expressions for the coefficients P_1, P_2, R_1, and R_2 are

$$P_1 = \frac{e}{e_w F}\left[a + \frac{L_w^2}{c_p R_v T^2}\right]\int_0^\infty frdm, \quad R_1 = \frac{e}{e_i F}\left[a + \frac{L_w L_i}{c_p R_v T^2}\right]\int_0^\infty frdm,$$
$$P_2 = \frac{e}{e_w F_i}\left[a + \frac{L_i L_w}{c_p R_v T^2}\right]\sum_i \int_0^\infty f_i C_i dm_i, \quad R_2 = \frac{e}{e_i F_i}\left[a + \frac{L_i^2}{c_p R_v T^2}\right]\sum_i \int_0^\infty f_i C_i dm_i. \tag{6.3.38}$$

and $a = \frac{1}{q(1+0.61q)}$. The equation system (6.3.37) describes the change of supersaturation during diffusional growth. The characteristic time scale of the supersaturation changes in warm and mixed-phase clouds, i.e., the phase relaxation time can be very small. It means that the changes of supersaturation should be taken into account when diffusional growth is calculated. So, equation system (6.3.37) should be solved together with the equations of diffusional growth. Numerically, however, these equations are solved in a sequence. The microphysical time step Δt in the model is chosen small enough, so coefficients P_1, P_2, R_1, and R_2 are considered constant during a single time step. In this case, Equation (6.3.37) has the analytical solution

$$S_w(t) = \alpha^{-1}\Big\{S_w(t_0)\Big(\beta\exp\{\gamma(t-t_0)\} + \gamma\exp\{-\beta(t-t_0)\}\Big)$$
$$-[P_1 S_w(t_0) + P_2 S_i(t_0)]\Big(\exp\{\gamma(t-t_0)\} - \exp\{-\beta(t-t_0)\}\Big)\Big\} \tag{6.3.39}$$

$$S_i(t) = \alpha^{-1}\Big\{S_i(t_0)\Big(\beta\exp\{\gamma(t-t_0)\} + \gamma\exp\{-\beta(t-t_0)\}\Big)$$
$$-[R_1 S_w(t_0) + R_2 S_i(t_0)]\Big(\exp\{\gamma(t-t_0)\} - \exp\{-\beta(t-t_0)\}\Big)\Big\}, \tag{6.3.40}$$

where $\alpha = \left[(P_1 - R_2)^2 + 4R_1P_2\right]^{1/2}$; $\beta = \frac{1}{2}(\alpha + P_1 + R_2)$; and $\gamma = \frac{1}{2}(\alpha - P_1 - R_2)$. These expressions allow analytical calculation of the supersaturation integrals over time step Δt: $\int_{t_0}^{t_0+\Delta t} S_w d\tau$ and $\int_{t_0}^{t_0+\Delta t} S_i d\tau$.

Then at each time step, a new mass of drops is calculated as

$$m^{2/3}(t_0+\Delta t) = m^{2/3}(t_0) + 2\left(\frac{4}{3}\pi\rho_w\right)^{2/3}\frac{1}{F}\int_{t_0}^{t_0+\Delta t} S_w d\tau \tag{6.3.41}$$

In cases of spherical ice particles, the changes of their mass by deposition or sublimation is described by Equation (6.3.41), where F is replaced by F_i and ρ_w by ρ_i. When ice particles are not spherical, Khain and Sednev (1996) calculate their mass changes using the following equation:

$$m_i(t_0+\Delta t) = m_i(t_0) + \frac{4\pi\rho_i C_i(t_0)}{F_i}\int_{t_0}^{t_0+\Delta t} S_i d\tau, \tag{6.3.42}$$

where C_i is taken at time $t = t_0$. Using the new mass values, size distributions are recalculated using remapping approaches described in Section 5.4. New size distributions are used during the next microphysical time step to calculate the new values of coefficients P_1, P_2, R_1 and R_2.

After calculating the new mass values of water and ice, the temperature and humidity changes are calculated using Equations (6.3.35) and (6.3.36). The new values of temperature and humidity allow calculating supersaturations at the end of each microphysical time step. In principle, these values should coincide with the solutions of Equations (6.3.35) and (6.3.36). However, since coefficients P_1, P_2, R_1, and R_2 are not exactly constant, the supersaturations calculated after applying the mass corrections (6.3.41) and (6.3.42) do not fully coincide with the solutions of Equations (6.3.35) and (6.3.36). To improve this agreement, an iterative procedure can be used, wherein the temperatures and the saturation values obtained after the mass correction are used to recalculate coefficients P_1, P_2, R_1, and R_2, as well as the recalculation of supersaturation integrals in

Equations (6.3.41) and (6.3.42). A few iterations lead to an agreement with the accuracy of 1% of the supersaturation magnitude.

Equations (6.3.35–6.3.37) are applied only at a microphysical time sub-step of the diffusion growth and deposition growth. As was discussed in Section 5.4, after a dynamic time sub-step used for solving the equations of advection the supersaturation (or subsaturation) can reach large values. During a microphysical time sub-step, the supersaturation rapidly decreases. As a result, supersaturation experiences significant fluctuations. In reality, all of these processes take place at the same time, so the positive tendency of supersaturation caused by the adiabatic cooling of the ascending air (the term $A_1 w$ in Equation (6.3.5)) is compensated, to a significant degree, by the second and third terms of this equation. For proper simulation of nucleation processes that depend on the supersaturation values (Sections 5.3 and 6.2), it is important to avoid numerically caused fluctuations of supersaturation values. Two ways can be proposed to decrease supersaturation fluctuations during a model time step: a) to reduce the time step itself and b) to take into account both the dynamically induced source and the sink of supersaturation within the same microphysical sub-step. In the WRF–SBM (Khain et al., 2010; Fan et al., 2013), the microphysical time step is separated into n_{cond} sub-steps, and all dynamically induced changes of supersaturations $\Delta S_{w,dyn}$ and $\Delta S_{i,dyn}$ (calculated as the difference in the values of supersaturations after advection and before advection) are also divided by n_{cond}. Thus, at the beginning of each sub-step supersaturations are updated by $\Delta S_{w,dyn}/n_{cond}$ and $\Delta S_{i,dyn}/n_{cond}$.

Another approach proposed by Khain et al. (2008) is described in Section 5.4 (see Equation (5.4.32) in respect to warm processes). Let us assume that due to the dynamic effects, the supersaturations over water and ice change at the rates $\left(\frac{\partial S_w}{\partial t}\right)_{dyn} = \frac{\Delta S_{w,dyn}}{\Delta t_{dyn}}$; $\left(\frac{\partial S_i}{\partial t}\right)_{dyn} = \frac{\Delta S_{i,dyn}}{\Delta t_{dyn}}$, where Δt_{dyn} is the dynamic time step used for advection and turbulent mixing. The changes of supersaturation at a microphysical time step due to diffusional growth and advection can be calculated using the following equations:

$$\frac{dS_w}{dt} = -P_1 S_w - P_2 S_i + \left(\frac{\partial S_w}{\partial t}\right)_{dyn},$$
$$\frac{dS_i}{dt} = -R_1 S_w - R_2 S_i + \left(\frac{\partial S_i}{\partial t}\right)_{dyn} \quad (6.3.43)$$

Assuming that time steps are small enough, the coefficients in Equation (6.3.43) can be considered as constants within a single time step. To solve the equations for diffusional growth, we need to know the integrals of the supersaturations over the microphysical time step. The corresponding expressions for the mixed-phase case are

$$\int_0^t S_w d\tau = \frac{c_{11}}{\gamma}[\exp(\gamma t) - 1] + \frac{c_{21}}{\beta}[\exp(-\beta t) + 1] + \frac{G_{31}}{G_2}t; \quad (6.3.44)$$

$$\int_0^t S_i d\tau = \frac{c_{12}}{\gamma}[\exp(\gamma t) - 1] + \frac{c_{22}}{\beta}[\exp(-\beta t) + 1] + \frac{G_{32}}{G_2}t, \quad (6.3.45)$$

where coefficients c and G are calculated as

$$c_{11} = \alpha^{-1}\left\{(\beta - P_1)S_w(0) - P_2 S_i(0) - \beta\frac{G_{31}}{G_2} + \left(\frac{\partial S_w}{\partial t}\right)_{dyn}\right\},$$
$$c_{12} = \alpha^{-1}\left\{(\gamma + P_1)S_w(0) + P_2 S_i(0) - \gamma\frac{G_{31}}{G_2} - \left(\frac{\partial S_w}{\partial t}\right)_{dyn}\right\},$$
$$c_{21} = \alpha^{-1}\left\{-R_1 S_w(0) + (\beta - R_2)S_i(0) - \beta\frac{G_{32}}{G_2} + \left(\frac{\partial S_i}{\partial t}\right)_{dyn}\right\},$$
$$c_{22} = \alpha^{-1}\left\{R_1 S_w(0) + (\gamma + R_2)S_i(0) - \gamma\frac{G_{32}}{G_2} - \left(\frac{\partial S_i}{\partial t}\right)_{dyn}\right\} \quad (6.3.46)$$

$$G_1 = P_1 + R_2; \; G_2 = P_1 R_2 - P_2 R_1;$$
$$G_{31} = R_2\left[\frac{\delta S_w}{\delta t}\right]_{dyn} - P_2\left[\frac{\delta S_i}{\delta t}\right]_{dyn} \quad (6.3.47)$$
$$G_{32} = P_1\left[\frac{\delta S_i}{\delta t}\right]_{dyn} - R_1\left[\frac{\delta S_w}{\delta t}\right]_{dyn}.$$

The approach remains valid in case the ventilation coefficient is included into the equations for diffusion growth and depositional growth (Section 6.4).

The method described has a disadvantage, since a solution of the two differential equations (6.3.37) requires calculation of determinant $G_2 = P_1 R_2 - P_2 R_1$. In case this determinant is close to zero, special techniques (representation of exponents by the Taylor series) are required to get a correct solution. To avoid this problem, a new method more computationally efficient is currently used in HUCM. This method takes into account the linear relationship between supersaturations over water and ice (Equation 6.3.2). By substituting Equation (6.3.2) into Equation (6.3.37), calculation of supersaturation over water reduces to a closed differential equation of the first order:

$$\frac{dS_w}{dt} = -(P_1 + R_2)S_w + P_2 - R_2 \qquad (6.3.48)$$

This equation has a simple solution:

$$S_w(t) = S_w(t_0)\exp[-(P_1 + R_2)(t - t_0)] + \frac{P_2 - R_2}{P_1 + R_2}\{1 - \exp[-(P_1 + R_2)(t - t_0)]\} \qquad (6.3.49)$$

Expression (6.3.49) contains $P_1 + R_2$ in the denominator of the second term on the right-hand side. In some cases, $P_1 + R_2$ become very small and the value of this term becomes uncertain. In these cases, if $(P_1 + R_2)(t - t_0) \ll 1$ the second exponential term on the right-hand side of (6.3.49) is expanded into the Taylor series yielding the equation that does not contain this uncertainty:

$$S_w(t) = S_w(t_0)\exp[-(P_1 + R_2)(t - t_0)] + (P_2 - R_2)(t - t_0)\left(1 - \frac{(P_1 + R_2)(t - t_0)}{2}\right) \qquad (6.3.50)$$

Then supersaturation over ice is calculated using Equation (6.3.2). Solutions (6.3.49) and (6.3.50) allow analytic calculation of integrals of supersaturation over a microphysical time step and calculation increase/decrease of particles mass using expressions (6.3.41) and (6.3.42).

At $T_C > 0°C$, the ice phase is not stable and the concept of the saturation over ice and corresponding concept of supersaturation over ice lose the sense since equilibrium state does not exist. At the same time, at $T_C > 0°C$ ice particles may fall long distances within the atmosphere. In this case two methods can be applied depending on the sophistication of the melting scheme. According to the simpler method, ice particles are assumed to be covered with a water film and evaporation/condensation is calculated similar to that for water drops. The more sophisticated and more precise method implies calculating the temperature on a particle surface via heat balance of falling particles during melting. Heat and moisture fluxes from the surface of such particles are assumed to be proportional to the temperature and moisture differences between a particle surface and the surrounding air (Section 6.6).

Bulk Parameterization Schemes

In the presence of the liquid phase, the supersaturation adjustment to zero is assumed to be a result of droplet diffusion growth. In most bulk schemes, supersaturation over ice is not changed during a single microphysical time step. In some mixed-phase bulk schemes, the adjustment is performed to water saturation for warm temperatures (above $\sim -5°C$) and to ice saturation for cold temperatures (below $\sim -20°C$), and linear interpolation is used for intermediate temperatures (Grabowski, 2015). The first approach implements empirical formulas for deposition rates. For instance, for deposition growth of ice crystals Reisner et al. (1998) used the formula

$$\left(\frac{\partial q_{iw}}{\partial t}\right)_{dep} = \min\left\{\frac{q - q_{si}}{q_S - q_{si}} a_1 \cdot \langle m_i \rangle^{a_2} N_i/\rho;\ \frac{q - q_{si}}{\Delta t_{microphys}}\right\} \qquad (6.3.51)$$

where q, q_S, and q_{si} are water vapor mixing ratio and saturation mixing ratios over water and ice, respectively; N_i is the ice crystal concentration, $\langle m_i \rangle = \rho q_{iw}/N_i$ is the mean mass of ice crystals, and a_1 and a_2 are the temperature-dependent parameters. Utilization of growth rates calculated using formulas such as Equation (6.3.51) may result in obtaining negative values of supersaturation over ice, which is physically impossible. Therefore, the growth rate is restricted by the limiting condition in Equation (6.3.51) to avoid obtaining negative values of q_{iw}.

The second approach is based on equations for depositional growth and assumes all of the ice particles to be spheres of different diameters d. We illustrate this approach by an example presented by Cotton et al. (1986) for depositional growth of snow (aggregates), obeying the Marshall–Palmer size distribution (Table 2.1.5):

$$N(d) = \frac{N_T}{d_m}\exp(-d/d_m), \qquad (6.3.52)$$

where N_T is the total aggregate concentration and d_m is the characteristic diameter of the aggregate population. It is assumed that $d_m = 0.33$ cm. The aggregate density is taken in the form suggested by Passarelli and Srivastava (1978):

$$\rho_{snow} = \beta_1 d^{-0.6}, \qquad (6.3.53)$$

where $\beta_1 = 0.015\ \mathrm{g\,m^{-2.4}}$. The change of aggregate mass due to vapor deposition is written as

$$\left(\frac{\partial q_{snow}}{\partial t}\right)_{dep} = \int_0^\infty \frac{1}{\rho}\frac{dm_{snow}}{dt} N(d)d(d), \qquad (6.3.54)$$

where $\frac{dm_{snow}}{dt}$ is given by Equation (6.3.9). The electrical capacitance of spherical particles $C_i = \frac{d}{2}$. Substituting

Equations (6.3.9), (6.3.52), and (6.3.53) into Equation (6.3.54) and taking into account the ventilation coefficient $F_v(Re)$, one gets

$$\left(\frac{\partial q_{snow}}{\delta t}\right)_{dep} = 4.03 \cdot q_{snow} d_m^{-1.4} \beta_1^{-1} S_i \left(\frac{\rho_w R_v T}{e_i D} + \frac{\rho_w L_i^2}{R_v T^2 k_a}\right)^{-1} \times F_v(Re) \quad (6.3.55)$$

Walko et al. (1995) and Meyers et al. (1997) used the general Gamma distribution (Table 2.1.4) instead of Marshall–Palmer distribution (6.3.52) for all types of ice particles.

Dealing with sublimation in bulk parameterization schemes causes a certain problem, since the smallest ice hydrometeors disappear completely into vapor during a single time step. In two-moment schemes such as that used in RAMS, the parameterization suggested by Harrington et al. (1995) is used to decrease the number concentration resulting from evaporation/sublimation. The relative decrease in concentration $N_f = \frac{\Delta N_i}{N_i}$ depends on the decrease in the relative mixing ratio $R_f = \frac{\Delta q_{iw}}{q_{iw}}$ and on the shape distribution parameter v, as well as on the type of the ice hydrometeor (on the crystal habit) as $N_f = F_e(R_f, v)$ (Meyers et al., 1997). Here q_{iw} and N_i are mass mixing ratio and the total number concentration of a given species and Δq_{iw} and ΔN_i are the changes in these quantities due to evaporation/sublimation. Formulation of the analytical function F_e is nontrivial, therefore Meyers et al. (1997) calculated the rates of the relative decrease in the number concentration using a bin representation of the distribution function. The rates are tabulated for *a priori* prescribed parameters of size distribution.

As was mentioned, the electrical capacitance of nonspherical crystals may be substantially higher than that of spheres. In most bulk parameterization schemes, the shapes of ice hydrometeors are assumed spherical, so the effect of nonsphericity on the depositional growth is neglected. No distinction is made among ice crystal types, so bulk schemes typically consider only one type of ice crystals called cloud ice or pristine ice crystals. An attempt to take into account the effect of differential growth of ice crystals of different shapes within the bulk-parameterization approach was made by Meyers et al. (1997). The authors adjusted the growth rate of ice particles, assuming that the type of ice crystals was determined by the environmental temperature. A method to take into account transformation of ice particle shapes in bulk parameterization schemes (in particular for simulation of mixed-phase stratiform clouds) was proposed by Sulia et al. (2011), Harrington et al. (2013), and Harrington and Sulia (2013).

6.4 Motion of Ice Particles

6.4.1 The Complexity of Ice Particle Motion

Motion of ice particles is in many aspects much more complicated than that of drops.

- Ice particles usually have complicated nonspherical shapes. The habit diagrams described in Section 2.4 contain several tens of basic ice crystal forms. Other types of ice particles (aggregates, graupel, and hail) also often have nonspherical shapes (Section 2.4, Wang, 2002). The motion of nonspherical particles is a complicated combination of translation and 3D rotation in space.
- Two ice particles having the same mass and belonging to the same type may have different aspect ratios influencing their motion. The dependences of the aspect ratio of crystals on the mean volume radius, shown in Section 6.1, Figure 6.1.5, are averaged dependences with high dispersion.
- The wide range of Reynolds numbers *Re* characterizing the motion of ice particles determines the qualitative differences in their motion. In calm air, small ice particles with *Re* smaller than forty fall down in a steady manner with larger cross sections oriented horizontally. The flow field around these particles is steady and does not change with time. The fall of larger ice particles is nonsteady, as they can tumble or rotate. For example, large plates may fall in a zigzag manner while ice columns rotate. A nonsteady motion is reflected in the complicated flow pattern behind a falling particle. An example of a computer simulation of a nonsteady fall of a short cylinder is illustrated in **Figure 6.4.1.**
- The bulk density of ice particles belonging to the same type can change within a wide range. The range is especially wide for snowflakes (Figure 6.1.6). The variability of the densities contributes to dispersion of particle terminal velocities as well as to other effects related to the particle inertia.
- Snow aggregates and low-density graupel are porous particles. The porosity creates a flow of air through a moving particle, increasing its sedimentation velocity and decreasing the effective drag coefficient. The porosity also affects the collection rate. Some porosity effects are taken into account in cloud models (Hashino, 2007).
- Ice particles often contain some fraction of liquid water that can appear as a result of melting. Ice particles can also get liquid water via their collisions with super-cooled water drops. These

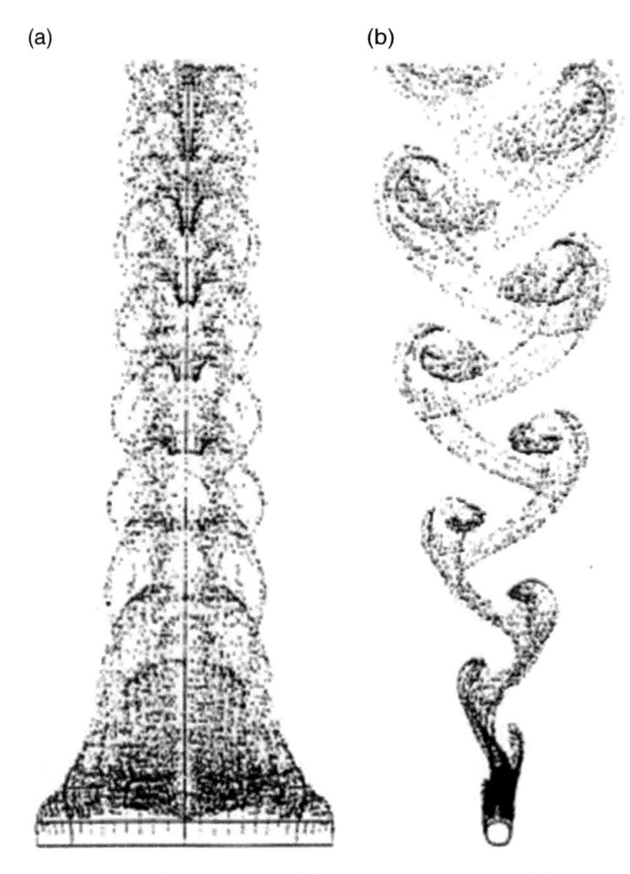

Figure 6.4.1 Computed particle streaks in a perturbed flow past the short cylinder at $Re = 70$. (a) Broadside view; (b) end view (from Wang, 2002; courtesy of Elsevier).

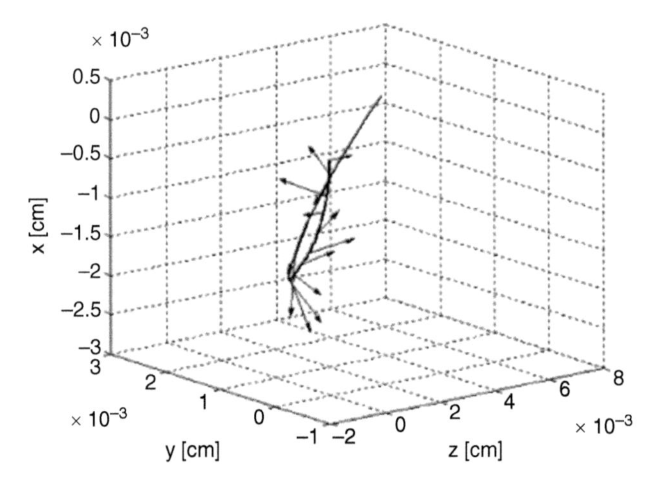

Figure 6.4.2 3D trajectories of an inertial prolate symmetric spheroid with a large semiaxis $a = 25$ μm and the aspect ratio of $\gamma = 24$ in a random realization of a turbulent strain tensor. Arrows show the orientation of the spheroid. Gravity is directed downward along x-axis. $t = 0.15$ s (from Pinsky et al., 2004; courtesy of Elsevier).

collisions are not accompanied by freezing of all the accreted water as, for instance, in the course of wet growth of hail. Depending on the ice particle porosity, water can be soaked inside the particle like in wet graupel, or form a water film on the particle surface. Since water within a particle affects its shape and density, the existence of water impacts the particle's motion. Falling of large ice particles covered with a water film can be accompanied by shedding of the skin that affects the fall velocity.

- In a turbulent medium, the motion of ice particles becomes even more complicated. 3D-rotation motions, permanently changing the orientation of particles, precession, etc., characterize this motion. An example of a spheroidal particle trajectory in a 3D shear flow is shown in **Figure 6.4.2.**

Similar to the case of drops, two main forces affect an ice particle motion, namely, the gravity force and the drag force, so the equation for the translation motion of an ice particle can be written in the form close Equation (5.5.1). The drag force in the case of a non-spherical particle is proportional to the geometrical cross-section S that is perpendicular to the relative velocity between the particle and the surrounding air. The proportionality factor is the drag coefficient $C_d(Re)$, which is a function of the Reynolds number Re (Section 5.5). The Reynolds number Re associated with the motion of a nonspherical particle is defined as $Re = \frac{L\rho}{\mu}|\mathbf{V} - \mathbf{u}|$, where L is the characteristic size of the particle in the direction perpendicular to the relative velocity $\mathbf{V} - \mathbf{u}$. For example, Wang (2002) determined the characteristic size of falling particles as $L = a$ **(Figure 6.4.3).** Sometimes characteristic size is defined as $2a$.

Different methods are used to analyze the motion of nonspherical ice particles. For small particles characterized by small Re that have simple forms such as a circular cylinder, a spheroid, and a disk, there are analytical solutions presented in Pruppacher and Klett (1997). The motion of particles with Re exceeding 40–100 becomes unstable, and to analyze it the numerical solution of the non-stationary Navier-Stokes equation with appropriate boundary conditions is used. The cases of a circular cylinder, a hexagonal plate and a broad-branch crystal are described by Wang (2002). An example of a calculated flow pattern for a nonstationary falling cylinder is shown in Figure 6.4.1. Investigations of falling particles of different shapes are also conducted using laboratory measurements in wind tunnels.

These measurements allow us to determine dependences within a wide range of *Re*. The results of such measurements and their comparison with numerical calculation are shown in **Figure 6.4.4**.

The motion of nonspherical ice particles in a turbulent shear flow is described by two coupled nonlinear equations, namely, the translation equation and the rotation equation. For example, the kinematic equations for a prolate spheroid moving in a 3D shear flow can be written as (Broday et al., 1998)

$$m\frac{d\mathbf{V}}{dt} = -\mu\mathbf{R}(\mathbf{V}-\mathbf{u}) + m\left(1-\frac{\rho}{\rho_w}\right)\mathbf{g} \tag{6.4.1}$$

$$\mathbf{I}\frac{d\boldsymbol{\omega}}{dt} - \mathbf{I}\boldsymbol{\omega}\times\boldsymbol{\omega} = \mathbf{Q}(\mathfrak{R}\times\mathbf{u}) - \boldsymbol{\Omega}\cdot\boldsymbol{\omega} \tag{6.4.2}$$

Here, μ is the dynamic viscosity, $\mathbf{u}$ and $\mathbf{V}$ are the fluid velocity, and the velocity of the particle's center of mass, respectively; $\boldsymbol{\omega}$ is an angular velocity vector; $\mathbf{R}$, $\mathbf{Q}$, and $\boldsymbol{\Omega}$ are the particle's hydrodynamic resistance tensors of the translational motion and the rotational motion in still air; m is the particle's mass and $\mathbf{I}$ is the particle's inertia tensor. Equation (6.4.1) expresses the second Newton law, according to which particle acceleration is determined by the sum of two forces: the gravity force (the second term on the right-hand side of Equation (6.4.1)) and the drag force (the first term on the right-hand side of Equation (6.4.1)). Expressions for the hydrodynamic resistance tensors and the spheroidal rotor-like differential operator $\mathfrak{R}$ are given in some studies (e.g., in Happel and Brenner, 1983). The resistance tensor $\mathbf{R}$ of a spheroid depends on its orientation, therefore Equations (6.4.1) and (6.4.2) are related to each other. Equation (6.4.2) was used to analyze the orientation change of small spheroids (Broday et al., 1998; Gavze et al., 2012, 2016). Knowledge on changes in a particle's orientation is highly important for a correct formulation of the collision equation for ice particles.

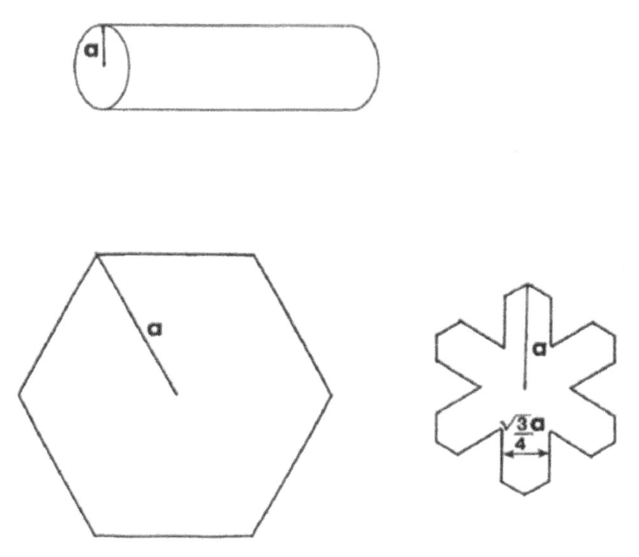

Figure 6.4.3 Three types of ice crystals and their characteristic geometric dimensions (from Wang, 2002; courtesy of Elsevier).

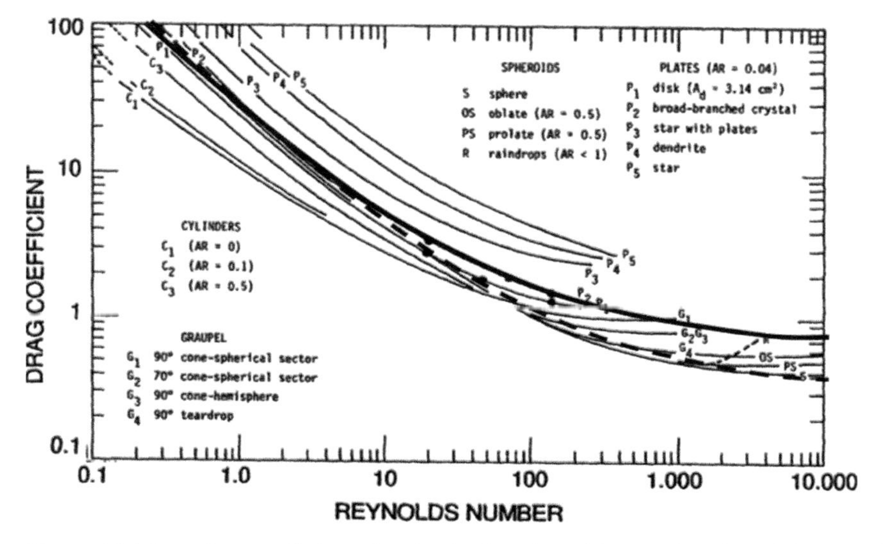

Figure 6.4.4 Drag-force coefficient of variously shaped particles as a function of the Reynolds number based on laboratory studies (from Beard, 1980, with changes). Thick solid and dashed lines are the theoretical curves calculated using Equation (6.4.7) with coefficients $C_0 = 0.292$ and $\delta_0 \approx 9.06$ (Abraham, 1970) and $C_0 = 0.6$ and $\delta_0 = 5.83$ (Mitchel, 1996).

The entire diversity of ice particle motions is too vast to be completely taken into account in cloud models. Only several average characteristics of ice particles movement are used to simulate sedimentation, ice collisions, and other processes related to ice particles' motion. The most important characteristic is the terminal fall velocity and its dependence on the particle's mass and geometrical characteristics.

6.4.2 Terminal Fall Velocity of Ice Particles

The terminal velocity is a major characteristic of an ice particle. It determines the residence time of particles within clouds and affects their deposition/sublimation rates. In most cloud models, the terminal fall velocity is a basic parameter for calculation of rates of collisions between ice particles as well as between ice particles and water drops, collisional breakup, secondary ice formation, and other processes. The terminal velocity affects the production of precipitating particles and, therefore, the rate and the amount of ice precipitation. Liquid precipitation also depends on the ice flux because in mixed-phase clouds liquid precipitation is a result of the melting of falling ice particles.

Since ice particles, even those belonging to the same hydrometeor type, have substantially different shapes, their terminal velocity values vary within a wide range. **Figure 6.4.5** shows the dependences of the terminal velocity of ice crystals on their sizes. One can see high variability of fall velocities of ice crystals of different types. Plates fall two times faster than stellar crystals. The terminal fall velocity of ice crystals is typically smaller than that of water drops of the same size (see Figure 5.5.1). This can be explained by the fact that the bulk density of ice crystals is lower than water density (1,000 kg/m^3), and larger cross sections of ice crystals determines a higher drag force.

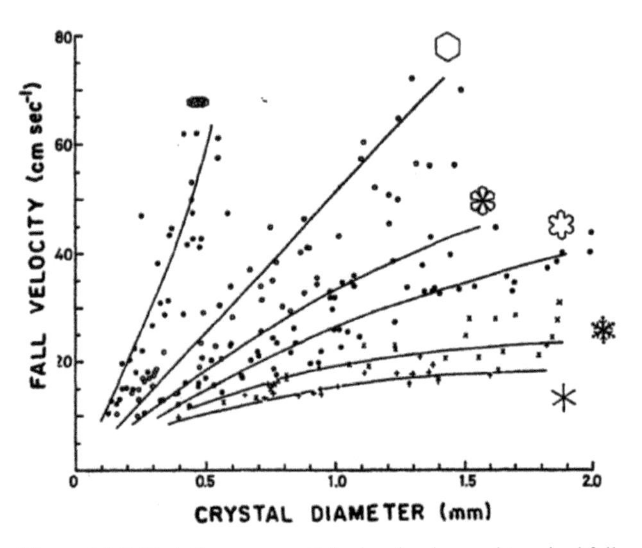

Figure 6.4.5 Variation over particle sizes in observed terminal fall velocities (symbols) and velocities computed from the drag data (lines) for ice crystals of various shapes at $T_C = -10°\text{C}$ and $p =$ 1,000 mb (from Kajikawa, 1972; courtesy of JMS).

It is often assumed that the drag force is proportional to squared fall velocity of a particles: $F_d = \frac{1}{2}\rho V_g^2 S C_d$, where C_d is the drag coefficient and S is the particle's area projection normal to the flow (here, projection in the horizontal plane). Knowing the dependence of the drag coefficient on the Reynolds number $C_d(Re) = C_d\left(\frac{r\rho}{\mu} V_g\right)$ the terminal velocity can be evaluated from the transcendental equation

$$V_g = \left[\frac{2mg}{\rho S C_d(Re)}\right]^{1/2}, \tag{6.4.3}$$

where ρ and μ are the density and the viscosity of the air, respectively, r is the linear characteristic size (the equivalent radius) of the particle in the horizontal plane, and m is the mass of the particle. To calculate the fall velocity using Equation (6.4.3), it is necessary to evaluate the particle's projection area S as well as the $C_d - Re$ relationship.

The basic theory of calculating the fall velocity was developed by Abraham (1970) (see also Böhm, 1992a). When a particle moves within a viscous fluid, the fluid creates a BL of depth δ around the particle. Outside of the BL, friction force is negligible. The depth δ determines a dividing line between potential frictionless flow outside the BL and rotational flow within the BL. In this case, the effective particle's projected area is defined as

$$S_{eff} = \pi(r+\delta)^2 = \pi r^2\left(1+\frac{\delta}{r}\right)^2, \tag{6.4.4}$$

where r is the radius of a sphere having the same projection area as the particle under consideration. The drag force F_d can now be written as

$$F_d = \frac{1}{2}\rho V_g^2 S_{eff} C_0, \tag{6.4.5}$$

where C_0 is the limiting value of the drag coefficient appropriate to describe the drag force affecting the particle with projection area S_{eff}. According to the BL theory, the ratio of the BL depth δ to the particle's characteristic radius r is proportional to $Re^{-1/2}$ (Tomotika, 1935; Batchelor, 1967; Abraham, 1970). That is,

$$\frac{\delta}{r} = \delta_0 Re^{-1/2} \tag{6.4.6}$$

Equations (6.4.3–6.4.6) lead to the following relationship between the drag coefficient and the Reynolds number:

$$C_d = C_0\left(1+\frac{\delta}{r}\right)^2 = C_0\left(1+\frac{\delta_0}{\sqrt{Re}}\right)^2 \tag{6.4.7}$$

For smooth rigid spheres with $Re < 10^4$, $C_0 = 0.292$, and $\delta_0 \approx 9.06$ (Abraham, 1970). The $C_d - Re$ relationship (6.4.7) is suitable for particles of different shapes (Mitchell, 1996). (See the thick black lines in Figure 6.4.4 plotted using formula (6.4.7).

The $C_d - Re$ relationships are often calculated using the Best number (or the Davis number), defined as

$$X = Re^2 C_d(Re) = \frac{8mg\rho}{\pi\mu^2} \tag{6.4.8}$$

The advantage of using the Best number is that for spherical particles it depends only on the particle mass and the environmental air parameters and, therefore, can be easily calculated if the particle mass is known. From Equation (6.4.7) and the expression for the Best number one can get the $Re - X$ relationship in the form proposed by Böhm (1989, 1992a):

$$Re = \frac{\delta_0^2}{4}\left[\left(1+\frac{4\sqrt{X}}{\delta_0^2\sqrt{C_0}}\right)^{1/2} - 1\right]^2 \tag{6.4.9}$$

The $Re - X$ relationship (6.4.9) indicates an excellent agreement with experimental results for rigid spheres and considered as a universal expression for parameterization of the terminal fall velocity of particles of different types and shapes (Böhm, 1989; 1992a; Mitchell et al., 1996; Heymsfield and Westbrook, 2010).

Figure 6.4.6 shows the $Re - X$ relationship derived by Mitchell (1996) from observations of hydrometeors of different types. In Figure 6.4.6, $S_r = \frac{4S}{\pi D^2}$ is the ratio of the particle's projected area S to the area of a circumscribing circle with diameter D, which is the maximum dimension of the particle's projection normal to the direction of fall. One can see that dispersion of the $Re - X$ dependencies for different ice particles is much lower than the dispersion of $C_d - Re$ relationships (Figure 6.4.4). It is another advantage of applying the $Re - X$ relationships in calculation of the fall velocities.

Despite the fact that Equation (6.4.9) is universal and is suitable for any particle shape, the values of δ_0 and C_0 are dependent on the hydrometeor type. Böhm (1989, 1992a) attributes this dependence to the surface roughness effects. Based on estimations of the drag coefficient for hailstones, Böhm (1989) found that $C_0 = 0.6$ and $\delta_0 = 5.83$. Figure 6.4.6 shows that the $Re - X$

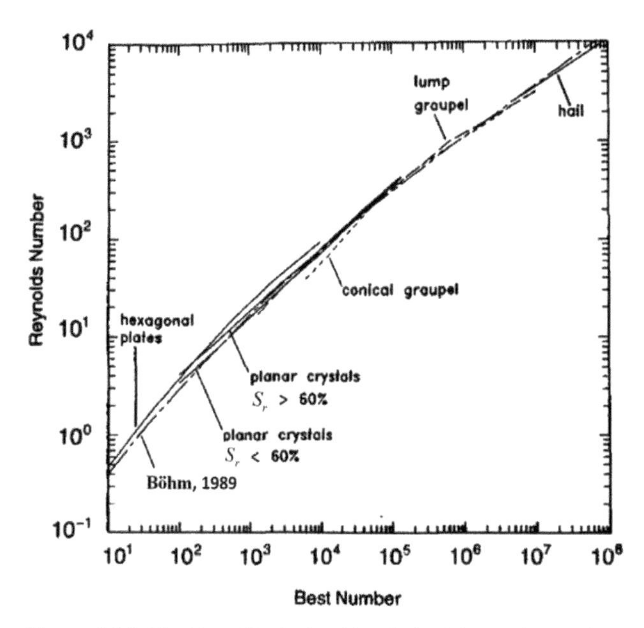

Figure 6.4.6 The $Re - X$ relationships derived from observations of hydrometeors of different types. The $Re - X$ relationship (6.4.9) with $C_0 = 0.6$ and $\delta_0 = 5.83$ is denoted by a dashed-dotted line (from Mitchell, 1996; © American Meteorological Society; used with permission).

relationship (6.4.9) for these values of δ_0 and C_0 is in a very good agreement with observed data for $Re < 10^4$. At $Re > 10^3$, the BL around a falling particle becomes turbulent and the drag coefficient increases. Böhm (1992a) took the turbulent effects into account by increasing C_0 by a factor of 1.6. Knight and Heymsfield (1983) showed that a factor of 1.3 provides even better agreement with the experimental data. The curve of the $Re - X$ relationship is plotted in Figure 6.4.6 only for laminar BL, since the turbulent correction cannot be distinguished in the logarithmic scale used in the figure.

The $Re - X$ relationship (6.4.9) corresponds to a variable drag coefficient depending on Re $C_d(Re)$. Based on several observational studies showing a linear $Re - X$ relationship in the log log space, Mitchell (1996) represented the $Re - X$ relationship in the power-law form:

$$Re = aX^b \tag{6.4.10}$$

Equation (6.4.10) enables us to express the terminal fall velocity as (Mitchell, 1996)

$$V_g = a\frac{\nu}{D}\left(\frac{2mgD^2}{\rho\nu^2 S}\right)^b, \tag{6.4.11}$$

where D is equivalent particle diameter and m is particle mass. This well known power law relationship describes the dependence of the fall velocity on the particle's geometrical sizes. In particular, at small Re when $b = 1$, Equation (6.4.11) reduces to the Stokes regime, where V_g is proportional to squared particle radius (Equation 5.5.6).

Mitchell (1996) approximated the $Re - X$ relationship (6.4.9) by four piecewise linear dependencies in the log–log space covering a wide range of the Reynolds numbers $5 \times 10^{-4} \leq Re \leq 10^4$. These $Re - X$ relationships are presented in **Table 6.4.1.** Rasmussen and Heymsfield (1987) also divided the entire range of the Re into four subranges and presented the $Re - X$ relationships for hydrometeors (mostly hail and graupel), as shown in **Table 6.4.2. Figure 6.4.7** shows dependencies of the fall velocity of ice crystals of different types on their maximum dimension, both derived from observations and calculated using Equation (6.4.9) and the $Re - X$ relationships presented in Table 6.4.1. The dependencies completely coincide, indicating a universality of Equation (6.4.9). A good agreement with observed dependencies is obvious.

There are several studies further developing the approach proposed by Mitchell (1996). Khvorostyanov and Curry (2002) approximated Equation (6.4.9) by a power law with continuous variable coefficients. Mitchell and Heymsfield (2005) and Khvorostyanov and Curry (2005b) included a correction of the drag coefficient, taking into account the turbulence effect at high Re. These results do not differ from those obtained by Mitchell (1996) over the range of $1 < Re < 10^3$. Detailed comparison between the calculated and the observational data shows that Equation (6.4.9) containing the coefficients used by Mitchell (1996) yields excellent results for comparatively large S_r (thin circular discs, hexagonal plates, branched crystals) (Heymsfield and Westbrook, 2010). At the same time, applying Equation (6.4.9) leads to overestimation of V_g for more tenuous and more complex crystals at small Re. The overestimation can reach 40–100% for low Re and S_r below ~30%. Heymsfield and Westbrook (2010) derived the $C_d - Re$ relationships for nonspherical particles such as crystals, aggregates, and graupel, combining the results of available laboratory measurements and in situ obtained data (**Figure 6.4.8**). One can see that Mitchell (1996), Khvorostyanov and Curry (2005b) and Mitchell and Heymsfield (2005) obtained excellent results within a wide range of Re, but the drag coefficients were underestimated (i.e., the fall velocity is overestimated) at very low Re.

To achieve a better approximation of the observed data for different ice particle shapes, Heymsfield and Westbrook (2010) improved Mitchell's (1996) approach. They used a modified drag coefficient $C_d^* = C_d S_r^{0.5}$ and the modified Best number $X^* = Re^2 C_d^* = \frac{8mg\rho}{\pi S_r^{0.5}\mu}$, where S_r is, as earlier, the ratio of the particle's projected area S to the area of a circumscribing circle with diameter D. The Reynolds number is calculated using Equation (6.4.9), where X^* is used instead of X and $C_0 = 0.35$ and $\delta_0 \approx 8.0$. According to the data obtained for deep ice clouds during the Tropical Rainfall Measuring Mission (TRMM) Kwajalein field program, the relationship between the area ratio and the particle diameter is $S_r = 0.29D^{-0.18}$ (Heymsfield et al., 2002), and in the deep cirrus layers $S_r = 0.18D^{-0.17}$ (Heymsfield and

Table 6.4.1 The $Re - X$ relationships according to Mitchell (1996); $W = \log_{10}X$.

Range of the Re	Range of the Best Numbers	Function $\log_{10} Re = f(W)$
1	$0.01 < X \leq 10$	$-1.3571 + 0.970W$
2	$10 < X \leq 585$	$-0.2183 + 0.831W$
3	$585 < X \leq 1.56 \times 10^5$	$-0.6836 + 0.638W$
4	$1.56 \times 10^5 < X < 10^8$	$0.036 + 0.499W$

Table 6.4.2 The $Re - X$ relationships for hydrometeors with different Re; $W = \log_{10}X$ (from Rasmussen and Heymsfield, 1987; © American Meteorological Society; used with permission).

Range of the Re	Range of the Best Numbers	Function $\log_{10} Re = f(W)$
$1 < Re \leq 12.2$	$73 < X < 562$	$1.7095 + 1.33438W - 0.11591W^2$
$12.2 < Re \leq 30$	$562 < X < 1.83 \times 10^3$	$-1.81391 + 1.34671W - 0.12427W^2 + 0.0063W^3$
$30 < Re \leq 2.4 \times 10^4$	$1.83 \times 10^3 < X < 3.46 \times 10^8$	$-0.348 + 0.5536W$
$2.4 \times 10^4 < Re \leq 3 \times 10^5$	$3.46 \times 10^8 < X < 5.4 \times 10^{10}$	$0.1109 + 0.5W$

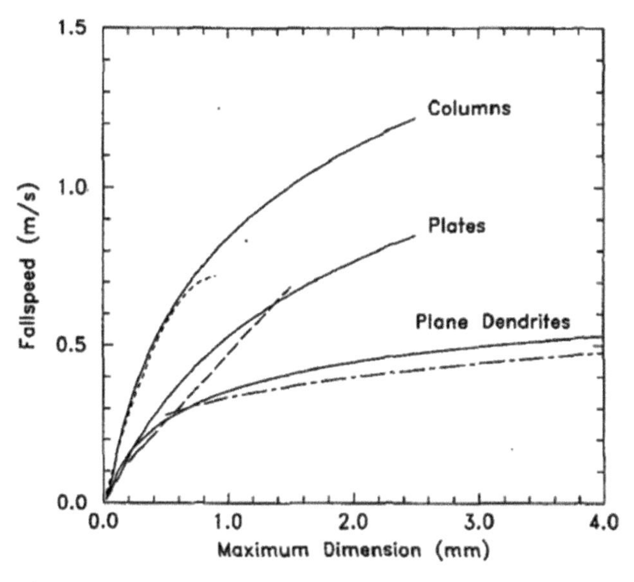

Figure 6.4.7 Dependencies of the fall velocity of ice crystals of different types on crystal's maximum dimensions, derived from observations (dashed lines) and calculated using Equation (6.4.9) and the $Re - X$ relationships presented in Table 6.4.1 (solid lines). The dependencies calculated using Equation (6.4.9) and Table 6.4.1 coincide. Empirically obtained expressions are plotted over the actual range of measurements (from Mitchell, 1996; © American Meteorological Society; used with permission).

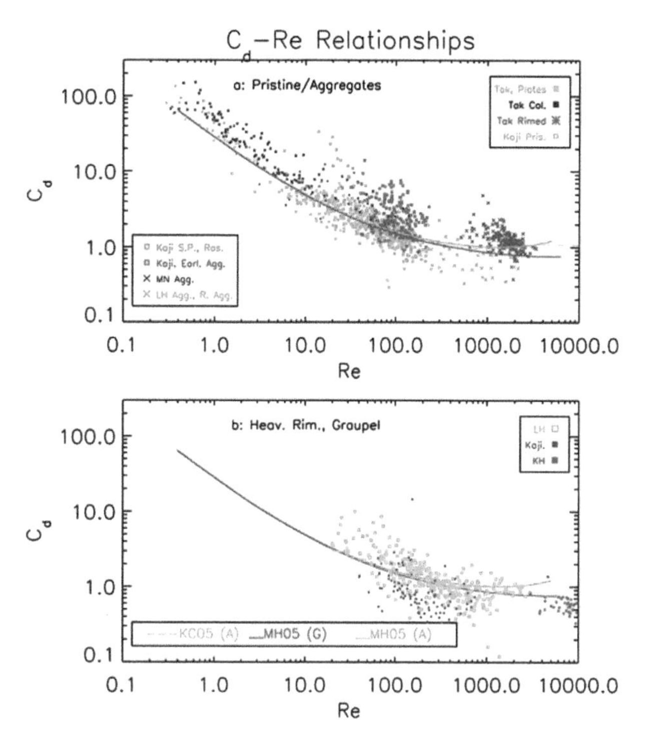

Figure 6.4.8 The $C_d - Re$ relationships derived from different measurements. (a) Pristine and lightly rimed particles; (b) heavily rimed particles, aggregates, and graupel. Dots denote different measurements data: Tak (Takahashi and Fukuta, 1988) and Takahashi et al. (1991); Kaji (Kajikawa, 1982); LH (Locatelli and Hobbs, 1974); and KH (Knight and Heymsfield, 1983). Curves: KC05 (Khvorostyanov and Curry, 2005) and MH05 (Mitchell and Heymsfield, 2005b) (from Heymsfield and Westbrook, 2010; © American Meteorological Society; used with permission).

Miloshevich, 2003). This approach allowed reducing the errors in the reproduction of the terminal velocity values down to less than 25%.

Utilization of the $Re - X$ relationship (6.4.9), and parameterization applied by Mitchell (1996) (Table 6.4.1) with corrections proposed by Heymsfield and Westbrook (2010), allows precise calculation of the drag coefficients using Equation (6.4.8) and of the fall velocity of particles of any type and shape using Equation (6.4.3).

6.4.3 Approximating the Terminal Velocity Relationships

Another approach to determination of the terminal fall velocities is based on empirical or semiempirical relationships between V_g and the mass or different geometrical characteristics of a particle. The dependencies of the terminal velocity on ice particle size or mass are usually described by power laws. Equation (6.4.11) can serve as the physical basis of this law. The advantage of the power law approximations is that the corresponding formulas for V_g can be easily used in bulk-parameterization schemes where particle size distributions are assumed to be exponential or the Gamma distributions. The empirical dependencies are usually approximated by the general expressions for V_g in the forms

$$V_g = Ad^B \tag{6.4.12a}$$

$$V_g = \alpha m^\beta \tag{6.4.12b}$$

In Equation (6.4.12a), d is usually assumed to be the mean diameter of an ice particle. Sometimes, instead of the mean diameter the largest dimension of the particle is used. The coefficients A, B, α, and β are assumed different for different types of hydrometeors. The most complete list of these parameterizations and conditions of their applicability was presented by Pruppacher and Klett (1997). **Tables 6.4.3–6.4.5** show the dependences (6.4.12a) for different types of hydrometeors. The dependence similar to Equation (6.4.12a) was obtained for hailstones:

Table 6.4.3 The terminal fall velocity-particle diameter relationships for unrimed planar snow crystals of various types; d in cm (from Heymsfield and Kajikawa, 1987; © American Meteorological Society; used with permission).

Crystal Type	V_g, cm s^{-1}	Validity Range, mm
Hexagonal plates (Pla)	$297d^{0.86}$	0.3–1.5
Broad branches (Plb)	$190d^{0.81}$	0.4–1.6
Dendrites (Plc)	$103d^{0.62}$	0.5–2.8
Dendrites (Pld)	$58d^{0.55}$	0.4–2.4
Dendrites (Ple)	$55d^{0.48}$	0.6–5.3

Table 6.4.4 The terminal fall velocity–particle mass relationships for columnar crystals and crystal aggregates observed on Cascade Mts. (750–1,500 m, Washington) (from Pruppacher and Klett, 1997 and Locatelly and Hobbs, 1974; courtesy of © John Wiley & Sons, Inc.).

Crystal Aggregates	V_g, cm s^{-1}	Validity Range, mm
Needle (Nla), mass in g	$155m^{0.271}$	
Hollow columns (Clf), mass in g	$253m^{0.271}$	
Aggregates of dendrites, mass in mg	$1.1m^{0.08}$	2–10
Aggregates of densely rimed dendrites, mass in mg	$1.3m^{0.15}$	2–12
Aggregates of unrimed plates, bullets and columns, mass in mg	$1.2m^{0.07}$	0.2–3.0
Aggregates of unrimed side plates, mass in mg	$1.2m^{0.14}$	0.5–4.0

$$V_g = 12.43 d_{max}^{0.5}, \tag{6.4.13}$$

where $d_{\max}$ is the maximum particle size in cm and V_g is in m/s. This formula shows that the terminal velocity of large hailstones is very high (of order of 30 m/s). Formation of such hailstones is typically associated with strong updrafts in cumulus clouds (thunderstorms, super-cell storms). **Figures 6.4.9a–c** show the dependences presented in Tables 6.4.3–6.4.5.

The relationships shown in Tables 6.4.3–6.4.5 and in Figure 6.4.9 can be directly applied in SBM cloud models together with the mass–size relationships described in Section 6.1. As an example, **Figure 6.4.10** shows the dependencies of the basic fall velocities on ice

Table 6.4.5 The terminal fall velocity-particle diameter relations for rimed crystals and graupel particles of various types; d in cm (from Heymsfield and Kajikawa, 1987; © American Meteorological Society; used with permission).

Particle Type	V_g, cm s^{-1}	Validity Range, mm
Rimed plates (R1c)	$92d^{0.27}$	0.8–2.7
Rimed stellar (R1d)	$79d^{0.36}$	0.7–5.3
Densely rimed plate (R2a)	$92d^{0.73}$	0.7–2.2
Densely rimed stellar (R2b)	$162d^{0.53}$	1.1–4.7
Rimed spatial branches (R2c)	$75d^{0.24}$	0.3–6.2
Jump graupel (R4b)	$733d^{0.89}$	0.4–9.0
Conical graupel (R4c)	$590d^{0.76}$	0.8–8.6

particle melted radii (radius of a drop obtained from a totally melted ice particle) derived from the empirical power laws and used in HUCM in the forty-three bins version. These velocities are further corrected to take into account rimed mass fractions, liquid water mass fractions, and hail growth mode (wet or dry).

Analysis of observations (Heymsfield et al., 2007a and 2007b; 2010) showed that coefficients A and B in Equation (6.4.12a) depend on temperature. **Table 6.4.6** shows the coefficients in the dependencies $A(T_C) = C_0 \exp(C_1 T_C)$ and $B = C_2 + C_3 T_C$ formulated by Equation (6.4.12a) and the corresponding values of coefficients α and β in Equation (6.1.1) ($m = \alpha D^{\beta}$), as derived from analysis of in situ measurements of ice crystals. These dependences are valid for particles with diameter exceeding 100 μm. The values of coefficients are given for $\beta = 1.75$ and β depending on temperature. The terminal velocity values calculated using coefficients A, B, α, and β, presented in Table 6.4.6, correspond to the pressure level of 1,000 hPa. For other levels, the terminal velocity should be multiplied by factor of $\left(\frac{\rho_0}{\rho}\right)^{0.54}$, where ρ is the air density at a given pressure level. In some bulk parameterization schemes (Ferrier, 1994; Ferrier et al., 1995; Milbrandt and Yau, 2005a, 2005b), the basic formula for the terminal velocity of particles of different types is given as follows:

$$V_{gi}(d) = \left(\frac{\rho_0}{\rho}\right)^{0.5} a_i d^{b_i} \exp(-f_i d) \tag{6.4.14}$$

The parameters of this dependence for different types of hydrometeors are presented in **Table 6.4.7.**

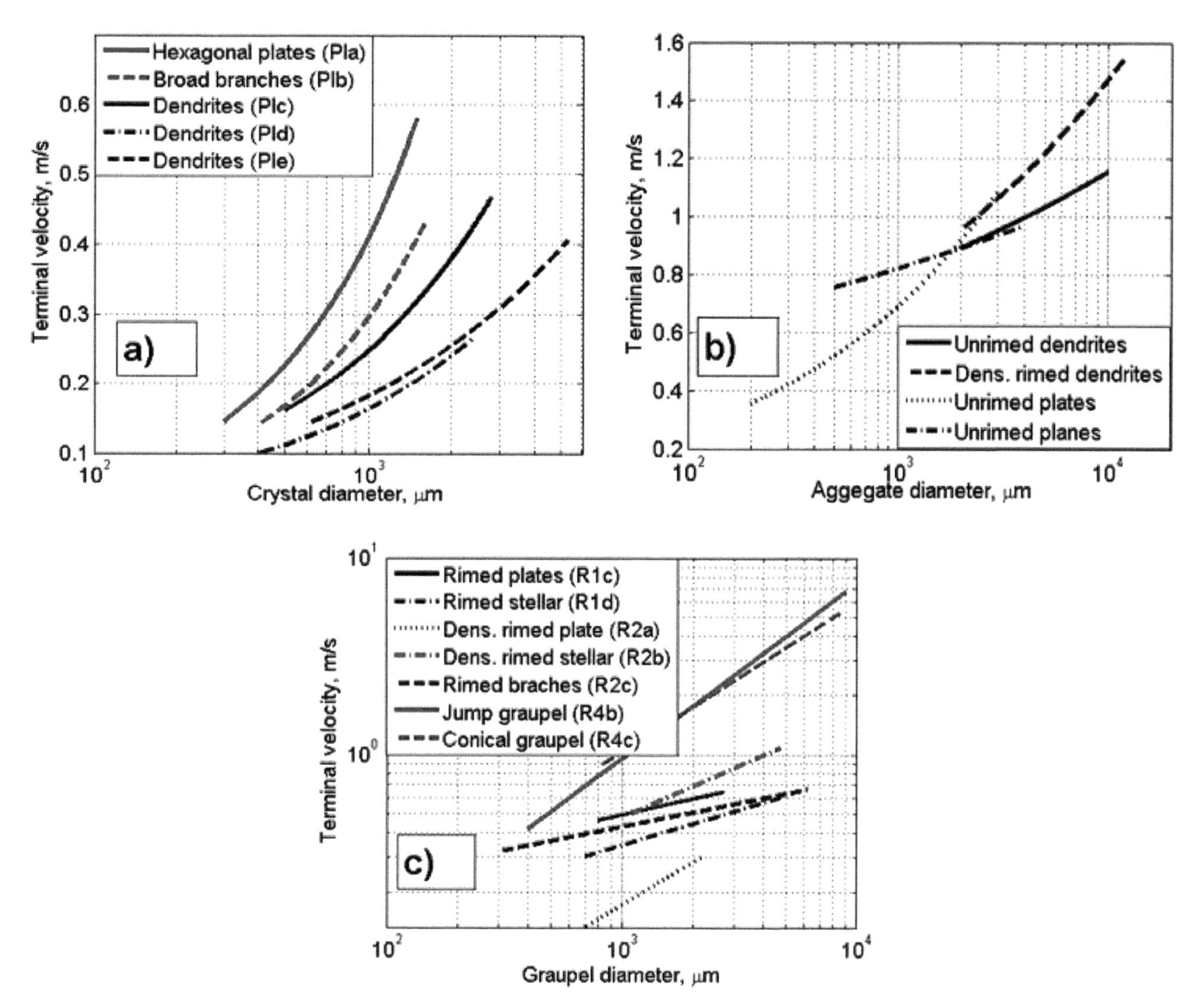

Figure 6.4.9 The relationships between the terminal fall velocity and particle diameter for unrimed planar snow crystals (a), crystal aggregates (b), and graupel (c) according to Tables 6.4.3–6.4.5.

In cloud modeling it is important to know the particle mass-size dependencies and their relationship with the terminal fall velocity. Mitchell (1996) proposed using the area–dimensional power law relationship:

$$S = \gamma D^{\sigma}, \tag{6.4.15}$$

where D is maximum dimension and S is the projected area. Substitution of this relationship and Equation (6.1.1) ($m = \alpha D^{\beta}$) into Equation (6.4.8) leads to the following expression for the Best number:

$$X = \frac{2\alpha g \rho}{\gamma \mu^2} D^{\beta+2-\sigma} \tag{6.4.16}$$

Although this parameterization is quite simple, it contains detailed information on a particle structure. Substituting Equations (6.4.15) and (6.1.1) into Equation (6.4.11), we can express the fall velocity V_g as

$$V_g = av\left(\frac{2\alpha g}{\rho \gamma v^2}\right)^b D^{b(\beta+2-\sigma)-1} \tag{6.4.17}$$

Equation (6.4.17) is similar to Equation (6.4.12a), however, Equation (6.4.17) shows in detail how the fall velocity depends on particle geometry parameters. Thus, geometrical parameters of ice particles can be used for determination of the particle fall velocity. Equation (6.4.17) provides the values that agree with observations both of ice crystals and rimed particles (e.g., rimed crystals and lump graupel) (Mitchell, 1996).

Morrison and Grabowski (2008), Morrison and Milbrandt (2015), and Morrison et al. (2015) calculated

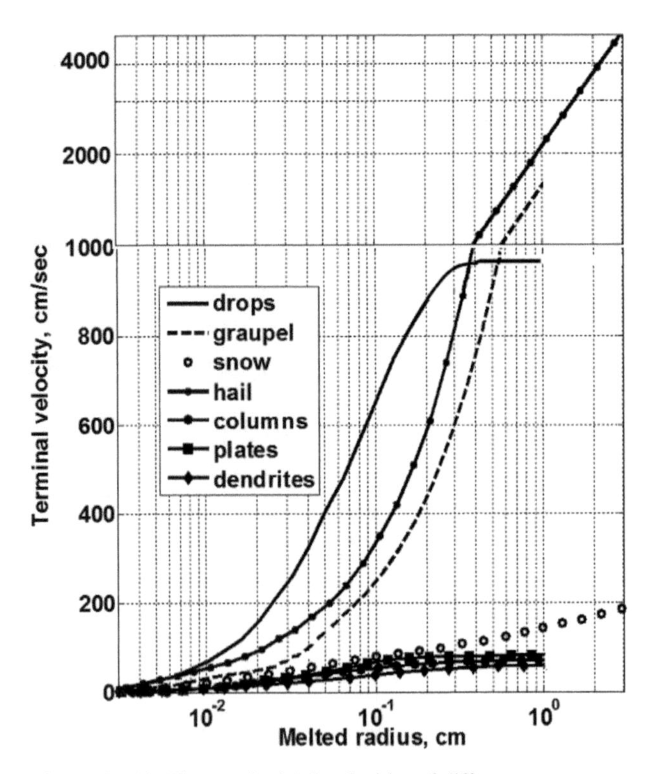

Figure 6.4.10 The terminal fall velocities of different hydrometeors as a function of their melted radii, $p = 1{,}000$ mb. Linear scale is replaced by logarithmic scale at $V_g > 1{,}000$ cm s^{-1} (from Khain et al., 2004, with changes; © American Meteorological Society; used with permission).

the rimed mass of nonspherical ice particles. The fall velocities of these particles are calculated as a function of the bulk rimed mass fraction by a linear interpolation between the value for graupel (the rimed fraction is equal to one) and for unrimed ice (the rimed fraction is equal to zero).

6.4.4 Fall Velocity of Particles Containing a Liquid–Water Fraction

In case hail or graupel particles collect significant mass of water drops, not all of the water freezes even at $T_C <$ 0°C since the particle surface heats due to release of the fusion heat. Also, when ice particles melt, particles containing both ice and liquid water arise. The appearance of liquid water changes a particle's shape and density, affecting its fall velocity.

Graupel bulk density is lower than that of pure ice due to the existence of air inclusions inside graupel. Water soaking into graupel increases its bulk density and decreases its size. This in turn leads to an increase in the fall velocity. After the soaking stage, or if the initial

Table 6.4.6 The values of coefficients in expressions for A and B (Equation 6.4.12a). Coefficient A is in cm/s. T_C is temperature in °C (from Heymsfield et al., 2007b; © American Meteorological Society; used with permission).

	C_0	C_1	C_2	C_3
Synoptic				
$\beta = 1.75$	113	−0.0120	0.127	−0.0102
$\beta = 1.86 + 0.004\ T_c$	131	−0.0138	0.185	−0.0084
Crystal face				
$\beta = 1.75$	182	−0.0040	0.207	−0.0060
$\beta = 1.84 + 0.0029\ T_c$	200	0.0004	0.244	−0.0049

Table 6.4.7 Parameters of Equation (6.4.14) for hydrometeors of different types (from Milbrandt and Yau, 2005b; © American Meteorological Society; used with permission).

Category	a_i, m$^{1-b_i}$ s^{-1}	b_i	f_i, m^{-1}
Rain	4,854.00	1.0	195
Ice	71.34	0.6635	0
Snow	8.996	0.42	0
Graupel	19.30	0.37	0
Hail	206.89	0.6384	0

density of the ice particle is close to that of pure ice, melting leads to formation of a water skin on the particle surface. The skin makes the surface smooth, which should increase the fall velocity. However, the effect of the water skin is more complicated than just smoothing the surface. According to laboratory experiments by Rasmussen et al. (1984a, 1984b), after the ice particle is fully soaked with meltwater, a torus of water near the equator of a solid ice sphere is formed. The torus increases the cross-section area that leads to decreasing fall velocity.

The fall velocity of melting snow is often calculated using some kind of interpolation between the fall velocity of dry snow $V_{gs,dry}$ and that of raindrops of the same mass V_{g_rd} (e.g., Phillips et al., 2007):

$$V_{gs} = V_{gs,dry} + \left(V_{g\_rd} - V_{gs,dry}\right) Y(f_w), \quad (6.4.18)$$

where Y is an empirical function of the liquid–water fraction f_w determined in the laboratory experiments in a wind tunnel (Mitra et al., 1990).

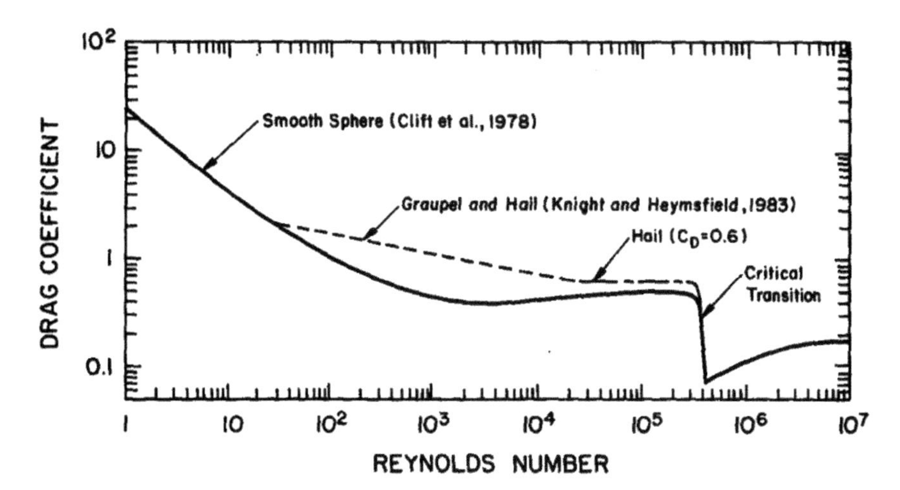

Figure 6.4.11 The $C_d - Re$ relationships for graupel and hail (dashed line) and for a smooth sphere (solid line). Dashed line corresponds to the $Re - X$ relationships given in Table 6.4.6 (from Rasmussen and Heymsfield, 1987; © American Meteorological Society; used with permission).

The $Re - X$ relationships used for particles covered with the water film are often assumed to be the same as in Table 6.4.2. However, to calculate fall velocity of spherical particles covered by water film, the $C_d - Re$ relationships should be taken as for a smooth sphere (**Figure 6.4.11**). Then the fall velocity is calculated using Equation (6.4.3). Laboratory experiments by Rasmussen et al. (1984b) and Rasmussen and Heymsfield (1987) showed that the fall velocity of melted particles depends on the Reynolds number of particles and the stage of melting that determines the roughness and the shape of the particle surface (Section 6.6). **Table 6.4.8** describes the algorithms for calculation of drag coefficient C_d for high-density hailstones and low-density hailstones (graupel) and the terminal velocity of these particles. Typically, once C_d is determined, the terminal velocity is calculated using Equation (6.4.3). In some cases, however, the terminal velocity is calculated from the Re determined in measurements.

Rasmussen and Heymsfield (1987) considered several stages of melting of graupel and hail (Section 6.6). At the first "just wet" stage, the particles' surface becomes wet. The next "soaking" stage takes place only for low-density hailstones (graupel) when the melted water is soaked inside of the particle. The soaking stage ends with the "just soaked" stage when the water film is formed on the particle's surface. At $5 \times 10^3 < Re < 2.5 \times 10^4$, a regime of equilibrium mass of water on the surface is established. The Reynolds number found in the laboratory measurements at this stage can be expressed as $Re = 4800 + 4831.5 m_i$, where m_i is the mass of the ice core. Accordingly, the formulas for the terminal velocities in stage five contain this expression for Re. At the soaking stage, the graupel fall velocity is determined using Equation (6.4.3), where either "dry" or "wet" Reynolds numbers are used. The dry Reynolds number is calculated under the hypothesis that all the water is soaked, so the graupel radius is equal to the radius of its ice part $r_{p\_dry} = \left(\frac{4}{3}\pi\rho_{bulk}\right)^{-1/3} m_i^{1/3}$, where ρ_{bulk} is the bulk density of dry graupel. The wet Reynolds number is determined from the $Re - X$ relationships (Table 6.4.2). The wet Reynolds number is used if the dry Reynolds number is below 4×10^3. For higher values of the dry Re, smoothness of the particle surface is irrelevant, as the Reynolds number and the drag coefficient are assumed to be unchanged, being equivalent to those of the dry unmelted particle.

Phillips et al. (2007) suggested a particle to be at the fully soaked stage if the volume of liquid water becomes larger than the volume of the air spaces of initial dry particle. At this stage the terminal velocity is determined by interpolation between the just-soaked value $V_{g,just\_soak}$ and the hypothetical equilibrium value, $V_{g,eq}$, corresponding to maximum amount of meltwater m_{w_cr}:

$$V_{g,\,fully\_soak} = V_{g,\,just\_soak} + \frac{m_w - m_{w\_soak}}{m_{w\_cr}}\left(V_{g,eq} - V_{g,\,just\_soak}\right), \tag{6.4.19}$$

where the mass of the meltwater varies within the range of $m_{w\_cr} \geq m_w \geq m_{w\_soak}$. m_{w_soak} is the mass of water soaked inside the graupel (see Section 6.6 for more detail). The hypothetical equilibrium value of the

Table 6.4.8 Calculation of the drag coefficient C_d and the terminal velocity of melting high-density hailstones and low-density hailstones (graupel) (from Rasmussen and Heymsfield, 1987, with changes; © American Meteorological Society; used with permission).

Number of Stage	Stage of Melting	Wet High-Density Hailstones	Graupel or Low-Density Hailstones
1	Just wet	$Re > 4{,}000$: C_d is calculated as for dry hailstones. $Re < 4{,}000$: C_d is calculated using the $C_d - Re$ relations for wet surface (smooth sphere) V_g calculated from Equation (6.4.3)	The same as for high-density hailstones
2	Soaking of water	N/A	As at stage(1)
3	Just soaked	N/A	If $\rho_i > 0.8$, $C_d - Re$ relationship is calculated as for dry graupel. If $\rho_i \leq 0.8$, C_d is calculated using the $C_d - Re$ relationship for wet surface (smooth sphere)
4	Transition from "just soaked" to equilibrium mass of water on the surface	The fall velocity is calculated by a linear interpolation between the terminal velocities at stages (1) and (5)	The fall velocity is calculated by a linear interpolation between the terminal velocities at stages (3) and (5)
5	Equilibrium mass of water on the surface	$Re < 5{,}000$: the terminal velocity is equal to that of raindrops; $5 \times 10^3 < Re < 2.5 \times 10^4$: $V_g = 0.15 \times (4{,}800 + 4{,}831.5m_i)/(2r)$, where m_i is the mass of the ice core. At $Re > 2.5 \times 10^4$: the terminal velocity is determined from Equation (6.4.3) using $C_d = 0.6$.	The same as for high-density hailstones, but the sum of f the ice core mass and the soaked, mass (m_s) is used instead of m_i

fall velocity is determined according to Table 6.4.8 (stage five).

6.4.5 The Ventilation Effect

Like in case of diffusional drop growth, the rate of deposition/sublimation of ice particles depends on their velocity relative to the air. This ventilation effect is described by the ventilation coefficient F_v defined by Equation (5.5.10) and showing the increase in the rates of sublimation or deposition of falling particles as compared with those of unmovable particles. The values of ventilation coefficient for ice particles differs from those for drops because of difference in material of surface (water vs. ice) and difference in shapes. The ventilation coefficient is used not only for calculation of rates of deposition/sublimation, but also to estimate the freezing/melting time (Section 6.6), as well as to analyze the scavenging process and to describe liquid water shedding from falling graupel and hail. Similar to the case of drops (see Section 5.5.4), the ventilation coefficient is parameterized using a combination of two numbers, namely, the Reynolds number Re and the Schmidt number Sc ($Sc = \frac{\nu}{D}$, where ν and D are the kinematic viscosity and diffusivity of water vapor, respectively). F_v is represented more precisely as a function of $Z = Sc^{1/3} \times Re^{1/2}$. Parameterization formulas suggested by different authors for different crystal types are presented in **Table 6.4.9**. These formulas are used in cloud models. The ventilation coefficient for graupel and hail, including particles covered with the water skin, is calculated as suggested by Phillips et al. (2007):

$$F_v = \begin{cases} 2\left(1.00 + 0.108Z^2\right) & Z < 1.4 \\ 2(0.78 + 0.308Z) & Z \geq 1.4 \end{cases} \quad \text{for} \quad Re < 250 \tag{6.4.20a}$$

$$F_v = \begin{cases} 1.00 + 0.108Z^2 & Z < 1.4 \\ 0.78 + 0.308Z & Z \geq 1.4 \end{cases} \quad \text{for} \quad 250 < Re < 6{,}000 \tag{6.4.20b}$$

$$F_v = 0.76Z/2 \quad \text{for} \quad 6{,}000 < Re < 20{,}000 \tag{6.4.20c}$$

$$F_v = \left(0.57 + 9 \cdot 10^{-6} Re\right)Z/2 \quad \text{for} \quad Re > 20{,}000 \tag{6.4.20d}$$

Table 6.4.9 Parameterization formulas for the ventilation coefficient for ice crystals of different type.

Any crystal shape (Hall and Pruppacher, 1976)

$$F_v = \begin{cases} 1.00 + 0.14Z^2 & Z < 1.0 \\ 0.86 + 0.28Z & Z \geq 1.0 \end{cases}$$

$Sc = 0.63; Z < 10$

Columnar ice crystal (Wang, 2002)

$$F_v = 1 + \sum_{k=1}^{4} A_k \left(\frac{Z}{4}\right)^k$$

$A_1 = -0.00668; A_2 = 2.39402; A_3 = 0.73409; A_4 = -0.73911$

$0.2 \leq Re \leq 20,\ Sc = 0.63$

Hexagonal plates (Wang, 2002)

$$F_v = 1 + \sum_{k=1}^{4} A_k \left(\frac{Z}{10}\right)^k$$

$A_1 = -0.06042; A_2 = 2.79820; A_3 = -0.31933; A_4 = -0.06247$

$1.0 \leq Re \leq 120,\ Sc = 0.63$

Broad-branch crystals (Wang, 2002)

$$F_v = 1 + \sum_{k=1}^{2} A_k \left(\frac{Z}{10}\right)^k$$

$A_1 = 0.35463; A_2 = 3.55338$

$1.0 \leq Re \leq 120,\ Sc = 0.63$

Oblate ice spheroid (Pitter et al., 1974)

$$F_v = \begin{cases} 1 + 0.142Z^2 + 0.054Z^4 \ln\left(0.893Z^2\right), & Z < 0.71 \\ 0.937 + 0.178Z, & Z \geq 0.71 \end{cases}$$

$1.0 \leq Re \leq 20,\ Sc = 0.71$

The dependencies of $F_v(Re) - 1$ on Re taken from Table 6.4.9 and Equations (6.4.20) are shown in **Figure 6.4.12**. The dependencies, calculated for water drops using Equation (5.5.12), are also shown for comparison.

6.4.6 Numerical Treatment of Particle Sedimentation in Cloud Models

Ice particle sedimentation in cloud models is represented similarly to that of drops (Section 5.5.3). In spectral bin microphysical models, sedimentation is described by equations similar to Equation (5.5.8), but applied to all hydrometeor types. Vertical sedimentation of particles is calculated separately for each bin by solving the advection equation using methods described in Section 4.2. The dependences of the fall velocity on particle mass for different hydrometeor types allow an accurate representation of size sorting, i.e., gravitation-induced separation of particles of different types and sizes in the vertical direction (Figure 5.5.4), which is important for an accurate description of collisions within clouds. Together with particles belonging to different bins, values characterizing their properties (such as the liquid–water mixing ratio, the rimed–water mixing ratio) are sediment using the same fall velocities as for corresponding particles.

In bulk-parameterization schemes, sedimentation of particle concentrations and mixing rations of different hydrometeors is calculated. These quantities are different moments of particle size distribution. The fall velocity of different moments is calculated using equations similar to those for raindrops (Equation 5.5.9), but applied for all hydrometeor types:

$$\frac{\partial M_i^k}{\partial t} = \frac{\partial \overline{V}_{g,i} M_i^k}{\partial z}, \tag{6.4.21}$$

where index i denotes the type of hydrometeor, $N_i = M_i^0 = \int_0^\infty f_i(m)dm$ is the zero moment representing the particle number concentration; $M_i^1 = \int_0^\infty m_i f_i(m)dm$ is the first moment representing the particle mass content. In Equation (6.4.21), $\overline{V}_{g,i}$ is the weighted (averaged) terminal fall velocity of particles belonging to the i-th type. As in cases of raindrops, the mass-averaged terminal fall velocity $\overline{V}_{g\_m,i}$ (Equation (3.2.13)) is used in the equation for the first moment. In the equation for the zero moment, the concentration-averaged terminal fall velocity $\overline{V}_{g,i}$ is used (Equation (3.2.14)). In most bulk parameterization schemes, ice particle size distributions are prescribed in the form of the Gamma distribution, and therefore, the fall velocities of ice particles $\overline{V}_{g\_m,i}$ and $\overline{V}_{g,i}$ can be analytically expressed via the Gamma distribution parameters.

Using the parameters presented in Table 6.4.6, Milbrandt and Yau (2005a) calculated both the mass- and concentration-weighted fall velocities. The ratio $\overline{V}_{g\_m,i}/\overline{V}_{g,i}$ is plotted in **Figure 6.4.13** as a function of the slope parameter α_i of the Gamma distribution (Section 2.1) for different types of hydrometeors. As expected, $\overline{V}_{g\_m,i}/\overline{V}_{g,i} > 1$. The physical meaning of this inequality is clear: the mass of hydrometeors is concentrated in large particles that fall fast. At the same time, the number concentration is determined by smaller particles whose fall velocity is lower. Due to this reason, the ratio $\overline{V}_{g\_m,i}/\overline{V}_{g,i}$ is the largest for raindrops and hail, especially at small values of the shape parameter.

Reproduction of sedimentation and size sorting in bulk-parameterization schemes that use one or few

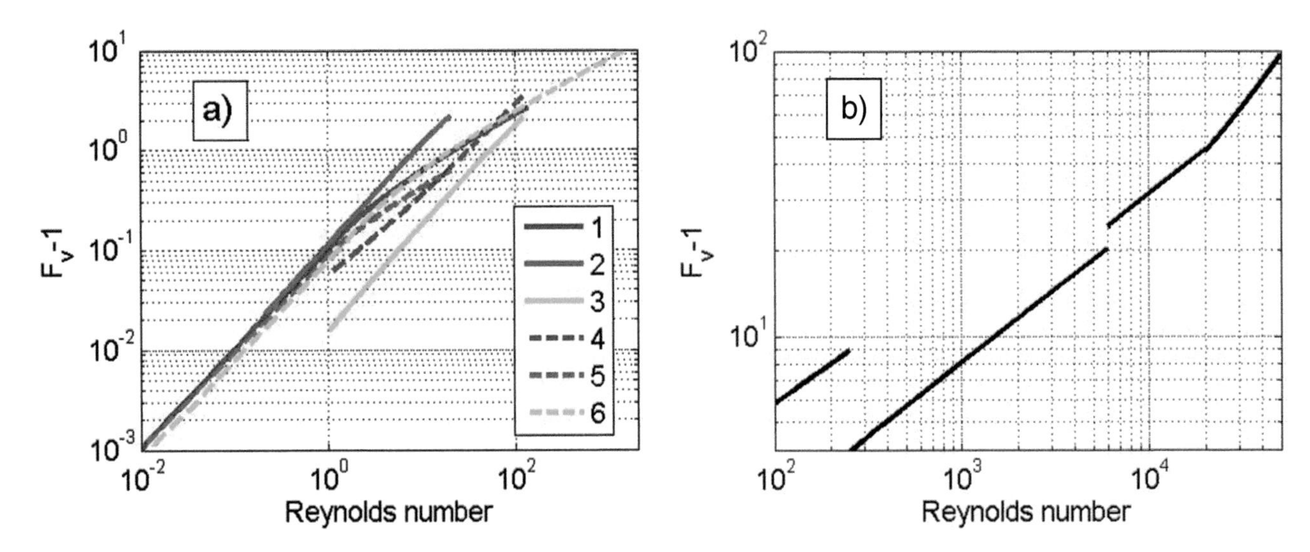

Figure 6.4.12 The dependencies of $F_v(Re) - 1$ on Re. **(a)**: 1-any crystal shape, $Sc = 0.63$ (Hall and Pruppacher, 1976); 2-columnar ice crystal, $Sc = 0.63$ (Wang, 2002); 3-hexagonal plates, $Sc = 0.63$ (Wang, 2002); 4-broad-branch crystals, $Sc = 0.63$ (Wang, 2002); 5-oblate ice spheroid, $Sc = 0.72$ (Pitter et al., 1974) and 6-water drops, $Sc = 0.63$ (Pruppacher and Klett, 1997); (b): graupel and hail according to Equation (6.4.11), $Sc = 0.63$.

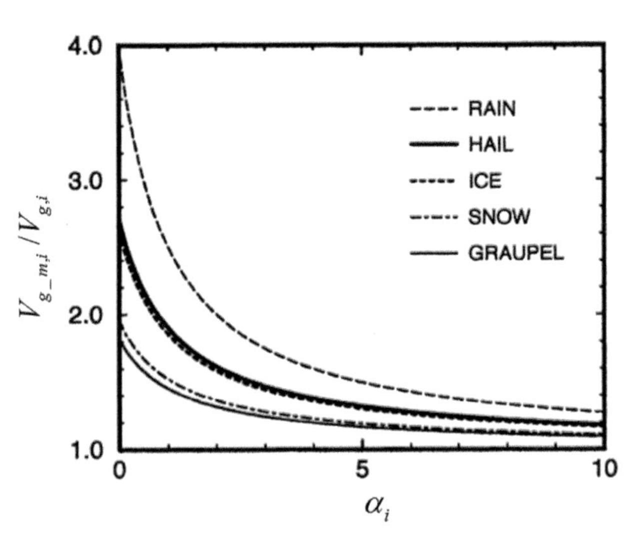

Figure 6.4.13 Dependence of $\overline{V}_{g\_m,i}/\overline{V}_{g,i}$ on the shape parameter α_i for different types of hydrometeors. Particle size distribution is taken in the form $f_i(d) = N_{i0}d^{\alpha_i}e^{-\lambda_i d}$, where N_{i0} is intercept, α_i is the shape parameter, and λ_i is the slope parameter (from Milbrandt and Yau, 2005a; © American Meteorological Society; used with permission).

moments of the SD encounters significant problems (e.g., Milbrandt and Yau, 2005a, 2005b; Wacker and Lupkes, 2009; Milbrandt and McTaggart-Cowan, 2010; Morrison, 2012) of physical and numerical nature. Milbrant and Yau (2005a) compared the solutions of Equation (6.4.21) for bulk-parameterization schemes to the solution obtained by the bin method that is considered as a reference. Simulations were performed using a 1D model. All of the microphysical processes except sedimentation were switched off. At time $t = 0$, hail was located within the layer between 8 and 10 km, as shown in **Figure 6.4.14**. The vertical distribution of the mass content was assumed sinusoidal with maximum of 1 g m^{-3} at $z = 9$ km. Several bulk schemes were compared: a one-moment scheme; two two-moment schemes where the initial size distributions were given in the form of the Gamma distribution $f(d) = N_0 d^{\alpha} e^{-\lambda d}$ with the shape parameter $\alpha = 3$; and two two-moment schemes where the initial SD was also given in the form of the Gamma distribution with the shape parameter set as a function of the mean mass diameter $D_m = \left(\frac{6}{\pi\rho_i}\right)^{1/3}\left(M^1/M^0\right)^{1/3}$. Equation (6.4.21) was solved using the upstream difference scheme (Section 4.2). Figure 6.4.14 shows the vertical profiles of the mass content (the first moment) and the concentration (the zero moment). The averaged fall velocities $\overline{V}_{g\_m,i}$ and $\overline{V}_{g,i}$ were calculated analytically using parameters presented in Table 6.4.4. The profiles were plotted each five min. The vertical profiles of the concentration and the mass content obtained in the bulk schemes were compared with those calculated using the SBM approach where the size distribution was divided over multiple bins, each sediment separately with its own fall velocity. One can see that the results of the one-moment bulk scheme are not satisfactory: at $t = 20$

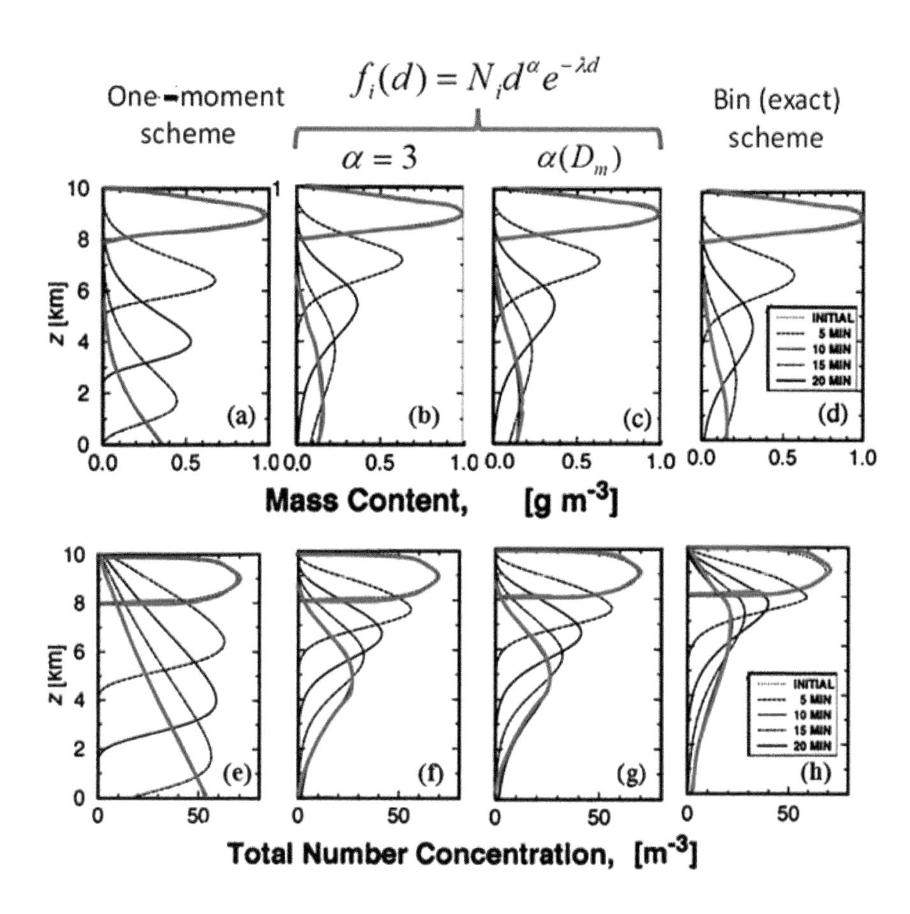

Figure 6.4.14 Vertical profiles of the mass content and the concentration in simulations with a 1D model (1st column); a two-moment scheme for $\alpha = 3$ (2nd column), a two-moment scheme with the shape parameter depending on the mean mass particle diameter D_m (3d column); and the profiles calculated using the bin approach (reference) (4th column) (from Milbrant and Yau, 2005a; © American Meteorological Society; used with permission).

min, the maximum of particle concentration is located at the surface, while in the SBM approach this maximum is located at $z = 7$ km. The concentration profiles produced by the two-moment bulk schemes are closer to the reference solution. The best agreement reached at the shape parameter value of $\alpha = 3$ (under the particular initial conditions and the parameters of the 1D model). However, SD shapes in mixed-phase clouds vary significantly, and consequently, the shape parameter should also vary within a wide range. Therefore, the choice $\alpha = 3$ as the parameter that produces the best agreement with the exact solution would not be very productive from a practical point of view. This is especially true if one takes into account that surface precipitation depends on the size distribution parameters, as shown in **Figure 6.4.15**. Also, one can see that the one-moment scheme indicates the worse result: precipitation is very intense but short, which is in agreement with the conceptual scheme plotted in Figure 5.5.5.

Further attempts to improve the representation of sedimentation and size sorting in bulk schemes were made by Milbrandt and McTaggart-Cowan (2010). In this study, vertical profiles of different quantities simulated using a 1D parcel model and obtained using different bulk schemes were compared with the profiles obtained using a bin scheme. In the simulations ice particles were initially located within the layer between 6,500 m and 8,000 m (**Figure 6.4.16).** The mass content was 0.5 $\mathrm{g\,m^{-3}}$ and the concentration was chosen 3 $\mathrm{L^{-1}}$. The initial particle size distribution was assumed exponential, i.e., the initial shape parameter value was $\alpha = 0$. The profiles of the zero, the first and the second moments (the second moment is directly related to the radar reflectivity), and the mean mass diameter D_m were calculated using different bulk-parameterization schemes and compared with the profiles obtained using the bin model assumed to produce the reference solution. In the three-moment scheme, the additional prognostic equation for the second moment (the radar reflectivity) was solved. The evolution of the prognostic moments was determined by solving Equation (6.4.21) by means of the box–Lagrangian scheme (Kato, 1995). As the Courant number in their study was substantially less than unity, the scheme produced results similar to those obtained by the upstream difference method. For the one-moment simulations, the closure assumptions are $N_{i0} = 8 \times 10^6$ m^{-4} and $\alpha = 0$ for all of the time points. In the two-moment scheme, the single

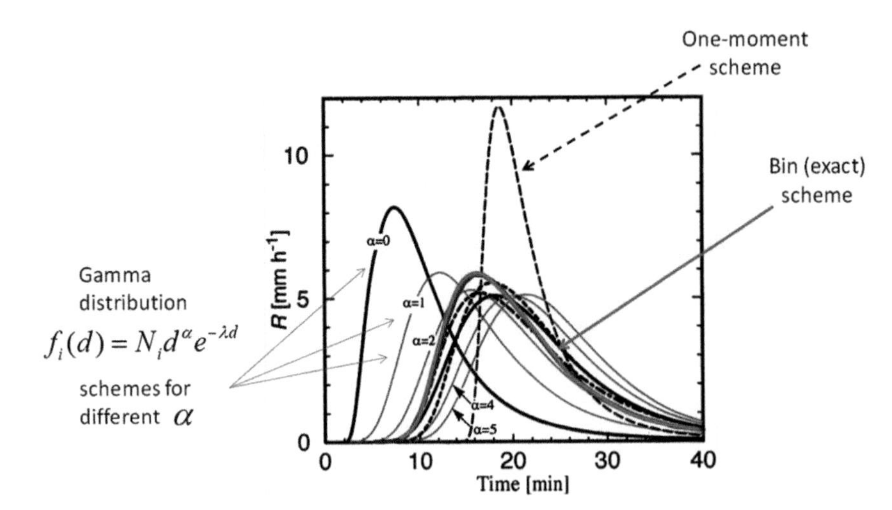

Figure 6.4.15 Dependence of precipitation rate on the shape parameter in simulations illustrated in Figure 6.4.14 (from Milbrandt and Yau, 2005a; © American Meteorological Society; used with permission).

constraint (the closure assumption) is $\alpha = 0$. Other parameters are determined from the values of the moments. In the three-moment scheme, all the SD parameters were determined using the values of the three moments, so this scheme does not require any closure assumptions.

The analysis showed that the profiles calculated using the one-moment scheme significantly differ from the reference solution. This scheme is not able to simulate size sorting, as the mean mass particle diameters are unrealistically small and practically do not depend on height. The profiles produced by the two-moment scheme are more realistic, while the errors remain significant, especially in the reproduction of the first and the second moments. Size sorting is significantly stronger than that in the reference solution. The simulations showed that the errors of the two-moment scheme depend on the choice of the moments to be used to simulate sedimentation. Implementation of the three-moment scheme considerably reduced the errors. Therefore, Milbrandt and Yau (2005b) and Milbrandt and McTaggart-Cowan (2010) recommended implementing the three-moment bulk-parameterization scheme containing the equation for the radar reflectivity in addition to the equations for the concentration and for the mass content.

Taking into account that most bulk-parameterization schemes are two-moment ones, Milbrandt and McTaggart-Cowan (2010) also proposed a method to improve the reproduction of sedimentation in the two-moment bulk schemes. The slope parameter α is parameterized as a function of the ratio of moments. As a result, the value of the shape parameter used to calculate the fall velocity differed from that used in other bulk-parameterization schemes. This method turned out to be highly effective in mitigating the problems associated with excessive size sorting in a standard two-moment scheme with fixed α.

Figures 6.4.14 and 6.4.16 show a significant stretching of the moment profiles in the vertical direction, which increases with time. In case a bin microphysics scheme is used, this stretching is a natural result of the difference in the fall velocities between particles having different masses. In bulk parameterization, this stretching cannot be fully attributed to the dependence of the average fall velocity on height. In case of a constant fall velocity, Equation (6.4.21) is reduced to the classical equation of advection

$$\frac{\partial M_i^k}{\partial t} = \begin{cases} -\overline{V}_{g,i}\dfrac{\partial M_i^k}{\partial z}, & k=0 \\ -\overline{V}_{g\_m,i}\dfrac{\partial M_i^k}{\partial z}, & k=1 \end{cases}.$$

The minus sign is used since the fall velocities are positive values directed downward. The analytical solutions of the last equations are $M_i^0(z,t) = M_i^0(z - \overline{V}_{g,i}t)$ and $M_i^1(z,t) = M_i^1(z - \overline{V}_{g\_m,i}t)$. According to these solutions, the initial profile of the moment is translated downward without any change to its shape. The stretching of the vertical profiles of the moments seen in Figures 6.4.14 and 6.4.16 can be partially attributed to the high computational viscosity of the upstream difference scheme used for simulation of sedimentation. The computational viscosity coefficient of the upstream finite-difference scheme can be written as $K_{num} = \frac{1}{2}\overline{V}_g\Delta z\left(1 - \frac{\overline{V}_g\Delta t}{\Delta z}\right)$ (Section 4.2). The low Courant number $\frac{\overline{V}_g\Delta t}{\Delta z}$ typically used in the simulations indicates a significant effect of the computational viscosity on the vertical profiles of the concentration and the mass of hydrometeors, and therefore, on the value of the mean mass diameter. Such stretching causes an overlap of the vertical profiles of different moments that allows us to get reasonable averaged particle sizes. However, the mean particle size turns out to be dependent on the properties of the numerical scheme used. Since $\overline{V}_{g\_m,i} > \overline{V}_{g,i}$, it can

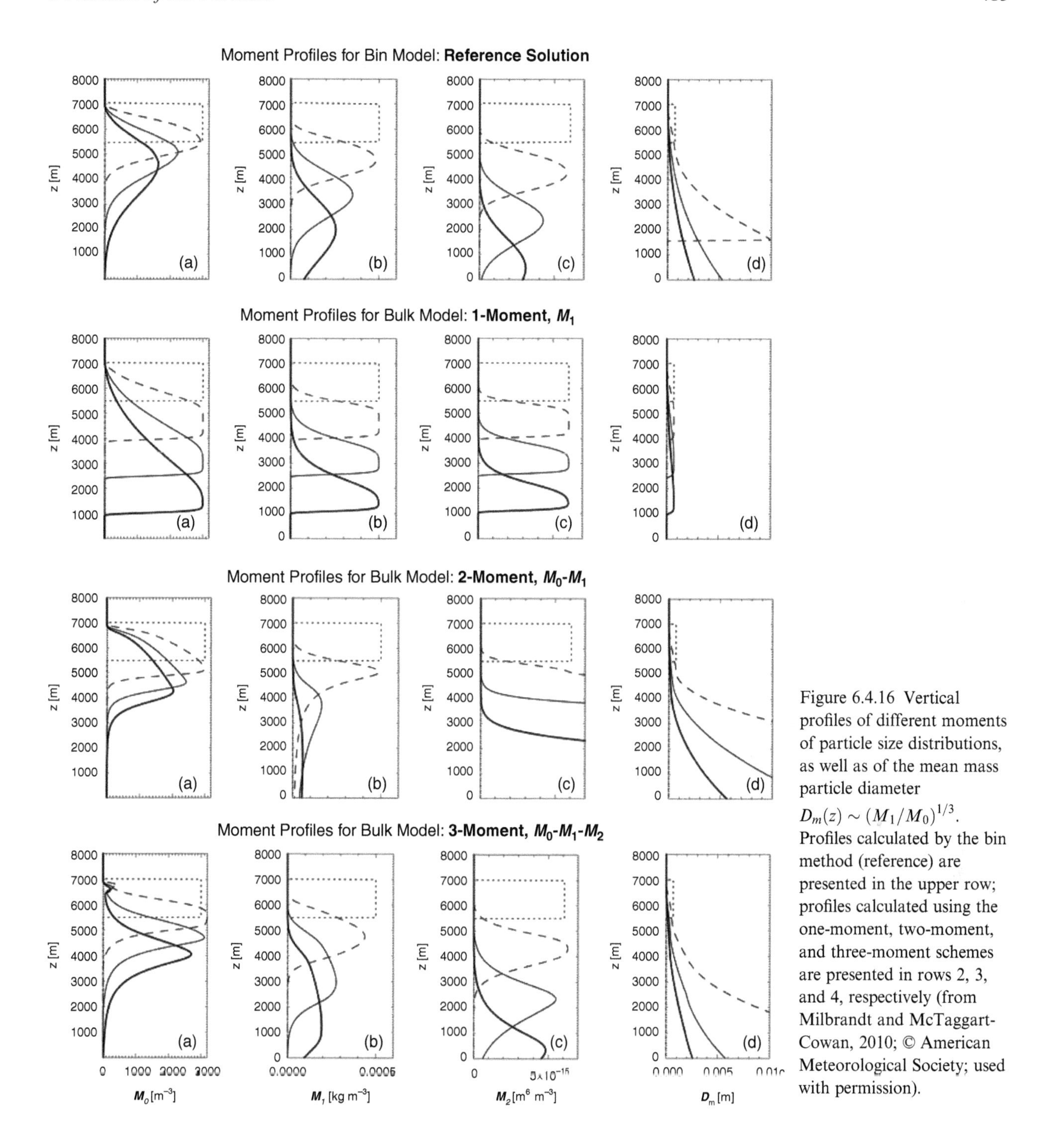

Figure 6.4.16 Vertical profiles of different moments of particle size distributions, as well as of the mean mass particle diameter $D_m(z) \sim (M_1/M_0)^{1/3}$. Profiles calculated by the bin method (reference) are presented in the upper row; profiles calculated using the one-moment, two-moment, and three-moment schemes are presented in rows 2, 3, and 4, respectively (from Milbrandt and McTaggart-Cowan, 2010; © American Meteorological Society; used with permission).

occur that at low heights the concentration of particles is equal to zero at nonzero mass contents, while at higher levels the mass content is zero at a nonzero concentration. In this case, the calculated values of the mean volume particle diameter become unrealistic. The accepted approach in bulk schemes is shifting these values into the reasonable range by means of the corresponding correction of the particle concentration values.

Another method of treating sedimentation in bulk schemes is known as the bin-emulating, or hybrid method. The particle size distribution given in the form of the Gamma distribution is discretized into several tens of bins before each time sub-step of the sedimentation procedure. Then, a sedimentation algorithm is applied to each bin. After sedimentation, the values of the mass and the concentration of particles are calculated and are used to restore the Gamma distribution. To decrease the computational time, bin-emulating schemes use lookup tables to calculate the fall distances of particles belonging to a particular bin. The validity of this approach has been a subject of debate: while Feingold et al. (1998) saw a substantial advantage of the bin-emulating schemes, Morrison (2012) showed that such schemes are similar to the traditional schemes with sedimentation of separate PSD moments. Morrison concluded that representation of sedimentation in two- and three-moment bulk schemes can be improved only if one takes into account the effects of changes in the size distribution shape (e.g., the shape parameters of the SD) on the sedimentation (e.g., Milbrandt and Yau, 2005a; Wacker and Lupkes, 2009; Milbrandt and McTaggart-Cowan, 2010).

Another method for improving the representation of ice particle sedimentation in mixed-phase clouds was proposed by Milbrandt and Morrison (2013), who included an additional prognostic equation for the bulk graupel volume mixing ratio B_g, so that the density of graupel at any grid point and at any time instance could be diagnosed from the relationship $\rho_g = q_g/B_g$, where q_g is the graupel mass mixing ratio. The changes of B_g were caused largely by riming. The bulk density of graupel recalculated at each time step was used for calculation of sedimentation velocity. The scheme was tested in idealized simulations of a convective mesoscale system. It was shown that the method is capable of reproducing a wide range of graupel densities and to improve the spatial distribution of graupel and simulation of surface precipitation.

6.4.7 Orientation of Ice Particles

Orientation of nonspherical ice particles is an important characteristic of hydrometeors, determining the particle's area projection normal to the flow, which affects the fall velocity, radiation, and optical cloud properties. Orientation of ice particles also impacts the rate of particle collisions, aggregation, ice splintering, and the radiative transfer in clouds as well as the optical and the radar cloud properties. Polarimetric characteristics of optical and radar signals used in remote sensing of atmosphere are directly related to orientation of ice particles. Data about the orientation characteristics is vital for

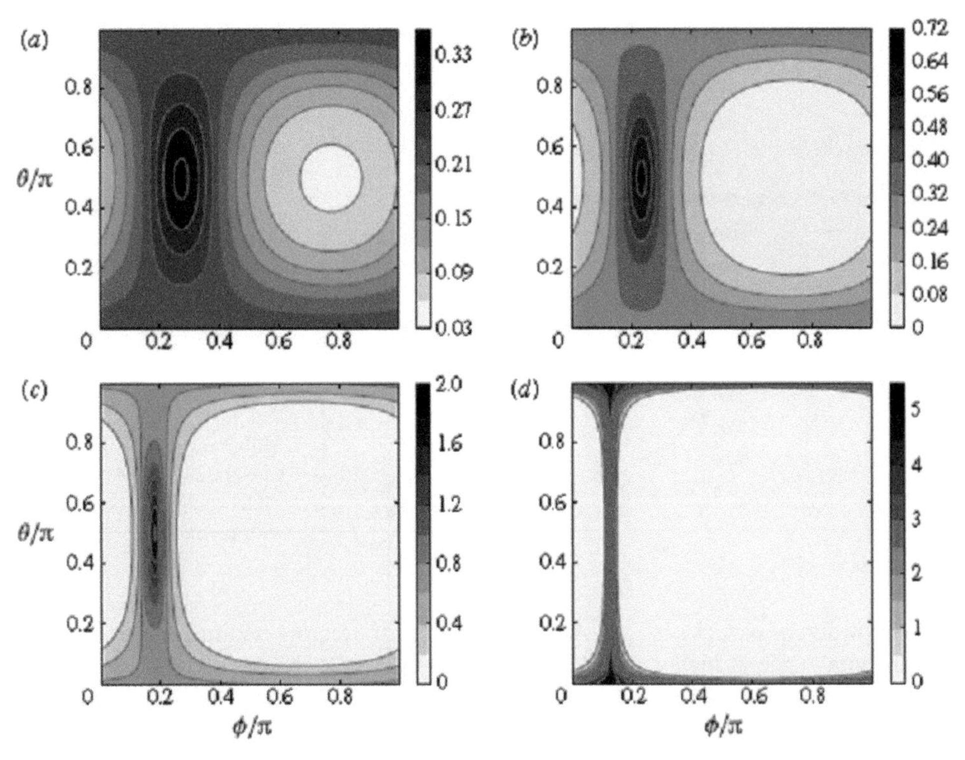

Figure 6.4.17 An example of the probability distribution function of orientation angles ϕ and θ of a prolate ellipsoid at different time instances t. The aspect ratio of the ellipsoid is equal to five. (a) $t = 0.5$; (b) $t = 1$; (c) $t = 2$; and (d) $t = 5$. The time is normalized on the Kolmogorov time scale (from Gavza et al., 2012).

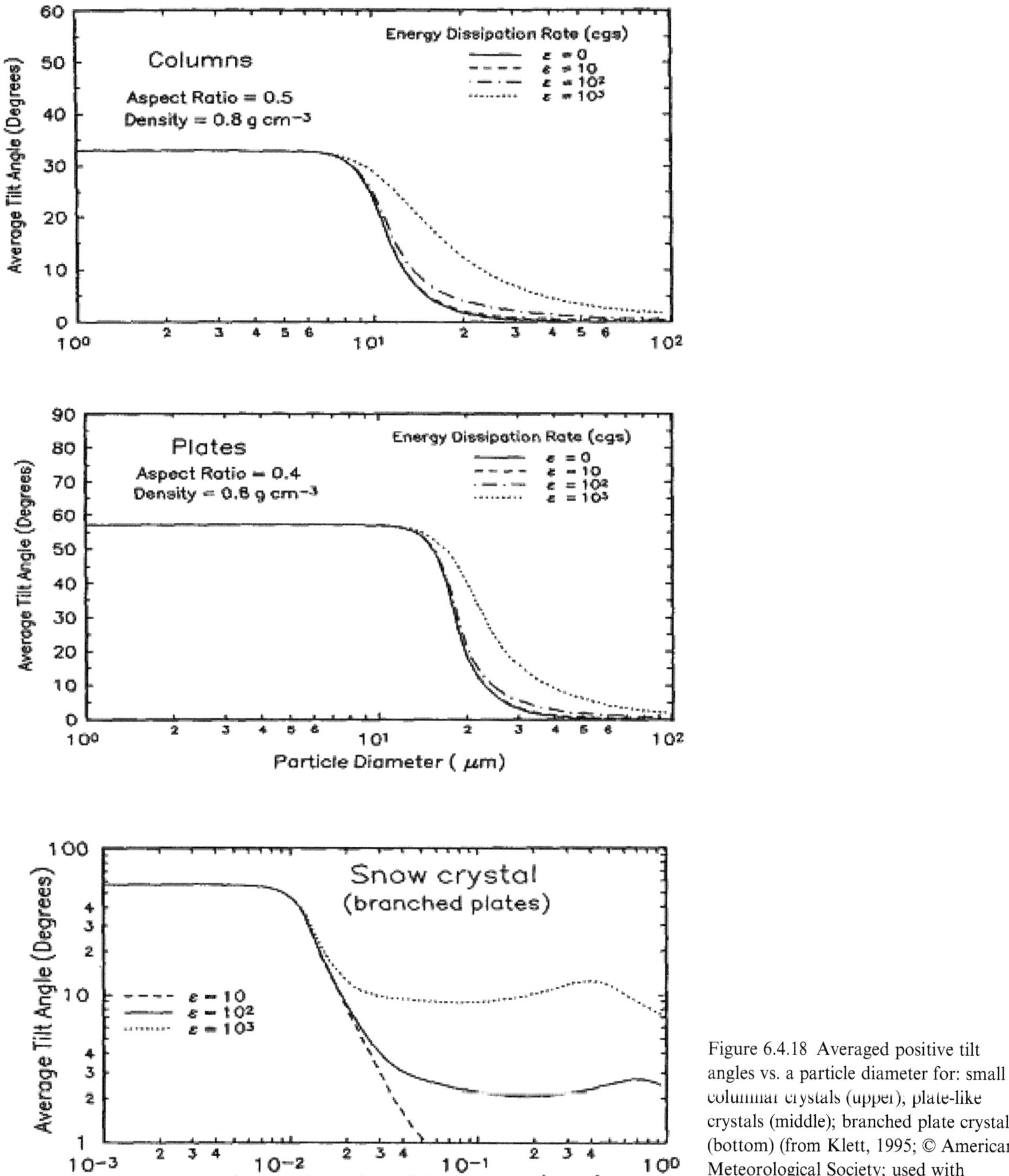

Figure 6.4.18 Averaged positive tilt angles vs. a particle diameter for: small columnar crystals (upper), plate-like crystals (middle); branched plate crystals (bottom) (from Klett, 1995; © American Meteorological Society; used with permission).

interpretation of various radar and lidar measurements in clouds; therefore, modeling ice particle orientation is a compulsory element of radiative cloud models.

Analysis of particle orientation in the atmosphere is a highly complicated theoretical task. As was mentioned, in calm air small ice particles fall in a stationary manner with a lager cross section oriented horizontally, so the particles have the preferential orientation. Existence of turbulent shears, even of a simple vertical shear, can lead to flipping or tumbling of the particle (Broday et al., 1998). Spheroidal particles also have an additional drift toward regions characterized by higher local velocities in the direction opposite to gravity. The drift rate depends on the particle shape being affected by its aspect ratio and its inertia.

According to Böhm (1992a), small particles with $Re < 10$ are randomly oriented. When $10 < Re < 200$, planar ice crystals fall with the axis of symmetry in their orientation directed parallel to the flow (vertical orientation), while for columnar ice crystals the axis of symmetry is perpendicular to the flow (horizontal orientation). At higher Reynolds numbers at $200 < Re < 3{,}000$, planar ice crystals tend to wobble. The wobble angle characterizing the deviation from the horizontal orientation is supposed to be below 20°. At the same time, graupel that are more spherical particles tend to tumble. At $Re > 3{,}000$, both graupel and hail tumble. Columnar crystals do not reach nonstationary regimes, as they have lower *Re*.

Analytical analysis of small spheroid motion in a 3D linear shear flow was conducted by Gavze et al. (2012, 2016). They analyzed the so-called Jeffery equation, which is a consequence of Equation (6.4.2). Gavze et al. (2012, 2016) showed the existence of a limit particle orientation that in the $\phi - \theta$ plane (where ϕ and θ are two angles characterizing orientation) is seen as a stable point or a stable limit cycle. This preferential limit orientation depends on the turbulent shear tensor $\frac{\partial u_i}{\partial x_j}$ and the aspect ratio of the ice particle. However, the time needed to reach the stable state is large enough, typically larger than the Kolmogorov time scale (Section 3.3), so one can consider the orientation as a random value and characterize ice particles in clouds by their orientation distribution (PDF). The PDF tends to the boundary value with time, as shown in **Figure 6.4.17**. The figure demonstrates large difficulties on the way to parameterize the orientation effects in cloud models. In other recent studies (Pumir and Wilkinson, 2011; Siewert et al., 2014) orientation distributions of ellipsoids were obtained. Pumir and Wilkinson (2011) also evaluated the characteristic time of particle alignment to stationary state as 5–15 of the Kolmogorov time scales.

Klett (1995) investigated random oscillations of falling inertial ice particles around stable preferable orientation arising in turbulent flow. These oscillations are described by the following scalar linear stochastic equation:

$$I\frac{d^2\theta}{dt^2} + R\frac{d\theta}{dt} + 2Q\theta = \Gamma, \qquad (6.4.22)$$

where θ is the angle between the vertical direction and the symmetry axis of the spheroid, $R = 8\pi l^3\eta$, $Q = \frac{87}{40}\pi\rho l^3 V_g^2|e|$, $2l$ is the characteristic length of the spheroid, and e is its eccentricity. Γ represents a random torque caused by turbulence and I is moment of inertia for the vertical tilting. **Figure 6.4.18** shows the dependence of the oscillation amplitude on the particle's diameter at different turbulent intensity values characterized by different turbulent dissipation rates ε typical of clouds of different types (Section 3.3). Increasing ε increases the size range of ice particles experiencing strong oscillations. The figures demonstrate that there exists a size range of particles that experience strong oscillations. This range is of the order of 10 – 20 μm for crystals of simple shapes. The oscillations of snow-like crystals with sizes of several hundred μm are rapidly damped due to the high inertia.

6.5 Collisions and Coalescence in Mixed-Phase Clouds and Ice Clouds

6.5.1 Specific Features of Particle Collisions and Coalescence in Mixed-Phase and Ice Clouds

Precipitation particles in mixed-phase clouds and ice clouds form as a result of collisions between ice particles as well as between ice particles and water drops. There are many common features between collisions of liquid drops and collisions with participation of ice particles:

a) the rate of collisions is determined by the collision kernel that is the product of the collision efficiency and the swept volume;
b) the collision efficiency is determined by hydrodynamic interaction of approaching particles;
c) collisions may lead to collisional breakup of particles;
d) the changes of PSD due to the collision process are described by stochastic collection equations (cf. Equation 5.6.5);
e) not all collisions lead to merging (aggregation) of particles. The fraction of collisions that results in merging is characterized by the coalescence efficiency (also known as the sticking efficiency). The product of collision efficiency and coalescence efficiency, known as the collection efficiency, characterizes the process of collection between colliding particles.

At the same time, collision processes in mixed-phase clouds significantly differ from those in warm clouds. (Next, we will use the term "collection" to refer to any collision outcome that involves merging of the two colliding particles.) The main specific features of collisions/coalescence in mixed-phase clouds are the following:

- Collision and collection kernels depend not only on a particle mass, but also on a particle type (e.g., aggregates and graupel of the same mass have different abilities to collide with other particles).
- The fall velocities of ice particles of the same mass are different due to differences in particle shapes, complicated trajectories, and variable spatial orientation (Section 6.4). Thus, unlike drops of the same mass, particles of the same mass may collide even in a quiescent environment.
- In contrast to drops, the composition of ice particles can be nonuniform. Processes such as melting, freezing, wet growth of hail, and others lead to the appearance of cloud hydrometeors that contain both ice and liquid fractions as well as fractions of different densities (e.g., fractions of rimed ice embedded into an aggregate of low bulk density). The composition of these particles changes with time and space, affecting their fall velocity and their ability to collect other particles. As a result of interparticle collisions, a particle type may change.
- In contrast to inter-drop collisions, collisions between particles in mixed-phase clouds are not symmetric. For instance, large aggregate-small drop collision rate differs significantly from that of large drop-small aggregate, in spite of the fact that in both cases the resulting particles are of the same mass. One of the reasons for this asymmetry is the difference in the relative velocities between a large ice particle and a small drop, as opposed to the case of a large drop and a small ice particle. Another reason is the difference in the hydrodynamic interaction in drop collector-small drop pairs and drop collector-small ice particle pairs.
- Collision efficiencies depend not only on the mass of particles, but also on their shape, which causes a significant variability of the measured collision efficiencies even for particles of the same mass.
- Coalescence efficiencies depend on the properties of ice particle surfaces, their shape, and on the temperature.
- Collection (merging) of hydrometeors of different types can lead to formation of a hydrometeor belonging either to the type of the collector or to a type that neither of the colliding particles belongs to. The resulting particle typically differs from the colliding particles in its density and shape.
- A particle's conversion to a different type can be a lengthy process of multiple collisions. For instance, an aggregate should collect many small drops to become graupel. It means that to analyze the result of a collision one should know the particle's history.
- The rules used in cloud models for simulating hydrometeor-type conversions caused by collision-coalescence processes are complicated and, to a certain extent, subjective. The rules applied in SBM schemes are sometimes different from those in bulk-parameterization schemes. In SBM, the rules are applied to collection of particles of a particular size belonging to a certain mass bin, while in bulk schemes these rules are applied to the entire spectrum of a particular hydrometeor type.
- Concentration of ice particles is typically much lower than that of drops. Therefore, the stochastic collection equations used for description of ice particle collisions should be averaged over a much larger volume than in the case of water drop collisions.

All of these factors make the analysis of collisions in mixed-phase clouds and ice clouds much more complicated and uncertain than in the case of water drops.

6.5.2 Conversion of Particles Due to Collision Coalescence in Mixed-Phase Cloud Models

A qualitative analysis of formation of different ice particle types is presented in Section 2.4. There are different kinds of collisions between hydrometeors in mixed-phase clouds, and several processes involved in these collisions. Hereafter, the two types of interacting hydrometeors are denoted as X and Y. Collisions between X and Y hydrometeors resulting in their collection can lead either to formation of hydrometeors of type X or of type Y, or produce particles of a third type: Z. The main possible outcomes of such collisions and of the corresponding processes are presented in **Table 6.5.1.**

Self-collection of ice crystals leads to formation of aggregates (Figure 2.4.7). Descriptions of the riming of ice crystals and aggregates in cloud models is a more complicated task than self-collection of ice crystals. Just one collision of a comparatively large water drop with an ice crystal or aggregate instantly leads to formation of an ice particle of comparatively large bulk density (i.e., graupel or freezing drop) due to freezing. At the same time, collisions of ice crystals or aggregates with small water drops lead to drop freezing on the ice surface (Figure 2.4.9) and gradual increase in the bulk

Table 6.5.1 Main kinds of possible collection outcomes and examples of the corresponding processes.

Kind of Collection and Outcomes	Name of Process	Examples of Processes
X + X =>X	Self-collection (aggregation)	snow + snow =>snow
X + X =>Z	Autoconversion	dendrites + dendrites =>snow
X + Y =>X	Accretion	Snow + ice crystal =>snow
X + Y =>X	Riming in case *Y* is a water drop	Snow + water =>snow
X + Y =>Z	Autoconversion in case *X* and *Y* are ice crystals	dendrites + plates =>snow
X + Y =>Z	Drop freezing by collisions with ice. *X* is a water drop; *Y* is an ice crystal or snow. Mass of *X* is larger than that of *Y*	raindrop + ice crystal =>graupel

density of the rimed aggregates. Such collisions are treated differently in models depending on the level of the model's sophistication.

Khain and Sednev (1996) assumed that collisions of ice particles and water drops led to formation of ice particles of the same type if the mass of the colliding ice particle was larger than that of the drop, or collision produced graupel (or hail) if the mass of the water drop was larger than that of the collected ice particle. This approach does not take into account that conversion of aggregates to graupel can be a gradual process. The riming rate depends on LWC in the surrounding of an ice particle. It means that the mass of small drops collected by an ice crystal or a snowflake during one time step might be large enough to assign the resulting particle to graupel. Accordingly, Khain et al. (2004) converted snowflakes to graupel if LWC exceeded a threshold value. The threshold value of LWC was chosen in such a way that during one time step the mass of a snowflake increased by riming about twice.

Accumulation of the rimed mass often takes multiple time steps. Therefore, Khain et al. (2011, 2012) introduced a new parameter, namely, the rimed mass within aggregates, which accumulated in the course of riming. Using the rimed fraction, a snowflake bulk density was recalculated at each time step. A snowflake was converted into graupel when its bulk density exceeded the threshold value of 0.2 $\mathrm{g\,cm^{-3}}$.

Riming of graupel increases its size and bulk density. Khain et al. (2004, 2008) assumed that graupel–water drop collisions led to hail formation if the environmental supercooled water content exceeded the threshold value of about 1–3 g $\mathrm{cm^{-3}}$. Such high supercooled LWC allowed a significant growth of the graupel mass due to riming during one time step. In studies carried out by Khain (2009) and Noppel et al. (2010), graupel belonging to a particular mass bin was converted to hail in two cases: if the wet growth had started or if the graupel diameter exceeded one cm. (Section 6.7). The wet growth started when the latent heat release caused by freezing of the collected drops partially prevented further freezing of collected water. The criterion for wet growth onset at freezing temperatures was derived by Blahak, who evaluated the surface temperature of graupel and hail particles (Blahak, 2008; Khain et al., 2011). The wet growth is described in detail in Section 6.7.

The probability of collection depends on the properties of the surfaces of colliding ice particles. For instance, wet growth changes the properties of hail surface, making it possible to collect ice particles. Collisions of graupel (or hail) with snow and ice crystals may lead rather to a breakup of snow and ice crystals and to formation of small ice splinters or fragments (Yano and Phillips, 2011; Phillips et al., 2017a, 2017b). Ice splinters also form as a result of collisions of graupel and hail with water drops, as well as collisions between graupel (hail) and aggregates. This process is known as a secondary ice production by ice multiplication (see Section 6.8).

The main rules of type conversion of ice particles following collisions and coalescence, applied in SBM models, are presented in **Table 6.5.2**. The utilization of algorithms describing conversion of particles from one type into another depending on the particle size is a significant advantage of SBM schemes compared to bulk schemes, where the conversion is often performed regardless of the particle size. These rules are somehow subjective and may eventually be refined in the course of further model development with increasing model sophistication and new knowledge of the processes involved.

The conversion rules used in bulk schemes are presented in **Table 6.5.3,** where cloud droplets (with radii below ~30–40 μm) and raindrops are treated as different

Table 6.5.2 Conversion of hydrometeors in different SBM models.

Hydrometeor Type	Drops	Ice Crystals	Snow	Graupel	Hail
Drops	R96, Y00, KS96, K04, K12: drops	R96, KS96, K04, K12: crystals if $m_d < m_c$, otherwise graupel	R96, Y00: graupel if $m_d > m_s$, otherwise snow. K04: graupel if $m_d > m_s$ or if $q_l > q_{l\_thesh}$, otherwise snow. K12: graupel if density of resulting particle $\rho_p > 0.2\, gcm^{-3}$, otherwise snow with rimed fraction	R96, KS96; K04: graupel; K12: graupel with liquid water fraction	K04: hail; K12: hail with liquid water fraction
Ice crystals		R96, Y00, KS96, K04, K12: snow	R96; Y00, KS96, K04, K12: snow	R96, Y00: graupel. K04: no collection; K12: no collection in dry growth, graupel in wet growth.	K12: no collection in dry growth, hail in wet growth.
Snow			R96, Y00, KS96, K04, K12: snow	R96: graupel. KS96: snow if $m_g < m_s$ otherwise graupel. K04: no collection K12: no collection in dry growth, graupel in wet growth.	K04; no collection K12: no collection in dry growth, hail in wet growth.
Graupel				KS96: graupel K04: no collection	KS96: hail K04: no collection; K12: no collection in dry growth, hail in wet growth
Hail					KS96: hail. K12: no collection in dry growth, hail in wet growth

hydrometeors. In most bulk-parameterization schemes, collisions between ice crystals are neglected. Formation of snow (aggregates) is assumed to occur during deposition growth of ice crystals (Section 6.3). In RAMS (Walko et al., 1995), collisions of snow and graupel lead to production of graupel, while collisions between snow and hail or between graupel and hail lead to production of hail. When considering ice-cloud drop collisions, Meyers et al. (1997) (also using RAMS) determined the fractions of the mass retained in the type of ice collector and a newly formed ice particle type according to the proportion between the masses of the ice collector and of the collected drop. This proportion was different for different types of collectors. Only a few bulk-parameterization schemes take hail into consideration (e.g., Noppel et al., 2010).

As follows from this short review, the criteria applied in models for particle type conversions are to a certain extent qualitative and subjective, being dependent on the model sophistication and the modeler's experience. Significant experimental and theoretical efforts are required to make these criteria quantitative.

6.5.3 Stochastic Collection Equations in SBM Models

Collisions in mixed-phase clouds are described by a set of stochastic collection equations (SCEs) formulated for each hydrometeor type. The particular form of the equation system depends on the rules of collisions applied by a particular model (Table 6.5.2). SCE are strictly valid under certain conditions listed in Section 5.6. (Equations 5.6.6–5.6.9). The most important condition is that the number of particles of the *x-th* type within the air volume under consideration has to be high enough, i.e., inequalities $\Delta x \Delta y \Delta z \int_0^\infty f_x(m)dm \gg 1$

Table 6.5.3 Conversion of hydrometeors in different bulk-parameterization schemes.

Hydrometeor Type	Cloud Droplets	Raindrops	Ice Crystals	Snow	Graupel	Hail
Cloud droplets	Droplets in case of selfcollection; raindrops in case of autoconversion	Raindrops	SB06: crystals, and possible conversion into graupel	SB06: crystals, and possible conversion into graupel	SB06: graupel	N10: hail
Raindrops	raindrops	raindrops	graupel	graupel	graupel	N10: hail
Ice crystals			C86, M05, B94: no collisions SB06: snow	snow	W95: graupel. SB06: no collection.	W95: hail
Snow (aggregates)				snow	S86; W95; SB06: graupel.	W95: hail
Graupel					MY05: No collection in dry growth, graupel in wet growth	W95: hail. N10, MY05: No collection in dry growth, hail in wet growth
Hail						N10: hail. MY05: No collection in dry growth, hail in wet growth

or a stricter condition $\Delta x \Delta y \Delta z \Delta m_i f_x(m_i) \gg 1$ should be valid. Since the concentration of ice particles is typically by three orders of magnitude lower than that of drops, the minimal size of volumes considered should be by about an order of magnitude larger than the volumes required for water drops collisions.

We present two examples of SCE used in HUCM in the version described by Khain and Sednev et al. (1996) and Khain et al. (2004) to simulate production of aggregates (snowflakes) and graupel.

In the first example, the SD function of snowflakes $f_s(m,t)$ is calculated using the following equation:

$$\frac{\partial f_s(m,t)}{\partial t} = \text{Gain}_s - \text{Loss}_s, \quad (6.5.1)$$

where index "s" denotes snowflakes, Gain_s is the sum of the terms which increase $f_s(m,t)$ as a result of collisions of particles of masses m' and $m-m'$ and Loss_s is the sum of the terms which decrease $f_s(m,t)$ due to collection of snowflakes of mass m and any other particles of mass m'. There are two ways to obtain snowflakes as a result of collisions:

a) collisions of particles belonging to the same hydrometeor type (e.g., ice crystals and ice crystals, snow and snow; in this case, the Gain integral is written similarly to the case of collisions between water drops (Section 5.5);

b) collisions of particles belonging to different hydrometeor types, when the collision kernels are not symmetric (e.g., snow and ice crystals), i.e., the result of collisions of particles of masses m' and $m-m'$ at $m' > m-m'$ is not the same as that at $m' < m-m'$.

We describe construction of expressions for the Gain integrals following Khain and Sednev (1996). Snowflakes form as a result of collisions of hydrometeors of different types, which can be separated into three groups. The first group includes collisions between ice crystals belonging to the same type, namely, plates-plates, dendrites-dendrites, columnar crystals-columnar crystals and snow-snow: $\sum_{x=1}^{4} G_{xx} = G_{pp} + G_{dd} + G_{cc} + G_{ss}$. Here indices p, d, c and s denote columns, dendrites, plate-type crystals and snowflakes, respectively. The Gain integral for this group can be written as

$$\sum_{x=1}^{4} G_{xx} = \sum_{x=1}^{4} \int_0^{m/2} f_x(m-m') K_{xx} f_x(m') dm' \quad (6.5.2)$$

The second group includes collisions between ice crystals belonging to different crystal types, namely, dendrites-columns, dendrites-plates and plates-columns, as well as between all types of crystals and snowflakes. The corresponding Gain integrals are denoted as G_{xy} if the mass of a type $\boldsymbol{X}$ hydrometeor is larger than that of type $\boldsymbol{Y}$ hydrometeor, and as G_{yx} in the reverse case. The sum $(G_{xy}+G_{yx})$ describes the contribution of collisions between hydrometeors $\boldsymbol{X}$ and $\boldsymbol{Y}$ to particle masses ranging from zero to m. For instance, the sum $(G_{sd}+G_{ds})$ describes production of snow by collisions between snow and dendrites, both in case when the snow mass is larger (the first term) and when the dendrite mass is larger (the second term). The sum of the Gain integrals describing collisions between crystals of different types (three terms) and between crystals and snow (three terms) can be written in the form:

$$\sum_{\substack{x,y=1\\x\neq y}}^{6}(G_{xy}+G_{yx})=\sum_{\substack{x,y=1\\x\neq y}}^{6}\int_0^m f_x(m-m')K_{xy}f_y(m')dm' \tag{6.5.3}$$

The upper limit of the integrals in Equation (6.5.3) is equal to m, which takes into account the asymmetry of collisions between $\boldsymbol{X}$ and $\boldsymbol{Y}$ hydrometeors.

The third group includes collisions of large snowflakes and smaller drops (Table 6.5.2). The corresponding Gain integral $G_{xy}=G_{sl}$ (index l denotes liquid drops) can be written as

$$G_{xy}=\int_0^{m/2} f_x(m-m')K_{xy}f_y(m')dm' \tag{6.5.4}$$

In Equation (6.5.4), the snowflake mass $m-m'$ should be larger than m' of a liquid drop. The total Gain integral of snowflakes Gain_s is the sum of the terms in Equations (6.5.2–6.5.4).

The Loss integral is written as:

$$\text{Loss}_s=f_s(m)\cdot\sum_{x}^{Y_{max}}\int_0^{\infty}K_{sx}f_x(m')dm' \tag{6.5.5}$$

where x denotes the types of hydrometeors that snowflakes may collect to produce particles of mass larger than m. In Equations (6.5.2–6.5.5), $K_{xy}(m_1,m_2)$ is the collection kernel matrix describing the probability of collection between particles with masses m_1 and m_2, belonging to hydrometeor types X and Y, respectively.

The second example is the SCE describing graupel formation. Like snow particles, graupel forms as the result of collisions of hydrometeors belonging to several types. The first group of collisions describes the riming rate of ice crystals and snowflakes by collisions with liquid drops of masses larger than the ice particles. This group can be expressed as sum $\sum_{y=1}^{4}G_{xy}=G_{ls}+G_{lp}+G_{ld}+G_{lc}$ that can be represented by the sum of integrals:

$$\sum_{y=1}^{4}G_{xy}=\sum_{y=1}^{4}\int_0^{m/2}f_l(m-m')K_{ly}f_y(m')dm' \tag{6.5.6}$$

Another type of collisions leading to formation of graupel represents collisions between graupel and liquid drops with mass larger or smaller than that of the graupel particle. The corresponding Gain integral can be written as:

$$G_{xy}=G_{gl}+G_{lg}=\int_0^m f_g(m-m')K_{gl}f_l(m')dm' \tag{6.5.7}$$

where index g denotes graupel. The upper limit of integration m indicates that collisions of graupel with drops of mass ranging from zero to m are considered.

In the HUCM version described by Khain et al. (2011), collection between graupel and snowflakes is allowed only in case when graupel has liquid water on its surface. The corresponding Gain integral can be written as:

$$G_{gs}=\int_0^{m/2}f_g(m-m')K_{gs}f_s(m')dm' \tag{6.5.8}$$

The total Gain integral Gain_g is the sum of the terms in Equations (6.5.6–6.5.8). The Loss integral for graupel Loss_g has a form analogous to Equation (6.5.5). SCE for collisions of other types of hydrometeors can be written in a similar way using the rules concerning hydrometeors' conversions formulated in Table 6.5.2.

Methods of numerical solution applied for SCE in SBM are discussed in Section 5.7. It should be mentioned that the method proposed by Berry and Reinhard (1974) becomes quite unstable in case collisions between particles of different types are treated. Riming, freezing and sedimentation lead to significant changes of particle size distributions (PSD), especially in the distribution tails. Utilization of high-order polynomials in the Berry and Reinhard method to approximate and to interpolate the PSD often leads to unrealistic and even to negative PSD values. In this case, even a simple linear interpolation (a polynomial of the first order) yields more realistic PSD. In contrast, methods developed by Kovetz and Olund (1969) and by Bott (1998) demonstrate high stability.

6.5.4 Collection Kernels for Different Hydrometeors: Theory and Representation in SBM Models

The collection kernel (known also as the aggregation kernel in case of ice particles) describing the probability of collection (aggregarion) of ice particles, as well as of ice particles and liquid drops is defined similarly to the collection kernel between drops (Equation 5.6.3). The expression is $K_g(m_1, m_2) = V_{swept} \cdot E \cdot E_{coal}$, where $V_{swept} = S|V_{g2} - V_{g1}|$ is the swept volume, E is the collision efficiency and E_{coal} is the coalescence efficiency. The product $E \cdot E_{coal}$ is known as the collection (aggregation) efficiency. The swept volume is determined by the product of the geometrical cross-section S and the absolute value of the relative velocity between colliding particles. There are several uncertainties in calculation of collection kernels in mixed-phase clouds caused by two main reasons. First, the geometrical cross-section is not well-defined if a particle has a complicated non-spherical shape. The problem becomes even more complicated when both colliding particles are non-spherical, and the swept volume depends on their mutual orientation. Second, there is a significant uncertainty concerning the values of the collision efficiency and the coalescence efficiency.

Calculation of Collision Cross Sections

With a few exceptions, collision cross-sections of ice particles are calculated in cloud models under the assumption that the colliding particles are "effective spheres" with a circular cross-section. An effect of non-spherical shapes of ice particles is taken into account implicitly by using the fall velocities empirically determined for non-spherical particles. The geometrical cross-section is calculated as $\pi(r_i + r_j)^2$ where r_i and r_j are the "effective" radii of the colliding ice particles or the maximum radii of their the cross-sections. In HUCM SBM scheme (Khain and Sednev, 1995, 1996) it is assumed that there is a preferential ice crystal orientation with the maximum cross-section perpendicular to the direction of particles downfall. The cross-sections are calculated on the basis of geometrical considerations. For instance, the geometrical cross-sections for collisions of columnar crystals with other hydrometeors are determined as shown in **Figure 6.5.1**. Panel (a) shows the geometrical cross-section in case a large columnar crystal collects small particles with a circular cross-section of radius r (drops, plates and dendrites). The cross-section is calculated as a product $(L + 2r)(d + 2r)$ where L is the length and d is the width of the columnar crystal. Panel (b) shows the geometrical cross-section in case a large columnar crystal with dimensions L_1 and d_1 collects small columnar crystals with dimensions L_2 and d_2. The geometrical cross-section is calculated as a product

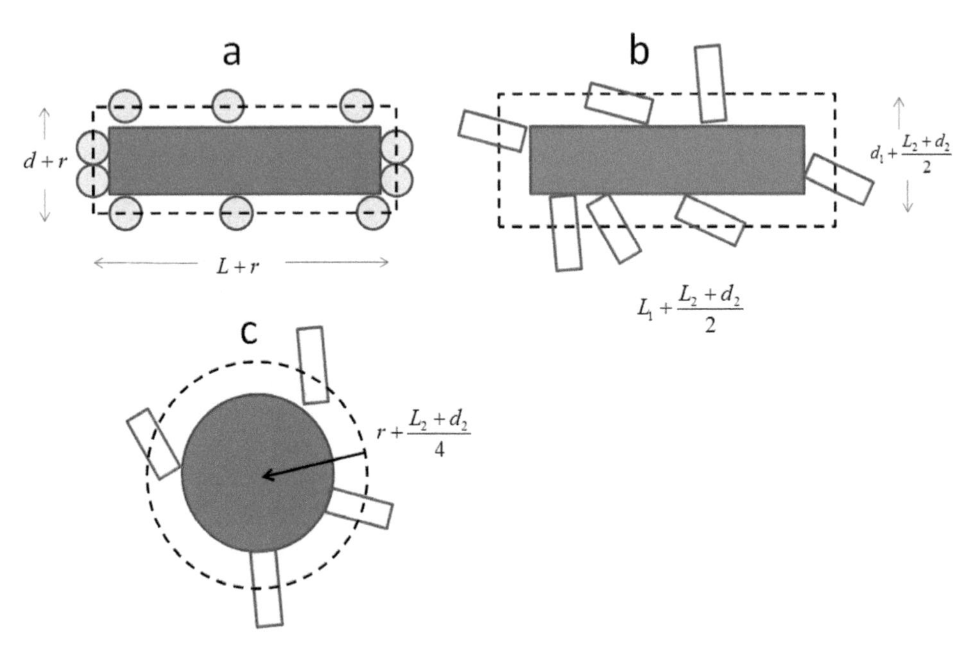

Figure 6.5.1 Geometrical cross-sections for collisions of columnar crystals with other hydrometeors as used in HUCM.

$\left(L_1 + \frac{L_2+d_2}{2}\right)\left(d_1 + \frac{L_2+d_2}{2}\right)$. Panel (c) shows the geometrical cross-section in case a large particle with a circular cross-section collects small columnar crystals. The geometrical cross-section is calculated as $\pi\left(r_1 + \frac{L_2+d_2}{4}\right)^2$. The cross-sections (b) and (c) are obtained by averaging the possible relative orientations of the two particles.

Collision Efficiencies between Large Ice Crystals and Small Water Drops

Collisions between ice particles and drops are asymmetric, i.e., the collision rates, as well as the type of the resulting hydrometeor, depend on the types of the collector and the collected particle (ice or drop). Therefore, the two types of collection are often considered separately: a) collisions of large ice crystals with small drops and b) collisions of large drops with small crystals. Wang and Ji (1992) calculated collision efficiency of plates, branch-type crystals and columns with drops of different sizes, but having fall velocities lower than those of ice crystals. The velocity field induced by the crystals was calculated numerically by means of the Navier-Stokes equations. The calculations were carried out at a fixed orientation of the crystals. Then the drop trajectories were calculated within this velocity field using the superposition method (Section 5.6). The dependencies of the collision efficiency between ice crystals of different types and drops of different radii are shown in **Figure 6.5.2**. The geometrical parameters of crystals corresponding to the dependencies shown in Figure 6.5.2 are shown in **Table 6.5.4**. One can see that the collision efficiency between ice crystals and small droplets is low; then it increases up to 0.5–0.9 with the increasing drop size but goes down to zero for drops whose fall velocity is equal to that of the ice crystal.

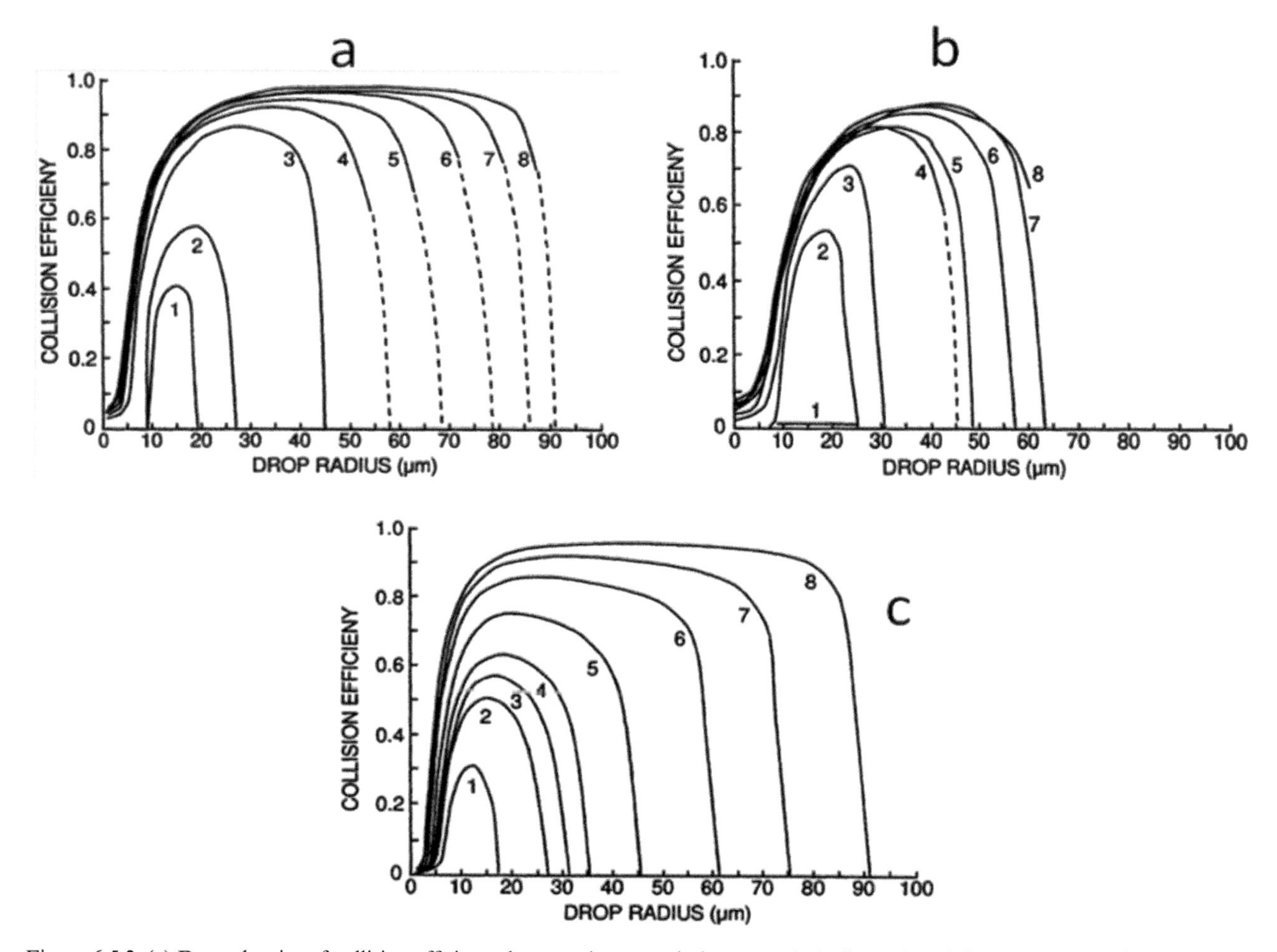

Figure 6.5.2 (a) Dependencies of collision efficiency between hexagonal plate crystals (collectors) and drops on drop radius. The geometrical parameters of the crystals are shown in Table 6.5.4. (b) The same as in (a), but for branched crystals. **(c)** The same as in (a), but for hexagonal columns (from Wang and Ji, 1992, with changes; © American Meteorological Society; used with permission).

Table 6.5.4 The geometrical parameters of crystals corresponding to the dependencies shown in Figure 6.5.2.

Number of Curve in Figure 6.5.2	Hexagonal Plate		Branched Crystals		Hexagonal Columns	
	d, μm	*h*, μm	*d*, μm	*h*, μm	*L*, μm	*d*, μm
1	160	18	200	15	67.1	47
2	226	20	250	18	93.2	65.4
3	506	32	700	32	112.6	73.2
4	716.3	37	1,000	40	128.3	83
5	947	41	1,500	50	237.4	106.8
6	1,240	45	2,000	60	514.9	154.4
7	1,500	48	2,500	65	1,067	213.4
8	1,700	49	3,100	73	2,440	292.8

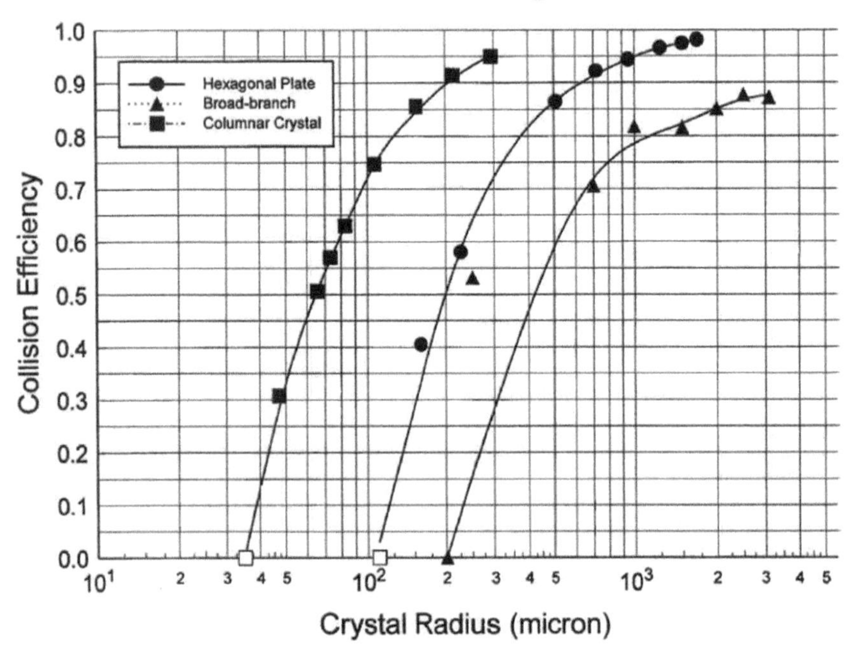

Figure 6.5.3 Dependence of collision efficiency between ice crystals and water drops on crystal sizes for different types of crystals. The curves are obtained by interpolation of values denoted by symbols. The cutoff riming ice crystal radii are clearly determined (from Wang, 2002; permission of Elsevier).

There is a minimum threshold size of ice crystals below which they cannot collect small water drops. Wang, 2002 determined this threshold radius for crystals of different sizes. This was done by plotting the maximum collision efficiency (the peak value of each curve in Figure 6.5.2 as a function of the corresponding crystal size for each crystal habit, as shown in **Figure 6.5.3**. The points where the extrapolated curves intersect the x-axis (i.e., points where collision efficiency is equal to zero) indicate the cutoff crystal size. The riming cutoff size is about 35 μm for columnar ice crystals, 110 μm for hexagonal plates and 200 μm for branched crystals. Ice crystals of sizes less than a few hundred of microns are unable to collide with droplets of radii of 5 – 10 μm. The threshold radius was theoretically derived by Pinsky et al. (2000). The threshold sizes below which ice crystals cannot collide with small drops are determined by the properties of particle hydrodynamic interaction. A falling particle induces a field of pressure perturbations around it. In front of the falling particle the pressure is higher than further away in its surrounding. At the same time, the pressure behind the particle is lower than in its surrounding. The pressure induced by the particle-collector tends to push away particles located in front of the collector (in the case of vertical motion, below the collector). As a result, a small collecting particle moves with the air flow around the collector. This effect determines the collision efficiency. Collisions take place because the collecting particle has inertia causing it to deviate from the air flow (Section 5.6). In contrast, low pressure behind the falling particle fosters the attraction of its counterpart. The magnitude of repulsion and attraction forces depend on the shape and the fall velocity of the particles. Pinsky et al. (2000) showed that collisions between drops are possible for any drop sizes. At the same time, the decrease of the collector particle fall velocity (for instance, due to lower particle density) increases the time during which the repulsion force is active, making collisions with small drops impossible. This indicates a significant difference in the hydrodynamic interaction in pairs containing an ice particle and those in drop pairs.

In case the collision efficiency value is known, the collision kernels can be calculated as a product of the geometrical collision kernel and the collision efficiency, as in case of drop-drop collisions (Section 5.6). The geometrical collision kernel is calculated using the terminal velocity values and expressions for the geometrical cross sections.

Collision Efficiencies and Collision Kernels of Large Drops with Small Ice Crystals

Experimental data concerning collisions of large drops with small ice crystals is scarce. The particles formed as a result of these collisions are typically assigned to graupel. In the latest HUCM version, such collisions lead to formation of freezing drops. Lew and Pruppacher (1983) calculated collision efficiency of drops of radii ranging from 100 μm to 600 μm colliding with columnar crystals of length ranging from 15 μm to 600 μm. The calculations were performed for two pressure values (500 mb and 900 mb) and three ice crystal orientations (**Figure 6.5.4a**). The dependence of collision efficiency on the drop size (Figure 6.6.4b) was found to be weak, collision efficiency is close to unity for ice crystals with length $L > 200$ μm. For smaller crystals, collision efficiency reaches its maximum in orientation (1) (Figure 6.5.4a) and increased as the pressure decreased (air density decreased).

Lew et al. (1985) calculated collision efficiency between drops and planar crystals, the diameter of the crystals ranging from 10 μm to 100 μm and the density of 0.9 g cm^{-3}. The orientation angle α was assumed

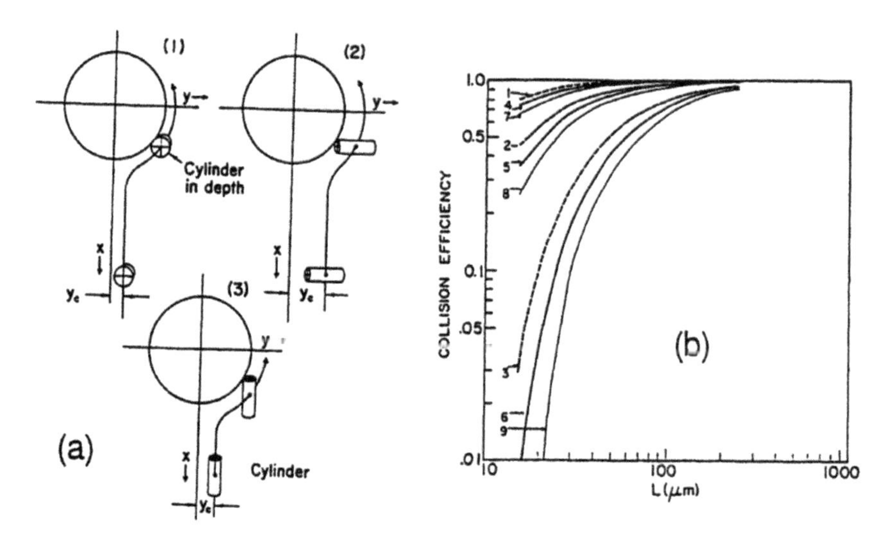

Figure 6.5.4 (a) Orientations of columnar ice crystals chosen for calculations. (b) collision efficiency as a function of a crystal length for three ice crystal orientations. Curves 1,2 and 3 are plotted for crystal density of 0.92 g cm^{-3}, $p = 900$ mb and $T_C = -6$°C; Curves 4,5 and 6 are plotted for crystal density of 0.5 g cm^{-3}, $p = 500$ mb and $T_C = -10$°C; curves 7,8 and 9 are plotted for ice crystal density of 0.5 g cm^{-3}, $p = 900$ mb and $T_C = -6$°C. The drop radius is 416 μm (from Lew and Pruppacher, 1983; © American Meteorological Society; used with permission).

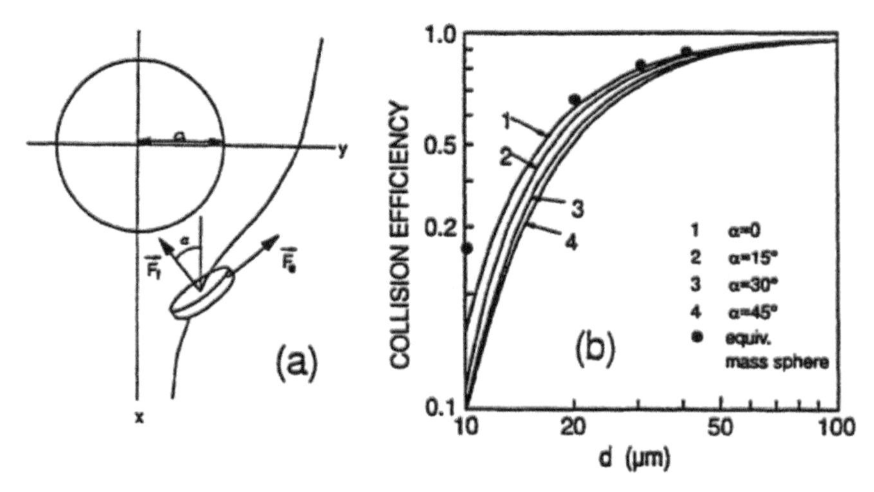

Figure 6.5.5 (a) The geometry of drop-plate ice crystal relative motion. (b) collision efficiency as a function of the ice crystal diameter at different α. The drop radius is 530 μm, $p = 400$ mb and $T_C = -12°\text{C}$. The collision efficiency between a drop collector and crystals approximated by "equivalent" spheres of the same mass and density is presented for comparison (from Lew et al. 1985; © American Meteorological Society; used with permission).

constant during the hydrodynamic interaction (**Figure 6.5.5a**). The drop radii varied from 530 μm to 760 μm. The dependence of the collision efficiency on the ice crystal diameter for a drop of 530 μm radius is shown in Figure 6.5.5b. The collision efficiency is close to unity for ice crystals of diameters exceeding 50 – 60 μm. For smaller crystals, the collision efficiency is lower than that of the effective sphere. This difference is especially significant in case of ice crystals of diameters below 10 μm. For a given orientation, the collision efficiency decreases with decreasing ice crystal density. The calculation of collision kernels was performed similar to that for drops. The fall velocities were calculated as described in Section 6.4.

Collision Efficiencies and Collision Kernels between Graupel (Hail) and Water Drops

Graupel particles form as a result of riming of ice crystals and of snow. After formation, graupel grows by riming until reaching sizes large enough to be assigned to hail. Graupel is one of the major hydrometeors responsible for precipitation in mixed-phase convective clouds, thus it is important to calculate the rate of its formation and growth as accurately as possible. In multiple studies, the collision efficiencies between graupel and water drops are assumed equal to those between a drop and another drop whose mass or size are the same as of graupel (e.g., Beheng, 1978; Johnson, 1987; Khain and Sednev, 1996; Khain et al., 1999; Ovchinnikov and Kogan, 2000). Ovchinnikov and Kogan (2000) assume that the graupel-drop collision kernel is equal to 80% of the drop-drop collision kernel.

The assumption that graupel-drop collision efficiencies are equal to or close to those of inter-drop collisions is based on an intuitive presumption that collision efficiency is determined solely by the shapes of colliding particles. However, a detailed analysis shows that collision efficiency depends also on the relative velocity of particles tending to collide (e.g., Pinsky et al., 2000). As the densities and, consequently, the terminal velocities of graupel differ from those of water drops of the same mass or size, graupel-graupel and graupel-drop collision efficiency differs from that of inter-drop collisions. To calculate water drop-graupel collision efficiencies, Khain et al. (2001a) used a superposition method described in Section 5.6. The graupel-drop collision efficiencies for different graupel radii and densities at the 750-mb pressure level are presented in **Figure 6.5.6**. As can be inferred from Figure 6.5.6, there is a cutoff size below which graupel is unable to collect water drops falling at a lower velocity than graupel. This cutoff size decreases with increase in graupel density. In addition, there exists a minimum drop size below which drops cannot be captured by graupel. Graupel-drop collision efficiencies increase as the drop size increases from zero to its maximum value, but then sharply decrease to zero when the drop terminal velocity approaches the graupel terminal velocity. As soon as the drop terminal velocity exceeds that of graupel (i.e., when the drop becomes a collector), the collision efficiency leaps to

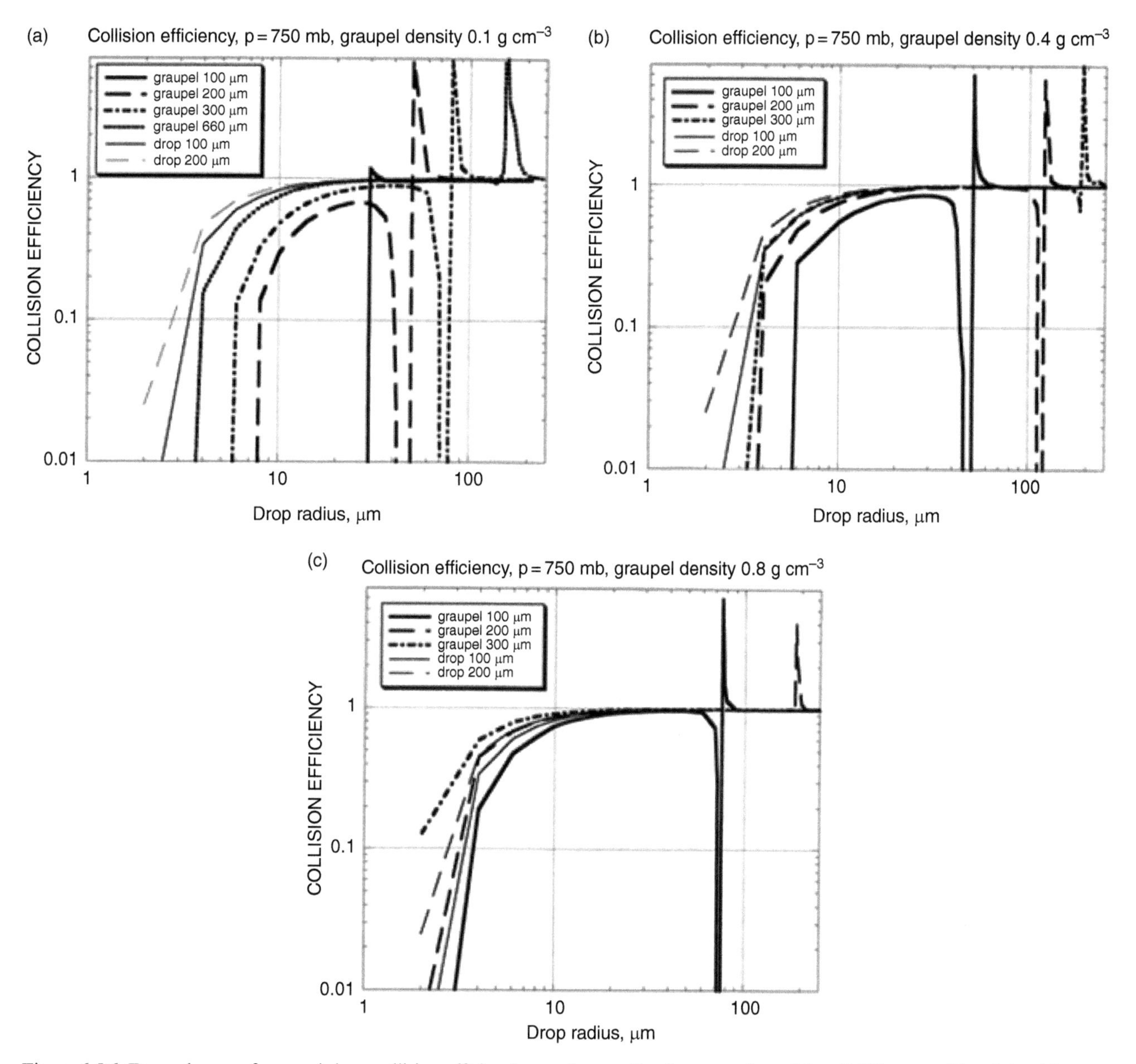

Figure 6.5.6 Dependences of graupel-drop collision efficiencies on drop radius for graupel particles of different radii and densities: (a) 0.1 g cm^{-3}, (b) 0.4 g cm^{-3} and (c) 0.8 g cm^{-3}. The collision efficiencies of drop-drop collisions are shown for comparison (from Khain et al., 2001a; © American Meteorological Society; used with permission).

values significantly exceeding unity, and then decreases rapidly to a value close to unity as the drop size grows. The leap of the collision efficiency to values larger than unity takes place for graupel of any density. Since the range of drop sizes at which collision efficiencies exceed unity is narrow, the mass grids of relatively low resolutions, used for hydrometeors in cloud models, do not resolve such peaks.

In collisions with drops, the collision efficiency of graupel increases with graupel density. Figure 6.5.6 shows that graupel–drop collision efficiencies for the low- and medium-density graupel are significantly lower than those for drop collectors of the same size. For small graupel, there is a significant difference in the collision efficiencies between drop-high-density graupel collisions and drop-drop collisions. For example, collision efficiency of a 100-μm-radius graupel and a 4-μm-radius drop is about twice as low as that between water drops of the same sizes. The collision efficiency of 0.8 g cm^{-3} density graupel particles with water drops is much closer

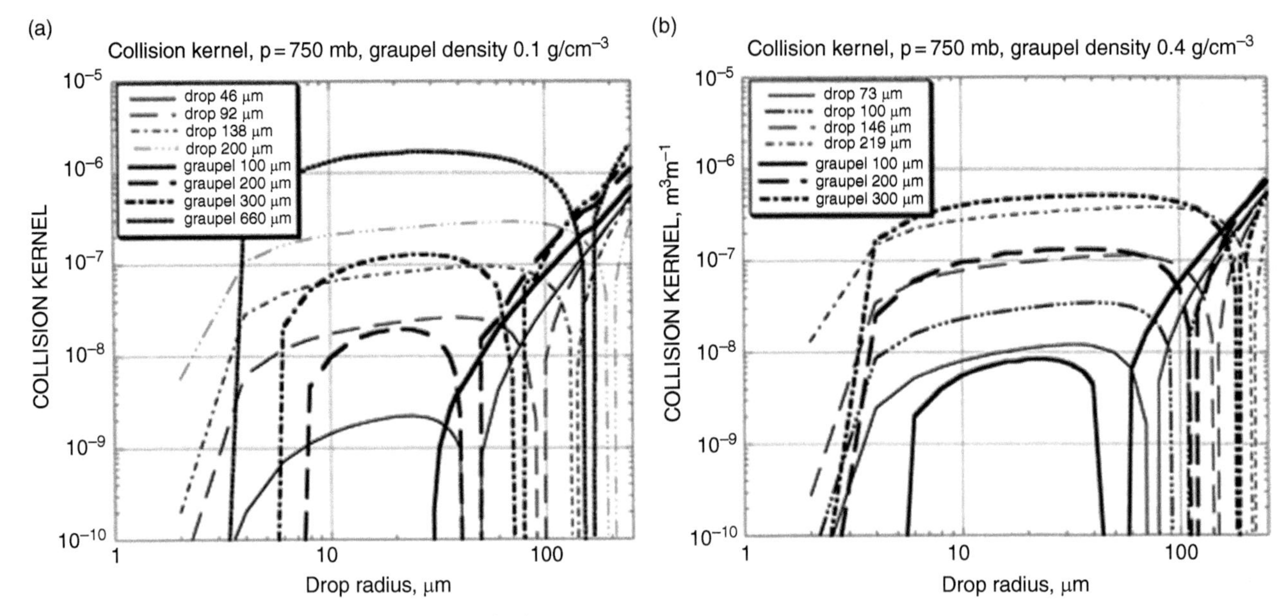

Figure 6.5.7 (a) Graupel-drop collision kernels (m^3s^{-1}) of low-density graupel with the radii of 100 μm, 200 μm, 300 μm and 660 μm as functions of the drop radii. Collision kernels of drops with radii of 46 μm, 92 μm and 138 μm having the same mass as graupel with radii of 100 μm, 200 μm and 300 μm, respectively) with drops of different radii are presented as well. The collision kernel for a 200-μm-radius drop is presented for comparison. (b) Collision kernels ($m^3\ s^{-1}$) of medium-density graupel with radii of 100 μm, 200 μm and 300 μm and drops as a function of drop radii. Drop-drop collision kernels for drops with radii of 73 μm, 146 μm, and 219 μm which have the same mass as graupel with radii of 100 μm, 200 μm and 300 μm, respectively, as well as of a 100-μm-radius drop are presented for comparison (from Khain et al., 2001a; © American Meteorological Society; used with permission).

to the collision efficiencies of inter-drop collisions than to that of low- and medium-density graupel.

Therefore, there are threshold values of graupel sizes and drop sizes below which collisions between them are impossible. As was shown by Khain et al. (2001b), the inability of graupel and ice crystals to collect small droplets is an important factor accounting for the existence of small supercooled droplets at the upper cloud levels alongside with a significant amount of graupel and ice crystals. This is a quite significant finding for the microphysics of deep convective clouds, explaining an extremely high concentration of ice crystals (of several hundred particles per cm^3) in anvils of deep convective cloud (Rosenfeld and Woodley, 2000; Heymsfield et al., 2009). These ice crystals are formed by homogeneous freezing of small droplets ascending in convective updraft.

The rate of graupel growth by riming is determined by the values of the collision kernel. **Figure 6.5.7** shows graupel-drop collision kernels of low- and medium-density graupel particles as a function of the drop radii. The collision kernels of a drop-drop pair where one of the drops has the same mass as graupel are presented for comparison. When graupel radii are below about 150 μm, graupel-drop collision kernels are smaller than those of drop-drop collisions. For graupel of radii above 200 – 300 μm, graupel-drop collision kernels become of the same order or even higher than the collision kernels of drop pairs where the mass of the drop collector is equal to that of the graupel. The ability of graupel to collect the smallest water droplets (of radius below 4 μm) is lower than that of drop collectors. When water drops serve as collectors, graupel-drop collision kernels are larger than those of drops of the same mass and/or the same size. Since the collision efficiencies in this case are close to unity both in graupel-drop and drop-drop collisions, the difference in the collision kernels is caused by higher drop-graupel relative velocities as compared to the drop-drop relative velocities. Graupel-drop collision efficiencies and kernels at $p = 750$ mb for graupel of low, medium and high density are presented in **Tables C1–C6** (**Appendix C**) that are recommended for application in cloud models.

In Section 5.6, two factors responsible for an increase of collision kernels between drops with height are discussed, namely: an increase in the relative fall velocity between drops and an increase in the collision efficiency. In mixed-phase deep convective clouds, graupel forms at the height of 4–12 km where the air density is

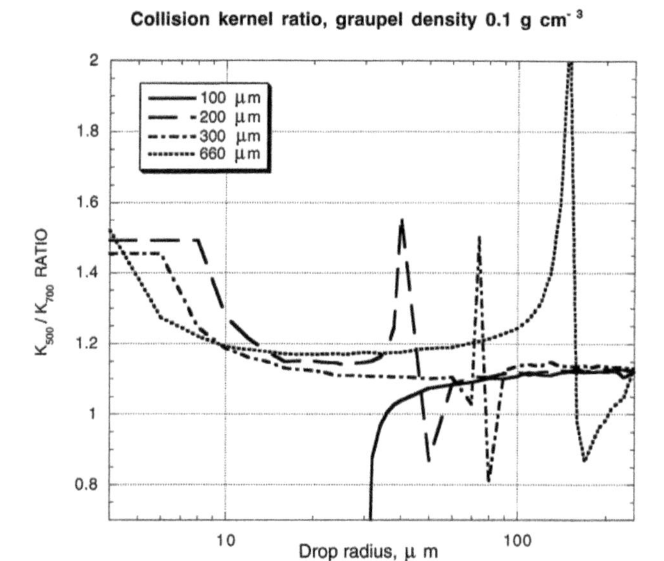

Figure 6.5.8 Ratio K_{500}/K_{750} of graupel-drop collision kernels calculated at the levels of 500 mb and 750 mb for low-density graupel (from Khain et al., 2001a; © American Meteorological Society; used with permission).

significantly lower than at the surface. Graupel particles arise above the freezing level, thus the riming process can be efficient up to levels of several kilometers above the freezing level. The ratio of graupel-drop collision kernels calculated at 500 mb (K_{500}) and 750 mb (K_{750}) for low-density graupel is shown in **Figure 6.5.8**. A significant increase (up to factor of 1.5) in the collision kernel takes place when collisions with the smallest drops occur. Peaks of this ratio take place at drop sizes corresponding to the leaps in the collision efficiencies when the terminal velocities of the drop and the graupel are equal. Due to the existence of the minimum size below which drops cannot be collected by small graupel, the ratio of the collision kernels for small drops cannot be determined. The collision kernel ratio for medium- and high-density graupel has similar features: K_{500}/K_{750} is about 1.4 for drop radii below 10 µm. This ratio abruptly decreases to 1.1–1.15 with the increasing size of the captured drop. The 15%–20% increase in the collision kernel for larger drops is determined by the increasing swept volume.

Collision efficiencies between hail and liquid drops can be assumed equal to those of dense graupel. Therefore, hail-drop collision kernels can be calculated using Table C3 from Appendix C with the fall velocities of graupel replaced for those of hail. Coalescence efficiencies for graupel-water drop and hail-water drop collisions are typically assumed equal to unity.

Collision Efficiencies and Collision Kernels between Ice Crystals and Aggregates

Collision efficiencies between spheroidal ice crystals were calculated by Bohm (1992b, c; 1999). Using simplified theory of boundary layer around solid body he obtained semi-empirical analytical solution for collision efficiency between two spheroidal particles. The collision efficiency depends on particle fall velocities, thickness of boundary layer, sizes and shapes of particles to be collide and other parameters. Since the particles of the same masses and shapes have different orientations and therefore different fall velocities, Bohm (1992c) introduced statistical distribution of particle fall velocity of the same mass and calculated dependences of average collision efficiency on particle equivalent radii.

Figure 6.5.9 shows the collision efficiency and collision kernels between branched planar ice crystals obtained using Bohm's theory. One can see that the collision efficiency between crystals with radii below 100 µm is less than 0.05, and rapidly increases with the increase in the size of the collector. However, if the collected crystal is small, the collision efficiency remains significantly below unity. The figure shows that the results depend on a way of calculation of relative fall velocity mostly when the sizes of colliding particles are close.

In HUCM, the collision efficiencies between aggregates and ice crystals, as well as between aggregates are assumed equal to those between effective spheres of corresponding particle densities (idealized shapes of ice particles). These collision efficiencies are calculated using the superposition method (e.g., Khain et al., 2001a). Collision kernels for ice crystals are calculated taking into account the correction proposed by Bohm (1992c).

6.5.5 Coalescence Efficiencies between Ice Particles

Collection of ice crystals triggers formation of aggregates, thus contributing to precipitation (especially in winter time). Not all collisions lead to particle collections. Mechanisms leading to collection of ice crystals and creating aggregates at low temperatures are not well known. There are many examples when relatively large elements of ice crystals and aggregates get "hooked" by narrow necks, for instance, when ice crystals have a T-shape. Aggregates often represent chains of ice crystals. The forces that keep these elements together are currently not known, yet it is assumed that these forces are of electric nature (e.g., Connolly et al., 2005), but no attempts have been made so far to quantitatively evaluate them.

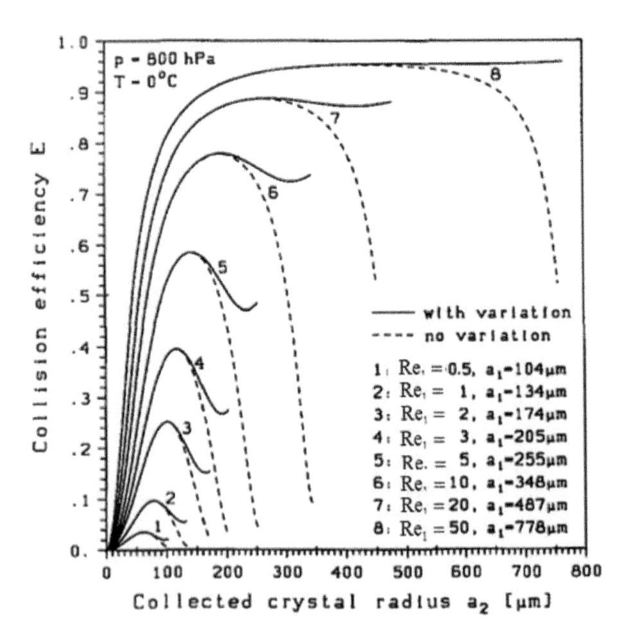

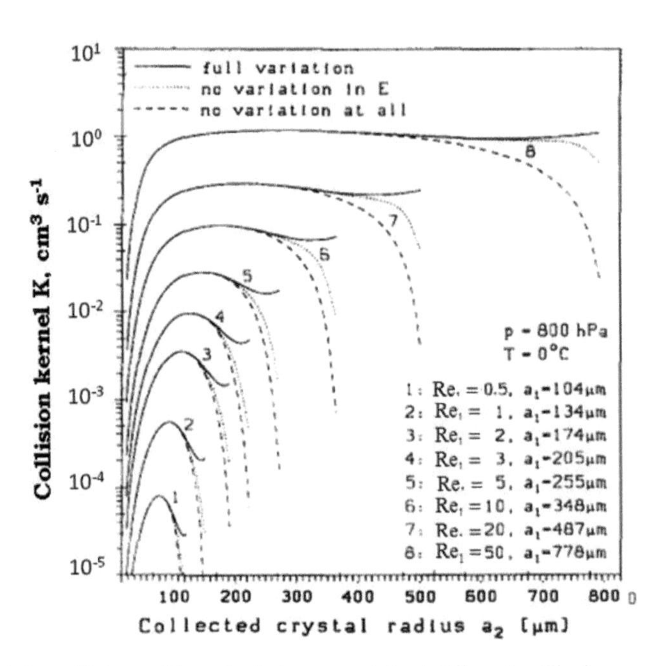

Figure 6.5.9 Collision efficiencies (left) and collision kernels (right) for collisions of branched planar crystals at different radii of collectors and collected crystals. The Reynolds numbers and the corresponding radii of collectors are presented in the legends. Solid and dashed lines show the values calculated using different ways of calculation of relative fall velocity. The dashed curves show calculations with the mean fall velocities used. The axial ratio of crystals is equal to 0.05 and the bulk density is 0.6 $g cm^{-3}$ (from Bohm, 1992c; courtesy of Elsevier).

It is known that coalescence (sticking) depends on the air temperature. Kajikawa and Heymsfield (1989) evaluated E_{coal} at $T_C = -35°C$ to $-40°C$ as ~0.05. According to the experimental data presented by Hostler et al. (1957), there is an obvious dependence of coalescence efficiency on water vapor pressure over ice. Khain and Sednev (1996) parameterize the dependence of coalescence efficiency on temperature and water vapor pressure as following

$$E_{coal} = \min\left[1,\ \frac{e}{e_i}\max\{0, a + bT_C + cT_C^2 + dT_C^3\}\right] \quad (6.5.9)$$

where e and e_i are the water vapor pressure and its saturated value over ice, respectively, $a = 0.883$, $b = 0.093$, $c = 0.00348$ and $d = 4.52 \times 10^{-5}$. Function E_{coal} shows a good agreement with the experimental data obtained by Rogers (1973) in field studies of snowflakes. According to Equation (6.5.9), the coalescence efficiency approaches the value of $0.883\frac{e}{e_i}$ at 0°C. Equation (6.5.9) shows that E_{coal} monotonically increases with increasing T_C. This formula does not take into account the ice particle type and size.

Mitchell (1988) found that E_{coal} reaches its maximum within a "dendritic zone" within the temperature range of $-13°C - 17°C$ at high humidity where dendrites is the main crystal habit. Within this range, aggregation rate increases, which is not taken into account by Equation (6.5.9). Connolly et al. (2012) also found the maximum of coalescence efficiency at $-15°C$. Thus, coalescence efficiency depends of temperature, on humidity and on the shape and texture of the ice crystal.

Phillips et al. (2015) made an attempt to theoretically formulate coalescence efficiency between ice particles. They treated the coalescence efficiency as the probability of two colliding particles getting stuck after collision. If K_0 is the initial collisional kinetic energy between two particles with masses m_{small} and m_{big}, it is calculated as (see also Equation 5.6.15):

$$K_0 = 0.5 \times \frac{m_{small} m_{big}}{m_{small} + m_{big}} \left(V_{big} - V_{small}\right)^2 \quad (6.5.10)$$

Suppose K_1 is the collisional kinetic energy between these two ice particles after collision. Let us denote the energy converted to heat and the noise caused by the particle's inelastic deformation as K_{loss}, and the work to be done to separate the particles after the impact as ΔS. According to the energy conservation law

$$K_0 = K_1 + K_{loss} + \Delta S \quad (6.5.11)$$

The rebound takes place at $K_1 > 0$. So, coalescence takes place at $K_0 \leq K_{loss} + \Delta S$.

If one of the particles is substantially smaller than the other, $K_{loss} = K_0(1 - q^2)$ where q is the coefficient of restitution, defined as the ratio of the relative velocities of the colliding particles after and before the impact (Wall et al., 1990). The coefficient of restitution depends on properties of the material of the particles, such as stiffness. Substituting the expression for K_{loss} to the last inequality leads to the condition at which coalescence takes place: $K_0 \leq \Delta S/q^2$. Accordingly, Phillips et al. (2015) define the collection efficiency as the probability of $K_0 < \Delta S/q^2$:

$$E_{coal} = P(K_0 q^2 < \Delta S) \quad (6.5.12)$$

Assuming that ΔS is proportional to the entire surface area of the smaller particle α, Phillips et al. (2015) showed that the collection efficiency can be written in the form

$$E_{coal} = \exp\left(-\frac{\beta(T, RH_i, \ldots)K_0}{\alpha}\right), \quad (6.5.13)$$

where

$$\beta(T, RH_i, \ldots) = \chi(RH_i) d\left[c - \frac{b}{b + (T - T_1)^2}\right] \bigg/ \xi V_{imp}^{\kappa} \quad (6.5.14)$$

where RH_i is the relative humidity with respect to ice. The factor $\chi(RH_i)$ is unity for $RH_i > 1$. Otherwise, $\chi(RH_i)$ has a larger value required to treat the smoothness of ice surfaces during sublimation. Quantity ξ describes the effect of the number of component ice crystals (monomers) on the coalescence efficiency. The coalescence efficiency is assumed to monotonically increase with increasing number of monomers. An empirical power law is proposed: $\xi = 1 + \omega\left[(n_j n_k)^{\gamma} - 1\right]$, where n_k and n_j are the numbers of monomers in colliding particles of type k and j, and ω and γ are empiric constants. In case of collision of ordinary ice crystals, the number of monomers in the colliding pair is $n_k = n_j = 1$ and $\xi = 1$.

The surface area of contact is assumed to be proportional to V_{imp}^{κ} (with κ close to unity) where V_{imp} is the impact velocity of the smaller particle at the surface of the larger one. Although this velocity is determined by both the difference in the gravitational settling velocities and the velocities of particle rotation, Rasumussen and Heymsfield (1985) and Phillips et al. (2015) calculate V_{imp} as the difference in the fall velocities of the colliding particles. The minimum value of V_{imp} was assumed

Table 6.5.5 Parameters of Equation (6.5.13) for parameterization of graupel-ice crystal and snow-ice crystal coalescence efficiency (from Phillips et al., 2015; © American Meteorological Society; used with permission).

Parameter	Graupel Collisions	Snow Collisions
b (°C^2)	0	16.98
c	1	1.023
d ($kg^{-1}s^{1/2}m^{3/2}$)	20,654	3,981
T_1 (°C)	–	−15
γ	1.	1.3
κ	1.5	1,5
$\chi(RH_i > 100\%)$	13	1
$\chi(RH_i < 100\%)$	0.1	3
ω		0.05

equal to 5 cm/s. The factor $d \times \left[c - \frac{b}{b+(T-T_1)^2}\right]$ takes into account the dependence of the coalescence efficiency on temperature. Using available laboratory measurements and observations, Phillips et al. (2015) proposed setting the values of the parameters in Equation (6.5.13) as presented in **Table 6.5.5**.

Figure 6.5.10 shows the temperature dependences of coalescence efficiency for snow-crystal collisions predicted by the parameterization used by Phillips et al. (2015) and obtained in measurements by Mitchel (1988) and Connolly et al. (2012). One can see that the parameterization used by Phillips et al. (2015) is able to describe the available observed data, in particular, the maximum of coalescence efficiency at $T = -15$°C.

There are no observations or laboratory experiments concerning graupel-graupel or graupel-hail coalescence efficiency. Phillips et al. (2015) recommend to use graupel-graupel or graupel-hail coalescence efficiencies set by order of magnitude less than those of graupel-ice crystals. Another option widely used in numerical modeling is to assume zero coalescence efficiencies between graupel-graupel and graupel-hail. In recent versions of HUCM (Khain et al., 2011; Ilotoviz et al., 2016), the temperature dependence of coalescence efficiency for all ice aggregation is used as shown by the dashed line in Figure 6.5.10. In this approach, coalescence efficiency does not depend on particle size or type. The coalescence efficiencies between graupel-graupel, graupel-hail, hail-snow, hail-hail, hail-ice crystals are lower than those between ice crystals or

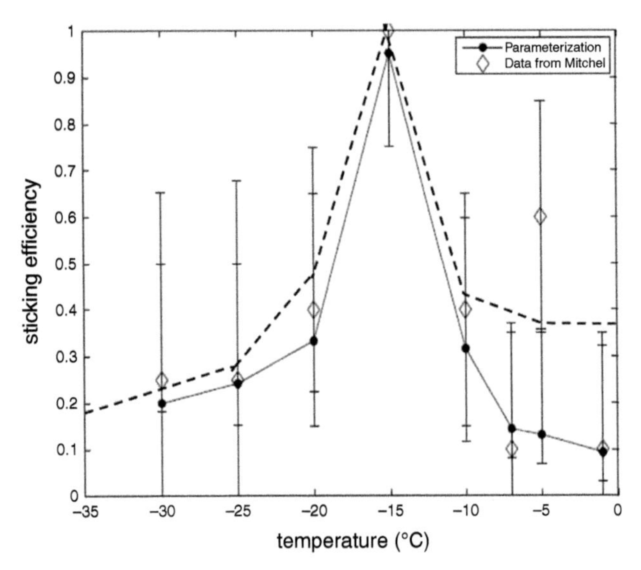

Figure 6.5.10 Temperature dependence of coalescence efficiency for snow-crystal collisions obtained using the Phillips et al. (2015) parameterization on the basis of experimental data obtained by Connolly et al. (2012) (black circles). Open diamonds represent data obtained by Mitchell (1988). The error bars mark the standard deviation of E_{coal} over the size distributions given by Connolly et al. (2012). The collection efficiency between ice crystals used in HUCM by Khain et al. (2011) (dashed line) is presented for comparison. (from Phillips et al., 2015, with changes; © American Meteorological Society; used with permission).

between snowflakes. The latter are often assumed equal to zero as in Khain et al. (2011).

6.5.6 Conversion of Aggregates into Graupel by Gradual Riming in SBM Models

The stochastic equation for collisions presented in Section 6.5.3 describes transformation of particle types regardless the history of a particle's evolution. However, particle riming is often a gradual process in which the bulk density of riming particles increases during a comparatively long time until it becomes similar to graupel density. Then, these rimed ice crystals or snowflakes should be converted to graupel. The idea to characterize hydrometeors by their bulk density continuously increasing by riming was elaborated by Morrison and Grabowski (2010), Morrison and Milbrandt (2015) and Morrison et al. (2015). In these studies, the bulk density of snow increases by riming and as it reaches large values the particles (initially unrimed snow) are formally assigned to graupel.

In model studies by Khain et al. (2011), Noppel et al. (2010) and Benmoshe et al. (2012) calculation of the rimed mass within snowflakes of different mass is included into simulations. Advection and sedimentation of the rimed mass are calculated using the same equations as for advection and sedimentation of snow itself. In the Lagrangian approach, this is equivalent to advection and sedimentation of particles with particular conservative properties. In case of collisions, the resulting aggregate has the mass equal to the sum of the masses of the colliding particles. A new rimed mass is equal to the sum of their rimed masses. A schematic illustration of rimed mass accumulation and conversion of heavily rimed snowflakes into graupel is illustrated in **Figure 6.5.11**. The rimed mass in snowflakes is calculated in all mass bins. The smallest snowflakes are typically rimed only weakly, while larger snowflakes collect water drops more efficiently. It is clear that the riming mass depends on the presence or absence of supercooled water in a snowflake's surrounding. At each time step, a new bulk snowflake density is calculated as

$$\rho_s = \frac{m_s}{Vol_s} = \frac{m_s}{\dfrac{m_{rime}}{\rho_i} + \dfrac{m_s - m_{rime}}{\rho_0}} \tag{6.5.15}$$

where Vol_s is the volume of a snowflake, m_s is the snowflake mass, m_{rime} is its rimed mass, ρ_i is the density of the rimed mass, which ranges from about 0.3 g cm^{-3} to 0.92 g cm^{-3} in different studies. ρ_0 is the density of the initial (unrimed) snowflake of mass m_s. The dependence of the density of pure unrimed snowflakes on their mass is presented in Figure 6.1.6. It is assumed that conversion of heavily rimed snowflakes into graupel takes place when the bulk density of a snowflake exceeds $\rho_{cr} = 0.2\ \text{g cm}^{-3}$.

Riming of graupel accompanied by recalculation of graupel density was also included in the two-moment bulk scheme suggested by Milbrandt and Morrison (2013).

6.5.7 Description of Collision and Collection in Bulk Microphysics Models

Numerous collision schemes have been developed for bulk parameterization models (e.g., Cotton et al., 1982, 1986; Lin et al., 1983; Rutledge and Hobbs, 1984; Murakami, 1990; Walko et al., 1995; Reisner et al., 1998; Thompson et al., 2004; Morrison et al., 2005; Seifert and Beheng, 2006; Li et al, 2008). Many modern schemes contain elements applied in earlier studies. There are several common features of bulk schemes for collisions:

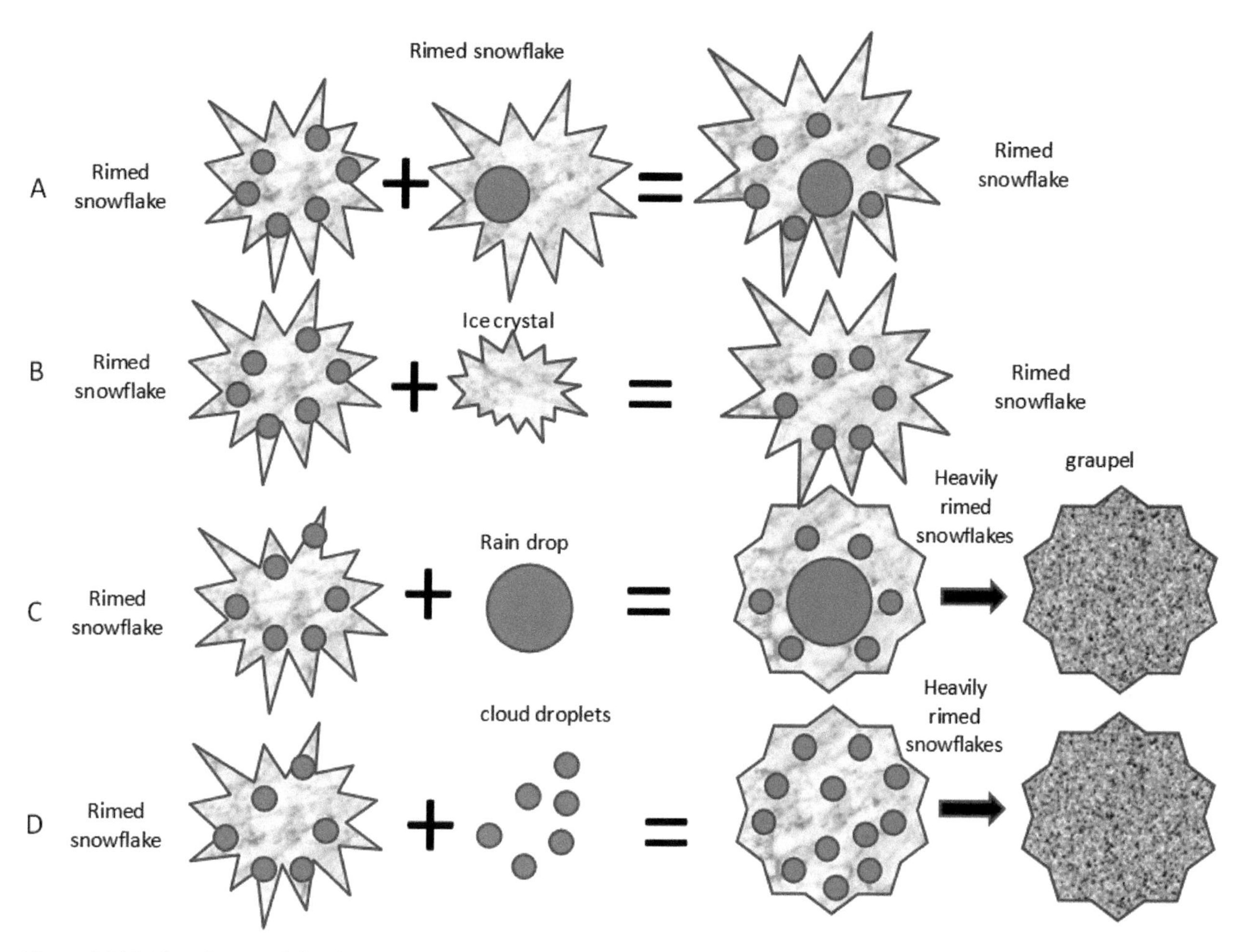

Figure 6.5.11 The schemes of rimed mass accumulation and conversion of heavily rimed snowflakes into graupel by collection of snow by snow (A), snow by crystals (B), snow by rain drops (C) and snow by cloud droplets (D), applied in HUCM. Blue areas within snowflakes depict rimed mass with density of 0.9 g cm^{-3}.

- Ice particles are considered as equivalent spheres having the Gamma or the Marshall-Parmer size distributions.
- Growth of particles by collisions and coalescence are often described using the equation of continuous growth (Equation 5.6.4) according to which particle collectors do not collide with each other and collect smaller particles independently.
- Since the collection equations are written for the PSD moments (mass contents in one-moment schemes, mass contents and concentrations in two-moment schemes), the parametrization schemes operate with values averaged over the corresponding distributions. For instance, averaged collection efficiencies between hydrometeors of different types are used. As a result, collection efficiencies used in parameterizations are independent of particle size. The same is true about fall velocities which are integrated over mass distribution or PSD (Section 6.4).

In a sense, the treatment of collisions in bulk parameterization schemes is more complicated than that in spectral bin microphysical schemes. Indeed, in SBM the collision procedure is reduced to solving the stochastic collection equations which are quite similar for all hydrometeors regardless their type. The main difference in treating different hydrometeors is in utilization of different collection kernels which is a quite simple and routine procedure. In contrast, complicated changes of PSD caused by collisions are described in bulk schemes using only one or two moments of PSD. This creates complicated physical and mathematical problems.

Bulk parameterization schemes describe the evolution of PSD moments. The moment of order k is defined as $M_x^k = \int_0^\infty m^k f_x(m)dm$, where $f_x(m)$ is the size distribution function, $k = 0$ corresponds to the particle concentration and $k = 1$ corresponds to the particle mass (Section 2.1). As shown in Table 6.5.1, all collections can be divided into several types.

Collection of Type $X + Y \to X$ and $X + Y \to Z$
Collection of particles of type X and type Y can lead either to formation of particles of type Z ($X + Y \to Z$) or of type X ($X + Y \to X$). An example of $X + Y \to Z$ collection is that between snow and raindrops, resulting in graupel formation. An example of $X + Y \to X$ collection is that between graupel and drops, leading to graupel growth (riming) (Tables 6.5.1, 6.5.3). Time evolution of PSD moments (collecting particles M_x^k, collected particles M_y^k and the resulting hydrometeor M_z^k) can be written as following (Seifert and Beheng, 2006):

$$\left(\frac{\partial M_x^k}{\partial t}\right)_{x+y->z} = -\int_0^\infty\int_0^\infty K_{xy} f_x(m_x) f_y(m_y) m_{xy}^k dm_x dm_y \tag{6.5.16}$$

$$\left(\frac{\partial M_y^k}{\partial t}\right)_{x+y->z} = -\int_0^\infty\int_0^\infty K_{xy} f_x(m_x) f_y(m_y) m_{yy}^k dm_x dm_y \tag{6.5.17}$$

$$\left(\frac{\partial M_z^k}{\partial t}\right)_{x+y->z} = \int_0^\infty\int_0^\infty K_{xy} f_x(m_x) f_y(m_y) (m_x+m_y)_y^k dm_x dm_y \tag{6.5.18}$$

where the collection kernel is calculated similar to the case of drop collisions (Section 5.6)

$$K_{xy} = \frac{\pi}{4}\left[D_x(m_x) + D_y(m_y)\right]^2 \left|V_x(m_x) - V_y(m_y)\right| E_{xy} \tag{6.5.19}$$

where D_x and D_y are diameters of the particles. Integrals (6.5.16) and (6.5.17) describe a decrease in concentration ($k = 0$) and mass content ($k = 1$) of hydrometeors of types X and Y; integral (6.5.18) represents the rate of increase in the concentration ($k = 0$) and of the mass ($k = 1$) of type Z hydrometeors. Equations (6.5.16–6.5.17) represent the Loss integrals in the stochastic collection equations written for the corresponding moments, while Equation (6.5.18) represents the Gain integral in those equations. One can see that an increase in the mass content of Z-type hydrometers is equal to the sum of mass losses of X and Y hydrometeors, which is consistent with the mass conservation law. An increase in the concentration of Z-type hydrometers is equal to the decrease in the concentration of particles belonging to either X or Y hydrometeors. Concentrations of X or Y decrease by the same value. The equality $\Delta M_z^0 = -\Delta M_x^0 = -\Delta M_y^0$ means that only binary collisions are taken into account.

Most of the existing bulk schemes, especially two-moment ones, are based on calculation of integrals (6.5.16–6.5.18). The differences between the bulk schemes are related to different treatments of these integrals, implying several simplifications since the integrals cannot be solved analytically. Below we describe the main simplifications used in calculation of the integrals (6.5.16–6.5.18).

Collection Efficiencies between Ice Particles
In most bulk schemes, the collection efficiencies are not represented as a product of collision and coalescence efficiencies. This simplification is related to the fact that aggregation between ice crystals (which of small size) is typically not considered. At the same time, collision efficiencies between large snow particles and ice crystals and other snow particles can be assumed equal to unity. In this case, the collection efficiency becomes equal to the coalescence efficiency. Another simplification is replacing the collection efficiencies E_{xy} by value $\overline{E}_{xy}$ standing for an averaged, or "effective" collection efficiency between hydrometeors of types X and Y. The value of $\overline{E}_{xy}$ can be removed from the integrals. The averaged collection efficiency $\overline{E}_{xy}$ does not depend on a particle size but only on the environment temperature. The formulas used for parameterization of the collection efficiency between ice particles of various types applied by different authors are presented in **Table 6.5.6**. The dependences of collection efficiency on temperature are presented in **Figure 6.5.12**. One can see significant differences in the collection efficiency values. E.g., at $T_C = 0°C$ formula (F1) yields $\overline{E}_{ii} = 1$ and formula (F2) yields $\overline{E}_{ii} = 0.2$.

Puppacher and Klett (1997) show that porosity of aggregates substantially increases collection efficiency. It is also known that snow forms efficiently at temperatures of about $-12°C$ when the concentration of dendrites of high porosity reaches its maximum. This phenomenon can be in part accounted for by a comparatively high collection efficiency between dendrites. Using the observed data, Passarelli (1978) evaluated the averaged collection efficiency for aggregating stellar and dendritic crystals as 1.4 ± 0.6 within the temperature range

Table 6.5.6 Expressions for collection efficiencies used in bulk parameterization schemes.

Author	Equation	NN formula
Lin et al. (1983)	$\bar{E}_{ii} = \bar{E}_{si} = \exp(0.025T_c)$	(F1)
Hallgren and Hosler (1960); Cotton et al. (1986)	$\bar{E}_{ii} = \min\{10^{0.035T_c-0.7}, 0.2\}$	(F2)
Ferrier et al. (1995); Milbrandt and Yau (2005b)	$\bar{E}_{ii} = 0.005\exp(0.1T_c)$	(F3)
Ferrier et al. (1995); Milbrandt and Yau (2005b)	$\bar{E}_{si} = 0.05\exp(0.1T_c)$	(F4)
Lin et al. (1983); Ferrier et al. (1995); Seifert and Beheng (2006)	$\bar{E}_{hi\_dry} = 0.1$ $\bar{E}_{hc\_dry} = \exp(0.09T_c)$ $\bar{E}_{hi\_wet} = \bar{E}_{hs\_wet} = 1.0$	(F5)
Milbrandt and Yau (2005b)	$\bar{E}_{hi\_dry} = \bar{E}_{hs\_dry} =$ $\bar{E}_{gi\_dry} = \bar{E}_{gs\_dry} = 0.01\exp(0.1T_c)$	(F6)
Ferrier et al. (1995); Milbrandt and Yau (2005); Seifert and Beheng (2006)	$\bar{E}_{gi\_wet} = \bar{E}_{gs\_wet} = 1$	(F7)
Milbrandt and Yau (2005)	$\bar{E}_{xl} = \bar{E}_{gl} = E_{hl} = \exp\left[-8.68\times10^{-7}\overline{D}_l^{-1.6}\overline{D}_x\right]$	(F8)
Milbrandt and Yau (2005b)	$\bar{E}_{ci} = \bar{E}_{ri} = \bar{E}_{is} = \bar{E}_{cs} =$ $\bar{E}_{rs} = \bar{E}_{rg} = \bar{E}_{rh} = 1$ $\bar{E}_{ss} = \bar{E}_{ii} = \bar{E}_{gh} = \bar{E}_{hh} = 0$	(F9)

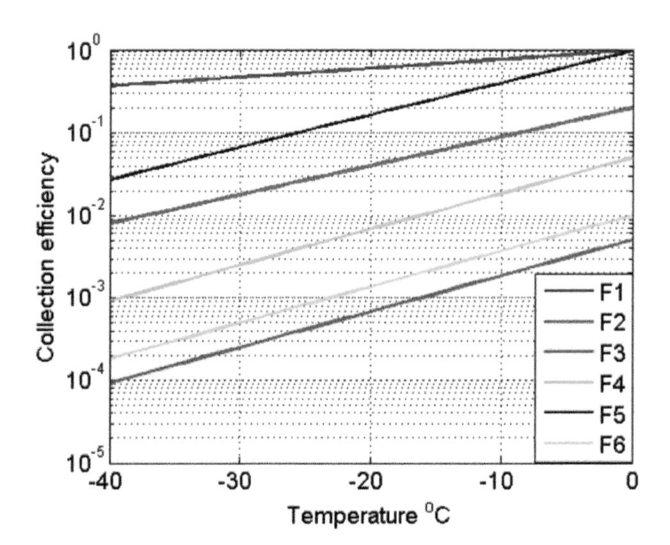

Figure 6.5.12 Temperature dependencies of collection efficiency applied by different authors: Notations: F1–F6 correspond to the equation numbers in Table 6.5.6.

of −15°C to 12°C. The high collection efficiency indicates an intense production of aggregates within this temperature range. In most bulk schemes, the collision efficiency between ice particles is assumed equal to one, making it equal to the coalescence efficiency.

Difference of Fall Velocity

Another simplification is that the difference in fall velocities of particles X and Y needed for calculation of collection kernels is also replaced by some kind of an average value $\overline{|\Delta V_{xy}|}$. This value can be removed from the integral (6.5.18). As a result, Equation (6.5.18) is replaced for the following expression:

$$\left(\frac{\partial M_z^k}{\partial t}\right)_{x+y->z} = \frac{\pi}{4}\overline{|\Delta V_{xy}|}\overline{E}_{xy}\int_0^\infty\int_0^\infty f_x(m_x)f_y(m_y)(m_x+m_y)_y^k(D_x+D_y)^2 dm_x dm_y \quad (6.5.20)$$

There are different formulas for calculating $\overline{|\Delta V_{xy}|}$ (**Table 6.5.7**). To simplify calculations, it is often assumed that the fall velocity of particle collectors is much higher than that of collecting particles, so the velocity of collected particles can be neglected. For instance, the fall velocity of ice particles during riming is often assumed to be much larger than that of collected drops (e.g., Cotton et al., 1986). In this case, $\overline{V}_y \ll \overline{V}_x$ and one can use approximated formula (F1) from Table 6.5.7. Such simplification is typically used in one-moment bulk schemes, where $\overline{V}_x$ and $\overline{V}_y$ are mass-weighed fall velocities. Constants C_n, C_1, C_2 and C_3 in Table 6.5.7 depend on the parameters of PSD (Straka, 2009).

The velocity of collected particles cannot always be neglected, so formula (F1) can be applied only in a limited number of cases. In some schemes (e.g., Wisner et al., 1972; Cotton et al., 1986) it is

Table 6.5.7 Expressions for calculation of $\overline{|\Delta V_{xy}|}$ used by different authors (from Khain et. al., 2015; courtesy of © John Wiley & Sons, Inc.).

Number of Formula	References	**Expressions for** $\overline{|\Delta V_{xy}|}$
(F1)	Cotton et al. (1986)	$\overline{V}_x$
(F2)	Wisner et al. (1972); Cotton et al. (1986)	$\left|\overline{V}_x - \overline{V}_y\right|$
(F3)	Murakami (1990); Milbrandt and Yau (2005a,b); Morrison et al. (2005)	$\sqrt{\left(\overline{V}_x - \overline{V}_y\right)^2 + 0.04\overline{V}_x\overline{V}_y}$
(F4)	Mizuno (1990)	$\sqrt{\left(1.2\overline{V}_x - 0.95\overline{V}_y\right)^2 + 0.08\overline{V}_x\overline{V}_y}$
(F5)	Murakami (1990); Mizuno (1990); Seifert and Beheng (2006)	$\left[\frac{1}{F_n}\int_0^\infty\int_0^\infty \left[V_x(m_x) - V_y(m_y)\right]^2 D_x^2 D_y^2 f_x(m_x) f_y(m_y) m_y dm_x dm_y\right]^{1/2}$ where $F_n = \int_0^\infty\int_0^\infty D_x^2 D_y^2 f_x(m_x) f_y(m_y) m_y dm_x dm_y = C_n N_x N_y D_x^2(\overline{m}_x) D_y^2(\overline{m}_y)\overline{m}_x$ C_n is a constant.
(F6)	Flatau et al. (1989); Ferner (1994)	$\left[\frac{1}{F_n}\int_0^\infty\int_0^\infty \left[V_x(m_x) - V_y(m_y)\right]^2 (D_x + D_y)^2 f_x(m_x) f_y(m_y) m_x dm_x dm_y\right]^{1/2} = \left\{\frac{1}{F_n} N_x N_y \overline{m}_x \left[C_1\overline{V}_x^2 - 2C_2\overline{V}_x\overline{V}_y + C_3\overline{V}_y^2\right]\right\}^{1/2}$ where $F_n = \int_0^\infty\int_0^\infty \left(D_x(\overline{m}_x) + D_y(\overline{m}_y)\right)^2 f_x(m_x) f_y(m_y) m_x dm_x dm_y$

assumed that $\overline{|\Delta V_{xy}|}$ is equal to the absolute value of the difference between the mass-weighed or concentration-weighed fall velocities ((F2) in Table 6.5.7). This expression used in many bulk schemes is not only a mathematically crude approximation, but it leads to a physical paradox of vanishing self-collisions. According to this formula, there are no collisions between hydrometeors belonging to the same type. Since large snowflakes have an average fall velocity close to that of small raindrops, utilization of this expression may also erroneously prevent snow-raindrop collisions, in spite of the fact that for particular snow-raindrop pairs the fall velocities of the colliding particles are different and collisions take place. To avoid this problem some bulk schemes (e.g., Milbrandt and Yau, 2005a,b; Morrison et al., 2005) use the formula proposed by Murakami (1990) (Equation (F3) in Table 6.5.7). The coefficient 0.04 characterizes the value of dispersion of the fall velocities. Mizuno (1990) proposed a similar approximation of $\overline{|\Delta V_{xy}|}$ (Equation (F4) in Table 6.5.7). To get a better approximation of the collision rate, Seifert and Beheng (2006) use RMS values instead of the mean absolute values (Equation (F5) in Table 6.5.7). Flatau et al. (1989) and Ferrier (1994) calculate $\overline{|\Delta V_{xy}|}$ using Equation (F6) having the same meaning as Equation (F5). The advantage of the two latter parameterizations is taking into account the specific forms of PSD obtained in recent model calculations.

In case of $X + Y \rightarrow X$ collections, the rate of increase of the k-th moment of type X hydrometeors is described by the integral

$$\left(\frac{\partial M_x^k}{\partial t}\right)_{x+y->x} = \int_0^\infty\int_0^\infty K_{xy} f_x(m_x) f_y(m_y) \left[(m_x + m_y)^k - m_{xy}^k\right] dm_x dm_y \tag{6.5.21}$$

The decrease in the concentration and the mass content of type Y hydrometeors due to their collection by X-type particles is described by Equation (6.5.17). In this class of collections, there is no change in the concentration of hydrometeors of type X, because the concentration of collectors does not change. This is clearly seen from Equation (6.5.21) if $k = 0$. The decrease of the mass content of collected particles can be calculated from the mass conservation equation:

$$\left(\frac{\partial q_y}{\partial t}\right)_{x+y->x} = -\left(\frac{\partial q_x}{\partial t}\right)_{x+y->x} \tag{6.5.22}$$

Substitution of the Gamma distributions into equations of type (6.5.20–6.5.21) leads to quite complicated expressions. Examples of these expressions are presented in **Appendix 6.5.8**. In most expressions the rates of the moment changes are proportional to the products of the mass content of one hydrometeor and the number concentration of the other one (Equation A12 in the Appendix 6.5.8) or to the products of concentrations of the interacting hydrometeors (Equation A13).

Simpler parameterizations formulas can be obtained using the Marshall-Palmer distributions typically applied in one-moment bulk schemes. For example, Cotton et al. (1986) propose the formula for snow-cloud droplets collision leading to riming of snow (of $X + Y \to X$) that follows from the assumption of the Marshall-Palmer distribution (Equation 2.1.29) for snow in the form $f(D) = \frac{N}{\overline{D}_s}\exp\left(-\frac{D}{\overline{D}_s}\right)$. The characteristic diameter of snowflakes $\overline{D}_s = 0.33$ cm. Neglecting the terminal velocity of small drops, Cotton et al. (1986) used the simplified equation (F1) from Table 6.5.7 for $\overline{|\Delta V_{xy}|}$. They also used the expression for snow particles fall velocity $V_{g,s} = \left(\frac{4g\rho_s}{3\rho C_D}\right)^{1/2} D^{0.2}$, where $C_D = 1.3$ and the dependence of snow density on the particle diameter $\rho_s = \beta_1 D^{-0.6}$ with $\beta_1 = 0.015\ \mathrm{gm}^{-2.4}$. As a result, the following equation for snow mass content rate was obtained

$$\left(\frac{\partial q_s}{\partial t}\right)_{s+l->s} = 2.42\left[\frac{g\rho}{3C_D\beta_1}\right]^{1/2} \overline{E}_{sl}\overline{D}_s^{-0.2} q_l q_s \tag{6.5.23}$$

where $\overline{E}_{sl}$ is the averaged collection efficiency between snowflakes and droplets.

An example of a parameterization formula describing collisions of $X + Y \to Z$ type in a one-moment scheme is the formula for snow-raindrops collisions leading to formation of graupel (Figure 6.5.11c). Assuming the Marshall-Palmer distribution in Equations (6.5.16–6.5.18) and using Equation (F2) from Table 6.5.7, the rates of decrease in the rain water content and of snowflake content due to raindrop-snow collisions are written as suggested by Cotton et al. (1986):

$$\left(\frac{\partial q_r}{\partial t}\right)_{r+s->g} = -\rho\left|\overline{V}_r - \overline{V}_s\right|\overline{E}_{sr} \times \left(\frac{10\overline{D}_r^2 + 4\overline{D}_r\overline{D}_s + \overline{D}_s^2}{\beta_1 \overline{D}_s^{2.4}}\right) q_r q_s \tag{6.5.24}$$

$$\left(\frac{\partial q_s}{\partial t}\right)_{r+s->g} = -0.25\rho\left|\overline{V}_r - \overline{V}_s\right|\overline{E}_{sr} \times \left(\frac{\frac{1}{4}\overline{D}_r^2 + 0.85\overline{D}_r\overline{D}_s + 1.87\overline{D}_s^2}{\rho_w \frac{\overline{D}_r^3}{8}}\right) q_r q_{snow} \tag{6.5.25}$$

The value $\overline{D}_r$ in Equations (6.5.24–6.5.25) is a primarily given characteristic diameter of raindrops. Increase in the graupel content is determined by mass conservation:

$$\left(\frac{\partial q_g}{\partial t}\right)_{r+s->g} = -\left[\left(\frac{\partial q_r}{\partial t}\right)_{r+s->g} + \left(\frac{\partial q_s}{\partial t}\right)_{r+s->g}\right] \tag{6.5.26}$$

One can see that all the rates of mass content changes are proportional to the product of the mass contents of hydrometeors of the types participating in collisions.

Besides conversion of snow into graupel described by Equations (6.5.24–6.5.26) and illustrated in Figure 6.5.11c, snow can convert into graupel by riming with small cloud droplets that are present in large amounts (Figure 6.5.11d). To parameterize this mechanism, Cotton et al. (1986) assume that snow-droplets collisions lead to production of new graupel if the riming rate of snowflakes is larger than that of graupel of the same mass. Accordingly, Cotton et al. (1986) calculate the rate of graupel production due to collection of small cloud droplets by snow as:

$$\left(\frac{\partial q_g}{\partial t}\right)_{s+l->g} = \max\left\{\left[\left(\frac{\partial q_s}{\partial t}\right)_{s+l->s} - \left(\frac{\partial q_{g=s}}{\partial t}\right)_{g+l->g}, 0.0\right]\right\} \tag{6.5.27}$$

where the second term on the right-hand side is the "imaginary" riming rate of graupel of the same mixing ratio as snow, i.e. $q_{g=s} = q_s$. At present riming and transfer of mass to graupel/hail in RAMS is performed via computations of hydrometeor internal energy (Walko et al. 1995).

Interactions of Type $X + X \rightarrow Z$
In bulk schemes, the most serious difficulties arise in the description of collisions and collection of particles belonging to the same type, namely the processes of self-collection and auto-conversion (Table 6.5.1). Similar problems in case of collisions between cloud droplets are described in Section 5.7. In auto-conversion, collection between hydrometeors of type X leads to formation of hydrometeors of type Z. In ice microphysics the process of auto-conversion is represented by collection between ice crystals leading to snow (aggregates) production. There are several approaches used in bulk schemes to calculate the aggregation rate of ice crystals to produce snow. In a widely used approach (e.g., Lin et al., 1983) the aggregation rate is described by an empirical formula written following the concept proposed by Kessler (1969) to simulate the rate of raindrop formation by cloud droplet collisions:

$$\left(\frac{\partial q_s}{\partial t}\right)_{c+c->s} = 0.001 \cdot \overline{E}_{ii}\left(q_c - q_{0,c}\right) \qquad (6.5.28)$$

where q_c is the ice crystal content and $q_{0,c}$ is a threshold value of ice crystal content. Lin et al. (1983) use $q_{0,c} = 10^{-3}$ kg/kg. The expression for $\overline{E}_{ii}$ is given in Table 6.5.6 (Equation (F1)). The conversion rate, therefore, is a linear function of the ice crystal mass content.

Cotton et al. (1986) derived the rate of auto-conversion of ice crystals into snow from a simple model of collection rate in a homogeneous population of ice crystals, according to which the rate of ice crystal concentration is proportional to the square of concentration N_i:

$$\frac{dN_i}{dt} = -\overline{K}_{ii}N_i^2 \qquad (6.5.29)$$

where $\overline{K}_{ii}$ is the collection kernel. For simplicity, a mono-disperse population of crystals is assumed. The collection kernel is written to take into account the differences in the properties of equal-sized crystals. The kernel has a form:

$$\overline{K}_{ii} = \frac{\pi \overline{D}_i^2}{6} C\overline{V}_{gi}\overline{E}_{ii} \qquad (6.5.30)$$

where $C\overline{V}_{gi}$ is the average relative velocity between ice crystals, assumed to be proportional to the averaged fall velocity of ice crystals $\overline{V}_{gi}$ with the proportion coefficient $C = 0.25$ and $\overline{D}_i$ is characteristic value of crystal diameter. Using Equations (6.5.29) and (6.5.30), the rate of auto-conversion can be written as

$$\left(\frac{\partial q_s}{\partial t}\right)_{c+c->s} = -\frac{q_i}{N_i}\frac{dN_i}{dt} = \frac{\pi \overline{D}_i^2}{6} C\overline{V}_{gi}\overline{E}_{ii}N_i q_i \qquad (6.5.31)$$

In contrast to Equation (6.5.28), Equation (6.5.31) reflects a strongly nonlinear dependence of the conversion rate on the characteristics of ice crystal population.

In several recent parameterizations (e.g., Harrington et al., 1995; Morrison et al., 2005) collection of ice crystals is neglected. Auto-conversion of ice crystals to snow is parameterized in terms of the vapor diffusion growth rate. Ice crystals growing by diffusion up to the diameters exceeding a threshold diameter (e.g., 125 μm, assumed by Harrington et al., 1995) are converted to snow. The parameterization formulas describe the flux of ice crystals through this threshold value in the particle sizes coordinates. Assigning of large crystals to a certain type of aggregates is widely used not only in bulk-schemes, but also in SBM schemes (e.g., Hall, 1980). This assumption is an obvious simplification because primary crystals may reach sizes of a few thousands of μm (Pruppacher and Klett, 1997).

Interactions of Type $X + X \rightarrow X$
Self-collection, when collisions between hydrometeors of the same type lead to formation of hydrometeors of the same type (Table 6.5.1) is typically considered only for snow. In one-moment schemes self-collection is not taken into account, because the mass content of snow does not change during this process. In two-moment and multi-moment schemes self-collection must be taken into account because it decreases concentration and increases the size of particles. Ferrier (1994) uses look up-tables of self-collection rates. These rates are derived from the solution of the SCE for snowflakes. Seifert and Beheng (2006) applied expressions A12 and A13 (Appendix 6.5.8) to calculate self-collection rates.

Application of Bin-Emulating Schemes
In bin-emulating bulk schemes (like RAMS), size distributions in the collision integrals Equations (6.5.16–6.5.18) are written in the normalized form, so that the products of total concentrations of hydrometeors of X and Y types are taken out from the integral. The remaining integrals are calculated numerically. To reduce the computational cost during model runtime, Walko et al. (1995) calculated a large number of solutions for the integral and combined the results in 3D look-up tables. Two of the table dimensions are the characteristic diameters of PSD for the two types of colliding particles. The third dimension is the values of the integrals for particle pairs that can be combined from the seven types of liquid and ice particles used in RAMS.

Other Criteria for Particle Conversions in Bulk Schemes As was mentioned in Section 6.5.1, the conditions for conversion of ice crystals and snow into graupel by collection of small drops are set rather arbitrarily (Tables 6.5.2, 6.5.3). According to Seifert and Beheng (2006), a rimed ice crystal is converted into a graupel particle as soon as the collected mass fills up its enveloping sphere of a diameter equal to the maximum size of the ice crystal. The critical rime mass needed for conversion of an ice crystal into graupel is expressed as

$$m_{c,rime} = \alpha_0 \rho_w \left(\frac{\pi}{6} \overline{D}_i^3 - \frac{\overline{m}_i}{\rho_i} \right) \tag{6.5.32}$$

where $\overline{D}_i$ is the mean diameter of ice crystal. The expression in the brackets is the difference between the volume of a sphere of diameter $\overline{D}_i$ and the volume occupied by ice. $\alpha_0 = 0.68$ is a so called space-filling coefficient characterizing the fraction of volume of the enveloping sphere that can be filled with water. Knowing the riming rate, i.e., the rate of decrease in the ice crystal content $\left(\frac{\partial q_i}{\partial t}\right)_{i+l->g}$ one can evaluate the characteristic time τ_c needed to fill up the enveloping sphere as:

$$\tau_c = -\frac{\overline{m}_{c,rime}}{\left(\frac{\partial \overline{m}_i}{\partial t}\right)_{i+l->g}} = -\frac{\alpha_0 \rho_w N_i \left(\frac{\pi}{6} \overline{D}_i^3 - \frac{\overline{m}_i}{\rho_i} \right)}{\left(\frac{\partial q_i}{\partial t}\right)_{i+l->g}} \tag{6.5.33}$$

Seifert and Beheng (2006) parameterize the rate of conversion from ice crystals into graupel, occurring due to riming with small cloud droplets, as the ratio between the ice content and the characteristic conversion time:

$$\left(\frac{\partial q_g}{\partial t}\right)_{i+l->g} = \frac{q_i}{\tau_c} = -\alpha_0 \left\{ \frac{\rho_w}{\rho_i} \left(\frac{\frac{\pi}{6}\rho_i \overline{D}_i^3}{\overline{m}_i} - 1 \right) \right\}^{-1} \left(\frac{\partial q_i}{\partial t}\right)_{i+l->i} \tag{6.5.34}$$

The changes in the number concentration of graupel are determined from the expression:

$$\left(\frac{\partial N_g}{\partial t}\right)_{i+l->g} = \frac{1}{\overline{m}_i} \left(\frac{\partial q_g}{\partial t}\right)_{i+l->g} \tag{6.5.35}$$

To suppress the premature formation of very small graupel particles, the conversion takes place only if the ice particle mean diameter $\overline{D}_i$ exceeds a certain threshold chosen as 500 μm. The same procedure is applied to the conversion of snow into graupel due to riming, but with the filling coefficient value of $\alpha_{0\_s} = 0.01$. The utilization of the characteristic riming time is used in several other bulk parameterizations, e.g., Risner et al. (1998) and Ferrier (1994).

The survey presented above does not cover all the methods and approaches used in bulk parameterization schemes for description of collisions. A more detailed list of formulas can be found in the book by Straka (2009). However, the present survey demonstrates a high variability of the approaches used to calculate different collisions and conversions in bulk schemes. This variability typically leads to different results in ice transformations, so a detailed comparison with observations and with the corresponding bin procedures is required to evaluate the validity of a certain assumption.

6.5.8 Appendix. An Example of Representation of Collection in a Two-Moment Bulk Scheme

As an example, we briefly describe representation of collisions of type $X + Y \rightarrow X$ in the advanced two-moment bulk parameterization scheme developed by Seifert and Beheng (2006). The parameterization are based on the master equations (6.5.16) and (6.5.18) written for zero moment N (concentration) and first moments q (content) as

$$\left(\frac{\partial N_x}{\partial t}\right)_{x+y->x} = 0 \tag{A1}$$

$$\left(\frac{\partial q_x}{\partial t}\right)_{x+y->x} = \int_0^\infty \int_0^\infty K_{xy} f_x(m_x) f_y(m_y) m_y dm_x dm_y \tag{A2}$$

$$\left(\frac{\partial N_y}{\partial t}\right)_{x+y->x} = -\int_0^\infty \int_0^\infty K_{xy} f_x(m_x) f_y(m_y) dm_x dm_y \tag{A3}$$

$$\left(\frac{\partial q_y}{\partial t}\right)_{x+y->x} = -\left(\frac{\partial q_x}{\partial t}\right)_{x+y->x} = -\int_0^\infty \int_0^\infty K_{xy} f_x(m_x) f_y(m_y) m_y dm_x dm_y \tag{A4}$$

In Equations (A1–A4), $K_{xy}(m_x, m_y)$ is the collection kernel and $f_x(m_x)$ and $f_y(m_y)$ are particle size distributions (PSD) represented in the form of the General Gamma distribution with given parameters

$$f_x(m) = A_x m^{\nu_x} \exp(-\lambda_x m^{\mu_x}) \tag{A5}$$

$$f_y(m) = A_y m^{\nu_y} \exp(-\lambda_y m^{\mu_y}) \tag{A6}$$

Then the following simplifications are used in this parameterization scheme:

- The simplified equation of the collection kernel is used instead of Equation (6.5.19)

$$K^*_{xy} = \frac{\pi}{4}\left[D_x(m_x) + D_y(m_y)\right]^2 \overline{\left|V_x - V_y\right|}\,\overline{E}_{xy} \tag{A7}$$

- The mean collection efficiency $\overline{E}_{xy}$ depends solely on mean particle masses $\overline{m}_x$ and $\overline{m}_y$.
- The mean velocity difference $\overline{\left|V_x - V_y\right|}$ is found analytically by means of Equation (F5) from Table 6.5.7, under assumption that the velocity–mass relationships are power laws with constant parameters

$$V_x(m) = \alpha_x m^{\beta_x} \left(\frac{\rho(0)}{\rho(z)}\right)^{\gamma_x} \tag{A8}$$

$$V_y(m) = \alpha_y m^{\beta_y} \left(\frac{\rho(0)}{\rho(z)}\right)^{\gamma_y} \tag{A9}$$

- Power law relationships are assumed between diameters and masses of the particles

$$D_x(m) = a_x m^{b_x} \tag{A10}$$

$$D_y(m) = a_y m^{b_y} \tag{A11}$$

These simplifications allowed to obtain equations for the rates of moments by analytical integration of Equations (A2, A3). The obtained equations are

$$\left(\frac{\partial q_x}{\partial t}\right)_{x+y->x} = \frac{\pi}{4}\overline{E}_{xy} N_{0x} q_y \left[\delta^0_x D^2_x(\overline{m}_x) + \delta^1_{xy} D_x(\overline{m}_x) D_y(\overline{m}_y) + \delta^1_y D^2_y(\overline{m}_y)\right] \times \left[\vartheta^0_x V^2_x(\overline{m}_x) - \vartheta^1_{xy} V_x(\overline{m}_x) V_y(\overline{m}_y) + \vartheta^0_y V^2_y(\overline{m}_y)\right]^{1/2} \tag{A12}$$

$$\left(\frac{\partial N_y}{\partial t}\right)_{x+y->x} = -\frac{\pi}{4}\overline{E}_{xy} N_{0x} N_{0y} \left[\delta^0_x D^2_x(\overline{m}_x) + \delta^0_{xy} D_x(\overline{m}_x) D_y(\overline{m}_y) + \delta^0_y D^2_y(\overline{m}_y)\right] \times \left[\vartheta^0_x V^2_x(\overline{m}_x) - \vartheta^0_{xy} V_x(\overline{m}_x) V_y(\overline{m}_y) + \vartheta^0_y V^2_y(\overline{m}_y)\right]^{1/2} \tag{A13}$$

In Equations (A12–A13) $D_x(\overline{m}_x)$ and $D_y(\overline{m}_y)$ are diameters of particles having mean masses $\overline{m}_x$ and $\overline{m}_y$, and δ and θ are dimensionless parameters depending on PSD parameters, as well as on parameters the diameter-mass relationships. The expressions for these parameters are given in **Table A6.5.1**. Note that $\delta^k_{yx} \neq \delta^k_{xy}$ and $\vartheta^k_{yx} \neq \vartheta^k_{xy}$, since collisions of type $X + Y \to X$ are non-symmetric with respect to particle classes X and Y. Asymmetry is also seen in Equation (A12) indicating that the rate of collector mass content is proportional to the product of the concentration of collecting particles and the mass content of collected particles. Equations for other types of collisions can be obtained analogously.

Table A6.5.1 Formulas for parameters in equations for rates of moments (A12–A13)

$$\delta^k_y = \frac{\Gamma\left(\frac{2b_y+\nu_y+1+k}{\mu_y}\right)}{\Gamma\left(\frac{\nu_y+1}{\mu_y}\right)}\left[\frac{\Gamma\left(\frac{\nu_y+1}{\mu_y}\right)}{\Gamma\left(\frac{\nu_y+2}{\mu_y}\right)}\right]^{2b_y+k}$$

$$\delta^k_{yx} = 2\frac{\Gamma\left(\frac{b_x+\nu_x+1+k}{\mu_x}\right)}{\Gamma\left(\frac{\nu_x+1}{\mu_x}\right)}\frac{\Gamma\left(\frac{b_y+\nu_y+1}{\mu_y}\right)}{\Gamma\left(\frac{\nu_y+1}{\mu_y}\right)}\left[\frac{\Gamma\left(\frac{\nu_x+1}{\mu_x}\right)}{\Gamma\left(\frac{\nu_x+2}{\mu_x}\right)}\right]^{b_x+k}\left[\frac{\Gamma\left(\frac{\nu_y+1}{\mu_y}\right)}{\Gamma\left(\frac{\nu_y+2}{\mu_y}\right)}\right]^{b_y}$$

$$\vartheta^k_y = \frac{\Gamma\left(\frac{2\beta_y+2b_y+\nu_y+1+k}{\mu_y}\right)}{\Gamma\left(\frac{2b_y+\nu_y+1+k}{\mu_y}\right)}\left[\frac{\Gamma\left(\frac{\nu_y+1}{\mu_y}\right)}{\Gamma\left(\frac{\nu_y+2}{\mu_y}\right)}\right]^{2\beta_y}$$

$$\vartheta^k_{yx} = 2\frac{\Gamma\left(\frac{\beta_x+b_x+\nu_x+1+k}{\mu_x}\right)}{\Gamma\left(\frac{b_x+\nu_x+1+k}{\mu_x}\right)}\frac{\Gamma\left(\frac{\beta_y+b_y+\nu_y+1}{\mu_y}\right)}{\Gamma\left(\frac{b_y+\nu_y+1}{\mu_y}\right)}\left[\frac{\Gamma\left(\frac{\nu_x+1}{\mu_x}\right)}{\Gamma\left(\frac{\nu_x+2}{\mu_x}\right)}\right]^{\beta_x}\left[\frac{\Gamma\left(\frac{\nu_y+1}{\mu_y}\right)}{\Gamma\left(\frac{\nu_y+2}{\mu_y}\right)}\right]^{\beta_y}$$

6.6 Melting and Freezing

Freezing of drops and melting of ice particles are important microphysical processes in mixed-phase clouds, which affect the microphysics, dynamics and radiative cloud properties. Freezing and melting are the processes of phase transitions accompanied by release or absorption of latent heat. The height level with $T_C = 0°C$ is often referred to as the melting/ freezing level. We will use both terms depending on the process under consideration. The freezing level height varies depending on the latitude and the season. In Tropics, the freezing level is located at about 4 km. Since deep convective clouds reach heights of 10–12 km, freezing and melting are efficient in Tropics as well as in midlatitudes.

6.6.1 Melting

The Role of Melting

Melting affects both the microphysics and the dynamics of clouds and cloud ensembles. The major part of surface precipitation including precipitation in Tropics forms by melting of snow, graupel or hail. Melting determines whether precipitation at the surface will be pure liquid, mixed-phase or ice. Melting also controls the size of raindrops and ice particles falling down to the surface. Cooling caused by melting may lead to destabilization of the atmosphere (Findeisen, 1940; Moore and Stewart, 1985; Rauber et al., 2000; Marwitz, 1983; Szeto et al., 1988; Szeto and Stewart, 1997). Descent of air driven by melting-induced cooling fosters formation of the sub-cloud cold pool beneath squall-type systems (Liu et al., 1997) and, as a consequence, formation of new clouds, spatially organized convection (e.g., Tao et al., 1991) and rain-bands (Barth and Parsons, 1996). The cold pool determines circulations within a storm and promotes formation of new cells. Melting-induced downdrafts intensify the low-level horizontal convergence in warm frontal precipitation systems, accelerating surface frontogenesis and invigorating updrafts (Szeto and Stewart, 1997).

The rates of drop freezing and ice particle melting are determined by particle thermodynamics, in particular, by the rates of the heat outflow (in case of freezing) or heat supply (in case of melting) at the ice-water interface within particles. Smaller ice particles crossing the melting level melt away quicker than larger ones (Willis and Heymsfield, 1989). The rate of heat transfer is proportional to the surface area of a particle. Since smaller ice particles have higher surface area-to-mass ratio than larger particles, the supply of latent heat of melting to their ice cores is more efficient. Therefore, during warm seasons only large hail can fall through the warm air by distances exceeding several kilometers and finally reach the ground.

The structure of cloud particles within the melting layer is quite complicated. Structurally, stratiform clouds differ from convective clouds due to differences in their microphysical structures above the melting level. Investigation of the microphysical structure of the melting layer is an important issue in Radar Meteorology. An important radar feature of melting layer in stratiform clouds is the existence of an enhanced reflectivity layer referred to as the radar *bright band* (e.g., Bauer et al., 2000). Within the bright band the radar reflectivity is high because melting snowflakes being covered with a water film reflect radio waves more effectively than dry particles of the same sizes (Battan, 1973; Klaassen, 1988). Melting determines the polarimetric radar signatures, e.g., the increasing differential radar reflectivity within the melting layer, indicating the existence of non-spherical particles (e.g., Fabry and Szyrmer, 1999).

The Thermodynamics of Melting Particles

Melting particles contain both ice and liquid water fractions. Typically, to describe the evolution of a particle structure it is necessary to use the mass budget equation and two heat budget equations, one for the particle surface and the other for the ice-water interface. The heat budget equations determine the relationship between the liquid component and the ice component of a particle, as well as the melting rate. The heat budget equation for the particle surface is used to calculate the particle surface temperature.

The evolution of a spherical hail particle structure during melting is schematically illustrated in **Figure 6.6.1**. In laboratory experiments, Rasmussen et al. (1984a,b) found that when the liquid water film is thin there are no motions within it, so the heat flux through the skin is determined by diffusion. At melting stages when the water film becomes thick enough, semi-turbulent internal circulation of the melt water arises. This circulation leads to formation of a well-mixed water layer and immediately eliminates the temperature gradient within the water film. Thus laboratory experiments indicate the existence of two regimes of the melting process.

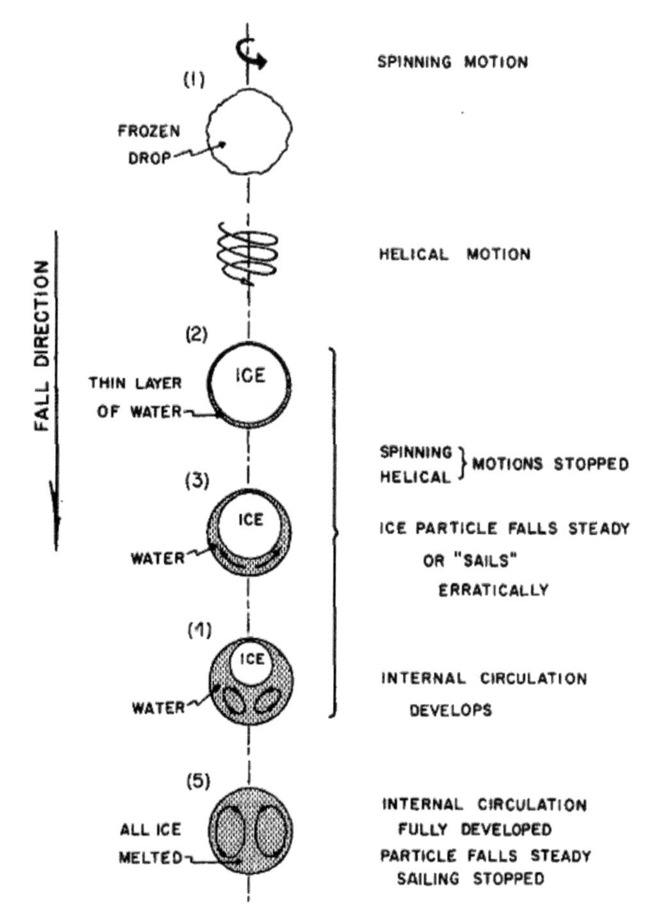

Figure 6.6.1 Schematic illustration of the actual melting behavior of a spherical ice particle (from Rasmussen and Pruppacher, 1982; © American Meteorological Society; used with permission).

Regime 1: Diffusion Heat Flux within the Water Film

The classical (and the simplest) model of melting of a spherical hail particle with the density of pure ice is illustrated in **Figure 6.6.2**. A hailstone with radius r is falling down through the air at constant humidity and temperature $T_C > 0°\text{C}$. The particle melts from the outside and at certain time t is covered with a water film of depth $r - r_i$, where r_i is the radius of the particle's ice core. The thickness of the water film is assumed to be uniform. This simplified model also assumes that r is constant in spite of the evaporation, and neither shedding of water nor particle collisions take place. The temperature of ice at the water-ice interface is equal to $T_0 = 0°\text{C}$, the surface temperature is T_s. The heat flux within the water film is directed from outside toward the ice-water interface.

At the stationary state of the spherical symmetric hailstone, the temperature changes linearly within the water film, so the heat flux through the water-ice interface surface of $4\pi r_i^2$ square can be written as $F = \frac{4\pi r r_i k_w(T_0 - T_s)}{r - r_i}$, where k_w is the heat conductivity of water (Mason, 1956; Rasmussen and Pruppacher, 1982). This heat flux during time dt results in melting of an ice mass dm_i located within a concentric shell of dr_i thickness (Figure 6.6.2), so $dm_i = 4\pi\rho_i r_i^2 dr_i$. As a result, the melting rate is described by the expression (Pruppacher and Klett, 1997):

$$L_m \frac{dm_i}{dt} = \frac{4\pi r r_i k_w(T_0 - T_s)}{r - r_i} \tag{6.6.1}$$

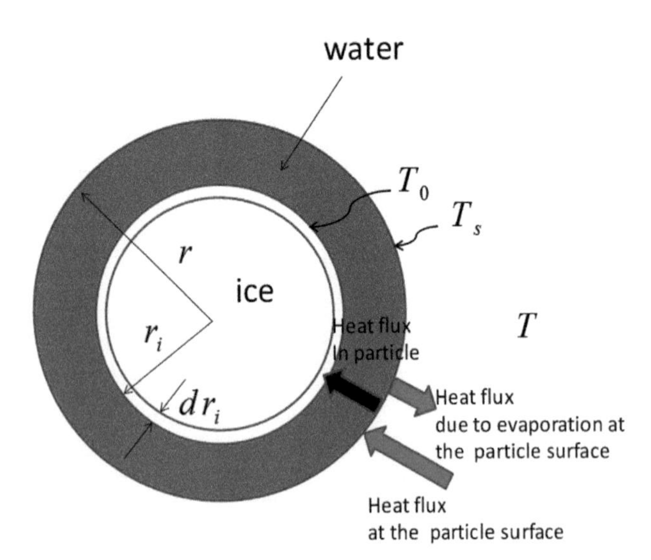

Figure 6.6.2 Scheme of heat fluxes during melting of an idealized hailstone.

Equation (6.6.1) represents the heat balance at the ice-water interface. In the stationary state, the heat flux through the water film is independent on the radial distance from the particle center. It allows to formulate the second equation that represents the heat balance on the particle surface. According to this equation, the heat flux within the water film is equal to the difference between the heat flux from the environment air toward the hailstone expressed as $4\pi r k_a(T - T_s)F_v$ and the heat flux caused by the evaporation from the hailstone surface expressed as $4\pi r \frac{DL_w}{R_v}(\rho_v - \rho_{vs})F_v$. In these expressions, $\rho_v = RH \cdot \rho_{vs}$ is the water vapor density remote from the particle, and ρ_{vs} is the saturating water vapor density at the particle surface temperature T_s, RH is the relative humidity of the environment air, D is the diffusivity of water vapor and F_v is the ventilation coefficient responsible for the increase of surface fluxes due to particle motion relative to the environment air (Section 5.5). As a result, the heat balance on the particle surface can be written as:

$$\frac{4\pi r r_i k_w(T_0 - T_s)}{r - r_i} = -4\pi r F_v \left[k_a(T - T_s) + \frac{DL_w}{R_v}(\rho_v - \rho_{vs}) \right] \tag{6.6.2a}$$

Using the state equation (3.1.25), Equation (6.6.2a) can be written in a more convenient form:

$$\frac{4\pi r r_i k_w(T_0 - T_s)}{r - r_i} = -4\pi r F_v \times \left[k_a(T - T_s) + \frac{DL_w}{R_v}\left(\frac{e}{T} - \frac{e_w}{T_s}\right) \right] \tag{6.6.2b}$$

Combining Equations (6.6.1) and (6.6.2b) one can get the expression for the melting rate:

$$\frac{dm_i}{dt} = -\frac{4\pi r}{L_m} F_v \left[k_a(T - T_s) + \frac{DL_w}{R_v}\left(\frac{e}{T} - \frac{e_w}{T_s}\right) \right] \tag{6.6.3}$$

Thus, the thermodynamics of melting is determined by two equations Equations (6.6.2b) and (6.6.3) that are used to determine the surface temperature T_s, to calculate the melting rate, i.e., to determine the time change of mass m_i (and the radius r_i) of the ice core.

Equation (6.6.3) allows to express the condition for the environment temperature necessary for the melting onset. Assuming that just prior to the melting onset $\frac{dm_i}{dt} = 0$ and $T_s = T_0$, Equations (6.6.2a) and (6.6.3) yield

$$T > T_{cr} = T_0 + \frac{DL_w}{k_a R_v}(\rho_{vs} - RH \cdot \rho_{vs}) \tag{6.6.4}$$

Equation (6.6.4) shows that melting starts at $T = T_0 = 0°\text{C}$ only in the saturating environment when $RH = 1$.

At lower relative humidity, the evaporation from the surface cools the particle, preventing melting at $T = 0°C$. Evaluations show that at the relative humidity of 50%, melting starts only at 4°C (Pruppacher and Klett, 1997).

Regime 2: Melting in the Presence of Water Circulation within the Water Film

Rasmussen et al. (1984a,b) found that semi-turbulent internal circulation of the melt water immediately eliminates the temperature gradient within the water film. It means that the temperature of melt water can be set, as a first approximation, equal to the temperature of the ice-water interface, i.e., to T_0. To parameterize the effect of this internal circulation on the melting rate, T_s in Equation (6.6.3) should be replaced by T_0. As a result, the melting equation is written in the form

$$\frac{dm_i}{dt} = -\frac{4\pi r F_v}{L_m}\left[k_a(T - T_0) + \frac{DL_w}{R_v}\left(\frac{e}{T} - \frac{e_w(T_0)}{T_0}\right)\right] \tag{6.6.5}$$

Equation (6.6.5) provides a good agreement with the observations. Condition $T_s = T_0$ is used instead of Equation (6.6.2) in order to determine the surface temperature. So, in the presence of the internal water circulation Equation (6.6.5) fully determines the melting rate in Regime 2.

Both melting regimes described above are idealized and agree with observations only within a certain range of geometrical parameters and only when melting of a dense hail particle is considered. There is a substantial difference in the melting processes of graupel and of hail. A schematic diagram presenting these differences is shown in **Figure 6.6.3**. Graupel (sometimes referred to as low-density hailstones) contain air volumes inside. When melting starts, melt water first soaks into the particle, so the surface remains dry. Soaking decreases, the particle size and increases the particle density. The water film forms when the water on the surface cannot soak inside any longer. Due to the air flow stress upon the surface of a falling melting particle, the water film is shifted toward the particle equator, and a torus of melt water forms. The torus grows with time since the melt water is advected into it from the lower half of the ice particle. The growing torus expands the cross-section area, decreasing the fall velocity as described in Section 6.4. The tours forming on large particles become thick enough to trigger shedding of water.

Rasmussen et al. (1984b) revealed seven melting modes in the laboratory experiments. A schematic of melting modes of a spherical hailstone of initial 2-cm diameter is shown in **Figure 6.6.4**. As melting starts

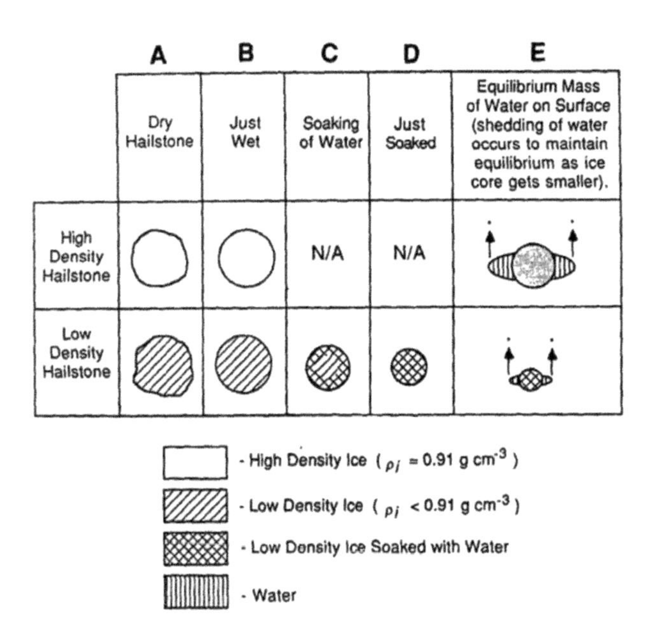

Figure 6.6.3 Schematic diagram showing the stages of melting experienced by ice particles of high and low density. The left-most panel shows a dry particle; panels progressively toward the right show the stages occurring with continuing melting. The density of the particles is given with reference to the initial ice density. Columns D and E represent hailstones with density values between those of ice and water. ρ_{bulk} is the bulk density of a melting particle (from Rasmussen and Heymsfield, 1987; © American Meteorological Society; used with permission).

(Mode 1) a ring of water forms near the particle equator. After about 20% of the particle mass has melted, the torus of melt water starts to shed drops with diameters of order of 1.5 mm (Mode 2). Since melting takes place mainly on the lower half of the particle, the ice core shape becomes oblate. Shedding leads to a decrease of the particle mass, and the fall velocity reduces. As a result, the torus moves upstream (Mode 3). The torus becomes wider and the drops sheared off become larger (up to ~3 mm in diameter). At the Reynolds numbers of Re~1.4×10^4, the particle's shape becomes complicated with sinusoidal deviations from its equilibrium position. As shedding continues, the particle size decreases, the torus thickens and moves upstream (Mode 4). At Re~8.5×10^3, the ice core is almost completely embedded in the melt water, and the particle shape approaches to that of a raindrop. It is important that perturbations of the melt water are damped out by the ice core. As a result, in contrast to a pure raindrop, a particle consisting mostly of water but containing an ice core may have the maximum diameters of up to 1.4 cm. As the ice core decreases further, the damping effect decreases triggering

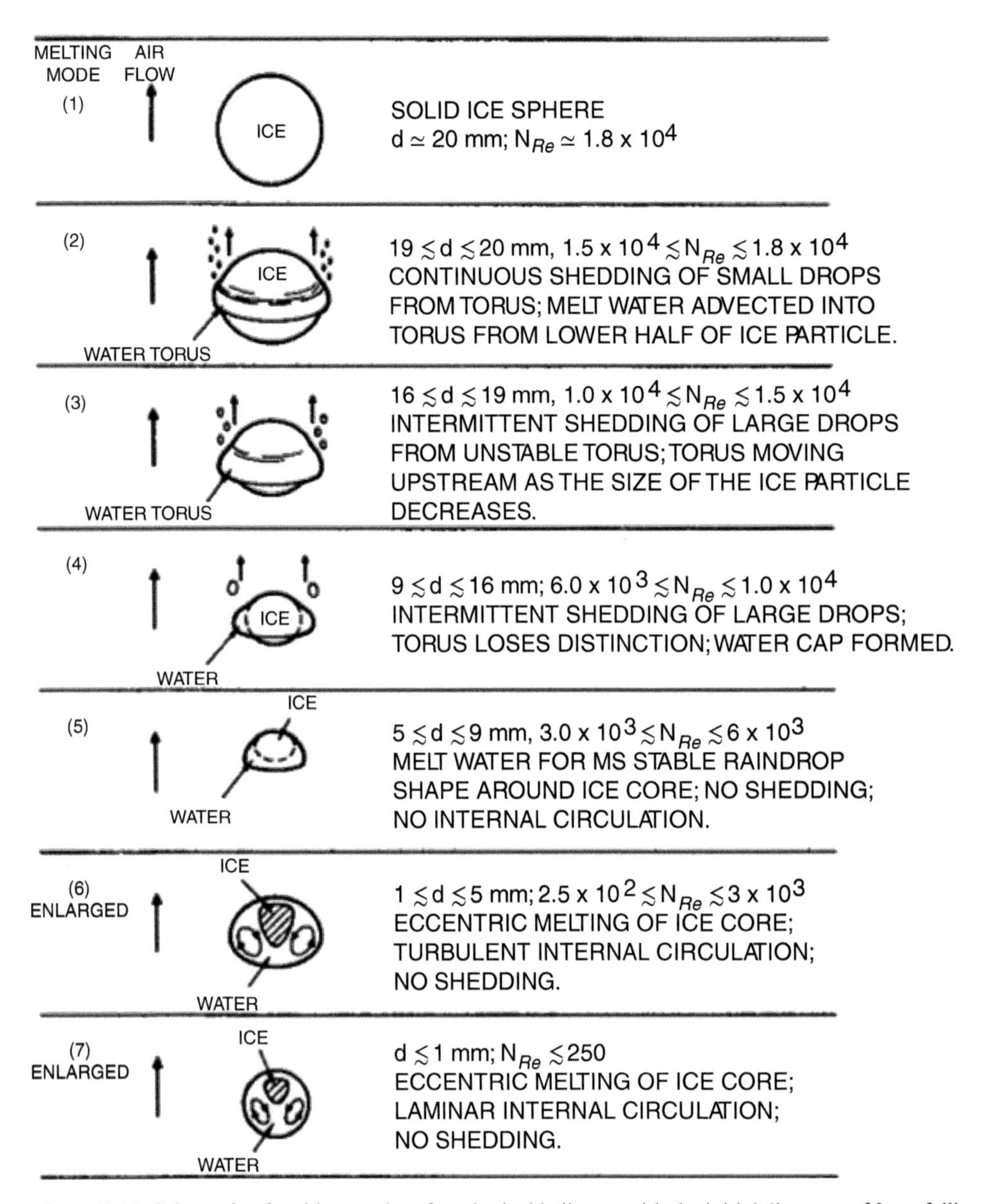

Figure 6.6.4 Schematic of melting modes of a spherical hailstone with the initial diameter of 2 cm falling in the calm air. In figure N_{Re} is the Reynolds number (from Rasmussen et al., 1984b; © American Meteorological Society; used with permission).

shedding of large drops with diameters of about 4.5 mm. These shed drops may sometimes collide with the parent particle, producing 300–400 μm diameter drops emerging from the melting particle. The shedding is possible when the particle diameter is larger than ~9 mm ($Re \sim 6 \times 10^3$). At smaller diameters raindrops are stable, and no shedding is possible (Mode 5). Intense internal water circulation exists till the particle diameter reaches 5 mm (Mode 6). For particles of smaller size, the internal circulation becomes laminar (mode 7).

According to the laboratory data, in stationary state there is a maximum mass m_w of melt water that can exist on the ice core of mass m_i. The equation relating the mass of the ice core m_i to the maximal mass m_w of melt water on its surface has a form (masses in g):

$$m_w = 0.268 + 0.1389 m_i \qquad (6.6.6)$$

The equilibrium linear dependence (6.6.6) is shown in **Figure 6.6.5**. If the mass of melt water exceeds maximum value, shedding of the excess water occurs. One can see that the mass of raindrop remaining after the

Table 6.6.1 Empirical and theoretical formulas for melting rates (from Rasmussen and Heymsfield, 1987; © American Meteorological Society; used with permission).

Mode of Melting, Figure 6.6.4	Reynolds Number	**Equation for the Melting Rate**, $\frac{dm_i}{dt}$	No.
7	$Re < 250$	$-\frac{4\pi r F_v}{L_m}\left[k_a(T - T_0) + \frac{DL_w}{R_v}\left(\frac{e}{T} - \frac{e_w}{T_0}\right)\right]$	T1
6	$2.5 \cdot 10^2 \le Re \le 3 \cdot 10^3$	As in mode 7, but with the ventilation coefficient twice lower	
5	$3 \cdot 10^3 \le Re \le 6 \cdot 10^3$	$-\frac{4\pi r F_v}{L_m}\left[k_a(T - T_s) + \frac{DL_w}{R_v}\left(\frac{e}{T} - \frac{e_w}{T_s}\right)\right]$	T3
1–4	$6 \cdot 10^3 \le Re \le 2 \cdot 10^4$	$-0.76 \cdot 2 \cdot \frac{2\pi r_i Re^{1/2}}{L_m}\left[\Pr^{1/3} \cdot k_a(T - T_0) + Sc^{1/3}\frac{DL_w}{R_v}\left(\frac{e}{T} - \frac{e_w}{T_0}\right)\right]$	T4
Very large particles	$Re \ge 2 \cdot 10^4$	$-\chi \cdot 2 \cdot \frac{2\pi r_i Re^{1/2}}{L_m}\left[\Pr^{1/3} \cdot k_a(T - T_0) + Sc^{1/3}\frac{DL_w}{R_v}\left(\frac{e}{T} - \frac{e_w}{T_0}\right)\right]$, $\chi = 0.57 + 9.0 \cdot 10^{-6} Re$	T5

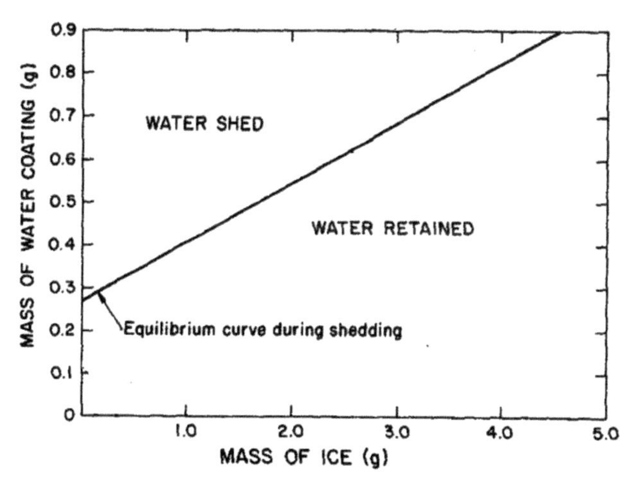

Figure 6.6.5 Experimentally found relationship between the maximum mass amount of melt water which can coat an ice core before shedding and the mass of a spherical ice core during melting or wet growth (from Rasmussen and Heymsfield, 1987; © American Meteorological Society; used with permission).

hailstone melting is 0.27 g, which corresponds to a raindrop diameter of ~1 cm. Equation (6.6.6) is widely used in numerical modeling of hail shedding (e.g., Ryzhkov et al., 2011).

Equation (6.6.5) describes well the melting of particles with sizes below about 5 mm. In these particles, the internal circulation of melt water is intense, so the condition $T_s = T_0$ is valid with a high precision. In melting Mode 5, no internal circulation occurs. Accordingly, in this case $T_s \neq T_0$, so the surface temperature should be determined from the equation for heat budget on the particle surface. Empirical and theoretical formulas for melting rates are presented in **Table 6.6.1**. Equation (T1) represents the melting stages in which the melt water film completely coats the ice core and no shedding occurs. This formula coincides with Equation (6.6.5). Equation (T3) represents melting Mode 5 where no internal circulation takes place. To find the surface temperature T_s, it is necessary to use equation of heat balance on the particle surface (6.6.2b). T_s can be found from this equation by iterations. For $Re > 2 \times 10^4$, heat transfer rapidly grows with increasing Reynolds number, which is essential for hailstone growth.

Formation of a melt water torus and transformation of the dry rough surface to a smooth one affect the fall velocity of melting particles. The formulas for calculation of fall velocity of meting particles are given in Section 6.4. Melting affects the dependence of the drag coefficient on the Reynolds number.

The schemes in Figures 6.6.1–6.6.4 illustrate melting of spherical particles of relatively high ice density. Melting of non-spherical particles, and especially of porous particles such as snowflakes, is much more complicated. **Figure 6.6.6** shows a photo of meltion snowflake. One can see that melting takes place within the entire particle volume. Laboratory experiments carried out by Mitra et al. (1990) showed that snowflakes start melting from their tips, and first liquid moves to the particle center being forced by the surface tension and the capillary forces. The branches remain liquid-water free at this stage. Then the central zone of a snowflake starts melting. At the next stage, a distortion of the branch structure takes place. Thus, snow melts much faster than graupel and hail. When snowflakes fall from clouds they typically fully melt within the layer with the temperature range of 0–5°C. Mitra et al. (1990) did not

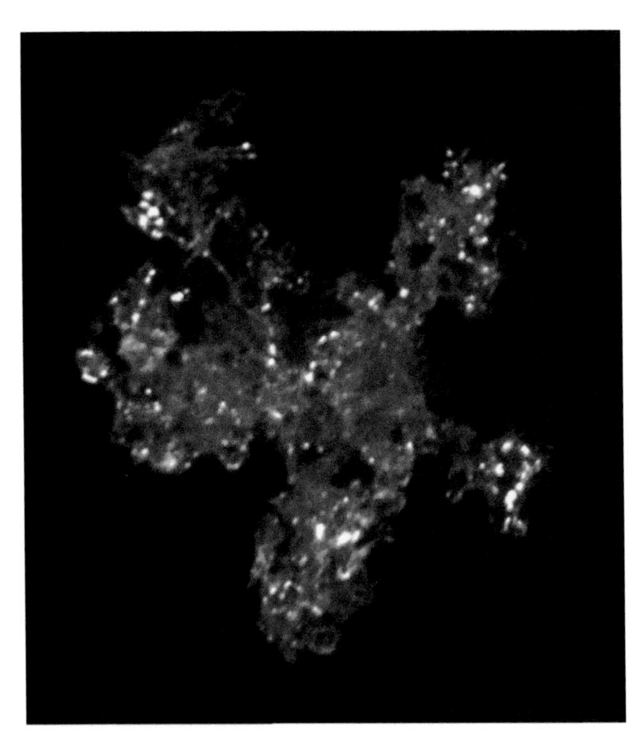

Figure 6.6.6 A pictures of melting snowflake. Permission of Prof. Thimothy Garrett at the University of Utah.

observe shedding during snow melting. Melting of snow in a dry environment can lead to a significant drop of the air temperature.

Melting Accompanied by Accretion

Equations for melting rate in Table 6.6.1 consider the ice particles as a solitary and neglect the heat influx caused by inter-particle collisions. In clouds containing liquid water, the accretion of water at positive temperatures provides the melting particle with an additional source of heat. Under the assumption that water accreted by a hailstone remains in contact with it and the equilibrium at the melting temperature is attained, the rate of the heat supply to the hailstone due to the accretion can be expressed as $c_w\left(\frac{dm_w}{dt}\right)_{accr}(T - T_0)$. Here $\left(\frac{dm_w}{dt}\right)_{accr}$ is the growth rate of the particle mass caused by accretion of liquid drops. In this expression, the temperature of the collected drops is assumed equal to the environmental temperature. As a result, instead of Equation (6.6.5) one obtains the following equation

$$\frac{dm_i}{dt} = -\frac{4\pi r F_v}{L_m}\left[k_a(T - T_0) + \frac{DL_w}{R_v}\left(\frac{e}{T} - \frac{e_w}{T_0}\right)\right] - \frac{c_w}{L_m}\left(\frac{dm_w}{dt}\right)_{accr}(T - T_0) \tag{6.6.7}$$

The minus before the second term on the right-hand side of Equation (6.6.7) indicates that the additional heat influx caused by accretion of relatively warm drops accelerates the melting. It is noteworthy that Equation (6.6.7) describes only the heat budget. The change of a melting particle mass caused by accretion should be determined as described in Section 6.5.

6.6.2 Description of Melting and Shedding in Cloud Models

In cloud models, melting is described at different levels of sophistication. The simplest approach is assumption of instantaneous melting when any ice particle is converted to a liquid drop of the same mass immediately below the freezing level (e.g., Khain and Sednev, 1996; Li et al., 2008). In most models, particles emerging due to collisions between ice particles and raindrops at $T_C > 0$ are assigned to raindrops without further treatment of the melting process (e.g., Cotton et al., 1986).

Melting in SBM Models

There are a few melting algorithms applied in the SBM models. Yin et al. (2000) solve Equation (6.6.5) for each size category of melting graupel at each time step. Then, the mass of melted water is assumed to be totally shed, thus, melting graupel does not contain melt water. Snowflakes and ice crystals are assumed to melt instantaneously as soon as the environmental temperature exceeds zero.

In the WRF/SBM model applied by Fan et al. (2012), Wang et al. (2013) and some other authors, the melting procedure includes the characteristic melting time $\tau_{melt,i}(r)$ which depends on the characteristic particle radius. Index "i" denotes the type of a hydrometeor. The values of size distribution of the i-th hydrometeor exponentially decrease during melting as

$$f_i(r,t) = f_i(r,0)\exp\left[-t/\tau_{melt,i}(r)\right] \tag{6.6.8}$$

The melted mass is transferred to the corresponding bin in the DSD. This procedure describes the main effects of melting, i.e., larger particles have a larger melting time scale and melt slower than smaller ones. At the same time, the parameterization expressed by Equation (6.6.8) significantly simplifies the process and does not describe the melting history of a particle. The fact that melting particles contain both ice and liquid water fractions is neglected, so the liquid fraction continuously grows without converting into drops or all liquid is shed immediately. In reality, the conversion of melted water into drops should take place either by partial shedding, or when the ice mass fraction gets below a certain minimum value. The approach based on Equation (6.6.8) resembles those applied in bulk parameterization models described below.

A more sophisticated approach was applied in HUCM (Phillips et al., 2007) and implemented into the Weather Research Forecatsing model (WRF). First step in this approach is calculation of the surface temperature at the time of the melting onset using Equation (6.6.4). Then Equation (6.6.5) is applied to calculate the melting rate of different types of ice hydrometeors. The differences between hydrometeor types are expressed by using different ventilation coefficients in Equation (6.6.5). For non-spherical particles, the surface fluxes are calculated using the capacitance of dry particles. Using the total mass of particles m and mass of ice m_i, the mass m_w of the melted water is calculated for each particle. HUCM deals with the distribution of melt water over the size spectrum of melting particles, so in each bin the particles are characterized by a certain liquid water fraction that changes as a result of melting, collisions, etc. This melt water is advected, diffused and settled which is simulated by means of equations similar to those used for size distributions of melting particles themselves. For each melting particle, the LWF is calculated as $f_w = m_w/m$. So, the value of LWF is known both before and after the melting procedure. In case $f_w > 0$, the ice mass decreases as described by Equation (6.6.7) and illustrated in Figure 6.6.2.

A melting snowflake is assumed to be an oblate spheroid of a variable aspect ratio γ_s. The capacitance of a dry particle C_{dry} is calculated as discussed in Section 6.3 (Table 6.3.1). The actual capacitance of a snowflake containing LWF $f_w > 0$ is calculated by a linear interpolation:

$$C = C_{dry}(1 - f_w) + C_w f_w \tag{6.6.9}$$

where C_w is the capacitance of a drop with the mass equal to that of the snowflake. The aspect ratio of a melting snowflake is also determined by interpolation as $\gamma_s = \gamma_{s,dry} + (\gamma_d - \gamma_{s,dry})f_w$, where γ_d is the aspect ratio of a liquid drop of a mass equal to that of a melting particle and $\gamma_{s,dry}$ is the aspect ratio of a dry snowflake. If $f_w = 0$, the particle mass changes by deposition/ sublimation as described in Section 6.3. If $f_w > 0$, water evaporates from the particle surface, so the particle mass decreases at the rate:

$$\frac{dm}{dt} = \frac{dm_w}{dt} = -\frac{4\pi r F_v D L_w}{L_m R_v}\left(\frac{e}{T} - \frac{e_w}{T_0}\right) \tag{6.6.10}$$

To simplify the treatment of melting of ice crystals, crystals are converted in HUCM to snowflakes at positive temperatures. Phillips et al. (2007) consider a more complicated approach to melting of ice crystals. The dry aspect ratios of crystals, $\gamma_i = h/d$ or $\gamma_i = L/d$, used in the approach are prescribed in the melting scheme from the empirical formulas presented in Table 6.1.3, and the dry capacitances are determined using formulas listed Table 6.3.1. The ventilation coefficient values for columns, plates and dendrites are calculated using the expressions given in Table 6.4.9.

Melting of Graupel and Hail

The ventilation coefficient values are calculated using Equation (6.4.20). The melting rate is calculated according to formulas presented in Table 6.6.1. To use these formulas, it is necessary to know the radius of a particle and its fall velocity. The fall velocity is necessary to calculate the ventilation coefficient. In HUCM, the fall velocities determined in the course of melting are also used to calculate the sedimentation rate as well as the collision kernels for melting particles. The capacitance of graupel and hail is calculated under assumption that they are spherical, i.e., it is equal to their radii. The decrease in the particle mass due to evaporation is calculated using Equation (6.6.10).

Expressions for the particle radius (and therefore, for the fall velocity) depend on the melting stage (Figure 6.6.4). The melting of graupel begins with the *soaking stage* (**Figure 6.6.7**). It is assumed that melt water first fills the volume occupied by the air. The soakable volume in a graupel is equal to $V_{soak} = V_{gr} - \frac{m}{\rho_i}$, where $V_{gr} = m/\rho_{bulk}$ and m are the volume and the mass of the graupel particle, respectively, ρ_{bulk} is the bulk density of a dry particle and $\rho_i = 920\ \text{kg/m}^3$ is the density of pure ice. The volume of melt water is $V_w = m_w/\rho_w$. During melting, both volumes V_{soak} and V_{gr} change. The melting stage is determined by the relationship between V_{soak} and V_w. During the soaking stage ($V_w < V_{soak}$) the melt water is located only inside the ice core. The entire particle's radius during the soaking stage is equal to the radius of a graupel of mass equal to $m - m_w$: $r = r_{i\_core} = \left(\frac{m - m_w}{\frac{4}{3}\pi\rho_{bulk}}\right)^{1/3}$. During the *fully soaked stage* ($V_w > V_{soak}$), the melt water accumulates on the exterior of the ice core and the entire particle's radius is $r = \left\{\frac{V_{i\_core} + V_w - V_{soak}}{4\pi/3}\right\}^{1/3}$. The maximum mass of the melt water that can be accumulated on the exterior of the particle is determined by Equation (6.6.6), where the mass $m_i + m_{soak}$ is used instead of m_i. Here $m_{soak} = \rho_w V_{soak}$ is the water mass within the soakable volume. If $m_w > m_{w\_cr}$ shedding occurs and new liquid drops emerge. The radius of a particle at the equilibrium stage when the water mass

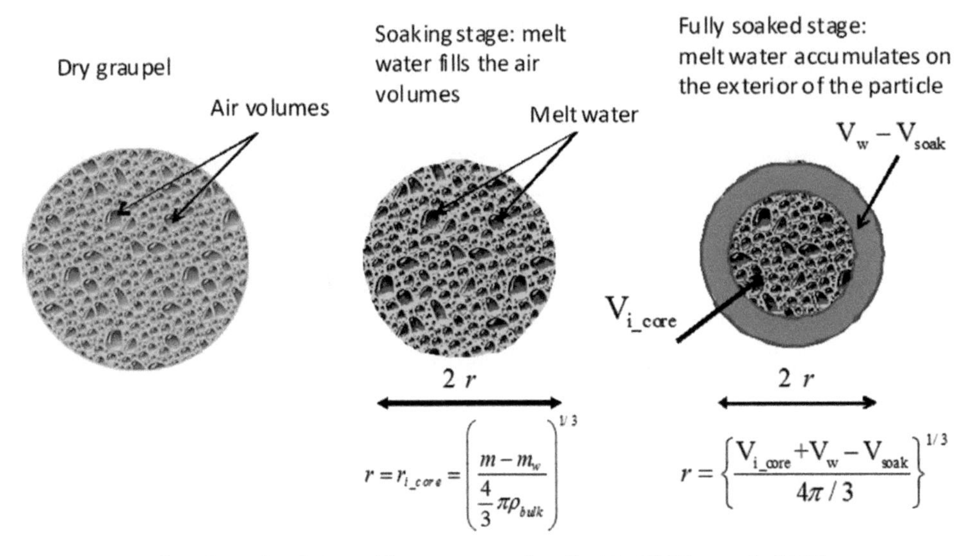

Figure 6.6.7 The schematic of the melting process as described by Phillips et al. (2007).

within the particle is equal to its maximum possible value m_{w_cr} is $r = \left[\frac{V_{i\_core} + m_{w\_cr}/\rho_w}{4\pi/3}\right]^{1/3}$.

Shedding
The size of shed drops depends on the particle's Reynolds number. According to the laboratory measurements carried out by Rasmussen and Heymsfield (1987), at $Re > 2.5 \times 10^4$ all the exterior melt water is shed resulting in formation of raindrops of 1.5 mm in diameter. At $1.5 \times 10^4 < Re < 2.5 \times 10^4$, the shedding of such raindrops is continuous, and the mass of the exterior melt water is maintained at the critical equilibrium value. At $Re < 1.5 \times 10^4$, there is an intermittent shedding of raindrops of 3 mm ($Re > 10^4$) or 4.5 mm ($Re < 10^4$) in diameter, reducing the mass of the exterior melt water in each shedding event, so finally this mass becomes less than m_{w_cr}. The mass of the melt water that is shed in any such event cannot exceed the prescribed fraction (50%) of m_{w_cr}.

Application of the Melting Algorithm in HUCM
We illustrate the effects of melting on the microphysics and dynamics of deep convective clouds using, by way of example, the clouds observed during the LBA–SMOCC campaign on 1900 UTC October 1, 2002 (10°S, 62°W) and 1900 UTC October 4, 2002 (10°S, 67°W) (Andreae et al., 2004). The clouds were simulated under different CCN concentrations (green-ocean, smoky and pyro-clouds) (Khain et al., 2004, 2005). The specific feature of the meteorological situations is a low relative humidity (35–40%) within the lower troposphere. The results obtained using the detailed melting scheme (Phillips et al., 2007) are compared to simulations where an instantaneous melting at the freezing level was implemented. Simulations of smoky and especially pyro-clouds indicate that the detailed melting scheme invigorates convection. The effects of applying the detailed melting scheme on the structure of smoky clouds are illustrated in **Figure 6.6.8** that compares the fields of the total ice content (the sum of all ice species) and of the rainwater content at t = 3,900 s in case of the detailed melting scheme (left panels) and the instantaneous melting scheme (right panels). One can see that the total ice content is larger and ice penetrates to higher levels if the detailed melting procedure is used. In both cases, the precipitation is caused mainly by melting, while the spatial distribution of rainwater content (RWC) is quite different. In the instantaneous melting, RWC reaches its maximum just below the freezing level and decreases downward due to strong evaporation. In the detailed melting, formation of raindrops increases downward due to melting of the largest graupel and hail. As a result, RWC increases downward reaching its maximum at the surface. In the detailed melting, the surface precipitation is much more intense than in the instantaneous melting.

Figure 6.6.9 shows that the detailed melting scheme results in formation of a more vigorous cloud penetrating higher levels. The vertical updrafts remain intense (>15 m s^{-1}) up to z = 12 km, while in the instantaneous melting the updrafts at upper levels do not exceed 5 m s^{-1}. The most intense downdrafts reach vertical

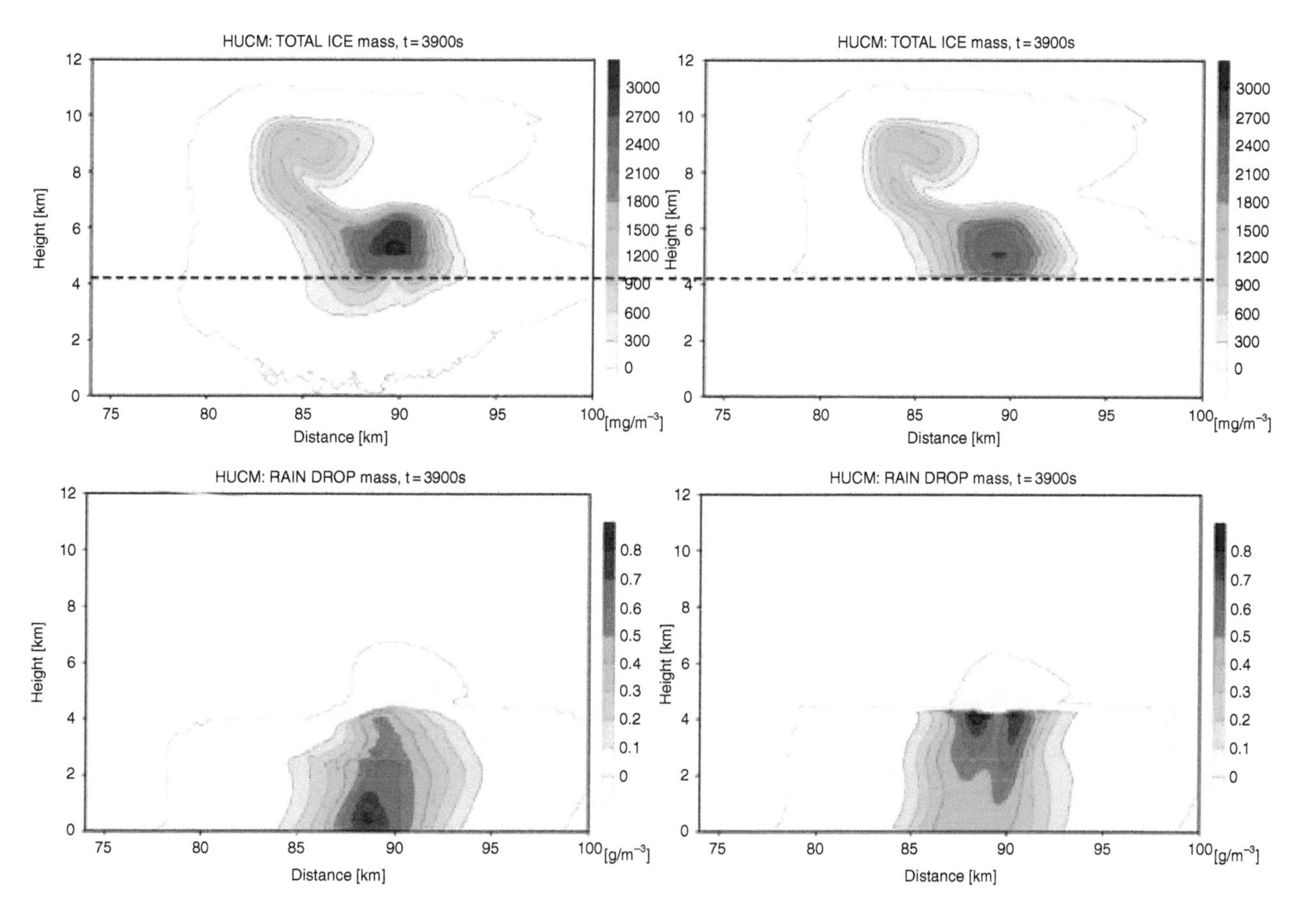

Figure 6.6.8 Total ice water content (upper row) and RWC (bottom row) in case of the detailed melting (left) and the instantaneous melting (right) for the smoky cloud observed in LBA–SMOCC. The height of the melting level is indicated with dashed line (from Phillips et al., 2007; © American Meteorological Society; used with permission).

speeds of 10 m s^{-1} in the instantaneous melting but attain 15 m s^{-1} in the detailed melting.

Figure 6.6.10 displays the size distributions of snow, graupel and hail, as well as of liquid water fraction f_w within these hydrometeors at different height levels in the pyro-cloud simulation. The distributions are sampled from the cross-sections through regions of penetration of the hydrometeor species downward below the freezing level. One can see that smaller particles are the first to melt, so at lower levels only the largest ones remain. Snow melts faster than graupel and hail. Hail particles are larger than graupel. In the simulations, the typical hail diameter is 3–5 mm. Snow melts more uniformly than graupel or hail, so the size distribution of snow is wider than those of graupel and hail.

Description of Melting in Bulk Parameterization Models
Since the melting rate strongly depends on particle size, parameterization of melting represents a difficult problem in bulk parameterization schemes. Most parameterizations of melting are based on the thermodynamic equation of type (6.6.5). Rutledge and Hobbs (1983, 1984) apply equation $\frac{dm_i}{dt} = -\frac{4\pi r F_v}{L_m} k_a(T - T_0)$ to melting of snow, neglecting the heat flux on the particle surface caused by evaporation. Multiplying this expression by the size distribution factor taken in the form of the Marshall-Palmer distribution, and integrating over the entire size range, Rutledge and Hobbs (1983) got the following expression for the melting rate of snow mass content

$$\left(\frac{dq_s}{dt}\right)_{melt} = -\frac{2\pi N_{0s}}{L_m} k_a(T - T_0) \times \Phi(\overline{F}_v, \lambda_s, B), \tag{6.6.11}$$

where N_{0s} is the intercept parameter in snowflake size distribution, $\Phi(\overline{F}_v, \lambda_s, B)$ is a function of the averaged ventilation coefficient $\overline{F}_v$, the slope parameter λ_s and the power of the particle diameter in the expression for the

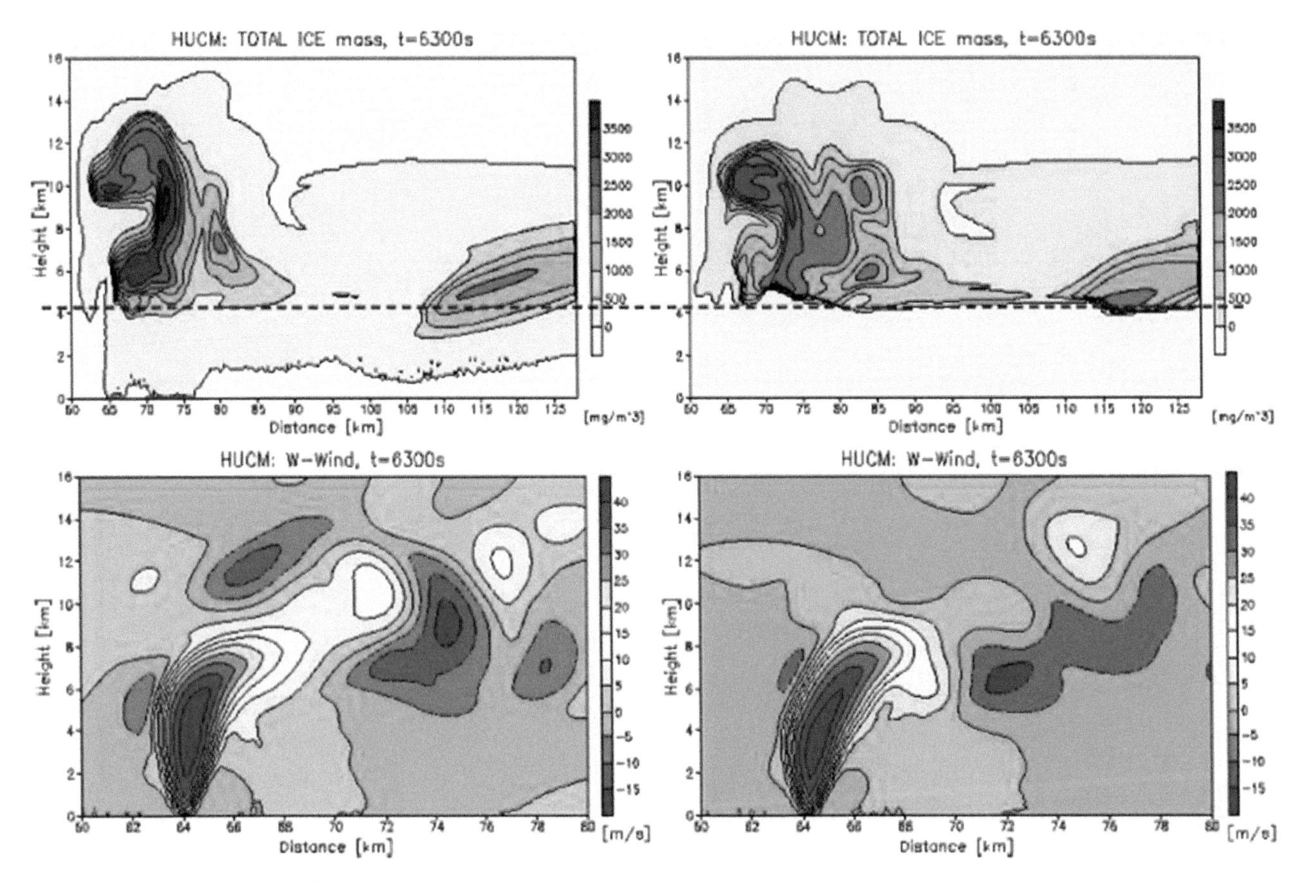

Figure 6.6.9 Fields of the total ice content (upper row) and the vertical velocity (bottom row) in a pyro-cloud at t = 6300 s in simulations with the detailed melting (left) and the instantaneous melting (right). The height of the freezing level is indicated with the dashed line (from Phillips et al., 2007; © American Meteorological Society; used with permission).

fall velocity B (Section 6.4, Equation 6.4.12a). Evaporation from snow surface at $T_C > 0°\text{C}$ is calculated using an expression similar to Equation (6.6.10). This expression is similar to that used for calculation of snow sublimation, the difference being that the particle surface is assumed to be wet, so the saturating water vapor pressure over water is assumed. Rutledge and Hobbs (1984) used similar expressions to describe graupel melting. This approach is used in several bulk schemes (e.g., Morrison et al., 2005) that assume an instantaneous melting of ice crystals.

Wisner et al. (1972), Lin et al. (1983) and Milbrandt and Yau (2005a) calculated the melting rate of snow, graupel and hail using the equation for heat balance, where different thermodynamic effects are taken into account, namely the effects of cooling caused by melting and of heating caused by a sensible heat flux upon the particle surface, as well as by the latent heat of condensation/evaporation at the particle surface. The heating from collected cloud droplets and raindrops (whose temperature is assumed equal to the environment temperature which is higher than the temperature of particle surface) is also taken into account. The increase in the mass of snow, graupel and hail particles due to collisions within the melting layer is calculated using the equation for continuous growth. The authors integrate Equation (6.6.7) over the hydrometeor spectra. The final expressions depend on the parameters of the particle size distribution and on the expression for the ventilation coefficient.

The approach proposed by Seifert and Beheng (2006) is based on Equation (6.6.5) expressing the characteristic melting time as:

$$\tau_{melt} = m\left(\frac{dm}{dt}\right)^{-1} = m\left(\frac{4\pi r F_v}{L_m}\left[k_a(T - T_0) + \frac{DL_w}{R_v}\left(\frac{e}{T} - \frac{e_w}{T_0}\right)\right]\right)^{-1} \tag{6.6.12}$$

After the characteristic melting time is evaluated, Equation (6.6.5) is used to write down the equation

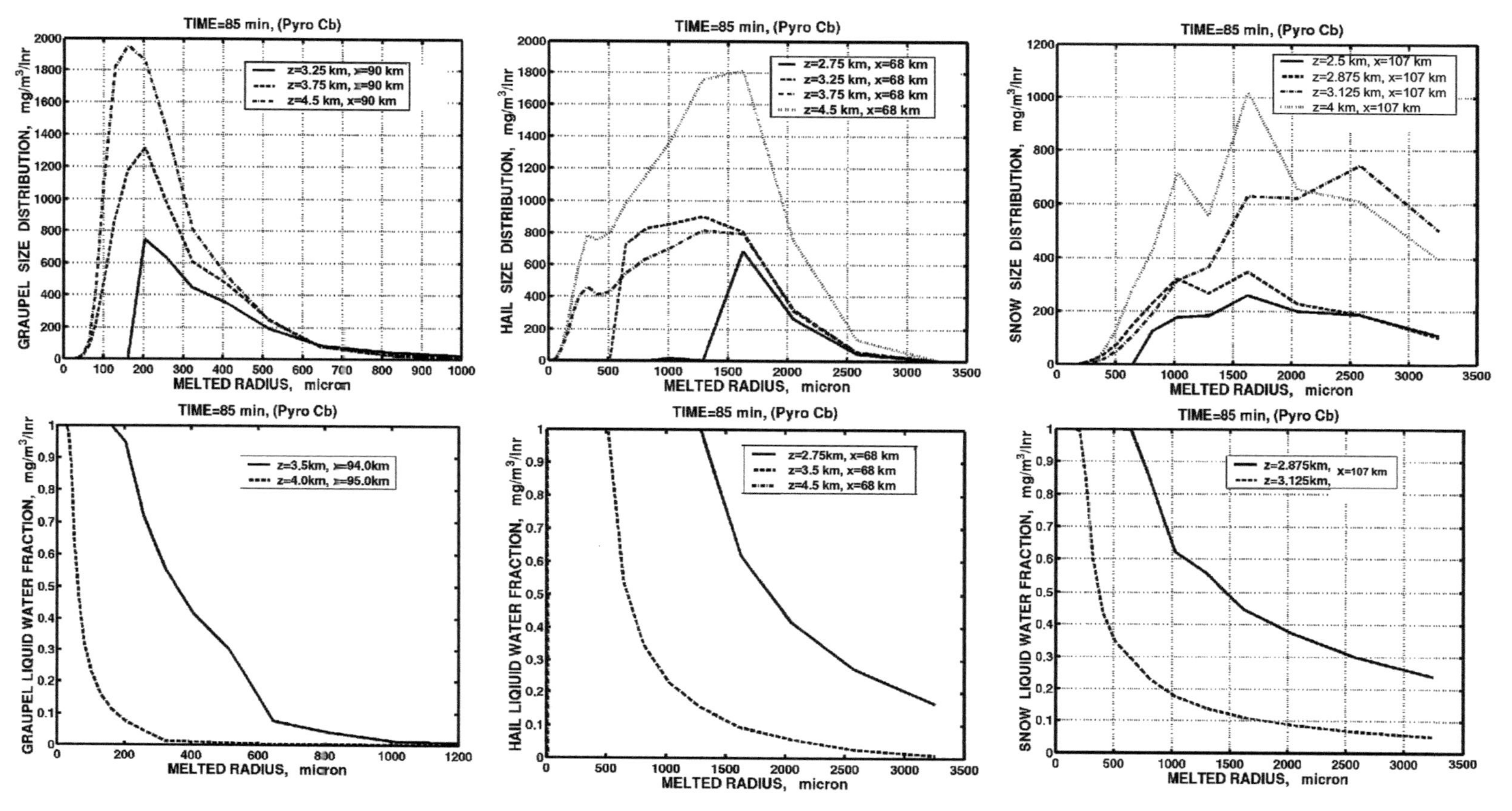

Figure 6.6.10 Upper row: Size distributions of graupel (left column), hail/frozen drops (middle) and snow (right column) and distributions of liquid water fraction, f_w (bottom row), in the corresponding mass bins at different levels. $f_w = 0$ corresponds to a totally frozen particle, $f_w = 1$ corresponds to a totally melted particle (from Phillips et al., 2007; © American Meteorological Society; used with permission).

for PSD moments, using a procedure typical of bulk parameterization schemes. The equation of change of the k-th moment due to melting is written in the form:

$$\left(\frac{\partial M^k}{\partial t}\right)_{melt} = -\int_0^\infty \frac{m^k f(m)}{\tau_{melt}} dm$$
$$= -\frac{4\pi \overline{F}_v}{L_m}\left[k_a(T - T_0) + \frac{DL_w}{R_v}\left(\frac{e}{T} - \frac{e_w}{T_0}\right)\right]$$
$$\times \int_0^\infty rF_v(m) f(m) m^{k-1} dm \qquad (6.6.13)$$

where $r \sim m^{1/3}$ is particle radius. The integral in Equation (6.6.13) is approximated as a product of averaged values: $\int_0^\infty rF_v(m)f(m)m^{k-1}dm \approx A \cdot \overline{r}N\overline{m}^{k-1}\overline{F}_v$, where N is the particle concentration and $\overline{F}_v$ is the averaged ventilation coefficient depending on the Reynolds number calculated using the mass-averaged velocity of the hydrometeors. Assuming the diameter-mass and the velocity-mass relations in the forms (6.1.1) and (6.4.12b), respectively, Seifert and Beheng (2006) got analytical expressions for averaged ventilation coefficients. This approach is applied to describe melting of all types of ice particles, i.e., graupel, snow and ice crystals. Evaporation of water from the surface of melting particles is calculated as evaporation of raindrops, but assuming the surface temperature equal to zero. Seifert and Beheng (2006) do not consider shedding.

Ferrier (1994) assumes that shedding of water from melting precipitating particles takes place when the mass fraction of water accreted during one time step exceeds 50% of the total mass of hydrometeors of a particular type. The excess over this maximum mass of the melt water is shed and added to the mass of raindrops. The changes of number concentrations of precipitation ice are calculated under the assumption that the PSD slopes do not change during melting.

Meyers et al. (1997) convert snow, that lost all the melted mass, to graupel. To determine the loss of the number concentration of snow, an explicit parcel bin model was used. As the liquid water fraction in graupel reaches 30%, graupel is converted to hail. Melting of hail is described using the lookup tables calculated by means of a parcel bin microphysics model. Meyers et al. (1997) calculate shedding of hail using Equation (6.6.5) in which masses of water and ice are interpreted as integrated over the corresponding spectra. Any excess of hydrometeor mass over the maximum possible value is shed. As was motioned above, the melt water includes the soaked water not available for shedding, therefore it is not feasible to treat this effect within a bulk microphysical framework. So Meyers et al. (1997) approximate all the melt water as a given fraction of the total mass content, available for shedding. The diameter of shed drops is assumed equal to 1 mm. Knowing the mass of the shed drops and their size, the concentration of shed drops is calculated.

6.6.3 Freezing

At sub-zero temperatures exceeding −37°C, hydrometeors in clouds can contain both ice and liquid water fractions. Liquid water within such particles freezes with time. The freezing rate is determined by the rate of dissipation of the latent heat to the ambient air. In the course of freezing, the shape, the size and the surface properties of precipitation particles change. As a result, freezing affects the collection rates and the fall velocities of particles. If surfaces of graupel and hail are wet, their ability to collect ice crystals increases. The presence of water in particles affects the radar reflectivity, the differential reflectivity and other radar properties of clouds (Kumjian et al., 2012). Detection of wet growth using polarimetric radars is important to improve prediction of a severe hail storms that can damage crops and urban areas. The intensity of freezing and wet growth of drops depends on the aerosol concentrations, so these processes are part of the cloud-aerosol interaction.

Freezing of the liquid phase at negative temperatures takes place in the course of several processes, among them freezing of drops after activation of immersion or of contact ice nuclei (see Section 6.2). These types of freezing can be also regarded as *nucleation freezing*. Nucleation freezing can take place either very rapidly or gradually, depending on the surrounding temperature and the drop size. In case of slow freezing, drops may collect other drops thus affecting the freezing time. Drop freezing can take place on the surface of larger ice particles such as snowflakes, graupel of hail, caused by their collisions with supercooled liquid drops. This process is known as riming of ice particles.

As was mentioned in Section 6.2, drop freezing consists of two stages. At the first stage, an ice germ forms inside a drop either on IN or spontaneously. This stage is very short and can be considered adiabatic. The second stage includes a comparatively long thermodynamic process of ice growth within a drop until it freezes totally. At this stage, an ice shell forms on the exterior encasing the liquid, while freezing proceeds inwards as the ice shell thickens. In most cloud models, drop freezing is treated as instantaneous. The second stage of freezing that in real clouds can be as long as several minutes is typically ignored. Instead, after the first stage drops are immediately converted to graupel or hail. However, in some cases a detailed simulation of the second stage of freezing is required. For instance, in hazardous conditions of *freezing rain*, an accurate calculation of the total time of drop freezing is extremely important to provide real-time information to aviation services, ground transportation etc. Freezing rain may arise during intrusion of cold air into the boundary layer which was initially located within warm air. Ice particles falling from above first melt fully or partially in the warm layer and then fall through the cold air of the boundary layer. Knowing whether the surface precipitation in this situation will be liquid, ice or a mixture of both, is crucial.

There is a distinct asymmetry between freezing and melting, so both processes are best treated separately in microphysical cloud models. Freezing generally proceeds from the outside inwards, with the exception of the initial ice nucleation that can happen anywhere inside a drop. Freezing from the outside inwards happens because the latent heat from freezing must be conducted from the ice-liquid interface (at about 0°C) to the particle surface and then to the air, so the surface must be supercooled. On the contrary, in melting which absorbs the latent heat the outer surface of the particle must be slightly warmed above 0°C to conduct the heat into the particle down to the ice-liquid interface. Thus, freezing characterized by ice on the outside is physically different from melting characterized by liquid on the outside. Melting is not merely an "inversion" of freezing. Both processes begin on the particle surface and propagate inside, but their heat fluxes are oppositely directed.

Nucleation freezing was investigated in the wind tunnel experiments (Johnson and Hallett, 1968; Pitter and Pruppacher, 1973). Drops tumble while freezing, which most often results in a radially symmetric loss of latent heat of freezing (Pruppacher and Klett, 1997). For this reason, drop freezing can be mathematically formulated (at least in the first approximation) as a spherically symmetric problem. To write the heat balance equation (similarly to that for melting) describing freezing at the second stage, it is necessary to describe the first stage in more detail since the result of the first stage (e.g., the mass of ice in the drop) serves as an initial condition for the second stage. According to Pruppacher and Klett (1997), at the first (adiabatic) freezing stage the interaction of the drop with its environment can be neglected, so most of heat released by freezing causes a quick increase in the drop temperature to about 0°C. The mass of ice formed at the first stage can be evaluated by means of the heat balance equation

$$m_i L_m = (c_w m_w + c_i m_i)\Delta T \qquad (6.6.14)$$

where m_i is the mass of ice within a drop of mass m and the remaining mass of water is $m_w = m - m_i$ and ΔT is the increase in temperature in the drop during the first stage. In Equation (6.2.14) $c_w m_w \gg c_i m_i$. So the ratio between m_i and m_w can be approximately evaluated as $\frac{m_i}{m_w} = \frac{c_w}{L_m}\Delta T \approx \frac{\Delta T}{80}$. For example, in case $\Delta T = 10°C$ at the first stage, only about 1/8 of the liquid water mass freezes. It is typically assumed that the ice formed at the first stage is homogeneously distributed over the drop volume. However, Equation (6.6.14) is an approximation, while in reality a particle remains slightly supercooled after the first stage. According to Hallett (1964), the temperature of the ice-water interface at the second stage is about $-0.2°C$.

The approach based on derivation of the heat balance at the second stage of drop freezing is somehow similar to that used in case of ice particle melting. However, as shown below, the solution of the balance equation for freezing is more complicated than that for melting. The scheme of fluxes needed to derive the heat balance equation is shown in **Figure 6.6.11**.

As in case of ice particle melting, the temperature of the ice-water interface is assumed equal to $T_0 = 0°C$.

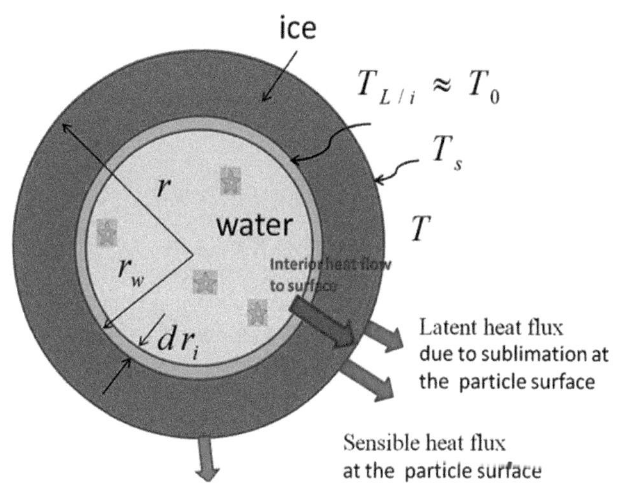

Figure 6.6.11 Schematic geometry of freezing of an idealized spherical drop and the heat fluxes forming the heat budget. Asterisks indicate that after the first stage of freezing the internal volume contains a small fraction of ice assumed to be homogeneously distributed over a sphere of radius r_w. $T_{L/i}$ is the temperature of the water-ice interface (from Phillips et al., 2015, with changes; © American Meteorological Society; used with permission).

For the quasi-steady state, the release of the latent heat of freezing is compensated by the heat loss due to the heat flux directed outward (Pruppacher and Klett, 1997):

$$L_m \frac{dm_w}{dt} = 4\pi\rho_w L_m r_w^2 \left(1 - \frac{c_w}{L_m}\Delta T\right)\frac{dr_w}{dt} = \frac{4\pi r r_w k_i (T_0 - T_s)}{r - r_w} \quad (6.6.15)$$

where k_i is the heat conductivity of ice. Equation (6.6.15) is the heat budget equation at the ice-water interface. The multiplier $\left(1 - \frac{c_w}{L_m}\Delta T\right)$ takes into account the fact that after the first stage of freezing the mass of water within the internal volume of radius r_w is equal to $m_w = \frac{4}{3}\pi\rho_w r_w^3\left(1 - \frac{c_w}{L_m}\Delta T\right)$. Accordingly, the mass of water within any volume $4\pi r_w^2 dr_w$ is equal to $dm_w = 4\pi\rho_w r_w^2\left(1 - \frac{c_w}{L_m}\Delta T\right)dr_w$. Since typically $\frac{c_w}{L_m}\Delta T \ll 1$, the interior volume can be considered as a volume of pure water. In the absence of accretion, the heat flux through the ice shell is equal to the sum of two fluxes, namely: the heat loss by conduction to the environment air $4\pi r k_a (T_s - T)F_v$ and the flux due to sublimation $4\pi D L_i(\rho_{vi} - \rho_v)F_v$ (Figure 6.6.11). Here, like in Equation (6.6.2), ρ_v is water vapor density far away from the drop, and ρ_{vi} is the saturating (over ice) water vapor density at saturating water temperature T_s; D is the diffusivity of water vapor and F_v is the ventilation coefficient. So the second thermodynamic equation for drop freezing expressing the heat balance on its surface can be written as

$$\frac{4\pi r r_w k_i (T_0 - T_s)}{r - r_w} = 4\pi r k_a (T_s - T)F_v + 4\pi r D L_i(\rho_{vi} - \rho_v)F_v \quad (6.6.16)$$

Equations (6.6.15) and (6.6.16) are similar to Equations (6.6.1) and (6.6.2), respectively. The difference is in the direction of the heat flux through the surface ice shell. Freezing leads to particle heating, so the particle is warmer than the environmental air and the heat flux is directed outside. While evaporation on the surface of a melting particle hinders its melting, the sublimation from the surface of a freezing drop accelerates its freezing. However, in contrast to Equation (6.6.5), the surface temperature in a freezing drop is not equal to T_0, but should rather be determined using the equation for heat balance at the drop surface. Pruppacher and Klett (1997) express the difference $\rho_{vi}(T_s) - \rho_v(T)$ in Equation (6.6.16) as

$$\rho_{vi} - \rho_v = (1 - RH)\rho_{vi}(T) + [T_s - T]\overline{\frac{d\rho_v}{dT}} \quad (6.6.17)$$

This expression describes the expansion of the differences between the water vapor densities (taken at different temperatures) into the Taylor series. The value $\overline{\frac{d\rho_v}{dT}}$ indicates the mean slope of the dependence $\rho_v(T)$ within the temperature range between T_s and T. Analysis of Equation (6.6.16) shows that the freezing rate in wet air (high RH) is slower than in dry air. In contrast, the melting rate increases in wet air.

Upon elimination of T_s from Equations (6.6.16) and (6.6.17), the freezing equation can be written for the ratio $y = \frac{r_w}{r}$:

$$3t_0 \frac{dy}{dt} = -\left[(1 - \alpha)y^2 + \alpha y\right]^{-1} \quad (6.6.18)$$

Parameters t_0 and α in Equation (6.6.18) are determined as

$$t_0 = \frac{\rho_w L_m r^2 [1 - (T_0 - T)c_w/L_m]}{3(T_0 - T)F_v\left(k_a + L_i D\overline{d\rho_v/dT}\right)},$$

$$\alpha = \frac{F_v\left(k_a + L_i D\overline{d\rho_v/dT}\right)}{k_i} \quad (6.6.19)$$

Integrating Equation (6.6.18) from 1 to 0 and taking into account that $\alpha \ll 1$, one can get the following equation for the time of complete freezing t_f

$$t_f = t_0\left(1 + \frac{\alpha}{2}\right) \approx t_0 \quad (6.6.20)$$

The dependence of the freezing time on drop size and temperature is presented in **Figure 6.6.12**. One can see that the time of complete drop freezing increases with its radius. Freezing time for most drops at $T_C < -5°C$ does not exceed 100 s and at $T_C < -15°C$ does not exceed 10 s. Cloud droplets freeze within a time period shorter than 1 s. For most purposes of cloud simulations, these durations can be considered negligible. For this reason, many models do not include the second stage of drop freezing. Inclusion of water drop accretion can, however, significantly change the freezing time.

Note that Figure 6.6.11 presents only a simplified geometric scheme of drop freezing, which is the basis for deriving the governing heat balance equations. As shown in **Figure 6.6.13**, the interface between ice and water has a very complicated shape. Therefore, at the

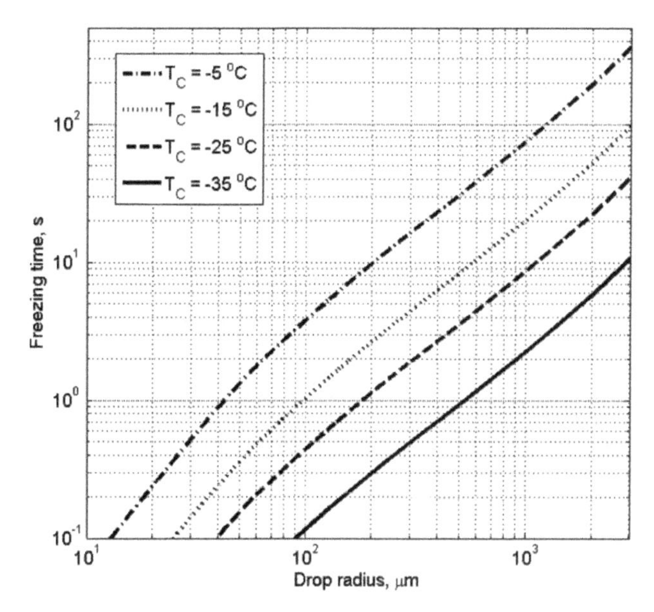

Figure 6.6.12 Dependence of the freezing time on the drop size and surrounding temperature.

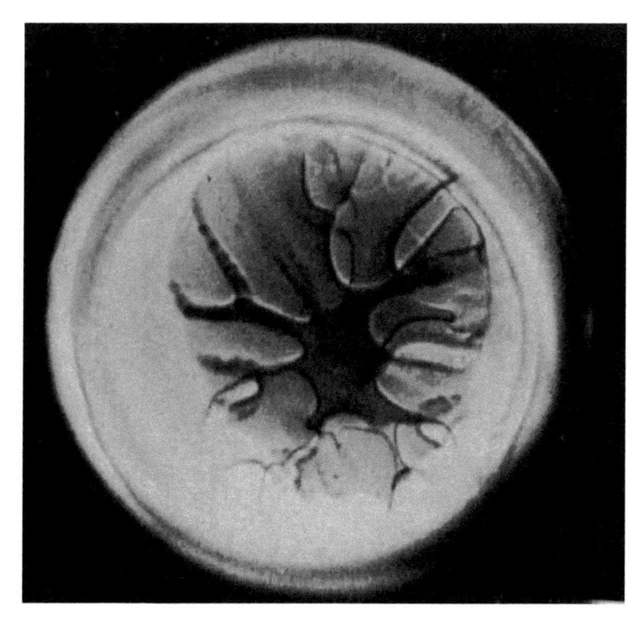

Figure 6.6.13 Photograph of a 1-mm drop undergoing freezing. The outer ice shell has already formed, and the dendritic growth is evident in the drop core. The photograph was taken 10 seconds after nucleation, and complete freezing took a few seconds (from Johnson and Hallett, 1968; courtesy of © John Wiley & Sons, Inc.).

second stage dendridic growth of ice (fingering) into the inner core takes place, so the interior volume is rather a mixture of water and ice than pure water. Therefore, a more sophisticated approach should be used to describe drop freezing more accurately.

6.6.4 Representation of Drop Freezing in Cloud Models with Spectral Bin Microphysics

With few exceptions, a detailed description of the second stage of drop freezing is not included in cloud models. Kumjian et al. (2012) and Ilotoviz et al. (2016) applied the time dependent freezing scheme described by Equations (6.6.15) and (6.6.16) to calculate process of freezing and hail formation. They also calculated polarimetric radar parameters such as radar reflectivity, differential radar reflectivity and others. It was found that in strong convective updrafts freezing drops contain the liquid phase during comparatively long time period producing a significant differential reflectivity.

Equations (6.6.15) and (6.6.16) describe a simplified case when the liquid water freezes only due to the fluxes of heat and moisture upon the particle surface. At the same time, drops during the second stage of freezing collect supercooled liquid droplets (accretion) (Figure 6.6.11). The accreted liquid slows down the raindrops freezing. Freezing of the accreted water leads to a growth of the outer ice shell. In the presence of accretion, two regimes of drop growth may take place. If all the accreted liquid water freezes instantaneously upon contact with the particle surface, the regime is referred to as *dry growth*.

If the accreted liquid water cannot freeze immediately so a fraction of accreted water remains unfrozen, the regime of growth is referred to as *wet growth*. During wet growth, the loss of heat by a drop due to the sensible and the latent surface fluxes cannot compensate the heating caused by the latent heat of freezing. In both dry and wet regimes, the surface temperature of freezing drops increases due to the latent heat of freezing. However, in the dry regime the surface temperature T_s remains lower than T_0.

Calculation of freezing rate begins with determination of the surface temperature under the assumption of the dry freezing mode. The surface temperature is obtained from the heat balance equation written for the drop surface. Quasi-equilibrium is assumed, which means that the heat loss via the sensible and the latent heat fluxes upon the surface, as well as the heat loss needed to warm the accreted water are all assumed to be balanced. This balance is achieved between the latent heat of the accreted water freezing and by the conductive heat flow through the ice shell toward the interior water. This budget equation has the form:

$$\underbrace{4\pi r F_v k_a (T_s - T)}_{\text{sensible heat at surface}} + \underbrace{4\pi r F_v \frac{DL_i}{R_v}\left(\frac{e_i}{T_s} - \frac{e}{T}\right)}_{\text{latent heat of sublimation at surface}} + \underbrace{c_w (T_s - T)\left(\frac{dm_w}{dt}\right)_{accr}}_{\text{heat loss due to warming accreted liquid from } T \text{ to } T_s} = \underbrace{4\pi r \frac{r_w k_i (T_0 - T_s)}{r - r_w}}_{\text{heat flow through shell of ice}} + \underbrace{L_m \left(\frac{dm_w}{dt}\right)_{accr}}_{\text{latent heat relesed by freezing of accreted water}} \tag{6.6.21}$$

where k_i is the thermal conductivity of the ice shell. In HUCM simulations, the accretion rate was calculated using the equation for continuous growth of freezing drops, that is similar to Equation (5.6.4): $\left(\frac{dm_w}{dt}\right)_{accr} = \frac{4\pi\rho_w}{3}\int_0^\infty r'^3 f_r(r') K_{g\_col}(r, r') dr'$, where r is the radius of the freezing drop. The surface temperature of freezing drops depends on drop size. Typically, drops of larger sizes have higher T_s, because large drops collect more water drops. Equation (6.6.21) is written under the assumption that all the accreted liquid freezes. The freezing rate of the interior water $\left(\frac{dm_w}{dt}\right)_{\text{int}}$ (i.e., the rate of the interior water mass decrease) is determined by the conductive heat flux from the interior water upon the drop surface: $\left(\frac{dm_w}{dt}\right)_{\text{int}} = 4\pi r \frac{r_w k_i (T_0 - T_s)}{L_m (r - r_w)}$. The last term on the right-hand side of Equation (6.6.21) $L_m \left(\frac{dm_w}{dt}\right)_{accr}$ describes the heat released by freezing of the accreted water.

Equation (6.6.21) is solved iteratively with respect to T_s. In case $T_s \leq T_0$, the assumption that the freezing drop grows in the dry mode is confirmed. In this case, the increase in the ice mass (equal to the decrease in the mass of all the water including both the accreted and the interior water) within a drop is determined from the heat budget for the entire drop:

$$L_m \frac{dm_w}{dt} = 4\pi r F_v \left[k_a (T_s - T) + \frac{DL_i}{R_v}\left(\frac{e_i}{T_s} - \frac{e}{T}\right)\right] + c_w (T_s - T)\left(\frac{dm_w}{dt}\right)_{accr} \tag{6.6.22}$$

Equation (6.6.22) is an analog of the balance equation (6.6.7) for melting particles that collect liquid drops. If the solution of Equation (6.6.21) corresponds to positive T_s, it means that not all the accreted water freezes, i.e., the freezing drops grow in the wet regime. Polarimetric radar observations of a multi-cell storm, performed by Bruning et al. (2007) show that freezing drops remain water-coated (i.e., undergo wet growth) due to the abundance of supercooled water within a vigorous updraft. This wet growth must control the internal freezing by temporarily halting it. Simulating wet growth, it is necessary to calculate the fraction of the accreted water that freezes on the surface of the freezing drop.

HUCM implements a special procedure for simulating wet growth of freezing drops. Following Pruppacher and Klett (1997), Phillips et al. (2015) assume that the mass of accreted water that freezes on the drop surface is enough to heat this surface to the temperature of $T_0 = 273.15\,K$. The rest of the accreted water remains on the surface. Internal freezing of the liquid trapped inside the shell is assumed to cease during wet growth due to the lack of a radial temperature gradient within the ice shell that could conduct the latent heat released from internal freezing through the ice to the particle's surface. Therefore, the freezing rate is calculated from an equation similar to Equation (6.6.22), by replacing T_s for T_0:

$$L_m \frac{dm_w}{dt} = 4\pi r F_v \left[k_a (T_0 - T) + \frac{DL_i}{R_v}\left(\frac{e_i}{T_0} - \frac{e}{T}\right)\right] + c_w (T_0 - T)\left(\frac{dm_w}{dt}\right)_{accr} \tag{6.6.23}$$

This equation is solved under the limiting condition proposed by Garcia-Garcia and List (1992) upon which the mass that freezes during one time step is not less than that needed to increase the surface temperature to T_0: $\Delta m_w = \Delta m_i \geq \frac{c_w}{L_m}(T_0 - T)\left(\frac{dm_w}{dt}\right)_{accr}\Delta t$. The algorithm of calculations is schematically presented in **Figure 6.6.14**. This algorithm is simplified since it does not trace the time evolution of the exterior water. However, the calculation of the exterior liquid water is important, as the existence of the exterior liquid water means that there is no freezing of interior water. So the exterior liquid water delays drop freezing. The exterior liquid also affects the fall velocity of a drop because it determines the roughness of the freezing drop surface.

The algorithm of time-dependent freezing was implemented in HUCM with spectral bin microphysics according to the scheme shown in Figure 6.6.14. A separate type of hydrometeor, namely freezing drops, was introduced. The size distribution of freezing drops

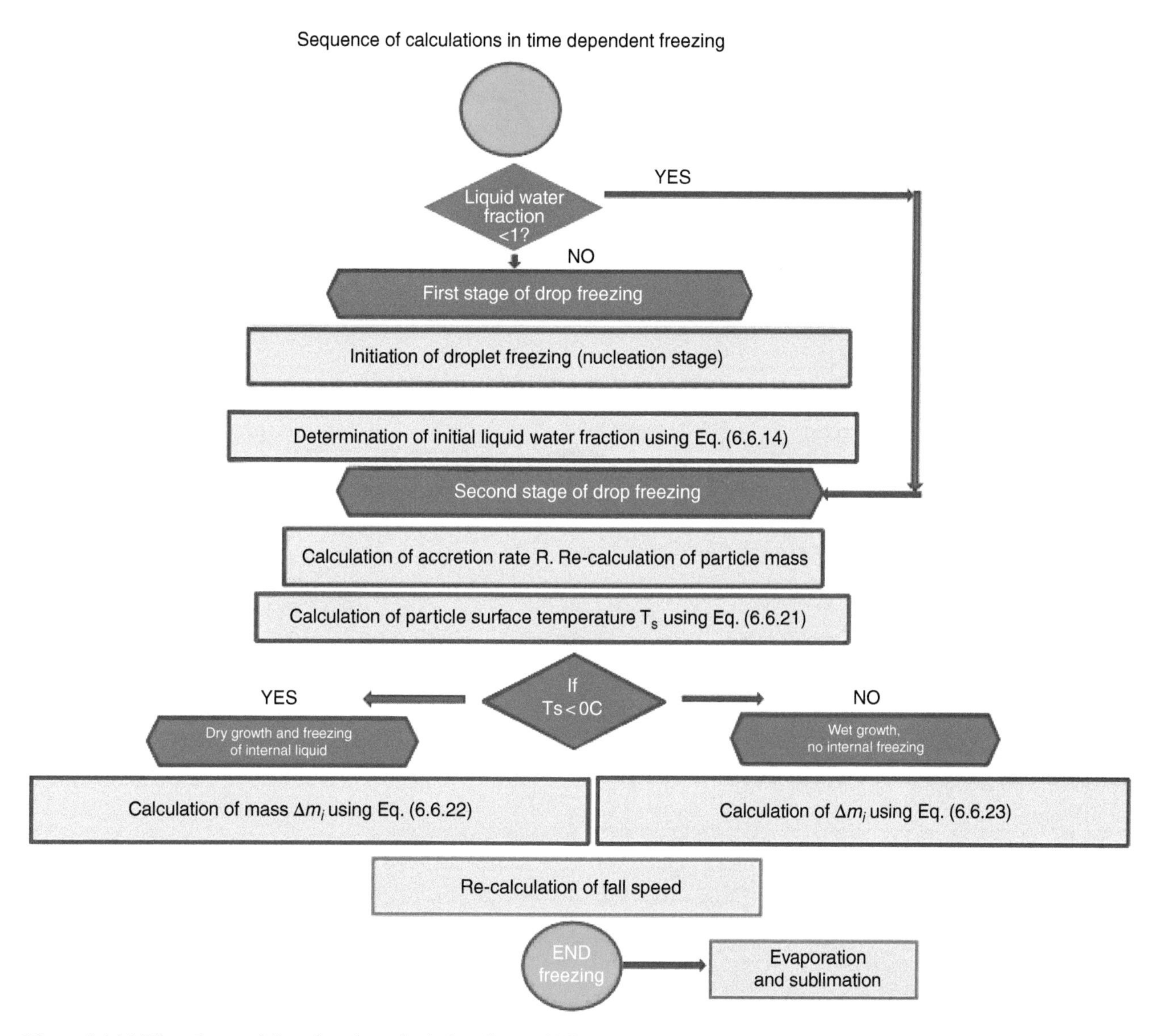

Figure 6.6.14 The scheme of drop freezing calculation (from Phillips et al., 2015; © American Meteorological Society; used with permission).

is calculated on the mass grid containing 43 bins, as for other hydrometeors. It was assumed that freezing drops growing by dry growth can collect only supercooled water droplets. In case of wet growth, the freezing drops were allowed to collide with other hydrometeors at collection efficiency equal to one. If in a collision event a freezing drop is larger than its counterpart, the resulting particle was assigned to the class of freezing drops. Otherwise, the resulting particle was assigned to the type of the counterpart. Since freezing drops contain both ice and liquid water, the mass of the liquid water in each bin is advected, diffused and settle similarly to the freezing drops themselves. When at negative temperatures the liquid water fraction due to freezing becomes smaller than a certain threshold value (say, 1%), freezing drops are converted to hail. At positive temperatures, freezing drops are converted to the melting hail, since melting occurs on the drop shell as in hail.

Formation and evolution of freezing drops, graupel and hail was investigated in simulations of a thunderstorm observed in Oklahoma on February 2, 2009. The meteorological conditions of this storm are described by Ryzhkov et al. (2011). In the simulations, the computational area of HUCM was 120 km × 19 km with the grid resolution of 300 m × 100 m. **Figure 6.6.15** shows

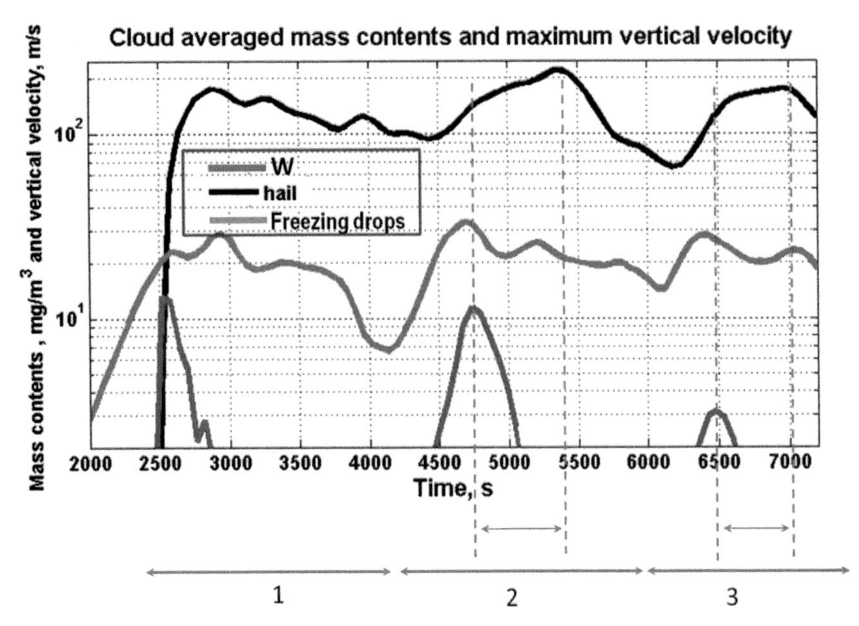

Figure 6.6.15 Time dependencies of cloud-averaged mass contents of freezing drops and hail, and the maximum values of the vertical velocity within the layer below 6 km (from Phillips et al., 2015; © American Meteorological Society; used with permission).

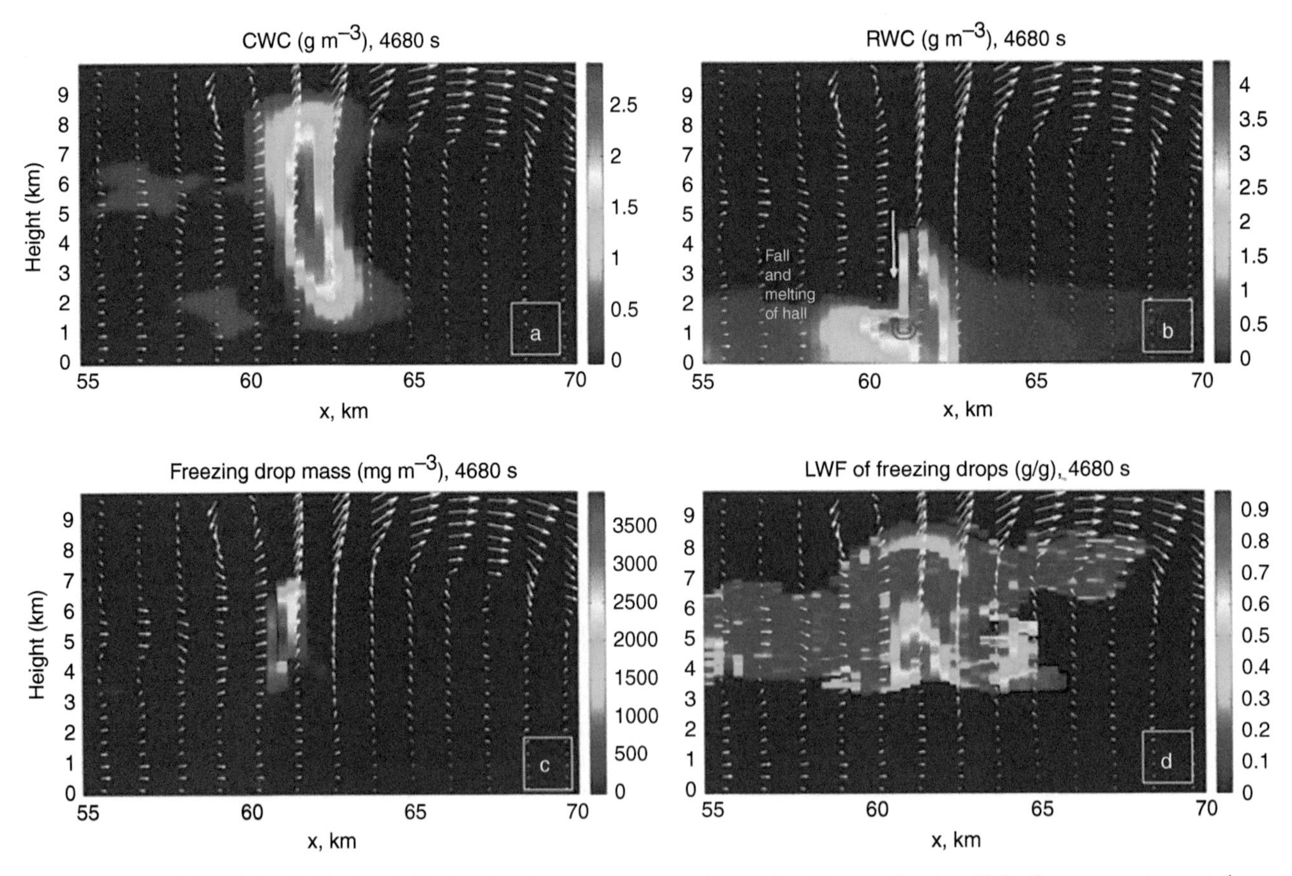

Figure 6.6.16 Fields of CWC (a), RWC (b), freezing drop mass content (c) and liquid water fraction (d) in the mature storm at the stage of a new cell development. Arrows show zones of hail fall, and the direction of movement of raindrops formed by hail melting (from Phillips et al., 2015; © American Meteorological Society; used with permission).

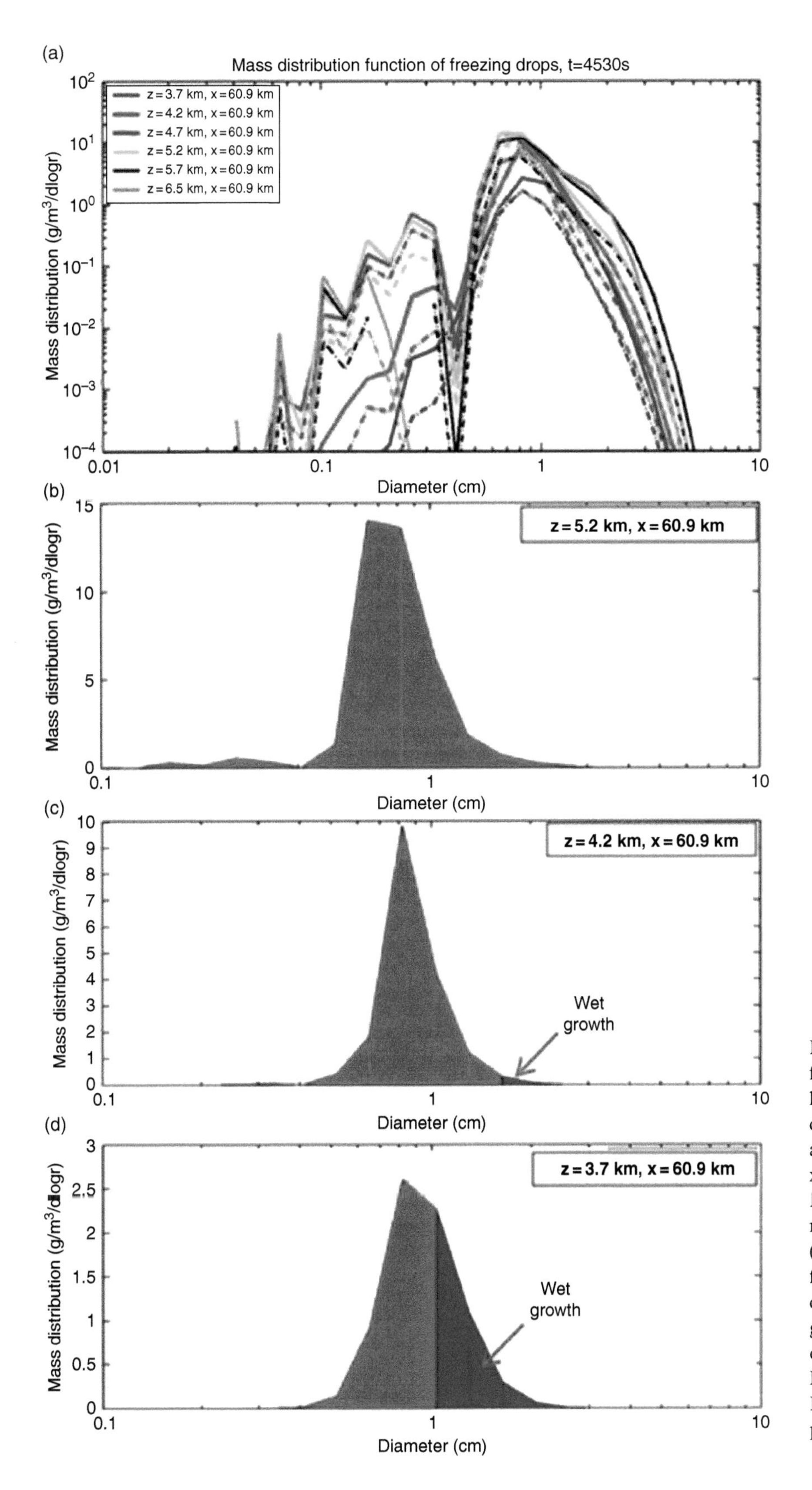

Figure 6.6.17 (a) Mass distributions of freezing drops (solid lines) and of liquid water mass within the freezing drops (dashed lines) at different altitudes at horizontal distance x = 60.9 km. (see spatial scale in Figure 6.6.16) during the stage of a new convective cell development. (b)–(d) The same distributions of freezing drops in a linear scale at different heights. The size range of dry growth is marked red and size range of the wet growth is marked blue (from Phillips et. al., 2015; © American Meteorological Society; used with permission).

the time dependencies of cloud-averaged mass contents of freezing drops and hail, as well as the maximum values of the vertical velocity within the layer below 6 km. One can see that the mass of freezing drops is typically substantially lower than that of hail or graupel. Figure 6.6.15 shows three periods of the vertical velocity increase, indicating the emergence of new developing convective cells, accompanied by hail formation. The mass content of freezing drops increases during the developing stage of convective cells when the maximum vertical velocity is located at comparatively low levels. High velocity leads to ascending of a significant amount of cloud droplets and raindrops. The largest raindrops freeze forming first freezing drops. It may take a few minutes for the freezing drops to convert to hail, thus starting the subsequent hail growth. As a result, the maximum of the hail mass and size is reached several minutes after freezing drops start forming. The emergence of freezing drops may be a precursor of an intense hail.

Figure 6.6.16 shows the fields of CWC, RWC, freezing drop mass content and LWF of freezing drops at the mature stage of the storm during the development of a new strong convective cell. Freezing drops form at about 4.0–4.5 km altitude due to nucleation freezing of raindrops (panels b and c). Complete freezing of liquid within freezing drops takes place at the altitude of about 7 km, where freezing drops are converted to hail. So, due to high CWC (panel a) accretion is intense and significant amount of liquid within the freezing drops remains up to 6.5 km (panel d). Due to collection of small cloud droplets (since CWC is high), freezing drops are comparatively large (**Figure 6.6.17**), so the radar reflectivity during the cell development exceeds 50 dBz. LWF exceeds 0.4 at the altitude of 4–5.5 km. Figure 6.6.17 shows the mass distributions of freezing drops and the distribution of liquid water at different heights within a column of the maximum mass content of freezing drops. One can see that the mean diameter of freezing drops is about 0.8 cm, but there is a small amount of freezing drops having diameters of up to 2–3 cm. The existence of those large freezing drops reflects the fact that drops containing both water and ice are more stable than pure water raindrops of the same size that can be destroyed by breakup. Mass distributions of freezing drops are often bi-modal indicating the difference in drop histories. The largest freezing drops (0.5–3 cm) in Figure 6.6.17a formed in the course of recirculation (the flow looks like a vortex in Figure 6.6.16b). It is natural that freezing drops of all sizes contain liquid water (according to their definition). However, the existence of liquid water does not reveal the regime of their growth since the water may be interior. Figures 6.6.17b–d show the same mass distributions (taken at the instance when the mass content reaches its maximum), with the display of growth modes marked different colors. The growth mode was determined according to the existence of the exterior liquid. One can see that at warm subzero temperatures (z = 3.7 km, $T_C = -2°\text{C}$, CWC ~ 0.5 g m^{-3}) drops with a diameter exceeding 1 cm grow by wet growth (Figure 6.6.17d). At colder higher levels, exterior water freezes, and above approximately 4.5 km ($T_C = -6°\text{C}$, CWC ~ 1 g m^{-3}) freezing drops grow by dry growth (Figures 6.6.17b,c). The boundary between the colors indicates the critical drop diameter D_{onset} separating freezing drops growing by dry growth from larger freezing drops growing by wet growth. The value of this diameter increases with decreasing ambient temperature and CWC. The accretion rate, in turn, depends on the freezing drop size and the CWC in the environment.

6.7 Dry and Wet Growth of Graupel and Hail

Hailstones may reach several centimeters in diameter and fall at velocities of several tens of m/s. This makes hail a dangerous natural phenomenon, which explains the special interest of researchers in the processes of hail formation and growth. Many studies and field experiments have been carried out in search of methods for preventing large hail formation. Success of hail prevention programs depends on how well the mechanisms of hail formation and growth are understood.

6.7.1 Regimes of Graupel and Hail Growth

Hail forms as a result of raindrops freezing (Rasmussen and Heymsfield, 1987), or emerge from graupel by accretion of supercooled liquid drops. The main difference between graupel and hail is in their densities. In some studies, separation between graupel and hail is made basing only on particle sizes, i.e., particles larger than 1 cm in diameter are assigned to hail. Processes of liquid freezing at the surface of graupel and hail are similar in many aspects. So, discussing the accretion growth we will use the terms (graupel or hail) used in corresponding original studies. The cases when the differences between the growth of graupel and hail are principal will be specially stipulated.

The rate of hail growth is determined by three processes: accretion, shedding and, to a less extent, by deposition/sublimation. The latent heat release during freezing of accreted drops is the main heat source that increases particle temperature. If the environmental temperature is low enough and the rate of liquid drops accretion (often called hail riming) is not high enough, all accreted water freezes actually immediately upon a contact with a hail particle. This regime of hail growth is known as *dry growth* (Section 6.6). The surface temperature of a hail particle during dry growth remains negative. With increasing supercooled drop content and hail size, the accretion rate increases, and freezing of the accreted drops gradually increases the temperature of the hail particle's surface. If the accretion rate becomes high enough and the environmental temperature is relatively high, the release of the latent heat of freezing results in increasing the particle surface temperature up to 0°C and above it. In this case, not all the accreted water freezes, so liquid water remains at the hail particle surface. This regime of hail growth is known as *wet growth*. Typically, the surface of a hail particle growing in the wet growth regime is covered with a liquid water film. The critical value of liquid water content (LWC) needed for triggering the wet growth is known as the Schumann-Ludlam Limit (SLL) (Schumann, 1938; Ludlam, 1958). The SLL depends on a particle's size and the environmental temperature. Typically, wet growth starts when the hail size exceeds a certain critical value (about 1–2 cm in diameter), depending on the environmental temperature. The wet growth regime is usually observed in polarimetric radar observations of a vigorous deep convection when supercooled water is abundant (Bruning et al., 2007).

There is a substantial difference between the structure of freezing drops during accretion and the structure of hail particles. Freezing drops are initially liquid and after the first stage of freezing the temperature of the liquid is close to zero. Freezing takes place from outside toward the drop center, forming an ice shell near the particle surface (Section 6.6.). Hail (and graupel) does not initially contain liquid; it arises only in case of a high accretion rate.

During dry growth, riming leads to growth of a spongy ice layer on hail surface. The spongy ice consists of a dense ice framework whose capillaries are filled with air and water. The density of the sponge depends on the impact velocity of drops at the surface of particles and on the temperature of this surface. The velocity at which drops impact the hail (or graupel) surface differs from the relative velocity between these particles and drops. The impact velocity is determined by both the difference in the gravitational settling velocities and the velocity of hail (graupel) rotation. Rasmussen and Heymsfield (1985) express the impact velocity as a function of the terminal velocity and the Stokes number of the riming graupel. At low impact speed and low environmental temperatures, drops freeze keeping the shape of hemispheres or truncated spheres. In this case, the sponge contains significant air volumes and its density is low (from 0.1 up to 0.3 g cm^{-3}). At high impact speed and warm temperatures, the drops freeze taking the shape of flattened particles, and the density of the sponge increases. Heymsfield and Pflaum (1985) proposed the following empirical expression for the density of sponge formed by collisions between graupel of radius r_g and drops of radius r_d:

$$\rho_{sponge}(r_g, r_d) = \begin{cases} AY^B, & \text{for } T_s \le -5°\text{C or} \\ & T_s > -5°\text{C}, Y < 1.6 \\ \exp\sum_{k=1}^{4} B_k Y^{k-1}, & \text{for } T_s > -5°\text{C}, Y > 1.6 \end{cases} \quad (6.7.1)$$

where $Y = \frac{r_d U_{imp}(r_g, r_d)}{T_s}$, $A = 0.3$, $B = 0.44$, $B_1 = -0.03115$, $B_2 = -1.703$, $B_3 = 0.9116$ and $B_4 = -0.1224$. Here the sponge density is in g cm^{-3}, the particle radii are in μm and U_{imp} is the impact velocity in m s^{-1}. Laboratory measurements (Rasmussen and Heymsfield, 1987) showed that the sponge density is closely related to the ice fraction $I_f = \frac{m_{wf}}{m_{wf}+m_{wl}}$, where m_{wf} and m_{wl} are the masses of the frozen and the liquid water accreted by a graupel particle, respectively. The relation is:

$$\rho_{sponge} = (1 - 0.08 I_f) I_f \quad (6.7.2)$$

where ρ_{sponge} is in g cm^{-3}. In Equation (6.7.2), the ice fraction I_f depends on the LWC q_w in the surrounding air. The empirical expression for the ice fraction is:

$$I_f = \begin{cases} I_{f0} + \frac{1 - I_{f0}}{1 + K_I(q_w - q_{w\_cr})} - 0.2 & \text{if } q_w \ge q_{w\_cr} \\ 1, & \text{if } q_w < q_{w\_cr} \end{cases} \quad (6.7.3)$$

where $I_{f0} = 0.25$, $K_I = 0.1798$ (in g^{-1} m^{-3}) and q_{w_cr} = 2 g m^{-3}. Equations (6.7.1–6.7.3) are empiric relationships and the units of the constants are chosen in order to get the correct magnitudes of ρ_{sponge} if it is measured in g cm^{-3}. Equations (6.7.1–6.7.3) are efficient for parameterization of graupel and hail growth in microphysical schemes.

Density of the sponge is an important parameter that determines the type of wet growth. There are two types of wet growth: *spongy wet growth* and *high-density wet growth* (Pruppacher and Klett, 1997). In case of wet growth of graupel having a skeletal framework like a sponge with the bulk density of 450–750 kg m^{-3}, the unfrozen liquid first fills the spaces inside this skeletal structure. This kind of wet growth is known as spongy wet growth and is similar to that at the soaking stage of graupel melting (Figure 6.6.7). The remaining liquid (if it exists) is located on the particle surface forming a bulb near the particle's equator.

When a hailstone is covered with a water film and grows in the high-density wet growth regime, the sponge layer is formed by dendritic growth within the water film (**Figure 6.7.1**). The figure shows that the ice sponge (IS) layer consists of two sub-layers: the layer with dendritic growth (D) and the transition zone of thickness δ between D and the water skin (WS). As a result, in case of the high-density wet growth the density of the growing sponge layer varies between the density of pure ice and that of pure water. List (2014a) evaluates the averaged value of this density as 0.958 g cm^{-3}. The transition from one growth regime to another causes formation of a layered hailstone structure with layers of different density and transparency.

Shedding of water takes place only from hail. Shedding from hail surface is similar to that during hail melting (Section 6.6.1) and can produce a significant number of liquid drops, some of which can serve as potential hail embryos. Lesins and List (1986) suppose that shedding may trigger intense rain. According to Rasmussen and Heymsfield (1987), concentration of shed drops can be 10^3 times higher than that of

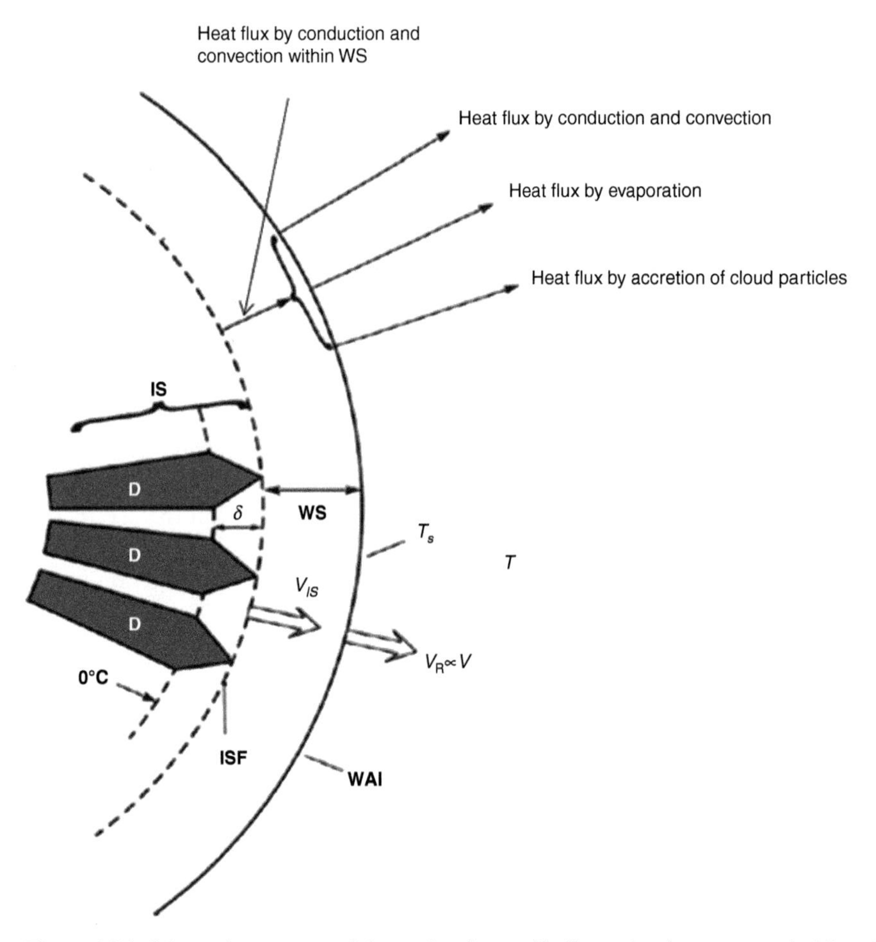

Figure 6.7.1 Schematic structure of the surface layer of hail growing by wet growth. Notations: Dendritic ice growth (D) in the water skin (WS), sponge transition zone thickness δ, ice sponge (IS), ice sponge front (ISF), water-air interface (WAI), surface temperature T_s, air temperature T, ice front speed V_{IS}, radial growth speed V_R and the hailstone fall speed V (adapted from List, 2014a, with changes; © American Meteorological Society; used with permission).

hailstones themselves. Observations show that shedding takes place only in case of intense wet growth. Moreover, the hail diameter must exceed 0.5–0.8 cm for shedding to occur. The maximum mass of water that can contain hailstone before shedding is determined by Equation (6.6.6). Based on their laboratory observations, Lesins and List (1986) presented a more detailed classification of hail growth types. In particular, they introduced the spongy-no-shedding, spongy-shedding, soaked-shedding and dry shedding regimes. The separation of these regimes on the air temperature-LWC plane is presented in **Figure 6.7.2**. The real picture of hail evolution may be more complicated. For instance, dry growth may occur after periods of wet growth along the trajectory of a particular particle in the cloud, owing to the variability in ambient conditions along it. In case wet growth is replaced by dry growth, a particle growing in the dry regime may contain internal liquid remaining after a previous period of wet growth. During dry growth, the loss of heat from the particle surface to the air may be sufficient to freeze both the liquid accreted at the surface and the internally soaked liquid, partially or totally. Freezing of the internal liquid during dry growth may take significant time and may be even incomplete (Pflaum, 1980). The reason is that any freezing can only progress as fast as its latent heat is dissipated into the air.

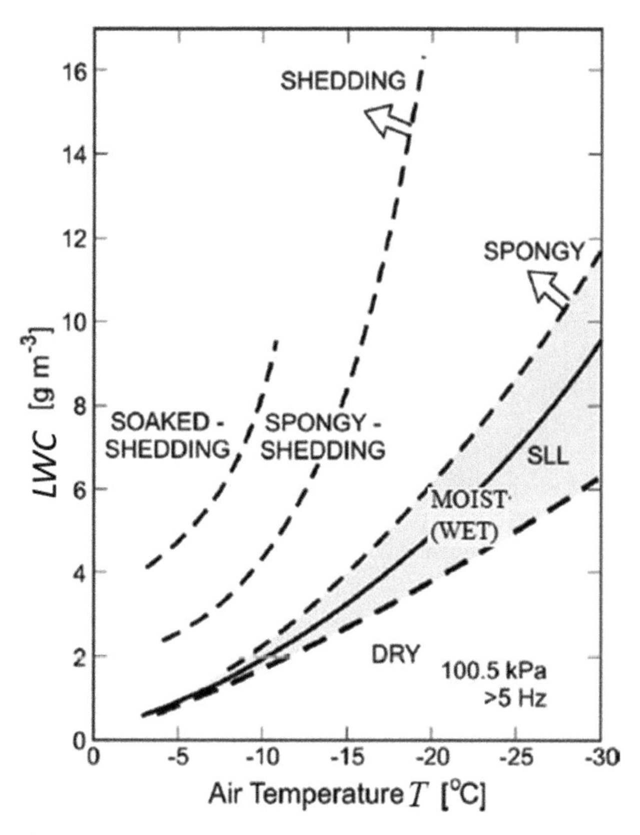

Figure 6.7.2 The five observed hailstone growth regimes shown by LWC-air temperature diagram. SLL is the Schumann-Ludlam Limit, the dividing line between dry growth and growth accompanied by shedding; W_t denotes LWC (from List, 2014b; © American Meteorological Society; used with permission).

6.7.2 Theories of Wet Growth Onset

The theory developed by Schumann (1938) and Ludlam (1958) describes the thermodynamics of hailstone growth under the assumption that the growing particle remains solid and all the water that does not freeze during wet growth is shed immediately. However, laboratory studies indicate that the shedding is quite rare. Moreover, the assumption that a hail particle is impervious was found incorrect, due to formation of the sponge layer, as discussed above. Despite many attempts, it remains a difficult theoretical problem to take into account the presence of a liquid film on the hailstone surface. There are two main theoretical approaches to describe hailstone growth. The first one is based on the equation of heat balance on hailstone surface in the absence of the water film. This approach is aimed at finding the critical liquid water content corresponding to the Schumann-Ludlam limit. Actually, the SLL theory does not consider the behavior of the water film after its formation.

In some studies, it is shown that formation of the water film does not automatically mean that wet growth begins. Several authors (Kachurin, 1962; Zhekamuhov, 1982; Zakinyan, 2008) analyzed the stability of the film. Kachurin (1962) found a theoretical equilibrium film thickness h_b, meaning that a film thinner than h_b is unstable and its thickness decreases with time. In this case, liquid freezes on the particle surface without forming a stable film, and the radius of the ice-water interface grows faster than the particle radius. This condition means that $V_{IS} > V_R$ (Figure 6.7.1). However, if the film is thicker than h_b, it does not disappear but begins growing and a transparent ice structure forms. In this case, the particle radius grows faster than the radius of the ice-water interface, i.e., $V_{IS} < V_R$. Kachurin's theory does not take into account the heat transfer and mass transfer on the film surface, so the film surface temperature is assumed to be constant. Zakinyan (2008) extended the study by Kachirin's approach by including surface fluxes into the analysis. He considers two stages of the film evolution: at the first stage, a film with a certain initial thickness forms spontaneously; at the second stage, the film grows and crystallizes. A new expression was found for the equilibrium film thickness h_b that somehow differs from that proposed in the Kachurin's theory. According to evaluations of Zakinyan (2008), for a hailstone of radius of 1 cm the initial thickness of the water film should exceed 1.7 mm to trigger the wet growth regime.

6.7.3 Theories of Wet Growth

Most theories of graupel and hail growth consider the heat and the mass budgets assuming a uniform depth of the water skin and a spherical shape of graupel or hail. At the same time, the large rimed particles are usually spheroidal in shape and become more spheroidal toward larger sizes. The axis ratio of hailstones declines down to about 0.6 with the size increasing toward a few cm (Knight, 1986). According to List (2014a), about 70% of hailstones with characteristic sizes of 2–5 cm are spheroids. The spheroidal shape of hail leads to inhomoheneity of the fluxes of the accreted water, resulting in the inhomogeneity of the surface temperature which can reach several degrees. However, in most theoretical models graupel or hail growing by wet growth are approximated by spheres covered with a homogeneous water film of a uniform temperature (Musil, 1970; Ferrier, 1994; Pruppacher and Klett, 1997; Zakinyan, 2008; List, 2014a,b). This assumption simplifies the theoretical analysis but makes it less general. The exception is the study by Phillips et al. (2014) where an attempt was made to take into account the inhomogeneous structure of growing hail and graupel, caused by a non-uniform distribution of the riming rate over a particle surface. Below, we discuss the approaches to description of the wet growth of graupel and hail under the assumption of a spherical symmetry of the hailstone and describe the results obtained in studies that took into account a non-uniform particle surface.

Theories of Spherically Symmetric Particle Growth

The change of a hail particle mass is equal to the sum of changes caused by accretion, deposition/sublimation (i.e., diffusion) and shedding:

$$\frac{dm}{dt} = \left(\frac{dm}{dt}\right)_{accr} + \left(\frac{dm}{dt}\right)_{diff} - \left(\frac{dm}{dt}\right)_{shed} \tag{6.7.4}$$

The thermodynamic equations of hail growth resemble those for drop freezing Equations (6.6.21) and (6.6.23). However, there are several important differences. The accretion is the main source of hail particle mass. The release of the latent heat of freezing of the accreted water is the main heat source determining the dry-to-wet growth transition. It means that the term describing the accretion rate plays an important role in the mass and the heat budgets and thus cannot be neglected. In practice, calculations of hailstone growth are performed under the primary assumption that hail grows in the dry growth regime. The heat budget for a growing spherical hail particle can then be written as

$$\underbrace{L_m\left(\frac{dm_w}{dt}\right)_{accr}}_{\substack{\text{latent heat relesed by}\\ \text{freezing of accretion}}} = \underbrace{4\pi r F_v k_a (T_s - T)}_{\substack{\text{sensible heat at}\\ \text{the surface}}} + \underbrace{4\pi r F_v \frac{D L_i}{R_v}\left(\frac{e_i}{T_s} - \frac{e}{T}\right)}_{\substack{\text{latent heat at sublimation}\\ \text{at surface}}} + \underbrace{c_w (T_s - T)\left(\frac{dm_w}{dt}\right)_{accr}}_{\substack{\text{heat loss due to warming}\\ \text{accreted liquid from } T \text{ to } T_s}} \tag{6.7.5}$$

where D is the coefficient of molecular diffusion of water vapor. This equation resembles Equations (6.6.21) and (6.6.22), the main difference being that the term $\frac{dm_w}{dt}$ at the left-hand side of Equation (6.6.22) includes the internal liquid, while Equation (6.7.5) deals with the freezing rate of the accreted liquid. Equation (6.7.5) is used to calculate the surface temperature T_s by the iteration method. In case $T_s < T_0$, the assumption concerning the dry growth is justified and the changes of the hail particle mass are calculated by the equation for mass budget (6.7.4), where the shedding term is omitted.

As in case of freezing drops, if the solution of Equation (6.7.5) shows that $T_s > T_0$ (which is unrealistic) the wet growth takes place. Then T_s should be recalculated using the heat balance equation. In case of wet growth, this equation is more complicated than Equation (6.7.5) since it takes into account the effects of accretion of water, which freezes only partially, in the presence of collection of ice particles and shedding. Since the surface of a particle growing by wet growth is wet, L_i should be replaced by L_w in the second term on the right-hand side of Equation (6.7.5). As a result, the balance equation can be written as:

$$\underbrace{I_f L_m R}_{\substack{\text{Heating during freezing}\\ \text{of accreted water}\\ \text{(heat source)}}} = \underbrace{4\pi r F_v k_a (T_s - T)}_{\substack{\text{Cooling due to heat}\\ \text{conductivity}}} + \underbrace{4\pi r F_v \frac{DL_w}{R_v}\left(\frac{e_w}{T_s} - \frac{e}{T}\right)}_{\substack{\text{Cooling due to water vapor}\\ \text{sublimation}}} + \underbrace{c_w (T_s - T) R}_{\substack{\text{Heat required to increase temprature}\\ \text{of accreted water from } T \text{ to } T_s}}$$
$$+ \underbrace{c_i (T - T_s)\left(\frac{dm}{dt}\right)_{col\_ice}}_{\text{Heat change due to collisions with ice}} + \underbrace{c_w E_{shed}(T_w - T_0) R}_{\substack{\text{Extraction of heat due to}\\ \text{shedding}}} \tag{6.7.6}$$

The multiplier $R = \left(\frac{dm_w}{dt}\right)_{accr}$ in Equation (6.7.6) indicates the accretion rate of liquid, the multiplier $\left(\frac{dm}{dt}\right)_{col\_ice}$ is the rate of collection of ice particles by the hailstone. The term on the left-hand side shows the heat source due to freezing of fraction I_f of the accreted liquid. The third and the last terms on the right-hand side show the heat needed to increase the temperature of the accreted liquid up to the particle surface temperature and the heat of shed water, respectively. In Equation (6.7.6), the frozen fraction I_f is defined as the ratio $I_f = \left(\frac{dm}{dt}\right)_{freez} / \left(\frac{dm}{dt}\right)_{accr}$. This fraction needs to be determined.

There are several approaches to calculating the frozen fraction I_f. The first approach is similar to that used in case of wet growth of freezing drops (Section 6.6). The freezing rate is calculated from the heat balance equation (6.7.6) under the assumption that the latent heat release caused by freezing of fraction I_f of the accreted liquid increases the particle surface temperature to T_0. It is assumed that any excess liquid is shed. In addition, L_w in the second term on the right-hand side of Equation (6.7.6) should be replaced by L_i. The rate of freezing needed to heat the surface to temperature T_0 is calculated as (Straka, 2009):

$$\left(\frac{dm}{dt}\right)_{freez} = \frac{4\pi r F_v k_a (T_0 - T) + 4\pi r F_v \frac{DL_i}{R_v}\left(\frac{e_w}{T_0} - \frac{e}{T}\right) + c_i (T_0 - T)\left(\frac{dm}{dt}\right)_{col\_ice}}{L_m + c_w (T - T_0)} \tag{6.7.7}$$

The rate of the increase of the liquid mass at the hail surface is then calculated as $(1 - I_f)R$. Under certain conditions, this water can be shed as described in Section 6.6.1.

The approach described by Equation (6.7.7) has some physical inconsistencies. First, it suggests that during the onset of wet growth the freezing rate remains equal to that at the instance of the transition from dry to wet growth at the surface temperature of 0°C. At the same time, liquid film dramatically affects all the heat fluxes as soon as it starts to form. Indeed, the surface temperature at wet growth is by several degrees below 0°C (List, 2014a,b). Substantial supercooling of particle surface is a necessary condition for the heat flux to be directed outward from the ice sponge front to the particle surface. The temperature of the ice sponge-liquid interface (ISF, Figure 6.7.1) is typically negative and decreases with increasing hailstone size. Second, Phillips et al. (2014) showed that the formation of the water skin leads to substantial changes in all the heat fluxes. For instance, in case of wet growth there is evaporation on the particle surface, but not sublimation as follows from Equation (6.7.7).

A different approach was used by List (2014a,b) who summarized the empirical and theoretical results of his previous investigations of hail wet growth and re-formulated the heat balance equation. The proposed equation system consists of the heat balance equations for an entire particle and for its surface. These equations contain two unknowns: the surface temperature and the ice fraction. To close the system, List (2014a,b) assumes that a) the speed of the increase of hailstone size V_R is equal to the speed of the ice sponge-liquid interface (ISF) (ice sponge front) V_{IS} (Figure 6.7.1) and b) there is a functional dependence of V_{IS} on the temperature at the ISF. Since the radial growth of a hailstone is fully determined by the accretion rate and the sponge density, this dependence allows to relate the surface temperature to the radial growth speed V_R and to the fall speed of a hailstone. Thus, this dependence provides an additional relationship that allows to find the surface temperature. The second assumption made by List (2014a,b) is that the temperature at the sponge-liquid interface is equal to zero. These assumptions allow to find the temperature gradient within the liquid film. In turn, using the temperature gradient allows to solve the equation system and to find both the surface temperature of the particle and the ice fraction. This approach, however, does not consider the internal structure of a hailstone.

Theory of Nonuniform Wet Growth of Hail Particles
There are only a small number of theoretical studies that attempted reproducing laboratory results about the structure of hailstones growing by wet growth. To take into account the observed inhomogeneity of hail particle surface wetness and temperature, Phillips et al. (2014) proposed a "2-part model" of hail wet growth and compared the simulation results with the laboratory data. As soon as wet growth begins, the volume of each hail particle is assumed to be divided into two parts a) a "wet part" covered with a liquid film and b) a "dry part" with an icy surface. The geometry of a hail particle during wet growth is presented in **Figure 6.7.3**. Assuming σ_d is the fraction of the dry surface of a hail particle, while the remaining part, i.e., $1 - \sigma_d$ is wet, Phillips et al. (2014) suggested that σ_d is related to the liquid water fraction (LWF) f_w in a simple linear fashion:

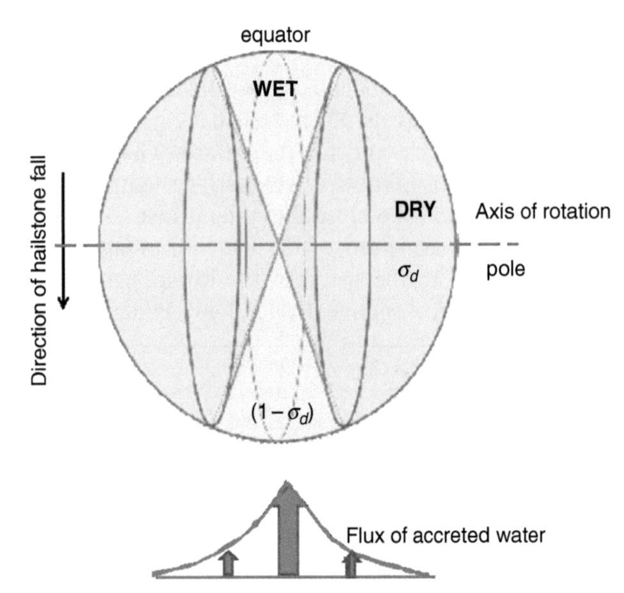

Figure 6.7.3 Cross section of a hail particle illustrating a schematic separation of a hailstone growing by wet growth into a wet part and a dry part. While falling, the particle rotates a nearly horizontal minor axis. The larger axis is directed vertically. The flux of accreted water onto the particle surface is shown by arrows (adapted from Phillips et al., 2014; © American Meteorological Society; used with permission).

$$\sigma_d = \begin{cases} \min\{1 - C_1 f_w,\ 0.95\}, & \text{if } f_w > 0 \\ 1, & \text{if } f_w = 0 \end{cases} \tag{6.7.8}$$

where, according to their evaluations, $C_1 \approx 5$. The spheroidal shape of hail creates a symmetric gyration in a free fall with rotation around an almost horizontal minor axis. The frequency of this rotation is typically of about 10 Hz. (Pruppacher and Klett, 1997; Straka 2009). As a result, accretion and liquid accumulation take place along an almost vertically oriented equator (Figure 6.7.3). Hence, hail "equatorial zone" receives most of the accreted condensate and becomes wet. This structure agrees with the laboratory experiments carried out by List (1959), Macklin (1961), Mossop and Kidder (1962), Kidder and Carte (1964), Lesins and List (1986) and Garcia-Garcia and List (1992), which show that water forms a vertical torus around the particle equator during wet growth, while the poles remain dry and icy. The particle structure shown in Figure 6.7.3 is typically observed in comparatively large hailstones. In contrast, smaller hail and graupel tumble and grow symmetrically.

The structure of the non-spherical hail particle differs from that in the case of melting of spherical ice particles. In melting, the torus of the exterior liquid forms within the cross-section perpendicular to the particle fall velocity (Section 6.6). This occurs because at positive temperatures melting takes place over the entire surface, and the surface water is shifted toward the horizontal axis under the pressure of the environment air flow. At negative temperatures, freezing of the accreted liquid hinders the shift of the liquid from the point of drop impact where the flux of accreted water is maximum (Figure 6.7.3).

Phillips et al. (2014) attempted taking into account the effect of the internal structure of hailstone on its growth. The surface temperature is calculated using the evaluated growth rate of a hailstone and the density of sponge, predicting the thickness of the liquid film and the temperature of the ice-liquid interface. The particle structure presented in **Figure 6.7.4** shows two vertical cross-sections: the equatorial one and the one perpendicular to it. One can see on the left panel that the hailstone volume is divided into two parts: a "dry" part having the relative surface σ_d, and a "wet" part having the relative surface of $1 - \sigma_d$ (Figure 6.7.3). The dry part consists of ice of bulk density equal to ρ_{bulk}. Besides, the hailstone has an inner core of an equivalent radius r_{core}. This core forms during the early stage of a hailstone history when it grows by dry growth and its structure is assumed homogeneous. The properties of

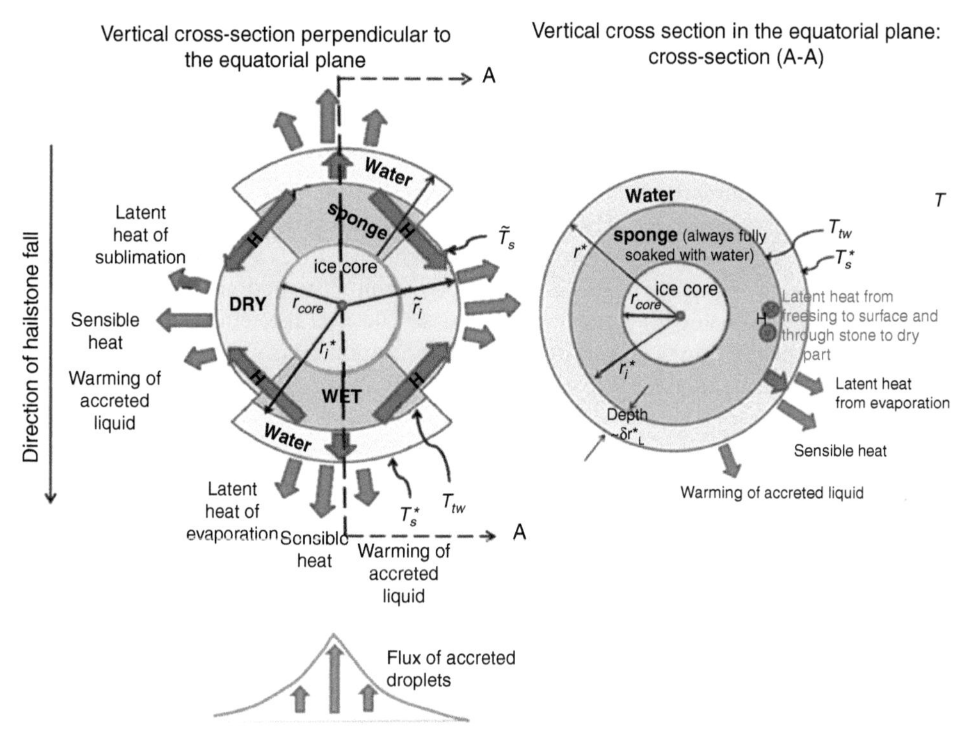

Figure 6.7.4 Schematic diagram of two vertical cross-sections of a hail particle during wet growth: the equatorial plane (right) and the perpendicular to the equatorial plane (left). Fluxes of heat featuring in the wet component heat budget directed outside (blue) and inside (red) of the particle are shown. The vertical cross-section bisects the two spherical sectors of the dry part (blue shading; left panel). All the parameters related to the wet part are denoted by symbol "*." All the parameters related to the dry part are denoted by symbol "~" (adapted from Phillips et al., 2014; © American Meteorological Society; used with permission).

the dry part and the ice core are identical. In the wet part, the spongy surface ice is formed by the dendritic propagation of ice into the surface water (Figure 6.7.1). The external radius of the sponge layer is r^*. In addition to the sponge layer, the wet part contains the exterior liquid film located within the radii range of $r^* > r > r_i^*$. The liquid mass in the film is a minor portion of the particle liquid since during the soaking stage the majority of the liquid is accumulated in the interior of the particle. After the sponge has reached its maximum volume possible in the wet component, the soaking process is extended to the dry component and to the core that may get soaked. Only when both the dry and the wet components are fully soaked, the extra liquid can accumulate on the exterior of the wet component. The liquid film is then thickened, extending outwards in a wedge-like manner. It is clear that both the assumed geometry of the wet and the dry components and the sequence of the processes are simplified. In reality, multiple mobile liquid skins are formed on the surface of a particle, especially at the onset of wet growth. Nevertheless, Phillips et al. (2014) demonstrated that the simplified geometry yields results quite close to laboratory observations.

In general, the approach to developing the heat balance equations in a 2-part model is the following. As shown in Figure 6.7.4, a hail particle is considered consisting of concentric shells, so different quantities characterizing a hailstone are assumed to vary only at a distance to the particle center. The inner shells form a homogeneous core. In the outer shells, the wet part is assumed to be a fraction $(1 - \sigma_d)$ of an entire shell. The inhomogeneity of the hailstone structure is merely represented by the contrast between the wet and the dry parts. Both parts are thermally coupled by the heat conduction through the interface between the parts of the particle. The interaction between the two parts affects the value of σ_d, the wet component surface temperature and the freezing rate.

Calculations of hailstone freezing rate are performed in several steps. First, the condition required for the onset of wet growth is checked. It is assumed that the growth process starts in the dry regime, and the corresponding hailstone surface temperature is calculated. At this stage, the structure of the hailstone is assumed to be azimuthally homogeneous. If the surface temperature of the hailstone $\hat{T}_s$ is found to be lower than the empirical critical threshold, T_{cr}, the assumption regarding the dry growth regime is justified and the surface temperature of the hailstone is assumed equal to $\hat{T}_s$. All the accreted water freezes determining the rate of the hailstone growth. Phillips et al. (2014) calculate T_{cr} using a simple fit of results obtained in laboratory observations by Garcia-Garcia and List (1992) and Levi and Lubart (1998) as

$$T_{cr} = T_0 - \min\left[\Delta T/2,\ 3°\text{C}\right] \tag{6.7.9}$$

Equation (6.7.9) takes into account that wet growth starts when the surface temperature is below T_0. The temperature $\hat{T}_s$ is evaluated from the heat budget equation

$$RL_m = 4\pi r F_v \chi \left[k_a\left(\hat{T}_s - T\right) + \frac{DL_i}{R_v}\left(\frac{e_w(\hat{T}_s)}{\hat{T}_s} - \frac{e}{T}\right)\right] + \left(\hat{T}_s - T\right)\left[c_w R + c_i \left(\frac{dm}{dt}\right)_{col\_ice}\right] \tag{6.7.10}$$

The accretion rate R of a hailstone having the effective radius r due to collection of liquid drops is calculated as

$$R \approx \int_0^\infty m_d \pi (r + r_d)^2 E \Delta V_g f(m_d) dm_d \tag{6.7.11}$$

where E is the collection efficiency, ΔV_g is the difference between fall velocities of the hailstone and the collected drops and $f(m_d)$ is DSD. Equation (6.7.10) differs from Equation (6.7.7) in two aspects: a) the surface temperature is not equal to T_0 and b) a correction factor χ is included to parameterize the enhancement of the surface area of a spheroidal particle as compared to a spherical particle. χ is the ratio of capacitances of the actual spheroidal particle and the spherical particles of radius r. The value of χ is calculated as $\chi = \frac{\varepsilon}{s^{1/3}\text{asin}(\varepsilon)}$ where s is the axial ratio and $\varepsilon = \sqrt{1 - s^2}$ is the eccentricity of the spheroidal particle. For spherical particles $\chi = s = 1$.

Equation (6.7.10) is solved iteratively. $\hat{T}_s$ depends on a particle size, as well as on the ambient temperature and on the LWC. Since a particle for which $\hat{T}_s > T_{cr}$ grows in the wet growth regime, there is the boundary radius r_{onset} corresponding to T_{cr}, so hailstones with radii exceeding r_{onset} grow in the wet growth regime at a given environmental LWC. In this case it is necessary to use the heat balance equations separately for each part of the hailstone.

The Rate of Freezing on the Dry Component

The heat balance of the dry component during wet growth differs from that in case of pure dry growth due to thermodynamic interaction between the dry and the wet parts. The average temperature of the icy surface of the dry component is assumed to be equal to the critical value:

$$\tilde{T}_s = T_{cr} \tag{6.7.12}$$

Condition (6.7.12) imposes a strong limitation on the rate of accretion rate $\tilde{R}$ on the dry part. As a result, it is lower than that on the wet part. The freezing rate on the dry component is obtained using the heat budget of the dry component surface:

$$L_m\tilde{R} = -H + 4\pi\tilde{r}_i F_v \chi \sigma_d \left[k_a\left(\tilde{T}_s - T\right) + \frac{DL_i}{R_v}\left(\frac{e_w(\tilde{T}_s)}{\tilde{T}_s} - \frac{e}{T}\right)\right] + \left(\tilde{T}_s - T\right)c_w\tilde{R} \tag{6.7.13}$$

where $\tilde{R} = \frac{d\tilde{m}_i}{dt}$. To apply Equation (6.7.13), it is necessary to know the heat flux H through the interior of the hailstone into its dry component (Figure 6.7.4). This heat flux is assumed to be proportional to the difference in temperatures of the dry and the wet components.

The Rate of Freezing on the Wet Component

To calculate the rate of freezing and to determine the geometry of the layers of the wet component of a hailstone (Figure 6.7.4), it is necessary to calculate the surface temperature, as well as the temperature T_{iw} of the liquid at the sponge-liquid interface. For raindrop freezing, the approximation T_{iw} = 0°C can be used (Hallett, 1964). As regards large hail, the supercooling temperature T_{iw} at the ice front can be as high as 0.5–1°C (List, 2014a). So the approximation T_{iw} = 0°C is too crude for hail, and thus T_{iw} should be calculated. The equation system used for this calculation consists of three heat balance equations: for particle surface, for sponge-liquid interface and for the entire wet component. The procedure involves estimating the change in thickness of the ice sponge layer in order to establish the radial speed of the ice-sponge front V_{IS} as well as temperature T_{iw} at the water-ice interface.

The first balance equation is written for the outer surface of the water skin and used to calculate the temperature T_s^* of the liquid surface:

$$\underbrace{\frac{k_w A_L (T_{iw} - T_s^*)}{\delta r_L^*}}_{\text{heat conducted through liquid skin}} = \underbrace{4\pi r^* F_v (1-\sigma_d)\chi C_3 \left[k_a (T_s^* - T) + \frac{D L_w}{R_v}\left(\frac{e_w(T_s^*)}{T_s^*} - \frac{e}{T}\right)\right]}_{\text{sensible and latent heat flux from particle surface to ambient air}}$$
$$+ \underbrace{(R_1^* c_w + I^* c_i)(T_s^* - T) + S c_w (T_w - T)}_{\text{loss of heat due to warning of accreted and shed condensate}} \quad (6.7.14)$$

In Equation (6.7.14), S is the rate of shedding assumed to be proportional to the accretion rate (Garcia-Garcia and List, 1992):

$$S = \begin{cases} R \times \min\left\{1 - \left[1 + K_E (q_w - q_{w\_cr}\right]^{-1}, 1\right\}, & \text{if } q_w \geq q_{w\_cr} \\ 0, & \text{if } q_w < q_{w\_cr} \end{cases} \quad (6.7.15)$$

where $K_E = 100 - 4\Delta T$ m^{3}kg^{-1}. Shedding is allowed if the LWC exceeds the critical value q_{w_cr}. According to Levi and Lubart (1998), $q_{w\_cr} = 10^{-3}\left(0.928 + 0.059\Delta T + \frac{3.227+0.21\Delta T}{200r}\right)$ kg m^{-3}, where r is the radius of the hailstone in meters and $\Delta T = T_0 - T$ is the supercooling temperature of the environment air. The effect of a spheroidal shape on the fluxes is taken into account as proposed by Knight (1986) and Lesins and List (1986). The net rate of the accretion of liquid onto the wet component is $R_1^* = (R - \tilde{R}) - S$, where $\tilde{R}$ is the rate of freezing on the dry component. The rate of accretion of ice by the wet component is $I^* = I - \tilde{I} \approx I$. An important parameter determining the effect of shedding on the hailstone thermodynamics is the temperature T_w of shed water. This temperature depends on the time during which the liquid water remained on the particle surface before being shed. While being on the surface, the liquid water is warmed deviating from the environmental temperature toward T_s^*. Following List et al. (2014a), it is assumed that the time prior to shedding is long enough, so $T_w - T_s^*$. A_L and $\delta r_L^* = r - r_w$ in the Equation (6.7.14) are the outer area and the depth of the liquid film, while k_w is the thermal conductivity of the liquid. $C_3 \approx 1.5$ is a dimensionless factor parameterizing the surface roughness (Garcia-Garcia and List, 1992; Aufdermauer and Joss, 1967; Schuepp and List, 1969a,b). The term on the left-hand side of Equation (6.7.14) is the diffusional flux of heat through the liquid film into the air. This flux is caused by the latent heat released of freezing. $T_{iw} - T_s^*$ is the difference in the temperature across the film. The heat flux H through the interior of the hailstone does not appear in Equation (6.7.14) because this flux is caused by the latent heat of freezing at the sponge-liquid interface. It does not pass through the film and, therefore, does not affect the heat balance on the particle surface.

The second equation is for the heat budget of the entire volume of the wet component, used to calculate the frozen mass in the wet component. The budget includes the latent heat from freezing and its conduction both through the outer liquid layer to the surface and through the interior of the hailstone into its dry components:

$$L_m \frac{dm_i^*}{dt} = \frac{k_w A_L (T_{iw} - T_s^*)}{\delta r_L^*} + H \quad (6.7.16)$$

In calculations, the surface temperature T_s^* of the liquid skin, found at the first step, is used. The heat flux H is directed through the interior of the hail particle, almost tangentially beneath the surface, from the ice front of the wet component toward the outer surfaces of the dry components. The flux is assumed to be proportional to $k_p(T_{iw} - \tilde{T}_s)$, where k_p is the averaged thermal conductivity of the hailstone. Since the internal structure of hail is complicated, k_p is calculated as a weighted average of the thermal conductivity of ice, liquid and air for the entire particle. Knowing m_i^* and the sponge density enables to calculate the radius r_i^* of the outer surface of the ice sponge. The theory proposed by Phillips et al. (2014) allows, therefore, to calculate the velocity of the ice-sponge interface V_{IS} as

$$V_{IS} \approx \frac{dr_i^*}{dt} \quad (6.7.17)$$

Expression (6.7.17) demonstrates the improvement of the wet growth theory as compared to that developed by List (1990, 2014a,b) and Lozowski (1991), who assumed that V_{IS} is equal to the rate of the particle radius growth.

The third equation allows to calculate the temperature T_{iw} at the sponge ice-liquid interface. Phillips et al. (2014) use the empirical relationship between V_{IS} and T_{iw} (List, 1990):

$$T_0 - T_{iw} = \left(\frac{V_{IS}\left[\mathrm{ms}^{-1}\right]}{0.00028}\right)^{1/2.39} \tag{6.7.18}$$

The value T_{iw} together with the surface temperature determines the temperature gradient across the outer liquid layer. This gradient determines the ice fraction and the rate of the latent heat dissipation into the air.

Technically, the equations for the dry and the wet components are solved simultaneously as a single set using the method of iterations. At each iteration, the masses of accreted liquid that become frozen in the dry $\Delta\tilde{m}_i$ and the wet Δm_i^* components of the hailstone are calculated, as well as the liquid fraction and σ_d. After convergence to solution of T_{iw}, the total mass of ice acquired per unit time is calculated as

$$\Delta m_i = \Delta\tilde{m}_i + \Delta m_i^* \tag{6.7.19}$$

Finally, both temperatures T_s^* and $\tilde{T}$ are used to alter m_w and m_i due to evaporation and sublimation from the wet and the dry components.

Comparison with Laboratory Observations

Figure 6.7.5 (upper row) shows the prediction of the hail surface temperature during hail growth, performed by means of the two-part model proposed by Phillips et al. (2014). The values of the calculated surface temperature averaged over the equatorial band of ±30° were compared with the laboratory observations made by Garcia-Garcia and List (1992). The average error in the equatorial surface temperature for all the values of the LWC observed for different hailstones is less than about 0.4°C within the environmental temperature range from −6°C to −20°C. The predicted difference between the surface temperatures averaged over the equatorial wet band and the dry component is also consistent with the experimental results obtained by (List et al., 1995). The dry component surface temperature was predicted to be lower than the equatorial average, as expected. The average absolute error in the predicted ice mass fraction is less than about 0.04 for all the experiments within the temperature range from −6°C to −20°C (Figure 6.7.5, bottom row).

A spherically symmetric model with parameters adjusted to maximize the agreement with the observations proved not very accurate. First, the one-component model predicts the surface temperature by 2–3°C higher in comparison to the observations. The deviation from observations is especially pronounced at intermediate LWC values close to the wet growth onset. The errors are especially high at colder ambient air temperatures at which the wet growth is less probable and the particle surface should be drier. Second, the critical LWC value required for the onset of the appreciable liquid film on the surface simulated by the one-component model is by about 100–200% larger compared to the observations; at small and intermediate LWC, the ice fraction decreases much sharper with respect to the environment LWC than in the observations. At the highest LWC values, the ice fraction simulated by the spherically uniform model is typically lower than that observed in the laboratory experiments.

This behavior of the one-component model can be explained as follows. Any wet film tends to slow the freezing process, partly by thermally insulating the freezing region in the sponge and cooling the particle's surface, thus inhibiting the heat loss to the air. Thus, the liquid film thickness controls the growth of the hailstone. Close to the onset of wet growth, the artificial spreading of any initial liquid film over the entire particle in the one-component model makes the film unrealistically thin and liable to freezing due to the insufficient thermal insulation at too warm surface temperature (Figure 6.7.5, upper panels). As a result, the effective onset of the wet growth occurs in the one-component model at LWC values higher than those in the observations. In the two-part model at LWC close to the Schumann-Ludlam Limit, the liquid film covers only a part of the entire surface, which leads to a slower decrease of the ice fraction which is in agreement with the laboratory measurements. It is noteworthy that if the wet part occupies over 90% of the particle surface, the hail properties predicted by the one-component model eventually converge toward those predicted by the two-component model. Thus, the spherically symmetric hail growth model produces realistic results in case when either LWC is quite high or when the time period of hailstone growing by wet growth is long enough for all the surface of the hailstone to become wet. The time period needed to reach such stage can exceed one-two minutes. During this time, the hailstone may fall down by more than 1 km and the environment conditions may change significantly. In this situation, the convergence of the one-component model results to those obtained by the two-part model may not be achieved.

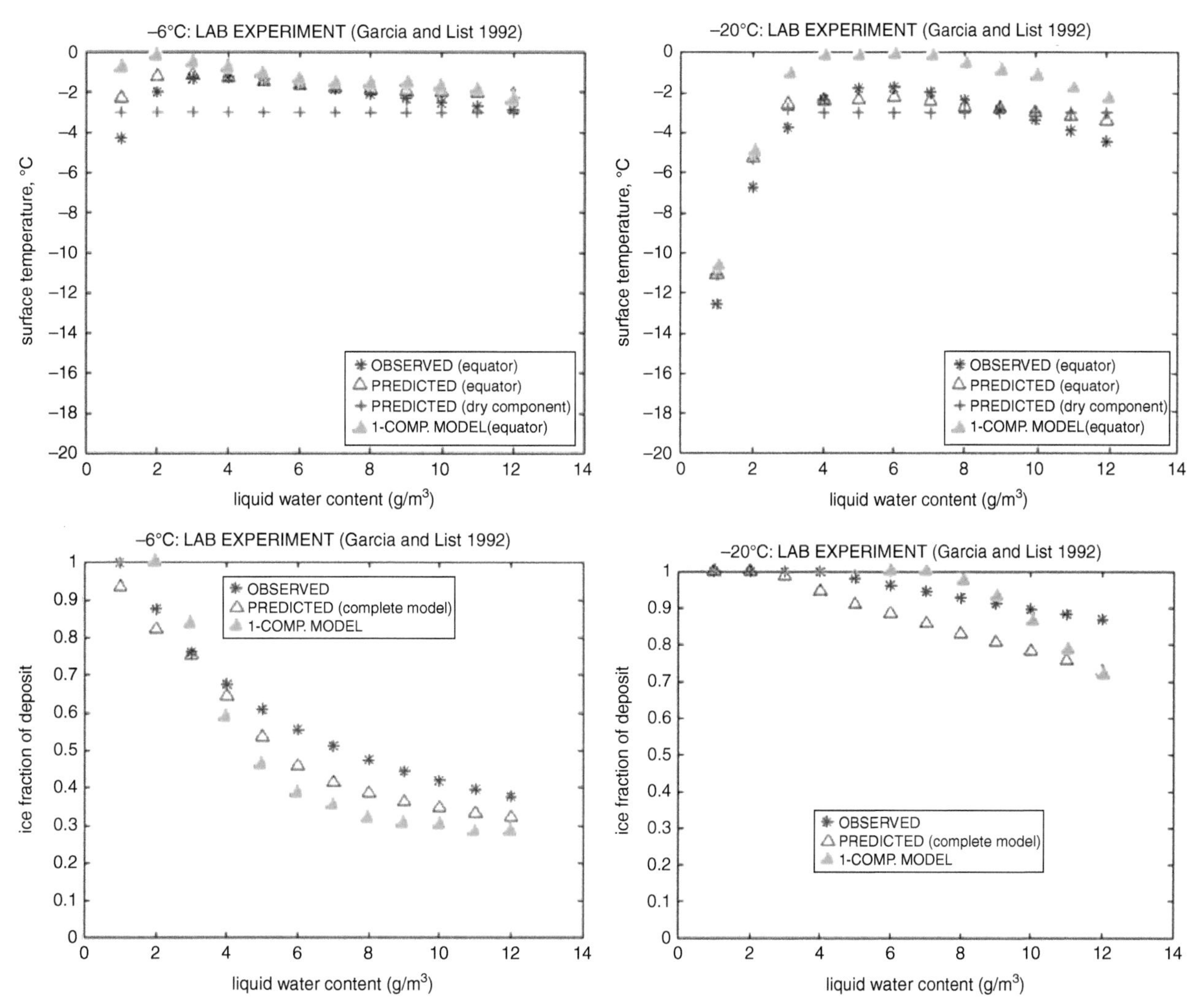

Figure 6.7.5 Surface temperatures averaged over the equatorial band of ±30° (upper row) and the ice mass fraction (bottom row), predicted by the two-component hail model (red triangles) and observed (blue asterisks). The predicted average temperature of the surface of the dry component (red '+' symbols) is also shown (upper row). The results obtained using a spherically symmetric hailstone model (the one-component model) are denoted by green triangles. The initial diameter of the hail particle is 2 cm. The left and the right panels show the air temperatures inside the tunnel (−6°C and −20°C, respectively) (from Phillips et al., 2014, with changes; © American Meteorological Society; used with permission).

6.7.4 Representation of the Dry and the Wet Growth of Hail and Graupel in Cloud Models

Spectral Bin Microphysics Models

A detailed description of the wet growth of hail requires calculation of time- and space-evolution of size distributions and of the liquid water mass at each grid point. This approach is too complicated for most Eulerian cloud models. Many studies investigated the growth of individual hailstones moving within a given thermodynamic and microphysical environment with a given amount of environmental supercooled water (e.g., Musil, 1970; Dennis and Musil, 1973; Buikov and Kuzmenko, 1978; Rassumussen and Heymsfield, 1987; Johnson and Rassumussen, 1992). In simplified 1D SBM models (e.g., Danielsen et al., 1972; Danielsen, 1975) hail growth was simulated at different initial DSD values. In most models, the description of the wet growth of hail is much simpler than it is described in Section 6.7.3.

Khain et al. (2011) calculated graupel diameter D_{onset} that corresponds to the onset of the wet growth. All graupel particles with diameters larger than D_{onset} were

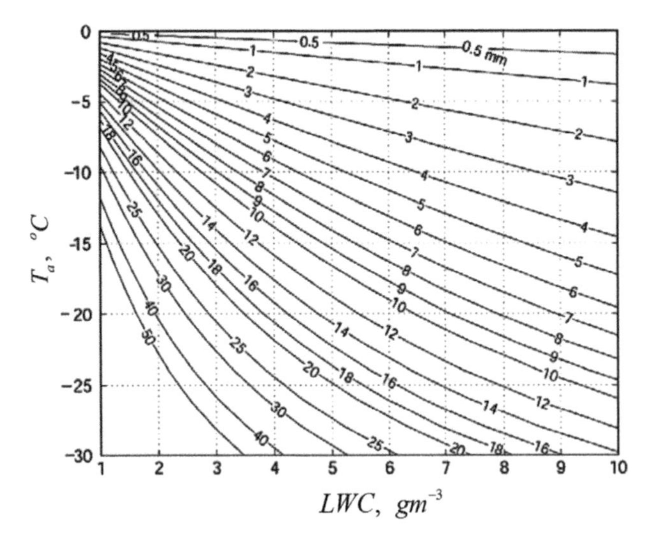

Figure 6.7.6 The critical diameter D_{onset} (in mm) of graupel, corresponding to the onset of the wet growth as a function of T and of q_w at fixed values of $p = 600$ hPa and $q_i = 1.0$ g m^{-3} (from Khain et al., 2011; courtesy of Elsevier).

converted to hail. The value of D_{onset} was calculated using the approach proposed by Blahak (2008), where the surface temperature T_s of graupel or hail is determined using the heat balance equation similar to Equation (6.7.5). To facilitate the application of this apparoach in HUCM, a lookup table of D_{onset} as a function of 4 parameters $D_{onset}(T_C, p, q_w, q_i)$ was calculated, using the diameter-fall speed relation (Figure 6.4.7). **Figure 6.7.6** shows diameter D_{onset} as a function of T and of q_w at fixed values of $p = 600$ hPa and $q_i = 1.0$ g m^{-3}. D_{onset} decreases while q_w increases and T_C decreases.

The processes of graupel conversion to hail due to the onset of the wet growth is illustrated by **Figure 6.7.7**, where the radar reflectivity fields calculated for graupel and hail are shown as obtained in a simulated hailstorm in Villingen-Schwenningen, southwest Germany, on 28.06.2006. One can see that in the presence of large supercooled liquid water content, formation of large hail starts at the level of 8–8.5 km. Above this level, the high reflectivity comes from graupel (Figure 6.7.7, the upper panel).

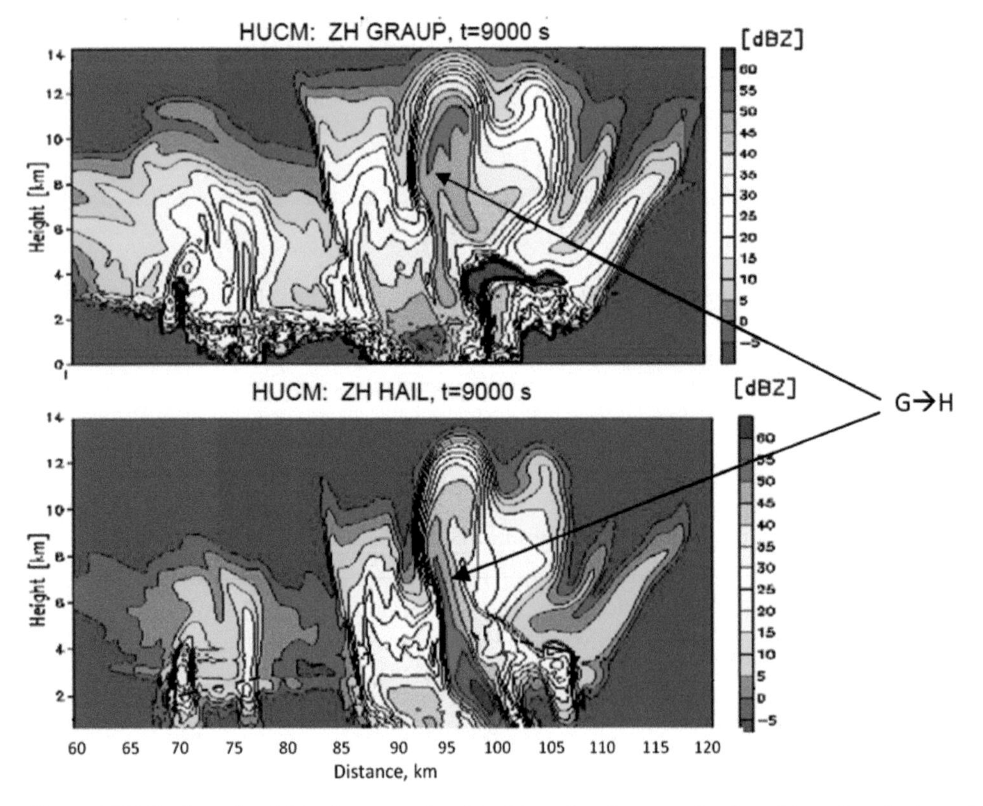

Figure 6.7.7 Radar reflectivities for graupel and hail in a thunderstorm developed in polluted atmosphere. The CCN concentration is 3000 cm^{-3}. Zones of transition of graupel to hail (the onset of the wet growth) are marked as G→H (from Khain et al., 2011, with changes; courtesy of Elsevier).

The approach applied by Khain et al. (2011) is rather simplified. The liquid water is not treated in graupel or hail at negative temperatures. Accordingly, no shedding at negative temperatures is taken into account. The wet growth of graupel is not considered. Calculation of D_{onset} is required only to convert graupel to hail. Besides, D_{onset} is calculated under the assumption of spherical symmetry of graupel and hail particles. To improve the representation of the wet growth of hail in HUCM, the scheme of time-dependent dry and wet growth of graupel and hail developed by Phillips et al. (2014) (Section 6.7.3) was implemented. This scheme allows to calculate the time evolution of liquid water within hail and graupel. Size distributions of liquid water in graupel and hail were implemented accordingly, so that particles belonging to each mass bin were characterized by a certain liquid water fraction. The liquid water mass in each bin of these hydrometeors was advected, diffused and settled together with the total particle masses. At each time step, the balance equations were solved to describe the freezing of accreted water. **Figure 6.7.8** shows the fields of CWC, the hail mass content and the liquid water fraction in hail in the hailstorm simulation of the in Villingen-Schwenningen hailstorm mentioned above. The figure shows the fields during growing stage and during the decay of a new convective cell during the mature stage of the storm evolution. Significant masses of hail form due to an intense accretion of cloud droplets within strong updrafts, whose velocities exceed 40 m/s at time instances of 4,680 s and 4,800 s. Zones of a very high liquid water fraction at 2 km level are the zones of

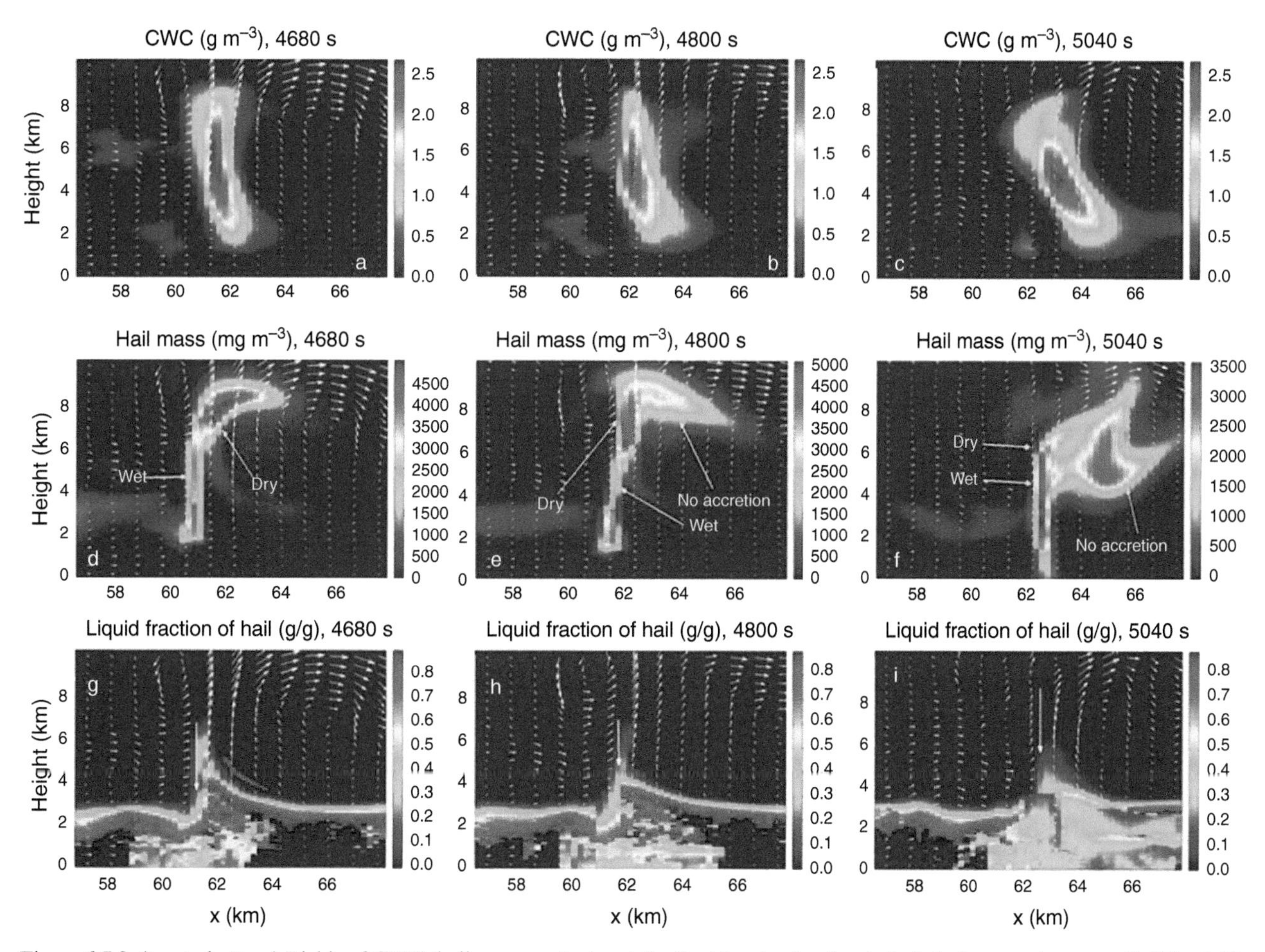

Figure 6.7.8 (top to bottom) Fields of CWC, hail mass content and the liquid water fraction in hail during growing stage (4,680 s, left) and decay (5,040 s, right) of a convective cell, after the mature stage (4,800 s, middle) of storm evolution. Zones where large hail falls are marked by white arrows. Zones of the wet and the dry growth are shown by yellow arrows. Some part of melting hail penetrates to updraft, as shown by the red arrow in panel g (from Phillips et al., 2015, with changes; © American Meteorological Society; used with permission).

intense melting of hail. Small hail particles melt within a layer of about 1.5–2 km below the freezing layer located at the height of 3.4 km. At the growing stage of the convective cell, the vertical velocities are high at low heights and some part of melting hail is involved into the zone of updrafts as shown by the red arrow in panel g) of Figure 6.7.8. The liquid water freezes in these ascending particles and the LWF in hail decreases during this ascent. The particles grow by wet growth up to comparatively large sizes of a few cm (see size distributions in **Figure 6.7.9**). Liquid water within hail exists up to heights of 6 km (−23°C). The large hail particles fall down being slightly shifted from the zone of the maximum updraft. These large hailstones also grow in the wet growth regime, although their LWF is small (<0.3). At the decay stage of the cell, large hail falls down to the surface (t = 5,100s, Figure 6.7.8f). One can see that the zones of high hail mass contents reach 10 km height. Within the large zones where the CWC is low or equal to zero, hail grows in the dry growth regime.

Figure 6.7.9 shows mass distributions of hail and of the liquid water in hailstones in the column of maximum mass content of hail at times t = 4,680 s (left) and 4,980 s (right). One can see that the peak of mass distribution occurs at hail diameters of about 1 cm. The largest hailstones reach 2–3 cm in diameter. Just above the freezing level, all hail particles contain unfrozen water. At the stage of the cell growing (Figure 6.7.9a), the vertical velocities are high, and even relatively large hail particles ascend. Due to accretion of supercooled drops, the mass of hailstones increases with height. The minimum diameter of hailstones containing liquid water in a strong updraft at the growing stage is about

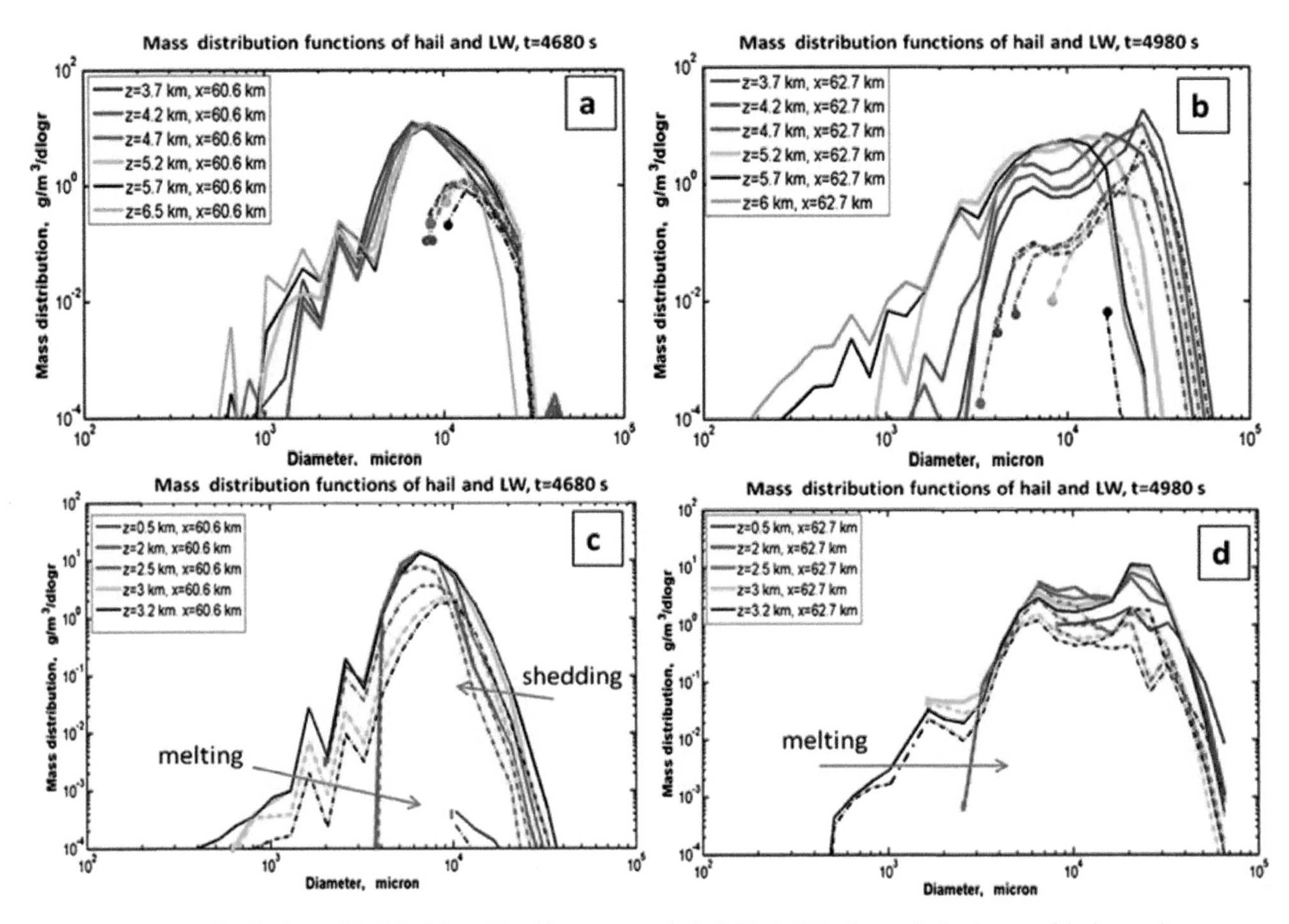

Figure 6.7.9 Mass distributions of hail (solid) and liquid water mass in hail (dashed) in the vertical columns with the maximum hail mass content at t = 4,680s (a), (c) and at t = 4,980 s (b),(d), above the freezing level (top) and below the freezing level (bottom). Small circles show the critical hail diameter D_{onset} at which the onset of wet growth begins. Arrows show the shift of the mass distribution branches due to melting and shedding. Various colors indicate different altitudes (from Phillips et al., 2015; © American Meteorological Society; used with permission).

1 cm. At the decaying stage (Figure 6.7.9b), large hail falls down, and the total mass and the maximum size of hail increase downward by collection of cloud drops. The maximum diameter of hail at the freezing level is than about 5–6 cm. Figures 6.7.9c and 6.7.9d show the mass distributions of hail below the melting level at the growing and the decay stages of a convective cell evolution. Note that the height of freezing level changes, being about 3.4–3.5 km within the convective zone and 3.3 km outside it.

During the growing stage of the cell evolution, the mass distribution of hail falling into the melting layer rapidly narrows due to two related processes: melting that increases the minimum size of hail particles, and shedding that reduces the largest size of hailstones. Consequently, near the ground (0.5 km, at 19°C) only a few hail particles with diameters of 1–2 cm remain. Hail does not reach the ground by this time. Later at the decay stage of the cell, the hail particles get larger and collect CWC more efficiently. As a result, despite the shedding, the largest hail particles grow as they fall. In agreement with the observations, the diameters of hailstones falling to the ground range from 0.8 to 6 cm (Figure 6.7.9d).

Figure 6.7.10 shows hail mass distributions at different heights at the growing and the decay stages of the convective cell plotted for the same atmosphereic columns as in Figure 6.7.9. The size ranges corresponding to the dry growth and the wet growth regimes are displayed in different colors. In strong updrafts, smaller hailstones grow by dry growth, while larger hailstones are in the wet growth regime. Because of the high CWC in the updrafts, the largest hail grows in the wet regime within the column up to about 6 km (−17°C). The diameter separating particles in the wet and the dry growth, D_{onset}, is about 1 cm and increases with height comparatively slowly. The weakness of this dependence of D_{onset} on height below 5.7 km (−14°C) can be attributed to the competition of two factors: whereas a decrease in the ambient air temperature tends to inhibit wet growth and to increase D_{onset}, the CWC increasing with height promotes the wet growth and tends to decrease D_{onset}. However, above 6 km the CWC decreases with height, and all the hail particles at this level are in the dry growth. Generally, D_{onset} is a function of both the temperature and the CWC. During the decaying stage, the mass distribution of falling hail contains more big hail stones. Due to the lower values of CWC, D_{onset} at z = 5.7 km during the decay stage is larger than at the growing stage of the convective cell. D_{onset} rapidly decreases downward with increase in temperature. Due to the larger size of hail during the decay stage and the decrease in D_{onset}, all the hail particles are in the wet growth regime below 4.2 km (at −6°C). These results show again that the onset of the wet growth substantially depends on particle size, temperature and environmental CWC. The wet growth regime begins when hail falls through a large mass of supercooled drops. Figure 6.7.10 shows that despite the complicated trajectories of hail particles, that are different for hail of different sizes, the mass distributions reveal only one boundary size separating the two regimes of hail growth.

Bulk Parameterization Schemes

In most bulk parameterization schemes, hail is not considered as a separate hydrometeor. Typically, graupel and hail are combined into one type and treated as high-density hydrometeors (usually referred to as "graupel"). The recent version of the two-moment bulk scheme by Noppel et al. (2010) regards hail as a separate hydrometeor. Hail formation is determined using the approach developed by Blahak (2008) described above. In bulk parameterization schemes proposed by Lin et al. (1983) and Milbrandt and Yau (2005b), the equation of wet growth is obtained using Equation (6.7.8) integrated over the hail size distribution. The accreted water that cannot freeze is assumed to be shed and its mass is added to the mass of raindrops. The collection efficiency for hail growing by the wet growth is assumed to be equal to one. Since rimed water consists of both cloud droplets and raindrops, the shedding in these schemes efficiently transforms cloud water to rain within the temperature range from 0°C to −10°C.

To improve the representation of hail using bulk parameterization schemes, Loftus et al. (2014) developed a scheme where hail evolution was described using three PSD moments, while the other hydrometeors were described using a two-moment scheme. This triple-moment hail bulk microphysics scheme (3MHAIL) has been implemented into the RAMS. This bulk parameterization is assigned to bin-emulating schemes (see Section 7.2 for detail). Utilization of additional hydrometeor type (hail) substantially improved the general quality of the bulk scheme in reproduction of precipitation, which stressed an important role of hail in formation of precipitation from deep mixed-phase clouds. This scheme proved to be more sensitive to aerosols as compared to the two-moment bin-emulating scheme used in RAMS (Cotton et al., 2003). In addition, the implementation of the third moment for hail description allowed reproduction of big hail of several cm in diameter (see Section 7.2). The necessity

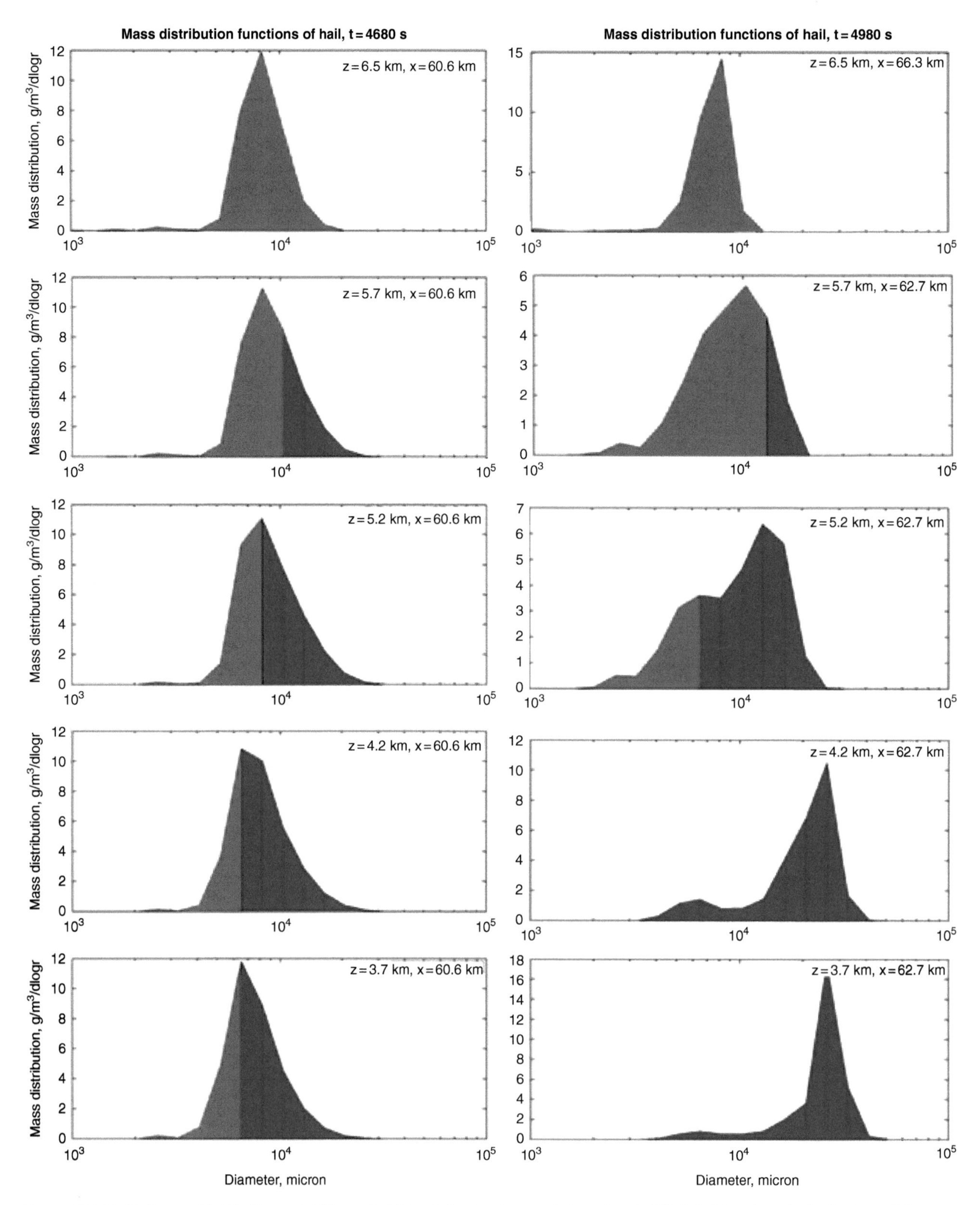

Figure 6.7.10 Hail mass distributions at different heights at the development stage (t = 4,680 s) (left) and at the beginning of the decay stage (t = 4,980 s) (right) of a convective cell evolution. The mass distributions of hail are plotted for the column with the maximum hail mass content. The size ranges corresponding to the dry and the wet growth regimes are colored red and blue, respectively. The particle diameter corresponding to the boundary between the zones is D_{onset} (from Phillips et al., 2015; © American Meteorological Society; used with permission).

to use 3-moment bulk schemes to simulate big hail was also stressed by Milbrandt and Yau (2006). Lang et al. (2014) and Tao et al. (2015) improved the microphysics of the Goddard Cumulus Ensemble (GCE) model by adding a fourth ice class (frozen drops/hail). This new 4ICE scheme was tested in simulations of an intense continental squall line and of a moderate, less organized continental mesoscale rain event. The results obtained provided a much better agreement with observational data than the two previous 3ICE versions of the Goddard microphysics where either graupel or hail were included.

6.8 Ice Multiplication and Its Representation in Cloud Models

Pristine ice nucleation, i.e., formation of ice crystals on IN, is described in Section 6.2. Observed concentration of ice crystals is sometimes much higher than concentration of IN. Ice crystals are observed in clouds at temperatures too high to explain their formation by activation of IN. While pristine ice crystals growing by diffusion growth in laboratory conditions have regular shape, a significant fraction of ice crystals in clouds, and sometimes most of ice crystals, have irregular shapes. This fact indicates the existence of other mechanisms of ice crystal formation and evolution than just IN activation. **Figure 6.8.1** shows images of ice crystals of irregular shapes. Ice crystals shown in the left panel supposedly form as a result of collisions of crystals or aggregation with other ice particles, graupel or hail. Crystals shown in the right panel of Figure 6.8.1 formed as a result of freezing of supercooled drops.

The process of ice generation that is not directly related to IN activation is known as *ice multiplication,* or *secondary ice production*. Ice multiplication processes were discovered and investigated in laboratory experiments. Several mechanisms of ice multiplication have been found.

6.8.1 Splintering During Liquid Drop-Ice Collisions

There is ice multiplication caused by riming and subsequent drop splintering i.e., a large ice particle collects water drops which may eject splinters while freezing. The process of splintering caused by riming is

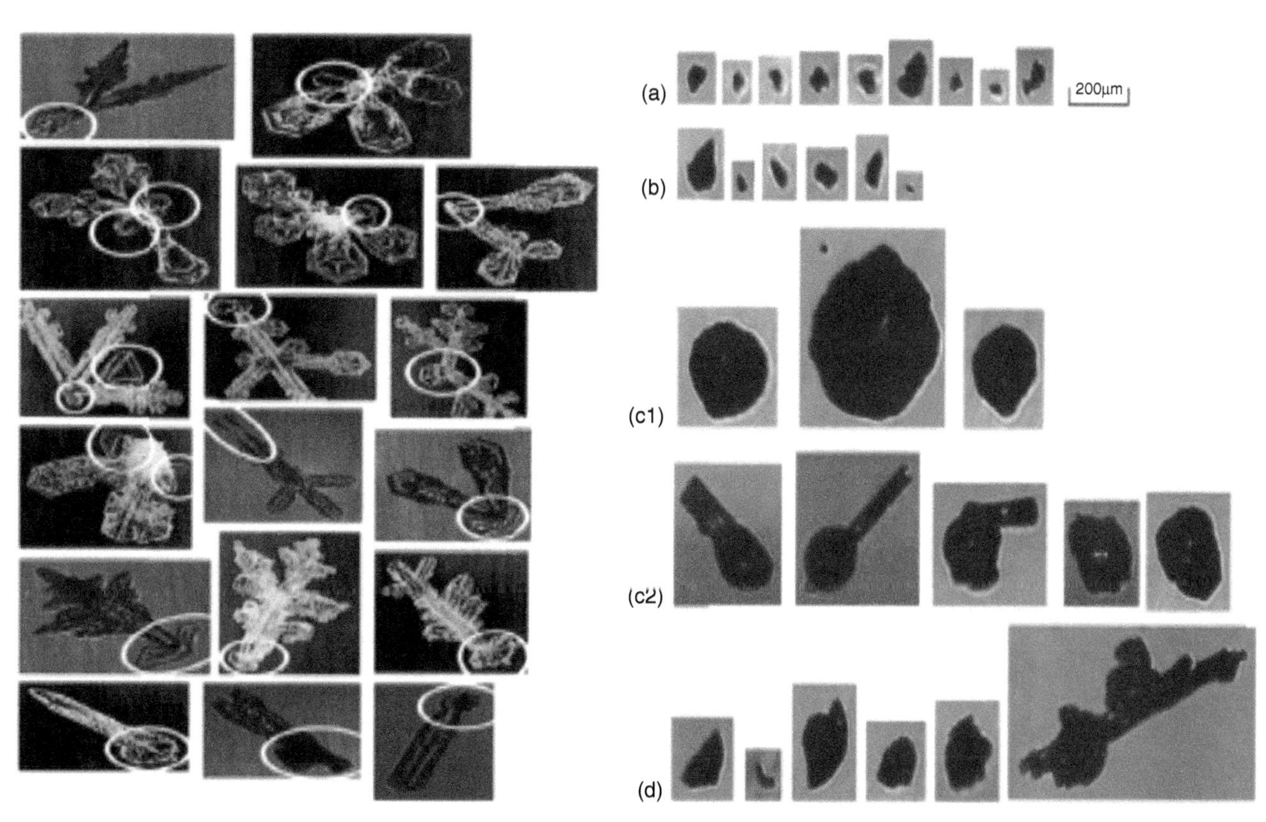

Figure 6.8.1 Left: Fragmented ice crystals with fewer than 6 original branches observed in Arctic clouds (from Schwarzenboeck et al., 2009; with permission from Elsevier). Right: ice crystal fragments with rounded elements formed due to fragmentation during raindrop freezing observed in stratiform clouds (from Rangno, 2008; © American Meteorological Society; used with permission).

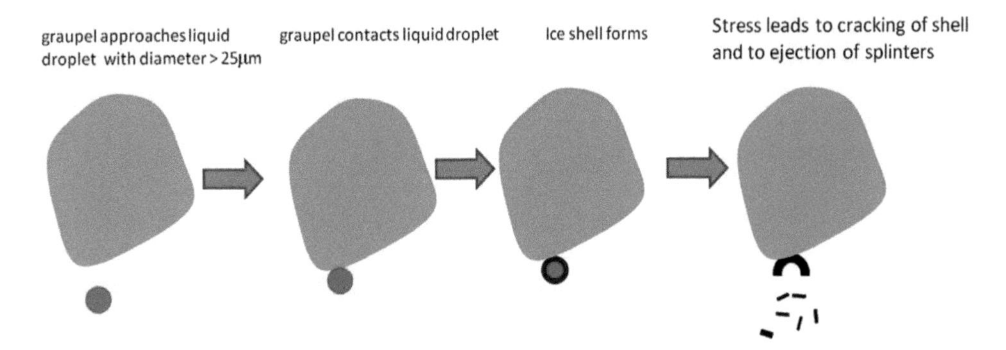

Figure 6.8.2 The process of splintering caused by riming. Graupel (or hail) particles are marked gray, drops are marked red.

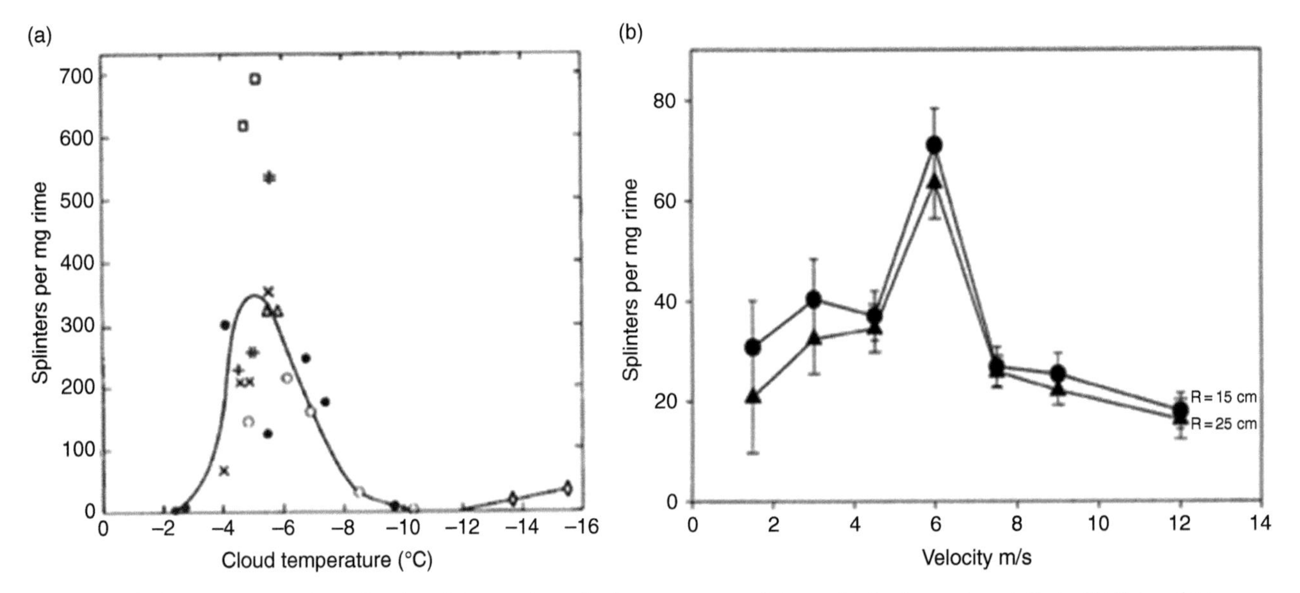

Figure 6.8.3 a) Temperature dependencies of the number of splinters per mg of rime, obtained experimentally by Hallet and Mossop (1974) (reprinted by permission from Macmillan Publishers Ltd.). b) Dependencies of the number of splinters per mg of rime on the velocity of the rimer, obtained experimentally by Saunders and Hosseini (2001), at the rimer temperature ranging from −4°C to −5°C. Two rods with lengths of 15 cm and 20 cm were used as rimers (courtesy of Willey and Sons. Ltd.).

schematically shown in **Figure 6.8.2**. The figure shows a scheme of the most known process of ice splintering by riming called the *Hallett-Mossop* (H-M) *mechanism* found in laboratory experiments by Hallett and Mossop (1974). In their laboratory, experiment drops collided with a cold solid cylindrical rod playing the role of a rimer. The splintering takes place within a narrow range of temperatures from −3°C to −8°C. The diameter of colliding drops which produce splinters by freezing should exceed 24 μm. When a drop freezes upon contact with the surface of graupel or hail, the increase in the volume of frozen water leads to formation of high surface stresses between the frozen water and the ice-collector. According to laboratory measurements, one of each 250 collisions leads to formation of one ice splinter. **Figure 6.8.3a** shows dependencies of the number of splinters per mg of rime on the temperature of the rimer, obtained in the laboratory experiments by Hallet and Mossop (1974).

Mossop (1978) found that the rate of ice splinters production depends not only upon concentration of large drops (>24 μm in diameter), but also upon the concentration of small droplets (<3 μm in diameter). The reason why the H-M mechanism is efficient within the temperature range from −3°C to −8°C is explained as follows. At temperatures warmer than −3°C no ice shell is formed around drops. At temperatures lower than −8°C, the ice shell is too strong and does not break under the internal pressure (stress). Saunders and Hosseini (2001) found that the intensity of the H-M mechanism depends on the velocity of the graupel pellet as it accretes supercooled drops. Within the range of 1.5–12 m s^{-1}, the maximum secondary ice particle ejection occurs at 6 m s^{-1} at the rate of 70 splinters per one mg of the accreted rime (Figure 6.8.3b).

The H-M mechanism is parameterized in SBM HUCM, where the number of graupel and hail collisions with liquid drops is calculated at each time step. The maximum production rate takes place at −5°C and linearly decreases to zero at −3°C and at −8°C.

To parameterize the H-M mechanism in bulk schemes, the concentration of splinters is expressed via the accreted mass of liquid drops, i.e., via the rime mass. According to Hallett and Mossop (1974), approximately 350 splinters are produced per every 1 mg of rime accreted on graupel particles at −5°C. Cotton et al. (1986) parameterize the rate of crystal production due to the H-M mechanism as the number of crystals per 1 g of the accreted water by equation:

$$P_{HM} = 3.5 \times 10^5 \left(\frac{\partial m_g}{\partial t}\right)_{g+l->g} F_{HM}(T)\left[\mathrm{g}^{-1}\right] \quad (6.8.1)$$

where $\left(\frac{\partial m_g}{\partial t}\right)_{g+l->g}$ is the riming rate of an ice crystal or a graupel particle (Section 6.5). Function $F_{HM}(T)$ is described by the following formula:

$$F_{HM}(T) = \begin{cases} 0, & 270.16\,\mathrm{K} < T \\ \dfrac{T-268.16}{2}, & 270.16 \geq T > 268.16\,\mathrm{K} \\ \dfrac{T-268.16}{3}, & 268.16 \geq T > 265.16\,\mathrm{K} \\ 0, & T \leq 265.16\,\mathrm{K} \end{cases} \quad (6.8.2)$$

Reisner et al. (1998) parameterize production of ice crystal mass by the H-M mechanism as:

$$\left(\frac{\partial q_i}{\partial t}\right)_{HM} = 3.5 \times 10^8 F_{HM1}(T) \left\{\left(\frac{\partial q_i}{\partial t}\right)_{s+l->s} + \left(\frac{\partial q_g}{\partial t}\right)_{g+l->g}\right\} m_{i0} \quad [\mathrm{kg/s}] \quad (6.8.3)$$

where m_{i0} is the mass of an ice splinter and $F_{HM1}(T)$ describes the dependence of ice multiplication rate on temperature. There are other parameterization formulas that describe secondary ice crystal generation by the H-M mechanism (e.g., Meyers et al., 1997).

The role of the H-M mechanism in the microphysics of mixed-phase clouds is different for different cloud types and depends on the presence of graupel, hail and of comparatively large drops within the temperature range of −3°C to −8°C. For instance, if the cloud base is cold and the aerosol concentration is high, drops are too small. Hence the H-M mechanism is too slow and cannot cause formation of ice crystals in strong cloud updrafts. Hobbs and Rangno (1990) argue that the H-M mechanism takes too long to account for the rapid glaciation in polar maritime clouds. At the same time, the H-M mechanism is efficient in tropical maritime deep convective clouds where graupel and frozen drops form not far above the freezing level, and drops are large enough.

6.8.2 Secondary Ice Production by Collisional Breakup of Ice

Recent studies show that formation of ice fragments during ice-ice collisions is probably the main mechanism of secondary ice production. Very high ice crystal concentrations forming as a result of graupel-graupel collisions were observed in laboratory measurements by Takahashi et al. (1995) who used two rimed ice spheres of 2 cm in diameter. However, these ice spheres were much larger than real graupel particles in clouds. This difference should be taken into account when applying these laboratory results to real clouds where graupel diameter is around 4 mm.

Intense fragmentation of snow and large ice crystals due to collisions with graupel was observed in Arctic clouds (Vardiman, 1978). According to Schwarzenboeck et al. (2009), 20–80% of ice crystals observed in Arctic clouds are naturally fragmented, formed supposedly due to collisions of ice crystals with graupel. A peak of fragmentation rate observed at −15°C can be attributed to the fact that dendrites tend to form and grow at this temperature. Branch ice crystals like dendrites can be easily broken by collisions with graupel or hail. According to Yano and Phillips (2011), tiny fragile branches that grow on the rimed surfaces may snap during graupel-graupel collisions. The major feature of the mechanical breakup is its potential occurrence at any subzero temperature or at comparatively low humidity.

A theory of secondary ice production by ice-ice collisions was proposed by Phillips et al. (2017a). This theory is based on the energy conservation idea, which provides a universally applicable constraint of the total numbers of fragments. The energy budget is written as in Equation (6.5.10) ($K_0 = K_1 + K_{loss} + \Delta S$), where K_0

is the initial kinetic energy of collision given by Equation (6.5.13) and K_{loss} is the energy loss by heat flux, noise and inelastic deformation. K_{loss} converts to other types of energy and determines how much energy remains for formation of fragments. It is assumed that a fraction c_2 of this energy is available for breaking branches: $\Delta K_{loss} = c_2 K_{loss} \approx c_2 K_0 (1 - q^2)$, where q is the coefficient of restitution that is the ratio of the relative velocities between the particles after and before breakup. ΔK_{loss} is assumed to be equally distributed among the branches located in the region of the particle contact, so all of them are impacted by the same energy available for potential breaking or deforming.

The number N_f of the ice fragments resulting from a collision is defined as the product of the number of breakable branches N_{cont} in the region of contact, and the probability of any such branch to be broken due to the impact:

$$N_f = N_{cont} P\left(\frac{\Delta K_{loss}}{N_{cont}} \geq G_{crit}\right) \tag{6.8.4}$$

where G_{crit} is the work required to break a branch of the particle. N_{cont} is assumed to be proportional to the area of small particle S_{small} (that experiences breakup) and to the number density of branches n_{branch}:

$$N_{cont} = c_1 S_{small} n_{branch} \tag{6.8.5}$$

where c_1 is a constant. Phillips et al. (2016) assume that the energy needed to break a branch is proportional to its cross-section area. Assuming that the cross-section area is proportional to the square of the branch's width, w_{branch}, they obtained that characteristic branch's width $w_{branch} \sim c_3 G_{crit}^{\gamma}$, where $\gamma \approx 1/2$ and c_3 is a constant. According to Polycarpou and Etsion (1999), the distribution of branch widths obeys the exponential law:

$$\begin{aligned} p(w_{branch}) &= \frac{1}{\langle w_{branch} \rangle} \exp\left(-w_{branch}/\langle w_{branch} \rangle\right) \\ &= \lambda \exp\left(-\lambda w_{branch}\right) \end{aligned} \tag{6.8.6}$$

where $\lambda = \langle w_{branch} \rangle^{-1}$. Since w_{branch} is a measure of G_{crit}, the probability that a particle will not be broken is equal to the probability that the branch width is below a critical value w_{branch_cr}, i.e.,

$$\begin{aligned} w_{branch\_cr} &= c_3 \left(\frac{\Delta K_{loss}}{N_{cont}}\right)^{\gamma} = c_3 \left(\frac{c_2 K_0 (1 - q^2)}{N_{cont}}\right)^{\gamma} \\ &= \Omega \left(\frac{K_0}{S_{small}}\right)^{\gamma} \end{aligned} \tag{6.8.7}$$

where $\Omega = c_3 \left(\frac{c_2(1-q^2)}{c_1 n_{branch}}\right)^{\gamma}$ is a function of temperature and ice particle properties, including the rimed mass fraction F_{rime}. Hence, using Equation (6.8.6) one can obtain the probability of collisional breakup of a branch:

$$\begin{aligned} P\left(W_{crit} \leq \frac{\Delta K_{loss}}{N_{cont}}\right)^{\gamma} &= P(w_{branch} < w_{branch\_cr}) \\ &= \lambda \int_0^{w_{branch\_cr}} \exp(-\lambda w_{branch}) dw_{branch} \\ &= 1 - \exp\left(-\Omega\lambda \left[\frac{K_0}{S_{small}}\right]^{\gamma}\right) \end{aligned} \tag{6.8.8}$$

Equations (6.8.4–6.8.7) lead to the final expression for the most probable number of fragments broken per one collision:

$$\begin{aligned} N_f = {} & S_{small} A(T, F_{rime}, D_{ice}) \\ & \left(1 - \exp\left[-\left(\frac{CK_0}{S_{small} A(T, F_{rime}, D_{ice})}\right)^{\gamma}\right]\right) \end{aligned} \tag{6.8.9}$$

where $A(T, F_{rime}, D_{ice}) = c_1 n_{branch}$ characterizes the number density of ice crystal branches per unit area. $C = \frac{c_2(1-q^2)\Gamma(1+\gamma^{-1})c_3^{-1/\gamma}}{\langle G_{crit} \rangle}$ is a constant that can be referred to as the "branch-fragility coefficient," and D_{ice} is the maximum size of the smaller ice particle in a colliding pair. The kinetic energy K_0 between two particles before collision is calculated using Equation (6.5.10). The parameterization expressed by Equation (6.8.9) allows to calculate production of ice crystals (fragments) by collisions between graupel or hail with ice crystals or snow, as well as between graupel and hail particles themselves. The values of constants in Equation (6.8.9) are chosen using observations made by Vardiman (1978) and Takahashi et al. (1995) and are presented in **Table 6.8.1**.

Parameterization (6.8.9) was used by Phillips et al. (2017b) to simulate deep convective clouds observed during the Severe Thunderstorm Electrification and Precipitation Study (STEPS) on June 19, 2000 (Lang et al., 2004). Two numerical models were used: a 3D cloud model with bin-emulating microphysics (Phillips et al., 2013) and HUCM with spectral bin microphysics (Ilotoviz et al., 2016). Both models yielded similar results indicating that collisional ice breakup is the major source of ice crystals in the convective zone below the level of homogeneous freezing. **Figure 6.8.4** shows the time dependences of accumulated secondary

Table 6.8.1 Parameters of Equation (6.8.9) in the ice breakup scheme (from Phillips et al., 2017a, with changes).

Participants of Collisions	Graupel of Diameters $0.5 < D_{ice} < 5$mm and Large Graupel	Hail and Hail	Rimed Crystal/Snow (Rimed Fraction $F_{rime} < 0.5$) and Larger Graupel/Hail	
			Dendrites $-12 > T_C > -17$ °C	Spatial Planar $-40 > T_c > -17$ °C $-9 > T_C > -12$ °C
$A\ [m^{-2}]$	$\frac{a_0}{3} + \max\left(\frac{2a_0}{3} - \frac{a_0}{9}\lvert T_C - T_0\rvert, 0\right) > -17°\text{C}$		$1.41{\cdot}10^6\left(1 + 10^2 F_{rime}^2\right)$	$1.58{\cdot}10^7\left(1 + 10^2 F_{rime}^2\right)$
	$a_0 = 3.78 \cdot 10^4\left(1 + \frac{0.0079}{D_{ice}^{1.5}}\right)[\text{J}]$	$a_0 = 4.35 \cdot 10^5[\text{J}]$	$\times\left(1 + \frac{3.98{\cdot}10^{-5}}{D_{ice}^{1.5}}\right)$	$\times\left(1 + \frac{1.33{\cdot}10^{-4}}{D_{ice}^{1.5}}\right)$
$C\ [J]$	2.21×10^4	3.31×10^5	1.08×10^4	2.48×10^4
γ	0.3	0.54	$0.5–0.25F_{rime}$	$0.5–0.25F_{rime}$
T_c °C	−15		–	–

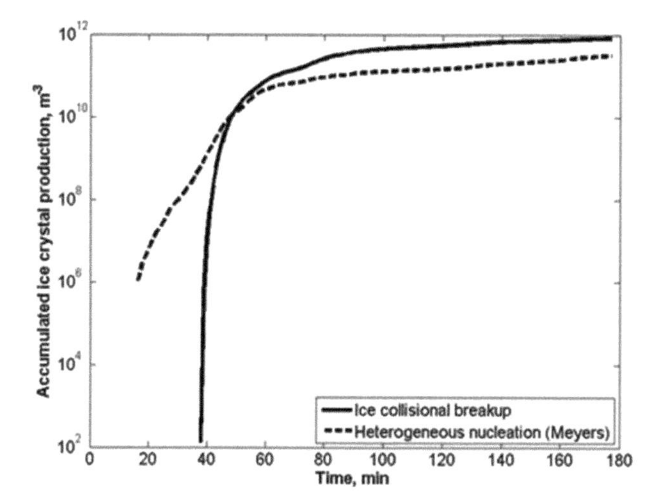

Figure 6.8.4 Time dependences of accumulated ice crystal production due to ice collision breakup and due to primary nucleation, obtained in simulations using HUCM.

ice (ice fragments) production in the entire cloud due to collisional breakup and due to primary heterogeneous nucleation, obtained in simulations using HUCM. It was assumed that breakup of snow leads to production of ice fragments which also can be assign to aggregates (snow). The primary ice nucleation in HUCM is described by parameterization proposed by Meyers et al. (1992). One can see that during cloud development when the cloud does not contain enough snow, graupel and hail, primary ice nucleation is the dominating process of ice crystal production. At the mature stage of cloud development, secondary ice production due to collisional breakup becomes dominating. Ice crystals formed in cloud anvils by homogeneous freezing, whose concentration is much higher than in the cloud below the homogeneous freezing level, are not included in Figure 6.8.4.

Figure 6.8.5 shows comparative contribution of different processes involved in secondary ice production at the mature stage of deep convective clouds, as follows from the simulations with HUCM. One can see that primary ice nucleation is responsible for about 5% of ice crystals. The most efficient process of secondary ice production is breakup of snow by collisions with graupel and hail (86%). Ice fragments produced as a result of snow-snow collisions are about 8% of the total ice fragments. The Hallet-Mossop mechanism produces less than 1% of secondary ice concentration. The diagram in Figure 6.8.5 was obtained in simulations that did not include mechanisms of secondary ice production by drop freezing. Note that the comparative contribution of different processes to concentration of small ice particles is calculated for conditions of STEPS, characterized by the cold cloud base, lack of warm rain, significant amount of snow and by negligible amount of hail particles. Under other conditions the comparative contribution of different processes to the secondary ice production can be different. In particular, in deep convective clouds with warm cloud base, the contribution of the Hallet-Mossop mechanism to production of secondary ice can increase.

6.8.3 Production of Ice Crystals by Drop Freezing

Ice splintering may take place during drop freezing. The shapes of frozen drops observed after freezing are shown in **Figure 6.8.6**. Appearance of the shapes according to scenarios b–f indicates that ice splintering took place during freezing. Modes of splintering production by drop freezing can be separated into two

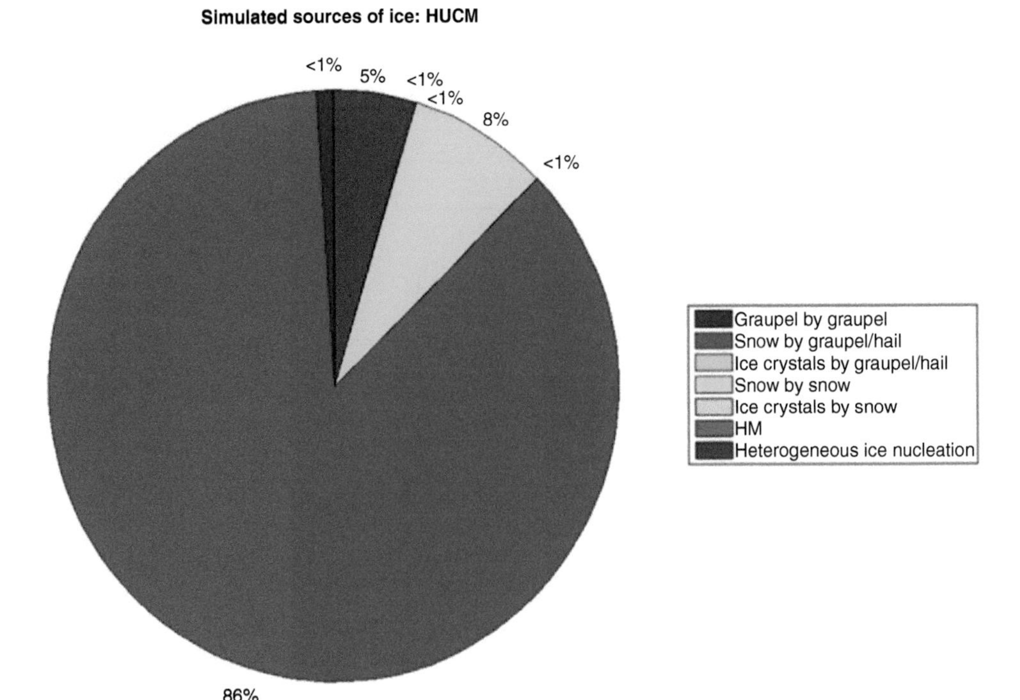

Figure 6.8.5 Comparative contribution of different processes in ice production to the convective cores of mature deep convective clouds, as follows from simulations with HUCM. Segments of different colors show comparative contributions of collisions graupel and graupel, breakup of snow by collisions with graupel and hail, breakup of ice crystals by collisions with graupel and hail; by snow-snow collisions, by ice crystals-snow collisions, due to the HM mechanism; and by heterogeneous ice nucleation. Ice crystals generated in the cloud anvil by homogeneous drop freezing are not included into the diagram. The averaging is performed over 150 min of storm evolution.

main groups characterized by different ice splintering shapes and sizes. The first group includes modes of drop freezing-fracturing-fragmentation (Koenig, 1963, 1965, 1966; Rangno, 2008; Leisner et al. 2014, Lawson et al., 2015). According to this mechanism illustrated in **Figure 6.8.7A**, large supercooled drops freeze and break into ice fragments. Examples of ice particles forming as a result of such drop explosions are shown in Figure 6.8.6a–d.

The second group includes modes of large supercooled drops producing spicules. The spicules eject bubbles that break into small ice fragments (Mason and Maybank, 1960; Koenig 1963, 1965, 1966; Hobbs and Alkezweeny, 1968; Brownscombe and Thorndyke, 1968; Takahashi and Yamashita, 1969, 1970; Gagin, 1972; Pruppacher and Schlamp, 1975; Kolomeychuk et al., 1975; Korolev et al. 2007a; Rangno, 2008; Leisner et al., 2014). This mechanism was observed in many laboratory experiments (e.g., Mason and Maybank, 1960; Gagin, 1972; and Leisner et al., 2014) and is illustrated in Figure 6.8.7B. In both groups of drops producing splinters during freezing, ice splinters arise because the outer shell of the drop freezes first, while the slushy interior freezes more slowly, causing internal pressure. This increased pressure leads either to drop breakup as shown in Figure 6.8.7A, or forces a spicule to form through the exterior shell as shown in Figure 6.8.7B. The exact mechanism of the splintering is not known. It is possible that freezing is accompanied by dendritic growth within the internal liquid creates internal stresses (Figure 6.6.13). Another possible mechanism of creation of stresses is non-uniform cooling of drops during their motion relative to air.

Figure 6.8.8 shows examples of frozen drops suspended on fine wires, as well as CPI (Cloud Particle Imager) images in regions with supercooled drizzle and CPI images captured in updrafts during Ice in Clouds Experiment–Tropical (ICE-T) (Heymsfield and Willis, 2014). The CPI images of frozen water drops with spicules are very similar to those found in the laboratory and shown in Figure 6.8.7. The spicules formed within a few milliseconds after the initial freezing of 80-mm drops. Korolev (2014, personal communication)

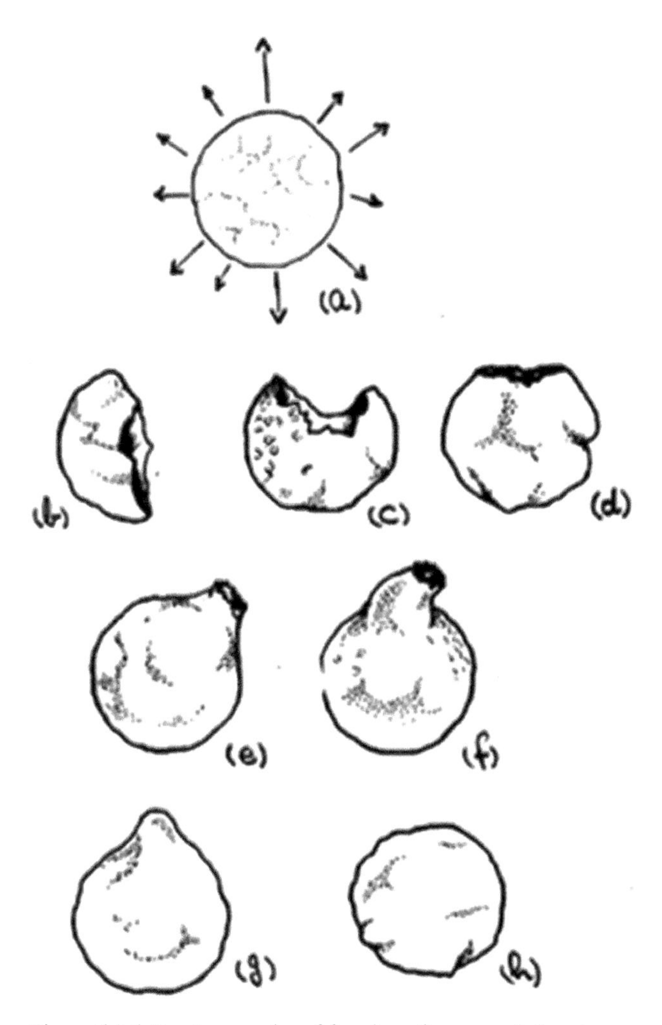

Figure 6.8.6 Rupture modes of freezing of supercooled water drops freely suspended in a wind tunnel airstream. Freezing drop totally ruptured by soap-bubble-type explosion (a); freezing drop ruptured into two almost equal halves (b); freezing drop ruptured into one small and one large fragment (c) and (d); frozen drop with a broken-off spike (e) and (f); frozen drop with a knob (g) and an irregular-shaped frozen drop (h) (from Pruppacher and Schlamp, 1975; courtesy of © John Wiley & Sons, Inc.).

reports that millimeter-diameter drops took 30–40 s to emit a spicule. The CPI image in the upper-right corner of Figure 6.8.8. resembles the laboratory image of a drop with a slushy center shown at 699.8 ms in Figure 6.8.7A. However, high-speed video photography does not capture all of the smallest ice particles that are ejected. Thus, even though laboratory studies estimate that only 10% of freezing drops (80-mm in diameter) produce splinters, the number of ice fragments can potentially be much larger.

Figure 6.8.9 shows images of hydrometeors and PSD measured during aircraft penetrations to the strong convective updrafts during the field experiment ICE-T. The penetrations were performed within the temperature range of $-12°C$ to $-20°C$. This temperature range was found to be in the region of rapid cloud glaciation. As seen in Figure 6.8.9, liquid drops with diameters exceeding about 300 μm were not observed in the rapid glaciation region. The ice images resemble those observed in the laboratory (Figure 6.8.7). The peaks of the DSD and ice PSD are at the diameters of about 20 μm and 100 μm, respectively. One can also see a significant amount of small ice crystals with sizes of 20–50 μm. Lawson et al. (2015) found rapid cloud glaciation in cloud updrafts. The high rate of glaciation can be seen from **Table 6.8.2**. showing the rapid decrease in the LWC and an increase in IWC in cloud updraft within the temperature range of $T_C = -8°C$ to 20°C.

Lawson et al. (2015) attribute the fast glaciation to splinter production during drop freezing. If the drop-freezing secondary ice production process is active, frequent liquid drop-ice collisions accompanied by drop freezing may potentially lead to releasing more ice splinters. Thus, a cascading process can develop that results in rapid glaciation within cloud updrafts. Immersion drop freezing, as well as other mechanisms considered in Sections 6.2 and 6.3 can also contribute to cloud glaciation.

Despite multiple laboratory experiments showing formation of splinters and drop cracking during drop freezing, the general theory of this effect has not been developed. This fact can be attributed to the high diversity of the experimental results as regards of both the number of splinters and the conditions at which such splintering and drop breakup take place. Mason and Maybank (1960) found that the maximum number of splinters is formed by slightly supercooled drops. The averaged number of splinters produced by freezing of a drop of 1 mm in diameter was evaluated as 50 at $T_C = 0°C$. In laboratory experiments by Takahashi and Yamashita (1970) and Leisner et al. (2014) the maximum production of ice splinters by drop freezing was found at $-15°C$ to $-16°C$. In different experiments, the number of ice splinters produced by a freezing drops ranges from about 1 to 50 with the typical value of about 10. Nearly all the experimentalists note, however, that drop freezing leads to production of possibly large number of ice crystals of undetectable size. Kolomeychuk et al. (1975) found substantial dependence of the splinter production rate on the surrounding air humidity.

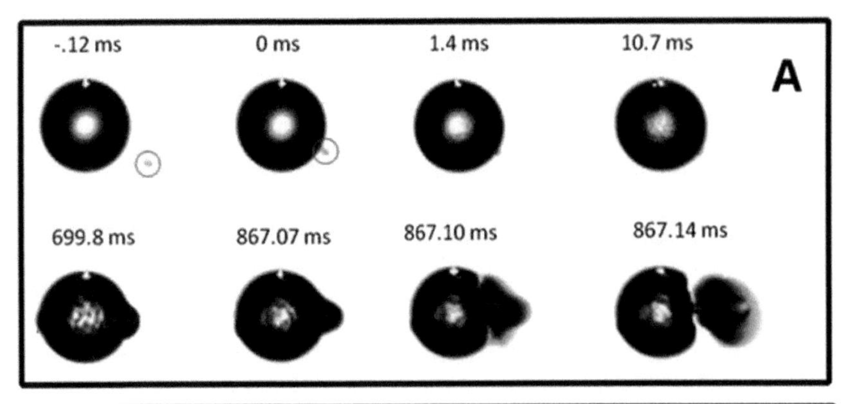

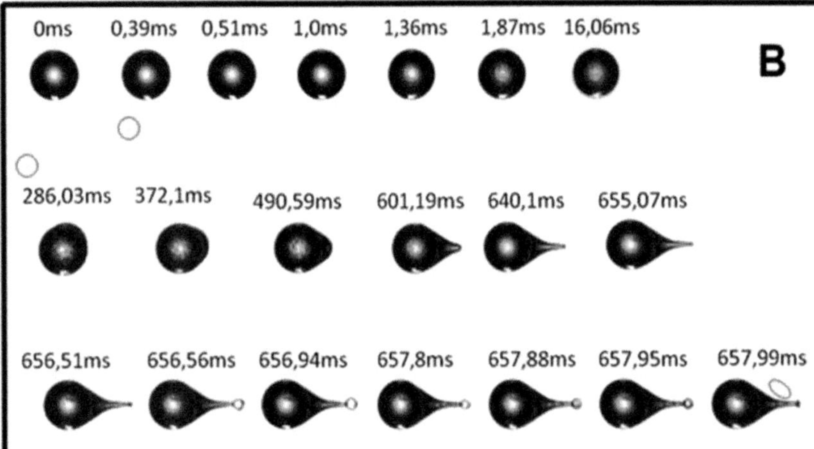

Figure 6.8.7 (A) High-speed photography (200,000 frames per second) of breakup of an 8-μm-diameter electrostatically suspended drop. (B) A spicule emitting bubbles from a 80-μm-diameter suspended drop at −10°C. Red circles denote an ice-forming nucleus entering the drop and leading to its freezing. Red ellipse in (B) at 657.99 ms indicates fragments from a burst bubble (from Lawson et al., 2015; adapted with permission from T. Leisner).

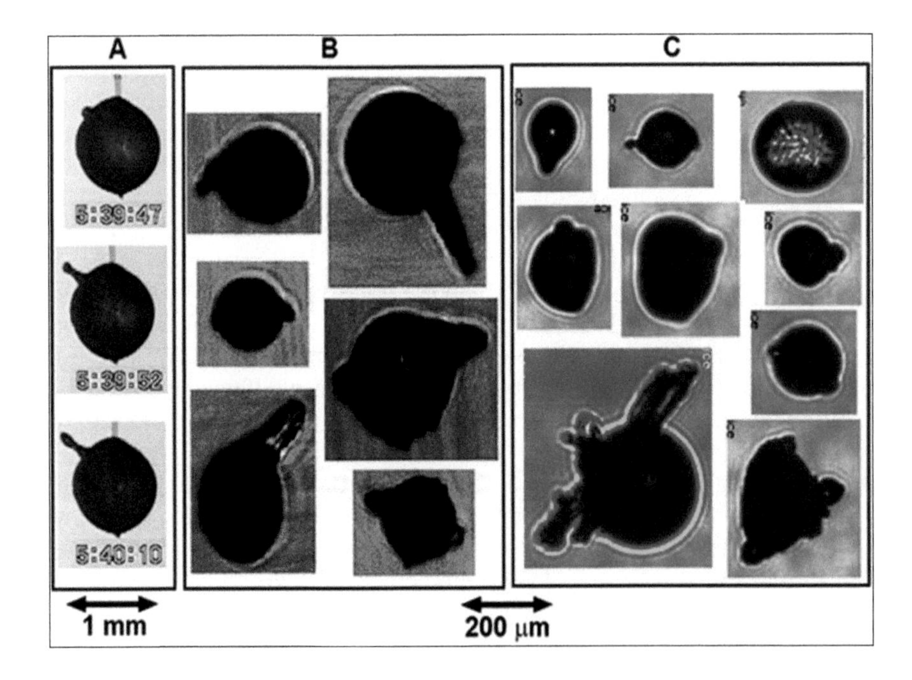

Figure 6.8.8 (A) Example of a spicule being emitted from a millimeter-diameter drop suspended from a wire and freezing at −10°C; (B) CPI images of drops with protrusions in freezing drizzle (Korolev et al., 2007a) and (C) CPI images of drops with protrusions in ICE-T clouds (Lawson et al., 2015; © American Meteorological Society; used with permission).

Table 6.8.2 The microphysical parameters in cloud updrafts measured by Lawson et al. (2015).

Temperature, °C	LWC, g m^{-3}	Droplet Concentration, cm^{-3}	IWC, g m^{-3}	Ice Crystal Concentration, L^{-1}
−8	1.9	34	0.001	20–30
−12	0.33	34	0.9	390
−20	0.05	5	2.8	500

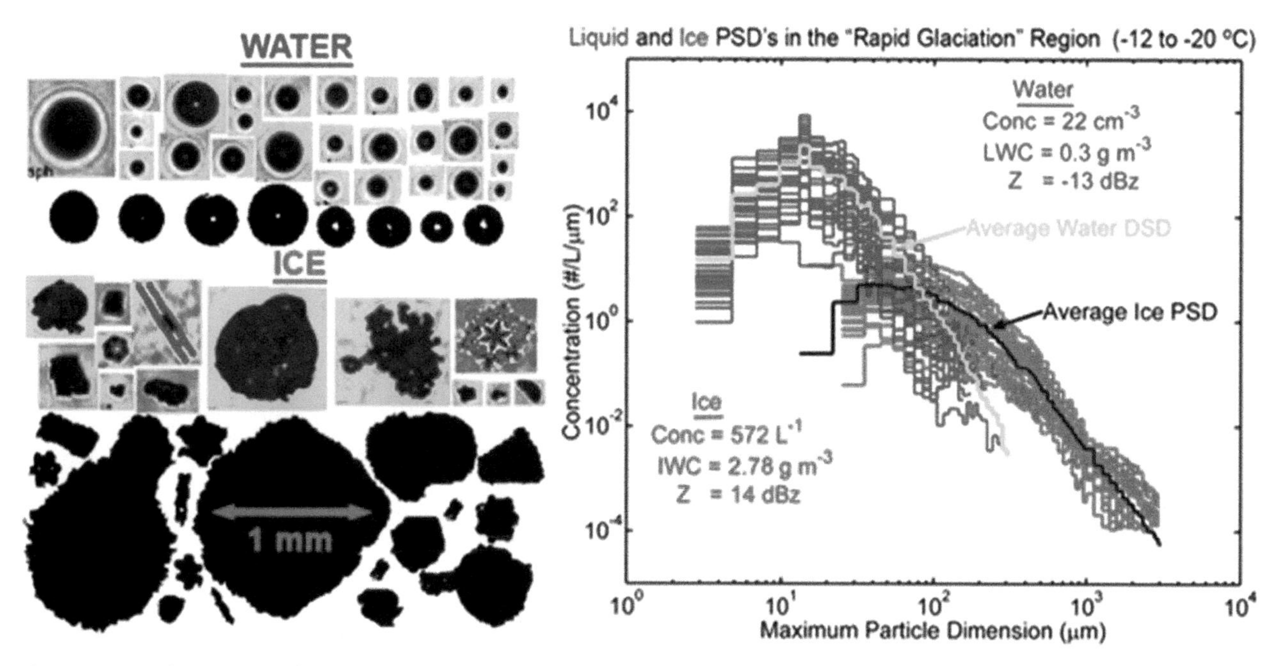

Figure 6.8.9 Left: examples of CPI and 2D-S images. Right: PSD of drops (blue) and ice particles (red) measured in the ICE-T experiment during aircraft penetrations to the updrafts of developing deep convective clouds at the levels of the rapid cloud glaciation (from Lawson et al., 2015; © American Meteorological Society; used with permission).

While in some studies no strong dependence of the number of ice splinters on drop size were reported (Mason and Maybank, 1960), many studies reported a strong increase in the number of ice splinters with an increase in drop size (Takahashi and Yamashita, 1969; Kolomeychuk et al., 1975; Lawson et al., 2015). Lawson et al. (2015) used a 1D bin cloud model to simulate ice multiplication by drop freezing. It was found that in case all ice fragments are formed via splintering during drop freezing, the number of ice fragments per a frozen drop can be estimated as $N_f = 2.5 \times 10^{-11} D^4$, where D is the drop diameter in μm.

Comparative contribution of splinter production due to collisional breakup of ice particles and drop freezing depends of cloud type. The first mechanism dominates, supposedly, in polluted clouds, where concentration of raindrops is low, but formation of graupel and hail is quite efficient. Intense ice production due to drop freezing is typically observed in deep Tropical convective clouds, where concentration of raindrops is high, but production of hail is not efficient.

The in situ observed shapes of ice particles and frozen drops and laboratory experiments conducted lately using high-speed photography indicate that secondary ice production is most likely the dominating mechanism of ice production in clouds below the homogeneous freezing level. The microphysical processes in convective clouds are highly non-linear. Multiple quantitative laboratory and in situ measurements, as well analysis using advanced microphysics models are needed to determine the role of secondary ice production in microphysics of mixed-phase clouds and in precipitation from such clouds.

References

Abraham, F.F., 1970: Functional dependence of drag coefficient of a sphere on Reynolds number. *Phys. Fluids*, **13**, 2194–2195.

1974: *Homogeneous Nucleation Theory*. Academic Press, New York, p. 263.

Al-Naimi, R., and C.P.R. Saunders, 1985a: Ice nucleus measurements: Effect of site location and weather. *Tellus*, **37B**, 296–303.

1985b: Measurements of natural deposition and condensation–freezing ice nuclei with a continuous flow chamber. *Atmos. Environ.*, **19**, 1871–1882.

Alheit, R. R., A. I. Flossmann, and H. R. Pruppacher, 1990: A theoretical study of the wet removal of atmospheric pollutants. Part 4: The uptake and redistribution of aerosol particles captured through nucleation and impaction scavenging by growing cloud drops and ice particles. *J. Atmos. Sci.*, **47**, 870–887.

Andreae, M.O., D. Rosenfeld, P. Artaxo, A.A. Costa, G.P. Frank, K.M. Longo, and M.A.F. Silva-Dias, 2004. Smoking rain clouds over the Amazon. *Science*, **303**, 1337–1342.

Aufdermaur, A. N., and J. Joss, 1967: A wind tunnel investigation on the local heat transfer from a sphere, including the influence of turbulence and roughness. *Z. Angew. Math. Phys.*, **18,** 852–866.

Barahona, D., 2012: On the ice nucleation spectrum. *Atmos. Chem. Phys.*, **12**, 3733–3752.

Barahona, D., and A. Nenes, 2008: Parameterization of cirrus formation in large scale models: Homogenous nucleation. *J. Geophys. Res.*, **113**, doi:10.1029/2007JD009355.

2009: Parameterizing the competition between homogeneous and heterogeneous freezing in cirrus cloud formation–monodisperse ice nuclei. *Atmos. Chem. Phys.*, **9**, 369–381.

Barth, M.C., and D.B. Parsons, 1996: Microphysical processes associated with intense frontal rainbands and the effect of evaporation and melting on frontal dynamics. *J. Atmos. Sci.*, **53,** 1569–1586.

Batchelor, G.K., 1967: *An Introduction to Fluid Dynamics*. Cambrige University Press, p. 615.

Battan, L.J., 1973: *Radar Observation of the Atmosphere*. University of Chicago Press, p. 324.

Bauer, P., A. Khain, A. Pokrovsky, R. Meneghini, C. Kummerow, F. Marzano, and J.P.V. Poiares Baptista, 2000: Combined cloud-microwave radiative transfer modeling of stratiform rainfall. *J. Atmos. Sci.*, **57,** 1082–1104.

Beard, K.V., 1976: Terminal velocity and shape of cloud and precipitation drops aloft. *J. Atmos. Sci.*, **33**, 851–864.

1980: The effects of altitude and electrical force on the terminal velocity of hydrometeors. *J. Atmos. Sci.*, **37**, 1363–1374.

1992: Ice initiation in warm-base convective clouds: An assessment of microphysical mechanisms. *Atmos. Res.*, **28**, 125–152.

Beheng, K., 1978: Numerical simulation of graupel development. *J. Atmos. Sci.*, **35**, 683–689.

Benmoshe, N., A. Khain, M. Pinsky, and A. Pokrovsky, 2012: Turbulent effects on cloud microstructure and precipitation of deep convective clouds as seen from simulations with a 2-D spectral microphysics cloud model. *J. Geophys Res.*, **117**, D06220.

Bergeron, T., 1935: On the physics of clouds and precipitation. *Proc. Ve Assemblée Générale de l'Union Géodésique et Geophysique Internationale*, Lisbon, Portugal, International Union of Geodesy and Geophysics, 156–180.

Berry, E.X., and R.J. Reinhardt, 1974: An analysis of cloud drop growth by collection: Part 1. Double distributions. *J. Atmos. Sci.*, **31**, 1814–1824.

Bigg, E.K., 1953: The formation of atmospheric ice crystals by the freezing of droplets. *Q. J. Royal Meteorol. Soc.*, **79**, 510–519.

Blahak, U., 2008: Towards a better representation of high density ice particles in a state-of-the-art two-moment bulk microphysical scheme. Extended Abstract, 15th International Conference on Clouds and Precipitation, Cancun, Mexico, July 7–11.

Blanchard, D.C., 1957: *The Supercooling, Freezing, and Melting of Giant Waterdrops at Terminal Velocity in Air. Artificial Stimulation of Rain*. Pergamon, NY, pp. 233–249.

Böhm, H.P., 1989: A general equation for the terminal fall speed of solid hydrometeors. *J. Atmos. Sci.*, **46,** 2419–2427.

Böhm, J.P., 1992a: A general hydrodynamic theory for mixed-phase microphysics. Part I: Drag and fall speed of hydrometeors. *Atmos. Res.*, **27**, 253–274.

1992b: A general hydrodynamic theory for mixed-phase microphysics. Part II: Collision kernels for coalescence. *Atmos. Res.*, **27**, 275–290.

1992c: A general hydrodynamic theory for mixed-phase microphysics. Part III: Riming and aggregation. *Atmos. Res.*, **28**, 103–123.

1999: Revision and clarification of "A general hydrodynamic theory for mixed-phase microphysics." *Atmos. Res.*, **52**, 167–176.

Bott, A., 1998: A flux method for the numerical solution of the stochastic collection equation. *J. Atmos. Sci.*, **55**, 2284–2293.

Broadley, S.L., B.J. Murray, R.J. Herbert, J.D. Atkinson, S. Dobbie, T.L. Malkin, E. Condliffe, and L. Neve, 2012: Immersion mode heterogeneous ice nucleation by an illite rich powder representative of atmospheric mineral dust. *Atmos. Chem. Phys.*, **12**, 287–307.

Broday, D., M. Fichman, M. Shapiro, and C. Gutfinger, 1998: Motion of spheroidal particles in vertical shear flows. *Phys. Fluids*, **10,** 86–100.

Brown, P.R.A., and P.N. Francis, 1995: Improved measurements of the ice water content in cirrus using a total-water probe. *J. Atmos. Oceanic Technol.*, **12**, 410–414.

Brownscombe, J.L., and N.S.C. Thorndyke, 1968: Freezing and shattering of water droplets in free fall. *Nature*, **220**, 687–689.

Bruning, E.C., W.D. Rust, T.J. Schuur, D.R. MacGorman, P.R. Krehbiel, and W. Rison, 2007: Electrical and polarimetric radar observations of a multicell storm in TELEX. *Mon. Wea. Rev.*, **135**, 2525–2544.

Buikov, M.B., and A.G. Kuzmenko, 1978: Hail growth in supercell hail clouds. *Meteorology and Hydrology*, N11, 60–69.

Cantrell, W., and A. Heymsfield, 2005: Production of ice in tropospheric clouds. A review. *Bull. Amer. Met. Soc.*, 795–807, DOI:10.1175/BAMS-86-6-795.

Carrió, G.G., and W.R. Cotton, 2010: Investigations of aerosol impacts on hurricanes: Virtual seeding flights. *Atmos. Chem. Phys. Discuss.*, **10**, 22,437–22,467.

Carrió, G.G., S.C. van den Heever, and W.R. Cotton, 2007: Impacts of nucleating aerosol on anvil-cirrus clouds: A modeling study. *Atmos. Res.*, **84**, 111–131.

Carstens, J.C., and J.J. Martin, 1982: In-cloud scavenging by thermophoresis, diffusiophoresis, and Brownian diffusion. *J. Atmos.Sci.*, **39,** 1124–1129.

Chisnell, R.F., and J. Latham, 1976: Ice particle multiplication in cumulus clouds. *Q. J. Royal Meteorol. Soc.*, **102**, 133–156.

Clift, R., J.R. Grace, and M.E. Weber, 1978: *Babbles, Drops and Particles*. Akademic Press, p. 380.

Connolly, P.J., C. Emersic, and P.R. Field, 2012: A laboratory investigation into the aggregation efficiency of small ice crystals. *Atmos. Chem. Phys.*, **12**, 2055–2076.

Connolly, P.J., C.P.R. Saunders, M.W. Gallagher, K.N. Bower, M.J. Flynn, T.W. Choularton, J. Whiteway, and R.P. Lawson, 2005: Aircraft observations of the influence of electric fields on the aggregation of ice crystals. *Q. J. Royal Meteorol. Soc.*, **131**, 1695–1712.

Cooper, W.A., 1980: A method of detecting contact ice nuclei using filter samples. 8th Int. Conf. Cloud. Phys. Clermont-Ferrand, France. pp. 665–668.

1986: Ice initiation in natural clouds. Precipitation enhancement. A scientific challenge. *Meteorol. Mon.*, **21**, 28.

Cotton, W.R. et al., 2003: RAMS 2001: Current status and future directions. *Meteorol. Atmos. Phys.*, **82**, 5–29.

Cotton, W.R., M.A. Stephens, T. Nehrkorn, and G.J. Tripoli, 1982: The Colorado State University three-dimensional cloud mesoscale model – 1982. Part II: An ice-phase parameterization. *J. Rech. Atmos.*, **16**, 295–320.

Cotton, W.R., G. Tripoli, R.M Rauber, and E.A. Mulvihill, 1986: Numerical simulation of the effects of varying ice crystal nucleation rates and aggregation processes on orographic snowfall. *J. Climate Appl. Meteorol.*, **25**, 1658–1680.

Curry, J.A., and V.I. Khvorostyanov, 2010: Assessment of parameterizations of ice nucleation. *Atmos. Chem. Phys. Discuss.*, **10**, 2669–2710.

2012: Assessment of some parameterizations of heterogeneous ice nucleation in cloud and climate models. *Atmos. Chem. Phys.*, **12**, 1151–1172.

Danielsen, E.F., 1975: A review of hail growth by stochastic collection in a cumulonimbus model. *Pageoph*, **113**, 1019–1034.

Danielsen, E.F., R. Bleck, and D.A. Morris, 1972: Hail growth in a cumulus model. *J. Atmos. Sci.*, **29**, 135–155.

DeMott, P.J., D.J. Cziczo, A.J. Prenni, D.M. Murphy, S.M. Kreidenweis, D.S. Thomson, R. Borys, and D.C. Rogers, 2003a: Measurements of the concentration and composition of nuclei for cirrus formation. *Proc. Natl. Acad. Sci. U.S.A.*, **100** (25), 14,655–14,660.

DeMott, P.J., M.P. Meyers, and W.R. Cotton, 1994: Parameterization and impact of ice initiation processes relevant to numerical model simulations of cirrus clouds. *J. Atmos. Sci.*, **51**, 77–90.

DeMott, P.J., O. Möhler, O. Stetzer, G. Vali, Z. Levin, M. Petters, M. Murakami, T. Leisner, U. Bundke, H. Klein, Z. Kanji, R. Cotton, H. Jones, S. Benz, M. Brinkmann, D. Rzesanke, H. Saathoff, M. Nicolet, A. Saito, B. Nillius, H. Bingemer, J. Abbatt, K. Ardon, E. Ganor, D.G. Georgakopoulos, and C. Saunders, 2011: Resurgence in ice nucleation research. *Bull. Amer. Meteorol. Soc.*, **92**, 1623–1635.

DeMott, P.J., A.J. Prenni, X. Liu, S.M. Kreidenweis, M.D. Petters, C.H. Twohy, M.S. Richardson, T. Eidhammer, and D.C. Rogers,.2010: Predicting global atmospheric ice nuclei distributions and their impacts on climate. *Proc. Natl. Acad. Sci.*, **107**, 11,217–11,222.

DeMott, P.J., and D.C. Rogers, 1990: Freezing nucleation rates of dilute solution droplets measured between -30° and -40°C in laboratory simulations of natural clouds. *J. Atmos. Sci.*, **47**, 1056–1064.

DeMott, P.J., K. Sassen, M.R. Poellot, D. Baumgardner, D.C. Rogers, S.D. Brooks, A.J. Prenni, and S.M. Kreidenweis, 2003b: African dust aerosols as atmospheric ice nuclei. *Geophys. Res. Lett.*, **30**(14), 1732.

Dennis, A.S., and D.J. Musil, 1973: Calculations of hailstone growth and trajectories in a simple cloud model. *J. Atmos. Sci.*, **30**, 278–288.

Deshler, T., 1982: Contact ice nucleation by submicron atmospheric aerosols. PhD Dissertation Dept. Phys. Astron, Univ. Wyoming, Laramie, p. 197.

Deshler, T., and G. Vali, 1992: Atmospheric concentrations of submicron contact-freezing nuclei. *J. Atmos. Sci.*, **49**, 773–784.

Diehl, K., S. Matthias-Maser, S.K. Mitra, and R. Jaenicke, 2001: Wind tunnel studies of the ice nucleating ability of leaf litter and pollen in the immersion and contact mode. Proc. EGS XXVI General Assembly, Nice, France, European Geophysical Society, Geophysical Research Abstracts, Vol. 3, 5512.

2002: The ice nucleating ability of pollen. Part II: Laboratory studies in immersion and contact freezing modes. *Atmos. Res.*, 61, 125–133.

Diehl, K., and S. Mitra, 1998: A laboratory study of the effects of a kerosene-burner exhaust on ice nucleation and the evaporation rate of ice crystals. *Atmos. Environ.*, 32, 3145–3151.

Diehl, K., and S. Wurzler, 2004: Heterogeneous drop freezing in the immersion mode: Model calculations considering soluble and insoluble particles in the drops. *J. Atmos. Sci.*, **61**, 2063–2072.

Duft, D., and T. Leisner, 2004: Laboratory evidence for volume-dominated nucleation of ice in supercooled water microdroplets. *Atmos. Chem. Phys. Discuss.*, **4**, 3077–3088.

Durant, A.J., and R.A. Shaw, 2005: Evaporation freezing by contact nucleation inside-out. *Geophys. Res. Lett.*, **32**, L20814.

Eidhammer, T., P.J. DeMott, and S.M. Kreidenweis, 2009: A comparison of heterogeneous ice nucleation parameterization using a parcel model framework. *J. Geophys. Res.*, **114**, D06202.

Fabry, F., and W. Szyrmer, 1999: Modeling of the melting layer. Part II: Electromagnetic. *J. Atmos. Sci.*, **56**, 3593–3600.

Fan, J., L.R. Leung, Z. Li, H. Morrison, H. Chen, Y. Zhou, Y. Qian, and Y. Wang, 2012: Aerosol impacts on clouds and precipitation in eastern China: Results from bin and bulk microphysics. *J. Geophys. Res.*, **117**, D00K36, doi: 10.1029/2011JD016537.

Fan, J., L.R. Leunga, D. Rosenfeld, Q. Chena, Z. Lid, J. Zhang, and H. Yan, 2013: Microphysical effects determine macrophysical response for aerosol impacts on deep convective clouds, *PNAS*, E4581–E4590.

Feingold, G., R.L. Walko, B. Stevens, and W.R. Cotton, 1998: Simulations of marine stratocumulus using a new microphysical parameterization scheme. *Atmos. Res.*, **47–48**, 505–528.

Ferrier, B.S., 1994: A two-moment multiple phase four-class bulk ice scheme. Part 1: Description. *J. Atmos. Sci.*, **51**, 249–280.

Ferrier, B.S., W.-K. Tao, and J. Simpson, 1995: A two-moment multiple phase four-class bulk ice scheme. Part II: Simulations of convective storms in different large-scale environments and comparisons with other bulk parameterizations. *J. Atmos. Sci.*, **52**, 1001–1033.

Findeisen, W., 1938: Kolloid-meteorologische Vorgänge bei Neiderschlags-bildung. *Meteorol. Z.*, **55**, 121–133.

1940: The formation of the 0°C-isothermal layer and fractocumulus under nimbostratus. *Meteorol. Z.*, **6**, 882–888.

Flatau, P., G.J. Tripoli, J. Verlinde, and W.R. Cotton, 1989: *The CSU_RAMS Cloud Microphysics Module: General Theory and Code Documentation*. Colorado State University, Department of Atmospheric Science, paper 451, p. 88.

Flatau, P.J., R.L. Walko, and W.R. Cotton, 1992: Polynomial fits to saturation vapor pressure. *J. Appl. Meteorol.*, **31**, 1507–1513.

Fletcher, N.H., 1962: *The Physics of Rainclouds*, Cambridge University Press, p. 390.

1969: Active sites and ice crystal nucleation. *J. Atmos. Sci.*, **26**, 1266–1271.

Flossmann, A.I., and H.R. Pruppacher, 1988: A theoretical study of the wet removal of atmospheric pollutants. Part III: The uptake, redistribution, and deposition of (NH4)2SO4 particles by a convective cloud using a two-dimensional cloud dynamics model. *J. Atmos. Sci.*, **45**, 1857–1871.

Gagin, A., 1972: The effect of supersaturation on the ice crystal production by natural aerosols. *J. Rech. Atmos*, **6**, 175–185.

Garcia-Garcia, F., and R. List, 1992: Laboratory measurements and parameterizations of supercoled water skin temperatures and bulk properties of gyrating hailstones. *J. Atmos. Sci.*, **49**, 2058–2072.

Gavze, E., M. Pinsky, and A. Khain, 2012: The orientations of prolate ellipsoids in linear shear flows. *J. Fluid Mech.*, **690**, 51–93.

2016: The orientation dynamics of small prolate and oblate spheroids in linear shear flows. *Int. J. Mult. Flow*, **83**, 103–114.

Gierens, K., 2003: On the transition between heterogeneous and homogeneous freezing. *Atmos. Chem. Phys.*, **3**, 437–446.

Grabowski, W.W., 2015: Untangling microphysical impacts on deep convection applying a novel modeling methodology. *J. Atmos. Sci.*, **72**, 2446–2464.

Hagen, D., R. Anderson, and J. Kassner Jr., 1981: Homogeneous condensation-freezing nucleation rate measurements for small water droplets in an expansion cloud chamber. *J. Atmos. Sci.*, **38**, 1236–1243.

Hall, W.D., 1980: A detailed microphysical model within a two dimensional framework: Model description and preliminary results. *J. Atmos. Sci.*, **37**, 2486–2507.

Hall, W.D., and H.R. Pruppacher, 1976: The survival of ice crystals falling from cirrus clouds in subsaturated air. *J. Atmos. Sci.*, **33**, 1995–2006.

Hallett, J., 1964: Experimental studies of the crystallization of supercooled water. *J. Atmos. Sci.*, **21**, 671–682.

Hallett, J., and S.C. Mossop, 1974: Production of secondary ice crystals during the riming process. *Nature*, **249**, 26–28.

Hallgren, R.E., and C.L. Hosler, 1960: Preliminary results on the aggregation of ice crystals. *Geophys. Monogr. Am. Geophys. Union*, **5**, 267–263.

Happel, J., and H. Brenner, 1983: *Low Reynolds Number Hydrodynamics*. Noordhoff International Publishing.

Harrington, J.Y., M.P. Meyers, R.L. Walko, and W.R. Cotton, 1995: Parameterization of ice crystal conversion processes in cirrus clouds using double-moment basis functions. Part I: Basic formulation and one-dimensional tests. *J. Atmos. Sci.*, **52**, 4344–4366.

Harrington, J.Y., and K. Sulia, 2013: A method for adaptive habit prediction in bulk microphysical models. Part II:

Parcel model corroboration. *J. Atmos. Sci.*, **70**, 365–376.

Harrington, J.Y., K. Sulia, and H. Morisson, 2013: A method for adaptive habit prediction in bulk microphysical models. Part I: Theoretical development. *J. Atmos. Sci.*, **70**, 349–364.

Hashino, T., 2007: Explicit simulation of ice particle habits in a numerical weather prediction model. University Wisconsin-Madison. Dissertation.

Heymsfield, A., 1972: Ice crystal terminal velocities. *J. Atmos. Sci.*, **29**, 1348–1357.

1975: Cirrus uncinus generating cells and the evolution of cirriform clouds. Part 111: Numerical computations of the growth of the ice phase. *J. Atmos. Sci.*, **32**, 820–830.

Heymsfield, A.J., 2007: On measurements of small ice particles in clouds. *Geophys. Res. Lett.*, **34**, L23812, doi:10.1029/2007GL030951.

Heymsfield, A.J., A. Bansemer, G. Heymsfield, and A.O. Fierro, 2009: Microphysics of maritime tropical convective updrafts at temperatures from −20°C to −60°C. *J. Atmos. Sci.*, **66**, 3530–3565.

Heymsfield, A.J., A. Bansemer, and C.H. Twohy, 2007a: Refinements to ice particle mass dimensional and terminal velocity relationships for ice clouds. Part I: Temperature dependence. *J. Atmos. Sci.*, **64**, 1047–1067.

Heymsfield, A.J., M. Kajikawa, 1987: An improved approach to calculating terminal velocities of plate-like crystals and graupel. *J. Atmos. Sci.*, **44**, 1088–1099.

Heymsfield, A.J., S. Lewis, A. Bansemer, J. Iaquinta, L.M. Miloshevich, M. Kajikawa, C. Twohy, and M.R. Poellot, 2002: A general approach for deriving the properties of cirrus and stratiform ice cloud properties. *J. Atmos. Sci.*, **59**, 3–29.

Heymsfield, A.J., and L.M. Milosevich, 1993: Homogeneous ice nucleation and supercooled liquid water in orographic wave clouds. *J. Atmos.Sci*, **50**, 2335–2353.

1995: Relative humidity and temperature influences on cirrus formation and evolution: Observations from wave clouds and FIRE-II. *J. Atmos. Sci.*, **52**, 4302–4303.

Heymsfield, A.J., and L.M. Miloshevich, 2003: Parameterizations for the cross-sectional area and extinction of cirrus and stratiform ice cloud particles. *J. Atmos. Sci.*, **60**, 936–956.

Heymsfield, A.J., and J.C. Pflaum, 1985: A quantativ assessment of the accuracy of techniques for calculating graupel growth. *J. Atmos. Sci.*, **42**, 2264–2274.

Heymsfield, A.J., and R.M. Sabin, 1989: Cirrus crystal nucleation by homogeneous freezing of solution droplets. *J. Atmos. Sci.*, **46**, 2252–2264.

Heymsfield, A.J., C. Schmitt, A. Bansemer, and C.H. Twohy, 2010: Improved representation of ice particle masses based on observations in natural clouds. *J. Atmos. Sci.*, **67**, 3303–3318.

Heymsfield, A.J., G.-J. Van Zadelhoff, D.P. Donovan, F. Fabry, R.J. Hogan, and A.J. Illingworth, 2007b: Refinements to ice particle mass dimensional and terminal velocity relationships for ice clouds. Part II: Evaluation and parameterizations of ensemble ice particle sedimentation velocities. *J. Atmos. Sci.*, **64**, 1068–1088.

Heymsfield, A.J., and C.D. Westbrook, 2010: Advances in the estimation of ice particle fall speeds using laboratory and field measurements. *J. Atmos. Sci.*, **67**, 2469–2482.

Heymsfield, A., and P. Willis, 2014: Cloud conditions favoring secondary ice particle production in tropical maritime convection. *J. Atmos. Sci.*, **71**, 4500–4526.

Hobbs, P.V., 1969: Ice multiplication in clouds. *J. Atmos. Sci.*, **26**, 315–318.

Hobbs, P.V., and A.J. Alkezweeny, 1968: The fragmentation of freezing water droplets in free fall. *J. Atmos. Sci.*, **25**, 881–888.

Hobbs, P.V., and A.L. Rangno, 1990: Rapid development of high ice particle concentrations in small polarmaritime cumuliformclouds. *J. Atmos. Sci.*, **47**, 2710–2722.

Hoffer, T.E., 1961: A laboratory investigation of droplet freezing. *J. Meteorol.*, **18**, 766–778.

Hoose, C., and O. Mohler, 2012: Heterogeneous ice nucleation on atmospheric aerosols: A review of results from laboratory experiments. *Atmos. Chem. Phys. Discuss.*, **12**, 12,531–12,621.

Hosler, C.L., D.C. Jensen, and L. Goldshlak, 1957: On the aggregation of ice crystals to form snow. *J. Meteorol.*, **14**, 415–420.

Huffman, P.J., 1973: Supersaturation spectra of AgI and natural ice nuclei. *J. Appl. Meteorol.*, **12**, 1080–1082.

Huffman, P.J., and G. Vali, 1973: The effect of vapor depletion on ice nucleus measurements with membrane filters. *J. Appl. Meteorol.*, **12**, 1018–1024.

Ilotoviz, E., N. Benmoshe, A.P. Khain, V.T.J. Phillips, and A.V. Ryzhkov, 2016: Effect of aerosols on freezing drops, hail and precipitation in a mid-latitude storm. *J. Atmos. Sci.*, **73**, 109–144.

Jeffery, C., and P. Austin, 1997: Homogeneous nucleation of supercooled water: Results from a new equation of state. *J. Geophys. Res.*, **102**, 25 269–25 279.

Johnson, D., and R.M. Rasmussen, 1992: Hail growth hysteresis. *J. Atmos. Sci.*, **49**, 2525–2532.

Johnson, D.A., and J. Hallett, 1968: Freezing and shattering of supercooled water drops. *Q. J. Royal Meteorol. Soc.*, **94**, 468–482.

Johnson, D.B., 1987: On the relative efficiency of coalescence and riming. *J. Atmos. Sci.*, **44**, 1671–1680.

Kachurin, L.G. 1962: To the theory of aircraft icing. *Izv. Akad. Nauk SSSR*, Ser. Geofiz. N 6, 38–46.

Kajikawa, M., 1972: Measurements of falling velocity of individual snow crystals. *J. Meteorol. Soc. Japan*, **50**, 577–583.

1982: Observation of the falling motion of early snow flakes. Part I: Relationship between the free-fall

pattern and the number and shape of component snow crystals. *J. Meteorol. Soc. Japan*, **60**, 797–803.
Kajikawa, M., and A.J. Heymsfield, 1989: Aggregation of ice crystals in cirrus. *J. Atmos. Sci.*, **46**, 3108–3121.
Karcher, B., and U. Lohmann, 2003: A parameterization of cirrus cloud formation: Heterogeneous freezing. *J. Geophys. Res.*, **108** (D14), 4402.
Kato, T., 1995: A box-Lagrangian rain-drop scheme. *J. Meteorol. Soc. Japan*, **73**, 241–245.
Kessler, E., 1969: On the distribution and continuity of water substance in atmospheric circulations. *Meteorol. Monogr.* 32. Boston: American Meteorological Society.
Khain, A.P., 2009: Effects of aerosols on precipitation: A critical review. *Environ. Res. Lett.*, **4**, 015004.
Khain, A.P., K.D. Beheng, A. Heymsfield, A. Korolev, S.O. Krichak, Z. Levin, M. Pinsky, V. Phillips, T. Prabhakaran, A. Teller, S.C. van den Heever, and J.-I. Yano, 2015: Representation of microphysical processes in cloud resolving models: Spectral (bin) microphysics versus bulk parameterization. *Rev. Geophys.*, **53**, 247–322.
Khain, A.P., N. BenMoshe, and A. Pokrovsky, 2008: Factors determining the impact of aerosols on surface precipitation from clouds: An attempt of classification. *J. Atmos. Sci.*, **65**, 1721–1748.
Khain, A.P., B. Lynn, and J. Dudhia, 2010: Aerosol effects on intensity of landfalling hurricanes as seen from simulations with WRF model with spectral bin microphysics. *J. Atmos. Sci.*, **67**, 365–384.
Khain, A.P., M. Ovchinnikov, M. Pinsky, A. Pokrovsky, and H. Krugliak, 2000: Notes on the state-of-the-art numerical modeling of cloud microphysics. *Atmos. Res.*, **55**, 159–224.
Khain, A.P., V. Phillips, N. Benmoshe, and A. Pokrovsky, 2012: The role of small soluble aerosols in the microphysics of deep maritime clouds. *J. Atmos. Sci.*, **69**, 2787–2807.
Khain, A., M. Pinsky, M. Shapiro, and A. Pokrovsky, 2001a: Collision rate of small graupel and water drops. *J. Atmos. Sci.*, **58**, 2571–2595.
Khain, A.P., A. Pokrovsky, M. Pinsky, A. Seifert, and V. Phillips, 2004: Effects of atmospheric aerosols on deep convective clouds as seen from simulations using a spectral microphysics mixed-phase cumulus cloud model. Part 1: Model description. *J. Atmos. Sci.*, **61**, 2963–2982.
Khain, A.P., A. Pokrovsky, and I. Sednev, 1999: Some effects of cloud–aerosol interaction on cloud microphysics structure and precipitation formation: Numerical experiments with a spectral microphysics cloud ensemble model. *Atmos. Res.*, **52**, 195–220.
Khain, A.P., D. Rosenfeld, and A. Pokrovsky, 2001b: Simulation of deep convective clouds with sustained supercooled liquid water down to –37.5 C using a spectral microphysics model. *Geoph. Res. Let.*, **28**, 3887–3890.
Khain, A., D. Rosenfeld, and A. Pokrovsky, 2005: Aerosol impact on the dynamics and microphysics of convective clouds. *Q. J. Royal Meteorol. Soc.*, **131**, 2639–2663.
Khain, A.P., D. Rosenfeld, A. Pokrovsky, U. Blahak, and A. Ryzhkov, 2011: The role of CCN in precipitation and hail in a mid-latitude storm as seen in simulations using a spectral (bin) microphysics model in a 2D dynamic frame. *Atmos. Res.*, 99, 129–146.
Khain, A.P., and I. Sednev, 1995: Simulation of hydrometeor size spectra evolution by water-water, ice-water and ice-ice interactions. *Atmos. Res.*, **36**, 107–138.
1996: Simulation of precipitation formation in the Eastern Mediterranean coastal zone using a spectral microphysics cloud ensemble model. *Atmos. Res.*, **43**, 77–110.
Khvorostyanov, V.I., and J.A. Curry, 2002: Terminal velocities of droplets and crystals: Power laws with continuous parameters over the size spectrum. *J. Atmos. Sci.*, **59**, 1872–1884.
2004: The theory of ice nucleation by heterogeneous freezing of deliquescent mixed CCN. Part 1: Critical radius, energy and nucleation rate. *J. Atmos. Sci.*, **61**, 2676–2691.
2005a: The theory of ice nucleation by heterogeneous freezing of deliquescent mixed CCN. Part 2: Parcel model simulation. *J. Atmos. Sci.*, **62**, 261–285.
2005b: Fall velocities of hydrometeors in the atmosphere: Refinements to a continuous analytical power law. *J. Atmos. Sci.*, **62**, 4343–4357.
Khvorostyanov, V.I., A.P. Khain, and E.A. Kogteva, 1989: Two dimensional nonstationary microphysical model of a three-phase convective cloud and evaluation of the effects of seeding by a crystallizing agent. *Sov. Meteorol. Hydrol.*, **5**, 33–45.
Khvorostyanov, V., and K. Sassen, 1998: Towards the theory of homogeneous nucleation and its parameterization for cloud models. *Geophys. Res. Lett.*, **25**, 3155–3158.
Kidder, R.E., and A.E. Carte, 1964: Structures of artificial hailstones. *J. Rech. Atmos.*, **1**, 169–181.
Klaassen, W., 1988: Radar observations and simulation of the melting layer of precipitation. *J. Atmos. Sci.*, **45**, 3741–3753.
Klett, J., 1995: Orientation model for particles in turbulence. *J. Atmos. Sci.*, **52**, 2276–2285.
Knight, C.A., 1979: Ice nucleation in the atmosphere. *Adv. Coll. Int. Sci.*, **10**, 369–395.
Knight, N.C., and A.J. Heymsfield, 1983: Measurement and interpretation of hailstone density and terminal velocity. *J. Atmos. Sci.*, **40**, 1510–1516.
Knight, N.C., 1986: Hailstone shape factor and its relation to radar interpretation of hail. *J. Clim. Appl. Met.*, **25**, 1956–1958.
Koenig, L.R., 1963: The glaciating behavior of small cumulonimbus clouds. *J. Atmos. Sci.*, **20**, 29–47.
1965: Drop freezing through drop breakup. *J. Atmos. Sci.*, **22**, 448–451.

1966: Numerical test of the validity of the drop-freezing / splintering hypothesis of cloud glaciation. *J. Atmos. Sci.*, **23**, 726–740.

Kolomeychuk, R.J., D.C. McKay, and J.V. Iribarne, 1975: The fragmentation and electrification of freezing drops. *J. Atmos. Sci.*, **32**, 974–979.

Korolev, A.V., 2007a: Reconstruction of the sizes of spherical particles from their shadow images. Part I: Theoretical considerations. *J. Atmos. Oceanic Technol.*, **24**, 376–389.

2007b: Limitations of the Wegener–Bergeron–Findeisen mechanism in the evolution of mixed-phase clouds. *J. Atmos. Sci.*, **64**, 3372–3375.

Korolev, A.V., and P.R. Field, 2008: The effect of dynamics on mixed-phase clouds: Theoretical considerations. *J. Atmos. Sci.*, **65**, 66–86.

Korolev, A.V., and G.A. Isaac, 2003: Phase transformation of mixed-phase clouds. *Q. J. Royal Meteorol. Soc.*, **129**, 19–38.

2005: Shattering during sampling by OAPs and HVPS. Part I: Snow Particles. *J. Atmos. Ocean Tech.*, **22**, 528–542.

Korolev, A.V., and I. Mazin, 2003: Supersaturation of water vapor in clouds. *J. Atmos. Sci.*, **60**, 2957–2974.

Kovetz, A.V., and B. Olund, 1969: The effect of coalescence and condensation on rain formation in a cloud of finite vertical extent. *J. Atmos. Sci.*, **26**, 1060–1065.

Kramer, B., O. Hubner, H. Vortisch, L. Woste, T. Leisner, M. Schwell, E. Ruhl, and H. Baumgartel, 1999: Homogeneous nucleation rates of supercooled water measured in single levitated microdroplets. *J. Chem. Phys.*, **111**, 6521–6527.

Kumjian, M.R., S.M. Ganson, and A.V. Ryzhkov, 2012: Freezing of raindrops in deep convective updrafts: A microphysical and polarimetric model. *J. Atmos. Sci.*, **69**, 3471–3490.

Lang, S.E., W.-K. Tao, J.D. Chern, D. Wu, and X. Li, 2014: Benefits of a fourth ice class in the simulated radar reflectivities of convective systems using a bulk microphysics scheme. *J. Atmos. Sci.*, **71**, 3583–3611.

Lang, T.J. et al., 2004: The severe thunderstorm electrification and precipitation study. *Bull. Am. Meteorol. Soc.*, **85**, 1107–1125.

Lawson, R.P., S. Woods, and H. Morrison, 2015: The microphysics of ice and precipitation development in tropical cumulus clouds. *J. Atmos. Sci.*, **72**, 2429–2445.

Leisner, T., T. Pander, P. Handmann, and A. Kiselev, 2014: Secondary ice processes upon heterogeneous freezing of cloud droplets. 14th Conf. on Cloud Physics and Atmospheric Radiation, Boston, MA, Amer. Meteorol. Soc., 2.3.

Lesins, G.B., and R. List, 1986: Sponginess and drop shedding of gyrating hailstones in a pressure-controlled icing wind tunnel. *J. Atmos. Sci.*, **43**, 2813–2825.

Levi, L., and L. Lubart, 1998: Modelled spongy growth and shedding process for spheroidal hailstones. *Atmos. Res.*, **47–48**, 59–68.

Lew, J.K., D.E. Kingsmill, and D.C. Montague, 1985: A theoretical study of the collision efficiency of small planar ice crystals colliding with large supercooled drops. *J. Atmos. Sci.*, **42**, 857–862.

Lew, J.K., and H.R. Pruppacher, 1983: A theoretical determination of the capture efficiency of small columnar ice crystals by large cloud drops. *J. Atmos. Sci.*, **40**, 139–145.

Li, G., Y. Wang, and R. Zhang, 2008: Implementation of a two-moment bulk microphysics scheme to the WRF model to investigate aerosol-cloud interaction. *J. Geophys. Res.*, **113**, D15211.

Lin, R.-F., D. O'C. Starr, P.J. DeMott, R. Cotton, K. Sassen, E. Jensen, B. Kärcher, and X. Liu, 2002: Cirrus parcel model comparison project. Phase 1: The critical components to simulate cirrus initiation explicitly. *J. Atmos. Sci.*, **59**, 2305–2329.

Lin, Y.-L., R.D. Farley, and H.D. Orville, 1983: Bulk parameterization of the snow field in a cloud model. *J. Climate Appl. Meteorol.*, **22**, 1065–1092.

List, R., 1959: Wachstrum von Eis-Wassergemischen im Hagel-versuchskanal. *Helv. Phys. Acta.*, **32**, 293–296.

1990: Physics of supercooling of thin water skins covering gyrating hailstones. *J. Atmos. Sci.*, **47**, 1919–1925.

2014a: New hailstone physics. Part I: Heat and mass transfer (HMT) and growth. *J. Atmos. Sci.*, **71**, 1508–1520.

2014b: New hailstone physics. Part II: Results. *J. Atmos. Sci.*, **71** (6), 2114–2129.

List, R., B.J.W. Greenan, and F. Garca-Garca, 1995: Surface temperature variations of gyrating hailstones and effects of pressure-temperature coupling on growth. *Atmos. Res.*, **38**, 161–175.

Litvinov, I.V., 1974: *The Structure of Atmospheric Precipitation*. Gidrometizdat, p. 154 (in Russian).

Liu, C., M.W. Moncrieff, and E.J. Zipser, 1997: Dynamical influence of microphysics in tropical squall lines: A numerical study. *Mon. Wea. Rev.*, **125**, 2193–2210.

Liu, X., and J. Penner, 2005: Ice nucleation parameterization for global models. *Meteorol. Z.*, **14**, 499–514.

Locatelli, J.D., and P.V. Hobbs, 1974: Fall speeds and masses of solid precipitation particles. *J. Geophys. Res.*, **79**, 2185–2197.

Loftus, A.M., W.R. Cotton, and G.G. Carrió, 2014: A triple-moment hail bulk microphysics scheme. Part I: Description and initial evaluation. *Atmos. Res.*, **148**, 35–57.

Lozowski, E.P., 1991: Comments on "Physics of supercooling of thin water skins covering gyrating hailstones." *J. Atmos. Sci.*, **48**, 1606–1608.

Ludlam, F.H., 1958: The hail problem. *Nubila*, **1**, 13.

Lynn, B. and A.P. Khain, 2007: Utilization of spectral bin microphysics and bulk parameterization schemes to simulate the cloud structure and precipitation in a mesoscale rain event. *J. Geophys. Res.*, **112**, D22205.

Macklin, W.C., 1961: Accretion in mixed clouds. *Q. J. Royal Meteorol. Soc.*, **87**, 413–424.

Magono, C., and C.W. Lee, 1966: Meteorological classification of natural snow crystals. *J. Fac. Sci.*, Hokkaido Univ., Ser. 2, 7, 321–335.

Mamouri, R.E., and A. Ansmann, 2015: Potential of polarization lidar to provide profiles of CCN- and INP-relevant aerosol parameters. *Atmos. Chem. Phys. Discuss.*, **15**, 34,149–34,204.

Marcolli, C., S. Gedamke, T. Peter, and B. Zobrist, 2007: Efficiency of immersion mode ice nucleation on surrogates of mineral dust. *Atmos. Chem. Phys.*, **7**, 5081–5091.

Marwitz, J.D., 1983: The kinematics of orographic airflow during Sierra storms. *J. Atmos. Sci.*, **40**, 1218–1227.

Mason, B.J., and J. Maybank, 1960: The fragmentation and electrification of freezing water drops. *Quarterly Journal of the Royal Meteorological Society*, **86** (368), 176–185.

Mason, N.J., 1956: On the melting of hailstones. *Q. J. Royal Meteorol. Soc.*, **82**, 209–216.

Meyers, M.P., P.J. DeMott, W.R. Cotton, 1992: New primary ice-nucleation parameterizations in an explicit cloud model. *J. Appl. Meteorol.*, **31**, 708–721.

Meyers, M.P., R.L. Walko, J.Y. Harrington, and W.R. Cotton, 1997: New Rams cloud microphyisics parameterization: Part II. The two-moment scheme. *Atmos. Res.*, **45**, 3–39.

Milbrandt, J.A., and R. McTaggart-Cowan, 2010: Sedimentation-induced errors in bulk microphysics schemes. *J. Atmos. Sci.*, **67**, 3931–3948.

Milbrandt, J.A., and H. Morrison, 2013: Prediction of graupel density in a bulk microphysics scheme. *J. Atmos. Sci.*, **70**, 410–429.

Milbrandt, J.A., and M.K. Yau, 2005a: A Multimoment Bulk Microphysics Parameterization. Part I: Analysis of the Role of the Spectral Shape Parameter. *J. Atmos. Sci.*, **62**, 3051–3064.

2005b: A multimoment bulk microphysics parameterization. Part 2: A proposed three-moment closure and scheme description. *J. Atmos. Sci.*, **62**, 3065–3081.

2006: A multimoment bulk microphysics parameterization. Part III: Control simulation of a hailstorm. *J. Atmos. Sci.*, **63**, 3114–3136.

Mitchell, D.L., 1988: Evolution of snow-size spectra in cyclonic storms. Part 1: Snow growth by vapor deposition and aggregation. *J. Atmos. Sci.*, **45**, 3431–3451.

1991: Evolution of snow-size spectra in cyclonic storms. Part II: Deviations from the exponential form. *J. Atmos. Sci.*, **48**, 1885–1899.

1996: Use of mass- and area-dimensional power laws for determining precipitation particle terminal velocities. *J. Atmos. Sci.*, **53**, 1710–1723.

Mitchell, D.L., and A.J. Heymsfield, 2005: Refinements in the treatment of ice particle terminal velocities, highlighting aggregates. *J. Atmos. Sci.*, **62**, 1637–1644.

Mitchell, D.L., R. Zhang, and R.L. Pitter, 1990: Mass-dimensional relationships for ice particles and the influence of riming on snowfall rates. *J. Appl. Meteorol.*, **29**, 153–163.

Mitra, S.K., O. Vohl, M. Ahr, and H.R. Pruppacher, 1990: A wind tunnel and theoretical study of the melting behavior of atmospheric ice particles. Part IV: Experiment and theory for snow flakes. *J. Atmos. Sci.*, **47**, 584–591.

Mizuno, H., 1990: Parameterization of the accretion process between different precipitation elements. *J. Meteorol. Soc. Japan*, **68**, 395–398.

Moore, G.W., and R.E. Stewart, 1985: The coupling between melting and convective air motions in stratiform clouds. *J. Geophys. Res.*, **90**, 10,659–10,666.

Morrison, H., 2012: On the numerical treatment of hydrometeor sedimentation in bulk and hybrid bulk–bin microphysics schemes. *Mon. Wea. Rev.*, **140**, 1572–1588.

Morrison, H., J.A. Curry, and V.I. Khvorostyanov, 2005: A New double-moment microphysics parameterization for application in cloud and climate models. Part I: Description. *J. Atmos. Sci.*, **62**, 1665–1677.

Morrison, H., and W.W. Grabowski, 2008: A novel approach for representing ice microphysics in models: Description and tests using a kinematic framework. *J. Atmos. Sci.*, **65**, 1528–1548.

2010: An improved representation of rimed snow and conversion to graupel in a multicomponent bin microphysics scheme. *J. Atmos. Sci.*, **67**, 1337–1360.

Morrison, H., and J.A. Milbrandt, 2015: Parameterization of cloud microphysics based on the prediction of bulk ice particle properties. Part I: Scheme description and idealized tests. *J. Atmos. Sci.*, **72**, 287–311.

Morrison, H., J.A. Milbrandt, G. Bryan, K. Ikeda, S.A. Tessendorf, and G. Thompson, 2015: A new approach for parameterizing microphysics based on prediction of multiple ice particle properties. Part 2: Case study comparison with observations and other schemes. *J. Atmos. Sci.*, **72**, 311–339.

Mossop, S.C., and Kidder, R.E., 1962: Artificial hailstones. *Bull. Obs. Puy. De Dom.*, **2**, 65–79.

Murakami, M., 1990: Numerical modeling of dynamical and microphysical evolution of an isolated convective cloud – The 19 July 1981 CCOPE cloud. *J. Meteorol. Soc. Japan*, **68**, 107–128.

Musil, D.J., 1970: Computer modeling of hailstone growth in feeder clouds. *J. Atmos. Sci.*, **27**, 474–482.

Noppel, H., A. Pokrovsky, B. Lynn, A.P. Khain, and K.D. Beheng, 2010: On precipitation enhancement due to a spatial shift of precipitation caused by introducing small aerosols: Numerical modeling. *J. Geophys. Res.*, **115**, D18212.

Ovtchinnikov, M., 1998: An investigation of ice production mechanisms using a 3-D cloud model with explicit microphysics. Cooperative Institute for Mesoscale Meteorological Studies, Norman, Oklahoma 73019, Report 107, p. 128.

Ovchinnikov, M., and Y. Kogan, 2000: An investigation of ice production mechanisms in small cumuliform clouds using a 3D model with explicit microphysics. Part I: Model description. *J. Atmos. Sci.*, **57**, 2989–3003.

Passarelli, R.E., 1978: Theoretical and observational study of snow-size spectra and snowflake aggregation efficiencies. *J. Atmos. Sci.*, **35**, 882–889.

Passarelli, R.E., and R.C. Srivastava, 1978: A new aspect of snowflake aggregation theory. *J. Atmos. Sci.*, **36**, 484–493.

Pflaum, J.C., 1980: Hail formation via microphysical recycling. *J. Atmos. Sci.*, **37**, 160–173.

Phillips, V.T.J., A.M. Blyth, P.R.A. Brown, T.W. Choularton, and J. Latham, 2001: The glaciation of a cumulus cloud over New Mexico. *Q. J. Royal Meteorol. Soc.*, 127, 1513–1534.

Philips, V.T.J., P.J. DeMott, and C. Andronache, 2008: An empirical parameterization of heterogeneous ice nucleation for multiple chemical species of aerosol. *J. Atmos. Sci.*, **65**, 2757–2783.

Phillips, V.T.J., P.J. DeMott, C. Andronache, K.A. Pratt, K.A. Prather, R. Sabramanian, and C. Twohy, 2013: Improvements to an empirical parameterization of heterogeneous ice nucleation and its comparison with observations. *J. Atmos. Sci.*, **70**, 378–409.

Phillips, V.T.J., A. Khain, N. Benmoshe, and E. Ilotovich, 2014: Theory of time-dependent freezing. Part I: Description of scheme for wet growth of hail. *J. Atmos. Sci.*, **71** (12), 133–163.

Phillips, V., A.P. Khain, N. Benmoshe, E. Ilotoviz, and A. Ryzhkov, 2015: Theory of time-dependent freezing. Part II: Scheme for freezing raindrops and simulations by a cloud model with spectral bin microphysics. *J. Atmos. Sci.*, **72**, 262–286.

Phillips, V., A.P. Khain, and A. Pokrovsky, 2007: The influence of melting on the dynamics and precipitation production in maritime and continental storm clouds. *J. Atmos. Sci.*, **64**, 338–359.

Phillips, V.T.J., S.C. Sherwood, C. Andronache, A. Bansemer, W.C. Conant, P.J. DeMott, R.C. Flagan, A. Heymsfield, H. Jonsson, M. Poellot, T.A. Rissman, J.H. Seinfeld, T. Vanreken, V. Varutbangkul, and J.C. Wilson, 2005: Anvil glaciation in a deep cumulus updraft over Florida simulated with an explicit microphysics model. I: The impact of various nucleation processes. *Q. J. Royal Meteorol. Soc.*, **131**, 2019–2046.

Phillips, V.T.J., J.-I. Yano, and A. Khain, 2017a: Ice Multiplication by Break-up in Ice-Ice Collisions. Part 1: Theoretical Formulation. *J. Atmos. Sci.*, **74**, 1705–1719.

Phillips, V.T.J., J.-I. Yano, M. Formenton, E. Ilotoviz, V. Kanawade, I. Kudzotsa, J. Sun, A. Bansemer, A.G. Detwiler, A. Khain, and S. Tessendorf, 2017b: Ice Multiplication by Breakup in Ice-Ice Collisions. Part 2: Numerical simulations. *J. Atmos. Sci.*, **74**, 2789–2811.

Pinsky, M., A. Khain, and A. Korolev, 2014: Analytical investigation of glaciating time in mixed-phase adiabatic cloud volumes. *J. Atmos. Sci.*, **71**, 4143–4157.

2015: Phase transformations in an ascending adiabatic mixed-phase cloud volume. *J. Geophys. Res., Atmos.*, **120**, 1329–1353.

Pinsky, M., A. Khain, and M. Shapiro, 2000: Stochastic effects of cloud droplet hydrodynamic interaction in a turbulent flow. *Atmos. Res.*, **53**, 131–169.

Pinsky, M., M. Shapiro, A. Khain, and H. Wirzberger, 2004: A statistical model of strains in homogeneous and isotropic turbulence. *Physica D.*, **191**, 297–313.

Pitter, R.L., and H.R. Pruppacher, 1973: A wind tunnel investigation of freezing of small water drops falling at terminal velocity in air. *Q. J. Royal Meteorol. Soc.*, 99, 540–550.

Pitter, R.L., H.R. Pruppacher, and A.E. Hamielec, 1974: A numerical study of the effect of forced convection on mass transport from a thin oblate spheroid of ice in air. *J. Atmos. Sci.*, **31**, 1058–1066.

Polycarpou, A.A., and I. Etsion, 1999: Analytical approximations in modeling contacting rough surfaces. *ASME J. Tribol.*, **121**, 234–239.

Prenni, A.J., P.J. DeMott, C. Twohy, M.R. Poellot, S.M. Kreidenweis, D.C. Rogers, S.D. Brooks, M.S. Richardson, and A.J. Heymsfield, 2007a: Examinations of ice formation processes in Florida cumuli using ice nuclei measurements of anvil ice crystal particle residues. *J. Geophys. Res.*, **112**, D10221, doi:10.1029/2006JD007549.

Prenni, A.J., J.Y. Harrington, M. Tjernstrom, P.J. DeMott, A. Avramov, C.N. Long, S.M. Kreidenweis, P.Q. Olsson, and J. Verlinde, 2007b: Can ice-nucleating aerosols affect arctic seasonal climate? *Bull. Am. Meteorol. Soc.*, **88** (4), 541–550.

Pruppacher, H.R., and J.D. Klett, 1997: *Microphysics of Clouds and Precipitation*, 2nd edition. Oxford University Press, p. 963.

Pruppacher, H.R., and R.J. Schlamp, 1975: A wind tunnel investigation on ice multiplication by freezing of waterdrops falling at terminal velocity in air. *J. Geophys. Res.*, **80**, 380–386.

Pumir, A., M. Wilkinson, 2011: Orientation statistics of small particles in turbulence. *New J. Phys.*, **13**, 093030.

Rangno, A.L., 2008: Fragmentation of freezing drops in shallow maritime frontal clouds. *J. Atmos. Sci.*, **65**, 1455–1466.

Rasmussen, D.H., 1982: Thermodynamic and nucleation phenomena. A set of experimental observations. *J. Crystal Growth*, **56**, 5646.

Rasmussen, R.M., I. Geresdi, G. Thompson, K. Manning, and E. Karplus, 2002: Freezing drizzle formation in stably stratified layer clouds: The role of radiative cooling of cloud droplets, cloud condensation nuclei, and ice initiation. *J. Atmos. Sci.*, **59**, 837–860.

Rasumussen, R.M., and A.J. Heymsfield, 1985: A generalized form for impact velocities used to determine graupel accretional densities. *J. Atmos. Sci.*, 42 (21), 2275–2279.

1987: Melting and shedding of graupel and hail. Part I: Model physics. *J. Atmos. Sci.*, **44**, 2754–2763.

Rasmussen, R.M., V. Levizzani, and H.R. Pruppacher, 1984a: A wind tunnel and theoretical study of the melting behavior of atmospheric ice particles. Part II: A theoretical study for frozen drops of radius < 500 μm. *J. Atmos. Sci.*, **41**, 374–380.

1984b: A wind tunnel and theoretical study of the melting behavior of atmospheric ice particles. Part III: Experiment and theory for spherical ice particles of radius > 500 μm. *J. Atmos. Sci.*, **41**, 381.

Rasmussen, R.M., and H.R. Pruppacher, 1982: A wind tunnel and theoretical study of the melting behavior of atmospheric ice particles. I: A wind tunnel study of frozen drops of radius < 500 μm. *J. Atmos. Sci.*, **39**, 152–158.

Rauber, R.M., L.S. Olthoff, M.K. Ramamurthy, and K.E. Kunkel, 2000: The relative importance of warm rain and melting processes in freezing precipitation events. *J. Appl. Meteorol.*, **39**, 1185–1195.

Reisin, T., Z. Levin, and S. Tzivion, 1996: Rain production in convective clouds as simulated in an axisymmetric model with detailed microphysics. Part I: Description of the model. *J. Atmos. Sci.*, **54**, 497–519.

Reisner, J., R.M. Rasmussen, and R.T. Bruintjes, 1998: Explicit forecasting of supercooled liquid water in winter storms using the MM5 mesoscale model. *Q. J. Royal Meteorol. Soc.*, **124**, 1071–1107.

Respondek, P.S., A.I. Flossman, R.R. Alheit, and H.R. Pruppacher, 1995: A theoretical study of the wet removal of atmospheric pollutants. Part V: The uptake, redistribution, and deposition of $(NH_4)_2SO_4$ by a convective cloud containing ice. *J. Atmos. Sci.*, **52**, 2121–2132.

Richardson, M.S., P.J. DeMott, S.M. Kreidenweis, D.J. Cziczo, E.J. Dunlea, J.L. Jimenez, D.S. Thomson, L.L. Ashbaugh, R.D. Borys, D.L. Westphal, G.S. Casuccio, and T.L. Lersch, 2007: Measurements of heterogeneous ice nuclei in the western United States in springtime and their relation to aerosol characteristics. *J. Geophys. Res.*, **112**, D02209.

Rogers, D.C., 1973: The aggregation of natural ice crystals. MS thesis, Dept. Atmos. Res., University of Wyoming, p. 86.

1982: Field and Laboratory Studies of Ice Nucleation in Winter Orographic Clouds, PhD Dissertation. Dept. Phys. Astron, Univ. Wyoming, p. 161.

Rosenfeld, D., R. Chemke, P. DeMott, R.C. Sullivan, R. Rasmussen, F. McDonough, J. Comstock, B. Schmid, J. Tomlinson, H. Jonsson, K. Suski, A. Cazorla, and K. Prather, 2013: The common occurrence of highly supercooled drizzle and rain near the coastal regions of the western United States. *J. Geophys. Res. Atmos.*, **118**, 1–15.

Rosenfeld, D., R. Chemke, K. Prather, K. Suski, J.M. Comstock, B. Schmid, J. Tomlinson, and H. Jonsson, 2014: Polluting of winter convective clouds upon transition from ocean inland over central California: Contrasting case studies. *Atmos. Res.*, **135–136**, 112–127.

Rosenfeld, D., and W.L. Woodley, 2000: Deep convective clouds with sustained highly supercooled liquid water until −37.5°C. *Nature*, **405**, 440–442.

Rutledge, S.A., and P.V. Hobbs, 1983: The mesoscale and microscale structure and organization of clouds and precipitation in midlatitude cyclones. Viii: A model for the "seeder-feeder" process in warm-frontal rainbands. *J. Atmos. Sci.*, **40**, 1185–1206.

1984: The mesoscale and microscale structure and organization of clouds and precipitation in midlatitude cyclones. XII: A diagnostic modeling study of precipitation development in narrow cold-frontal rainbands. *J. Atmos. Sci.*, **41**, 2949–2972.

Ryzhkov, A., M. Pinsky, A. Pokrovsky, and A. Khain, 2011: Polarimetric Radar Observation Operator for a Cloud Model with Spectral Microphysics. *J. Appl. Meteorol. Climatol.*, **50**, 873–894.

Sassen, K., and S. Benson, 2000: Ice nucleation in cirrus clouds. A model study of the homogeneous and heterogeneous nucleation modes. *Geophys. Res. Lett.*, **27**, 521–524.

Sassen, K., and G.C. Dodd, 1988: Homogeneous nucleation rate for highly supercooled cirrus cloud droplets. *J. Atmos. Sci.*, **45**, 1357–1369.

1989: Haze particle nucleation simulation in cirrus clouds, and application for numerical and lidar studies. *J. Atmos. Sci.*, **46**, 3005–3014.

Sastry, S., 2005: Water: Ins and outs of ice nucleation. *Nature*, **438**, 746–747.

Saunders, C.P.R., and A.S. Hosseini, 2001: A laboratory study of the effect of velocity on Hallett-Mossop ice crystal multiplication. *Atmos. Res.*, **59–60**, 3–14.

Schuepp, P.H., and R. List, 1969a: Mass transfer of rough hailstone models in flows of various turbulence levels. *J. Appl. Meteorol.*, **8**, 254–263.

1969b: Influence of molecular properties of the fluid on simulation of the total heat and mass transfer of solid precipitation particles. *J. Appl. Meteorol.*, **8**, 743–746.

Schumann, T.E.W., 1938: The theory of hailstone formation. *Q. J. Royal Meteorol. Soc.*, **64**, 3–21.

Schwarzenboeck, A., V. Shcherbakov, R. Lefevre, J.-F. Gayet, Y. Pointin, and C. Duroure, 2009: Indications for stellar-crystal fragmentation in Arctic clouds. *Atmos. Res.*, **92**, 220–228.

Seifert, A., and K.D. Beheng, 2006: A two-moment cloud microphysics parameterization for mixed-phase clouds. Part 1: Model description. *Meteorol. Atmos. Phys.*, **92**, 45–66.

Shaw, R.A., A.J. Durant, and Y. Mi, 2005: Heterogeneous surface crystallization observed in undercooled water. *J. Phys. Chem. B.*, **109**, 9865–9868.

Shaw, R., and D. Lamb, 1999: Homogeneous freezing of evaporating cloud droplets. *Geophys. Res. Lett.*, **26**, 1181–1184.

Siewert, C., R.P.J. Kunnen, M. Meinke, and W. Schroder, 2014: Orientation statistics and settling velocity of ellipsoids in decaying turbulence. *Atmos. Res.*, 142, 45–56.

Slinn, W.G.N., and J.M. Hales, "Phoretics Effects in Scavenging," Precipitation Scavenging, AEC Symposium Series 22, Washington, DC, 1970.

Stith, J.L., V. Ramanathan, W.A. Cooper, G.C. Roberts, P.J. DeMott, G. Carmichael, C.D. Hatch, B. Adhikary, C.H. Twohy, D.C. Rogers, D. Baumgardner, A.J. Prenni, T. Campos, R. Gao, J. Anderson, and Y. Feng, 2009: An overview of aircraft observations from the Pacific Dust Experiment campaign. *J. Geophys. Res.*, **114**, D05207, doi:10.1029/2008JD010924.

Stockel, P., 2001: Homogene Nukleation in levitierten Tropfchen aus stark unterkhltem H2O und D2O. PhD thesis. Free University of Berlin, p. 197.

Straka, J.M., 2009: *Cloud and Precipitation Microphysics.* Cambridge University Press, p. 392.

Sulia, K.J., and J.Y. Harrington, 2011: Ice aspect ratio influences on mixed-phase clouds: Impacts on phase partitioning in parcel models. *J. Geop. Res.*, **116**, D21309.

Szeto, K.K., and R.E. Stewart, 1997: Effects of melting on frontogenesis. *J. Atmos. Sci.*, **54**, 689–702.

Szeto, K.K., R.E. Stewart, and C.A. Lin, 1988: Mesoscale circulations forced by the melting of snow in the atmosphere. Part II: Application to meteorological features. *J. Atmos. Sci.*, **45**, 1642–1650.

Takahashi, C., and A. Yamashita, 1969: Deformation and fragmentation of freezing water drops in free fall. *J. Meteorol. Soc. Japan*, **47**, 431–434.

1970: Shattering of frozen water drops in a supercooled cloud. *J. Meteorol. Soc. Japan*, **48**, 373–376.

Takahashi, T., 1978: Riming electrification as a charge generation mechanism in thunderstorms. *J Atmos. Sci.*, **35**, 1536–1548.

Takahashi, T., and G. Wakahama, 1991: Vapor diffusional growth of free-falling snow crystals between −3 and −23 °C. *J. Meteorol. Soc. Japan*, **69**, 15–30.

Takahashi, T., and N. Fukuta, 1988: Supercooled cloud tunnel studies on the growth of snow crystals between −4 and −20 °C. *J. Meteorol. Soc. Japan*, **66**, 841–855.

Takahashi, T., Y. Nagao, and Y. Kushiyama, 1995: Possible high ice particle production during graupel–graupel collisions. *J. Atmos. Sci.*, **52**, 4523–4527.

Tao, W.-K., J. Simpson, and S.-T. Soong, 1991: Numerical simulation of a subtropical squall line over the Taiwan Strait. *Mon. Wea. Rev.*, **119**, 2699–2723.

Tao, W.-K., D. Wu, S. Lang, J.-D. Chern, C. Peters-Lidard, A. Fridlind, and T. Matsui, 2015: High-resolution NU-WRF simulations of a deep convective-precipitation system during MC3E: Further improvements and comparisons between Goddard microphysics schemes and observations. *J. Geophys. Res. Atmos.*, **121**, 1278–1305, doi:10.1002/2015JD023986.

Thompson, G., R.M. Rasmussen, and K. Manning, 2004: Explicit forecasts of winter precipitation using an improved bulk microphysics scheme. Part I: Description and sensitivity analysis. *Mon. Weather Rev.*, **132**, 519–542.

Tinsley, B.A., 2004: Contact ice nucleation near cloud tops due to electroscavenging, Ice Initiation Workshop, NCAR June 7–9.

Tinsley, B.A., R.P. Rohrbauch, and M. Hei, 2001: Electroscavenging in clouds with broad droplet size distributions and weak electrification. *Atmos. Res.*, **59–60**, 115–135.

Tomotika, S., 1935: The laminar boundary layer on the surface of a sphere in a uniform stream. Rep. and memo., Great Britain Aeronautical Research Commettee, 1678, H.M.S.O., London, UK, p. 14.

Vali, G., 1975: Remarks on the mechanism of atmospheric ice nucleation. Proc. 8th Int. Conf. on Nucleation, Leningrad, Sept. 23–29, I.I. Gaivoronsky ed., Gidrometeoizdat, pp. 265–269.

1976: Contact freezing nucleation measured by the DFC instrument. Preprints Third Internat. Workshop on Ice Nucleous Measurements, Laramie, Univ. of Wyoming, pp. 159–178.

1994: Freezing rate due to heterogeneous nucleation. *J. Atmos. Sci.*, **51**, 1843–1856.

1999: ICE NUCLEATION THEORY. A TUTORIAL FOR PRESENTATION AT THE NCAR/ASP 1999 SUMMER COLLOQUIUM, <vali@uwyo.edu>; www-das.uwyo.edu/~vali, p. 22.

Vardiman, L., 1978: The generation of secondary ice particles in clouds by crystal–crystal collision. *J. Atmos. Sci.*, **35**, 2168–2180.

Vohl, O., S.K. Mitra, K. Diehl, H. Huber, S.C. Wurzler, K.l. Kratz, and H.R. Pruppacher, 2001: A wind tunnel study of turbulence effects on the scavenging of aerosol particles by water drops. *J. Atmos. Sci.*, **58**, 3064–3072.

Wacker, U., and C. Lupkes, 2009: On the selection of prognostic moments in parameterization schemes for drop sedimentation. *Tellus*, **61A**, 498–511.

Walko, R., W.R. Cotton, M.P. Meyers, and J.Y. Harrington, 1995: New RAMS cloud microphysics parameterization. Part I: The single-moment scheme. *Atmos. Res.*, **38**, 29–62.

Wall, S., W. John, H. Wang, and S.L. Goren, 1990: Measurements of kinetic energy loss for particles impacting surfaces. *Aerosol Sci. Technol.*, **12**, 926–946.

Wang, P.K., and W. Ji, 1992: Numerical simulation of three-dimensional unsteady flow past ice crystals. *J. Atmos. Sci.*, **54**, 2261–2274.

Wang, P.K., 2002: *Ice Microdynamics*. Academic Press, p. 273.

Wang, Y., J. Fan, R. Zhang, L. Leung, and C. Franklin, 2013: Improving bulk microphysics parameterizations in simulations of aerosol effects. *J. Geophys. Res. Atm.*, **118**, 1–19.

Wegener, A., 1911: *Thermodynamik der Atmosphäre*. J. A. Barth (in German), p. 331.

Weil, J.C., R.P. Lawson, and A.R. Rodi, 1993: Relative dispersion of ice crystals in seeded cumuli. *J. Appl. Meteorol.*, **32**, 1055–1073.

Willis, P.T., and A.J. Heymsfield, 1989: Structure of the melting layer in mesoscale convective system stratiform precipitation. *J. Atmos. Sci.*, **46**, 2008–2025.

Wisner, C., H.D. Orville, and C. Myers, 1972: A numerical model of a hail-bearing cloud. *J. Atmos. Sci.*, **29**, 1160–1181.

Wood, S., M. Baker, and B. Swanson, 2002: Instrument for studies of homogeneous and heterogeneous ice nucleation in free-falling supercooled water droplets. *Rev. Sci. Instrum.*, **73**, 3988–3996.

Wurzler, S., and A. Bott, 2000: Numerical simulations of cloud microphysics and drop freezing as function drop contamination. *J. Aerosol Sci.*, **31**, S152–S153.

Yano, J.-I., and V.T.J. Phillips, 2011: Ice–ice collisions: An ice multiplication process in atmospheric clouds. *J. Atmos. Sci.*, **68**, 322–333.

Yin, Y., Z. Levin, T. Reisin, and S. Tzivion, 2000: The effects of giant cloud condensational nuclei on the development of precipitation in convective clouds: A numerical study. *Atmos. Res.*, **53**, 91–116.

Young, K.C., 1974: The role of contact nucleation in ice phase initiation in clouds. *J. Atmos. Sci.*, **31**, 768–776.

Zakinyan, R.G., 2008: On the theory of hailstone growth. *Atmospheric and Oceanic Physics*, **44**, (2), 207–212.

Zhekamukhov, M.K., 1982: *Some Problems of Formation of Hailstone Structure*. Gidrometeoizdat, Moscow (in Russian), p. 256.

7 Modeling: A Powerful Tool for Cloud Investigation

One of the most efficient ways to investigate atmospheric processes, and often the sole method available to do so, is numerical modeling. Atmospheric motions have a very broad spectrum of scales, ranging from 1 mm to thousands of kilometers (see Section 3.4). At one end of the spectrum there are DNS models, able to resolve the smallest turbulent vortices; at the other end there are models of General Circulation of the atmosphere (GCM), capable of describing large-scale atmospheric motions across the globe. These models are also called Global Circulation Models. Each model has its own specific features, and is typically designed to solve a particular set of problems. We shall now discuss models that deal with clouds and their effects.

7.1 Characteristics of State-of-the-Art Cloud and Cloud-Resolving Models

7.1.1 Convection in Large-Scale Atmospheric Models

Limited Area Models (LAM) and GCM have grid spacing that exceeds 10–20 km and do not resolve clouds. Thus, convection in such models is a subgrid process (Section 3.4). Traditional methods of convective parameterization are usually used to describe convective heating/cooling (Emanuel and Raymond, 1993). The goal of these parameterizations is to describe the overall effect of subgrid cumulus convection on the large spatial scales explicitly resolved by the models. These parameterizations are based largely on empirical or semiempirical mass and energy budget considerations (Plant and Yano, 2015). Convective parameterizations use simple representations of clouds in the form of plumes transporting mass and water vapor upward. The mass flux in clouds, as well as the distribution of clouds with respect to their size, are determined by one or another closure assumption and the assumed geometry of the cloud model. Examples are a cylinder with a constant radius (Arakawa and Shubert, 1974); thermals or bubbles (Anthes, 1977); and two coaxial cylinders with entrainment–detrainment rates that vary with cloud size (Tiedtke, 1983). These parameterizations describe cloud microphysics in a very crude manner.

Within LAM and GCM, convectively induced heating/cooling effects and cloud-induced radiative effects are treated by two independent parameterization schemes, despite the fact that both types of effects are caused by the same clouds. The convective parameterizations do not allow for a determination of cloud coverage. Instead, the cloud coverage is empirically related to the horizontally averaged relative humidity. These simplifications are responsible for the high diversity of results generated by the different models.

Several studies have addressed the methods of convective parameterization used in LAM and GCM (Arakawa, 2004; Yano et al., 2005, 2013; Plant, 2010; De Rooy et al., 2013; Plant and Yano, 2015). The need to improve the representation of convection in large-scale models led to the idea of the "superparameterization" of small and mesoscale processes in large-scale models and GCM (Grabowski and Smolarkiewicz, 1999; Khairoutdinov and Randall, 2001; Khairoutdinov et al., 2005; Grabowski, 2001, 2003, 2006). Superparameterization entails explicitly reproducing clouds on high-resolution 2D meshes implemented within grid columns of GCM. Convective heating rate, precipitation rate, and other parameters characterizing the combined effect of clouds on large-scale processes are obtained by horizontal averaging over these 2D grids and then transferring them to the GCM. In this way, traditional parameterization of convection is replaced by explicitly calculated heating/cooling profiles. The rapid and continuous increase in computing capabilities has made it feasible to utilize cloud-resolving grid spacing in regional numerical weather predictions, or within nested meshes of GCM. A serious drawback of the current superparameterization schemes, however, is the utilization of 2D grids with fixed orientation for description of convection. Such schemes have difficulty simulating convective processes accompanied by formation of such coherent structures as squall lines, supercell storms, and tornadoes.

7.1.2 Cloud-Resolving Models and Atmospheric Modeling Systems

Utilization of grid spacing of less than about 2–3 km permits the resolution of individual clouds, at least the largest ones. Models with such a better resolution are

CRM. Explicit simulations of clouds allow for the handling of all processes associated with clouds, including thermodynamics and radiation, as well as microphysical processes.

State-of-the-art CRM can be divided into several groups. In the first group there are the multipurpose models known as atmospheric system models. Among such models are the WRF model (Skamarock et al., 2008), the RAMS (Cotton et al., 2003), and the COSMO (Baldauf et al., 2011). These models allow simulations of both small-scale processes such as single clouds, and mesoscale processes such as thunderstorms, squall lines, and precipitation events across regions. Some of the models are designed for both atmospheric research and operational forecasting needs.

A striking example of such a model is the WRF, developed at the National Center for Atmospheric Research (NCAR). The WRF model allows researchers to perform simulations using either real data or idealized atmospheric conditions. It provides a flexible and robust platform for operational forecasting, while offering advances in physics, numerical approaches, and data assimilation. This model has a large global community of users (more than 23,000 registered users in more than 150 countries). The principal components of the WRF system are depicted in **Figure 7.1.1**. The WRF Software Framework (WSF) provides infrastructure that accommodates dynamics solvers, physics packages that interface with the solvers, programs for initialization, data assimilation WRF–Var (Variational) and description of Atmospheric Chemistry, and WRF–Chem. The Advanced Research WRF solver (ARW) that performs an integration of the model equation system, and the physics packages that describe different physical processes such as microphysics in clouds, processes in the boundary layer, and radiation constitute the main apparatus of WRF software infrastructure. The main characteristics of the ARW are presented in **Table 7.1.1**. The model physics includes cumulus parameterization, microphysics, surface physics, physics of the boundary layer, radiation, and atmospheric chemistry. The main characteristics of the WRF Physical Package are described next.

Cumulus parameterization. Cumulus parameterization schemes are used in mesoscale and large-scale models that do not resolve clouds. The package includes convective adjustment and mass-flux schemes. In convective adjustment schemes, convection is thought to nudge vertical profiles of temperature and humidity to a neutral state. In mass-flux schemes, the integral effect of convective clouds is determined by construction of cloud ensembles consisting of clouds of different heights. The distribution of clouds with respect to height is determined using elementary cloud models, and applying a closure hypothesis to analytically describe the relationship between large-scale processes and cumulus-cloud convection (for more details, see Emanuel and Raymond, 1993; Pielke and Pearce, 1994). The WRF package includes Kain–Fritsch's (Kain, 2004) mass flux schemes, the Grell–Devenyi scheme, and the scheme known as Grell–3 (Grell and Devenyi, 2002). The convective adjustment is represented by the Betts–Miller–Janjic scheme (Janjic, 2000).

Microphysics. WRF microphysics include microphysical schemes of varying complexity, ranging from simplified physics suitable for numerical weather prediction requiring high computational efficiency to sophisticated mixed-phase physics suitable for time-consuming scientific investigations. The package includes the option to use different bulk-parameterization schemes (e.g., Kessler, Purdue Lin, WSM3, WSM5, WSM6, Eta GCP,

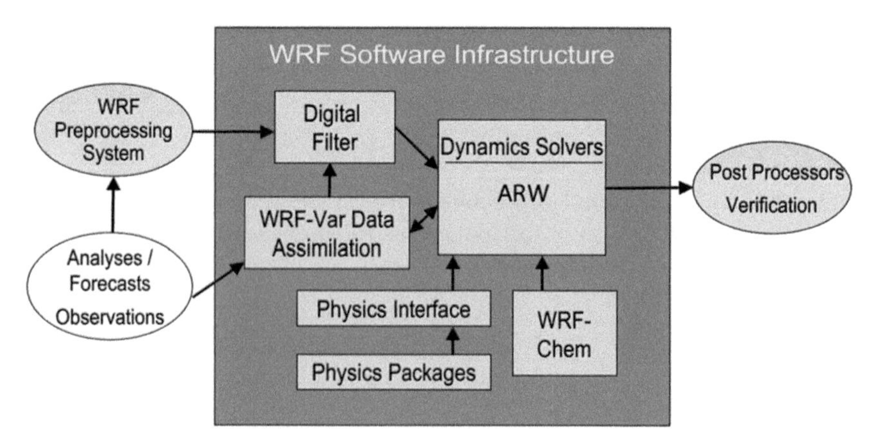

Figure 7.1.1 The principal components of the WRF system (from Skamarock et al., 2008; courtesy of University Corporation for Atmospheric Research).

Table 7.1.1 The main characteristics of the ARW system (from Skamarock et al., 2008; courtesy of University Corporation for Atmospheric Research).

Characteristic	Description
Prognostic variables	Horizontal velocity components u and v in Cartesian coordinate, vertical velocity w, perturbation of potential temperature, perturbation of geopotential, perturbation surface pressure of dry air. Optionally, turbulent kinetic energy, any number of scalars such as water-vapor mixing ratio, rain/snow-mixing ratio, cloud water/ice-mixing ratio, and chemical species and tracers. In the WRF/SBM version, size distribution functions of drops, ice crystals, snow (aggregates), graupel, and hail (optionally). Size distribution of cloud condensational nuclei.
Equation system	Fully compressible, Euler nonhydrostatic with a run-time hydrostatic option available. The equations are written in the flux form (see Chapter 4). Conservative for scalar variables.
Vertical coordinate	Terrain-following, dry hydrostatic-pressure vertical coordinate η is defined as $\eta = (p_h - p_{h<\$\$\$>})/(p_{h<\$\$\$>} - p_{h<\$\$\$>})$, where p_h is the hydrostatic component of the pressure, p_{hs} and p_{ht} are pressures along the surface and top boundaries, respectively. Top of the model is a constant pressure surface. Vertical grid stretching is permitted.
Horizontal grid	Arakawa C–grid staggering (see Sections 4.4 and 4.5).
Time integration	Time-split integration using a 2nd- or 3rd-order Runge–Kutta scheme with smaller time step for acoustic and gravity-wave modes (see Section 4.5). Variable time step capability.
Lateral boundary conditions	Periodic, open, symmetric, and specified options available.
Top boundary conditions	Gravity waves absorbing (diffusion Rayleigh damping or implicit Rayleigh damping for vertical velocity) (see Section 4.5). Constant pressure level at top boundary along a material surface. Rigid lid option.
Bottom boundary conditions	Physical boundary conditions (heat and moisture fluxes, wind stresses, etc.) are calculated using boundary layer and soil model schemes. The utilization of free-slip conditions is possible.
Earth's rotation	Coriolis terms are included.
Mapping to sphere	Four options for map projections for real data simulations: polar stereographic, Lambert conformal, Mercator, and latitude-longitude (allowing rotated pole). Curvature terms are included.
Nesting	One-way interactive, two-way interactive and moving nests (see Section 4.5). Multiple levels and integer ratios.
Nudging	Grid (analysis) and observation nudging capabilities available.
Numerical schemes for advection	Options for 2nd–6th order of approximation of advection in horizontal and vertical.
Model filters and turbulent mixing	Subgrid scale turbulence formulation in both coordinate and physical space. Divergence damping, external-mode filtering (see Chapters 3, 4).
WRF software framework	Highly modular codes, separation of scientific codes from parallelization, distributed and shared memory, vector and scalar

Thompson, Goddard, and Morrison 2-Moment). The number of parameterization schemes continuously increases. These schemes are described in Section 7.2. The SBM scheme implemented into the WRF allows calculation of size distribution functions of liquid and ice hydrometeors, as well as size distribution functions of cloud condensational nuclei (Khain et al., 2004a).

Surface physics. The WRF contains a number of schemes to determine surface fluxes. It includes different multilayer land surface models, ranging from a simple thermal model to full vegetation and soil-moisture models, including snow cover and sea ice. The surface layer schemes based on the Monin–Obuhov similarity theory calculate friction velocities and exchange coefficients that enable the calculation of surface heat and moisture fluxes in the land-surface models, and surface stress in the planetary boundary layer scheme. Over water surfaces, the surface fluxes and surface diagnostic fields are computed. The Land-Surface Models (LSM) use atmospheric information from the surface-layer scheme, radiative forcing from the radiation scheme, and precipitation forcing from the microphysics and convective schemes, together with internal information on the land's state variables and land surface properties, to provide heat and moisture fluxes over land points and sea ice points. These fluxes play the role of boundary conditions at the atmosphere–land or atmosphere–sea interfaces. Also, there is an option to use one of four land surface models. For simulation of ocean mixing under the influence of hurricanes, the ocean mixed-layer model is included. In the mixed layer, the prognostic variables are the layer depth, vector of horizontal current, and mean temperature taken to be the sea surface temperature (SST).

Atmospheric radiation physics includes longwave and shortwave radiation schemes with multiple spectral bands and a simple shortwave scheme suitable for climate and weather applications. Cloud effects and surface fluxes are included.

Chemical processes are calculated in a WRF–Chem module (Figure 7.1.1). This includes a description of processes such as dry deposition, coupled with the soil/vegetation scheme, biogenic emissions, anthropogenic emissions, gas-phase chemical reactions, photolysis, aerosol chemistry, and tracer transport. More details about the WRF model can be found in the study by Skamarock et al., (2008)

Other cloud resolving atmospheric modeling systems are in some respects (e.g., a non-hydrostatic equation system, nested grid structure) structurally similar to the WRF. For instance, RAMS, developed at Colorado State University, may be configured to cover an area as large as a planetary hemisphere for simulating mesoscale and large scale atmospheric systems, but may also be used for simulations of tornadoes and boundary layer eddies, as well as motions of small scale like turbulent flows over buildings. Two-way interactive grid nesting in RAMS allows the resolution of coherent-like atmospheric systems such as thunderstorms, while simultaneously modeling the large scale environment of the systems on a coarser grid.

While most cloud resolving atmospheric modeling systems are not as universally applicable as the WRF, they do have their own specific features and preferred areas of application. For instance, RAMS uses bin-emulating microphysics (see Section 7.2), which is considered more accurate than many other bulk-parameterization schemes. This computationally time-consuming microphysical scheme renders RAMS unsuitable as a weather forecast model. Nonetheless, RAMS is widely used in research dedicated to cloud–aerosol interaction at different scales (see, for example, Saleeby and Cotton, 2004, 2008; Saleeby and van den Heever, 2013).

Regional weather forecasts in the German Weather Service typically rely on the COSMO model. COSMO is also used for climate limited-area modeling. The research version of COSMO contains an advanced bulk microphysics parameterization scheme (Section 7.2) and is used to investigate problems such as the effects of aerosols on precipitation (Seifert et al., 2012).

7.1.3 Cloud Models

Cloud resolving atmospheric modeling systems allow simulation of mesoscale cloud-related phenomena at scales of hundreds and thousands of kilometers as well as the crude simulation of isolated clouds. There are models that are specifically designed for the simulation of isolated clouds and cloud ensembles. These models are often referred to as cloud models. Examples are the Goddard Cumulus Ensemble (GCE) model developed at NASA Goddard Space Center (Tao et al., 2003), the System for Atmospheric Modeling (SAM), the HUCM, and the cloud model of Tel Aviv University (TAU model). Accurate descriptions of microphysical processes are the hallmark of cloud models. At the same time, the descriptions of other processes are typically simpler than in WRF and RAMS. For instance, most of these models use the z-coordinate system, and do not take into account topography. The 3D SAM uses periodic lateral boundary conditions. The lateral boundary conditions of the GCE model can be chosen to be cyclic, open, or mixed (cyclic in one lateral boundary and open in another). In the HUCM, open lateral conditions (see Chapter 4) are applied.

The grid spacing of these models in different simulations varies from a few tens of meters to a few kilometers. The GCE uses a stretched vertical coordinate with height increments increasing from 20 m near the surface to 1,150 m at the upper levels. Time steps used in cloud models are of a few seconds duration; the duration of process simulation is several hours. Such models are widely used for simulation of single clouds and cloud systems documented in field experiments and in situ measurements (Ovchinnikov et al., 2011, 2014; Fan et al., 2009a, 2009b, 2011; Kogan et al., 2012; Khain et al., 2013), and in the investigation of the fine effects of cloud–aerosol interaction in warm and mixed-phase convective and stratocumulus clouds (Benmoshe et al., 2012; Benmoshe and Khain, 2014; Kumjian et al., 2014; Ovchinnikov et al., 2014).

Simulations of convection in these models are usually performed using one of two approaches. In the first approach, cloud systems develop under the influence of large-scale flow. Continuous large-scale forcing is applied by varying one or more of the following: atmospheric instability, wind speed and direction, surface temperature, and radiation during long-term simulations. Such convection might be termed "continuously forced convection." In the second approach, there is no dynamical large-scale forcing, and clouds and cloud systems develop in the wake of initial convective instability given by profiles of temperature, humidity, and background wind. Such convection might be termed "self-forced convection."

The GCE model is available in 2D and 3D versions. The HUCM is a 2D model. There are two dynamical versions of the TAU cloud model: 2D axisymmetric and 2D slab symmetric. Given the typical availability of computer resources, 2D models more readily permit the use of the very high model resolution needed to simulate fine features of cloud evolution and precipitation formation, such as high spatial and time variability of cloud structure, and hydrometeor recirculation in updrafts and downdrafts. These models also allow the utilization of complicated microphysical schemes in which size distributions of many hydrometeors are explicitly calculated, and each cloud particle – in addition to major parameters such as mass and size – is characterized by aspect ratio, capacitance, liquid-water fraction in ice particles, rimed mass within snow, and so forth.

SBM schemes have been developed in only a very few scientific centers (see Section 7.2). Rich in microphysics, the HUCM and TAU cloud models have turned out to be laboratory tools for the development, modification, and testing of spectral bin microphysics vs. observations.

At present, there exists a high level of international collaboration for developing and utilizing atmospheric system models. Leading meteorological centers such as NCAR and NCEP have developed dynamical and numerical models such as WRF, principal components of which are shown in Figure 7.1.1. Each such model provides a dynamical platform or skeleton, including an advanced dynamical core. Such systems are made up of modules containing descriptions of different physical processes. The microphysical schemes implemented into WRF, for instance, were developed in the United States, Germany, Korea, Israel, and other countries. The SBM packages developed for the HUCM and TAU models are used in the WRF model (Khain et al., 2010; Lynn et al., 2014; Sarkadi et al., 2016), RAMS (Lebo and Seinfeld, 2011), the GCE model (Tao et al., 2003; Iguchi et al., 2012a, 2012b, 2014), and the SAM model (Fan et al., 2009a, 2009b, 2011; Ovchinnikov et al., 2011, 2014; Wang et al., 2011).

7.1.4 LES and LEM Cloud Models

Cloud models with grid spacing of tens of meters and smaller are referred to as Large Eddy Simulation (LES) models. Typically, LES models are used to simulate stratocumulus clouds in the atmospheric BL, where vertical mixing is determined by vortices with scales of the BL thickness (large eddies) (Stevens et al., 2005). Such high resolution is necessary to accurately represent microphysical processes that take place at very short spatial scales, such as drizzle within a shallow cloud layer with low vertical velocities and supersaturations. Drizzle-formation investigation demands a very accurate description of diffusion growth, collisions, and aerosol effects. In 3D LES models, velocities are calculated using the Navier–Stokes equations (Section 3.2) and model dynamics is coupled with microphysics and radiation schemes. Such models are often used for the determination of spatial scales of drizzle zones and investigations of instabilities in the BL, as well as effects of aerosols on the dynamics of Sc and Cu (e.g., Wang and Feingold, 2009a, 2009b; Feingold et al., 2010; Kazil et al., 2011; Wang et al., 2011; Yamaguchi and Feingold, 2013; Dagan et al., 2016). Feingold et al. (2010) used an LES model of the cloud system to demonstrate the mechanism of open-cellular system oscillation. The model domain size used was 60 km × 60 km × 1.5 km. The grid spacing was 300 m in the horizontal direction and 30 m in vertical direction, and the time step was 3 s. Feingold et al. (2010) examined how precipitating clouds produce an open-cellular cloud pattern that oscillates between different, weakly stable states. The oscillations are a result of precipitation causing downward motion and outflow from clouds that were previously positively buoyant. The evaporating precipitation drives air down to Earth's surface, where the air flows diverge and collide with the outflows of neighboring precipitating cells. These colliding outflows form surface convergence zones and new clouds. In turn, the newly formed clouds produce precipitation and new colliding outflow patterns that are displaced from the previous ones. As successive cycles of this kind unfold, convergence zones alternate with divergence zones and new cloud patterns emerge to replace old ones.

The Eulerian approach, when the values of different quantities are calculated in a given point of a finite difference grid, does not allow for an accurate description of drop diffusion growth, a fact that inevitably leads to artificial drop spectrum broadening (see Section 5.4). To reduce errors in the representation of

DSD evolution, trajectory ensemble models (TEM) have been developed (Stevens et al., 1996; Feingold et al., 1998; Harrington et al., 2000; Cooper et al., 2011; Erlick et al., 2005). A typical TEM consists of two components, namely, an LES model used to create a "background" field of velocity, temperature, and humidity; and a model of a Lagrangian air parcel moving within this background field. TEM yield highly accurate microphysical descriptions in parcel models. Such models use movable mass grids, in which the mass of bins changes according to the diffusion growth equation, thus eliminating artificial DSD broadening. This approach allows for the simulation of DSD formation in parcels with different histories, and also for investigating the mechanisms of DSD variability. In TEM, cloud parcels are isolated and far apart. Hence, the application of sedimentation schemes is inconsistent with the assumptions.

Exceptionally accurate descriptions of microphysical processes have been achieved using the LEM of boundary layer (Pinsky et al., 2008; Magariz-Ronen et al., 2014, 2016) (see also Sections 5.10 and 5.11). We present the model description in **Appendix 7.1.5**. In the model, about 2,000–3,000 Lagrangian adjacent parcels moving within a turbulent-like flow cover the entire computational area. The time-dependent turbulent-like velocity field generated by the model reproduces observed turbulent intensity and correlation properties. In these studies, diffusion growth and collisions of drops, as well as drop sedimentation from one parcel to another, were calculated at mass grids containing 500 mass bins. Turbulent mixing of thermodynamic and microphysical quantities between neighboring parcels was also modeled. Since drop sedimentation is calculated within the Eulerian framework, the model is referred to as a hybrid Lagrangian–Eulerian model of stratocumulus clouds.

There is another type of models referred to as the Lagrangian Cloud Models (LCM). LCM aim at a description of the processes of DSD formation, paying special attention to the role of mixing in these processes (Andrejczuk et al., 2008, 2009, 2010; Shima et al., 2009; Riechelmann et al., 2012). The leading idea of LCM is to calculate growth of droplets moving within the velocity field by following the trajectory of each droplet. Taking into account a huge number of droplets in real clouds, the sample of real droplets is replaced by a sample of super droplets which number much lower than that of droplets (Shima et al., 2009). Each super droplet represents a large number of real droplets of equal size and location. It is assumed that droplets belonging to the same super droplet will behave in an identical manner regarding growth, sedimentation, and movement. The radii of super droplets changed in the course of condensation/evaporation processes are calculated using Equation (5.1.14), taking into account curvature and chemical terms. The process of collisions in LCM is also described by the collision of super droplets. A special Monte-Carlo-type algorithm was developed for this purpose (Shima et al., 2009). The LCM allows calculating droplet size distribution on spatial scales of the order of tens to 100 m. It has been demonstrated that LCM can be successfully used in the investigation of the microphysical processes of cloud–aerosol interaction in warm stratocumulus clouds. These potentially powerful models are currently at the development stage, and further research is required to properly take into account processes related to collisions, droplet nucleation, and the formation of raindrops.

Many 1D models of parcels or plumes exist. Most of these are used for simulation of particular microphysical processes such as nucleation, diffusional growth, collisions, and sedimentation. Some 1D models describe a complex of microphysical processes that end with the formation of precipitating particles. These models can be considered dynamically simple cloud models. Examples are warm microphysics models (e.g., Pinsky and Khain, 2002; see Section 5.11) and mixed-phase microphysics models (e.g., Heymsfield and Miloshevich, 1993). 1D models with advanced microphysics are often used as a benchmark for different parameterization schemes.

7.1.5 Appendix. Brief Description of a Hybrid Lagrangian–Eulerian Model of Warm Stratocumulus Clouds (Pinsky et al., 2008)

The 2D LEM consists of two main parts: a dynamical part determining velocity field, and a microphysical part.

Model dynamics. The velocity field in the LEM caused by different types of eddies and small-scale turbulence is represented by the sum of a great number of harmonics, including those representing large eddies with the scales of the BL depth. The mean horizontal velocity $v_0(z)$ is also taken into account. The expressions for the vertical $w(x,z,t)$ and the horizontal $v(x,z,t)$ velocity components are as follows:

$$w(x,z,t) = f(z)\sum_{m=1}^{M}\sum_{n=1}^{N} C_n D_m \left[a_{mn}(t)\sin\frac{2\pi mx}{L} + b_{mn}(t)\cos\frac{2\pi mx}{L}\right]\sin\frac{\pi nz}{H} \quad \text{(A1)}$$

$$v(x,z,t) = \sum_{m=1}^{M}\sum_{n=1}^{N} C_n D_m \frac{n}{m}\frac{L}{2H}$$
$$\left[a_{mn}(t)\cos\frac{2\pi mx}{L} - b_{mn}(t)\sin\frac{2\pi mx}{L}\right]\cos\frac{\pi nz}{H}$$
$$\times\left[\frac{df(z)}{dz}\sin\frac{\pi nz}{H} + \frac{\pi n}{H}f(z)\cos\frac{\pi nz}{H}\right] + v_0(z) \quad \text{(A2)}$$

where M and N are the numbers of harmonics in the horizontal and the vertical directions, respectively; L and H are the sizes of the computational area in the horizontal and vertical directions, respectively. Random fluctuations of the velocity field in time are described by independent stationary auto-regression series of the first-order $a_{mn}(t)$ and $b_{mn}(t)$. Auto-regression represents a discrete version of the Langevin equation widely used to describe turbulent diffusion. The coefficients D_m are normalized to obey the condition $\sum_{m=1}^{M} D_m^2 = 1$. The velocity field (Equations A1 and A2) obeys zero boundary conditions at upper and low boundaries

$$w(x,0,t) = w(x,H,t) = 0 \quad \text{(A3)}$$

and periodical boundary conditions at lateral boundaries

$$\begin{aligned} w(0,z,t) &= w(L,z,t) \\ v(0,z,t) &= v(L,z,t) \end{aligned} \quad \text{(A4)}$$

The mean values of velocity components are equal to $\langle w\rangle = 0$, $\langle v\rangle = v_0(z)$. The velocity variations are expressed as

$$\langle w^2\rangle = f^2(z)\sum_{n=1}^{N} C_n^2 \sin^2\frac{\pi nz}{H} \quad \text{(A5)}$$

$$\left\langle (v-v_0)^2\right\rangle = \left(\frac{L}{2\pi}\right)^2 \sum_{m=1}^{M}\left(\frac{D_m}{m}\right)^2 \sum_{n=1}^{N} C_n^2$$
$$\times\left[\frac{df(z)}{dz}\sin\frac{\pi nz}{H} + \frac{\pi n}{H}f(z)\cos\frac{2\pi nz}{H}\right]^2 \quad \text{(A6)}$$

The mean values and variations depend only on the vertical coordinate z.

The correlation functions of velocity components along the horizontal direction are given as

$$B_w(\Delta x, z) = f^2(z)\left[\sum_{n=1}^{N} C_n \sin\frac{\pi nz}{H}\right]^2 \sum_{m=1}^{M} D_m^2 \cos\frac{2\pi m\Delta x}{L} \quad \text{(A7)}$$

$$B_v(\Delta x, z) = \left(\frac{L}{2H}\right)^2 \sum_{n=1}^{N} C_n^2 \left[\frac{df(z)}{dz}\sin\frac{\pi nz}{H} + \frac{\pi n}{H}f(z)\cos\frac{\pi nz}{H}\right]^2$$
$$\times \sum_{m=1}^{M}\left(\frac{D_m}{m}\right)^2 \cos\frac{2\pi m\Delta x}{L} \quad \text{(A8)}$$

The correlation functions depend only on $\Delta x = x_2 - x_1$, indicating that velocity fields are statistically homogeneous in the horizontal direction. The cross-correlation functions of the velocity components along the horizontal direction also depend only on $\Delta x = x_2 - x_1$:

$$B_{vw}(x_1, x_2, z) = B_{vw}(\Delta x, z);\quad B_{vw}(0,z) = 0 \quad \text{(A9)}$$

As a result, correlations between two velocity components in one and the same point are equal to zero. Thus, the velocity field (Equations A1 and A2) is stationary and statistically uniform in the horizontal direction. The field obeys ergodic properties, namely, averaging with respect to realizations is equivalent to averaging over a long enough time period.

The model allows turning of the dynamics to the observed data. The velocity field (Equations A1 and A2) contains three groups of parameters to be calculated from observations: coefficients $C_n, n = 1\ldots N$, $D_m, m = 1\ldots M$, and the characteristic correlation time. The function $f(z)$ is used to tune the profile of the vertical velocity variation to the observed profile $\langle w_0^2\rangle^{1/2}$. First, the coefficients C_n are calculated as

$$C_n = \frac{1}{H}\int_{-H}^{H} \langle w_0^2\rangle^{1/2} \sin\frac{2\pi nz}{H} dz \quad \text{(A10)}$$

Then the function $f(z)$ is calculated as

$$f(z) = \frac{\sum_{n=1}^{N} C_n \sin\frac{\pi nz}{H}}{\left(\sum_{n=1}^{N} C_n^2 \sin^2\frac{\pi nz}{H}\right)^{1/2}} \quad \text{(A11)}$$

The correlation properties of the vertical velocity in the horizontal direction are determined by the lateral correlation function $B_w(\Delta x, z)$ or the lateral structure function $D_w(\Delta x) = \left\langle [w(x_1) - w(x_2)]^2\right\rangle$ of the vertical velocity in the horizontal direction (see Section 3.3). Using the correlation or the structure functions obtained from observations, coefficients D_m are calculated as

$$D_m(z) = \frac{1}{L} \int_{-L/2}^{L/2} \frac{B_w(\Delta x)}{B_w(0)} \cos \frac{2\pi m x}{L} dx$$
$$= \frac{1}{L} \int_{-L/2}^{L/2} \left[1 - \frac{D_w(\Delta x)}{D_w(\infty)} \right] \cos \frac{2\pi m x}{L} dx \quad \text{(A12)}$$

The characteristic time scales of the velocity fluctuations of different spatial scales are also subject to adaptation. The characteristic time is assumed to meet the Kolmogorov relationship for each harmonic:

$$\gamma_n = \varepsilon^{-1/3} l^{2/3} = \varepsilon^{-1/3} (H/n)^{2/3} \quad \text{(A13)}$$

where ε is the dissipation rate (see Section 3.3). So, three observed quantities are required to generate time-dependent turbulent-like velocity fields, namely, observed profile $\langle w_0^2 \rangle^{1/2}$, observed lateral structure function $D_w(\Delta x)$, and turbulent dissipation rate ε. Dynamical parameters used in different simulations are given in **Table A7.1.5**.

The LEM Microphysics

Diffusional growth. Full equation for diffusional drop growth (Equation 5.1.12) that includes "curvature" and chemical terms is used. Each particle is an aqueous solution of salt, depending on the chemical composition of CCN. The growth of particles of all sizes is calculated using the same equation, irrespective of whether the particle is a drop or a haze.

Collision of drops is calculated by solving the stochastic collision equation (Equation 5.6.5). The collision droplet growth is calculated using a collision efficiency table with a high 1-μm resolution in droplet radii (**Appendix B, Tables B1, B2, B3**), Pinsky et al., 2001).

Sedimentation of drops from parcels located above to neighboring adjacent parcels located below is taken into account. To calculate changes of DSD as a result of sedimentation, the entire computational area is covered by a grid with resolution of 1 m (hurricane BL) or 5 m (Sc). The interfaces between parcels are determined using this grid (see Section 5.5.3 and Figure 5.5.6).

Table A7.1.5 Dynamical parameters used in simulations using the LEM.

Parameter	Value, object of simulation
Characteristic size of air parcels	6 m (fog, hurricane BL), 25–40 m (Sc)
Number of parcels	1,200–9,000 in different simulations
Length of the area L	300–600 m (hurricane BL), 2,550–20,000 m (Sc)
Height of the area H	400 m (hurricane BL), 850–1250 m (Sc)
Maximum r.m.s vertical velocity fluctuation	0–3.5 m/s
Number of harmonics M, N	50–500
Lifetime of harmonics	1–1,000 s
Time period of updating of velocity field	0.1 s
Turbulent dissipation rate	10–2,000 cm^2/s^3 (hurricane BL), 5–10 $cm^2\ s^{-3}$ (Sc)

Table B7.1.5 Model parameters.

Number of mass bins	500
Range of cloud particles radii, μm	0.01–3,000 μm
Time steps for diffusion growth	0.01 s
Time steps for collisions, mixing, and sedimentation	1.0 s
Chemical composition	NaCl, option to use another chemical agent
Initial data	Size distributions of dry aerosol, vertical profiles of temperature and humidity. Initially all parcels are subsaturated and contain only wet aerosols (haze particles)

Turbulent mixing between Largangian parcels is described following Pinsky et al. (2010) and Magaritz-Ronen et al. (2014) (see Section 5.10.2). The mixing represents an extension of the K-theory to the case of nonconservative values, such as drop-size distributions. The latent heat release during the mixing process is taken into account.

Surface fluxes of heat and moisture are calculated with the help of aerodynamic formulas using the wind velocity and temperature gradients in the lowest 10 m-depth layer. The main physical and microphysical parameters used in simulations of marine Sc and sea-spray induced fog are presented in **Table B7.1.5**.

7.2 Two Methodologies in Cloud Microphysics: Bulk Schemes and Bin Schemes

7.2.1 Main Definitions

Microphysical processes in cloud-resolving models are described and simulated using two main approaches: bulk microphysics parameterization (bulk parameterizations or bulk schemes) and SBM. In this section, we summarize specific features of both approaches and compare the results of cloud-resolving models that utilize them. Implementations of the two approaches in modeling of individual microphysical processes are discussed in Chapters 5 and 6. Both approaches describe the same microphysical processes and the outputs of the microphysical schemes are similar (e.g., in each approach cloud and rain-water content, cloud ice content, precipitation rates, etc. are calculated). However, the methodologies underlying these approaches are quite different, since they were initially developed to solve different theoretical and applied problems.

The major goal of bulk parameterization was to replace the traditional schemes of convective parameterization in cloud-resolving mesoscale and large-scale models by more accurate schemes treating microphysical processes (Section 7.1). Bulk-parameterization schemes aim at representing the most general microphysical cloud properties using a semiempirical description of PSD $f_k(m)$ (index k corresponds to particular hydrometeor type). This approach is assumed to be computationally efficient. Kessler (1969) developed the first bulk-parameterization scheme allowing simulation of warm cloud microphysics and cloud evolution in numerical models (Section 5.7).

The computational efficiency of bulk-parameterization schemes stems from implementation of microphysical equations for a few PSD moments rather than for PSD of different hydrometeor types themselves (e.g., cloud droplets, raindrops, ice crystals, aggregates, graupel, hail). The moments of PSD M^i (i is order of a moment) are defined in Section 2.1 (Equation 2.1.9). The schemes that use only one moment (typically, it is the mass content of hydrometeors $i = 1$) are known as one-moment or single-moment schemes, while the schemes using two moments (typically, number concentrations $i = 0$ and mass contents $i = 1$) are known as two-moment schemes. Less frequently, three-moment bulk schemes are used, in which the variables are number densities, mass content, and radar reflectivity ($i = 2$). The system of equations for the PSD moments is not closed, since the equations for i-th moment M^i include terms with a higher-order moment M^{i+1} (Seifert and Beheng, 2001). The closure problem is circumvented by representing PSD in the form of specific mathematical functions that are completely determined by only a few parameters and relations between their moments are known. A four-parameter Gamma distribution is typically used as the master function for PSD (Section 2.1).

The first bulk-parameterization scheme described only warm microphysical processes. Starting from the schemes developed by Lin et al. (1983) and Rutledge and Hobbs (1984), all bulk parameterizations describe both warm and ice processes and have been often used in different mesoscale models with spatial resolutions of several kilometers. Today, more than twenty different bulk-parameterization schemes are available. Simplified bulk-parameterization schemes were also implemented in climate models (e.g., Boucher and Lohmann, 1995; Lohmann and Feichter, 1997; Ghan et al., 2001). Despite the significant variety of components, all bulk-parameterization schemes share a basic assumption concerning the PSD shape. Even in bulk-parameterization schemes containing more than ten hydrometeor types, the PSD of particles belonging to each hydrometeor type are approximated by exponential or the Gamma distributions, or much more rarely by the lognormal distribution. The parameters of these distributions are chosen based on the type of the cloud system being simulated.

The second main approach in microphysical description is SBM, also referred to as explicit microphysics, bin microphysics, and bin-resolving or size-resolving microphysics. SBM aims at simulating microphysical and precipitation processes in clouds of different types as accurately as possible. Calculation of PSD by solving explicit microphysical equations is fundamental to bin microphysics. Hence, no a priori information about the PSD shape is required or assumed. Instead, PSD are defined and calculated on a finite-difference mass grid containing several tens to several hundred mass bins. Contemporary bin-microphysics schemes differ substantially from the early generations (Clark, 1973; Takahashi, 1976; Kogan et al., 1984; Tzivion et al., 1987). The difference mainly lies in the level of specification in descriptions of microphysical processes. The dramatic increase in computing power has made it possible to explicitly describe cloud microstructure using all the knowledge previously accumulated in cloud physics. SBM schemes have not only improved the representation of separate microphysical processes, but have also been successfully implemented in cloud models and cloud-resolving mesoscale models for simulation of a wide range of meteorological phenomena (Section 7.2.4).

Equations used in bin microphysics do not depend on particular meteorological situations, which makes them universal, i.e., the same scheme can be used without any modification to simulate different atmospheric phenomena from stratiform arctic clouds to tropical cyclones. The number of equations to be solved in bin microphysics schemes is proportional to the number of bins and the number of hydrometeor types (including aerosols) and, therefore, computational requirements are substantially higher than in bulk-parameterization schemes. The bin-microphysics approach requires five to fifty times more computer time than contemporary bulk-parameterization schemes.

In addition, there are so-called hybrid schemes that actually represent modifications of bulk parameterization combined with some SBM features. Onishi and Takahashi (2011) developed a scheme describing warm microphysical processes via the SBM approach and processes related to ice formation and evolution via bulk parameterization. This combined method enabled them to achieve higher accuracy in drop-formation treatment and deal with significant uncertainties regarding ice-formation treatment.

Another example of a hybrid scheme is the bin-emulating approach similar to that used in the RAMS (e.g., Meyers et al., 1997; Cotton et al., 2003; Carrió et al., 2007), as well as by Heymsfield and Sabin (1989) and Feingold and Heymsfield (1992). To calculate the rates of microphysical processes in the bin-emulating procedure, the Gamma distributions are discretized into bins. Then, the rates of various microphysical processes are calculated in the way similar to that in the SBM approach using this discretization. In the calculation the rates of the microphysical processes obtained in offline calculations with a parcel SBM model for a wide range of atmospheric conditions are widely used. The results obtained for individual microphysical processes (e.g., collisions, sedimentation) are tabulated in lookup tables incorporated within the bulk-microphysical module of the 3D RAMS model. Finally, the masses and concentrations calculated after each microphysical time step in each bin are recombined to construct new Gamma distributions to be used at the next advection substep.

The choice of an appropriate microphysical approach depends on the objectives of a particular study and requirements imposed on its scope (for instance, the rate of calculation, the computer power, etc.).

7.2.2 Spectral Bin Microphysics (SBM)

The kinetic equations for PSD of the k-th hydrometeor type $f_k(m)$ used in SBM models can be written in the form (see Equation 3.2.15)

$$\frac{\partial \rho f_k}{\partial t} + \frac{\partial \rho u f_k}{\partial x} + \frac{\partial \rho v f_k}{\partial y} + \frac{\partial \rho (w - V_t(m)) f_k}{\partial z}$$
$$= \rho \left\{ \left[\frac{\delta f_k}{\delta t}\right]_{nucl} + \left[\frac{\delta f_k}{\delta t}\right]_{c/e} + \left[\frac{\delta f_k}{\delta t}\right]_{d/s} + \left[\frac{\delta f_k}{\delta t}\right]_{f/m} + \left[\frac{\delta f_k}{\delta t}\right]_{col} \right\}$$
$$+ \cdots + \frac{\partial}{\partial x_j}\left(K \frac{\partial}{\partial x_j} \rho f_k \right) \qquad (7.2.1)$$

where u, v, and w are components of wind speed and V_t is the fall velocity that depends on the mass and type of the hydrometeor, on the characteristics of particle shape, and on the air density ρ. The terms on the right-hand side of Equation (7.2.1) determine the rates of different microphysical processes such as nucleation $[\bullet]_{nucl}$, condensation/evaporation $[\bullet]_{c/e}$, deposition/sublimation $[\bullet]_{d/s}$, freezing/melting $[\bullet]_{f/m}$, collisions $[\bullet]_{col}$, etc. The last term determines the changes in PSD caused by turbulent mixing with the turbulent coefficient K. The methods used to calculate the rates of microphysical processes are discussed in Chapters 5 and 6. Statistically significant PSD require a sufficient number of particles. Concentrations of large hydrometeors such as raindrops and hail are low, so determination of PSD requires counting particles within volumes of several cubic meters and more. Consequently, Equation (7.2.1) is a result of spatial averaging over corresponding volumes. This averaging is consistent with a description of mixing using the K-theory, which assumes subgrid fluxes to be proportional to gradients of model-resolvable values.

Two main schemes are used to compute the evolution of PSD described by Equation (7.2.1), BM and MMM. In publications, BM is often referred to as SBM.

Bin microphysics (BM). This method dates back to the classic studies by Berry and Reinhardt (1974a, 1974b, 1974c) and defines PSD on the logarithmic equidistance mass grid containing several tens of bins (Section 2.1). Instead of PSD $f(m)$ the function

$g(\ln r) = 3m^2 f(m)$ is applied. The conservation equation for function g on the mass grid is written as

$$\frac{\partial g_{i,k}}{\partial t} + \frac{\partial u g_{i,k}}{\partial x} + \frac{\partial v g_{i,k}}{\partial y} + \frac{\partial (w - V_{t,k}(m_i)) g_{i,k}}{\partial z} = \left[\frac{\delta g_{i,k}}{\delta t}\right]_{nucl} + \left[\frac{\delta g_{i,k}}{\delta t}\right]_{c/e} + \left[\frac{\delta g_{i,k}}{\delta t}\right]_{d/s} + \left[\frac{\delta g_{i,k}}{\delta t}\right]_{f/m} + \left[\frac{\delta g_{i,k}}{\delta t}\right]_{col} + \cdots + \frac{\partial}{\partial x_j}\left(K_\theta \frac{\partial}{\partial x_j} g_{i,k}\right), \qquad (7.2.2)$$

where i is the bin number and k denotes hydrometeor type. Actually, representation of PSD on the mass grid is similar to representation of continuous meteorological variables on finite-difference grids.

The main particle characteristic in BM schemes is particle mass. Actual particles are characterized by density, capacitance, shape, salinity, fall velocity, charge, etc., which can be different for different particles of the same mass. In the advanced BM schemes, ice particles are also characterized by rimed or liquid water fraction (e.g., Benmoshe et al., 2012; Phillips et al., 2014, 2015; Ilotoviz et al., 2016). The growth rates of particles of the same mass (and even of the same hydrometeor type) may differ due to different shape or different salinity. Strictly speaking, particles of the same mass should be further categorized by other parameters, and multidimensional PSD should be used. Several studies use 2D size distributions. For instance, to investigate the effects of aerosols on drop growth in an Sc boundary layer, Bott (2000) used 2D distributions by introducing categorization of drops with respect to their solute concentration. To investigate the effects of drop charge on drop collisions, Khain et al. (2004b) introduced categorization with respect to drop charge. However, utilization of multidimensional PSD is too complicated both mathematically and computationally. The typical approach is to use averaged values of the mentioned parameters for each mass bin. For instance, various fall velocities of snowflakes of the same mass are replaced by an averaged fall velocity of such snowflakes. Under this simplification, all particle parameters such as bulk density, equivalent radius (or diameter), shape, fall velocity, and other parameters are expressed via their mass using empirical relationships (Sections 6.1, 6.4). As a result, the PSD of a particular hydrometeor type is 1D, being dependent only on a particle mass. In this case, the number of equations for PSD is equal to the number of bins multiplied by the number of hydrometeor types.

The advantage of BM is its simplicity and amenability to modifications and novel implementations. For example, BM allows an easy implementation of new types of hydrometeors, as well as of new parameters of cloud particles. Besides, BM allows utilization of any type of collision kernels, including kernels in a turbulent flow that vary randomly in space and time. These advantages explain the high popularity of BM in different versions of the HUCM, as well as in mesoscale cloud-resolving models such as WRF, SAM, and RAMS (**Tables 7.2.1** and **7.2.2**).

The microphysical method of moments (MMM). The MMM approach is rooted in works by Young (1975) and Enukashvili (1980). Tzivion et al. (1987, 1989), Feingold et al. (1988), and Reizin et al. (1996) further developed MMM and implemented it in axisymmetric and slab-symmetric models. In MMM, the axis of mass is separated into categories (Section 2.1). The boundaries of the categories form a logarithmic equidistance grid. The PSD is assumed to be continuous within each category, for instance, having a certain linear mass dependence. PSD can be discontinuous at the boundaries of categories. In each category, the PSD is characterized by several moments. Tzivion et al. (1987, 1989) used two moments to describe PSD within each mass category, namely, the mass content and number concentration. So, the equations for each category are formulated not for PSD, as in the BM, but for its moments defined for each category (Section 2.1.3).

MMM uses the same assumption as the BM, namely, that properties of all particles belonging to the same category are similar. The form of the equations is similar to that of Equation (7.2.2). To simplify the calculations, Tzivion et al. (1987) used the Low and List (1982a, 1982b) collision kernel that allowed a preliminary calculation of collision integrals. The number of equations in MMM is equal to the product of the number of hydrometeor types, the number of categories, and the number of moments used for each category. The MMM algorithms for description of diffusion growth and collisions are discussed in Chapter 5. This MMM scheme has already been used in investigations of squall lines, orographic clouds, deep tropical convection, Mediterranean convective clouds, marine stratocumulus clouds, and warm cumulus using LES.

Properties and Applications of SBM Models

The properties and areas of applications of some models using SBM schemes are presented in Tables 7.2.1 and 7.2.2. They illustrate a continuous advancement of the bin-microphysics approach in two directions: improving the description of different microphysical processes such as melting, freezing, and ice nucleation, and broadening the area of SBM application from modeling single clouds to modeling TC and synoptic processes. Recently, SBM models have been used to solve various remote-sensing problems.

Table 7.2.1 Properties of different models using SBM schemes.

Authors	Type of SBM, name of model	Classes of hydrometeors	Number of bins or categories	Specific features of models
Young (1974, 1975)	MMM	5 classes: liquid water, freezing water, ice crystals, snowflakes, graupel	45	Parcel framework, dynamics are prescribed; number density of particles in each category is linearly proportional to their radius. Description of all microphysical processes is highly simplified.
Khvorostyanov et al. (1989, 1995)	BM	Drops and ice; small ice is assigned to ice crystals, large particles to graupel	31	2D; 10 bins of 2 μm and 21 bins with logarithmically increasing increments. Only one size distribution is used for all ice types.
Hall (1980)	BM	Drops and ice; small ice is assigned to ice crystals, large particles to graupel		Axi-symmetric, two size distributions (drops and ice). Density of ice particles increases with their size. Only one size distribution is used for all ice types.
Tzivion et al. (1987); Reisin et al. (1996)	MMM, TAU	Drops, ice crystals, snow and graupel	34	Axi-symmetric geometry, utilization of different size distributions for different hydrometeors, analytical calculation of supersaturation and diffusion growth, immediate melting at $T_C > 0$.
Kogan (1991)	BM	Drops and CCN	30 for drops, 19 for CCN	First 3D cloud model with bin microphysics, logarithmically equidistant mass grid.
Kogan et al. (1995)	BM; LES	Drops and CCN	25 for droplets, 19 for CCN	3D, logarithmically equidistant mass grid. The range of drop radius 1–256 μm, application of a method decreasing numerical DSD broadening
Ackerman et al. (1995)	BM, DHARMA	Drops and CCN	50	Logarithmically equidistant grid. Keeping track of the volume of dissolved CCN allowing to conserve solute mass. Maximum drop radius 500 μm. The radiative term is included in the droplet condensation equation.
Khain and Sednev (1996)	BM, HUCM	Drops, 3 types of ice crystals, snow, graupel, hail, CCN	33	2D, logarithmically equidistant mass grid. Each hydrometeor type is described by a separate size distribution. CCN budget included. Immediate melting at $T_C > 0$. Analytical calculation of supersaturation and diffusional growth. Solving SCE using Berry and Reinhard scheme (1974a,b).
Yin et al. (2000)	MMM, TAU	Drops, ice crystals, snow and graupel	36	2D slab symmetric geometry. Microphysics as in Tzivion et al., 1987. Time-dependent melting of graupel is included. Liquid water shed immediately forming two equal drops.

Khain et al. (2004); Phillips et al. (2007b)	BM, HUCM	Drops, 3 types of ice crystals, snow, graupel, hail, liquid mass in snow, graupel and hail, CCN	33	As in Khain and Sednev, 1996. Improved description of collisions (Bott, 1998). Collisional breakup and height-dependent collision kernels between drops and between drops and graupel are introduced. Liquid water mass within snow, graupel and hail is calculated to describe time-dependent melting.
Muhlbauer et al. (2010); Hashino and Tripoli (2007, 2008, 2011)	MMM, UWNMS	Drops, plates, rimed crystals, rimed aggregates, graupel	30 for drops, 20 for ice phase	Only one mass grid is used for all ice types. Contribution of pristine ice crystals dominates in the low-mass category, graupel contribute largely to category of large particle mass.
Khain et al. (2008)	BM, HUCM	Drops, 3 types of ice crystals, snow, graupel, hail, rimed mass in snow, CCN. Liquid water in snow, graupel and hail at $T_C > 0$	33	As in Khain et al., 2004. Improved scheme of diffusion growth, size distribution of rimed mass in snow are implemented. Snow density is calculated.
Lynn and Khain (2005, 2007)	BM, MM5/BM	Drops, 3 types of ice crystals, snow, graupel, hail, CCN	33	3D nested grid. First application of bin microphysics to simulate a mesosale phenonenon. Immediate melting is assumed.
Pinsky et al. (2008); Magaritz et al. (2009)	BM, LEM	Drops, haze, CCN	500	Lagrangian-Eulerian model of Sc clouds, drizzle formation with direct calculation of aerosol particles' growth. Utilization of movable mass grids to calculate diffusion growth of drops and haze particles. Solving full equation for diffusion growth including curvature and chemical effects.
Fan et al. (2009)	BM, SAM/BM	Drops, 3 types of ice crystals, snow, graupel, hail, CCN, IN	33	As in Khain et al., 2004. An aerosol-dependent and a temperature- and supersaturation-dependent ice nucleation scheme is implemented. IN size distribution are treated prognostically.
Khain et al. (2010)	BM, WRF/BM	Drops, snow, graupel/hail and CCN	33	3D nested grid. Bin microphysics is used to simulate hurricanes, spatial and time inhomogeneity of CCN concentration is taken into account.
Lebo and Seinfeld (2011)	MMM WRF/ MMM	Drops, ice crystals, snow and graupel	36	3D, Long (1974) collection kernel is used for collisions between drops. For ice-ice, ice-snow, ice-graupel, snow-graupel, snow-snow, liquid-ice, liquid-snow, liquid-graupel, graupel-graupel collisions the gravitational collection kernel is used.
Khain et al. (2011)	BM, HUCM	Drops, 3 types of ice crystals, snow, graupel, hail, rimed mass in snow, CCN. Liquid water in snow, graupel and hail at $T_C > 0$	43	As in Khain et al., 2008. Improved description of graupel-hail conversion. Recalculation of snow density by riming. Calculation of polarimetric parameters is included.

Table 7.2.1 (cont.)

Authors	Type of SBM, name of model	Classes of hydrometeors	Number of bins or categories	Specific features of models
Benmoshe et al. (2012, 2013)	BM, HUCM	Drops, 3 types of ice crystals, snow, graupel, hail, CCN. Liquid water in snow, graupel and hail at $T_C > 0$	43	As in Khain et al., 2011. Calculation of turbulence intensity and turbulent collision kernels are implemented.
Iguchi et al. (2012)	BM, JMA-NHM	Drops, 3 types of ice crystals, snow, graupel, hail, CCN	33	3D model, The ice nucleation rate was updated following Cotton et al., (1986).
Geresdi et al. (2014)	MMM	Drops, snowflakes and graupel	36	A one-dimensional kinematic model framework. Introduction of a version of time dependent melting.
Fan et al. (2014)	BM WRF/BM	Drops, 3 types of ice crystals, snow, graupel, hail, CCN, IN	33	As in Khain et al, 2004 and Fan et al., 2009. Improvement of homogeneous freezing. Implementation of simplified procedure of time-dependent melting, melted water is shed immediately.
Phillips et al. (2014, 2015); Ilotoviz et al. (2016)	BM, HUCM	Drops, 3 types of ice crystals, snow, graupel, hail, freezing drops, CCN. Liquid water in snow, graupel and hail at $T_C > 0$. Rimed mass in snow at $T_C < 0°C$. Liquid water mass in freezing drops and in hail	43	Based on Khain et al., 2011. Included: detailed time-dependent freezing procedure; calculation of liquid water in snow, in graupel and hail at $T_C > 0°C$; calculation of rimed mass in snow and liquid water mass in freezing drops and in hail at $T_C < 0°C$. Dry and wet growth regimes of freezing drops and hail are simulated.
Iguchi et al. (2014)	BM WRF/BM	Drops, 3 types of ice crystals, snow, graupel, hail, CCN.	33	3D mesoscale, as Khain et al., 2011.
Magaritz-Ronen et al. (2014, 2016)	BM, LEM	Drops, CCN, haze	500	As in Pinsky et al., 2008. Mixing between cloud parcels and mixing with environment is implemented.
Phillips et al. (2016, 2017a,b)	HUCM	Drops, 3 types of ice crystals, snow, graupel, hail, freezing drops, CCN,	43	As in Ilotoviz et al. 2016. Detailed calculation of sticking efficiency based on energetic balance is included. Ice collisional breakup of different ice hydrometeors is calculated.
Sarkadi et al. (2016)	WRF/MMM	Drops, one type of ice crystals, snowflakes and graupel. Calculation of liquid water in snow and graupel and rimed mass in snow	36	Time dependent melting of snow and graupel. Conversion of rimed snow to graupel. Shedding of liqud from graupel if liquid water fraction >0.25.

* Notations: HUCM – Hebrew University Cloud model; JMA–NHM – Japan Meteorological Agency Non-Hydrostatic Model; LEM – Lagrangian–Eulerian model of a Sc cloud; UWNMS – The University of Wisconsin Non-Hydrostatic Model; TAU – Tel Aviv University; DHARMA – Distributed Hydrodynamic Aerosol and Radiative Modeling Application

Table 7.2.2 Areas of investigation using SBM schemes.

Authors	Type of SBM, name of model	Application
Khvorostyanov et al. (1989, 1995)	BM	Rain-enhancement simulations (glaciogenic seeding).
Hall (1980)	BM	Test of methodic in bin microphysics.
Tzivion et al. (1987)	MMM, TAU	Rain enhancement simulations (CCN seeding).
Kogan (1991)	BM	Rain formation in an isolated Cu.
Khairoutdinov and Kogan (2000)	BM, LES	Simulation of drizzle formation, parameterization of autoconversion rate.
Ackerman et al. (1995, 2009)	BM, DHARMA	Drizzle formation in Sc.
Khain and Sednev (1996)	BM HUCM	Effects of breeze during winter time in the Eastern Mediterranean.
Reisin et al. (1996)	MMM	Rain enhancement (CCN seeding).
Yin et al. (2000)	MMM	Rain enhancement (CCN seeding).
Wurzler et al. (2000)	MMM	Modification of mineral dust panicles by cloud processing and subsequent effects on drop-size distributions.
Khain et al. (2004)	BM	Aerosol effects of microphysics and precipitation from isolated clouds.
Lynn et al. (2007)	BM	Effects of aerosols on precipitation from orographic clouds.
Khain and Lynn (2009)	BM	Simulation of a super-cell storm in clean and dirty atmosphere.
Li et al. (2009a,b); Khain et al. (2009)	BM	Simulation of squall line and aerosol effects.
Khain et al. (2008a)	BM	Aerosol effects of microphysics and precipitation of cloud ensembles. Effects of environment conditions. Simulation of pyro-clouds.
Lynn and Khain (2007); Khain and Lynn (2009)	BM	Simulation of a super cell storm in clean and dirty atmosphere.
Pinsky et al. (2008); Magaritz et al. (2009); Magaritz-Ronen et al. (2014, 2016)	BM, LEM	Simulation of drizzle formation in stratocumulus clouds. Effect of mixing on DSD formation.
Shpund et al. (2011, 2012, 2014)	BM, LEM	Simulation of sea spray effects on microphysics of the boundary layer under strong wind.
Teller et al. (2012)	MMM	The effects of mineral dust particles, aerosol regeneration, and ice nucleation parameterizations on clouds and precipitation.
Khain et al. (2010, 2015); Lynn et al. (2014)	BM	Simulation of aerosol effects on TC intensity.
Khain et al. (2011); Ilotoviz et al. (2016); Phillips et al. (2014, 2015)	BM	Hail formation, aerosol effects on hail, calculation of polarimetric signatures from hail storms.
Lebo and Seinfeld (2011)	MMM	Aerosol-induced convective invigoration.
Benmoshe et al. (2012); Benmoshe and Khain (2014)	BM	Analysis of effects of turbulence on cloud microphysics and rain formation.
Khain et al. (2013)	BM	Analysis of first rain-drop formation.
Noppel et al. (2010a)	BM	Effects of seeding of the spatial distribution of precipitation.
Fan et al. (2009a)	BM	Effects of wind shear on cloud–aerosol interaction.
Fan et al. (2009b); Ovchinnikov et al. (2011, 2014)	BM	Simulation of mixed-phase stratiform clouds.
Fan et al. (2012a, 2012b, 2013)	BM	Simulation of aerosol indirect effects over large areas.
Han et al. (2012)	BM	Effects of urban on precipitation.

Table 7.2.2 (cont.)

Authors	Type of SBM, name of model	Application
Iguchi et al. (2012a, 2012b, 2014); Suzuki et al. (2010, 2011)	BM	Utilization of SBM for remote-sensing problems (satellites).
Kumjian et al. (2014)	BM	Utilization of SBM model for interpretation of dual-polarimetric radar signatures.
Geresdi et al. (2014)	MMM	Investigation of melting process.
Sarkadi et al. (2016)	MMM	Simulation of orographic clouds.
Phillips et al. (2017a,b)	BM	Investigation of the role of secondary ice production in cloud microphysics and precipitation.

7.2.3 Modern Bulk-Microphysics Parameterization

Multiplying Equation (7.2.1) by m^i and integrating over PSD, one can get the equations for the i-th PSD moment of the hydrometeor of the k-th type $M_k^i = \int_0^\infty m^i f_k(m)dm$:

$$\frac{\partial \rho M_k^i}{\partial t} + \frac{\partial \rho u M_k^i}{\partial x} + \frac{\partial \rho v M_k^i}{\partial y} + \frac{\partial \rho \left(w - \overline{V}_{t,k}^i\right) M_k^i}{\partial z} = \rho \left\{ \left[\frac{\delta M_k^i}{\delta t}\right]_{nucl} + \left[\frac{\delta M_k^i}{\delta t}\right]_{c/e} + \left[\frac{\delta M_k^i}{\delta t}\right]_{d/s} + \left(\frac{\delta M_k^i}{\delta t}\right)_{f/m} + \left[\frac{\delta M_k^i}{\delta t}\right]_{col} \right\} + \cdots + \frac{\partial}{\partial x_j}\left(K \frac{\partial}{\partial x_j} \rho M_k^i\right) \quad (7.2.3)$$

where $\overline{V}_{t,k}^i = \frac{1}{M^i}\int_0^\infty m^i f(m) V_{t,k}(m)dm$ is the averaged fall velocity of the i-th moment of the PSD. Equation (7.2.3) is a generalization of Equations (3.5.5) and (3.5.6), written for mixing ratios of hydrometeors. The physical meaning of the terms in Equation (7.2.3) is similar to that in Equation (7.2.1), but all of the rates of microphysical processes are written for the PSD moments. The number of Equation (7.2.3) is equal to the number of moments used in a particular bulk-parameterization scheme, multiplied by the number of hydrometeor types. Thus, the number of equations in bulk-parameterization schemes is typically by an order of magnitude lower than that in SBM.

Modern bulk-parameterization schemes are far more sophisticated than those developed a decade ago, as they contain more types of hydrometeors and more moments used to describe the PSD of the hydrometeors (**Table 7.2.3**). One can see that the number of microphysical variables predicted by bulk-parameterization schemes increased over time, from two in the Kessler (1969) scheme to seventeen in the scheme by Loftus et al. (2014) and Loftus and Cotton (2014a, 2014b), and eighteen in Milbrandt and Yau (2005, 2006).

A new bulk scheme by Morrison and Milbrandt (2015) may contain both one- or two-moment descriptions of liquid drops. Ice particles are described by means of a single hydrometeor type with a three-parameter Gamma size distribution and characterized by four prognostic mixing-ratio variables: the total ice mixing ratio, the ice number concentration, the ice mixing ratio from rime growth, and the bulk rime volume. These characteristics are chosen because they can be treated as conserved prognostic variables. These parameters change via vapor deposition, aggregation, and riming (dry and wet growth). Using these prognostic variables, several important predicted properties are derived, namely, the rime-mass fraction, bulk density, the mean particle size, and the fall velocity. Ice particles at a particular grid point and a particular time instance can be diagnosed, if necessary, using the above set of characteristics. An advantage of such approach is that it eliminates, to some extent, uncertainties related to transition of particles from one type to another. A disadvantage is that in real clouds ice particles of entirely different structures, such as low density snowflakes and high density hail, can exist at the same spatial point.

Later, Milbrandt and Morrison (2016) introduced multiple ice particle types in a 1D kinematic model. In this study, several types of ice particles are considered. These types are initially determined by the mean particle diameter. In the course of model integration, the ice types are identified by the rimed fractions and the

Table 7.2.3 The main characteristics of widely used bulk-parameterization schemes.

Authors	Name/comments	Type of hydrometeors							
		DROPS $N\ q\ Z$	DRIZZLE $N\ q\ Z$	RAIN $N\ q\ Z$	ICE $N\ q\ Z$	AGGR $N\ q\ Z$	SNOW $N\ q\ Z$	GRAUP $N\ q\ Z$	HAIL $N\ q\ Z$
Kessler (1969)	1M/first warm rain bulk parameterization, included in the microphysical WRF package.	- X -		- X -					
Lin et al. (1983)	1M/hail is treated as a high density hydrometeor.	- X -		- X -	- X -		- X -		- X -
Rutledge and Hobbs (1984)	1M	- X -		- X -	- X -		- X -	- X -	
Chen and Sun (2002)	Purdue Lin scheme (used in WRF). Based on Lin et al. (1983) and Rutledge and Hobbs (1984) studies.	- X -		- X -	- X -		- X -	- X -	
Cotton et al. (1986)	1M/first bin parameterization used in RAMS. Ice-multiplication, melting, and shedding are included.	- X -		- X -	- X -	- X -		- X -	- X -
Tao et al. (1989)	1M, 3ICE. Used in WRF and in NASA Goddard scheme, based on Lin et al. (1983) and Rutledge and Hobbs (1984) schemes. A new procedure of ice-water saturation adjustment is included.	- X -		- X -	- X -		- X -	- X -	
Murakami (1990)	1M/snow includes singly snow crystals and aggregates, many algorithms are similar to those in Lin et al. (1983) and Cotton et al. (1986). Improved approach for calculation of relative fall velocity between different hydrometeors.	- X -		- X -	- X -		- X -	- X -	

Table 7.2.3 (cont.)

Authors	Name/comments	Type of hydrometeors							
		DROPS $N\ q\ Z$	DRIZZLE $N\ q\ Z$	RAIN $N\ q\ Z$	ICE $N\ q\ Z$	AGGR $N\ q\ Z$	SNOW $N\ q\ Z$	GRAUP $N\ q\ Z$	HAIL $N\ q\ Z$
Verlinde et al. (1990)	1M/used in RAMS, similar to Cotton et al. (1986) but with analytical expressions for integrals of collisions. Application of lookup tables.	- X -		- X -	- X -	- X -		- X -	- X -
Ferrier (1994)	2M/Ice and precipitation particle concentrations are predicted.	- X -		- X -	XX -		XX -	XX -	XX -
Walko et al. (1995)	1M/used in RAMS. Ice crystals are separated into small pristine crystals and large pristine crystals attributed to snow. Hail and graupel are of different density.	- X -		- X -	- X -	- X -	- X -	- X -	- X -
Meyers et al. (1997)	2M/used in RAMS. Ice and precipitation particle concentrations are predicted.	- X -		XX -	XX -	XX -	XX -	XX -	XX -
Reisner et al. (1998)	2M/NCAR/Penn State Mesoscale Model Version 5 (MM5).	- X -		XX -	XX -		XX -	XX -	
Cohard and Pinty (2000)	2M/warm microphysics. Implementation of some analytical expressions for rates.	XX -		XX -					
Seifert and Beheng (2001)	2M/warm processes, analytical formulas for autoconversion, accretion and self-collection.	XX -		XX -					

Cotton et al. (2003)	2M/used in RAMS. Based on schemes by Verlinde et al. (1990) and Walko et al. (1995). Bin-emulating bulk scheme. Ice is categorized into pristine ice, large pristine ice (snow) that can be rimed, aggregates, graupel, and hail.	- X -		XX -	XX -		XX -	XX -	XX -	XX -
Hong et al. (2004); Hong and Lim (2006)	1M/used in WRF, 3-class (WSM3) scheme. This scheme is efficient at grid spacing between the mesoscale and cloud-resolving.	- X -		- X -	- X -					
Lim and Hong (2005)	1M/used in WRF, 5-class scheme WSM5, a gradual melting of snow is allowed. This scheme is efficient at grid spacing between the mesoscale and cloud-resolving grid.	- X -		- X -	- X -			- X -		
Hong and Lim (2006)	1M/used in WRF, 6-class scheme WSM6, includes graupel. The scheme is the most suitable for cloud-resolving grids.	- X -		- X -	- X -			- X -	- X -	
Skamarock et al. (2008)	Eta Grid-scale Cloud and precipitation scheme (2001), EGCP01 or the Eta Ferrier scheme, used in WRF; the mean size of precipitating ice is temperature dependent.	- X -		- X -	- X -			- X - Precipitation ice		
Milbrandt and Yau (2006)	3M/implementation of dependencies between parameters of Gamma distributions, simulation of hail.	XXX		XXX	XXX			XXX	XXX	XXX

Table 7.2.3 (cont.)

Authors	Name/comments	Type of hydrometeors							
		DROPS $N\ q\ Z$	DRIZZLE $N\ q\ Z$	RAIN $N\ q\ Z$	ICE $N\ q\ Z$	AGGR $N\ q\ Z$	SNOW $N\ q\ Z$	GRAUP $N\ q\ Z$	HAIL $N\ q\ Z$
Thompson et al. (2004) (as it is in WRF)	1M for all hydrometeors, except ice./Ice is described using a 2M scheme. Empirical relation between parameters of Gamma distributions, and between the parameters and environmental conditions. A scheme is built for high-resolution simulations. It includes ice number concentrations.	- X -		- X -	XX-		- X -	- X -	
Thompson et al. (2008)	As Thompson et al. (2004), but two-moment approach is used for cloud ice and rain.	- X -		XX -	XX-		- X -	- X -	
Thompson and Eidhammer (2014)	2M scheme for cloud water, rain, ice and 1M for snow and graupel./Variable density snow is used. Prediction of the aerosol number concentration.	XX -		XX -	XX-		- X -	- X -	
Morrison et al. (2005a, 2009a)	2M, used in WRF. Improved calculation of supersaturation is included. Concentration of CCN is prescribed.	XX -		XX -	XX -		XX -	XX -	
Morrison and Gettelman (2008)	2M for cloud water and ice/ used in the NCAR Community Atmosphere Model (CAM). Developed for GCM. Explicitly predicts mass and number-mixing ratios of cloud droplets and ice crystals. Precipitation is calculated diagnostically in neglecting horizontal advection of precipitating particles.	XX -		Diagnostic	XX -		Diagnostic	Diagnostic	

Lin and Hong (2010)	2M for droplets and rain/ used in WRF. A prognostic treatment of cloud condensation nuclei (CCN) is introduced.	XX -		XX -	- X -	- X -	- X -		
Lin and Colle (2011)	1M/5-class hydrometeor Stony Brook University Y. Lin scheme; SBU_YLIN, the projected area, mass, and fall velocity of precipitating ice depends on riming intensity and temperature.	- X -		- X -	X	- X - Precipitating ice			
Saleeby and Cotton (2004)	2M/used in RAMS, bin-emulating scheme. Drizzle is included as a separate hydrometeor. Fully interactive with prognostic CCN and IN aerosol schemes.	XX -	XX -	XX -	XX -	XX -	XX -	XX -	XX -
Seifert and Beheng (2006)	2M/The rates of collisions are calibrated versus a bin scheme. The collision rate depends on the stage of cloud evolution.	XX -		XX -	XX -	XX -	XX -	XX -	
Mansell et al. (2010)	2M, NSSL-7-Class/Graupel density is recalculated. Conversion of graupel to hail takes place when wet growth begins.	XX -		XX -	XX -	XX -	XX -	XX -	XX -
Saleeby and Van den Heever (2013)	Same as Saleeby and Cotton (2004), except modified prognostic aerosol schemes. IN are parameterized using DeMott et al. (2010) scheme. Many aerosol types and a variety of aerosol processes included.	XX -	XX -	XX -	XX -	XX -	XX -	XX -	XX -
Loftus et al. (2014)	RAMS, 3M for HAIL/Same as Saleeby and Van den Heever (2013), except that a triple-moment scheme has been included for hail.	XX -	XX -	XX -	XX -	XX -	XX -	XX -	XXX

Table 7.2.3 (cont.)

Authors	Name/comments	Type of hydrometeors							
		DROPS $N\ q\ Z$	DRIZZLE $N\ q\ Z$	RAIN $N\ q\ Z$	ICE $N\ q\ Z$	AGGR $N\ q\ Z$	SNOW $N\ q\ Z$	GRAUP $N\ q\ Z$	HAIL $N\ q\ Z$
Phillips et al. (2007a, 2008, 2009, 2013); Formenton et al. (2013a,b)	Bin-emulating for coagulation of snow, rain, and graupel/hail; supersaturation in clouds is predicted; bulk treatment of six aerosol species is applied.	XX -		- X -	XX -		- X -	- X -	
Lang et al. (2014); Tao et al. (2015)	1M, 4ICE the same as Tao et al. (1989), but with additional implementation of hail.	- X -		- X -	- X -		- X -	- X -	- X -
Morrison and Milbrandt (2015); Morrison et al. (2015)	The Predicted Particle Properties (P3) scheme represents all ice hydrometeors using a single “free” category, in which the bulk properties evolve smoothly through changes in the prognostic variables, allowing for the representation of any type of ice particle.	- X - or XX -		XX-	Only one hydrometeor type with predicted bulk properties				
Milbrandt and Morrison (2016)	Same as Morrison and Milbrandt, 2015, but with introduction of multiple ice types.	- X - or XX -		XX-	Several types of ice hydrometeors are implemented. These types are identified by mean–mass diameter and bulk density				

fall velocity, which are maximum for graupel-like particles. At each time step, the distance between the ice types is evaluated using the characteristic parameters. If the difference in the mean–mass diameters is below 150 μm and the difference in the bulk densities is below 100 kg m^{-3}, the ice types are merged into single type. The resulting ice type is determined by summing each prognostic variable and assigning the obtained values to one of the ice types and setting these values at zero for the other types.

From the computational point of view, treatment of each ice particle characteristic is equivalent to treatment of a particular hydrometeor type in other bulk schemes. Treatment of three particle ice types in the Milbrandt and Morrison (2016) scheme, each of them being described by four prognostic characteristics, is equal to the treatment of twelve prognostic variables, which is comparable with the number of major variables used in two- or three-moment schemes.

Fit of Observed PSD by the Gamma Distribution

Most bulk-parameterization schemes use the Gamma distribution as a master function for the approximation of PSD of different cloud hydrometeors. The properties of the Gamma distribution are discussed in Section 2.1. A major advantage of using Gamma distributions in microphysical applications is the ability to derive the rates of microphysical processes in a straightforward manner.

The Gamma distribution $f(m) = N_0 m^\nu \exp(-\lambda m^\mu)$ (see Equation 2.1.30) is equal to zero at $m = 0$, has one maximum, and tends to zero when $m \to \infty$. As a rule, observed DSD have two modes: for cloud droplets with radii below 20–25 μm centered at $r \approx 10-15$ μm, and for raindrops with radii sometimes as large as 3–4 mm (Pruppacher and Klett, 1997). These modes are separated by a pronounced minimum within the radii range of 25–60 μm. Size distributions containing both small cloud droplets and raindrops obviously do not obey single Gamma distribution. To avoid this problem, all bulk-parameterization schemes distinguish between small cloud droplets with a distribution parameterized by a Gamma distribution, and raindrops whose distribution is approximated usually, although not always, by exponential functions. Thus, liquid drops are represented by two types of hydrometeors: cloud droplets and raindrops. Drops with a radius of about 30 μm is the distinction value separating cloud droplets from raindrops. **Figure 7.2.1** shows examples of general Gamma distributions used for approximation of DSD of different widths in bulk-parameterization schemes. As can be seen in multiple examples of DSD measured in situ (Figures 2.3.8–2.3.10; 5.9.17 and 5.9.18), many DSD have a complicated shape and contain two or more modes. The shapes of DSD averaged over lengths of ten or more kilometers are often closer to the shapes of the Gamma distributions, shown

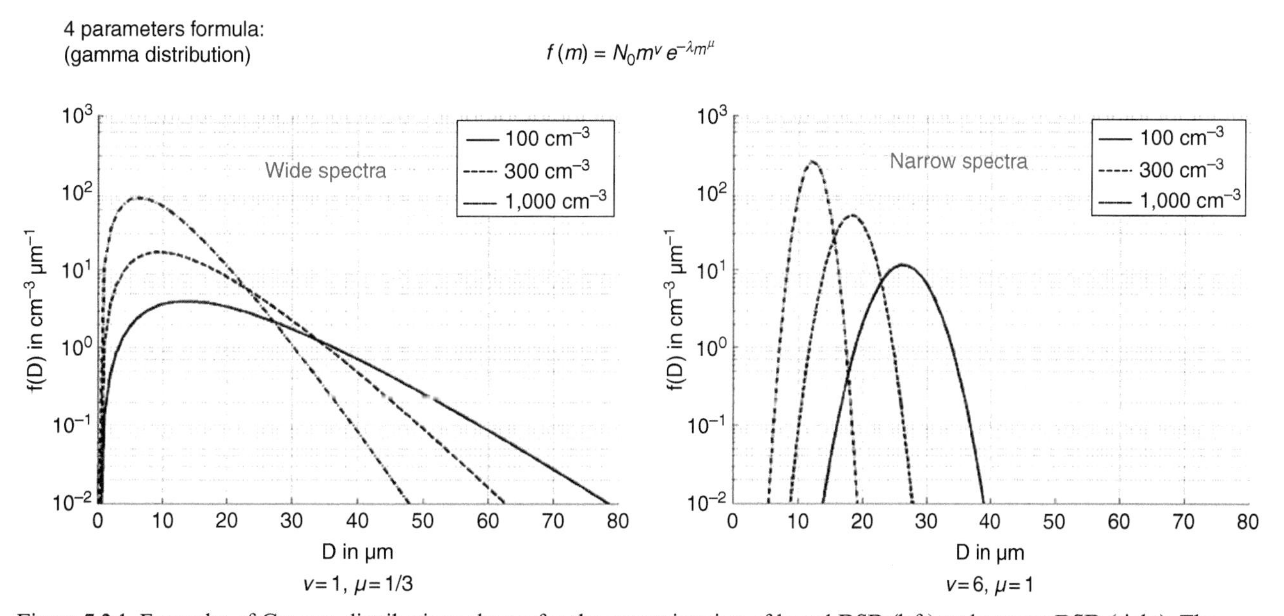

Figure 7.2.1 Examples of Gamma distributions chosen for the approximation of broad DSD (left) and narrow DSD (right). The distributions are plotted for different parameters ν and μ. The values of the intercept parameter N_0 were chosen in order to get desirable concentrations of droplets, indicated in the boxes. Solid, dash-dotted, and dotted lines denote DSD for maritime, intermediate, and continental conditions, respectively (from Noppel et al., 2010b; courtesy of © John Wiley & Sons, Inc.).

in Figure 7.2.1. Note that the rates of the microphysical processes are determined by local DSD and not by spatial-averaged DSD.

Several attempts have been made to evaluate the possibility of approximating in situ measured DSD by the Gamma distributions, as well as lognormal and exponential distributions (e.g., Costa et al., 2000). The exponential distribution proved unsuitable for approximation of most observed DSDs. The Gamma distributions typically approximate the observed DSD better than exponential distributions. However, the DSD shapes are quite variable. DSD shapes depend on the aerosol type, so the DSDs are different for maritime, coastal, continental, and urban aerosols. Besides, significant variability in the values of the parameters was observed in clouds belonging to the same type, and even between different zones within clouds. Accordingly, to approximate the observed DSD, different parameters of the Gamma distribution should be used. **Figure 7.2.2** shows histograms characterizing occurrence of the shape parameter and scale diameter in all the Gamma distribution fits. It was also found that for more than 10% of cases no fit was possible. Costa et al. (2000) conclude that such variability of DSD shapes imposes strict limitations on bulk microphysical modeling.

Analyses of observed data performed by Tampieri and Tomasi (1978), and later by Dooley (2008) showed that the DSD shape in clouds evolves with time and height, so all the parameters of the Gamma distributions should be changed in a consistent way to preserve the approximation of DSD. This conclusion is illustrated in **Figure 7.2.3**, showing the intercept parameter N_0 vs. slope parameter λ scattering diagram acquired by the Gamma distribution fits to the observed DSD.

Geoffroy et al. (2010) attempted to relate the values of parameters of lognormal and Gamma distributions with the values of DSD moments using the DSD measured in stratocumulus and small cumulus clouds. The dispersion in the values of parameters was found to be very large, as each microphysical process affects the values of the parameters differently. The researchers tried to decrease the scatter by treating different microphysical processes separately. However, on the basis of observations it was impossible to isolate a process that could be considered the dominating one. It was concluded that bulk schemes have serious limitations due to extremely high variability of observed DSD shapes and the resulting problems relate to choosing the master functions and their parameters. Illingworth and Blackman (2002) and Handwerker and Straub (2011) used radar reflectivity–rain rate relationships to evaluate parameters of the Gamma distributions used to approximate raindrop-size distributions. In these studies, a strong variability of the parameters was also reported. Igel and van den Heever (2017) found that parameters of the Gamma distributions approximating DSD in small cumulus clouds are different in zones of condensation and evaporation.

As regards ice particles, their size distributions with radii above about 50–100 μm measured by 2D imaging probes were approximated by exponentials (Sekhon and Srivastava, 1970), bimodals (Mitchell et al., 1996; Yuter et al., 2006; Lawson et al., 2006), the Gaussian size distributions (Delanoë et al., 2005), the sum of an exponential and Gamma (Field et al., 2005), and lognormal size distributions (Tian et al., 2010) and by Gamma (Heymsfield et al., 2013). In most studies, the DSD were obtained using drops registered over long traverses of several hundred kilometers and over different clouds. As follows from numerical simulations using SBM models, the local PSD of aggregates, graupel, and hail at spatial scales of a few hundred meters can have the Gamma, bimodal, or more complicated shapes (e.g., Phillips et al., 2015; Ilotoviz et al., 2016). Ovchinnikov et al. (2014) showed that an accurate representation of liquid–ice separation and the longevity of mixed-phase clouds can be reached only when the parameters of the Gamma distributions are properly chosen.

Strictly speaking, four equations must be solved to determine the four parameters of the Gamma distribution in Equation (2.1.30). Accordingly, four-moment bulk-parameterization schemes should be used. However, the state-of-the-art bulk schemes use no more than three PSD moments (Table 7.2.3). So, in one-, two-, or three-moment bulk-parameterization schemes, three, two, and one parameters, respectively, are either fixed a priori or are determined using additional semiempirical relationships. The analysis of observed PSD shows that fixing parameters does not yield a good approximation of PSD by the Gamma distribution.

Three major approaches have been suggested to improve PSD approximation by the Gamma distribution in bulk-parameterization schemes. The first approach, widely applied in two-moment bulk schemes, entails the implementation of dependencies between parameters of the Gamma distributions, as shown in Figure 7.2.3. Similar relations between the parameters were used in a bulk-microphysics scheme by Thompson et al. (2004), who derived them from bin-microphysics simulations. To describe a raindrop-size distribution, a relationship between N_0 and the rain-mass mixing ratio is used. For graupel with a Gamma size distribution, a relation between the intercept parameter N_0 and the slope parameter λ is used, while the shape parameter is fixed. In the recent version of this bulk parameterization

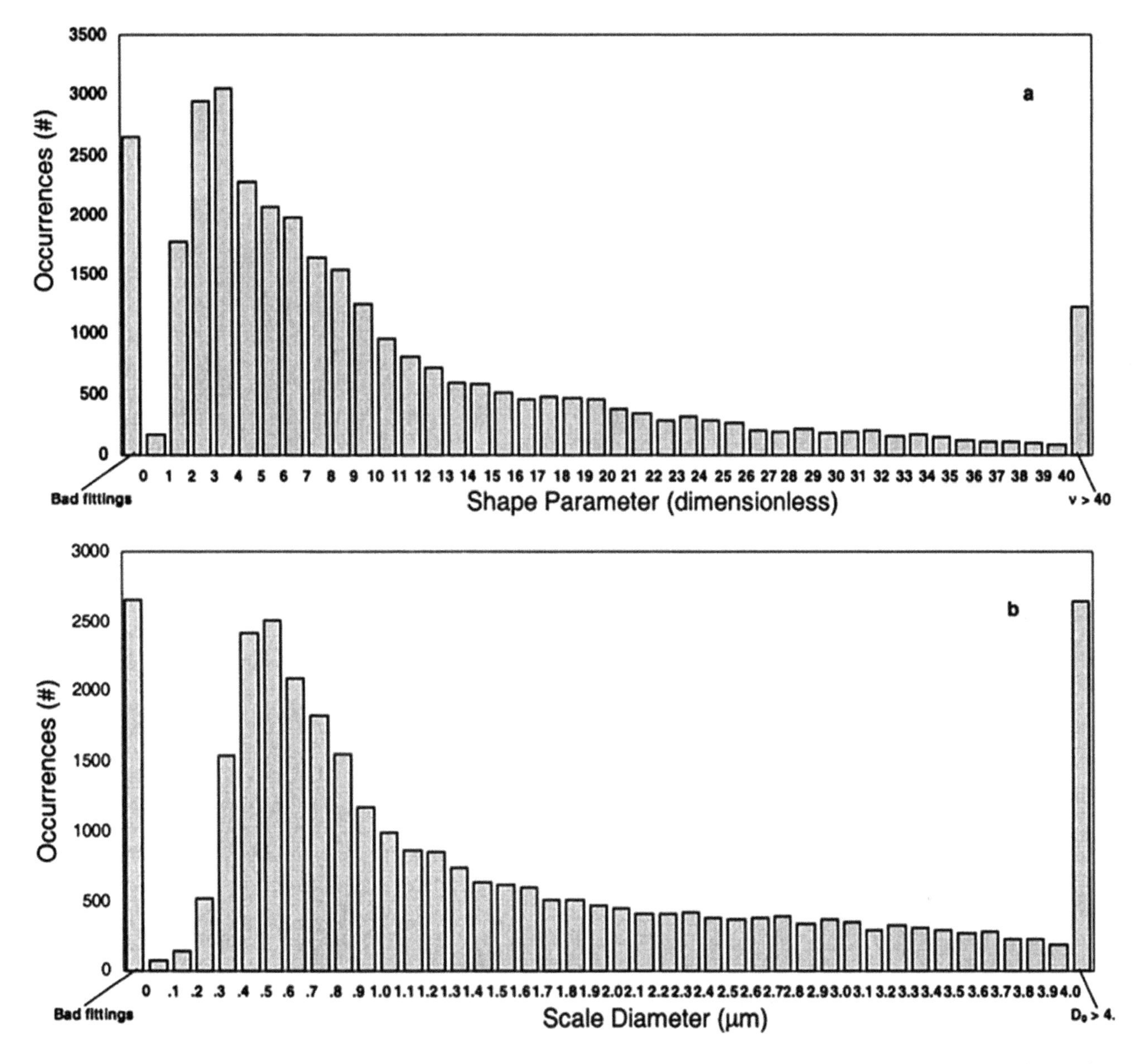

Figure 7.2.2 Occurrences of the shape parameter (upper panel) and the scale diameter (bottom panel) in all the Gamma distribution fits. DSD were measured in shallow cumuli, Northeast Brazil (adapted from Costa et al., 2000; courtesy of Elsevier).

(G. Thompson, personal communication), the graupel intercept parameter depends on the supercooled liquid-water content and graupel content, as shown in **Figure 7.2.4**.

To treat snow, Thompson et al. (2004) used a relation between N_0 and temperature. Formenton et al. (2013a, 2013b) applied a similar idea by deriving an iterative numerical solution for shape, slope, and the intercept parameters of a Gamma size distribution in a one-moment scheme. These parameters are diagnosed from a given snow- mass mixing ratio at every grid-point for each time step. Formenton et al. (2013a, 2013b) used an observed relation between the parameters of the Gamma distributions, obtained by Heymsfield et al. (2002) in their analysis of aircraft data from various field campaigns. However, utilization of relationships between the parameters of Gamma distribution still keeps the PSD unimodal, which may substantially mis-represent the cloud microphysical structure and the associated microphysical processes.

The second approach to improving the PSD representation involves implementation of additional

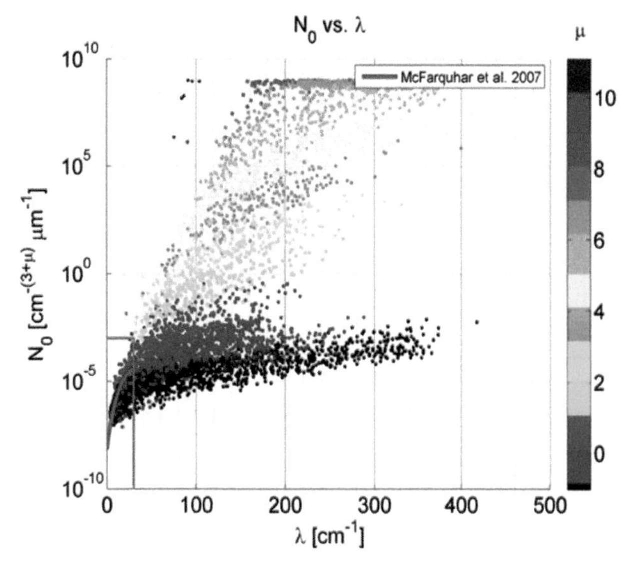

Figure 7.2.3 Scatter plots of the parameters of the Gamma distributions used to fit the observed DSD. Parameters N_0 and λ calculated using 1the 0-s Gamma distribution fits to the observed DSD. Solid red line marks the best fit line for the variation of N_0 vs. λ, obtained by McFarquhar et al. (2007) (from Dooley, 2008; permission of McFarquhar).

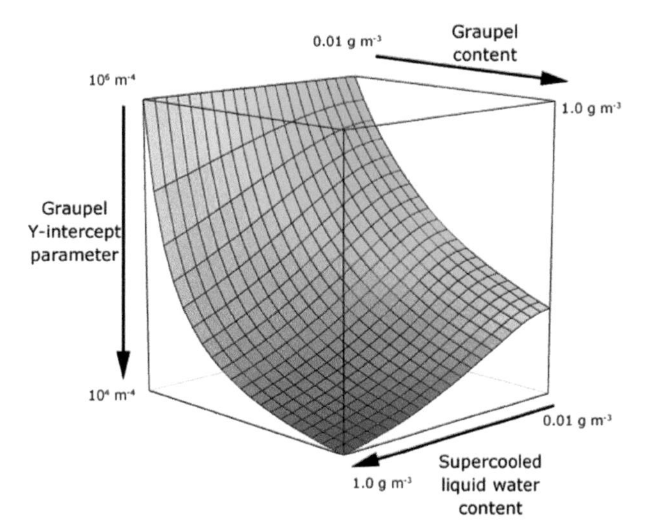

Figure 7.2.4 Dependence of the intercept parameter of graupel PSD on supercooled liquid-water content and graupel content in the updated Thompson bulk-parameterization scheme (G. Thompson, personal communication, 2014).

hydrometeor types described by their master function. This approach alongside the Gamma distributions is applied in RAMS, where a third liquid-water drizzle mode was introduced (Saleeby and Cotton, 2004). The drizzle mode that includes drops with diameters of 50–100 μm is intermediate between mode of cloud droplets (2–50 μm in diameter) and the mode of raindrops (diameter exceeding 100 μm). Implementation of the drizzle mode was found to slow down rain formation, thus enabling them to achieve a better agreement of simulated precipitation rates with observations. The drizzle mode was also implemented by Sant et al. (2013) in simulations of drizzle formation in a warm stratocumulus cloud. Thompson et al. (2008) approximated the snow PSD using the sum of two Gamma distributions centered at different radii, which provided more degrees of freedom in PSD simulations.

Finally, the third approach entails utilization of three-moment bulk-parameterization schemes, thereby enabling calculation of three parameters in Equations (2.1.30). This approach allowed Milbrandt and Yau (2006) and Loftus and Cotton (2014a) to successfully simulate large hail with a broad size distribution, previously unachievable via two-moment schemes.

7.2.4 Comparison of Results Obtained by Means of SBM and Bulk-Parameterization Schemes

Here we present several examples of comparisons between bulk and SBM schemes in simulations of different meteorological phenomena. In some examples, responses of different microphysical schemes to variations in aerosol concentration are compared. The ability to represent aerosol effects on clouds and on precipitation is now considered a measure of a model's capabilities in investigating local and global climatic changes.

SBM numerical models made it possible to simulate the observed relations between aerosol concentration and droplet concentration (Segal and Khain, 2006; Ghan et al., 2011; Pinsky et al., 2012). Two-moment bulk-parameterization schemes calculate mass contents and number concentrations, i.e., they enable evaluation of the mean volume radius of droplets, decreasing with increasing CCN concentration. The similarity in the response of bulk models and SBM to changes in the aerosol concentration would demonstrate an ability of the bulk-parameterization schemes to take into account aerosol effects in cloud and cloud-resolving models. Several examples are presented to illustrate the capabilities of bulk parameterizations and SBM in simulating different atmospheric phenomena of various spatial and time scales.

1D Models

Shipway and Hill (2012) compared the rates of diffusional growth, collisions, sedimentation, and surface precipitation between bulk-microphysical schemes and the MMM method developed at Tel Aviv University (Tzivion et al., 1987). The comparison was performed by means of a 1D kinematic model where the velocity field evolving over time was prescribed. The set of the bulk-parameterization schemes including three one-moment schemes (1M), three two-moment schemes (2M), and one three-moment scheme (3M) was used to calculate the warm microphysical processes. The results obtained using the MMM method with thirty-four drop-mass categories were considered the benchmark. The schemes differed significantly in the resulting surface precipitation rates, rain onset timing, and accumulated surface precipitation. The early precipitation onset was shown to be a persistent feature of the 1M scheme (**Figure 7.2.5**). Results obtained via the 2M scheme were closer to the benchmark, with the exception of the peak of precipitation rate, which was larger than that in the MMM. The best agreement with the MMM results was obtained using the 3M. Figure 7.2.5 shows that bulk schemes produce sharp and narrow precipitation peaks compared to a weaker but much wider precipitation distribution simulated by the SBM schemes. The reasons of such differences were discussed in Section 5.5.3.

Mesoscale rain event. Lynn et al. (2005a, 2005b) were apparently the first to use the SBM scheme in a mesoscale model (the fifth-generation Pennsylvania State University–NCAR Mesoscale Model – MM5 (Grell et al., 1994) for simulation of a large convective-system evolution. The rain event over Florida on July 27, 1991, observed during the Convection and Precipitation Electrification Experiment (CaPE), was simulated by means of a 3-km resolution grid for a comparatively long period of time and over a significant area. Both low

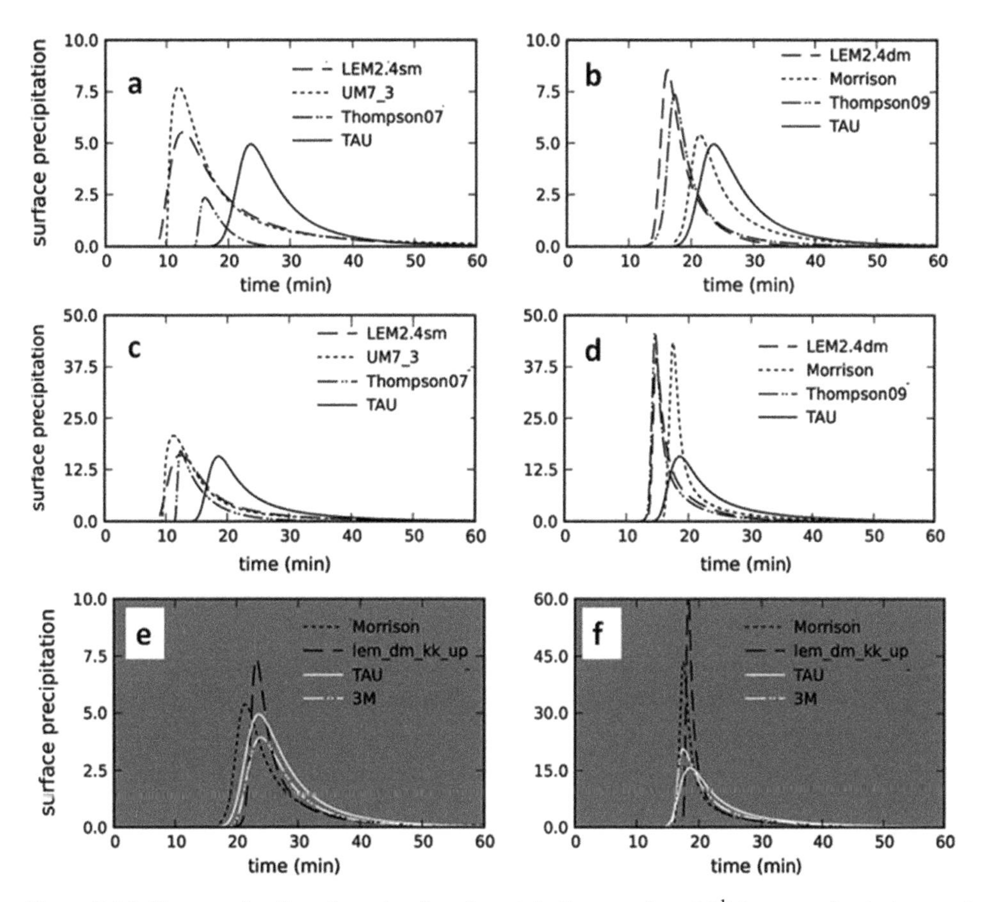

Figure 7.2.5 Comparative time dependencies of precipitation rate (mm h^{-1}) between simulations performed by means of a kinematic model using different bulk-parameterization schemes and the MMM scheme. Panels (a), (b), and (e) illustrate simulations with the vertical velocity of 2 m/s. Panels (c), (d), and (f) illustrate simulations with the vertical velocity of 3 m/s. Panels (a) and (c) show the results of the 1M: LEM2.4sm (Swann, 1998), Thompson07 (modified Thompson et al., 2004), and UM7_3 (Tripoli and Cotton, 1980). Panels (b) and (d) show the results of the 2M: schemes LEM2.4dm (Swann, 1998), Thompson09 (modified Thompson et al., 2004), and Morrison (Morrison et al., 2011). Panels (e) and (d) and (f) show results of the 3M scheme (Shipway and Hill, 2012) (from Shipway and Hill, 2012, with changes; courtesy of © John Wiley & Sons, Inc.).

and high CCN concentrations were used. The so-called Fast SBM (FSBM) version was used, in which cloud hydrometeors are described using three size distribution functions for drops, low density ice (aggregates), and high density ice (hail or graupel). The results of FSBM were compared to those obtained by means of different one-moment bulk-parameterization schemes. While the SBM reproduced observed large areas of stratiform rain, the bulk schemes tended to underestimate the area of weak and stratiform rain (**Figure 7.2.6**). **Figure 7.2.7** shows time dependencies of the area average and maximum rain amounts. FSBM produces results closer to observations. The one-moment bulk models produced unrealistically large rain rates within a comparatively narrow line of cumulus clouds. As in the previous example, the possible cause of these results might be the incorrect representation of precipitation sedimentation (Section 5.5.3).

These results illustrate the substantial limitations of one-moment bulk-parameterization schemes in simulations of the structure of individual clouds, cloud ensembles, and precipitation rates. Lynn and Khain (2007) simulated the same rain event over Florida on July 27, using other bulk schemes: Thompson scheme (Thompson et al., 2006), the Reisner–Thompson scheme (Reisner2), and the two-moment scheme described by Seifert et al. (2006) (Table 7.2.3). The results obtained are quite similar to those shown in

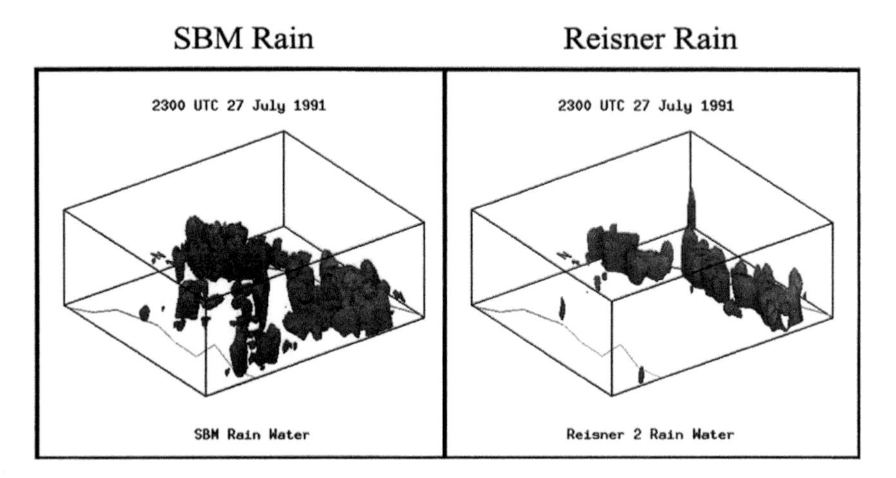

Figure 7.2.6 3D structure of rainwater content in FSBM and in bulk Reisner2 scheme (Table 7.2.3) at 2,300 UTC in simulations of the rain event over Florida on July 27. Solid line denotes the land–sea boundary (from Lynn et al., 2005b; courtesy of © American Meteorological Society. Used with permission).

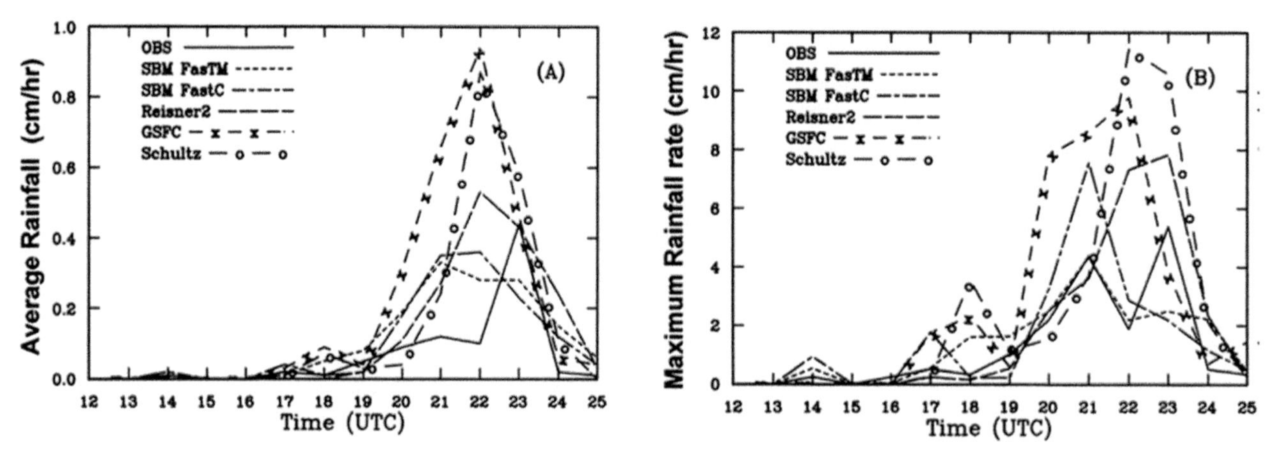

Figure 7.2.7 Average (A) and maximum (B) rainfall rates obtained from observations and model simulations. Notations: Reisner2 (Reisner et al., 1998), GSFC (Goddard Space Flight Center) (Tao et al., 2003), and Schultz (Schultz, 1995) (from Lynn et al., 2005b; courtesy of © American Meteorological Society. Used with permission).

Figure 7.2.7: all of the bulk schemes overestimated averaged and maximum rain rates by a factor of two to three, as compared to the observations. The two-moment schemes predicted rain rates better than the one-moment schemes. The SBM overestimated the maximum rain rate by about 20%.

Cumulus and Stratocumulus Clouds
Morrison and Grabowski (2007) used a 2D kinematic model to compare the microphysical structure of warm stratocumulus and cumulus clouds simulated using the Morrison et al. (2005a) two-moment bulk scheme with a bin scheme under different aerosol loadings. Three different parameterizations for the coalescence process developed by Beheng (1994), Seifert and Beheng (2001), and Khairoutdinov and Kogan (2000) were tested. Simulations revealed differences in horizontally averaged values of RWC, as well as in the mean raindrop diameters. It is interesting that the difference in the accumulated rain produced by different schemes was comparatively low (up to 20%). This result suggests that accumulated rain is determined, to a large extent, by such atmospheric properties as environmental humidity, as well as by the fact that most water vapor in excess of the saturated value is condensed. At the same time, precipitation rates obtained by different schemes may differ substantially. Morrison and Grabowski (2007) showed that results of one-moment schemes were much worse than those of two-moment schemes. They found that in order to adjust the results of one-moment schemes to those of two-moment ones, the value of the intercept parameter N_0 should be changed with height by five orders of magnitude. This strong sensitivity of the simulation results to the value of the intercept parameter makes one-moment schemes unsuitable for cloud-resolving models to be used in regional and global climate simulations.

A detailed comparison of bulk and bin approaches in simulations of structure and precipitation of continental clouds (Texas clouds) and tropical maritime clouds (GATE 1974) was done by Seifert et al. (2006). The two-moment scheme developed by Seifert and Beheng (2006a) was implemented into the HUCM, thus both microphysical schemes were compared using the same dynamic framework of the HUCM. Simulations were performed for two aerosol loadings: a high CCN concentration of 1,260 cm^{-3} and a low CCN concentration of 100 cm^{-3}. It was found that an accurate representation of the warm phase autoconversion process is the most important factor in achieving a reasonable agreement between drop concentration and the mass contents in bulk parameterization and SBM. To obtain comparable results, the parameters of various PSD in the bulk scheme were calibrated. In spite of differences between the PSD in the bin and the bulk simulations, the precipitation rates and accumulated precipitations were relatively close, especially at high CCN concentrations **(Figure 7.2.8)**. The substantial difference in the accumulated rain amounts from maritime clouds at low CCN concentrations seen in Figure 7.2.8 (right panel) is related to the fact that the bulk scheme does not take into account a decrease in aerosol concentration due to nucleation scavenging. Saleeby and van den Heever (2013) also showed that taking into account transport

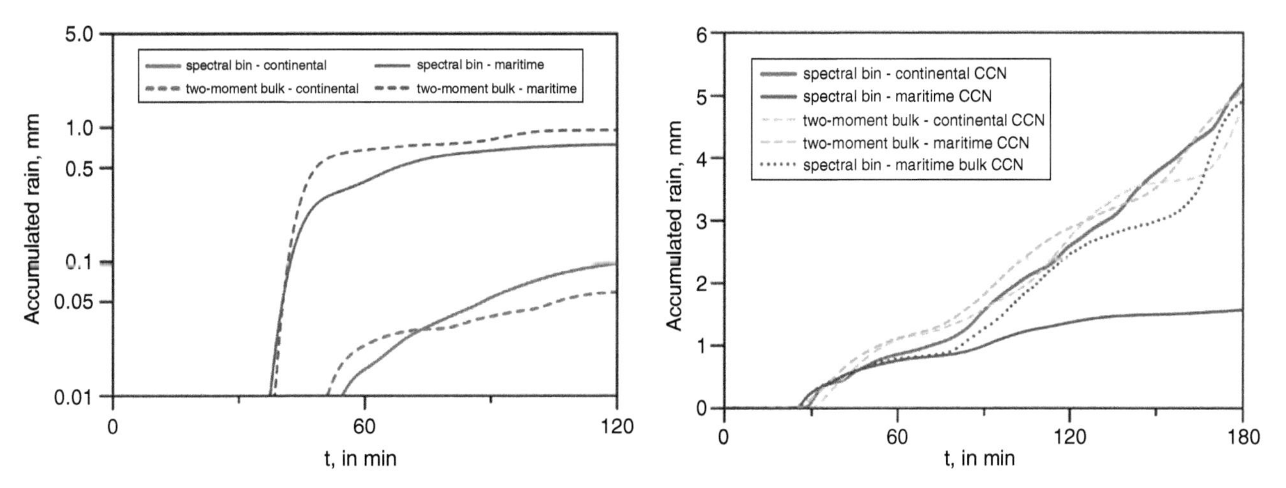

Figure 7.2.8 Time dependencies of grid-averaged accumulated surface precipitation from the four Texas cloud simulations (left) and the 5 GATE simulations (right) obtained using a bin scheme and a two-moment bulk scheme. High (continental) and low (maritime) CCN concentrations were used in simulations (from Seifert et al., 2006; courtesy of Elsevier).

and nucleation scavenging of aerosols in bulk-parameterization schemes yields substantially better results.

Fan et al. (2012a, 2012b) and Wang et al. (2013) reported a significant improvement in a two- moment bulk parameterization scheme when nucleation aerosol scavenging was taken into account. Sensitivity experiments performed by these authors using four different types of autoconversion schemes revealed that the saturation adjustment employed in calculating condensation/evaporation in the bulk scheme is the main factor responsible for the errors in predicting cloud water content. Thus, an explicit calculation of diffusion growth with predicted supersaturation is a promising way to improve bulk microphysics schemes. It was also found that rain evaporation in bulk schemes occurs too fast as compared to SBM (Shipway and Hill, 2012).

Squall Lines

A squall line is a good example of a mesoscale system in which convective clouds form a line of several hundred kilometers in length (Figure 1.2.15). The front of a squall line represents an elongated thunderstorm characterized by heavy rain, hail, and lightning. The width of the leading updraft core of a squall line typically ranges from ten to several tens of kilometers. A wide zone of light precipitation caused largely by aggregates forms behind the convective zone and reaches a few hundred kilometers in width. Squall lines represent a typical severe-weather phenomenon occurring over both land and sea. Over the sea, squall lines are the dominant type of mesoscale convective systems. The lifetime of squall lines can be as long as 10 h, as compared to about 1 h for an isolated cloud.

Li et al. (2009a, 2009b) simulated a squall line typical of continental conditions observed during the Preliminary Regional Experiment for Storm-Scale Operational and Research Meteorology 1985 (PRE-STORM) (Zhang et al., 1989; Braun and House, 1997). Simulations were performed using a 2D anelastic version of the GCE with two types of microphysical schemes, namely: the one-moment bulk microphysical scheme by Lin et al. (1983) with prognostic equations for mixing ratios of cloud water, rain, ice, snow, and graupel/hail, and SBM (Khain et al., 2004; Phillips et al., 2007b). The model resolution in the zone of interest was 1 km, the time step was 6 s. **Figure 7.2.9** shows the surface rainfall time–domain plots obtained in these simulations. Figure 7.2.9 shows that the squall line simulated by SBM has a much larger trailing stratiform area than the squall line simulated using the bulk parameterization. This result agrees well with those shown in Figure 7.2.6. The SBM indicates a much larger contribution of light rain to the total surface rainfall. During the mature stage, about 20% of the total surface rainfall was stratiform in the SBM simulation, compared with only 7% in the bulk simulation. According to observations by Johnson and Hamilton (1988), about 29% of the surface rain came from the stratiform region. This means that the SBM describes separation of rain into convective and stratiform modes better than the one-moment bulk-parameterization scheme.

Another significant feature of the simulated structures of a squall line was that the bulk scheme yielded high rainfall-rate streaks extending from the leading convection rather deep into the stratiform region, while such streaks were absent in the SBM simulations and in the observations. The appearance of the multicell

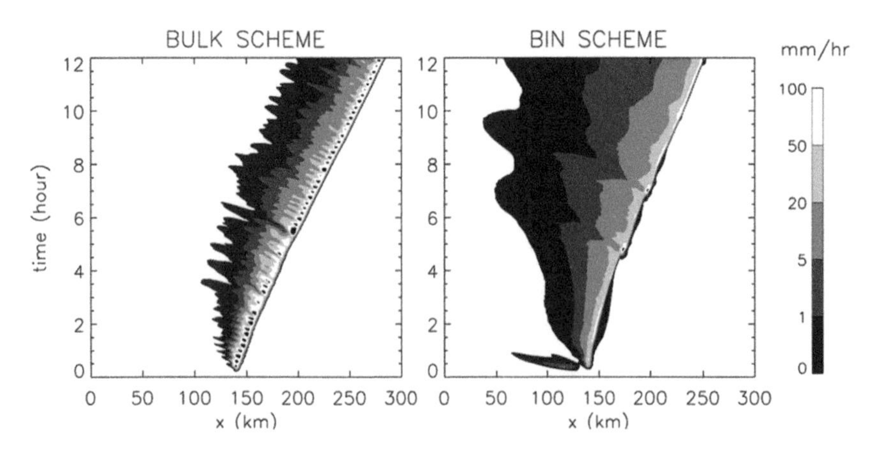

Figure 7.2.9 The time-domain diagram of surface rainfall obtained by the bulk scheme (left) and the bin scheme (right) (adopted from Li et al., 2009a; courtesy of © American Meteorological Society. Used with permission).

structure of the squall line in the bulk parameterization simulations was attributed by Li et al. (2009a, 2009b) to errors in the treatment of the evaporation rate (i.e., to cooling at low levels) as well as to errors in values of the fall velocity of graupel. Tuning the evaporation rate and the graupel fall velocity enabled them to obtain results closer to those of the SBM.

Figure 7.2.10 shows the vertical profiles of the heat budget items related to different microphysical processes. The budgets were calculated for the entire computational area. Bulk parameterization yields to higher condensation and stronger drop evaporation, yet similar rates of freezing and melting as compared to the SBM. In general, despite great differences in the local precipitation rates and spatial precipitation distributions, the accumulated rain is nearly the same: the area-averaged rain rates are 2.7 mm/h in bulk parameterization and 2.4 mm/h in SBM. The large spikes in both deposition and sublimation at around 10 km in the bulk simulation seen in Figure 7.2.10 are noteworthy. These spikes are produced artificially by utilization of the saturation adjustment in the bulk scheme. Although these two spikes on the whole cancel each other out and have little effect on the total energy budget, the heating and cooling they cause can substantially affect the structure of a squall line in the bulk-parameterization.

A squall line was also simulated by Morrison et al. (2009a), who used a two-moment bulk parameterization scheme they developed earlier (Morrison et al., 2005b; Morrison and Grabowski, 2007) and implemented into the WRF. This model includes prognostic variables for the mixing ratio and the number concentration of graupel. The simulation results were compared to those obtained using a single-moment scheme. A Hovmöller plot of the surface rainfall rate for the two-moment and one-moment simulations is presented in **Figure 7.2.11**. The main finding of the study was that the two-moment scheme produced a much wider and more prominent region of trailing stratiform precipitation as compared to the single-moment scheme. This difference was attributed to the fact that the rain evaporation rate in the stratiform region in the two-moment scheme was higher as compared to the one-moment scheme. The difference in the rain-evaporation rates reflects the difference in the raindrop distributions expressed in the values of the intercept parameter and the slope parameter. In the two-moment scheme where the intercept parameter was calculated, it ranged from 10^5 to 10^7 m^{-4} in the stratiform region, and from 10^7 to 10^9 m^{-4} in the convective region. In contrast, in the single-moment scheme, the value of this parameter was set constant at 10^7 m^{-4}. Larger values of the intercept parameter in the convective region are related to higher collision rates, while rain in the stratiform region was primarily produced by melting snow. Morrison et al. (2009a) failed in choosing parameters of raindrops' PSD in their 1M scheme to reproduce the results of the 2M scheme. They also showed that the spatial-precipitation distribution strongly depends on the method of calculating parameters determining the PSD shape. Baldauf et al. (2011) carried out multiple simulations of squall lines by means

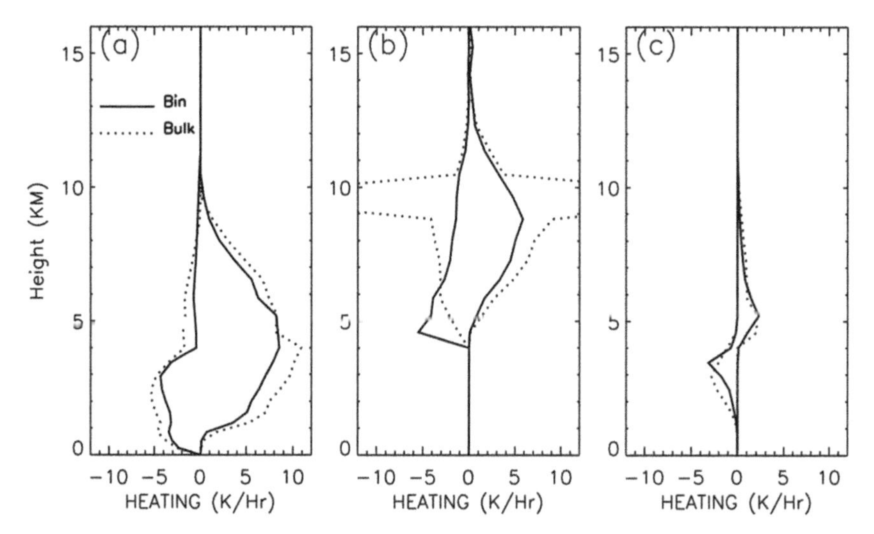

Figure 7.2.10 Vertical profiles of air heating/cooling due to different microphysical processes simulated by the bulk scheme (dashed lines) and the bin scheme (solid lines): (a) condensation and evaporation, (b) deposition and sublimation, and (c) melting and freezing (adopted from Li et al., 2009a; courtesy of © American Meteorological Society. Used with permission).

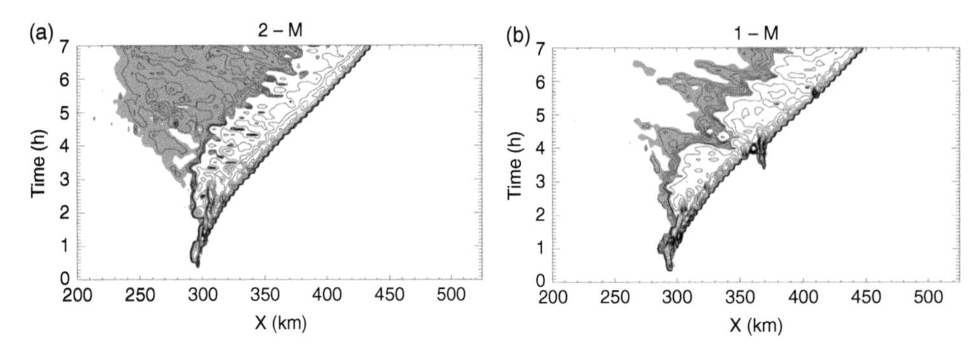

Figure 7.2.11 Hovmöller plot of the surface rainfall rate for the two-moment (a) and the one-moment (b) simulations. The contour interval is every 1 mm h^{-1} for rates between 0 and 5 mm h^{-1} and every 10 mm h^{-1} for rates exceeding 10 mm h^{-1}. To highlight the stratiform rain-precipitation region, moderate precipitation rates between 0.5 and 5 mm h^{-1} are shaded gray (from Morrison et al., 2009a; courtesy of © American Meteorological Society. Used with permission).

of the COSMO weather forecast model and also found an advantage of a more sophisticated two-moment scheme in simulations of strong squall lines.

Khain et al. (2009) simulated the same squall line as Morrison et al. (2009a), but using the WRF with two microphysical schemes: a SBM scheme and a two-moment bulk-parameterization scheme (Thompson et al., 2004, 2008) (Table 7.2.3). Simulations were carried out at different CCN and, correspondingly, at different cloud droplet concentrations. The bulk scheme produced substantially lower CWC at low CCN concentration and substantially lower RWC at high CCN concentration as compared to the SBM results. Similar differences in the simulated structure of squall lines were reported by Li et al. (2009a, 2009b), where another bulk scheme was used.

All the results discussed support the conclusion that two-moment bulk-parameterization schemes surpass single-moment ones in predicting the spatial structure of rain in squall lines, but the spatial distribution depends heavily on parameters determining the DSD shape.

Supercell Storms

Khain and Lynn (2009) simulated a supercell storm using a 2-km resolution WRF/SBM model (Fast SBM) and the WRF with the Thompson (2004) bulk-parameterization schemes. The computational area was 252 km × 252 km and the maximum time step was 10 s. The simulations were performed for both clean and polluted conditions, as well as at moderate relative humidity (more typical of the Great Plains) and at higher relative humidity (more typical of the southern Gulf Coast). The difference between the relative humidity values was about 10%. The maximum vertical velocities in the SBM simulations range from 25 to 40 m/s. The Thompson scheme produced maximum vertical velocities ranging from 45 to 65 m/s. Both schemes produced two branches of precipitation, but in the SBM simulations the right-hand branch dominates, as compared to the Thompson simulations, where the left-hand branch dominates (**Figure 7.2.12**). Analysis shows that these differences in precipitation structures are caused by the differences in the vertical velocities produced by the schemes, which cause the hydrometeors to ascend by different altitudes with different directions of the background flow. The bulk-parameterization scheme produced twice as much accumulated rain compared with the SBM. An increase in the humidity by several percent in both schemes increases the precipitation amount by a factor of two.

A study investigating the same supercell storm was carried out by Lebo and Seinfeld (2011). The results obtained using the MMM scheme (Reisin et al., 1996) were compared with those obtained by means of the two-moment bulk scheme developed by Morrisson et al. (2005a). The cumulative surface rain was similar in the SBM and the MMM at both high and low relative humidity. The accumulated rain amount produced by the bulk scheme was twice as large as that in MMM. Results of MMM were similar to those of the WRF/BM. Both Khain and Lynn (2009) and Lebo and Seinfeld (2011) found that the responses of the accumulated rain amount to changes in CCN concentration in the bulk schemes were opposite to those simulated by the SBM schemes. Lebo and Seinfeld (2011) explained the different responses by the different impacts of the microphysical processes on the latent heat release, and subsequently on the vertical velocity.

Hailstorms

Hailstorms pose a serious threat to agriculture and property across the globe. The hail-induced damage

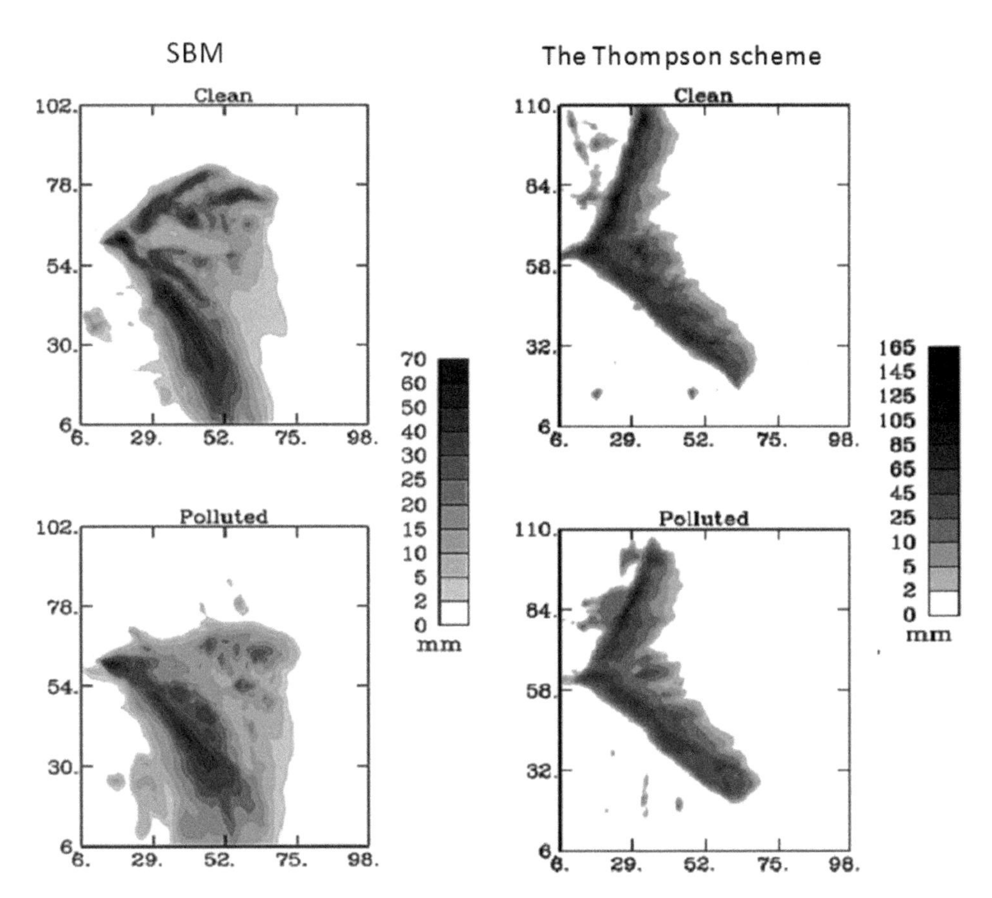

Figure 7.2.12 Accumulated surface rain simulated using the SBM and the Thompson scheme at 240th min of simulation at relatively high humidity in clean air (upper row) and in polluted air (bottom row). Gray scales applied for illustration of results obtained using SBM and the bulk scheme are different (from Khain and Lynn, 2009; courtesy of © John Wiley & Sons, Inc.).

rapidly increases with hail size. Most bulk schemes do not include hail, although it is a major hydrometeor determining precipitation in deep convective clouds (Khain et al., 2011). Representation of hail in some bulk schemes was improved: Noppel et al. (2010b) included large hail as a new hydrometeor type into the 2M bulk scheme of the COSMO model; Lang et al. (2014) and Tao et al. (2015) included hail as a fourth hydrometeor type into the Goddard Flight Center cloud model; Loftus et al. (2014) and Loftus and Cotton. (2014a, 2014b) developed a 3MHAIL bulk parameterization in which hail was described using three PSD moments, while the other hydrometeors were described by only two moments. The 3MHAIL parameterization was included into RAMS (Table 7.2.3).

Since hail grows largely by accretion of supercooled water, the mass of supercooled water would be expected to substantially affect the size of hail particles. Simulations of large hail require an accurate reproduction of the wide PSD tail. To illustrate this issue, we compare the results obtained by two advanced microphysical models used to simulate the hail storm in Villingen-Schwenningen, southwest Germany, on June 28, 2006. This storm produced hailstones with diameters up to 5 cm and the radar reflectivity of up to 70 DBz. Measurements showed that CCN concentration was very high, up to 6,000 cm^{-3}. The storm was simulated using the 2D SBM HUCM (Khain et al., 2011; Kumjian et al., 2014; Ilotoviz et al., 2016) and the 3D COSMO with the two-moment bulk microphysical scheme (Seifert and Beheng 2006a, 2006b), and a parameterization described by Noppel et al. (2010b). The models are discussed in Section 7.1. To investigate effects of aerosols on hail characteristics, the simulations were performed under high continental CCN concentrations and low CCN concentrations typical of maritime air masses.

Although both models proved able to simulate the hailstorm, the parameters of the simulations turned out to be quite different. The radar-reflectivity field simulated by HUCM with high CCN concentration agrees well with the observations, as the maximum radar reflectivity reaches 70 DBz and the high reflectivity values reach altitudes of 10–11 km. The SBM-simulated maximum reflectivity at CCN concentration was substantially higher than at low CCN concentration. The diameter of hailstones at high CCN concentration reaches 5 cm, which was in agreement with the observations. In clean air, the diameter of hail particles was typically below 2 cm. The hail shaft at the surface was much higher for the polluted air than in the clean air. The appearance of a negligibly small hail shaft in the clean air was related to melting of comparatively small hail particles. In contrast, the COSMO with bulk parameterization predicts higher values of the radar reflectivity and larger areas of high radar reflectivity in the clean air. In the simulations with bulk parameterization, the radar reflectivity rapidly decreases with height above 6 km. The accumulated rain, the hail shaft, and the size of hailstones were substantially larger in the clean air.

Actually, SBM and two-moment schemes simulate different mechanisms of hail growth. In SBM at high CCN concentration, hail grows largely by the accretion of supercooled water, while in a two-moment bulk scheme hail forms largely by freezing of raindrops at comparatively low distances above the freezing level. Accordingly, the SBM and the bulk models responded differently to changes in aerosol concentration.

Loftus and Cotton (2014a, 2014b) used the 3MHAIL bulk parameterization to simulate the hail storm over northwest Kansas on June 29, 2000, during the Severe Thunderstorm and Electrification and Precipitation Study (STEPS). The 3MHAIL results were compared with those of two different two-moment schemes: the bin-emulating scheme (REG2M) used in RAMS (Cotton et al., 2003) and a simplification of the 3MHAIL scheme. All of the bulk schemes tested by Loftus and Cotton, (2014b) managed to simulate formation of strong hail storms with maximum velocities up to 40 m/s. In all simulations, the integral mass of hail in cloud volume is larger at lower CCN concentrations. At the same time, the two-moment bulk schemes were capable of simulating the observed high radar reflectivity at altitudes of 9–11 km. Large hail in the bulk simulations did not exceed 2 cm in diameter. **Figure 7.2.13** shows surface-accumulated amounts of hail and rain at the end of the simulations (210 minutes) using REG2M and 3MHAIL schemes. In Figure 7.2.13, accumulated rain amounts are also presented. One can see that the surface precipitation simulated by 3MHAIL consists largely of hailstones, while the two-moment scheme REG2M does not produce hail at the surface. Total (rain + hail) precipitation in the 3MHAIL simulation is substantially higher.

A comparison of the calculated radar reflectivity, polarimetric parameters, and hail size with observations indicates that the SBM and 3MHAIL simulated the microphysics of the hail storm much better than the two-moment schemes. Both the SBM and 3MHAIL produced hailstones with diameters of several centimeters at high CCN concentration, and predicted an increase in hail precipitation at the surface level in the polluted air. Simulations of a hail storm using a three-moment bulk scheme (Milbrandt and Yau, 2006) confirmed the ability of such schemes to simulate large hail stones. The results of the studies discussed show that simulation of phenomena such as large hail requires an accurate description of the PDS tail, which can be reached by the utilization of at least three moments of hail PSD to simulate hailstones of several centimeters in diameter.

Hurricanes

Hurricane intensity depends on the rate and the spatial distribution of latent heat release in clouds. We will illustrate the effects of microphysical parameterizations on the intensity of a simulated TC using Hurricane Irene (which moved northward along the U.S. coast during the second half of August, 2011) as an example. Simulations were performed using WRF/SBM. The model geometry contained several nested grids with the resolution of the finest grid over the area of 500 km × 500 km of 1 km. This grid moved together with the TC center. TC–ocean coupling was taken into account. Two aerosol scenarios were considered. In the first scenario, the aerosol concentration was assumed to be low (100 cm^{-3} at $S_w = 1\%$) over the entire computational area, including North America and the Atlantic. In the second scenario, the concentration of aerosols over land was increased to ~2,000 cm^{-3}. Simulations with low CCN concentrations are denoted as "Mar," simulations with high CCN concentrations over land are denoted as "COMP," and simulations with bulk schemes using high CCN concentrations (or low droplet concentrations) over the entire area are denoted as "Con." The model denoted as Tho–Aerosol Aware (Thompson and Eidhammer, 2014) predicts the number concentration of aerosols. Similarly to SBM, the Tho–Aerosol Aware scheme was used in two versions denoted Tho–A_MAR and Tho–A_COMP. The microphysical schemes are generally described in Table 7.2.3.

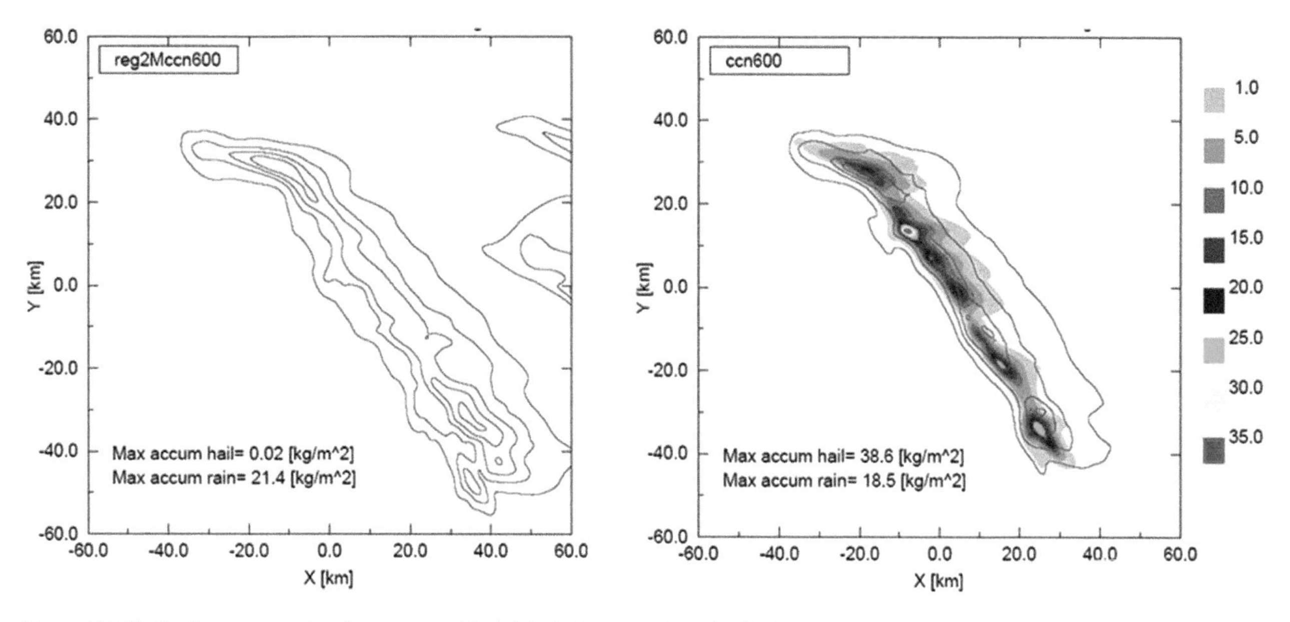

Figure 7.2.13 Surface-accumulated amounts of hail (shaded contours) and rain (blue contours) at end of simulation (210 minutes) by REG2M (left) and 3MHAIL (right). Contour values are 1, 5, 10, 15, 20, 25, 30, and 35 kg·m^{-2}. The CCN concentration is 600 cm^{-3} (from Loftus et al., 2014b, with changes; courtesy of © Elsevier).

Figure 7.2.14a–c shows time dependencies of the minimum pressure and the maximum wind speed in observations as well as in different simulations using SBM and bulk schemes. The upper row shows time dependencies of the minimum pressure (left) and the maximum wind velocity (middle and right) in simulations carried out with WRF with bin microphysics and using the Morrison and Thompson schemes allowing changes of aerosol or droplet concentrations. A_COMP produced the best agreement with observations. Among the bulk schemes, the Thompson and Morrison schemes and especially the two-moment Tho–A_COMP produce the best agreement with observations. However, Figure 7.2.14a–c shows that the TC intensities of TC simulated by these bulk schemes are not sensitive to aerosols (in contrast to the SBM). This problem is discussed in Section 7.3.

A specific feature of Hurricane Irene was a 40-h time shift between the maximum wind speed and the minimum surface pressure. This effect is related to the increase in the TC size caused by aerosols that penetrated to it and fostered formation of secondary eyewall at a larger distance from the TC center (Khain et al., 2016; Lynn et al., 2016). Figure 7.2.14 (upper row) shows that only SBM that takes into account the effect of continental CCN was able to simulate this time shift.

Figure 7.2.14 (bottom row) shows the time dependencies calculated in all the simulations with different bulk schemes. The high spread in the simulated data suggests a substantial sensitivity of simulated TC intensity to the treatment of cloud microphysics. The variability of the minimum surface pressure and the maximum wind speed predicted by different bulk schemes is very high (up to 40 mb and 60 kt (30 m/s), respectively). Tao et al. (2011) also reported a high variability of model TC intensities simulated by different bulk schemes. Such high variability indicates that TC intensity is extremely sensitive to the value and spatial distribution of latent heat release that is determined differently by different bulk schemes. None of the bulk schemes predicted the observed time shift between the maximum wind speed and the minimum surface pressure.

Figure 7.2.15 shows the dependencies of the maximum wind speed on the minimum surface pressure in the 1-km resolution simulations of Irene with different bulk schemes and with WRF/BM. The observed dependence is also presented. One can see that the WRF/SBM simulates the observed dependence well. Among the bulk schemes, the two-moment Thompson scheme (Thompson and Eidhammer, 2014) produces the best agreement with the observations, while other schemes show substantial differences with the observed data. All of the bulk schemes predicted the minimum surface pressure and the maximum wind speed at the same time instance. (Figure 7.2.15). The SBM predicts the maximum wind speed and the minimum surface

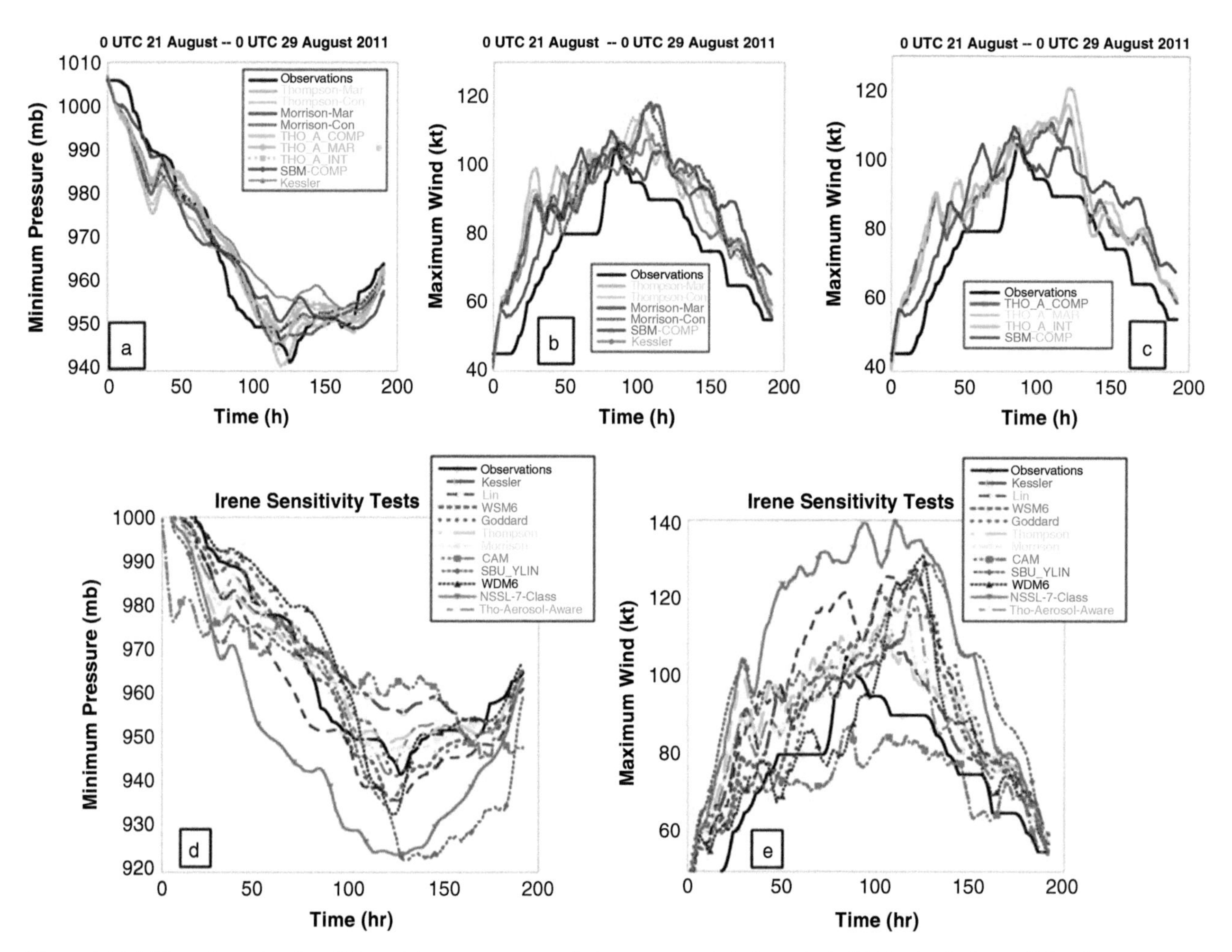

Figure 7.2.14 Time dependencies of the minimum pressure and the maximum wind speeds in WRF simulations of Hurricane Irene (2011). Upper row: time dependencies derived from observations, and simulated by WRF/SB and by the Morrison and Thompson schemes. The dependencies obtained in simulations using the simplest warm-rain Kessler parameterization are also presented. Bottom row: time dependencies in WRF simulations using different bulk schemes. Notations: simulations with low CCN (or droplet) concentrations are denoted as "MAR." Simulations with high CCN (or droplet) concentrations are denoted as "CON." The scheme denoted as "Thompson" is described by Thompson et al. (2004). The scheme denoted as "Morrison" is described by Morrison et al. (2005b). The scheme denoted as Tho–Aerosol Aware is described in Thompson and Eidhammer (2014) and was used in two versions denoted as Tho–A_MAR and Tho–A_COMP (from Khain et al., 2016, with changes; courtesy of © Elsevier).

pressure at different time instances, which is seen by two circles on the corresponding curve. The SBM results agree well with the observations. The successful simulation of the maximum wind speed–minimum surface pressure dependencies by the WRF/BM confirms the importance of high-quality microphysical schemes in predicting TC intensity.

Simulation of Cloud Radar Properties

Multiple studies investigated the microphysical structure of clouds using satellites or ground-based radars as well as using cloud models of various complexity, which are able to calculate radiation and radar parameters from dynamically simple 1D SBM models (e.g., Kumjian et al., 2012; Ryzhkov et al., 2013a, 2013b) to mesoscale and large-scale ones (e.g., Suzuki et al., 2011; Alexandrov et al., 2012, Kumjian et al., 2014, Carlin et al., 2016; see also Table 7.2.2). By way of example, one of these studies is discussed next.

Iguchi et al. (2012a) simulated three precipitation events using the Japan Meteorological Agency Nonhydrostatic Model (JMA-NHM). These events were also observed by a ship-borne radar and space-borne W-band cloud radars of the CloudSat satellite. The

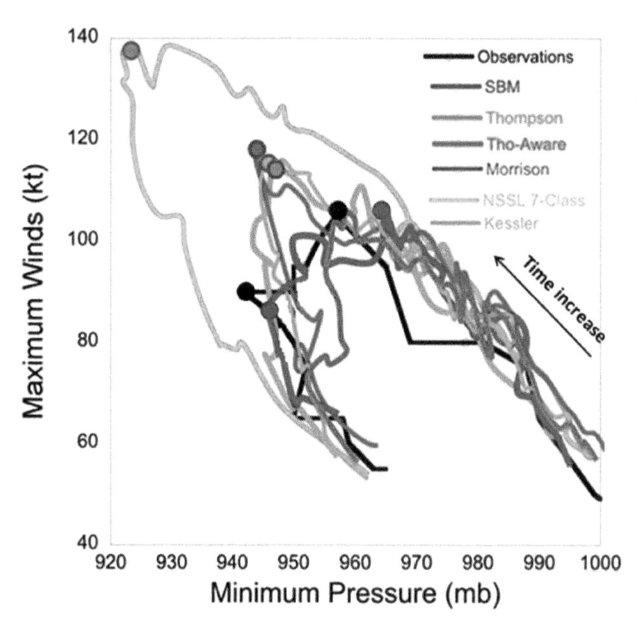

Figure 7.2.15 The dependencies of the maximum wind speed on the minimum surface pressure calculated at the 1-km resolution simulations of Irene using the WRF model with the TC–ocean coupling taken into account. The dependencies are plotted for different microphysical schemes. The dependence calculated using the observed data is presented, as well. Circles denote the points of the minimum surface pressure and the maximum wind speed. In bulk schemes the circles coincide, which indicates that these schemes predict minimum pressure and maximum wing at the same time instances.

SBM (Khain et al., 2004a) and a single-moment bulk microphysics scheme (Lin et al., 1983) were employed for comparison. **Figure 7.2.16** shows the Contoured Frequency Altitude Diagrams (CFADs) (normalized reflectivity–height histograms) obtained during the period from 1200 UTC May 22 to 1200 UTC May 23, 2001. One can see that the SBM simulates the observed vertical distribution of the radar reflectivity much better than the bulk scheme. The SBM simulates the decrease in the reflectivity with height down to −30 DBz at the altitude of 11 km. At the same time, bulk scheme substantially overestimates the reflectivity at high levels. Iguchi et al. (2012a) found that it was necessary to take into account the increase of the fall velocity of snow due to riming (panel c). This example shows that comparison of simulated and observed radar data can be used to improve a model's microphysics.

Microstructure of Stratiform Arctic Clouds

The sensitivity of phase composition of Arctic stratiform clouds to IN concentration varies between models. Model intercomparison studies (e.g., Morrison et al., 2009b, 2011; Muhlbauer et al., 2010) reveal a dramatic dispersion in prediction of the liquid-ice partitioning under the same environment conditions between different models. Ovchinnikov et al. (2014) conducted intercomparison of seven LES models with different microphysical parameterizations. The models were used to simulate Arctic clouds observed during Indirect and Semi-Direct Aerosol Campaign (ISDAC) (McFarquhar et al., 2011). Among the models used for intercomparison, only two models used SBM, namely SAM–SBM and DHARMA–bin (Distributed Hydrodynamic Aerosol and Radiative Modeling Application) (Avramov and Harrington, 2010) (Table 7.2.1). Although the major uncertainty concerning ice nucleation was excluded (concentration was set at the observed value), significant differences between models were found in depositional growth rates, precipitation fluxes, and the values of the liquid-water pass (LWP) and the ice water pass (IWP). These differences are related to the differences in representations of diffusion growth/deposition and evaporation/sublimation that are the basic processes in mixed-phase stratiform clouds. The dispersion of the results is illustrated in **Figure 7.2.17**, which compares the time dependencies of LWP and IWP obtained using LES models with different microphysics in the set of simulations with target ice crystal concentration of 4 L^{-1}. One can see a wide spread of the LWP values. Regarding the IWP, a stationary state was reached. The values of the IWP in the SBM simulations were found to be substantially larger than in the bulk ones. Interestingly, different SBM schemes produced similar results. Ovchinnikov et al. (2014) found that bulk schemes provide better agreement with bin schemes when ice-size spectra are approximated by the Gamma distributions with widths comparable to that predicted by the bin schemes. The results show that an accurate simulation of PSD of ice crystals and drops is required to simulate the main microphysical parameters of mixed-phase stratocumulus clouds.

Large-Scale and Long-Term Convection

The distribution of clouds with respect to their heights at long time scales is determined by long-term forcing such as radiation forcing and aerosol forcing.

Observational studies show that over large regions and on time scales of months, cloud top, cloud cover, and precipitation rates are higher in polluted environments (Niu and Li, 2012; Bell et al., 2008; Koren et al., 2012). At the same time, idealized modeling performed using bulk schemes does not show any sensitivity of cloud fraction to CCN (Van den Heever et al., 2011; Khairoutdinov and Yang, 2013).

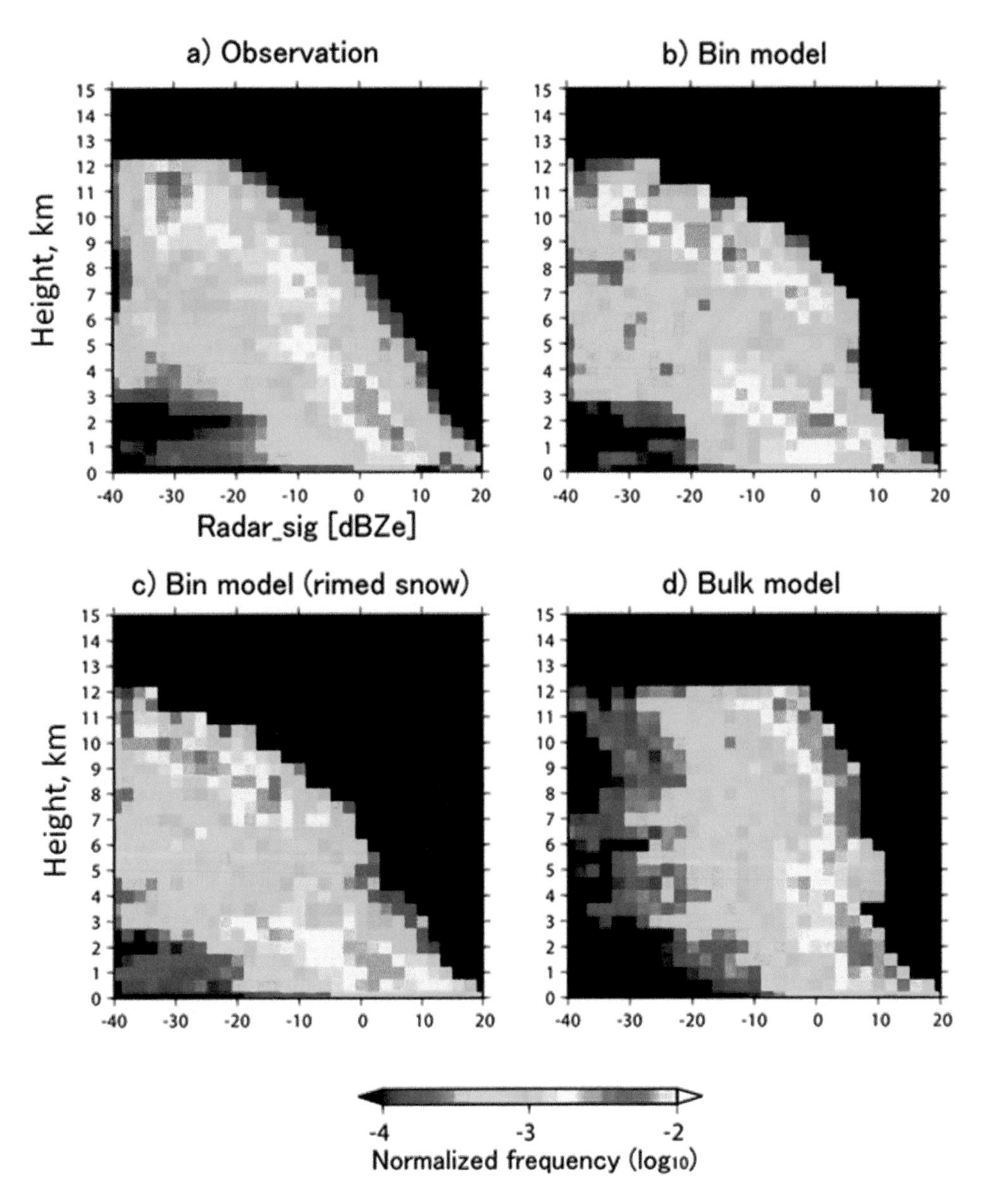

Figure 7.2.16 Normalized dBZe-height histograms constructed on the basis of the radar measurement (a) and the simulations by the bin (control) (b), the bin (rimed snow) (c), and the bulk model (d) during the period of 1200 UTC May 22 to 1200 UTC May 23, 2001 (from Iguchi et al., 2012a; courtesy of © American Meteorological Society. Used with permission).

Using WRF/BM, Fan et al. (2014) conducted 3D month-long cloud-resolving simulations of summertime convection in three regions: the tropical western Pacific, southeastern China, and the U.S. Southern Great Plains to represent tropical, midlatitude coastal, and midlatitude inland summer convective clouds, respectively. Computation areas were covered by nested grids with the size of the outermost grid of 1,400 km and of inmost grids of about 600 km × 600 km. The grid spacing of inmost grids was 2-km. Two simulations were performed using two CCN concentrations that differed by a factor of six to represent clean vs. polluted environments. The relatively long time scale allowed Fan et al. (2014) to calculate the radiative properties of clouds in a quasi-stationary state. The WRF/SBM simulations captured the observed macrophysical and microphysical properties of summer convective clouds and precipitation in the tropics and midlatitudes. This study showed that an increase CCN concentration leads to the increase in the area and the depth of anvils of

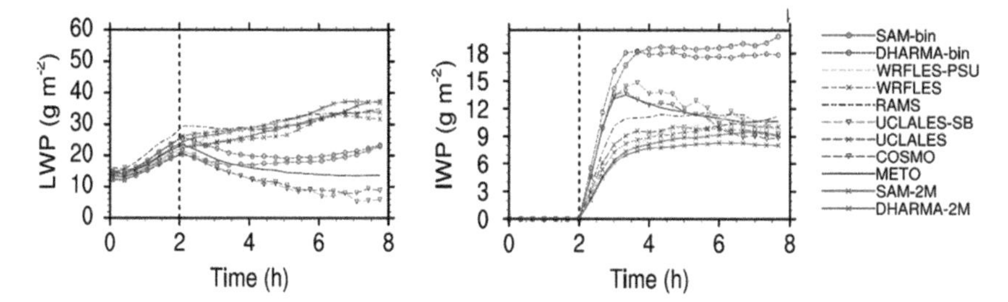

Figure 7.2.17 Time dependencies of LWP and IWP obtained using different models. Ice crystal concentration was 4 L^{-1}. Titles of the models: SAM_SBM (PNNL); SAM–2M (bulk) (PNNL); DHARMA–bin; DHARMA–2M(bulk) (NASA GISS); UCLA–LES–SU (Stockholm Univ); UCLA–LES (NASA Langley); COSMO (Karlsruhe Univ.); MetOffice (Met. Office UK); RAMS (Penn State); WRF_LES (NOAA); WRF–LES–PSU (Penn State). SAM–2M, DHARMA–2M, UCLA–LES, and WRF–LES employ the same bulk microphysics scheme (from Ovchinnikov et al., 2014; courtesy of © American Meteorological Society. Used with permission).

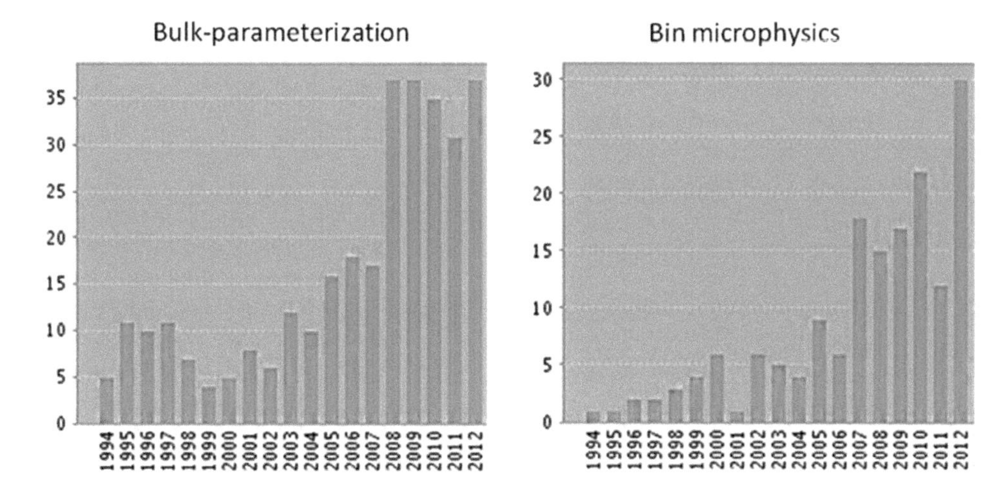

Figure 7.2.18 Annual number of publications found in "Web Science" via the key words "bulk parameterization" and "bin microphysics" (from Khain et al., 2015; courtesy of © John Wiley & Sons, Inc.).

deep convective clouds in agreement with observations. The increase in the cloud cover is explained in the study by the fact that the increase in the CCN concentration leads to formation of smaller ice crystals with lower fall velocities.

The decrease in the sedimentation velocity of ice crystals with increasing CCN concentration is the key factor responsible for slow cloud dissipation and increased cloud fraction, cloud top height, and cloud thickness. Fan et al. (2014) believed that bulk schemes cannot inherently simulate such effects because they do not calculate a distribution of particle size-resolved fall speeds within each model grid box. As a result, the evaluations of the CCN effect on the radiative balance were also found to be different. This example indicates the extreme importance of an accurate representation of sedimentation velocity and sublimation rates, not only for large precipitating particles, but also for small ice crystals affecting the radiative balance of the atmosphere.

7.2.5 Some Statistics on SBM and Bulk Model Application

Figure 7.2.18 shows the annual numbers of publications found via the key words "bulk parameterization" and "bin microphysics" since 1994. Needless to say, the number of related publications is much larger than that shown in Figure 7.2.18, since many titles do not contain these key words. One can see a surge in publications reporting utilization of both bulk-parameterization and SBM schemes in 2007–2008. Although the share of SBM is lower, it has been continuously increasing, perhaps indicating the increased application of SBM in mesoscale cloud-resolving models.

7.3 Effects of Aerosols on the Structure and Microphysics of Clouds and Cloud Systems

7.3.1 Major Aerosol Effects

The atmosphere contains a great number of suspended aerosols. It was assumed in climatic research and, in particular, GCM that the main effect of aerosols is that on the radiation budget of the atmosphere. There are numerous mechanisms by means of which aerosols affect different items of the budgets of atmospheric heat and radiation. First, aerosols directly affect the radiation budget of the atmosphere via their impact on reflection, scattering, and extinction of solar and infrared radiation. Effect of this type is called *direct aerosol effects*.

Clouds play a crucial role in the radiation balance of the atmosphere (see Section 1.1). Changes in their microphysical structure lead to changes in reflection, and to changes in the radiation balance as a whole. Twomey (1974, 1977) focused on the effect of aerosols on radiation via their impact on droplet size and concentration. An increase in aerosol concentrations increases concentration of droplets and decreases their size. As a result, for the same liquid-water path, the reflectance of droplet water surfaces increases, which means an increase in cloud albedo. Since aerosols affect radiation indirectly, via their effects on droplet size, this effect is called the *first indirect aerosol effect*. Beginning with these studies, many different mechanisms by means of which aerosols affect radiational cloud properties were reported.

Aerosols affecting DSD also change the liquid-water path by affecting precipitation formation and thus the liquid water content, cloud lifetime, and cloud coverage (Albrecht, 1989). This is another mechanism of aerosol impact on the radiative properties of clouds, known as the *second indirect aerosol effect*. An important component of the second indirect effect is the effects of aerosols on precipitation (including drizzle) in clouds. The onset of precipitation dramatically changes all cloud radiation characteristics. The direct and indirect aerosol effects as they were formulated at the Intergovernmental Panel on Climate Change (IPCC) are schematically illustrated in **Figure 7.3.1**. The scheme in Figure 7.3.1 includes the aerosol effects on radiative properties only via warm-cloud microphysics. At the same time, aerosols also affect liquid-ice partitioning and glaciation time (Section 6.3). Since liquid clouds and ice clouds have different radiative properties, the first and second indirect aerosol effects include processes of cloud glaciation and precipitation formation in mixed-phase clouds.

Despite significant efforts, many uncertainties still remain in explaining the second indirect effect. These uncertainties hinder the investigation of many fundamental problems, including the global atmospheric circulation and climate. According to the early hypotheses, an increase in aerosol concentration leads to suppression of precipitation formation and to an increase in the liquid-water path (Albrecht, 1989). Subsequent studies showed that the picture is more complicated, since many dynamic and thermodynamic processes are involved in precipitation formation and in determination of surface precipitation.

The role of clouds is not limited to their effects on the radiative balance of the atmosphere. Clouds and precipitation determine the hydrological cycle of the atmosphere (Section 1.1) and local and global precipitation regimes. The latent heat release in clouds is the main energy source of atmospheric motions. Thus, a special branch of cloud physics deals with the investigation of aerosol effects on precipitation formation and on rain amounts. This inquiry was triggered by the need to evaluate the effects of the continuously increasing production of anthropogenic aerosols on local and global precipitation regimes.

Processes related to the first and second indirect aerosol effects, as well as to aerosol effects on the thermodynamics of the atmosphere via their effects on clouds, are called *cloud-aerosol interaction*. Cloud-aerosol interaction also includes effects of clouds on concentration and size distribution of aerosols. Strictly speaking, the term "cloud-aerosol interaction" is not precise, because aerosol particles, i.e., CCN and IN are actually cloud particles, so cloud droplets and ice crystals are just the result of the diffusion growth of aerosols. However, we will use the term "cloud-aerosol interaction" as it is widely accepted. Aerosols affect all microphysical processes in clouds. The concentration and size distribution of CCN, together with the vertical velocity at cloud base, determine droplet concentration and, consequently, droplet size, height of first raindrop formation, rate of raindrop formation, etc. CCN and IN affect the microstructure of mixed-phase clouds (Chapters 5 and 6).

Aerosol effects are complicated even on the level of separate microphysical processes. In clouds, these numerous processes interact in mutual feedback, making the prediction of the net effects of change in aerosol characteristics quite challenging. For instance, an increase in CCN concentration increases the

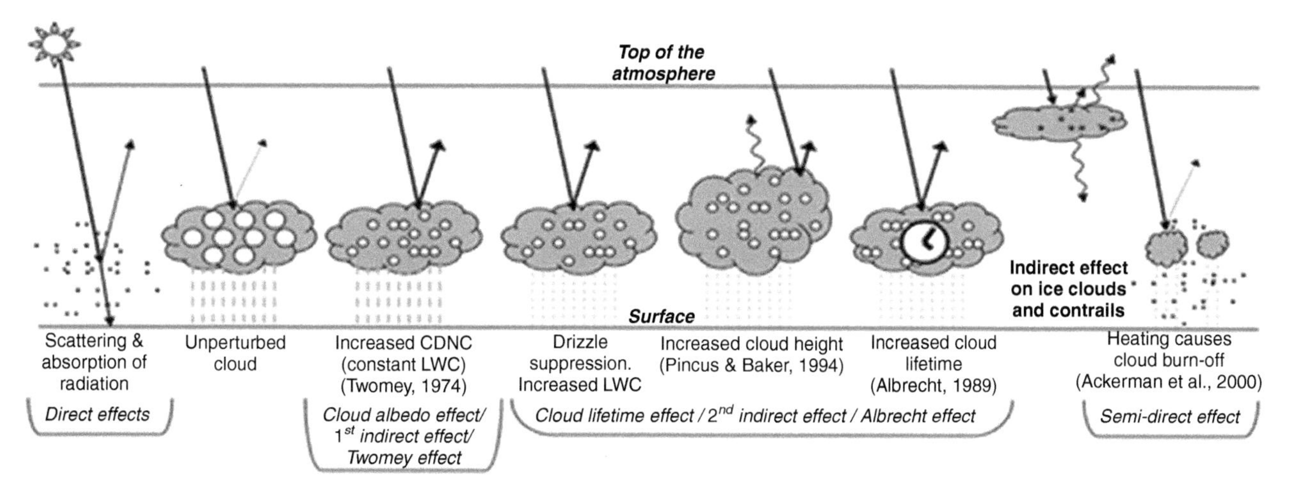

Figure 7.3.1 Various mechanisms proposed to explain how aerosols affect the radiation properties of clouds via affecting cloud microphysics and dynamics (adapted from Tao et al., 2012; courtesy of © John Wiley & Sons, Inc.).

concentration of droplets and decreases their size (Sections 5.2, 5.3). Since the collision kernels of small droplets are smaller than those of larger ones, smaller droplet sizes are associated with lower rates of collision. On the other hand, the rate of collisions is proportional to the square of the droplet concentration (Sections 5.6, 5.8), so an increase in CCN concentration should increase the collision rate. Thus, it is a complicated task to predict the net effect of aerosols on the collision process. Cloud models with most advanced microphysical schemes are required to investigate the effects of aerosols on the microphysical properties of clouds.

New aerosol effects on cloud microphysics have been found in the past decade. In this section, we will try to classify aerosol effects on the microphysics and dynamics of clouds, as well as on cloud-related phenomena (e.g., storms, hurricanes). The main focus will be on the effects of aerosols on precipitation. Three main characteristics of aerosols affect cloud microphysics and dynamics: aerosol concentration, aerosol sizes characterized by a size distribution, and the ratio of soluble and non-soluble fractions in aerosols. The physical mechanisms of aerosol impact on precipitation and dynamics of clouds and cloud systems depend on spatial scales and cloud types.

7.3.2 Effects of Aerosols on Surface Precipitation from Single Clouds

The main microphysical effect of aerosols in single deep convective clouds is that on droplet concentration, namely, an increase in CCN concentration increases droplet concentration (Section 5.3). Different CCN concentrations make clouds microphysically maritime when the CCN concentration is low, and microphysically continental when the CCN concentration is high. When droplet concentration is low, the supersaturation over water in the clouds is high, and droplets rapidly form raindrops at comparatively small distances above cloud base. These droplets grow fast, giving rise to the onset of warm rain (Section 5.11). New droplets are efficiently collected by raindrops, which shortens the residential time of droplets in clouds.

When droplet concentration is high, the supersaturation over water is low; droplets grow slowly and reach altitudes of several kilometers above cloud base before the first raindrops are formed. Thus, an increase in the CCN concentration increases the residential time of droplets in clouds. When the CCN concentration is high, raindrops form at negative temperatures (e.g., −10–15°C), so a substantial fraction of these raindrops freeze. The increase in the residential time of droplets within clouds and the intensification of freezing with increasing CCN concentrations leads to an increase in the latent heat release, which in turn increases the vertical velocity and sometimes cloud-top height and the area of cloud anvils. This dynamical effect, especially pronounced for convective clouds with a warm cloud base and high altitude of the freezing levels, is known as aerosol-induced convection invigoration. Aerosol-induced convection invigoration has been reported in many observational and numerical studies (Khain, 2009; Rosenfeld et al., 2008; Tao et al., 2012).

A schematic illustration of the microphysical and dynamical effects of CCN in deep convective clouds with a warm cloud base is presented in **Figure 7.3.2.**

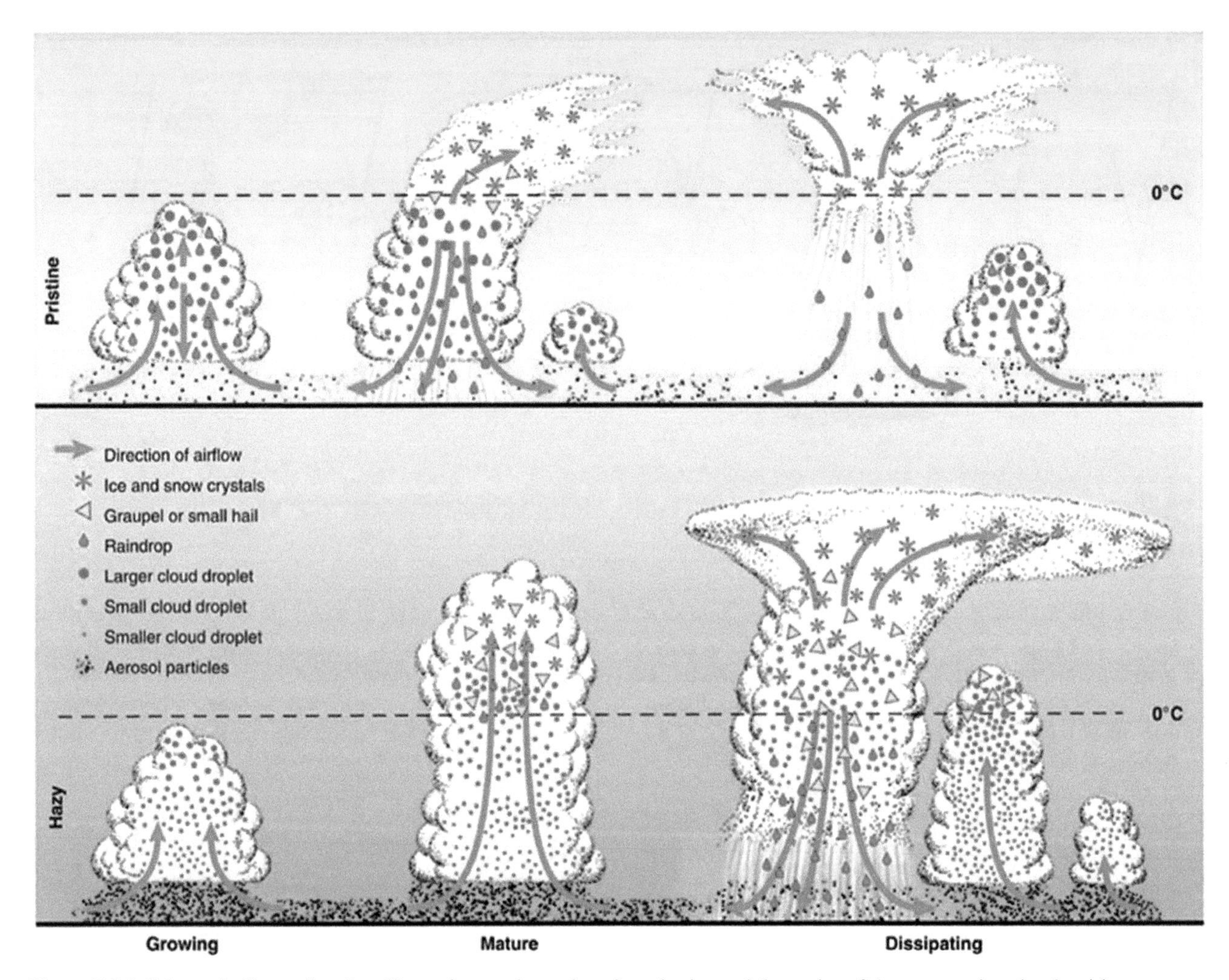

Figure 7.3.2 Schematic illustrating the effects of aerosols on the microphysics and dynamics of deepconvective clouds with a warm cloud base and high freezing level. The plots correspond to clean (top) and polluted (bottom) environments (adapted from Rosenfeld et al., 2008; published with permission of the American Association for the Advancement of Science).

The scheme shows a delay in warm rain formation in polluted atmospheres as compared to the clean air, and precipitation intensification at the mixed-phase mature stage in polluted clouds, as well as a higher cloud top and a larger area of cloud anvils in clouds developing in polluted air.

To simulate microphysical and dynamical aerosol effects, advanced microphysical models are required. A model of this kind was used by Fan et al. (2013), who simulated the evolution of clouds under different thermodynamical and aerosol conditions. The typical structure of clean and polluted deep convective clouds at the mature and dissipation stages as seen from these simulations is shown in **Figure 7.3.3**. In the polluted cloud, convective cores detrain larger amounts of cloud hydrometeors of much smaller sizes as compared to the clean environment. The decrease in particle size associated with high aerosol concentrations leads to a large expansion and much slower dissipation of the cloud anvils. This results in higher cloud top height and greater cloud depth at the mature stage. This is a consequence of lower fall velocities of ice particles formed in polluted air. Fan et al. (2013) found that the changes of the cloud anvil areas caused by an increase in CCN concentrations substantially affect different items of the radiation balance. Via their effect on the microphysical structure of each convective cloud, aerosols affect the rates of heating and cooling in the entire cloud region. **Table 7.3.1** summarizes observations of cloud and cloud system features continental environments with high aerosol and low aerosol concentration.

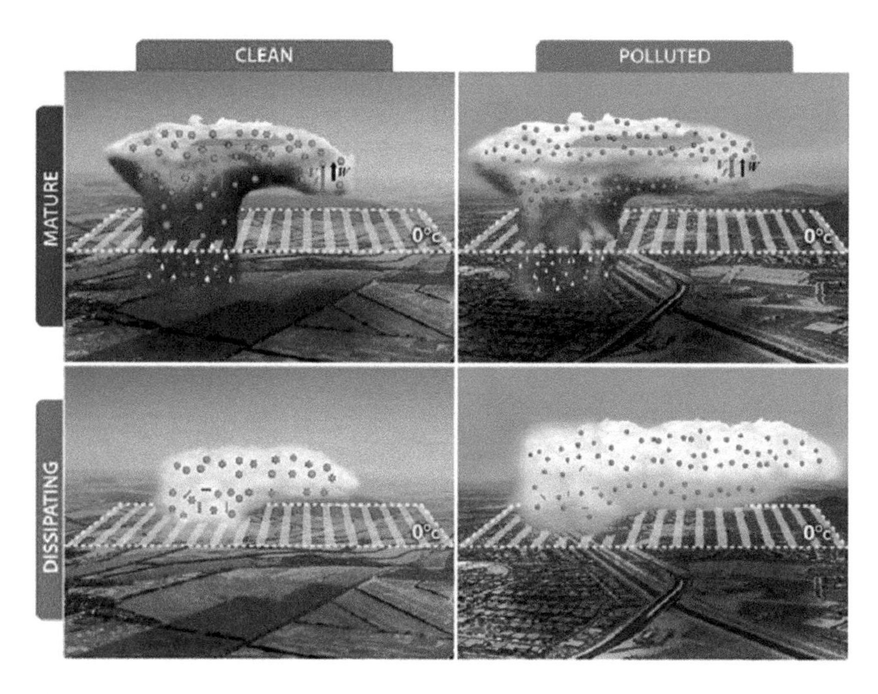

Figure 7.3.3 Schematic illustration of the differences in cloud top height, cloud fractions, and cloud thickness for deep convective clouds in clean and polluted environments. Red dots denote cloud droplets, light-blue dots denote raindrops, and blue shapes denote ice particles (from Fan et al., 2013).

Mass Hydrometeor Budget
One of most complicated problems regarding aerosol effects on cloud microphysics is the effects of aerosols on surface precipitation. To understand the mechanisms underlying these effects, it is necessary to consider the mass and energetic budgets of clouds. Clouds of different types respond differently to changes in aerosol concentration. Moreover, this response may be different for clouds of the same type depending on environmental conditions.

In a stationary situation, when the influx and the outflux of hydrometeor mass into/from a column of the atmosphere are small, the hydrometeor mass budgets in the column can be written as

$$P \approx G - L \tag{7.3.1}$$

where P is the precipitation rate and G is the generation rate of hydrometeor mass due to condensation or deposition and $L = E + S$ where L is the loss rate of the hydrometeor mass due to evaporation E and sublimation S. Precipitation efficiency PE is defined as the ratio (Khain, 2009):

$$PE \approx \frac{P}{G} = \frac{G - L}{G} \tag{7.3.2}$$

The PE is a function of time. Sometimes, precipitation efficiency is defined as the ratio of the precipitation amount to the vapor flux at the cloud base (e.g., Sui et al., 2007). This definition is thought to have come from traditional convection parameterizations assuming that water vapor ascends within clouds only, and totally condensates reaching the cloud top. When this assumption is valid, this definition is similar to that given in Equation (7.3.2).

Aerosols change all of the terms of the mass balance equation (Equation 7.3.1). Changes in aerosols lead to changes in the rate of generation by ΔG, and to changes in the loss of hydrometeor mass by ΔL. The net changes in precipitation caused by aerosols are

$$\Delta P \approx \Delta G - \Delta L \tag{7.3.3}$$

Consequently, the relative change in precipitation efficiency is

$$\frac{\Delta PE}{PE} \approx \frac{\Delta P}{P} - \frac{\Delta G}{G} \tag{7.3.4}$$

We illustrate aerosol effects on the mass and heat budgets using as example deep convective clouds typical of maritime tropical conditions observed during Experiment GATE-74, 261 day (Khain et al., 2008b; Khain, 2009). Simulations of tropical clouds were performed for the CCN concentrations (at 1% of supersaturation) of 100 cm^{-3} (M-run) and 2,500 cm^{-3} (M-c run). The simulations were performed using the

Table 7.3.1 Key observational studies identifying the differences in microphysical properties, cloud characteristics, thermodynamics, and dynamics associated with clouds and cloud systems developed in polluted and clean environments (from Tao et al., 2012, with additions; courtesy of John Wiley & Sons, Inc.©).

Properties	Increase in CCN concentration leads to:	References
Cloud droplet size; DSD width	Decrease in droplet size, narrowing DSD	Squires (1958), Radke et al. (1989), Ferek et al. (2000), Rosenfeld and Lensky (1998), Rosenfeld (1999, 2000), Rosenfeld et al. (2001), Rosenfeld and Woodley (2000), Andreae et al. (2004), Koren et al. (2005), Yuan et al. (2008), Freud et al. (2008), Freud and Rosenfeld (2012), Prabha et al. (2011, 2012)
Warm rain process	Delay and sometimes total suppression of warm rain formation	Squires (1958), Radke et al. (1989), Albretch (1989), Rosenfeld (1999, 2000), Rosenfeld and Woodley (2000), Rosenfeld and Ulbrich (2003), Andreae et al. (2004), Lin et al. (2006), Givati and Rosenfeld (2004), Li et al. (2011a), Freud et al. (2008), Freud and Rosenfeld (2012), Prabha et al. (2011, 2012)
The height of first raindrop formation, the level of first radar echo	Increase	Freud and Rosenfeld (2012)
Cold rain process	Intensification of ice processes	Rosenfeld and Woodley (2000), Orville et al. (2001), Williams et al. (2002), Andreae et al. (2004), Lin et al. (2006), Bell et al. (2008)
Mixed-phase region	Increase in the depth of mixed-phase region in Cu	Rosenfeld and Lensky (1998), Williams et al. (2002), Andreae et al. (2004), Koren et al. (2005, 2008, 2010a, 2010b), Lin et al. (2006), Li et al. (2011a), Niu and Li (2011)
Anvils of deep convective clouds	Increase in the area and depth of the anvils	Koren et al. (2010b)
Turbulence	Increase	Benmoshe et al. (2012, 2014)
Lightning	Enhanced (downwind side); higher flash maximum	Williams et al. (2002), Orville et al. (2001), Steiger et al. (2002), Steiger and Orville (2003), Yuan et al. (2011), Khain et al. (2008b), Cohen and Khain (2009)

HUCM with the computational area of 170 km × 17 km and a resolution of 350 m in the horizontal direction and 125 m in the vertical direction. To simplify the explanation of aerosol effects on the mass budget and the heat budget, we first present the vertical profiles of the mass-budget items (**Figure 7.3.4**). The figure shows the profiles of horizontally averaged moistening caused by evaporation and sublimation, and drying resulting from condensation and deposition for the 4-h period in the simulations of tropical clouds (M-run). The vertically integrated positive rates of evaporation and sublimation leading to the air moistening determine precipitation loss $L = E + S$ in Equation (7.3.1). The vertically integrated rate of the air moisture decreases due to condensation C and deposition D (the negative components in the air-moisture budget in Figure 7.3.1) corresponds to generation of hydrometeor mass $G = -(C + D)$. One can see that condensation of water vapor on droplets is the main source of the hydrometeor mass. It is important that the vertical profiles of condensation and evaporation rates, as well as of deposition and sublimation rates, are not mirror images. For instance, the condensation rate is substantial up to heights of 9 km, while the evaporation rate is concentrated largely below the melting level (4 km). The sum of deposition and condensation determines the generation of G, while the sum of sublimation and evaporation determines the loss of the condensate mass L. According to Equation (7.3.1), $G - L$ determines precipitation.

The net-aerosol effect on the heat and moisture budgets components can be seen in **Figure 7.3.5**, presenting the vertical profiles of horizontally averaged moistening/drying (left panel) and heating/cooling (right panel) over the 4-h period for M and M-c clouds. The increase in the CCN concentrations leads to an

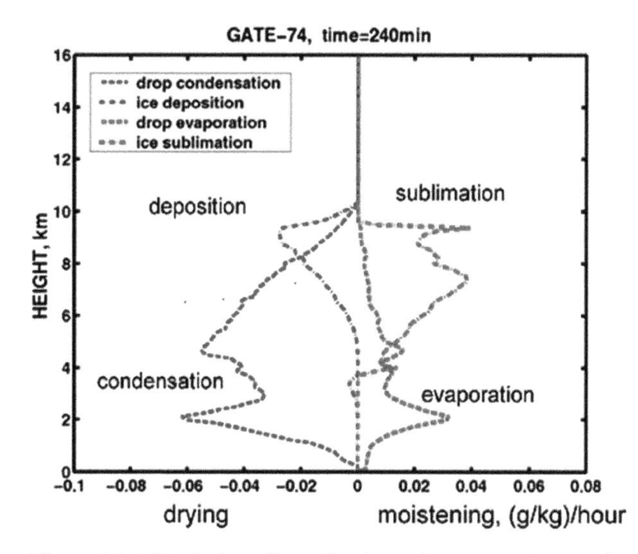

Figure 7.3.4 Vertical profiles of horizontally averaged rates of moistening and drying of the atmosphere for a period of 4 h in the M-run (adopted from Khain, 2009; courtesy of © IOP Publishing CC BY–NC–SA).

increase in both heating and cooling (as well as drying and moistening), i.e., increasing both the generation of hydrometeor mass ($\Delta G > 0$) and its loss ($\Delta L > 0$). This conclusion seems to be relevant for clouds and cloud systems, regardless of cloud type and environmental conditions (e.g., Khain et al., 2005, 2008b, 2011). The increase in the condensate mass with aerosol loading, which indicates an increase in atmospheric heating by the latent heat release, is the consequence of microphysical aerosol effects illustrated in Figure 7.3.2. A higher condensate loss by sublimation and evaporation in cases of larger aerosol loading ($\Delta L > 0$) is attributed to the fact that particles in polluted clouds are usually smaller than those in clouds forming in clean air since they fall from higher levels. Besides, in the presence of the vertical shear of the background flow, the particles in polluted clouds tend to fall outside of cloud updrafts through comparatively dry air, which increases the sublimation and evaporation. In contrast, raindrops at low CCN concentration form at low levels within humid cloudy air, so the precipitation loss is relatively low.

The effect of aerosols on precipitation is determined by the relation between ΔG and ΔL (Equation 7.3.3). If an increase in aerosol concentration leads to the condition $\Delta L - \Delta G > 0$, a decrease in precipitation takes

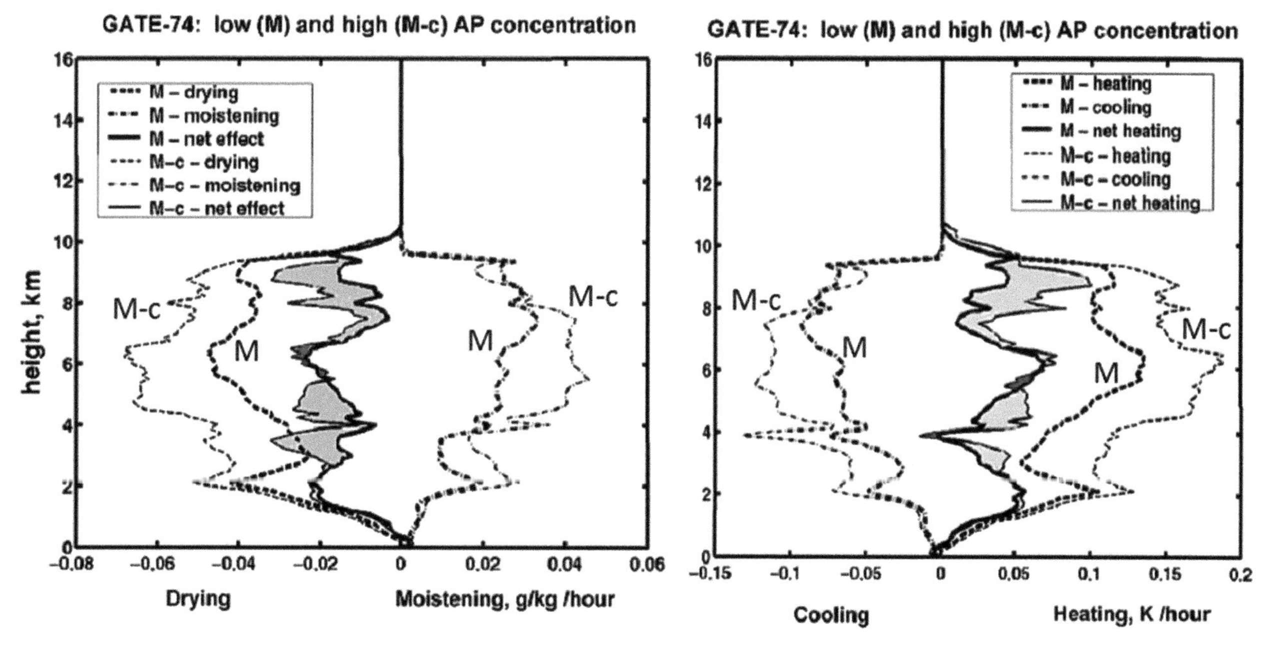

Figure 7.3.5 Vertical profiles of horizontally averaged atmospheric moistening/drying (left panel) and heating/cooling (right panel) over a 4-h period in simulations of tropical maritime convective clouds. Areas marked gray show zones where larger heating and drying take place in polluted clouds; areas marked blue denote zones where larger heating and drying take place in non-polluted clouds. A larger area of gray zones indicates that an increase in aerosol concentrations leads to an increase in precipitation (adopted from Khain, 2009; courtesy of © IOP Publishing CC BY–NC–SA).

place. Vice versa, if an increase in aerosol concentration leads to the condition $\Delta L - \Delta G < 0$, an increase in precipitation takes place. In Figure 7.3.5 the difference $\Delta L - \Delta G$ is represented by an area between the profiles indicating moistening and drying. The areas marked gray show zones where $\Delta G - \Delta L > 0$; the areas marked blue denote zones where $\Delta G - \Delta L < 0$. For the tropical clouds, the difference $\Delta G - \Delta L$ integrated over the vertical atmospheric column is positive ($\Delta G - \Delta L > 0$), i.e., an increase in the CCN concentrations leads to an increase in precipitation from these clouds. The condition $\Delta G - \Delta L > 0$ means that the heating of the atmosphere by condensation and deposition is larger than the cooling due to sublimation and evaporation (Figure 7.3.5, right panel). An increase in precipitation ($\Delta P > 0$) does not necessarily imply that PE also increases. Typically, in dirty clouds $\frac{\Delta G}{G} > \frac{\Delta P}{P}$, i.e., precipitation efficiency decreases with an increase in CCN concentration.

There are at least three factors that affect the relation between ΔL and ΔG, namely: cloud type, air humidity, and wind shear. We first consider warm stratocumulus and small cumulus clouds.

Stratocumulus and Warm Cumulus Clouds

Many studies have shown that precipitation from small cumulus and stratocumulus clouds is suppressed in polluted air (e.g., Albrecht, 1989; Rosenfeld, 1999; Cheng et al., 2007). Simulations with a trajectory ensemble model of stratocumulus clouds show that an increase in the CCN concentration from 200 to 600 cm^{-3} totally prevents drizzle formation in stratocumulus clouds under thermodynamic conditions of the DYCOMS-II field experiment (**Figure 7.3.6**). This strong response can be explained as follows. An increase in AP concentration does not significantly affect the generation of hydrometeors, but substantially decreases droplet size, making droplets unable to trigger collisions. As a result, droplets evaporate in downdrafts, which increases the condensate loss ΔL. Small cumulus clouds undergo intense mixing with the environment, which additionally increases ΔL by evaporation. Thus, an increase in CCN concentrations in these clouds leads to $\Delta L > \Delta G$.

Deep Convective Clouds

Available observations and numerical simulations show that there is no definite answer to the question

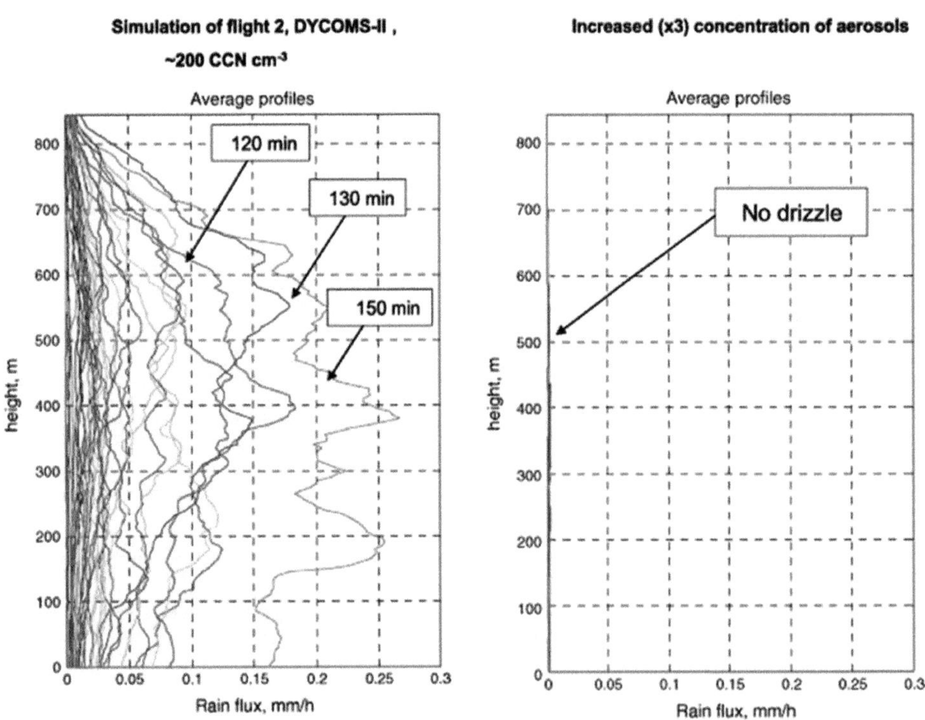

Figure 7.3.6 Vertical profiles of drizzle flux at different time instances in simulations of stratocumulus clouds observed in the DYCOMS-II field experiment (research flight RF02) with CCN concentration of 200 cm^{-3} (left) and 600 cm^{-3} (right). Vertical profiles of the fluxes are plotted in different colors with a time increment of 5 min (adopted from Khain, 2009; courtesy of © IOP Publishing CC BY–NC–SA).

of whether higher CCN concentrations increase or decrease precipitation from deep convective clouds. While an increase in CCN concentrations always increases the generation of hydrometeor mass, the condition $\Delta G > \Delta L$ is not obvious, since the loss of hydrometeor mass at high CCN concentration can be very large. An increase in CCN concentrations leads to a decrease in particle size and to their higher elevation, therefore, the loss of hydrometeor mass via evaporation and sublimation increases. The increase in the loss ΔL depends on the relative humidity of the surrounding air. At high environmental humidity, ΔL is comparatively low, and $\Delta L < \Delta G$, which corresponds to an increase in precipitation. Such a precipitation increase from maritime deep convective clouds caused by an increase in CCN concentrations was reported in many numerical studies (Khain, 2009; Tao et al., 2012). The precipitation increase from deep clouds over the tropics was also derived from observations (Koren et al., 2012). In contrast, in a dry environment ΔL is high and $\Delta L > \Delta G$, which corresponds to a decrease in precipitation caused by an increase in aerosol loading. Several simulations found that an increase in CCN concentrations increases surface precipitation from deep convective clouds at high RH, and decreases surface precipitation in a dry environment (RH < 50–60% in the middle atmosphere) (Khain and Pokrovsky, 2004; Khain, 2009). Similar interrelations were obtained by Khain and Lynn (2009).

Any factor favorable to evaporation and sublimation processes contributes to a decrease in precipitation. For instance, wind shear increases detrainment of hydrometeors, transporting hydrometeors away from cloud updrafts and into areas of relatively low humidity (Khain et al., 2005; Fan et al., 2009a). An example of the combined effect of CCN concentration and wind shear is presented in **Figure 7.3.7**. This figure shows precipitation rates obtained in simulations of a deep convective cloud using WRF/SBM (Fan et al., 2009a) under thermodynamic conditions typical of northern Australia close to Darwin, during the period of November, 2005–February, 2006. One can also see that precipitation in humid air is by an order of magnitude larger than in dry air. It is also clear that an increase in the wind shear leads to a dramatic decrease of precipitation, both in dry and moist environment. This indicates that precipitation amount is highly sensitive to thermodynamic factors. Aerosols is only one (and not the strongest) of several factors affecting precipitation amounts. Yet, it has been a focus of close attention of cloud research since the persistent anthropogenic emission of aerosols creates significant trends in rainfall regimes over different regions at climatic time scales.

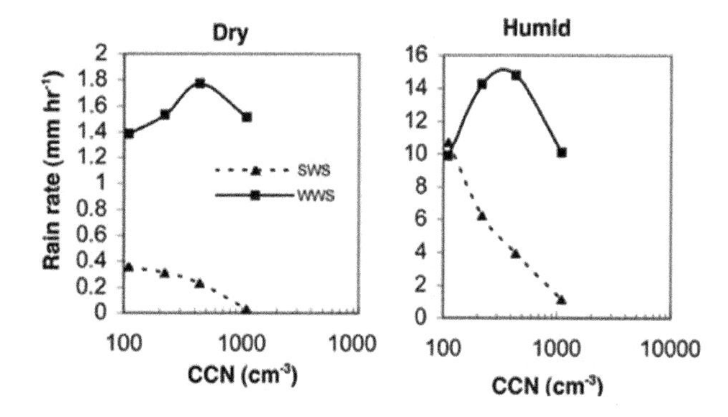

Figure 7.3.7 Averaged surface precipitation rates versus CCN concentrations under dry and humid conditions. Within the layer 500–900 mb, the values of RH in dry conditions differ by 20% from those in humid conditions. Simulations were performed both with strong wind shear (SWS) of about 13/7 m/s/km and weak wind shear (WWS) of about 5/7 m/s/km (from Fan et al., 2009; courtesy of © John Wiley & Sons, Inc.).

Midlatitude Storms

The CCN effects shown in Figure 7.3.2 are related largely to clouds with a warm cloud base, developing in a warm environment with a high freezing level, such as deep tropical convective clouds. The increase in CCN concentrations in such clouds leads to an increase in the altitude of raindrop formation, i.e., raindrops form above the freezing level. These raindrops tend to freeze, delaying warm rain formation and intensifying the ice processes. This is an important component of the convective invigoration discussed previously.

In deep convective clouds or convective storms forming in midlatitudes, the cloud base is often located in the vicinity of the freezing level. Therefore, raindrops form at negative temperatures both at low and high CCN concentrations. Freezing just above the freezing level more often takes place at low CCN concentrations, as large raindrops freeze more efficiently (Section 6.2). Numerical simulations show that CCN dramatically affect cloud microphysics and precipitation in midlatitude, deep convective clouds, and also storms (Khain et al., 2011; Kumjian et al., 2014; Ilotoviz et al., 2016). By way of example, we consider the results of simulation of a midlatitude hailstorm in Villingen-Schwenningen, southwest Germany, on June 28, 2006. This hailstorm produced hailstones with diameters of up to 5–6 cm, and the radar reflectivity exceeding 60 dBz up to the 10 km level. The storm developed in a fairly polluted atmosphere with the CCN concentration of several thousand particles per cm^3. The simulations were conducted using HUCM, at different CCN concentrations from 100 to 5,000 cm^{-3}.

The CCN effects on cloud microphysics are illustrated in **Figure 7.3.8**, which shows vertical profiles of maximum values of different hydrometeor concentrations and mass contents under different aerosol loadings at the developing stage of the storm. Effects of CCN on cloud microphysics and dynamics of mixed-phase deep convective clouds and storms are determined to a large extent by their effect on supercooled CWC. If the CCN concentration is high, droplets are small and ascend to high levels (panels a, c), forming a significant mass of supercooled water aloft (panel c). The concentration of raindrops is low, the first raindrops form at levels of about 5 km, and the RWC is low (panels b, d). On the other hand, if the CCN concentration is low, the concentration of droplets and CWC are low (panels a, c). However, because droplets are larger in case of low CCN concentration, their collisions are efficient and lead to faster and more intense raindrop production. As a result, the mass and concentration of raindrops just above the freezing level turns out to be much larger in cases of low CCN concentration (panels b, d).

These differences in warm microphysics largely determine the difference in the fields of ice particles. The main differences are:

- Since freezing of raindrops is much more probable than that of cloud droplets, the mass of FD is larger in cases of low CCN concentration at the developing stage of a storm (panel e);
- Total freezing of liquid within FD leads to the production of a significant hail-mass content around the altitude of 6 km in cases of low CCN concentration (panel f). On the other hand, the formation of the first hail in cases of high CCN concentration takes place at higher levels (panel f);
- the Hallet and Mossop mechanism (Section 6.8) leads to formation of ice crystals within the layer with temperatures from −3 to −8°C (altitude range from 4 to 4.8 km) in cases of low CCN concentration (panel h). Collisions of these crystals lead to snow formation at these levels (panel g). The lack of FD, graupel, and hail as well as droplets with diameters exceeding 24 μm at these levels in cases of high CCN concentration makes the Hallet-Mossop mechanism inefficient at the developing stage. In this case, ice crystals form at high levels of 9 km and higher (panel h).

The rates of production of different hydrometeors and their vertical distributions vary during the storm evolution. However, several basic differences between microphysical structures at different aerosol concentrations remain throughout the simulation period. For instance, RWC and raindrop-number concentration above the freezing level in cases of low CCN concentration remains larger than in cases of high CCN concentration, while the supercooled CWC remains larger in cases of high CCN concentration.

The differences in the mass of supercooled droplets between clean and polluted conditions lead to different mechanisms of hail formation and to different hail parameters. The difference is illustrated in **Figure 7.3.9**, showing the fields of the mean-volume radius of hail particles at high aerosol concentrations and at low CCN concentrations. At high CCN concentration (Figure 7.3.9, left), hail particles fall near the cloud edge and penetrate cloud updrafts. These particles grow within updrafts with high CWC to large sizes by accretion of supercooled droplets. Large hail particles ascend in strong updrafts up to heights of 8 km. Analysis shows that these hail particles grow in the wet growth regime. If the CCN concentration is low (Figure 7.3.9, right), the supercooled CWC in updrafts is also low, and raindrops forming near the freezing level freeze, being of comparatively small sizes. No substantial growth by accretion takes place.

Since small hail completely melts below the melting layer, the hail shaft at the surface is substantially smaller in clean atmosphere than in polluted atmosphere (**Figure 7.3.10**, left). One can see that the accumulated hail at the surface reaches its maximum at CCN concentration of 3,000 cm^{-3}. The hail shaft in clean air is negligibly small, as compared to the high CCN concentration case. Figure 7.3.10 (right) shows the time dependence of cumulative rain at the surface in these simulations. Precipitation begins earlier at low CCN concentration, but accumulated rain reaches its maximum at high CCN concentrations (2,000–3,000 cm^{-3}). Accumulated rain increases rapidly with the increase in CCN concentration from 100 cm^{-3} to about 500 cm^{-3}. Precipitation is less sensitive to higher CCN concentrations.

The decrease in the size of hail particles falling at the surface at CCN concentrations exceeding 3,000 cm^{-3} (Figure 7.3.10) can be attributed to formation of a significant amount of small ice crystals by homogeneous freezing of small droplets ascending the levels of ~9.5 km. These crystals spread in cloud anvils over a large area and sublimate, increasing the loss of the condensate mass.

The increase in surface precipitation due to an increase in CCN concentration can be understood by analyzing the heat and the water vapor mass budgets plotted in **Figure 7.3.11** for two CCN concentrations: 100 cm^{-3} denoted M-run and 3,000 cm^{-3} denoted

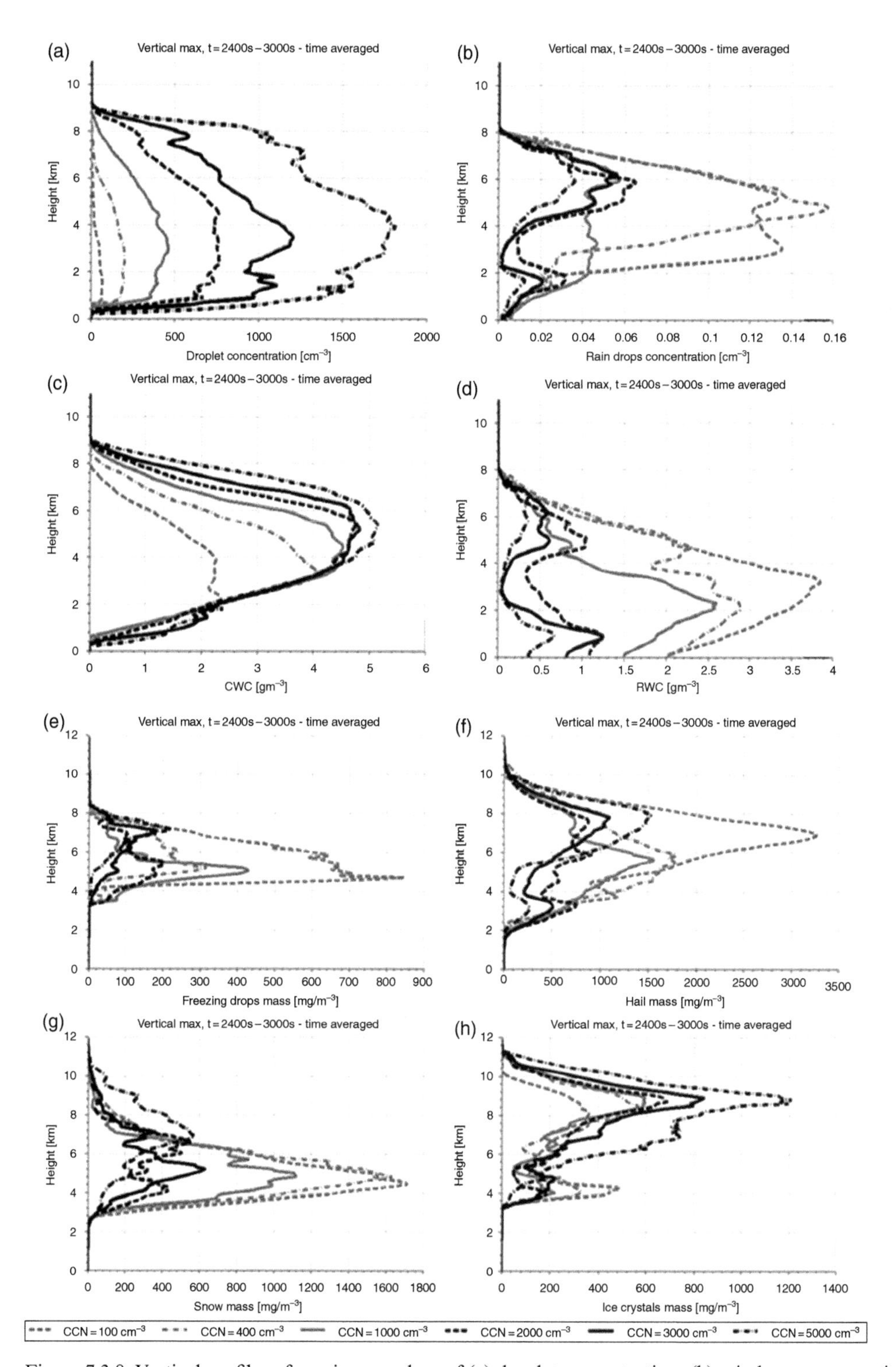

Figure 7.3.8 Vertical profiles of maximum values of (a) droplet concentration, (b) raindrop concentration, (c) cloud water content (CWC), (d) rain water content (RWC), and mass (e) contents of freeing drops, (f) hail, (g) snow, and (h) total mass of ice crystals at the developing stage of storm (deep convective cell). The profiles are obtained by averaging over the time period of 2,400–3,000 s (from Ilotoviz et al., 2016; courtesy of © American Meteorological Society. Used with permission).

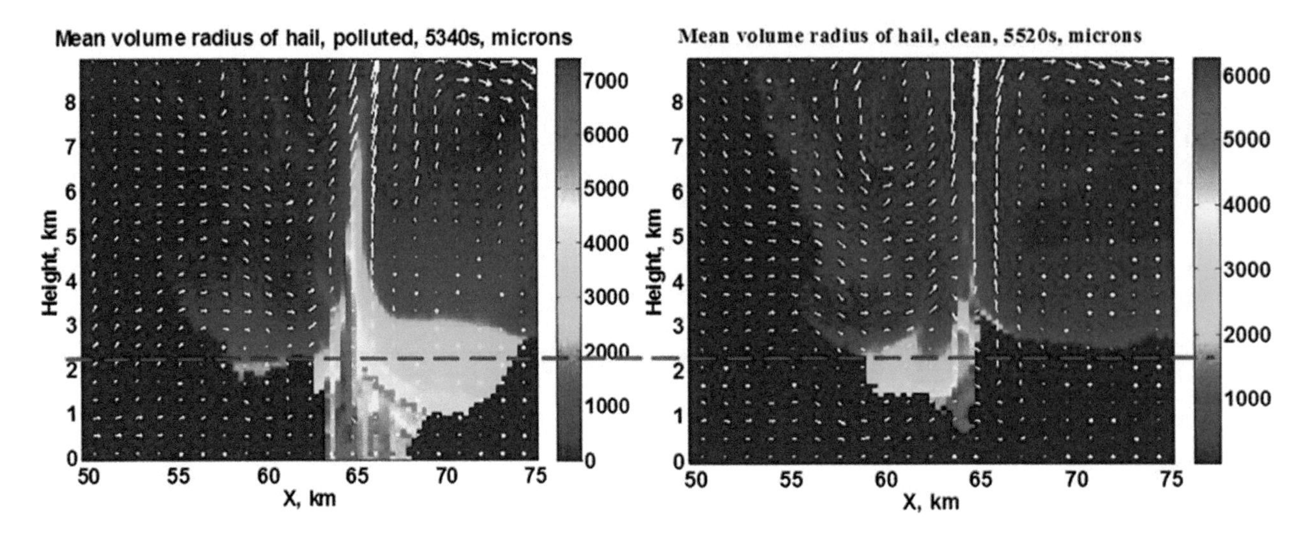

Figure 7.3.9 Fields of mean volume radii of hail at high CCN concentrations (3,000 cm^{-3}) (left) and low CCN concentrations (100 cm^{-3}) (right) at the mature stage of storm development. The dashed red line denotes the freezing level of 273.15°C (from Ilotoviz et al., 2016, with changes; courtesy of © American Meteorological Society. Used with permission).

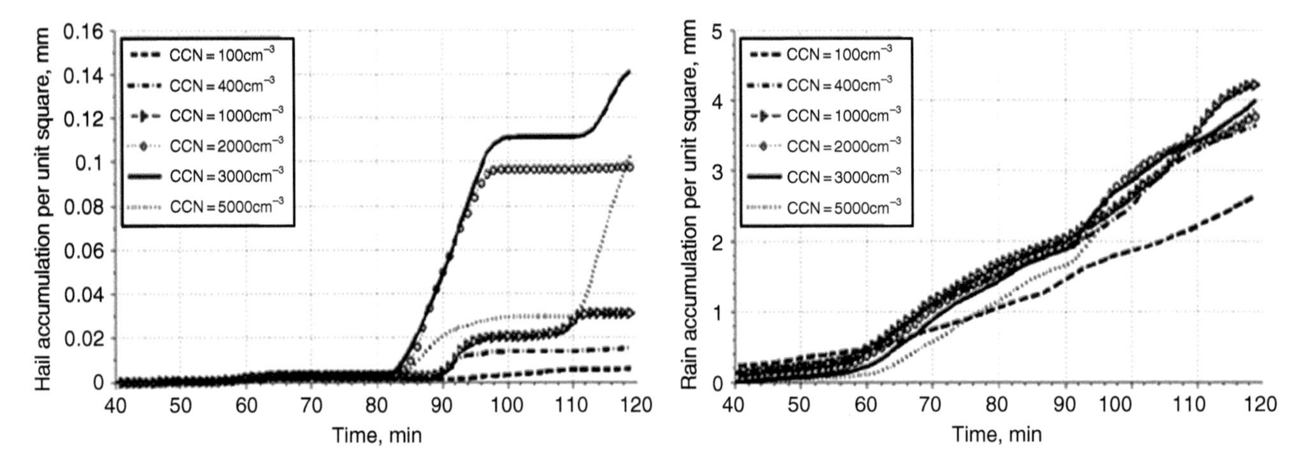

Figure 7.3.10 Time dependence of accumulated surface hail precipitation (left) and accumulated rain (right) in simulations with different CCN concentrations (from Ilotoviz et al., 2016; courtesy of © American Meteorological Society. Used with permission).

C-run. Comparison of Figures 7.3.11 and 7.3.5 indicates a significant difference between aerosol effects in tropical deep convective clouds and midlatitude storms. Under tropical conditions, an increase in CCN concentration increases both the condensate generation, leading to convective invigoration, and the condensate loss, decreasing the precipitation efficiency. In midlatitude storms, an increase in CCN concentration does not increase the condensate generation significantly. Actually, all of the droplets ascend above freezing level and freeze due to different microphysical processes. At the same time, aerosols foster formation of large hail. Due to the high fall velocity, large hail efficiently collects supercooled cloud droplets during recirculation and does not sublimate during its fall, which decreases the precipitation loss. Thus, small aerosols triggering formation of effective collectors such as large hail increase the precipitation efficiency in deep mixed-phase storms.

Figure 7.3.12 shows the time dependence of the precipitation efficiency in different deep convective clouds and hailstorms with different CCN concentrations. One can see that the precipitation efficiency of maritime clouds observed in GATE-74, as well as clouds forming in clean air over Brazil (BO-100), ranges from 0.3 to 0.4, which means that 30–40% of the condensate fall

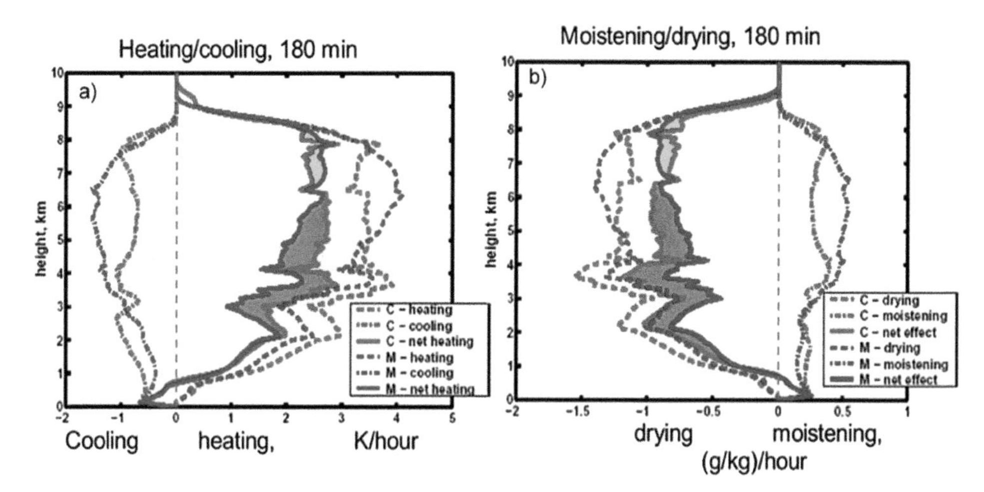

Figure 7.3.11 Vertical profiles of horizontally averaged (a) heating/cooling and (b) moistening/drying in a hail storm in southern Germany calculated for low (blue curves) and high (red curves) aerosol concentrations. Areas marked pink denote zones where net heating and drying are higher in the high aerosol concentration case. Areas marked light blue denote zones where net heating and drying are higher in the low aerosol concentration case (adapted from Khain et al., 2009; courtesy of © IOP Publishing CC BY–NC–SA).

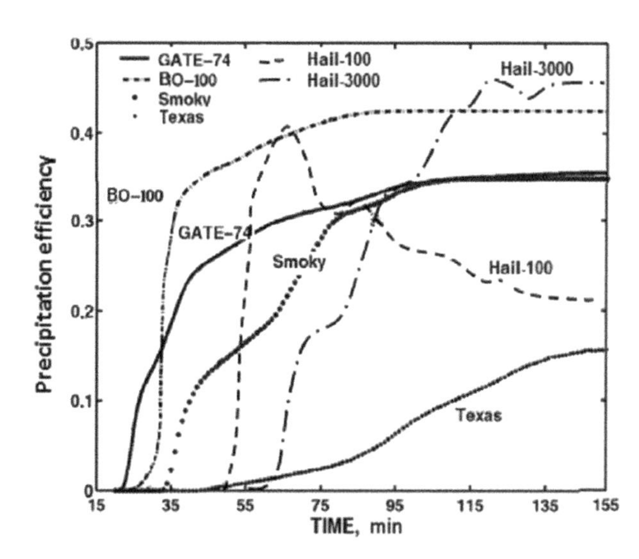

Figure 7.3.12 Time dependence of precipitation efficiency in clean and polluted air in the simulation of different deep convective clouds. GATE–74: maritime clouds observed in the field experiment in tropical Atlantic, CCN concentration of 100 cm^{-3}; BO–100: Blue-ocean clouds observed in LBA–SMOCC field experiment in Brazil under very clean conditions, CCN concentration of 100 cm^{-3}; Smoky: clouds observed in LBA–SMOCC field experiment in Brazil under polluted conditions, CCN concentration of 3,000 cm^{-3}; Texas: deep convective clouds observed during summertime in Texas, CCN concentration of 3,000 cm^{-3}; Hail–100 (CCN concentration of 100 cm^{-3}) and Hail-3,000 (CCN concentration of 3,000 cm^{-3}) are clouds in the midlatitude hail storm in Villingen-Schwenningen, southwest Germany, observed on June 28, 2006 (from Khain et al., 2009, with changes; courtesy of © IOP Publishing CC BY-NC-SA).

to the surface as precipitation and 60–70% of the condensate evaporate. The precipitation efficiency in tropical clouds developing in polluted conditions is typically lower than in unpolluted clouds, especially in clouds developing in very dry air, such as Texas clouds. The precipitation efficiency of Texas clouds developing at high CCN concentration is about 0.1. The precipitation efficiency of clouds in the midlatitude hail storm depends on CCN concentration. One can see that at $t \sim 90$ min, when large hail forms, the precipitation efficiency in the polluted case (Hail-3000) becomes larger than in clean air (Hail-100) and exceeds the precipitation efficiency of maritime tropical clouds. Hence, hail (as well as large graupel) plays a very important role in precipitation increase in polluted air due to increasing precipitation efficiency of midlatitude storms.

7.3.3 Effects of Aerosols on Surface Precipitation from Convective Systems

Under certain thermodynamic conditions, downdrafts created by primary clouds lead to formation of secondary clouds or squall lines (Rotunno et al., 1988). Squall lines are a major type of convective structure over the ocean. Khain et al. (2004a, 2005) and Lynn et al. (2005a) found that an increase in aerosol concentration intensifies secondary clouds and fosters formation of squall lines. By way of example, **Figure 7.3.13** shows the fields of radar reflectivity from clouds developing at low CCN concentration (left panels) and at high CCN concentration (right panels), as calculated using the

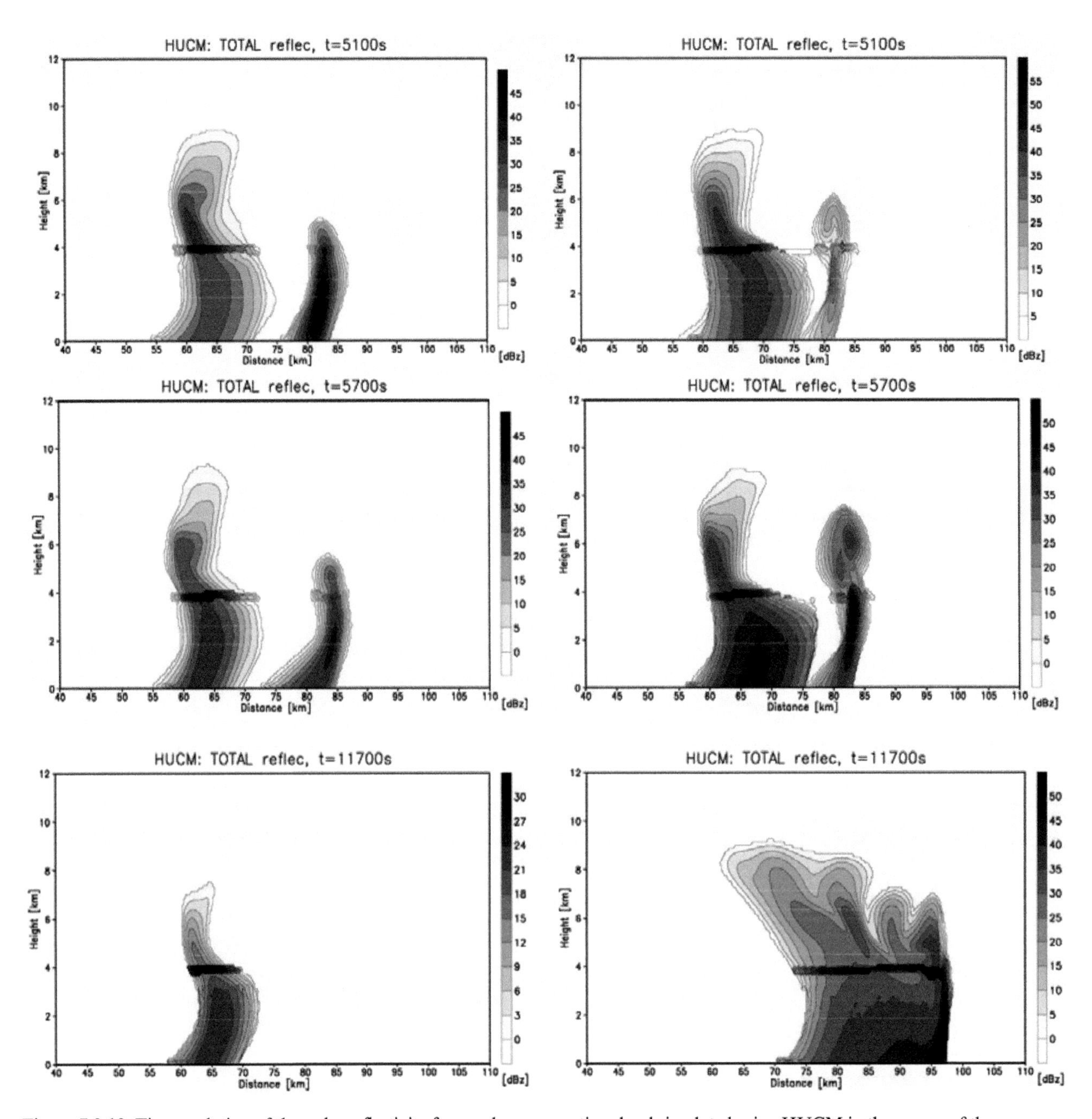

Figure 7.3.13 Time evolution of the radar reflectivity from a deep convective cloud simulated using HUCM in the course of the PRE–STORM sounding. Left panels correspond to low CCN concentration. Right panels show results obtained at high CCN concentration (from Khain et al., 2005, with changes; courtesy of © John Wiley & Sons, Inc.).

HUCM under the PRE–STORM field experiment (Oklahoma region, USA, dry environment) sounding. Formation of a squall line in the latter case is seen in the right panels. Seifert et al. (2006) and van den Heever and Cotton (2007) also demonstrated the significant role of aerosols in secondary cloud development, as well as in the associated surface precipitation.

As discussed, an increase in aerosol concentrations in clouds with warm base leads to extra cooling of the air in cloud downdrafts due to increased melting and evaporation. In the presence of a favorable wind shear, ice particles and small drops that are detrained from the zone of updrafts fall downwind into comparatively dry air. Evaporation increases the downdrafts as well as the air convergence in the boundary layer, creating a cool pool. On the one hand, an increase in evaporation decreases precipitation from individual clouds. On the other hand, more evaporation fosters formation of secondary clouds, which in turn increases precipitation when aggregated over the scale of the cloud system. Whether one of these two effects would dominate depends on several thermodynamic and dynamic factors, such as environmental air humidity, wind shear, instability of the boundary layer, etc.

Fan et al. (2007b) and Tao et al. (2007) focused on the role of air humidity in formation of secondary clouds and squall lines. The simulations were performed using the 2D Goddard Flight Center Cloud Model with SBM. It was shown that an increase in aerosol concentration at high humidity intensifies convection and leads to stronger evaporative cooling as compared to the clean-air conditions. **Figure 7.3.14** shows the vertical profiles of the evaporation rate in the simulations of squall lines observed in two field experiments: TOGA CORE (Central Pacific, humid environment) and PRE–STORM. The simulations were performed for low- and high AP concentrations. One can see that evaporation rates are higher in the dryer atmosphere (PRE–STORM). The difference in evaporation rates between the polluted and clean cases is larger in moist maritime air (TOGA COARE). Increased cooling intensifies downdrafts and leads to an intensification of maritime squall lines and an increased condensate generation. As a result, higher aerosol loading increases the condensate generation to a larger extent than the condensate mass loss, which leads to rain enhancement in the Pacific squall line. The simulations indicate some aerosol-induced decrease in precipitation in the Oklahoma case (dry environment), and no significant effect in the Florida squall line.

Lee et al. (2005, 2008b) simulated the evolution of cloud ensembles using the WRF model with two-moment bulk parameterization for different wind-shear magnitudes. They found that an increase in aerosol loading leads to an increase in precipitation at high humidity, strong wind shear, and instability. The strong shear and the presence of aerosols lead to self-organization of deep convection, intensification of deep clouds, and decaying of weak clouds. The factors causing this aerosol-induced redistribution of clouds over height include the convective invigoration that takes place mostly in deep convective clouds, and downdrafts caused by large clouds that tend to suppress small clouds.

Seifert and Beheng (2006a) used a 3D model with two-moment bulk-parameterization microphysics to simulate supercell and multicell storms. They found that while an increase in aerosol loading decreases precipitation from ordinary single cells and supercell storms, it leads to precipitation enhancement in multicell cloud systems.

7.3.4 Effects of Aerosols on the Spatial Shift of Surface Precipitation

Now we consider the effects of aerosols on orographic clouds, on precipitation in urban areas, and on precipitation related to breeze circulation.

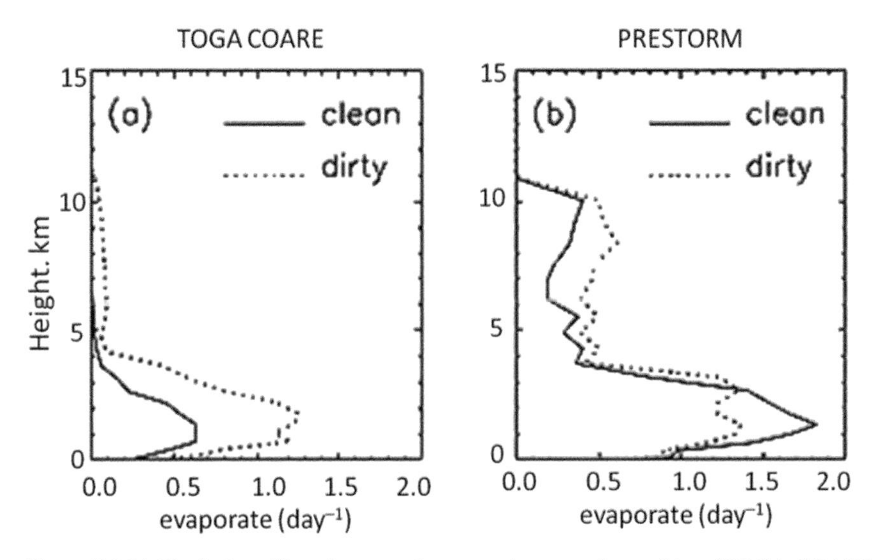

Figure 7.3.14 Vertical profiles of averaged evaporation rates in maritime (TOGA COARE) (left) and continental (PRE–STORM) (right) squall lines (adopted from Tao et al., 2007; courtesy of © John Wiley & Sons, Inc.).

Orographic Clouds
Givati and Rosenfeld (2004) and Jirak and Cotton (2006) reported a 30% decrease of precipitation in the Sierra Nevada Mountains over the mountain slopes located downwind from urban areas during a few past decades. They assumed that the decrease in precipitation was caused by an increase in CCN concentration caused by developing of the urban zone. Lynn et al. (2007) simulated orographic clouds in this region using the 2D version of the WRF model with SBM. The simulations were carried out using maritime (clean-air) and continental (polluted-air) aerosols. **Figure 7.3.15** shows the spatial distribution of accumulated surface precipitation toward the end of the simulations at low and high AP concentrations. The results were obtained at the relative humidity typical of the region. The increase in aerosol loading leads to a decrease in precipitation and to a spatial shift of precipitation downwind. The accumulated surface precipitation decreased by about 30%, from 0.44 mm in clean air to 0.32 mm in polluted air. There are several factors causing the decrease in precipitation. Additional aerosol loading leads to warm rain suppression and to generation of a significant amount of snow particles with small sedimentation velocities. Snow is advected downwind, which increases precipitation over the mountain peak. These ice crystals and snow particles sublimate on the downwind side of the highest mountain peak due to very low humidity on the downwind slope of the mountain ridge. Due to the sublimation of ice crystals and snow, the simulation with continental aerosols produced less precipitation over the whole mountain slope. An increase in air humidity coinciding with an increase in CCN concentrations reduces the loss of precipitating cloud mass, while leading to an increase in precipitation.

Urban Effects
Observational studies show that more clouds, lightning, and precipitation occur over and downwind of urban areas (e.g., Shepherd et al., 2002; Inoue and Kimura, 2004; Mote et al., 2007). Several causes of urban-induced or urban-modified convective phenomena have been suggested, including the urban heat island, increased urban surface roughness, and increased urban aerosols (e.g., Baik et al., 2001; Shepherd, 2005; van den Heever and Cotton, 2007). Several observations (Bornstein and Lin, 2000) and numerical studies (Baik et al., 2001; Rozoff et al., 2003) indicate that the urban heat island plays an important role in initiating convective thunderstorms and precipitation on the downwind side of urban areas, whereas upwind convergence induced by increased urban surface roughness is not strong enough to initiate moist convection.

Han et al. (2012) investigated the impact of urban aerosols on clouds and precipitation using HUCM. Extensive numerical experiments with various CCN concentrations were performed under different environmental moisture conditions. To take into account the urban heat island and urban air pollution, the heating in the urban area was assumed to be low level, and the aerosol concentration to be higher than that in the

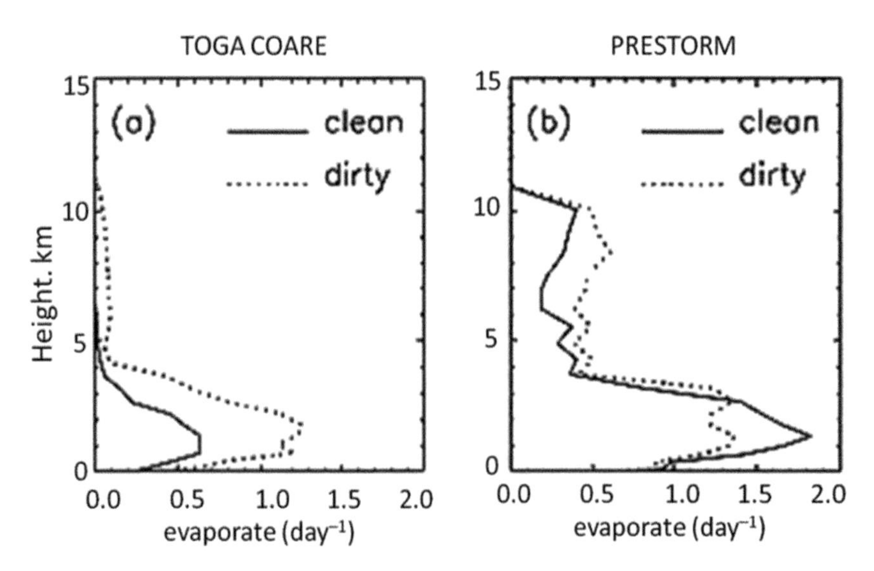

Figure 7.3.15 The spatial distribution of accumulated surface precipitation from orographic clouds toward the end of simulations at low and high AP concentrations. Grey curve denotes the topography; the distance is from the foot of the mountain (adopted from Lynn et al., 2007; courtesy of © John Wiley & Sons, Inc.).

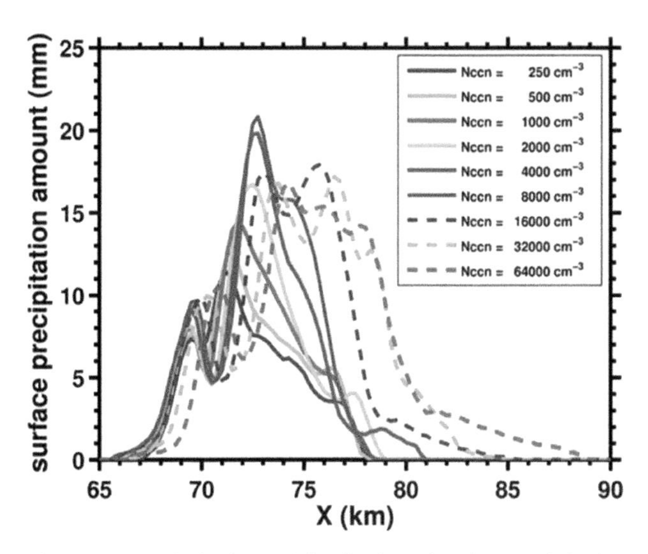

Figure 7.3.16 The horizontal distribution of surface precipitation amount summed over the area where the surface precipitation rate resulting from the deep convective cloud system exceeds 5 mm. Simulations were performed using various CCN concentrations for the water vapor mixing ratio within the boundary layer of 18 g/kg. The urban center is located at the point of x = 50 km (from Han et al., 2012; courtesy of © American Meteorological Society. Used with permission).

surrounding rural area. Simulation results show that a low-level updraft induced by the urban heat island leads to formation of a low-level cloud, and later a deep convective cloud downwind of the urban area. When the aerosol concentration is high, warm rain is suppressed; ice crystals and snow, having lower fall velocity than liquid drops, are shifted downwind. The onset of precipitation is delayed and ice particles are advected downwind (**Figure 7.3.16**). A higher aerosol concentration generally leads to development of a stronger convective cloud, to increase in hail size and to precipitation enhancement downwind of the urban area in all moisture environments considered.

Precipitation Shift in a Coastal Zone

As an example of a precipitation shift in a coastal zone, we consider the effects of breeze in the Eastern Mediterranean. Precipitation in the Eastern Mediterranean takes place during the cold season when the sea-surface temperature is by 5–10°C higher than the land-surface temperature. This temperature difference leads to formation of a land-breeze-like circulation that interacts with the dominating westerlies and leads to intense cloud formation over the sea ~10–20 km from the coastal line. As a result, most of the precipitation falls onto the sea without reaching the land. To investigate the possibility of shifting the rainfall from sea to land, numerical simulations were performed using the 2D HUCM and the 3D WRF, both operating with spectral bin microphysics, as well as the 3D COSMO model applying a two-moment bulk parameterization for cloud microphysics. In the numerical simulations, the concentration of small CCN over the sea in the boundary layer below cloud base was increased in order to investigate effects of CCN on the spatial precipitation distribution. Such increase in the CCN concentration below cloud base of developing clouds can be reached by seeding of developing clouds below their cloud base. The results of all the simulations indicate that an increase in small aerosols concentration delays raindrop formation and fosters formation of extra ice particles with a low settling velocity. This ice is advected inland by the background wind. As a result, precipitation over land increases by 15–20% at the expense of precipitation over sea. **Figure 7.3.17** shows the fields of accumulated rain calculated for a rain event on January 20–21, 2007, at both low (natural) and high (like after seeding) aerosol concentrations. The results were obtained using the WRF/SBM. One can see that the area covered with precipitation within the coastal zone is larger at high AP concentration. Such result was also obtained using both HUCM and COSMO. At the same time, precipitation over the sea decreased. So, increase in the CCN concentration over the sea led to a shift of some precipitation fraction from the sea to the land. The effect of aerosols on total precipitation over the computational area depends on the air humidity. At lower humidity, evaporation increases with increasing CCN concentration, and total precipitation may decrease.

As was discussed, the impact of aerosols on precipitation strongly depends on the air humidity. The air humidity is a major factor determining the sign of the difference $\Delta G - \Delta L$ in Equation (7.3.3). The analysis of results obtained in many observational and numerical studies leads to the classification table shown in **Figure 7.3.18**. As seen from Figure 7.3.18, precipitation from deep convective clouds developing in extremely dry atmosphere at RH ~ 35% (like in case of the Texas clouds) may decrease with increasing CCN concentration. At the same time, precipitation from dynamically similar clouds developing in wet air (RH 80–90%) increases with increasing CCN concentration. Rain from small cumuli and stratocumulus clouds evaporates efficiently even at high humidity. As a result, precipitation from these clouds decreases or totally disappears in polluted air. The impact of aerosols on precipitation amounts in individual clouds may be different from that in cloud ensembles. The increase in precipitation may take place due to formation of secondary clouds or

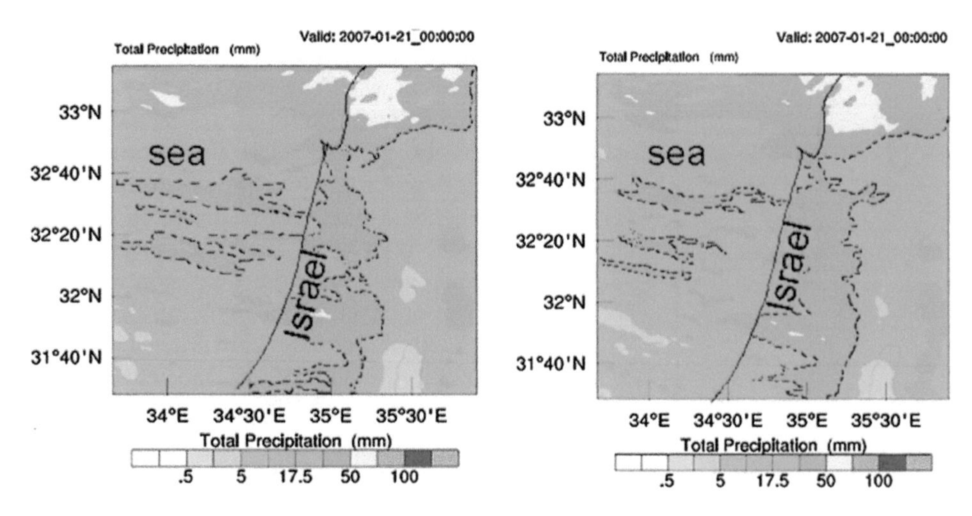

Figure 7.3.17 Fields of accumulated rain calculated for a rain event on January 20–21, 2007, at low aerosol concentration (left) and high aerosol concentration over the sea (right). Dashed lines denote the areas of intense rain. The area covered with precipitation within the coastal zone (dark green) is larger at high AP concentration (from Noppel et al., 2010a; courtesy of © John Wiley & Sons, Inc.).

squall lines. As was discussed, the wind shear plays the dual role: it decreases precipitation from individual clouds, but promotes formation of secondary clouds, squall lines, and convective mesoscale systems.

7.3.5 The Role of Giant CCN in Cloud Microphysics and Precipitation

Multiple studies explored the effect of giant CCN (GCCN) with dry radii exceeding a few microns on precipitation. Some researchers consider the existence of GCCN to be a necessary condition for formation of raindrops in convective clouds and drizzle drops in stratocumulus clouds (Feingold et al., 1999; Jensen and Lee, 2008). In simulations with very narrow DSD, collisions between cloud droplets were inefficient, and each particle that grew on a GCCN functioned as a drop collector. In other studies, where DSD was not negligibly narrow and collisions between cloud droplets were substantial, raindrops formed without GCCN in the aerosol size spectrum (Section 5.7.3; Figure 5.7.10). At the same time, the presence of GCCN may accelerate formation of raindrops or drizzle drops.

GCCN are thought to contribute to rain formation through two mechanisms. First, particles forming on GCCN grow rapidly near the cloud base, decreasing the supersaturation maximum and thereby decreasing the concentration of nucleated droplets. This competition effect was analyzed in Section 5.3.2. By the second mechanism, particles growing on GCCN first reach sizes allowing them to trigger intense collisions, i.e., accelerate formation of warm rain. Some observational and numerical studies questioned the effect of GCCN on precipitation onset and precipitation amount. At low CCN concentrations, raindrops form quickly even in the absence of GCCN. At high CCN concentrations, GCCN accelerate rain formation if their concentration is comparatively high (Section 5.7). Note that at high CCN concentrations, raindrops in deep convective clouds may form in course of drop recirculation when drops descending along the cloud edges penetrate cloud updrafts and collect smaller drops. In mixed-phase clouds raindrops form by the melting of ice particles.

Using HUCM, Khain et al. (2012) found a synergetic effect of the smallest CCN (with radii below 0.01 μm) and GCCN on the microphysics of a deep convective cloud. If the concentration of GCCN is high enough, these GCCN accelerate formation of warm rain by acceleration of drop collisions. Drops falling leads to unloading and, consequently, to an increase in buoyancy force and to an increase in updraft. These two factors, the increase in updraft and the decrease in droplet concentration due to collisions, lead to an increase in supersaturation and to in-cloud nucleation of droplets on the smallest CCN that were not activated at the cloud base. As a result, new droplets appear in clouds at heights of 6–8 km and efficiently participate in the ice processes. Thus, the existence of the GCCN fosters the in-cloud nucleation and contributes to formation of significant ice crystal concentrations (several tens to hundred per cm^3) in anvils of tropical clouds. This synergetic effect of GCCN and the smallest CCN

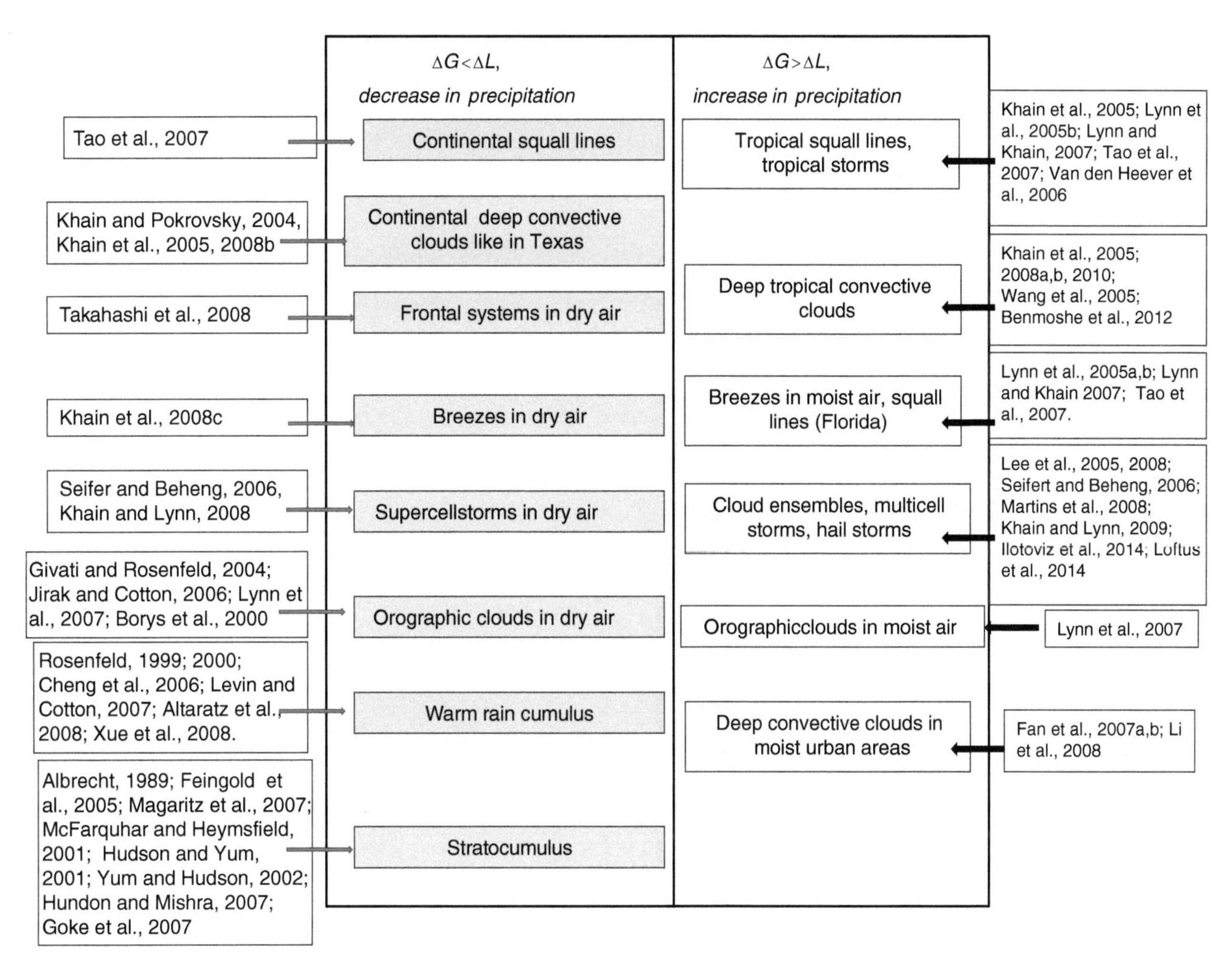

Figure 7.3.18 The classification scheme of aerosol effects on precipitation, obtained by analysis of observed data and numerical simulations. In cases when a decrease of precipitation was reported ($\Delta G < \Delta L$), the relative humidity was typically low. In cases when an increase in precipitation was reported ($\Delta G > \Delta L$), the meteorological situation was typically characterized by high humidity.

leads to intensification of lightning (including lightning in hurricanes eyewalls), that arises only in case of coexistence of supercooled water, graupel, and crystals at low temperatures.

Khain et al. (2012) investigated the synergetic effect of GCCN and small CCN in simulations using HUCM. Parameters of aerosols used in these simulations are shown in **Table 7.3.2**. **Figure 7.3.19** shows the time dependence of precipitation in the simulations. One can see that including GCCN in the CCN spectrum that does not contain the smallest CCN (E100_NS_G) increases accumulated rain only slightly. Including the smallest CCN in E100_S increases precipitation. However, the maximum accumulated rain was obtained in simulation E100_SG, where both the giant and the smallest CCN were present. Simulations with high CCN concentrations show an increase in precipitation with increasing CCN concentrations.

7.3.6 Effects of Aerosols on Microphysics and Intensity of Tropical Cyclones

Several dynamical and thermodynamic factors affect TC) intensity: SST, vertical wind shear, vorticity of the steering current, air humidity, the Coriolis force, and others. It was found that aerosols affect TC intensity by invigorating convection within a TC (e.g., Khain et al., 2008a, 2010; Rosenfeld et al., 2012; Cotton et al., 2007; Lynn et al., 2016). TC structure, pressure gradients, and maximum wind depend on the relation between the convection intensity in TC center and at TC periphery. TC approaching land often absorb a

Table 7.3.2 Parameters of the initial CCN spectra in the simulations (from Khain et al., 2012; American Meteorological Society; used with permission).

Expt	N_0 (cm^{-3})	k	Minimum CCN radius (μm)	Maximum CCN radius (μm)	Concentration of CCN with radius > 1 μm (cm^{-3})
E100_S	100	0.9	0.003	2	0.2
E100_NS	100	0.9	0.0125	2	0.2
E100_SG	100	0.9	0.003	2	0.6
E100_NS_G	100	0.9	0.0125	2	0.6
E3500_S	3500	0.9	0.006	2	7.0
E3500_NS	3500	0.9	0.015	2	7.0

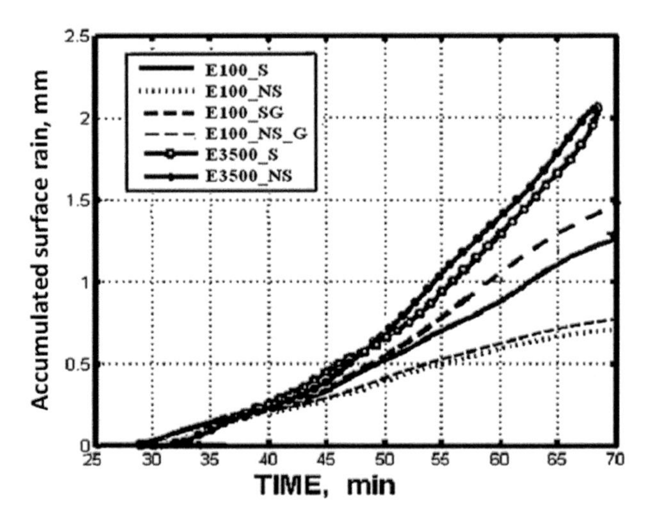

Figure 7.3.19 Time dependencies of space-averaged accumulated rain at the surface in simulations with and without GCCN and with and without the smallest CCN. Conditions of the simulations are given in Table 7.3.2 (from Khain et al., 2012; courtesy of © American Meteorological Society. Used with permission).

significant aerosol mass from continents into the TC periphery. As a result, convection in landfalling TC is invigorated at the TC periphery. The aerosol-induced changes in TC structure and intensity caused by convection invigoration at TC periphery are schematically illustrated in **Figure 7.3.20**. The figure shows a vertical cross section through the hurricane center, located at point zero (left-bottom point in the figure). The convective zone closest to the TC center is the TC eyewall, where huge latent heat is released in deep convective clouds. At the TC periphery, convective clouds form rain bands approaching the TC center (like squall lines).

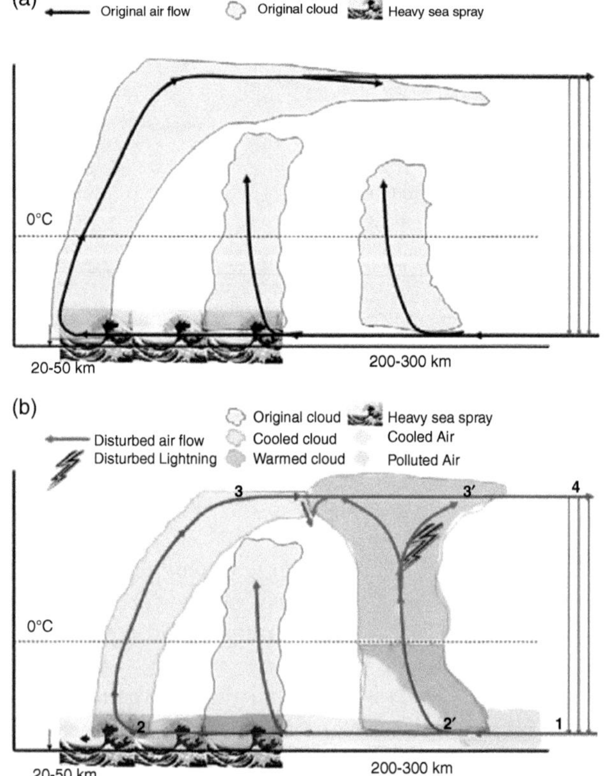

Figure 7.3.20 Conceptual model of aerosol impacts on tropical cyclones. (a) TC structure with low CCN concentrations over the entire TC area, typical over the open sea. (b) The structure of a landfalling TC, when low CCN concentrations in the TC inner zone are accompanied by high CCN concentrations at the TC periphery. The closure of the circulation system with subsiding air far away from the TC is indicated (blue lines) (from Rosenfeld et al., 2012; courtesy of © American Meteorological Society. Used with permission).

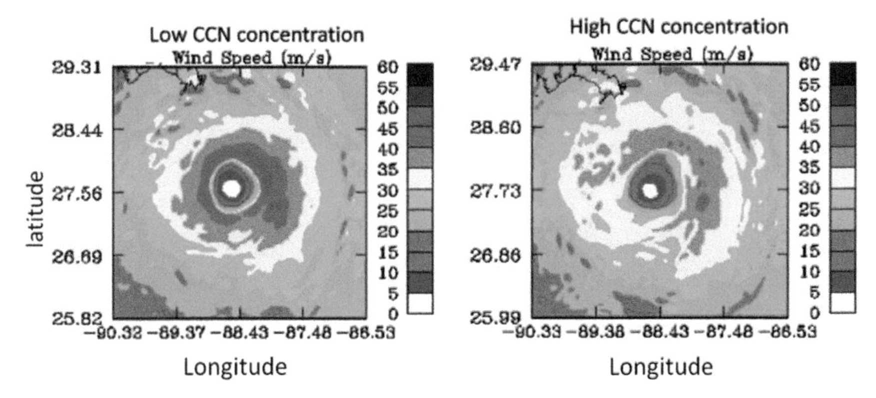

Figure 7.3.21 The maximum wind speed in a simulation of Hurricane Katrina (2005) with low CCN concentration (left) and when effect of continental aerosols on convection at the hurricane periphery was taken into account. Time is 2200 UTC Aug. 28 (from Khain et al., 2010, with changes; courtesy of © American Meteorological Society. Used with permission).

Aerosol-induced intensification of the outer rain bands causes a certain fraction of air to ascend within them (trajectory 2'–3'), but not in the eyewall (trajectory 2–3) (Figure 7.3.20b). This decreases the inflow toward the eyewall and, consequently, decreases the convection within the TC inner core. Intensification of rainbands also decreases the horizontal pressure gradient, which in turn leads to a decrease in the maximum wind.

Figure 7.3.21 shows the fields of maximum wind speed in simulations of Hurricane Katrina (2005) using WRF–SBM both at low CCN concentrations all over the computational area and when the effects of continental aerosols at the periphery convection are taken into account. One can see that aerosols penetrating the TC periphery lead to a substantial TC weakening. The radius of maximum winds increases. At the same time, aerosol penetration leads to an increase in the TC size determined by the radius at which wind speed is equal to 30 ms^{-1}.

TCs may be affected by aerosols not only during their landfall, but also over open oceans when TC cross wide bands of the Saharan dust moving over the Atlantic Ocean. In these cases, aerosols can penetrate the eyewall and intensify the convection in the inner core. The effects of aerosols penetrating the TC inner core as well as the periphery during the TC landfall were investigated by Lynn et al. (2016), who performed simulations of Hurricane Irene (2011). Hurricane Irene (Aug. 22–28, 2011) crossed a wide band of the Saharan dust and then moved northward along the eastern coast of the United States. The simulations were performed using WRF/BM with the resolution of the finest innermost grid of 1 km. The fields of different hydrometeors in two simulations were compared: MAR with the CCN concentration assumed equal to 100 cm^{-3} all over the entire computational domain, and DUST with the CCN concentration initially set equal to 2,000 cm^{-3} in the latitudinal zone from 10 to 30 N over the Atlantic Ocean (representing the Saharan dust) and over continents. In other areas over the ocean, CCN concentration of 100 cm^{-3} was assumed. The difference in TC intensity in MAR and DUST was due exclusively to aerosol effects on deep convective clouds.

The evolution of cloud microphysical structure determining changes in TC intensity is illustrated in **Figures 7.3.22** and **7.3.23**. A major effect of CCN on cloud microphysical structure is an increase in the cloud water mixing ratio of supercooled droplets. During August 25, Hurricane Irene crossed the dust band and a significant amount of CCN penetrated the TC eyewall. The aerosol effects on deep convective clouds in the TC inner core are fully in agreement with the scheme shown in Figure 7.3.2. The high CCN concentration in the DUST run led to convection invigoration and to formation of a significant number of supercooled cloud droplets ascending to the altitude of homogeneous freezing (~9–10 km) (Figure 7.3.22B, D, F). In contrast, in MAR the CWC is negligible above 2 km (panels A, C). This difference indicates the difference in the rates of latent heat release between MAR and DUST that leads to differences in TC intensity. Indeed, in DUST the radius of the eyewall decreased during Aug. 25 (compare panels B and D), which is indicative of TC intensification and deepening. Later on, the concentration of aerosols at the TC periphery increased, which led to an intensification of convection at the TC periphery and to an increase in the radius of the eyewall (panel F). This increase in the eyewall radius is indicative of TC weakening. In contrast, in MAR the radius of the eyewall continued decreasing until Aug. 26 (panel E), which indicates continuous TC deepening.

The difference in the CWC fields leads to other changes in the microphysical structure of clouds. As seen in Figure 7.3.23, the maximum values of the snow-mixing ratio are larger in MAR than in DUST. This is a typical property of deep convective clouds developing

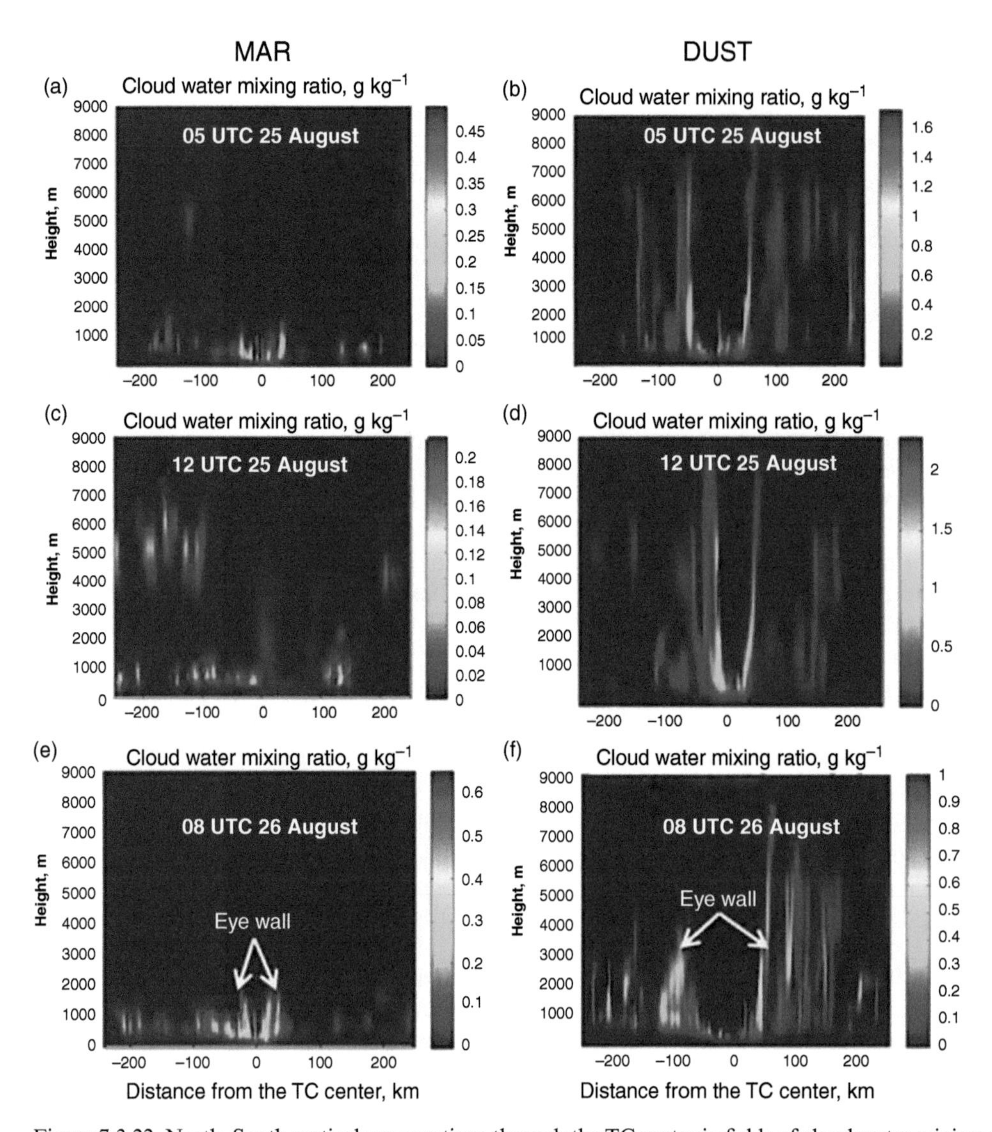

Figure 7.3.22 North–South vertical cross sections through the TC center in fields of cloud water mixing ratio in MAR (left) and DUST (right) simulations. The fields of cloud-water mixing ratio are shown for three time instances, as shown in the figure. Grid spacing is 1 km. Freezing level is 3.8 km (from Lynn et al., 2016; courtesy of © American Meteorological Society. Used with permission).

in clean air. Since they contain much less supercooled water than clouds forming in polluted air, production of graupel by riming of snow is relatively inefficient. Thus, these clouds have relatively more snow and less graupel. When hurricane Irene moved along the coast in the course of the simulations, the aerosol-induced intensification of the rainbands at the TC periphery on Aug. 26 in DUST led to much more snow at the northern periphery of the hurricane than in MAR. Aerosols at the TC periphery lead to an increase in the number of the rainbands (compare panels C and D in Figure 7.3.23).

Figure 7.3.24 demonstrates that taking the aerosol effects into account substantially improves results of TC intensity simulations. The effect of aerosols on the TC intensity turns out as important as the effect of sea-surface temperature and wind shear. Figure 7.3.24 also shows that the aerosol effect on the TC intensity depends on the aerosol location. When Irene crossed the Saharan dust band (experiment DUST, 50–70 h), aerosols intensified convection in the TC eyewall, fostering the TC intensification. At the same time, if aerosols intensify convection of the TC periphery ($t > 100$ h) TC weaken. Simulations also show that the Saharan dust can foster the development of tropical depressions.

Sea spray arising at the ocean surface at strong winds is a specific feature of hurricanes. The size range of sea-spray drops is quite wide: from about 0.03 μm to several hundred of μm in radius. Spray at the ocean surface

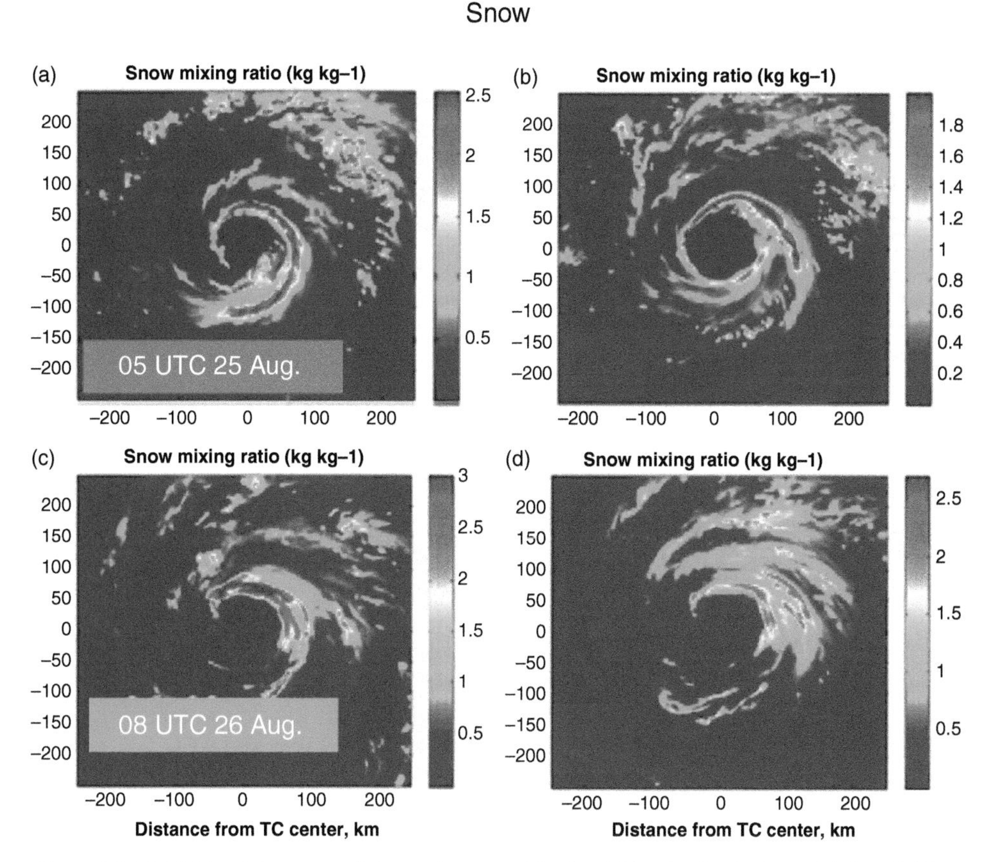

Figure 7.3.23 Fields of the snow mixing ratio at an altitude of 7 km in simulations with low CCN concentrations (MAR, left column) and with high CCN concentrations (DUST, right column) at two time instances: 05 UTC Aug. 25 and 08 UTC Aug. 26, 2011 (from Lynn et al., 2016; courtesy of © American Meteorological Society. Used with permission).

decreases atmosphere–ocean friction (the drag coefficient) (Andreas et al., 2004; Bao et al., 2011). The spray drops also affect the thermodynamics and microphysics of the BL. The spray particles have high salinity and can grow in the boundary layer at subsaturation conditions. Large eddies always existing in the boundary layer of TC transport the spray to cloud base of deep convective clouds and affect sensible and latent surface fluxes (Shpund et al., 2011, 2012, 2014). At last, Shpund et al. (2016) showed that sea spray affects microphysics and dynamics of deep convective clouds. The sea spray droplets entering the deep clouds increase droplet concentration and decrease droplet effective radius. As a result, the deep convective clouds in the eyewall of hurricanes have properties of both maritime clouds (i.e., raindrops form very fast) and continental clouds (high droplet concentration and small size of most droplets). It was found that sea spray ascending in cloud updrafts increases cloud intensity and cloud-top height in the TC eyewall, fostering TC intensification.

7.3.7 Effects of IN

The impact of IN on deep convective clouds is quite different from that on mixed-phase stratiform clouds. Only a few studies investigated the effects of an increasing IN concentration on precipitation from deep convective clouds. Fan et al. (2010) used a 3D model SAM/BM to examine the effects of both CCN and IN on properties of anvils of deep convective clouds, and water vapor content in the tropical tropopause layer. Simulations of deep convective clouds show that CCN effects on anvil microphysical properties, size, and lifetime are much more pronounced than IN effects. The main effect is related to homogeneous freezing of cloud droplets, resulting in formation of ice crystals whose concentration substantially exceeds the concentration of IN. The sensitivity study performed by Fan et al. (2010) shows that IN can increase ice number concentrations in cloud anvils, but have little effect on convective strength and precipitation. As shown in Section 6.8, concentration of ice crystals produced by mechanisms of ice multiplication can be substantially

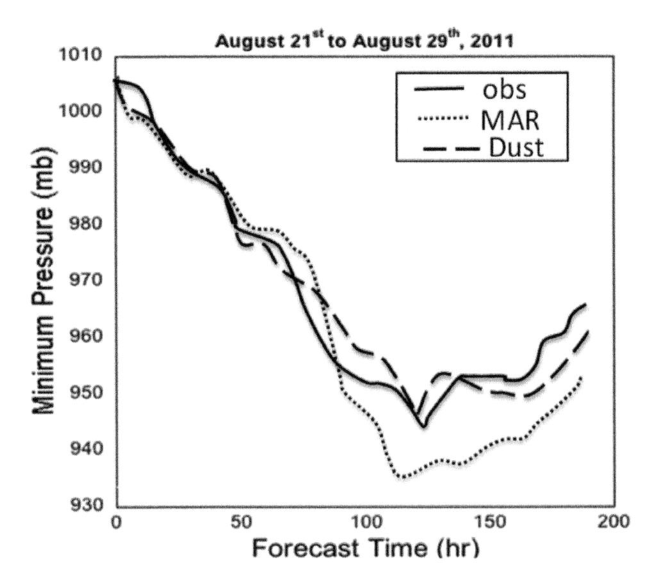

Figure 7.3.24 Time dependences of surface minimum pressure in simulations of Hurricane Irene (2011) in two simulations: low (maritime) aerosols everywhere (MAR) and high CCN concentration in zones of the Saharan dust band and over the land (DUST). Observed time dependence of minimum pressure is presented, as well (from Lynn et al., 2016; courtesy of © American Meteorological Society. Used with permission).

larger than the concentration of ice crystals formed due to primary ice nucleation. This fact can explain relatively low sensitivity of microphysics of mixed-phase deep convective clouds to concentration of IN. The main role of IN is the triggering of ice processes: formation of the first ice crystals, and immersion drop freezing.

The concentration of IN is much more important in stratiform and orographic winter clouds, where the temperature is warm enough to prevent formation of ice crystals by homogeneous droplet freezing. According to the theory (Section 6.3), an increase in ice crystal concentrations, caused by an increase in IN concentrations in such clouds increases the mass of ice crystals and shortens the glaciation time. According to Fan et al. (2014), who used WRF/BM, an increase in IN concentrations can lead to a 10–20% increase in precipitation from winter orographic clouds.

7.3.8 Effects of Aerosols on Large-Scale and Long-Term Processes

On large spatial and time scales, indirect effects of aerosols become more uncertain. The reason is that the GCM are not cloud resolving, and the convective parameterizations used in GCM do not describe properly either cloud microphysics or cloud dynamics. Convective parameterizations do not allow one to calculate cloud coverage in general and, in particular, the total area of the cloud anvils. Therefore, these parameterizations do not take into account the effects of aerosols on cloud properties and cloud size. Recently, efforts have been made to introduce a highly simplified two-moment bulk-parameterization scheme for convective clouds in GCM. Ming et al. (2007) implemented a calculation of cloud droplet number concentration into the convective parameterization of the GCM. Lohmann (2008) tested a two-moment scheme for convective clouds in the atmospheric general circulation model (ECHAM) developed at the Max Planck Institute for Meteorology. Song and Zhang (2011) tested a two-moment scheme including four hydrometeor types of convective clouds, into a single-column climate model. Obviously, the task is still very challenging due to the subgrid nature of convection in current climate models.

Using WRF/BM, Fan et al. (2013) conducted 3D month-long cloud-resolving simulations of summertime convection in three regions: the tropical western Pacific (TWP), southeastern China (SEC), and the U.S. southern Great Plains (SGP) to represent tropical, midlatitude coastal, and midlatitude inland summer convective clouds, respectively. Over TWP and SEC, the atmosphere was generally humid with weak vertical wind shears, but SGP was relatively dry with strong vertical wind shears during most of the simulation period. Clouds at SGP were generally organized by frontal systems. Computation areas were designed using a nested grid system with the linear sizes of outermost and internal meshes of 1,400 and 600 km, respectively. The grid spacing of the internal grid was 2 km. Duration of calculations allowed the authors to clarify the climatic properties of convection in these regions. In the simulations, the radiative properties of the clouds and the variations of cloud coverage were calculated at two concentrations of CCN, that differed by a factor of six to represent the clean and the polluted environments.

Figure 7.3.25 shows the vertical profiles of cloud occurrence frequencies during the one-month period from simulations and observations as well as of cloud fractions averaged over the one-month simulation period for TWP, SGP, and SEC regions (both in clean and polluted air). Figure 7.3.25 shows that the long-term simulations using WRF/BM reproduce several fine climatic features of cloudiness in the different regions. In particular, the SBM simulations show the existence of two major zones of cloud occurrence maxima, namely, Sc in the boundary layer and anvils of deep convective clouds. The model reproduces well the maximum cloud coverage caused by anvils of deep convective clouds. At last, the model shows the increase of this coverage in a polluted atmosphere in agreement with observations.

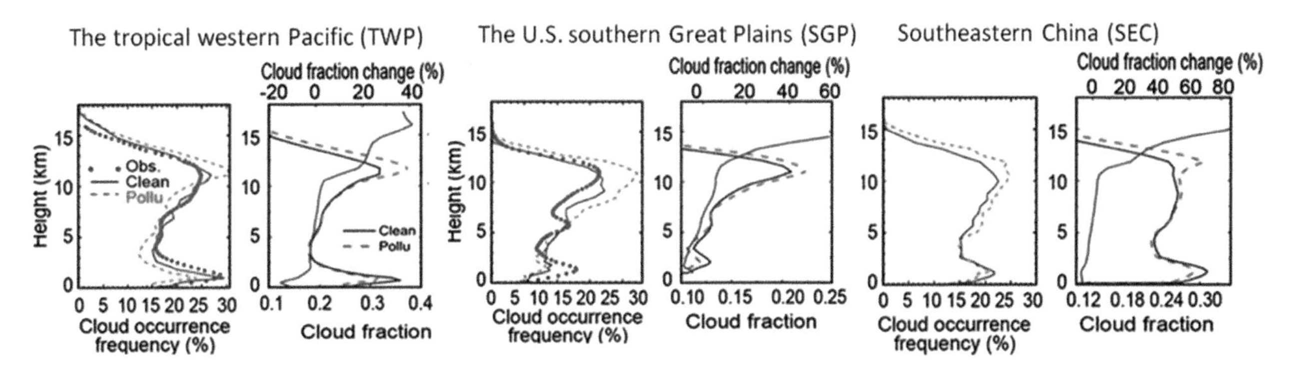

Figure 7.3.25 Vertical profiles of cloud fraction for TWP (left), SGP (middle), and SEC (right) regions. The left columns show the cloud occurrence frequencies during the one-month period from simulation for clean air (Clean; black) and polluted air (Pollu; dashed red) and observations (Obs; dotted blue). The right column in each panel shows the vertical profiles of cloud fractions averaged over the one-month simulation period. The percentage cloud fraction changes are shown by blue lines with the secondary axis in the same plots (from Fan et al., 2013, with changes; courtesy of PNAS).

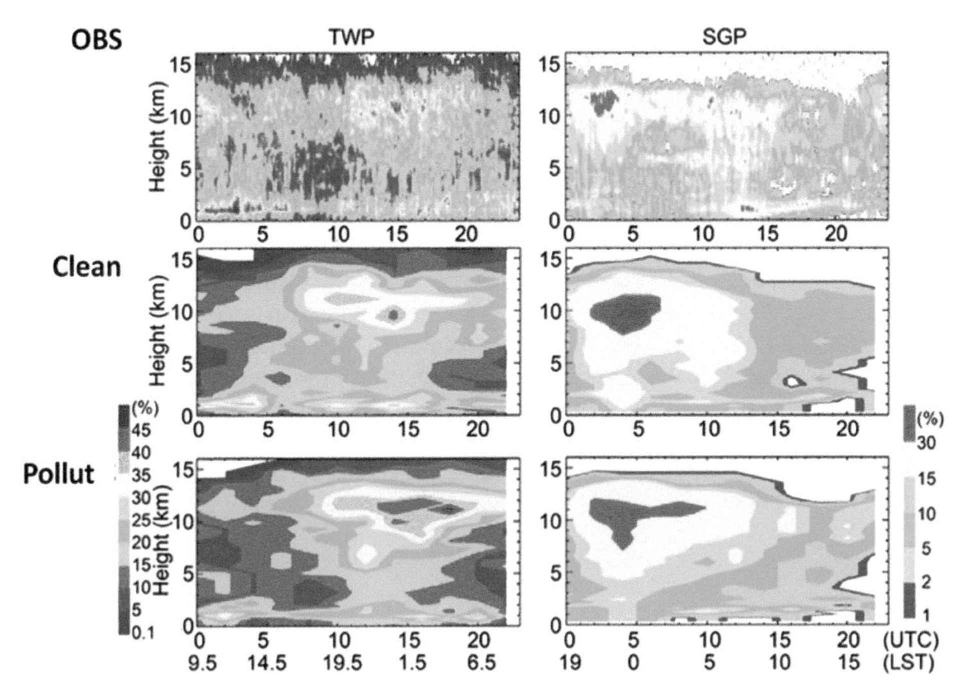

Figure 7.3.26 Comparison of diurnal cycle of cloud-occurrence frequency with observations averaged over the one-month simulation period for TWP and SGP areas. Observations (upper row); Clean: results of simulations for clean air (middle row); Pollut: results of simulations for pollute air (bottom row). Both Coordinated Universal Time (UTC) and Local Standard Time (LST) are shown on the x axis (from Fan et al., 2013, with changes; courtesy of PNAS).

Figure 7.3.26 compares a diurnal cycle of cloud-occurrence frequency with observations averaged over the one-month simulation period for TWP and SGP areas. Calculations were performed at different CCN concentrations. The figures show that an increase in CCN concentration causes a dramatic increase in cloud cover and cloud thickness, as well as in the cloud top height. Analysis showed that this response of deep convective clouds to an increasing CCN concentration was caused mostly by the induction of higher concentrations of long-lasting ice particles in the cloud anvils. These ice particles of relatively small size formed due to homogeneous freezing. The dynamic invigoration of convection is responsible only for ~30% of the effect.

According to Fan et al. (2013), the overall indirect aerosol effect is atmospheric radiative warming

(3–5 W·m^{-2}) and surface cooling (−5 to −8 W·m^{-2}). This impact is quite substantial and can increase the stability of the atmosphere. The most significant change in precipitation caused by aerosols was found to be redistribution of rain. The frequency of light rain is reduced, while heavy rain in the tropics becomes more frequent in polluted environments.

Several studies discuss the global effects of aerosols. These effects are caused by huge desert areas such as the Sahara that produce dust transported over enormous areas. Lau et al. (2009) used a GCM to show that the Saharan dust and biomass burning have a significant impact on the climate and water cycles of the North Atlantic. It was found that radiation heating of the Saharan dust and intensification of convection over Eastern Africa leads to the east–west circulation with subsidence over the Caribbean region, accompanied by a subsidence induced suppression of convection (see scheme in **Figure 7.3.27**). At global scales, aerosols may affect the intensity of circulation and precipitation over large areas. This effect has multiple outcomes: an increase in the moisture transport from the eastern Atlantic and the Gulf of Guinea to West Africa; enhanced rainfall over the Sahel and the Intertropical Convergence Zone (ITCZ) off the coast of West Africa; subsidence and suppressed cloudiness in the central Atlantic and Caribbean and the Gulf of Guinea; and cooling of West African Monson land and the upper ocean in the eastern Pacific underneath the dust plume.

Some observations reported that events such as tornados, hailstorms, and lightning activity over large areas such as the United States have a weekly midweek peak (e.g., Bell et al., 2008; Kim et al., 2010). Rosenfeld and Bell (2011) relate the peak to the time-coinciding peak in anthropogenic aerosol concentration over the eastern United States during summertime. These statistically significant variations over large areas require numerical modeling using advanced microphysical schemes.

7.3.9 Effects of Clouds on Aerosol Size Distribution

Cloud–aerosol interaction is a two-way interaction. Clouds also affect concentration, size distribution, and production of atmospheric aerosols.

As an example of effects of clouds on the aerosol concentration in the atmosphere, we present **Figure 7.3.28**, showing fields of aerosol concentration in simulations of hurricane Irene (Aug. 22–28, 2011) in the case of low aerosol concentration (left panel) and in case of high aerosol concentration over the land and in the Saharan dust band (right panel). Simulations were performed using WRF/SB (Lynn et al., 2016). One can see a significant decrease in the aerosol concentration in the TC region. The right panel shows that Irene crossed the band of Saharan dust, leaving a strip of clean air. Cloud clusters, storms, and hurricanes clean huge areas over the ocean and land by nucleation scavenging. At the same time, atmospheric motions advect aerosols on large distances and vertical velocities transport a significant amount of aerosols to the upper atmosphere. Evaporation of drops and sublimation of ice particles leads to release of aerosol particles into the atmosphere at height levels that can significantly differ from the levels of drop nucleation.

There are various mechanisms by means of which clouds affect concentration of aerosols. The most efficient mechanism of decrease in concentration of soluble AP is *nucleation scavenging*. The soluble AP playing the role of

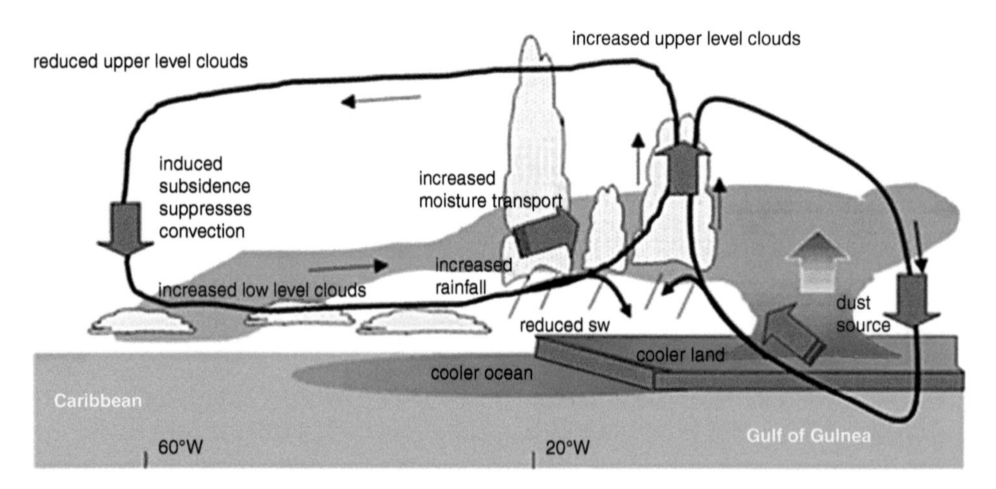

Figure 7.3.27 Schematic showing key features in the latitude domain of 5°N–15°N, associated with the "elevated heat pump" mechanism by radiative heating caused by the Saharan dust (adapted from Tao et al., 2012; courtesy of © John Wiley & Sons, Inc.).

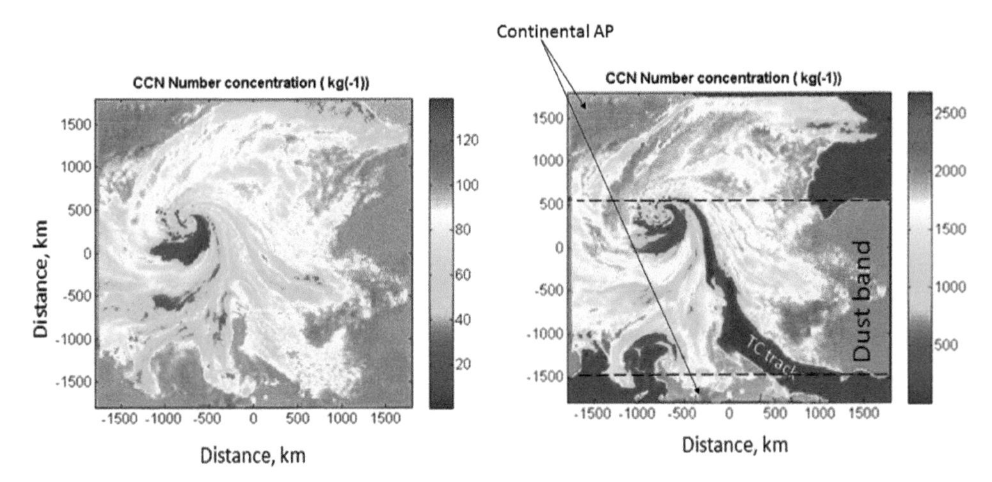

Figure 7.3.28 Fields of aerosol concentration in the boundary layer in simulations of hurricane Irene (Aug. 22–28, 2011) using WRF/SB (Lynn et al., 2016) in the case of low aerosol concentration (left panel) and in the case of high aerosol concentration over the land and in the Saharan dust band (right panel). Color scales in the panels are different.

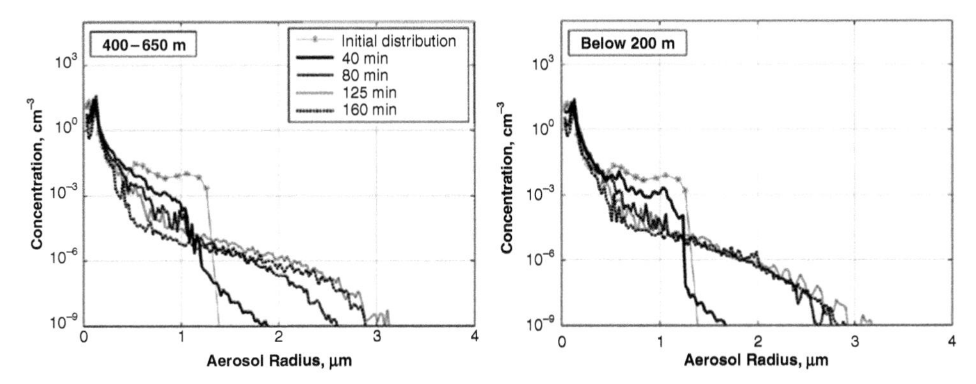

Figure 7.3.29 Aerosol size distributions (outside and inside of drops) averaged within cloud layer (left) and below cloud (right) at different time instances. The initial aerosol size distribution is shown, as well. Sum of values in the bins equals the averaged aerosol concentration (from Magaritz et al., 2010; courtesy of © Elsevier).

CCN are activated and give rise to droplet formation. Collision of these droplets leads to formation of precipitation particles. Precipitation particles fall to the surface, decreasing concentration of CCN in the atmosphere.

Processes of collection of aerosols, thermophoresis, and diffusiophoresis (Section 6.2) also strongly influence elimination of aerosol from clouds.

Magaritz et al. (2010) investigated effects of clouds on aerosol size and drop salinity in a stratiform cloud. In the way of example, the Sc observed in the research flight RF07 of DYCOMS-II field experiment was simulated with the LEM. Cloud was located within the layer at 400–800 m. Initial aerosols (CCN) size distribution is shown in **Figure 7.3.29**. The maximum aerosol radius was 1.3 μm. In the simulation, drizzle formed at about 60 min and fell to the surface. Figure 7.3.29 shows aerosol size distributions (outside and inside of drops) averaged within cloud layer (400–600 m) and below cloud ($z < 200$ m) at different time instances. One can see that drop collisions lead to the appearance of aerosol particles (if AP is dissolved within a drop, its size is evaluated by the size of particle that could be obtained by complete drop evaporation) with radii about three times as large as the maximum CCN size in the initial CCN spectrum. The largest aerosols are located in the largest drops (drizzle), which collected maximum amount of droplets. Concentration of small CCN decreases also, as a result of collection of small droplets (containing small CCN) by drizzle drops.

Since drizzle falls to the surface, mass content of aerosols in the boundary layer decreases, as shown in **Figure 7.3.30**. This figure shows that at $t < 60$ min, drizzle does not reach the surface and aerosol mass content (including aerosol inside of drops) remains constant. Since drizzle flux was small in RF07, the decrease in the aerosol mass (and also concentration) was comparatively small. Since the drizzle drops contain aerosols of maximum size, the drizzle fall to the surface leads to decrease in the concentration of largest aerosols. Thus, precipitation from stratocumulus clouds eliminates first of all of the largest aerosol particles.

One of the important topics of atmospheric chemistry is the analysis of salinity of drops, especially salinity of raindrops. Acid rains can cause serious damage to the environment. **Figure 7.3.31** shows dependence of salinity of drops on drop sizes simulated by Magaritz et al.

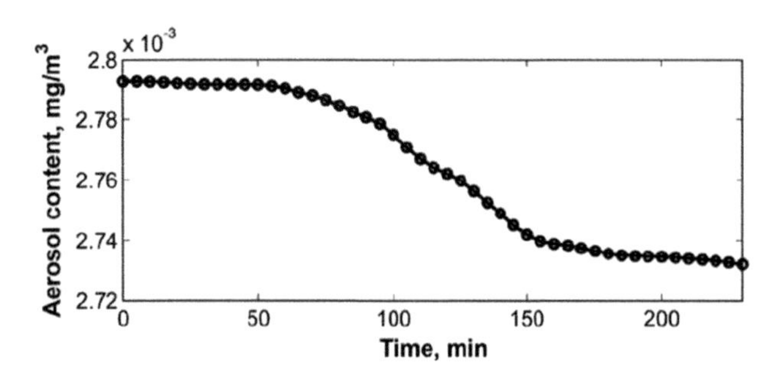

Figure 7.3.30 Time dependence of aerosol-mass content over the computational area in simulation of the Sc observed during the research flight RF07, as simulated by LEM (from Magaritz et al., 2010; courtesy of © Elsevier).

(2010) for a slightly drizzling maritime stratocumulus cloud. Salinity of a drop is determined as ratio of aerosol mass in the drop to the total drop mass. Figure 7.3.31 indicates the existence of three size ranges, in which the slopes of the dependencies are different: a) range of haze particles, b) range of diffusion growth, and c) range where collisions are dominating. The salinity of haze particles is maximum, because the mass of pure water in these particles is minimal. Magaritz et al. (2010) showed that within this zone salinity decreases as r^{-1}. This dependence follows from the Kohler theory. Diffusion growth of drops with the radii range from 1 μm to about 20 μm leads to strong decrease in drop salinity. The dependence of salinity on drop radius during diffusional drop growth is close to r^{-3}. This dependence is related to the fact that during diffusion growth mass of pure water within drops increases, while the mass of salt (aerosol mass) remains unchanged. Note that the sizes of droplets arising on AP of different size are different, and larger droplets grow slower. Accordingly, salinity of larger droplets decreases with time slower, and at the stage when effect of collisions is relatively unimportant, turns out to be larger than smaller droplets. It leads to a peak of salinity at radii of ~20 μm (Figure 7.3.31, left). In the top layers the salinity of drops is lower than in the lower levels (especially below cloud base).

The beginning of collisions between the largest droplets (within the peak of salinity) and smaller droplets decreases salinity of these largest droplets. At the same

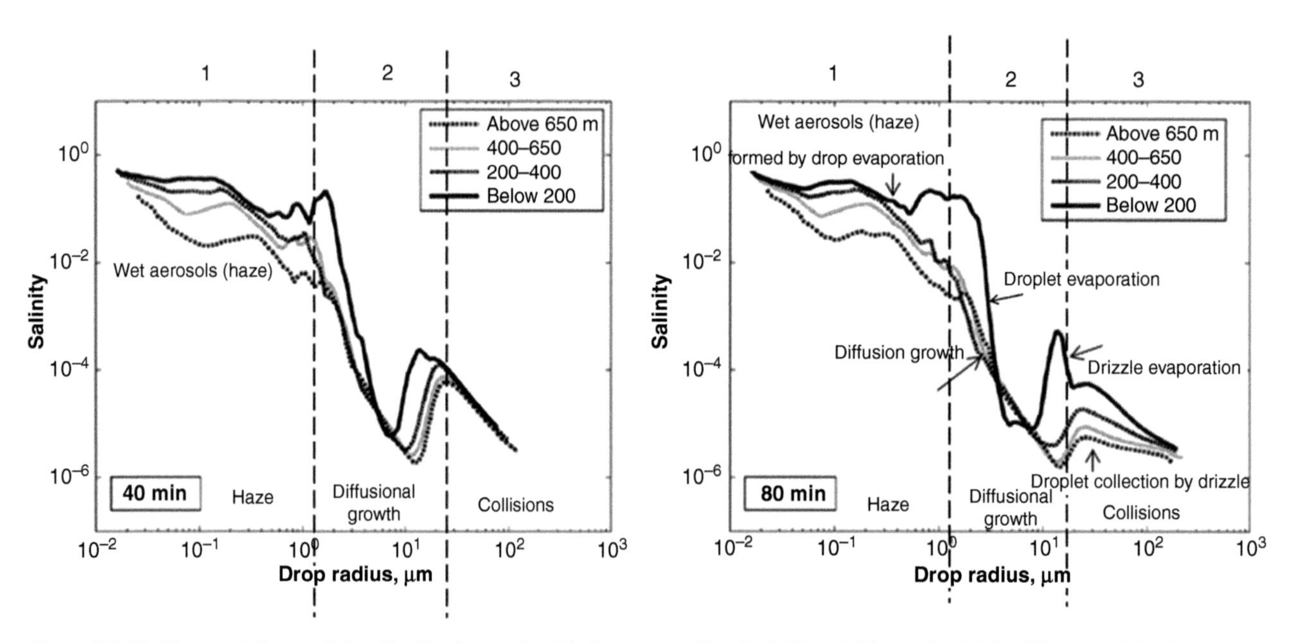

Figure 7.3.31 Haze and drop salinity distribution at the drizzle stage at 40 min (left) and 80 min (right) in different height layers. Numbers and vertical dashed lines denote three size ranges with different slopes of the dependencies of drop salinity on the radius (from Magaritz et al., 2010, with changes; courtesy of © Elsevier).

time, collisions of drizzle drops with small cloud droplets, which have comparatively high salinity, stops decreasing salinity with the further increase in drop size, so that drizzle drops are characterized by nearly the same salinity (Figure 7.3.31, right). Note that dependencies of drop salinity on drop size can be somehow different in cases of drop condensation and evaporation. This difference is caused by the existence of in-cloud nucleation in updrafts, different amounts of aerosols in ascending and descending drops. Below cloud base evaporation is the dominating process, so changes in the upper curve in Figure 7.3.31 result from evaporation of drizzle drops. In the process of evaporation, the mass of the aerosol does not change and there is a decrease only in the water mass, leading to higher salinity.

7.4 Some Advances in Modeling Clouds and Cloud-Related Phenomena

During the past decade, a serious advancement has been made in modeling individual clouds and cloud-related phenomena. This progress has been possible due to significant developments in different fields of studies. First of all, there is development of CRM that are used now in many areas of research and weather and climate forecast. The quality of microphysical bulk-parameterization schemes has increased permanently. The rapid growth of computational power allows us to apply higher model resolutions that enable us to improve model forecast skills and allows simulation of local small scale phenomena. Sections 7.2 and 7.3 present many examples of modern simulations, among them applications of advanced SBM models that explicitly calculate PSD of different hydrometeors and their higher moments.

Significant achievements have been made in evaluation of aerosol effects on precipitation formation. Since aerosols affect clouds via their effects on PSD, SBM models have advantage in this issue. SBM models made it possible to explain formation of high ice crystal concentration in Cb cloud anvils by mechanism of homogeneous freezing of droplets formed due to in-cloud nucleation of the smallest CCN.

Another important result was the performance of long-term simulations of aerosol effects on the radiation balance of cloud systems within synoptic scale regions (see Section 7.3; Fan et al., 2013). It was found that an increase in aerosol concentration tends to enlarge the cloud cover (the area of cloud anvils) due to the formation of smaller-sized ice crystals of high concentration that ascend higher than in the case of low aerosol concentration and spread over a larger area without any significant sedimentation. The overall aerosol indirect effect, as follows from the simulations, is manifested in radiative atmospheric warming of 3–5 W·m^{-2} and surface cooling of 5–8 W·m^{-2}.

Impressive findings have been made in the vitally important field of forecasting TC intensity. The simulations of TC evolution using WRF/SBM discovered high sensitivity of TC intensity to aerosols within the TC area (Section 7.3). Thus, a new factor was found that affects TC intensity in addition to SST and wind shear.

Better representation of PSD and their high moments in models enabled us to accelerate research and make steady progress in several related areas such as development of new bulk schemes for weather forecasting, remote sensing, weather modification, etc. In this section, we present examples of major contributions to a deeper understanding of fine microphysical processes in clouds that became possible due to the application of advanced numerical models. The focus is placed on the ability of advanced models to simulate PSD of different hydrometeors and to give answers to such long-pending problems as drizzle formation, effects of turbulence on formation and amount of rain in convective clouds, and the mechanism of large hail formation. The distinctive feature of advanced simulations is their good agreement with observations in "fine" aspects, for instance, correct reproduction of PSD and their moments. Such an agreement indicates that the model is able to reproduce realistic results, and, therefore, can be used for investigation of clouds and for parameterizations of the corresponding processes to be applied in large-scale models.

7.4.1 Drizzle Formation in Warm Stratocumulus Clouds

Maritime Sc cover enormous areas of Earth's surface and determine the radiative balance of the atmosphere. Formation of drizzle dramatically decreases the surface covered by Sc clouds and leads to a decrease in Earth's albedo. Formation of drizzle is a fine microphysical phenomenon. Drizzle drops with radii of about 100 μm form within the narrow cloud layer of ~300 m depth. In cumulus clouds, drops of such size typically form at distances substantially exceeding 300 m above cloud base even in maritime clouds with low CCN concentration (Arabas et al., 2009; Freud et al., 2008). In convective clouds, supersaturation is typically larger and turbulence is more intense than in Sc. In addition, convective clouds are not limited by inversion from above, so there is enough time for droplets to grow by diffusion and collisions up to raindrop sizes during their ascent. A relevant simulation of formation of wide DSD containing large droplets that would allow triggering efficient collisions within a limited time interval is a comprehensive task requiring an accurate description of diffusion growth, collisions, mixing, and sedimentation.

Warm stratocumulus clouds were investigated numerically using LES of different levels of complexity to describe cloud microphysical processes (Feingold et al., 1996; Khairoutdinov and Kogan, 1999; Kogan and Kogan, 2001; Stevens et al., 2003, 2005; Ackerman et al., 2009). Khairoutdinov and Kogan (1999) and Kogan and Kogan (2001) developed the Cooperative Institute for Mesoscale Meteorological Studies (CIMMS) LES model with the SBM microphysics. All dynamic and microphysical components of the model were carefully tested. The model used a high-order positive definite advection scheme, as well as the Berry and Reinhard scheme of collisions (Section 5.7). Aerosols were described using a separate size distribution function containing seventeen mass bins. The simulations were performed at the vertical resolution of 25 m. To eliminate the artificial spectrum broadening during diffusion growth, a variation-optimization method for the drop spectrum remapping in the semi-Lagrangian condensation/evaporation calculations was used (Liu et al., 1997). This model was the first to reproduce the vertical profiles of important thermodynamic variables such as LWC, the virtual liquid water potential temperature, and DSD. **Figure 7.4.1** shows an example of the DSD measured during the ASTEX flight A209 and simulated DSD at three different levels (top, middle, and bottom regions of the cloud layer). The simulated DSD are in good agreement with observations both for cloud droplets sampled by the FSSP and for drizzle drops sampled by the 2-DC probe.

Upon comparison of the model results with data obtained in multiple measurements conducted over the Atlantic and Pacific, it was concluded that the model-derived DSD can be used as a data source for parameterizations. Khairoutdinov and Kogan (2000) and Kogan and Kogan (2001) derived parameterization formulas for rates of different microphysical processes, as well as for calculation of the effective radius (**Table 7.4.1**). These formulas are now used in large-scale models to calculate the integral microphysical variables such as CWC and droplet concentration (Randall et al., 2003). The parameterizations were verified by Wood (2000) on the basis of observations from eleven flights carried out in different drizzling Sc. The in situ data were collected during different seasons at different locations in the Atlantic around the United Kingdom. The results of the study showed that the drizzle parameterization, derived using CIMMS LES with SBM provided the most accurate drizzle predictions. The development of these parameterizations is an example of a significant achievement made owing to advanced cloud models.

Another successful step in developing parameterizations using the SBM model was parameterization of the droplet effective radius for drizzling marine Sc (Kogan and Kogan, 2001). Cloud droplet effective radius r_{eff} is an important parameter needed for calculations of cloud optical properties. The satellite-derived annually averaged values of r_{eff} vary from 6 to 17 μm depending on cloud type and location. On the basis of GCM simulations, Slingo (1990) showed that a reduction of r_{eff} by 1–2 μm can introduce changes in Earth's

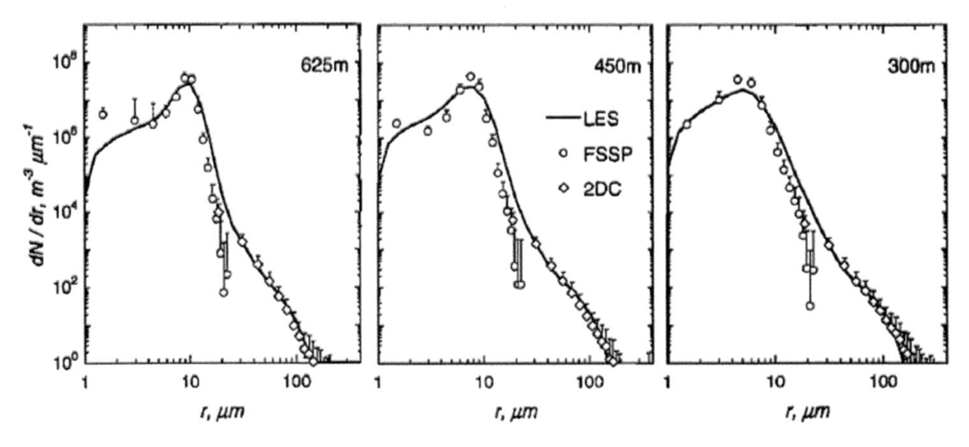

Figure 7.4.1 Simulated DSD (solid lines) and DSD measured during the ASTEX flight A209 (symbols) at three different levels within a stratocumulus cloud layer. The standard deviation of the measured values is indicated by bars (from Kogan and Kogan, 2001; courtesy of © American Meteorological Society. Used with permission).

Table 7.4.1 Parameterization of microphysical process rates according to Khairoutdinov and Kogan (2000) and Kogan and Kogan (2001).

Process	Parameterization expression	Comments
Condensation/ evaporation	$\left(\frac{\partial q_r}{\partial t}\right)_{cond} = 3\frac{C_{ev}}{F(T,p)}\left(\frac{4\pi\rho_w}{3\rho_a}\right)^{2/3} q_r^{1/3} N_r^{2/3} S_w$	Index "r" denotes drizzle drops $C_{ev} = 0.86$; q_r is in kg kg^{-1}.
Autoconversion	$\left(\frac{\partial q_r}{\partial t}\right)_{auto} = 1350 q_c^{2.47} N_c^{-1.75} S_w$ $\left(\frac{\partial q_r}{\partial t}\right)_{auto} = 4.1\times10^{-15} r_v^{5.67}$ $\left(\frac{\partial N_r}{\partial t}\right)_{auto} = \left(\frac{\partial q_r}{\partial t}\right)_{auto}\left(\frac{4\pi\rho_w}{3\rho_a} r_0^3\right)^{-1}$	Index "c" denotes cloud droplets;, q_r in $kg\ kg^{-1}$, N_c is in cm^{-3}; the mean volume radius r_v in μm. r_0 is the minimum radius of drizzle drops assumed equal to 25 μm.
Accretion	$\left(\frac{\partial q_r}{\partial t}\right)_{accr} = 67(q_c q_r)^{1.15}$ $\left(\frac{\partial N_c}{\partial t}\right)_{accr} = \frac{\left(\frac{\partial q_r}{\partial t}\right)_{auto} + \left(\frac{\partial q_r}{\partial t}\right)_{accr}}{\left(\frac{4\pi\rho_w}{3\rho_a} r_v^3\right)}$	
Sedimentation	$V_{N_r} = 0.007 r_v - 0.1;$ $V_{q_r} = 0.012 r_v = 0.2$	V_{N_r} and V_{q_r} are concentration and mass averaged fall velocities, respectively; r_v is the mean volume radius of the drizzle drops in μm; terminal fall velocities are in ms^{-1}
Parameterization for effective radius in non-precipitating Sc	$r_{eff} = 74.7\, N^{-0.38} q_w^{0.28}$	Units: N in cm^{-3}; q_w in gm^{-3}, Z in dBZ. Regimes of moderate and heavy drizzle are separated according to the value of the mean volume radius:
Light and moderate drizzle (~0.2 mm/day)	$r_{eff} = 35.1\, N^{-0.38} q_w^{0.26} (Z+50)^{0.2}$	$r_v < 16$ μm-moderate drizzle; $r_v > 16$ μm-heavy drizzle;
Heavy drizzle (~0.2 mm/day)	$r_{eff} = 3.4\, N^{-0.3} q_w^{0.1} (Z+50)^{0.71}$	

radiative budget comparable to the global warming effect from doubling the CO_2 content. Thus, a 10% error in calculating r_{eff}, which is of the order of 1.5 μm, can hardly be ignored (Slingo, 1990). The value of the effective radius is also important to characterize the rate of collisions and precipitation formation, since it characterizes the shape of DSD.

Using airborne observations in non-precipitating stratocumulus, Johnson et al. (1992), Martin et al. (1994), Gultepe et al. (1996), Reid et al. (1999), and McFarquar and Heymsfield (2001) demonstrated that r_{eff} can be evaluated using a simple relationship between the mean volume radius and the cloud droplet effective radius:

$$r_{eff} = \alpha r_v \quad (7.4.1)$$

Values of α depend on the shape of DSD and vary from $\alpha = 1.08 \pm 0.02$ for pristine clouds down to $\alpha = 1.11 \pm 0.03$ for polluted clouds and $\alpha = 1.19 \pm 0.07$ for drizzling clouds (Martin et al., 1994; Johnson et al., 1992; McFarquar and Heymsfield, 2001). Using CIMMS, Kogan and Kogan (2001) proposed a parameterization of r_{eff} that is valid not only for the non-drizzling case, but also in moderate and heavy drizzle conditions. To derive parameterization for the drizzling case, the third parameter – namely, radar reflectivity – was included in the list of governing parameters alongside droplet concentration and CWC. The parameterization expressions were obtained using statistical analysis of model simulations at different drizzle rates. The expressions obtained are also presented in Table 7.4.1.

The problem of drizzle formation can be formulated as follows: does drizzle form at any location within Sc due to favorable fluctuations of some parameters, or the source and location where drizzle forms can be predicted? Several factors favoring drizzle formation have been investigated. In idealized simulations, Feingold et al. (1996) showed the importance of cloud thickness and the residential time of drops within cloud. In more detail, the importance of the residential time was investigated by Kogan (2006), who used the CIMMS LES model to calculate a great number of Lagrangian trajectories of idealized parcels playing the role of passive tracers. It was shown that air cycling is an essential feature of drizzling stratocumulus cloud dynamics. According to the observation data, drizzle rate is proportional to the cube of cloud thickness and inversely proportional to droplet concentration (van Zanten et al., 2005).

The problem of drizzle formation was addressed in a set of the studies (Pinsky et al., 2008; Magaritz et al., 2009; Magaritz-Ronen et al., 2014; Magaritz-Ronen et al., 2016). In these studies, a hybrid LEM of the maritime BL was used. Sc clouds observed in two research flights RF01 and RF07 in the field experiment DYCOMP II were simulated. The model is described in detail in Section 7.1. There was only a slight difference in the properties of RF01 and RF07. Nevertheless, RF01 was non-drizzling, while a light drizzle formed in RF07. The LEM was able to distinguish between these cases and to reproduce both the horizontally averaged profiles of the main microphysical variables and particular microphysical characteristics (**Figure 7.4.2** and **Table 7.4.2**). Comparison with data measured in observations indicates that the model reproduces quite fine microphysical properties of real Sc (Table 7.4.2).

In addition, the model realistically reproduces spatial variability of the main thermodynamic and microphysical quantities (Figure 5.10.4), as well as the relationship between r_{eff} and r_v measured in observations (Magaritz-Ronen et al., 2016).

Another confirmation of the high accuracy of the model can be seen in **Figure 7.4.3**, showing the *Z*–LWC scatter diagrams plotted for the entire period of simulations (0–400 min) of RF01 (a) and RF07 (b). Simulated by LEM *Z*–LWC scattering diagrams are compared with diagrams presented by Krasnov and Russchenberg (2002) and Khain et al. (2008d) (panel (c)), designed using the measured DSD in different Sc clouds (non-drizzling and drizzling, including heavy drizzling). One can see that dots in the scatter diagrams are concentrated along three lines (Figure 7.4.3b) representing three stages of cloud evolution: diffusional growth (1), intense collisions (2), and drizzle regime when high reflectivity forms below cloud base (3). The superposition of these three zones results in the scattering diagram in Figure 7.4.3c. The agreement between *Z*–LWC dependence scattering diagrams simulated by the model (panel b) and the observations data (panel c) demonstrates the model's ability to reproduce very fine microphysical features of real clouds, which makes it applicable for investigation of drizzle-formation processes.

The diagrams allow us to analyze the stages of cloud evolution leading to drizzle formation. At the stage of diffusion growth (line 1), the radar reflectivity increases as LWC increases, with the slope of relationship in log–log coordinates equal to two. This slope forms because for relatively narrow DSD radar reflectivity (the 6-th moment of DSD) is proportional to square of the LWC

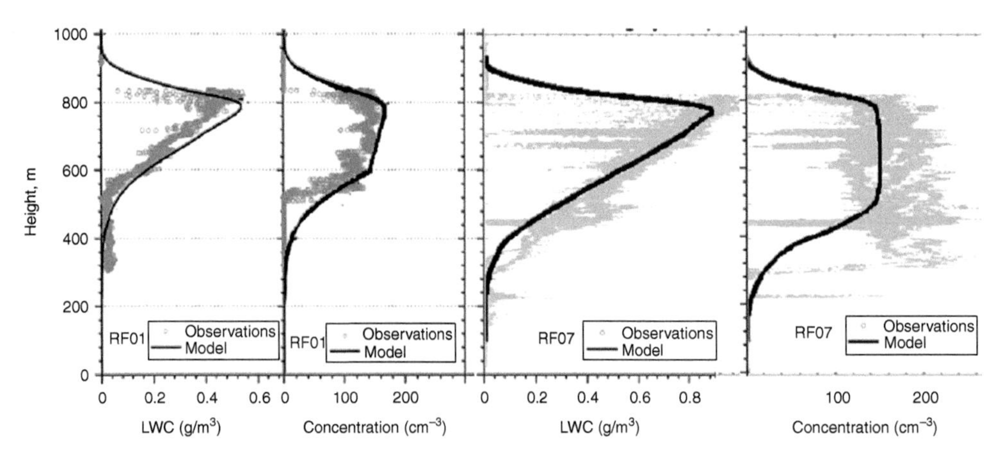

Figure 7.4.2 Vertical profiles of horizontally averaged LWC and droplet concentration obtained in simulations (solid lines). Light-gray symbols denote in situ observations. Two left panels are related to RF01 (non-drizzling) and two right panels are related to RF07 (weak drizzle) (from Magaritz-Ronen et al., 2014; courtesy of © American Meteorological Society. Used with permission).

Table 7.4.2 Comparison of calculated values with data measured during RF01 and RF07 (from Pinsky et al., 2008, with changes).

	RF01	No-drizzle case	RF07	Drizzle case
Cloud base (m)*	585	530–600	310	350–450
$\overline{q}_e$(g kg^{-1})	9	9	10	10
LWC$_{max}$ (g kg^{-1})	0.5	0.6	0.8	0.8
Drizzle flux (mm day^{-1})**	Below detection level	Below detection level	0.6 (±0.18)	0.5
Droplet concentration (cm^{-3})	~150	~190	~150	~160
The range of effective radii at 820 m (μm)		8–12	10–14	10–14
Mean effective radius of DSD near the surface (μm)		75	100	100
Maximum effective radius near the surface (μm)		100	160	200
Maximum radar reflectivity (dBZ)	−12	−10	10–12	10–12

* The evaluation of the cloud-base height was carried out using profiles of LWC; the vertical profiles of the LWC in the cloud layer were extrapolated from above to the level where the extrapolated value was equal to zero.
** Model value is an average over the entire run; the drizzling detection level (threshold) was about 0.03 mm day^{-1}.

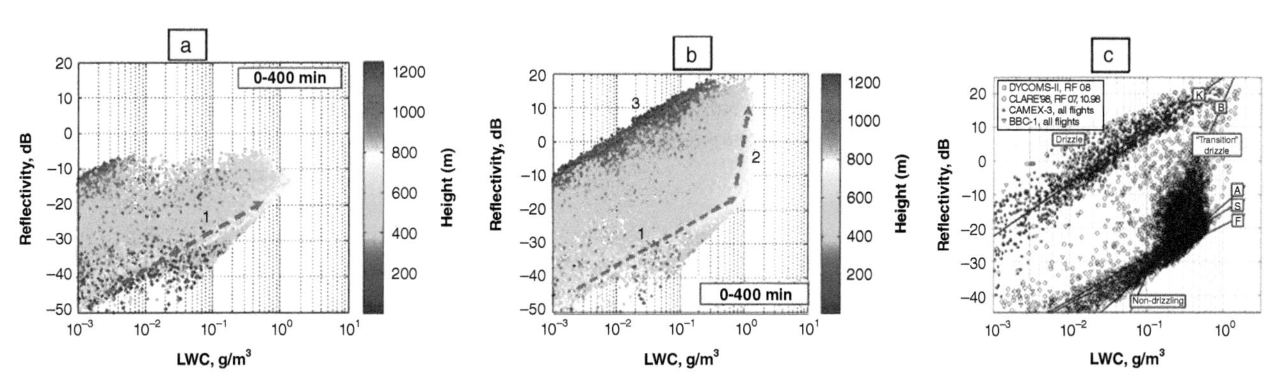

Figure 7.4.3 The Z–LWC scattering diagrams for clouds simulated in (a) RF01, (b) in RF07, and (c) calculated using measured DSD in Sc clouds in different field experiments. Each point in diagrams (a) and (b) denotes location of a separate parcel marked every 5 min. Color scale in these diagrams denotes the height of parcels above the surface. Dashed lines indicate different stages of cloud development in the simulations. Lines A, S, and F in diagram (c) show different approximations of Z–LWC dependence using in situ measurements (diagram (c) is taken from Krasnov and Russchenberg, 2002 and Khain et al., 2008d; courtesy of © American Meteorological Society. Used with permission).

(the third moment of DSD). The stage of intense collisions corresponds increasing radar reflectivity at nearly the same LWC (line 2). One can see that the RF01 cloud does not pass the stage of intense collisions, and, as a result, does not produce drizzle. Triggering efficient collisions takes place when LWC reaches about 1 g/m^3 and the maximum radar reflectivity reaches −20 to −15 dBZ. Since LWC increases with height, intense collisions and first drizzle drops formation begin near cloud top. Drizzle drops formed at this stage fall down to parcels containing lower LWC and to non-cloud parcels below the cloud base. As a result, in parcels below cloud base high radar reflectivity occurs at small LWC. Line 3 corresponding to drizzle drops is parallel to Line 1.

The parcels where first drizzle drops form have LWC ~1 g m^{-3}. This value exceeds the maximum horizontally averaged value of 0.8 g m^{-3} and is close to the adiabatic value. Analysis of the calculated statistics of the values of effective radius shows that r_{eff} in parcels near cloud top is 11–12 μm, which agrees with the measurements (Van Zanten et al., 2005). However, drizzle forms first in parcels having the largest LWC (lucky parcels). The maximum LWC in a lucky parcel near the cloud top can be reached if the LCL of this parcel is lower than that of other neighboring parcels. Accordingly, humidity in such lucky parcels should be larger than in all neighboring parcels. Statistical analysis showed that lucky parcels spend significant time near the surface getting water vapor from the ocean surface and then ascend within a comparatively wide updraft zone of a large eddy (Magaritz-Ronen et al., 2016). The significant width of the updraft zone allows ascending lucky parcels to remain undiluted.

Figure 7.4.4 illustrates the evolution of the microphysical parameters and DSD of a single lucky parcel that is the first to produce drizzle drops (of radius exceeding 40 μm). This parcel ascends from the cloud base up to 800 m during 13 min, has low LCL, and DSD with a large effective radius near the cloud top. The formation of drizzle-size drops substantially accelerates at $t = 160$ min when $r_{eff} = 11$ μm and LWC reaches 1 g m^{-3}. An elongated tail of largest drops of the DSD forms and develops due to active collisions (panel b). Toward $t = 166$ min, the parcel contains small drizzle drops of radii as large as 40 μm that fall down due to sedimentation. DSD shown in Figure 7.4.4 is quite similar to the measured DSD (Van Zanten, 2005).

Magaritz-Ronen et al. (2016) revealed the dual impact of turbulent mixing between parcels on drizzle formation in Sc clouds. On the one hand, mixing decreases the maximum values of LWC in ascending parcels, which retards drizzle formation. On the other hand, the mixing leads to homogenization of cloud structure, elimination of zones of decreased humidity formed by parcels entraining from the inversion zone. This prevents drizzle evaporation within the zones of decreased humidity and fosters the drizzle growth by collisions during sedimentation. Drizzle formation in the LEM is highly sensitive to aerosols, so an increase in aerosol concentration can prevent drizzle formation (see Section 7.3).

7.4.2 Rain Formation in Warm Cumulus Clouds

Formation of DSD and raindrops was discussed in Section 5.11.3. The microphysical processes were discussed using results of a 1D model, so comparison with observations could be only qualitative. Next, we present results obtained in multidimensional cloud models that demonstrate agreement of simulated microstructures with observations as well as enable calculating surface precipitation.

Numerous simulations show that SBM cloud models provide accurate reproduction of DSD in convective clouds. In simulations using the Tel Aviv University cloud model, Levin et al. (1991) reported a good agreement between the calculated raindrop size distribution and that measured at the top and bottom of Mt Rigi in Switzerland. Similarly, the HUCM SBM model accurately simulated DSD and their changes with height at different aerosol loading values. **Figure 7.4.5** compares simulated DSD with in situ measured DSD during the LBA-SMOCC 2002 field experiment in the Amazon region. Developing clouds with different aerosol concentrations were classified into Blue Ocean (BO) clouds (clean atmosphere), Green Ocean (GO) clouds with intermediate aerosol concentrations and Smoky clouds (S) developing in a very polluted atmosphere with CCN concentrations of about 10^4 cm^{-3} (Rissler et al., 2006).

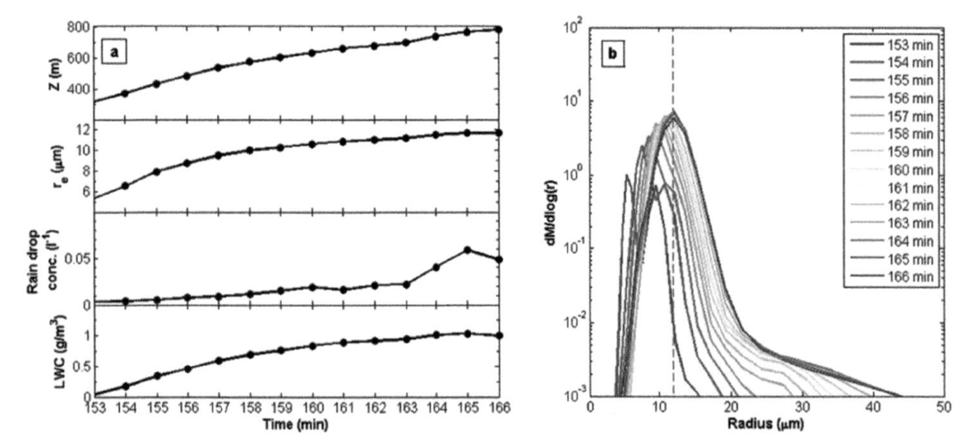

Figure 7.4.4 Time evolution of the microphysical parameters (a) and mass distribution (b) in the single lucky parcel. Drizzle drops are defined as drops of radii exceeding 40 μm (from Magaritz-Ronen et al., 2016; courtesy of ACP).

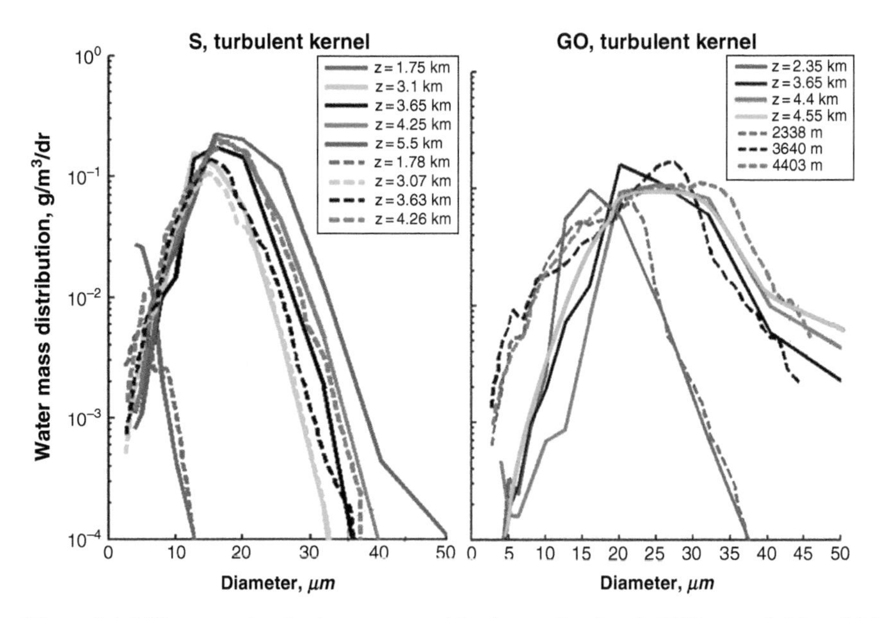

Figure 7.4.5 The mass distributions measured in situ on October 5, 2002 up to height of 4.2 km (dashed lines) and calculated using HUCM (solid lines). Left: developing S clouds. Right: developing GO clouds. The CCN concentration at 1% of supersaturation is 5,000 cm^{-3} in S and 1,000 cm^{-3} in GO (from Benmoshe et al., 2012; courtesy of © John Wiley & Sons, Inc.).

Figure 7.4.5 shows DSD at different heights in GO and S clouds during the earlier stages of cloud development, before the formation of the first raindrops. These DSD were obtained by averaging the observed DSD over the entire cloud traverses. Figure 7.4.5 also shows DSD obtained in simulations with the effects of turbulence on cloud-droplet collisions taken into account. The values of ε and Re_λ were calculated using the equation for turbulent kinetic energy at each time step and at each model grid point (Section 5.11.4). Figure 7.4.5 demonstrates that the model reproduces the DSD evolution in clouds quite well. The DSD in GO are wider than in S and centered at 12–14 μm, as compared with 10 μm in S, which agrees with the observations. Such correct reproduction of DSD temporal and height variations at different aerosol concentrations creates a solid basis for investigation of rain formation in simulated clouds.

The effective radius r_{eff} is an important microphysical characteristic of clouds, indicating their ability to produce raindrops. Measurements show that raindrop formation takes place if r_{eff} exceeds 14 μm in cumulus clouds developing in comparatively clean air, and 10–11 μm in clouds developing in polluted atmosphere (Freud et al., 2008; Freud and Rosenfeld, 2012). **Figure 7.4.6** shows the vertical profiles of the effective radius, r_{eff} in tops of developing BO, GO, and S clouds. The vertical profiles of r_{eff} obtained using in situ observations in tops of multiple developing convective clouds during the LBA–SMOCC are presented, as well. The vertical arrows show the values of r_{eff} corresponding to formation of first raindrops. The model results are in good agreement with observations: first raindrops form at $r_{eff} = 14$ μm in BO, at $r_{eff} = 12$ μm in GO, and at $r_{eff} = 10.5-11$ μm in highly polluted S. Also the heights of first raindrop formation marked by horizontal lines in Figure 7.4.6 are close, with the observed precision of about 100–200 m.

Khain et al. (2013) modeled the process of warm rain formation in deep convective clouds observed during the CAIPEEX (2009) field experiment over India. The simulations of deep convective clouds were performed using the SAM with the SBM. The model successfully simulated the dependence of first raindrop formation height on CCN concentration. In addition, vertical profile of the effective radius $r_{eff}(z)$ in developing clouds was obtained and found in good agreement with the values measured in situ (see Figure 5.10.10). The relationship between effective and mean-volume radii calculated using the model results have a form $r_{eff}(z) = \alpha r_v$ with $\alpha = 1.08$ for non-precipitating clouds and $\alpha = 1.17$ for the onset of precipitation stage (**Figure 7.4.7**). The same relationship was established by Freud and Rosenfeld (2012) on the basis of data obtained in numerous field campaigns in non-precipitating developing convective clouds in different geographical regions.

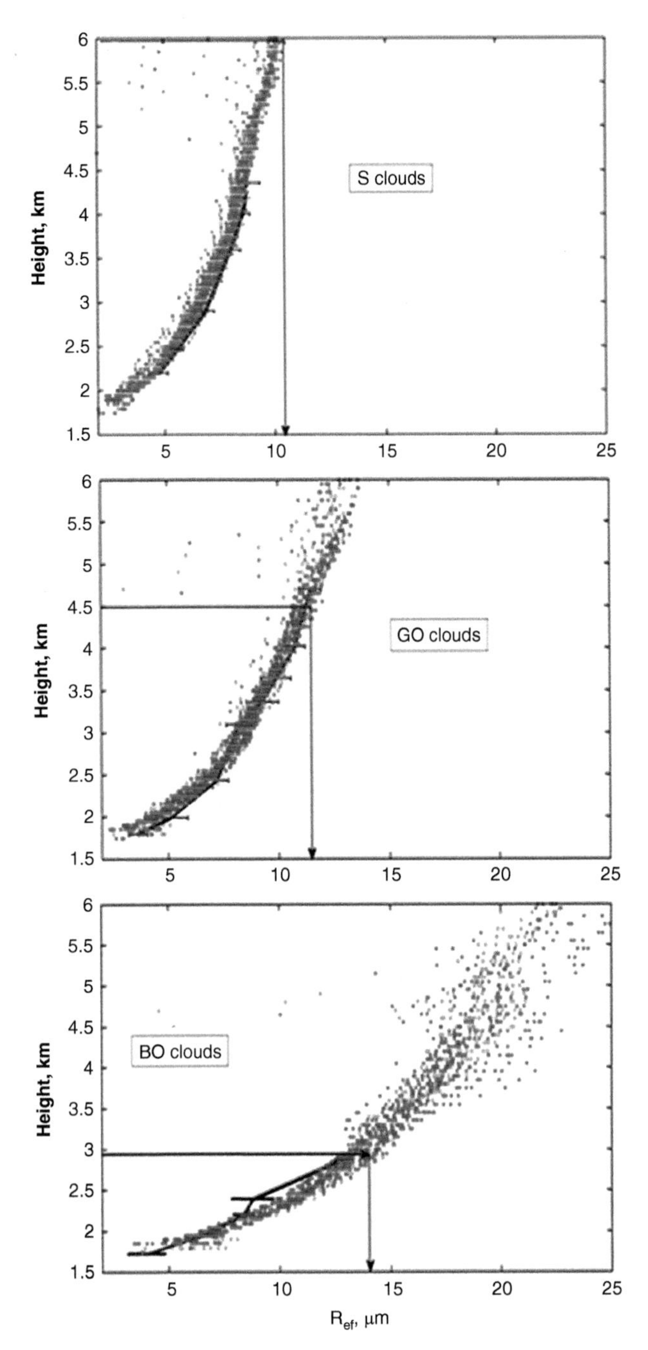

Figure 7.4.6 Height dependence of the effective radius in S (top), GO (middle), and BO (bottom) clouds. Solid lines show the averaged observed data. Simulations with gravitation collision kernel are marked by blue circles; red asterisks denote simulations with turbulent collision kernels. The dots located at the same height denote values of effective radius in different grid points in cloud at this altitude. The values of critical values of effective radius and levels of first rain formation obtained in the simulations are plotted by the arrows (from Benmoshe et al., 2012; courtesy of © John Wiley & Sons, Inc.).

Thus, the reliability of the simulations is confirmed by the successful reproduction of the DSD shapes, as well as of the vertical profiles of $r_{eff}(z)$ for different CCN concentrations, and of the relationship between the effective radii and the mean-volume radii determined by high moments of DSD. An important conclusion made as a result of these studies is that the first raindrops form in undiluted (adiabatic) or slightly diluted cloud volumes near the cloud top where turbulence intensity reaches its maximum. The effective drop radius in these volumes is 12–14 μm.

The realistic reproduction of microphysical cloud structure using turbulent collision kernels allowed scientists to evaluate the impact of turbulence on precipitation formation. Benmoshe et al. (2012) showed that taking into account the turbulent effects enables one to lower the level of first raindrop formation by a few hundred meters and accelerate rain onset by several minutes (i.e., by 30% as compared to simulations using gravitational). Lee and Baik (2016) simulated a rain event over the Korean Peninsula by means of WRF/BM following the approach developed by Benmoshe et al. (2012) to describe turbulence effects (see Section 5.8). They showed that taking into account turbulent effects on collisions improved the prediction of surface precipitation. Comparison of precipitation rates over two regions of the Korean Peninsula in simulations with turbulence-induced collision enhancement and without it is presented in **Figure 7.4.8**. One can see that turbulence increases the rain rate and accelerates the rain onset not only in a particular cloud, but in cloud ensembles where clouds are at different stages of their evolution.

Elaboration of the role of turbulence in rain formation is an important accomplishment made possible due to the cloud SBM model.

The effects of turbulence on surface precipitation turned out to be not crucial for several reasons. First, turbulence in clouds is intense only in a comparatively low fraction of clouds. Second, more importantly, turbulence intensifies collisions of cloud droplets accelerating the formation of first raindrops. However, as soon as raindrops formed, further raindrop growth is determined largely by gravitational collisions and the role of turbulence decreases.

7.4.3 Simulation of Deep Mixed-Phase Clouds

Today's mixed-phase and ice microphysics is far from being complete. Nevertheless, despite all gaps and uncertainties, the advanced microphysical models are

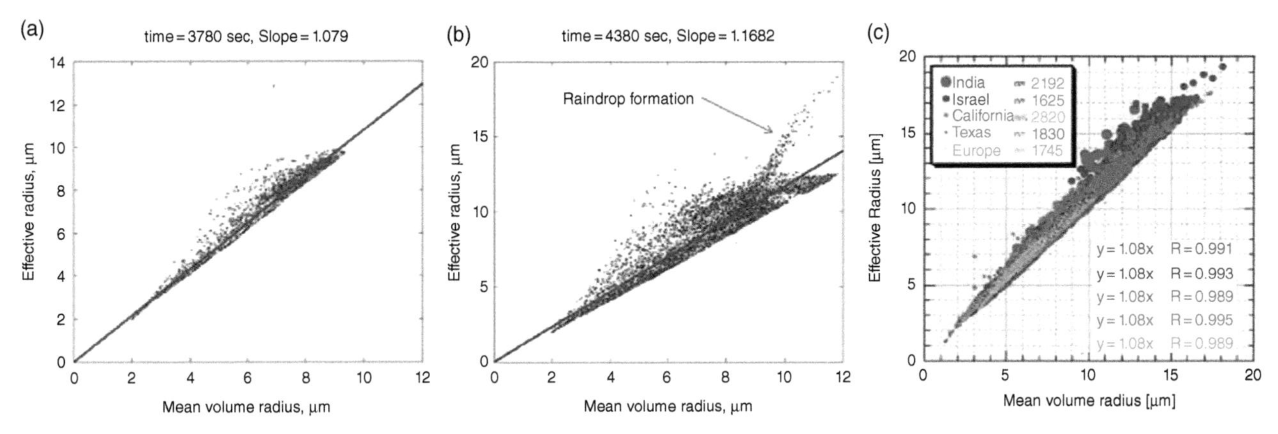

Figure 7.4.7 (a) The relationships between the effective radii and the mean-volume radii in a numerically simulated deep convective cloud at the non-precipitating development stage; (b) several minutes after raindrop formation. The solid straight lines show the approximation of the relationship by linear dependences obtained using the least root square method; (c) the relationship between the effective radius and the mean volume radius obtained for 1 Hz-averaged DSD measured in various locations and clouds of different types by different cloud-droplet probes. The color marking denotes different field campaigns and location data: CAIPEEX (red); the Israeli rain-enhancement program (blue); Suppression of Precipitation Experiment in California (purple); the Southern Plains Experiment in Cloud seeding of Thunderstorms for Rainfall Augmentation (green); European Integrated Project on Aerosol Cloud Climate and Air Quality interactions performed over the Netherlands and the North Sea (gray). Numbers in the legend denote the number of measurements that were used to calculate the linear best fit for each location (from Khain et al., 2013; courtesy of © John Wiley & Sons, Inc.).

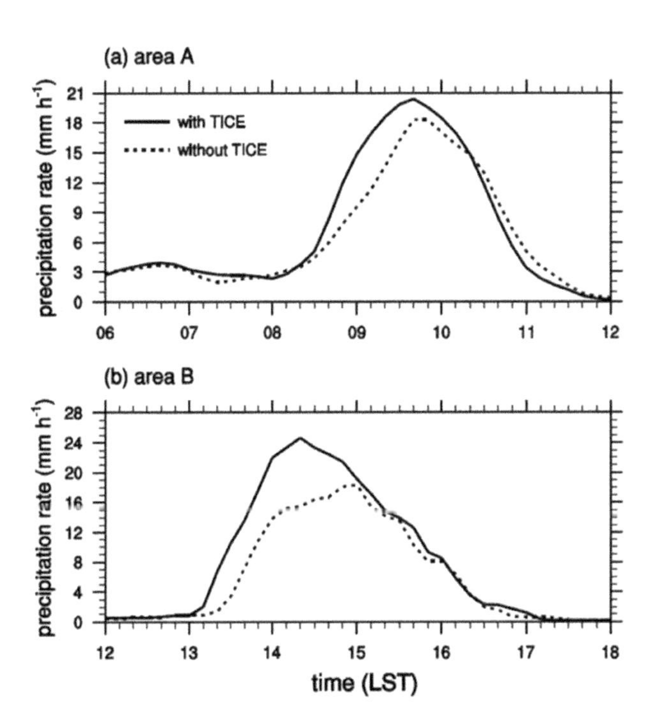

Figure 7.4.8 Time dependence of precipitation rate over two regions of the Korean Peninsula in simulations with turbulence-induced collision enhancement (TICE) and without it (from Lee and Baik, 2016, personal communication).

able to reproduce important observed characteristics of mixed-phase clouds. To justify the HUCM results, the height dependencies of radar reflectivity (Z) in a typical hail storm simulated for the Oklahoma conditions were compared with the dependencies derived from radar data and more than 3,000 surface hail reports obtained in 2010–2014 from the Severe Hazards Analysis and Verification Experiment (SHAVE; Ortega et al., 2016). Vertical profiles of Z in SHAVE were plotted for the following surface precipitation types: no hail, small hail diameter $D < 2.5$ cm, large hail with $2.5 \text{ cm} < D < 5$ cm, and giant hail with $D > 5$ cm. The data were segregated into six classes (corresponding to six height layers) with respect to the altitude h of $T_w = 0°C$ wet-bulb temperature: $z < h - 3$ km (Class 1), $h - 3 \text{ km} < z < h - 2$ km (Class 2), $h - 2 \text{ km} < z < h - 1$ km (Class 3), $h - 1 \text{ km} < z < h$ km (Class 4), $h < z < z(T_w = -25°C)$ (Class 5), and $z > z(T_w = -25°C)$ (Class 6). The distributions of Z were calculated for each separate height layer.

Figure 7.4.9 shows the vertical profiles and PDF of radar reflectivity from the model (left panel) and from observations (right panel). The vertical lines in each PDF box (that is, the left edge of the box, the vertical line near the center of the box, and the right edge of the box) represent the 25th, 50th (i.e., median), and 75th

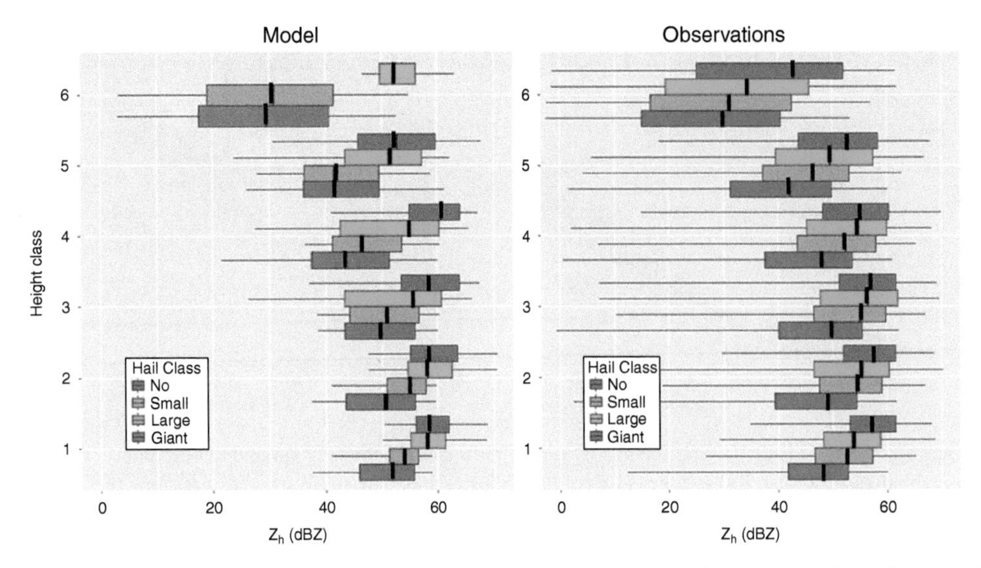

Figure 7.4.9 Distributions of Z_H in different height layers of hail size (giant, large, and small are marked purple, blue, and green, respectively) and no hail (red) from model (left) and radar observations (right). The whiskers mark the 95th percentile, the boxes mark the interquartile range, and the vertical line marks the median value. Layers 1–4 are warm (with $T_w \geq 0°C$) and layers 5 and 6 are cold ($T_w < 0°C$) (Ilotoviz et al., 2017).

percentiles of the distribution, respectively; the whiskers of the plots extend to the 2.5 and 97.5 percentiles of the distribution. All these percentiles are based on the total dataset (from model or observations). Height classes 1–4 represent areas of the atmosphere below the 0 °C wet-bulb temperature height (an approximation for the start of the melting layer). Height classes 5–6 represent heights above freezing level. A good agreement of simulations with the observations is seen. The reflectivity values are highest in big hail cases with maximum around ~65 dBZ in both (model and observations). The giant hail is excluded from the statistics due to a small statistic sample. Minimum Z takes place in no-hail cases. In all classes maximum Z depend on height relatively weakly.

Figure 7.4.10 shows scattering diagrams of radar-reflectivity-rain water content (Z vs. RWC) obtained in simulations using HUCM of a midlatitude Cb (hail storm) with hail stones of diameter up to 4–5 cm (Ilotoviz et al., 2017). An empirical Z–RWC dependence $RWC = 1.74 \times 10^{-3}Z^{0.640}$ obtained deep convection (no or small hail) in Oklahoma is designated by the black curve in Figure 7.4.10. The red line in the panel represents the results of a 1D SBM melting-hail model (Ryzhkov et al., 2013a). Note that RWC and Z are formed as the result of many microphysical processes: diffusion growth, deposition, collisions, riming, melting, shedding, spontaneous, and collision breakups of raindrops. Therefore, a good agreement of simulated scattering diagrams with the observations indicates realness of simulation of not only warm processes, but also the main ice microphysical processes in the model. The model reproduces with good precision many other empirical dependencies between microphysical and radar variables, thus indicating that it is realistic enough to analyze observed phenomena, including the formation of big hail.

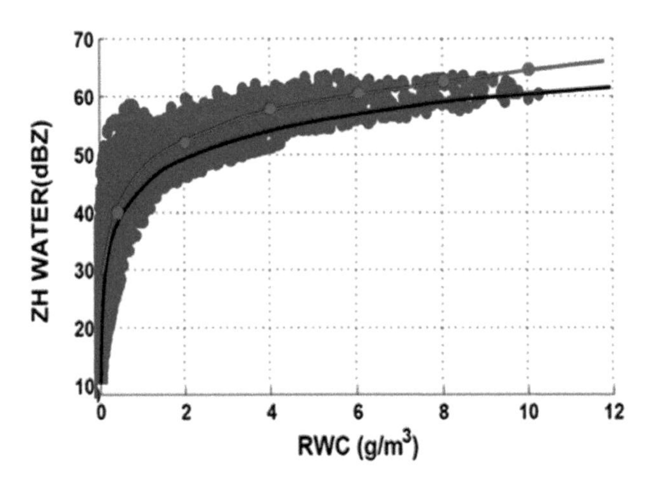

Figure 7.4.10 Z–RWC scattering diagrams in the HUCM simulations of a midlatitude hail storm. The black line shows the dependence $RWC = 1.74 \times 10^{-3}Z^{0.640}$ derived from observation data on hail storms in Oklahoma (Carlin et al., 2015) where Z is in mm^6/m^3. The red curve represents the results of the hail model developed by Ryzhkov et al. (2013a) (Ilotoviz et al., 2017).

Mechanisms of Big Hailstone Formation

Insufficient knowledge about the mechanisms of big hailstone formation makes it difficult to forecast this dangerous natural hazard. Observations show that big hail events are often characterized by a specific structure of the radar reflectivity field. The radar reflectivity in Cb clouds (thunderstorms) often exceeds 50–60 dBZ up to altitudes of 10 km, and the maximum reflectivity sometimes exceeds 70 dBZ at heights of 4–5 km. Zones of high reflectivity, and especially of differential reflectivity indicating the existence of large nonspherical raindrops, typically coincide with zones of strong updrafts with vertical velocities exceeding several tens of m s^{-1}. The mechanism leading to formation of large nonspherical particles in cloud updrafts was analyzed in simulations of a hail storm under high CCN concentration (3,000 cm^{-3}) (Kumjian et al., 2014; Ilotoviz et al., 2016, 2017). **Figure 7.4.11** shows the field of radar reflectivity, as well as MD of drops, freezing drops, and hail. The MD in the left column are plotted for the cloud edge with a weak updraft, where hail mass content is maximum (x = 60.3 km). The MD in the cloud core with the maximum updraft are plotted in the right column (x = 60.9 km). The radar reflectivity exceeds 50 dBZ up to altitudes of 10 km, with the maximum exceeding 60 dBZ within the layer of 4–5 km. The field of radar reflectivity in Figure 7.4.11 agrees with observations (Noppel et al., 2010b).

The MD for drops in the cloud updraft clearly shows the existence of two modes: cloud droplets and raindrops. A significant mass of large raindrops exists in the updrafts already at the altitude of 3 km. Hail particles of 1 cm in diameter also exist in updrafts at 3-km height. Neither raindrops nor hail can fall down to this

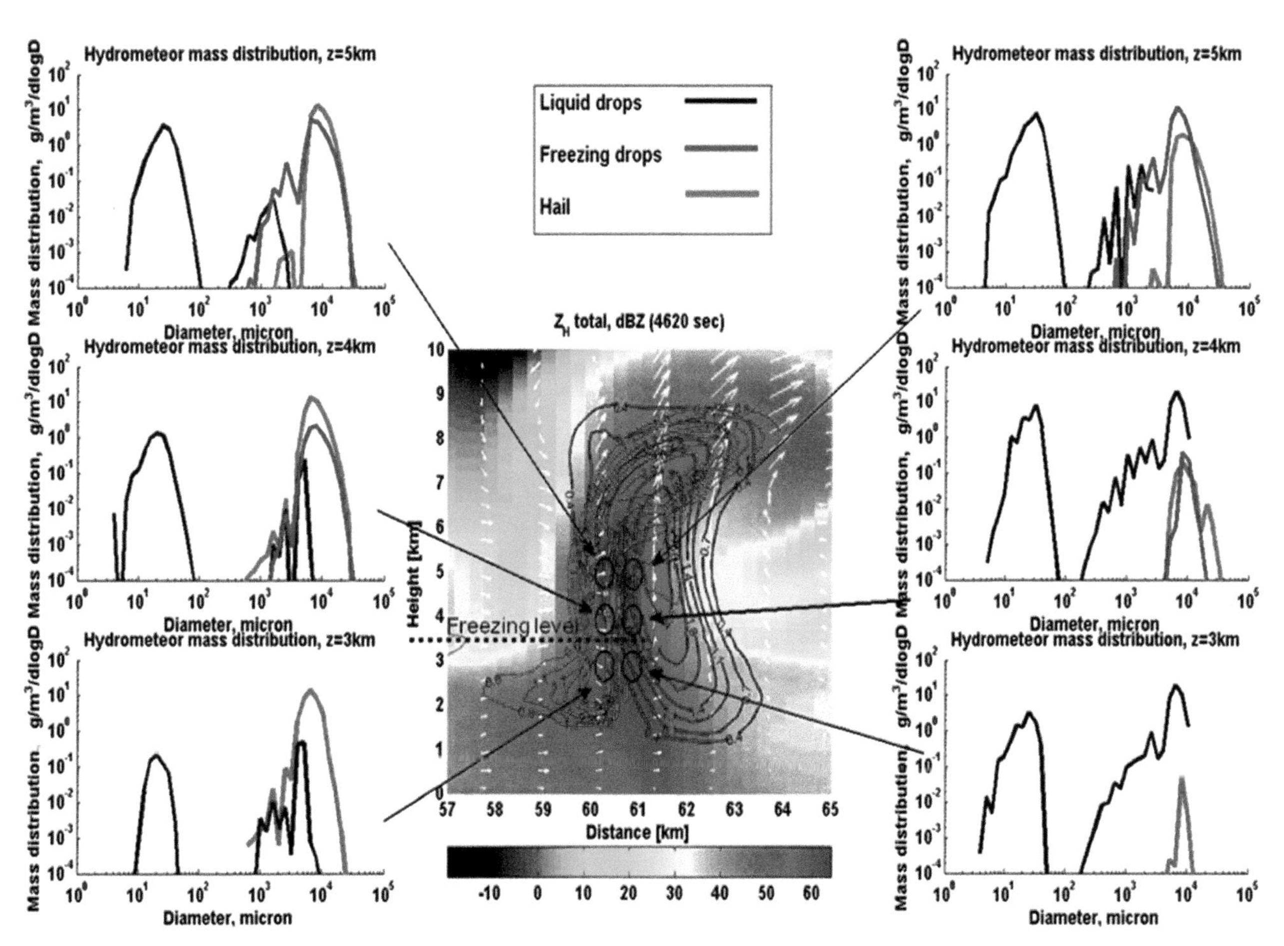

Figure 7.4.11 Microphysical structure of the storm at the mature stage as simulated in the model. Center panel: the field of the radar reflectivity (dBZ). Overlaid are the contours of the cloud-water content (blue) and mass content of hail (green) in g m^{-3}. Left panels: PSD at different heights in the cloud within the column of maximum hail-mass content (x = 60.3 km). Right panels: PSD at different heights in the cloud within the column of maximum updraft (x = 60.9 km). The arrows and circles indicate the location of volumes with the corresponding PSD (from Ilotoviz et al., 2016; courtesy of © American Meteorological Society. Used with permission).

level from above where the vertical velocities are very high. An analysis of the MD in the updraft at z = 4 km and at z = 5 km where $w = 15$ and $20\ \mathrm{m\,s^{-1}}$, respectively, shows that the mass contents and the sizes of hail particles and freezing drops increase upward. At z = 5 km, large raindrops disappear, being converted into freezing drops and then to hail particles whose maximum diameter reaches ~3.3 cm. Some fraction of large hailstones reaches high levels, accounting for the high reflectivity. MD in downdraft at the cloud edge (left panels) substantially differ from that in the updraft. Both freezing drops and hail fall down along the cloud edge. Shedding from large hail and melting of the smallest hail particles and other ice particles below melting level increases the RWC. The maximum size of hail falling along the cloud edge changes only slightly within the layer between altitudes of 4–5 km, and slightly decreases downward, apparently due to shedding. The comparison of hail size distributions in updrafts and in downdrafts shows that hail grows largely by accretion in cloud updrafts.

The mechanism of formation of big hail, large raindrops, and freezing drops in cloud updrafts is further illustrated in **Figure 7.4.12**. This figure shows the fields of vertical velocity and CWC at the mature stage of Cb evolution overlaid with ten-minute backward trajectories calculated for tracer particles having different terminal velocities. These tracer particles can be hail or raindrops with corresponding fall velocities. The trajectories originate in cloud-edge zones with weak updrafts or downdrafts. The trajectories were chosen so that their initial points were located within areas of significant hail-water content below the melting level. One can see that particles having comparatively large fall velocities enter the updraft zone and start ascending. Therefore, melting hail particles participate in the recycling process. Only the largest particles with fall velocity of $15\ \mathrm{m\ s^{-1}}$ fall to the ground. It can be concluded that the dominant mechanism of hail growth in midlatitude continental Cb (thunderstorms) is accretion in the cloud updrafts during the recycling process. Thus, the high radar reflectivity in the thunderstorms up to 9–10 km altitude, as well as formation of large hail, are the result of recycling of large raindrops and hail. During the recycling, raindrops may convert to hail and hail may melt, at least partially, to form raindrops. Thus, the utilization of advanced SBM cloud models allowed researchers to reveal mechanism of hail growth in Cb.

Hail Size Distributions

Data of hail size distributions are scarce since it is dangerous to carry out in situ measurements within hail storms with vertical velocities exceeding 40 m/s. It seems established, however, that the hail is distributed

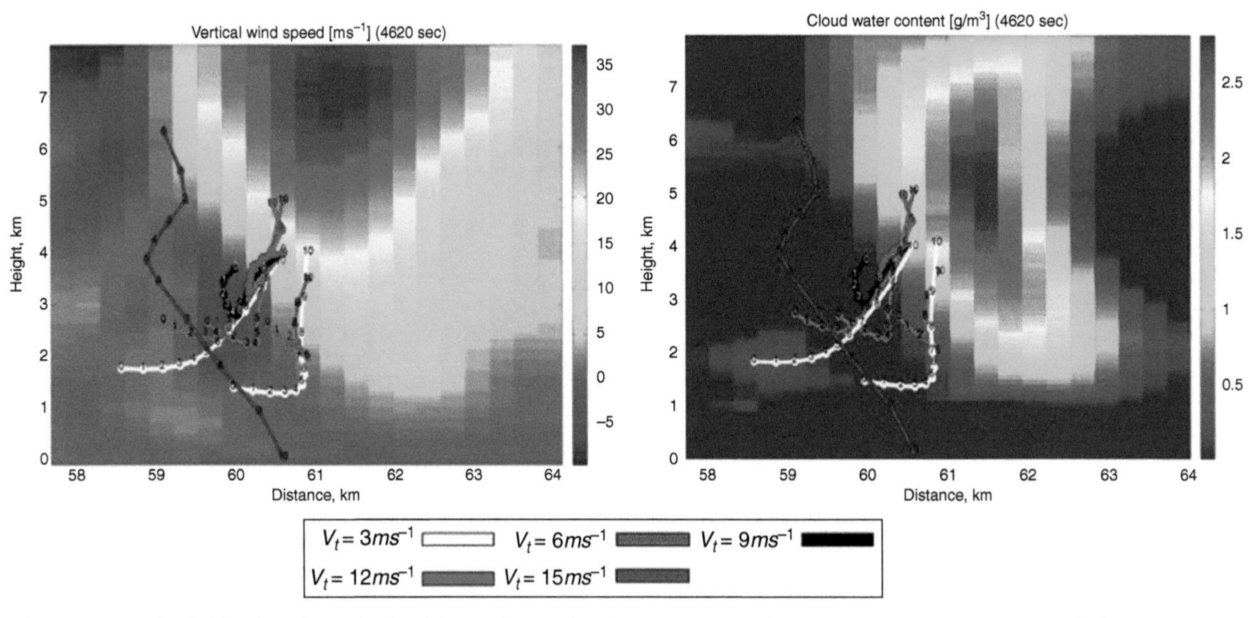

Figure 7.4.12 The fields of vertical velocity (left panel) and cloud-water content (right panel) at the mature stage of Cb evolution (t = 4,620 s). The ten-minute backward trajectories of tracer particles with different prescribed terminal velocities are also shown in the panels. The fall velocities of tracer particles were chosen within the range from $3\ \mathrm{m\,s^{-1}}$ to $15\ \mathrm{m\,s^{-1}}$. Small circles on the trajectories indicate particles' locations at one-min intervals (from Ilotoviz et al., 2016; courtesy of © American Meteorological Society. Used with permission).

nonuniformly over clouds and the sizes of falling hail change with cloud evolution. The processes of precipitation formation (including hail shaft) are determined by local instantaneous size distributions, which can substantially differ from those obtained by averaging over a span of time. Therefore, results obtained using cloud models with advanced microphysics are especially useful for understanding the physical mechanism of hail formation.

Hail number concentration distribution functions calculated below the freezing level for high (left) and low (right) CCN concentrations at different heights are shown in **Figure 7.4.13**. The figure illustrates hail size distributions at time instances when hail-mass content near the surface was maximum. Cheng et al. (1985) proposed the following empirical formula for hail size distributions near the surface:

$$N(D) = C\Lambda^{4.11} \exp(-\Lambda D) \quad (7.4.2)$$

where D is the hail diameter. If $N(D)$ is measured in $m^{-3}{\cdot}mm^{-1}$, Λ varies between 0.1 and 1.0 mm^{-1} and factor C varies between 60 and 300. These dependencies for $C = 300$, $\Lambda = 0.4\ mm^{-1}$, and $0.6\ mm^{-1}$ are presented in Figure 7.4.13a and 7.4.13b. One can see that in the polluted case (panel a) the calculated size distribution near the surface (z = 0.1 km) agrees quite well with the exponential distribution at $\Lambda = 0.4\ mm^{-1}$. At the same time, the size distribution calculated for low CCN concentration typical of maritime conditions (panel b) is better approximated at $\Lambda = 0.6\ mm^{-1}$. Hail stones grow larger in a polluted atmosphere. Analysis of Figures 7.4.13a and 7.4.13b shows that Λ increases with height. The height dependence can be used in bulk schemes to improve hail parameterizations.

7.4.4 Simulation of Mixed-Phase Stratiform Clouds

During the past decade, significant progress has been reached in understanding phase coexistence and competition in mixed-phase stratiform clouds owing to analytical and semi-analytical studies (Section 6.3). The analytical results were obtained under simplifying assumptions, for instance, of adiabaticity of cloud volumes, monodisperse PSD, and others. These studies allowed better qualitative understanding of the mechanisms of ice–liquid water coexistence. These results are especially important since it is still difficult to make quantitative predictions of microphysical and dynamical structure of mixed-phase Sc due to large uncertainties regarding production of ice particles (Sections 6.2). Under such situation, numerical simulations with cloud models actually play the role of laboratory experiments helping to describe quantitatively the rates of different microphysical processes. Such simulations were performed in several studies (Fan et al., 2009b, 2011; Yang et al., 2013; Ovchinnikov et al., 2014) using LES models of mixed-phase stratiform clouds and SBM. Fan et al. (2011) simulated two types of Arctic clouds observed during the ISDAC and M–PACE field campaigns using a 3D cloud-resolving model SAM with SBM. Single-layer clouds and multilayer mixed-phase clouds were

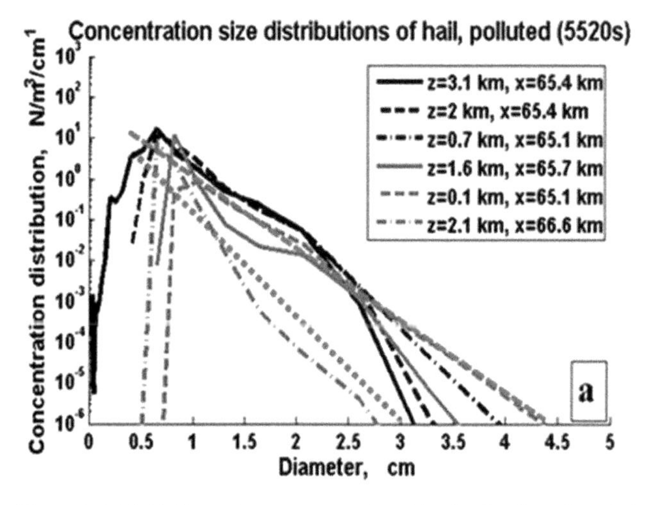

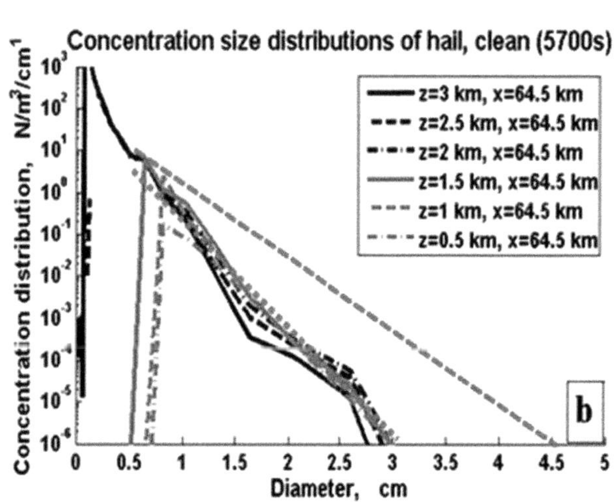

Figure 7.4.13 Hail number concentration distribution functions below the freezing level in simulations of a midlatitude Cb (hail storm) at the stage when hail size reaches its maximum in cases of high aerosol concentration (panel a) and low aerosol concentration (panel b). The distributions are plotted at different altitudes in zones of maximum hail-water content. Empirical distributions (Equation 7.4.2) with $C = 300$ and $\Lambda = 0.4$ mm and $\Lambda = 0.6$ mm are shown by red dashed and dotted lines, respectively (from Ilotoviz et al., 2016; courtesy of © American Meteorological Society. Used with permission).

observed in ISDAC and M–PACE, respectively. The goal of simulations was to get a better understanding of formation of mixed-phase structure in Sc.

Figure 7.4.14 compares the model vertical profiles of LWC, of IWC, droplet number concentrations N_w, and ice particle number concentrations N_i, with the corresponding aircraft observation data obtained in ISDAC, Flight 31 during 0:00–3:30 A.M. on April 27, 2008. The measurements of LWC made with the King probe are also plotted for comparison (red line). The simulated LWC and N_w agree very well with the observed values. The agreement of the modeled IWC with observations could also be considered as good, especially taking into account significant uncertainties regarding IWC in measurements. The modeled PDF of the vertical velocity in general also agree well with the observations. The modeled PSD agrees reasonably well with the observed PSD within the radii range 100–500 μm. Below 100 μm, no reliable measurements of ice particles are available.

Figure 7.4.15 compares the modeled cloud with observations made in M–PACE. A significant discrepancy between the aircraft observations and the ground observations is distinctly seen, which complicates the comparison of the model results with observations. Figure 7.4.14 shows that, in general, the model captures the vertical changes of the cloud fraction. In agreement with observations, the model simulates the clouds containing three maxima in liquid content with ice crystals falling between the liquid layers. Figures 7.4.13 and 7.4.14 show that despite significant uncertainties, the model enables us to simulate major properties of mixed-phase Sc, which allowed us to find the most important factors that determine microphysical properties of the mixed-phase Sc such as PDF of vertical velocity, ice particle capacitance, and PSD shape of ice particles. This is an important model result. The fact that the results obtained using different SBM schemes produced similar Sc structures (Ovchinnikov et al., 2014; Section 7.2) is quite encouraging and substantially increases the model reliability.

Yang et al. (2013) showed that the IWC in clouds observed in ISDAC has a power law relationship with the ice number concentration with the slope close to 2.5. To check whether such a dependence is reproduced by an advanced model, simulations were performed using LES models with SBM. The simulations were carried out under the simplified assumption that ice particles are produced by drop freezing, so glaciation occurs with a constant freezing probability for any drop, regardless of its size and location. The results show that the models can serve as an efficient tool to investigate the ice-nucleation process. The results also demonstrate a strong relationship between the microphysics of mixed-phase Sc and cloud dynamics. The observed 2.5-power law is fundamentally tied to crystal growth in updrafts. This law reflects a supposedly physically based connection between cloud dynamics, ice-nucleation cloud-microphysical properties that influence precipitation, cloud lifetime, and cloud optical

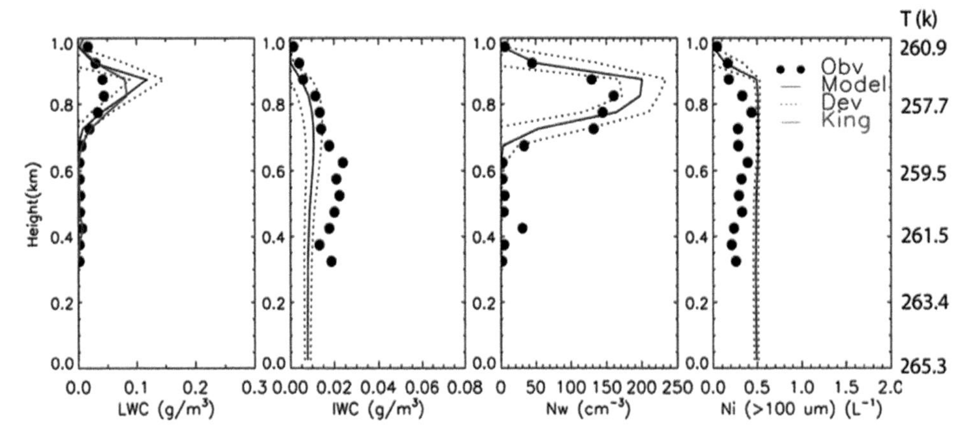

Figure 7.4.14 Comparisons of the vertical profiles of the mean LWC, IWC, N_w, and N_i (IWC and N_i considered only for ice particles of maximum dimension exceeding 100 μm). Black dots refer to the mean observed values at given heights within ±50 m. Solid and dotted lines denote the mean modeled values and the mean ± one-standard deviation, respectively. Red line represents the LWC measured with the King Probe ("King"). LWC and N_w were measured using the FSSP probe-sizing particles of 3 μm to 47 μm in radius. IWC and N_i were calculated from the composite size distribution measured by the Optical Array Probe, the 2D cloud probe (2DC), and the precipitation probe (2DP). The PSD were calculated from the 2DC and 2DP measurements (from Fan et al., 2011; courtesy of © John Wiley & Sons, Inc.).

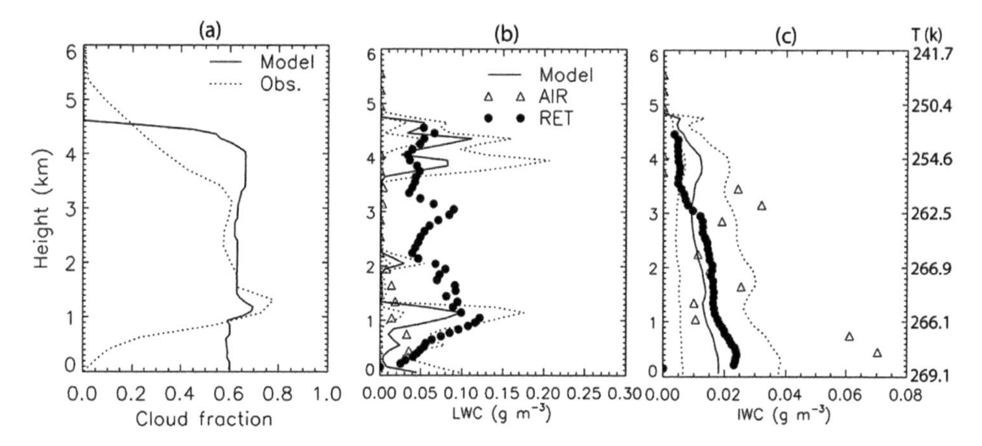

Figure 7.4.15 (a) Comparison of the vertical profile of the modeled cloud fraction and the corresponding profile derived from the ARM Best Estimated Climate Product. Comparison of modeled LWC (b) and IWC (c) with the aircraft measurements ("AIR") and the retrieved ground-based measurements ("RET") from multi-sensors (radar, lidar, microwave radiometer, radio sounds). Dashed lines in panels (b) and (c) denote one standard deviation from the model results. Cloud boundaries are defined by the threshold value of cloud-ice mixing ratio exceeding 10^{-4} g kg^{-1} (from Fan et al., 2011; courtesy of © John Wiley & Sons, Inc.).

properties. Further investigations are required in this direction.

7.4.5 Simulations of Hurricanes

First simulations of hurricanes using SBM models were performed by (Khain et al., 2010). Due to high time consumption, a simplified version of SBM was implemented including only three size distributions (aerosols; ice crystal+snowflakes, and graupel). Simulations were performed at 3-km and 1-km spatial resolution (Lynn et al., 2014, 2016; Khain et al., 2016; Section 7.3). Simulations of hurricane development and motion performed with SBM using complex triple-nested grid systems over the area with linear sizes of several thousands of kilometers represent a significant achievement in investigation of synoptic-scale cloud-related phenomena. Using SBM enables us to improve predictions of both the minimum pressure and the maximum wind speed in a hurricane. Also, only SBM allowed reproducing the effects of aerosols on TC structure (Figures 7.3.21–7.3.24). Next we present two examples of achievements reached in TC simulation using SBM models.

Simulation of TC Structure

Figure 7.4.16 compares SSMI/S 91 GHz color composite images with the fields of the total mass content of ice crystals and aggregates in simulations of hurricane Irene, carried out using the SBM model. The maps are plotted near the time of the lowest pressure, at the height of the cirrus shield (~11 km). The structure and time evolution of the cirrus shield in the simulation shows a good spatial agreement with the observations (panel b). Indeed, a significant area of cirrus clouds at high altitudes over the northern periphery was found both in the observations and in the SBM simulation. The same is true about several rainbands occurring in the interior of Irene to the east from the center on Aug. 26 (see also Figure 7.3.23). Panel (c) shows that utilization of a smaller grid spacing describes the same structure, but with more details. This example shows that implementing detailed microphysics likely allows for predicting not only TC intensity, but also TC structure. More simulations are required to get statistical evaluations of potential improvements that can be achieved using the SBM schemes.

Simulation of Lightning Activity

Adequate description of the microphysical structure of hurricanes allowed for explaining and predicting lightning intensity in the TC zone. It is known that lightning in clouds is the result of charge separation during collisions between graupel and ice crystals in the presence of liquid water (Takahashi, 1978; Saunders, 1993). Thus, to forecast the intensity and location of lightning, it is necessary to forecast the emergence of the appropriate cloud microstructure. The ability of WRF/SBM to simulate lightning in hurricanes is illustrated by **Figure 7.4.17**. Figure 7.4.17 (left) shows zones of intense lightning in hurricane Katrina (2005) approaching land. Lightning takes place both within the TC eyewall and in the rainbands located within the 100–300 km-radius

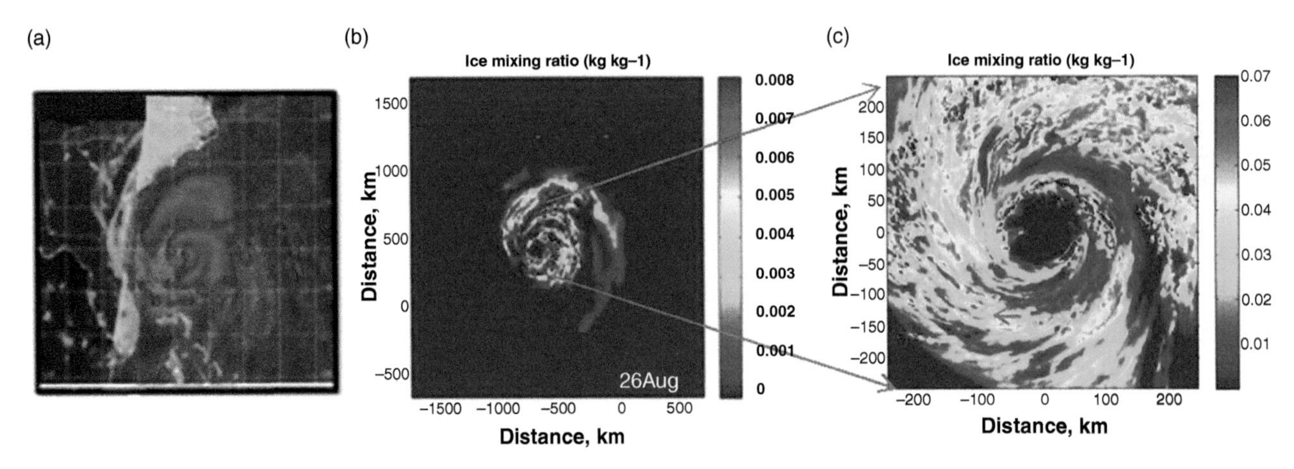

Figure 7.4.16 a) SSMI/S 91 GHz color composite images of hurricane Irene at 1040 UTC on August 26, 2011. b) The field of ice crystal and aggregate contents in the cirrus shield in the simulations. The model resolution is 3 km. c) The same as in panel b), but at grid resolution of 1 km (from Lynn et al., 2016, with changes; courtesy of © American Meteorological Society. Used with permission).

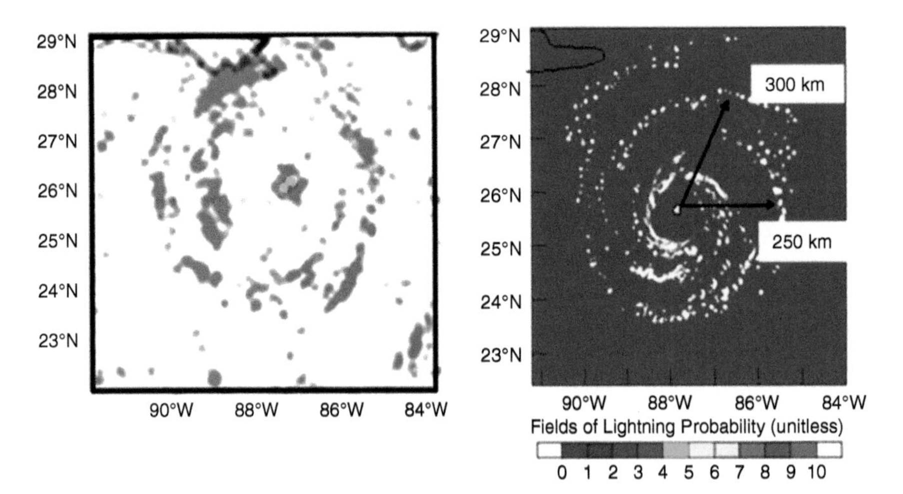

Figure 7.4.17 Left: lightning in Katrina (2005) on Aug. 25, 2005 at 17:30–18:30; zones of lightning are marked by red dots; the TC eye is marked green (from Shao et al., 2005). Right: fields of the lightning potential calculated for Katrina with intrusion of continental aerosols taken into account (from Khain et al., 2008a, with changes; courtesy of © American Meteorological Society. Used with permission).

ring around the hurricane center. This lightning pattern remains during the entire period of Katrina's presence in Mexican Bay. Such spatial structure of lightning is typical of land-falling hurricanes (Molinari et al., 1999). Khain et al. (2008a) showed that conditions favorable for charge separation arise due to the effects of small soluble aerosols (CCN) penetrating the TC periphery from the continent. These aerosols lead to formation of a large amount of small droplets reaching the high levels in clouds, as well as to formation of graupel and ice crystals. Charge separation takes place in clouds at low temperatures due to collisions between graupel and ice crystals. An important condition for charge separation is the presence of supercooled water (Saunders et al., 1993; Saunders, 2008). Figure 7.4.17 (right) shows the fields of lightning potential (determined as a product of mass contents of ice crystals, graupel, and supercooled water at low CCN concentration and intrusion of continental aerosols taken into account. In such aerosol conditions the amount of supercooled water as well as of graupel grows, increasing the intensity of lightning events. Figure 7.4.17 shows that continental aerosols lead to formation of lightning bands of 150–300 km in radius, which is in agreement with observations.

7.5 Conclusions and Perspectives of Cloud Modeling

In this book we have discussed different aspects of cloud microphysics and its representation in modern cloud models, as well as the significant advancement achieved in this field over the past decades. The book outlines different methods used in investigation of microphysical processes, including SBM schemes and bulk-parameterization schemes, and discusses both the advantages and the drawbacks of both approaches. We have also explained how expressions for rates of different microphysical processes are used in cloud models and cloud-resolving models.

The SBM schemes have an obvious advantage for research where the knowledge of the PSD shapes is important. The ability of SBM models to calculate PSD makes them a powerful tool for investigation and parameterization of cloud microphysical processes. Application of bin-microphysical models made it possible to discover such effects as aerosol-induced convective invigoration, an increase in hail size in polluted clouds, and the outstripping formation of rain in undiluted cloud volumes and in zones of maximum turbulence. Modern cloud models are able to represent the microphysical structure of different types of clouds and cloud-related phenomena. Microphysical packages that had been used earlier only in LES cloud models at scales of a few kilometers have become applicable in mesoscale cloud-resolving models with computational areas of a synoptic scale.

Substantial progress has also been made in the development of bulk-parameterization schemes, which are now widely used in cloud-resolving models. Utilization of observation data and comparison with SBM allowed relating the parameters of Gamma distributions to environment conditions and to find relationships between these parameters, thus substantially improving the representation of PSD in the bulk schemes. Implementation of variable density of ice hydrometeors in models allowed description of the main properties of ice particles of different types. Single-moment bulk models are replaced by two- and three-moment schemes. These improvements allowed advanced bulk schemes to produce results comparable with those obtained with SBM schemes, at least regarding representation of the general cloud structure. As a result, traditional convective parameterization schemes are being rapidly replaced by microphysical schemes in mesoscale and large-scale models able to explicitly resolve clouds.

There are multiple applications of SBM and bulk-parameterization models. First of all, using the models provides a better understanding of cloud-related phenomena across a wide range of scales from individual clouds to global circulation. Weather forecasting as well as simulations of climate and climate changes are among the most important applications of cloud-resolving models. Significant improvement of model results can be reached by combining cloud models with satellite, radar, and lidar measurements. Significant steps forward have already been taken. The combining allows development of new remote sensing schemes, to perform a better evaluation of energy and radiation budgets of the atmosphere at different scales up to the global ones.

Advanced SBM models today are elaborate enough to allow reliable testing of weather modification hypotheses. The models promote development in both the traditional areas of weather modification (hail suppression, fog elimination, and rain enhancement) and novel methods that can be used in geo-engineering, such as weakening of hurricane intensity and changing the cloud cover over specific regions to prevent undesirable climate changes. In a pioneering study, Carrió and Cotton (2011) performed a series of multigrid cloud-resolving simulations using RAMS to examine the response of a simulated hurricane to the targeted insertion of CCN.

Many problems are to be solved to improve cloud-modeling results and to make them more reliable. Three main directions can be outlined for further development of cloud models and cloud-resolving models.

7.5.1 Gaining a Deeper Understanding of Microphysical Processes via Observations

To improve microphysical schemes, we need a deeper knowledge of cloud microphysical processes. Despite significant progress, the understanding of effects of entrainment and mixing in warm and mixed-phase clouds requires significant improvement. At present, many mechanisms of ice crystal formation including primary and secondary ice production are not known even qualitatively. Gaining this knowledge is crucial, since concentration of ice crystals, their shapes, and vertical velocities determine the phase structure of clouds. It is highly important to understand ice formation in stratocumulus clouds that make a major contribution to the radiative balance of the atmosphere. Other understudied phenomena are collisions and coalescence between ice crystals and ice particles of different kinds. The lack of knowledge of collision and collection efficiencies lead to the problems in simulation of mixed-phase clouds, cirrus clouds, convective mesoscale systems, and squall lines. Effects of ice crystals porosity on the aggregate formation still have not been studied. Ice multiplication caused by drop freezing and ice collisional breakup require quantitative description.

Another example of a gap in the knowledge is related to multiple aerosol effects on cloud microphysics and radiative cloud properties. Since most microphysical processes depend on size, chemical composition, and concentration of aerosols, these aerosol parameters should be adequately treated in cloud models. The investigation of aerosol effects requires application of bin-microphysics models allowing taking into account effects of aerosols on particle size distributions.

The advanced knowledge should be obtained both from theoretical studies and experimental data. The first source of experimental data is laboratory experiments that can provide new information on ice nucleation, including primary and secondary ice production, immersion drop freezing, particle collisions, breakup of drops and aggregates, clustering of particles in turbulent flows, aerosol properties, etc.

Another important data source is *in situ* observations. It is necessary to develop more precise devices that allow measurements at higher frequency. An example of efforts in this direction is a new device for measuring concentration of ice crystals that is not subjected to effects of shattering. This device developed and implemented made it possible to correct the measured crystal concentrations by factors of 10^2–10^3 and provided a much better agreement between model results and experimental data (Korolev and Field, 2015). The application of such probes allows the isolation of artificial mechanisms of ice formation from real processes of ice multiplication. Another example of an advanced experimental device is described by Beals et al. (2015) and Bodenschatz (2015), who reported direct discovery of droplet concentration inhomogeneity in clouds using a sophisticated holographic imaging system that allowed them to measure both the locations and the sizes of micrometer-size water droplets in three dimensions.

Weather radars including Doppler and polarimetric radars opened a new era in remote sensing of microphysical cloud structure. The polarimetric radars measure signals in two perpendicular planes and yield data that enable evaluation of the level of particle sphericity, as well as distinguishing between particles phases (liquid vs. ice). The combination of different measured polarimetric signatures and advanced cloud models allows both a better interpretation of radar signals and a better understanding of microphysical processes in clou are ds. An example of an efficient radar-model coupling used for development of advanced remote sensing algorithms is presented by Carlin et al. (2016), where the analysis of microphysical cloud structure is used to design novel parameterizations and improve short-range severe weather forecasts.

Satellite measurements already play a highly important role in remote sensing and weather forecast. Satellites are used for investigation of clouds, retrieval of different hydrometeors parameters, determination of cloud optical depth, evaluation of latent heat release needed in assimilation algorithms, and estimation of precipitation and DSD parameters. The quality of satellite-obtained data has been continuously improving, and new information is obtained using multiwavelength retrievals. Satellite observations cover the entire globe and are thus the most useful method for investigating oceanic regions not covered by other measurement networks. Satellite observations are important in investigation of climate changes.

In addition, satellites are used for retrieval of multiple thermodynamical and microphysical values used in cloud-resolving models. In particular, it is possible to measure certain DSD parameters, such as the effective radius needed for calculations of cloud radiation properties. Using the recent methods of estimation of cloud-base vertical velocity, one can evaluate concentration and other properties of aerosols (Rosenfeld et al., 2014). Monitoring of areas of open and closed cells allows the evaluation of the effects of aerosols on radiation cloud properties at global scales (Goren and Rosenfeld, 2015; Feingold et al., 2015).

Lidar data contribute significantly to measurements of optical properties of BL and aerosol concentration in BL, and are efficiently incorporated in cloud models.

A number of observational studies that were initially focused on investigation of cloud microphysical processes actually triggered further development of microphysical schemes in cloud and cloud-resolving models. Studies by Rosenfeld and Gutman (1994) and Freud and Rosenfeld (2012) who found the critical effective radius at which first raindrops form, gave an incentive to investigations of the mechanism of raindrop formation in cumulus clouds. Realistic reproduction of the actual values of the effective radius and raindrop formation at observed altitudes became a quality criterion for microphysical schemes. Li et al. (2010) demonstrated the necessity of taking into account the rimed mass in aggregates and its effect on fall velocity. Studies by Ryzhkov et al. (2013a, 2013b), Snyder et al. (2015),

Carlin et al. (2016) and others stimulated investigations of hail formation and time-dependent freezing and other microphysical processes.

7.5.2 Improving Representation of Physical Processes in SBM Cloud Models

Independent description of different processes that are physically closely related is a serious drawback of today's cloud models and cloud-resolving models. For instance, droplet activation and drop freezing are caused by the same aerosols containing both soluble and insoluble fractions. But treatment of these processes in most models today is independent: droplet nucleation algorithms interpret aerosols as soluble CCN particles, while algorithms of primary ice production consider aerosols as IN. Regarding drop freezing, it is usually described using semiempirical formulas that do not explicitly include any information about aerosols. A similar gap exists between descriptions of radiative properties and microphysical cloud properties. When radiative quantities are calculated, the microphysical structure of clouds is often assumed to be typical of clouds of a certain type, which is not consistent with the microphysical structure of particular clouds simulated by the model.

Further development of cloud models should include explicit aerosol budget calculations that take into account transformation of dry aerosols into wet aerosols (haze) and further aerosol conversion into aerosol inside drops and ice particles, as well as the consequent aerosol transport together with drops and ice particles, and aerosol release during drop evaporation. At present, there are only a few warm-microphysics models that take into account the history of soluble aerosols within drops. No models so far would describe the history of IN within ice particles.

Future modifications should include more precise descriptions of ice microphysics, first of all, production of ice particles, their growth, and collisions. As soon as more grounded expressions for the rates of these processes are determined, the SBM models will incorporate these new rates comparatively easily.

7.5.3 Modernization of Bulk Parameterization Schemes

The potential of bulk parameterization schemes is far from being exhausted. Several directions of further development include:

- Implementation of more accurate rates of microphysical processes obtained by SBM models.
- Utilization of two-moment or three-moment schemes that could dramatically improve representation of particle sedimentation and size sorting as compared to one-moment schemes.
- Excluding the saturation adjustment approach to achieve better representation of diffusional growth.
- Improvement of calculation of particle sedimentation.
- Better description of autoconversion and accretion processes and representation of melting and freezing of hydrometeors implementing variable particle density.
- Optimization of choice of the moments.
- Achieving higher resolutions and smaller time steps. Using different sets of governing parameters of bulk parameterization schemes that are more appropriate for different geographical regions (sea, land, urban, etc.). This fine-tuning of the parameters can be achieved by comparison with SBM models.

There are reasons to believe that the ongoing growth of computer power will allow, in the near future, the development and use of cloud-resolving global circulation models, at least with bulk parameterization schemes, as well as the use of SBM within large geographical regions where higher resolution will be applied. We believe that it will improve weather forecasting skills at different scales, including forecasting of precipitation, prediction of dangerous atmospheric phenomena, and evaluation of climatic changes.

References

Ackerman, A.S., P.V. Hobbs, and O.B. Toon, 1995: A model for particle microphysics, turbulent mixing, and radiative transfer in the stratocumulus-topped marine boundary layer and comparisons with measurements. *J. Atmos. Sci.*, **52**, 1204–1236.

Ackerman, A.S., M.C. van Zanten, B. Stevens, V. Savic-Jovcic, C.S. Bretherton, et al., 2009: Large-eddy simulations of a drizzling, stratocumulus-topped marine boundary layer. *Mon. Weather Rev.*, **137**, 1083–1110.

Albrecht, B., 1989: Aerosols, cloud microphysics and fractional cloudiness. *Science*, **245**, 1227–1230.

Alexandrov, M.D., B. Cairns, C. Emde, A.S. Ackerman, and B. van Diedenhoven, 2012: Accuracy assessments of cloud droplet size retrievals from polarized reflectance measurements by the research scanning polarimeter. *Remote Sens. Environ.*, **125**, 92–111.

Altaratz O., I. Koren, T. Reisin, A. Kostinski, G. Feingold, et al., 2008: Aerosols' influence on the interplay between condensation, evaporation and rain in warm cumulus cloud. *Atmos. Chem. Phys.* **8**, 15–24.

Andreae, M.O., D. Rosenfeld, P. Artaxo, A.A. Costa, G.P. Frank, et al., 2004: Smoking rain clouds over the Amazon, *Science*. **303**, 1337–1342.

Andreas, E.L., 2004: Spray stress revisited. *J. Phys. Oceanogr.*, **34**, 1429–1440.

Andrejczuk, M., W.W. Grabowski, S.P. Malinowski, and P.K. Smolarkiewicz, 2009: Numerical simulation of cloud–clear air interfacial mixing: Homogeneous versus inhomogeneous mixing. *J. Atmos. Sci.*, **66**, 2493–2500.

Andrejczuk, M., W.W. Grabowski, J. Reisner, and A. Gadian, 2010: Cloud-aerosol interactions for boundary layer stratocumulus in the Lagrangian Cloud Model. *J. Geophys. Res.*, **115**, D22214.

Andrejczuk, M., J.M. Reisner, B. Henson, M.K. Dubey, and C.A. Jeffery, 2008: The potential impacts of pollution on a nondrizzling stratus deck: Does aerosol number matter more than type? *J. Geophys. Res.*, **113**, D19204.

Anthes, T.A., 1977: A cumulus parameterization scheme utilizing a one dimensional cloud model. *Mon. Weather Rev.*, **105**, 270–286.

Arabas, S., H. Pawlowska, and W.W. Grabowski, 2009: Effective radius and droplet spectral width from in-situ aircraft observations in trade-wind cumuli during RICO. *Geophys. Res. Lett.*, **36**, L11803, doi:10.1029/2009GL038257.

Arakawa, A., 2004: The cumulus parameterization problem: Past, present, and future. *J. Climate*, **17**, 2493–2522.

Arakawa, A., and W.H. Shubert, 1974: Interaction of a cumulus cloud ensemble with the large-scale environment. Part. 1. *J. Atmos. Sci.*, **31**, 674–701.

Avramov, A., and J.Y. Harrington, 2010: Influence of parameterized ice habit on simulated mixed phase Arctic clouds. *J. Geophys. Res.*, **115**, D03205, doi:10.1029/2009JD012108.

Baik, J.-J., Y.-H. Kim, and H.-Y. Chun, 2001: Dry and moist convection forced by an urban heat island. *J. Appl. Meteorol.*, **40**, 1462–1475.

Bao, J.W, C.W. Fairall, S.A. Michelson, and L. Bianco, 2011: Parameterizations of sea-spray impact on the air-sea momentum and heat fluxes. *Mon. Weather Rev.*, **139**, 3781–3797.

Baldauf, M., A. Seifert, J. Forstner, D. Majewski, M. Raschendorfer, and T. Reinhardt, 2011: Operational convective-scale numerical weather prediction with the COSMO model: Description and sensitivities. *Mon. Weather Rev.*, **139**, 3887–3905.

Beals, M.J., J.P. Fugal, R.A. Shaw, J. Lu, S.M. Spuler, and J.L. Stith, 2015: Holographic measurements of inhomogeneous cloud mixing at the centimeter scale. *Science*, **350**, 89–90.

Beheng, K.D., 1994: A parameterization of warm cloud microphysical conversion processes. *Atmos. Res.*, **33**, 193–206.

Bell, T.L., D. Rosenfel, K.-M. Kim, J.-M. Yoo, M.-I. Lee, and M. Hahnenberger, 2008: Midweek increase in U.S. summer rain and storm heights suggests air pollution invigorates rainstorms. *J. Geophys. Res.*, **113** (D2), D02209.

Benmoshe, N., and A.P. Khain, 2014: The effects of turbulence on the microphysics of mixed-phase deep convective clouds investigated with a 2-D cloud model with spectral bin microphysics. *J. Geophys. Res.*, **119**, 207–221, doi:10.1002/2013JD020118.

Benmoshe, N., A. Khain, M. Pinsky, and A. Pokrovsky, 2012: Turbulent effects on cloud microstructure and precipitation of deep convective clouds as seen from simulations with a 2-D spectral microphysics cloud model. *J. Geophys. Res.*, **117**, D06220, doi:10.1029/2011JD016603.

Berry, E.X., and R.L. Reinhardt, 1974a: An analysis of cloud droplet growth by collection: Part I. Double distributions. *J. Atmos. Sci.*, **31**, 1814–1824.

Berry, E.X., and R.L., Reinhardt, 1974b: An analysis of cloud drop growth by collection: Part II. Single initial distributions. *J. Atmos. Sci.*, **31**, 1825–1831.

1974c: An analysis of cloud drop growth by collection: part III. Accretion and selfcollection. *J. Atmos. Sci.*, **31**, 2118–2126.

Bodenschatz, E., 2015: Clouds resolved. *Science*, **350**, 40–41, DOI: 10.1126/science.aad1386.

Bornstein, R., and Q. Lin, 2000: Urban heat islands and summertime convective thunderstorms in Atlanta: Three case studies. *Atmos. Environ.*, **34**, 507–516.

Borys R.D., D.H. Lowenthal, and D.L. Mitchell, 2000: The relationships among cloud microphysics, chemistry, and precipitation rate in cold mountain clouds. *Atmos. Environ.*, **34**, 2593–2602.

Bott, A., 1998: A flux method for the numerical solution of the stochastic collection equation. *J. Atmos. Sci.*, **55**, 2284–2293.

2000: A numerical model of the cloud-topped planetary boundary-layer: Influence of the physico-chemical properties of aerosol particles on the effective radius of stratiform clouds. *Atmos. Res.*, **53**, 15–27.

Boucher, O., and Lohmann, U., 1995: The sulfate-CCN-cloud albedo effect, a sensitivity study with two general circulation models. *Tellus*, **47B**, 281–300.

Braun, S.A., and R. House Jr., 1997: The evolution of the 10–11 June 1985 PRE-STORM squall line: Initiation, development of rear inflow, and dissipation. *Mon. Weather Rev.*, **125**, 478–504.

Carlin, J., A. Ryzhkov, J. Snyder, and A. Khain, 2016: Hydrometeor mixing ratio retrievals for storm-scale radar data assimilation: Utility of current relations and potential benefits of polarimetry. *Mon. Weather Rev.*, **144**, 2981–3001. doi:10.1175/MWR-D-15-0423.1.

Carrió, G.G., and W.R. Cotton, 2011: Investigations of aerosol impacts on hurricanes: Virtual seeding flights. *Atmos. Chem. Phys.*, **11**, 2557–2567.

Carrió, G.G., S.C. van den Heever, and W.R. Cotton, 2007: Impacts of nucleating aerosol on anvil-cirrus clouds: A modeling study. *Atmos. Res.*, **84**, 111–131.

Chen, S.-H., and W.-Y. Sun, 2002: A one-dimensional time dependent cloud model. *J. Meteorol. Soc. Japan*, **80**, 99–118.

Cheng, C.-T., W.-C. Wang, and J.-P. Chen, 2007: A modeling study of aerosol impacts on cloud microphysics and radiative properties. *Q. J. Royal Meteorol. Soc.*, **133**, 283–297.

Cheng, L., M. English, and R. Wong, 1985: Hailstone size distributions and their relationship to storm thermodynamics. *J. Climate Appl. Meteorol.*, **24**, 1059–1067.

Clark, T.L., 1973: Numerical modeling of the dynamics and microphysics of warm cumulus convection. *J. Atmos. Sci.*, **30**, 857–878.

Cohard, J.-M., and J.P. Pinty, 2000: A comprehensive two-moment warm microphysical bulk model scheme: I: Description and tests. *Q. J. Royal Meteorol. Soc.*, **126**, 1815–1842.

Cohen, N., and A.P. Khain, 2009: Aerosol effects on lightning and intensity of landfalling hurricanes. In *Hurricanes and Climate Change*, J.B. Elsner and T.H. Jagger, eds. New York: Springer, pp. 189 212.

Cooper, W.A., S.G. Lasher-Trapp, and A.M. Blyth, 2011: Initiation of coalescence in a cumulus cloud: A beneficial influence of entrainment and mixing. *Atmos. Chem. Phys. Discuss.*, **11**, 10557–10613.

Costa, A., A.C.J. de Oliveira, J.C.P. de Oliveira, and A.J.C. Sampaio, 2000: Microphysical observations of warm cumulus clouds in Ceara′, Brazil. *Atmos. Res.*, **54**, 167–199.

Cotton, W.R., R.A. Pielke Sr., R.L. Walko, G.E. Liston, C.J. Tremback, H. Jiang, R.L. McAnnelly, J.Y. Harrington, M.E. Nicholls, G.G. Carrió, and J.P. McFadden, 2003: RAMS 2001: Current status and future directions. *Meteorol. Atmos. Phys.*, **82**, 5–29.

Cotton, W.R., G.J. Tripoli, R.M. Rauber, and E.A. Mulvihill, 1986: Numerical simulation of the effects of varying ice crystal nucleation rates and aggregation processes on orographic snowfall. *J. Clim. Appl. Meteorol.*, **25**, 1658–1680.

Cotton, W.R., H. Zhang, G.M. McFarquhar, and S.M. Saleeby, 2007: Should we consider polluting hurricanes to reduce their intensity? *J. Weather Modif.*, **39**, 70–73.

Dagan, G., I. Koren, O. Altaratz, and R.H. Heiblum, 2016: Aerosol effect on the evolution of the thermodynamic properties of warm convective cloud fields. *Science. Rep.*, **6**, 38769, doi:10.1038/srep38769.

Delanoë, J., A. Protat, J. Testud, D. Bouniol, A.J., Heymsfield, A. Bansemer, P.R.A. Brown, and R.M. Forbes, 2005: Statistical properties of the normalized ice particle size distribution. *J. Geophys. Res.*, **110**, D10201, doi:10.1029/2004JD005405.

DeMott, P.J., A.J. Prenni, X. Liu, S M. Kreidenweis, M.D. Petters, C.H. Twohy, M.S. Richardson, T. Eidhammer, and D.C. Rogers, 2010: Predicting global atmospheric ice nuclei distributions and their impacts on climate. *Proc. Natl. Acad. Sci.*, **107**(25), 11217–11222.

De Rooy, W.C., P. Bechtold, K. Frohlich, C. Hohenegger, H. Jonker, D. Mironov, A.P. Siebesma, J. Teixeiraf, and J.-I. Yano, 2013: Entrainment and detrainment in cumulus convection: An overview. *Q. J. Royal Meteorol. Soc.*, **139**, 1–19.

Dooley, A.L., 2008: Ice microphysics observations in tropical cyclones from NAMMA. M.S. Thesis, University of Illinois at Urbana-Champaign, p. 65.

Emanuel, K.A., and D.J. Raymond, 1993: The representation of cumulus convection in numerical models. *Meteorol. Monogr.*, **24**, (46), 246.

Enukashvili, I.M., 1980: A numerical method for integrating the kinetic equation of coalescence and breakup of cloud droplets. *J. Atmos. Sci.*, **37**, 2521–2534.

Erlick, C., A. Khain, M. Pinsky, and Y. Segal, 2005: The effect of wind velocity fluctuations on drop spectrum broadening in stratocumulus clouds. *Atmos. Res.*, **75**, 15–45.

Fan, J., J.M. Comstock, and M. Ovchinnikov, 2010: The cloud condensation nuclei and ice nuclei effects on tropical anvil characteristics and water vapor of the tropical tropopause layer. *Environ. Res. Lett.*, **5**, 044005, doi:10.1088/1748-9326/5/4/044005.

Fan, J., S. Ghan, M. Ovchinnikov, X. Liu, P.J. Rasch, and A. Korolev, 2011: Representation of Arctic mixed-phase clouds and the Wegener–Bergeron–Findeisen process in climate models: Perspectives from a cloud-resolving study. *J. Geophys. Res.*, **116**, D00T07.

Fan, J., L.R. Leung, P.J. DeMott, J.M. Comstock, B. Singh, D. Rosenfeld, J.M. Tomlinson, A. White, K.A. Prather, P. Minnis, J.K. Ayers, and Q. Min, 2014: Aerosol impacts on California winter clouds and precipitation during CalWater 2011: Local pollution versus long-range transported dust. *Atmos. Chem. Phys.*, **14**, 81–101.

Fan, J., L.R. Leung, Z. Li, H. Morrison, H. Chen, Y. Zhou, Y. Qian, and Y. Wang, 2012a: Aerosol impacts on clouds and precipitation in eastern China: Results from bin and bulk microphysics. *J. Geophys. Res.*, **117**, D00K36, doi: 10.1029/2011JD016537.

Fan, J., L.R. Leung, D. Rosenfeld, Q. Chena, Z. Lid, J. Zhang, and H. Yan, 2013: Microphysical effects determine macrophysical response for aerosol impacts on deep convective clouds. *PNAS*, November 11, E4581–E4590.

Fan, J., M. Ovtchinnikov, J.M. Comstock, S.A. McFarlane, and A. Khain, 2009a: Ice formation in Arctic mixed-phase clouds: Insights from a 3-D cloud resolving model with size-resolved aerosol and cloud microphysics. *J. Geophys. Res.*, **114**, D04205.

Fan, J., D. Rosenfeld, Y. Ding, L.R. Leung, and Z. Li, 2012b: Potential aerosol indirect effects on atmospheric circulation and radiative forcing through deep convection. *Geophys. Res. Lett.*, **39**, L09806, doi:10.1029/2012GL051851.

Fan, J., T. Yuan, J.M. Comstock, S. Ghan, A.P. Khain, L.R. Leung, Z. Li, V.J. Martins, and M. Ovchinnikov, 2009b: Dominant role by vertical wind shear in regulating aerosol effects on deep convective clouds. *J. Geophys. Res.*, **114**, D22206, doi:10.1029/2009JD012352.

Fan, J., R. Zhang, G. Li, and W.-K. Tao, 2007a: Effects of aerosols and relative humidity on cumulus clouds.

J. Geophys. Res., **112**, (D14) D14204, doi:10.1029/2006JD008136.

Fan, J., R. Zhang, G. Li, W.-K. Tao, and X. Li, 2007b: Simulations of cumulus clouds using a spectral microphysics cloud resolving model. *J. Geophys. Res.*, **112**, D04201.

Feingold, G., W.R. Cotton, S.M. Kreidenweis, and J. Davis, 1999: The impact of giant cloud condensation nuclei on drizzle formation in stratocumulus: Implications for cloud radiative properties. *J. Atmos. Sci.*, **56**, 4100–4117.

Feingold, G., and A.J. Heymsfield, 1992: Parameterizations of condensational growth of droplets for use in general circulation models. *J. Atmos. Sci.*, **49**, 2325–2342.

Feingold, G., H. Jiang, and J.Y. Harrington, 2005: On smoke suppression of clouds in Amazonia. *Geophys. Res. Lett.*, **32**, L02804.

Feingold, G., I. Koren, H. Wang, H. Xue, and W.A. Brewer, 2010: Precipitation-generated oscillations in open cellular cloud fields. *Nature*, **466**, doi:10.1038/nature09314.

Feingold, G., I. Koren, T. Yamaguchi, and J. Kazil, 2015: On the reversibility of transitions between closed and open cellular convection. *Atmos. Chem. Phys.*, **15**, 7351–7367.

Feingold, G., S.M. Kreidenweis, and Y.P. Zhang, 1998: Stratocumulus processing of gases and cloud condensation nuclei. 1. Trajectory ensemble model. *J. Geophys. Res.-Atmos.*, **103**, 19527–19542.

Feingold, G., B. Stevens, W.R. Cotton, and A.S. Frisch, 1996: The relationship between drop in-cloud residence time and drizzle production in numerically simulated stratocumulus cloud. *J. Atmos. Sci.*, **53**, 1108–1121.

Feingold, G., S. Tzivion, and Z. Levin, 1988: The evolution of raindrop spectra with altitude. 1: Solution to the stochastic collection/breakup equation using the method of moments. *J. Atmos. Sci.*, **45**, 3387–3399.

Ferek, R.J., T. Garrett, P.V. Hobbs, S. Strader, D. Johnson, et al., 2000: Drizzle suppression in ship tracks. *J. Atmos. Sci.*, **57**, 2705–2728.

Ferrier, B.S., 1994: A double-moment multiple-phase four-class bulk ice scheme. Part I: Description. *J. Atmos. Sci.*, **51**, 249–280.

Field, P.R., R.J. Hogan, P.R.A. Brown, A.J. Illingworth, T.W. Choularton, and R.J. Cotton, 2005: Parameterization of ice particle size distribution for mid-latitude stratiform cloud. *Q. J. Royal Meteorol. Soc.*, **131**, 1997–2017.

Formenton, M., G. Panegrossi, D. Casella, S. Dietrich, A. Mugnai, P. Sanò, F. Di Paola, H.-D. Betz, C. Price, and Y. Yair, 2013a: Using a cloud electrification model to study relationships between lightning activity and cloud microphysical structure. *Nat. Hazards Earth Syst. Sci.*, **13**, 1085–1104.

Formenton, M., V.T.J. Phillips, and B. Lienert, 2013b: A new snow microphysics parameterization applied to a cloud electrification model: Framework and preliminary results, 93rd AMS Annual Meeting, Austin, Tex., Jan. 6–10.

Freud, E., and D. Rosenfeld, 2012: Linear relation between convective cloud drop number concentration and depth for rain initiation. *J. Geophys. Res.*, **117**, D02207, doi:10.1029/2011JD016457.

Freud, E., D. Rosenfeld, M.O. Andreae, A.A. Costa, and P. Artaxo, 2008: Robust relations between CCN and the vertical evolution of cloud drop size distribution in deep convective clouds. *Atmos. Chem. Phys.*, **8**, 1661–1675.

Geoffroy, O., J.-L. Brenguier, and F. Burnet, 2010: Parametric representation of the cloud droplet spectra for LES warm bulk microphysical schemes. *Atmos. Chem. Phys.*, **10**, 4835–4848.

Geresdi, I., N. Sarkadi, and G. Thompson, 2014: Effect of the accretion by water drops on the melting of snowflakes. *Atmos. Res.*, 149, 96–110.

Ghan, S.J., H. Abdul-Razzak, A. Nenes, Y. Ming, X. Liu, M. Ovchinnikov, B. Shipway, N. Meskhidze, J. Xu, and X. Shi, 2011: Droplet nucleation: Physically based parameterizations and comparative evaluation. *J. Adv. Model. Earth Syst.*, **3**, M10001, doi:10.1029/2011MS000074.

Ghan, S.J., R.C. Easter, J. Hudson, and F.-M. Bŕeon, 2001: Evaluation of aerosol indirect radiative forcing in MIRAGE. *J. Geophys. Res.*, **106**, 5317–5334.

Givati, A., and D. Rosenfeld, 2004: Quantifying precipitation suppression due to air pollution. *J. Appl. Meteorol.*, **43**, 1038–1056.

Goke, S, H.T. Ochs III, and R.M. Raube, 2007: Radar analysis of precipitation initiation in maritime versus continental clouds near the Florida coast: Inferences concerning the role of CCN and giant nuclei. *J. Atmos. Sci.*, **64**, 3695–3707.

Goren, T., and D. Rosenfeld, 2015: Extensive closed cell marine stratocumulus downwind of Europe – A large aerosol cloud mediated radiative effect or forcing? *J. Geophys. Res. Atmos.*, **120**(12), 6098–6116.

Grabowski, W.W., 2001: Coupling cloud processes with the large-scale dynamics using the Cloud-Resolving Convection Parameterization (CRCP). *J. Atmos. Sci.*, **58**, 978–997.

2003: Impact of cloud microphysics on convective-radiative quasi-equilibrium revealed by cloud-resolving convection parameterization (CRCP). *J. Climate*, **16**, 3463–3475.

2006: Comments on "Preliminary tests of multiscale modeling with a two-dimensional framework: Sensitivity to coupling methods" by Jung and Arakawa. *Mon. Weather Rev.*, **134**, 2021–2026.

Grabowski, W.W., and P.K. Smolarkiewicz, 1999: CRCP: A cloud resolving convection parameterization for modeling the Tropical convecting atmosphere. *Physica D.*, **133**, 171–178.

Grell, G.A., and D. Devenyi, 2002: A generalized approach to parameterizing convection combining ensemble and data assimilation techniques. *Geophys. Res. Lett.*, **29**(14), 1693.

Grell, G., J. Dudhia, and D.R. Stauffer, 1994: A description of the fifth generation Penn State/NCAR Mesoscale Model (MM5). NCAR Tech. Note NCAR/TN-398 +STR, p. 121.

Gultepe, I., G.A. Isaac, W.R. Leaitch, and C.M. Banic, 1996: Parameterizations of marine stratus microphysic based on in situ observations: Implications for GCMs. *J. Climate*, **9**(2), 345–357.

Hall, W.D., 1980: A detailed microphysical model within a two-dimensional dynamic framework: Model description and preliminary results. *J. Atmos. Sci.*, **37**, 2486–2507.

Han, J.-G., J.-J. Baik, and A.P. Khain, 2012: A numerical study of urban aerosol impacts on clouds and precipitation. *J. Atmos. Sci.*, **69**, 504–520.

Handwerker, J., and W. Straub, 2011: Optimal determination of parameters for Gamma-type drop size distributions based on moments. *J. Atmos. Oceanic Technol.*, **28**, 513–529.

Harrington, J.Y., G. Feingold, and W.R. Cotton, 2000: Radiative impacts on the growth of a population of drops within simulated Summertime Arctic stratus. *J. Atmos. Sci.*, **57**, 766–785.

Hashino, T., and G.J. Tripoli, 2007: The Spectral Ice Habit Prediction System (SHIPS). Part I: Model description and simulation of the vapor deposition process. *J. Atmos. Sci.*, **64**, 2210–2237.

2008: The Spectral Ice Habit Prediction System (SHIPS). Part II: Simulation of nucleation and depositional growth of polycrystals. *J. Atmos. Sci.*, **65**, 3071–3094, doi:10.1175/2008JAS2615.1.

2011: The Spectral Ice Habit Prediction System (SHIPS). Part IV: Box model simulations of the habit-dependent aggregation process. *J. Atmos. Sci.*, **68**, 1142–1161, doi:10.1175/2011JAS3667.1.

Heymsfield, A.J., A. Bansemer, P.R. Field, S.L. Durden, J.L. Stith, J.E. Dye, W. Hall, and C.A. Grainger, 2002: Observations and parameterizations of particle size distributions in deep tropical cirrus and stratiform precipitating clouds: Results from in situ observations in TRMM field campaigns. *J. Atmos. Sci.*, **59**, 3457–3491.

Heymsfield, A.J., and L.M. Miloshevich, 1993: Homogeneous ice nucleation and supercooled liquid water in orographic wave clouds. *J. Atmos. Sci.*, **50**, 2335–2353.

Heymsfield, A.J., and R.M. Sabin, 1989: Cirrus crystal nucleation by homogeneous freezing of solution droplets. *J. Atmos. Sci.*, **46**, 2252–2264.

Heymsfield, A., C. Schmitt, and A. Bansemer, 2013: Ice cloud particle size distributions and pressure-dependent terminal velocities from in situ observations at temperatures from 0 to −86 C. *J. Atmos. Sci.*, **70**, 4123–4154.

Hong, S.-Y., J. Dudhia, and S.-H. Chen, 2004: A revised approach to ice microphysics processes for the bulk parameterization of clouds and precipitation. *Mon. Weather Rev.*, **132**, 103–120.

Hong, S.-Y., and J.-O.J. Lim, 2006: The WRF single-moment 6-Class microphysics scheme (WSM6). *J. Korean Meteorol. Soc.*, **42**, 2, 129–151.

Hudson, J.G., and S. Mishra, 2007: Relationships between CCN and cloud microphysics variations in clean maritime air. *Geophys. Res. Lett.*, **34**, L16804.

Hudson, J.G., and S.S. Yum, 2001: Maritime-continental drizzle contrasts in small cumuli. *J. Atmos. Sci.*, **58**, 915–26.

Igel, A.L., and S.C. van den Heever, 2017: The Role of the Gamma Function Shape Parameter in Determining Differences between Condensation Rates in Bin and Bulk Microphysics Schemes. *Atmos. Chem. Phys.*, **17**, 4599–4609.

Iguchi, T., T. Matsui, J.J. Shi, W.-K. Tao, A.P. Khain, A. Hou, R. Cifelli, A. Heymsfield, and A. Tokay, 2012a: Numerical analysis using WRF-SBM for the cloud microphysical structures in the C3VP field campaign: Impacts of supercooled droplets and resultant riming on snow microphysics. *J. Geophys. Res.*, **117**, D23206, doi:10.1029/2012JD018101.

Iguchi, T., T. Matsui, W.-K. Tao, A.P. Khain, V.T.J. Phillips, C. Kidd, T. L'Ecuyer, S.A. Braun, and A. Hou, 2014: WRF-SBM simulations of melting layer structure in mixed-phase precipitation events observed during LPVEx. *J. Appl. Meteorol. Climatol.*, **53**, 2710–2731.

Iguchi, T., T. Nakajima, A. Khain, K. Saito, T. Takemura, H. Okamoto, T. Nishizawa, and W.-K. Tao, 2012b: Evaluation of cloud microphysics in JMA-NHM simulations using bin or bulk microphysical schemes through comparison with cloud radar observations. *J. Atmos. Sci.*, **69**, 2566–2586.

Illingworth, A.J., and T.M. Blackman, 2002: The need to represent raindrop size spectra as normalized Gamma distributions for the interpretation of polarization radar observations. *J. Appl. Meteorol.*, **41**, 286–297.

Ilotoviz, E., A.P. Khain, V. Phillips, N. Benmoshe, and A. Ryzhkov, 2016: Effect of aerosols on freezing drops, hail and precipitation in a mid-latitude storm. *J. Atmos. Sci.*, **73**, 1, 109–144.

Ilotoviz, E., A. Khain, A. Ryzhkov, and J.C. Snyder, 2017: Relationship between hail microphysics and Zdr columns. *J. Atmos. Sci.* (in press).

Inoue, T., and F. Kimura, 2004: Urban effects on low-level clouds around the Tokyo metropolitan area on clear summer days. *Geophys. Res. Lett.*, **31**, L05103.

Janjic, Z.I., 2000: Comments on "Development and evaluation of a convection scheme for use in climate models." *J. Atmos. Sci.*, **57**, 3686.

Jensen, J.B., and S. Lee, 2008: Giant sea-salt aerosols and warm rain formation in marine stratocumulus. *J. Atmos. Sci.*, **65**, 3678–3694.

Jirak, I.L., and W.R. Cotton, 2006: Effect of air pollution on precipitation along the front range of the Rocky Mountains. *J. Appl. Meteorol. Climatol.*, **45**, 236–245.

Johnson, D.W., P.R.A. Brown, G.M. Martin, and S.J. Moss, 1992: Recent measurement campaigns at the U.K. Meteorological research flight to improve numerical cloud parameterizations, in Proceedings of WMO Workshop on Cloud Microphysics and Applications to Global Change, pp. 257–262, World Meteorol. Organ., Geneva, Switzerland.

Johnson, R.H., and P.J. Hamilton, 1988: The relationship of surface pressure features to the precipitation and airflow structure of an intense midlatitude squall line. *Mon. Weather Rev.*, **16**, 1444–1472.

Kain, J.S., 2004: The Kain–Fritsch convective parameterization: An update. *J. Appl. Meteorol.*, **43**, 170–181.

Kazil, J., H. Wang, G. Feingold, A.D. Clarke, J.R. Snider, and A.R. Bandy, 2011: Chemical and aerosol processes in the transition from closed to open cells during VOCALS-REx. *Atmos. Chem. Phys.*, **11**, 7491–7514.

Kessler, E., 1969: On the distribution and continuity of water substance in atmospheric circulations. *Meteorol. Monogr.*, 10, *Amer. Meteorol. Soc.*, **10**, #32.

Khain, A.P., 2009: Notes on state-of-the-art investigations of aerosol effects on precipitation: A critical review. *Environ. Res. Lett.*, **4**, 015004.

Khain, A.P., V. Arkhipov, M. Pinsky, Y. Feldman, and Y. Ryabov, 2004a: Rain enhancement and fog elimination by seeding with charged droplets. Pt. 1. Theory and numerical simulations. *J. Appl. Meteorol.*, **43**, 1513–1529.

Khain, A.P., K.D. Beheng, A. Heymsfield, A. Korolev, S.O. Krichak, Z. Levin, M. Pinsky, V. Phillips, T. Prabhakaran, A. Teller, S.C. van den Heever, and J.-I. Yano, 2015: Representation of microphysical processes in cloud resolving models: Spectral (bin) microphysics versus bulk parameterization. *Rev. Geophys.*, **53**, 247–322.

Khain, A.P., N. Benmoshe, and A. Pokrovsky, 2008a. Factors determining the impact of aerosols on surface precipitation from clouds: An attempt of classification. *J. Atmos. Sci.*, **65**, 1721–1748.

Khain, A.P., N. Cohen, B. Lynn, and A. Pokrovsky, 2008b: Possible aerosol effects on lightning activity and structure of hurricanes. *J. Atmos. Sci.*, **65**, 3652–3667.

Khain A.P. and B. Lynn, 2009: Simulation of a super cell storm in clean and dirty atmosphere. *J. Geophys. Res.*, **114**, D19209, DOI: 10.1029/2009JD011827.

Khain, A.P., L.R. Leung, B. Lynn, and S. Ghan, 2009: Effects of aerosols on the dynamics and microphysics of squall lines simulated by spectral bin and bulk parameterization schemes. *J. Geophys. Res.*, DOI: 10.1029/2009JD011902.

Khain, A., B. Lynn, and J. Dudhia, 2010: Aerosol effects on intensity of landfalling hurricanes as seen from simulations with the WRF model with spectral bin microphysics. *J. Atmos. Sci.*, **67**, 365–384.

Khain, A., B. Lynn, and J. Shpund, 2016: High resolution WRF simulations of hurricane Irene: Sensitivity to aerosols and choice of microphysical schemes. *Atmos. Res.*, **167**, 129–145.

Khain, A.P., V. Phillips, N. Benmoshe, and A. Pokrovsky, 2012: The role of small soluble aerosols in the microphysics of deep maritime clouds. *J. Atmos. Sci.*, **69**, 2787–2807.

Khain, A.P., M. Pinsky, L. Magariz, O. Krasnov, and H.W.J. Russchenberg, 2008c: Combined observational and model investigations of the Z-LWC relationship in stratocumulus clouds. *J. Appl. Meteorol.*, **47**, 591–606.

Khain, A.P., and A. Pokrovsky, 2004: Effects of atmospheric aerosols on deep convective clouds as seen from simulations using a spectral microphysics mixed-phase cumulus cloud model. Part 2: Sensitivity study. *J. Atmos. Sci.*, **61**, 2983–3001.

Khain, A.P, A. Pokrovsky, U. Blahak, and D. Rosenfeld, 2008d: Is the dependence of warm and ice precipitation on the aerosol concentration monotonic? 15th Int. Conf. on Clouds and Precipitation, Cancun, July.

Khain, A.P., A. Pokrovsky, M. Pinsky, A. Seifert, and V. Philips, 2004b: Simulation of effects of atmospheric aerosols on deep turbulent convective clouds by using a spectral microphysics mixed-phase cumulus cloud model. Part I: Model description and possible applications. *J. Atmos. Sci.*, **61**, 2963–2982.

Khain, A., T.V. Prabha, N. Benmoshe, G. Pandithurai, and M. Ovchinnikov, 2013: The mechanism of first raindrops formation in deep convective clouds. *J. Geophys. Res.*, **118**, 9123–9140.

Khain, A., D. Rosenfeld, and A. Pokrovsky, 2005: Aerosol impact on the dynamics and microphysics of convective clouds. *Q. J. Royal. Meteorol. Soc.*, **131**, 2639–2663.

Khain, A.P., D. Rosenfeld, A. Pokrovsky, U. Blahak, and A. Ryzhkov, 2011: The role of CCN in precipitation and hail in a mid-latitude storm as seen in simulations using a spectral (bin) microphysics model in a 2D dynamic frame. *Atmos. Res.*, **99**, 129–146.

Khain, A.P., and I. Sednev, 1996: Simulation of precipitation formation in the eastern Mediterranean coastal zone using a spectral microphysics cloud ensemble model. *Atmos. Res.*, **43**, 77–110.

Khairoutdinov, M.F., and Y.L. Kogan, 1999: A large eddy simulation model with explicit microphysics: Validation against aircraft observations of a stratocumulus-topped boundary layer. *J. Atmos. Sci.*, **56**, 2115–2131.

2000: A new cloud physics parameterization in a large-eddy simulation model of marine stratocumulus. *Mon. Weather Rev.*, **128**(1), 229–243.

Khairoutdinov, M.F., and D.A. Randall, 2001: A cloud resolving model as a cloud parameterization in the NCAR Community Climate System Model: Preliminary Results. *Geophys. Res. Lett.*, **28**, 3617–3620.

Khairoutdinov, M.F., D.A. Randall, and C. DeMotte, 2005: Simulations of the atmospheric general circulation using a cloud-resolving model as a superparameterization of physical processes. *J. Atmos. Sci.*, **62**, 2136–2154.

Khairoutdinov, M.F., and C.E. Yang, 2013: Cloud-resolving modeling of aerosol indirect effects in idealized radiative-convective equilibrium with interactive and fixed sea surface temperature. *Atmos. Chem. Phys.*, **13** 8, 4133–4144.

Khvorostyanov, V.I., A.P. Khain, N. Chrekasova, and E.L. Kogteva, 1995: A two-dimensional model of dynamic cloud seeding. *Soviet Meteorology and Hydrology*, **9**, 68–84.

Khvorostyanov, V.I., A.P. Khain, and E.L. Kogteva, 1989: A two-dimensional non stationary microphysical model of a three-phase convective cloud and evaluation of the effects of seeding by a crystallizing agent. *Soviet Meteorology and Hydrology*, **5**, 33–45.

Kim, K.-Y., R.J. Park, K.-R. Kim, and H. Na, 2010: Weekend effect: Anthropogenic or natural?, *Geophys. Res. Lett.*, **37**, L09808, doi:10.1029/2010GL043233.

Kogan, Y.L., 1991: The simulation of a convective cloud in a 3-D model with explicit microphysics. Part I: Model description and sensitivity experiments. *J. Atmos. Sci.*, **48**, 1160–1189.

2006: Large-eddy simulation of air parcels in stratocumulus clouds: Time scales and spatial variability. *J. Atmos. Sci.*, **63**, 952–967.

Kogan, Y.L., M.P. Khairoutdinov, D.K. Lilly, Z.N. Kogan, and Q. Liu, 1995: Modeling of stratocumulus cloud layers in a large eddy simulation model with explicit microphysics. *J. Atmos. Sci.*, **52**, 2923–2940.

Kogan, Z.N., and Y.L. Kogan, 2001: Parameterization of drop effective radius for drizzling marine stratocumulus. *J. Geophys. Res.*, **106**, 9757–9764.

Kogan, Y., I.P. Mazin, B.N. Sergeev, and V.I. Khvorostyanov, 1984: *Numerical Cloud Modeling*. Gidrometeoizdat, Moscow, p. 183.

Kogan, Y.L., D.B. Mechem, and K. Choi, 2012: Effects of sea-salt aerosols on precipitation in simulations of shallow cumulus. *J. Atmos. Sci.*, **69**, 463–483.

Koren, I., O. Altaratz, L.A. Remer, G. Feingold, J.V. Martins, and R.H. Heiblum, 2012: Aerosol-induced intensification of rain from the tropics to the mid-latitudes. *Nat. Geosci.*, **5**(2), 118–122.

Koren, I., G. Feingold, and L.A. Remer, 2010a: The invigoration of deep convective clouds over the Atlantic: Aerosol effect, meteorology or retrieval artifact?. *Atmos. Chem. Phys.*, **10**, 8855–8872.

Koren, I., Y.J. Kaufman, D. Rosenfeld, L.A. Remer, and Y. Rudich, 2005: Aerosol invigoration and restructuring of Atlantic convective clouds. *Geophys. Res. Lett.*, **32**, L14828, doi: 10.1029/2005GL023187.

Koren, I., J.V. Martins, L.A. Remer, and H. Afargan, 2008: Smoke invigoration versus inhibition of clouds over the Amazon. *Science*, **321**, 946–949, doi: 10.1126/science.1159185.

Koren, I., L.A. Remer, O. Altaratz, J.V. Martins, and A. David, 2010b: Aerosol-induced changes of convective cloud anvils produce strong climate warming. *Atmos. Chem. Phys.*, **10**, 5001–5010.

Korolev, A., and P.R. Field, 2015: Assessment of the performance of the inter-arrival time algorithm to identify ice shattering artifacts in cloud particle probe measurements. *Atmos. Meas. Tech.*, **8**, 761–777.

Krasnov, O., and H. Russchenberg, 2002: An enhanced algorithm for the retrieval of liquid water cloud properties from simultaneous radar and lidar measurements. Part I: The basic analysis of in situ measured drop size spectra. *European Conference on Radar Meteorology (ERAD)*, **1**, 173–178.

Kumjian, M.R., S.M. Ganson, and A.V. Ryzhkov, 2012: Raindrop freezing in deep convective updrafts: A microphysical and polarimetric model. *J. Atmos. Sci.*, **69**, 3471–3490.

Kumjian, M.R., A.P. Khain, N. Benmoshe, E. Ilotoviz, A.V. Ryzhkov, and V.T.J. Phillips, 2014: The anatomy and physics of ZDR columns: Investigating a polarimetric radar signature with a spectral bin microphysical model. *J. Appl. Meteorol. Climatol.*, **53**, 1820–1843.

Lang, S.E., W.-K. Lang, J.-D. Tao, D. Chern, Wu, and X. Li, 2014: Benefits of a fourth ice class in the simulated radar reflectivities of convective systems using a bulk microphysics scheme. *J. Atmos. Sci.*, **71**, 3583–3611.

Lau, K.M., K.M. Kim, Y.C. Sad, and G.K. Walker, 2009: A GCM study of the response of the atmospheric water cycle of West Africa and the Atlantic to Saharan dust radiative forcing. *Ann. Geophys.*, **27**, 4023–4037.

Lawson, R.P., B. Baker, B. Pilson, and Q. Mo, 2006: In situ observations of the microphysical properties of wave, cirrus, and anvil clouds. Part II: Cirrus clouds. *J. Atmos. Sci.*, **63**, 3186–3203.

Lebo, Z.J., and J.H. Seinfeld, 2011: Theoretical basis for convective invigoration due to increased aerosol concentration. *Atmos. Chem. Phys.*, **11**, 5407–5429.

Lee, H., and J.-J. Baik, 2016: Effects of turbulence-induced collision enhancement on heavy precipitation: The 21 September 2010 case over the Korean Peninsula. *J. Geophys. Res.*, **121**, Issue 20, 12,319–12,342.

Lee, S.-S., L.J. Donner, and V.T.J. Phillips, 2005: Impact of aerosols on deep convection. *Q. J. Roy. Meteorol. Soc.*, **4** (8).

Lee, S.S., L.J. Donner, V.T.J. Phillips, and Y. Ming, 2008a: Examination of aerosol effects on precipitation in deep convective clouds during the 1997 ARM summer experiment. *Q. J. Roy. Meteorol. Soc.*, **134**, 1201–1220.

2008b: The dependence of aerosol effects on clouds and precipitation on cloud-system organization, shear and stability. *J. Geophys. Res.*, **113**, D16202, doi:10.1029/2007JD009224.

Levin, Z., and W.R. Cotton, 2007: Aerosol pollution impact on precipitation: A scientific review WMO/IUGG Report.

Levin, Z., G. Feingold, S. Tzivion, and A. Waldvogel, 1991: The Evolution of raindrop spectra: Comparison between modeled and observed spectra along a mountain slope in Switzerland. *J. Appl. Meteorol.*, **30**, 893–900.

Li, Z., F. Niu, J. Fan, Y. Liu, D. Rosenfeld, and Y. Ding, 2011: The long-term impacts of aerosols on the vertical development of clouds and precipitation. *Nat. Geosci.*, **4**, 888–894, doi: 10.1038/ngeo1313.

Li X., C.-H. Sui, and K.-N. Lau, 2002: Precipitation efficiency in the tropical deep convection: A 2D cloud resolving model study. *J. Meteorol. Soc. Jpn*, **80**, 205–212.

Li, X., W.-K. Tao, A.P. Khain, J. Simpson, and D.E. Johnson, 2009a: Sensitivity of a cloud-resolving model to bulk and explicit bin microphysical schemes. Part I: Validation with a PRE-STORM case. *J. Atmos. Sci.*, **66**, 3–21.

2009b: Sensitivity of a cloud-resolving model to bulk and explicit bin microphysical schemes. Part II: Cloud microphysics and storm dynamics interactions. *J. Atmos. Sci.*, **66**, 22–40.

Li, X., W.-K. Tao, T. Matsui, C. Liu, and H. Masunaga, 2010: Improving a spectral bin microphysics scheme using TRMM satellite observations. *Q. J. Royal Meteorol. Soc.*, **136**, 382–389, DOI:10.1002/qj.569.

Lim, J.-O., J. Lim, and S.-Y. Hong, 2005: Effects of bulk ice microphysics on the simulated monsoonal precipitation over east Asia. *J. Geophys. Res.*, 110 (D24), 166–06181.

Lin, K.-S., and S.-Y. Hong, 2010: Development of an effective double-moment cloud microphysics scheme with prognostic cloud condensation nuclei (CCN) for weather and climate models. *Mon. Weather Rev.*, **138**, 1587–1612.

Lin, Y., and B.A. Colle, 2011: A new bulk microphysical scheme that includes riming intensity and temperature-dependent ice characteristics. *Mon. Weather Rev.*, **139**, 1013–1035.

Lin, Y.-L., R.D. Farley, and H.D. Orville, 1983: Bulk parameterization of the snow field in a cloud model. *J. Climate Appl. Meteorol.*, **22**, 1065–1092.

Liu, Q.-F., Y.L. Kogan, D.K. Lilly, and M.P. Khairoutdinov, 1997: Variational optimization method for calculation of cloud drop growth in an Eulerian drop-size framework. *J. Atmos. Sci.*, **54**, 2493–2504.

Loftus, A.M., and W.R. Cotton, 2014a: A triple-moment hail bulk microphysics scheme. Part II: Verification and comparison with two moment bulk microphysics. *Atmos. Res.*, **150**, 97–128.

2014b: Examination of CCN impacts on hail in a simulated supercell storm with a triple-moment hail bulk microphysics. *Atmos. Res.*, **147–148**, 183–204.

Loftus, A.M, W.R. Cotton, and G.G. Carrió, 2014: A triple-moment hail bulk microphysics scheme. Part I: Description and initial evaluation. *Atmos. Res.*, **149**, 35–57.

Lohmann, U., 2008: Global anthropogenic aerosol effects on convective clouds in ECHAM-HAM. *Atmos. Chem. Phys.*, **8**, 2115–2131.

Lohmann, U., and J. Feichter, 1997: Impact of sulfate aerosols on albedo and lifetime of clouds: A sensitivity study with the ECHAM GCM. *J. Geophys. Res.*, **102**, 13,685–13,700.

Long, A., 1974: Solutions to the droplet collection equation for polynomial kernels. *J. Atmos. Sci.*, **31**, 1040–1052.

Low, T.B., and R. List, 1982a: Collision coalescence and breakup of raindrops: Part I. Experimentally established coalescence efficiencies and fragments size distribution in breakup. *J. Atmos. Sci.*, **39**, 1591–1606.

1982b: Collision coalescence and breakup of raindrops: Part II. Parameterization of fragment size distributions in breakup. *J. Atmos. Sci.*, **39**, 1607–1618.

Lynn, B., and A.P. Khain, 2007: Utilization of spectral bin microphysics and bulk parameterization schemes to simulate the cloud structure and precipitation in a mesoscale rain event. *J. Geophys. Res.*, **112**, D22205.

Lynn, B H., A.P. Khain, J.W. Bao, S.A. Michelson, T. Yuan, G. Kelman, and N. Benmoshe, 2014: The sensitivity of the WRF-simulated hurricane Irene to physics configuration. Abstract at the 94-th AMS conference, Atlanta, February 2014.

Lynn, B.H., A.P. Khain, J.W. Bao, S.A. Michelson, T. Yuan, G. Kelman, D. Rosenfeld, J. Shpund, and N. Benmoshe, 2016: The sensitivity of hurricane Irene to aerosols and ocean coupling: Simulations with WRF spectral bin microphysics. *J. Atmos. Sci.*, **73**, 467–486.

Lynn, B., A.P. Khain, J. Dudhia, D. Rosenfeld, A. Pokrovsky, and A. Seifert, 2005a: Spectral (bin) microphysics coupled with a mesoscale model (MM5). Part 1. Model description and first results. *Mon. Weather Rev.*, **133**, 44–58.

2005b: Spectral (bin) microphysics coupled with a mesoscale model (MM5). Part 2: Simulation of a CaPe rain event with squall line. *Mon. Weather Rev.*, **133**, 59–71.

Lynn, B., A.P. Khain, D. Rosenfeld, and W.L. Woodley, 2007: Effects of aerosols on precipitation from orographic clouds. *J. Geophys. Res.*, **112**, D10225, doi:10.1029/2006JD007537.

Magaritz, L, M. Pinsky, and A. Khain, 2007: Drizzle formation in stratocumulus clouds Abstracts of IUGG (Perugia, July).

2010: Effects of stratocumulus clouds on aerosols in the maritime boundary layer. *Atmos. Res.*, **97**, 498–512.

Magaritz, L., M. Pinsky, A.P. Khain, and O. Krasnov, 2009: Investigation of droplet size distributions and drizzle formation using a new trajectory ensemble model. Part 2: Lucky parcels in non-mixing limit. *J. Atmos. Sci.*, **66**, 781–805.

Magaritz-Ronen, L., M. Pinsky, and A. Khain, 2014: Effects of turbulent mixing on the structure and macroscopic properties of stratocumulus clouds, demonstrated by a Lagrangian trajectory model. *J. Atmos. Sci.*, **71**, 1843–1862.

2016: Drizzle formation in stratocumulus clouds: Effects of turbulent mixing. *Atmos. Chem. Phys.*, **15**, 1849–1862.

Mansell, E.R., C.L. Ziegler, and E.C. Bruning, 2010: Simulated electrification of a small thunderstorm with two-moment bulk microphysics. *J. Atmos. Sci.*, **67**, 171–194.

Martin, G.M., D.W. Johnson, and A. Spice, 1994: The measurement and parameterization of effective radius of droplets in warm stratocumulus clouds. *J. Atmos. Sci.*, **51**, 1823–1842.

Martins, J.A., M.A.F. Silva Dias, and F.L.T. Goncalves, 2009: Impact of biomass burning aerosols on precipitation in the Amazon: A modelling case study *J. Geophys. Res.* **114**, D02207, doi: 10.1029/2007JD009587.

McFarquhar, G.M., S. Ghan, J. Verlinde, A. Korolev, J.W. Strapp, et al., 2011: Indirect and Semi-Direct Aerosol Campaign (ISDAC): The impact of Arctic aerosols on clouds. *Bull. Am. Meteorol. Soc.*, **92**, 183–201.

McFarquhar, G.M., M.S. Timlin, R.M. Rauber, B.F. Jewett, J.A. Grim, and D.P. Jorgensen, 2007: Vertical variability of cloud hydrometeors in the stratiform region of mesoscale convective systems and bow echoes. *Mon. Weather Rev.*, **135**, 3405–3428.

McFarquhar, G.M., and A.J. Heymsfield, 2001: Parameterizations of INDOEX microphysical measurements and calculations of cloud susceptibility: Applications for climate studies. *J. Geophys. Res.*, **106**, 28675–28698.

Meyers, M.P., R.L. Walko, J.Y. Harrington, and W.R. Cotton, 1997: New RAMS cloud microphysics parameterization. Part II: The two-moment scheme. *Atmos. Res.*, **45**, 3–39.

Milbrandt, J., and H. Morrison, 2016: Parameterization of cloud microphysics based on the prediction of bulk ice particle properties. Part III: Introduction of multiple free categories. *J. Atmos. Sci.*, **73**, 975–995.

Milbrandt, J.A., and M.K. Yau, 2005: A multi-moment bulk microphysics parameterization. Part II: A proposed three-moment closure and scheme description. *J. Atmos. Sci.*, **62**, 3065–3081.

2006: A multi-moment bulk microphysics parameterization. Part III: Control simulation of a hailstorm. *J. Atmos. Sci.*, **63**, 3114–3136.

Ming, Y., V. Ramaswamy, L.J. Donner, V.T.J. Phillips, S.A. Klein, P.A. Ginoux, and L.W. Horowitz, 2007: Modeling the interactions between aerosols and liquid water clouds with a self-consistent cloud scheme in a general circulation model. *J. Atmos. Sci.*, **64**, 1189–1209.

Mitchell, D.L., S.K. Chai, Y. Liu, A.J. Heymsfield, and Y. Dong, 1996: Modeling cirrus clouds. Part I: Treatment of bimodal size spectra and case study analysis. *J. Atmos. Sci.*, **53**, 2952–2966.

Molinari, J., P. Moore, and V. Idone, 1999: Convective structure of hurricanes as revealed by lightning locations. *Mon. Weather Rev.*, **127**, 520–534.

Morrison, H., J.A. Curry, and V.I. Khvorostyanov, 2005a: A new double-moment microphysics parameterization for application in cloud and climate models. Part I: Description. *J. Atmos. Sci.*, **62**(6), 1665–1677.

Morrison, H., and A. Gettelman, 2008: A new two-moment bulk stratiform cloud microphysics scheme in the Community Atmosphere Model, version 3 (CAM3). Part I: Description and numerical tests. *J. Climate*, **21**, 3642–3658.

Morrison, H., and W.W. Grabowski, 2007: Comparison of bulk and bin warm-rain microphysics models using a kinematic framework. *J. Atmos. Sci.*, **64**(8), 2839–2861.

Morrison, H., and J.A. Milbrandt, 2015: Parameterization of ice microphysics based on the prediction of bulk particle properties. Part 1: Scheme description and idealized tests. *J. Atmos. Sci.*, **72**, 287–311.

Morrison, H., J.A. Milbrandt, G. Bryan, K. Ikeda, S.A. Tessendorf, and G. Thompson, 2015: Parameterizing of cloud microphysics based on prediction of bulk ice particle properties. Part 2: Case study comparison with observations and other schemes. *J. Atmos. Sci.*, **72**, 312–339.

Morrison, H., M.D. Shupe, J.O. Pinto, and J.A. Curry, 2005b: Possible roles of ice nucleation mode and ice nuclei depletion in the extended lifetime of Arctic mixed-phase clouds. *Geophys. Res. Lett.*, **32**, L18801, doi:10.1029/2005GL023614.

Morrison, H., G. Thompson, and V. Tatarskii, 2009a: Impact of cloud microphysics on the development of trailing stratiform precipitation in a simulated squall line: Comparison of one- and two-moment schemes. *Mon. Weather Rev.*, **137**, 991–1007.

Morrison, H., R.B. McCoy, S.A. Klein, S. Xie, Y. Luo, et al., 2009b: Intercomparison of model simulations of mixed-phase clouds observed during the ARM Mixed-Phase Arctic Cloud Experiment. II: Multilayered cloud. *Q. J. Royal Meteorol. Soc.*, **135**, 1003–1019.

Morrison, H., P. Zuidema, A.S. Ackerman, A. Avramov, G. DeBoer, J. Fan, A.M. Fridlind, T. Hashino, J.Y. Harrington, Y. Luo, M. Ovchinnikov, and B. Shipway, 2011: Inter-comparison of cloud model simulations of

Arctic mixed-phase boundary layer clouds observed during SHEBA/FIRE-ACE. *J. Adv. Model. Earth Syst.*, **3**, M06003.

Mote, T.L., M.C. Lacke, and J.M. Shepherd, 2007: Radar signatures of the urban effect on precipitation distribution: A case study for Atlanta, Georgia. *Geophys. Res. Lett.*, **34**, L20710.

Muhlbauer, A., T. Hashino, L. Xue, A. Teller, U. Lohmann, R. Rasmussen, I. Geresdi, and Z. Pan, 2010: Intercomparison of aerosol-cloud-precipitation interactions in stratiform orographic mixed-phase clouds. *Atmos. Chem. Phys.*, **10**, 8173–8196.

Murakami, M., 1990: Numerical modeling of dynamical and microphysical evolution of an isolated convective cloud – The 19 July 1981 CCOPE cloud. *J. Meteorol. Soc. Jpn*, **68**, 107–128.

Niu, F., and Z. Li, 2011: Cloud invigoration and suppression by aerosols over the tropical region based on satellite observations. *Atmos. Chem. Phys. Discuss.*, **11**, 5003–5017, doi: 10.5194/acpd-11-5003-2011.

2012: Systematic variations of cloud top temperature and precipitation rate with aerosols over the global tropics. *Atmos. Chem. Phys.*, **12**(18), 8491–8498.

Noppel, H., U. Blahak, A. Seifert, and K.D. Beheng, 2010a: Simulations of a hailstorm and the impact of CCN using an advanced two-moment cloud microphysical scheme. *Atmos. Res.*, **96**, 286–301.

Noppel, H., A. Pokrovsky, B. Lynn, A.P. Khain, and K.D. Beheng, 2010b: A spatial shift of precipitation from the sea to the land caused by introducing submicron soluble aerosols: Numerical modeling. *J. Geophys. Res.*, **115**, D18212, doi: 10.1029/2009JD012645.

Onishi, R.Y.O., and K. Takahashi, 2011: A warm-bin–cold-bulk hybrid cloud microphysical model. *J. Atmos. Sci.*, **69**, 1474–1497.

Ortega, K.L., J.M. Krause, and A.V. Ryzhkov, 2016: Polarimetric radar characteristics of melting hail: Part III: Validation of the algorithm for hail size discrimination. *J. Appl. Meteorol. Climatol.*, **55**, 829–848.

Orville, R.E., R. Zhang, J.N. Gammon, D. Collins, B. Ely, and S. Steiger, 2001: Enhancement of cloud-to-ground lightening over Houston, Texas. *Geophys. Res. Lett.*, **28**(13), 2597–2600, doi: 10.1029/2001GL012990.

Ovchinnikov, M., A.S. Ackerman, A. Avramov, A. Cheng, J. Fan, A.M. Fridlind, S. Ghan, J. Harrington, C. Hoose, A. Korolev, G.M. McFarquhar, H. Morrison, M. Paukert, J. Savre, B.J. Shipway, M.D. Shupe, A. Solomon, and K. Sulia, 2014: Intercomparison of large-eddy simulations of Arctic mixed-phase clouds: Importance of ice size distribution assumptions. *J. Appl. Meteorol. Clim.*, **6**, 223–248.

Ovchinnikov, M., A. Korolev, and J. Fan, 2011: Effects of ice number concentration on dynamics of a shallow mixed-phase stratiform cloud. *J. Geophys. Res.*, **116**, D00T06.

Phillips, V.T.J., C. Andronache, B. Christner, C.E. Morris, D.C. Sands, A. Bansemer, A. Lauer, C. McNaughton, and C. Seman, 2009: Potential impacts from biological aerosols on ensembles of continental clouds simulated numerically. *Biogeosciences*, **6**, 1–28.

Phillips, V.T.J., P.J. DeMott, and C. Andronache, 2008: An empirical parameterization of heterogeneous ice nucleation for multiple chemical species of aerosol. *J. Atmos. Sci.*, **65**, 2757–2783.

Phillips, V.T.J., P.J. DeMott, C. Andronache, K.A. Pratt, K.A. Prather, R. Subramanian, and C. Twohy, 2013: Improvements to an empirical parameterization of heterogeneous ice nucleation and its comparison with observations. *J. Atmos. Sci.*, **70**, 378–409.

Phillips, V.T.J., L.J. Donner, and S.T. Garner, 2007a: Nucleation process in deep convection simulated by a cloud-system-resolving model with double-moment bulk microphysics. *J. Atmos. Sci.*, **64**, 738–761.

Phillips, V.T.J., A. Khain, N. Benmoshe, A. Ryzhkov, and E. Ilotovich, 2014: Theory of time-dependent freezing and its application in a cloud model with spectral bin microphysics. Part I. Wet growth of hail. *J. Atmos. Sci.*, **71**, 4527–4557.

2015: Theory of time-dependent freezing and its application in a cloud model with spectral bin microphysics. II: Freezing raindrops and simulations. *J. Atmos. Sci.*, **72**, 262–286.

Phillips, V., Khain, A. P., and A. Pokrovsky 2007b: The influence of melting on the dynamics and precipitation production in maritime and continental storm-clouds. *J. Atmos. Sci.*, **64**, 338–359.

Phillips, V.T.J., J.-I. Yano, M. Formenton, E. Ilotoviz, V. Kanawade, I. Kudzotsa, J. Sun, Aaron Bansemer, A.G. Detwiler, A. Khain and S.A. Tessendorf, 2017b: Ice multiplication by break-up in ice-ice collisions. Part 2: Numerical simulations. *J. Atmos. Sci.*, **74**, 2789–2811, doi: 10.1175/JAS-D-16-0223.1.

Phillips, V.T.J, J.-I. Yano, and A. Khain, 2017a: Ice multiplication by break-up in ice-ice collisions. Part 1: Theoretical formulation. *J. Atmos. Sci.*, **74**(6), 1705–1719.

Pielke, R.A., and R.P. Pearce, 1994: Mesoscale modelling of the atmosphere. *Meteorol. Manogr.*, **25** (47), 168.

Pinsky, M., and A.P. Khain, 2002: Effects of in-cloud nucleation and turbulence on droplet spectrum formation in cumulus clouds. *Q. J. Roy. Meteorol. Soc.*, **128**, 1–33.

Pinsky, M., A. Khain, and L. Magaritz, 2010: Representing turbulent mixing of non-conservative values in Eulerian and Lagrangian cloud models. *Q. J. Roy. Meteorol. Soc.*, **136**, 1228–1242.

Pinsky, M., A. Khain, L. Magaritz, and A. Sterkin, 2008: Simulation of droplet size distributions and drizzle formation using a new trajectory ensemble model of cloud topped boundary layer. Part 1: Model description and first results in non-mixing limit. *J. Atmos. Sci.*, **65**, 2064–2086.

Pinsky, M., A. Khain, and M. Shapiro, 2001: Collision efficiency of drops in a wide range of Reynolds numbers: Effects of pressure on spectrum evolution. *J. Atmos. Sci.*, **58**, 742–764.

Pinsky, M., I. Mazin, A. Korolev, and A. Khain, 2012: Analytical estimation of droplet concentration at cloud base. *J. Geophys. Res.*, **117**, D18211.

Plant, R.S., 2010: A review of the theoretical basis for bulk mass flux convective parameterization. *Atmos. Chem. Phys.*, **10**, 3529–3544.

Plant, R.S., and J.-I. Yano, 2015: *Parameterization of Atmospheric Convection* (In 2 Volumes). *Volume 1: Theoretical Background and Formulation; Volume 2: Current Issues and New Theories*. Under editing by Plant, R.S. and J.-I. Yano. Cambridge University Press.

Prabha, T.V., A. Khain, R.S. Maheshkumar, G. Pandithurai, J.R. Kulkarni, M. Konwar, and B.N. Goswami, 2011: Microphysics of premonsoon and monsoon clouds as seen from in situ measurements during CAIPEEX, *J. Atmos. Sci.*, **68**, 1882–1901.

Prabha V.T., S. Patade, G. Pandithurai, A. Khain, D. Axisa, P. Pradeep Kumar, R.S. Maheshkumar, J.R. Kulkarni, and B.N. Goswami, 2012: Spectral width of premonsoon and monsoon clouds over Indo-Gangetic valley during CAIPEEX, *J. Geophys. Res.*, **117**, D20205, doi: 10.1029/2011JD016837.

Pruppacher, H.R., and J.D. Klett, 1997: *Microphysics of Clouds and Precipitation*, 2nd edition. Oxford University Press, p. 963.

Radke, L.F., J.A. Coakley Jr., and M.D. King, 1989: Direct and remote sensing observations of the effects of ships on clouds. *Science*, **246**, 1146–1149, doi: 10.1126/science.246.4934.1146.

Randall, D., M. Khairoutdinov, A. Arakawa, and W. Grabowski, 2003: Breaking the cloud parameterization deadlock. *Bull. Am. Meteor. Soc.*, **84**, 1547–1564.

Reid, J.S., P.V. Hobbs, A.L. Rangno, and D.A. Hegg, 1999: Relationships between cloud droplet effective radius, liquid water content, and droplet concentration for warm clouds in Brazil embedded in biomass smoke. *J. Geophys. Res.*, **104**, 6145–6153.

Reisin, T., Z. Levin, and S. Tzvion, 1996: Rain production in convective clouds as simulated in an axisymmetric model with detailed microphysics. Part I: Description of the model. *J. Atmos. Sci.*, **53**, 497–519.

Reisner, R., R.M. Rasmussen, and R.T. Bruintjes, 1998: Explicit forecasting of supercooled liquid water in winter storms using the MM5 mesoscale model. *Q. J. Royal Meteorol. Soc.*, **124**, 1071–1107.

Riechelmann, T., Y. Noh, and S. Raasch, 2012: A new method for large-eddy simulations of clouds with Lagrangian droplets including the effects of turbulent collision. *New J. Phys.*, **14**, 065008.

Rissler, J., A. Vestin, E. Swietlicki, G. Fisch, J. Zhou, P. Artaxo, and M.O. Andreae, 2006: Size distribution and hygroscopic properties of aerosol particles from dry-season biomass burning in Amazonia. *Atmos. Chem. Phys.*, **6**, 471–491.

Rosenfeld, D., 1999: TRMM observed first direct evidence of smoke from forest fires inhibiting rainfall. *Geophys. Res. Lett.*, **26**, 3105–3108, doi:10.1029/1999GL006066.

2000: Suppression of rain and snow by urban and industrial air pollution. *Science*, **287**, 1793–1796.

Rosenfeld, D., and T.L. Bell, 2011: Why do tornados and hailstorms rest on weekends? *J. Geophys. Res.*, **116**, D20211, doi:10.1029/2011JD016214.

Rosenfeld, D., B. Fischman, Y. Zheng, T. Goren, and D. Giguzin, 2014: Combined satellite and radar retrievals of drop concentration and CCN at convective cloud base. *Geophys. Res. Lett.*, **41**(9), 3259–3265.

Rosenfeld, D., and G. Gutman, 1994: Retrieving microphysical properties near the tops of potential rain clouds by multispectral analysis of AVHRR data. *Atmos. Res.*, **34**, 259–283.

Rosenfeld, D., U. Lohmann, G.B. Raga, C.D. O'Dowd, M. Kulmala, S. Fuzzi, A. Reissell, and M.O. Andreae, 2008: Flood or drought: How do aerosols affect precipitation? *Science*, **321**, 1309–1313.

Rosenfeld, D., W.L. Woodley, A. Khain, W.R. Cotton, G. Carrió, I. Ginis, and J.H. Golden, 2012: Aerosol effects on microstructure and intensity of tropical cyclones. *Bul. Am. Meteorol. Soc.*, **July**, 987–1001.

Rotunno, R., J.B. Klemp, and M.L. Weisman, 1988: A theory for strong, long-lived squall lines. *J. Atmos. Sci.*, **45**, 463–485.

Rozoff, C.M., W.R. Cotton, and J.O. Adegoke, 2003: Simulation of St. Louis, Missouri, land use impacts on thunderstorms. *J. Appl. Meteorol.*, **42**, 716–738.

Rutledge, S.A., and P.V. Hobbs, 1984: The mesoscale and microscale structure and organization of clouds and precipitation in midlatitude cyclones. XII: A diagnostic modeling study of precipitation development in narrow cold-frontal rainbands. *J. Atmos. Sci.*, **41**, 2949–2972.

Ryzhkov, A.V., M. Kumjian, and S. Ganson, 2013a: Polarimetric radar characteristics of melting hail, Part 1: Theoretical simulations using spectral microphysical modeling. *J. Appl. Meteorol. Climatol.*, **52**, 2849–2869.

Ryzhkov, A.V., M. Kumjian, S. Ganson, and P. Zhang, 2013b: Polarimetric radar characteristics of melting hail, Part II: Practical implications. *J. Appl. Meteorol. Climatol.*, **52**, 2871–2886.

Saleeby, S.M., and W.R. Cotton, 2004: A Large-Droplet Mode and Prognostic Number Concentration of Cloud Droplets in the Colorado State University Regional Atmospheric Modeling System (RAMS). Part I: Module Descriptions and Supercell Test Simulations. *J. Appl. Meteorol.*, **43**, 182–195.

2008: A binned approach to cloud-droplet riming implemented in a bulk microphysics model. *J. Appl. Meteorol. Climatol.*, **47**, 694–703.

Saleeby, S.M., and S.C. Van den Heever, 2013: Developments in the CSU-RAMS aerosol model: Emissions, nucleation, regeneration, deposition, and radiation. *J. Appl. Meteorol. Climatol.*, **52**, 2601–2622.

Sant, V., U. Lohmann, and A. Seifert, 2013: Performance of a tri-class parameterization for the collision–coalescence process in shallow clouds. *J. Atmos. Sci.*, **70**, 1744–1767.

Sarkadi, N., I. Geresdi, and G. Thompson, 2016: Numerical simulation of precipitation formation in the case orographically induced convective cloud: Comparison of the results of bin and bulk microphysical schemes. *Atmos. Res.*, **180**, 241–261.

Saunders, C.P.R., 1993: A review of thunderstorm electrification processes. *J. Appl. Meteorol.*, **32**, 642–655.

2008: *Lightning: Principles, Instruments and Applications*. Springer, Amsterdam.

Schultz, P., 1995: An explicit cloud physics parameterization for operational numerical weather prediction. *Mon. Weather Rev.*, **123**, 3331–3343.

Segal, Y., and A.P. Khain, 2006: Dependence of droplet concentration on aerosol conditions in different cloud types: Application to droplet concentration parameterization of aerosol conditions. *J. Geophys. Res.*, **111**, D15204, doi:10.1029/2005JD006561.

Seifert, A., and K.D. Beheng, 2001: A double-moment parameterization for simulating autoconversion, accretion and selfcollection. *Atmos. Res.*, **59–60**, 265–281.

2006a: A two-moment cloud microphysics parameterization for mixed-phase clouds. Part 1: Model description. *Meteorol. Atmos. Phys.*, **92**, 45–66.

2006b: A two-moment cloud microphysics parameterization for mixed-phase clouds. Part 2: Maritime vs. continental deep convective storms. *Meteorol. Atmos. Phys.*, **92**, 67–82.

Seifert, A., A. Khain, A. Pokrovsky, and K.D. Beheng, 2006: A comparison of spectral bin and two-moment bulk mixed-phase cloud microphysics. *Atmos. Res.*, **80**(1), 46–66.

Seifert, A., C. Köhler, and K.D. Beheng, 2012: Aerosol-cloud-precipitation effects over Germany as simulated by a convective-scale numerical weather prediction model. *Atmos. Chem. Phys.*, **12**, 709–725.

Sekhon, R.S., and R.C. Srivastava, 1970: Snow size spectra and radar reflectivity. *J. Atmos. Sci.*, **27**, 299–307.

Shao, X.M., J. Harlin, M. Stock, M. Stanley, A. Regan, K. Wiens, T. Hamlin, M. Pongratz, D. Suszcynsky, and T. Light, 2005: Katrina and Rita were lit up with lightning. *EOS*, **86**(42), 398–399.

Shepherd, J.M., 2005: A review of current investigations of urban-induced rainfall and recommendations for the future. *Earth Interactions*, **9**. http://EarthInteractions.org.

Shepherd, J.M., H. Pierce, and A.J. Negri, 2002: Rainfall modification by major urban areas: Observations from spaceborne rain radar on the TRMM satellite. *J. Appl. Meteorol.*, **41**, 689–701.

Shima, S., K. Kusano, A. Kawano, T. Sugiyama, and S. Kawahara, 2009: The super-droplet method for the numerical simulation of clouds and precipitation: A particle-based and probabilistic microphysics model coupled with a non-hydrostatic model. *Q. J. Royal Meteorol. Soc.*, **135**, 1307–1320.

Shipway, B.J., and A.A. Hill, 2012: Diagnosis of systematic differences between multiple parameterizations of warm rain microphysics using a kinematic framework. *Q. J. Royal Meteorol. Soc.*, **138**, 2196–2211.

Shpund, J., A. Khain, and D. Rosenfeld, 2016: The Effects of Sea-Spray on Deep Mixed-Phase Convective Cloud under strong wind conditions. ICCP2016, Manchester, July 25–29.

Shpund, J., M. Pinsky, and A. Khain, 2011: Microphysical structure of the marine boundary layer under strong wind and spray formation as seen from simulations using a 2D explicit microphysical model. Part I: The impact of large eddies. *J. Atmos. Sci.*, **68**, 2366–2384.

Shpund, J., J.A. Zhang, M. Pinsky, and A. Khain, 2012: Microphysical structure of the marine atmospheric mixed layer under strong wind and sea spray formation as seen from a 2-D Explicit Microphysical Model Part II: The role of sea spray. *J. Atmos. Sci.*, **69**, 3501–3514.

2014: Microphysical structure of the marine boundary layer under strong wind and sea spray formation as seen from a 2D Explicit Microphysical Model. Part III: Parameterization of height-dependent droplet size distribution. *J. Atmos. Sci.*, **71**, 1914–1934.

Skamarock, W.C., J.B. Klemp, J. Dudhia, D.O. Gill, D.M. Barker, M.G. Duda, X.Y. Huang, W. Wang, and J.G. Powers, 2008. A description of the advanced research WRF Version 3. NCAR/TN–475+STR, NCAR Technical note, National Center for Atmospheric Research Boulder, Colorado, USA, p. 113.

Slingo, A., 1990: Sensitivity of the earth's radiation budget to changes in low clouds. *Nature*, **349**, 49–52.

Snyder, J.C., A.V. Ryzhkov, M.R. Kumjian, A.P. Khain, and J. Picca, 2015: A Zdr column detection algorithm to examine convective storm updrafts. *Weather and Forecast.*, **30**, 1819–1844.

Song, X., and G.J. Zhang, 2011: Microphysics parameterization for convective clouds in a global climate model: Description and single-column model tests. *J. Geophys. Res.*, **116**, D02201, doi:10.1029/2010JD014833.

Squires, P., 1958: The microstructure and colloidal stability of warm clouds. *Tellus*, **10**, 256–271.

Steiger, S.M., and R.E. Orville, 2003: Cloud-to-ground lightning enhancement over southern Louisiana. *Geophys. Res. Lett.*, **30**(19), 1975, doi:10.1029/2003GL017923.

Steiger, S.M., R.E. Orville, and G. Huffines, 2002: Cloud-to-ground lightning characteristics over Houston, Texas: 1989–2000. *J. Geophys. Res.*, **107**(D11), 4117, doi: 10.1029/2001JD001142.

Stevens, B., D.H. Lenschow, G. Vali, H. Gerber, A. Bandy, et al., 2003: Dynamics and chemistry of maritime stratocumulus-DYCOMS-II. *Bull. Am. Meteorol. Soc.*, **84**, 579–593.

Stevens, B., and Coauthors, 2005: Evaluation of large-eddy simulations via observations of nocturnal marine stratocumulus. *Mon. Weather Rev.*, **133**, 1443–1462.

Stevens, B., G. Feingold, W.R. Cotton, and R.L. Walko, 1996: Elements of the Microphysical Structure of Numerically Simulated Nonprecipitating Stratocumulus. *J. Atmos. Sci.*, **53**, 980–1006.

Sui, C.-S., X. Li,. and M.-J. Yang, 2007: On the definition of the precipitation efficiency. *J. Atmos. Sci.*, **64**, 4506–4513.

Suzuki, K., T. Nakajima, T.Y. Nakajima, and A.P. Khain, 2010: A Study of microphysical mechanisms for correlation patterns between droplet radius and optical thickness of warm clouds with a spectral bin microphysics cloud model. *J. Atmos. Sci.*, **67**, 1126–1141.

Suzuki, K., G.L. Stephens, S.C. van den Heever, and T.Y. Nakajima, 2011: Diagnosis of the warm rain process in cloud-resolving models using joint CloudSat and MODIS observations. *J. Atmos. Sci.*, **68**, 2655–2670.

Swann, H., 1998: Sensitivity to the representation of precipitating ice in CRM simulations of deep convection. *Atmos. Res.*, **48**, 415–435.

Takahashi, T., 1976: Hail in an axisymmetric cloud model. *J. Atmos. Sci.*, **33**, 1579–1601.

1978: Riming electrification as a charge generation mechanism in thunderstorms. *J. Atmos. Sci.*, **35**, 1536–1548.

Takahashi, I., T. Nakajima, A. Khain, K. Saito, T. Takemura, and K. Suzuki, 2008: A study of the cloud microphysical properties influenced by aerosols in an East Asia region using a meso-scale model coupled with a bin microphysics for clouds. *J. Geophys. Res.*, **113**, D14215.

Tampieri, F., and C. Tomasi, 1978: Size distribution models of fog and cloud droplets in terms of the modified Gamma function. *Tellus* XXVIII, **4**, 333–347.

Tao, W.-K., J.-P. Chen, Z. Li, C. Wang, and C. Zhang, 2012: Impact of aerosols on convective clouds and precipitation. *Rev. Geophys.*, **50**, RG2001, doi:10.1029/2011RG000369.

Tao, W.-K., X. Li, A. Khain, T. Matsui, S. Lang, and J. Simpson, 2007: The role of atmospheric aerosol concentration on deep convective precipitation: Cloud-resolving model simulations. *J. Geophys. Res.*, **112**, D24S18.

Tao, W.-K., J. Simpson, D. Baker, S. Braun, M.-D. Chou, B. Ferrier, D. Johnson, A. Khain, S. Lang, B. Lynn, C.-L. Shie, D. Starr, C.-H. Sui, Y. Wang, and P. Wetzel, 2003: Microphysics, radiation and surface processes in the Goddard Cumulus Ensemble (GCE) model. *Meteorol. Atmos. Phys.*, **82**, 97–137.

Tao, W.-K., J. Simpson, and M. McCumber, 1989: An ice-water saturation adjustment. *Mon. Weather Rev.*, **117**, 231–235.

Tao, W.-K., J.J. Shi, S.S. Chen, S. Lang, P.-L. Lin, S.-Y. Hong, C. Peters-Lidard, and A. Hou, 2011: The impact of microphysical schemes on intensity and track of hurricane. *Asia-Pacific J. Atmos. Sci.*, **47**, 1–16.

Tao, W.-K., D. Wu, S. Lang, J.Chern, C. Peters-Lidard, A. Fridlind, and T. Matsui, 2015: High-resolution NU-WRF simulations of a deep convective-precipitation system during MC3E: Further improvements and comparisons between Goddard microphysics schemes and observations. *J. Geophys. Res. Atmos.*, **121**, 1278–1305, doi:10.1002/2015JD023986.

Teller, A., L. Xue, and Z. Levin, 2012: The effects of mineral dust particles, aerosol regeneration and ice nucleation parameterizations on clouds and precipitation, *Atmos. Chem. Phys.*, **12**, 9303–9320.

Thompson, G., and T. Eidhammer, 2014: A study of aerosol impacts on clouds and precipitation development in a large winter cyclone. *J. Atmos. Sci.*, **71**, 3636–3658.

Thompson, G., P.R. Field, W.D. Hall, and R. Rasmussen, 2006: A new bulk microphysical parameterization for WRF (& MM5). WRF Conference, Natl. Cent. for Atmos. Res., Boulder, Colorado, June.

Thompson, G., P.R. Field, R.M. Rasmussen, and W.D. Hall, 2008: Explicit forecasts of winter precipitation using an improved bulk microphysics scheme. Part II: Implementation of a new snow parameterization. *Mon. Weather Rev.*, **136**, 5095–5115.

Thompson, G., R.M. Rasmussen, and K. Manning, 2004: Explicit forecasts of winter precipitation using an improved bulk microphysics scheme. Part I: Description and sensitivity analysis. *Mon. Weather Rev.*, **132**, 519–542.

Tian, L., G.M. Heymsfield, L. Li, A.J. Heymsfield, A. Bansemer, C.H. Twohy, R.C. Srivastava, 2010: A study of cirrus ice particle size distribution using TC4 observations. *J. Atmos. Sci.*, **67**, 195–216.

Tiedtke, M., ECMWF, 1983: The sensitivity of the time-mean large-scale flow to cumulus convection in the ECMWF model. Workshop on Convection in Large-Scale Numerical Models. Nov. 28–Dec. 1, pp. 297–316.

Tripoli, G.J., and W.R. Cotton, 1980. A numerical investigation of several factors contributing to the observed variable density of deep convection over south Florida. *J. Appl. Meteorol.*, **19**, 1037–1063

Twomey, S., 1974: Pollution and the planetary albedo. *Atmos. Environ.*, **8**, 1251–1256.

1977: The influence of pollution on the shortwave albedo of clouds. *J. Atmos. Sci.*, **34**, 1149–1152.

Tzivion, S., G. Feingold, and Z. Levin, 1987: An efficient numerical solution to the stochastic collection equation. *J. Atmos. Sci.*, **44**, 3139–3149.

1989: The evolution of raindrop spectra II: Collisional collection/breakup and evaporation in a rainshaft. *J. Atmos. Sci.*, **46**, 3312–3327.

Van den Heever, S.C., and W.R. Cotton, 2007: Urban aerosol impacts on downwind convective storms. *J. Appl. Meteorol. Climatol.*, **46**, 828–850.

Van den Heever, S.C., G.G. Carrió, W.R. Cotton, P.J. Demott, and A.J. Prenni, 2006: Impacts of nucleating aerosol on Florida storms. Part I: Mesoscale simulations. *J. Atmos. Sci.*, **63**, 1752–1775.

Van den Heever, S.C., G.L. Stephens, and N.B. Wood, 2011: Aerosol indirect effects on tropical convection characteristics under conditions of radiative-convective equilibrium. *J. Atmos. Sci.*, **68**, 699–718.

Van Zanten, M.C., B. Stevens, G. Vali, and D.H. Lenschow, 2005: Observations of drizzle in nocturnal marine stratocumulus. *J. Atmos. Sci.*, **62**, 88–106.

Verlinde, J., P.J. Flatau, and W.R. Cotton, 1990: Analytical solutions to the collection growth equation: Comparison with approximate methods and application to cloud micro-physics parameterization schemes. *J. Atmos. Sci.*, **47**, 2871–2880.

Walko, R.L., W.R. Cotton, M.P. Meyers, and J.Y. Harrington, 1995: New RAMS cloud microphysics parameterization Part I: The single-moment scheme. *Atmos. Res.*, **38**, 29–62.

Wang, C., 2005: A modeling study of the response of tropical deep convection to the increase of cloud condensational nuclei concentration: 1. Dynamics and microphysics. *J. Geophys. Res.*, **110**, D21211.

Wang, H., and G. Feingold, 2009a: Modeling mesoscale cellular structures and drizzle in marine stratocumulus. Part 1: Impact of drizzle on the formation and evolution of open cells. *J. Atmos. Sci.*, **66**, 3237–3256.

2009b: Modelling mesoscale cellular structures and drizzle in marine stratocumulus. Part II: The microphysics and dynamics of the boundary region between open and closed cells. *J. Atmos. Sci.*, **66**, 3257–3275.

Wang, M., S. Ghan, M. Ovchinnikov, X. Liu, R. Easter, E. Kassianov, Y. Qian, R. Marchand, and H. Morrison, 2011: Aerosol indirect effects in a multi-scale aerosol-climate model PNNL-MMF. *Atmos. Chem. Phys.*, **11**, 5431–5455.

Wang, Y., J. Fan, R. Zhang, L. Leung, and C. Franklin, 2013: Improving bulk microphysics parameterization in simulation of aerosol effects. *J. Geophys. Res.*, **118**, 1–19.

Williams, E., D. Rosenfeld, N. Madden, J. Gerlach, N. Gears, et al., 2002: Contrasting convective regimes over the Amazon: Implications for cloud electrification, *J. Geophys. Res.*, **107**(D20), 8082, doi: 10.1029/2001JD000380.

Wood, R., 2000: Parametrization of the effect of drizzle upon the droplet effective radius in stratocumulus clouds. *Q. J. Royal Meteorol. Soc.*, **126**, 3309–3324.

Wurzler, S., T.G. Reisin, and Z. Levin, 2000: Modification of mineral dust particles by cloud processing and subsequent effects on drop size distributions. *J. Geophys. Res. Atmos.*, **105**, 4501–4512.

Xue, H., G. Feingold, and B. Stevens, 2008: Aerosol effect on clouds, precipitation, and the organization of shallow cumulus clouds. *J. Atmos. Sci.*, **65**, 392–406.

Yamaguchi, T., and G. Feingold, 2013: On the size distribution of cloud holes in stratocumulus and their relationship to cloud-top entrainment. *Geophys. Res. Lett.*, **40**, 2450–2454.

Yang, F., M. Ovchinnikov, and R.A. Shaw, 2013: Minimalist model of ice microphysics in mixed-phase stratiform clouds. *Geophys. Res. Lett.*, **40**, 3756–3760, doi: 10.1002/grl.50700.

Yano, J.I., M. Bister, Z. Fuchs, L. Gerard, V. Phillips, S. Barkidija, and J.M. Piriou, 2013: Phenomenology of convection-parameterization closure. *Atmos. Phys. Chem.*, **13**, 4111–4131.

Yano, J.-I., J.-L. Redelsperger, F. Guichard, and P. Bechtold, 2005: Mode decomposition as a methodology for developing convective-scale representations in global models. *Q. J. Roy. Meterol. Soc.*, **131**, 2313–2336.

Yin, Y., Z. Levin, T. Reisin, and S. Tzivion, 2000: Seeding convective clouds with hygroscopic flares: Numerical simulations using a cloud model with detailed microphysics. *J. Appl. Meteorol.*, **39**, 1460–1472.

Young, K.C., 1974: A numerical simulation of wintertime, orographic precipitation: Part I. Description of model microphysics and numerical techniques. *J. Atmos. Sci.*, **31**, 1735–1748.

1975: The evolution of drop spectra due to condensation, coalescence and breakup. *J. Atmos. Sci.*, **32**, 965–973.

Yuan, T., Z. Li, R. Zhang, and J. Fan, 2008: Increase of cloud droplet size with aerosol optical depth: An observation and modeling study. *J. Geophys. Res.*, **113**, D04201, doi: 10.1029/2007JD008632.

Yuan, T., L.A. Remer, K.E. Pickering, and H. Yu. 2011: Observational evidence of aerosol enhancement of lightning activity and convective invigoration. *Geophys. Res. Lett.*, **38**, L04701, doi: 10.1029/2010GL046052.

Yum S.S. and J.G. Hudson, 2002: Maritime/continental microphysical contrasts in stratus. *Tellus*. **B 54**, 61–73.

Yuter, S.E., D. Kingsmill, L.B. Nance, and M. Löffler-Mang, 2006: Observations of precipitation size and fall speed characteristics within coexisting rain and wet snow. *J. Appl. Meteorol. Clim.*, **45**, 1450–1464.

Zhang, D.-L., K. Gao, and D.B. Parsons, 1989: Numerical simulation of an intense squall line during 10–11 June 1985 PRE-STORM, Part 1: Model verification, *Mon. Weather Rev.*, **117**, 960–994.

Appendix A Tensors

Tensors allow one to represent numerous equations in compact form that simplify writing and transformation of these equations. In 3D space the first-rank tensor is the vector, having 3 components $\mathbf{u} = u_i$, $i = 1,2,3$. The second-rank tensor A_{ij}, $i,j = 1,2,3$ is 3×3 square matrix, having 9 components. The third-rank tensor $B_{ijk}, i, j, k = 1, 2, 3$ is $3 \times 3 \times 3$ cubic matrix, having 27 components. The main rule of tensor algebra is summation with respect to repeating indexes. For example, scalar multiplication of two vectors a_i and b_i is written as $a_i b_i = a_1 b_1 + a_2 b_2 + a_3 b_3$.

Elementary Tensors

Two elementary tensors are widely used in equations written in tensor form: (a) Kronecker symbol δ_{ij} is unit symmetric second-rank tensor components of which $\delta_{11} = \delta_{22} = \delta_{33} = 1$ and other components are equal to zero; (b) Levi-Civita symbol ε_{ijk} is a unit third-rank tensor anti-symmetric with respect to each pair of indexes, which components are $\varepsilon_{123} = \varepsilon_{312} = \varepsilon_{231} = 1$, $\varepsilon_{132} = \varepsilon_{321} = \varepsilon_{213} = -1$, and other components are equal to zero.

Table A.1 Differential operators in tensor form

Operator	Divergence	Gradient	Rotor			Laplace
Vector form	$div(\mathbf{u}) = \nabla \cdot \mathbf{u}$	$grad(\Phi) = \nabla\Phi$	$rot(\mathbf{u}) = \nabla \times \mathbf{u}$	$(\mathbf{u} \cdot \nabla)\Phi$	$(\mathbf{v} \cdot \nabla)\mathbf{u}$	$\Delta\mathbf{u} = \nabla^2\mathbf{u}$
Tensor form	$\frac{\partial u_i}{\partial x_i}$	$\frac{\partial \Phi}{\partial x_i}$	$\varepsilon_{ijk} \frac{\partial u_k}{\partial x_j}$	$u_j \frac{\partial \Phi}{\partial x_j}$	$v_j \frac{\partial u_i}{\partial x_j}$	$\frac{\partial^2 u_i}{\partial x_j^2}$

Appendix B Collision Efficiency between Drops and Turbulent Enhancement Factor

Table B1 The collision efficiencies of drops with radii below 40 μm at the 1000-mb level (from Pinsky et al., 2001; © American Meteorological Society. Used with permission).

2 μm	3 μm	4 μm	5 μm	6 μm	7 μm	8 μm	9 μm	10 μm	11 μm	12 μm	13 μm	14 μm	15 μm	16 μm	17 μm	18 μm	19 μm	20 μm	21 μm	
0.02	0.0144	0.0105	0.0079	0.0063	0.0051	0.004	0.0033	0.0029	0.0023	0.0021	0.0018	0.0016	0.0013	0.0012	0.001	0.0009	0.0009	0.0008	0.0008	**1 μm**
	0.0266	0.0229	0.019	0.0159	0.0135	0.0113	0.0097	0.0082	0.0069	0.006	0.0054	0.0045	0.004	0.0038	0.0033	0.0029	0.0027	0.0025	0.0023	**2 μm**
		0.0292	0.026	0.0229	0.0201	0.0174	0.0154	0.0131	0.0113	0.0101	0.009	0.0079	0.0072	0.0066	0.006	0.0054	0.0051	0.0048	0.0043	**3 μm**
			0.0306	0.0279	0.0247	0.0223	0.0195	0.0174	0.0154	0.014	0.0126	0.0113	0.0105	0.0097	0.0093	0.0086	0.0082	0.0079	0.0079	**4 μm**
				0.0306	0.0279	0.0253	0.0229	0.0206	0.0184	0.0174	0.0159	0.0149	0.0144	0.014	0.0135	0.0135	0.0135	0.0135	0.014	**5 μm**
					0.0306	0.0285	0.026	0.0229	0.0212	0.0201	0.0195	0.019	0.019	0.019	0.0195	0.0201	0.0212	0.0229	0.0253	**6 μm**
						0.0306	0.0279	0.0247	0.0229	0.0223	0.0217	0.0223	0.0229	0.0241	0.0266	0.0292	0.0333	0.0393	0.0474	**7 μm**
							0.0306	0.026	0.0241	0.0235	0.0235	0.0241	0.026	0.0292	0.0333	0.0401	0.0491	0.062	0.0764	**8 μm**
								0.0266	0.0247	0.0235	0.0235	0.0253	0.0279	0.0326	0.0393	0.05	0.066	0.0841	0.1058	**9 μm**
									0.0279	0.0253	0.0247	0.026	0.0285	0.0341	0.0433	0.0582	0.0786	0.1032	0.1313	**10 μm**
										0.0272	0.0253	0.0253	0.0285	0.0341	0.0441	0.061	0.0853	0.1162	0.1488	**11 μm**
											0.0266	0.0253	0.0272	0.0319	0.0417	0.0582	0.0864	0.1202	0.158	**12 μm**
												0.0266	0.026	0.0292	0.037	0.0527	0.0797	0.1175	0.1611	**13 μm**
													0.0266	0.0272	0.0326	0.0441	0.068	0.107	0.1534	**14 μm**
														0.0272	0.0285	0.0363	0.0545	0.0899	0.1384	**15 μm**
															0.0279	0.0312	0.0425	0.069	0.1162	**16 μm**
																0.0292	0.0341	0.06	0.0876	**17 μm**
																	0.0306	0.0385	0.061	**18 μm**
																		0.0333	0.0441	**19 μm**
																			0.0363	**20 μm**

22 μm	23 μm	24 μm	25 μm	26 μm	27 μm	28 μm	29 μm	30 μm	31 μm	32 μm	33 μm	34 μm	35 μm	36 μm	37 μm	38 μm	39 μm	40 μm	
0.0007	0.0007	0.0007	0.0006	0.0006	0.0006	0.0006	0.0006	0.0006	0.0007	0.0007	0.0007	0.0007	0.0007	0.0008	0.0008	0.0008	0.0008	0.0008	**1 μm**
0.0021	0.0021	0.0019	0.0019	0.0019	0.0018	0.0018	0.0018	0.0018	0.0019	0.0019	0.0019	0.0021	0.0021	0.0021	0.0023	0.0023	0.0023	0.0023	**2 μm**
0.0043	0.004	0.0038	0.0038	0.0038	0.0038	0.0038	0.004	0.004	0.0043	0.0043	0.0045	0.0048	0.0051	0.0054	0.0057	0.006	0.0063	0.0069	**3 μm**
0.0076	0.0076	0.0076	0.0076	0.0079	0.0082	0.0086	0.0093	0.0101	0.0109	0.0118	0.0135	0.0149	0.0174	0.0195	0.0229	0.0272	0.0319	0.0378	**4 μm**
0.0144	0.0148	0.0159	0.0174	0.0195	0.0217	0.0253	0.0292	0.0348	0.0417	0.05	0.0591	0.069	0.0808	0.0922	0.1045	0.1175	0.1299	0.1428	**5 μm**
0.0285	0.0333	0.0393	0.0457	0.0545	0.065	0.0755	0.0676	0.1007	0.1146	0.1299	0.1443	0.1595	0.1755	0.1923	0.206	0.2243	0.2394	0.2549	**6 μm**
0.0572	0.069	0.0819	0.0958	0.1109	0.1271	0.1428	0.1611	0.1772	0.1957	0.2134	0.2318	0.249	0.2669	0.2853	0.3023	0.3197	0.3354	0.3514	**7 μm**
0.0934	0.1122	0.1313	0.1503	0.1706	0.1906	0.2098	0.2299	0.249	0.2689	0.2895	0.3088	0.3286	0.3468	0.3654	0.3821	0.4017	0.4167	0.4345	**8 μm**
0.1299	0.1534	0.1772	0.2009	0.2225	0.2471	0.2689	0.2916	0.3131	0.3331	0.3537	0.3749	0.3943	0.4142	0.4319	0.4501	0.4686	0.4848	0.5013	**9 μm**
0.1595	0.1872	0.2152	0.2432	0.2689	0.2938	0.3197	0.3422	0.3654	0.3894	0.4092	0.4319	0.4501	0.4713	0.4903	0.5096	0.5265	0.5436	0.5581	**10 μm**
0.1821	0.2152	0.2471	0.2771	0.3066	0.3354	0.3607	3.387	0.4117	0.4345	0.4554	0.4794	0.4985	0.518	0.5379	0.5552	0.5728	0.5877	0.6058	**11 μm**
0.1974	0.2337	0.2709	0.3044	0.3354	0.3678	0.3943	0.4217	0.4475	0.4713	0.4958	0.518	0.5379	0.5581	0.5787	0.5937	0.6119	0.6273	0.6428	**12 μm**
0.2044	0.2451	0.2853	0.3241	0.3584	0.3918	0.4217	0.4501	0.4794	0.504	0.5265	0.5494	0.5728	0.5907	0.6088	0.6273	0.648	0.6617	0.6745	**13 μm**
0.2027	0.249	0.2938	0.3354	0.3749	0.4092	0.4423	0.474	0.5013	0.5293	0.5552	0.5758	0.5997	0.618	0.6366	0.6554	0.6713	0.6874	0.7004	**14 μm**
0.1923	0.2451	0.2938	0.3399	0.3821	0.4217	0.458	0.4903	0.5208	0.5494	0.5758	0.5997	0.6211	0.6428	0.6617	0.6777	0.6939	0.7103	0.7235	**15 μm**
0.1739	0.2318	0.2874	0.3376	0.3845	0.4268	0.466	0.5013	0.535	0.564	0.5907	0.6149	0.6397	0.6586	0.6777	0.6972	0.7136	0.7268	0.7436	**16 μm**
0.1458	0.2098	0.2709	0.3286	0.3797	0.4268	0.4713	0.5096	0.5436	0.5758	0.6027	0.6303	0.6522	0.6745	0.6939	0.7136	0.7302	0.7436	0.7571	**17 μm**
0.1109	0.1772	0.2451	0.3109	0.3678	0.4217	0.4686	0.5096	0.5465	0.5817	0.6119	0.6397	0.6649	0.6874	0.707	0.7235	0.7402	0.7571	0.7708	**18 μm**
0.0753	0.1384	0.2116	0.2833	0.3491	0.4092	0.4606	0.5068	0.5465	0.5847	0.6149	0.646	0.6713	0.6939	0.7169	0.7335	0.7503	0.7673	0.7811	**19 μm**
0.0509	0.0946	0.1674	0.2471	0.3197	0.387	0.4449	0.4985	0.5436	0.5817	0.618	0.6491	0.6745	0.7004	0.7235	0.7436	0.7605	0.7742	0.7915	**20 μm**
0.0409	0.061	0.1162	0.1992	0.2833	0.3584	0.4243	0.4821	0.5322	0.5758	0.6149	0.6491	0.6777	0.7037	0.7268	0.7469	0.7673	0.7811	0.795	**21 μm**
	0.0466	0.0732	0.1428	0.2318	0.3175	0.3943	0.4606	0.518	0.5669	0.6088	0.646	0.6777	0.7037	0.7302	0.7503	0.7708	0.788	0.8019	**22 μm**
		0.0536	0.0899	0.1723	0.2669	0.3537	0.4319	0.4958	0.5523	0.5967	0.6397	0.6745	0.7037	0.7302	0.7537	0.7742	0.7915	0.8054	**23 μm**
			0.062	0.1096	0.2027	0.3023	0.3918	0.466	0.5293	0.5817	0.6273	0.6681	0.7004	0.7268	0.7537	0.7742	0.7915	0.8089	**24 μm**
				0.0721	0.1327	0.2375	0.3399	0.4268	0.5013	0.561	0.6119	0.6554	0.6939	0.7235	0.7503	0.7742	0.7915	0.8089	**25 μm**
					0.0853	0.1595	0.2709	0.3749	0.4633	0.535	0.5937	0.6428	0.6842	0.7169	0.7469	0.7708	0.7915	0.8089	**26 μm**
						0.1007	0.1889	0.3088	0.4117	0.4958	0.5669	0.6211	0.6681	0.707	0.7402	0.7673	0.788	0.8089	**27 μm**
							0.1175	0.2225	0.3445	0.4475	0.5293	0.5967	0.6491	0.6939	0.7302	0.7605	0.7845	0.8054	**28 μm**
								0.137	0.2569	0.3821	0.4821	0.584	0.6273	0.6777	0.7169	0.7503	0.7776	0.8019	**29 μm**
									0.1611	0.2916	0.4192	0.518	0.5937	0.6522	0.7004	0.7402	0.7708	0.795	**30 μm**
										0.1855	0.3286	0.4554	0.5494	0.6242	0.6809	0.7235	0.7605	0.788	**31 μm**
											0.2134	0.3654	0.4903	0.5817	0.6491	0.7037	0.7469	0.7776	**32 μm**
												0.2413	0.4017	0.5236	0.6119	0.6777	0.7268	0.7639	**33 μm**
													0.273	0.4345	0.5552	0.6366	0.7004	0.7469	**34 μm**
														0.3023	0.4686	0.5847	0.6649	0.7202	**35 μm**
															0.3331	0.5013	0.6119	0.6874	**36 μm**
																0.3631	0.5293	0.6366	**37 μm**
																	0.3918	0.5552	**38 μm**
																		0.4192	**39 μm**

Table B2 The collision efficiencies of small drops and raindrops at the 1000-mb level (from Pinsky et al., 2001; © American Meteorological Society. Used with permission).

	50 μm	60 μm	70 μm	80 μm	90 μm	100 μm	110 μm	120 μm	130 μm	140 μm	150 μm
1 μm	0.0009	0.0008	0.0006	0.0005	0.0004	0.0004	0.0003	0.0003	0.0003	0.0003	0.0003
2 μm	0.0027	0.0025	0.0023	0.0021	0.0019	0.0018	0.0018	0.0018	0.0018	0.0019	0.0021
3 μm	0.0122	0.0223	0.0401	0.0601	0.0786	0.0946	0.1109	0.1243	0.1384	0.1518	0.1642
4 μm	0.1148	0.1772	0.2188	0.249	0.273	0.2916	0.3088	0.3241	0.3376	0.3514	0.3631
5 μm	0.2589	0.3308	0.3773	0.4067	0.4294	0.4475	0.4633	0.4767	0.4903	0.5013	0.5124
6 μm	0.3821	0.4554	0.4985	0.5265	0.5465	0.564	0.5787	0.5907	0.5997	0.6119	0.6211
7 μm	0.4794	0.5523	0.5907	0.618	0.6366	0.6491	0.6617	0.6713	0.6809	0.6907	0.6972
8 μm	0.5581	0.6273	0.6617	0.6842	0.7004	0.7136	0.7235	0.7335	0.7402	0.7469	0.7571
9 μm	0.6211	0.6842	0.7169	0.7368	0.7503	0.7639	0.7708	0.7776	0.7845	0.7915	0.7984
10 μm	0.6713	0.7302	0.7605	0.7776	0.7915	0.8019	0.8089	0.816	0.8195	0.8266	0.8302
11 μm	0.7136	0.7673	0.795	0.8089	0.8231	0.8302	0.8373	0.8409	0.8481	0.8517	0.8553
12 μm	0.7469	0.795	0.8195	0.8373	0.8445	0.8517	0.8589	0.8661	0.8698	0.8734	0.8771
13 μm	0.7742	0.8195	0.8445	0.8553	0.8661	0.8734	0.8771	0.8807	0.8881	0.8918	0.8918
14 μm	0.7984	0.8409	0.8625	0.8734	0.8807	0.8881	0.8918	0.8955	0.8992	0.9029	0.9066
15 μm	0.816	0.8553	0.8771	0.8881	0.8955	0.8992	0.9066	0.9103	0.9103	0.914	0.9178
16 μm	0.8302	0.8698	0.8881	0.8992	0.9066	0.9103	0.914	0.9178	0.9215	0.9253	0.9253
17 μm	0.8445	0.8807	0.8992	0.9103	0.914	0.9215	0.9253	0.9253	0.929	0.9328	0.9328
18 μm	0.8553	0.8918	0.9066	0.9178	0.9215	0.929	0.9328	0.9328	0.9366	0.9404	0.9404
19 μm	0.8661	0.8992	0.914	0.9253	0.929	0.9328	0.9366	0.9404	0.9404	0.9442	0.948
20 μm	0.8734	0.9066	0.9215	0.929	0.9366	0.9404	0.9442	0.9442	0.948	0.948	0.9518
21 μm	0.8807	0.914	0.929	0.9366	0.9404	0.9442	0.948	0.948	0.9518	0.9518	0.9556
22 μm	0.8881	0.9178	0.9328	0.9404	0.9442	0.948	0.9518	0.9518	0.9556	0.9556	0.9594
23 μm	0.8918	0.9215	0.9366	0.9442	0.948	0.9518	0.9556	0.9556	0.9594	0.9594	0.9632
24 μm	0.8955	0.929	0.9404	0.948	0.9518	0.9556	0.9594	0.9594	0.9632	0.9632	0.9632
25 μm	0.8992	0.9328	0.9442	0.9518	0.9556	0.9594	0.9594	0.9632	0.9632	0.9671	0.9671
26 μm	0.9029	0.9328	0.948	0.9556	0.9594	0.9594	0.9632	0.9632	0.9671	0.9671	0.9709
27 μm	0.9066	0.9366	0.948	0.9556	0.9594	0.9632	0.9671	0.9671	0.9671	0.9709	0.9709
28 μm	0.9103	0.9404	0.9518	0.9594	0.9632	0.9671	0.9671	0.9709	0.9709	0.9709	0.9748
29 μm	0.9103	0.9404	0.9556	0.9594	0.9632	0.9671	0.9709	0.9709	0.9709	0.9748	0.9748
30 μm	0.9103	0.9442	0.9556	0.9632	0.9671	0.9671	0.9709	0.9709	0.9748	0.9748	0.9748
31 μm	0.914	0.9442	0.9556	0.9632	0.9671	0.9709	0.9709	0.9748	0.9748	0.9748	0.9786
32 μm	0.914	0.948	0.9594	0.9632	0.9671	0.9709	0.9748	0.9748	0.9748	0.9786	0.9786
33 μm	0.914	0.948	0.9594	0.9671	0.9709	0.9709	0.9748	0.9748	0.9786	0.9786	0.9786
34 μm	0.914	0.948	0.9594	0.9671	0.9709	0.9748	0.9748	0.9786	0.9786	0.9786	0.9786
35 μm	0.914	0.948	0.9632	0.9671	0.9709	0.9748	0.9786	0.9786	0.9786	0.9786	0.9825
36 μm	0.9103	0.948	0.9632	0.9709	0.9748	0.9748	0.9786	0.9786	0.9786	0.9825	0.9825
37 μm	0.9103	0.9518	0.9632	0.9709	0.9748	0.9748	0.9786	0.9786	0.9825	0.9825	0.9825
38 μm	0.9066	0.9518	0.9632	0.9709	0.9748	0.9786	0.9786	0.0786	0.9825	0.9825	0.9825
39 μm	0.9029	0.9518	0.9632	0.9709	0.9748	0.9786	0.9786	0.9825	0.9825	0.9825	0.9825

160 μm	170 μm	180 μm	190 μm	200 μm	210 μm	220 μm	230 μm	240 μm	250 μm	
0.0003	0.0003	0.0002	0.0002	0.0002	0.0002	0.0002	0.0002	0.0002	0.0002	**1 μm**
0.0023	0.0027	0.0035	0.0048	0.0072	0.0113	0.0184	0.0299	0.0425	0.0572	**2 μm**
0.1772	0.1889	0.2027	0.217	0.2318	0.2451	0.2609	0.2771	0.2959	0.3153	**3 μm**
0.3749	0.387	0.4017	0.4142	0.4294	0.4423	0.4554	0.4713	0.4903	0.5096	**4 μm**
0.5236	0.535	0.5465	0.5581	0.5698	0.5817	0.5937	0.6088	0.6211	0.6397	**5 μm**
0.6303	0.6397	0.6491	0.6586	0.6681	0.6777	0.6874	0.7004	0.7136	0.7268	**6 μm**
0.707	0.7136	0.7202	0.7302	0.7402	0.7469	0.7571	0.7673	0.7776	0.788	**7 μm**
0.7605	0.7673	0.7742	0.7811	0.7915	0.795	0.8054	0.8125	0.8231	0.8302	**8 μm**
0.8054	0.8089	0.816	0.8231	0.8266	0.8337	0.8409	0.8481	0.8553	0.8625	**9 μm**
0.8373	0.8409	0.8445	0.8517	0.8553	0.8625	0.8661	0.8734	0.8807	0.8844	**10 μm**
0.8625	0.8661	0.8698	0.8734	0.8771	0.8844	0.8881	0.8918	0.8992	0.9029	**11 μm**
0.8807	0.8844	0.8881	0.8918	0.8955	0.8992	0.9029	0.9066	0.914	0.9178	**12 μm**
0.8955	0.8992	0.9029	0.9066	0.9103	0.914	0.9178	0.9215	0.9253	0.929	**13 μm**
0.9103	0.9103	0.914	0.9178	0.9215	0.9253	0.9253	0.929	0.9328	0.9366	**14 μm**
0.9215	0.9215	0.9253	0.929	0.929	0.9328	0.9366	0.9404	0.9404	0.9442	**15 μm**
0.929	0.9328	0.9328	0.9366	0.9366	0.9404	0.9442	0.9442	0.948	0.9518	**16 μm**
0.9366	0.9366	0.9404	0.9442	0.9442	0.948	0.948	0.9518	0.9556	0.9556	**17 μm**
0.9442	0.9442	0.948	0.948	0.9518	0.9518	0.9556	0.9556	0.9594	0.9594	**18 μm**
0.948	0.948	0.9518	0.9518	0.9556	0.9556	0.9594	0.9594	0.9632	0.9632	**19 μm**
0.9518	0.9556	0.9556	0.9556	0.9594	0.9594	0.9632	0.9632	0.9671	0.9671	**20 μm**
0.9556	0.9594	0.9594	0.9594	0.9632	0.9632	0.9671	0.9671	0.9671	0.9709	**21 μm**
0.9594	0.9594	0.9632	0.9632	0.9671	0.9671	0.9671	0.9709	0.9709	0.9748	**22 μm**
0.9632	0.9632	0.9671	0.9671	0.9671	0.9709	0.9709	0.9709	0.9748	0.9748	**23 μm**
0.9671	0.9671	0.9671	0.9709	0.9709	0.9709	0.9709	0.9748	0.9748	0.9786	**24 μm**
0.9671	0.9709	0.9709	0.9709	0.9709	0.9748	0.9748	0.9748	0.9786	0.9786	**25 μm**
0.9709	0.9709	0.9709	0.9748	0.9748	0.9748	0.9748	0.9786	0.9786	0.9786	**26 μm**
0.9709	0.9748	0.9748	0.9748	0.9748	0.9786	0.9786	0.9786	0.9786	0.9825	**27 μm**
0.9748	0.9748	0.9748	0.9748	0.9786	0.9786	0.9786	0.9786	0.9825	0.9825	**28 μm**
0.9748	0.9748	0.9786	0.9786	0.9786	0.9786	0.9825	0.9825	0.9825	0.9825	**29 μm**
0.9786	0.9786	0.9786	0.9786	0.9786	0.9825	0.9825	0.9825	0.9825	0.9864	**30 μm**
0.9786	0.9786	0.9786	0.9786	0.9825	0.9825	0.9825	0.9825	0.9864	0.9864	**31 μm**
0.9786	0.9786	0.9825	0.9825	0.9825	0.9825	0.9825	0.9864	0.9864	0.9864	**32 μm**
0.9786	0.9825	0.9825	0.9825	0.9825	0.9825	0.9864	0.9864	0.9864	0.9864	**33 μm**
0.9825	0.9825	0.9825	0.9825	0.9825	0.9864	0.9864	0.9864	0.9864	0.9864	**34 μm**
0.9825	0.9825	0.9825	0.9825	0.9864	0.9864	0.9864	0.9864	0.9864	0.9864	**35 μm**
0.9825	0.9825	0.9825	0.9864	0.9864	0.9864	0.9864	0.9864	0.9864	0.9903	**36 μm**
0.9825	0.9825	0.9864	0.9864	0.9864	0.9864	0.9864	0.9864	0.9903	0.9903	**37 μm**
0.9825	0.9864	0.9864	0.9864	0.9864	0.9864	0.9864	0.9903	0.9903	0.9903	**38 μm**
0.9864	0.9864	0.9864	0.9864	0.9864	0.9864	0.9903	0.9903	0.9903	0.9903	**39 μm**

Table B3 The collision efficiencies of comparable size drops at p = 1000 mb (from Pinsky et al., 2001; © American Meteorological Society. Used with permission).

R===>	50 µm	60 µm	70 µm	80 µm	90 µm	100 µm	110 µm	120 µm	130 µm	140 µm	150 µm
R-1 µm	0.5948	0.6984	0.8262	0.9907	1.1958	1.4446	1.7477	2.0967	2.5051	2.9701	3.4636
R-2 µm	0.7107	0.7661	0.8103	0.8666	0.9361	1.0201	1.1201	1.241	1.3784	1.5339	1.7016
R-2 µm	0.7763	0.8182	0.8449	0.8721	0.908	0.9475	0.9995	1.0589	1.1294	1.2118	1.3005
R-4 µm	0.8155	0.8503	0.8693	0.8885	0.9108	0.9333	0.9647	0.9995	1.0439	1.0924	1.1482
R-5 µm	0.8422	0.8748	0.8885	0.9024	0.9164	0.9333	0.9532	0.9762	1.0054	1.038	1.0741
R-6 µm	0.8611	0.8913	0.9052	0.9136	0.9248	0.9361	0.9503	0.9676	0.9878	1.0113	1.038
R-7 µm	0.8748	0.9052	0.9164	0.9248	0.9333	0.9418	0.9532	0.9647	0.9791	0.9966	1.0172
R-8 µm	0.8858	0.9136	0.9248	0.9305	0.9389	0.9446	0.9532	0.9647	0.9762	0.9878	1.0024
R-9 µm	0.8913	0.922	0.9305	0.9361	0.9418	0.9503	0.9561	0.9647	0.9733	0.982	0.9966
R-10 µm	0.8996	0.9276	0.9361	0.9418	0.9475	0.9532	0.9589	0.9647	0.9704	0.9791	0.9907
R-11 µm	0.9024	0.9333	0.9418	0.9475	0.9503	0.9561	0.9589	0.9647	0.9704	0.9791	0.9878
R-12 µm	0.908	0.9361	0.9446	0.9503	0.9532	0.9589	0.9618	0.9676	0.9733	0.9791	0.9849
R-13 µm	0.908	0.9389	0.9475	0.9532	0.9561	0.9618	0.9647	0.9676	0.9733	0.9791	0.9849
R-14 µm	0.9108	0.9418	0.9503	0.9561	0.9589	0.9618	0.9647	0.9704	0.9733	0.9791	0.982
R-15 µm	0.9136	0.9446	0.9532	0.9589	0.9618	0.9647	0.9676	0.9704	0.9733	0.9791	0.982
R-16 µm	0.9136	0.9446	0.9561	0.9618	0.9647	0.9676	0.9676	0.9704	0.9762	0.9791	0.982
R-17 µm	0.9136	0.9475	0.9589	0.9618	0.9647	0.9676	0.9704	0.9733	0.9762	0.9791	0.982
R-18 µm	0.9136	0.9475	0.9589	0.9647	0.9676	0.9704	0.9704	0.9733	0.9762	0.9791	0.982
R-19 µm	0.9136	0.9503	0.9618	0.9647	0.9676	0.9704	0.9733	0.9733	0.9762	0.9791	0.982
R-20 µm	0.9108	0.9503	0.9618	0.9676	0.9704	0.9704	0.9733	0.9762	0.9791	0.9791	0.982
R-21 µm	0.9108	0.9503	0.9618	0.9676	0.9704	0.9733	0.9733	0.9762	0.9791	0.9791	0.982
R-22 µm	0.908	0.9503	0.9647	0.9676	0.9704	0.9733	0.9762	0.9762	0.9791	0.982	0.982
R-23 µm	0.9052	0.9503	0.9647	0.9704	0.9733	0.9733	0.9762	0.9791	0.9791	0.982	0.982
R-24 µm	0.9052	0.9503	0.9647	0.9704	0.9733	0.9762	0.9762	0.9791	0.9791	0.982	0.9849
R-25 µm	0.8996	0.9503	0.9647	0.9704	0.9733	0.9762	0.9791	0.9791	0.982	0.982	0.9849
R-26 µm	0.8968	0.9475	0.9647	0.9704	0.9733	0.9762	0.9791	0.9791	0.982	0.982	0.9849
R-27 µm	0.8913	0.9475	0.9647	0.9733	0.9762	0.9762	0.9791	0.9791	0.982	0.982	0.9849
R-28 µm	0.8885	0.9475	0.9647	0.9733	0.9762	0.9791	0.9791	0.982	0.982	0.9849	0.9849
R-29 µm	0.8803	0.9446	0.9647	0.9733	0.9762	0.9791	0.9791	0.982	0.982	0.9849	0.9849
R-30 µm	0.8748	0.9418	0.9647	0.9733	0.9762	0.9791	0.9791	0.982	0.982	0.9849	0.9849
R-31 µm	0.8666	0.9418	0.9647	0.9733	0.9762	0.9791	0.982	0.982	0.982	0.9849	0.9849
R-32 µm	0.8557	0.9389	0.9647	0.9733	0.9762	0.9791	0.982	0.982	0.9849	0.9849	0.9849
R-33 µm	0.8449	0.9361	0.9647	0.9733	0.9791	0.9791	0.982	0.982	0.9849	0.9849	0.9849
R-34 µm	0.8315	0.9333	0.9618	0.9733	0.9791	0.9791	0.982	0.982	0.9849	0.9849	0.9878
R-35 µm	0.8155	0.9305	0.9618	0.9733	0.9791	0.982	0.982	0.9849	0.9849	0.9849	0.9878
R-36 µm	0.7971	0.9276	0.9618	0.9733	0.9791	0.982	0.982	0.9849	0.9849	0.9849	0.9878
R-37 µm	0.7738	0.922	0.9589	0.9733	0.9791	0.982	0.982	0.9849	0.9849	0.9849	0.9878
R-38 µm	0.7482	0.9192	0.9589	0.9733	0.9791	0.982	0.982	0.9849	0.9849	0.9878	0.9878

160 μm	170 μm	180 μm	190 μm	200 μm	210 μm	220 μm	230 μm	240 μm	250 μm	<===R
4.001	4.6022	5.3195	6.1322	7.0259	7.9555	9.0059	10.3541	11.9308	13.9237	**R-1 μm**
1.8819	2.0755	2.3053	2.5611	2.8353	3.108	3.4039	3.7586	4.1488	4.5771	**R-2 μm**
1.3922	1.4979	1.6148	1.7516	1.898	2.0461	2.1997	2.386	2.5893	2.8107	**R-3 μm**
1.2054	1.2706	1.3443	1.4305	1.5231	1.6148	1.7169	1.834	1.9631	2.1052	**R-4 μm**
1.1139	1.1576	1.2086	1.2673	1.3341	1.3992	1.4693	1.5465	1.641	1.74	**R-5 μm**
1.065	1.0985	1.1357	1.1766	1.2248	1.2739	1.324	1.3853	1.4516	1.5267	**R-6 μm**
1.038	1.062	1.0893	1.1232	1.1576	1.1958	1.2345	1.2805	1.3341	1.3922	**R-7 μm**
1.0201	1.038	1.0589	1.0863	1.1139	1.145	1.1766	1.2118	1.2541	1.3005	**R-8 μm**
1.0083	1.0231	1.0409	1.062	1.0832	1.1077	1.1325	1.1639	1.199	1.2345	**R-9 μm**
1.0024	1.0142	1.029	1.0439	1.062	1.0832	1.1047	1.1294	1.1576	1.1894	**R-10 μm**
0.9966	1.0054	1.0172	1.032	1.0469	1.065	1.0832	1.1016	1.1263	1.1513	**R-11 μm**
0.9937	1.0024	1.0113	1.0231	1.035	1.0499	1.065	1.0832	1.1016	1.1263	**R-12 μm**
0.9907	0.9966	1.0054	1.0172	1.026	1.0409	1.0529	1.068	1.0832	1.1047	**R-13 μm**
0.9878	0.9937	1.0024	1.0113	1.0201	1.032	1.0439	1.0559	1.071	1.0863	**R-14 μm**
0.9878	0.9937	0.9995	1.0083	1.0142	1.026	1.035	1.0469	1.0589	1.0741	**R-15 μm**
0.9878	0.9907	0.9966	1.0024	1.0113	1.0201	1.029	1.038	1.0499	1.062	**R-16 μm**
0.9849	0.9907	0.9966	1.0024	1.0083	1.0142	1.0231	1.032	1.0409	1.0529	**R-17 μm**
0.9849	0.9907	0.9937	0.9995	1.0054	1.0113	1.0201	1.026	1.035	1.0469	**R-18 μm**
0.9849	0.9878	0.9937	0.9966	1.0024	1.0083	1.0142	1.0231	1.032	1.0409	**R-19 μm**
0.9849	0.9878	0.9937	0.9966	1.0024	1.0054	1.0113	1.0201	1.026	1.035	**R-20 μm**
0.9849	0.9878	0.9907	0.9966	0.9995	1.0054	1.0113	1.0172	1.0231	1.029	**R-21 μm**
0.9849	0.9878	0.9907	0.9937	0.9995	1.0024	1.0083	1.0142	1.0201	1.026	**R-22 μm**
0.9849	0.9878	0.9907	0.9937	0.9966	1.0024	1.0054	1.0113	1.0172	1.0231	**R-23 μm**
0.9849	0.9878	0.9907	0.9937	0.9966	0.9995	1.0054	1.0083	1.0142	1.0201	**R-24 μm**
0.9849	0.9878	0.9907	0.9937	0.9966	0.9995	1.0024	1.0083	1.0113	1.0172	**R-25 μm**
0.9849	0.9878	0.9907	0.9937	0.9966	0.9995	1.0024	1.0054	1.0113	1.0172	**R-26 μm**
0.9849	0.9878	0.9907	0.9937	0.9966	0.9995	1.0024	1.0054	1.0083	1.0142	**R-27 μm**
0.9878	0.9878	0.9907	0.9937	0.9937	0.9966	0.9995	1.0024	1.0083	1.0113	**R-28 μm**
0.9878	0.9878	0.9907	0.9907	0.9937	0.9966	0.9995	1.0024	1.0054	1.0113	**R-29 μm**
0.9878	0.9878	0.9907	0.9907	0.9937	0.9966	0.9995	1.0024	1.0054	1.0083	**R-30 μm**
0.9878	0.9878	0.9907	0.9907	0.9937	0.9966	0.9995	1.0024	1.0054	1.0083	**R-31 μm**
0.9878	0.9878	0.9907	0.9907	0.9937	0.9966	0.9995	0.9995	1.0024	1.0083	**R-32 μm**
0.9878	0.9878	0.9907	0.9907	0.9937	0.9966	0.9966	0.9995	1.0024	1.0054	**R-33 μm**
0.9878	0.9878	0.9907	0.9907	0.9937	0.9966	0.9966	0.9995	1.0024	1.0054	**R-34 μm**
0.9878	0.9878	0.9907	0.9907	0.9937	0.9937	0.9966	0.9995	1.0024	1.0054	**R-35 μm**
0.9878	0.9878	0.9907	0.9907	0.9937	0.9937	0.9966	0.9995	1.0024	1.0054	**R-36 μm**
0.9878	0.9878	0.9907	0.9907	0.9937	0.9937	0.9966	0.9995	0.9995	1.0024	**R-37 μm**
0.9878	0.9907	0.9907	0.9907	0.9937	0.9937	0.9966	0.9995	0.9995	1.0024	**R-38 μm**

Table B4 Turbulent enhancement factor for $\varepsilon = 0.001\ m^2s^{-3}$, $Re_\lambda = 5 \times 10^3$ and $p = 1000$ mb (The conditions are suitable for stratocumulus clouds) (from Pinsky et al., 2008; © American Meteorological Society. Used with permission).

	1 μm	**2 μm**	**3 μm**	**4 μm**	**5 μm**	**6 μm**	**7 μm**	**8 μm**	**9 μm**	**10 μm**	**11 μm**	**12 μm**	**13 μm**	**14 μm**	**15 μm**	**16 μm**	**17 μm**	**18 μm**	**19 μm**	**20 μm**	**21 μm**
1 μm	1.595	1.719	1.43	1.363	1.366	1.39	1.387	1.404	1.451	1.539	1.559	1.635	1.671	1.722	1.736	1.872	1.951	2.038	2.02	2.169	2.184
2 μm	1.719	1.595	1.471	1.223	1.177	1.151	1.146	1.148	1.13	1.14	1.149	1.149	1.142	1.153	1.149	1.162	1.163	1.21	1.233	1.211	1.254
3 μm	1.43	1.471	1.499	1.427	1.198	1.137	1.12	1.113	1.111	1.107	1.11	1.089	1.109	1.095	1.095	1.083	1.085	1.094	1.081	1.079	1.09
4 μm	1.363	1.223	1.427	1.416	1.404	1.187	1.13	1.12	1.096	1.094	1.101	1.089	1.092	1.101	1.088	1.087	1.072	1.082	1.044	1.047	1.033
5 μm	1.366	1.177	1.198	1.404	1.399	1.394	1.182	1.122	1.114	1.094	1.095	1.086	1.082	1.09	1.1077	1.072	1.076	1.069	1.07	1.056	1.045
6 μm	1.39	1.151	1.137	1.187	1.394	1.392	1.39	1.171	1.125	1.107	1.088	1.086	1.079	1.069	1.059	1.044	1.028	1.012	0.998	0.978	0.97
7 μm	1.387	1.146	1.12	1.13	1.182	1.39	1.394	1.398	1.176	1.127	1.103	1.097	1.086	1.071	1.064	1.056	1.05	1.035	1.008	0.961	0.946
8 μm	1.404	1.148	1.113	1.12	1.122	1.171	1.398	1.393	1.388	1.185	1.128	1.098	1.089	1.082	1.082	1.069	1.055	1.019	1.009	0.99	0.968
9 μm	1.451	1.13	1.111	1.096	1.114	1.125	1.176	1.388	1.435	1.483	1.192	1.124	1.091	1.089	1.086	1.072	1.045	1.043	1.025	1.017	0.997
10 μm	1.539	1.14	1.107	1.094	1.094	1.107	1.127	1.185	1.483	1.467	1.452	1.174	1.124	1.096	1.089	1.082	1.071	1.069	1.04	1.025	1.026
11 μm	1.559	1.149	1.11	1.101	1.095	1.088	1.103	1.128	1.192	1.452	1.452	1.453	1.172	1.11	1.091	1.087	1.082	1.08	1.066	1.046	1.04
12 μm	1.635	1.149	1.089	1.089	1.086	1.086	1.097	1.098	1.124	1.174	1.453	1.456	1.46	1.167	1.115	1.09	1.078	1.084	1.08	1.068	1.057
13 μm	1.671	1.142	1.109	1.092	1.082	1.079	1.086	1.089	1.091	1.124	1.172	1.46	1.464	1.468	1.169	1.109	1.093	1.093	1.081	1.081	1.073
14 μm	1.722	1.153	1.095	1.101	1.09	1.069	1.071	1.082	1.089	1.096	1.11	1.167	1.468	1.462	1.455	1.162	1.103	1.087	1.082	1.076	1.08
15 μm	1.736	1.149	1.095	1.088	1.077	1.059	1.064	1.082	1.086	1.089	1.091	1.115	1.169	1.455	1.457	1.459	1.16	1.102	1.09	1.089	1.083
16 μm	1.872	1.162	1.083	1.087	1.072	1.044	1.056	1.069	1.072	1.082	1.087	1.09	1.109	1.162	1.459	1.469	1.479	1.147	1.109	1.09	1.083
17 μm	1.951	1.163	1.085	1.072	1.076	1.028	1.05	1.055	1.045	1.071	1.082	1.078	10.93	1.103	1.16	1.479	1.477	1.476	1.155	1.105	1.088
18 μm	2.038	1.21	1.094	1.082	1.069	1.012	1.035	1.019	1.043	1.069	1.08	1.084	1.093	1.087	1.102	1.147	1.476	1.48	1.484	1.153	1.103
19 μm	2.02	1.233	1.081	1.044	1.07	0.998	1.008	1.009	1.025	1.04	1.066	1.08	1.081	1.082	1.09	1.109	1.155	1.484	1.5	1.516	1.144
20 μm	2.169	1.211	1.079	1.047	1.056	0.978	0.961	0.99	1.017	1.025	1.046	1.068	1.081	1.076	1.089	1.09	1.105	1.153	1.516	1.521	1.525
21 μm	2.184	1.254	1.09	1.033	1.045	0.97	0.946	0.968	0.997	1.026	1.04	1.057	1.073	1.08	1.083	1.083	1.088	1.103	1.144	1.525	1.521

Table B5 Turbulent enhancement factor for $\varepsilon = 0.02\ m^2s^{-3}$, $Re_\lambda = 2 \times 10^4$ and $p = 1000$ mb (The conditions are suitable for cumulus clouds) (from Pinsky et al., 2008; © American Meteorological Society. Used with permission).

	1 μm	**2 μm**	**3 μm**	**4 μm**	**5 μm**	**6 μm**	**7 μm**	**8 μm**	**9 μm**	**10 μm**	**11 μm**	**12 μm**	**13 μm**	**14 μm**	**15 μm**	**16 μm**	**17 μm**	**18 μm**	**19 μm**	**20 μm**	**21 μm**
1 μm	4.53	5.075	3.859	3.602	3.64	3.788	3.864	4.018	4.22	4.646	4.809	5.17	5.398	5.729	5.91	6.513	6.853	7.291	7.355	7.961	8.146
2 μm	5.075	4.53	3.985	2.666	2.3	2.137	2.063	2.046	2.028	2.056	2.102	2.155	2.187	2.261	2.311	2.405	2.426	2.578	2.686	2.683	2.844
3 μm	3.859	3.985	3.867	3.748	2.437	2	1.831	1.729	1.664	1.64	1.632	1.6	1.625	1.61	1.624	1.62	1.626	1.667	1.668	1.684	1.725
4 μm	3.602	2.666	3.748	3.698	3.649	2.342	1.903	1.715	1.579	1.522	1.494	1.457	1.449	1.447	1.444	1.458	1.458	1.511	1.518	1.569	1.607
5 μm	3.64	2.3	2.437	3.649	3.633	3.617	2.305	1.844	1.649	1.527	1.468	1.426	1.41	1.427	1.427	1.452	1.502	1.557	1.631	1.693	1.777
6 μm	3.788	2.137	2	2.342	3.617	3.616	3.616	2.279	1.812	1.619	1.492	1.439	1.418	1.429	1.465	1.514	1.575	1.641	1.714	1.763	1.812
7 μm	3.864	2.063	1.831	1.903	2.305	3.616	3.631	3.647	2.281	1.838	1.596	1.498	1.458	1.464	1.514	1.58	1.65	1.699	1.723	1.69	1.64
8 μm	4.018	2.046	1.729	1.715	1.844	2.279	3.647	3.643	3.638	2.39	1.851	1.601	1.509	1.51	1.569	1.632	1.699	1.71	1.658	1.582	1.509
9 μm	4.22	2.028	1.664	1.579	1.649	1.812	2.281	3.638	3.916	4.193	2.488	1.871	1.612	1.562	1.604	1.666	1.711	1.686	1.614	1.528	1.45
10 μm	4.646	2.056	1.64	1.522	1.527	1.619	1.838	2.39	4.193	4.117	4.04	2.417	1.845	1.632	1.619	1.685	1.731	1.686	1.584	1.491	1.427
11 μm	4.809	2.102	1.632	1.494	1.468	1.492	1.596	1.851	2.488	4.04	4.079	4.119	2.428	1.833	1.667	1.696	1.751	1.701	1.594	1.489	1.424
12 μm	5.17	2.155	1.6	1.457	1.426	1.439	1.498	1.601	1.871	2.417	4.119	4.162	4.205	2.425	1.852	1.72	1.756	1.732	1.622	1.511	1.431
13 μm	5.398	2.187	1.625	1.449	1.41	1.418	1.458	1.509	1.612	1.845	2.428	4.205	4.249	4.293	2.443	1.867	1.785	1.789	1.68	1.548	1.455
14 μm	5.729	2.261	1.61	1.447	1.427	1.429	1.464	1.51	1.562	1.632	1.833	2.425	4.293	4.313	4.332	2.45	1.899	1.83	1.758	1.601	1.488
15 μm	5.91	2.311	1.624	1.444	1.427	1.465	1.514	1.569	1.604	1.619	1.667	1.852	2.443	4.332	4.378	4.425	2.466	1.945	1.856	1.704	1.544
16 μm	6.513	2.405	1.62	1.458	1.452	1.514	1.58	1.632	1.666	1.685	1.696	1.72	1.867	2.45	4.425	4.504	4.582	2.48	2.015	1.845	1.639
17 μm	6.853	2.426	1.626	1.458	1.502	1.575	1.65	1.699	1.711	1.731	1.751	1.756	1.785	1.899	2.466	4.582	4.64	4.697	2.545	2.062	1.806
18 μm	7.291	2.578	1.667	1.511	1.557	1.641	1.699	1.71	1.686	1.686	1.701	1.732	1.789	1.83	1.945	2.48	4.697	4.794	4.891	2.612	2.085
19 μm	7.355	2.686	1.668	1.518	1.631	1.714	1.723	1.658	1.614	1.584	1.594	1.622	1.68	1.758	1.856	2.015	2.545	4.891	5.034	5.178	2.677
20 μm	7.961	2.683	1.684	1.569	1.693	1.763	1.69	1.582	1.528	1.491	1.489	1.511	1.548	1.601	1.704	1.845	2.062	2.612	5.178	5.31	5.441
21 μm	8.146	2.844	1.725	1.607	1.777	1.812	1.64	1.509	1.45	1.427	1.424	1.431	1.455	1.488	1.544	1.639	1.806	2.085	2.677	5.441	5.31

Table B6 Turbulent enhancement factor for $\varepsilon = 0.1\ m^2s^{-3}$, $Re_\lambda = 2 \times 10^4$ and $p = 1000$ mb (The conditions are suitable for cumulonimbus clouds) (from Pinsky et al., 2008; © American Meteorological Society. Used with permission).

	1 μm	**2 μm**	**3 μm**	**4 μm**	**5 μm**	**6 μm**	**7 μm**	**8 μm**	**9 μm**	**10 μm**	**11 μm**	**12 μm**	**13 μm**	**14 μm**	**15 μm**	**16 μm**	**17 μm**	**18 μm**	**19 μm**	**20 μm**	**21 μm**
1 μm	9.509	10.72	7.719	7.312	7.438	7.774	8.02	8.403	8.937	9.804	10.33	11.12	11.66	12.51	12.91	14.25	15.06	16.07	16.16	17.51	17.96
2 μm	10.72	9.509	8.301	5.267	4.49	4.116	3.975	3.932	3.938	4.02	4.163	4.311	4.412	4.641	4.855	5.155	5.373	5.88	6.354	6.667	7.399
3 μm	7.719	8.301	8.051	7.801	4.771	3.817	3.431	3.236	3.17	3.225	3.374	3.516	3.845	4.15	4.605	5.085	5.704	6.524	7.364	8.358	9.568
4 μm	7.312	5.267	7.801	7.691	7.581	4.578	3.627	3.276	3.139	3.251	3.515	3.847	4.382	5.014	5.732	6.621	7.537	8.811	9.893	11.35	12.74
5 μm	7.438	4.49	4.771	7.581	7.54	7.498	4.526	3.591	3.369	3.46	3.824	4.338	5.016	5.874	6.718	7.681	8.765	9.816	10.94	11.95	13.02
6 μm	7.774	4.116	3.817	4.578	7.498	7.49	7.481	4.519	3.725	3.723	4.052	4.683	5.451	6.331	7.252	8.129	8.936	9.614	10.13	10.34	10.37
7 μm	8.02	3.975	3.431	3.627	4.526	7.481	7.52	7.56	4.63	4.096	4.268	4.894	5.715	6.581	7.475	8.184	8.664	8.771	8.554	7.921	7.238
8 μm	8.403	3.932	3.236	3.276	3.591	4.519	7.56	7.545	7.529	5.057	4.579	4.991	5.805	6.727	7.581	8.078	8.234	7.829	7.059	6.219	5.503
9 μm	8.937	3.938	3.17	3.139	3.369	3.725	4.63	7.529	8.157	8.785	5.539	5.222	5.826	6.775	7.607	8.003	7.823	7.087	6.169	5.33	4.669
10 μm	9.804	4.02	3.225	3.251	3.46	3.723	4.096	5.057	8.785	8.638	8.492	5.716	5.752	6.556	7.433	7.899	7.654	6.733	5.66	4.826	4.247
11 μm	10.33	4.163	3.374	3.515	3.824	4.052	4.268	4.579	5.539	8.492	8.615	8.738	6.173	6.367	7.245	7.896	7.738	6.696	5.534	4.63	4.047
12 μm	11.12	4.311	3.516	3.847	4.338	4.683	4.894	4.991	5.222	5.716	8.738	8.891	9.043	6.656	7.082	7.808	7.929	6.964	5.654	4.646	3.977
13 μm	11.66	4.412	3.845	4.382	5.016	5.451	5.715	5.805	5.826	5.752	6.173	9.043	9.223	9.404	7.275	7.676	8.198	7.596	6.084	4.838	4.046
14 μm	12.51	4.641	4.15	5.014	5.874	6.331	6.581	6.727	6.775	6.556	6.367	6.656	9.404	9.544	9.684	7.871	8.24	8.233	6.845	5.246	4.238
15 μm	12.91	4.855	4.605	5.732	6.718	7.252	7.475	7.581	7.607	7.433	7.245	7.082	7.275	9.684	9.919	10.15	8.533	8.723	7.98	6.078	4.63
16 μm	14.25	5.155	5.085	6.621	7.681	8.129	8.184	8.078	8.003	7.899	7.896	7.808	7.676	7.871	10.15	10.48	10.81	9.152	9.134	7.438	5.375
17 μm	15.06	5.373	5.704	7.537	8.765	8.936	8.664	8.234	7.823	7.654	7.738	7.929	8.198	8.24	8.533	10.18	11.12	11.44	9.96	9.243	6.777
18 μm	16.07	5.88	6.524	8.811	9.816	9.614	8.771	7.829	7.087	6.733	6.696	6.964	7.596	8.233	8.723	9.152	11.44	11.9	12.36	10.69	9.092
19 μm	16.16	6.354	7.364	9.893	10.94	10.13	8.554	7.059	6.169	5.66	5.534	5.654	6.084	6.845	7.98	9.134	9.96	12.36	12.99	13.62	11.33
20 μm	17.51	6.667	8.358	11.35	11.95	10.34	7.921	6.219	5.33	4.826	4.63	4.646	4.838	5.246	6.078	7.438	9.243	10.69	13.62	14.25	14.88
21 μm	17.96	7.399	9.568	12.74	13.02	10.37	7.238	5.503	4.669	4.247	4.047	3.977	4.046	4.238	4.63	5.375	6.777	9.092	11.33	14.88	14.25

Table B7 DNS results of g_{12}. The values in the parentheses are the corresponding statistical uncertainties (± one standard deviation) (from Ayala et al., 2008, Courtesy © IOP Publishing & Deutsche Physikalische Gesellschaft)

a_1 (μm)	a_2 (μm)	ϵ 10 $cm^2 s^{-3}$			ϵ 100 $cm^2 s^{-3}$			ϵ 400 $cm^2 s^{-3}$		
		R_λ			R_λ			R_λ		
		23.4	43.0	72.4	23.4	43.0	72.4	23.4	43.0	72.4
10	10	1.488	1.000	1.000	1.562	0.953	1.433	1.213	1.639	0.945
		(0.352)	(0.000)	(0.000)	(0.411)	(0.405)	(0.630)	(0.232)	(0.432)	(0.252)
10	20	1.026	1.006	0.945	1.041	1.042	1.104	1.060	1.076	1.043
		(0.029)	(0.033)	(0.076)	(0.024)	(0.026)	(0.054)	(0.021)	(0.031)	(0.033)
10	30	0.973	1.000	0.987	0.997	1.010	1.010	1.012	1.015	1.040
		(0.024)	(0.028)	(0.078)	(0.010)	(0.013)	(0.037)	(0.006)	(0.008)	(0.021)
10	40	1.015	1.004	1.010	1.010	0.986	0.944	1.009	1.047	1.047
		(0.034)	(0.039)	(0.104)	(0.022)	(0.033)	(0.036)	(0.017)	(0.040)	(0.044)
10	50	0.977	0.984	1.122	1.019	1.033	1.045	0.975	0.994	1.045
		(0.031)	(0.044)	(0.097)	(0.021)	(0.038)	(0.042)	(0.022)	(0.039)	(0.049)
10	60	0.994	1.063	1.038	0.991	0.968	0.974	0.995	1.005	1.019
		(0.039)	(0.044)	(0.003)	(0.017)	(0.046)	(0.050)	(0.008)	(0.031)	(0.044)
20	20	1.552	1.067	1.407	1.209	1.936	1.814	3.179	5.032	5.087
		(0.310)	(0.202)	(0.571)	(0.103)	(0.231)	(0.193)	(0.157)	(0.324)	(0.332)
20	30	1.054	0.985	0.971	1.037	1.003	1.024	1.032	1.127	1.125
		(0.046)	(0.042)	(0.093)	(0.044)	(0.022)	(0.051)	(0.041)	(0.043)	(0.046)
20	40	0.999	1.002	0.884	1.071	0.930	0.974	1.004	1.018	1.038
		(0.050)	(0.056)	(0.081)	(0.042)	(0.051)	(0.058)	(0.029)	(0.052)	(0.057)
20	50	1.046	0.946	0.991	1.081	0.984	0.985	0.970	1.087	1.048
		(0.053)	(0.066)	(0.135)	(0.033)	(0.057)	(0.085)	(0.027)	(0.056)	(0.063)
20	60	1.139	1.091	1.055	1.028	1.006	1.001	1.007	1.057	0.959
		(0.079)	(0.059)	(0.140)	(0.026)	(0.068)	(0.097)	(0.021)	(0.045)	(0.077)

Table B7 (cont.)

a_1 (μm)	a_2 (μm)	ϵ 10 cm^2 s^{-3}			100 cm^2 s^{-3}			400 cm^2 s^{-3}		
		R_λ			R_λ			R_λ		
		23.4	43.0	72.4	23.4	43.0	72.4	23.4	43.0	72.4
30	30	1.671	1.382	0.879	2.313	3.854	3.227	8.184	13.256	16.824
		(0.331)	(0.248)	(0.356)	(0.183)	(0.295)	(0.493)	(0.240)	(0.406)	(0.591)
30	40	0.881	0.927	1.222	0.979	1.024	0.979	1.108	1.053	1.114
		(0.060)	(0.065)	(0.108)	(0.036)	(0.069)	(0.059)	(0.032)	(0.057)	(0.073)
30	50	0.969	1.030	1.077	0.993	1.073	0.934	1.024	1.048	1.089
		(0.063)	(0.086)	(0.118)	(0.040)	(0.081)	(0.074)	(0.022)	(0.048)	(0.092)
30	60	1.020	0.915	1.018	0.969	0.993	1.054	0.989	1.011	0.996
		(0.098)	(0.100)	(0.116)	(0.032)	(0.067)	(0.112)	(0.018)	(0.039)	(0.073)
40	40	1.284	0.916	2.240	3.690	7.163	7.913	8.701	14.019	15.711
		(0.240)	(0.235)	(0.534)	(0.192)	(0.476)	(0.549)	(0.252)	(0.550)	(0.641)
40	50	1.144	0.962	0.939	1.179	1.244	1.206	1.168	1.132	0.984
		(0.074)	(0.085)	(0.085)	(0.033)	(0.070)	(0.108)	(0.021)	(0.038)	(0.077)
40	60	1.019	0.992	1.267	1.089	1.057	1.051	1.063	1.061	0.997
		(0.105)	(0.111)	(0.140)	(0.030)	(0.059)	(0.120)	(0.017)	(0.035)	(0.064)
50	50	0.892	1.577	2.526	2.974	8.006	10.781	7.254	12.180	14.491
		(0.207)	(0.342)	(0.554)	(0.158)	(0.524)	(0.760)	(0.211)	(0.470)	(0.720)
50	60	1.112	1.214	1.016	1.330	1.250	0.959	1.429	1.230	1.191
		(0.093)	(0.091)	(0.103)	(0.028)	(0.056)	(0.093)	(0.014)	(0.035)	(0.060)
60	60	1.194	1.268	2.568	2.733	5.812	9.566	5.647	9.373	13.721
		(0.274)	(0.258)	(0.543)	(0.105)	(0.441)	(0.802)	(0.153)	(0.360)	(0.689)

Appendix C Graupel–Drop Collision Efficiency and Kernel

Table C1 Graupel–drop collision efficiencies for 0.1 g cm^{-3} density graupel. The upper row shows radii of graupel in μm. The left column shows radii of drops in μm (from Khain et al., 2001; © American Meteorological Society. Used with permission).

	100	120	140	160	180	200	220	240	260	280	300	320	340	360
2	0	0	0	0	0	0	0	0	0	0	0	0	0	0
4	0	0	0	0	0	0	0	0	0	0	0	0	0	0
6	0	0	0	0	0	0	0	0.03	0.07	0.1	0.13	0.16	0.18	0.2
8	0	0	0	0	0.08	0.14	0.19	0.23	0.27	0.3	0.33	0.36	0.38	0.4
10	0	0	0	0.14	0.23	0.29	0.34	0.38	0.42	0.45	0.48	0.5	0.52	0.54
12	0	0	0.09	0.25	0.34	0.4	0.46	0.5	0.53	0.56	0.58	0.61	0.62	0.64
14	0	0	0.15	0.33	0.42	0.49	0.54	0.58	0.61	0.64	0.66	0.68	0.7	0.71
16	0	0	0.17	0.38	0.49	0.56	0.61	0.64	0.67	0.7	0.72	0.74	0.75	0.76
18	0	0	0.15	0.41	0.53	0.6	0.65	0.69	0.72	0.74	0.76	0.78	0.79	0.8
20	0	0	0.08	0.42	0.56	0.63	0.69	0.72	0.75	0.77	0.79	0.81	0.82	0.83
22	0	0	0	0.41	0.57	0.66	0.71	0.75	0.78	0.8	0.82	0.83	0.84	0.85
24	0	0	0	0.37	0.57	0.67	0.73	0.77	0.8	0.82	0.84	0.85	0.86	0.87
26	0	0	0	0.28	0.56	0.68	0.74	0.78	0.81	0.83	0.85	0.87	0.88	0.88
28	0	0	0	0.09	0.53	0.67	0.75	0.79	0.83	0.85	0.86	0.88	0.89	0.9
30	1.18	0	0	0	0.47	0.66	0.75	0.8	0.83	0.86	0.87	0.88	0.9	0.9
32	1.05	0	0	0	0.37	0.64	0.74	0.8	0.84	0.86	0.88	0.89	0.9	0.91
34	0.99	1.37	0	0	0.17	0.6	0.73	0.8	0.84	0.87	0.88	0.9	0.91	0.92
36	0.96	1.18	2.47	0	0	0.53	0.72	0.8	0.84	0.87	0.89	0.9	0.91	0.92
38	0.94	1.08	1.64	0	0	0.41	0.68	0.79	0.84	0.87	0.89	0.91	0.92	0.93
40	0.94	1.03	1.31	4	0	0.18	0.64	0.77	0.83	0.87	0.89	0.91	0.92	0.93
50	0.95	0.98	1.01	1.09	1.39	6.79	0	0.45	0.73	0.83	0.88	0.9	0.92	0.93
60	0.97	0.98	0.99	1.01	1.04	1.14	1.61	16	0	0.61	0.79	0.87	0.9	0.92
70	0.97	0.98	0.99	1	1.01	1.03	1.07	1.23	2.37	0	0.19	0.71	0.83	0.89
80	0.97	0.98	0.99	0.99	1	1.01	1.02	1.05	1.12	1.42	6.92	0	0.48	0.77
90	0.97	0.98	0.99	0.99	0.99	1	1.01	1.01	1.03	1.08	1.21	1.88	16	0
100	0.97	0.98	0.99	0.99	0.99	1	1	1.01	1.01	1.03	1.05	1.13	1.35	2.88
110	0.97	0.98	0.99	0.99	0.99	0.99	1	1	1.01	1.01	1.02	1.04	1.08	1.19
120	0.97	0.98	0.99	0.99	0.99	0.99	1	1	1	1.01	1.01	1.02	1.03	1.06
130	0.97	0.98	0.99	0.99	0.99	0.99	1	1	1	1	1.01	1.01	1.01	1.03
140	0.97	0.98	0.99	0.99	0.99	0.99	1	1	1	1	1	1.01	1.01	1.01
150	0.97	0.98	0.99	0.99	0.99	0.99	1	1	1	1	1	1	1.01	1.01
160	0.97	0.98	0.99	0.99	0.99	0.99	1	1	1	1	1	1	1	1.01
170	0.98	0.99	0.99	0.99	0.99	0.99	1	1	1	1	1	1	1	1
180	0.98	0.99	0.99	0.99	0.99	0.99	1	1	1	1	1	1	1	1
190	0.98	0.99	0.99	0.99	0.99	0.99	1	1	1	1	1	1	1	1
200	0.98	0.99	0.99	0.99	0.99	0.99	1	1	1	1	1	1	1	1
210	0.98	0.99	0.99	0.99	0.99	0.99	1	1	1	1	1	1	1	1
220	0.98	0.99	0.99	0.99	0.99	0.99	1	1	1	1	1	1	1	1
230	0.98	0.99	0.99	0.99	0.99	1	1	1	1	1	1	1	1	1
240	0.98	0.99	0.99	0.99	0.99	1	1	1	1	1	1	1	1	1
250	0.99	0.99	0.99	0.99	0.99	1	1	1	1	1	1	1	1	1

380	400	420	440	460	480	500	520	540	560	580	600	620	640	660
0	0	0	0	0	0	0	0	0	0	0	0	0	0	0
0	0	0.01	0.02	0.03	0.05	0.06	0.08	0.09	0.1	0.11	0.12	0.13	0.15	0.15
0.22	0.24	0.26	0.28	0.29	0.31	0.33	0.34	0.36	0.37	0.39	0.4	0.42	0.43	0.44
0.42	0.44	0.46	0.47	0.49	0.5	0.52	0.53	0.55	0.56	0.58	0.59	0.6	0.62	0.63
0.56	0.58	0.59	0.61	0.62	0.63	0.65	0.66	0.67	0.68	0.69	0.71	0.72	0.73	0.74
0.66	0.67	0.68	0.7	0.71	0.72	0.73	0.74	0.75	0.76	0.77	0.78	0.79	0.8	0.82
0.73	0.74	0.75	0.76	0.77	0.78	0.79	0.8	0.81	0.82	0.82	0.83	0.84	0.85	0.86
0.77	0.79	0.79	0.81	0.81	0.82	0.83	0.84	0.84	0.85	0.86	0.87	0.87	0.88	0.89
0.81	0.82	0.83	0.84	0.84	0.85	0.86	0.87	0.87	0.88	0.88	0.89	0.9	0.91	0.91
0.84	0.85	0.86	0.86	0.87	0.88	0.88	0.89	0.89	0.9	0.9	0.91	0.91	0.92	0.93
0.86	0.87	0.88	0.88	0.89	0.89	0.9	0.9	0.91	0.91	0.92	0.93	0.93	0.93	0.94
0.88	0.88	0.89	0.9	0.9	0.91	0.91	0.92	0.92	0.93	0.93	0.93	0.94	0.94	0.95
0.89	0.9	0.9	0.91	0.91	0.92	0.92	0.93	0.93	0.93	0.94	0.94	0.95	0.95	0.96
0.9	0.91	0.91	0.92	0.92	0.93	0.93	0.93	0.94	0.94	0.94	0.95	0.95	0.96	0.96
0.91	0.92	0.92	0.93	0.93	0.93	0.94	0.94	0.94	0.95	0.95	0.96	0.96	0.96	0.96
0.92	0.92	0.93	0.93	0.94	0.94	0.94	0.95	0.95	0.95	0.96	0.96	0.96	0.96	0.97
0.92	0.93	0.93	0.94	0.94	0.94	0.95	0.95	0.95	0.96	0.96	0.96	0.96	0.97	0.97
0.93	0.93	0.94	0.94	0.94	0.95	0.95	0.96	0.96	0.96	0.96	0.97	0.97	0.97	0.97
0.93	0.94	0.94	0.94	0.95	0.95	0.96	0.96	0.96	0.96	0.97	0.97	0.97	0.97	0.97
0.93	0.94	0.94	0.95	0.95	0.96	0.96	0.96	0.96	0.96	0.97	0.97	0.97	0.97	0.98
0.94	0.95	0.95	0.96	0.96	0.96	0.97	0.97	0.97	0.97	0.97	0.98	0.98	0.98	0.98
0.94	0.94	0.95	0.96	0.96	0.97	0.97	0.97	0.97	0.97	0.98	0.98	0.98	0.98	0.99
0.92	0.93	0.94	0.95	0.96	0.96	0.97	0.97	0.97	0.98	0.98	0.98	0.98	0.99	0.99
0.86	0.9	0.93	0.94	0.95	0.96	0.96	0.97	0.97	0.97	0.98	0.98	0.98	0.99	0.99
0.61	0.8	0.87	0.91	0.93	0.94	0.96	0.96	0.97	0.97	0.97	0.98	0.98	0.98	0.99
0	0.04	0.67	0.82	0.88	0.91	0.94	0.95	0.96	0.96	0.97	0.97	0.98	0.98	0.99
1.56	4.9	0	0.26	0.7	0.82	0.88	0.91	0.93	0.95	0.96	0.96	0.97	0.97	0.98
1.12	1.26	1.77	8.24	0	0.3	0.69	0.81	0.87	0.91	0.93	0.94	0.96	0.97	0.97
1.05	1.08	1.16	1.33	1.91	6.61	0	0.19	0.62	0.77	0.86	0.9	0.93	0.94	0.96
1.02	1.04	1.06	1.1	1.18	1.35	1.8	3.35	16	0	0.52	0.72	0.83	0.89	0.93
1.01	1.02	1.03	1.05	1.07	1.11	1.19	1.3	1.55	2.03	3.45	16	1.79	1.17	1.16
1.01	1.01	1.02	1.03	1.03	1.05	1.07	1.11	1.16	1.23	1.34	1.5	1.72	2.05	2.43
1.01	1.01	1.01	1.01	1.02	1.03	1.04	1.05	1.07	1.09	1.13	1.16	1.21	1.26	1.31
1	1.01	1.01	1.01	1.01	1.02	1.02	1.03	1.04	1.05	1.06	1.08	1.09	1.11	1.13
1	1	1.01	1.01	1.01	1.01	1.01	1.02	1.02	1.03	1.04	1.04	1.05	1.06	1.07
1	1	1	1.01	1.01	1.01	1.01	1.01	1.01	1.02	1.02	1.03	1.03	1.04	1.04
1	1	1	1	1	1.01	1.01	1.01	1.01	1.01	1.01	1.02	1.02	1.02	1.03
1	1	1	1	1	1	1.01	1.01	1.01	1.01	1.01	1.01	1.01	1.02	1.02
1	1	1	1	1	1	1	1.01	1.01	1.01	1.01	1.01	1.01	1.01	1.01
1	1	1	1	1	1	1	1	1.01	1.01	1.01	1.01	1.01	1.01	1.01
1	1	1	1	1	1	1	1	1	1.01	1.01	1.01	1.01	1.01	1.01

Table C2 Graupel–drop collision efficiencies for 0.4 g cm^{-3} density graupel. The upper row shows radii of graupel in μm. The left column shows radii of drops in μm (from Khain et al., 2001; © American Meteorological Society. Used with permission).

	100	120	140	160	180	200	220	240	260	280	300	320	340	360	380	400
2	0	0	0	0	0	0	0	0	0	0	0	0	0	0	0	0
4	0	0.02	0.07	0.12	0.16	0.2	0.24	0.27	0.3	0.33	0.35	0.38	0.4	0.42	0.45	0.47
6	0.22	0.29	0.35	0.4	0.44	0.48	0.52	0.54	0.57	0.59	0.61	0.63	0.65	0.67	0.69	0.71
8	0.41	0.48	0.54	0.58	0.62	0.65	0.67	0.7	0.72	0.73	0.75	0.76	0.78	0.79	0.81	0.82
10	0.55	0.61	0.66	0.7	0.73	0.75	0.77	0.79	0.8	0.81	0.83	0.84	0.85	0.86	0.87	0.88
12	0.64	0.7	0.74	0.77	0.8	0.82	0.83	0.84	0.86	0.86	0.87	0.88	0.89	0.9	0.91	0.92
14	0.71	0.76	0.79	0.82	0.84	0.86	0.87	0.88	0.89	0.9	0.9	0.91	0.92	0.93	0.93	0.94
16	0.75	0.8	0.83	0.86	0.87	0.89	0.9	0.9	0.91	0.92	0.93	0.93	0.94	0.94	0.95	0.95
18	0.78	0.83	0.86	0.88	0.9	0.91	0.91	0.92	0.93	0.93	0.94	0.94	0.95	0.95	0.96	0.96
20	0.81	0.85	0.88	0.9	0.91	0.92	0.93	0.94	0.94	0.94	0.95	0.95	0.96	0.96	0.96	0.97
22	0.83	0.87	0.9	0.91	0.93	0.93	0.94	0.94	0.95	0.95	0.96	0.96	0.96	0.97	0.97	0.97
24	0.84	0.88	0.91	0.93	0.94	0.94	0.95	0.95	0.96	0.96	0.96	0.97	0.97	0.97	0.97	0.98
26	0.84	0.89	0.92	0.93	0.94	0.95	0.96	0.96	0.96	0.97	0.97	0.97	0.97	0.97	0.98	0.98
28	0.85	0.9	0.93	0.94	0.95	0.96	0.96	0.96	0.97	0.97	0.97	0.97	0.97	0.98	0.98	0.98
30	0.85	0.9	0.93	0.94	0.95	0.96	0.96	0.97	0.97	0.97	0.97	0.98	0.98	0.98	0.98	0.99
32	0.84	0.91	0.93	0.95	0.96	0.96	0.97	0.97	0.97	0.97	0.98	0.98	0.98	0.98	0.98	0.99
34	0.84	0.91	0.94	0.95	0.96	0.97	0.97	0.97	0.97	0.98	0.98	0.98	0.98	0.98	0.99	0.99
36	0.82	0.91	0.94	0.96	0.96	0.97	0.97	0.97	0.98	0.98	0.98	0.98	0.98	0.99	0.99	0.99
38	0.79	0.91	0.94	0.96	0.97	0.97	0.97	0.98	0.98	0.98	0.98	0.98	0.99	0.99	0.99	0.99
40	0.74	0.9	0.94	0.96	0.97	0.97	0.97	0.98	0.98	0.98	0.98	0.99	0.99	0.99	0.99	0.99
50	0	0.81	0.93	0.96	0.97	0.98	0.98	0.98	0.99	0.99	0.99	0.99	0.99	0.99	0.99	0.99
60	1.03	0	0.87	0.95	0.97	0.98	0.98	0.99	0.99	0.99	0.99	0.99	0.99	0.99	0.99	0.99
70	0.99	1.09	0	0.92	0.96	0.98	0.98	0.99	0.99	0.99	0.99	0.99	0.99	0.99	0.99	0.99
80	0.99	1	1.28	0.7	0.94	0.97	0.98	0.99	0.99	0.99	0.99	0.99	0.99	0.99	0.99	1
90	0.99	0.99	1.02	3.85	0.87	0.96	0.98	0.99	0.99	0.99	0.99	0.99	0.99	0.99	0.99	1
100	0.99	0.99	1	1.07	0	0.93	0.97	0.98	0.99	0.99	0.99	0.99	0.99	0.99	1	1
110	0.99	0.99	0.99	1.01	1.27	0.73	0.95	0.97	0.99	0.99	0.99	0.99	0.99	0.99	1	1
120	0.99	0.99	0.99	1	1.03	5.54	0.88	0.96	0.98	0.99	0.99	0.99	0.99	0.99	1	1
130	0.99	0.99	0.99	1	1.01	1.11	0	0.93	0.97	0.98	0.99	0.99	0.99	0.99	0.99	1
140	0.99	0.99	0.99	0.99	1	1.02	1.4	0.75	0.96	0.98	0.99	0.99	0.99	0.99	0.99	1
150	0.99	0.99	0.99	0.99	1	1.01	1.05	5.46	0.88	0.97	0.98	0.99	0.99	0.99	0.99	1
160	0.99	0.99	0.99	0.99	1	1	1.01	1.14	0	0.93	0.97	0.99	0.99	0.99	0.99	1
170	0.99	0.99	0.99	0.99	1	1	1.01	1.03	1.4	0.71	0.95	0.98	0.99	0.99	0.99	1
180	0.99	0.99	0.99	1	1	1	1	1.01	1.07	2.44	0.86	0.96	0.98	0.99	0.99	0.99
190	0.99	0.99	0.99	1	1	1	1	1.01	1.02	1.14	10.4	0.91	0.97	0.99	0.99	0.99
200	0.99	0.99	1	1	1	1	1	1	1.01	1.04	1.23	16	0.94	0.98	0.99	1
210	0.99	0.99	1	1	1	1	1	1	1.01	1.02	1.06	1.33	16	1.01	1	1
220	0.99	0.99	1	1	1	1	1	1	1	1.01	1.03	1.08	1.35	6.16	1.33	1.07
230	0.99	0.99	1	1	1	1	1	1	1	1.01	1.01	1.03	1.09	1.29	2.31	16
240	0.99	1	1	1	1	1	1	1	1	1	1.01	1.02	1.04	1.09	1.21	1.52
250	0.99	1	1	1	1	1	1	1	1	1	1.01	1.01	1.02	1.04	1.08	1.14

Table C3 Graupel–drop collision efficiencies for 0.8 g cm^{-3} density graupel. The upper row shows radii of graupel in μm. The left column shows radii of drops in μm (from Khain et al., 2001; © American Meteorological Society. Used with permission).

	100	120	140	160	180	200	220	240	260	280	300	320
2	0	0	0	0	0	0.01	0.03	0.06	0.08	0.1	0.13	0.15
4	0.19	0.26	0.32	0.37	0.41	0.45	0.48	0.51	0.54	0.57	0.59	0.62
6	0.47	0.54	0.59	0.63	0.66	0.68	0.71	0.73	0.75	0.77	0.79	0.81
8	0.64	0.69	0.73	0.76	0.78	0.8	0.82	0.83	0.84	0.86	0.87	0.89
10	0.75	0.78	0.81	0.83	0.85	0.86	0.88	0.88	0.9	0.91	0.92	0.93
12	0.81	0.84	0.86	0.88	0.89	0.9	0.91	0.92	0.93	0.93	0.94	0.95
14	0.85	0.88	0.9	0.91	0.92	0.93	0.93	0.94	0.94	0.95	0.96	0.96
16	0.88	0.9	0.92	0.93	0.94	0.94	0.95	0.95	0.96	0.96	0.96	0.97
18	0.9	0.92	0.93	0.94	0.95	0.95	0.96	0.96	0.96	0.97	0.97	0.97
20	0.92	0.93	0.94	0.95	0.96	0.96	0.96	0.97	0.97	0.97	0.98	0.98
22	0.93	0.94	0.95	0.96	0.96	0.97	0.97	0.97	0.97	0.98	0.98	0.98
24	0.94	0.95	0.96	0.97	0.97	0.97	0.97	0.98	0.98	0.98	0.98	0.99
26	0.94	0.96	0.96	0.97	0.97	0.97	0.98	0.98	0.98	0.98	0.99	0.99
28	0.95	0.96	0.97	0.97	0.97	0.98	0.98	0.98	0.98	0.99	0.99	0.99
30	0.95	0.96	0.97	0.97	0.98	0.98	0.98	0.99	0.99	0.99	0.99	0.99
32	0.96	0.97	0.97	0.98	0.98	0.98	0.99	0.99	0.99	0.99	0.99	0.99
34	0.96	0.97	0.98	0.98	0.98	0.99	0.99	0.99	0.99	0.99	0.99	0.99
36	0.96	0.97	0.98	0.98	0.99	0.99	0.99	0.99	0.99	0.99	0.99	0.99
38	0.96	0.97	0.98	0.98	0.99	0.99	0.99	0.99	0.99	0.99	0.99	0.99
40	0.96	0.97	0.98	0.99	0.99	0.99	0.99	0.99	0.99	0.99	0.99	0.99
50	0.96	0.98	0.99	0.99	0.99	0.99	0.99	0.99	0.99	0.99	0.99	1
60	0.94	0.98	0.99	0.99	0.99	0.99	0.99	0.99	0.99	0.99	1	1
70	0.74	0.97	0.99	0.99	0.99	0.99	0.99	0.99	1	1	1	1
80	1.16	0.95	0.98	0.99	0.99	0.99	0.99	1	1	1	1	1
90	0.99	0.83	0.98	0.99	0.99	0.99	0.99	1	1	1	1	1
100	0.99	1.36	0.97	0.99	0.99	0.99	0.99	1	1	1	1	1
110	0.99	1	0.93	0.99	0.99	0.99	0.99	1	1	1	1	1
120	0.99	0.99	0	0.98	0.99	0.99	0.99	1	1	1	1	1
130	0.99	0.99	1.02	0.97	0.99	0.99	0.99	1	1	1	1	1
140	0.99	0.99	0.99	0.91	0.99	0.99	0.99	1	1	1	1	1
150	0.99	0.99	0.99	1.15	0.98	0.99	0.99	1	1	1	1	1
160	0.99	0.99	0.99	1	0.99	0.99	0.99	1	1	1	1	1
170	0.99	0.99	0.99	0.99	3.59	0.99	0.99	1	1	1	1	1
180	0.99	0.99	0.99	0.99	1.03	1.01	0.99	1	1	1	1	1
190	0.99	0.99	0.99	0.99	1	2.14	1	1	1	1	1	1
200	0.99	0.99	0.99	0.99	1	1.12	1.02	1	1	1	1	1
210	0.99	1	1	0.99	1	1.01	1.42	1.01	1	1	1	1
220	0.99	1	1	1	1	1	1.34	1.04	1.01	1	1	1
230	1	1	1	1	1	1	1.04	1.52	1.02	1.01	1	1
240	1	1	1	1	1	1	1.01	1.49	1.09	1.01	1.01	1
250	1	1	1	1	1	1	1	1.06	3.75	1.06	1.02	1

Table C4 Graupel–drop collision kernels (m^3s^{-1}) for 0.1 g cm^{-3} density graupel. The upper row shows radii of graupel in μm. The left column shows radii of drops in μm (from Khain et al., 2001; © American Meteorological Society. Used with permission).

	100	120	140	160	180	200	220
2	0.00E+00	0.00E+00	0.00E+00	0.00E+00	0.00E+00	0.00E+00	0.00E+00
4	0.00E+00	0.00E+00	0.00E+00	0.00E+00	0.00E+00	0.00E+00	0.00E+00
6	0.00E+00	0.00E+00	0.00E+00	0.00E+00	0.00E+00	0.00E+00	0.00E+00
8	0.00E+00	0.00E+00	0.00E+00	0.00E+00	1.72E−09	4.63E−09	9.10E−09
10	0.00E+00	0.00E+00	0.00E+00	2.09E−09	5.16E−09	9.79E−09	1.66E−08
12	0.00E+00	0.00E+00	8.42E−10	3.69E−09	7.75E−09	1.36E−08	2.21E−08
14	0.00E+00	0.00E+00	1.38E−09	4.78E−09	9.48E−09	1.64E−08	2.61E−08
16	0.00E+00	0.00E+00	1.54E−09	5.40E−09	1.07E−08	1.83E−08	2.90E−08
18	0.00E+00	0.00E+00	1.32E−09	5.61E−09	1.13E−08	1.95E−08	3.08E−08
20	0.00E+00	0.00E+00	6.43E−10	5.48E−09	1.15E−08	1.99E−08	3.18E−08
22	0.00E+00	0.00E+00	0.00E+00	5.00E−09	1.12E−08	2.01E−08	3.22E−08
24	0.00E+00	0.00E+00	0.00E+00	4.16E−09	1.07E−08	1.96E−08	3.20E−08
26	0.00E+00	0.00E+00	0.00E+00	2.87E−09	9.78E−09	1.89E−08	3.15E−08
28	0.00E+00	0.00E+00	0.00E+00	8.62E−10	8.51E−09	1.78E−08	3.03E−08
30	2.31E−10	0.00E+00	0.00E+00	0.00E+00	6.90E−09	1.63E−08	2.90E−08
32	1.01E−09	0.00E+00	0.00E+00	0.00E+00	4.80E−09	1.45E−08	2.72E−08
34	1.78E−09	1.00E−09	0.00E+00	0.00E+00	1.85E−09	1.23E−08	2.51E−08
36	2.60E−09	2.23E−09	6.60E−10	0.00E+00	0.00E+00	9.78E−09	2.27E−08
38	3.52E−09	3.40E−09	2.97E−09	0.00E+00	0.00E+00	6.61E−09	1.98E−08
40	4.53E−09	4.64E−09	4.55E−09	2.79E−09	0.00E+00	2.36E−09	1.67E−08
50	1.10E−08	1.25E−08	1.37E−08	1.37E−08	1.26E−08	1.52E−08	0.00E+00
60	1.99E−08	2.32E−08	2.63E−08	2.80E−08	2.76E−08	2.49E−08	2.16E−08
70	3.10E−08	3.67E−08	4.22E−08	4.62E−08	4.79E−08	4.68E−08	4.25E−08
80	4.45E−08	5.28E−08	6.11E−08	6.79E−08	7.23E−08	7.38E−08	7.17E−08
90	6.08E−08	7.18E−08	8.29E−08	9.31E−08	1.00E−07	1.05E−07	1.06E−07
100	7.96E−08	9.38E−08	1.08E−07	1.21E−07	1.33E−07	1.41E−07	1.45E−07
110	1.01E−07	1.19E−07	1.37E−07	1.54E−07	1.68E−07	1.80E−07	1.88E−07
120	1.26E−07	1.47E−07	1.70E−07	1.90E−07	2.08E−07	2.23E−07	2.35E−07
130	1.54E−07	1.79E−07	2.05E−07	2.30E−07	2.52E−07	2.71E−07	2.88E−07
140	1.85E−07	2.14E−07	2.45E−07	2.73E−07	3.01E−07	3.24E−07	3.45E−07
150	2.20E−07	2.53E−07	2.89E−07	3.21E−07	3.53E−07	3.81E−07	4.06E−07
160	2.58E−07	2.96E−07	3.36E−07	3.74E−07	4.11E−07	4.43E−07	4.73E−07
170	3.01E−07	3.44E−07	3.88E−07	4.31E−07	4.72E−07	5.09E−07	5.45E−07
180	3.47E−07	3.96E−07	4.45E−07	4.94E−07	5.39E−07	5.81E−07	6.22E−07
190	3.98E−07	4.51E−07	5.06E−07	5.60E−07	6.11E−07	6.58E−07	7.04E−07
200	4.52E−07	5.11E−07	5.72E−07	6.32E−07	6.88E−07	7.40E−07	7.92E−07
210	5.11E−07	5.76E−07	6.42E−07	7.08E−07	7.70E−07	8.28E−07	8.86E−07
220	5.77E−07	6.46E−07	7.18E−07	7.90E−07	8.58E−07	9.22E−07	9.86E−07
230	6.45E−07	7.21E−07	7.99E−07	8.78E−07	9.51E−07	1.03E−06	1.09E−06
240	7.19E−07	8.04E−07	8.89E−07	9.71E−07	1.05E−06	1.13E−06	1.20E−06
250	8.01E−07	8.89E−07	9.81E−07	1.07E−06	1.16E−06	1.24E−06	1.32E−06

240	260	280	300	320	340	360	380
0.00E+00	0.00E+00	0.00E+00	0.00E+00	0.00E+00	0.00E+00	0.00E+00	0.00E+00
0.00E+00	0.00E+00	0.00E+00	0.00E+00	0.00E+00	0.00E+00	0.00E+00	0.00E+00
2.21E−09	6.31E−09	1.22E−08	2.05E−08	3.12E−08	4.41E−08	5.99E−08	7.97E−08
1.58E−08	2.45E−08	3.64E−08	5.16E−08	7.06E−08	9.27E−08	1.20E−07	1.51E−07
2.62E−08	3.87E−08	5.49E−08	7.51E−08	9.99E−08	1.29E−07	1.63E−07	2.02E−07
3.40E−08	4.92E−08	6.85E−08	9.23E−08	1.21E−07	1.55E−07	1.94E−07	2.38E−07
3.95E−08	5.64E−08	7.80E−08	1.05E−07	1.36E−07	1.73E−07	2.16E−07	2.64E−07
4.34E−08	6.19E−08	8.48E−08	1.14E−07	1.47E−07	1.86E−07	2.31E−07	2.82E−07
4.61E−08	6.56E−08	8.99E−08	1.19E−07	1.55E−07	1.96E−07	2.42E−07	2.95E−07
4.76E−08	6.79E−08	9.28E−08	1.24E−07	1.60E−07	2.02E−07	2.50E−07	3.04E−07
4.85E−08	6.91E−08	9.51E−08	1.27E−07	1.64E−07	2.07E−07	2.55E−07	3.11E−07
4.87E−08	6.97E−08	9.59E−08	1.28E−07	1.66E−07	2.09E−07	2.59E−07	3.14E−07
4.83E−08	6.95E−08	9.59E−08	1.28E−07	1.67E−07	2.11E−07	2.61E−07	3.17E−07
4.74E−08	6.90E−08	9.57E−08	1.28E−07	1.67E−07	2.11E−07	2.62E−07	3.19E−07
4.61E−08	6.77E−08	9.43E−08	1.27E−07	1.66E−07	2.10E−07	2.61E−07	3.18E−07
4.43E−08	6.58E−08	9.27E−08	1.26E−07	1.64E−07	2.09E−07	2.60E−07	3.17E−07
4.22E−08	6.37E−08	9.05E−08	1.23E−07	1.62E−07	2.07E−07	2.58E−07	3.14E−07
3.98E−08	6.11E−08	8.79E−08	1.21E−07	1.59E−07	2.04E−07	2.55E−07	3.12E−07
3.69E−08	5.81E−08	8.48E−08	1.18E−07	1.56E−07	2.01E−07	2.51E−07	3.09E−07
3.37E−08	5.48E−08	8.14E−08	1.14E−07	1.52E−07	1.96E−07	2.48E−07	3.04E−07
1.08E−08	3.27E−08	5.85E−08	8.99E−08	1.27E−07	1.70E−07	2.19E−07	2.74E−07
1.20E−09	0.00E+00	2.51E−08	5.65E−08	9.24E−08	1.33E−07	1.80E−07	2.34E−07
3.57E−08	3.29E−08	0.00E+00	6.30E−09	4.74E−08	8.27E−08	1.33E−07	1.83E−07
6.55E−08	5.65E−08	4.66E−08	6.34E−08	0.00E+00	2.64E−08	7.40E−08	1.23E−07
1.02E−07	9.50E−08	8.36E−08	6.92E−08	5.89E−08	2.85E−09	0.00E+00	4.87E−08
1.44E−07	1.40E−07	1.31E−07	1.17E−07	9.96E−08	8.27E−08	7.94E−08	0.00E+00
1.91E−07	1.90E−07	1.84E−07	1.73E−07	1.57E−07	1.38E−07	1.17E−07	9.94E−08
2.42E−07	2.45E−07	2.43E−07	2.35E−07	2.23E−07	2.07E−07	1.86E−07	1.62E−07
2.98E−07	3.05E−07	3.07E−07	3.04E−07	2.95E−07	2.82E−07	2.64E−07	2.43E−07
3.60E−07	3.70E−07	3.76E−07	3.77E−07	3.74E−07	3.65E−07	3.51E−07	3.33E−07
4.26E−07	4.41E−07	4.51E−07	4.58E−07	4.57E−07	4.54E−07	4.43E−07	4.31E−07
4.97E−07	5.17E−07	5.32E−07	5.42E−07	5.48E−07	5.48E−07	5.45E−07	5.35E−07
5.74E−07	5.99E−07	6.19E−07	6.33E−07	6.42E−07	6.50E−07	6.50E−07	6.48E−07
6.56E−07	6.86E−07	7.11E−07	7.30E−07	7.45E−07	7.59E−07	7.64E−07	7.65E−07
7.43E−07	7.79E−07	8.09E−07	8.34E−07	8.54E−07	8.70E−07	8.85E−07	8.92E−07
8.37E−07	8.78E−07	9.13E−07	9.44E−07	9.69E−07	9.91E−07	1.01E−06	1.03E−06
9.36E−07	9.83E−07	1.02E−06	1.06E−06	1.09E−06	1.12E−06	1.14E−06	1.17E−06
1.04E−06	1.09E−06	1.14E−06	1.18E−06	1.22E−06	1.25E−06	1.28E−06	1.31E−06
1.15E−06	1.21E−06	1.27E−06	1.31E−06	1.36E−06	1.40E−06	1.43E−06	1.49E−06
1.27E−06	1.34E−06	1.40E−06	1.45E−06	1.50E−06	1.55E−06	1.59E−06	1.63E−06
1.40E−06	1.47E−06	1.53E−06	1.59E−06	1.65E−06	1.70E−06	1.75E−06	1.80E−06

Table C4 (cont.) The left column shows radii of drops in μm.

	400	420	440	460	480	500	520
2	0.00E+00	0.00E+00	0.00E+00	0.00E+00	0.00E+00	0.00E+00	0.00E+00
4	0.00E+00	3.42E−09	1.19E−08	2.31E−08	3.85E−08	5.38E−08	7.20E−08
6	1.03E−07	1.30E−07	1.61E−07	1.96E−07	2.36E−07	2.81E−07	3.29E−07
8	1.87E−07	2.29E−07	2.75E−07	3.27E−07	3.83E−07	4.47E−07	5.14E−07
10	2.47E−07	2.98E−07	3.54E−07	4.16E−07	4.83E−07	5.60E−07	6.35E−07
12	2.89E−07	3.46E−07	4.09E−07	4.77E−07	5.51E−07	6.34E−07	7.19E−07
14	3.18E−07	3.80E−07	4.47E−07	5.19E−07	5.99E−07	6.86E−07	7.77E−07
16	3.40E−07	4.03E−07	4.74E−07	5.50E−07	6.31E−07	7.23E−07	8.15E−07
18	3.55E−07	4.20E−07	4.92E−07	5.71E−07	6.56E−07	7.47E−07	8.43E−07
20	3.65E−07	4.32E−07	5.06E−07	5.87E−07	6.74E−07	7.65E−07	8.63E−07
22	3.73E−07	4.41E−07	5.15E−07	5.98E−07	6.84E−07	7.80E−07	8.75E−07
24	3.77E−07	4.47E−07	5.22E−07	6.06E−07	6.93E−07	7.87E−07	8.87E−07
26	3.81E−07	4.50E−07	5.28E−07	6.11E−07	6.98E−07	7.93E−07	8.91E−07
28	3.81E−07	4.52E−07	5.29E−07	6.12E−07	7.00E−07	7.95E−07	8.94E−07
30	3.82E−07	4.52E−07	5.29E−07	6.13E−07	7.01E−07	7.97E−07	8.96E−07
32	3.80E−07	4.52E−07	5.29E−07	6.13E−07	7.02E−07	7.98E−07	8.97E−07
34	3.79E−07	4.49E−07	5.26E−07	6.10E−07	6.99E−07	7.95E−07	8.94E−07
36	3.76E−07	4.46E−07	5.23E−07	6.07E−07	6.95E−07	7.91E−07	8.90E−07
38	3.72E−07	4.42E−07	5.19E−07	6.03E−07	6.91E−07	7.87E−07	8.86E−07
40	3.68E−07	4.38E−07	5.14E−07	5.98E−07	6.86E−07	7.82E−07	8.77E−07
50	3.36E−07	4.04E−07	4.79E−07	5.60E−07	6.46E−07	7.40E−07	8.32E−07
60	2.92E−07	3.58E−07	4.31E−07	5.10E−07	5.92E−07	6.82E−07	7.70E−07
70	2.39E−07	3.02E−07	3.70E−07	4.46E−07	5.24E−07	6.09E−07	6.95E−07
80	1.77E−07	2.36E−07	3.02E−07	3.73E−07	4.48E−07	5.28E−07	6.08E−07
90	1.04E−07	1.61E−07	2.23E−07	2.91E−07	3.61E−07	4.37E−07	5.14E−07
100	2.33E−09	7.12E−08	1.33E−07	1.99E−07	2.65E−07	3.37E−07	4.08E−07
110	1.17E−07	0.00E+00	1.91E−08	9.11E−08	1.56E−07	2.25E−07	2.91E−07
120	1.39E−07	1.22E−07	1.84E−07	0.00E+00	2.48E−08	9.79E−08	1.63E−07
130	2.17E−07	1.92E−07	1.66E−07	1.50E−07	2.01E−07	0.00E+00	1.43E−08
140	3.12E−07	2.85E−07	2.58E−07	2.28E−07	2.05E−07	1.88E−07	1.98E−07
150	4.13E−07	3.91E−07	3.66E−07	3.37E−07	3.10E−07	2.83E−07	2.61E−07
160	5.22E−07	5.06E−07	4.85E−07	4.59E−07	4.35E−07	4.08E−07	3.87E−07
170	6.39E−07	6.27E−07	6.11E−07	5.93E−07	5.73E−07	5.49E−07	5.32E−07
180	7.65E−07	7.57E−07	7.48E−07	7.34E−07	7.20E−07	7.01E−07	6.89E−07
190	8.95E−07	8.96E−07	8.90E−07	8.83E−07	8.74E−07	8.63E−07	8.57E−07
200	1.03E−06	1.04E−06	1.04E−06	1.04E−06	1.04E−06	1.03E−06	1.04E−06
210	1.18E−06	1.19E−06	1.20E−06	1.20E−06	1.21E−06	1.21E−06	1.22E−06
220	1.34E−06	1.35E−06	1.37E−06	1.38E−06	1.39E−06	1.40E−06	1.42E−06
230	1.50E−06	1.52E−06	1.54E−06	1.56E−06	1.58E−06	1.60E−06	1.63E−06
240	1.67E−06	1.70E−06	1.73E−06	1.75E−06	1.78E−06	1.81E−06	1.84E−06
250	1.85E−06	1.89E−06	1.92E−06	1.95E−06	1.99E−06	2.02E−06	2.06E−06

540	560	580	600	620	640	660
0.00E+00	0.00E+00	0.00E+00	0.00E+00	0.00E+00	0.00E+00	0.00E+00
9.32E−08	1.17E−07	1.45E−07	1.75E−07	2.06E−07	2.42E−07	2.76E−07
3.84E−07	4.40E−07	5.30E−07	5.70E−07	6.42E−07	7.19E−07	7.96E−07
5.87E−07	6.66E−07	7.51E−07	8.36E−07	9.31E−07	1.03E−06	1.13E−06
7.20E−07	8.11E−07	9.08E−07	1.01E−06	1.12E−06	1.22E−06	1.34E−06
8.11E−07	9.07E−07	1.01E−06	1.12E−06	1.23E−06	1.35E−06	1.47E−06
8.72E−07	9.75E−07	1.08E−06	1.19E−06	1.31E−06	1.43E−06	1.55E−06
9.15E−07	1.02E−06	1.13E−06	1.24E−06	1.36E−06	1.49E−06	1.61E−06
9.46E−07	1.05E−06	1.16E−06	1.28E−06	1.40E−06	1.52E−06	1.65E−06
9.64E−07	1.07E−06	1.18E−06	1.30E−06	1.42E−06	1.54E−06	1.67E−06
9.82E−07	1.09E−06	1.20E−06	1.32E−06	1.44E−06	1.56E−06	1.69E−06
9.91E−07	1.10E−06	1.21E−06	1.33E−06	1.45E−06	1.57E−06	1.70E 06
9.95E−07	1.10E−06	1.22E−06	1.33E−06	1.46E−06	1.58E−06	1.70E−06
9.99E−07	1.11E−06	1.22E−06	1.34E−06	1.46E−06	1.58E−06	1.70E−06
1.00E−06	1.11E−06	1.22E−06	1.34E−06	1.45E−06	1.58E−06	1.70E−06
9.99E−07	1.11E−06	1.22E−06	1.34E−06	1.46E−06	1.57E−06	1.70E−06
9.96E−07	1.10E−06	1.22E−06	1.33E−06	1.45E−06	1.57E−06	1.69E−06
9.92E−07	1.10E−06	1.21E−06	1.33E−06	1.44E−06	1.56E−06	1.69E−06
9.87E−07	1.10E−06	1.21E−06	1.32E−06	1.44E−06	1.55E−06	1.67E−06
9.82E−07	1.09E−06	1.20E−06	1.31E−06	1.43E−06	1.55E−06	1.66E−06
9.34E−07	1.03E−06	1.14E−06	1.25E−06	1.36E−06	1.48E−06	1.996−06
8.68E−07	9.64E−07	1.07E−06	1.17E−06	1.28E−06	1.38E−06	1.49E−06
7.88E−07	8.82E−07	9.78E−07	1.08E−06	1.17E−06	1.28E−06	1.37E−06
6.95E−07	7.83E−07	8.76E−07	9.65E−07	1.06E−06	1.16E−06	1.24E−06
5.94E−07	6.76E−07	7.62E−07	8.48E−07	9.35E−07	1.02E−06	1.10E−06
4.83E−07	5.59E−07	6.41E−07	7.18E−07	7.98E−07	8.78E−07	9.54E−07
3.62E−07	4.33E−07	5.06E−07	5.77E−07	6.51E−07	7.22E−07	7.92E−07
2.30E−07	2.95E−07	3.65E−07	4.28E−07	4.94E−07	5.60E−07	6.20E−07
8.12E−08	1.44E−07	2.09E−07	2.68E−07	3.28E−07	3.86E−07	4.93E−07
1.50E−07	0.00E+00	4.62E−08	9.87E−08	1.52E−07	2.05E−07	2.51E−07
2.43E−07	2.73E−07	2.51E−07	5.37E−07	9.47E−09	4.96E−08	8.44E−08
3.63E−07	3.45E−07	3.28E−07	3.18E−07	3.13E−07	3.15E−07	3.22E−07
5.10E−07	4.93E−07	4.77E−07	4.64E−07	4.57E 07	4.52E−07	4.54E−07
6.73E−07	6.26E−07	6.48E−07	6.43E−07	6.38E−07	6.36E−07	6.44E−07
8.46E−07	8.43E−07	8.37E−07	8.34E−07	8.37E−07	8.42E−07	8.57E−07
1.03E−06	1.03E−06	1.03E−06	1.04E−06	1.05E−06	1.07E−06	1.09E−06
1.23E−06	1.24E−06	1.24E−06	1.26E−06	1.28E−06	1.30E−06	1.33E−06
1.43E−06	1.45E−06	1.46E−06	1.49E−06	1.52E−06	1.55E−06	1.59E−06
1.64E−06	1.67E−06	1.70E−06	1.73E−06	1.76E−06	1.81E−06	1.86E−06
1.87E−06	1.90E−06	1.94E−06	1.98E−06	2.03E−06	2.08E−06	2.14E−06
2.10E−06	2.15E−06	2.19E−06	2.24E−06	2.29E−06	2.35E−06	2.43E−06

Table C5 Graupel–drop collision kernels (m^3s^{-1}) for 0.4 g cm^{-3} density graupel. The upper row shows radii of graupel in μm. The left column shows radii of drops in μm (from Khain et al., 2001; © American Meteorological Society. Used with permission).

	100	120	140	160	180	200	220	240
2	0.00E+00	0.00E+00	0.00E+00	0.00E+00	0.00E+00	0.00E+00	0.00E+00	0.00E+00
4	0.00E+00	2.78E−10	2.20E−09	6.31E−09	1.36E−08	2.51E−08	4.07E−08	6.24E−08
6	2.13E−09	5.33E−09	1.15E−08	2.20E−08	3.77E−08	6.04E−08	8.96E−08	1.26E−07
8	4.13E−09	9.19E−09	1.82E−08	3.26E−08	5.36E−08	8.25E−08	1.19E−07	1.65E−07
10	5.26E−09	1.18E−08	2.27E−08	3.97E−08	6.37E−08	9.67E−08	1.38E−07	1.89E−07
12	6.64E−09	1.37E−08	2.58E−08	4.46E−08	7.09E−08	1.06E−07	1.50E−07	2.05E−07
14	7.38E−09	1.51E−08	2.80E−08	4.80E−08	7.56E−08	1.13E−07	1.59E−07	2.16E−07
16	7.91E−09	1.60E−08	2.79E−08	5.04E−08	7.93E−08	1.18E−07	1.66E−07	2.23E−07
18	8.23E−09	1.67E−08	3.09E−08	5.23E−08	8.21E−08	1.22E−07	1.70E−07	2.30E−07
20	8.43E−09	1.71E−08	3.17E−08	5.38E−08	8.44E−08	1.25E−07	1.75E−07	2.36E−07
22	8.51E−09	1.74E−08	3.23E−08	5.49E−08	8.60E−08	1.27E−07	1.78E−07	2.39E−07
24	8.46E−09	1.75E−08	3.27E−08	5.57E−08	8.74E−08	1.29E−07	1.80E−07	2.42E−07
26	8.32E−09	1.75E−08	3.30E−08	5.62E−08	8.84E−08	1.30E−07	1.83E−07	2.46E−07
28	8.07E−09	1.74E−08	3.30E−08	5.66E−08	8.88E−08	1.32E−07	1.84E−07	2.48E−07
30	7.73E−09	1.71E−08	3.29E−08	5.66E−08	8.92E−08	1.32E−07	1.85E−07	2.49E−07
32	7.30E−09	1.67E−08	3.26E−08	5.65E−08	8.94E−08	1.33E−07	1.86E−07	2.51E−07
34	6.78E−09	1.62E−08	3.22E−08	5.63E−08	8.95E−08	1.34E−07	1.87E−07	2.52E−07
36	6.13E−09	1.56E−08	3.17E−08	5.60E−08	8.94E−08	1.33E−07	1.87E−07	2.53E−07
38	5.36E−09	1.49E−08	3.10E−08	5.54E−08	8.92E−08	1.33E−07	1.88E−07	2.54E−07
40	4.46E−09	1.41E−08	3.02E−08	5.48E−08	8.85E−08	1.33E−07	1.87E−07	2.53E−07
50	0.00E+00	7.96E−09	2.43E−08	4.91E−08	8.34E−08	1.29E−07	1.85E−07	2.51E−07
60	6.63E−09	0.00E+00	1.51E−08	4.01E−08	7.46E−08	1.20E−07	1.76E−07	2.44E−07
70	1.58E−08	9.85E−09	0.00E+00	2.76E−08	6.22E−08	1.08E−07	1.64E−07	2.32E−07
80	2.75E−08	2.24E−08	1.10E−08	1.50E−08	4.62E−08	9.19E−08	1.48E−07	2.16E−07
90	4.19E−08	3.83E−08	2.66E−08	1.29E−08	2.59E−08	7.21E−08	1.28E−07	1.96E−07
100	5.90E−08	5.69E−08	4.64E−08	2.66E−08	0.00E+00	4.82E−08	1.04E−07	1.71E−07
110	7.88E−08	7.87E−08	6.97E−08	5.03E−08	2.35E−08	1.85E−08	7.60E−08	1.42E−07
120	1.02E−07	1.04E−07	9.65E−08	7.81E−08	4.97E−08	2.85E−08	4.26E−08	1.10E−07
130	1.27E−07	1.32E−07	1.27E−07	1.10E−07	8.19E−08	4.38E−08	0.00E+00	7.20E−08
140	1.57E−07	1.64E−07	1.61E−07	1.45E−07	1.18E−07	7.97E−08	3.74E−08	2.72E−08
150	1.89E−07	1.99E−07	1.98E−07	1.85E−07	1.60E−07	1.21E−07	7.44E−08	4.78E−08
160	2.25E−07	2.39E−07	2.40E−07	2.28E−07	2.05E−07	1.68E−07	1.21E−07	6.67E−08
170	2.66E−07	2.82E−07	2.85E−07	2.76E−07	2.55E−07	2.19E−07	1.73E−07	1.17E−07
180	3.10E−07	3.29E−07	3.35E−07	3.29E−07	3.09E−07	2.75E−07	2.30E−07	1.75E−07
190	3.58E−07	3.80E−07	3.89E−07	3.86E−07	3.68E−07	3.35E−07	2.92E−07	2.38E−07
200	4.10E−07	4.36E−07	4.50E−07	4.47E−07	4.32E−07	4.01E−07	3.60E−07	3.07E−07
210	4.66E−07	4.96E−07	5.13E−07	5.14E−07	5.01E−07	4.72E−07	4.33E−07	3.83E−07
220	5.28E−07	5.61E−07	5.82E−07	5.85E−07	5.74E−07	5.48E−07	5.11E−07	4.62E−07
230	5.93E−07	6.31E−07	6.55E−07	6.62E−07	6.54E−07	6.30E−07	5.95E−07	5.48E−07
240	6.64E−07	7.08E−07	7.34E−07	7.44E−07	7.38E−07	7.17E−07	6.84E−07	6.40E−07
250	7.40E−07	7.89E−07	8.19E−07	8.31E−07	8.29E−07	8.10E−07	7.80E−07	7.38E−07

260	280	300	320	340	360	380	400
0.00E+00	0.00E+00	4.00E−12	1.35E−11	3.14E−11	6.06E−11	2.49E−10	2.55E−09
9.02E−08	1.24E−07	1.66E−07	2.16E−07	2.73E−07	3.36E−07	4.08E−07	4.89E−07
1.73E−07	2.27E−07	2.91E−07	3.64E−07	4.45E−07	5.37E−07	6.36E−07	7.43E−07
2.21E−07	2.86E−07	3.61E−07	4.46E−07	5.39E−07	6.42E−07	7.51E−07	8.68E−07
2.50E−07	3.21E−07	4.03E−07	4.93E−07	5.93E−07	7.00E−07	8.17E−07	9.39E−07
2.70E−07	3.44E−07	4.30E−07	5.24E−07	6.29E−07	7.38E−07	8.55E−07	9.82E−07
2.83E−07	3.60E−07	4.48E−07	5.46E−07	6.52E−07	7.65E−07	8.85E−07	1.01E−06
2.92E−07	3.72E−07	4.63E−07	5.61E−07	6.70E−07	7.83E−07	9.05E−07	1.03E−06
3.01E−07	3.82E−07	4.72E−07	5.74E−07	6.83E−07	7.97E−07	9.18E−07	1.05E−06
3.07E−07	3.89E−07	4.81E−07	5.83E−07	6.92E−07	8.08E−07	9.30E−07	1.06E−06
3.11E−07	3.94E−07	4.88E−07	5.91E−07	7.01E−07	8.19E−07	9.42E−07	1.07E−06
3.16E−07	4.00E−07	4.94E−07	5.98E−07	7.08E−07	8.26E−07	9.50E−07	1.08E−06
3.20E−07	4.05E−07	4.99E−07	6.03E−07	7.13E−07	8.32E−07	9.57E−07	1.08E−06
3.22E−07	4.08E−07	5.03E−07	6.08E−07	7.19E−07	8.38E−07	9.60E−07	1.09E−06
3.25E−07	4.10E−07	5.07E−07	6.13E−07	7.24E−07	8.44E−07	9.66E−07	1.10E−06
3.27E−07	4.13E−07	5.10E−07	6.14E−07	7.29E−07	8.46E−07	9.69E−07	1.10E−06
3.28E−07	4.15E−07	5.11E−07	6.18E−07	7.31E−07	8.48E−07	9.74E−07	1.10E−06
3.29E−07	4.16E−07	5.14E−07	6.19E−07	7.32E−07	8.53E−07	9.76E−07	1.11E−06
3.30E−07	4.18E−07	5.15E−07	6.20E−07	7.36E−07	8.54E−07	9.77E−07	1.11E−06
3.31E−07	4.18E−07	5.15E−07	6.23E−07	7.36E−07	8.55E−07	9.81E−07	1.11E−06
3.29E−07	4.18E−07	5.18E−07	6.25E−07	7.39E−07	8.57E−07	9.83E−07	1.11E−06
3.22E−07	4.12E−07	5.11E−07	6.18E−07	7.34E−07	8.52E−07	9.74E−07	1.10E−06
3.11E−07	4.00E−07	4.99E−07	6.08E−07	7.21E−07	8.38E−07	9.58E−07	1.08E−06
2.95E−07	3.84E−07	4.84E−07	5.90E−07	7.02E−07	8.18E−07	9.36E−07	1.06E−06
2.74E−07	3.63E−07	4.62E−07	5.67E−07	6.78E−07	7.92E−07	9.08E−07	1.03E−06
2.49E−07	3.38E−07	4.34E−07	5.40E−07	6.50E−07	7.61E−07	8.79E−07	9.95E−07
2.20E−07	3.08E−07	4.04E−07	5.08E−07	6.16E−07	7.26E−07	8.40E−07	9.53E−07
1.87E−07	2.74E−07	3.69E−07	4.72E−07	5.77E−07	6.85E−07	7.96E−07	9.06E−07
1.49E−07	2.35E−07	3.30E−07	4.29E−07	5.34E−07	6.39E−07	7.44E−07	8.53E−07
1.06E−07	1.91E−07	2.84E−07	3.83E−07	4.86E−07	5.88E−07	6.90E−07	7.95E−07
5.68E−08	1.43E−07	2.34E−07	3.31E−07	4.31E−07	5.32E−07	6.30E−07	7.31E−07
0.00E+00	8.85E−08	1.79E−07	2.75E−07	3.72E−07	4.70E−07	5.65E−07	6.62E−07
6.04E−08	2.62E−08	1.18E−07	2.12E−07	3.07E−07	4.01E−07	4.94E−07	5.86E−07
1.11E−07	6.31E−08	5.08E−08	1.44E−07	2.37E−07	3.28E−07	4.17E−07	5.03E−07
1.74E−07	1.07E−07	1.15E−07	6.93E−08	1.60E−07	2.48E−07	3.32E−07	4.15E−07
2.45E−07	1.75E−07	1.07E−07	4.31E−08	7.82E−08	1.62E−07	2.43E−07	3.22E−07
3.21E−07	2.52E−07	1.79E−07	1.16E−07	8.41E−08	7.45E−08	1.49E−07	2.22E−07
4.03E−07	3.36E−07	2.63E−07	1.93E−07	1.35E−07	1.50E−07	6.17E−08	1.21E−07
4.92E−07	4.26E−07	3.56E−07	2.85E−07	2.19E−07	1.67E−07	1.46E−07	2.77E−08
5.87E−07	5.22E−07	4.54E−07	3.86E−07	3.19E−07	2.63E−07	2.17E−07	1.88E−07
6.85E−07	6.26E−07	5.61E−07	4.94E−07	4.31E−07	3.74E−07	3.27E−07	2.89E−07

Table C6 Graupel–drop collision kernels (m^3s^{-1}) for 0.8 g cm^{-3} density graupel. The upper row shows radii of graupel in μm. The left column shows radii of drops in μm (from Khain et al., 2001; © American Meteorological Society. Used with permission).

	100	120	140	160	180	200	220	240	260	280	300	320
2	1.95E−12	5.47E−12	1.67E−11	4.16E−11	1.37E−10	1.21E−09	8.49E−09	2.11E−08	3.89E−08	6.18E−08	9.12E−08	1.25E−07
4	3.34E−09	9.12E−09	1.99E−08	3.75E−08	6.28E−08	9.72E−08	1.42E−07	1.97E−07	2.63E−07	3.40E−07	4.27E−07	5.24E−07
6	8.59E−09	1.94E−08	3.74E−08	6.49E−08	1.02E−07	1.51E−07	2.12E−07	2.85E−07	3.70E−07	4.66E−07	5.37E−07	6.90E−07
8	1.20E−08	2.56E−08	4.76E−08	8.03E−08	1.24E−07	1.80E−07	2.49E−07	3.31E−07	4.24E−07	5.29E−07	6.43E−07	7.65E−07
10	1.43E−08	2.97E−08	5.42E−08	8.98E−08	1.38E−07	1.97E−07	2.72E−07	3.57E−07	4.55E−07	5.65E−07	6.83E−07	8.06E−07
12	1.59E−08	3.26E−08	5.87E−08	9.67E−08	1.47E−07	2.10E−07	2.86E−07	3.76E−07	4.76E−07	5.88E−07	7.07E−07	8.34E−07
14	1.72E−08	3.47E−08	6.21E−08	1.02E−07	1.53E−07	2.18E−07	2.97E−07	3.88E−07	4.92E−07	6.04E−07	7.26E−07	8.55E−07
16	1.81E−08	3.64E−08	6.48E−08	1.06E−07	1.59E−07	2.25E−07	3.06E−07	3.99E−07	5.03E−07	6.17E−07	7.39E−07	8.70E−07
18	1.89E−08	3.78E−08	6.69E−08	1.09E−07	1.63E−07	2.31E−07	3.12E−07	4.07E−07	5.13E−07	6.28E−07	7.52E−07	8.81E−07
20	1.94E−08	3.88E−08	6.88E−08	1.11E−07	1.67E−07	2.35E−07	3.18E−07	4.15E−07	5.22E−07	6.39E−07	7.64E−07	8.91E−07
22	1.99E−08	3.98E−08	7.03E−08	1.14E−07	1.70E−07	2.40E−07	3.25E−07	4.22E−07	5.29E−07	6.48E−07	7.74E−07	9.02E−07
24	2.02E−08	4.06E−08	7.17E−08	1.16E−07	1.73E−07	2.44E−07	3.29E−07	4.28E−07	5.36E−07	6.56E−07	7.80E−07	9.12E−07
26	2.05E−08	4.11E−08	7.28E−08	1.18E−07	1.76E−07	2.47E−07	3.34E−07	4.32E−07	5.43E−07	6.61E−07	7.89E−07	9.18E−07
28	2.06E−08	4.18E−08	7.38E−08	1.20E−07	1.78E−07	2.51E−07	3.38E−07	4.38E−07	5.47E−07	6.69E−07	7.94E−07	9.28E−07
30	2.07E−08	4.20E−08	7.47E−08	1.21E−07	1.80E−07	2.54E−07	3.41E−07	4.43E−07	5.54E−07	6.73E−07	8.03E−07	9.33E−07
32	2.07E−08	4.24E−08	7.55E−08	1.22E−07	1.82E−07	2.56E−07	3.45E−07	4.46E−07	5.58E−07	6.81E−07	8.08E−07	9.39E−07
34	2.06E−08	4.26E−08	7.62E−08	1.23E−07	1.84E−07	2.59E−07	3.48E−07	4.50E−07	5.64E−07	6.85E−07	8.12E−07	9.48E−07
36	2.04E−08	4.26E−08	7.66E−08	1.24E−07	1.86E−07	2.61E−07	3.50E−07	4.54E−07	5.67E−07	6.89E−07	8.17E−07	9.52E−07
38	2.02E−08	4.27E−08	7.68E−08	1.25E−07	1.87E−07	2.63E−07	3.54E−07	4.57E−07	5.71E−07	6.93E−07	8.24E−07	9.57E−07
40	1.99E−08	4.26E−08	7.73E−08	1.26E−07	1.88E−07	2.65E−07	3.56E−07	4.60E−07	5.74E−07	6.99E−07	8.29E−07	9.61E−07
50	1.69E−08	4.07E−08	7.69E−08	1.28E−07	1.92E−07	2.71E−07	3.66E−07	4.73E−07	5.90E−07	7.15E−07	8.45E−07	9.83E−07
60	1.19E−08	3.64E−08	7.38E−08	1.26E−07	1.93E−07	2.75E−07	3.71E−07	4.80E−07	5.99E−07	7.25E−07	8.61E−07	9.95E−07
70	4.12E−09	2.97E−08	6.83E−08	1.22E−07	1.91E−07	2.75E−07	3.73E−07	4.84E−07	6.07E−07	7.35E−07	8.68E−07	1.00E−06
80	4.34E−09	2.05E−08	5.99E−08	1.16E−07	1.86E−07	2.71E−07	3.72E−07	4.86E−07	6.08E−07	7.37E−07	8.71E−07	1.01E−06
90	1.53E−08	8.24E−09	4.90E−08	1.06E−07	1.78E−07	2.65E−07	3.67E−07	4.83E−07	6.06E−07	7.36E−07	8.70E−07	1.00E−06
100	2.94E−08	5.97E−09	3.52E−08	9.31E−08	1.67E−07	2.55E−07	3.59E−07	4.77E−07	6.01E−07	7.31E−07	8.64E−07	9.97E−07
110	4.63E−08	2.14E−08	1.83E−08	7.78E−08	1.52E−07	2.42E−07	3.48E−07	4.66E−07	5.92E−07	7.22E−07	8.55E−07	9.87E−07
120	6.58E−08	4.11E−08	0.00E+00	5.89E−08	1.35E−07	2.26E−07	3.33E−07	4.53E−07	5.79E−07	7.10E−07	8.42E−07	9.73E−07
130	8.83E−08	6.40E−08	2.31E−08	3.70E−08	1.14E−07	2.07E−07	3.15E−07	4.36E−07	5.62E−07	6.93E−07	8.25E−07	9.55E−07
140	1.14E−07	9.00E−08	4.85E−08	1.18E−08	8.97E−08	1.84E−07	2.93E−07	4.15E−07	5.42E−07	6.72E−07	8.04E−07	9.32E−07
150	1.43E−07	1.20E−07	7.77E−08	1.84E−08	6.19E−08	1.57E−07	2.67E−07	3.90E−07	5.17E−07	6.48E−07	7.78E−07	9.04E−07
160	1.76E−07	1.53E−07	1.11E−07	4.88E−08	3.09E−08	1.26E−07	2.37E−07	3.61E−07	4.88E−07	6.18E−07	7.48E−07	8.72E−07
170	2.12E−07	1.90E−07	1.48E−07	8.47E−08	1.61E−08	9.13E−08	2.04E−07	3.27E−07	4.55E−07	5.85E−07	7.13E−07	8.35E−07

180	2.52E−07	2.30E−07	1.89E−07	1.25E−07	4.56E−08	5.35E−08	1.66E−07	2.90E−07	4.18E−07	5.47E−07	6.74E−07	7.94E−07
190	2.95E−07	2.75E−07	2.34E−07	1.69E−07	8.86E−08	2.11E−08	1.24E−07	2.48E−07	3.75E−07	5.04E−07	6.29E−07	7.47E−07
200	3.43E−07	3.24E−07	2.83E−07	2.18E−07	1.37E−07	4.26E−08	7.83E−08	2.01E−07	3.28E−07	4.56E−07	5.80E−07	6.95E−07
210	3.95E−07	3.79E−07	3.38E−07	2.72E−07	1.90E−07	9.19E−08	3.49E−08	1.50E−07	2.76E−07	4.03E−07	5.25E−07	6.40E−07
220	4.52E−07	4.36E−07	3.96E−07	3.31E−07	2.48E−07	1.48E−07	4.33E−08	9.62E−08	2.21E−07	3.46E−07	4.66E−07	5.77E−07
230	5.15E−07	4.99E−07	4.59E−07	3.94E−07	3.10E−07	2.10E−07	9.79E−08	4.66E−08	1.60E−07	2.83E−07	4.00E−07	5.07E−07
240	5.81E−07	5.67E−07	5.27E−07	4.62E−07	3.78E−07	2.78E−07	1.63E−07	5.45E−08	9.74E−08	2.15E−07	3.30E−07	4.31E−07
250	6.52E−07	6.39E−07	6.00E−07	5.35E−07	4.51E−07	3.50E−07	2.35E−07	1.16E−07	5.78E−08	1.45E−07	2.55E−07	3.50E−07

References

Ayala O., B. Rosa, L.-P. Wang, and W.W. Grabowski, 2008: Effects of turbulence on the geometric collision rate of sedimenting droplets: Part 1. Results from direct numerical simulation. *New J. Physics*, **10**, 075015.

Khain A., M. Pinsky, M. Shapiro and A. Pokrovsky, 2001: Collision rate of small graupel and water drops. *J. Atmos. Sci.*, **58**, 2571–2595.

Pinsky M., A. P. Khain, and M. Shapiro, 2001: Collision efficiency of drops in a wide range of Reynolds numbers: Effects of pressure on spectrum evolution. *J. Atmos. Sci.* **58**, 742–764.

Pinsky, M., Khain, A., Krugliak, H., 2008: Collisions of cloud droplets in a turbulent flow. Part 5: application of detailed tables of turbulent collision rate enhancement to simulation of droplet spectra evolution. *J. Atmos. Sci.*, **65**, 357–374.

Index

κ–ε theory with 1.5 order closure, 97

Acceleration, 68, 351
  gravitational, 68, 178
  Lagrangian, 92
Accretion, 58
Activation spectrum, 160
Adiabatic air parcel, 74
Adiabatic liquid water content, 283
Adiabatic process, 69
Advection, 77, 80, 86, 111
  -diffusion equation, 80
  equation, 126, 129
  sub-step, 189, 199, 203
Aerosol, 3
  activation, 331
  direct effect, 536
  first indirect effect, 536
  mass content, 562
  particles, 29
  second indirect effect, 536
Aerosols, 8, 30, 82
  background, 30
  desert dust storm, 31
  maritime, 30
  polar, 31
  remote continental, 31
  rural, 31
  urban, 31
Aggregates, 3–4, 8
Aliasing, 103
  error, 103
Amplification factor, 127, 138
Amplitude errors, 128
Approximation accuracy, 124
Artificial spectrum broadening, 242, 314
Aspect ratio, 21, 39
  of crystals, 397
  of drop, 39, 447
  of hail, 58
  of ice particle, 416
Atmospheric boundary layer, 26
Autoconversion, 81, 247
Available convective potential energy, 75
Averaging operators, 103

Barometric formula, 68
Batchelor formula, 219
Binary collisions, 229
Bin-emulating procedure, 211, 506
Biomass burning aerosol particles, 35, 204
Boltzmann constant, 355
Boundary conditions, 85
  cyclic, 144
  Neumann, 143
  periodic, 144
  radiation, 145
Boundary layer, 4–5
Boussinesq approximation, 84
Bright band, 50
Brownian diffusion, 222
Brunt-Vaisala frequency, 120
Bubbles, 278
Budget equation, 79
Bulk density, 20–21

Centered finite difference, 122–123
Characteristic mixing time, 277
Chemical composition of aerosols, 158
Chemical potential, 152
Chemistry term, 181
Clausius-Clapeyron equation, 71
Closure problem, 249
Cloud anvils, 55
Cloud base, 4
Cloud condensational nuclei, 8, 33
Cloud condensational nuclei (CCN)
    efficiency, 160–161
Cloud streets, 13, 15
Cloud turbulence, 42, 98
Cloud water content, 81, 274, 526
Cloud-aerosol interaction, 19
Clouds, 1–2
  altocumulus, 6
  altostratus, 6
  arctic, 366, 477, 479
  cirrus, 6
  continental, 45–46, 49, 62, 176, 186
  convective, 7, 42, 45, 95, 166, 169, 183, 198,
    294–296, 301
  cumulonimbus, 8
  cumulus, 7
  cumulus congestus, 8
  deep convective, 2, 535
  maritime, 46, 48, 62, 160–161, 168–169,
    175–176, 185–186, 205, 245–246, 279,
    284, 318–319, 322, 324, 328
  mixed-phase, 8, 51
  nimbostratus, 6
  of vertical development, 4
  orographic, 9
  pyro, 47–48, 198
  smoky, 47, 201
  stratiform, 4, 6
  stratocumulus, 5–6, 8, 26, 44–45, 98, 120,
    158, 169, 252, 302, 328, 381, 501, 542
  streets, 115
Cloud-topped boundary layer (BL), 85
Clustering effect, 219
Coalescence, 63
Coalescence efficiency, 224
Coalescence energy, 228
Coefficient, 207
  of computational viscosity, 129
  correlation, 2
  drag, 207
  friction, 135–136
  of molecular diffusion, 80, 155
  of molecular heat conductivity, 80
  osmotic, 154
  thermodiffusion, 88
  turbulent, 89, 108
  turbulent viscosity, 109, 118
  of variation, 24
  ventilation, 167, 211
Coherent structure, 91
Collection efficiency, 224, 234, 252
Collection kernel, 224, 249
Collection-breakup equation, 231, 245
Collector, 222
Collision, 3, 8, 13
Collision cross section, 262
Collision efficiency, 222–223, 226
Collision kernel, 223, 225
  cylindrical, 224
  Golovin, 225
  hydrodynamical, 224, 235
  Long, 154, 225
  spherical, 259
Collision kinetic energy, 227, 232
Collision rate, 40, 48
Collisional breakup, 151, 215
Columnar crystals, 52, 350
Concentration, 4, 13, 15
  of aerosols, 315, 373
  of cloud condensational nuclei, 42, 82, 151,
    160
  of drops, 27, 200, 209
  of ice particles, 63, 375
  preferential, 219, 221, 265
Concentration-averaged fall velocity, 81, 110,
    203
Conservation law, 17, 68, 133
Conservative thermodynamic quantities, 72
Continuity equation, 68, 79
Continuous growth equation, 224, 238
Convection, 6, 15, 50, 64
Convection invigoration, 537, 554–555
Convective available potential energy
    (CAPE), 77–78
Convective cell, 5, 7, 13
  closed, 13, 584
  open, 13, 586
Convective motion, 100, 102–104
Coriolis parameter, 83, 139
Correlation function, 91–92
Courant number, 127, 129, 131
Courant-Friedrichs-Lewy condition (CFL), 128
Covariance, 90
Critical Reynolds number, 114–115
Critical Richardson number, 87
Critical supersaturation, 157–158
Crystal habit, 51, 53–54, 397
Curvature term, 154, 156, 167

Dalton law, 68, 70
Damköhler number, 282–283, 285
Deliquesce humidity, 19
Dendrites, 20–22, 52
Density, 9, 20
  of aerosol particle, 154
  of dry air, 68
  of ice, 51, 351
  of water, 20
  of water vapor, 3, 19
  of wet air, xv, 84, 142, 208
Deposition, 13–14, 23
Depositional growth, 51, 378
Dew point, 4, 76
Diagnostic equation, 166
Differential activation spectrum, 164
Diffusion flux, 80–81, 107
Diffusion operator, 277
Diffusional growth, 19, 26–27
Diffusional growth sub-step, 199
Diffusion-collision stage of cloud development, 210, 342, 481
Diffusiophoresis, 375
Diffusivity of water vapor, 168, 212, 408
Dilution, 33, 278–279
Direct numerical simulation, 85, 144, 179
Disdrometer, 49, 66
Dispersion of size distribution, 24, 28–29
Dispersion relationship, 137
Dissipation sub-range of turbulence, 89, 93
Distribution, 1, 4, 16
  bimodal, 60, 65
  equilibrium raindrops, 245, 333
  exponential, 24, 28
  Gamma, 28
  general Gamma, 28–29
  lognormal, 29, 176, 231
  Marshall-Palmer, 50, 57, 245
  mass, 23, 26
  number, 23, 30
  Poisson, 265, 267
  size, 17, 24–25
  two-dimensional, 27, 332
Doppler radar, 64, 98, 121
Drizzle, 2
Drop clustering, 218, 257
Drop inertia, 207, 216–217
Drop motion equation, 207, 226
Drop oscillations, 213, 341
Drop relaxation time, 187, 207, 209
Drop size distribution, 42, 49, 64
Droplet spectrum broadening, 43
Dry adiabatic gradient, 70, 74
Dry adiabatic lapse rate, 74, 120, 183
Dynamic equations, 85, 106, 109–110
Dynamic viscosity, 83, 207, 374

Electrical capacitance, 378, 393, 396
Embedded convection, 98
Energy cascade, 88–89
Enhancement factor, 222, 253, 264
Ensemble averaging, 103
Entity-type entrainment mixing theory, 311
Entrainment, 182–183, 278
Entrainment parameter, 298, 313
Equation for the turbulent kinetic energy, 112
Equilibrium size, 19, 155, 158
Equilibrium state, 219, 280, 283
Equivalent particles, 20, 22
Equivalent potential temperature, 73–74, 80
Equivalent shapes, 20–21
Exner function, 70, 142–143
Exponential function, 63, 91, 203
External turbulent scale, 88, 93–94

Fast Fourier transform (FFT), 143
Ferrel cell, 117
Filament, 12, 94, 231
Filling algorithms, 135
Finite difference approach, 102
Finite difference quotient, 123
Finite difference representation, 123, 129, 134
Finite spatial difference, 122
Finite-difference grid, 102–103
First law of thermodynamics, 17, 68
First rain drops, 243, 569
Flatness, 90, 94, 98
Force, 4, 7, 68
  added mass, 207
  Archimedes, 207
  Basset, 207
  buoyancy, 4
  centrifugal, 216, 218
  Coriolis, 83–84
  drag, 207
  friction, 7, 75, 82
  gravity, 68, 83, 207
  inertial, 84, 86
Fourier series, 123, 127, 143
Fourier transform, 91–92
Free convection, 77, 114
Free-molecular regime, 167
Freezing, 3–4
  condensation, 346, 365, 375
  contact, 346, 375
  deliquescent-heterogeneous, 346
  heterogeneous, 346, 486, 490
  homogeneous, 55, 345, 356
  immersion, 346, 361, 368
  nucleation, 357, 363, 365
Freezing level, 58, 61, 345
Freezing rain, 344, 453
Friction equation, 135, 148, 206
Frozen drops, 20, 51–52, 346
Frozen flows, 104
Frozen turbulence hypothesis, 104
Fully developed turbulence, 332, 337

Gain integral, 224, 230
Geometrical cross section, 223, 425
Giant cloud condensational nuclei, 343, 496
Giant particles, 19, 66
Glaciation, 339, 347, 382
Glaciation time, 391–393, 536
Glaciogenic seeding, 19, 511
Graupel, 3, 9, 20
Gravitational collisions, 223, 259, 272
Gravity waves, 75, 87, 112
Grazing trajectories, 223
Grid point, 25–26, 102
Grid spacing, 94, 102
Grid spacing lengths, 105

Habit diagram, 52, 63, 397
Hadley cell, 117
Hail, 3, 8–9
  dry growth, 20, 59, 419
  high-density wet growth, 347
  spongy wet growth, 462
  wet growth, 20, 59, 344
Hailstones, 8, 58–59
Hailstorm, 8, 62, 344
Hallett-Mossop mechanism, 478
Haze particles, 19, 33
Heat conductivity, 109, 115
  of air, 347, 349
  of ice, 349, 454
  of water, 442
Homogeneous and isotropic turbulence, 88, 92–93
Homogeneous mixing, 280, 282
Humidity, 3–5
  absolute, 70, 166, 188
  relative, 4, 8–9, 34
Hurricane, 15, 25, 65
Hydrodynamic drop interaction, 222, 259, 262
Hydrometeors, 3–4, 9, 17
Hyperbolic equation, 137

Ice crystals, 3, 5–7
Ice multiplication, 344, 346, 418, 477, 479
Ice nuclei, 19, 333
Ice water content, 55, 64, 81, 377
Ice-water potential temperature, xv
Impact velocity, 431, 461
Inertia gravity waves, 137, 139
Inertial sub-range of turbulence, 89, 92, 218
Initial conditions, 30, 122, 166
In-situ measurements, 15, 183
Instability, 4, 7, 13
  baroclinic, 119
  barotropic, 118
  cloud boundary, 279
  cloud-environment interface, 120, 279, 324
  convective, 74–75
  hydrodynamic, 87, 119, 213
  nonlinear, 131–132
  Rayleigh-Benard, 117
Integral length, 168
Integral turbulent scale, 93
Intermittency, 94, 98
Internal liquid, 463, 482
Internal turbulent scale, 88
Isobaric process, 69
Isochoric process, 69
Isothermal process, 69

Karman vortex street, 114
Kelvin formula, 156
Kelvin's law, 152
Kinematic viscosity, 83, 87, 115
Kinetic energy, 3, 78, 85
Kinetic equation, 27, 64
Knudsen number, xiii, 374
Köhler curve, 155–156
Köhler theory, 158, 193
Kolmogorov law, 89, 97, 218
Kolmogorov microscale, 89, 254, 258
Kolmogorov time scale, xv, 258, 416
Kronecker symbol, xv, 81, 595
K-theory, 107–108, 112

Lagrange interpolation formula, 237
Lagrangian approach, 134, 370
Lagrangian correlation function, 91

rid, 25,

on, 93
n, 93, 219

349, 351
0, 417
tion, 23, 243, 274
50, 318
sedimentation velocity, 81
, 15
r, 50, 441, 450
onvective system (MCS), 11, 13
4, 15–16
ating-direction implicit, 203
y and Reinhardt, 235–237
x, 134–135
inite-difference, 80, 102, 203
flux, 211, 237
implicit, 203
of moments, 26–27, 202
point-based, 235
spectral moment, 235, 239–240
splitting-up, 122, 143, 203
superposition, 226, 263
von Neumann, 127
Mid-latitude storm, 65, 489–490, 543
Mixing, 4–5, 36, 66
conservative, 151, 345
extremely inhomogeneous, 280, 282–283
homogeneous, 280, 282
in-cloud, 305, 328
inhomogeneous, 280, 282–284
mechanical, 282
nonconservative, 151
passive, 300
turbulent, 4, 109–110
Mixing diagram, 282–283
Mixing length, 107
Mixing operator, 276, 278, 299
Mixing ratio, 70–72
of ice water, 110, 354, 385
of liquid water, 73, 110, 112
water vapor, 112, 152, 165

Mode
accumulation, 30–31
acoustic, 142
Aitken, 31, 158
coarse, 30
lognormal, 176
nucleation, 30
Model, 2
anelastic, 142
bulk parameterization, 210, 579
diffusion-evaporation, 16–17, 284, 286, 289
Eulerian cloud, 471
explicit mixing parcel, 312
fully compressible, 142
global circulation, 367, 497, 558
hybrid, 257
kinematic of turbulence, 257, 269
Lagrangian, 26
Lagrangian parcel, 311
large-eddy simulation (LES), 64, 302, 501
mesoscale, 126, 143, 149
non-adiabatic parcel, 298
non-hydrostatic, 142, 150
numerical, 21, 24
spectral bin microphysics, 82, 192
state-of-the-art numerical cloud, 271
statistical Langevin-type, 258
statistical of turbulence, 121, 258
trajectory ensemble, 65–66, 149
Moist adiabatic gradient, 74–75, 319
Moist adiabatic lapse rate, 74
Molecular diffusion, 80, 109, 115
Molecular weight, 68
Molecular weight of aerosol salt, 154
Molecular weight of water, 70
Moment of the size distribution, 23
Monodisperse spectra, 169
Motion equations, 68, 79, 82
Moving averaging, 103
Multi-fractal structure, 94

Navier–Stokes equation, 83, 85–86
Nesting grid, 144, 146
Neutral atmosphere, 84
N-LWC, 166
Non-resolvable scales, 123
Nontrivial solution, 138
Nucleation, 8, 30–31
deposition, 346, 360–361
droplet, 8, 64, 136
heterogeneous, 155–156, 359
homogeneous, 155, 345
in-cloud, 66, 158, 174
Nucleation procedure, 188, 199, 204
Numerical dispersion, 131, 139
Numerical viscosity, 128–129

Optical extinction, 24, 28–29
Order of approximation accuracy, 124–125
Oscillations of a cloud parcel, 174

Particle size distribution (PSD), 158
of hail, 63
of ice particles, 62, 64, 66
of snow, 63
Phase errors, 130–131
Phase relaxation time, 148, 168
Phase speed, 131, 137
Phase transition, 16, 73–74

Phase velocity, 145–146
Planck constant, 356
Poisson noise, 23, 25, 277
Polarimetric radar, 63, 66, 441
Polydisperse spectra, 66
Potential evaporation parameter, 286, 293
Potential temperature, 69–70, 73
Prandtl concept, 107
Prandtl number, 109, 112, 375
Precipitation, 3, 6
Pressure, 15, 64, 68
of dry air, 68, 187, 499
of moist air, 71, 188
Primary ice crystals, 51
Probability density function (PDF), 276
Problem of the equation closure, 106
Prognostic equation, 113, 411, 414
Psychrometric correction, 206

Quasi-compressible approximation, 143
Quasi-geostrophic motion, 139
Quasi-hydrostatic approximation, 165, 188
Quasi-static equation, 68, 83
Quasi-stationary turbulence, 88
Quasi-steady supersaturation, 168–170
Quasi-stochastic collection equation, 225

Radar reflectivity, 24, 26
Radial distribution function (RDF), 265
Radiative balance of atmosphere, 19
Radiative cooling, 182, 493
Radiative forcing, 51, 67, 500
Radiative heating, 73, 80, 560
Radius, 6, 8, 20
of correlation, 93
critical, 157–158
driving, 321
of drop, 210
of dry aerosol particle, 31, 154
effective, 24, 28–29
equivalent, 21, 23
of ice particle, 25, 379
of insoluble fraction of aerosol particle, 368
mean, 168
mean volume, 21, 194, 250
modal, 363
of wet particle, 158
Rainwater content (RWC), 81, 545
Rainfall, 50, 63, 65
Random field, 90
Random process, 90–91
Random variable, 90–91, 271
Rankine vortex, 217
Raoult's law, 153
Rayleigh damping, 135, 146, 499
Rayleigh number, 88, 115–116
Recycling, 311, 493, 574
Relative humidity (RH), 283, 358
Relaxation time, 148, 168
Relaxation zone, 147
Remapping, 26, 200
Residence time, 34–36
Resolvable scales, 103, 105, 132
Reynolds number, 84–86
Reynolds stress tensor, 82, 106
Richardson energetic cascade, 94
Richardson law, 89
Richardson number, 87

Riming, 20, 51
Roll vortices, 114–115

Saturation adjustment hypothesis, 199
Saturation mixing ratio, 74, 78
Saturation vapor pressure, 71, 73
  over ice, 376
  over water, 71, 376
Scavenging, 64, 373, 375
  collisional, 373
  nucleation, 525
Scheme, 3, 5, 8
  Adams–Bashforth, 126
  Arakawa, 140
  backward, 125–126
  box-Lagrangian, 209, 411
  bulk microphysical parameterization, 81
  conditionally stable, 128–129
  Euler, 125
  Euler-backward, 125
  explicit, 125, 128, 134
  forward in time and upstream advection, 126
  forward-backward, 143
  Houn, 125–126
  implicit, 125, 128
  iterative, 125–126
  Kovetz–Olund, 200–201
  leapfrog, 126, 129
  Matsuno, 125
  one-moment, 27, 242, 410
  positive definite, 134
  Runge–Kutta iterative, 125–126
  semi-implicit, 142
  three-level, 125–126
  three-moment, 142, 411
  trapezoidal, 125–126
  two-level, 125–126
  two-moment, 154, 242, 397
  unstable, 128–129
  upstream difference, 209, 410, 412
Schmidt number, 212, 408
Schumann–Ludlam limit, 461, 463
Sea spray, 33, 511, 557
Secondary ice crystals, 51, 64, 488
Secondary ice formation, 345–346, 400
Secondary ice production, 346, 418, 477
Self-collection, 250–251, 274
  of droplets, 247, 250, 274
  of raindrops, 250
Self-organization, 11, 119, 324
Semi-analytic solution, 205
Sensitivity study, 332, 336, 557
Shear production term, 112, 325
Shedding, 60, 345, 347
Similarity, 84–85, 114
Size sorting, 209–210
Skewness, 90, 94
Skew-T diagram, 75, 78
Slope parameter, 161–162

Small-scale intermittency, 94
Smoluchowski equation, 224
Snowflakes, 8, 20
Soaking stage, 406–407, 447
Solubility of aerosol water solution, 154
Soluble fraction, 19, 30
Sound wave, 128, 142–143
Spatial averaging, 44–45, 90
Specific gas constant, 68, 70
  of dry air, 68
  of water vapor, 70
Specific heat capacity, 69
  at constant pressure, 69
  at constant volume, 69
Specific humidity, 5, 70, 73
Spectral density function, 91–92
Spectrum of turbulent kinetic energy, 96
Spheroid, 20, 53, 352
  oblate, 21, 352, 378
  prolate, 20–21, 378
Splintering, 51
Sponge ice, 470
Spontaneous breakup, 15, 20, 40
Squall line, 13
Stability of finite difference scheme, 137
Staggered grids, 139–141
State equation, 68–69
Static energy, 70, 74
Stochastic breakup equation (SBE), 241
Stochastic collection equation (SCE), 224
Stochastic condensation theory, 179, 301
Stokes droplets, 208
Stokes number, 218–219, 221
Stokes' law of friction, 82
Stress tensor, 82, 106
Structure function, 91–93
Subgrid process, 105, 111, 117
Sublimation, 26, 73, 80
Superadiabatic droplets, 281, 309
Supercell storm, 143, 497, 528
Supercooled water drops, 346, 483, 489
Supersaturation, 30, 47, 52
  over ice, 354, 365, 370
  over water, 165, 344, 358
Supersaturation maximum near cloud base, 169–170, 178
Surface tension, 152
  of ice-air interface, 348
  of water-air interface, 348
  of water-ice interface, 355
Swept volume, 121, 222–223

Taylor hypothesis, 92
Taylor microscale, 96, 260, 324
Taylor microscale Reynolds number, 96, 324
Taylor series, 74, 124
Temperature gradient, 19, 70–71
Temperature lapse rate, 183
Tephigram, 75–77
Terminal velocity, 64, 207–208

  of
  of g
  of h
  of ic
  of v
Ther
Ther
The
Time
Time
Tran
Tran
Triple
Tropic
Turbul
Turbul
Turbule
Turbulen
Turbulent
Turbulent
Turbulent fr
Turbulent he
Turbulent kin
Turbulent Pran
Turbulent shear
Turbulent transpo
Turbulent viscosity,
Turbulent vortices, 8
Twomey's formula, 16

Uncentered finite differe
Universal gas constant, 68
Universal profile of supersa

Van't Hoff factor, 153
Variation, 2, 15, 30
Velocity divergence, 13
Velocity gradient tensor, 91
Ventilation effect, 151–152, 211
Virtual temperature, 70–71, 78
Viscous sub-range, 93
Vortex rings, 279

Warm microphysical processes, 7, 151, 344
Warm rain formation, 243, 538
Water balance equation, 166
Water drops, 1, 20, 51
Water film, 345–347
Water potential temperature, 80, 121
Water vapor pressure, 70–71
Water-ice interface, 346, 354, 442
Wave number, 89, 92, 105
Wavelength, 88, 98, 102
Weber number, 227, 229
Wegener-Bergeron-Findeizen process, 347
Wet convection, 97
Wettability parameter, 359–360
Wetting, 346, 359
Wind-tunnel measurements, 41, 66
Winter precipitation, 51, 495, 593